Interpr

Micromorp .cal
Features of ɔoils and
Regoliths

Second Edition

Interpretation of Micromorphological Features of Soils and Regoliths

Second Edition

Edited by

Georges Stoops Ghent University, Ghent, Belgium

Vera Marcelino Ghent University, Ghent, Belgium

Florias Mees Royal Museum for Central Africa, Tervuren, Belgium

ELSEVIER

Elsevier
Radarweg 29, PO Box 211, 1000 AE Amsterdam, Netherlands
The Boulevard, Langford Lane, Kidlington, Oxford OX5 1GB, United Kingdom
50 Hampshire Street, 5th Floor, Cambridge, MA 02139, United States

Library of Congress Cataloging-in-Publication Data
A catalog record for this book is available from the Library of Congress

British Library Cataloguing in Publication Data
A catalogue record for this book is available from the British Library

ISBN: 978-0-444-63522-8

For information on all Elsevier publications visit our website at
https://www.elsevier.com/books-and-journals

Working together
to grow libraries in
developing countries

www.elsevier.com • www.bookaid.org

Publisher: Candice Janco
Acquisition Editor: Marisa LaFleur
Editorial Project Manager: Emily Thomson
Production Project Manager: Anitha Sivaraj
Designer: Greg Harris

Typeset by TNQ Technologies

Contents

List of Contributors

W. Paul Adderley   University of Stirling, Stirling, Scotland, United Kingdom

José Aguilar   University of Granada, Granada, Spain

Octavio Artieda   Extremadura University, Plasencia, Spain

Maria Bronnikova   Institute of Geography, Russian Academy of Sciences, Moscow, Russia

Matthew G. Canti   Historic England, Eastney, United Kingdom

Marie-Agnès Courty   Universidad Rovira y Virgili, Tarragona, Spain

Mauro Cremaschi   Università degli Studi di Milano, Milano, Italy

Donald A. Davidson   University of Stirling, Stirling, Scotland, United Kingdom

Yannick Devos   Université Libre de Bruxelles, Brussels, Belgium

Nicolas Durand   Mines Paris-Tech, Centre de Géosciences, Fontainebleau, France

William R. Effland   Natural Resources Conservation Service, U.S. Department of Agriculture, Beltsville, MD, United States

Nicolas Fedoroff   Agrotech Paris, Paris, France (Deceased)

Catherine A. Fox   Agriculture and Agri-Food Canada, Harrow, ON, Canada

Maria Gerasimova   Moscow Lomonosov University, Moscow, Russia;   Dokuchaev Soil Science Institute, Moscow, Russia

Paul Goldberg   Boston University, Boston, MA, United States;   University of Wollongong, Wollongong, NSW, Australia

Zhengtang Guo   Chinese Academy of Sciences, Beijing, China

Ma. del Carmen Gutiérrez-Castorena Colegio de Postgraduados, Campus Montecillo, Texcoco, Mexico

Kristin Ismail-Meyer University of Basel, Basel, Switzerland

Danuta Kaczorek Warsaw University of Life Sciences, Warsaw, Poland; Leibniz-Centre for Agricultural Landscape Research, Müncheberg, Germany

Panagiotis Karkanas American School of Classical Studies, Athens, Greece

Maja J. Kooistra International Soil Reference and Information Centre and Kooistra Micromorphological Services, Rhenen, The Netherlands (Retired)

Irina Kovda Dokuchaev Soil Science Institute, Moscow, Russia; Institute of Geography, Russian Academy of Sciences, Moscow, Russia

Peter Kühn Eberhard Karls Universität Tübingen, Tübingen, Germany

Frans Kwaad University of Amsterdam, Purmerend, The Netherlands (Retired)

Marina Lebedeva Dokuchaev Soil Science Institute, Moscow, Russia

David L. Lindbo United States Department of Agriculture, Washington, DC, United States

Richard I. Macphail University College London, London, United Kingdom

Vera Marcelino Ghent University, Ghent, Belgium

Florias Mees Royal Museum for Central Africa, Tervuren, Belgium

Ahmet R. Mermut Harran University, Sanliurfa, Turkey

Rienk Miedema Wageningen University, Wageningen, The Netherlands (Retired)

H. Curtis Monger USDA-Natural Resources Conservation Service, Lincoln, NE, United States

Herman Mücher University of Amsterdam, Amsterdam, The Netherlands (Deceased)

Marcello Pagliai   Centro di Ricerca per l'Agrobiologia e la Pedologia, Florence, Italy

Ákos Petó   Szent István University, Gödölló, Hungary

Rosa M. Poch   University of Lleida, Lleida, Spain

Mirjam M. Pulleman   Wageningen University, Wageningen, The Netherlands

Dominique Righi   Université de Poitiers, Poitiers, France (Deceased)

Carlos E.G.R. Schaefer   Federal University of Viçosa, Viçosa, Minas Gerais, Brazil

Sergey Sedov   Universidad Nacional Autónoma de México, Mexico City, México

Sergei Shoba   Lomonosov Moscow State University, Moscow, Russia

Felipe N.B. Simas   Federal University of Viçosa, Viçosa, Minas Gerais, Brazil

Ian A. Simpson   University of Stirling, Stirling, Scotland, United Kingdom

Mark H. Stolt   University of Rhode Island, Kingston, RI, United States

Georges Stoops   Ghent University, Ghent, Belgium

Luca Trombino   Università degli Studi di Milano, Milano, Italy

Tatiana V. Tursina   Dokuchaev Soil Science Institute, Moscow, Russia

Eric Van Ranst   Ghent University, Ghent, Belgium

Henk van Steijn   University of Utrecht, Wijk bij Duurstede, The Netherlands (Retired)

Brigitte Van Vliet-Lanoë   CNRS Geosciences Océan, University of Brest, Plouzané, France

Michael J. Vepraskas   North Carolina State University, Raleigh, NC, United States

Eric P. Verrecchia   University of Lausanne, Lausanne, Switzerland

Luc Vrydaghs   Université Libre de Bruxelles, Brussels, Belgium

Amanda J. Williams   SWCA Environmental Consultants, Las Vegas, NV, United States

Clare A. Wilson   University of Stirling, Stirling, Scotland, United Kingdom

Michael A. Wilson   USDA-NRCS, Lincoln, NE, United States

Siti Zauyah   University Putra Malaysia, Serdang, Malaysia (Retired)

Andrea Zerboni   Università degli Studi di Milano, Milano, Italy

Preface to the First Edition

During his more than 30 years of teaching experience in the field of soil micromorphology, the senior Editor was confronted with an always returning question from students and alumni: where to find information on the interpretation of micromorphological data. Apart from a few reviews, often quite outdated by now, this information is dispersed in numerous research papers in various national and international journals and congress proceedings. To provide students and researchers with an up-to-date and comprehensive tool, he took the initiative at the end of his career to prepare, with two colleagues, a textbook on the significance of micromorphological features, with the help of international specialists.

The relationship between micromorphological characteristics and genesis, classification or physical, chemical and mineralogical properties of soils and regoliths is a vast and complex issue, requiring much further systematic research. The Editors therefore consider this book as a first approximation towards an inventory of micromorphological knowledge, and they are aware that it is surely not as complete as it could be. They hope that the gaps in our knowledge and understanding of the fabric of soils and regoliths, as pointed out in the various chapters of the book, will stimulate young researchers to contribute to further progress in the field of micromorphology.

Since the early days of micromorphology, different descriptive systems and terminologies have been used. To avoid terminology-related confusion, authors were asked to use the concepts and terminology proposed by Bullock et al. (1985) and Stoops (2003), and where necessary to 'translate' other terms to this system.

An effort was made to cover the different aspects of micromorphological interpretation, except in the field of quantitative micromorphology, also known as micromorphometry. A general overview of the use of micromorphology and its interpretation is presented in the first two chapters, the second serving as an open key to the next chapters. The approach of the last two chapters, dealing with archaeology and palaeosoils is different. These chapters do not systematically treat observed features, but rather explain general ideas, illustrated with many case studies.

The amount of work done by the authors should not be underestimated. For some topics, little published information is available (e.g., saprolite fabric). For others the amount of literature data is overwhelming (e.g., clay illuviation), often scattered over publications coming from various disciplines such as soil science, geology and engineering (e.g., carbonates, laterites). Moreover, the available information is often contradictory, making it necessary to be critical while reviewing published interpretations.

It would not have been possible to reach the high scientific standard of the various chapters without the greatly appreciated assistance of many dedicated referees. All manuscripts were reviewed by at least two specialists, whose constructive and critical comments have contributed to the final quality of this book. The Editors thank all referees for their excellent work. The following persons have assisted the Editors by reviewing one or more manuscripts: P. Adderley, J. Alexandre, A.M. Alonso Zarza, J. Arocena, I.E. Brochier, B. Buck, P. Bullock (†), A. Bustillo-Revuelta, P. Buurman, M. Canti, J. Catt, D. Davidson, S. Driese, D. Dubroeucq, M.H. Farpoor, N. Fedoroff, E.A. FitzPatrick, D. Gabriels, R. Joeckel, A. Jongmans,

D. Kaczorek, M. Kooistra, P. Kühn, R. Langohr, K. Milek, A. Mindszenty, S. Mooney, H. Morrás, H. Mücher, D. Nahon, D.J. Nash, E. Padmanabhan, M. Pagliai, R.M. Poch, J. Poesen, C. Schaefer, I. Simpson, A. Singer, O. Spaargaren, M. Stolt, V. Tagulian, L. Trombino, A. Tsatskin, J. van der Meer, B. Van Vliet, M. Wieder, L. Wilding, D. Yaalon and S. Zauyah.

The Editors are much indebted to Prof. Dr. P. Van den haute, Head of the Mineralogy and Petrology unit of the Department of Geology and Soil Science, formerly the Laboratory for Mineralogy, Petrology and Micropedology, who provided the facilities to carry out editorial work for this book within his division.

Much of the research and teaching experience that led to the realisation of this book is the result of years of activity at the International Training Centre for Post-Graduate Soil Scientists of the Ghent University, where micromorphology was an important discipline since the early 1960s.

Georges Stoops
Vera Marcelino
Florias Mees

References

Bullock, P., Fedoroff, N., Jongerius, A., Stoops, G., Tursina, T. & Babel, U., 1985. Handbook for Soil Thin Section Description. Waine Research Publications, Wolverhampton, 152 p.

Stoops, G., 2003. Guidelines for Analysis and Description of Soil and Regolith Thin Sections. Soil Science Society of America, Madison, 184 p.

Preface to the Second Edition

The first edition of *Interpretation of Micromorphological Features of Soils and Sediments*, published in 2010, was warmly welcomed by a large audience of scientists using micromorphology in their research, such as soil scientists, Quaternary geologists and an increasing number of archaeologists. It was the first systematic review of the interpretation of micromorphological features to be published, and it appears to have become widely used.

When asked a few years ago by the Publisher to consider the preparation of a second edition, the Editors saw an opportunity to upgrade and update the existing texts, by incorporating information contained in more recent publications and in overlooked older literature.

For the new edition, revision by the original authors was preferred. However, some authors passed away, and others were no longer able to contribute because of changes in their professional career. Some chapters were therefore updated by only one or two of the original authors. For other chapters, new authors or coauthors had to be invited, giving other experienced scientists, often of a younger generation, the opportunity to contribute to this book.

The chapter by Fedoroff and co-authors on palaeosoils remained unchanged, but a new additional chapter on palaeosoils, using a different perspective, was added. Also the chapters on faunal activity and saprolites have remained unchanged. The chapter on siliceous features was split into two chapters: one dealing with pedogenic siliceous features, and the other dealing with biogenic silica. A new chapter about the significance of groundmass characteristics was added, as a subject that is mostly neglected in the literature, despite its potential relevance in micromorphological studies.

The total number of authors increased from 46 in the first edition to 58 in the present one, and the number of references increased from about 2400 to 3500.

The scientific quality of a book depends not only on authors and editors but also for a large part on the efforts of referees. The Editors therefore acknowledge the important input of an international team of referees who contributed to the quality of the second edition: F. Berna, H. Breuning-Madsen, M. Bronnikova, B. Buck, J. Deák, C. Doronsorro, S. Driese, V. Felde, M. Francis, C. French, R. Gilkes, M. Joeckel, F. Khormali, R. Langohr, R. Miedema, A. Mindszenty, H.C. Monger, H. Morrás, C. Nicosia, E. Padmanabhan, B. Sageidet, P. Sanborn, J. Shamshuddin, A. Solé-Benet, F. Sulas, S. Ullyott, L. Uzarowicz, K. Vancampenhout, E. Van Ranst, and Z. Zagórski.

The Editors thank Prof. Dr. Johan De Grave, Head of the Mineralogy and Petrology unit of the Department of Geology, Ghent University, for providing the facilities to carry out the editorial work for this book.

Georges Stoops
Vera Marcelino
Florias Mees

List of Abbreviations

BLF   Blue-light fluorescence microscopy
CPL   Circular polarized light
EDS   Energy-dispersive spectroscopy
OIL   Oblique incident light
PPL   Plane-polarized light
SEM   Scanning electron microscopy
TEM   Transmission electron microscopy
UVF   Ultraviolet fluorescence microscopy
WDS   Wavelength-dipersive spectroscopy
XPL   Cross-polarized light

Chapter 1

Micromorphology as a Tool in Soil and Regolith Studies

Georges Stoops

GHENT UNIVERSITY, GHENT, BELGIUM

CHAPTER OUTLINE

1. Introduction

The aim of micropedology is to contribute to solving problems related to the genesis, classification and management of soils, including soil characterisation in palaeopedology and archaeology. The interpretation of features observed in thin sections is the most important part of this type of research, based on an objective detailed analysis and description.

Attempts to understand soil genesis by the study of thin sections were already made in the beginning of the 20th century by Delage and Lagatu (1904), Lacroix (1913), Agafonoff (1929, 1935/1936, 1936) and Allen (1930), but it was only after the publication of the book *Micropedology* (Kubiëna, 1938) that a real start was made. The use of micromorphology was promoted further by Kubiëna's later books on soil genesis and classification (Kubiëna, 1948, 1953; see also Stoops, 2009a). New methodologies to prepare large thin sections with cold-setting resins (Borchert, 1961; Altemüller, 1962;

Jongerius & Heintzberger, 1962) allowed the study of microstructures and other relatively large features. In the course of the following years, much progress was made in methods of thin section description (Brewer, 1964; FitzPatrick, 1984, 1993; Bullock et al., 1985; Stoops, 2003) and quantitative analyses (Kubiëna, 1967; Jongerius, 1974; Mermut & Norton, 1992; Terribile et al., 1997; see also Stoops, 2009b). The interpretation of micromorphological features of various soil and regolith materials advanced as well, but no general overview of these achievements had been compiled before publication of the first edition of this book.

The evolution of micromorphological publications, from 1900 until 2000, was analysed by Stoops (2014) based on bibliometric data. In the following paragraphs, an update until 2015, using the same methodology as described in the original paper and completed with many overlooked references, is presented. After publication of the book by Kubiëna (1938), the number of publications continuously increased, especially since 1956, reaching a maximum in the period 1986-1990 (Fig. 1) and decreasing thereafter slightly for various reasons (Stoops, 2014).

Besides changes in the number of micromorphological publications, also the fields of research that they represent strongly varied since its inception (Fig. 2). Before 1940, publications on methodology and concepts were dominant, followed by soil genesis. Soil genesis became the main application after 1940, reaching a maximum in the period 1956-1960 (80%) and still accounting for more than 50% until 1996-2005. During and after this period, a gradual increase in the number of publications on archaeological

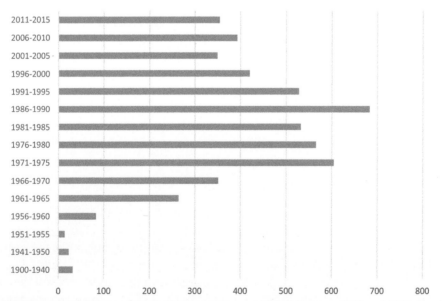

FIGURE 1 Total number of papers on soil micromorphology, published between 1900 and 2015. *Based on an extended and updated version of the dataset compiled by Stoops (2014).*

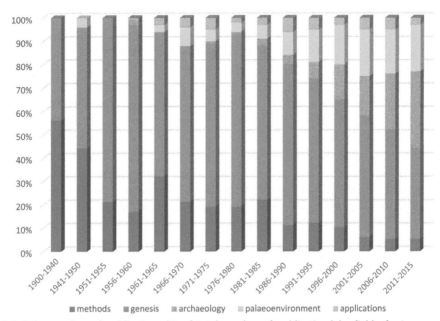

FIGURE 2 Relative number of publications (% of total number of publications) by field of micromorphological research: methods (thin section preparation, ultramicroscopy, analytical methods, concepts, terminology), genesis (soil development, classification, mineral weathering, neoformation), archaeology, palaeoenvironment (palaeopedology, geomorphology, quaternary geology) and applications (agronomy, soil mechanics). *Based on an extended and updated version of the dataset compiled by Stoops (2014).*

and palaeopedological topics was observed, and from 2011 onwards, these fields of research together account for more than 50% of all micromorphological publications.

With regard to the language of publication, it was only after 1970 that English definitely passed the threshold of 50%, which has now increased to more than 90%. Before 1970, many papers were written in German (>50% in the period 1941-1950) and also in French, Spanish and Russian, besides other languages such as Czech, Dutch, Hungarian, Portuguese and Polish, representing the countries where micromorphology was strongly developed as field of research. The loss of this linguistic diversity represents a real impoverishment, for example, in the creation and formulation of concepts.

Several difficulties are encountered when making a synthesis of existing literature data on micromorphological characteristics of materials and horizons in relation to their genesis and classification (Stoops, 2001):

- Most published micromorphological descriptions are unavoidably incomplete, mentioning only the most relevant, striking or 'diagnostic' features, or only those that the author was able to recognise. The most complete descriptions can be found in reports and dissertations.

- The micromorphological terminology used is frequently not sufficiently precise, in some cases partly consisting of an author's own descriptive system without well-defined concepts. The use of a standardised and uniform terminology is therefore very important. The influential system of Brewer (1964) is still in use by some authors, but more widely accepted modern terminology (Bullock et al., 1985; Stoops, 2003) is now recommended. The latter has been translated into more than 15 languages, producing lists of terms that have been made available online.
- Soils or materials studied in thin sections are often not sufficiently precisely identified (e.g., 'a brown soil') or they are classified according to often poorly accessible national systems. The many changes through time in classification criteria used in the different versions of the FAO-WRB and USDA systems also strongly hinder correlation between soil type and micromorphological characteristics, based on papers published over periods of several decades.
- Horizons or layers that are not considered as diagnostic or interesting are generally not described. For instance, practically no micromorphological information is available on cambic horizons occurring together with other diagnostic horizons in a same profile, or on aspects of surface horizons other than their microstructure characteristics.
- In several publications only interpretations are given, not supported by detailed descriptions, and therefore making it impossible to use the data for comparison or for database construction.

The general significance of certain micromorphological features can be explained rather easily, for example when relating the presence of clay coatings to the process of clay illuviation. For many other features, the interpretation is less straightforward. In many cases the combined occurrence of features, or the absence of a certain feature, is more diagnostic than the presence of individual features. For instance, fine granular microstructures and an undifferentiated b-fabric are found in oxic, andic and spodic materials. If the field context is poorly known, the presence and abundance of volcanic components, weatherable minerals and fine organic matter allow differentiation between these materials.

Interpretation of micromorphological characteristics is based on comparison between the recognised features and field observations, laboratory analysis results, and results of experimental studies. In the case of palaeosoils, comparing features of buried or relict soils with those in present-day soils is evidently a must. In archaeology, the study of modern analogues is also useful.

For a full understanding of soil and regolith thin sections, sufficient knowledge of polarised light microscopy and optical mineralogy, as well as a basic knowledge of petrography, is needed. This knowledge is often missing among certain groups of researchers such as archaeologists (see Shahack-Gross, 2015) and soil scientists with a background in agricultural studies.

For all these reasons it is not possible to relate, in a simple way, micromorphological characteristics with genesis, classification or physical, chemical and mineralogical properties. Much systematic research is therefore still needed.

2. Soil and Regolith Genesis

Micromorphology has been applied mainly in view of disentangling processes of soil and regolith formation.

Before publication of the first edition of this book (Stoops et al., 2010), review papers on the micromorphological characteristics of different soil types were available in volumes edited by Bullock and Murphy (1983) and Douglas and Thompson (1985). Other reviews are scarce, and they are scattered over various scientific journals and congress proceedings. Many of them suffer from the fact that their authors are acquainted only with the literature published in their own language or country. There is also a considerable risk that in future reviews only publications that are registered or available online will be taken into consideration. As a result, much useful information can be lost. Early Russian knowledge on the interpretation of soil micromorphology was summarised by Parfenova and Yarilova (1965) and widely used after translation, whereas no translation is available of the more recent volume by Romeshkevich and Gerasimova (1982). The English-language book by Gerasimova et al. (1996) on Russia and adjacent countries has not been widely distributed. For the work of Chinese micromorphologists, no review is available.

The interpretation of micromorphological features has been based essentially on comparison between thin section observations, macromorphological features and laboratory data. The most probable hypothesis is then accepted, and in many cases gradually considered as a proven fact. Interpretation is also often based on the authority of an author, as it is anyway also the case in many other scientific disciplines. Especially in the early days of micromorphology, the opinion of Kubiëna was often taken for granted, although it generally lacked supporting laboratory analyses results.

In several cases the genesis of specific features or fabrics remains controversial, such as the origin of the small rounded aggregates in Oxisols (see Marcelino et al., 2010, 2018), Andosols (see Sedov et al., 2010; Stoops et al., 2018) and spodic horizons (see Wilson & Righi, 2010; Van Ranst et al., 2018).

Interpretation can also be based on experimental work in the laboratory or in the field (see Section 7), or on field observations. A good example of the latter is the correlation between determinations of hydromorphic conditions in the field and the resulting micromorphological features (Veneman et al., 1976).

Regolith genesis as a result of weathering has been studied in thin sections by petrographers since the first part of the 20th century (e.g., Harrison, 1933). In the 1950s, pedologists also began using this technique to study weathering, saprolite formation and pedoplasmation, especially in Africa by ORSTOM (IRD) researchers (see also Stoops &

Schaefer, 2010, 2018; Zauyah et al., 2010, 2018). This resulted in numerous publications in journals and congress proceedings, besides a large number of unpublished reports and dissertations. Much of this material has been summarised in books by Nahon (1991), Tardy (1993) and especially Delvigne (1998).

3. Soil and Regolith Classification

In the genetic soil classification system of Kubiëna (1948, 1953), micromorphology was important to distinguish soil types such as Braunlehm and Terra Rossa or humus layers such as Mor and Anmoor. In modern soil classification systems only the one of FitzPatrick (2005) considers micromorphological features as important criteria for the identification of some horizons. In WRB (IUSS Working Group WRB, 2015), micro-morphological characteristics are used to a limited extent in the identification of horizons.

One difficulty in relating micromorphological data to classification criteria is that in most soil classification systems (e.g., WRB, Soil Taxonomy), diagnostic horizons or diagnostic materials are defined not only on a morphological basis but also on the basis of chemical or physical characteristics. Moreover, some horizons are defined not only by their own characteristics but also by differences with other parts of the same profile. Other horizons comprise a very broad and heterogeneous group of materials, such as the mollic and cambic horizons in Soil Taxonomy (Soil Survey Staff, 2014). Also, non-morphological properties, such as climate, often play a dominant role, keying out soils before their morphology is considered (e.g., Aridisols). It is worth mentioning that various international commissions on soil classification (e.g., ICOMAND, ICOMID, ICOMOX) intensively used micromorphology in their discussions, but these micro-morphological data were never published in peer-reviewed journals.

In few cases, direct information on soil type is obtained from thin section observations. For example, the presence of fine clay coatings points to an argillic horizon, whereas other types of coatings may indicate natric or agric horizons. However, not all argillic horizons clearly show the occurrence of clay illuviation in thin sections (e.g., most Red Mediterranean soils), and not all clay illuviation features are the result of vertical transport from an E horizon to a Bt horizon (see Kühn et al., 2010, 2018). Similar considerations are possible for other horizons such as the gypsic (see Poch et al., 2010, 2018) and calcic (see Durand et al., 2010, 2018). Other horizons, such as the oxic, display a characteristic combination of clearly recognisable micromorphological features (see Marcelino et al., 2010, 2018), but not all materials with these characteristics key out as oxic. On the other hand, some features observed in thin sections are not directly related to specific diagnostic horizons, such as features related to gley phenomena (see Lindbo et al., 2010; Vepraskas et al., 2018), stress movements, bioturbation or eluviation, but they may be important for soil classification at other levels (e.g., moisture regime, vertic and vermic characteristics).

For some diagnostic horizons practically no published micromorphological data are available, as is the case for the sombric horizon, although a micromorphological identification criterion is included in the description of this horizon in Soil Taxonomy.

Several materials, considered as parent materials in Soil Taxonomy (Soil Survey Staff, 2014) and WRB (IUSS Working Group WRB, 2015), are considered as part of the regolith profile by other authors with a more naturalistic and holistic approach, especially in the tropics where deep regoliths are common.

For regolith materials, no internationally accepted classification system exists. The proposal of Buol (1994) for saprolites and regoliths, although based on reflections of an international working group, is hardly used, and the system does not contain micromorphological criteria. The latter is also the case for the nomenclature system that was more recently proposed by Juilleret et al. (2016). Also for the subdivision of individual regolith types, such as laterites, tephra and colluvial sediments, no micromorphological criteria were used.

4. Palaeopedology, Quaternary Geology and Archaeology

Climatic changes during the Quaternary, characterised by an alternation of colder and warmer periods, have been recorded by fabric changes in soils or sediments. Micromorphological methods have been applied, especially in the study of soils from Eurasia and North America, to observe imprints of former cold periods, such as freeze-thaw features (see Van Vliet-Lanoë, 2010; Van Vliet-Lanoë & Fox, 2018). Micromorphological studies of glacial sediments started already in the 1960s (Korina & Faustova, 1964). Later, they have been promoted mainly by van der Meer and Menzies (e.g., van der Meer, 1987, 1993, 1996; Menzies, 1998; Menzies et al., 2010; van der Meer & Menzies, 2011) and soon became a discipline on its own. Changes in climate and vegetation also resulted in formation of surface sediments, such as colluvium and solifluction layers that were studied with micromorphological techniques (see also Mücher et al., 2010, 2018).

In arid and semiarid areas, climate changes are mainly recorded in the fabric and composition of calcareous (see Durand et al., 2010, 2018) and gypsic (see Poch et al., 2010, 2018) horizons, or deposits of soluble salts (see Mees & Tursina, 2010, 2018). Their micromorphological study contributed definitely to the understanding of past climate and environmental change.

The use of micromorphology to disentangle the complex fabric of palaeosoils was already mentioned by Kubiëna (1956). It is evident that interpretation of micromorphological features of palaeosoils is only possible based on a thorough knowledge and understanding of the characteristics of present-day soils. This is complicated by the fact that few soils are the result of a single pedogenic cycle, which implies that most frequently several cycles have to be considered (see Fedoroff et al., 2010, 2018; Cremaschi et al., 2018). Early micromorphological research on fossil (buried) soils was published, for instance, by Morozova (1964), Smolikova (1967, 1978), Ferrari and Magaldi (1968), Dalrymple (1972) and Bullock and Murphy (1979) (see also Mücher & Morozova, 1983). Especially in the field of loess stratigraphy a huge amount of micromorphological research has been done (e.g., Smolikova, 1958, 1967, 1972, 1978; Bronger, 1969/1970, 1972; Kemp & Derbyshire, 1998; Kemp, 1999, 2001). The amount of micromorphological data on pre-Quaternary soils is still limited, and practically no information is available on diagenetic features developed after burial.

Today, it is difficult to imagine geoarchaeology without micromorphology. Although the use of micromorphological techniques in archaeology dates already from the 1950s (e.g., Cornwall, 1958), only after the late 1980s have archaeologists become a very active group in micromorphology (see Courty et al., 1989; Davidson et al., 1992; Stoops, 2014; see also Nicosia & Stoops, 2017). Archaeologists apply micromorphology mainly in two fields:

- The study of soils and related materials, such as those resulting from forest clearing, ancient cultivation methods or compaction through trampling by domesticated animals, as well as buried soils, infills and earthy structures (see Adderley et al., 2010, 2018); this type of research is closely related to the study of natural soils, and existing micromorphological concepts and terms are adequate for the description of these materials.
- The interpretation of archaeological layers and the identification of archaeological materials, such as excrements, mortars, burned materials, ashes and metal slag (see Karkanas & Goldberg, 2010, 2018; Macphail & Goldberg, 2010, 2018); here special features and materials are encountered, not completely covered by the standard micromorphological concepts.

5. Soil Management and Other Applications

The contributions of micromorphology in the field of soil and land management are mainly indirect. By providing a better understanding of processes in the soil, micromorphology helps improving or choosing good management practices. Most direct applications are based on quantitative studies of porosity. They often deal with compaction by heavy equipment or trampling. Monitoring seal and crust formation by micromorphological methods has been successful (see Pagliai & Stoops, 2010; Williams et al., 2018). Other examples of qualitative research are studies of changes in structure (Borchert, 1961, 1964, 1972), root penetration (Stewart et al., 1994), silting of drainage systems (Houot & Berthelin, 1987; Herrero Isern et al., 1989) and slurry application (Sveistrup et al., 1995).

Forensic studies using micromorphological techniques are still rare, mainly because their confidential nature generally does not permit publication. However, micromorphology can become an important tool for this science, once the results of this type of investigation become more accessible, as for instance for detecting clandestine graves (Goldberg & Macphail, 2006).

6. Correlations Between Micromorphology and Physical Data

Relatively few efforts have been made to correlate quantitative or semiquantitative micromorphological data with data of chemical or physical measurements. Most work has been done on the comparison between porosity measurements in thin sections and soil physical porosity data. These studies include the pioneering works of Jongerius (Jongerius et al., 1972; Jongerius, 1974), and later those of Bouma et al. (1979), comparing saturated flow in clayey soils with micromorphometric data, and also various studies by Pagliai on soil structure changes (see Pagliai, 2008). With respect to soil structure, qualitative correlations have since long been made between transformations of porosity patterns and management techniques (e.g., Borchert, 1964, 1972).

The second-most quantified type of features are clay coatings/infillings and their fragments, which can be correlated with relative clay contents determined by textural analyses (e.g., Grossman, 1964; Miedema & Slager, 1972; see also Kühn et al., 2010, 2018). For this application, Murphy and Kemp (1984) presented nomograms to help correlating the volumetric and weight increase of fine clay. Efforts have also been made to correlate b-fabrics with plasticity and the amount of easily dispersible clay in oxic materials (Embrechts & Stoops, 1986; Zainol & Stoops, 1986; see also Stoops & Mees, 2018) and in Vertisols (see Kovda & Mermut, 2010, 2018).

Good correlations between micromorphological and physical data may become important in future, to link micropedology with applied soil science, employing methods used in pedometrics. This will require, however, a standardisation of methods and parameters used by different laboratories to obtain results that can be compared, as results depend strongly on methodology (e.g., Marcelino et al., 2007).

7. Monitoring Experimental Work and Analyses

Micromorphology is a powerful tool to monitor experimental studies aimed at simulating and understanding processes of soil and regolith genesis. The fact that small changes may become visible in thin sections before they can be detected by physical or chemical analyses makes it a valuable method. However, no systematic research has been done yet in this field and relative few data have been published. Experimental studies were done for instance on the development of striated b-fabrics by shearing (Morgenstern & Tchalenko, 1967; Jim, 1990; see also Stoops & Mees, 2018), freeze-thaw features (Frese & Czeratzki, 1957; Coutard & Van Vliet-Lanoë, 1994), colluvial sedimentation (Mücher & De Ploey, 1977, 1984, 1990), faunal activity (Jeanson, 1964; Bal, 1977; Eschenbrenner, 1987), formation of clay coatings (Brewer & Haldane, 1957;

Dalrymple, 1972; Theocharopoulos & Dalrymple, 1987), pyrite formation (Rabenhorst, 1990) and aggregate stability (Breuer, 1997). Experimental work has been exceptionally important for the study of the formation of surface seals and crusts, which is largely based on combined field and laboratory experiments (see Pagliai & Stoops, 2010; Williams et al., 2018).

The precipitation and crystallisation of carbonates, gypsum and highly soluble salts have commonly been studied in thin sections or with the stereomicroscope. Early examples are studies of calcite precipitation by Wells (1965) and Aguilar et al. (1978). Many laboratory experiments have been done for gypsum crystallisation to understand the significance of its various crystal habits (see Poch et al., 2010, 2018).

In archaeology, efforts have been made to support interpretations by comparing micromorphological features for excavated and experimental materials (e.g., Mathieu & Stoops, 1972; Canti, 2003; see also Macphail & Goldberg, 2010, 2018; Nicosia & Stoops, 2017).

The role of micromorphology in checking and monitoring physical or chemical data is not sufficiently recognised and appreciated. Thin section analyses of undisturbed samples allow, for instance, an assessment of the results of grain-size analyses, by showing the presence of pseudosands (e.g., Pede & Langohr, 1983) or by revealing the real grain size of minerals that can become fragmented during laboratory treatments. Micromorphology is also one of the few techniques that allow lithogenic and pedogenic occurrences of a specific mineral, such as calcite or gypsum, to be distinguished, whereas chemical or mineralogical analyses only yield information on the presence and abundance of these constituents.

Thin section studies provide information on the composition of the fraction coarser than 2 mm, which is commonly removed before soil analyses are performed (see also Stoops & Mees, 2018). This information is, for instance, important when pedochemical balance calculations are envisaged. For example, manganese and coprecipitated elements accumulated in nodules larger than 2 mm will generally escape detection during routine analyses. Micromorphology can also help in avoiding errors in the interpretation of mineralogical data. For instance, in strongly weathered tropical soils, routine mineralogical analyses may show the presence of weatherable minerals, thus excluding an Oxisol classification for the profile. In several cases, thin sections show that these minerals are totally enclosed by iron oxides and thus no longer take part in soil processes (Baert & Van Ranst, 1997). Micromorphology has also been used to control the validity of various specific analyses of soil and regolith materials. An excellent example is that of palynology, where thin section studies can help to find out whether pollen grains are *in situ* or reworked (Van Mourik, 1999, 2003; Kooistra & Kooistra, 2003).

8. Evolutions in Analytical Methods

Concepts and terminology that are used in micromorphological studies seem at present to be rather stable, and no major changes have been proposed over the last years, but considerable progress has been made in analytical techniques that can be applied to components in thin sections. In general, soil micromorphologists did not take much advantage of this progress, except in archaeological studies. FTIR-microscopy (or μ-FTIR) was used, for instance, to analyse burned wood and bone (Berna & Goldberg, 2007; Berna et al., 2012), and organic remains in middens (Shillito, 2009). In studies of soil development, this methodology could be very useful for analysis of organic or amorphous phases, for instance, in spodic horizons, organic surface horizons or soils showing mineral weathering. Also other methods of thin section microanalysis, such as μ-XRD, μ-XRF and LA-IRMS, have been rarely used in soil micromorphology, except for archaeological soil materials (e.g., Bruneau et al., 2002; Berthold & Mentzer, 2017).

Although the methodology for obtaining powders by microdrilling in thin sections or corresponding impregnated blocks is known in micromorphology since several decades (van Oort et al., 1994; Denaix et al., 1999; Jongmans et al., 1999), it has only rarely been used. A recent example is its use in archaeology for stable isotope analysis (Mentzer & Quade, 2013; Mentzer, 2017).

Developments in imaging techniques to improve image quality for thin section documentation at the mesoscale have been recently reported (Carpentier & Vandermeulen, 2016).

The application of micro-CT scans in micromorphology was up to now mainly restricted to porosity studies (e.g., Elliot and Heck, 2007; Bendle et al., 2015) but could be a help to obtain 3D images of the distribution of some components.

9. Conclusions

The role of micromorphology in genetic studies and classification of soils and regoliths is clear, but up to now systematic and structured information on interpretation of micromorphological features is missing. This also applies for the study of palaeosoils, as the same type of information is requested in this field. Much important micromorphological data that would be useful to extend the possibilities of interpretation exist but were never published in journals. Greater accessibility of this information, hidden in reports, excursion guides and dissertations, would therefore be important for the further progress of micromorphology.

Correlations between micromorphological data, both qualitative and quantitative, and chemical, physical, mineralogical and land use data are scarce. Relevant correlations could enhance the possibilities of using micromorphological analyses as a routine method, making it a tool in pedometrics.

References

Adderley, W.P., Wilson, C.A., Simpson I.A. & Davidson, D.A., 2010. Anthropogenic features. In Stoops, G., Marcelino, V. & Mees, F. (eds.), Interpretation of Micromorphological Features of Soils and Regoliths. Elsevier, Amsterdam, pp. 569-588.

Adderley, W.P., Wilson, C.A., Simpson, I.A. & Davidson, D.A., 2018. Anthropogenic features. In Stoops, G., Marcelino, V. & Mees, F. (eds.), Interpretation of Micromorphological Features of Soils and Regoliths. Second Edition. Elsevier, Amsterdam, pp. 753-777.

Agafonoff, V., 1929. Sur quelques sols rouges et Bienhoa de l'Indochine. Revue de Botanique Appliquée et d'Agriculture Coloniale 9, 89-90.

Agafonoff, V., 1935/1936. Sols types de Tunisie. Annales du Service Botanique et Agronomique de Tunisie 12/13, 43-413.

Agafonoff, V., 1936. Les sols de France au Point de Vue Pédologique. Dunod, Paris, 154 p.

Aguilar J., Ramos, A. & Ruiz, F., 1978. Contribution to the study of the crystals formation process by bacteria. In Delgado, M. (ed.), Soil Micromorphology. University of Granada, pp. 139-168.

Allen, V.T., 1930. Petrographic studies bearing on the genesis and morphology of Illinois Soils. Proceedings of Second International Congress on Soil Science, Volume 5. Selkolkhozgiz, Moscow, 113-117.

Altemüller, H.J., 1962. Verbesserung der Einbettungs- und Schleiftechnik bei der Herstellung von Bodendünnschliffen mit VESTOPAL. In Altemüller, H.J. & Frese, H. (eds.), Arbeiten aus dem Gebiet der Mikromorphologie des Bodens. Verlag Chemie, Weinheim, pp. 230-243.

Baert, G. & Van Ranst, E., 1997. Total reserve in bases as an alternative for weatherable mineral content in soil classification: a micromorphological investigation. In Shoba, S., Gerasimova, M. & Miedema, R. (eds.), Soil Micromorphology: Studies on Soil Diversity, Diagnostics, Dynamics. Moscow, Wageningen, pp. 41-51.

Bal, L., 1977. The formation of carbonate nodules and intercallary crystals in the soil by the earthworm *Lumbricus rubellus*. Pedobiologia 17, 102-106.

Bendle, J.M, Palmer, A.P. & Carr, S.J., 2015. A comparison of micro-CT and thin section analysis of Late glacial glaciolacustrine varves from Glen Roy, Scotland. Quaternary Science Reviews 114, 61-77.

Berna, F. & Goldberg, P., 2007. Assessing Paleolithic pyrotechnology and associated hominin behavior in Israel. Israel Journal of Earth Science 56, 107-121.

Berna, F., Goldberg, P., Horwitz, L.K., Brink, J., Holt, S., Bamford, M. & Chazan, M., 2012. Microstratigraphic evidence of *in situ* fire in the Acheulean strata of Wonderwerk Cave, Northern Cape province, South Africa. Proceedings of the National Academy of Sciences 109, E1215-E1220.

Berthold, C. & Mentzer, S.M., 2017. X-ray microdiffraction. In Nicosia, C. & Stoops, G. (eds), Archaeological Soil and Sediment Micromorphology. John Wiley & Sons Ltd, Chichester, pp. 417-429.

Borchert, H., 1961. Einfluss der Bodenerosion auf die Bodenstruktur und Methoden zu ihrer Kennzeichnung. Geologisches Jahrbuch 78, 439-502.

Borchert, H., 1964. Mikromorphologische Beobachtungen an meliorierten Böden. In Jongerius, A. (ed.), Soil Micromorphology. Elsevier, Amsterdam, pp. 459-466.

Borchert, H., 1972. Mehrjährige mikroskopische Beobachtungen von Hohlraum-veränderungen meliorierter Böden und deren zugehörige physikalische Kennwerte. In Kowalinski, S. & Drozd, J. (eds.), Soil Micromorphology. Panstwowe Wydawnicto Naukowe, Warsawa, pp. 649-660.

Bouma, J., Jongerius, A. & Schoonderbeek, D., 1979. Calculation of saturated hydraulic conductivity of some pedal clay soils using micromorphometric data. Soil Science Society of America Journal 43, 261-264.

Breuer, J., 1997. Dünnschliffe von Aggregatsäulen aus Perkolationsversuchen. In Stahr, K. (ed.), Mikromorphologische Methoden in der Bodenkunde. Hohenheimer Bodenkundliche Hefte 40, 105-116.

Brewer, R., 1964. Fabric and Mineral Analysis of Soils. John Wiley and Sons, New York, 470 p.

Brewer, R. & Haldane, A.D., 1957. Preliminary experiments in the development of clay orientations in soils. Soil Science 84, 301-309.

Bronger, A., 1969/1970. Zur Mikromorphogenese und zum Tonmineralbestand quartärer Lößböden in Südbaden. Geoderma 3, 281-320.

Bronger, A., 1972. Zur Mikromorphologie und Genese von Paläoböden auf Löss im Karpatenbecken. In Kowalinski, S. & Drozd, J. (eds.), Soil Micromorphology. Panstwowe Wydawnicto Naukowe, Warsawa, pp. 607-616.

Bruneau, P.M.C., Ostle, N., Davidson, D.A., Grieve, I.C. & Fallick, A., 2002. Determination of rhizosphere 13C pulse signals in soil thin sections by Laser Ablation Isotope Ratio Mass Spectromery (LA-IRMS). Rapid Communications in Mass Spectrometry 16, 2190-2194.

Bullock, P. & Murphy, C.P., 1979. Evolution of a paleo-argillic brown earth (paleudalf) from Oxfordshire, England. Geoderma 22, 225-252.

Bullock, P. & Murphy, C.P. (eds.), 1983. Soil Micromorphology. AB Academic Publishers, Berkhamsted, 705 p.

Bullock, P., Fedoroff, N., Jongerius, A., Stoops, G., Tursina, T. & Babel, U., 1985. Handbook for Soil Thin Section Description. Waine Research Publications, Wolverhampton, 152 p.

Buol, S.W., 1994. Saprolite-regolith taxonomy - an approximation. In Whole Regolith Pedology. Soil Science Society of America Special Publication 34, 119-133.

Canti, M.G., 2003. Aspects of the chemical and microscopic characteristics of plant ashes found in archaeological soils. Catena 54, 339-361.

Carpentier, F. & Vandermeulen, B., 2016. High-resolution photography for soil micromorphology slide documentation. Geoarchaeology 31, 603-607.

Cornwall, I.W., 1958. Soils for the Archaeologist. Phoenix House, London, 230 p.

Courty, M.A., Goldberg, P. & Macphail, R.I., 1989. Soils and Micromorphology in Archaeology. Cambridge Manuals in Archaeology. Cambridge University Press, Cambridge, 344 p.

Coutard, J.P. & Van Vliet-Lanoë, B., 1994. Cryoexpulsion et cryoreptation en milieu limono-argileux. Expérimentation en laboratoire. Biuletyn Peryglacjalny 33, 5-20.

Cremaschi, M., Trombino, L. & Zerboni, A., 2018. Palaeosoils and relict soils, a systematic review. In Stoops, G., Marcelino, V. & Mees, F. (eds.), Interpretation of Micromorphological Features of Soils and Regoliths. Second Edition. Elsevier, Amsterdam, pp. 863-894.

Dalrymple, J.B., 1972. Experimental micropedological investigations of iron oxide-clay complexes and their interpretation with respect to the soil fabrics of Paleosols. In Kowalinski, S. & Drozd, J. (eds.), Soil Micromorphology. Panstwowe Wydawnicto Naukowe, Warsawa, pp. 583-594.

Davidson, D.A., Carter, S.P. & Quine, T.A., 1992. An evaluation of micromorphology as an aid to archaeological interpretation. Geoarchaeology 7, 55-65.

Delage, A. & Lagatu, H., 1904. Sur la constitution de la terre arable. Comptes Rendus de l'Académie des Sciences 109, 1043-1044.

Delvigne, J.E., 1998. Atlas of Micromorphology of Mineral Alteration and Weathering. Canadian Mineralogist, Special Publication 3, 495 p.

Denaix, L., Van Oort, F., Pernes, M. & Jongmans, A.G., 1999. Transmission X-Ray diffraction of undisturbed soil microfabrics obtained by microdrilling in thin sections. Clays and Clay Minerals 47, 637-646.

Douglas, L.A. & Thompson, M.L. (eds.), 1985. Soil Micromorphology and Soil Classification. Soil Science Society of America Special Publication 15, SSSA, Madison, 216 p.

Durand, N., Monger, H.C. & Canti, M.G., 2010. Calcium carbonate features. In Stoops, G., Marcelino, V. & Mees, F. (eds.), Interpretation of Micromorphological Features of Soils and Regoliths. Elsevier, Amsterdam, pp. 149-194.

Durand, N., Monger, H.C. & Canti, M.G., 2018. Calcium carbonate features. In Stoops, G., Marcelino, V. & Mees, F. (eds.), Interpretation of Micromorphological Features of Soils and Regoliths. Second Edition. Elsevier, Amsterdam, pp. 205-258.

Elliot, T.R. & Heck, R.J., 2007. A comparison of optical and X-ray CT techniques for void analysis in soil thin sections. Geoderma 141, 60-70.

Embrechts, J. & Stoops, G., 1986. Relations between microscopical features and analytical characteristics of a soil catena in a humid tropical climate. Pedologie 36, 315-328.

Eschenbrenner, V., 1987. Les Glébules des Sols de Côte d'Ivoire. Nature et Origine en Milieu Ferrallitique. Modalités de leur Concentration. Rôle des Termites. PhD Dissertation, Université de Bourgogne, Dijon, 498 + 282 p.

Fedoroff, N., Courty, M.A. & Zhengtang Guo, 2010. Palaeosols and relict soils. In Stoops, G., Marcelino, V. & Mees, F. (eds.), Interpretation of Micromorphological Features of Soils and Regoliths, Elsevier, Amsterdam, pp. 623-662.

Fedoroff, N., Courty, M.A. & Zhengtang Guo, 2018. Palaeosoils and relict soils, a conceptual approach. In Stoops, G., Marcelino, V. & Mees, F. (eds.), Interpretation of Micromorphological Features of Soils and Regoliths. Second Edition. Elsevier, Amsterdam, pp. 821-862.

Ferrari, G.A. & Magaldi, D., 1968. I paleosuoli di Collecchio (Parma) ed il loro significato (Quarternario continentale padono) − Nota 1. L'Ateneo Parmense. Acta Naturalia 4, 57-92.

FitzPatrick, E.A., 1984. Micromorphology of Soils. Chapman and Hall, London, 433 p.

FitzPatrick, E.A., 1993. Soil Microscopy and Micromorphology. John Wiley & Sons, Chichester, 304 p.

FitzPatrick, E.A., 2005. Horizon Identification − Reference Point System. University of Aberdeen. [CD ROM].

Frese, H. & Czeratzki, W., 1957. Strukturbilding im Ackerboden. Die Umschau in Wissenschaft und Technik 16, 495-498.

Gerasimova, M.I., Gubin, S.V. & Shoba, S.A., 1996. Soils of Russia and Adjacent Countries: Geography and Micromorphology. Moscow State University & Wageningen Agricultural University, 204 p.

Goldberg, P. & Macphail, R.I., 2006. Practical and Theoretical Geoarchaeology. Blackwell Publishing, Oxford, 455 p.

Grossman, R.B., 1964. Composite thin sections for estimation of clay film volume. Soil Science Society of America Proceedings 28, 132-133.

Harrison, J.B., 1933. The Katamorphism of Igneous Rocks under Humid Tropical Conditions. Imperial Bureau of Soil Science, Rothamsted Experimental Station, Harpenden, 79 p.

Herrero Isern, J., Rodriguez Ochoa, R. & Porta Casanellas, J., 1989. Colmatacion de Drenes en Suelos Afectados por Salinidad. Institución Fernando el Católico, Zaragoza, 133 p.

Houot, S. & Berthelin J., 1987. Dynamique du fer et de la formation du colmatage ferrique des drains dans des sols hydromorphes à amphigley (Acric Haplaquepts). In Fedoroff, N., Bresson, L.M., & Courty, M.A. (eds), Micromorphologie des Sols, Soil Micromorphology. AFES, Plaisir, pp. 345-352.

IUSS Working Group WRB, 2015. World Reference Base for Soil Resources 2014, Update 2015. World Soil Resources Reports No. 106. FAO, Rome, 192 p.

Jeanson, C., 1964. Micromorphologie et pédozoologie expérimentale: contribution à l'étude sur plaques minces de grandes dimensions d'un sol artificiel structuré par les Lombricides. In Jongerius, A. (ed.), Soil Micromorphology. Elsevier, Amsterdam, pp. 47-55.

Jim, C.Y., 1990. Stress, shear deformation and micromorphological clay orientation: a synthesis of various concepts. Catena 17, 431-447.

Jongerius, A., 1974. Recent developments in soil micromorphometry. In Rutherford, G.K. (ed.), Soil Microscopy. The Limestone Press, Kingston, Ontario, pp. 67-83.

Jongerius, A. & Heintzberger, G., 1962. The Preparation of Mammoth Sized Thin Sections. Soil Survey Papers, Nr. 1. Netherlands Soil Survey Institute, Wageningen, 37 p.

Jongerius, A., Schoonderbeek, D. & Jager, A., 1972. The application of the Quantimet 720 in soil micromorphometry. The Microscope 20, 243-254.

Jongmans, A.G., van Oort, F., Denaix, L. & Jaunet, A.M., 1999. Mineral micro- and nano-variability revealed by combined micromorphology and in situ submicroscopy. Catena 35, 259-279.

Juilleret, J., Dondeyne, S., Vancampenhout, K., Deckers, J. & Hissler, C., 2016. Mind the gap: a classification system for integrating the subsolum into soil surveys. Geoderma, 264, 332-339.

Karkanas, P. & Goldberg, P., 2010. Phosphatic features. In Stoops, G., Marcelino, V. & Mees, F. (eds.), Interpretation of Micromorphological Features of Soils and Regoliths. Elsevier, Amsterdam, pp. 521-541.

Karkanas, P. & Goldberg, P., 2018. Phosphatic features. In Stoops, G., Marcelino, V. & Mees, F. (eds.), Interpretation of Micromorphological Features of Soils and Regoliths. Second Edition. Elsevier, Amsterdam, pp. 323-346.

Kemp, R.A., 1999. Micromorphology of loess-paleosol sequences: a record of palaeoenvironmental change. Catena 35, 179-196.

Kemp, R.A., 2001. Pedogenic modification of loess: significance for palaeoclimatic reconstructions, Earth-Science Reviews 54, 145-156.

Kemp, R.A. & Derbyshire, E., 1998. The loess soils of China as records of climatic change. European Journal of Soil Science 49, 525-539.

Kooistra, M.J. & Kooistra, L.I., 2003. Integrated research in archaeology using soil micromorphology and palynology. Catena 54, 603-618.

Korina, N.A. & Faustova, M.A., 1964. Microfabric of modern and old moraines. In Jongerius, A. (ed.), Soil Micromorphology. Elsevier, Amsterdam, pp. 333-338.

Kovda, I. & Mermut, A., 2010. Vertic features. In Stoops, G., Marcelino, V. & Mees, F. (eds.), Interpretation of Micromorphological Features of Soils and Regoliths. Elsevier, Amsterdam, pp. 109-127.

Kovda, I. & Mermut, A., 2018. Vertic features. In Stoops, G., Marcelino, V. & Mees, F. (eds.), Interpretation of Micromorphological Features of Soils and Regoliths. Second Edition. Elsevier, Amsterdam, pp. 605-632.

Kubiëna, W.L., 1938. Micropedology. Collegiate Press, Ames, 242 p.

Kubiëna, W.L., 1948. Entwicklungslehre des Bodens. Springer-Verlag, Wien, 215 p.

Kubiëna, W.L., 1953. The Soils of Europe. Thomas Murby & Co., London, 317 p.

Kubiëna, W.L., 1956. Zur Mikromorfologie, Systematik und Entwicklung der rezenten und fossilen Lössboden. Eiszeitalter und Gegenwart 7, 102-112.

Kubiëna, W.L. (ed.), 1967. Die Mikromorphometrische Bodenanalyse. Ferdinant Enke Verlag, Stuttgart, 188 p.

Kühn, P., Aguilar, J. & Miedema, R., 2010. Textural features and related horizons. In Stoops, G., Marcelino, V. & Mees, F. (eds.), Interpretation of Micromorphological Features of Soils and Regoliths. Elsevier, Amsterdam, pp. 217-250.

Kühn, P., Aguilar, J., Miedema, R. & Bronnikova, M., 2018. Textural features and related horizons. In Stoops, G., Marcelino, V. & Mees, F. (eds.), Interpretation of Micromorphological Features of Soils and Regoliths. Second Edition. Elsevier, Amsterdam, pp. 377-423.

Lacroix, A., 1913. Les latérites de la Guinée et les produits d'altération qui leur sont associés. Nouvelles Archives du Musée National d'Histoire Naturelle, Paris, 5, 255-356.

Lindbo, D.L., Stolt, M.H. & Vepraskas, M.J., 2010. Redoximorphic features. In Stoops, G., Marcelino, V. & Mees, F. (eds.), Interpretation of Micromorphological Features of Soils and Regoliths. Elsevier, Amsterdam, pp. 129-147.

Macphail, R.I. & Goldberg, P., 2010. Archaeological materials. In Stoops, G., Marcelino, V. & Mees, F. (eds.), Interpretation of Micromorphological Features of Soils and Regoliths. Elsevier, Amsterdam, pp. 589-622.

Macphail, R.I. & Goldberg, P., 2018. Archaeological materials. In Stoops, G., Marcelino, V. & Mees, F. (eds.), Interpretation of Micromorphological Features of Soils and Regoliths. Second Edition. Elsevier, Amsterdam, pp. 779-819.

Marcelino, V., Cnudde, V., Vansteelandt, S. & Carò, F., 2007. An evaluation of 2D-image analysis techniques for measuring soil microporosity. European Journal of Soil Science 58, 133-140.

Marcelino, V., Stoops, G. & Schaefer, C.E.G.R., 2010. Oxic and related materials. In Stoops, G., Marcelino, V. & Mees, F. (eds.), Interpretation of Micromorphological Features of Soils and Regoliths. Elsevier, Amsterdam, pp. 305-327.

Marcelino, V., Schaefer, C.E.G.R. & Stoops, G., 2018. Oxic and related materials. In Stoops, G., Marcelino, V. & Mees, F. (eds.), Interpretation of Micromorphological Features of Soils and Regoliths. Second Edition. Elsevier, Amsterdam, pp. 663-689.

Mathieu, C. & Stoops, G., 1972. Observations pétrographiques sur la paroi d'un four à chaux carolingien creusé en sol limoneux. Archéologie Médiévale 2, 347-354.

Mees, F. & Tursina, T., 2010. Salt minerals in saline soils and salt crusts. In Stoops, G., Marcelino, V. & Mees, F. (eds.), Interpretation of Micromorphological Features of Soils and Regoliths. Elsevier, Amsterdam, pp. 441-469.

Mees, F. & Tursina, T.V., 2018. Salt Minerals in Saline Soils and Salt Crusts. In Stoops, G., Marcelino, V. & Mees, F. (eds.), Interpretation of Micromorphological Features of Soils and Regoliths. Second Edition. Elsevier, Amsterdam, pp. 289-321.

Mentzer, S.M., 2017. Micro XRF. In Nicosia, C. & Stoops, G. (eds), Archaeological Soil and Sediment Micromorphology. John Wiley & Sons Ltd, Chichester, pp. 431-440.

Mentzer, S.M. & Quade, J., 2013. Compositional and isotopic analytical methods in archaeological micromorphology Geoarchaeology 28, 87-97.

Menzies, J., 1998. Microstructures within subglacial diamictons. In Kostrzewski, A. (ed.), Relief and Deposits of Present-day and Pleistocene Glaciation in the Northern Hemisphere – Selected Problems. Adam Mickiewicz University Press, Geography Series 58, 153-166.

Menzies, J., van der Meer, J.J.M., Domack, E. & Wellner, J., 2010. Micromorphology as a tool in the detection, analyses and interpretation of (glacial) sediments and man-made materials. Proceedings of the Geologists' Association 121, 281-292.

Mermut, A.R. & Norton, L.D. (eds.), 1992. Digitization, Processing and Quantitative Interpretation of Image Analysis in Soil Science and Related Areas. Geoderma Special Issue 53, 179-418.

Miedema, R. & Slager, S., 1972. Micromorphological quantification of clay illuviation. Journal of Soil Science 23, 309-314.

Morgenstern, N.R. & Tchalenko, J.S., 1967. Microscopic structures in kaolin subjected to direct shear. Géotechnique 17, 309-328.

Morozova, T.D., 1964. The micromorphological method in paleopedology and paleogeography. In Jongerius, A. (ed.), Soil Micromorphology. Elsevier, Amsterdam, pp. 325-332.

Mücher, H.J. & De Ploey, J., 1977. Experimental and micromorphological investigation of erosion and repositon of loess by water. Earth Surface Processes and Landforms 2, 117-124.

Mücher, H.J. & De Ploey, J., 1984. Formation of afterflow silt loam deposits and structural modification due to drying under warm conditions: an experimental and micromorphological approach. Earth Surface Processes and Landforms 9, 523-531.

Mücher, H.J. & De Ploey, J., 1990. Sedimentary structures formed in eolian-deposited silt loams under simulated conditions on dry, moist and wet surfaces. In Douglas, L.A. (ed.), Soil Micromorphology: A Basic and Applied Science. Developments in Soil Science, Volume 19. Elsevier, Amsterdam. pp. 155-160.

Mücher, H.J. & Morozova, T.D., 1983. The application of soil micromorphology in Quaternary geology and geomorphology. In Bullock, P. & Murphy, C.P. (eds.), Soil Micromorphology. Volume 1. Techniques and Applications. AB Academic Publishers, Berkhamsted, pp. 151-194.

Mücher, H., van Steijn, H. & Kwaad, F., 2010. Colluvial and mass wasting deposits. In Stoops, G., Marcelino, V. & Mees, F. (eds.), Interpretation of Micromorphological Features of Soils and Regoliths. Elsevier, Amsterdam, pp. 37-48.

Mücher, H., van Steijn, H. & Kwaad, F., 2018. Colluvial and mass wasting deposits. In Stoops, G., Marcelino, V. & Mees, F. (eds.), Interpretation of Micromorphological Features of Soils and Regoliths. Second Edition. Elsevier, Amsterdam, pp. 21-36.

Murphy, C.P. & Kemp, R.A., 1984. The over-estimation of clay and the underestimation of pores in soil thin sections. Journal of Soil Science 35, 481-495.

Nahon, D.B., 1991. Introduction to the Petrology of Soils and Chemical Weathering. John Wiley & Sons Ltd., New York, 313 p.

Nicosia, C. & Stoops, G. (eds.), 2017. Archaeological Soil and Sediment Micromorphology. John Wiley & Sons Ltd, Chichester, 476 p.

Pagliai, M., 2008. Crusts, crusting. In Chesworth, W. (ed.), Encyclopedia of Soil Science. Springer, Heidelberg, pp. 171-179.

Pagliai, M. & Stoops, G., 2010. Physical and biological surface crusts and seals. In Stoops, G., Marcelino, V. & Mees, F. (eds.), Interpretation of Micromorphological Features of Soils and Regoliths. Elsevier, Amsterdam, pp. 419-440.

Parfenova, E.I. & Yarilova, E.A., 1965. Mineralogical Investigations in Soil Science. Israel Program for Scientific Translations, Jerusalem, 178 p.

Pede, K. & Langohr, R., 1983. Microscopic study of pseudo-particles in dispersed soil samples. In Bullock, P. & Murphy, C.P. (eds.), Soil Micromorphology. AB Academic Publishers, Berkhamsted, pp. 265-271.

Poch, R.M., Artieda, O., Herrero, J. & Lebedeva-Verba, M., 2010. Gypsic features. In Stoops, G., Marcelino, V. & Mees, F. (eds.), Interpretation of Micromorphological Features of Soils and Regoliths. Elsevier, Amsterdam, pp. 195-216.

Poch, R.M., Artieda, O. & Lebedeva, M., 2018. Gypsic features. In Stoops, G., Marcelino, V. & Mees, F. (eds.), Interpretation of Micromorphological Features of Soils and Regoliths. Second Edition. Elsevier, Amsterdam, pp. 259-287.

Rabenhorst, M.C., 1990. Micromorphology of induced iron sulfide formation in a Chesapeake Bay (USA) tidal marsh. In Douglas, L.A. (ed.), Micromorphology: A Basic and Applied Science. Developments in Soil Science, Volume 19. Elsevier, Amsterdam, pp. 303-310.

Romashkevich, A.I. & Gerasimova, M.I., 1982. Micromorphology and Diagnostics of Pedogenesis [*in Russian*]. Nauka, Moscow, 125 p.

Sedov, S., Stoops, G. & Shoba, S., 2010. Regoliths and soils on volcanic ash. In Stoops, G., Marcelino, V. & Mees, F. (eds.), Interpretation of Micromorphological Features of Soils and Regoliths. Elsevier, Amsterdam, pp. 275-303.

Shahack-Gross, R., 2015. Archaeological micromorphology self-evaluation exercise. Geoarchaeology 31, 49-57.

Shillito, L.M., Almond, M.J., Nicholson, J., Pantos, M. & Matthews, W., 2009. Rapid characterisation of archaeological midden components using FT-IR spectroscopy, SEM-EDX and micro-XRD. Spectrochimica Acta A 73, 133-139.

Smolikova, L., 1958. Zur Mikromorphologie der Bodenkomplexe in der Lössserie bei Svitávka in der Boskowitzer Furche (Mähre). Casopis pro Mineralogii a Geologii 13, 75-84.

Smolikova, L., 1967. Polygenese der fossilen Lössböden der Tschechoslowakei im Lichte mikromorphologischer Untersuchungen. Geoderma 1, 315-324.

Smolikova, L., 1972. Genesis of fossil soil types in the loess series of Czechoslovakia. Acta Universitatis Carolinae, Geographica 2, 45-58.

Smolikova, L., 1978. Bedeutung der Bodenmikromorphologie für die Datierung archäologischer Horizonte. In Delgado, M. (ed.), Soil Micromorphology. University of Granada, pp. 1199-1222.

Soil Survey Staff, 2014. Keys to Soil Taxonomy. 12th Edition. USDA-Natural Resources Conservation Service, Washington, 360 p.

Stewart, J.B., Moran, C.J. & McBratney, A.B., 1994. Measurement of root distribution from sections through undisturbed soil specimens. In Ringrose-Voase, A. & Humphreys, G.S. (eds.), Soil Micromorphology: Studies in Management and Genesis. Developments in Soil Science, Volume 22. Elsevier, Amsterdam, pp. 507-514.

Stoops, G., 2001. Micropedology. Methods and Applications. Provisional Lecture Notes. International Training Centre for Post-Graduate Soil Scientists, Ghent University, 141 p.

Stoops, G., 2003. Guidelines for Analysis and Description of Soil and Regolith Thin Sections. Soil Science Society of America, Madison, 184 p.

Stoops, G., 2009a. Evaluation of Kubiena's contribution to micropedology. at the occasion of the seventieth anniversary of his book 'Micropedology'. Eurasian Soil Science 42, 693-698.

Stoops, G., 2009b. Seventy years 'Micropedology' 1938-2008. The past and future. Journal of Mountain Science 6, 101-106.

Stoops, G., 2014. The 'fabric' of soil micromorphological research in the 20th century – a bibliometric analysis. Geoderma 2013, 193-202.

Stoops, G. & Mees, F., 2018. Groundmass composition and fabric. In Stoops, G., Marcelino, V. & Mees, F. (eds.), Interpretation of Micromorphological Features of Soils and Regoliths. Second Edition. Elsevier, Amsterdam, pp. 73-125.

Stoops, G. & Schaefer, C.E.G.R., 2010. Pedoplasmation: formation of soil material. In Stoops, G., Marcelino, V. & Mees, F. (eds.), Interpretation of Micromorphological Features of Soils and Regoliths, Elsevier, Amsterdam, pp. 69-79.

Stoops, G. & Schaefer, C.E.G.R., 2018. Pedoplasmation: formation of soil material. In Stoops, G., Marcelino, V. & Mees, F. (eds.), Interpretation of Micromorphological Features of Soils and Regoliths. Second Edition. Elsevier, Amsterdam, pp. 59-71.

Stoops, G., Marcelino, V. & Mees, F. (eds.), 2010. Interpretation of Micromorphological Features of Soils and Regoliths, Elsevier, Amsterdam, 720 p.

Stoops, G., Sedov, S. & Shoba, S., 2018. Regoliths and soils on volcanic ash. In Stoops, G., Marcelino, V. & Mees, F. (eds.), Interpretation of Micromorphological Features of Soils and Regoliths. Second Edition. Elsevier, Amsterdam, pp. 721-751.

Sveistrup, T., Marcelino, V. & Stoops, G., 1995. Effects of slurry application on the microstructure of the surface layers of soils from northern Norway. Norwegian Journal of Agricultural Sciences 9, 1-13.

Tardy, Y., 1993. Pétrologie des Latérites et des Sols Tropicaux. Masson, Paris, 459 p.

Terribile F., Wright, R. & FitzPatrick, E.A., 1997. Image analysis in soil micromorphology: from univariate approach to multivarlate solution. In Shoba, S., Gerasimova, M. & Miedema, R. (eds.), Soil Micromorphology: Studies on Soil Diversity, Diagnostics, Dynamics. Moscow, Wageningen, pp. 397-417.

Theocharopoulos, S.P. & Dalrymple, J.B., 1987. Experimental construction of illuviation cutans (channel argillans) with differing morphological and optical properties. In Fedoroff, N., Bresson, L.M. & Courty, M.A. (eds.), Micromorphologie des Sols, Soil Micromorphology. AFES, Plaisir, pp. 245-250.

van der Meer, J.J.M., 1987. Micromorphology of glacial sediments as a tool in distinguishing genetic varieties of till. Geological Survey of Finland Special Paper 3, pp. 77-89.

van der Meer, J.J.M., 1993. Microscopic evidence of subglacial deformation. Quaternary Science Reviews 12, 553-587.

van der Meer, J.J.M., 1996. Micromorphology. In Menzies, J. (ed.), Glacial Environments. Vol. II. Past Glacial Environments, Sediments, Forms and Techniques. Butterworth-Heineman, Oxford, pp. 335-355.

van der Meer, J.J.M. & Menzies, J., 2011. The micromorphology of unconsolitated sediment. Sedimentary Geology, 238, 213-232.

Van Mourik, J.M., 1999. The use of micromorphology in soil pollen analysis. The interpretation of the pollen content of slope deposits in Galicia, Spain. Catena 35, 239-258.

Van Mourik, J.M., 2003. Life cycle of pollen grains in mormoder humus forms of young acid forest soils: a micromorphological approach. Catena 54, 651-664.

van Oort, F., Jongmans, A.G. & Jaunet, A.M., 1994. The progression from optical light microscopy to transmission electron microscopy in the study of soils. Clay Minerals 29, 247-254.

Van Ranst, E., Wilson, M.A. & Righi, D., 2018. Spodic materials. In Stoops, G., Marcelino, V. & Mees, F. (eds.), Interpretation of Micromorphological Features of Soils and Regoliths. Second Edition. Elsevier, Amsterdam, pp. 633-662.

Van Vliet-Lanoë, B., 2010. Frost action. In Stoops, G., Marcelino, V. & Mees, F. (eds.), Interpretation of Micromorphological Features of Soils and Regoliths. Elsevier, Amsterdam, pp. 81-108.

Van Vliet-Lanoë, B. & Fox, C., 2018. Frost action. In Stoops, G., Marcelino, V. & Mees, F. (eds.), Interpretation of Micromorphological Features of Soils and Regoliths. Second Edition. Elsevier, Amsterdam, pp. 575-603.

Veneman, P.L.M., Vepraskas, M.J. & Bouma, J., 1976. The physical significance of soil mottling in a Wisconsin toposequence. Geoderma 15, 103-118.

Vepraskas, M.J., Lindbo, D.L. & Stolt, M.H., 2018. Redoximorphic features. In Stoops, G., Marcelino, V. & Mees, F. (eds.), Interpretation of Micromorphological Features of Soils and Regoliths. Second Edition. Elsevier, Amsterdam, pp. 425-445.

Wells, C.B., 1965. The formation of calcium carbonate concentrations. In Hallsworth, E.G. & Crawford, D. V. (eds.), Experimental Pedology. Butterworths, London, pp. 165-172.

Williams, A.J., Pagliai, M. & Stoops, G., 2018. Physical and biological surface crusts and seals. In Stoops, G., Marcelino, V. & Mees, F. (eds.), Interpretation of Micromorphological Features of Soils and Regoliths. Second Edition. Elsevier, Amsterdam, pp. 539-574.

Wilson, M.A. & Righi, D., 2010. Spodic materials. In Stoops, G., Marcelino, V. & Mees, F. (eds.), Interpretation of Micromorphological Features of Soils and Regoliths. Elsevier, Amsterdam, pp. 251-273.

Zainol, E. & Stoops, G., 1986. Relationship between plasticity and selected physico-chemical and micromorphological properties of some inland soils from Malaysia. Pedologie 36, 263-275.

Zauyah, S., Schaefer, C.E.G.R. & Simas, F.N.B., 2010. Saprolites. In Stoops, G., Marcelino, V. & Mees, F. (eds.), Interpretation of Micromorphological Features of Soils and Regoliths. Elsevier, Amsterdam, pp. 49-68.

Zauyah, S., Schaefer, C.E.G.R. & Simas, F.N.B., 2018. Saprolites. In Stoops, G., Marcelino, V. & Mees, F. (eds.), Interpretation of Micromorphological Features of Soils and Regoliths. Second Edition. Elsevier, Amsterdam, pp. 37-57.

Chapter 2

Colluvial and Mass Wasting Deposits

Herman Mücher[1], Henk van Steijn[2], Frans Kwaad[3]

[1]UNIVERSITY OF AMSTERDAM, AMSTERDAM, THE NETHERLANDS (DECEASED);
[2]UNIVERSITY OF UTRECHT, WIJK BIJ DUURSTEDE, THE NETHERLANDS (RETIRED);
[3]UNIVERSITY OF AMSTERDAM, PURMEREND, THE NETHERLANDS (RETIRED)

CHAPTER OUTLINE

1. Introduction

Slope deposits are sediments found on slopes or at the base or foot of slopes. Several processes are usually involved in the development of a given slope deposit, reflecting varying environmental conditions in time and space. Two main groups of formative processes can be distinguished: (i) slope wash, hill wash or rain wash, including soil erosion on agricultural land, and (ii) mass wasting.

The identification of various types of slope deposits is important in geomorphology, geology and palaeopedology. Slope deposits reflect the lithology but also the pedological history of the source area by the type of rock fragments and transported pedofeatures they contain. During geological prospection, they can provide valuable information on ore mineral sources.

Solifluction deposits and landslides (especially earth slides and earth flows) and debris-flow deposits (with high fine earth content) are related to the group of mass wasting processes. Other mass wasting deposits such as those produced by rock fall, (dry) grain flow and avalanching do not contain enough fine earth to allow a meaningful micromorphological study and will thus not be treated in this chapter.

In the absence of macroscopic indications, transported or displaced material on slopes can still be recognised by a number of micromorphological features. However, different slope processes may generate similar microfacies and it is therefore necessary to use macroscopic, microscopic and granulometric criteria for a reliable diagnosis of the studied sediments (Bertran & Texier, 1999). For instance, inverse graded bedding can occur in sieving crusts (see Valentin & Bresson, 1992) that are produced by intense rainfall on sandy soils along slopes (e.g., Puigdefábregas et al., 1999) and they may also be formed by kinematic sieving under dry conditions (e.g., Makse, 1999). Another example are vesicles, which can form in debris flows related to liquefaction (Bernard et al., 2009), but also in physical soil crusts as a result of air entrapment (e.g., Brown & Dunkerley, 1996; Bertran & Texier, 1999).

Moreover, post-depositional processes can deeply alter original microstructures, and thus severely complicate the analysis of slope deposits.

The micromorphology of slope deposits is a research theme that has received relatively little attention. Already in 1974 a plea was made for the study of the micromorphology of slope deposits (Mücher, 1974), which was repeated in 1983 (Mücher & Morozova, 1983). This did not generate much response at that time or later, as indicated by the review of the literature on the facies and microfacies of the main types of slope deposits by Bertran and Texier (1999). In recent years the interest for this subject remained limited. Among the ca. one hundred papers in the Proceedings of the 14th International Working Meeting on Soil Micromorphology (Poch et al., 2012), only one deals with the micromorphology of slope deposits (Trindade et al., 2012).

Macromorphological characteristics of various types of slope deposits are fairly well documented (e.g., Bertran et al., 1992, 1997; Hétu et al., 1995; van Steijn et al., 1995, 2002; Blikra & Nemec, 1998; Pawelec et al., 2015), whereas investigations on the associated micromorphological aspects are sparse and mainly concern solifluction deposits (e.g., Harris & Ellis, 1980; Harris, 1981; Van Vliet-Lanoë, 1985, 1987, 2010; Van Vliet-Lanoë & Fox, 2018).

Identification of colluvium and other sediments, in combination with knowledge of soil formation, is also important for understanding the geomorphic and palaeoenvironmental history of archaeological sites (Courty et al., 1989). Soil micromorphology is a reliable tool to characterise and distinguish between *in situ* soils, transported materials and materials affected by post-depositional processes, hence providing relevant information for the reconstruction of the palaeoclimatic conditions (Courty et al., 1989; Bertran, 2005; Pustovoytov et al., 2011; see also Macphail & Goldberg, 2018).

The contribution of tillage erosion to the formation of colluvial deposits by water action has often been documented (e.g., Govers et al., 1994; van Oost et al., 2000; van Muijsen et al., 2002) but, as far as we know, no micromorphological criteria are available to discriminate water from tillage erosion deposits. Altogether, micromorphological observations were rarely used in investigations of colluvial deposits (Kwaad & Mücher, 1977, 1979; Bolt et al., 1980).

In the present chapter, the scarce literature on the micromorphology of slope deposits and associated processes is reviewed. Available micromorphological data mostly concerns colluvial deposits on loess (e.g., Mücher, 1974; Mücher & Morozova, 1983). Information on the micromorphology of colluvium originating from materials other than loess will be mostly based on Bertran and Texier (1999). Only one micromorphological study of colluvial wedges along faults appears to exist (Miedema & Jongmans, 2002). A related study presents thin section observations for palaeoearthquake-related reworking of colluvium (Vanneste et al., 2008). Micromorphological data for most mass wasting deposits is also very limited. To date, the review of microfacies features of slope deposits by Bertran and Texier (1999) is the most comprehensive. In addition, Bertran (2004) gives a very extensive account on slope deposits with some micromorphological information for a limited number of depositional types.

Although focusing on cryogenic features, the work of Pawelec et al. (2015) covers a wider range of slope processes, environments and related deposits. These authors used micromorphological observations to identify sedimentary structures that could not be detected macroscopically in fine-grained sediments and in sediments transformed by soil processes. Examples of the use of thin sections for the study of a variety of sedimentary environments are also presented by van der Meer and Menzies (2011).

2. General Features of Translocated Material

The first question that arises when studying soil material on slopes is whether the material is *in situ* or translocated. A number of macroscopic features can be helpful in establishing the presence of translocated material on slopes such as layering, repetitions of the same sequence of soil horizons overlying each other, and stone lines. Another criterion in the case of landslides is an irregular topography (scarps, lobes, local depressions).

In the absence of these features, translocated or displaced material on slopes can still be recognised by the presence of certain micromorphological features lacking in the *in situ* material. Examples include fragments of pedofeatures (clay coatings, crusts), typic ferruginous nodules with sharp boundaries, horizontally oriented root fragments, rounded aggregates composed of material derived from other soil horizons, relicts of palaeosoil intervals, anthropogenic materials, sharply bounded and mostly (sub) rounded rock fragments, fresh rock fragments practically free of weathering rims and/or unknown in the area (Mücher, 1974; Kwaad & Mücher, 1977). However, the presence of only one of the abovementioned features is often not enough to indicate translocation.

For example, fragments of clay coatings incorporated in the groundmass can also arise from biological activity and frost action.

3. Colluvial Deposits

The term colluvium is often used when discussing slope deposits, but no consensus exists on its definition (e.g., Kleber, 2006). It has been used as a synonym for (i) slope deposits, regardless of the mode of formation of the deposits (Hilgard, 1892; Selby, 1985; Soil Survey Staff, 1993; Bertran, 2004; Osterkamp, 2008; Pietsch & Machado, 2014), (ii) slope deposits formed by mass wasting processes, excluding slope wash (Merrill, 1897; Blikra & Nemec, 1998), (iii) slope deposits formed by slope wash (hill wash, rain wash) (Mücher, 1974; Mücher et al., 1981), and (iv) slope deposits formed by slope wash on agricultural land (Leopold & Völkel, 2007).

In this paper, colluvium is used in the sense of slope deposits formed by slope wash. It consists of transported material found on slopes, being particularly prominent at the slope foot. In loess regions it generally occurs on gentle slopes, less than 7 degrees. It commonly rests upon a soil horizon *in situ* (for example the Bt horizon of a truncated Alfisol profile) and shows an abrupt and smooth, sometimes wavy boundary with the horizon below (Mücher, 1974). In the field, two types of colluvium are recognised, laminated and non-laminated or massive. Laminated deposits mostly occur in valley bottoms and at the end of rills in alluvial fans.

Modern colluvial deposits derived from materials other than loess, from various regions, were studied by Bertran and Texier (1999). These deposits are generally moderately sorted, with marked spatial grain-size variation over short distances and randomly or weakly oriented coarse fragments. The deposits are typically composed of laminated bands, alternating with non-laminated intervals.

Compostella et al. (2014) detected several phases of palaeosoil development and deposition of colluvium in the treeline zone in the Northern Apennines, Italy. Multiple colluvial layers containing material of different origins could be identified as part of dismantled soils using micromorphological evidence.

3.1   Laminated Colluvium

Laminated colluvium commonly reflects alternating processes of *in situ* structural soil crust formation and sedimentation of detached soil material, including crust fragments, due to raindrop impact and overland flow (see Pagliai & Stoops, 2010 and Williams et al., 2018).

In loess, laminated colluvium is composed of individual layers predominantly 1-2 mm thick, homogeneous in composition and more or less parallel to the soil surface. Field observations revealed that these deposits are formed during heavy rainstorms when rill erosion occurred on fallow arable land, or on arable land only partly covered by crops like maize in an early stage, which fail to protect the soil from rain splash impact (Mücher, 1974).

In thin sections, the alternating layers (Figs. 1 and 2) are composed of individual laminae with different size distributions of mineral grains and rounded aggregates and with different degrees of sorting (Mücher et al., 1981). The rounded aggregates result from disruption by running water of subsoil horizons, mostly Bt horizons, and are rather stable during transport in water. The layers contain few voids, mainly simple and compound packing voids and occasionally vesicles, which usually show a referred orientation pattern parallel to the soil surface.

Laboratory experiments on loess with the aim of studying the mode of redeposition of silt loam have shown that turbulent rainwash generally leads to the formation of poorly layered colluvial deposits with poorly sorted individual laminae (some containing rounded clods smaller than 3 mm); the ratio between the percentage by volume of laminae composed of grains between 10 and 50 μm and the percentage by volume of those composed of grains smaller than 30 μm is larger than one (Mücher & De Ploey, 1977; Mücher et al., 1981). Under conditions of flow without splash, for instance simulating afterflow that occurs for a short period after rainfall has ceased or meltwater flow, the finer fractions are preferentially deposited and the ratio mentioned above is equal to or smaller than one. These deposits are loosely packed, well layered with individual laminae displaying excellent sorting and mineral differentiation (Mücher et al., 1981; Mücher & Morozova, 1983).

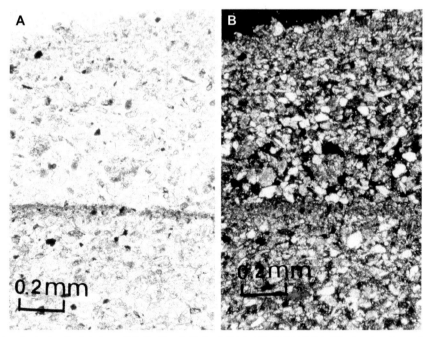

FIGURE 1 Laminated colluvial deposit (afterflow) derived from calcareous loess. The image shows, from top to bottom, a fine silt layer enriched with carbonates, a more porous layer with larger silt-sized mineral grains and few aggregates, a thin clayey layer and a layer with unsorted silt. (A) PPL; (B) XPL. *Images by H. Mücher.*

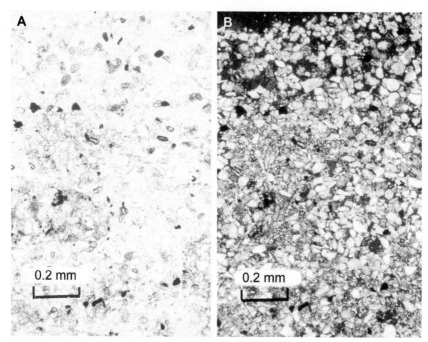

FIGURE 2 Poorly laminated colluvial deposit (rainwash), composed of tightly packed and moderately sorted mineral grains. (A) PPL; (B) XPL. *Images by H. Mücher.*

Laminated parts in colluvial deposits, originating from materials other than loess, appear in thin sections as formed by juxtaposed mineral grains and rounded to sub-angular peds, the relative amount of the latter depending on aggregate stability of the original material (Bertran & Texier, 1999). According to these authors, the laminated and sorted parts in colluvial sediments represent areas of rapid vertical or lateral particle accumulation, which can be extensive (e.g., downslope of large gullies). The c/f-related distribution pattern ranges from monic or enaulic to chitonic-gefuric, the latter corresponding to periods of slow water flow, which resulted in impregnations by fine material. Laminated intervals alternate with massive, poorly sorted deposits with chitonic to single-spaced porphyric c/f-related distribution pattern. These massive zones correspond either to former structural crusts (Valentin & Bresson, 1992), to deposits formed by overland flow in combination with splash (Mücher & De Ploey, 1977; cf. frequent scattered flakes of silt laminae), or to hyperconcentrated flow deposits.

Overland flow giving rise to colluvial deposits derived from the erosion of clayey soils are composed predominantly of clayey aggregates (Alberts et al., 1980). Colluvium can also be almost exclusively formed by mesofauna excrements, if derived from humic soils (Mücher et al., 1972).

3.2   Non-laminated Colluvium

In the field, non-laminated loess-derived colluvial sediments have a silty loam texture, are mainly massive but sometimes have a weakly developed angular blocky structure, with medium and fine pores.

In thin sections, the groundmass is mainly composed of silt-sized mineral grains and small amounts of clay (Fig. 3). Pedofeatures, such as anorthic iron oxide nodules, fragments of clay coatings from the Bt horizon, fragments of chert and other rock types not occurring in the area, coal and brick fragments are also often present (Mücher, 1974).

Laboratory experiments and microscopic observations on loess materials (Mücher & De Ploey, 1977; Mücher et al., 1981) revealed that non-laminated deposits are produced by splash, i.e., raindrop impact leading to displacement of material. In thin sections, these deposits show no lamination, no sorting and practically no mineral differentiation.

Field experiments showed that niveo-eolian deposits are not laminated (Dijkmans & Mücher, 1989).

3.3   Post-depositional Alterations

The main modification of colluvial deposits after deposition is soil formation.

Bolt et al. (1980) recognised four types of illuvial clay coatings micromorphologically: type 1 — (dirty) dark brown, speckled coatings of clay with weak continuous orientation;

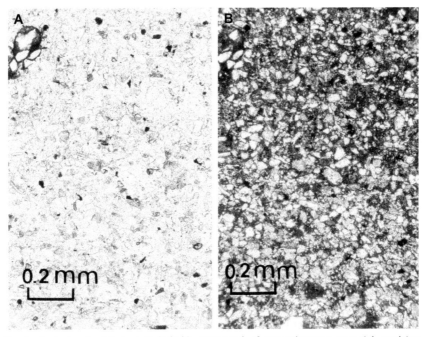

FIGURE 3 Non-laminated colluvial deposit (splash), composed of non-calcareous material overlying calcareous loess. (A) PPL; (B) XPL. *Images by H. Mücher.*

type 2 — (dirty) brown, speckled, impure clay coatings, with weak striated to weak strial b-fabric (matriargillans; Van Schuylenborgh et al., 1970) or coatings consisting of clay, silt and organic matter (agricutans; Jongerius, 1970); type 3 — compound clay coatings consisting of a combination of types 1 and 2; type 4 — yellow to yellowish brown, limpid clay coatings with a strong parallel orientation. These four types of clay coatings probably reflect increasing soil formation with increasing age of the deposit (see also Kühn et al., 2010, 2018). With time, the layered colluvia can be partly destroyed by tillage. However, small crust fragments are still recognisable under the microscope.

Miedema and Jongmans (2002) observed, in a colluvial wedge in Late Weichselian colluvial deposits, large greyish areas in which former coatings of strongly oriented clay show a decreased orientation of the clay and a greyish appearance, which they explained as a result of the so-called ferrolysis (see also Kühn et al., 2010, 2018). Such grey coatings also occurred around the outer edges of large fragments of the Bt horizon in the colluvial infilling of the wedge.

4. Mass Wasting Deposits

Micromorphological information was mainly found for solifluction, certain types of landslides (earth-slide and earth-flow deposits) and debris flows. In the field, mass wasting deposits may show crude or clear stratification, whereas traces of stratification may be completely absent in part of the sediments. Most of these deposits show a diamictic character. Some are accumulations of loosely stacked clasts, often with openwork arrangement.

4.1 Solifluction Deposits

Solifluction, the slow downslope flowing of saturated slope material, is generally related to periglacial environments (cf. French, 1996) but can be found in non-periglacial landscapes as well. It usually results from the combination of two mechanisms, namely frost creep and gelifluction (e.g., Van Vliet-Lanoë et al., 1984; Harris, 1987).

Solifluction deposits generally occur as lobes or sheets with fairly well-defined limits on slopes of a wide variety of steepness classes. Solifluction deposits are characterised by a well-developed preferred clast orientation perpendicular to contours (Brochu, 1978; Nelson, 1985; Bertran et al., 1997; Bertran & Texier, 1999) and by the presence of a platy structure with deformed features like folds or downslope overturned strata (Benedict, 1970; Van Vliet-Lanoë & Valadas, 1983; van Steijn et al., 1995). However, platy structures can develop also in other environments.

Microfabrics related to soli(geli)fluction are generally characterised by a preferred oblique orientation of elongated grains and rotated silt cappings (e.g., Huijzer, 1993).

Micromorphological features characteristic of periglacial solifluction deposits are further discussed by Van Vliet-Lanoë (2010) and Van Vliet-Lanoë & Fox (2018).

4.2   Landslides

Landslides are mass displacements along a basal shear plane when shear forces working along the slope overcome the resistance of the material. In many cases there are several internal slip planes above this basal shear plane. The moving mass may become brittle or ductile, depending on the material properties before and after failure, and this eventually determines to a certain degree the sedimentary properties of the deposit. (Ground)water plays an important role, both by inducing failure of the material and by affecting its behaviour after initial motion and at the end of the movement.

All kinds of combination of slabs and blocks, and coarse and fine fractions may be present in landslide deposits. However, a micromorphological characterisation is only relevant for earth-slide and earth-flow deposits, which have a relatively large mineral fraction <2 mm.

Micromorphological characterisation of subaqueous sliding and subglacial sediment deformation is not in the scope of this chapter (see van der Meer, 1993, 1996; van der Meer et al., 1993, 2003; Hiemstra et al., 2004).

Bertran and Texier (1999) studied the micromorphological characteristics of clay-rich landslides. Earth slide deposits typically show a heterogeneous material characterised by fragments of sedimentary material (sedimentary clasts) incorporated in a dense groundmass (brecciated fabric) with an open porphyric c/f-related distribution pattern and a great variety of deformation features. Areas with angular to subrounded sediment aggregates are juxtaposed to areas with deformed, boudinaged (elongated in one direction and having the shape of a sausage in cross section) and stretched clay lumps. The b-fabric ranges from undifferentiated to crystallitic depending on the composition of the material. Striated b-fabrics corresponding to shearing bands are locally observed. A brecciated fabric is characteristic of earth slides where deformation is weak (see Fig. 4).

The initial stages of deformation are usually marked by the occurrence of features due to rotation of rigid inclusions such as tails around gravels (galaxy or turbate fabrics) (e.g., van der Meer, 1997; Bertran & Texier, 1999). Individual slip planes in clay-rich sediments (see Fig. 5) are characterised in thin sections by a mono- or parallel striated b-fabric (Fabre et al., 2004).

Jaeger et al. (2014) and Jaeger and Menzies (2015) compared micromorphological features of crown, middle part and toe of landslides near Würzburg in Germany and observed systematic differences between those segments, discussed in terms of landslide development processes.

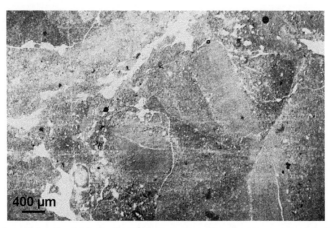

FIGURE 4 Landslide deposit (earth slide) developed from varved silts, showing a brecciated microstructure (PPL). *Reprinted from Bertran and Texier (1999) with permission from Elsevier.*

4.3   Debris-Flow Deposits

Debris flows, i.e., rapid mass movements of poorly sorted granular solids, water and air, leave elongated, narrow tracks of very heterogeneous material. Generally, levees occur along the track, which ultimately join to form a terminal lobe. Lenses of openwork, loose mineral and rock fragments (clasts) may alternate with fine earth-rich material, the latter either in clast-supported or matrix-supported arrangement. This applies to both parts of the debris-flow deposits. Sedimentary properties are discussed by

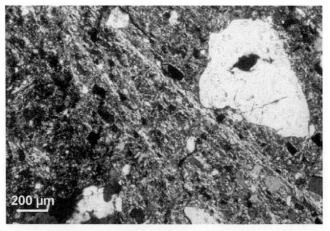

FIGURE 5 Slip plane showing parallel striated b-fabric in a Pleistocene landslide in clayey to sandy terrace deposits (Vallée de l'Isle, Aquitaine, France) (XPL). *From Fabre et al. (2004); reproduced with permission from the authors and the Association Française pour l'Étude du Quaternaire.*

van Steijn et al. (1995) and Bertran et al. (1997). Bertran and Texier (1999) describe some micromorphological features found in groundmass-rich, 'cohesive' debris-flow deposits. In particular, a homogeneous fine-grained groundmass, a porphyric c/f-related distribution pattern and an undifferentiated or crystallitic b-fabric are mentioned to be typical, and the occasional presence of intact to deformed vesicles has also been reported.

Lachniet et al. (2001) compared microstructures of non-glacial sediment-flow deposits and subglacial sediments, describing several comparable elements like folds, pressure shadows and reorientation of clasts around a 'core' stone, due to ductile deformation, or shears, faults and brecciation, caused by brittle deformation. Other features in common are fluid escape and injection structures, clast haloes and fissility. Also, Menzies and Zaniewski (2003) compared the micromorphological characteristics of a fresh debris-flow deposit with the poorly sorted Quaternary glacial deposits from which it derived and concluded they were both very similar. The most important difference between both deposits is the presence of 'tiled' structures produced by the downslope flow deceleration and dewatering resulting in transverse bands of clays developing in rhythmiclike patterns. 'Marbled' structures showing complex intercalations of differing matrix sediments, caused by intense deformation were also not found in the *in situ* Quaternary diamicton. In a beautifully illustrated paper, Phillips (2006) described micromorphological features near the base of a Pleistocene debris flow in central Scotland. The features observed include 'turbate' and 'galaxy' fabrics together with specific b-fabrics (often granostriated), several types of folds and faults, and hydrofractures. The phenomena found are interpreted as showing evidence of basal shearing, hydrofracturing, liquefaction and rotational deformation during the emplacement of the deposits. According to this author, the micromorphological features observed, such as galaxy and turbate structures, are comparable to those found in subglacial diamictons and thus not diagnostic by themselves.

More recently, Pleskot (2015) analyzed micromorphological features of moraine-derived debris-flow lobes on Svalbard to reconstruct flow properties that were not detected macroscopically.

4.4   Grain-Flow Deposits

(Dry) grain-flow deposits (cf. van Steijn et al., 1995) generally occur on steep taluses and consist mostly of loose, very poorly sorted angular aggregates. Near the bottom of the deposit, a fine-grained (fine sand to silt) interval is often found, indicating the slip plane along which the flow moved down. Bertran and Texier (1999) describe micromorphological features of such a slip plane found in Pleistocene talus deposits. The 'sole layers', as the slip-plane deposits are called by the authors, are characterised by 'scattered sand grains within a homogeneous fine matrix (open porphyric c/f-related distribution pattern) with a crystallitic b-fabric (randomly distributed small crystals of calcite)'. The layers involved were slightly indurated by secondary carbonates.

5. Conclusions

Micromorphological studies can play an important role in the recognition of transported materials and the identification of their mode of redeposition. However, still too little is known about the micromorphological characteristics of slope deposits. For mass wasting, micromorphological data are very limited and mainly regards solifluction deposits, earth slides and earth flows. For colluvial deposits, information is mainly available for laminated and non-laminated colluvium derived from loess. Another field where micromorphology could be a relevant tool is the role of tillage erosion, which has been related to the development of colluvial deposits in field experiments. A combination of field and laboratory research, including thin section studies, would contribute to a better understanding and knowledge of slope processes.

Acknowledgements

The authors gratefully acknowledge the following persons for providing publications: Dr. P. Bertran (CNRS, Talence, France), Dr. G. Govers (Katholieke Universiteit Leuven, Belgium), Dr. R.I. Macphail (Institute of Archaeology, London, England), and Dr. B. Van Vliet-Lanoë (Institut Universitaire Européen de la Mer, Plouzané, France).

References

Alberts, E.E., Moldenhauer W.C. & Foster, G.R., 1980. Soil aggregates and primary particles transported in rill and interrill flow. Soil Science Society of America Journal 44, 590-595.

Benedict, J.B., 1970. Downslope soil movement in a Colorado alpine region. Arctic and Alpine Research 2, 165-226.

Bernard, B., van Wyk de Vries, B. & Leyrit, H., 2009. Distinguishing volcanic debris avalanche deposits from their reworked products: the Perrier sequence (French Massif Central). Bulletin of Volcanology 71, 1041-1056.

Bertran, P. (ed.), 2004. Dépôts de Pente Continentaux – Dynamique et Faciès. Quaternaire, International Journal of the French Quaternary Association, Hors Série n° 1, 266 p.

Bertran, P., 2005. Stratigraphie du site de Peyrugues (Lot), une coupe de référence pour le dernier Pléniglaciaire en Aquitaine. Quaternaire 16, 25-44.

Bertran, P. & Texier, J.P., 1999. Facies and microfacies of slope deposits. Catena 35, 99-121.

Bertran, P., Coutard, J.P., Francou, B., Ozouf, J.C. & Texier, J.P., 1992. Nouvelles données sur l'origine du litage des grèzes. Implications paléoclimatiques. Géographie Physique et Quaternaire 46, 97-112.

Bertran, P., Hétu, B., Texier, J.P. & van Steijn, H., 1997. Fabric characteristics of subaerial slope deposits. Sedimentology 44, 1-16.

Blikra, L.H. & Nemec, W., 1998. Postglacial colluvium in western Norway: depositional processes, facies and paleoclimatic record. Sedimentology 45, 909-959.

Bolt, A.J.J., Mücher, H.J., Sevink, J. & Verstraten, J.M., 1980. A study on loess-derived colluvia in southern Limbourg (The Netherlands). Netherlands Journal of Agricultural Science 28, 110-126.

Brochu, M., 1978. Disposition des fragments rocheux dans les dépôts de solifluction, dans les éboulis de gravité et dans les dépôts fluviatiles: mesures dans l'Est de l'Arctique nord-américain et comparaison avec d'autres regions du globe. Biuletyn Peryglacjalny 27, 35-51.

Brown, K.J. & Dunkerley, D.L., 1996. The influence of hillslope gradient, regolith texture, stone size and stone position on the presence of a vesicular layer and related aspects of hillslope hydrologic processes: a case study from the Australian arid zone. Catena 26, 71-84.

Compostella, C., Mariani, G.S. & Trombino, L., 2014. Holocene environmental history at the treeline in the Northern Apennines, Italy: a micromorphological approach. The Holocene 24, 393-404.

Courty, M.A., Goldberg, P. & Macphail, R.I., 1989. Soils and Micromorphology in Archaeology. Cambridge Manuals in Archaeology. Cambridge University Press, Cambridge, 344 p.

Dijkmans, J.W.A. & Mücher, H.J., 1989. Niveo-aeolian sedimentation of loess and sand: an experimental and micromorphological approach. Earth Surface Processes and Landforms 14, 303-315.

Fabre, R., Clément, B. & Bertran, P., 2004. Glissements de terrain. In Bertran, P. (ed.), Dépôts de pente continentaux − dynamique et faciès. Quaternaire, Hors-série Numéro 1-2004, pp. 110-131.

French, H.M., 1996. The Periglacial Environment. 2nd Edition. Longman, Harlow, 341 p.

Govers, G., Vandaele, K., Desmet, P.J.J., Poesen, J. & Bunte, K., 1994. The role of soil tillage on soil redistribution on hillslopes. European Journal of Soil Science 45, 469-478.

Harris, C., 1981. Microstructures in solifluction sediments from South Wales and North Norway. Biuletyn Peryglacjalny 28, 221-226.

Harris, C., 1987. Mechanisms of mass movement in periglacial environments. In Anderson, M.G. & Richards, K.S. (eds.), Slope Stability. John Wiley & Sons, Chichester, pp. 531-559.

Harris, C. & Ellis, S., 1980. Micromorphology of soils in soliflucted materials, Okstindan, Northern Norway. Geoderma 23, 11-29.

Hétu, B., van Steijn, H. & Bertran, P., 1995. Le rôle des coulées de pierres sèches dans la genèse d'un certain type d'éboulis stratifiés. Permafrost and Periglacial Processes 6, 173-194.

Hiemstra, J.F., Zaniewski, K., Powell, R.D. & Cowan, E.A., 2004. Strain signatures of fjord sediment sliding: micro-scale examples from Yakutat Bay and Glacier Bay, Alaska, U.S.A. Journal of Sedimentary Research 74, 760-769.

Hilgard, E.W., 1892. A Report on the Relations of Soil to Climate. U.S. Department of Agriculture, Weather Bureau Bulletin 3, 59 p.

Huijzer, A.S., 1993. Cryogenic Microfabrics and Macrostructures: Interrelations, Processes, and Paleoenvironmental Significance. PhD Dissertation, Vrije Universiteit Amsterdam, 245 p.

Jaeger, D. & Menzies, J., 2015. Micromorphology in unconsolidated landslide sediments − investigating mass movement deposits at a different scale. Geophysical Research Abstracts 17, n° 2916, 2 p.

Jaeger, D., Menzies, J. & Terhorst, B., 2014. Micromorphology in landslide sediments − a different approach for investigating mass movement deposits. Abstracts of the International Conference on Analysis and Management of Changing Risks for Natural Hazards, Padua, Italy, Abstract n° AP17, 10 p.

Jongerius, A., 1970. Some morphological aspects of regrouping phenomena in Dutch soils. Geoderma 4, 311-331.

Kleber, A., 2006. 'Kolluvium' does not equal 'colluvium'. Zeitschrift für Geomorphologie 50, 541-542.

Kühn, P., Aguilar, J. & Miedema, R., 2010. Textural features and related horizons. In Stoops, G., Marcelino, V. & Mees, F. (eds.), Interpretation of Micromorphological Features of Soils and Regoliths. Elsevier, Amsterdam, pp. 217-250.

Kühn, P., Aguilar, J., Miedema, R. & Bronnikova, M., 2018. Textural features and related horizons. In Stoops, G., Marcelino, V. & Mees, F. (eds.), Interpretation of Micromorphological Features of Soils and Regoliths. Second Edition. Elsevier, Amsterdam, pp. 377-423.

Kwaad, F.J.P.M. & Mücher, H.J., 1977. The evolution of soils and slope deposits in the Luxembourg Ardennes near Wiltz. Geoderma 17, 1-37.

Kwaad, F.J.P.M. & Mücher, H.J., 1979. The formation and evolution of colluvium on arable land in northern Luxembourg. Geoderma 22, 173-192.

Lachniet, M.S., Larson, G.J., Lawson, D.E., Evenson, E.B. & Alley, R.B., 2001. Microstructures of sediment flow deposits and subglacial sediments: a comparison. Boreas 30, 254-262.

Leopold, M. & Völkel, J., 2007. Colluvium: definition, differentiation, and possible suitability for reconstructing Holocene climate data. Quaternary International 162/163, 133-140.

Macphail, R. & Goldberg, P., 2018. Archaeological materials. In Stoops, G., Marcelino, V. & Mees, F. (eds.), Interpretation of Micromorphological Features of Soils and Regoliths. Second Edition. Elsevier, Amsterdam, pp. 779-819.

Makse, H.A., 1999. Kinematic segregation of flowing grains in sandpiles. European Physical Journal B 7, 271-276.

Merrill, G.P., 1897. A Treatise on Rocks, Rock-Weathering and Soils. Macmillan Co., New York, 411 p.

Menzies, J. & Zaniewski, K., 2003. Microstructures within a modern debris flow deposit derived from Quaternary glacial diamicton − a comparative micromorphological study. Sedimentary Geology 157, 31-48.

Miedema, R. & Jongmans, T., 2002. Soil formation in Late Glacial Meuse sediments related to the Peel Boundary Fault activity. Geologie en Mijnbouw 81, 71-81.

Mücher, H.J., 1974. Micromorphology of slope deposits: the necessity of a classification. In Rutherford, G.K. (ed.), Soil Microscopy. The Limestone Press, Kingston, pp. 553-566.

Mücher, H.J. & De Ploey, J., 1977. Experimental and micromorphological investigation of erosion and redeposition of loess by water. Earth Surface Processes and Landforms 2, 117-124.

Mücher, H.J. & Morozova, T.D., 1983. The application of soil micromorphology in Quaternary geology and geomorphology. In Bullock, P. & Murphy, C.P. (eds.), Soil Micromorphology, Volume 1, Techniques and Applications. AB Academic Publishers, Berkhamsted, pp. 151-194.

Mücher, H.J., Carballas, T., Guitián Ojea, F., Jungerius, P.D., Kroonenberg, S.B. & Villar, M.C., 1972. Micromorphological analysis of effects of alternating phases of landscape stability and instability on two soil profiles in Galicia, N.W. Spain. Geoderma 8, 241-266.

Mücher, H.J., De Ploey, J. & Savat, J., 1981. Response of loess materials to simulated translocation by water: micromorphological observations. Earth Surface Processes and Landforms 6, 331-336.

Nelson, F.E., 1985. A preliminary investigation of solifluction macrofabrics. Catena 12, 23-33.

Osterkamp, W.R., 2008. Annotated Definitions of Selected Geomorphic Terms and Related Terms of Hydrology, Sedimentology, Soil Science, and Ecology. US Geological Survey Open-File Report 2008-1217, 49 p.

Pagliai, M. & Stoops, G., 2010. Physical and biological surface crusts and seals. In Stoops, G., Marcelino, V. & Mees, F. (eds.), Interpretation of Micromorphological Features of Soils and Regoliths. Elsevier, Amsterdam, pp. 419-440.

Pawelec, H., Drewnik, M., & Zyla, M., 2015. Paleoenvironmental interpretation based on macro- and microstructure analysis of Pleistocene slope covers: a case study from the Miechów Upland, Poland. Geomorphology 232, 145-163.

Phillips, E., 2006. Micromorphology of a debris flow deposit: evidence of basal shearing, hydrofracturing, liquefaction and rotational deformation during emplacement. Quaternary Science Reviews 25, 720-738.

Pietsch, D. & Machado, M.J., 2014. Colluvial deposits – proxies for climate change and cultural chronology. A case study from Tigray, Ethiopia. Zeitschrift für Geomorphologie, Supplementband 58, 119-136.

Pleskot, K., 2015. Sedimentological characteristics of debris flow deposits within ice-cored moraine of Ebbabreen, central Spitsbergen. Polish Polar Research 36, 125-144.

Poch, R.M., Casamitjana, M. & Francis, P.L. (eds.), 2012. Proceedings of the 14th International Working Meeting on Soil Micromorphology. Universitat de LLeida, Lleida, 387 p.

Puigdefábregas, J., Sole, A., Gutierrez, L., del Barrio, G. & Boer, M., 1999. Scales and processes of water and sediment redistribution in drylands: results from the Rambla Honda field site in Southeast Spain. Earth-Science Reviews 48, 39-70.

Pustovoytov, K., Deckers, K. & Goldberg, P., 2011. Genesis, age and archaeological significance of a pedosediment in the depression around Tell Mozan, Syria. Journal of Archaeological Science 38, 913-924.

Selby, M.J., 1985. Earth's Changing Surface. Clarendon Press, Oxford, 607 p.

Soil Survey Staff, 1993. Soil Survey Manual. USDA Handbook Number 18. U.S. Government Printing Office, Washington, 437 p.

Trindade, A., Vieira, G. & Schaefer, C., 2012. Micromorphology analysis of relict slope deposits of Serra da Estrela (Portugal): preliminary results. In Poch, R.M., Casamitjana, M. & Francis, P.L. (eds.), Proceedings of the 14th International Working Meeting on Soil Micromorphology. Universitat de Lleida, Lleida, pp. 266-269.

Valentin, C. & Bresson, L.M., 1992. Morphology, genesis and classification of surface crusts in loamy and sandy soils. Geoderma 55, 225-245.

van der Meer, J.J.M., 1993. Microscopic evidence of subglacial deformation. Quaternary Science Reviews 12, 553-587.

van der Meer, J.J.M., 1996. Micromorphology. In Menzies, J. (ed.), Glacial Environments. Vol. II. Past Glacial Environments: Sediments, Forms and Techniques. Butterworth-Heinemann, Oxford, pp. 335-355.

van der Meer, J.J.M., 1997. Particle and aggregate mobility in till: microscopic evidence of subglacial processes. Quaternary Science Reviews 16, 827-831.

van der Meer, J.J.M. & Menzies, J., 2011. The micromorphology of unconsolidated sediments. Sedimentary Geology 238, 213-232.

van der Meer, J.J.M., Mücher, H.J. & Höfle, H.C., 1993. Micromorphological observations on till samples from the Shackleton Range and North Victoria Land, Antarctica. Polarforschung 62, 57-65.

van der Meer, J.J.M., Menzies, J. & Rose, J., 2003. Subglacial till: the deforming glacier bed. Quaternary Science Reviews 22, 1659-1685.

van Muijsen, W., Govers, G. & van Oost, K., 2002. Identification of important factors in the process of tillage erosion: the case of mouldboard tillage. Soil & Tillage Research 6, 77-93.

Vanneste, K., Mees, F. & Verbeeck, K., 2008. Thin section analysis as a tool to aid identification of palaeoearthquakes on the 'slow', active Geleen Fault, Roer Valley Graben. Technophysics 453, 94-109.

van Oost, K., Govers, G. & Desmet, P., 2000. Evaluating the effects of changes in landscape structure on soil erosion by water and tillage. Landscape Ecology 15, 577.

Van Schuylenborgh, J., Slager, S. & Jongmans, A.G., 1970. On soil genesis in temperate humid climate. VIII. The formation of a 'Udalfic' Eutrochrept. Netherlands Journal of Agricultural Science 18, 207-214.

van Steijn, H., Bertran, P., Francou, B., Hétu B. & Texier, J.P., 1995. Review of models for genetical interpretation of stratified slope deposits. Permafrost and Periglacial Processes 6, 125-146.

van Steijn, H., Boelhouwers, J., Harris, S. & Hétu, B., 2002. Recent research on the nature, origin and climatic relations of blocky and stratified slope deposits. Progress in Physical Geography 26, 551-575.

Van Vliet-Lanoë, B., 1985. Frost effects in soils. In Boardman, J. (ed.), Soils and Quaternary Landscape Evolution. Wiley Publishers, London, pp. 117-158.

Van Vliet-Lanoë, B., 1987. Cryoreptation, gélifluxion et coulées boueuses: une dynamique de solifluxion continue en relation avec le drainage et la stabilité de l'agrégation cryogénique. In Pécsi, M. & French, H.M. (eds.), Loess and Periglacial Phenomena. Akadémiaì Kiadó, Budapest, pp. 203-226.

Van Vliet-Lanoë, B., 2010. Frost action. In Stoops, G., Marcelino, V. & Mees, F. (eds.), Interpretation of Micromorphological Features of Soils and Regoliths. Elsevier, Amsterdam, pp. 81-108.

Van Vliet-Lanoë, B. & Fox, C., 2018. Frost action. In Stoops, G., Marcelino, V. & Mees, F. (eds.), Interpretation of Micromorphological Features of Soils and Regoliths. Second Edition. Elsevier, Amsterdam, pp. 575-603.

Van Vliet-Lanoë, B. & Valadas, B., 1983. A propos des formations déplacées des versants cristallins des massifs anciens. Bulletin de l'Association Française pour l'Etude du Quaternaire 4, 153-160.

Van Vliet-Lanoë, B., Coutard, J.P. & Pissart, A., 1984. Structures caused by repeated freezing and thawing in various loamy sediments. A comparison of active, fossil and experimental data. Earth Surface Processes and Landforms 9, 553-565.

Williams, A.J., Pagliai, M. & Stoops, G., 2018. Physical and biological surface crusts and seals. In Stoops, G., Marcelino, V. & Mees, F. (eds.), Interpretation of Micromorphological Features of Soils and Regoliths. Second Edition. Elsevier, Amsterdam, pp. 539-574.

Chapter 3

Saprolites

Siti Zauyah[1], Carlos E.G.R. Schaefer[2], Felipe N.B. Simas[2]
[1]UNIVERSITY PUTRA MALAYSIA, SERDANG, MALAYSIA (RETIRED);
[2]FEDERAL UNIVERSITY OF VIÇOSA, VIÇOSA, MINAS GERAIS, BRAZIL

CHAPTER OUTLINE

1. Introduction

Saprolite is isovolumetrically weathered bedrock that retains the original lithic fabric (Stolt & Baker, 1994). Most tropical landscapes are underlain by these weathered mantles

Interpretation of Micromorphological Features of Soils and Regoliths. https://doi.org/10.1016/B978-0-444-63522-8.00003-6

of varying thickness. The saprolite usually reaches down to more than 50 m in humid areas on crystalline rocks, where they form an important part of the regolith. Due to long-term weathering under stable tropical conditions, most present-day humid tropical soils developed from deeply weathered saprolites rather than from the parent rock, and their physical and chemical properties are related to the nature of the saprolite. Hence, saprolites should be treated as parent materials of the majority of soils in the tropics and subtropics, especially those in acid igneous and metamorphic rock provinces (Stoops, 1989). In Europe and North America, most of the thick saprolite covers have during the Quaternary glaciations been removed by erosion (e.g., by rivers or glaciers) or covered (e.g., by loess). In these continents, present-day weathering rates are also far slower than the rate of pedoplasmation so that in general no saprolite mantle is formed.

Micromorphological and submicroscopic techniques have been used in the study of saprolites developed on various rock types, both in tropical and subtropical regions (Stoops, 1967; Fölster, 1971; Eswaran & Wong Chaw Bin, 1978; Santos et al., 1989; Zauyah & Stoops, 1990; Scholten et al., 1997; Schaefer et al., 2002) and in temperate and Mediterranean areas (Scarciglia et al., 2005; Jiménez-Espinosa et al., 2007). However, the number of micromorphological studies on saprolites is limited and in many cases they focus much more on mineral weathering than on fabric changes.

The aim of this chapter is to give information on micromorphological aspects of saprolite formation, illustrating especially changes in fabric and in the weathering sequence of minerals. For more detailed information on weathering of individual minerals, the reader is referred to excellent specialised volumes, such as those by Nahon (1991), Tardy (1993) and Delvigne (1998) (see also Stoops & Mees, 2018). The basic micromorphological terminology used here is that of Stoops (2003).

2. The Saprolite Profile

Most weathering profiles include a basal interval of compact altered rock, termed 'saprock' (Zauyah, 1986; Tonui et al., 2003), which is a term adopted from Meunier et al. (1983), and mostly corresponds to the so-called lower saprolite. The upper saprolite is less compact and, toward the surface, it gradually loses its original lithic fabric to form soil material, a process named pedoplasmation (see Stoops & Schaefer, 2010, 2018). Exceptionally, the lithic fabric is preserved by deposition of Al and Fe oxides in profiles where the saprolite is transformed directly to bauxitic or lateritic materials ('isoalterite').

Saprolites are long-term products of chemical weathering, which may be older than 20 Ma in most tropical areas, reaching depths exceeding 150 m in some places (Thomas, 1994). The formation of deep saprolites requires tectonic stability, as the rate of chemical weathering and downward progress of the weathering front must exceed the rate of erosion (Butt et al., 1997). The relief of the area should also allow sufficient leaching of the weathering products to occur. Deep saprolites have been observed not only in tropical regions but also in humid temperate regions, where they are generally considered to be palaeofeatures. When studying saprolites, one should be aware of the considerable three-dimensional variability of their nature, and the difficulties that arise from sampling materials at great depths, commonly exceeding 20 m.

In general, the type of bedrock influences the nature of the saprolite. Saprolites are commonly deep on quartz-rich lithologies such as granite, granitic gneiss, schists and sandstones, but they are shallow to nonexistent on quartz-free calcareous, basaltic and mafic lithologies.

3. General Micromorphological Features

The micromorphological features of saprolite profiles are the result of weathering of the primary minerals, and of processes such as biological activity, illuviation and neo-formation, including the accumulation of iron oxides and neoformed clays in voids. The resulting features are highly variable, both vertically and horizontally, as a result of differences in fabric and mineralogical composition of the parent rock, degree of weathering and hydrological conditions (Scholten et al., 1997), which can be further complicated by hydrothermal alteration that occurred before saprolite development (Evans & Bothner, 1993). Most descriptions of rock weathering and saprolites deal with mineral transformations, especially in rocks composed of large crystals of feldspar, mica, hornblende and quartz (e.g., Gilkes & Suddhiprakarn, 1979; Calvert et al., 1980; Rebertus et al., 1986; Buol & Weed, 1991; Robertson & Eggleton, 1991; Wilson, 2004). Pedofeatures often observed in tropical saprolites include clay coatings and infillings, Fe oxide coatings, hypocoatings and impregnative nodules, as well as excrement infillings. In arid areas, infillings composed of calcite and/or soluble salts (gypsum, thenardite, halite) may develop, which contributes to the physical weathering of rocks and minerals, a process called 'plasmasprengung' (Kerpen et al., 1960).

The fabric of the saprolite is related to the original lithic fabric, which is most obvious in the lower saprolite (Fig. 1). As the primary minerals weather, voids develop and secondary minerals are formed, which can cause the breakup of the lithic fabric in the upper saprolite (see also Stoops & Schaefer, 2010, 2018). Saprolites on gneiss, schists and phyllites may preserve the original schistosity.

Voids are common features in saprolites. They may be inherited from the rock, in which they can be formed by mechanical or thermal stress, partly controlled by the bedding or schistosity. Inter-, intra- and transmineral voids generally have a different genesis. These voids may increase in number and/or in size with increasing degree of weathering and can be filled with weathering products. The voids constitute the primary path for weathering solutions into the rock (Meunier, 1983). Void patterns have been suggested to be responsible for spheroidal weathering patterns (Bisdom, 1967a; Buss et al., 2008).

Mechanical fracturing in the saprolite may affect quartz grains (Curmi & Maurice, 1981; Zauyah, 1986). These fractures may become filled with illuvial clay (Fig. 2), derived

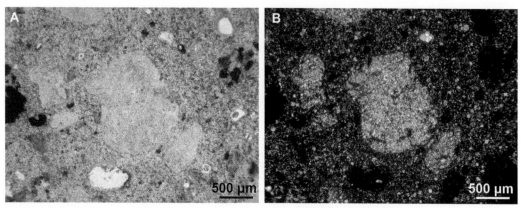

FIGURE 1 Feldspar phenocrysts stained with safranine (pink), in saprolite on porphyric diorite (Quenast, Belgium). The feldspar crystals are sericitised (alteration) and the pink staining indicates, in addition, formation of clay minerals by weathering. (A) PPL. (B) XPL. (Ghent University archive).

FIGURE 2 Selectively weathered perthite (p), with alternating weathered Na-feldspar laths and unaltered K-feldspar laths and biotite grains (b) in early stages of weathering and exfoliation, in a granite saprolite (deep weathering profile, Akagera Park, Rwanda). Illuvial orange limpid clay fills fissures and dissolution voids. (A) PPL. (B) XPL. (Ghent University archive).

from local weathering of biotite or feldspar grains (Chartres & Walker, 1987). Quartz is also affected by chemical weathering, as revealed by a surface morphology characterised by etch pits of varying size and density in SEM images (Eswaran & Stoops, 1979; Marcelino et al., 1999; Schulz & White, 1999).

Weathering products may be different depending on the climate at the time of weathering or exposure of the rock. For instance, weathering of feldspar in tropical saprolites results in *in situ* replacement by 1:1 clay minerals (halloysite and kaolinite) or in dissolution, whereas 2:1 phyllosilicates form in less humid conditions. Biotite is usually weathered to vermiculite and mixed layer clays in dry climates, whereas kaolinite booklets stained with iron oxides are typical alteromorphs under wetter conditions (Bisdom et al., 1982) (Fig. 2). Garnet shows only partial dissolution in dry climates, in contrast to almost complete transformation to iron oxides under wetter conditions (Embrechts & Stoops, 1982; Simas et al., 2005).

Weathering of pyroxenes in a Mediterranean climate (central Italy) shows iron oxides lining cleavage planes and fractures, as well as bleaching of the edges of crystals, as signs of early alteration (Certini et al., 2006). In humid tropical regions, weathering of pyroxenes proceeds from cracks that become coated by goethite, evolving to a boxwork structure by congruent dissolution of the mineral residues (Mulyanto et al., 1999). In drier climates, olivine and pyroxenes can be transformed to alteromorphs composed of expandable 2:1 clays and iron oxides of varying crystallinity (Chesworth et al., 2004), whereas in wetter climates (tropical to subtropical), kaolinite and iron oxides are formed (Delvigne et al., 1979).

Amphiboles (actinolite) in a Mediterranean climate show progressive replacement by a yellow brown trioctahedral vermiculite along fractures and cleavage cracks (Abreu & Vairinho, 1990). In the tropics, however, amphiboles first undergo alteration along cleavage planes which become filled with smectite (Verheye & Stoops, 1975) or iron oxides (Zauyah & Stoops, 1990), forming a boxwork structure by congruent dissolution of the remaining amphibole fragments. SEM images show that dissolution causes pronounced etching along cleavage planes, which commonly exhibit a dented appearance (Anand & Gilkes, 1984).

4. Saprolites on Specific Lithological Rock Types

4.1   Igneous Rocks

4.1.1   Granite

The main components of granite are quartz, alkali feldspar, plagioclase, biotite and muscovite, forming a rock with a granular or porphyric lithic fabric. The fabric of the saprolite has been described as rock-controlled (Stolt et al., 1991; Frazier & Graham, 2000), meaning that the distribution and orientation of mineral grains and void spaces is controlled by the structural and textural properties of the parent rock. In the lower saprolite, the microfabric may show inter-, intra- and transmineral planar voids resulting from mechanical weathering (Espino & Paneque, 1974; Chartres & Walker, 1987; Poetsch, 1990; Scarciglia et al., 2005), such as fractures in quartz and plagioclase grains, which can be filled by amorphous clay or sesquioxides (Fig. 2) (Rutherford, 1987; Evans & Bothner, 1993). In the upper saprolite, the fractures may break the crystals into smaller angular fragments.

A decreasing stability sequence, such as quartz > muscovite > alkali feldspar > biotite > plagioclase, is observed in various climatic regions, with varying types of weathering products (Espino & Paneque, 1974; Verheye & Stoops, 1975; Eswaran & Wong Chaw Bin, 1978; Gilkes et al., 1980; Curmi & Maurice, 1981; Melfi et al., 1983; Poetsch, 1990; Taboada & García, 1999a, 1999b; Frazier & Graham, 2000; Le Pera & Sorriso-Valvo, 2000; Sequeira Braga et al., 2002; Jiménez-Espinosa et al., 2007).

Pseudomorphic or alteromorphic weathering of plagioclases usually begins in the saprock, where they obtain a speckled appearance caused by inclusions of newly formed minerals. Some of these minerals (e.g., epidote, chlorite and sericite) can be the result of hydrothermal alteration rather than weathering. In the tropics, commonly reported weathering products are kaolinite, halloysite and gibbsite (Eswaran & Wong Chaw Bin, 1978; Gilkes et al., 1980). In temperate climates, illite, vermiculite and chlorite are observed (Sequeira Braga et al., 2002), although gibbsite has been reported as well (Bisdom, 1967b; Taboada & García, 1999a). Perthite and plagioclase are commonly sericitised, especially near the grain centres, which is believed to be due to hydrothermal alteration (Evans & Bothner, 1993). In a Mediterranean climate, feldspar is transformed to kaolinite and illite in densely jointed rocks and to interstratified illite/smectite in less fractured rock (Jiménez-Espinosa et al., 2007). Stoops and Dedecker (2006) report the formation of clay minerals oriented perpendicular to the walls of fissures in plagioclase in the saprock. Gibbsite alteromorphs after plagioclase have commonly been reported for saprolites of the subtropics (Felix-Henningsen et al., 1989) and the tropics (Delvigne, 1965; Delvigne & Martin, 1970). The most common weathering product of alkali feldspar is kaolinite, in various climatic conditions (Melfi et al., 1983).

The alteration of biotite is expressed by a loss of pleochroism and a decrease of interference colours (Fig. 2), indicating transformation to interstratified biotite-vermiculite and iron oxides along cleavage planes or crystal edges (Taboada & García, 1999b). Further weathering to kaolinite often results in exfoliation of the biotite grains (Verheye & Stoops, 1975; Curmi & Maurice, 1981; Bisdom et al., 1982). In semiarid regions of Brazil, mica breakdown is associated with smectite formation, whereas in more humid regions, interlayered mica-vermiculite, vermiculite and kaolinite are formed (Melfi et al., 1983). The subsequent pseudomorphic transformation of biotite to chlorite, smectite and ultimately to kaolinite was observed by Stoops and Dedecker (2006).

Muscovite shows a decrease in birefringence during weathering, resulting in grey yellow interference colours toward the edges while the centre remains unchanged. Exfoliation features, although much less common than in biotite, are usually observed in the upper saprolite (Curmi & Maurice, 1981; Rutherford, 1987). Dark brown and black stains are observed along some crystal edges.

Amphiboles in a tropical saprolite first form a fragile boxwork of smectitic clays (Stoops, 2003), and the amphibole remnants gradually disappear by congruent dissolution (Verheye & Stoops, 1975).

Large quartz grains often show fractures that may have infillings of clay, iron oxides or gibbsite, forming a feature termed runiquartz (Espino & Paneque, 1974; Eswaran et al., 1975; Eswaran & Wong Chaw Bin, 1978; Frazier & Graham, 2000). Various types of fracture patterns in weathered granitic rocks were described by Power et al. (1990).

Common pedofeatures observed for granitic saprolites are complete or incomplete infillings of oriented yellowish or reddish brown clay in intra- and transmineral voids (Espino & Paneque, 1974; Chartres & Walker, 1987; Rutherford, 1987; Felix-Henningsen et al., 1989). Ferruginous coatings of fissures are also common (Ismail, 1981). For the upper saprolite, impregnative iron oxide nodules and chalcedony nodules were reported by Stoops and Dedecker (2006).

4.1.2 Diorite

Weathering of dioritic rocks has been studied by several authors (Sousa & Eswaran, 1975; Eswaran et al., 1977; Ismail, 1981; Scholten et al., 1997; Thomas et al., 1999; Buss et al., 2008). In the reduction zone of the lower saprolite, dominated by smectite, green amphiboles show distinct pleochroism. In the middle saprolite, iron oxide coatings cover the mineral remnants and smectite is present as an intermediate phase. In the upper (kaolinitic) saprolite, the amphiboles have been completely dissolved and iron oxide pseudomorphs with a boxwork fabric remain (Scholten et al., 1997).

Biotite grains in a granodioritic saprolite show alteromorphic alteration to kaolinite and goethite, the latter occurring as framboids on crystal surfaces and edges (Sousa & Eswaran, 1975).

Feldspar shows pseudomorphic transformation to kaolinite and gibbsite (Ismail, 1981; Thomas et al., 1999) (see Fig. 1). SEM observations of the surface of altered feldspar from a dolerite saprolite reveal abundant elongated etch pits indicating congruent dissolution (Anand & Gilkes, 1984).

Spheroidal weathering producing corestones surrounded by concentric, partially weathered rinds or shells was studied for a saprolite on quartz diorite under a tropical climate (Buss et al., 2008). Disaggregation at the bedrock weathering zone is initiated by spheroidal macrocracking which is favoured by the oxidation of biotite, which is the earliest documented alteration process. Disintegration of the shells to saprolite is marked by extensive microcracking across multiple crystals and by complete weathering of plagioclase and hornblende. Plagioclase weathering starts, within the shells, with partial dissolution and precipitation of secondary phases in the interior of the crystals. Within the shells-protosaprolite zone, chlorite and plagioclase are highly weathered, resulting in a groundmass with low brightness in backscattered electron images. Quartz and hornblende show a high degree of microcracking but remain unaltered. As weathering progresses toward the protosaprolite and saprolite zones, plagioclase is entirely lost and pore space is filled mostly by gibbsite and, to a lesser extent, by kaolinite. Hornblende crystals become gradually smaller, with increasing dissolution features along cracks and cleavage planes, surrounded by gibbsite and kaolinite aggregates.

4.1.3 Gabbro

Saprolites on gabbro and metagabbro in a temperate climate were studied by Rice et al. (1985). Hornblende in both rocks is transformed to smectite and goethite, whereas feldspar is weathered to kaolinite. Chlorite is transformed to regularly interstratified chlorite-vermiculite, which alters in turn to randomly interstratified chlorite-vermiculite and smectite.

4.1.4 Andesitic Rocks

Andesitic rocks consist of large euhedral and anhedral phenocrysts of plagioclase, alkali feldspar, pyroxenes, biotite and quartz in a fine-grained groundmass of feldspar and quartz or set in an optically isotropic vitric matrix. The saprolite still shows the porphyritic texture of the rock (Fig. 3) (Gu et al., 2007) with some zigzag planar voids (Ibrahim, 1997).

A deep saprolite (20-30 m) on a feldspar-rich trachyandesite in SE Australia shows three distinct zones, corresponding to the lower, middle and upper saprolite (Tonui et al., 2003). The abundance of inter- and transmineral fissures increases from the lower toward the upper saprolite. Ferruginous coatings may line these fissures. In the lower saprolite, plagioclase starts to weather to smectite, kaolinite and sericite, and in the upper saprolite it is completely pseudomorphically replaced by a mixture of kaolinite booklets and mica, as shown by SEM observations. Alkali feldspar persists in the upper saprolite.

In the saprock zone of a dacite profile, plagioclase is weathered to kaolinite along cleavage planes and fractures, which is then completely pseudomorphically replaced by kaolinite in the saprolite (Gu et al., 2007). K-feldspar is also highly altered to illite. Halloysite, kaolinite and gibbsite are common alteration products of feldspar, as observed in SEM images. Biotite grains show alteration to chlorite and vermiculite, and they are no longer recognisable in the overlying saprolite.

In saprolites on trachydacite studied by Certini et al. (2006), pyroxenes are altered to kaolinite along cleavage planes and biotite shows a loss of pleochroism, the space along cleavage planes being filled with weakly birefringent material, assumed to be vermiculite, whereas the feldspar is still fresh. The optically isotropic vitric matrix is transformed to a mixture of small birefringent mineral grains and opaque iron oxides.

In temperate environments, pyroxenes still appear fresh while feldspar is partially altered to vermiculite (Drees et al., 2003). In the tropics, the ferromagnesian minerals are

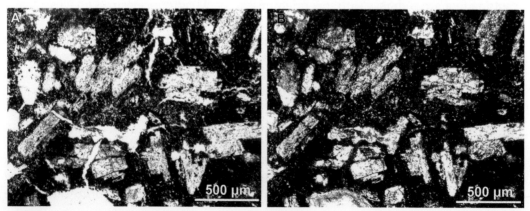

FIGURE 3 Iron oxides impregnating the groundmass and filling fissures in weathered feldspar grains, in an andesite saprolite (Jerantut, Pahang, Malaysia). (A) PPL. (B) XPL.

altered to such an extent that the opaque iron oxides that are formed dominate the groundmass (Ibrahim, 1997).

Mulyanto et al. (1999) and Mulyanto and Stoops (2003) studied spheroidal weathering of andesite boulders in tropical environments. Weathering of plagioclase (both pheno-crysts and small crystals) passes through successive phases from the centre toward the outer shell of the boulders. Amorphous material, sometimes mixed with kaolinite and/or halloysite, appears first, followed by a predominance of gibbsite, which in the outer shell finally disappears, leaving moldic voids. Pyroxene weathering starts along cleavage planes parallel to [001], which become filled by goethite. Due to congruent dissolution of the pyroxene, empty boxworks are formed that gradually become filled with fine-grained goethite, preserved in the outer shell as goethitic pseudomorphs. Sequential mineral formation in vacuoles (Mulyanto & Stoops, 2003) points to a weathering in distinct, separate phases.

In maritime Antarctica, weathering of andesite affected by hydrothermal minerali-sation of pyrite yields an optically isotropic yellowish upper saprolite with a porphyric c/f-related distribution pattern (Schaefer et al., 2008). The groundmass is a mixture of kaolinite and irregular interstratified clay minerals. Oxidation of sulfides results in intense chemical weathering and iron mobilisation with formation of secondary sulfates (jarosite, Na-jarosite) and amorphous iron oxide phases. Most andesitic clasts, with plagioclase as the main phenocryst type, are strongly altered, showing embayments with alteration products and illuvial, amorphous Fe-rich precipitates filling desiccation fractures. Coarse kaolinite fragments indicate intense weathering of plagioclase.

Common pedofeatures are iron oxide nodules and clay coatings covering the sides of voids. The saprolite shows speckled and undifferentiated b-fabrics. Gu et al. (2007) also reported isopachous amorphous silica cement in intergranular voids.

4.1.5 Basalt

Basalt generally contains mafic minerals (olivine, pyroxenes), feldspar (plagioclase) and accessory magnetite as phenocrysts, in a matrix of glass or fine-grained feldspar (Fig. 4).

In a cool temperate environment, Glassmann and Simonson (1985) observed an initial stage of pseudomorphic alteration of plagioclase (labradorite) to beidellite. The dissolution voids may also be filled by authigenic clays, such as beidellite, hydrated halloysite, chlorite and gibbsite (Fig. 5). Augite is completely altered, resulting in the formation of smectite or goethite pseudomorphs after augite crystals.

Horváth et al. (2000) observed the relic texture of basaltic rock in a saprolite. Moldic voids derived from olivine, pyroxene and feldspar phenocrysts are always filled by iron oxides. These infillings are commonly recognised in macropores and in intergranular pores. They have orange colour and are composed of thin goethitic-clayey zones parallel to the surface of the irregular voids. Gibbsite is closely related to the strongly intergrown feldspar crystals of the groundmass, coated by iron oxides. Macrocrystalline gibbsite can be seen along some cracks and in voids. In the intergranular space, trace amounts of clay coatings are present.

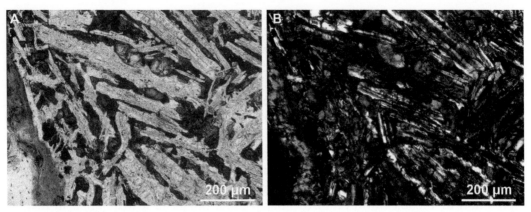

FIGURE 4 Fragmented and partly weathered feldspar laths, weathered pyroxenes releasing iron oxides and clay coating along the sides of a large void (left), in a shallow basalt saprolite (Isla Santa Cruz, Galapagos). (A) PPL. (B) XPL. (Ghent University archive).

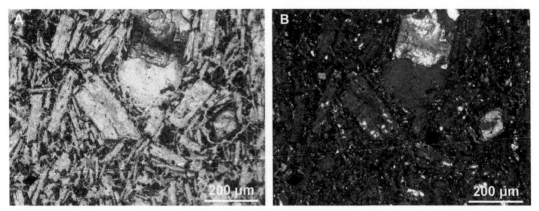

FIGURE 5 Feldspar grains, completely transformed to optically isotropic colloids, in a shallow basalt saprolite (Isla Santa Cruz, Galapagos). (A) PPL. (B) XPL. (Ghent University archive).

In a tropical saprolite studied by Eswaran (1972), the first component to disappear is volcanic glass, combined with an enrichment in iron. Clay coatings were observed in some voids.

Schaefer et al. (2008), studying soils derived from basalts in maritime Antarctica, observed the alteration of plagioclase phenocrysts to a dark brown smectitic groundmass, while pyroxene and magnetite were unaltered.

4.2   Metamorphic Rocks

4.2.1   Gneiss

Several studies describe saprolites developed on gneiss in different climatic conditions (Eswaran & Heng, 1976; Embrechts & Stoops, 1982; Zauyah, 1985; Stolt et al., 1991; Stoops et al., 1994; Bouchard et al., 1995; Kretzschmar et al., 1997; Schwarz, 1997;

Jolicoeur et al., 2000; Le Pera et al., 2001; Schaefer et al., 2002; Ibraimo et al., 2004; Simas et al., 2005; Muggler et al., 2007). Weathering of gneiss produces saprolites that can be very thick (>20 m), with abundant core boulders.

From a micromorphological and mineralogical point of view, two major stages of saprolite development can be recognised. In the first stage, corresponding to the lower saprolite or saprock, the morphological features of the original rock are preserved and physical weathering is the main process, resulting in microfracture development (Zauyah, 1985; Le Pera et al., 2001). Stolt et al. (1991) described the fabric as rock-controlled, based on the parallel orientation of the coarse material and voids in relation to the foliation. The parallel planar voids show iron oxide hypocoatings. The second stage involves further development of microcracks and progressive chemical attack of the minerals along compositional and microstructural discontinuities. Plagioclase shows edge pitting. Clay minerals and ferruginous products formed by alteration replace feldspar, biotite and garnet. The larger grains show weathering features as described by Delvigne (1998), their mineralogical composition differing according to climate or past climate. The groundmass of the saprolite can also contain small diffuse nodules composed of small iron oxide aggregates (Eswaran & Heng, 1976).

Biotite in saprolites developed on gneiss can show evidence of chloritisation (Zauyah, 1985). The usual parallel linear and pellicular alteration (Stolt et al., 1991) produces alteromorphs composed of hydroxyl-interlayered vermiculite, kaolinite and interstratified clays. Kaolinite and halloysite alteromorphs form in humid temperate climates (Kretzschmar et al., 1997), with or without mica-vermiculite intermediate phase (Schwarz, 1997; Jolicoeur et al., 2000). An arrangement of halloysite parallel to the cleavage planes of the biotite crystals was described by Eswaran and Heng (1976). The edges of biotite crystals are covered with tubular halloysite (Kretzschmar et al., 1997) and/or goethite aggregates (Zauyah, 1985; Schwarz, 1997; Le Pera et al., 2001).

In a humid temperate climate, feldspar may be partially altered to kaolinite or gibbsite (Stolt et al., 1991). In the humid tropics, it is completely altered to gibbsite (Simas et al., 2005) or it has an opaque, brownish appearance, due to replacement by a microcrystalline secondary product, probably a clay mineral.

Garnet weathering in gneiss starts with the formation of fissures, which are later filled by iron oxides released by weathering of biotite (Embrechts & Stoops, 1982), finally resulting in a boxwork fabric by congruent dissolution of the remaining garnet residues. Phyllosilicates occurring in cracks within garnet grains have also been reported (Bouchard et al., 1995). Complete replacement by macrocrystalline kaolinite booklets was observed in tropical conditions (Muggler et al., 2007).

Other features of saprolites on gneiss include the coexistence of gibbsite and primary minerals (Schwarz, 1997; Schaefer et al., 2002; Ibraimo et al., 2004; Simas et al., 2005), a predominance of gibbsitic lithorelict structures as alteromorphs, preserving the original cleavage patterns of feldspars (Delvigne, 1998) and the presence of ferruginous zones along pores, related to intense desilification of mafic minerals (Lacerda et al., 2000; Schaefer, 2001; Simas et al., 2005).

4.2.2   Schists

Saprolites on micaschist are deeper in the tropics (about 8 m; Zauyah, 1986) than in humid temperate conditions (3 m; Kretzschmar et al., 1997). The lower saprolite in the tropics is characterised by a parallel alignment of muscovite and biotite crystals in a matrix of interlocking quartz crystals and straight accommodated voids, parallel to the schistosity. Quartz, microcline and albite also occur as phenoblasts. The transmineral voids become wider toward the upper saprolite. The general weathering trend in schists is that biotite weathers first, followed by feldspar, while muscovite and quartz remain unchanged (Zauyah, 1986).

In both humid temperate and tropical environments, biotite grains exhibit expanded edges or exfoliation and are transformed to kaolinite pseudomorphs. Released iron accumulates as opaque and cryptocrystalline oxides on the edges of the altered biotite or along expanded cleavage planes. In temperate environments, only minor amounts of biotite-vermiculite mixed layers are observed in the clay and silt fractions (Kretzschmar et al., 1997). In the upper saprolite, pedoplasmation of these pseudomorphs is observed (Zauyah, 1987). The alteration pattern of muscovite is similar to that of biotite, i.e., pellicular and linear (Zauyah, 1987).

In the lower saprolite, two types of feldspar weathering (Zauyah, 1986) are observed (i) *in situ* replacement by gibbsite, and (ii) dissolution creating vacuoles within the grains and embayments along their boundaries. Moreover, the boundaries begin to lose their sharpness by weathering to tubular halloysite as shown by SEM studies.

Quartz phenoblasts ranging in size from 500 to 260 μm as well as polycrystalline quartz in the groundmass appear unaltered in the lower saprolite. In the upper saprolite, cracks appear in the larger quartz grains and the polycrystalline aggregates begin to separate (Fig. 6). The cracks are infused by clay or brown iron oxides.

SEM images of almandine grains in a saprolite on micaschist show etch pits, occasionally covered by goethite and hematite coatings (Graham et al., 1989).

Pedofeatures in saprolite on micaschist include iron segregation features parallel to the schistosity, clay coatings in voids and thin ferruginous coatings in transmineral fissures in the upper saprolite (Zauyah, 1986).

4.2.3   Amphibolites

The saprolite on an amphibole schist in the tropics comprises many core stones in a looser groundmass (Zauyah & Stoops, 1990). The slightly weathered centre of the cores is composed of alternating bands of epidote-clinozoisite and actinolite. The pale yellow weathering rims in the lower saprolite consist of booklets of vermiform kaolinite in a micromass of fine-grained kaolinite and some tubular halloysite (SEM). Parallel-oriented opaque particles mark the original schistosity. In the upper saprolite, the dark brown weathering rim consists of iron oxides forming schistosity-related networks, filled with fine-grained gibbsite and/or kaolinite stained by iron oxides. Based on morphological criteria, gibbsite is believed to be the alteration product of epidote and clinozoisite. Actinolite weathers faster than those two minerals. The brownish soft saprolite between

FIGURE 6 Fragmented polycrystalline quartz grain, in the upper part of a quartz micaschist saprolite (Ulu Kelang, Selangor, Malaysia) (XPL).

the core stones still shows a distinct relic schistosity due to the parallel alignments recognised for the goethite framework (SEM), filled with kaolinite booklets and more fine-grained kaolinite, gibbsite being rare. Evidence of faunal activity includes the presence of infillings with excrements and a crescent-like internal fabric, in planar voids that are parallel to the schistosity planes.

In cold temperate climate, a very thin saprolite layer on amphibolite (2 cm) (Protz et al., 1974) shows a lithic fabric with red and golden coatings along the sides of weathered biotite and other mineral grains and within voids. Vermiculite is the dominant clay mineral.

Other pedofeatures observed in saprolites on amphibolite include iron oxide coatings on transmineral voids.

4.2.4 Serpentinite

Only few studies discuss the micromorphology of saprolites developed on serpentinite (García et al., 1974; Zauyah, 1986; Lambiv Dzemua et al., 2011). The lower saprolite of a serpentinite profile in a Mediterranean climate shows abundant altered pyroxenes, serpentine, chlorite, plagioclase and chalcedony (García et al., 1974). Iron oxides are deposited between the serpentine grains. In the upper saprolite, serpentine is more weathered, and iron oxides accumulate in fissures. A few manganese oxide nodules are also observed.

In the lower saprolite of a profile on serpentinite, composed of antigorite, chrysotile, magnetite and chlorite, in a tropical area, the antigorite crystals remain unchanged (Zauyah, 1986). Magnetite grains show mainly pellicular alteration to iron oxides, and chlorite is partially altered to kaolinite. The lithic fabric is still preserved. Dense, incomplete, ferruginous infillings in planar voids associated with chlorite and chrysotile veins are common.

Replacement of serpentine by kaolinite, with preservation of the lithic fabric, was reported by Lambiv Dzemua et al. (2011). Iron oxide masses replace the dissolved magnesite grains (Fig. 7). Iron oxide hypocoatings and coatings are observed along planar voids and gradually form the only remaining material, as kaolinite disappears in the upper saprolite. Coatings of limpid authigenic clay occur in some voids in the lower saprolite.

4.2.5   Phyllites and Slates

Phyllites and slates are fine-grained metamorphic rocks that may produce very thick saprolite zones (Zauyah, 1986; Felix-Henningsen, 1994). In a tropical environment, the saprolite is composed of fine-grained quartz and sericite aligned according to the original schistosity (Zauyah, 1986; Stoops et al., 1990) (Fig. 8). A few muscovite, biotite and graphite grains and iron segregations may occur in the groundmass. Most planar voids are parallel to the foliation and are coated by iron oxides. Some channels and vughs can be present. Weathering of the minerals shows the normal stability trend, i.e., quartz > muscovite > biotite. Biotite loses its pleochroism and alters first to a mica-vermiculite interstratified clay mineral. Sericite weathers to kaolinite with a mosaic-like extinction pattern. Granular hematite becomes more common toward higher parts of the saprolite (Muggler et al., 2001).

Pedofeatures observed in the saprolite include clay coating in voids (Zauyah, 1986).

4.3   Sedimentary Rocks

4.3.1   Siliciclastic Rocks

Saprolites on shales, siltstones and slates in a tropical climate with savannah vegetation (Muggler et al., 2007) consist predominantly of quartz and fine-grained illite-type mica, with some accessory minerals such as ilmenite. The mica is transformed to hydroxy-Al interlayered vermiculite and kaolinite. Polycrystalline quartz grains are often fractured

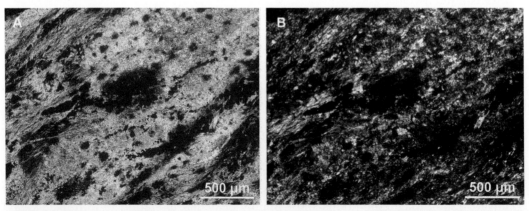

FIGURE 7 Kaolinite mass replacing serpentine and iron oxides precipitating in voids created by the dissolution of magnesite crystals (centre) and along planar voids, in a serpentinite saprolite (bottom of laterite profile, SE Cameroon). (A) PPL. (B) XPL. (Ghent University archive).

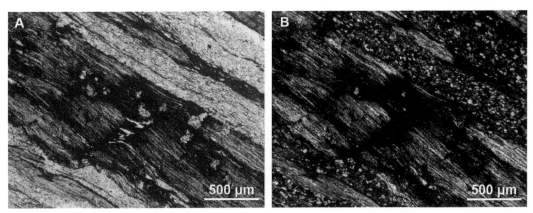

FIGURE 8 Mica layers disrupted and stained by iron oxides, in a sericite phyllite saprolite (bottom of laterite profile, Malakka, Malaysia). (A) PPL. (B) XPL. (Ghent University archive).

and separated into individual grains. Infillings of the fractures with red or dotted clay gives rise to runiquartz (Eswaran et al., 1975).

In a subtropical to tropical region, Islam et al. (2002) studied weathered samples of sedimentary rocks (sandstone, siltstone, shale and claystone), containing quartz, plagioclase, alkali feldspar, biotite, muscovite and lesser carbonates and epidote as the main primary minerals. Quartz starts to weather at the boundaries and along fracture planes, plagioclase shows alteration along cleavage and fracture planes, while the less weatherable alkali feldspar and muscovite began to alter at a later stage along boundaries and cleavage planes.

In saprolites on mudstone and shale in a tropical area, the related distribution pattern is open porphyric and the b-fabric is stipple-speckled (Sugandi, 2005). Pedofeatures observed in these saprolites are large iron oxide nodules with irregular shapes (mottles) and clay coatings along voids.

4.3.2 Calcareous Rocks

As a rule, saprolites on calcareous rocks (limestone, dolomite) show complete destruction of the original rock structure due to dissolution and subsequent collapse (Stoops et al., 1979; Stoops & Schaefer, 2010, 2018). Impure limestones such as marls may show complicated patterns.

Micromorphological observations of a limestone saprolite in a humid temperate climate (Lee et al., 1990; McKay et al., 2005) showed that the material was composed of thin (0.1-1 cm) undulating layers of limestone and shale/siltstone. Iron oxide coatings occurred along closely spaced planar voids parallel and perpendicular to the bedding, as well as in vughs and channels. Clay infillings were observed in vughs and channels. In the intervals derived from limestone layers, microcrystalline manganese oxide nodules were common.

5. Conclusions

Although there is a wealth of information on mineralogical and physicochemical aspects of weathering, relatively few papers deal with of fabric changes. The fabric of saprolites is by definition closely related to the rock fabric but also to the degree and type of weathering. No systematic study exists on differences in fabric developed under different hydrological conditions (below the water table, in the zone of groundwater fluctuations and in the aerated zone), although this might be of great importance.

The presence and development of inter- and transmineral voids plays an important role in the evolution of weathering, but this is not yet well understood (e.g., in the case of spheroidal weathering) and needs to be investigated more in detail. Micromorphometric studies of the porosity of saprolites are practically inexistent. They could contribute to a better understanding of water movement and storage in saprolites (Trindade et al., 2005).

A better understanding of the fabric of saprolites may be of importance in predicting risks of earth slides. In the tropics, saprolites may be quite rapidly transformed to a lateritic or bauxitic material, and therefore not recognised as such. Information on the fabric of saprolites in temperate regions is rather restricted.

References

Abreu, M.M. & Vairinho, M., 1990. Amphibole alteration to vermiculite in a weathering profile of gabbro-diorite. In Douglas, L.A. (ed.), Soil Micromorphology: A Basic and Applied Science. Developments in Soil Science, Volume 19. Elsevier, Amsterdam, pp. 493-500.

Anand, R.R. & Gilkes, R.J., 1984. Weathering of hornblende, plagioclase and chlorite in meta-dolerite, Australia. Geoderma 34, 261-280.

Bisdom, E.B.A., 1967a. The role of micro-crack systems in the spheroidal weathering of an intrusive granite in Galicia (NW Spain). Geologie en Mijnbouw 46, 333-340.

Bisdom, E.B.A., 1967b. Micromorphology of a weathered granite near the Ria de Arosa (NW Spain). Leidse Geologische Mededelingen 37, 33-67.

Bisdom, E.B.A., Stoops, G., Delvigne, J., Curmi, P. & Altemüller, H.J., 1982. Micromorphology of weathering biotite and its secondary products. Pedologie 32, 225-252.

Bouchard, M., Jolicoeur, S. & Pierre, P., 1995. Characteristics and significance of two pre-late-Wisconsinan weathering profiles (Adirondacks, USA and Miramichi Highlands, Canada). Geomorphology 12, 75-89.

Buol, S.W. & Weed, S.B., 1991. Saprolite-soil transformations in the Piedmont and Mountains of North Carolina. Geoderma 51, 15-28.

Buss, H.L., Sak, P.B., Webb, S.M. & Brantlet, S.L., 2008. Weathering of the Rio Blanco quartz diorite Luquillo Mountains, Puerto Rico: coupling, oxidation, dissolution and cracking. Geochimica et Cosmochimica Acta 72, 4488-4507.

Butt, C.R.M., Lintern, M.J. & Anand, R.R., 1997. Evolution of regoliths and landscapes in deeply weathered terrain-implications for geochemical exploration. In Gubins, A.G. (ed.), Proceedings of Exploration 97: Fourth Decennial International Conference on Mineral Exploration. GEO F/X Division of AG information Systems Ltd, Toronto, pp. 323-334.

Calvert, C.S., Buol, S.W. & Weed, S.B., 1980. Mineralogical characteristics and transformations of a vertical rock-saprolite-soil sequence in the North Carolina Piedmont: II. Feldspar alteration products – their transformations through the profile. Soil Science Society of America Journal 44, 1104-1112.

Certini, G., Wilson, M.J., Hillier, S.J., Fraser, A.R. & Delbos, E., 2006. Mineral weathering in trachydacitic-derived soils and saprolites involving formation of embryonic halloysite and gibbsite at Mt. Amiata, Central Italy. Geoderma 133, 173-190.

Chartres, C.J. & Walker, P.H., 1987. The development of a soil with textural contrast on granite in south eastern Australia. In Fedoroff, N., Bresson, L.M. & Courty, M.A. (eds.), Micromorphologie des Sols, Soil Micromorphology. AFES, Plaisir, pp. 125-130.

Chesworth, W., Dejou, J., Larroque, P. & Garcia Rodeja, E., 2004. Alteration of olivine in a basalt from central France. Catena 56, 21-30.

Curmi, P. & Maurice, F., 1981. Microscopic characterization of the weathering in a granitic saprolite. In Bisdom, E.B.A. (ed.), Submicroscopy of Soils and Weathered Rocks, Proceedings of the First Workshop of the International Working Group on Submicroscopy of Undisturbed Soil Materials. Pudoc, Wageningen, pp. 249-270.

Delvigne, J., 1965. Pedogenèse en Zone Tropicale. La Formation des Minéraux Secondaires en Milieu Ferralitique. Mémoires ORSTOM 13, 177 p.

Delvigne, J.E., 1998. Atlas of Micromorphology of Mineral Alteration and Weathering. Canadian Mineralogist, Special Publication 3, 495 p.

Delvigne, J.E. & Martin, H., 1970. Analyse à la microsonde électronique de l'altération d'un plagioclase en kaolinite par l'intermédiaire d'une phase amorphe. Cahiers ORSTOM, Série Géologie 2, 259-295.

Delvigne, J., Bisdom, E.B.A., Sleeman, J. & Stoops, G., 1979. Olivines, their pseudomorphs and secondary products. Pedologie 29, 247-309.

Drees, L.R., Wilding, L.P, Owens, P.R., Wu, B., Perotto, H. & Sierra, H., 2003. Steepland resources: characteristics, stability and micromorphology. Catena 54, 619-636.

Embrechts, J. & Stoops, G., 1982. Microscopical aspects of garnet weathering in a humid tropical environment. Journal of Soil Science 33, 535-545.

Espino, C. & Paneque, G., 1974. Weathering of granitic rocks and pedogenesis in the Sierra de Francia (Salamanca, Spain). In Rutherford, G.K. (ed.), Soil Microscopy. The Limestone Press, Kingston, pp. 366-382.

Eswaran, H., 1972. Micromorphological indicators of pedogenesis in some tropical soils derived from basalts from Nicaragua. Geoderma 7, 15-31.

Eswaran, H. & Heng, Y.Y., 1976. The weathering of biotite in a profile on gneiss in Malaysia. Geoderma 16, 9-20.

Eswaran, H. & Stoops, G., 1979. Surface textures of quartz in tropical soils. Soil Science Society of America Journal 43, 420-424.

Eswaran, H. & Wong Chaw Bin, 1978. A study of a deep weathering profile on granite in Peninsular Malaysia. I. Physico-chemical and micromorphological properties. Soil Science Society of America Journal 42, 144-149.

Eswaran, H., Sys, C. & Sousa, E.C., 1975. Plasma infusion. A pedological process of significance in the humid tropics. Anales de Edafología y Agrobiología 34, 665-674.

Eswaran, H., Stoops, G. & Sys, C., 1977. The micromorphology of gibbsite forms in soils. Journal of Soil Science 28, 136-143.

Evans, C.V. & Bothner, W.A., 1993. Genesis of altered Conway Granite (grus) in New Hampshire, USA. Geoderma 58, 201-218.

Felix-Henningsen, P., 1994. Mesozoic-Tertiary weathering and soil formation on slates of the Rhenish Massif, Germany. Catena 21, 229-242.

Felix-Henningsen, P., Zakosek, H. & Liu, L.W., 1989. Distribution and genesis of red and yellow soils in the central subtropics of southeast China. Catena 16, 73-89.

Fölster, H., 1971. Ferrallitische Böden aus Sauren Metamorphen Gesteine in den Feuchten und Wechselfeuchten Tropen Afrikas. Göttinger Bodenkundliche Berichte 20, 231 p.

Frazier, C.S. & Graham, R.C., 2000. Pedogenic transformation of fractured granitic bedrock, Southern California. Soil Science Society of America Journal 64, 2057-2069.

García, A., Aguilar, J. & Delgado, M., 1974. Micromorphological study of soils developed on serpentine rock from Sierra de Carratraca (Malaga, Spain). In Rutherford, G.K. (ed.), Soil Microscopy. The Limestone Press, Kingston, pp. 394-407.

Gilkes, R.J. & Suddhiprakarn, A., 1979. Biotite alteration in deeply weathered granite. I. Morphological, mineralogical and chemical properties. Clays and Clay Minerals 27, 349-360.

Gilkes, R.J., Suddhiprakarn, A. & Armitage, T.M., 1980. Scanning electron microscope morphology of deeply weathered granite. Clays and Clay Minerals 28, 29-34.

Glassman, J.R. & Simonson, G.H., 1985. Alteration of basalt in soils of Western Oregon. Soil Science Society of America Journal 49, 262-273.

Graham, R.C., Weed, S.B., Bowen, L.H. & Buol, S.B., 1989. Weathering of iron-bearing minerals in soils and saprolite on the North Carolina Blue Ridge Front: I. Sand-size primary minerals. Clays and Clay Minerals 37, 19-28.

Gu, S., Wan, G. & Mao, J., 2007. Sodic metasomatism in a dacite weathering profile in Pinxiang, Guangxi, China. Chinese Journal of Geochemistry 26, 434-438.

Horváth, Z., Varga, B. & Mindszenty, A., 2000. Micromorphological and chemical complexities of a lateritic profile from basalt (Jos Plateau, Central Nigeria). Chemical Geology 170, 81-93.

Ibrahim, S., 1997. Pedological Study and Classification of Soils Developed on Volcanic Rocks in Jerantut District, Pahang. MSc Dissertation, University Putra Malaysia, 204 p.

Ibraimo, M.M., Schaefer, C.E.G.R., Ker, J.C., Lani, J.L., Rolim-Neto, F.C., Albuquerque, M.A. & Miranda, V. J., 2004. Gênese e micromorfologia de solos sob vegetação xeromórfica (caatinga) na região dos Lagos (RJ). Revista Brasileira de Ciência do Solo 28, 695-712.

Islam, M.R., Stuart, R., Risto, A. & Vesa, P., 2002. Mineralogical changes during intense chemical weathering of sedimentary rocks in Bangladesh. Journal of Asian Earth Sciences 20, 889-901.

Ismail, A.B., 1981. Profile Development and Mineralogical Changes of Some Soils Derived from Plutonic Rocks in Peninsular Malaysia. MSc Dissertation, Ghent University, 141 p.

Jiménez-Espinosa, R., Vázquez, M. & Jiménez-Millán, J., 2007. Differential weathering of granitic stocks and landscape effects in a Mediterranean climate, Southern Iberian Massif (Spain). Catena 70, 243-252.

Jolicoeur, S., Ildefonse, P. & Bouchard, M., 2000. Kaolinite and gibbsite weathering within saprolites and soils of Central Virginia. Soil Science Society of America Journal 64, 1118-1129.

Kerpen, W., Gewehr, H. & Scharpenseel, H.W., 1960. Zur Kentniss der ariden Irrigationsböden des Sudan. II Teil. Mikroskopische Untersuchungen. Pedologie 10, 303-323.

Kretzschmar, R., Robarge, W.P, Amoozegar, A. & Vepraskas, M.J., 1997. Biotite alteration to halloysite and kaolinite in soil-saprolite profiles developed from mica schist and granite gneiss. Geoderma 75, 155-170.

Lacerda, M.P.C., Andrade, H. & Quéméneur, J.J.G., 2000. Micropedologia da alteração em perfis de solos com horizonte B textural na região de Lavras, MG. Revista Brasileira de Ciência do Solo 4, 829-841.

Lambiv Dzemua, L.G., Mees, F., Stoops, G. & Van Ranst, E., 2011. Micromorphology, mineralogy and geochemistry of lateritic weathering over serpentinite in south-east Cameroon. Journal of African Earth Sciences 60, 38-48.

Lee, S.Y., Phillips, D.H., Ammons, J.T. & Lietzke, D.A., 1990. A microscopic study of iron and manganese oxide distribution in soils from East Tennessee. In Douglas, L.A. (ed.), Soil Micromorphology: A Basic and Applied Science. Developments in Soil Science, Volume 19. Elsevier, Amsterdam, pp. 511-517.

Le Pera, E. & Sorriso-Valvo, M., 2000. Weathering and morphogenesis in a Mediterranean climate, Calabria, Italy. Geoderma 34, 251-270.

Le Pera, E., Critelli, S. & Sorriso-Valvo, M., 2001. Weathering of gneiss in Calabria, Southern Italy. Catena 42, 1-15.

Marcelino, V., Mussche, G. & Stoops, G., 1999. Surface morphology of quartz grains from tropical soils and its significance for assessing soil weathering. European Journal of Soil Science 50, 1-8.

McKay, L.D., Driese, S.G., Smith, K.H. & Vepraskas, M.J., 2005. Hydrogeology and pedology of saprolite formed from sedimentary rock, eastern Tennessee, USA. Geoderma 126, 27-45.

Melfi, A.J., Cerri, C.C., Kronberg, B.I., Fyfe, W.S. & McKinnon, B., 1983. Granitic weathering: a Brazilian study. Journal of Soil Science 34, 841-851.

Meunier, A., 1983. Micromorphological advances in rock weathering studies. In Bullock, P. & Murphy, C. P. (eds.), Soil Micromorphology, Volume 2, Soil Genesis. AB Academic Publishers, Berkhamsted, pp. 467-483.

Meunier, A., Velde, B., Dudoignon, P. & Beaufort, D., 1983. Identification of weathering and hydro-thermal alteration in acidic rock: petrology and mineralogy of clay minerals. Sciences Géologiques, Mémoire 72, 93-99.

Muggler, C.C., van Loef, J.J., Buurman, P. & van Doesburg, J.D.J., 2001. Mineralogical and (sub)microscopic aspects of iron oxides in polygenetic Oxisols from Minas Gerais, Brazil. Geoderma 100, 147-171.

Muggler, C.C., Buurman, P. & van Doesburg, J.D.J., 2007. Weathering trends and parent material characteristics of polygenetic Oxisols from Minas Gerais, Brazil. 1. Mineralogy. Geoderma 138, 39-48.

Mulyanto, B. & Stoops, G., 2003. Mineral neoformation in pore spaces during alteration and weathering of andesitic rocks in humid tropical Indonesia. Catena 54, 385-391.

Mulyanto, B., Stoops, G. & Van Ranst, E., 1999. Precipitation and dissolution of gibbsite during weathering of andesitic boulders in humid tropical west Java, Indonesia. Geoderma 89, 287-306.

Nahon, D.B., 1991. Introduction to the Petrology of Soils and Chemical Weathering. John Wiley & Sons, New York, 313 p.

Poetsch, T., 1990. Micromorphology studies of the weathering of granite in semi-arid lands, Northern Darfur, Republic of Sudan. In Douglas, L.A. (ed.), Soil Micromorphology: A Basic and Applied Science. Developments in Soil Science, Volume 19. Elsevier, Amsterdam, pp. 537-544.

Power, T., Smith, B.J. & Whalley, W.B., 1990. Fracture patterns and grain release in physically weathered granitic rocks. In Douglas, L.A. (ed.), Soil Micromorphology: A Basic and Applied Science. Developments in Soil Science, Volume 19. Elsevier, Amsterdam, pp. 545-550.

Protz, R., Gillespie, J.E. & Brewer, R., 1974. Micromorphology and genesis of four soils derived from different rocks in Peterborough County, Ontario, Canada. In Rutherford, G.K. (ed.), Soil Microscopy. The Limestone Press, Kingston, pp. 481-497.

Rebertus, R.A., Weed, S.B. & Buol, S.W., 1986. Transformations of biotite to kaolinite during saprolite-soil weathering. Soil Science Society of America Journal 50, 810-819.

Rice, T.J., Buol, S.W & Weed, S.B., 1985. Soil-saprolite profiles derived from mafic rocks in the North Carolina Piedmont. 1. Chemical, morphological and mineralogical characteristics and transformations. Soil Science Society of America Journal 49, 171-178.

Robertson, I.D.M. & Eggleton, R.A., 1991. Weathering of granitic muscovite to kaolinite and kaolinite to halloysite. Clays and Clay Minerals 39, 113-126.

Rutherford, G.K., 1987. Pedogenesis of two ultisols (Red earth soils) on granite in Belize, Central America. Geoderma 40, 225-236.

Santos, M.C.D., Mermut, A.R. & Ribeiro, M.R., 1989. Submicroscopy of clay microaggregates in an Oxisol from Pernambuco, Brazil. Soil Science Society of America Journal 53, 1895-1901.

Scarciglia, F., Le Pera, E. & Critelli, S., 2005. Weathering and pedogenesis in the Sila Grande Massif (Calabria, South Italy): from field scale to micromorphology. Catena 61, 1-29.

Schaefer, C.E.R., 2001. Brazilian latosols and their B horizon microstructure as long-term biotic constructs. Australian Journal of Soil Research 39, 909-926.

Schaefer, C.E.R., Ker, J.C., Gilkes, R.J., Campos, J.C., da Costa, L.M. & Saadi, A., 2002. Pedogenesis on the uplands of the Diamantina Plateau, Minas Gerais, Brazil: a chemical and micropedological study. Geoderma 107, 243-269.

Schaefer, C.E.G.R., Simas, F.N.B., Gilkes, R.J., Mathison, C., da Costa, L.M. & Albuquerque, M.A., 2008. Micromorphology and microchemistry of selected Cryosols from maritime Antarctica. Geoderma 144, 104-115.

Scholten, T., Felix-Henningsen, P. & Schotte, M., 1997. Geology, soils and saprolites of the Swaziland Middleveld. Soil Technology 11, 229-246.

Schulz, M.S. & White, A.F., 1999. Chemical weathering in a tropical watershed, Luquillo Mountains, Puerto Rico III: quartz dissolution rates. Geochimica et Cosmochimica Acta 63, 337-350.

Schwarz, T., 1997. Distribution and genesis of bauxite on the Mambilla Plateau, SE Nigeria. Applied Geochemistry 12, 119-131.

Sequeira Braga, M.A., Paquet, H. & Begonha, H., 2002. Weathering of granites in a temperate climate (NW Portugal): granitic saprolites and arenization. Catena 49, 41-56.

Simas, F.N.B., Schaefer, C.E.G.R., Fernandes Filho, E.I., Chagas, A.C. & Brandão, P.C., 2005. Chemistry, mineralogy and micropedology of highland soils on crystalline rocks of Serra da Mantiqueira, southeastern Brazil. Geoderma 125, 187-201.

Sousa, E.C. & Eswaran, H., 1975. Alteration of mica in the saprolite of a profile from Angola. A morphological study. Pedologie 25, 71-79.

Stolt, M.H. & Baker, J.C., 1994. Strategies for studying saprolite and saprolite genesis. In Creemens, D.L., Brown, R.B. & Huddleston, J.H. (eds.), Whole Regolith Pedology. Soil Science Society of America Special Publication 34. SSSA, Madison, pp. 1-20.

Stolt, M.H., Baker, J.C. & Simpson, T.W., 1991. Micromorphology of the soil-saprolite transition zone in Hapludults of Virginia. Soil Science Society of America Journal 55, 1067-1075.

Stoops, G., 1967. Le profil d'altération au Bas-Congo (Kinshasa). Sa description et sa genèse. Pedologie 17, 60-105.

Stoops, G., 1989. Relict properties in soils of humid tropical regions with special reference to Central Africa. In Bronger, A. & Catt, J.A. (eds.), Paleopedology. Nature and Application of Paleosols. Catena Supplement 16, pp. 95-106.

Stoops, G., 2003. Guidelines for Analysis and Description of Soil and Regolith Thin Sections. Soil Science Society of America, Madison, 184 p.

Stoops, G. & Dedecker, D., 2006. Microscopy of undisturbed sediments as a help in planning dredging operations. A case study from Thailand. International Conference 'Hubs, Harbours and Deltas in Southeast Asia: Multidisciplinary and Intercultural Perspective'. Royal Academy of Overseas Sciences, Brussels, pp. 193-211.

Stoops, G. & Mees, F., 2018. Groundmass composition and fabric. In Stoops, G., Marcelino, V. & Mees, F. (eds.), Interpretation of Micromorphological Features of Soils and Regoliths. Second Edition. Elsevier, Amsterdam, pp. 73-125.

Stoops, G. & Schaefer, C.E.G.R., 2010. Pedoplasmation: formation of soil material. In Stoops, G., Marcelino, V. & Mees, F. (eds.), Interpretation of Micromorphological Features of Soils and Regoliths. Elsevier, Amsterdam, pp 69-79.

Stoops, G. & Schaefer, C.E.G.R., 2018. Pedoplasmation: formation of soil material. In Stoops, G., Marcelino, V. & Mees, F. (eds.), Interpretation of Micromorphological Features of Soils and Regoliths. Second Edition. Elsevier, Amsterdam, pp. 59-71.

Stoops, G., Altemuller, H.J., Bisdom, E.B.A., Delvigne, J., Dobrosvolsky, V.V., FitzPatrick, E.A., Paneque, G. & Sleeman, J., 1979. Guidelines for the description of mineral alterations in soil micromorphology. Pedologie 29, 121-135.

Stoops, G., Shi Guang Chun & Zauyah, S., 1990. Combined micromorphological and mineralogical study of a laterite profile on graphite sericite phyllite from Malacca (Malaysia). Bulletin de la Société Belge de Géologie 99, 79-92.

Stoops, G., Marcelino, V., Zauyah, S. & Maas, A., 1994. Micromorphology of soils of the humid tropics. In Ringrose-Voase, A & Humphreys, G.S. (eds.), Soil Micromorphology: Studies in Management and Genesis. Developments in Soil Science, Volume 22. Elsevier, Amsterdam, pp. 1-15.

Sugandi, A., 2005. Characterisation and Classification of Some Common Soils in Tawau-Semporna Area, Sabah and their Suitability for Oil Palm. MSc Dissertation, University Putra Malaysia, 309 p.

Taboada, T. & García, C., 1999a. Smectite formation produced by weathering in a coarse granite saprolite in Galicia (NW Spain). Catena 35, 281-290.

Taboada, T. & García, C., 1999b. Pseudomorphic transformation of plagioclases during the weathering of granitic rocks in Galicia (NW Spain). Catena 35, 291-302.

Tardy, Y., 1993. Pétrologie des Latérites et des Sols Tropicaux. Masson, Paris, 459 p.

Thomas, M., 1994. Geomorphology in the Tropics. A Study of Weathering and Denudation in Low Latitudes. John Wiley, Chichester, 460 p.

Thomas, M., Thorp, M. & McAlister, J., 1999. Equatorial weathering, landform development and the formation of white sands in north western Kalimantan, Indonesia. Catena 36, 205-232.

Tonui, E., Eggleton, T. & Taylor, G., 2003. Micromorphology and chemical weathering of K-rich trachyandesite and an associated sedimentary cover (Parkes, SE Australia). Catena 53, 181-207.

Trindade, E. de S., Schaefer, C.E.G.R., Abrahão, W.A.P., Ribeiro Jr., E.S., Oliveira, D.M.F. & Teixeira, P.C., 2005. Crostas biológicas de saprólitos da região do Quadrilátero Ferrífero, MG: ciclagem biogeoquímica e micromorfológica. Geonomos 13, 37-45.

Verheye, W. & Stoops, G., 1975. Nature and evolution of soils developed on the granite complex in the subhumid tropics (Ivory Coast). II. Micromorphology and mineralogy. Pedologie 25, 40-55.

Wilson, M.J., 2004. Weathering of the primary rock-forming minerals: processes, products and rates. Clay Minerals 39, 233-266.

Zauyah, S., 1985. Weathering phenomena and the evolution of soils developed on quartz_mica schist and granitic gneiss. In Mokhtaruddin, A.M., Shamshuddin, J., Aminuddin, H. & Chow, W.T. (eds.), Advances in Soil Research in Malaysia. Universiti Pertanian Malaysia Press, pp. 231-242.

Zauyah, S., 1986. Characterization of Some Weathering Profiles on Metamorphic Rocks in Peninsular Malaysia. PhD Dissertation, Ghent University, 388 p.

Zauyah, S., 1987. The microfabric changes in a weathering profile over quartz-mica schist in Peninsular Malaysia. Pertanika 10, 283-288.

Zauyah, S. & Stoops, G., 1990. A study of the ferrallitic weathering of an amphibole schist in Peninsular Malaysia. Pertanika 13, 85-93.

Chapter 4

Pedoplasmation: Formation of Soil Material

Georges Stoops[1], Carlos E.G.R. Schaefer[2]

[1]*GHENT UNIVERSITY, GHENT, BELGIUM;*
[2]*FEDERAL UNIVERSITY OF VIÇOSA, VIÇOSA, MINAS GERAIS, BRAZIL*

CHAPTER OUTLINE

1. Introduction

In the general process of soil formation, the transformation of rock into soil parent material is followed by the vertical differentiation of the soil material into horizons (pedogenesis). In the first stage, the situation is different for loose sediments and for coherent rocks. The transformation of loose sediments to soil material is mainly a matter of pedoturbation, especially bioturbation (see also Kooistra & Pulleman, 2010, 2018). For coherent rocks, two discrete steps can be distinguished: (i) weathering, giving rise to a saprolite, which is reviewed by Zauyah et al. (2010, 2018), and (ii) homogenisation resulting in the transformation of the saprolite into soil material (pedoplasmation), which is the subject of the present chapter.

Flach et al. (1968) introduced the term 'pedoplasmation', which they defined as 'the transformation of weathered rock (saprolite) to soil B horizons'. In the case of deep tropical soils, it would have been more appropriate to speak more generally of transformation to loose soil parent material rather than to B horizons. It can then be considered as a type of proanisotropic pedoturbation, forming a layer of soil parent material over saprolite (Johnson et al., 1987, Johnson & Watson-Steger, 1990). Because transformation of isovolumetrically weathered rock fragments to soil material also happens in higher parts of the profile (e.g., in soils on colluvial material), we propose to extend the term pedoplasmation to a more general concept of '*in situ* transformation of lithic fabrics to soil fabrics'. Although the term 'plasma' (Kubiëna, 1938; Brewer, 1964) as such has been abandoned in micromorphology, the term pedoplasmation is still in use, meaning the *in situ* development of a structured, loose material in which the original fabric of the saprolite or sediment is no longer visible. Other terms have been used to refer to this process. Romashkevich (1964) described the transformation of 'weathering crust' to 'active soil material', and she later called this process 'soil formation' (Romashkevich, 1965; Romashkevich et al., 1977). Fölster (1971) used the word 'homogenisation', and Nahon (1991) used the general term 'pedoturbation'.

The boundary between saprolite and totally pedoplasmated material (the so-called pedoplasmation front) can be sharp (even represented in a single thin section) or gradual, as already noticed by Harrison (1933) in his classic work on various igneous rocks from Guyana. The thickness of the pedoplasmation zone can reach several decimetres or metres on coarse-grained rocks in the tropics, and the zone can be almost absent on fine-grained, quartz-free material. Pedoplasmation, therefore, is best studied by comparing sequential thin sections, from the saprolite upward. In arid zones, where weathering is slower than pedoturbation, no pedoplasmation zone is observed, except on palaeosaprolites formed during stages with other climatic conditions.

Pedoplasmation does not only take place at the upper boundary of the saprolite, but also in some preferential spots, such as along diaclases (Beaudou & Chatelin, 1979). According to Flach et al. (1968), the chemical and mineralogical composition of the material does not change during this process, although several physicochemical changes are noticed, such as an increase in clay content, linear extensibility and dispersibility of the soil, indicating that the process is not restricted to simple mixing. These authors considered pedoplasmation as an *in situ* rearrangement of constituents without input or output of materials. However, this isochemical and isomineral pedoplasmation is generally accompanied by an exchange of components, in suspension or in solution, vertically laterally or both. The most visible result of this exchange in thin sections is the presence of clay coatings (Beaudou & Chatelin, 1979) and the crystallisation of minerals in pores (e.g., gibbsite) (Stoops et al., 1990; Stoops, 1991) or in the groundmass (e.g., diffuse iron oxide nodules). The introduction of organic substances can also play an important role in the arrangement of components due to the development of organomineral complexes (Romashkevich et al., 1977). The newly formed minerals (e.g., clay minerals, gibbsite, calcite) in the created porosity are often different from those in the groundmass because other thermodynamic conditions prevailed there (Tardy & Novikoff, 1988). Kew et al. (2010) compared physical parameters of pedoplasmated zones with those of indurated or compacted subsoil materials of fluvial and aeolian origin. Based on SEM and microprobe analyses they concluded that those subsoil materials have a less porous micromass than saprolite that shows a certain degree of pedoplasmation, possibly due to closer packing of clay particles.

With few exceptions, no systematic study of the micromorphological aspects of pedoplasmation has been published so far. This chapter is therefore mainly based on the works of Nahon (1991), Stoops (1991) and the authors. Most other available publications discuss deeply weathered soils in the tropics (Fölster, 1971; Beaudou & Chatelin, 1979; Stoops, 1989; Stoops et al., 1990; Muggler & Buurman, 1997; Nunes et al., 2000; Schaefer, 2001; Muggler et al., 2007), and a few deal with palaeosoils in the Mediterranean region (e.g., Romashkevich et al., 1977; Martin-Garcia et al., 1999; Gerasimova & Gurov, 2012).

2. General Features

Pedoplasmation is expressed by changes in various micromorphological features. In most cases a gradual transition exists between a complete saprolite fabric, a continuous saprolite fabric with pedoplasmated spots, a continuous soil fabric with saprolite remains and finally a complete soil fabric (Fölster, 1971; Stoops et al., 1990; Stoops, 1991) (Fig. 1).

Sometimes pedoplasmation results in the formation of millimetre- or centimetre-sized parallel layers with alternating lithic fabrics and soil fabrics, as a result of repeated

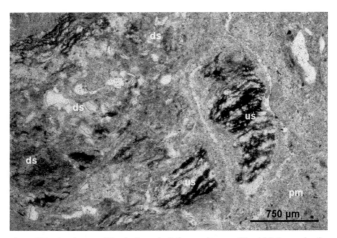

FIGURE 1 Fragments of undisturbed saprolite (us) and deformed saprolite (ds) in pedoplasmated material (pm). Note the gradual destruction of the fragment by further pedoplasmation. (PPL) (3 m depth, Malacca, Malaysia).

small-scale shears formed in the upper part of the saprolite (Curmi, 1979). At the contact with the overlying mass movement deposit, a breccia-like microstructure can be formed as a result of mechanical friction. Examples of this phenomenon were described by Altemüller (1960) in the C horizon of soils on marl along the contact with a solifluction deposit.

The lithic fabric is preserved for a longer time in spots with iron oxide impregnation than in their surroundings (Fölster, 1971). In the centre of ferruginous nodules in the lower part of laterite profiles, lithic fabrics are gradually transformed to soil fabrics as a result of continuous recrystallisation of kaolinite, whereby large kaolinite mesomorphs after biotite are converted to increasingly fine-grained kaolinite aggregates with random fabric (Muller & Bocquier, 1987). In most cases, changes occur over a very short distance between the saprolite and the overlying pedoplasmated zone. In saprolites developed from gneiss and granitic rocks, kaolinite pseudomorphs after biotite and feldspar are common. They are progressively incorporated into the soil by biological activity, but they survive as rare alteromorphic domains within the oxic Bw horizon, due to upward transport by termite activity (Pinto, 1971; Nunes et al., 2000; Schaefer, 2001).

When checking the use of the term pedoplasmation in earth science literature, it appears that it is often misused to describe simply pedoturbation, which can cause confusion.

3. Disappearance of the Original Rock Fabric

The preserved lithic fabric in a saprolite is formed by a mixture of relict minerals (e.g., quartz) and alteromorphs (e.g., clays after feldspar or biotite, goethite after olivine or garnet) (Nahon, 1991). As weathering progresses, two major features are observed in the pedoplasmation zone: loss of the outlines of the clay alteromorphs and fragmentation of the ferruginous or gibbsitic pseudomorphs and resistant primary minerals (Fig. 2A and B). Deformation of the fabric due to volume increase during weathering, such as pseudomorphic weathering of biotite to kaolinite with extension along the c-axis, has been observed in granite (Stoops, 1991) (Fig. 2C) and gneiss (Nunes et al., 2000). Deformation can also develop as a result of pressure exerted during crystal growth, for example during formation of calcite (Fig. 2D) or gypsum in arid environments or gibbsite in tropical areas.

Another observed feature is the loss of optical continuity between fragments of fractured coarse groundmass components (Nahon, 1991). An example is rotation and progressive separation of the fragments of fractured quartz grains (see Marcelino et al., 2018; Stoops & Marcelino, 2018). A special case is pedoplasmation of saprolites derived from sedimentary rocks or foliated metamorphic rocks, marked by the disturbance and eventual disappearance of the parallel arrangement of constituents, which is mainly a consequence of pedoturbation. SEM studies on weathered mica schist gravels revealed the importance of local clay illuviation in the disruption of the original schistosity (Martin-Garcia et al., 1999).

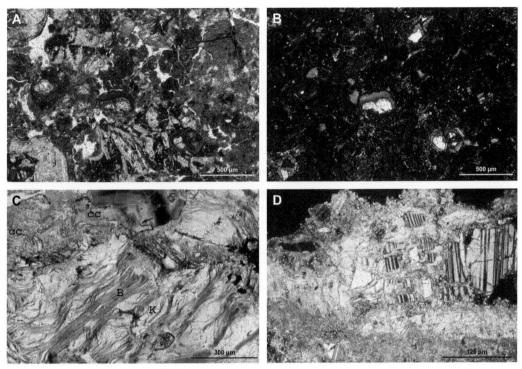

FIGURE 2 Deformation of individual minerals and fabrics. (A) Pedoplasmation of basalt saprolite: several pseudomorphs after lath-shaped feldspar are still visible, and olivine phenocrysts, partly altered to iddingsite, are more or less in their original position, but homogenised brown groundmass material is abundant (e.g., right). (PPL) (Isla Santa Cruz, Galápagos). (B) Idem in XPL. (C) Weathering biotite in a saprolite on granite, deforming the surrounding lithic fabric. The biotite (B) is locally transformed to vermiculite and to kaolinite (K), and the surrounding material includes limpid clay coatings (cc) (PPL) (Akagera National Park, Rwanda). (D) Deformation of a plagioclase grain by calcite crystal growth within cracks (XPL) (layer beneath a petrocalcic horizon, Fuertaventura, Canary Islands).

Overall, bioturbation is the main process contributing to the destruction of the saprolite fabric and further homogenisation, giving rise to voids (channels, chambers, vughs) (Fig. 3) and passage features (Fig. 4A). The latter seem to be less abundant on coarse-grained saprolites (e.g., on granite) than on fine-grained materials, probably because the size of the constituents exceeds the transport capabilities of the soil mesofauna (Stoops, 1991). Clusters of pellets formed by termite activity are commonly observed (e.g., Martins, 2007).

The collapse of moldic voids formed by congruent dissolution also plays a role during early stages of pedoplasmation, when part of the outlines of the dissolved minerals is still recognisable. This process is particularly important in limestone transformation, producing weathering profiles in which saprolite fabrics are virtually absent, as described in Section 6.

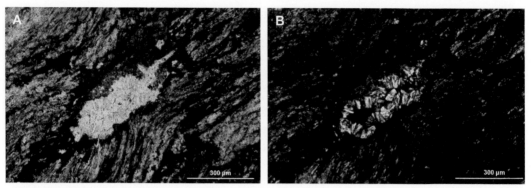

FIGURE 3 Development of porosity and passage features. (A) Channel partly lined by brownish pedoplasmated material and with an infilling of euhedral gibbsite crystals, in a graphite-sericite phyllite saprolite (PPL) (3 m depth, Malacca, Malaysia). (B) Idem in XPL.

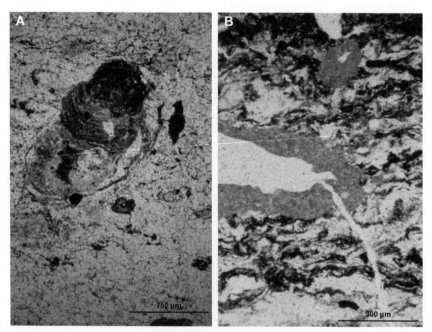

FIGURE 4 (A) Crescent infilling documenting the first stage of pedoplasmation and homogenisation by bioturbation in a saprolite on weathered graphite-sericite phyllite (PPL) (5 m depth, Malacca, Malaysia). (B) Coatings of pedoplasmated material derived from a higher layer in a kaolinite-hematite saprolite developed on sedimentary parent material (note horizontal banding) (PPL) (mottled clay, 14 m depth, Trombetas bauxite profile, Brazil; ISRIC laterite collection, Wageningen).

4. Development of Microstructure and Porosity

Whereas the saprolite is massive, a pedal microstructure gradually develops during pedoplasmation (Beaudou & Chatelin, 1979), accompanied by an increase in porosity (Hamdan et al., 2003). Increased development of structural units with decreasing ped size has been observed in well-drained deep B horizons of Oxisols (Muggler & Buurman, 1997). However, this seems not to be the case when hydromorphic conditions prevail, as for instance in mottled clays with a massive microstructure (Stoops et al., 1990). In this case, iron oxides are removed in parts of the saprolite, which results in a decrease in aggregate stability, an enhancement of the face-to-face arrangement of clay particles and a decrease in total porosity.

In saprolites, only moldic voids and contact voids (within the alteromorphs) are generally present. In the soil material, new voids are progressively formed by bioturbation (channels, chambers, vughs), by collapse of moldic voids formed by congruent dissolution (irregular vughs) and by shearing and drying (planar voids). This can give rise to local changes of mineral equilibrium conditions, resulting for instance in the precipitation of gibbsite in voids, whereas the groundmass remains kaolinitic (Fig. 3) (Tardy & Novikoff, 1988).

5. Other Changes

5.1 Changes in Grain-Size Distribution

The fragmentation of coarse constituents, mostly quartz (see Section 3), results in particle size changes from coarse/medium sand to fine sand/silt. In addition, decrease in silt content and increase in clay content are also typical features of pedoplasmation. This is observed in thin sections and in the field and also corroborated by the results of physical analyses. Grain-size distribution changes are also caused both by the comminution and dissolution of silt- and sand-sized kaolinite booklets and by the neoformation of clay minerals (Beaudou & Chatelin, 1979; Muggler & Buurman, 1997). Muggler et al. (2007) stress that comminution alone cannot explain the change and that an observed decrease in crystallinity of kaolinite indicates that a process of dissolution and neoformation must be active (see also Muller & Bocquier, 1987). The abrupt loss of kaolinite crystallinity in the pedoplasmation zone (e.g., Varajão et al., 2001; Melo et al., 2001, 2002) could be attributed to ingestion of soil by termites, in whose digestive track the pH is higher than 12, accounting for greater silicate dissolution (Schaefer, 2001).

5.2 Colour Homogenisation and Limpidity Changes

Saprolites are heterogeneous at the microscopic scale. Pedoplasmation gradually leads to homogenisation of the original material. Tropical saprolites generally show a juxtaposition of white zones (clay) and dark red or opaque zones rich in iron oxides (hematite, goethite), which after pedoplasmation form a uniform light reddish or orange material (Fig. 4B)

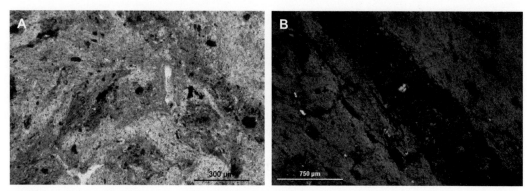

FIGURE 5 Changes of lithic fabric. (A) Heterogeneous, incompletely homogenised graphite-sericite phyllite saprolite (PPL) (2 m depth, Malacca, Malaysia). (B) Pedoturbated zone (centre, NW-SE oriented) with speckled b-fabric in a claystone saprolite with unistrial b-fabric (XPL) (Cameroon).

(Muggler & Buurman, 1997). An increase of the free iron content in the homogeneous material has been reported (Beaudou & Chatelin, 1979; Stoops et al., 1990).

Besides the well-expressed colour homogenisation, changes in limpidity are observed as well. Starting from a clustered distribution, a random distribution of dispersed fine opaque particles gradually appears (dotted or speckled limpidity). This feature is often overlooked in micromorphological studies. Total homogenisation is reached only after several cycles of pedoplasmation (Fölster, 1971; Stoops et al., 1990) (Fig. 5A). An indication for this in thin sections is the observation that the last generation of passage features is more homogeneous than earlier generations.

5.3   Changes of the b-Fabric

During pedoplasmation, the orientation of clay particles in the alteromorphs is destroyed, resulting in the appearance of other b-fabrics. Muggler and Buurman (1997) report the presence of undifferentiated b-fabrics or only weakly expressed other b-fabrics in soils developed from phyllites. In the case of saprolites derived from siltstones and shales, which originally have a strial b-fabric, the randomisation of the clay domains leads to the development of striated, speckled or even undifferentiated b-fabrics in the pedoplasmated material (Dasog et al., 1987; Stoops, unpublished data) (Fig. 5B). The type of b-fabric formed is most probably related to the active type of pedoturbation process, but no information is available on this issue.

6. Pedoplasmation on Carbonate Rocks

Pedoplasmation on carbonate rocks is a specific case, where a generalised dissolution results in the total collapse of the rare weathering residue. This mechanism was first described by Stoops and Mathieu (1970) for clay with chert (*argile à silex*), the weathering product of chert-bearing chalk, in a region with a loess cover and is illustrated in Figure 6. A three-dimensional network of fissures is first formed near the

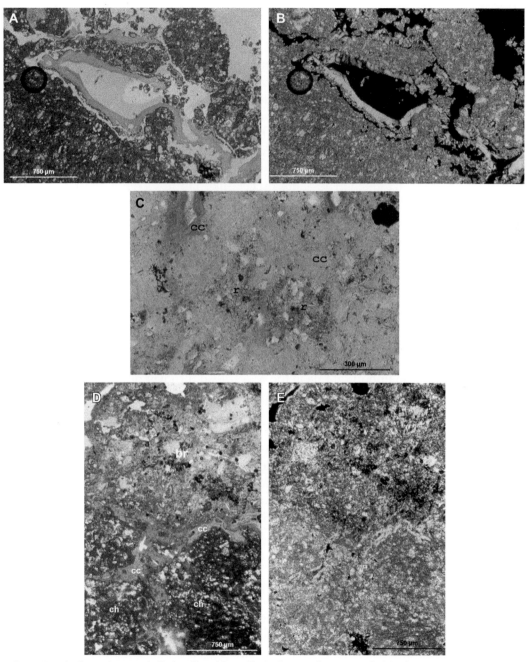

FIGURE 6 Pedoplasmation on chalk (profile with *argile à silex*, northern France). (A) Clay coatings in dissolution voids in chalk (PPL). (B) Idem in XPL. (C) Abundant clay coatings and infillings (cc), and chalk residues (r) (PPL). (D) Chalk substrate (ch) covered by layer with breccia-like fabric (br) consisting of fragmented clay coatings, and clay coatings and infillings (cc) in fissures in the chalk (PPL). (E) Idem in XPL.

surface of the parent rock, separating millimetre- or centimetre-sized blocks of chalk. The fissures become gradually more numerous and are widened by dissolution, and coatings and infillings of limpid fine clay develop (Fig. 6A and B). The chalk blocks gradually disappear, leaving only a small amount of insoluble weathering residue (clay, fine sand, chert) (Fig. 6C). Due to the collapse of the dissolution cavities, the clay coatings are fragmented and closely packed, forming a residue with a breccia-like fabric (Fig. 6D and E), which becomes more and more homogeneous toward the top. The illuvial clay is largely derived from the remobilisation of clay in the uppermost part of the alteration zone but may also partly originate from the overlying loess soil (Luvisol), as can be deduced from its mineralogical composition (Stoops & Mathieu, 1970; Buurman et al., 1985). The same breccia-like fabric of randomly packed fragments of clay coatings was observed in a similar context (Thorez et al., 1971; Buurman et al., 1985). Brewer et al. (1983) describe a similar fabric (papular fabric) on weathered pyroclastic material.

7. Conclusions

Pedoplasmation is not always apparent in a single thin section. A comparative study of sequential samples of undisturbed regolith is often needed and large thin sections are recommended for such studies. Pedoplasmation is the result of a combination of various physicochemical and biological processes, whose interaction is not well understood. Bioturbation plays an important role in the widespread turnover of soil materials and the structural changes observed at a microscopic scale, in the transition zone between the saprolite and the soil.

Most studies on pedoplasmation were performed for deeply weathered soils in the tropics and palaeosoils in the Mediterranean region. Research on the transition of weathered rock to soil material at shallower depths in temperate, cold and arid environments needs more attention. The transformation of unconsolidated sediments to soil parent material is poorly documented in the literature, except for the role of bioturbation, and research is also necessary in this field.

References

Altemüller, H.J., 1960. Mikromorphologische Untersuchungen an einigen Gipskeuperböden im Raum Iphofen. Bayerisches Wirtschaftliches Jahrbuch, Sonderheft 37, 70-85.

Beaudou, A.G. & Chatelin, Y., 1979. La pédoplasmation dans certain sols ferrallitiques rouges de savane en Afrique Centrale. Cahiers ORSTOM, Série Pédologie 17, 3-8.

Brewer, R., 1964. Fabric and Mineral Analysis of Soils. John Wiley and Sons, New York, 470 p.

Brewer, R., Sleeman, J.R. & Foster, R.C., 1983. The fabric of Australian soils. In Soils: An Australian Viewpoint. CSIRO, Melbourne & Academic Press, London, pp. 439-476.

Buurman, P., Jongmans, A.G., Broekhuizen, J. & Miedema, R., 1985. Genesis of the flint eluvium and related beds in South Limburg, The Netherlands. Geologie en Mijnbouw 64, 89-102.

Curmi, P., 1979. Genèse d'une structure litée à granulométrie hétérogène dans une arène granitique. Comptes Rendus de l'Académie des Sciences 288, 731-733.

Dasog, G.S., Acton, D.F. & Mermut, A.R., 1987. Genesis and classification of clay soils with vertic properties in Saskatchewan. Soil Science Society of America Journal 51, 1243-1250.

Flach, K.W., Cady, J.G. & Nettleton, W.D., 1968. Pedogenic alteration of highly weathered parent materials. Transactions of the 9th International Congress of Soil Science, Volume IV, Adelaide, pp. 343-351.

Fölster, H., 1971. Ferrallitische Böden aus Sauren Metamorphen Gesteine in den Feuchten und Wechselfeuchten Tropen Afrikas. Göttinger Bodenkundliche Berichte 20, 231 p.

Gerasimova, M.I. & Gurov, I.A., 2012. Micromorphology of zheltozems on hard sedimentary rocks and products of their decomposition: pedogenic and lithogenic features (with the Sochi arboretum as an example). Eurasian Soil Science 45, 22-32.

Hamdan, J., Pelp, M. & Ruhana, B., 2003. Weathering behaviour of a basaltic regolith from Pahang, Malaysia. Pertanika Journal of Tropical Agricultural Science 26, 79-88.

Harrison, J.B., 1933. The Katamorphism of Igneous Rocks under Humid Tropical Conditions. Imperial Bureau of Soil Science, Rothamsted Experimental Station, Harpenden, 79 p.

Johnson, D.L. & Watson-Steger, D., 1990. The soil-evolution model as a framework for evaluating pedoturbation in archaeological site formation. In Lasca, N.P. & Donahue, J. (eds.), Archaeological Geology of North America, Centennial Special Volume 4. Geological Society of America, Boulder, 541-560.

Johnson, D.L., Watson-Stegner, D., Johnson, D.N. & Schaetzl, R.J., 1987. Proisotropic and proanisotropic processes of pedoturbation. Soil Science 143, 278-292.

Kew, G.A., Gilkes R.J. & Evans D., 2010. Relationships between fabric, water retention, and strength of hard subsoils in the south of Western Australia. Australian Journal of Soil Research 48, 167-177.

Kooistra, M.J. & Pulleman, M.M., 2010. Features related to faunal activity. In Stoops, G., Marcelino, V. & Mees, F. (eds.), Interpretation of Micromorphological Features of Soils and Regoliths. Elsevier, Amsterdam, pp. 397-418.

Kooistra, M.J. & Pulleman, M.M., 2018. Features related to faunal activity. In Stoops, G., Marcelino, V. & Mees, F. (eds.), Interpretation of Micromorphological Features of Soils and Regoliths. Second Edition, Elsevier, Amsterdam, pp. 447-469.

Kubiëna, W.L., 1938. Micropedology. Collegiate Press, Ames, 242 p.

Marcelino, V., Schaefer, C.E.G.R. & Stoops, G., 2018. Oxic and related materials. In Stoops, G., Marcelino, V. & Mees, F. (eds.), Interpretation of Micromorphological Features of Soils and Regoliths. Second Edition. Elsevier, Amsterdam, pp. 663-689.

Martin-Garcia, J.M., Delgado, G., Párraga, J.F., Gámiz, E. & Delgado, R., 1999. Chemical, mineralogical and micromorphological study of coarse fragments in Mediterranean Red Soils. Geoderma 90, 23-47.

Martins, G.M., 2007. Efeitos da Ação de Cupins Sobre Propriedades de um Perfil de Solo em uma Vertente da Represa Billings, São Bernardo do Campo/SP. MSc Dissertation, Universidade de São Paulo, 88 p.

Melo, V.F., Singh, B., Schaefer, C.E.G.R., Novais, R.F. & Fontes, M.P.F., 2001. Chemical and mineralogical properties of kaolinite rich Brazilian soils. Soil Science Society of America Journal 65, 1324-1333.

Melo, V.F., Schaefer, C.E.G.R., Singh, B., Novais, R.F. & Fontes, M.P.F., 2002. Propriedades químicas e cristalográficas da caulinita e dos óxidos de ferro em sedimentos do Grupo Barreiras no município de Aracruz, estado do Espírito Santo. Revista Brasileira de Ciência do Solo 26, 53-64.

Muggler, C.C. & Buurman, P., 1997. Micromorphological aspects of polygenetic soils developed on phyllitic rocks in Minas Gerais, Brazil. In Shoba, S., Gerasimova, M. & Miedema, R. (eds.), Soil Micromorphology: Studies on Soil Diversity, Diagnostics, Dynamics. Moscow, Wageningen, pp. 129-138.

Muggler, C.C., Buurman, P. & van Doesburg, J.D.J., 2007. Weathering trends and parent material characteristics of polygenetic Oxisols from Minas Gerais, Brazil. 1. Mineralogy. Geoderma 138, 39-48.

Muller, J.P. & Bocquier, G., 1987. Textural and mineralogical relationships between ferruginous nodules and surrounding clayey matrices in a laterite from Cameroon. In Schultz, L.G., van Olphen, H. & Mumpton, F.A. (eds.), Proceedings of the International Clay Conference., Denver, pp. 186-194.

Nahon, D.B., 1991. Introduction to the Petrology of Soils and Chemical Weathering. John Wiley & Sons, New York, 313 p.

Nunes, W.A.G.A., Schaefer, C.E.G.R., Ker, J.C. & Fernandes Filho, E.I., 2000. Caracterização micropedológica de alguns solos da Zona da Mata Mineira. Revista Brasileira de Ciência do Solo 24, 103-115.

Pinto, O.C.B., 1971. Formation of Kaolinite from a Biotite Feldspar Gneiss in Four Strongly Weathered Soil Profiles from Minas Gerais, Brazil. MSc Dissertation, Purdue University, 133 p.

Romashkevich, A.I., 1964. Micromorphological indications of the processes associated with the formation of the krasnozems (red earths) and the red-coloured crust of weathering in the Transcaucasus. In Jongerius, A. (ed), Soil Micromorphology. Elsevier, Amsterdam, pp. 261-268.

Romashkevich, A.I., 1965. Micromorphological features of processes leading to the formation of red earths and the red weathering crust of the Black Sea coast of Caucasus. Soviet Soil Science 4, 407-415.

Romashkevich, A.I., Gerasimova, M.I. & Tursina, T.V., 1977. Evolution of microfabric in soils of humid environment. Problems of Soil Science. Contributions of Soviet Pedologists to the XI International Congress of Soil Science. USSR Academy of Sciences, Moscow, pp. 456-465.

Schaefer, C.E.R., 2001. Brazilian latosols and their B horizon microstructure as long-term biotic constructs. Australian Journal of Soil Research 39, 909-926.

Stoops, G., 1989. Relict properties in soils of humid tropical regions with special reference to Central Africa. In Bronger, A. & Catt, J.A. (eds.), Paleopedology. Nature and Application of Paleosols. Catena Supplement 16, pp. 95-106.

Stoops, G., 1991. The influence of the fauna on soil formation in the tropics. Micropedological aspects. Bulletin des Séances de l'Académie Royale des Sciences d'Outre-Mer 36, 461-469.

Stoops, G. & Marcelino, V., 2018. Lateritic and bauxitic materials. In Stoops, G., Marcelino, V. & Mees, F. (eds.), Interpretation of Micromorphological Features of Soils and Regoliths. Second Edition, Elsevier, Amsterdam, pp. 691-720.

Stoops, G. & Mathieu, C, 1970. Aspects micromorphologiques des argiles à silex de Thiérarche. Science du Sol 2, 103-116.

Stoops, G., Shi Guang Chun & Zauyah, S., 1990. Combined micromorphological and mineralogical study of a laterite profile on graphite sericite phyllite from Malacca (Malaysia). Bulletin de la Société Belge de Géologie 99, 79-92.

Tardy, Y. & Novikoff, A., 1988. Activité de l'eau et déplacement des équilibres gibbsite-kaolinite dans les profils latéritiques. Comptes Rendus de l'Académie des Sciences 306, 39-44.

Thorez, J., Bullock, P., Catt, J.A. & Weir, A.H., 1971. The petrography and origin of deposits silling solution pipes in the chalk near South Mimms, Hertfordshire. Geological Magazine 108, 413-423.

Varajão, A.F.D.C., Gilkes, R.J. & Hart, R.D., 2001. The relationships between kaolinite crystal properties and the origin of materials for a Brazilian kaolin deposit. Clays and Clay Minerals 49, 45-59.

Zauyah, S., Schaefer, C.E.G.R. & Simas, F.N.B., 2010. Saprolites. In Stoops, G., Marcelino, V. & Mees, F. (eds.), Interpretation of Micromorphological Features of Soils and Regoliths. Elsevier, Amsterdam, pp. 49-68.

Zauyah, S., Schaefer, C.E.G.R. & Simas, F.N.B., 2018. Saprolites. In Stoops, G., Marcelino, V. & Mees, F. (eds.), Interpretation of Micromorphological Features of Soils and Regoliths. Second Edition, Elsevier, Amsterdam, pp. 37-57.

Chapter 5

Groundmass Composition and Fabric

Georges Stoops[1], Florias Mees[2]

[1]GHENT UNIVERSITY, GHENT, BELGIUM;
[2]ROYAL MUSEUM FOR CENTRAL AFRICA, TERVUREN, BELGIUM

CHAPTER OUTLINE

Interpretation of Micromorphological Features of Soils and Regoliths. https://doi.org/10.1016/B978-0-444-63522-8.00005-X

1. Introduction

Most publications dealing with the interpretation of thin section or SEM observations of soils and regoliths focus on the presence and composition of pedofeatures. For specific soil types, much attention has also been given to microstructure development. In contrast, information about groundmass composition and fabric is relatively limited, with some exceptions (e.g., FitzPatrick, 1984, 1993). Nevertheless, the groundmass is commonly an important source of information about the nature of the parent material and about pedogenic processes. For some soil materials, such as deep soil horizons and weakly developed soils, it can be the only available source of micromorphological data. This is also the case for many soils transformed by human activities (cf. Adderley et al., 2018). The aim of this chapter is therefore to compile information that is useful for the interpretation of groundmass characteristics.

Traditionally, a coarse fraction and a fine fraction are distinguished for micromorphological descriptions. These two fractions have been called 'skeleton' and 'plasma' (Kubiëna, 1938), 'skeleton grains' and 'plasma' (Brewer, 1960, 1964b), 'f-members' and 'f-matrix' (Brewer & Pawluk, 1975; Brewer & Sleeman, 1988) and 'coarse material' and 'fine material' or 'micromass' (Stoops & Jongerius, 1975; Bullock et al., 1985; Stoops, 2003).The skeleton grain/plasma concept of Brewer (1964b) is based not only on grain size but also on mobility and solubility, whereby plasma is composed of material that is easily moved, reorganised or concentrated by soil-forming processes, including all material of colloidal size (<2 μm) and soluble compounds. In practice, the skeleton grain/plasma concept is difficult to apply systematically for various reasons, including its promotion of all particles that are larger than 2 μm to skeleton grain status, the difficult detection of weakly contrasting small particles, and the dependence of solubility on soil conditions. Moreover, only single mineral grains are considered as skeleton grains, whereas all compound grains, such as rock fragments, are classified as pedological features (lithorelicts). The coarse/fine concept, introduced by Stoops and Jongerius (1975), eliminates these problems by being based exclusively on size, with size limits that are defined by the user for each individual case. Using particle size as a criterion has been opposed by FitzPatrick (1984, 1993), referring to the continuity in grain size that is observed for many soils, but he nevertheless uses a similar concept ('matrix') that corresponds to a continuous fine fraction. It is clear that mixing various concepts, such as plasma, micromass and matrix, is erroneous and should be avoided, also because some terms have a different meaning in different systems.

In this chapter, coarse material, fine material and c/f-related distribution patterns are discussed under separate headings. The general outline is based in part on the review by Stoops (2015a, 2015b). For practical reasons, we use 10 μm as limit between the coarse and fine fractions, in view of standard thin section thickness and of the magnification that is reached by most optical microscopes.

As discussed in this chapter, the coarse fraction is mainly inherited from the parent material, and its thin section study can document various aspects that are not revealed by mineralogical or chemical analysis of bulk samples. The fine fraction can be either inherited or formed *in situ* by weathering of coarse components or by precipitation from soil solutions. Some information about its composition can be gained from thin section studies, and its b-fabric can provide a record of soil behaviour. Orientation and distribution patterns offer information about the nature and alteration of the parent material and about pedogenic processes.

2. Coarse Fraction

The study of the coarse material allows obtaining information about several important aspects of soil materials: (i) origin of the parent material, by providing information on the rock type from which it developed, (ii) profile homogeneity, by revealing vertical variations in the composition, nature and relative abundance of coarse components, (iii) contamination by aeolian material and other surface deposits, including anthropogenic material, by showing the presence of components that cannot be derived from the local rock or sediment substrate, and (iv) occurrence and degree of alteration, providing information about the type and duration of soil development, as well as about the nature of the parent material.

Some information concerning the coarse fraction can also be obtained through other methods, such as X-ray diffraction (XRD) analysis, grain mount study of the heavy mineral fraction and grain size analysis, but thin section studies can limit or eliminate the need to apply such other methods. In addition, thin section observations can have various advantages over other methods: (i) rock fragments can be identified, (ii) grain morphology and alteration patterns can be determined, (iii) different types of the same mineral species can be distinguished, based on characteristics such as colour and the presence of inclusions, (iv) amorphous inorganic components such as opal and volcanic glass, as well as organic constituents, can be detected, (v) certain components occurring in abundances below the detection limit of bulk sample analysis are easily recognised, and (vi) information can be obtained for units that are too thin or small for easy subsampling for other analyses. As an alternative optical method, grain mount analysis yields results that are more quantitative, but thin section studies have the advantage of allowing the study of coarser mineral grains, easier mineral identification due to standard section thickness, better recognition of internal features of mineral grains, and identification of rock fragments (Stoops & Van Driessche, 2007). The same identification possibilities are offered by thin sections of coarse fractions obtained by sieving, with the added advantage of yielding more representative results for the coarse fraction of clayey soil materials.

The coarse fraction is composed of mineral grains, rock fragments, coarse organic components, inorganic components of biological origin, and anthropogenic materials. Organic components are not considered in this review, and inorganic biogenic components and anthropogenic constituents are only briefly discussed (see Section 2.3), as they are treated *in extenso* in other chapters of this book.

In documenting the presence of specific minerals, thin section observations can provide information that is required for soil classification, including the absence of weatherable minerals as requirement for Oxisols and the presence of volcanic glass for vitric characteristics (Soil Survey Staff, 2014). In the same way, thin section studies can provide information about reserves of certain nutrients, as well as possible sources of toxic elements. For nutrient status, thin sections permit the description and identification of mineral grains and rock fragments that are larger than 2 mm, which allows to correct results of routine analyses in soil science, for which this fraction is systematically removed and neglected. The importance of considering the coarse fraction was already pointed out by Lelong (1967), and it has been confirmed by some later studies (e.g., Certini et al., 2004; De Léon-González et al., 2007; Sanborn, 2010; Small et al., 2014).

2.1   Mineral Grains

In this section, only minerals that are part of the parent material, or added to the profile by later incorporation of surface deposits, are described and discussed. Pedogenic minerals, ranging from halides to authigenic silicates and iron oxides, are covered in various other chapters of this book. Information about non-pedogenic occurrences of minerals that are mainly relevant as pedogenic phases (e.g., gypsum, pyrite) is also contained in other chapters. In principle, nearly all known minerals can be present as part of the coarse fraction of the groundmass, but only the most common rock-forming minerals will be discussed. In saprolites or soils on specific rock types or on ore deposits, various other minerals can obviously occur in important quantities.

It is not the aim of this chapter to provide exhaustive crystallographic and optical information about selected minerals or to present a key for their optical identification. Readers are referred to the many excellent manuals on optical mineralogy that have been published in several languages, including several well-illustrated atlases (e.g., MacKenzie & Guilford, 1980; MacKenzie & Adams, 1994; Melgarejo, 1997; Pichler & Schmitt-Riegraf, 1997; Perkins & Henke, 2004). Identification of minerals based on their optical properties requires a minimum knowledge of optical mineralogy, for which various handbooks are available (e.g., Parfenoff et al., 1970; Nesse, 2003). A short description of the optical characteristics, occurrence and alteration features of the most important rock-forming minerals occurring in the coarse fraction of soils is presented in the Annex of this chapter.

For soils developed on sedimentary parent materials, grain properties such as size distribution and degree of rounding can be determined and interpreted as described in handbooks on sediment petrology (e.g., Tucker, 2001; Boggs, 2009).

2.1.1   Mineralogical Composition and Internal Fabric

Mineral associations, as determined by thin section studies and other methods, will strongly differ between soils formed on contrasting parent materials, such as granite and basalt. For parent materials with similar mineralogical compositions, thin section observations can provide information that is otherwise unavailable. For instance, soils on granite and gneiss may have a similar mineralogical composition, but the shape and internal fabric of coarse mineral grains will be different, most prominently through the presence of quartz with undulose extinction in metamorphic rocks.

For parent material characterisation, including recognition of lithological discontinuities, micromorphological studies can identify differences that remain undetected by XRD analysis. Examples include the recognition of colour varieties (e.g., green vs brown biotite or hornblende), the presence of mineral, glass or fluid inclusions (e.g., basaltic glass in feldspar), and the presence of zoning or exsolution. Indications of weathering can be detected to distinguish vertical trends that are related to parent material heterogeneity or postdepositional alteration.

2.1.2   Alteration

Only few reviews of micromorphological aspects of mineral alteration have been published, including the book by Delvigne (1998) and a short paper by Delvigne and Stoops (1990).

The description of the type and degree of alteration of coarse mineral grains, and the nature of their alteration products, is important for the interpretation of the coarse fraction of the groundmass. The term alteration is used here in a wide sense, as proposed by Stoops et al. (1979), covering changes caused both by geological processes (e.g., burial diagenesis, deuteric reactions, hydrothermal activity) and by pedogenic processes. The latter type of alteration, resulting from interaction with atmospheric agents and organisms in earth-surface conditions, is generally called weathering. It is evident, but sometimes overlooked by micromorphologists, that changes requiring high temperatures or pressures cannot take place in soils and therefore cannot be interpreted as pedogenic.

Susceptibility to alteration is a reflection of mineral stability in soil environments, which is determined by many factors, including temperature, moisture conditions, pH, redox conditions, degree of leaching, and the nature, composition and grain size of the mineral. A discussion of these factors is beyond the scope of this chapter but can be found in other publications (e.g., Nahon, 1991). For intrinsic mineral stability, a sequence was already proposed by Goldich (1938), limited to minerals that are common in magmatic rocks. He proposed two parallel series, one from olivine over augite and hornblende to biotite, and the other from Ca-rich to Na-rich plagioclase, both continued by a series from potassium feldspar over muscovite to quartz. Related efforts for mineral stability classification are various weathering indexes, based on chemical composition and crystal structure (e.g., Reiche, 1950; Parker, 1970; Nesbitt & Young, 1982, 1984). These approaches are not always applicable because other factors may interfere. For instance, augite may be less stable than associated olivine in basalt because of its prominent cleavage (Brewer, 1964b; Stoops, 2013). Classifications of mineral stability that have been proposed in sedimentology are also not entirely suitable in pedology because they take physical stability during transport into account, which is not clearly relevant for soil environments (e.g., Boswell, 1941; Pettijohn et al., 2012).

Alteration of mineral grains can occur in the form of transformation to newly formed mineral phases, but it can also consist of congruent dissolution, without *in situ* formation of alteration products. In this case, with calcite and quartz dissolution as common examples, circumgranular voids (also called contact voids) are formed, separating the affected grain from surrounding or neighbouring components. This space can later be filled with other minerals, giving the false impression of mineral transformation or coupled dissolution-precipitation. In these and other contexts, formation of an enclosing mineral phase can contribute to dissolution of the surrounded grain through the effect of crystallisation pressure on solubility. After complete dissolution of a mineral grain, a mouldic void, preserving the original shape of the grain, can be left, if the surrounding material is sufficiently coherent. This has mostly been documented for minerals such as gypsum or calcite, but under tropical conditions it can also occur for less soluble minerals such as quartz. After their development, mouldic voids can be filled with authigenic minerals, resulting in the development of pseudomorphs after the dissolved phase. The same result can be reached by gradual dissolution and mineral formation, which often requires input of elements from external sources. Examples include goethite pseudomorphs after garnet, requiring Fe input, and gibbsite pseudomorphs after feldspar, requiring Al input (see Annex).

In the case of alteration through mineral transformation, a certain crystallographic relationship can exist between the original phase and its transformation product (topotactic replacement; e.g., vermiculite after biotite (see Fig. 1E and F), and iddingsite after olivine (see Fig. 5E and F)) (see Annex).

When a mineral is completely replaced by newly formed material, by dissolution-precipitation or transformation, an 'alteromorph' is formed, preserving the original boundaries of the weathered mineral. Alteration is described as *isomorphous* when the alteromorph has the same shape and size as the original mineral grain, *mesomorphous* when the shape has been preserved but the size has changed in one or two directions, and *katamorphous* when the shape has been completely lost (Stoops et al., 1979). More detailed subdivisions have been proposed by Delvigne (1998).

Polyphase or polygenic alteration is a common phenomenon. The original mineral is in that case altered to another product, which is, in turn, transformed to another phase. For instance, hornblende in igneous rocks can be altered to biotite by late-magmatic alteration, followed by biotite transformation to vermiculite and next to kaolinite in soil conditions, resulting in kaolinite alteromorphs after hornblende. If the first step of non-pedogenic alteration is not complete, two associated inherited mineral phases remain side by side, which may alter subsequently following different pathways, making interpretation quite difficult. Such complex situations can only be studied using microscopic techniques. More examples are given by Delvigne (1998). During pedoplasmation, most unstable alteromorphs are destroyed (see Stoops & Schaefer, 2018).

For the description of morphological types and degrees of alteration, the terminology of Bullock et al. (1985) and Stoops (2003), first proposed by Stoops et al. (1979), is available. Pellicular and irregular linear weathering patterns are commonly observed for minerals without well-expressed cleavage or fracture patterns (e.g., garnet, olivine), whereas parallel and cross-linear patterns point to minerals with prominent clear cleavage (e.g., pyroxenes, amphiboles, micas).

Besides chemical weathering, also physical weathering is visible in thin sections. The most important processes are fracturing by growth of ice or pedogenic minerals in cracks. The latter include salt weathering, which is mainly observed for arid environments, but similar processes can take place in tropical soils, for instance through gibbsite formation. Because the mineralogical and chemical composition of the grains does not change, this process cannot be detected by various methods, except thin section studies and grain-size analyses.

More information about mineral weathering is given in the chapter on saprolites (Zauyah et al., 2018). It should, however, be kept in mind that weathering in soils is often different from weathering in saprolites, where other conditions of permeability, pH or Eh and microbiological activity prevail.

2.2   Rock Fragments

Rock fragments are often a quantitatively important component of the coarse fraction of the groundmass, especially in young soils and weakly weathered materials. They can help to identify the parent material or the underlying rock substrate, and to recognise the addition of volcanic or aeolian material, as well as to assess the degree of weathering.

2.2.1   Identification and Occurrence

Identification of rock fragments is based on a combination of fabric characteristics and mineralogical composition.

The fabric of a rock, generally referred to as texture or structure by petrographers, is described in terms of absolute and relative grain size of the crystals, their shape and arrangement. For the three main categories of rocks that are distinguished, i.e., igneous, metamorphic and sedimentary, specific terms that are widely used to describe their texture are available.

Most igneous rocks and all metamorphic rocks are composed entirely of crystals (holocrystalline). Some volcanic rocks are composed mainly of volcanic glass (holohyaline or vitric) or composed of both glass and crystals (hypocrystalline). The distinction between vitric, hypocrystalline and holocrystalline is especially important when studying soils on tephra, as pedogenesis can be different for each type of material (e.g., Stoops, 2013). Igneous rocks almost always lack specific distribution and orientation patterns, except for occasional clusters or flow structures, whereas most common metamorphic rocks often show a clear parallel distribution and orientation pattern.

Clastic sedimentary rocks range in composition from claystone to conglomerate, but intergrades are most common, with an array of sand/clay ratios and distribution patterns. For calcareous sedimentary rocks, detailed systems of terminology have been developed for rock types and their components, including the widely used systems proposed by Dunham (1962) and Folk (1962). Diagenetic transformation of sedimentary deposits can involve the development of carbonate, silica or iron oxide cements as important components.

In describing the mineralogical composition of rock fragments, names of mineral species or groups are used. For igneous rocks, a distinction is commonly made between felsic minerals (quartz, feldspars and feldspathoids), with mostly light colours, and mafic minerals (e.g., amphiboles, pyroxenes, olivine, mica, opaque minerals), with mostly dark colours.

In naming igneous and metamorphic rocks, only terms that were defined and approved by nomenclature subcommissions of the International Union of Geological Sciences should be used (Le Maitre, 2002; Fettes & Desmons, 2007). Many rock types are part of a continuum, separated by quantitative limits that were agreed upon by convention. When the required quantitative data, in a statistically significant manner, cannot be obtained for rock fragments in thin sections, it is recommended to describe their fabric and composition rather than giving them a name. For igneous rocks, the field name categories proposed by Le Maitre (2002, p. 29, p. 39) can be used in this case.

Fragments of coarse-grained rocks can pose a specific problem, because the observed volume in relatively small thin section is typically not sufficiently large to be representative. Additional difficulties include optical identification of plagioclase composition for certain magmatic rocks, and uncertainty regarding rock fragment orientation that can hinder recognition of foliated metamorphic rocks. Moreover, identifications can be complicated by differential alteration, whereby the most unstable components are first destroyed. In this case, alteration products can provide some information about the nature of the altered phase.

Identification of rock fragments in thin sections requires a basic knowledge of petrography. Many useful atlases and manuals are available (e.g., MacKenzie & Guilford, 1980; MacKenzie et al., 1982; Adams et al., 1984; Yardley et al., 1990; MacKenzie & Adams, 1994; Melgarejo, 1997; Pichler & Schmitt-Riegraf, 1997; Adams & MacKenzie, 1998; Perkins & Henke, 2004). For the description of volcanic material in soils, information can be found in some review articles (e.g., De Paepe & Stoops, 2007; see also Stoops et al., 2018).

2.2.2   Alteration

Although alteration of rock fragments in soils may be quite different from that of the same rock in a saprolite, useful information on the topic can be found in some other chapters (Zauyah et al., 2018; Stoops et al., 2018), as well as in the book by Delvigne (1998).

2.3   Other Components

Besides mineral grains and rock fragments, some other types of mineral components can occur as part of the coarse fraction of the groundmass. Because they are covered in other chapters, only a brief overview is provided here, in the form of tables, both for mineral components of biological origin (Table 1) and for anthropogenic components (Table 2).

For mineral components of biological origin (Table 1), a distinction is made between remnants consisting of opal, calcium carbonates, hydroxylapatite and ceratine. In thin sections, opal is characterised by optical isotropy and relatively low relief. Calcium carbonates show high-order interference colours and negative to strongly positive relief. As example of hydroxylapatite occurrences only bone is considered here, composed of a mixture of hydroxylapatite and collagen. In thin sections, bone is characterised by light yellowish colours (PPL), first-order white to grey interference colours (XPL) and strong blue (UVF) or greenish (BLF) fluorescence. For calcareous remains, thin section identification is covered in various excellent handbooks (e.g., Adams & MacKenzie, 1998; Scholle & Ulmer-Scholle, 2003; Flügel, 2010).

Anthropogenic components are obviously mainly observed for archaeological materials and Technosols (e.g., Charzynski et al., 2013), but they can also be encountered in cultivated soils (e.g., plaggen soils, Amazonian Dark Earths), mass-transported materials (e.g., colluvium) and tsunami deposits (e.g., Bruins et al., 2008). Detailed descriptions of

Table 1   Mineral components of biological origin

Component	Main characteristics	Typical context	References
Opal (see also Gutiérrez-Castorena, 2018; Kaczorek et al., 2018)			
Phytoliths	Various forms, equant to columnar	Grassland areas, A horizons, volcanic soils	Vrydaghs et al. (2017)
Diatom frustules	Radial or bilateral symmetry	Marine sediments, wetland areas	Verleyen et al. (2017)
Sponge spicules	Elongated, with central axial canal	Marine sediments, wetland areas	Vrydaghs (2017)
Radiolaria skeletons	Various complex forms	Marine sediments	Verleyen et al. (2017)
Calcium Carbonates (see also Canti, 2017a; Durand et al., 2018)			
Bivalvia remains	Thick shells	Marine or lacustrine sediments	Canti (2017a, 2017b)
Ostracoda remains	Thin valves	Marine or lacustrine sediments	Canti (2017a, 2017b)
Gastropoda remains	Coiled shells	Marine or lacustrine sediments	Canti (2017a, 2017b)
Charophyta remains	Calcified oogonia and stems	Lacustrine sediments	—
Foraminifera remains	Tests with multiple chambers	Marine sediments	—
Egg shells	Shells composed of columnar crystals	Archaeological sites	Canti (2017c)
Apatite (see also Karkanas, 2017; Karkanas & Goldberg, 2018)			
Bone	Fibrous aspect, tubular pores	Archaeological sites, carnivore excrements	Villagran et al. (2017), Brönnimann et al. (2017)
Ceratine			
Antler, horn		Archaeological sites	Villagran et al. (2017)

Table 2   Anthropogenic components

Component	Main characteristics	Selected references
Mortar	Cryptocrystalline calcite micromass (rarely gypsum), rounded coarse grains	Stoops et al. (2017a, 2017b)
Plaster	Similar to mortar, commonly layered	Stoops et al. (2017a, 2017b)
Ceramic materials and soil affected by firing	Changed colour; less intense interference colours	Freestone (1995), Betancourt & Peterson (2009), Quinn (2013), Maritan (2017), Röpke & Dietl (2017)
Slags	Glass matrix, vesicles, skeletal fayalite crystals	Angelini et al. (2017)
Lithic artefacts	Chert, quartzite or volcanic glass composition, angular shape	Angelucci (2010, 2017)

See also Macphail and Goldberg (2018).

various types of anthropogenic coarse components can be found in several chapters of the volume edited by Nicosia and Stoops (2017), and some information is compiled in Table 2. An overview of anthropogenic materials, including their micromorphological aspects, is given by Howard and Orlicki (2016), and examples for urban contexts are presented by Mazurek et al. (2016).

3. Micromass

Whereas coarse material provides important information about parent material and degree of weathering, the composition and fabric of the micromass are a source of information about alteration, soil evolution, and mechanical behaviour. Several pedogenic changes affecting the micromass are recorded as matrix pedofeatures, which are not considered in this chapter.

3.1 Composition

Using a 10 μm c/f limit, the micromass appears uniform, as most particles are beyond the resolution that can be reached when a standard optical microscope and thin section thickness are used. Only contrasting small particles, such as opaque grains, and mineral grains with high interference colours, such as calcite or mica (sericite) particles, can be detected individually. For other micromass components, characteristics such as colour, limpidity and birefringence can be recognised for domains with strong parallel orientation of fine particles, if not masked by other components (e.g., iron oxides, organic material, carbonates).

Colour − Most types of clay have a greyish colour. Chlorite and glauconite, which are greenish, are rare exceptions. Such minerals can be dominant micromass components, as reported mainly for glauconite (De Coninck & Laruelle, 1964; Skiba et al., 2014; see also Annex). A reddish colour is always associated with the presence of fine hematite (see Marcelino et al., 2018; Stoops & Marcelino, 2018). Intensely reddish colours are already observed at hematite contents of 10% (Baert & Van Ranst, 1997). A yellow or yellowish brown colour seems associated with the presence of goethite in the clay fraction, but also ferrihydrite might be responsible. Brown to purplish brown colours can be caused by the presence of organic material. The staining effect of iron oxides is clearly demonstrated by selective dissolution treatment of uncovered thin sections (Arocena et al., 1990), and similar tests exist for organic matter (Babel, 1964; see also Stoops, 2003). Changes in colour of Bt horizons in a soil sequence, evaluated by digital measurements, reveal changes in parent material composition (Brzychcy et al., 2012).

Limpidity − The limpidity of the micromass may yield information about particle size. The finer the particle size, the more limpid the micromass will appear. Based on observations for illuvial clay coatings, complete limpidity is reached when the clay is finer than 0.2 μm. Limpidity is strongly reduced by the presence of contrasting small particles (e.g., organic material). A cloudy limpidity has been described for materials with high concentrations of cryptocrystalline gibbsite (Bennema et al., 1970; Marcelino et al., 2018; Stoops & Marcelino, 2018).

Birefringence — Low interference colours (first-order grey) generally point to kaolinite, chlorite or palygorskite-sepiolite compositions. Higher interference colours are only compatible with 2:1 phyllosilicate minerals, such as mica, vermiculite and smectite.

3.2   Fabric

Russian micromorphologists have a long tradition of describing the fabric of the fine material, as microtexture, using national sedimentological terminology (see Dobrovol'ski, 1983, 1991). In other parts of the world, micromass fabrics have received systematic attention only since the 1960s, when Brewer (1964a, 1964b) formulated concepts and terminology to describe the 'plasmic fabrics' of fine material (Table 3). Based on the work of Brewer and Russian authors, the concept of 'b-fabric' was introduced by Bullock et al. (1985). The earlier Russian concepts often seem inconsistent to non-Russian scientists, but this is mainly explained by the use of different expressions for the same term in translations.

Table 3   Types of b-fabrics and their equivalents in other description systems

b-Fabrics Bullock et al. (1985), Stoops (2003)	Plasmic fabrics Brewer (1964a, 1964b)	Microtextures Dobrovol'ski (1983)[a]
Undifferentiated	Isotic, undulic	Isotropic
Crystallitic	Crystic	—
Speckled	Asepic	Scaly
Stipple-speckled	—	Separate scaly
Mosaic-speckled	—	Coarse scaly
Striated	Sepic	Fibrous
Porostriated	Vosepic	—
Granostriated	Skelsepic	—
Monostriated	—	—
Parallel striated	Masepic	Flow fibrous
Cross-striated	Masepic, lattisepic	Cross fibrous
Random striated	Omnisepic	Random fibrous
Circular striated	—	—
Concentric striated	—	Concentric fibrous
Crescent striated	—	—
Strial	Strial	—
	Unistrial	
	Bistrial	

[a]Review of Russian literature.

The presence of optically anisotropic zones showing interference colours in the micromass of soils was already observed by Kubiëna (1938), who considered it to be a characteristic of dispersed ('peptised') clay, in contrast to flocculated ('pectised') fine material with optically isotropic behaviour. Zones showing interference colours were called 'birefringent streaks' (*Doppelbrechende Schlieren*) by Kubiëna (1948), without making a distinction between parallel-oriented clay within the groundmass and as part of illuvial clay coatings. This approach was used till the 'plasmic fabric' concept of Brewer (1964a, 1964b) was introduced, until this was in turn replaced by the b-fabric concept (Bullock et al., 1985).

Optically anisotropic zones are caused by parallel arrangement of clay-sized plate-shaped phyllosilicate particles (e.g., Mitchell, 1956; Minashina, 1958; Stephen, 1960). Clay particles are small (<2 µm equivalent diameter) relative to the standard thickness of a thin section (20-30 µm), resulting in superposition of many particles. If stacking of the individual particles is random, the effects of their optical properties on transmitted light will annihilate each other, resulting in a statistical isotropy. In zones where clay particles have a parallel arrangement, the contributions of individual particles will enhance each other and the zone will behave optically as a single crystal. In most cases, clay plates are organised in small aggregates of about 20 µm, called 'domains' by Aylmore and Quirk (1959) and 'pseudocrystals' in Russian literature. Such domains display optical anisotropy, visible as small dots showing interference colours. Grim (1962) supposed this was due to 'form birefringence', caused by parallel alignment of crystals in a medium with a different refractive index, but this was later disproven (Morgenstern & Tchalenko, 1967a). Mitchell (1956) obtained domain size values of 10-800 µm (mainly 15-150 µm) for various clays, both natural and remoulded.

Information about the submicroscopic fabric of the fine material of soils is scarce, although several studies were done on natural and remoulded clays in soil mechanics. TEM studies of microtome sections and replicas have demonstrated the presence of domains in red earths (Silva et al., 1965). A high degree of preferred orientation was proven for shear zones by TEM analysis by Foster and De (1971). Eswaran (1983) performed SEM studies of the fabric of the groundmass of various soils, concluding that it depends on mineralogical composition. Observations for a series of monomineralic micromass materials showed that (i) micromass materials composed of illite, chlorite or vermiculite comprise booklets with face-to-face arrangement, forming domains that can be stacked with a card-house fabric, (ii) kaolinite plates show subparallel packing, producing domains with parallel or card-house arrangement, (iii) smectitic material has a wavy appearance because of curling of particle edges, and it is characterised by face-to-face packing, with strong parallel orientation in slickensides, (iv) allophanic soils show a globular fabric consisting of spherical aggregates (5-500 µm), which also characterises allophane that encapsulates the other clay particles, and (v) halloysitic materials show a random arrangement of tubular particles. In the same study, oxic material is reported to display small aggregates consisting of very fine, randomly oriented kaolinite, except along the sides of the aggregates, where kaolinite particles are parallel oriented. This is in agreement with TEM observations for ultrathin sections by Bresson (1981), showing both edge-to-edge and edge-to-face configurations for kaolinite particles (0.05-0.5 µm) in ferralitic soils.

In the descriptive system of Bullock et al. (1985) and Stoops (2003), five main types of b-fabric are considered, based on the presence or absence and patterns of crystals and domains showing interference colours (undifferentiated, crystallitic, speckled, striated, strial), partly with several subtypes. In an image analysis study of b-fabrics, not all types could be identified, although they were recognised visually, and the possibility that some observed b-fabric types are in fact an optical illusion was considered (Zaniewski, 2001; Zaniewski & van der Meer, 2005).

3.2.1   Undifferentiated b-Fabrics

The b-fabric of the micromass is considered as undifferentiated when no interference colours appear in XPL, using sufficiently high magnification and light intensity. The absence of interference colours can have different reasons: (i) absence of optically anisotropic fine material, (ii) absence of oriented platy or fibrous particles, and (iii) masking of interference colours by other substances.

Absence of optically anisotropic fine material – Undifferentiated b-fabrics have been described by many authors for soils on volcanic material, especially Andisols (see Stoops et al., 2018). These soils are characterised by a micromass consisting of nanocrystalline materials (allophane, imogolite). An undifferentiated b-fabric can also be observed for siliceous duricrusts, when the micromass is dominated by colloidal silica or opal (Gutiérrez-Castorena, 2018). It also characterises spodic materials dominated by Al and Fe colloids or by pure organic fine material, such as dopplerite and monomorphic fine organic material (see Van Ranst et al., 2018).

Absence of oriented platy clay particles – As discussed before, random arrangement of platy clay particles results in apparent optical isotropy. Completely randomly arranged clay particles (flocculated) are exceptional in nature, occurring only along the sediment-water interface in certain recent deposits, especially at high salinity. Such deposits are commonly transformed to materials with parallel horizontal alignment of the clay particles upon burial (e.g., Moon, 1972). Randomly arranged clay is therefore not expected to be common in soils, unless external factors prohibit parallel orientation. The presence of large amounts of iron oxides or humic material is often considered to prevent alignment. A negative correlation between degree of clay particle alignment (birefringence ratio; see Section 5) and organic matter content was observed by Morgenstern and Tchalenko (1967a), who concluded that organic material hinders parallel clay orientation. In soils on volcanic ash, undifferentiated b-fabrics are observed for horizons dominated by halloysite, which is probably related to the tubular shape of halloysite particles (Eswaran, 1983).

Masking effect – In some cases an undifferentiated b-fabric is caused by the masking effect of the presence of other substances. This is most notably the case for Oxisols, due to the presence of abundant small hematite particles (Marcelino et al., 2018). This is clearly demonstrated by the appearance of weak striated b-fabrics following selective extraction of iron and manganese oxides in uncovered thin sections (Arocena et al., 1990; Stoops, 2003). Similarly, speckled and striated b-fabric can be recognised for iron-depletion features in oxic materials with undifferentiated b-fabric. In surface horizons, masking of interference colours by organic compounds may explain the presence of an undifferentiated b-fabric, but this has never been demonstrated.

3.2.2   Crystallitic b-Fabrics
By definition, a crystallitic b-fabric is caused by the abundant presence of small optically anisotropic particles. The components giving rise to this fabric are larger than clay size but smaller than the c/f limit that is used, otherwise the particles should be described as part of the coarse fraction.

The most common example is calcitic crystallitic b-fabrics, which occur mainly in soils of arid to subhumid regions (Durand et al., 2018). In most cases the calcite crystals are of pedogenic origin, but a detrital origin is common in soils on various types of sedimentary parent materials. The crystals almost always show a random distribution. In standard thin sections (20-30 µm), the crystals are often overlapping, but when the thickness of the section is reduced, the crystals can be observed as non-overlapping particles (see Stoops, 2003). Removal of calcite by acid treatment of uncovered thin sections can reveal a masked speckled or striated b-fabric (Wilding & Drees, 1990). Similar observations can be made for calcite depletion hypocoatings (e.g., Mees & Van Ranst, 2011; Mees, 2018).

Other examples include micaceous crystallitic b-fabrics, which can, for instance, be observed for young soils on marine clays. The relatively coarse mica particles may show various orientation patterns, described in the same way as for materials with striated b-fabrics. In bauxitic materials, gibbsitic crystallitic b-fabrics are common (see Stoops & Marcelino, 2018), and quartzic cystallitic b-fabrics have been described for silcretes with chalcedony or non-fibrous microcrystalline quartz as micromass material (see Gutiérrez-Castorena, 2018).

3.2.3   Speckled, Striated and Strial b-Fabrics
Speckled, striated and strial b-fabrics are the result of the presence of domains of parallel-arranged platy clay particles. Mitchell (1956), studying the influence of compaction on clay fabrics, and Morgenstern and Tchalenko (1967b), focussing on shear stress effects, were among the first to investigate experimentally the formation of zones with interference colours in clayey materials. Natural samples of sedimentary clay show greater amounts of randomly oriented particles than remoulded material, in which parallel orientations are more common (Mitchell, 1956). As soils can be considered as

materials remoulded repeatedly by pedoturbation processes (e.g., swelling-shrinking, bioturbation), domains of oriented clay can be expected to be omnipresent. Because dispersed clay (with parallel arrangement) always occupies a smaller volume than flocculated clay (with random orientation), the relative amount of oriented clay increases under pressure (Mitchell, 1956). Differences between dispersed and flocculated material are also expressed by b-fabrics of non-deformed material, as illustrated by strial b-fabrics obtained for clays settled in distilled water, in contrast to speckled b-fabrics that develop when phosphogypsum is added as flocculant (Southard et al., 1988).

3.2.3.1   Speckled b-Fabrics

When shear stress is weak relative to the shear strength of the material, no striated b-fabric will develop and a speckled b-fabric can be preserved. Speckled b-fabrics can therefore be dominant in soil materials that are not subjected to shrink-swell forces or in which vertical constraints are missing, as in surface horizons. The orientation of domains in materials with speckled b-fabrics has been identified as random (Hill, 1970) and shows a card-house arrangement, both before and after compaction (Bresson & Zambaux, 1990). Mosaic-speckled b-fabrics characterise the groundmass of soils and saprolites composed of relatively coarse clay aggregates without preferential orientation, which has been reported for materials dominated by minerals such as palygorskite (e.g., Stahr et al., 2000) and antigorite (e.g., Lambiv Dzemua et al., 2011).

3.2.3.2   Striated b-Fabrics

Striated b-fabrics are the result of parallel arrangement of clay domains (Brewer, 1964a, 1964b), forming striae. Clay particles are aligned parallel to planes of shearing in shear zones, but preferred orientation is also increased by compaction (Moon, 1972). As reviewed by Jim (1990), swelling pressure is exerted when the increase in volume is counteracted by lateral or vertical confinement. Stress develops when a difference in swelling exists between vertical and lateral directions. When this stress exceeds the shear strength of the soil, shear deformation occurs, which only takes place during the swelling stage. It results in two sets of shear planes at an angle of 45 degrees to the direction of the strain, reported as an angle of 35-55 degrees to the horizontal by Magaldi (1974). In thin sections, the apparent angle depends on the orientation of the section. With increasing bulk density of the material, the fabric is more strongly expressed (Jim, 1986). Using the method of Morgenstern and Tchalenko (1967a) to study remoulded Bt horizon material, Greene-Kelly and Mackney (1970) concluded that drying as such has no impact on the total degree of orientation within the micromass. The impact of earthquakes on clay arrangement in the micromass has been studied experimentally by Magaldi et al. (1994), showing only changes at submicroscopic level.

The nature of features related to shearing in soils was first studied using optical methods by Morgenstern and Tchalenko (1967b). An early TEM study, using ultrathin sections of experimentally deformed material, was performed by Foster and De (1971). They concluded that a high degree of preferred orientation occurs in shear-induced structures compared with the original material, and they observed that these striations have irregular boundaries and variable width, despite their rather regular appearance when studied using optical microscopy.

Striated b-fabrics are most commonly recognised for soil materials with a single-spaced porphyric to fine monic c/f-related distribution pattern, subject to wetting and drying. They are especially well expressed in soil materials subjected to vertic or cryogenic processes (Kovda & Mermut, 2018; Van Vliet-Lanoë & Fox, 2018). Blokhuis et al. (1970) concluded that no simple relationship exists between b-fabric type and soil classification category for soils with vertic properties.

Porostriated b-fabrics — In natural soils, porostriated b-fabrics are commonly observed for vertic materials (Kovda & Mermut, 2018), in which they occur along accommodating planar voids that correspond to slickensides in the field. SEM images of slickensides clearly show parallel orientation of clay particles (Stoops & De Mets, 1970; Eswaran, 1983). A study by Hill (1970) yielded orientation diagrams demonstrating a close parallelism between the orientation of planar voids and that of domains in zones with porostriated b-fabric, in contrast to random orientations in the remaining micromass. Porostriated b-fabrics are also observed along root channels in clayey soils (Blevins et al., 1970; Stoops, 2003). These observations partly pertain to channels with roots *in situ*, suggesting that the clay became oriented by lateral pressure exerted by the growing root. Porostriated b-fabrics can be created as artefacts when cutting moist clay samples with a spade or knife, or by inserting a Kubiëna box in the clay, so care should be taken in recording clay orientation along sample edges.

Granostriated b-fabric — In conditions of isotropic stress, such as during swelling without confinement, no b-fabrics related to shearing can develop, but orientation of clay particles can still take place around coarse mineral grains, crystals and hard nodules (Dalrymple & Jim, 1984; Jim, 1986). The presence of granostriated b-fabrics in the absence of poro-, mono- or random striation therefore points to isotropic stress. Granostriated b-fabrics also form in the presence of shear stress, as confirmed by many descriptions of b-fabrics of vertic materials (see Kovda & Mermut, 2018). Blokhuis et al. (1970) observed that the thickness of the zone showing granostriation increases with increasing size of the surrounded coarse component. The presence of a granostriated b-fabric around a nodule indicates that the latter behaves as a hard body. Stoops (1972) suggested that granostriated b-fabrics can be one of the factors in the development of concentric Fe/Mn nodules, related to variations in permeability at a microscale and explaining why rather diffuse concentric nodules are common in Vertisols.

Mono-, parallel-, cross- and random striated b-fabrics — Monostriated b-fabrics are often observed as a prolongation of porostriated zones along planar voids, suggesting that they developed in part along pores that are now closed. Random striated b-fabrics form as a result of several successive periods of shearing, which is characteristic of most vertic materials, unless masked by a calcitic crystallitic b-fabric. It is, however, not restricted to materials meeting the requirements for vertic properties in the main soil classification systems. In clayey soils with ustic or xeric moisture regimes, striated b-fabrics are often formed by shear deformation of illuvial clay coatings, which also resulted in closure of the pores along which they were deposited (see Kühn et al., 2018). Striated b-fabrics are also recognised for landslide deposits (see Mücher et al., 2018).

Circular and concentric striated b-fabrics – Circular striated b-fabrics are relatively rarely observed. One of the best illustrated cases is that of oxic materials (Marcelino et al., 2018), in which weak striation occurs around granules with undifferentiated b-fabric. Similar patterns have been recognised in SEM studies (Eswaran, 1983). In soil materials studied after removal of easily dispersible clay, circular striated b-fabrics are not anymore present (Embrechts & Stoops, 1987). Circular striated b-fabrics have also been observed for Vertisols (Field et al., 1997), as well as for thixotropic clay (Stoops, unpublished data). Bresson (1981) observed similar concentric fabrics for illitic clays, at a submicroscopic scale.

Crescent striated b-fabric – Crescent b-fabrics have often been observed for clayey tropical soils (Stoops, 2003; see also Marcelino et al., 2018), and they have also been reported for polder soils (Miedema, pers. comm.). They might be generated by small worms, such as nematodes.

Strial b-fabrics – Strial b-fabrics are typically inherited from clayey parent materials, deposited in calm conditions with good dispersion. The same fabric can also characterise saprolites derived from mudstones or shales. The only pedogenic occurrence of materials with strial b-fabrics are surface crusts (Williams et al., 2018). The presence of a strial b-fabric points to the absence of pedoturbation. Magaldi (1987) showed that the strial fabric of recent lake deposits was totally obliterated over a period of 100 years of soil development.

3.2.3.3 Comparative b-Fabric Studies

The type and development of b-fabrics is an essential criterion in several indexes created to measure the stage of soil development based on micromorphology (e.g., Dorronsoro, 1994; Magaldi & Tallini, 2000; Khormali et al., 2003). This was already implied by Cagauan and Uehara (1965), who suggested that strongly weathered tropical soils with an undifferentiated b-fabric are less fertile than those whose micromass shows interference colours. For B horizons of soils representing various stages of soil development on basalt in a tropical region, Eswaran (1972) recognised a trend in b-fabric expression. The speckled b-fabric that is observed for Entisols evolves to striated, with a maximum expression in Inceptisols, and gradually declines again to speckled in Ultisols, next becoming undifferentiated in Oxisols.

In another type of comparative b-fabric studies, Zainol and Stoops (1986) found that a relationship exists between the development of optical anisotropic domains and the plasticity index of soil material (Table 4). A relationship between soil plasticity and the development of clay orientation had earlier been observed by Brewer and Blackmore

Table 4 Relationship between plasticity index, b-fabric and soil type

Plasticity index (%)	Type of b-fabric	Soil order
5-7	Undifferentiated	Oxisol
5-12	Stipple speckled	Oxisol
5-17	Stipple speckled	Ultisol
12-18	Mosaic speckled	Ultisol
15-30	Random striated	Ultisol

After Zainol & Stoops (1986).

(1976), but these authors did not make a distinction between striated b-fabrics and illuvial clay coatings. A strong linear relationship between plasticity index and degree of b-fabric development, determined by point counting of domains of oriented clay, was recognised by Embrechts and Stoops (1987), for a soil sequence with ferralitic weathering. In the same study, materials with undifferentiated b-fabric were found to have a lower content of dispersible clay than those with other b-fabrics, which is probably related to a lack of parallel arrangement of the non-dispersable clay.

Overall, pedoplasmation and pedoturbation are important factors affecting b-fabrics (see Stoops & Schaefer, 2018).

4. Orientation and Distribution Patterns in the Groundmass

4.1 Basic Orientation and Distribution

In most soil materials, the basic and referred orientation and distribution patterns of groundmass constituents are random. If not, they can provide information about the nature of the rock substrate from which the soil is derived, as well as about geomorphological and pedogenic processes.

The arrangement of residual quartz grains in strongly weathered saprolites may help to distinguish between granite, gneiss or quartzite as original material. Another example are clusters of oriented orthoclase grains as remnants of perthitic intergrowths, from which the albite component has been removed through weathering (see Fig. 6A and B).

A banded distribution pattern can be observed for soils on sedimentary parent materials, or on derived saprolites, that have not yet been affected by pedoturbation. Sorting and layering are also observed for laminated colluvium (Mücher et al., 2018). Korina and Faustova (1964) published polar diagrams that clearly demonstrate preferred orientations of quartz and feldspar grains in moraine deposits, related to the glacier flow. Using the same methods, Hill (1981) and Sen and Mukherjee (1972) reconstructed the flow direction of fluvioglacial parent materials. A preferred oblique orientation of elongate grains occurs in solifluction deposits (Mücher et al., 2018).

Lafeber (1972), using polar diagrams for measurements obtained for horizontal thin sections, reported an orientation of platy particles that is parallel to the strike of the terrain slope and an orientation of elongated grains that is parallel to the dip of the slope.

Parallel arrangements of platy particles have also been observed for litter layers, surface crusts (Williams et al., 2018), frost-affected materials (Van Vliet-Lanoë & Fox, 2018), and surfaces affected by trampling (Rentzel et al., 2017).

Crescent arrangements, corresponding to passage features that are not sufficiently pore-related to qualify as infillings, were described for soils with vermic characteristics (e.g., Kooistra & Pulleman, 2018), for termite nests (Stoops, 1964), and for gypsic horizons (Barzanji & Stoops, 1974; Poch et al., 2018).

4.2   Related Distribution Between Coarse and Fine Particles

The mutual relation between coarse and fine components has been considered since the early days of soil micromorphology as an important soil characteristic. Kubiëna (1938) was the first to distinguish several types of arrangements, named 'elementary fabrics', whose definitions also take into account the state of the fine material (dispersed vs. flocculated). Brewer (1960, 1964b) distinguished four types of related distributions for plasma and skeleton grains (agglomeroplasmic, granular, intertextic, porphyroskelic), which was later replaced by the c/f-related distribution concept (Stoops & Jongerius, 1975; Bullock et al., 1985; Stoops, 2003). A system proposed by Eswaran and Baños (1976) comprises elements of microstructure, and so does the system proposed by Brewer and Pawluk (1975), Brewer (1979) and Brewer and Sleeman (1988). An overview of these systems is given by Jim (1988a). A different emphasis is used in sediment petrography, in which a distinction is mainly made between grain-supported and matrix-supported fabrics.

The c/f-related distribution pattern of the groundmass depends largely on clay content, with for instance porphyric patterns restricted to materials with high clay content, but silt/sand ratio also plays a role, especially for soils with intermediate clay contents (Dalrymple & Jim, 1984). The type of c/f-related distribution pattern can also be influenced by the type of cohesive forces (covalent vs ionic bonds) between the constituents (Chadwick & Nettleton, 1990). An overview of the occurrence of c/f-related distribution patterns in various materials is given elsewhere (Stoops et al., 2018).

In a study of changes in c/f-related distribution patterns and soil volume during soil development, Chadwick and Nettleton (1994) recognise a general evolution from coarse monic, over gefuric, chitonic or enaulic, to porphyric and finally open porphyric, whereby the increase in fine material is due either to weathering or to external input.

Enaulic c/f-related distribution patterns are particularly common in surface horizons and dominant in horizons with high organic matter contents, except when dominated by monomorphic fine material. In young soils, c/f-related distribution patterns may be the only micromorphological feature whose nature varies between different horizons or profiles (e.g., Stoops et al., 2001).

Chitonic c/f-related distribution patterns occur commonly in sandy soils when suspended matter settles on and around grains, as is the case in incipient spodic or cambic horizons. This type of c/f-related distribution pattern was obtained experimentally by Brewer and Haldane (1957) using clay suspensions in sand columns. When development increases, a combination of chitonic and gefuric is created. Not all occurrences of chitonic c/f-related distribution patterns should be attributed to illuviation, because the surface of detrital sand grains is commonly covered by very thin clay coatings, which can be continuous or discontinuous. This is prominently observed for deposits such as reddish dune sands.

5. Quantitative Analysis of the Groundmass

Two quantitative aspects of groundmass material can be determined: (i) abundance of components and fabrics, expressed as numbers per surface unit or as area or volume percentages, and (ii) orientation of grains and fabrics. Quantitative data for groundmass constituents have mainly been obtained as a by-product of the quantification of pedo-features and porosity, yielding information about c/f ratio, composition of the coarse fraction and the relative proportion of groundmass, pedofeatures and voids. Almost all published results were obtained through point counting. For quantification of the orientation of groundmass constituents, several attempts have been made to replace manual methods by image analysis techniques (e.g., Tovey et al., 1990, 1992a, 1992b; Tovey & Sokolov, 1997; Tovey & Wang, 1997). The older literature on this subject has been reviewed by Tovey (1974).

5.1   Coarse Fraction

Besides applying point counting, it has been tried by some authors to obtain quantitative data for the mineralogical composition of the coarse fraction through image analysis (e.g., Terribile & FitzPatrick, 1992, 1995; Terribile et al., 1997), but this approach is not yet clearly applicable for soils with complex mineral and rock fragment assemblages. The use of SEM-EDS mapping could be a solution, but its use is limited by the small field observed with SEM, the complexity of rock fragments, and the lack of direct mineralogical information.

Quantification of the orientation of coarse groundmass components is most easily achieved by determining the orientation of two-dimensional cross sections of plate-shaped (e.g., mica) or elongated (e.g., hornblende) particles within the plane of the thin section, which is done by rotating the graduated microscope stage and plotting the angle between the longitudinal axis and a reference point on a rosette or rose diagram (e.g., Hill, 1970). A more correct but complex method is measuring the orientation, in three dimensions, of the optical axes of mineral grains (e.g., quartz, feldspar, mica), making use of a universal stage (Fyodorov stage), and plotting them in stereographic projection (polar diagram). This second technique, commonly used in petrographical studies of metamorphic rocks, requires a good knowledge of optical mineralogy and stereographic projection theories (Lafeber, 1964, 1967, 1972). Examples of studies of this type for soil-related materials include the work of Korina and Faustova (1964) on moraine deposits, and that of Sen and Mukherjee (1972) and Hill (1981) on fluvioglacial sediments.

5.2   b-Fabric

Quantification of the b-fabric is a relatively difficult issue because many factors play a role. The basic orientation already has several components, including domain size, abundance (or occupied area), intensity, and birefringence. The referred orientation deals with the orientation of individual domains with respect to a reference feature, such as a void, a grain, or a horizontal or vertical line. For determination of relative area, expressed as percentage of the total micromass, the occurrence of extinction every 90 degrees when the microscope stage is turned is an additional problem. This can be overcome by the use of circular polarised light (see Stoops, 2003). The thickness of the thin section and the type of light source are important factors, making comparison between studies questionable. Photometric or image analysis methods for b-fabric quantification were developed by Jim (1988b), Zaniewski (2001) and Zaniewski and van der Meer (2005).

Quantitative analyses of b-fabrics have been performed since several decades, initially as part of studies in soil mechanics (Mitchell, 1956; Morgenstern et al., 1967) and later also by soil micromorphologists.

To determine the total degree of orientation of a given zone, Morgenstern and Tchalenko (1967a) proposed to use the ratio between minimum and maximum light intensity in XPL when the microscope stage is rotated ('birefringence ratio', β), measured with a photometer that is installed on the microscope. In this study, several classes were distinguished: $\beta = 1.0$ (random orientation), <1.0-0.9 (slight orientation), <0.9-0.5 (medium orientation), <0.5 (strong orientation) and 0.0 (perfect orientation). Applying this method for remoulded Bt horizons, Greene-Kelly and Mackney (1970) studied the impact of drying on the micromass.

Point counting of birefringent domains has often been used as an adequate method to determine their frequency, as done, for instance, by Mitchell (1956) in soil mechanics, and later also in soil micromorphology by Brewer and Blackmore (1976), Zainol and Stoops (1986) and Embrechts and Stoops (1987).

Stoops and Eswaran (1973) proposed a method using circular polarised light and a photometric cell mounted on a microscope, showing a clear difference between studied soil types. Overall, results are too much influenced by the light source and thin section thickness to allow comparison between results of different analysts.

The orientation pattern of domains can be measured by determining, for each individual domain, the extinction angle relative to a reference (e.g., soil surface, void wall), by rotating the microscope. A sufficient number of domains should be measured (e.g., 100, as proposed by Hill, 1970), and the results should be plotted on a rosette or rose diagram. As only orientations and no directions are measured, the upper part of the diagram is sufficient, the lower part being its mirror image. Analysis of the orientation of domains is generally hindered by a lack of information about orientation of the thin section relative to field conditions, for instance with regard to slope. Comparison between horizontal and vertical sections should be useful, but this has apparently not yet been performed.

6. Discussion and Conclusions

Although little attention has been given to the groundmass in reviews of micromorphological characteristics, important information is scattered over many publications in the fields of pedology, sedimentology and soil mechanics. Several systems for quantitative analysis of area and/or orientation of fabric components, including photometric and image analysing systems, have been proposed, but results of such analyses have only rarely been published.

Analysis of the coarse fraction requires a certain degree of knowledge of mineralogy, petrography and optical techniques, which is sometimes missing among users of soil thin sections (Shahack-Gross, 2015). This is also the case for the interpretation of alteration, for which a good knowledge of chemical reactions and balances is required. For some minerals, such as olivines, pyroxenes and feldspars, much information is available on their alteration, whereas for others, such as epidotes and amphiboles, information is limited.

The behaviour of the clay mass under stress is one of the fields in which much experimental work has been done, including optical and SEM studies. Nevertheless, several interpretations of b-fabrics were actually never proven and merit more attention.

Few attempts have been made to explain the genesis of some c/f-related distribution patterns, and more study is necessary.

Annex: Identification of Minerals

For each mineral, the following information is given:
- *Name, chemical formula* — both according to the November 2017 edition of the official list of minerals of the International Mineralogical Association; some minerals have been informally grouped under headings of non-valid but commonly used mineral names (in italics), rather than using IMA-approved terminology for hierarchical levels
- *Optical characteristics* — selected relevant properties, in plane-polarised light (colour, pleochroism, relief, shape, cleavage), cross-polarised light (interference colours at standard thin section thickness, extinction angle, optical orientation) and oblique incident light (colour, lustre), as well as autofluorescence in ultraviolet light and blue light; for numerical data the reader is referred to handbooks (e.g., Tröger, 1971; Fleischer et al., 1984)
- *Occurrence in rocks, occurrence in soils* — information about occurrence in rocks and soils, summarised from various handbooks of mineralogy, petrography and soil science (e.g., Tröger, 1969; Dixon & Weed, 1989; FitzPatrick, 1993)
- *Alteration* — information about non-pedogenic alteration is mainly based on standard handbooks of mineralogy and petrography (e.g., Tröger, 1969); information about pedogenic alteration is compiled from various publications (e.g., Delvigne & Stoops, 1990; FitzPatrick, 1993; Delvigne, 1998) and research articles (cited where relevant) (see also Zauyah et al., 2018)

Augite $((Ca,Mg,Fe)_2Si_2O_6)$

- *Optical characteristics* − colourless, pale greenish, pale purple (Ti-augite), weak or absent pleochroism, strong relief, short prismatic to xenomorphic, distinct cleavage parallel to c-axis, crossing at nearly right angles in cross sections perpendicular to the c-axis; second-order interference colours, oblique extinction (36-45 degrees), length fast; in volcanic rocks − common zonation, twinning, basaltic glass inclusions (Fig. 1A and B)
- *Occurrence in rocks* − in magmatic rocks, common in intermediate (e.g., diorites, andesites), basic (e.g., tholeiitic basalts, gabbros, dolerites) and mafic magmatic rocks; common in high-grade metamorphic rocks; also in derived recent marine and fluvial sediments; Ti-augite is common in basalt, often in association with olivine
- *Occurrence in soils* − absent in strongly weathered material
- *Non-pedogenic alteration* − highly susceptible to alteration; in magmatic rocks, pellicular transformation to hornblende, next to actinolite and chlorite; in metamorphic rocks, alteration to hornblende (uralitisation), next to chlorite, talc and epidote
- *Pedogenic alteration* − highly weatherable; initially altered by widening of cleavage planes, decolouration along these fissures, accompanied by lowering of interference colours; smectite formation within fissures, resulting in boxwork patterns, with denticulate augite remnants; in oxic material, possible complete replacement by goethite, without specific optical orientation (Eswaran, 1979)

Biotite

- *Main members* − **phlogopite** $(KMg_3(AlSi_3O_{10})(OH)_2)$ as Mg end-member, **annite** $(KFe_3^{2+}(AlSi_3O_{10})(OH)_2)$ as Fe end-member
- *Optical characteristics* − colourless to yellowish brown or green, with strong pleochroism, most pronounced colour for high-Fe members, low to moderate positive relief, plate-shaped crystals, clear cleavage parallel to plate faces; fourth-order interference colours, masked by mineral colour, extinction parallel to cleavage, bright spots remaining in extinction position (bird's-eye effect), length slow for cross sections perpendicular to plates; occasional uranium-bearing inclusions (e.g., zircon), surrounded by halo with darker colour and weaker pleochroism; oxidised form (oxibiotite) − yellow to dark red, with strong pleochroism, higher relief, higher interference colours
- *Occurrence in rocks* − in most intrusive magmatic rocks and coarse-foliated metamorphic rocks, ranging from undersaturated (with Mg-rich biotite) to saturated (with Fe-rich biotite); also in derived sedimentary rocks; rare in volcanic rocks (as oxibiotite)

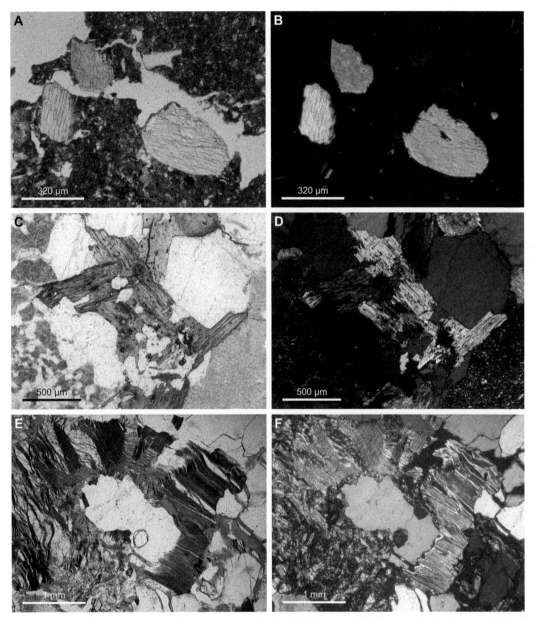

FIGURE 1 (A) Residual augite grains, showing strong relief and distinct parallel cleavage (Andosol, Rwanda) (PPL). (B) Idem in XPL. (C) Biotite, altered to chlorite, with lighter colours and absence of pleochroism of transformed parts (altered granite, Okertal, Germany) (PPL). (D) Idem in XPL, illustrating abnormal bluish interference colours of chlorite. (E) Biotite, altered to a vermiculite-like phase (right) and to kaolinite (e.g., upper left) (saprolite on granite, Akagera National Park, Rwanda) (PPL). (F) Idem in XPL, illustrating second-order interference colours of the vermiculite-like component and absence of bird's-eye effect.

- *Occurrence in soils* – in weakly to moderately weathered material
- *Alteration* – reviewed by Bisdom et al. (1982); some inclusions weather more easily than the biotite host, leaving dissolution cavities (Ghabru et al., 1987)
- *Non-pedogenic and pedogenic alteration to chlorite* – deuteric alteration with pellicular pattern; pedogenic alteration affecting the whole crystal (De Coninck et al., 1987; Stoops & Dedecker, 2006) (Fig. 1C and D)
- *Non-pedogenic and pedogenic alteration to vermiculite or interstratified phyllosilicates* – decrease in colour, pleochroism and relief, lowering of interference colours, disappearance of 'bird's-eye effect' (Fig. 1E and F); common subsequent alteration to kaolinite
- *Pedogenic alteration to kaolinite (less commonly hydrothermal)* – wedge-shaped penetration of kaolinite between biotite lamellae, causing exfoliation, resulting first in sheaf-like aspect, commonly with goethite formation along contact with soil micromass (Penven et al., 1981; Taboada & García, 1999a), kaolinite oriented either parallel (e.g., Wilson, 1966) or perpendicular (e.g., Seddoh & Robert, 1972) to the cleavage direction of the biotite host (SEM); expansion advances towards the centre, which can continue until a kaolinite booklet has formed, much thicker than the original biotite grain (up to 250%; Delvigne, 1998) (Fig. 2A and B); in well-drained tropical saprolites, associated goethite or hematite aggregates, especially along the sides of the pseudomorphs (Curmi & Maurice, 1981); kaolinite alteromorphs are quickly destroyed in the soil by pedoplasmation
- *Pedogenic alteration to gibbsite* – in strongly leaching conditions, with gibbsite crystals mainly oriented perpendicular to cleavage planes (Wilson, 1966, 1975; Sidhu & Gilkes, 1977); Al not necessarily derived from the biotite lattice
- *Pedogenic physical alteration* – subject to physical alteration, by frost (Bronger et al., 1976) or by crystal growth along cleavage planes (e.g., calcite, gypsum)

Calcite (Ca(CO$_3$))

- *Optical characteristics* – colourless, alternating negative to strongly positive relief when stage is rotated, distinct rhombohedral cleavage; high-order interference colours, polysynthetic twinning in crystals formed under pressure (e.g., marble); common ultraviolet and blue light fluorescence and cathodoluminescence; difficult distinction with aragonite and dolomite, except through staining tests or acid reaction (Stoops, 2003)
- *Occurrence in rocks* – in sedimentary rocks as a main constituent of limestone and other calcareous rocks, in various forms (skeletal debris, micritic material, diagenetic features); in magmatic rocks as principal constituent of carbonatites, occasionally as accessory phase resulting mostly from deuteric alteration; in metamorphic rocks as principal constituent of marble and as component of other metamorphic rocks derived from calcareous sediments, commonly with pressure twinning; possible distinction between calcite of different origins based on fluorescence or cathodoluminescence behaviour (e.g., Khormali et al., 2006)

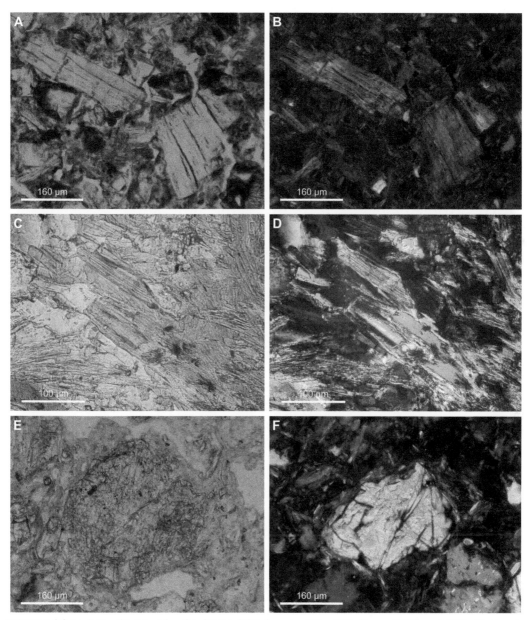

FIGURE 2 (A) Kaolinite alteromorphs after biotite (Ultisol on prasinite, Matadi area, DR Congo) (PPL). (B) Idem in XPL. (C) Pale green chlorite, weathered to yellowish brown 2:1 mixed-layer phyllosilicates (lower part of young soil on prasinite, Matadi area, DR Congo) (PPL). (D) Idem in XPL, illustrating the very low interference colours of chlorite and higher interference colours of the mixed-layer clay. (E) Epidote grain, showing high relief and light greenish colour (young soil on prasinite, Matadi area, DR Congo) (PPL). (F) Zoisite grain, showing abnormal blue interference colours (young soil on prasinite, Matadi area, DR Congo) (XPL).

- *Occurrence in soils* – non-pedogenic calcite as common inherited phase in soils of semiarid to arid regions, or in young soils on calcareous substrates; also as anthropogenic additions, including liming material; common as pedogenic phase (see Durand et al., 2018)
- *Alteration* – easily dissolved, especially at low pH; pellicular pattern of congruent dissolution, which may result in denticulate residues surrounded by void space; corrosion patterns (SEM) related to environmental conditions (Delmas & Berrier, 1987)

Chlorite

- *Main members* – **chamosite** $((Fe^{2+},Mg,Al,Fe^{3+})_6(Si,Al)_4O_{10}(OH,O))$, **clinochlore** $(Mg_5Al(AlSi_3O_{10})(OH)_8)$, **pennantite** $(Mn_5^{2+}Al(Si_3Al)O_{10}(OH))$
- *Optical characteristics* – pale green, pleochroic, moderate positive relief, plate-shaped crystals, clear cleavage parallel to plate faces; very low first-order interference colours, commonly abnormal bluish (Fe-rich chlorite) or brownish (Mg-rich chlorite), extinction parallel to cleavage, length fast for cross sections perpendicular to plates
- *Occurrence in rocks* – common constituent of low-grade metamorphic rocks; in magmatic rocks as product of deuteric alteration (e.g., of biotite) (see Fig. 1C and D); common as part of veins and infillings; in sedimentary rocks as component of unweathered fine-grained deposits
- *Occurrence in soils* – only in weakly weathered material; pedogenic chlorites are typically not observed in thin sections because of their fine grain size
- *Pedogenic alteration* – pellicular pattern, or along cleavage planes, characterised by a decrease in pleochroism and an increase in interference colours when transformed to vermiculite, smectite or interstratified phyllosilicates (Anand & Gilkes, 1984a; De Coninck et al., 1987) (Fig. 2C and D)

Enstatite-Ferrosilite

- *Main members* – **enstatite** $(Mg_2Si_2O_6)$, **ferrosilite** $(Fe_2^{2+}Si_2O_6)$; **hypersthene** as discredited name for occurrences with intermediate compositions $((Mg,Fe^{2+})_2Si_2O_6)$
- *Optical characteristics* – colourless (enstatite) to pale green or pale red with pleochroism (hypersthene), medium to strong positive relief, prismatic forms, distinct cleavage parallel to c-axis, crossing at nearly right angles in cross sections perpendicular to the c-axis; first-order interference colours, parallel extinction, length slow
- *Occurrence in rocks* – in intrusive magmatic rocks, ranging from granitic to ultramafic, common in norites, pyroxenites and peridotites; in some volcanic rocks (andesites, dacites); in certain high-grade metamorphic rocks (granulites); in derived sedimentary rocks

- *Occurrence in soils* − absent in strongly weathered material
- *Alteration* − easily weathered
- *Non-pedogenic alteration* − late-magmatic alteration to amphiboles or to chrysotile (serpentine) (alteromorphs are called bastite); deuteric alteration along cleavage planes to smectite (nontronite)
- *Pedogenic alteration* − congruent dissolution along cleavage planes, resulting in denticulate residues; also alteration to smectite, chlorite, ferrihydrite, goethite and hematite (FitzPatrick, 1993)

Epidote-Zoisite

- *Main members* − **epidote** $(Ca_2(Al_2Fe^{3+})[Si_2O_7][SiO_4]O(OH))$, **zoisite/clinozoisite** $(Ca_2Al_3[Si_2O_7][SiO_4]O(OH))$
- *Optical characteristics* − colourless to yellowish green with weak pleochroism (epidote), colourless (zoisite, clinozoisite), strong positive relief, mainly equant crystals (prismatic in vein deposits), perfect cleavage in one direction; end first-order to second-order interference colours (epidote), first-order grey with abnormal bluish tint (zoisite, clinozoisite) (Fig. 2E and F)
- *Occurrence in rocks* − typical components of low-grade metamorphic rocks, (greenschist facies) (in association with actinolite, albite, chlorite) and of metamorphic rocks derived from impure limestone; in magmatic rocks as hydrothermal alteration product of plagioclase, pyroxenes and amphiboles (saussuritisation) (Fig. 3A)
- *Occurrence in soils* − only in weakly weathered material
- *Pedogenic alteration* − scarce information; congruent dissolution with pellicular or irregular linear patterns, without associated deposition of specific reaction products

Garnet

- *Main members* − **almandine** $(Fe_3^{2+}Al_2(SiO_4)_3)$, **andradite** $(Ca_3Fe_2^{3+}(SiO_4)_3)$, **grossular** $(Ca_3Al_2(SiO_4)_3)$, **pyrope** $(Mg_3Al_2(SiO_4)_3)$, **spessartine** $(Mn_3^{2+}Al_2(SiO_4)_3)$
- *Optical characteristics* − colourless, strong to extreme positive relief, equant crystals, no cleavage; optically isotropic; no optical distinction between species
- *Occurrence in rocks* − in metamorphic rocks as common component, especially in micaschist, hornblende-bearing schist and gneiss (almandine), also in metamorphic rocks derived from impure limestone (andradite, grossular); in undersaturated magmatic rocks such as kimberlite (pyrope)
- *Occurrence in soils* − rare
- *Non-pedogenic alteration* − common deuteric alteration to talc, serpentine and chlorite
- *Pedogenic alteration in saprolites* − pellicular and irregular linear patterns (due to absence of cleavage), with development of goethite coatings with palisade fabric

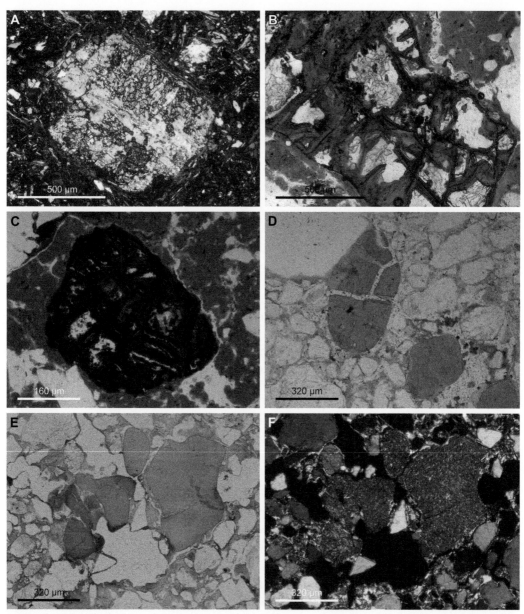

FIGURE 3 (A) Saussuritisation of plagioclase grain (young soil on prasinite, Matadi area, DR Congo) (XPL). (B) Garnet alteration, producing a boxwork pattern by goethite formation along cracks, separating isolated garnet remnants (oxic material derived from metamorphic rocks, Mont Febe, Cameroon) (PPL). (C) Goethite alteromorph after garnet, formed by complete congruent dissolution of garnet and infilling of boxwork cells by cryptocrystalline goethite (oxic material derived from metamorphic rocks, Mont Febe, Cameroon) (PPL). (D) Physical degradation of glauconite (Alfisol, Campine area, Belgium) (PPL). (E) Chemical degradation of glauconite through oxidation (brown colour), and formation of glauconitic micromass (Alfisol, Campine area, Belgium) (PPL). (F) Idem in XPL.

along cracks, resulting in boxwork system, followed by congruent dissolution of garnet residues, leaving cells that become filled with fine-grained goethite and less commonly with gibbsite (e.g., Embrechts & Stoops, 1982; Parisot et al., 1983; Velbel, 1984) (Fig. 3B); near the soil surface, the goethite pseudomorphs may be transformed to hematite or become physically degraded; the internal fabric of iron oxide pseudomorphs can be preserved during colluvial or alluvial transport, allowing detection of the former presence of garnet in source rocks of reworked material (Fig. 3C); cracks in garnet crystals can also be filled with authigenic phyllosilicates (Bouchard et al., 1995); in saprolites on specific manganese ore deposits, spessartine is altered to lithiophorite and next to cryptomelane, in the form of pseudomorphic replacement (Grandin & Perseil, 1977; Beauvais & Nahon, 1985)

- *Pedogenic alteration in soils* − alteration to microcrystalline goethite alteromorphs without visible boxwork fabric; complete replacement by kaolinite in tropical soils (Muggler et al., 2007)

Glauconite $((K,Na)(Fe^{3+},Al,Mg)_2(Si,Al)_4O_{10}(OH)_2)$

- *Optical characteristics* − green to yellowish green, pleochroic, low to moderate relief, as aggregates with xenotopic or fibrous fabric (mosaic-speckled b-fabric); first- to second-order interference colours, masked by mineral colour
- *Occurrence in rocks* − typically as pellets in shallow marine deposits, also as infillings of foraminifer tests and other bioclasts; also reworked and incorporated in younger sediments as allochthonous grains
- *Occurrence in soils* − mainly in weakly weathered soil materials derived from sandy marine deposits
- *Pedogenic alteration* − physical alteration can transform glauconite pellets to a green micromass (De Coninck & Laruelle, 1964; Skiba et al., 2014) (Fig. 3D); chemical alteration initially produces pellicular patterns (formation of brownish iron oxides, related to oxidation of Fe^{2+}), gradually affecting the entire grain (Fig. 3E and F), with gradual glauconite transformation to smectitic clay (Van Ranst & De Coninck, 1983; Skiba et al., 2014)

Hornblende

- *Main members* − **magnesiohornblende** $(Ca_2(Mg_4Al)(Si_7Al)O_{22}(OH)_2)$, **ferrohornblende** $(Ca_2(Fe_4^{2+}Al)(Si_7Al)O_{22}(OH)_2)$
- *Optical characteristics* − green to brown, strongly pleochroic, moderate to high positive relief, lath-shaped crystals, distinct cleavage parallel to c-axis, crossing at 60/120-degree angles in cross sections perpendicular to the c-axis; end first-order interference colours, oblique extinction (c. 20 degrees), length slow
- *Occurrence in rocks* − very common constituent of intrusive magmatic rocks (e.g., granite, diorite, syenite) and metamorphic rocks (e.g., schist, gneiss, amphibolite, associated with plagioclase); rare in volcanic rocks (e.g., andesite)

- *Occurrence in soils* – in weakly to moderately weathered material
- *Non-pedogenic alteration* – deuteric alteration to talc, chlorite, epidote, calcite and quartz
- *Pedogenic alteration* – congruent dissolution, with parallel or cross-linear patterns (parallel to cleavage), combined with pellicular patterns for isolated grains, with denticulate aspect of unaltered remnants (Anand & Gilkes, 1984a) (Fig. 4A and B); in initial phase, associated smectite formation in cracks, producing a boxwork pattern (Fig. 4C and D) (Verheye & Stoops, 1975)

Ilmenite ($Fe^{2+}Ti^{4+}O_3$)

- *Optical characteristics* – opaque, platy or anhedral; black metallic lustre in OIL
- *Occurrence in rocks* – in magmatic rocks as accessory component, more commonly in undersaturated rocks than in saturated rocks, with high concentrations in gabbro and anorthosite; in sedimentary rocks, locally as placer deposit
- *Occurrence in soils* – small quantities of single grains
- *Alteration* – more stable than magnetite; common pedogenic and non-pedogenic alteration to leucoxene, a microcrystalline mixture of various Ti minerals (opaque, cream colour in OIL)

Magnetite ($Fe^{2+}Fe_2^{3+}O_4$)

- *Optical characteristics* – opaque, commonly octahedral (rectangular or triangular cross sections), also clustered irregular grains, grey metallic lustre in OIL
- *Occurrence in rocks* – in magmatic rocks as accessory mineral, most abundant in gabbro/basalt and related rocks; in metamorphic rocks; in sedimentary rocks, locally as placer deposit
- *Occurrence in soils* – highly stable and therefore common in all soils, concentrated in strongly weathered soils; may occur as the only inherited phase following weathering of volcanic ash (Stoops, 2013; Stoops et al., 2018)
- *Deuteric alteration* – common deuteric alteration to hematite pseudomorphs (martite)
- *Pedogenic alteration* – in soils, slow transformation to ferrihydrite or goethite; in laterites, pellicular and linear alteration to porous hematite pseudomorphs, partly with characteristic lamellar martite fabric (Anand & Gilkes, 1984b)

Microcline ($K(AlSi_3O_8)$)

- *Optical characteristics* – colourless, low negative relief, mainly equant grains, perfect cleavage in one direction, imperfect in others; first-order grey interference colours, typical polysynthetic twinning in two perpendicular directions (Fig. 4E)
- *Occurrence in rocks* – in saturated leucocratic magmatic rocks (e.g., granite, granodiorite, syenite), especially in pegmatites; in some metamorphic rocks (gneiss); also in derived sedimentary rocks

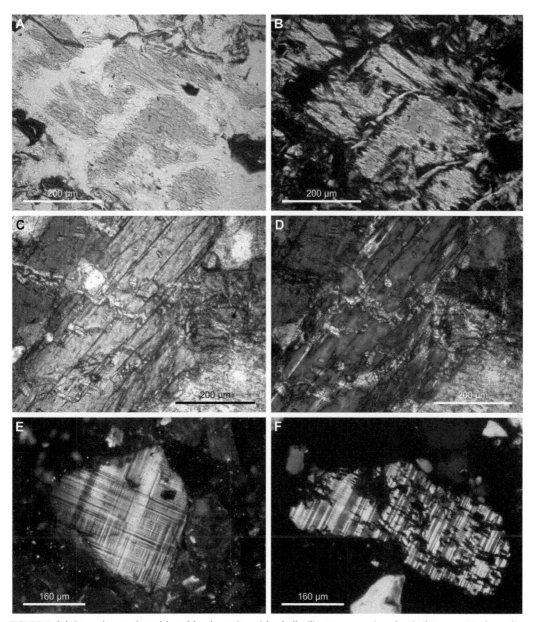

FIGURE 4 (A) Strongly weathered hornblende grain, with phyllosilicates separating denticulate remains (saprolite beneath Oxisol on granite, Ivory Coast) (PPL). (B) Idem in XPL. (C) Hornblende grain, showing early stages of weathering, in the form of fracturing and phyllosilicate formation (alteration products stained reddish by safranine) (saprolite beneath Oxisol on granite, Ivory Coast) (PPL). (D) Idem in XPL. (E) Unaltered microcline grain, showing characteristic cross-hatched pattern of polysynthetic twinning (young soil on prasinite, Matadi area, DR Congo) (XPL). (F) Weathered microcline grain, with iron oxide infillings (Oxisol on granite, Ivory Coast) (XPL).

- *Occurrence in soils* − relatively stable
- *Alteration* − pellicular and cavernous patterns (Fig. 4F); tubular weathering has been observed for microcline in Podzols (Jongmans et al., 1997)

Muscovite ($KAl_2(Si_3Al)O_{10}(OH)_2$)

- *Optical characteristics* − colourless, low to medium relief; plate-shaped crystals, clear cleavage parallel to plate faces; upper second-order interference colours, extinction parallel to cleavage, length slow for cross sections perpendicular to plates
- *Occurrence in rocks* − abundant in granitoid rocks (random orientation), micas-chists and gneisses (parallel orientation); common in certain sedimentary rocks, oriented parallel to bedding; fine-grained muscovite (and/or paragonite), called sericite, occurs as non-pedogenic alteration product of alkali feldspar (sericitisation)
- *Occurrence in soils* − because of its resistance to weathering, concentrated in strongly weathered soils and therefore allowed for oxic horizon criteria (Soil Survey Staff, 2014)
- *Alteration* − physical alteration by separation of lamellae (Pinto et al., 2015) or by interlamellar crystallisation of authigenic minerals (e.g., salt minerals, gibbsite)

Olivine

- *Main members* − **forsterite** ($Mg_2(SiO_4)$), **fayalite** ($Fe_2^{2+}(SiO_4)$)
- *Optical characteristics* − colourless, pale green (forsterite) or yellowish brown (fayalite), weak pleochroism, medium to strong relief, equant crystals, no distinct cleavage; second- to third-order interference colours (Fig. 5A and B); glass inclusions point to volcanic origin
- *Occurrence in rocks* − in undersaturated magmatic rocks, including occurrences as a dominant (xenotopic) constituent of mafic and ultramafic intrusive rocks (e.g., peridotite, dunite, harzburgite), also as (euhedral) phenocrysts in dolerites and basalts, often with zoning (Mg-rich in the centre, Fe-rich at the border); in metamorphic rocks, derived from dolomite-bearing rocks (forsterite); in sedimentary rocks, locally as placer deposit
- *Occurrence in soils* − in young or poorly weathered material on olivine-bearing rocks, including some serpentinites
- *Alteration* − see Delvigne et al. (1979) and Delvigne (1998); highly weatherable (cf. Goldich, 1938), but because of the lack of cleavage, it may be more stable in soils than augite (Stoops, 2013); pellicular and irregular linear alteration patterns
- *Alteration to serpentine* − common hydrothermal alteration to serpentine or talc in undersaturated rocks, often with exsolution of magnetite crystallites, producing serpentinites from which unaltered olivine remains may be liberated to form part of the coarse fraction of derived soils; serpentine development can be followed by iddingsitisation of the remaining olivine, leaving serpentine unaffected

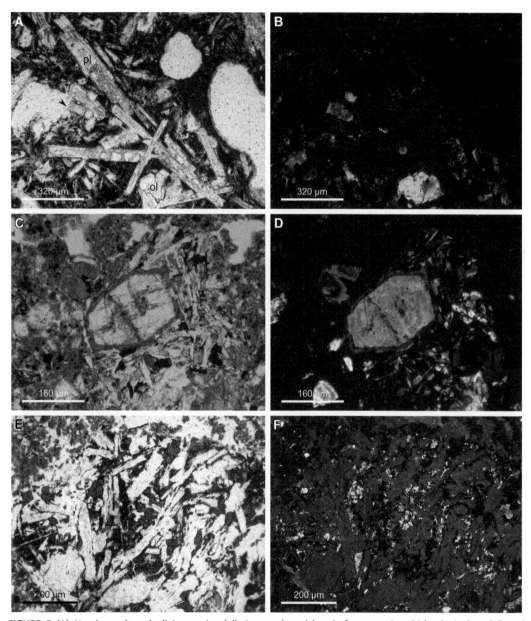

FIGURE 5 (A) Nearly unaltered olivine grains (ol), in weathered basalt fragment in which plagioclase (pl) and interstitial augite (*arrow*) are almost entirely weathered (volcanic ash soil, Isla Santa Cruz, Galápagos) (PPL). (B) Idem in XPL. (C) Pellicular alteration of olivine to iddingsite (volcanic ash soil, Isla Santa Cruz, Galápagos) (PPL). (D) Idem in XPL. (E) Iddingsite grains in weathered basalt (volcanic ash soil, Isla Santa Cruz, Galápagos) (PPL). (F) Idem in XPL, illustrating that the plagioclase is completely altered to optically isotropic colloidal material.

- *Alteration to 'chlorophaeite' (chlorite-related amorphous material)* – deuteric alteration, leaving denticulate residues (Delvigne, 1998)
- *Alteration to 'iddingsite' (mixture of cryptocrystalline goethite and phyllosilicates, such as smectite, chlorite and talc)* – common product of deuteric alteration, advancing inward from the sides along cracks (in intrusive magmatic rocks), or with pellicular patterns (in volcanic rocks) (Fig. 5C and D) (Delvigne et al., 1979), in both cases yielding partly denticulate remnants; can be followed by further olivine alteration with associated zeolite formation (Delvigne, 1998); iddingsite pseudomorphs are characterised by yellow to reddish brown colours, weak to distinct pleochroism, masked second- to third-order interference colours, optical single-crystal behaviour, with same extinction position as the olivine host, sharp to gradual boundaries, in contrast to goethite pseudomorphs; iddingsite rims can be covered by olivine, pointing to late-magmatic iddingsite formation (Sheppard, 1962; Ildefonse, 1983); common alteration of olivine residues to other minerals within the iddingsite pseudomorph; pedogenic alteration to iddingsite is not excluded, but as yet not proven; in soil conditions, iddingsite rims can protect the remaining olivine from pedogenic alteration; occasionally as only coarse component of soils on strongly weathered pyroclastic deposits (Stoops, 2013) (Fig. 5E and F)
- *Alteration to chlorite* – common pellicular or irregular linear late-magmatic alteration to chlorite in Mg-rich basic or ultramafic intrusive rocks; initially along cracks, forming coatings composed of chlorite plates oriented perpendicular to the sides, followed by development of sawtooth patterns as chlorite orientation becomes controlled by that of the olivine host, ultimately resulting in development of a complete pseudomorph (Baker & Haggerty, 1967); these alteromorphs evolve to more brownish aggregates through alteration and are easily destroyed by pedoplasmation
- *Alteration to 'bowlingite' (green material with variable composition, mainly Mg smectites and Mg chlorites)* – formed during final cooling stages of magma and in early stages of saprolite development; alteration along cracks, with partially radial fabric of the bowlingite deposits; possible alteration of bowlingite and olivine remains to iddingsite or goethite in higher parts of saprolite profiles (Eggleton et al., 1987); alteromorphs destroyed by pedoplasmation
- *Alteration to nontronite and beidellite* – in conditions with limited leaching, common pedogenic alteration to nontronite and beidellite (see Delvigne et al., 1979; Ildefonse, 1983; Delvigne & Stoops, 1990); formation of iron-rich smectite coatings along cracks, with smectite plates oriented perpendicular to the sides, leaving isolated remnants that can be altered to produce aggregates of randomly or radially oriented smectite; smectite alteromorphs changing to brown by goethite formation in higher parts of saprolite profiles, ultimately destroyed by pedoplasmation
- *Alteration to iron oxides* – in well-drained conditions in humid tropical environments, common pedogenic alteration to goethite, along fractures, producing boxwork patterns, with olivine remnants that undergo congruent dissolution, resulting

in an open boxwork system; rare preferential arrangement, with extinction parallel to the c-axis of olivine; in less well-drained conditions, cells of the boxwork system are initially filled with yellowish amorphous material, which is subsequently dissolved

Orthoclase (K(AlSi$_3$O$_8$))

- *Optical characteristics* – colourless, often cloudy, low negative relief, perfect cleavage in one direction, less perfect in the others; first-order grey interference colours, simple Carlsbad twinning; can occur as part of exsolution features composed of orthoclase and albite lamellae (perthite, antiperthite)
- *Occurrence in rocks* – in saturated intrusive rocks (e.g., granite)
- *Occurrence in soils* – almost absent in strongly weathered material but more stable than plagioclase, pyroxenes and amphiboles in soil environments (Taboada & García, 1999a, 1999b)
- *Pedogenic alteration* – along fractures, and with cross-linear patterns related to cleavage and twinning directions, followed by cavernous alteration (Delvigne et al., 1987); replaced by gibbsite in conditions with strong leaching; upon alteration of perthite or antiperthite, albite will disappear first, leaving a set of parallel orthoclase lamellae separated by weathering products (Fig. 6A and B)

Plagioclase

- *Main members* – **albite** (Na(AlSi$_3$O$_8$)), **anorthite** (Ca(Al$_2$Si$_2$O$_8$)), informal names such as oligoclase, andesine, labradorite and bytownite for intermediate compositions
- *Optical characteristics* – colourless, weak negative (albite) to weak positive (anorthite) relief, more or less equidimensional in intrusive magmatic rocks, lath-shaped in volcanic rocks (e.g., basalt, andesite) (see Fig. 5A and E); first-order grey interference colours; polysynthetic twinning (alternating light and dark parallel lamellae); in magmatic rocks, common zonation with anorthite-rich core and albite-rich rim, which influences weathering patterns (Taboada & García, 1999a, 1999b) (Fig. 6C and D)
- *Occurrence in rocks* – in almost all magmatic rocks, ranging from oversaturated to granitic; in micaschist and gneiss; in arkosic sedimentary rocks
- *Occurrence in soils* – common in young soils, almost absent in strongly weathered material
- *Alteration* – anorthite-rich plagioclase is more easily altered than albite-rich occurrences, which can be expressed by alteration of the Ca-rich core of zoned crystals; pellicular alteration mostly combined with alteration along fissures parallel to twinning planes
- *Non-pedogenic alteration* – forms of deuteric alteration include alteration to fine-grained muscovite or paragonite within plagioclase crystals (sericitisation),

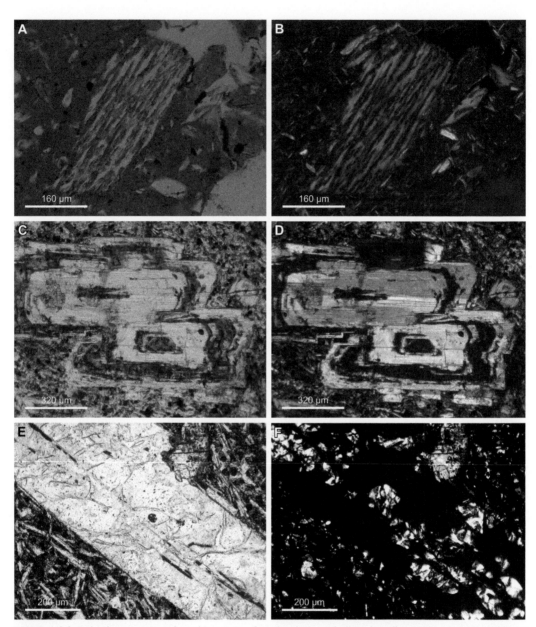

FIGURE 6 (A) Orthoclase remains of perthite grain after dissolution of albite lamellae (soil on granite, Akagera National Park, Rwanda) (PPL). (B) Idem in XPL. (C) Zoned plagioclase, with preferential alteration of Ca-rich zones (soil on andesite, western Java) (PPL). (D) Idem in XPL. (E) Plagioclase phenocrystal in basalt fragment, showing complex pattern of alteration (volcanic ash soil, Isla Santa Cruz, Galápagos) (PPL). (F) Idem in XPL, showing transformation of plagioclase to optically isotropic colloids, from which fan-like gibbsite aggregates have crystallised.

alteration to (clino)zoisite or epidote with granular aspect (saussuritisation) (see Fig. 3A), alteration to zeolites in the form of crystals with irregular shapes and distribution patterns within the plagioclase (Delvigne, 1998), and alteration to kaolinite, resulting in cloudy limpidity

- *Pedogenic alteration* – linear or banded, along with intersecting cleavage planes; in conditions with limited leaching, formation of amorphous colloids (Delvigne & Martin, 1970), which can evolve to halloysite (Eswaran & Wong Chaw Bin, 1978; Anand & Gilkes, 1984a), kaolinite or gibbsite (Gilkes et al., 1980) (Fig. 6E and F), depending on leaching conditions (Delvigne, 1965; Eswaran, 1979), whereby the newly formed clay domains show simultaneous extinction, without optical continuity with the plagioclase host (Cremeens et al., 1990), or perpendicular to fracture walls (Stoops & Dedecker, 2006); in conditions with strong leaching, gibbsite is formed directly, often resulting in complete alteromorphs; in arid soils, physical alteration can take place through growth of calcite in cracks (Stoops, 2003)

Quartz (SiO$_2$)

- *Optical characteristics* – colourless, very weak positive or negative relief, depending on impregnation medium; first-order grey interference colours, partly with wavy extinction and then biaxial; quartz of different origins can be distinguished by cathodoluminescence (blue violet for magmatic rocks, brown for metamorphic rocks, greenish brown for hydrothermal deposits) (Marshall, 1988)
- *Occurrence in rocks* – most common mineral in saturated intrusive magmatic rocks and metamorphic rocks, and in derived sedimentary rocks; rare in volcanic rocks; nearly absent in undersaturated and mafic to ultramafic magmatic rocks; microcrystalline quartz, partly with fibrous morphology (chalcedony), as a principal component of chert; **cristobalite** and **tridymite** are high-temperature SiO$_2$ polymorphs, occurring in cavities of rapidly cooled volcanic rocks (e.g., obsidian)
- *Occurrence in soils* – the most common mineral of the coarse fraction of most soils, and practically the only main mineral surviving strong weathering (e.g., in ferralitic soils); quartz grains are often composed of two or more crystals (fragments of rocks such as granite, gneiss and quartzite); wavy extinction points in most cases to metamorphic origin; for pedogenic quartz, see review by Gutiérrez-Castorena (2018)
- *Alteration* – fracturation caused by release of internal stress or by pressure of surrounding weathering minerals (Power et al., 1990); as no distinct cleavage is present, congruent dissolution of quartz follows a pellicular or irregular linear pattern; in ferralitic soils and lateritic materials, the irregular cracks are often filled with fine-grained hematite or gibbsite, resulting in 'runiquartz' development (Eswaran et al., 1975; Marcelino et al., 2018; Stoops & Marcelino, 2018); SEM studies of the surface of quartz grains can provide relevant information about the origin of the grains (e.g., Huang et al., 1990) or about degree and type of weathering (e.g., Marcelino & Stoops, 1996; Marcelino et al., 1999; Degórski & Kowalkowski, 2011)

Sanidine (K(AlSi$_3$O$_8$))

- *Optical characteristics* — colourless, low negative relief, perfect cleavage in one direction, less perfect in the others; first-order grey interference colours, no twinning.
- *Occurrence in rocks* — in saturated volcanic rocks
- *Occurrence in soils* — only in soils on volcanic material, almost absent in strongly weathered material
- *Alteration* — no available data

Serpentine

- *Main members* — **antigorite** (Mg$_3$Si$_2$O$_5$(OH)$_4$), **chrysotile** (Mg$_3$Si$_2$O$_5$(OH)$_4$)
- *Optical characteristics* — colourless to pale green, then weakly pleochroic, low positive relief, anhedral to fibrous (microcrystalline); first-order grey to yellow interference colours, parallel extinction, length slow
- *Occurrence in rocks* — principal constituent of serpentinite; low-temperature metamorphic or hydrothermal alteration product of olivine or Mg pyroxenes in saturated and ultramafic rocks (e.g., peridotite)
- *Occurrence in soils* — only as part of rock fragments in young soils on serpentinite, or in lower parts of saprolite profiles
- *Alteration* — no micromorphological descriptions are available; probably congruent dissolution

Tourmaline

- Main *members* — **dravite** (NaMg$_3$Al$_6$(Si$_6$O$_{18}$) (BO$_3$)$_3$(OH)$_3$(OH)), **elbaite** (Na(Al$_{1.5}$Li$_{1.5}$)Al$_6$(Si$_6$O$_{18}$)(BO$_3$)$_3$(OH)$_3$(OH)), **schorl** (NaFe$_3$$^{2+}Al_6$(Si$_6O_{18}$)(BO$_3$)$_3(OH)_3$(OH))
- *Optical characteristics* — grey, brown, green, blue, with strong pleochroism, prismatic, no cleavage; second- to third-order interference colours, generally masked by mineral colour, parallel extinction, length fast
- *Occurrence in rocks* — in intrusive magmatic rocks (pegmatites) and certain metamorphic rocks (micaschist, gneiss), as accessory mineral; very common as minor component in sands and sandstones, because of its high stability
- *Occurrence in soils* — commonly concentrated in strongly weathered material (e.g., oxic)
- *Alteration* — one of the most stable minerals, both chemically and physically; no evidence of alteration at microscopic scale

Tremolite-Actinolite

- *Main members* — **tremolite** (Ca$_2$(Mg$_{5.0-4.5}$Fe$_{0.0-0.5}$$^{2+}$)Si$_8O_{22}(OH)_2$), **actinolite** (Ca$_2$(Mg$_{4.5-2.5}Fe_{0.5-2.5}$$^{2+}$)Si$_8O_{22}(OH)_2$)

- *Optical characteristics* – colourless (tremolite) to green (actinolite), weak pleochroism, prismatic habit, high positive relief, distinct cleavage parallel to c-axis, crossing at 60/120-degree angles in cross sections perpendicular to the c-axis; first- to second-order interference colours, oblique extinction (10-20 degrees), length slow
- *Occurrence in rocks* – common in metamorphic rocks (e.g., gneiss, greenschist facies rocks), generally associated with calcite
- *Occurrence in soils* – in weakly to moderately weathered material
- *Non-pedogenic alteration* – late-magmatic and deuteric alteration to other amphiboles (uralitisation) or to antigorite or talc with associated magnetite
- *Pedogenic alteration* – pellicular, and advancing inward from the sides along cracks rather than along cleavage planes; formation of amorphous material, followed by smectite with parallel orientation, replaced by kaolinite in conditions with strong leaching (Delvigne, 1983); alteration of actinolite to vermiculite along fractures and cleavage planes, forming pseudomorphic large single crystals (Abreu & Vairinho, 1990); in conditions with strong leaching, formation of iron oxides along fractures, leaving denticulate residues, which can be separated from the iron oxide deposits by voids

Zircon (Zr(SiO$_4$))

- *Optical characteristics* – colourless, extreme positive relief, short prismatic crystals with pyramidal terminations, elliptic when transported, no clear cleavage; third-order interference colours, straight extinction, length slow (Fig. 7A and B)
- *Occurrence in rocks* – as accessory mineral in many intrusive magmatic rocks (e.g., granites, diorites, syenites); in sedimentary rocks, as almost omnipresent but minor component in sandy deposits; also in derived metamorphic rocks

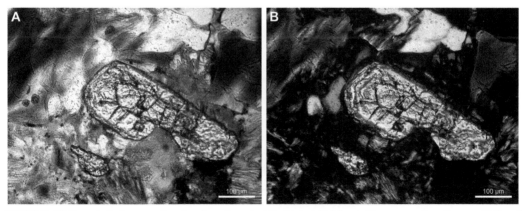

FIGURE 7 (A) Rounded zircon grain, showing characteristic high relief (saprolite on granite, Akagera National Park, Rwanda) (PPL). (B) Idem in XPL, showing typical high interference colours.

- *Occurrence in soils* – seldom observed in soil thin sections, except for highly weathered material (e.g., oxic), in which it is concentrated by residual enrichment
- *Alteration* – one of the most stable minerals, with no visible chemical alteration at microscopic scale

References

Abreu, M.M. & Vairinho, M., 1990. Amphibole alteration to vermiculite in a weathering profile of gabbro-diorite. In Douglas, L.A. (ed.), Soil Micromorphology: A Basic and Applied Science. Developments in Soil Science, Volume 19. Elsevier, Amsterdam, pp. 493-500.

Adams, A.E. & MacKenzie, W.S., 1998. A Colour Atlas of Carbonate Sediments and Rocks under the Microscope. Manson Publishing, London, 180 p.

Adams, A.E., MacKenzie, W.S. & Guilford, C., 1984. Atlas of Sedimentary Rocks under the Microscope. Longman, Harlow, 104 p.

Adderley, W.P., Wilson, C.A., Simpson I.A. & Davidson, D.A., 2018. Anthropogenic features. In Stoops, G., Marcelino, V. & Mees, F. (eds.), Interpretation of Micromorphological Features of Soils and Regoliths. Second Edition. Elsevier, Amsterdam, pp. 753-777.

Anand, R.R. & Gilkes, R.J., 1984a. Weathering of hornblende, plagioclase and chlorite in meta-dolerite, Australia. Geoderma 34, 261-280.

Anand, R.R. & Gilkes, R.J., 1984b. Mineralogical and chemical properties of weathered magnetite grains from lateritic saprolite. Journal of Soil Science 35, 559-567.

Angelini, I., Artioli, G. & Nicosia, C., 2017. Metals and metalworking residues. In Nicosia, C. & Stoops, G. (eds.), Archaeological Soil and Sediment Micromorphology. John Wiley & Sons Ltd, Chichester, pp. 213-222.

Angelucci, D.E., 2010. The recognition and description of knapped lithic artifacts in thin section. Geoarchaeology 25, 220-232.

Angelucci, D., 2017. Lithic artefacts. In Nicosia, C. & Stoops, G. (eds.), Archaeological Soil and Sediment Micromorphology. John Wiley & Sons Ltd, Chichester, pp. 223-230.

Arocena J.M., De Geyter, G., Landuydt, C. & Stoops, G., 1990. Study of the distribution and extraction of iron (and manganese) in soil thin sections. In Douglas, L.A. (ed.), Soil Micromorphology: A Basic and Applied Science. Developments in Soil Science, Volume 19. Elsevier, Amsterdam, pp. 621-626.

Aylmore, L.A.G. & Quirk, J.P., 1959. Swelling of clay-water systems. Nature 183, 1752-1753.

Babel, U., 1964. Chemische Reaktionen an Bodendünnschliffen. Leitz-Mitteilungen für Wissenschaft und Technik 3, 12-14.

Baert, G. & Van Ranst, E., 1997. Comparative micromorphological study of representative weathering profiles on different parent materials in the Lower Zaire. In Shoba, S., Gerasimova, M. & Miedema, R. (eds.), Soil Micromorphology: Studies on Soil Diversity, Diagnostics, Dynamics. Moscow, Wageningen, pp. 28-40.

Baker, I. & Haggerty, S.E., 1967. The alteration of olivine in basaltic and associated lavas. II: Intermediate and low temperature alteration. Contributions to Mineralogy and Petrology 16, 258-273.

Barzanji, A.F. & Stoops, G., 1974. Fabric and mineralogy of gypsum accumulations in some soils of Iraq. Transactions of the 10th International Congress of Soil Science, Volume VII, Moscow, pp. 271-277.

Beauvais, A. & Nahon, D., 1985. Nodules et pisolites de dégradation des profils d'altération manganésifères sous conditions latéritiques. Exemples de Côte d'Ivoire et du Gabon. Sciences Géologiques, Bulletin, 38, 359-381.

Bennema, J., Jongerius, A. & Lemos, R., 1970. Micromorphology of some oxic and argillic horizons in South Brazil in relation to weathering sequences. Geoderma 4, 333-355.

Betancourt, P.P. & Peterson, S.E., 2009. Thin-section Petrography of Ceramic Materials. Institute for Aegean Prehistory, Archaeological Excavation Manual 2. INSTAP Academic Press, Philadelphia, 27 p.

Bisdom, E.B.A., Stoops, G., Delvigne, J., Curmi, P. & Altemüller, H.J., 1982. Micromorphology of weathering biotite and its secondary products. Pedologie 32, 225-252.

Blevins, R.L., Holowaychuk, N. & Wilding, L.P., 1970. Micromorphology of soil fabric at tree root-soil interface. Soil Science Society of America Proceedings 34, 460-465.

Blokhuis, W.A., Slager, S. & Van Schagen, R.H., 1970. Plasmic fabric of two Sudan Vertisols. Geoderma 4, 127-137.

Boggs, S., 2009. Petrology of Sedimentary Rocks. Second Edition. Cambridge University Press, Cambridge, 600 p.

Boswell, P.G.H., 1941. Stability of minerals in sedimentary rocks. Nature 147, 734-737.

Bouchard, M., Jolicoeur, S. & Pierre, P., 1995. Characteristics and significance of two pre-late-Wisconsinan weathering profiles (Adirondacks, USA and Miramichi Highlands, Canada). Geomorphology 12, 75-89.

Bresson, L.M., 1981. Etude ultramicroscopique d'assemblages plasmiques sur lames ultraminces de sols réalisées par bombardement ionique. In Bisdom, E.BA. (ed.), Submicroscopy of Soils and Weathered Rocks, Proceedings of the First Workshop of the International Working Group on Submicroscopy of Undisturbed Soil Materials. Pudoc, Wageningen, pp. 173-190.

Bresson, L.M. & Zambaux, C., 1990. Micromorphological study of compaction induced by mechanical stress for a Dystrochreptic Fragiudalf. In Douglas, L.A. (ed.), Soil Micromorphology: A Basic and Applied Science. Developments in Soil Science, Volume 19. Elsevier, Amsterdam, pp. 33-42.

Brewer, R., 1960. The petrographic approach to the study of soils. Transactions of the 7th International Congress of Soil Science, Volume I, Madison, pp. 1-13.

Brewer, R., 1964a. Classification of plasmic fabrics of soil materials. In Jongerius, A. (ed.), Soil Micromorphology. Elsevier, Amsterdam, pp. 95-108.

Brewer, R., 1964b. Fabric and Mineral Analysis of Soils. John Wiley and Sons, New York, 470 p.

Brewer, R., 1979. Relationship between particle size, fabric and other factors in some Australian soils. Australian Journal of Soil Research 17, 29-41.

Brewer, R. & Blackmore, A.V., 1976. Subplasticity in Australian soils. II. Relationships between subplasticity rating, optically oriented clay, cementation and aggregate stability. Australian Journal of Soil Research 14, 237-248.

Brewer, R. & Haldane, A.D., 1957. Preliminary experiments in the development of clay orientations in soils. Soil Science 84, 301-309.

Brewer, R. & Pawluk, S., 1975. Investigations of some soils developed in hummocks of the Canadian Sub-Arctic and Southern-Arctic regions. 1. Morphology and micromorphology. Canadian Journal of Soil Science 55, 301-319.

Brewer, R. & Sleeman, J.R., 1988. Soil Structure and Fabric. CSIRO, Adelaide, 171 p.

Bronger, A., Kalk, E. & Schroeder, D., 1976. Ueber Glimmer und Feldspat Verwitterung sowie Entstehung und Umwandlung von Tonmineralien in rezenten und fossilen Lössböden. Geoderma 16, 21-54.

Brönnimann, D., Pümpin, C., Ismail-Meyer, K., Rentzel, P. & Égüez, N., 2017. Excrements of omnivores and carnivores. In Nicosia, C. & Stoops, G. (eds.), Archaeological Soil and Sediment Micromorphology. John Wiley & Sons Ltd, Chichester, pp. 67-81.

Bruins, H.J., MacGillivray, J.A., Synolakis, C.E., Benjamini, C., Keller, J., Kisch, H.J., Klügel, A. & van der Plicht, J., 2008. Geoarchaeological tsunami deposits at Palaikastro (Crete) and the Late Minoan IA eruption of Santorini. Journal of Archaeological Science 35, 191-212.

Brzychcy, S., Zagórski, Z., Sieczko, L. & Kaczorek, D., 2012. Analysis of groundmass colour as a tool for evaluating the extent of pedogenic processes in chromic soils. Roczniki Gleboznawcze 63, 3-7.

Bullock, P., Fedoroff, N., Jongerius, A., Stoops, G., Tursina, T. & Babel, U., 1985. Handbook for Soil Thin Section Description. Waine Research Publications, Wolverhampton, 152 p.

Cagauan, B. & Uehara, G., 1965. Soil anisotropy and its relation to aggregate stability. Soil Science Society of America Proceedings 29, 198-200.

Canti, M.G., 2017a. Mollusc shell. In Nicosia, C. & Stoops, G. (eds.), Archaeological Soil and Sediment Micromorphology. John Wiley & Sons Ltd, Chichester, pp. 43-46.

Canti, M.G., 2017b. Burnt carbonates. In Nicosia, C. & Stoops, G. (eds.), Archaeological Soil and Sediment Micromorphology. John Wiley & Sons Ltd, Chichester, pp. 181-188.

Canti, M.G., 2017c. Avian eggshell. In Nicosia, C. & Stoops, G. (eds.), Archaeological Soil and Sediment Micromorphology. John Wiley & Sons Ltd, Chichester, pp. 39-41.

Certini, G., Campbell, C.D. & Edwards, A.C., 2004. Rock fragments in soil support a different microbial community from the fine earth. Soil Biology and Biochemistry 36, 1119-1128.

Chadwick, O.A. & Nettleton, W.D., 1990. Micromorphologic evidence of adhesive and cohesive forces in soil cementation. In Douglas, L.A. (ed.), Soil Micromorphology: A Basic and Applied Science. Developments in Soil Science, Volume 19. Elsevier, Amsterdam, pp. 207-212.

Chadwick, O.A. & Nettleton, W.D., 1994. Quantitative relationships between net volume change and fabric properties during soil evolution. In Ringrose-Voase, A.J. & Humphreys, G.S. (eds.), Soil Micromorphology: Studies in Management and Genesis. Developments in Soil Science, Volume 22. Elsevier, Amsterdam, pp. 353-360.

Charzynski, P., Hulisz, P. & Bednack, R., 2013. Technogenic Soils of Poland. Polish Society of Soil Science, Torún, 358 p.

Cremeens, D.L., Darmody, R.G. & Mokma, D.L., 1990. Micromorphology of feldspar weathering in a lithic clast (semi-closed system) versus the associated S-matrix (open system) in a till Paleosol. In Douglas, L.A. (ed.), Soil Micromorphology: A Basic and Applied Science. Developments in Soil Science, Volume 19. Elsevier, Amsterdam, pp. 531-536.

Curmi, P. & Maurice, F., 1981. Caractérisation microscopique de l'altération dans une arène granitique à structure conservée. In Bisdom, E.B.A. (ed.), Submicroscopy of Soils and Weathered Rocks, Proceedings of the First Workshop of the International Working Group on Submicroscopy of Undisturbed Soil Materials. Pudoc, Wageningen, pp. 249-270.

Dalrymple, J.B. & Jim, C.Y., 1984. Experimental study of soil microfabrics induced by isotropic stresses of wetting and drying. Geoderma 34, 43-68.

De Coninck, F. & Laruelle, J., 1964. Soil development in sandy materials of the Belgian Campine. In Jongerius, A. (ed.), Soil Micromorphology. Elsevier, Amsterdam, pp. 169-188.

De Coninck, F., Stoops, G. & Van Ranst, E., 1987. Mineralogy and micromorphology of a soil toposequence near Matadi (Lower Zaire) on chloritic green rocks. In Rodriguez-Clemente, R. & Tardy, Y. (eds.), Geochemistry and Mineral Formation in the Earth Surface. Consejo Superior de Investigaciones Científicas, Madrid, pp. 157-174.

Degórski, M. & Kowalkowski, A., 2011. The use of SEM morphoscopy in researching the litho-pedogenetic environments evolution of Late Pleistocene and Holocene. Geographia Polonica, 84, Special Issue, Part 1, 17-38.

De León-González, F., Gutiérez-Castorena, M.C., González-Chávez, M.C.A. & Castillo-Juárez, H., 2007. Root-aggregation in a pumiceous sandy soil. Geoderma 142, 308-317.

Delmas, A.B. & Berrier, J., 1987. Les figures de corrosion de la calcite: typologie et séquences évolutives. In Fedoroff, N., Bresson, L.M. & Courty, M.A. (eds.), Micromorphologie des Sols, Soil Micromorphology. AFES, Plaisir, pp. 303-308.

Delvigne, J., 1965. Pedogenèse en Zone Tropicale. La Formation des Minéraux Secondaires en Milieu Ferralitique. Mémoires ORSTOM 13, 177 p.

Delvigne, J., 1983. Micromorphology of the alteration and weathering of pyroxenes in the Koua Bocca ultramafic intrusion, Ivory-Coast, western Africa. In Nahon, D. & Noack, Y. (eds.), Pétrologie des Altérations et des Sols. Volume II: Pétrologie des Séquences Naturelles. Sciences Géologiques, Mémoires, 72, 57-68.

Delvigne, J.E., 1998. Atlas of Micromorphology of Mineral Alteration and Weathering. Canadian Mineralogist, Special Publication 3, 495 p.

Delvigne, J. & Martin, H., 1970. Analyse à la microsonde électronique de l'altération d'un plagioclase en kaolinite par l'intermédiaire d'une phase amorphe. Cahiers ORSTOM, Série Géologie, 2, 259-295.

Delvigne, J. & Stoops, G., 1990. Morphology of mineral weathering and neoformation. I. Weathering of most common silicates. In Douglas, L.A. (ed.), Soil Micromorphology: A Basic and Applied Science. Developments in Soil Science, Volume 19. Elsevier, Amsterdam, pp. 471-482.

Delvigne, J., Bisdom, E.B.A., Sleeman, J. & Stoops, G., 1979. Olivines, their pseudomorphs and secondary products. Pedologie 29, 247-309.

Delvigne, J., Soubies, F. & Sardela, I., 1987. Micromorphologie des altérations supergènes de la lujaurite du massif alcalin de Pocos de Caldas (Minas Gerais, Brésil). In Fedoroff, N., Bresson, L.M. & Courty, M.A. (eds.), Micromorphologie des Sols, Soil Micromorphology. AFES, France, pp. 97-104.

De Paepe, P. & Stoops, G., 2007. A classification of tephra in volcanic soils. A tool for soil scientists. In Arnalds, O., Bartoli, F., Buurman, P., Oskarsson, H., Stoops, G. & Garcia-Rodeja, E. (eds.), Soils of Volcanic Regions in Europe. Springer-Verlag, Berlin, pp. 119-125.

Dixon, J.B. & Weed, S.B., 1989. Minerals in Soil Environments. Soil Science Society of America, Madison, 1244 p.

Dobrovol'ski, G.V. (ed.), 1983. A Methodological Manual of Soil Micromorphology [*in Russian*]. Izdatelstvo Moskovskogo Universiteta, 63 p.

Dobrovol'ski, G.V. (ed.), 1991. A Methodological Manual of Soil Micromorphology. International Training Centre for Post-Graduate Soil Scientists, Publication Series 3, Ghent, 63 p.

Dorronsoro, C., 1994. Micromorphological index for the evaluation of soil evolution in central Spain. Geoderma 61, 237-250.

Dunham, R.J., 1962. Classification of carbonate rocks according to depositional texture. In Ham, W.E. (ed.), Classification of Carbonate Rocks. American Association of Petroleum Geologists Memoir 1, pp. 108-121.

Durand, N., Monger, H.C., Canti, M.G. & Verrecchia, E., 2018. Calcium carbonate features. In Stoops, G., Marcelino, V. & Mees, F. (eds.), Interpretation of Micromorphological Features of Soils and Regoliths. Second Edition. Elsevier, Amsterdam, pp. 205-258.

Eggleton, R.A., Foudoulis, C. & Varkevisser, D., 1987. Weathering of basalt: changes in rock chemistry and mineralogy. Clay and Clay Minerals 35, 161-169.

Embrechts, J. & Stoops, G., 1982. Microscopical aspects of garnet weathering in a humid tropical environment. Journal of Soil Science 33, 535-545.

Embrechts, J. & Stoops, G., 1987. Microscopic identification and quantitative determination of micro-structure and potentially mobile clay in a soil catena in a humid tropical soil environment. In Fedoroff, N., Bresson, L.M. & Courty, M.A. (eds.), Micromorphologie des Sols, Soil Micromorphology. AFES, Plaisir, pp. 157-164.

Eswaran, H., 1972. Micromorphological indicators of pedogenesis in some tropical soils derived from basalts from Nicaragua. Geoderma 7, 15-31.

Eswaran, H., 1979. The alteration of plagioclases and augite under differing pedoenvironmental conditions. Journal of Soil Science 30, 547-555.

Eswaran, H., 1983. Characterization of domains with the scanning electron microscope. Pedologie 33, 41-54.

Eswaran, H. & Baños, C., 1976. Related distribution patterns in soils and their significance. Anales de Edafologia y Agrobiologia 35, 33-45.

Eswaran, H. & Bin, W.C., 1978. A study of a deep weathering profile on granite in Peninsular Malaysia. III. Alteration of feldspars. Soil Science Society of America Journal 42, 154-158.

Eswaran, H., Sys, C. & Sousa, E.C., 1975. Plasma infusion. A pedological process of significance in the humid tropics. Anales de Edafología y Agrobiología 34, 665-674.

Fettes, D. & Desmons, J., 2007. Metamorphic Rocks. A Classification and Glossary of Terms. Recommendations of International Union of Geological Sciences Subcommission on the Systematics of Metamorphic Rocks. Cambridge University Press, Cambridge, 244 p.

Field, D.J., Koppi, A.J. & Drees L.R., 1997. The characterisation of microaggregates in Vertisols using scanning electron microscopy (SEM) and thin section. In Shoba, S., Gerasimova, M. & Miedema, R. (eds.), Soil Micromorphology: Studies on Soil Diversity, Diagnostics and Dynamics. Moscow, Wageningen, pp. 80-86.

FitzPatrick, E.A., 1984. Micromorphology of Soils. Chapman and Hall, London, 433 p.

FitzPatrick, E.A., 1993. Soil Microscopy and Micromorphology. John Wiley & Sons, Chichester, 304 p.

Fleischer, M., Wilcox, R.E. & Matzko, J.J., 1984. Microscopic Determination of the Nonopaque Minerals. U.S. Geological Survey Bulletin 1627, 453 pp.

Flügel, E., 2010. Microfacies of Carbonate Rocks. Analysis, Interpretation and Application. Springer, Heidelberg, 984 p.

Folk, R.L., 1962. Spectral subdivision of limestone types. In Ham, W.E. (ed.) Classification of Carbonate Rocks. American Association of Petroleum Geologists Memoir 1, pp. 62-84.

Foster, R.H. & De, P.K., 1971. Optical and electron microscopic investigations of shear induced structures in lightly consolidated (soft) and heavily consolidated (hard) kaolinite. Clays and Clay Minerals 19, 31-47.

Freestone, I., 1995. Ceramic petrography. American Journal of Archaeology 99, 111-115.

Ghabru, S.K., Mermut, A.R. & St. Arnaud, R.J., 1987. The nature of weathered biotite in sand-sized fractions of Gray Luvisols (Boralfs) in Saskatchewan, Canada. Geoderma 40, 65-82.

Gilkes, R.J., Suddhiprakarn, A. & Armitage, T.M. 1980. Scanning electron microscope morphology of deeply weathered granite. Clays and Clay Minerals 28, 29-34.

Goldich, S.S., 1938. A study in rock weathering. Journal of Geology 46, 17-58.

Grandin, G. & Perseil, E.A., 1977. Le gisement de manganèse de Mokta (Côte d'Ivoire). Transformations minéralogiques des minerais par action météorique. Bulletin de la Société Géologique de France 19, 309-317.

Greene-Kelly, R. & Mackney, D., 1970. Preferred orientation of clay in soils, the effect of drying and wetting. In Osmond, D.A. & Bullock, P. (eds.), Micromorphological Techniques and Applications. Soil Survey Technical Monograph 2, pp. 43-52.

Grim, R.E., 1962. Applied Clay Mineralogy. McGraw-Hill, New York. 422 p.

Gutiérrez-Castorena, Ma. del C., 2018. Pedogenic siliceous features. In Stoops, G., Marcelino, V. & Mees, F. (eds.), Interpretation of Micromorphological Features of Soils and Regoliths. Second Edition. Elsevier, Amsterdam, pp. 127-155.

Hill, I.D., 1970. The use of orientation diagrams in describing plasmic fabrics in soil materials. Journal of Soil Science, 21, 184-187.

Hill, I.D., 1981. A method for the quantitative measurement of anisotropic plasma in soil thin sections. Journal of Soil Science 32, 461-464.

Howard, J.L. & Orlicki, K.M., 2016. Composition, micromorphology and distribution of microartifacts in anthropogenic soils, Detroit, Michigan, USA. Catena 138, 103-116.

Huang R.C., Pan, G.X. & Xiong, D.X., 1990. SEM observations of quartz sand surface structures as applied to parent materials of some albic soils of East China. In Douglas, L.A. (ed.), Soil Micromorphology: A Basic and Applied Science. Developments in Soil Science, Volume 19. Elsevier, Amsterdam, pp. 501-510.

Ildefonse, P., 1983. Altération préméteorique et météorique des olivines du basalte de Belbex (France). In Nahon, D. & Noack, Y. (eds.) 1983. Pétrologie des Altérations et des Sols. Volume II: Pétrologie des Séquences Naturelles. Sciences Géologiques, Mémoires, 72, 69-80.

Jim, C.Y., 1986. Experimental study of soil microfabrics induced by anisotropic stresses of confined swelling and shrinking. Geoderma 37, 91-112.

Jim, C.Y., 1988a. A classification of soil microfabrics. Geoderma 41, 315-325.

Jim, C.Y., 1988b. Microscopic-photometric measurement of polymodal clay orientation using circular-polarized light and interference colours. Applied Clay Science 3, 307-321.

Jim, C.Y., 1990. Stress, shear deformation and micromorphological clay orientation: a synthesis of various concepts. Catena 17, 431-447.

Jongmans, A.G., van Breemen, N., Lundstrom, U., van Hees, P.A.W., Finlay, R.D., Srinivasan, M., Unestam, T, Giesler, R., Melkerud, P.A. & Olsson, M., 1997. Rock-eating fungi. Nature 389, 682-683.

Kaczorek, D., Vrydaghs, L., Devos, Y., Petó, A. & Effland, W., 2018. Biogenic siliceous features. In Stoops, G., Marcelino, V. & Mees, F. (eds.), Interpretation of Micromorphological Features of Soils and Regoliths. Second Edition. Elsevier, Amsterdam, pp. 157-176.

Karkanas, P., 2017. Guano. In Nicosia, C. & Stoops, G. (eds.), Archaeological Soil and Sediment Micromorphology. John Wiley & Sons Ltd, Chichester, pp. 83-89.

Karkanas, P. & Goldberg, P., 2018. Phosphatic features. In Stoops, G., Marcelino, V. & Mees, F. (eds.), Interpretation of Micromorphological Features of Soils and Regoliths. Second Edition. Elsevier, Amsterdam, pp. 323-346.

Khormali, F., Abtahi, A., Mahmoodi, S. & Stoops, G., 2003. Argillic horizon development in calcareous soils of arid and semiarid regions of southern Iran. Catena 53, 273-301.

Khormali, F., Abtahi, A. & Stoops, G., 2006. Micromorphology of calcitic features in highly calcareous soils of Fars Province, Southern Iran. Geoderma 132, 31-46.

Kooistra, M.J. & Pulleman, M.M., 2018. Features related to faunal activity. In Stoops, G., Marcelino, V. & Mees, F. (eds.), Interpretation of Micromorphological Features of Soils and Regoliths. Second Edition. Elsevier, Amsterdam, pp. 447-469.

Korina, N.A. & Faustova, M.A., 1964. Microfabric of modern and old moraines. In Jongerius, A. (ed.), Soil Micromorphology. Elsevier, Amsterdam, pp. 333-338.

Kovda, I. & Mermut, A., 2018. Vertic features. In Stoops, G., Marcelino, V. & Mees, F. (eds.), Interpretation of Micromorphological Features of Soils and Regoliths. Second Edition. Elsevier, Amsterdam, pp. 605-632.

Kubiëna, W.L., 1938. Micropedology. Collegiate Press, Ames, 242 p.

Kubiëna, W.L., 1948. Entwicklungslehre des Bodens. Springer-Verlag, Wien, 215 p.

Kühn, P., Aguilar, J., Miedema, R. & Bronnikova, M., 2018. Textural Pedofeatures and Related Horizons. In Stoops, G., Marcelino, V. & Mees, F. (eds.), Interpretation of Micromorphological Features of Soils and Regoliths. Second Edition. Elsevier, Amsterdam, pp. 377-423.

Lafeber, D., 1964. Soil fabric and soil mechanics. In Jongerius, A. (ed.), Soil Micromorphology. Elsevier, Amsterdam, pp. 351-360.

Lafeber, D., 1967. The optical determination of spatial (three dimensional) orientation of platy clay minerals in soil thin sections. Geoderma 1, 359-370.

Lafeber, D., 1972. Micromorphometric techniques in engineering soil fabric analysis. In Kowalinski, S. & Drozd, J. (eds.), Soil Micromorphology. Panstwowe Wydawnictwo Naukowe, Warszawa, pp. 669-687.

Lelong, F., 1967. Sur les formations latéritiques de Guyane Française: 'manière d'être' de la kaolinite et de la gibbsite; origine des phyllites micacées. Comptes Rendus de l'Académie des Sciences 264, 2713-2716.

Lambiv Dzemua, L.G., Mees, F., Stoops, G. & Van Ranst, E., 2011. Micromorphology, mineralogy and geochemistry of lateritic weathering over serpentinite in south-east Cameroon. Journal of African Earth Sciences 60, 38-48.

Le Maitre, R.W., 2002. Igneous Rocks. A Classification and Glossary of Terms. Recommendations of International Union of Geological Sciences Subcommission on the Systematics of Igneous Rocks. Second Edition. Cambridge University Press, Cambridge, 236 p.

MacKenzie, W.S. & Adams, A.E., 1994. A Colour Atlas of Rocks and Minerals in Thin Section. Manson Publishing, London, 192 p.

MacKenzie, W.S. & Guilford, C., 1980. Atlas of Rock-Forming Minerals in Thin Sections. Longman, Harlow, 98 p.

MacKenzie, W.S., Donaldson, C.H. & Guilford, C., 1982. Atlas of Igneous Rocks and their Textures. Longman, Essex, 148 p.

Macphail, R.I. & Goldberg, P., 2018. Archaeological materials. In Stoops, G., Marcelino, V. & Mees, F. (eds.), Interpretation of Micromorphological Features of Soils and Regoliths. Second Edition. Elsevier, Amsterdam, pp. 779-819.

Magaldi, D., 1974. Caratteri a modalita dell orientamento delle argille nel'orizzonte B di alcuni suoli. Atti della Società Toscana di Scienze Naturali, Memorie, Serie A, 81, 152-166.

Magaldi, D., 1987. Degree of soil plasma orientation in relation to age in some hydromorphic soils of Tuscany. In Fedoroff, N., Bresson, L.M. & Courty, M.A. (eds.), Micromorphologie des Sols, Soil Micromorphology. AFES, Plaisir, pp. 605-610.

Magaldi, D. & Tallini, M., 2000. A micromorphological index of soil development for the Quaternary geology research. Catena 41, 261-276.

Magaldi, D., Rissoneand, P. & Totani, G., 1994. Effects of simulated seismic acceleration on silty clay soil horizons: a submicroscopic, micromorphological and geotechnical approach. In Ringrose-Voase, A.J. & Humphreys, G.S. (eds.), Soil Micromorphology: Studies in Management and Genesis. Developments in Soil Science, Volume 22. Elsevier, Amsterdam, pp. 719-728.

Marcelino, V. & Stoops, G., 1996. A weathering score for sandy soil materials based on the intensity of etching of quartz grains. European Journal of Soil Science 47, 7-12.

Marcelino, V., Mussche, G. & Stoops, G., 1999. Surface morphology of quartz grains from tropical soils and its significance for assessing soil weathering. European Journal of Soil Science 50, 1-8.

Marcelino, V., Schaefer, C.E.G.R. & Stoops, G., 2018. Oxic and related materials. In Stoops, G., Marcelino, V. & Mees, F. (eds.), Interpretation of Micromorphological Features of Soils and Regoliths. Second Edition. Elsevier, Amsterdam, pp. 663-689.

Marshall, D., 1988. Cathodoluminescence of geological materials. Unwin Hyman Ltd, London. 146 p.

Maritan, L., 2017. Ceramic materials. In Nicosia, C. & Stoops, G. (eds.), Archaeological Soil and Sediment Micromorphology. John Wiley & Sons Ltd, Chichester, pp. 205-212.

Mazurek, R., Kowalska, J., Gasiorek, M. & Setlak, M., 2016. Micromorphological and physico-chemical analyses of cultural layers in the urban soil of a medieval city - a case study from Krakow, Poland. Catena 141, 73-84.

Mees, F., 2018. Authigenic silicate minerals – sepiolite-palygorskite, zeolites and sodium silicates. In Stoops, G., Marcelino, V. & Mees, F. (eds.), Interpretation of Micromorphological Features of Soils and Regoliths. Second Edition. Elsevier, Amsterdam, pp. 177-203.

Mees, F. & Van Ranst, E., 2011. Micromorphology of sepiolite occurrences in recent lacustrine deposits affected by soil development. Soil Research 49, 547-557.

Melgarejo, J.C., 1997. Atlas de Asociaciones Minerales en Lámina Delgada. Edicions Universitat de Barcelona, Barcelona, 1076 p.

Minashina, N.G., 1958. Optically oriented clays in soils. Soviet Soil Science 4, 424-430.

Mitchell, J.K., 1956. The fabric of natural clays and its relation to engineering properties. Highway Research 35, 693-713.

Moon, C.F., 1972. The microstructure of clay sediments. Earth-Science Reviews 8, 303-321.

Morgenstern, N.R. & Tchalenko, J.S., 1967a. The optical determination of preferred orientation in clays and its application to the study of microstructure in consolidated kaolin. I. Proceedings of the Royal Society of London A 300, 218-234.

Morgenstern, N.R. & Tchalenko, J.S., 1967b. Microscopic structures in kaolin subjected to direct shear. Géotechnique 17, 309-328.

Morgenstern, N.R. & Tchalenko, J.S. & Skempton, A.W., 1967. The optical determination of preferred orientation in clays and its application to the study of microstructure in consolidated kaolin. II. Proceedings of the Royal Society of London 300, 235-250.

Mücher, H., van Steijn, H. & Kwaad, F., 2018. Colluvial and mass wasting deposits. In Stoops, G., Marcelino, V. & Mees, F. (eds.), Interpretation of Micromorphological Features of Soils and Regoliths. Second Edition. Elsevier, Amsterdam, pp. 21-36.

Muggler, C.C., Buurman, P. & van Doesburg, J.D.J., 2007. Weathering trends and parent material characteristics of polygenetic oxisols from Minas Gerais, Brazil. 1. Mineralogy. Geoderma 138, 39-48.

Nahon, D.B., 1991. Introduction to the Petrology of Soils and Chemical Weathering. John Wiley & Sons, New York, 313 p.

Nesbitt, H.W. & Young, G.M., 1982. Early Proterozoic climates and plate motions inferred from major element chemistry of lutite. Nature 299, 715-717.

Nesbitt, H.W. & Young, G.M., 1984. Prediction of some weathering trends of plutonic and volcanic rocks based on thermodynamic and kinetic considerations. Geochimica et Cosmochimica Acta 48, 1523-1534.

Nesse, W.D., 2003. Introduction to Optical Mineralogy. Third Edition. Oxford University Press, Oxford, 370 p.

Nicosia, C. & Stoops, G. (eds.), 2017. Archaeological Soil and Sediment Micromorphology. John Wiley & Sons Ltd, Chichester, 476 p.

Parfenoff, A., Pomerol, C. & Tourenq, J., 1970. Les Minéraux en Grains. Méthodes d'Etude et de Détermination. Masson, Paris, 578 p.

Parisot, J.C., Delvigne, J. & Groke, M.C.T., 1983. Petrographical aspects of the supergene weathering of garnet in the Serra dos Carajas (Para, Brazil). In Nahon, D. & Noack, Y. (eds.), Pétrologie des Altérations et des Sols. Volume II. Pétrologie des Séquences Naturelles. Sciences Géologiques, Mémoires 72, 141-148.

Parker, A., 1970. An index of weathering for silicate rocks. Geological Magazine 107, 501-504.

Penven, M.J., Fedoroff, N. & Robert, M., 1981. Altération météorique des biotites en Algérie. Geoderma 26, 287-309.

Perkins, D. & Henke, K.R., 2004. Minerals in Thin Sections. Second Edition. Prentice-Hall, Upper Saddle River, 163 p.

Pettijohn, F.J., Potter, P.E. & Siever, R., 2012. Sand and Sandstone. Second Edition. Springer-Verlag, Berlin, 553 p.

Pichler, H. & Schmitt-Riegraf, C., 1997. Rock-forming Minerals in Thin Sections. Chapman & Hall, London. 220 pp.

Pinto, L.C., Zinn, Y.L., de Mello, C.R., Owens, P.R. Norton, L.D. & Curi, N., 2015. Micromorphology and pedogenesis of mountainous Inceptisols in the Mantiqueira Range (MG). Ciência e Agrotecnologia 39, 455-462.

Poch, R.M., Artieda, O. & Lebedeva, M., 2018. Gypsic features. In Stoops, G., Marcelino, V. & Mees, F. (eds.), Interpretation of Micromorphological Features of Soils and Regoliths. Second Edition. Elsevier, Amsterdam, pp. 259-287.

Power, T., Smith, B.J. & Whalley, W.B., 1990. Fracture patterns and grain release in physically weathered granitic rocks. In Douglas, L.A. (ed.), Soil Micromorphology: A Basic and Applied Science. Developments in Soil Science, Volume 19. Elsevier, Amsterdam, pp. 545-550.

Quinn, P.S., 2013. Ceramic Petrography: The Interpretation of Archaeological Pottery & Related Artefacts in Thin Sections. Archaeopress, Oxford, 260 p.

Reiche, P., 1950. A Survey of Weathering Processes and Products. University of New Mexico Publications in Geology 3, 95 p.

Rentzel, P., Nicosia, C., Gebhardt, A., Brönnimann, D., Pümpin, C. & Ismail-Meyer, K., 2017. Trampling, poaching and the effect of traffic. In Nicosia, C. & Stoops, G. (eds.), Archaeological Soil and Sediment Micromorphology. John Wiley & Sons Ltd, Chichester, pp. 281-297.

Röpke, A. & Dietl, C., 2017. Burnt soils and sediments. In Nicosia, C. & Stoops, G. (eds.), Archaeological Soil and Sediment Micromorphology. John Wiley & Sons Ltd, Chichester, pp. 173-180.

Sanborn, P., 2010. Soil formation on supraglacial tephra deposits, Klutlan Glacier, Yukon Territory. Canadian Journal of Soil Science 90, 611-618.

Scholle, P.A. & Ulmer-Scholle, D.S., 2003. A Color Guide to the Petrography of Carbonate Rocks: Grains, Textures, Porosity, Diagenesis. American Association of Petroleum Geologists Memoir 77, 474 p.

Seddoh, F.K. & Robert, M., 1972. Intérêt de l'utilisation du microscope électronique à balayage pour l'étude des micas et de leur evolution (evolution expérimentale et dans le milieu naturel). Bulletin de la Société Française de Minéralogie et de Cristallographie 95, 75-83.

Sen, R. & Mukherjee, A.D., 1972. The relation of petrofabrics with directional orientation of mineral grains from soil parent materials. An example from Norway. Soil Science 113, 57-58.

Shahack-Gross, R., 2015. Archaeological micromorphology self-evaluation exercise. Geoarchaeology 31, 49-57.

Sheppard, R.A., 1962. Iddingsitization and recurrent crystallization of olivine in basalts from the Simcoe Mountains, Washington. American Journal of Science 260, 67-74.

Sidhu, P.S. & Gilkes, R.J., 1977. Mineralogy of soils developed on alluvium in the Indo-Gangetic Plain (India). Soil Science Society of America Journal 41, 1194-1202.

Silva, S.R., Spiers, V.M. & Gross, K.A., 1965. Soil Fabric Study with the Electron Microscope. A Progress Report. Australian Defence Scientific Service, Defence Standards Laboratories, Report 282, 13 p.

Skiba, M., Maj-Szeliga, K., Szymański, W. & Błachowski, A., 2014. Weathering of glauconite in soils of temperate climate as exemplified by a Luvisol profile from Góra Puławska, Poland. Geoderma 235/236, 212-226.

Small, S.J., Clinton, P.W., Allen, R.B. & Davis, M.R., 2014. New evidence indicates the coarse soil fraction is of greater relevance to plant nutrition than previously suggested. Plant and Soil 374, 371-379.

Southard, R.J., Shainberg, I. & Singer, M.J., 1988. Influence of electrolyte concentration on the micromorphology of artificial depositional crust. Soil Science 145, 278-288.

Soil Survey Staff, 2014. Keys to Soil Taxonomy. 12th Edition. USDA-Natural Resources Conservation Service, Washington, 360 p.

Stahr, K., Kühn, J., Trommler, J., Papenfuss, K.H., Zarei, M. & Singer, A., 2000. Palygorskite-cemented crusts (palycretes) in Southern Portugal. Australian Journal of Soil Research 38, 169-188.

Stephen, I., 1960. Clay orientation in soils. Science Progress 48, 322-331.

Stoops, G., 1964. Application of some pedological methods to the analysis of termite mounds. In Bouillon, A. (ed.), Etudes sur les Termites Africains. Université de Léopoldville, pp. 379-398.

Stoops, G., 1972. Micromorphology of some important soils of Isla Santa Cruz (Galápagos). In Kowalinski, S. & Drozd, J. (eds.), Soil Micromorphology. Panstwowe Wydawnicto Naukowe, Warszawa, pp. 407-420.

Stoops, G., 2003. Guidelines for Analysis and Description of Soil and Regolith Thin Sections. Soil Science Society of America, Madison, 184 p.

Stoops, G., 2013. A micromorphological evaluation of pedogenesis on Isla Santa Cruz (Galápagos). Spanish Journal of Soil Science 3, 14-37.

Stoops, G., 2015a. Análisis de la contextura de la masa basal y los rasgos edáficos del suelo. In Loaiza, J.C., Stoops, G., Poch, R. & Casamitjana, M. (eds.), Manual de Micromorfología de Suelos y Técnicas Complementarias. Fondo Editorial Pascual Bravo, Medellin, pp. 87-154.

Stoops, G., 2015b. Composition de la massa basal y de los edaforasgos. In Loaiza, J.C., Stoops, G., Poch, R. & Casamitjana, M. (eds.), Manual de Micromorfología de Suelos y Técnicas Complementarias. Fondo Editorial Pascual Bravo, Medellin, pp. 155-204.

Stoops, G. & Dedecker, D. 2006. Microscopy of undisturbed sediments as a help in planning dredging operations. A case study from Thailand. International Conference 'Hubs, Harbours and Deltas in Southeast Asia: Multidisciplinary and Intercultural Perspective'. Royal Academy of Overseas Sciences, Brussels, pp. 193-211.

Stoops, G. & De Mets, M., 1970. Scanning Elektronenmikroskopie toegepast in de bodemmikromorfologie. Natuurwetenschappelijk Tijdschrift 52, 10-16.

Stoops, G. & Eswaran, H., 1973. Application of a microphotometric analyser to micromorphological studies. A preliminary note. Pedologie, 23, 2, p. 141.

Stoops, G. & Jongerius, A., 1975. Proposal for a micromorphological classification of soil materials. I. A classification of the related distributions of fine and coarse particles. Geoderma 13, 189-199.

Stoops, G. & Marcelino, V., 2018. Lateritic and bauxitic materials. In Stoops, G., Marcelino, V. & Mees, F. (eds.), Interpretation of Micromorphological Features of Soils and Regoliths. Second Edition. Elsevier, Amsterdam, pp. 691-720.

Stoops, G. & Schaefer, C.E.G.R., 2018. Pedoplasmation: formation of soil material. In Stoops, G., Marcelino, V. & Mees, F. (eds.), Interpretation of Micromorphological Features of Soils and Regoliths, Second Edition. Elsevier, Amsterdam, pp. 59-71.

Stoops, G. & Van Driessche, A., 2007. Mineralogy of the sand fraction - results and problems. In Arnalds, Ó., Óskarsson, H, Bartoli, F., Buurman, P., Stoops, G. & García-Rodeja, E. (eds.), Soils of Volcanic Regions of Europe. Springer, Berlin, pp 141-153.

Stoops, G., Altemuller, H.J., Bisdom, E.B.A., Delvigne, J., Dobrosvolsky, V.V., FitzPatrick, E.A., Paneque, G. & Sleeman, J., 1979. Guidelines for the description of mineral alterations in soil micromorphology. Pedologie 29, 121-135.

Stoops, G., Van Ranst, E. & Verbeek, K., 2001. Pedology of soils within the spray zone of the Victoria Falls (Zimbabwe). Catena 46, 63-83.

Stoops, G., Canti, M.G. & Kapur, S., 2017a. Calcareous mortars, plasters and floors. In Nicosia, C. & Stoops, G. (eds.), Archaeological Soil and Sediment Micromorphology. John Wiley & Sons Ltd, Chichester, pp. 189-199.

Stoops, G., Tsatskin, A. & Canti, M.G., 2017b. Gypsic mortars and plasters. In Nicosia, C. & Stoops, G. (eds.), Archaeological Soil and Sediment Micromorphology. John Wiley & Sons Ltd, Chichester, pp. 201-204.

Stoops, G., Sedov, S. & Shoba, S., 2018. Regoliths and soils on volcanic ash. In Stoops, G., Marcelino, V. & Mees, F. (eds.), Interpretation of Micromorphological Features of Soils and Regoliths. Second Edition. Elsevier, Amsterdam, pp. 721-751.

Taboada, T. & García, C., 1999a. Smectite formation produced by weathering in a coarse granite saprolite in Galicia (NW Spain). Catena 35, 281-290.

Taboada, T. & García, C., 1999b. Pseudomorphic transformation of plagioclases during the weathering of granitic rocks in Galicia (NW Spain). Catena 35, 291-302.

Terribile, F. & FitzPatrick, E.A., 1992. The application of multilayer digital image processing techniques to the description of soil thin sections. Geoderma 55, 159-174.

Terribile, F. & FitzPatrick, E.A., 1995. The application of some image-analysis techniques to recognition of some micromorphological features. European Journal of Soil Science 46, 29-45.

Terribile F., Wright, R. & FitzPatrick, E.A., 1997. Image analysis in soil micromorphology: from univariate approach to multivariate solution. In Shoba, S., Gerasimova, M. & Miedema, R. (eds.), Soil Micromorphology: Studies on Soil Diversity, Diagnostics, Dynamics. Moscow, Wageningen, pp. 397-417.

Tovey, N.K., 1974. Some applications of electron microscopy to soil engineering. In Rutherford, G.K. (ed.), Soil Microscopy. The Limestone Press, Kingston, pp. 119-142.

Tovey, N.K. & Sokolov, V.N., 1997. Image analysis applications in soil micromorphology. In Shoba, S., Gerasimova, M. & Miedema, R. (eds.), Soil Micromorphology: Studies on Soil Diversity, Diagnostics, Dynamics. Moscow, Wageningen, pp. 345-356.

Tovey, N.K. & Wang, J.M., 1997. Quantitative three-dimensional orientation analysis using confocal microscopic studies on soil microfabric. In Shoba, S., Gerasimova, M. & Miedema, R. (eds.), Soil Micromorphology: Studies on Soil Diversity, Diagnostics, Dynamics. Moscow, Wageningen, pp. 427-435.

Tovey, N.K., Smart, P. & Hounslow, M.W., 1990. Quantitative orientation analysis of soil microfabric. In Douglas, L.A. (ed.), Soil Micromorphology: A Basic and Applied Science. Developments in Soil Science, Volume 19. Elsevier, Amsterdam, pp. 631-640.

Tovey, N.K., Smart, P. & Hounslow, M.W., 1992a. Automatic orientation mapping of some types of soil fabric. Geoderma 53, 179-200.

Tovey, N.K., Krinsley, D.H., Dent, D.L. & Corbett, W.M., 1992b. Techniques to quantitatively study the microfabric of soils. Geoderma 53, 217-235.

Tröger, W.E., 1969. Optische Bestimmung der Gesteinsbildenden Minerale. Teil 2 Textband. E. Schweizerbart'sche Verlagsbuchhandlung, Stuttgart, 822 p.

Tröger, W.E., 1971. Optische Bestimmung der Gesteinsbildenden Minerale. Teil 1 Bestimmungstabellen. E. Schweizerbart'sche Verlagsbuchhandlung, Stuttgart, 188 p.

Tucker, M.E., 2001. Sedimentary Petrology. An Introduction to the Origin of Sedimentary Rocks. Third Edition. Blackwell Science, Oxford, 272 p.

Van Ranst, E. & De Coninck, F., 1983. Evolution of glauconite in imperfectly drained sandy soils of the Belgian Campine. Zeitschrift für Pflanzenernährung und Bodenkunde 146, 415-426.

Van Ranst, E., Wilson, M.A. & Righi, D., 2018. Spodic materials. In Stoops, G., Marcelino, V. & Mees, F. (eds.), Interpretation of Micromorphological Features of Soils and Regoliths. Second Edition. Elsevier, Amsterdam, pp. 633-662.

Van Vliet-Lanoë, B. & Fox, C., 2018. Frost action. In Stoops, G. Marcelino, V. & Mees, F. (eds.), Interpretation of Micromorphological Features of Soils and Regoliths. Second Edition. Elsevier, Amsterdam, pp. 575-603.

Velbel, M.A., 1984. Natural weathering mechanisms of almandine garnet. Geology 12, 631-634.

Verheye, W. & Stoops, G., 1975. Nature and evolution of soils developed on the granite complex in the subhumid tropics (Ivory Coast). II. Micromorphology and mineralogy. Pedologie 25, 40-55.

Verleyen, E., Sabbe, K., Vyverman, W. & Nicosia, C., 2017. Siliceous microfossils from single-celled organisms: diatoms and crysophycean stomatocysts. In Nicosia, C. & Stoops, G. (eds.), Archaeological Soil and Sediment Micromorphology. John Wiley & Sons Ltd, Chichester, pp. 165-170.

Villagran, X.S., Huisman, D.J., Mentzer, S.M., Miller, C.E. & Jans, M.M., 2017. Bone and other skeletal tissues. In Nicosia, C. & Stoops, G. (eds.), Archaeological Soil and Sediment Micromorphology. John Wiley & Sons Ltd, Chichester, pp. 11-38.

Vrydaghs, L., 2017. Opal sponge spicules. In Nicosia, C. & Stoops, G. (eds.), Archaeological Soil and Sediment Micromorphology. John Wiley & Sons Ltd, Chichester, pp. 171-172.

Vrydaghs, L., Devos, Y. & Pető, Á., 2017. Opal phytoliths. In Nicosia, C. & Stoops, G. (eds.), Archaeological Soil and Sediment Micromorphology. John Wiley & Sons Ltd, Chichester, pp. 155-163.

Wilding, L.P. & Drees, L.R., 1990. Removal of carbonates from thin sections for microfabric interpretation. In Douglas, L.A. (ed.), Soil Micromorphology: A Basic and Applied Science. Developments in Soil Science, Volume 19. Elsevier, Amsterdam, pp. 613-620.

Williams, A.J., Pagliai, M. & Stoops, G., 2018. Physical and biological surface crusts and seals. In Stoops, G., Marcelino, V. & Mees, F. (eds.), Interpretation of Micromorphological Features of Soils and Regoliths. Second Edition. Elsevier, Amsterdam, pp. 539-574.

Wilson, M.J., 1966. The weathering of biotite in some Aberdeenshire soils. Mineralogical Magazine 35, 1080-1093.

Wilson, M.J., 1975. Chemical weathering of some primary rock-forming minerals. Soil Science 119, 349-355.

Yardley, B.W.D., MacKenzie, W.S. & Guilford, C., 1990. Atlas of Metamorphic Rocks and their Textures. Longman, Harlow, 120 p.

Zainol, E. & Stoops, G., 1986. Relationship between plasticity and selected physico-chemical and micromorphological properties of some inland soils from Malaysia. Pedologie 36, 263-275.

Zaniewski, K., 2001. Plasmic Fabric Analysis of Glacial Sediments Using Quantitative Image Analysis Methods and GIS Techniques. PhD Dissertation, University of Amsterdam, 220 p.

Zaniewski, K. & van der Meer, J.J.M., 2005. Quantification of plasmic fabric through image analysis. Catena 63, 109-127.

Zauyah, S., Schaefer, C.E.G.R. & Simas, F.N.B., 2018. Saprolites. In Stoops, G., Marcelino, V. & Mees, F. (eds.), Interpretation of Micromorphological Features of Soils and Regoliths. Second Edition. Elsevier, Amsterdam, pp. 37-57.

Chapter 6

Pedogenic Siliceous Features

Ma. del Carmen Gutiérrez-Castorena

COLEGIO DE POSTGRADUADOS, CAMPUS MONTECILLO, TEXCOCO, MEXICO

1. Introduction

Pedogenic siliceous features occur as crystalline forms, such as quartz, tridymite and cristobalite, and as poorly ordered or non-crystalline forms including opal-A and opal-CT. Silica may be concentrated in varying degrees through weathering, vertical and lateral translocation, and precipitation and dissolution, resulting in numerous macro-morphological forms expressed as silica-rich horizons, including cemented formations or duricrusts (e.g., silcretes and duripans) and non-cemented or poorly cemented materials (e.g., fragipans and sepiocretes).

Historically, micromorphological studies dealing with pedogenic silica started in the 1960s, focusing on arid and semiarid climatic regions and on areas with volcanic materials. McKeague and Cline (1963) provided the first comprehensive overview of siliceous features in soils, discussing their forms, origin and deposition through pedogenic

Interpretation of Micromorphological Features of Soils and Regoliths. https://doi.org/10.1016/B978-0-444-63522-8.00006-1

and biogenic processes, as well as their geographic distribution. Yarilova (1956) studied siliceous features for chernozems, Litchfield and Mabbut (1962) for duripans in Australia, Arnold (1963) for brunizems in the United States and von Buch (1967) for soils on volcanic ashes. Later, the petrography and mineralogy of silica in soils and palaeosoils was studied worldwide (see Table 2).

The aim of this chapter is to review micromorphological aspects of pedogenic silica accumulations in soils, palaeosoils and regoliths.

2. Forms of Silica

Siliceous features in soils and regoliths comprise both crystalline SiO_2 polymorphs (quartz, tridymite and cristobalite) and hydrated SiO_2 forms ($SiO_2 \cdot nH_2O$), with varying degrees of crystallinity, elemental impurities and water content (opal-A, opal-C, opal-CT). Comprehensive descriptions of silica polymorphs can be found in publications

Table 1 Physical and optical properties of silica minerals

Mineral and crystal system	Physical properties	Optical properties
Low quartz *Trigonal*	Colourless and transparent or white, vitreous lustre; mostly anhedral grains, seldom prismatic; conchoidal fracture; $D = 2.65$ g/cm³	$n_o = 1.5442$, $n_e = 1.5533$; whitish or greyish interference colours; U+; parallel extinction; length-slow; grains with undulating extinction are often biaxial
Chalcedony	Colourless to pale brown; translucent to semi-transparent nearly waxlike lustre; microcrystalline, fibrous habit	$n_o = 1.530$, $n_e = 1.543$; whitish or greyish interference colours; commonly anomalously B+; parallel extinction; length-fast or length-slow
Tridymite *Orthorhombic (pseudo-hexagonal)*	White or colourless; thin hexagonal tablets but mainly anhedral; $D = 2.26$ g/cm³	$n_\alpha = 1.469$, $n_\beta = 1.469$, $n_\gamma = 1.473$; $\Delta n = 0.004$; moderate negative relief; very weak whitish or greyish first order interference colours; B+
Cristobalite *Tetragonal (pseudo-isometric)*	Colourless, octahedral habit; spherical aggregates (lepispheres) of thin, intersecting plates or blades at submicroscopic scale; $D = 2.32$ g/cm³	$n_o = 1.484$, $n_e = 1.487$; very weak, greyish interference colours; U-
Opal (general)	Colourless to pale brown; vitreous lustre; transparent to nearly opaque; $D = 1.5$-2.3 g/cm³	$n = 1.40$-1.46; colourless, often pale grey or brown; isotropic
Opal-A	Inorganic forms: massive, botryoidal, globular, stalactitic, filamentous, pisolitic; spheres (<0.5 μm) on quartz grain surfaces. Distinctive biogenic forms: pseudomorphs of biologic cell or structure, unless fragmented	$n = 1.41$-1.47; isotropic
Opal-C	Colourless, $D = 2.3$ g/cm³	$n = 1.485$-1.487; isotropic
Opal-CT	Small (<10 μm) lepispheres; cryptocrystalline to fairly coarsely crystalline; $D = 2.20$-2.38 g/cm³	$n = 1.42$-1.44; often slightly birefringent

Adapted from Summerfield, 1983a, 1983b; Drees et al., 1989; Graetsch, 1994; Monger and Kelly, 2002; Neuendorf et al., 2007.

by Flörke et al. (1984), Miehe and Graetsch (1992), Graetsch (1994), and Monger and Kelly (2002) (Table 1).

As described for marine sediments, the generalised diagenetic sequence for silica is opal-A (siliceous biogenic ooze) → opal-A' (secondary amorphous silica) → opal-CT → reordered opal-CT → cryptocrystalline quartz or chalcedony → microcrystalline quartz (Williams et al., 1985). The main factors related to the diagenetic transformations (dissolution-reprecipitation) of non-crystalline opal-A to microcrystalline quartz are temperature, pH, ionic strength and degree of silica saturation (Iler, 1973, 1979; Williams et al., 1985). Burial depth, time, heat flow and host materials are other important factors (Stamatakis et al., 1991).

Because the optical determination of the various SiO_2 minerals is not always evident, a combination of other techniques (X-ray diffraction, electron microscopy, differential thermal analysis, infrared spectroscopy) is recommended. Using these techniques, several researchers determined the quantity, composition and distribution of silica in soils and regoliths (Smith & Whalley, 1982; Drees et al., 1989; Singh & Gilkes, 1993; Petrovic et al., 1996; Monger & Kelly, 2002; Mantha et al., 2012; Schmidht et al., 2013; Zhang & Moxon, 2014). Determination of the elemental ratio of Si to Al by EDS or WDS analysis (Boettinger & Southard, 1990; Chartres & Fitzgerald, 1990; Effland, 1990) or of oxygen isotope ratios may be necessary, for instance, to establish whether a component is inherited or pedogenic (Mizota et al., 1991). High-resolution petrographic and sub-microscopic investigations are relevant to evaluate the alteration of the various forms of silica (Thiry et al., 2014).

2.1 Quartz and Chalcedony

Quartz is the most widely recognised silica mineral in almost all soils and parent materials, as a dominant inherited component of the silt and sand fractions. Its habit in soils is mainly anhedral (Drees et al., 1989). Euhedral authigenic quartz crystals or overgrowths (Fig. 1) have been reported for earth-surface environments (Chafetz & Zhang,

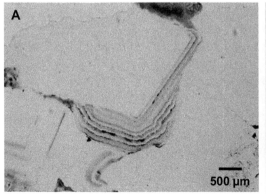 

FIGURE 1 Layered quartz overgrowth (silcrete from Nahe River Valley, Germany). (A) PPL. (B) XPL.

1998), including silcrete occurrences (Thiry, 1997, 1999). Quartz is precipitated at low temperature in soils or at water table levels. It has a higher solubility than quartz grains that are part of the groundmass, which have an ultimately magmatic origin, because pedogenic quartz typically has lower crystallinity and more structural defects (Thiry et al., 2014). The crystal size of pedogenic quartz ranges from crypto- to macrocrystalline.

Although quartz is considered to be a resistant mineral, several studies have shown that it is susceptible to weathering, with climate, time, organic acids, morphology and surface area as important controlling factors (Crook, 1968; Claisse, 1972; Krinsley & Doornkamp, 1973; Eswaran & Stoops, 1979; Bennett, 1991; Pell & Chivas, 1995; Scarciglia et al., 2005). In this regard, quartz grains show a combination of chemical dissolution and precipitation surface features. Characteristics of dissolution include etch patterning, triangular etch pits and solution features, whereas precipitation features include edge rounding, silica veneers and plate- and sheetlike structures. The degree of quartz weathering can be established by the frequency and size of etch pits (Marcelino & Stoops, 1996; Marcelino et al., 1999). For example, during long-term exposure to chemical weathering, quartz shows progressively deeper dissolution features: triangular etch pits, then an interconnected network of grooves and crevasses (Leneuf, 1973; Le Ribault, 1977; Tshidibi, 1985; Marcelino & Stoops, 1996; Marcelino et al., 1999; Moretti & Morras, 2013) and finally complete dissolution of the grain (Howard et al., 1996). In tropical regions, fractures are frequently filled with iron oxides and other compounds producing a feature called runiquartz (Eswaran et al., 1975; see Marcelino et al., 2018; Stoops & Marcelino, 2018).

Pedogenic quartz features such as infillings, coatings (e.g., Milnes et al., 1991; Thiry & Milnes, 1991; Thiry, 1992; Misik, 1996; Thiry & Ribet, 1999) or crystal intergrowths (e.g., Bustillo & Pérez-Jiménez, 2005; El Khoriby, 2005) were described for silcretes or palaeosoils (Srivastava & Sauer, 2014). Eswaran (1972) observed quartz nodules and coatings crystallised from soil solutions saturated with silica derived from weathering of basaltic rocks in Ultisols.

Chalcedony is a fibrous microcrystalline quartz variety (Fig. 2). Two varieties are recognised according to sign of elongation (length-slow or length-fast) (Folk & Pittman, 1971; Keene, 1983). It is most commonly length-fast, implying that the fibres are elongated perpendicular to the c-axis of quartz (e.g., Graetsch, 1994). Length-slow chalcedony, composed of fibres elongated parallel to the c-axis (e.g., Miehe et al., 1984), is more rare and has been related to conditions with high alkalinity (e.g., Folk & Pittman, 1971; Arbey, 1980; English, 2001; Bustillo & Alonso-Zarza, 2007), although this has to be done with caution (Keene, 1983), and it has also been related to evaporitic environments (Keene, 1983; Heaney & Post, 1992). Fluctuations in water composition can result in a succession of layers of length-fast and length-slow chalcedony in vughs (Summerfield, 1983b). Their crystal size seems to be controlled by the nature of silica supply.

Pedofeatures composed of chalcedony include coatings, which have been described for duripans (Flach et al., 1974; Oleschko, 1990; Rodas et al., 1994; Curlik & Forgac,

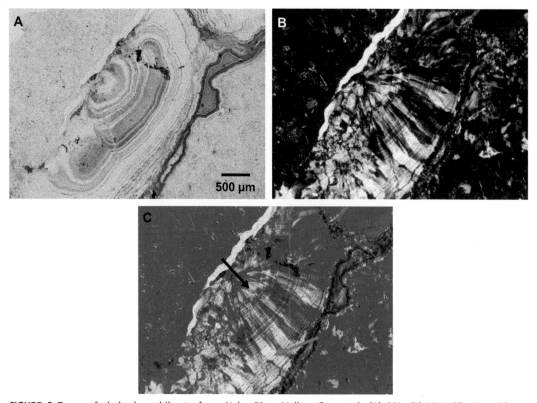

FIGURE 2 Types of chalcedony (silcrete from Nahe River Valley, Germany). (A) PPL. (B) XPL. (C) XPL with 1λ compensator plate (SE-NW), allowing recognition of length-fast chalcedony (*black arrow*) and length-slow chalcedony (*red arrow*).

1996; Thiry, 1999; Sauer et al., 2015), silcretes (Watts, 1978), silicified sandstone (e.g., Thiry & Ben Brahim, 1990), petroplinthite (Rellini et al., 2015), calcretes and silcretes (McCoy, 2011) and Ultisols (Moretti & Morras, 2013). The tendency of increasing crystal size towards vugh or channel centres suggests a declining rate of flow of silica-bearing solutions, in response to progressive void infilling and porosity reduction (Summerfield, 1983b). Other pedofeatures are loose continuous infillings with length-fast chalcedony in volcanic soils (Curlik & Forgac, 1996), nodules of length-slow chalcedony (Thiry & Milnes, 1991), loose discontinuous infillings of length-slow and length-fast chalcedony in silcretes (McCoy, 2011), microlaminar infillings (Thiry & Simon-Coinçon, 1996), and microlaminated bow-like infillings of length-slow and length-fast chalcedony with associated fine- and coarse-grained quartz in silcretes (Summerfield, 1983b), typic nodules (1-5 mm) of length-slow chalcedony in silcretes (Misik, 1996), concentric nodules of length-fast chalcedony in silcretes (Watts, 1978) and aggregates of silica spherulites (1-5 mm) composed of length-slow chalcedony, which is transformed to coarse-grained quartz with radial undulatory extinction along the outer edge (Misik, 1996).

2.2 Cristobalite and Tridymite

Cristobalite and tridymite are SiO_2 polymorphs in which SiO_4 tetrahedra are packed in a two-layer structure (tridymite) or in a three-layer structure (cristobalite) (Drees et al., 1989). The more open structure of cristobalite and tridymite in comparison with quartz allows a certain degree of incorporation of other elements into the crystal structure.

Cristobalite is usually reported, in thin section, as white to milky white, ranging from translucent to opaque in PPL (e.g., Howard, 1939; Usai & Dalrymple, 2003; Gutiérrez-Castorena et al., 2006). It occurs commonly as detrital grains in soils that formed on pyroclastic volcanic materials (Parfitt & Wilson, 1985, Mizota et al., 1987; Drees et al., 1989; Cronin et al., 1996; Molina-Ballesteros et al., 1997; Sommer et al., 2006). It has also been reported as part of the fine fraction of alluvial or lacustrine sediments (Reséndiz-Paz et al., 2013; Rivera-Vargas, 2013) and of soils around tailings piles (Duarte-Zaragoza et al., 2015).

Cristobalite forms several types of pedofeatures, such as coatings in andesitic lava (Nieuwenhuyse & Verburg, 2000), typic and amiboidal nodules in alluvial fans (Bustillo & Alonso-Zarza, 2003), and microlaminated coatings and fissure infillings in duripans (Gutiérrez-Castorena et al., 2006, 2007; Quantin, 2010). Transformation from opal to cristobalite in duripans on volcanic ash is related to dehydration during dry seasons (Elssas et al., 2000) or following artificial drying of lacustrine sediments with high amorphous silica content (Gutiérrez-Castorena et al., 2006).

Tridymite occurs less frequently in soils and is only rarely recognised as a pedofeature. It has been reported for siliceous calcretes (Vaniman et al., 1994), limestone (Verstraten & Sevink, 1979), chalk (Nørnberg et al., 1985) and silcretes (McCoy, 2011). Mulyanto and Stoops (2003) described thick coatings consisting of prismatic tridymite crystals oriented perpendicular to the pore walls in weathered andesite. The crystals are corroded, which is probably linked to hydrothermal processes at high pH.

2.3 Opal

Opal is a hydrated form of SiO_2 with a wide range of crystallinity, with water contents ranging from 1.5 wt% for opal-C to approximately 34 wt% for amorphous plant opals (Monger & Kelly, 2002). In transmitted light, the colour of opal-A ranges from colourless over light tan to various shades of brown to black (Drees et al., 1989); it is optically isotropic to weakly anisotropic in XPL (Figs. 3 and 4). In SEM images, opal-A coatings in duripans appear as bodies with a dense vitreous appearance, often containing small silica spheres (1 μm) (Sullivan, 1994), whereas opal-CT can appear as small lepispheres (<10 μm).

Opal-A is mainly found as phytoliths, diatoms, sponge spicules and radiolaria (Monger & Kelly, 2002; see Kaczorek et al., 2018). As non-biogenic precipitate, it forms in

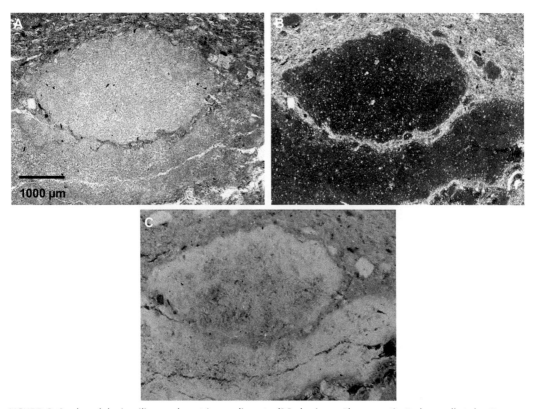

FIGURE 3 Opal nodule in siliceous lacustrine sediments (3Cg-horizon, Fluvaquentic Endoaquoll, Lake Texcoco, Mexico). (A) PPL. (B) XPL. (C) OIL.

various soils and young sediments when environmental conditions favour supersaturated Si levels in solution (Drees et al., 1989), and it can eventually be transformed to crystalline phases such as cristobalite and quartz (Thiry et al., 2014). Opal-A has been described as pendants, typic coatings and infillings, occurring in volcanic materials (Harden et al., 1991; Hessmann, 1992; Poetsch & Arikas, 1997; Gutiérrez-Castorena et al., 2007), on marine terraces (Moody & Graham, 1997), in saprolites (Chadwick et al., 1987, 1995; Milnes et al., 1991; Sauer et al., 2015), in silcretes (Thiry & Millot, 1987; Thiry & Milnes, 1991; Rodas et al., 1994; Thiry et al., 2006; Sauer et al., 2015), in duripans (Gutiérrez-Castorena et al., 2006), in tundra soils (Pustovoytov, 1998), and in Alfisols (Munk & Southard, 1993). Blank and Fosberg (1990) reported two types of opal in gravel coatings and pendants associated with carbonate accumulations in alluvial fan deposits: clear opal (clear to translucent and colourless or light brown to yellow, optically isotropic), and dirty opal (mottled, grey, translucent, optically isotropic, milky adamantine lustre in OIL). Opal-A has been observed as nodules in laminar duripans (Chadwick et al., 1987, 1995; Thiry et al., 2006) and palustrine deposits (Mees, 1999; Gutiérrez-Castorena et al., 2005, 2006), as coatings in duripans on volcanic

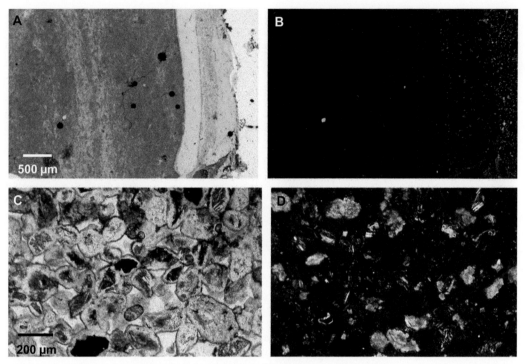

FIGURE 4 Opal (Petrocalcic Calciustoll, Atenco, Mexico). (A) Laminated opal-CT coating (PPL). (B) Same as (A) in XPL. (C) Typic opal-CT coatings and infillings (PPL). (D) Same as (C) in XPL.

material (Hessmann, 1992; Poetsch & Arikas, 1997; Gutiérrez-Castorena et al., 2007), as infillings in saprolites (Milnes et al., 1991) and as pendants in Palexeralfs (Munk & Southard, 1993).

Opal-A is often associated with other minerals, partly as laminar coatings, in which it is interlayered with micritic calcite (Blank & Fosberg, 1991b), ferruginous material (Kendrick & Graham, 2004), halloysite-smectite (Gutiérrez-Castorena et al., 2007) and unidentified phyllosilicates (Chadwick et al., 1987). Plet et al. (2012) described brown opal nodules and infillings in silcretes-calcretes.

Opal-CT has been described for sediments (Williams et al., 1985; Drees et al., 1989), groundwater silcretes (Thiry, 1999), petrified wood in volcanic ash soils (Dubroeucq et al., 1997) and duricrusts (Rodas et al., 1994). One example is coatings and infillings of opal-CT (Fig. 4) observed for volcanic soils in Mexico (Ma. del C. Gutiérrez-Castorena, unpublished data). Usai and Dalrymple (2003) described opal-C with accessory opal-A, opal-CT and well-ordered cristobalite as whitish or brownish coatings and infillings in a buried palaeosoil in Sardinia. In duripans that formed in volcanic ash, halloysite may be subjected to progressive transformation to opal-A or opal-C, through loss of aluminium during alteration of clay minerals (Elssas et al., 2000).

3. Pedogenic Silica and Types of Duricrust

Accumulation of silica in soils or regoliths is controlled by various environmental factors such as temperature, soil pH, oxidation-reduction potential, presence of cations, organic compounds and iron oxides, and the occurrence and activity of various silica-concentrating organisms (Brinkman, 1977; Iler, 1979; Sommer et al., 2006; Thiry et al., 2014). Other factors that affect silicification process include porosity, groundwater flow and changes in salinity of pore fluids that induce silica precipitation (Bustillo, 2010).

The solubility of silica is nearly constant between pH values of 2.0 and 8.5 (50-60 mg/L), but it increases markedly above pH 9, reaching several thousand mg/L at pH 11 (e.g., Drees et al., 1989). In addition, the solubility of silica is controlled by other factors such as crystallinity and degree of impurity. For example, highly crystalline minerals with only small amounts of impurities typically exhibit lower solubility (Monger & Kelly, 2002) than poorly ordered or non-crystalline forms with higher concentrations of impurities (Drees et al., 1989). Different sequences reported for duricrusts are opal-A → opal-CT (Jones & Segnit, 1971; Lynne & Campbell, 2004); opal-CT → microcrystalline quartz (Williams et al., 1985); opal-CT → quartz (Cady et al., 1996); and microcrystalline quartz → macrocrystalline quartz (Flörke et al., 1982, 1991; Moxon et al., 2006; Nash & Ullyott, 2007).

Pedogenic silica can cement soil particles, binding them at grain contact points (Monger & Kelly, 2002). Larger amounts of pedogenic silica might be required to harden sandy soils than clayey soils (Singh & Gilkes, 1993). The groundmass of the latter can be considered to be an integral part of the cement because it acts as a sorption surface (Chadwick & Nettleton, 1990). Pedogenic silica may also occur as nanocrystalline or crystalline pedofeatures (coatings, infillings and nodules) (Webb et al., 2013). During weathering, movement and deposition of silica results in development of cemented horizons in saprolites (Milnes et al., 1991). Depending on the type of parent material, different pedofeatures can form. For example, pale brown fibrous opal, overlain by limpid chalcedony, has been described for clayey materials, whereas infillings composed of fibrous opal-CT and massive limpid chalcedony occur in sandy sediments. When interstitial clay is absent, quartz coatings can form on quartz grains that are part of the groundmass.

Cemented silica horizons are classified as silcretes and duripans, and non-cemented horizons with silica-enrichment can qualify as fragipans. Differences between these types of horizons are related to illuvial processes, degree of crystallisation and other factors. Milnes et al. (1991) observed that pedogenic silcretes typically contain microcrystalline quartz, whereas duripans are characterised by opal-A as pedogenic silica phase.

Selected examples of occurrences of pedogenic silica in soils, sediments and regoliths are presented in Table 2. Micromorphological classifications exist for silcretes (Summerfield, 1983a; Nash & Ullyott, 2007; Ullyott et al., 2015), but no general classification that considers all types of silica-cemented horizons is available.

Table 2 Selected examples of occurrences of pedogenic siliceous features

Feature	Mineralogy	Occurrence	References
Typic coatings	Opal	Brunizems (Iowa, USA)	Arnold (1963)
Microlaminated coatings and nodules	Opal	Tertiary silcretes (Paris Basin and Central Massif, France)	Thiry & Millot (1987)
Concentric nodules	–	Haplic Durargids, Duric Natrargids (Nevada, USA)	Chadwick et al. (1987)
Clay and silica infillings and coatings	Opal	Typic Torripsamments (Saudi Arabia)	Khalifa et al. (1989)
Typic coatings, pendants	Opal	Alluvial fan deposits (California, USA)	Harden et al. (1991)
Pendants	–	Typic Palexeralfs (California, USA)	Munk & Southard (1993)
Coatings	Opal	Soddy Tundra soils (NE USSR)	Pustovoytov (1998)
Pendants	–	Typic Durixeralfs (California, USA)	Kendrick & Graham (2004)
Coatings, infillings	–	Fragipans (Gillan region, Iran)	Eghbal et al. (2012)
Typic coatings	Opal	Silcretes (Tanzania)	Fossum (2012)
Coatings	–	Typic Epiaquolls (California, USA)	Moody & Graham (1997)
Infillings	Opal	Duripans (Namaqualand, South Africa)	Francis et al. (2013)
Coatings and infillings	Opal-A	Haplic Durargids (New South Wales, Australia)	Sullivan (1994)
Lenticular and irregular typic nodules	–	Fluvaquentic Endoaquolls (Texcoco, México)	Gutiérrez-Castorena et al. (2006)
Orthic siliceous nodules, coatings and hypocoatings	Opal-A, opal-C	Calcareous pan margin deposits (Namibia)	Mees (2002)
Dense incomplete or incomplete infillings	Opal-A, opal-CT	Silcretes (Australia)	Singh et al. (1992)
Typic and concentric nodules	Opal, calcite	Duripan (Idaho, USA)	Blank & Fosberg (1991a, 1991b)
Calcite nodules embedded in opaline infillings	–	Haplic Durargids, Duric Natrargids (Nevada, USA)	Chadwick et al. (1987)
Nodules and coatings	–	Typic Calciargids (New Mexico, USA)	Gile et al. (1995)
Coatings and infillings	Opal-C, cristobalite	Buried palaeosoil (Alfisol) (Sardinia, Italy)	Usai & Dalrymple (2003)
Typic and concentric nodules and coatings	Opal-CT, chalcedony	Typic Durixerolls, Xerollic Durargids (Northern Great Basin, USA)	Chadwick et al. (1995)
Silica ovoids, opal nodules and fibres formed from coalescing small spheroids	Opal-CT, length-slow chalcedony	Miocene silcretes (Madrid Basin, Spain)	Bustillo & Bustillo (2000)
Nodules	Opal-CT	Alluvial deposits (Madrid Basin, Spain)	Bustillo & Alonso-Zarza (2003)

Feature	Mineralogy	Context (location)	Reference
Aggregates	Length-slow chalcedony transformed to coarse-grained quartz	Ancient silcretes (Western Carpathians)	Misik (1996)
Infillings and nodules	Length-slow chalcedony	Nodular to concretionary silcretes (Australia)	Watts (1978)
Pisoliths with layers of cryptocrystalline silica	Length-fast chalcedony	Fluvial silcretes (Botswana)	Shaw & Nash (1998)
Coatings	Chalcedony	Typic Durixeralfs (California, USA)	Flach et al. (1974)
Concentric banding infillings	–	Paleogene silcrete (Central Spain)	Rodas et al. (1994)
Infillings	–	Duripans on volcanic ash (Texcoco, Mexico)	Oleschko (1990)
Loose continuous infillings	Chalcedony, quartz	Silcretes (Southern Africa)	Summerfield (1983a)
Infillings	Quartz, chalcedony, opal-CT	Weathering and non-weathering silcrete profile types (Kalahari Basin, Southern Africa)	Summerfield (1983a)
Microlaminated coatings and nodules	Chalcedony, quartz	Tertiary silcretes (Paris Basin and Central Massif, France)	Thiry & Millot (1987)
Dense complete infillings	Chalcedony	Groundwater silcrete (South Australia)	Webb & Goulding (1998)
Infillings and laminar coatings	Quartz	Silcretes (Paris Basin and Central Massif, France)	Thiry (1992)
Infillings	Quartz	Silcretes (Paris Basin, France)	Thiry & Ribet (1999)
Fine-grained silica cements as isolated patches and vein- or caplike geopetal features	–	Silcretes (South Downs, Sussex, UK)	Ullyott & Nash (2006)
Coatings, infillings	Quartz	Groundwater silcretes (Southern UK)	Ullyott et al. (2015)
Coatings of prismatic crystals	Tridymite	Altered andesitic rocks (Indonesia)	Mulyanto & Stoops (2003)

3.1   Silcretes

Silcrete is a lithological term for a strongly indurated rock composed mainly of inherited quartz grains and siliceous cement (Singh et al., 1992). The term was first suggested by Lamplugh (1902) to describe a highly resistant and well-cemented near-surface crust formed by silica accumulation, cementing a pre-existing soil, sediment, rock or weathered material (Nash & Ullyott, 2007). As silcretes range widely in age, although they are mainly recognized for Cainozoic formations (Summerfield, 1983a; Monger & Kelly, 2002), they can be cemented by various SiO_2 polymorphs and other components (Fig. 5), and

FIGURE 5 Silcrete (Nahe River Valley, Germany) (XPL). (A) Dense complete infillings of macrocrystalline quartz (*yellow arrow*), chalcedony and non-fibrous microcrystalline quartz. (B) Detail of zone with chalcedony (*red arrow* in A). (C) Detail of zone with non-fibrous microcrystalline quartz (*blue arrow* in A).

they can present distinctive morphological and mineralogical features resulting from a succession of phases of silica dissolution and crystallisation (Thiry et al., 2006). Mainly, non-fibrous microcrystalline quartz has been reported (Khalaf, 1988; Milnes et al., 1991; Thiry & Simon-Coinçon, 1996), but opal and chalcedony with small amounts of non-fibrous quartz (Rodas et al., 1994) and opal-A to opal-CT (Curlik & Forgac, 1996; Lee & Gilkes, 2005) have also been observed, as well as disordered, length-fast chalcedony and nearly optically isotropic cryptocrystalline silica (Shaw & Nash, 1998). Singh et al. (1992) mention the presence of fine-grained kaolinite, zircon and anatase. The anatase (TiO_2) is used as an indication of the high degree of weathering of the parent material (Watts, 1980; Summerfield, 1983b; Nash et al., 1994). For calcrete-silcrete intergrades, Watts (1980) reports four different types of silica cement, with quartz replacing calcite in advanced stages of silcrete formation. Types of silica cement range from opaline silica in a thin (<5 µm) diffuse layer over a moderately thick layer of semi-opaque brown, length-slow chalcedony, to a rapidly alternating sequence of length-fast and length-slow chalcedony, followed by a quartz infilling. Detailed mineralogical and micromorphological studies show that silica can systematically change with decreasing depth from more soluble to less soluble phases: opal → microcrystalline quartz → euhedral quartz (Thiry et al., 2014), with different features at each stage.

Silcretes occur extensively in arid and semiarid regions of southern Africa (e.g., Summerfield, 1983a, 1983b) and Australia (e.g., Thiry & Milnes, 1991; Lee & Gilkes, 2005) (Table 2). They are typically more strongly cemented and presumably older than duripans (Monger & Kelly, 2002). Silcretes can be formed in various environmental conditions, with regard to pH and climate (Summerfield, 1979, 1983b). For instance, Wopfner (1978) reports quartz silcretes in kaolinised profiles formed in acid groundwater environments, as well as opal silcretes formed under alkaline hypersaline conditions. Most fabrics of silcretes reflect successive dissolution, precipitation, recrystallisation and reworking in soil conditions (Watts, 1978), and they also record the influence of parent material characteristics and geochemical environments during initial silica precipitation and postsilicification diagenesis (Summerfield, 1983a; Lee & Gilkes, 2005). Some silcretes result from the silicification of more or less weathered rocks, or by replacement of carbonates by microcrystalline quartz (Thiry & Ben Brahim, 1990).

Pedogenic silcretes can form in various environmental conditions with alternating wet and dry seasons, causing multiple episodes of water infiltration and percolation (Summerfield, 1979, 1983a; Thiry et al., 2006). Strong evaporation during dry periods results in an increase in pH, favouring alteration of silicates that are part of the groundmass, whereas in the beginning of the wet period, a drop in pH results in opal precipitation (Blanco et al., 2008). Non-pedogenic types of silcretes can form under the influence of phreatic conditions (groundwater silcrete), in alluvial sediment fills (drainage-line silcretes) and along lake margins (pan or lacustrine silcrete) (Ullyott et al., 2004; Nash & Ullyott, 2007; Ullyott & Nash, 2016).

Silcretes have been classified according to their morphology and mineralogical composition (Smale, 1973; Wopfner, 1978), genesis (Milnes & Thiry, 1992; Thiry, 1999), and petrographical or micromorphological features (Goudie, 1973; Wopfner, 1978; Summerfield, 1983a; Milnes & Thiry, 1992; Thiry, 1999; Webb et al., 2013; Thiry et al., 2014; Ullyott & Nash, 2016). Based on their genesis, chemical composition and the geomorphological setting, Nash and Ullyott (2007) divided silcretes into five categories, comprising two pedogenic types and three non-pedogenic types.

3.1.1 Pedogenic Silcretes

Silcretes show a complex fabric with a variety of features that overwrite the original fabric of the parent material (Thiry, 1999; Webb et al., 2013), as a result of a succession of phases of silica dissolution and recrystallisation during multiple episodes of water infiltration and percolation in alternately wet and dry conditions (Watts, 1978; Thiry & Millot, 1987; Thiry et al., 2006).

In the groundmass of silcretes, various polymorphs of silica occur. Distinguishing between inherited and pedogenetic silica polymorphs can be difficult. Microcrystalline quartz or chalcedony can give rise to a quartz-crystallitic b-fabric (Webb & Goulding, 1998; Nash & Hopkinson, 2004; Ullyott et al., 2015; Ullyott & Nash, 2016). When opal dominates, an undifferentiated b-fabric is observed (Smale, 1973; McCoy, 2011; Fossum, 2012). In silcretes-calcretes, a quartz-calcite-crystallitic b-fabric is common (Vaniman et al., 1994; Nash et al., 2004; Plet et al., 2012). During silcrete development, the c/f-related distribution pattern of the soil as a whole typically changes gradually from monic to chitonic or gefuric, followed by single- and double-spaced porphyric, and in the final stage, by open porphyric (Summerfield, 1983a, 1983b). Singh et al. (1992) report a close porphyric c/f-related distribution pattern and the occurrence of two cement types in the packing voids: (i) an opaque cement, pale yellow in reflected light, composed of microcrystalline quartz, anatase and zircon, and (ii) a translucent brown cement, white in reflected light, with a banded, gel-like massive texture, composed of aluminosilicates (kaolinite) that formed at a later stage. Coatings and infillings are the most common pedofeatures.

Opal coatings are common as overgrowths on quartz grains (Thiry et al., 2014), formed by dissolution-precipitation processes (Webb & Goulding, 1998; Thiry, 1999; Fossum, 2012; Sauer et al., 2015). Coatings and dense infillings composed of opal-A, opal-CT or gel-like aluminosilicates (Singh et al., 1992; Thiry, 1992), or loose infillings consisting of chalcedony or non-fibrous quartz (Summerfield, 1983a) are common as illuvial features. Thiry (1992) reports various illuvial pedofeatures that are typical of pedogenic silcretes, such as loose continuous infillings of quartz grains at the top of the silcrete, laminar coatings of microcrystalline quartz overlain by opal in the middle part, and laminar coatings and infillings of opal at the base. Thiry and Simon-Coinçon (1996) describe alternating layers of quartz grains cemented by microcrystalline quartz in silcretes formed in palaeoclimate conditions with a pronounced dry season. Compound coatings/infillings consisting of various silica types (e.g., opal, chalcedony, non-fibrous quartz) usually occur in subsurface horizons (Thiry & Milnes, 1991; Thiry, 1999), starting with amorphous opal and progressing through chalcedony to non-fibrous microcrystalline and coarser quartz towards the centre of voids (Thiry, 1978, 1988, 1992; Thiry & Millot, 1987; Rodas et al., 1994; Molina-Ballesteros et al., 1997). Typic chalcedony coatings are formed by illuviation and recrystallisation (Summerfield, 1982; Thiry & Milnes, 1991; Fossum, 2012; Thiry et al., 2014; Sauer et al., 2015).

Crescent coatings and infillings are often composed of laminae consisting of various silica phases, clay and TiO_2 (Summerfield, 1982; Callen, 1983; Thiry & Millot, 1987; Thiry, 1992, 1999; Webb & Goulding, 1998). Their origin is unclear, but the fact that they are related to macropores indicates that they are formed at sites of preferential flow of solutions (Nash & Ullyott, 2007; Ullyott & Nash, 2016). Biological activity can also be involved in the formation of these crescent coatings (McCoy, 2011). The type of silica is partly controlled by soil solution composition, whereas the TiO_2 phase is a residual product of mineral dissolution (Watts, 1980; Thiry, 1997; Thiry et al., 2014).

3.1.2 *Groundwater and Palustrine Silcretes*

In non-pedogenic silcretes, structures of the parent material are well preserved (Wopfner, 1983; Thiry & Milnes, 1991; Milnes & Thiry, 1992; Thiry et al., 2014). The c/f-related distribution pattern can be open, double-spaced or single-spaced porphyric, which has been described for cases where calcite or dolomite cements were replaced by silica (Watts, 1980; Bustillo & Alonso-Zarza, 2007; Thiry et al., 2015) or sepiolite (Francis et al., 2013). In groundwater silcretes, typic coatings and cappings (Tòfalo & Pazos, 2010; Dupuis et al., 2014) or dense complete infillings (Tófalo & Pazos, 2010) have been described to occur, composed of various types of authigenic silica such as opal-CT, cristobalite and tridymite (Wofpner, 1983), chalcedony and non-fibrous microcrystalline quartz (Thiry, 1999), and micro- and/or macrocrystalline quartz (Ullyott et al., 2004, 2015). Significant accumulations of TiO_2 are absent (Thiry, 1999; Thiry et al., 2014). Cappings composed of microcrystalline quartz are related to translocation of fine material and subsequent silicification (Ullyott et al., 2004, 2015). These features originate from phreatic fluctuations (Thiry & Milnes, 1991), climatic changes from wet to dry (Callen, 1983) or variations in pH (Shaw & Nash, 1998), causing silica precipitation.

In pan/lacustrine silcretes, lenticular and irregular opal nodules are common (Fig. 6). They are formed by changes in pore fluid salinity, inducing opal precipitation (Armenteros et al., 1995; Thiry, 1999; Ringrose et al., 2005; Gutiérrez-Castorena et al., 2006; Bustillo, 2010).

3.2 Duripans

A duripan is an indurated subsurface diagnostic soil horizon cemented by illuvial silica (Chartres & Fitzgerald, 1990; Singh & Gilkes, 1993), usually in the form of opal or microcrystalline quartz, and may or may not contain additional cementing agents such as calcite (Neuendorf et al., 2007; Soil Survey Staff, 2014). Weakly cemented to indurated

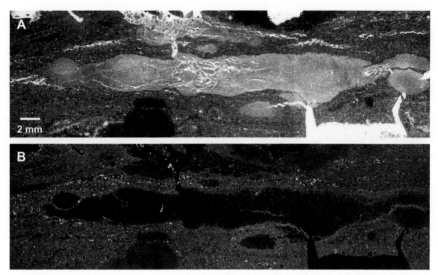

FIGURE 6 Lacustrine silcrete (3Lg-horizon, Lake Texcoco, Mexico). (A) Irregular lenticular brownish opal nodule (PPL). (B) Idem in XPL.

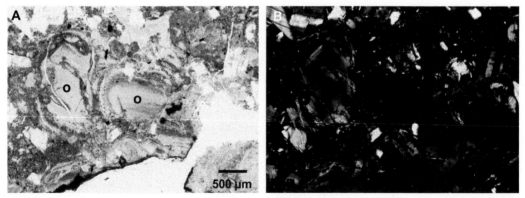

FIGURE 7 Duripan (6Btq-horizon, Miocene palaeosoil, Coatlinchan, Texcoco, Mexico). (A) Fragments of opal (o) coatings and infillings. (PPL). (B) Idem in XPL.

nodules of opal or microcrystalline quartz (durinodes) are the most common pedofeatures in duripans (Soil Survey Staff, 2014). Opal can be identified by its limpid, light yellow to dull grey aspect in PPL, and by its optical isotropy in XPL (Fig. 7) (Chadwick et al., 1987; Sullivan, 1994). This indurated horizon has various local names, such as red-brown hardpan in Australia (Litchfield & Mabbut, 1962), and tepetate or canganhua in Latin America, the latter two referring to duripans in soils on volcanic ash.

The formation of silica features is more related to the stage of soil development, parent material, degree of cementation, soil fabric and location within the horizon than to the source of silica (e.g., weathering of dust inputs or surface horizons) and hydrological aspects (e.g., lateral subsurface flow, groundwater depth, water movement) (Chadwick et al., 1987).

Soil fabrics influence the type of opaline silica precipitation. Chadwick et al. (1987) mention that, in loamy soils, opaline silica occurs as bridges between sand grains and as small (1-5 μm) nodules and intercalations that impregnate clay aggregates and sesquioxides. In loamy soils, opaline silica and clay occur as compound coatings, whereas in sandy soils opaline silica occurs as laminar coatings. Laminated opal infillings, concentric nodules and other composite features with chalcedony and calcite have been reported for alluvial fans (Blank & Fosberg, 1990, 1991a, 1991b) and for Mollisols and Aridisols (Chadwick et al., 1995). By the end of the process of duripan development, a porphyric c/f-related distribution pattern has formed (Chadwick et al., 1987). Amorphous silica deposits on sand grains in duripans were described by Chartres and Norton (1994) and Hollingsworth and Fitzpatrick (1994), both using submicroscopic techniques.

Silica cementation starts as small, diffuse, optically isotropic opal nodules, or as silica coatings on ped faces. In more strongly developed duripans, chalcedony appears as extremely thin (1.0 μm), weakly birefringent intercalations, throughout the groundmass and associated with clay coatings (Flach et al., 1969). In marine terraces, the following features appear successively with increasing degree of cementation (Moody & Graham, 1997): opal coatings and bridges on sand grains in weakly cemented material, then opal that partially fills packing voids and later complete infillings of those voids, resulting in a completely cemented material (Moody & Graham, 1997).

Complex silica pedofeatures can form when various processes occur at different stages, related to aeolian input, subaerial exposure, erosion and colonisation by cryptogamic organisms (Blank et al., 1998), or in response to climate fluctuations or bioturbation (Eghbal & Southard, 1993), such as laminated opal infillings, concentric nodules and other composite features containing chalcedony and calcite (Blank & Fosberg, 1990, 1991a, 1991b; Chadwick et al., 1995).

Within different layers of the duripan, different pedofeatures can be recognised. Boettinger and Southard (1990) report that opaline silica is present as distinct laminae (2-10 μm) in the laminar cap of a duripan, whereas spheroidal and ellipsoidal nodules (0.04-1.0 mm) occur within the calcite-dominated groundmass beneath the thin laminar cap.

In some regions, duripans on volcanic ash are important materials, with specific micromorphological characteristics. In duripans that developed in ash-flow tuff deposits, silica first precipitates as pendants, it then impregnates the groundmass and it finally precipitates around the clasts in the form of coatings (Kendrick & Graham, 2004).

All of those features eventually produce a plugged horizon with a laminar cap (Harden et al., 1991).

Cementing agents in silica-enriched soil horizons in volcanic ashes in various regions of Mexico (see also Sedov et al., 2010; Stoops et al., 2018) mainly consist of opaline silica (Hidalgo et al., 1992; Poetsch & Arikas, 1997; Poetsch, 2004), carbonates (Gutiérrez-Castorena & Ortiz-Solorio, 1992) and silica associated with illuvial clay and iron oxides (Acevedo-Sandoval et al., 2004). Opal-A is present as microlaminar brown yellow, optically isotropic coatings and infillings and as part of the groundmass with an undifferentiated b-fabric (Hessmann, 1992; Poetsch & Arikas, 1997; Gutiérrez-Castorena et al., 2007). Compound coatings and infillings composed of clay, opal-A, opal-CT and microcrystalline cristobalite have been described (Quantin, 2010). Nimlos (1989) reports silica films (up to 1 mm) at depths of up to 3 m, supporting the hypothesis of silica illuviation. These indurated horizons are the product of complex soil processes (eluviation-illuviation), redoximorphic conditions, clay degradation and progressive silicification (Quantin, 2010), although erosion/sedimentation processes have also been reported (Gutiérrez-Castorena et al., 2007). For similar formations in Ecuador and surrounding countries, silica cementation by colourless amorphous silica associated with clay coatings on planar voids was described by Creutzberg et al. (1990).

Duripans may occur in association with calcic or petrocalcic horizons. In this case, silica tends to be mutually exclusive with calcite: opaline silica impregnates the groundmass, whereas carbonates occur as infillings, at least in loamy soils (Chadwick et al., 1987). Silica and calcium carbonates can have a phreatic origin rather than being exclusively pedogenic (Monger & Adams, 1996). Laminar cappings and spheroidal to ellipsoidal nodules (0.04-1.0 mm) with crystallitic b-fabric (Boettinger & Southard, 1990) or lined by clay coatings (Monger & Adams, 1996) are recognized for these environments.

3.3   Fragipans

Fragipan is a subsurface diagnostic horizon that is compact but not or only weakly cemented, with evidence of pedogenesis (e.g., soil structure) and limited root penetration, a brittle consistency and redoximorphic features (Soil Survey Staff, 2014).

Several researchers have concluded that amorphous silica coatings (Harlan et al., 1977; Franzmeier et al., 1991) and illuvial or authigenic clay explain the brittle consistency of fragipans (Steinhardt et al., 1982; Lindbo & Veneman, 1989; Norton, 1994). In addition, iron and aluminium compounds can be present (Bridges & Bull, 1983; Brown & Mahler, 1988; Marsan & Torrent, 1989; Duncan & Franzmeier, 1999; Szymański et al., 2012). Weisenborn and Schaetzl (2005a, 2005b) report non-laminated or microlaminated typic and crescent coatings and infillings, composed of silty clay or silt, as well as dense infillings of Al compounds, Fe oxides, organic matter and SiO_2.

Chartres and Norton (1994) observed that silica and nanocrystalline Fe oxides (ferrihydrite), adhering to clay surfaces, are important elements in the cohesion of a fragipan. Mullins et al. (1990), McKyes et al. (1994) and Daniells (2012) demonstrated that silica coatings on mineral surfaces, under reduced humidity, promote reduction of porosity and hydraulic conductivity of soils.

Using SEM and EDS analysis of samples of E horizons, Weisenborn (2001) described dissolution pits on quartz grain surfaces. They have also been observed for duripans (Thompson et al., 1994).

Some silica-enriched horizons on volcanic ash that meet the requirements of fragipan (Gutiérrez-Castorena et al., 2007) show slightly birefringent, limpid, non-laminated, dense complete infillings of opal-CT, resulting in materials with chitonic and double-spaced porphyric c/f-related distribution patterns and undifferentiated b-fabric.

4. Conclusions

Pedogenic siliceous features in soils can occur in many different phases, ranging from amorphous opal-A over opal-C and opal-CT to chalcedony and non-fibrous micro-crystalline quartz. These materials are responsible for the formation of certain diagnostic soil horizons (fragipans, duripans), as well as silcretes and other types of silicified soil or regolith horizons, with a wide range of ages.

Fragipans, duripans and pedogenic silcretes differ in complexity and mineralogical composition. In fragipans, features are relatively simple (e.g., typic coatings, infillings, nodules) and mainly consist of opal. In duripans and silcretes, the features are more complex (e.g., typic and laminar coatings/infillings), and they consist of various silica phases (opal, opal-CT, cristobalite, chalcedony and non-fibrous microcrystalline quartz).

Systematic micromorphological studies of duricrusts are strongly recommended because current understanding of their development in various settings is still incomplete, and the proven potential of microscopic studies can still be more fully explored. As distinction between opal and fine-grained Si-Al colloids is almost impossible with current optical techniques, more research using other *in situ* analytical techniques is needed.

References

Acevedo-Sandoval, O.A., Ortiz-Hernández, L.E., Flores-Román, D., Velásquez-Rodríguez, A.S. & Flores-Castro, K., 2004. Physical and chemical characterization of hardened horizons (tepetate) in soils of volcanic origin of Mexico State. Agrociencia 37, 435-499.

Arbey, F., 1980. Les formes de la silice et l'identification des évaporites dans les formations silicifiées. Bulletin des Centres de Recherches Exploration-Production Elf-Aquitaine 4, 309-365.

Armenteros, I., Bustillo, M.A. & Blanco, J.A., 1995. Pedogenic and groundwater processes in a closed Miocene basin (northern Spain). Sedimentary Geology 99, 17-36.

Arnold, R.W., 1963. Silans in Some Brunizem Soils. PhD Dissertation, Iowa State University, 236 p.

Bennett, P.C., 1991. Quartz dissolution in organic-rich aqueous systems. Geochimica et Cosmochimica Acta 55, 1781-1797.

Blanco, J.A., Armenteros, I. & Huerta, P., 2008. Silcrete and alunite genesis in alluvial paleosols (late Cretaceous to early Palaeocene, Duero basin, Spain. Sedimentary Geology 211, 1-11.

Blank, R.R. & Fosberg, M.A., 1990. Micromorphology and classification of secondary calcium carbonate accumulations that surround or occur on the undersides of coarse fragments in Idaho (USA). In Douglas, L.A. (ed.), Soil Micromorphology: A Basic and Applied Science. Developments in Soil Science, Volume 19. Elsevier, Amsterdam, pp. 341-346.

Blank, R.R. & Fosberg, M.A., 1991a. Duripans of the Owyhee Plateau region of Idaho: genesis of opal and sepiolite. Soil Science 152, 116-133.

Blank, R.R. & Fosberg, M.A., 1991b. Duripans of Idaho, USA: in situ alteration of eolian dust (loess) to an opal-A/X-ray amorphous phase. Geoderma 48, 131-149.

Blank, R.R., Cochran, B. & Fosberg, M.A., 1998. Duripans of southwestern Idaho: polygenesis during the Quaternary deduced through micromorphology. Soil Science Society of America Journal 62, 701-709.

Boettinger, J.L. & Southard, R.J., 1990. Micromorphology and mineralogy of a calcareous duripan formed in granitic residuum, Mojave Desert, California, USA. In Douglas, L.A. (ed.), Soil Micromorphology: A Basic and Applied Science. Developments in Soil Science, Volume 19. Elsevier, Amsterdam, pp. 409-415.

Bridges, E.M. & Bull, P.A., 1983. The role of silica in the formation of compact and indurated horizons in the soils of south Wales. In Bullock, P. & Murphy, C.P. (eds.), Soil Micromorphology. Volume 2. Soil Genesis. AB Academic Publishers, Berkhamsted, pp. 605-614.

Brinkman, R., 1977. Surface-water gley soils in Bangladesh: genesis. Geoderma 17, 111-144.

Brown, T.H. & Mahler, R.L., 1988. Relations between soluble silica and plow pans in Palouse silt loam soils. Soil Science 145, 359-364.

Bustillo, M.A., 2010. Silicification of continental carbonates. In Alonso-Zarza, A.M. & Tanner, L.H. (eds.), Carbonates in Continental Settings: Processes, Facies and Applications. Developments in Sedimentology, Volume 61. Elsevier, Amsterdam, pp. 153-174.

Bustillo, M.A. & Alonso-Zarza, A.M., 2003. Transformaciones edáficas y diagenéticas de los depósitos aluviales distales del Mioceno de la Cuenca de Madrid, Area de Paracuellos de Jarama. Estudios Geológicos 59, 39-52.

Bustillo, M.A & Alonso-Zarza, A.M., 2007. Overlapping of pedogenesis and meteoric diagenesis in distal alluvial and shallow lacustrine deposits in the Madrid Miocene Basin, Spain. Sedimentary Geology 198, 255-271.

Bustillo, M.A. & Bustillo, M., 2000. Miocene silcretes in argillaceous playa deposits, Madrid Basin, Spain: petrological and geochemical features. Sedimentology 47, 1023-1037.

Bustillo, M.A. & Pérez-Jiménez, J.L., 2005. Características diferenciales y génesis de los niveles silíceos explotados en el yacimiento arquelógico de Casa Montero (Vicálvaro, Madrid). Geogaceta 38, 243-246.

Cady, S.L., Wenk, H.R. & Downing, K.H., 1996. HRTEM of microcrystalline opal in chert and porcelanite from the Monterrey Formation, California. American Mineralogist 81, 1380-1395.

Callen, R.A., 1983. Late Tertiary 'grey billy' and the age and origin of surficial silicifications (silcrete) in South Australia. Journal of the Geological Society of Australia 30, 393-410.

Chadwick, O.A. & Nettleton, W.D., 1990. Micromorphologic evidence of adhesive and cohesive forces in soil cementation. In Douglas, L.A. (ed.), Soil Micromorphology: A Basic and Applied Science. Developments in Soil Science, Volume 19. Elsevier, Amsterdam, pp. 207-212.

Chadwick, O.A., Hendricks, D.M. & Nettleton, W.D., 1987. Silica in duric soils. I. A depositional model. Soil Science Society of America Journal 51, 975-982.

Chadwick, O.A., Nettleton, W.D. & Staidl, G.J., 1995. Soil polygenesis as a function of Quaternary climate change, northern Great Basin, USA. Geoderma 68, 1-26.

Chafetz, H.S. & Zhang, J., 1998. Authigenic euhedral megaquartz crystals in a quaternary dolomite. Journal of Sedimentary Research 68, 994-1000.

Chartres, C.J. & Fitzgerald, J.D., 1990. Properties of siliceous cements in some Australian soils and saprolites. In Douglas, L.A. (ed.), Soil Micromorphology: A Basic and Applied Science. Developments in Soil Science, Volume 19. Elsevier, Amsterdam, pp. 199-205.

Chartres, C.J. & Norton, L.D., 1994. Micromorphological and chemical properties of Australian soils with hardsetting and duric horizons. In Ringrose-Voase, A. & Humphreys, G.S. (eds.), Soil Micromorphology: Studies in Management and Genesis. Developments in Soil Science, Volume 22. Elsevier, Amsterdam, pp. 825-834.

Claisse, G., 1972. Etude sur la solubilisation du quartz en voie d'altération. Cahiers ORSTOM, Série Pédologie 10, 97-122.

Creutzberg, D., Kauffman, J.H., Bridges, E.M. & Guillermo Del Posso, G.M., 1990. Micromorphology of 'cangahua': a cemented subsurface horizon in soils from Ecuador. In Douglas, L.A. (ed.), Soil Micromorphology: A Basic and Applied Science. Developments in Soil Science, Volume 19. Elsevier, Amsterdam, pp. 367-372.

Cronin, J.S., Neall, V.E. & Palmer, A.S., 1996. Investigation of an aggrading paleosol developed into andesitic ring-plan deposits, Ruapehu volcano, New Zealand. Geoderma 69, 119-135.

Crook, K., 1968. Weathering and roundness of quartz sand grains. Sedimentology 11, 171-182.

Curlik, J. & Forgac, J., 1996. Mineral forms and silica diagenesis in weathering silcretes of volcanic rocks in Slovakia. Geologica Carpathica 47, 107-118.

Daniells, I.G., 2012. Cohesive soils: a review. Soil Research 50, 349-359.

Drees, L.R., Wilding, L.P., Smeck, N.E. & Senkayi, A.L., 1989. Silica in soils: quartz and disordered silica polymorphs. In Dixon, J.B. & Weed, S.B. (eds.), Minerals in Soil Environments. Second Edition, SSSA Book Series No. 1, Soil Science Society of America, Madison, pp. 913-974.

Duarte Zaragoza, V.M., Gutiérrez Castorena, E.V., Gutiérrez Castorena, Ma. Del C., Carrillo González, R., Ortiz Solorio, C.A. & Trinidad Santos, A., 2015. Heavy metals contamination in soils around tailing heaps with various degrees of weathering in Zimapán, México. International Journal of Environmental Studies 72, 24-40.

Dubroeucq, D., Geissert, D. & Roger, P., 1997. Pine roots-induced petrocalcic horizons in volcanic ash soil of the Mexican Altiplano. In Zebrowski, C., Quantin, P. & Trujillo, G. (eds.), Suelos Volcánicos Endurecidos. Impressora Polar, Quito, pp. 98-106.

Duncan, M.M. & Franzmeier, D.P., 1999. Role of free silicon, aluminium, and iron in fragipan formation. Soil Science Society America Journal 63, 923-929.

Dupuis, C., Quesnel, F. & Baele, J.M., 2014. PETM related paleoweathering on the southern margins of the North Sea Basin. Examples from Mons Basin (Belgium) and the Dieppe-Hampshire Basin (Normandy, France). Societá Geologica Italiana 31, 68-69.

Effland, W.R., 1990. Genesis of clayey sediments and associated upland soils near the Upper Iowa River, Northeast Iowa. PhD Dissertation, Iowa State University, 223 p.

Eghbal, M.K. & Southard, R.J., 1993. Micromorphological evidence of polygenesis of three Aridisols, western Mojave Desert, California. Soil Science Society of America Journal 57, 1041-1050.

Eghbal, M.K., Givi, J., Torabi, H. & Miransari, M., 2012. Formation of soils with fragipan and plinthite in old beach deposits in the South of the Caspian Sea, Gilan province, Iran. Applied Clay Science 64, 44-52.

El Khoriby, E.M., 2005. Origin of the gypsum-rich silica nodules, Moghra Formation, Northwest Qattara depression, Western Desert, Egypt. Sedimentary Geology 177, 41-55.

Elssas, F., Dubroeucq, D. & Thiry, M., 2000. Diagenesis of silica minerals from clay minerals in volcanic soils of Mexico. Clay Minerals 35, 477-489.

English, P.M., 2001. Formation of analcime and moganite at Lake Lewis, central Australia: significance of groundwater evolution in diagenesis. Sedimentary Geology 143, 219-244.

Eswaran, H., 1972. Micromorphological indicators of pedogenesis in some tropical soils derived from basalts from Nicaragua. Geoderma 7, 15-31.

Eswaran, H. & Stoops G., 1979. Surface textures of quartz in tropical soil. Soil Science Society of America Journal 43, 420-424.

Eswaran, H., Sys, C. & Souza, E.C., 1975. Plasma infusion. A pedological process of significance in the humid tropics. Anales de Edafología y Agrobiología 34, 665-674.

Flach, K.W., Nettleton, W.D., Gile, L.H. & Cady, J.G., 1969. Pedocementation: induration by silica, carbonates, and sesquioxides in the Quaternary. Soil Science 107, 442-452.

Flach, K.W., Nettleton, W.D. & Nelson, R.E., 1974. The micromorphology of silica-cemented soil horizons in western North America. In Rutherford, G.K. (ed.), Soil Microscopy. The Limestone Press, Kingston, Ontario, pp. 715-729.

Flörke, O.W., Köhler-Herbertz, B., Langer, K. & Tönges, I., 1982. Water in microcrystalline quartz of volcanic origin: agates. Contributions to Mineralogy and Petrology 80, 324-333.

Flörke, O.W., Flörke, U. & Giese, U., 1984. Moganite: a new microcrystalline silica mineral. Neues Jahrbuch für Mineralogie, Abhandlungen 149, 325-326.

Flörke, O.W., Graetsch, B., Martin, B., Röller, K. & Wirth, R., 1991. Nomenclature of micro-and non-crystalline silica minerals, based on structure and microstructure. Neues Jahrbuch für Mineralogie, Abhandlungen 163, 19-42.

Folk, R.L. & Pittman, J.S., 1971. Length-slow chalcedony: a new testament for vanished evaporates. Journal of Sedimentary Research 41, 1045-1058.

Fossum, K., 2012. Sedimentology, Petrology and Geochemistry of the Kilimatinde Cement, Central-Tanzania. MSc Dissertation, University of Oslo, 128 p.

Francis, M.L., Ellis, F., Lambrechts, J.J.N. & Poch, R.M., 2013. A micromorphological view through a Namaqualand termitaria (Heuweltjie, a Mima-like mound). Catena 100, 57-73.

Franzmeier, D.P., Norton, L.D. & Steinhardt, G.C., 1991. Fragipan formation in loess of the midwestern United States. In Smeck, N.E. & Ciolkosz, E.J. (eds.), Fragipans: Their Occurrence, Classification, and Genesis. Soil Science Society of America, Inc., Madison, pp. 69-98.

Gile, L.H., Hawley, J.W., Grossman, R.B., Monger, H.C., Montoya, C.E. & Mack, G.H., 1995. Supplement to the Desert Project Guidebook, with Emphasis on Soil Micromorphology. New Mexico Bureau of Mines and Mineral Resources Bulletin 142, 96 p.

Goudie, A.S., 1973. Duricrust in Tropical and Subtropical Landscapes. Clarendon Press, Oxford, 174 p.

Graetsch, H., 1994. Structural characteristics of opaline and microcrystalline silica minerals. Reviews in Mineralogy 29, 209-232.

Gutiérrez-Castorena, Ma. del C. & Ortiz-Solorio C.A., 1992. Caracterización del tepetate blanco en Texcoco. Terra 10, 202-209.

Gutiérrez-Castorena, Ma. del C., Stoops, G., Ortiz Solorio, C.A. & López Avila, G., 2005. Amorphous silica materials in soils and sediments of the Ex-Lago de Texcoco, Mexico: an explanation for its subsidence. Catena 60, 205-226.

Gutiérrez-Castorena, Ma. del C., Stoops, G., Ortiz-Solorio, C.A. & Sánchez-Guzmán, P., 2006. Micromorphology of opaline features in soils on the sediments of the ex-Lago de Texcoco, México. Geoderma 132, 89-104.

Gutiérrez-Castorena, Ma. del C., Ortiz-Solorio, C.A. & Sánchez-Guzmán, P., 2007. Clay coating formation in tepetates from Texcoco, Mexico. Catena 71, 411-424.

Harden, J.W., Taylor, E.M., Reheis, M.C. & McFadden, L.D., 1991. Calcic, gypsic and siliceous soil chronosequences in arid and semiarid environments. In Nettleton, W.D. (ed.), Occurrence, Characteristics and Genesis of Carbonate, Gypsum and Silica Accumulations in Soils. Soil Science Society of America Special Publication 26. SSSA, Madison, pp. 1-16.

Harlan, P.W., Franzmeier, D.P. & Roth, C.B., 1977. Soil formation on loess in southwestern Indiana. II. Distribution of clay and free oxides and fragipan formation. Soil Science Society of America 42, 99-103.

Heaney, P.J. & Post, J.E., 1992. The widespread distribution of a novel silica polymorph in microcrystalline quartz varieties. Science 255, 441-443.

Hessmann, R., 1992. Micromorphological investigations on 'tepetate' formation in the 'toba' sediments of the state of Tlaxcala (Mexico). Terra 10, 145-150.

Hidalgo, M.C., Quantin, P. & Zebrowski, C., 1992. La cementación de los tepetates: estudio de la silicificación. Terra 10, 192-201.

Hollingsworth, L.D. & Fitzpatrick, R.W., 1994. Nature and origin of a duripan in a Durixeralf-Duraqualf toposequence: micromorphological aspects. In Ringrose-Voase, A. & Humphreys, G.S. (eds.), Soil Micromorphology: Studies in Management and Genesis. Developments in Soil Science, Volume 22. Elsevier, Amsterdam, pp. 835-844.

Howard, A.D., 1939. Cristobalite in southwestern Yellowstone Park. American Mineralogist 24, 485-491.

Howard, J.L., Amos, D.F. & Daniels, W.L., 1996. Micromorphology and dissolution of quartz sand in some exceptionally ancient soils. Sedimentary Geology 105, 51-62.

Iler, R.K., 1973. Effect of adsorbed alumina on the solubility of amorphous silica in water. Journal of Colloid and Interface Science 43, 399-408.

Iler, R.K., 1979. The Chemistry of Silica. Solubility, Polymerization, Colloid and Surface Properties, and Biochemistry. John Wiley & Sons, New York, 866 p.

Jones, J.B. & Segnit, E.R., 1971. The nature of opal I. Nomenclature and constituent phases. Journal of the Geological Society of Australia 18, 57-68.

Kaczorek, D., Vrydaghs, L., Devos, Y., Pető, Á. & Effland, W.R., 2018. Biogenic siliceous features. In Stoops, G., Marcelino, V. & Mees, F. (eds.), Interpretation of Micromorphological Features of Soils and Regoliths. Second Edition. Elsevier, Amsterdam, pp. 157-176.

Keene, J.B., 1983. Chalcedonic quartz and occurrence of quartzine (length-slow chalcedony) in pelagic sediments. Sedimentology 30, 449-454.

Kendrick, K.J. & Graham, R.C., 2004. Pedogenic silica accumulation in chronosequence soils, southern California. Soil Science Society of America Journal 68, 1295-1303.

Khalaf, F.I., 1988. Petrography and diagenesis of silcrete from Kuwait, Arabian Gulf. Journal of Sedimentary Petrology 58, 1014-1022.

Khalifa, E.M., Reda, M. & Al-Awajy, M.H., 1989. Changes in soil fabrics of Torripsamments under irrigated date palms, Saudi Arabia. Geoderma 44, 307-317.

Krinsley, D.H. & Doornkamp, J.C., 1973. Atlas of Quartz Sand Surface Textures. Cambridge University Press, London, 91 p.

Lamplugh, G.W., 1902. Calcrete. Geological Magazine 9, 575.

Lee, S.Y. & Gilkes, R.J., 2005. Groundwater geochemistry and composition of hardpans in southwestern Australian regolith. Geoderma 126, 59-84.

Leneuf, N., 1973. Observations stéréoscopiques sur les figures de corrosion du quartz dans certaines formations superficielles. Cahiers ORSTOM, Série Pédologie 11, 43-51.

Le Ribault, L., 1977. L'Exoscopie des Quartz. Editions Masson, Paris, 200 p.

Lindbo, D.L. & Veneman, P.L.M., 1989. Fragipans in the northeastern United States. In Smeck, N.E. & Ciolkosz, E.J. (eds.), Fragipans: Their Occurrence, Classification and Genesis. Soil Science Society of America Special Publication 24. SSSA, Madison, pp. 11-31.

Litchfield, W.H. & Mabbut, J.A., 1962. Hardpan in soils of semi-arid Western Australia. Journal of Soil Science 13, 148-159.

Lynne, B.Y. & Campbell, K.A., 2004. Morphologic and mineralogic transitions from opal-A to opal-CT in low-temperature siliceous sinter diagenesis, Taupo Volcanic Zone, New Zealand. Journal of Sedimentary Research 74, 561-579.

Mantha, N.M., Schindler, M., Murayama, M. & Hochella, M.F., 2012. Silica- and sulfate-bearing rock coatings in smelter areas: products of chemical weathering and atmospheric pollution. I. Formation and mineralogical composition. Geochimica et Cosmochimica Acta 85, 254-274.

Marcelino, V. & Stoops, G., 1996. A weathering score for sandy soil materials based on the intensity of etching of quartz grains. European Journal of Soil Science 47, 7-12.

Marcelino, V., Mussche, G. & Stoops G., 1999. Surface morphology of quartz grains from tropical soils and its significance for assessing soil weathering. European Journal of Soil Science 50, 1-8.

Marcelino, V., Schaefer, C.E.G.R. & Stoops, G., 2018. Oxic and related materials. In Stoops, G., Marcelino, V. & Mees, F. (eds.), Interpretation of Micromorphological Features of Soils and Regoliths. Second Edition. Elsevier, Amsterdam, pp. 663-689.

Marsan, F.A. & Torrent, J., 1989. Fragipan bonding by silica and iron oxides in a soil from northwestern Italy. Soil Science Society of America Journal 53, 1140-1145.

McCoy, Z., 2011. The distribution and origin of silcrete in the Ogallala formation, Garza County, Texas. MSc Dissertation, Texas Tech University, 148 p.

McKeague, J.A. & Cline, M.J., 1963. Silica in soils. Advances in Agronomy 15, 339-396.

McKyes, E., Nyamugafata, P. & Nyamapfene, K.W., 1994. Characterization of cohesion, friction and sensitivity of two hardsetting soils from Zimbabwe. Soil Tillage Research 29,357-366.

Mees, F., 1999. The unsuitability of calcite spherulites as indicators for subaerial exposure. Journal of Arid Environments 42, 149-154.

Mees, F., 2002. The nature of calcareous deposits along pan margins in eastern central Namibia. Earth Surface Processes and Landforms 27, 719-735.

Miehe, G. & Graetsch, H., 1992. Crystal structure of moganite: a new structure type for silica. European Journal of Mineralogy 4, 691-706.

Miehe, G., Graetsch, H. & Flörke, O.W., 1984. Crystal structure and growth fabric of length-fast chalcedony. Physics and Chemistry of Minerals 10, 197-199.

Milnes, A.R. & Thiry, M., 1992. Silcretes. In Martini, I.P. & Chesworth, W. (eds.). Weathering, Soils and Paleosols. Developments in Earth Surface Processes, Volume 2. Elsevier, Amsterdam, pp. 349-378.

Milnes, A.R., Wright, M.J. & Thiry, M., 1991. Silica accumulations in saprolites and soils in South Australia. In Nettleton, W.D. (ed.), Occurrence, Characteristics and Genesis of Carbonate, Gypsum and Silica Accumulations in Soils. Soil Science Society of America Special Publication 26. SSSA, Madison, pp. 121-149.

Misik, M., 1996. Silica spherulites and fossil silcretes in carbonate rocks of the Western Carpathians. Geologica Carpathica 47, 91-105.

Mizota. C., Toh, N. & Matsuhisa, Y., 1987. Origin of cristobalite in soils derived from volcanic ash in temperate and tropical regions. Geoderma 39, 323-330.

Mizota, C., Itoh, M., Kusakabe, M. & Noto, M., 1991. Oxygen isotope ratios of opaline silica and plant opal in three recent volcanic ash soils. Geoderma 50, 211-217.

Molina-Ballesteros, E., García-Talegón, J. & Hernández, M.A., 1997. Palaeoweathering profiles developed on the Iberian hercynian basement and their relationship to the oldest Tertiary surface in central and western Spain. In Widdowson, M. (ed.), Palaeosurfaces: Recognition, Reconstruction and Paleoenvironmental Interpretation. Special Publication 120. Geological Society Publishing House, London, pp. 175-185.

Monger, H.C. & Adams, H.P., 1996. Micromorphology of calcite-silica deposits, Yucca Mountain, Nevada. Soil Science Society of America Journal 60, 519-530.

Monger, H.C. & Kelly, E.F., 2002. Silica minerals. In Dixon, J.B. & Schulze, D.G. (eds.), Soil Mineralogy with Environmental Applications. SSSA Book Series, No. 7. Soil Science Society of America, Madison, pp. 611-636.

Moretti, L.M. & Morras, H.J.M., 2013. New microscopic evidences of the autochthony of the ferrallitic pedological mantle in the Misiones Province, Argentina. Latin American Journal of Sedimentology and Basin Analysis 20, 129-142.

Moody, L.E. & Graham, R.C., 1997. Silica-cemented terrace edges, central California coast. Soil Science Society of America Journal 61, 1723-1729.

Moxon, T., Nelson, D.R. & Zhang, M., 2006. Agate recrystallisation: evidence from samples found in Archaean and Proterozoic host rocks, Western Australia. Australian Journal of Earth Sciences 53, 235-248.

Mullins, C.E., MacLeod, D.A., Northcote, K.H., Tisdall, J.M. & Young, I.M., 1990. Hardsetting Soils: behavior, occurrence and, management. Advances in Soil Science 11, 37-108.

Mulyanto, B. & Stoops, G., 2003. Mineral neoformation in pore spaces during alteration and weathering of andesitic rocks in humid tropical Indonesia. Catena 54, 385-391.

Munk, L.P. & Southard, R.J., 1993. Pedogenic implications of opaline pendants in some California Late-Pleistocene Palexeralfs. Soil Science Society of America Journal 57, 149-154.

Nash, D.J. & Hopkinson, L., 2004. A reconnaissance laser Raman and Fourier transform infrared survey of silcretes from the Kalahari Desert, Botswana. Earth Surface Processes and Landforms 29, 1541-1558.

Nash, D. J. & Ullyott, J.S., 2007. Silcrete. In Nash, D. & McLaren, S.J. (eds), Geochemical Sediments and Landscapes. Blackwell, Oxford, pp. 95-143.

Nash, D.J. Thomas, D.S.G. & Shaw, O.A., 1994. Siliceous duricrusts as paleoclimatic indicators: evidence from the Kalahari desert of Botswana. Palaeogeography, Palaeoclimatology, Palaeoecology 112, 279-295.

Nash, D.J., McLaren, S.J. & Webb, J.A., 2004. Petrology, geochemistry and environmental significance of silcrete-calcrete intergrade duricrusts at Kang Pan and Tswaane, Central Kalahari, Botswana. Earth Surface Processes and Landforms. 29, 1559-1586.

Neuendorf, K.K.E., Mehl, J.P. & Jackson, J.A. (eds.), 2007. Glossary of Geology, 5th Edition. American Geological Institute, Alexandria, 779 p.

Nieuwenhuyse, A. & Verburg, P.S.J., 2000. Mineralogy of a soil chronosequence on andesitic lava in humid tropical Costa Rica. Geoderma 98, 61-82.

Nimlos, T.J., 1989. The density and strength of Mexican tepetate (duric materials). Soil Science 147, 23-27.

Nørnberg, P., Dalsgaard, K. & Skammelsen, E., 1985. Morphology and composition of three mollisol profiles over chalk, Denmark. Geoderma 36, 317-342.

Norton, L.D., 1994. Micromorphology of silica cementation in soils. In Ringrose-Voase, A. & Humphreys, G.S. (eds.), Soil Micromorphology: Studies in Management and Genesis. Developments in Soil Science, Volume 22. Elsevier, Amsterdam, pp. 811-824.

Oleschko, K., 1990. Cementing agents morphology and its relation to the nature of 'Tepetates'. In Douglas, L.A. (ed.) Soil Micromorphology: A Basic and Applied Science. Developments in Soil Science, Volume 19. Elsevier, Amsterdam, pp. 381-386.

Parfitt, R.L. & Wilson, A.D., 1985. Estimation of allophane and halloysite in three sequences of volcanic soils, New Zealand. Catena 7, 1-8.

Pell, S.D. & Chivas, A.R., 1995. Surface features of sand grains from the Australian continental Dunefield. Palaeogeography, Palaeoclimatology, Palaeoecology 113, 119-132.

Petrovic, I., Heaney, P.J. & Navrotsky, A., 1996. Thermochemistry of the new silica polymorph moganite. Physics and Chemistry of Minerals 23, 119-126.

Plet, C., Bustillo, M.A. & Alonso-Zarza, A.M., 2012. Calcrete-silcrete duricrust from distal-alluvial fan deposits (Madrid Basin, Torrijos area, Toledo, Spain). Geogaceta 52, 85-88.

Poetsch, T., 2004. Forms and dynamics of silica gel in a tuff-dominated soil complex: results of micromorphological studies in the central highlands of Mexico. Revista Mexicana de Ciencias Geológicas 21, 195-201.

Poetsch, T. & Arikas, K., 1997. The micromorphological appearance of free silica in some soils of volcanic origin in central Mexico. In Zebrowski, C., Quantin, P. & Trujillo, G. (eds.), Suelos Volcánicos Endurecidos. Impressora Polar, Quito, pp. 56-64.

Pustovoytov, K.E., 1998. Opal cutans in the profile of mountain tundra sod soil in the extreme northeast of the USSR. Moscow University Soil Science Bulletin 45, 12-17.

Quantin, P., 2010. Genesis of petroduric and petrocalcic horizons in Latinoamerica volcanic soils. Geophysical Research Abstracts 12, 1.

Rellini, I., Trombino, L., Carbone, C. & Firpo, M., 2015. Petroplinthite formation in a pedosedimentary sequence along a northern Mediterrean coast: from micromorphology to lanscape evolution. Journal of Soils and Sediments 15, 1311-1328.

Reséndiz-Paz, Ma. de la L., Gutiérrez-Castorena, Ma. del C., Gutiérrez-Castorena, E.V., Ortiz-Solorio, C.A., Cajuste-Bontempts, L. & Sánchez-Guzmán, P., 2013. Local soil knowledge and management of Anthrosols: a case study in Teoloyucan, Mexico. Geoderma 193-194, 41-51.

Ringrose, S., Huntsman-Mapila, P., Kampunzu, A.B., Downey, W., Coetzee, S., Vink, B., Matheson, W. & Vanderpost, C., 2005. Sedimentological and geochemical evidence for palaeo-environmental change in the Makgadikgadi subbasin, in relation to the MOZ rift depression, Botswana. Palaeogeography, Palaeoclimatology, Palaeoecology 217, 265-287.

Rivera-Vargas, G., 2013. Relación de los minerales amorfos con metales pesados en Antrosoles irrigados con aguas residuales. MSc Dissertation, Colegio de Postgraduados, Texcoco, 63 p.

Rodas, M., Luque, F.J., Mas, R. & Garzón, M.G., 1994. Calcretes, palycretes and silcretes in the palaeogene detrital sedimetns of the Duero and Tajo basins, Central Spain. Clay Minerals 29, 273-285.

Sauer, D., Stein, C., Glatzel, S., Kühn, J., Zarei, M. & Stahr, K., 2015. Duricrust in soils of the Alentejo (southern Portugal) — types, distribution, genesis and time of their formation. Journal of Soils and Sediments 15, 1437-1453.

Scarciglia, F., Le Pera E. & Critelli, S., 2005. Weathering and pedogenesis in the Sila Grande Massif (Calibria, South Italy): from field scale to micromorphology. Catena 61, 1-29.

Schmidht, P., Slodczyk, A., Léa, V., Davidson, A., Puaud, S. & Sciau, P., 2013. A comparative study of the thermal behaviour of length-fast chalcedony, length-slow chalcedony (quartzine) and moganite. Physics and Chemistry of Minerals 40, 331-340.

Sedov, S., Stoops, G. & Shoba, S., 2010. Regoliths and soils on volcanic ash. In Stoops, G., Marcelino, V. & Mees, F. (eds.), Interpretation of Micromorphological Features of Soils and Regoliths. Elsevier, Amsterdam, pp. 275-303.

Shaw, P.A. & Nash, D.J., 1998. Dual mechanisms for the formation of fluvial silcretes in the distal reaches of the Okavango Delta Fan, Botswana. Earth Surface Processes and Landforms 23, 705-714.

Singh, B. & Gilkes, R.J., 1993. The recognition of amorphous silica in indurated soil profiles. Clay Minerals 28, 461-474.

Singh, B., Gilkes, R.J. & Butt, C.R., 1992. An electron optical investigation of aluminosilicate cement in silcretes. Clays and Clay Minerals 40, 707-721.

Smale, D., 1973. Silcretes and associated silica diagenesis in southern Africa. Journal of Sedimentary Petrology 43, 1077-1089.

Smith, B.J. & Whalley, W.B., 1982. Observations on the composition and mineralogy of an Algerian duricrust complex. Geoderma 28, 285-311.

Soil Survey Staff, 2014. Keys to Soil Taxonomy. 12th Edition. USDA-Natural Resources Conservation Service, Washington, 360 p.

Sommer, M., Kaczorek, D., Kuzyakov, Y. & Breuer, J., 2006. Silicon pools and fluxes in soils and landscapes — a review. Journal of Plant Nutrition and Soil Science 169, 310-329.

Srivastava, P. & Sauer, D., 2014. Thin-section analysis of lithified paleosols from Dagshai Formation of the Himalayan Foreland: identification of paleopedogenic features and diagenetic overprinting and implications for paleoenvironmental reconstruction. Catena 112, 86-98.

Stamatakis, M.G., Kanaris-Sotiriou, R. & Spears, A., 1991. Authigenic silica polymorphs and the geochemistry of Pliocene siliceous swamp sediments of the Aridea volcanic province, Greece. Canadian Mineralogist 29, 587-598.

Steinhardt, G.C., Franzmeier, D.P. & Norton L.D., 1982. Silica associated with fragipan and non-fragipan horizons. Soil Science Society of America Journal 46, 656-657.

Stoops, G. & Marcelino, V., 2018. Lateritic and bauxitic materials. In Stoops, G., Marcelino, V. & Mees, F. (eds.), Interpretation of Micromorphological Features of Soils and Regoliths. Second Edition. Elsevier, Amsterdam, pp. 691-720.

Stoops, G., Sedov, S. & Shoba, S., 2018. Regoliths and soils on volcanic ash. In Stoops, G., Marcelino, V. & Mees, F. (eds.), Interpretation of Micromorphological Features of Soils and Regoliths. Second Edition. Elsevier, Amsterdam, pp. 721-751.

Sullivan, L.A., 1994. Micromorphology and composition of silica accumulations in a hardpan. In Ringrose-Voase, A. & Humphreys, G.S. (eds.), Soil Micromorphology: Studies in Management and Genesis. Developments in Soil Science, Volume 22. Elsevier, Amsterdam, pp. 845-854.

Summerfield, M.A., 1979. Origin and palaeoenvironmental interpretation of sarsens. Nature 281, 137-139.

Summerfield, M.A., 1982. Distribution, nature and probable genesis of silcrete in arid and semi-arid southern Africa. In Yaalon, D.H. (ed.), Aridic Soils and Geomorphic Processes. Catena Supplement 1, 37-65.

Summerfield, M.A., 1983a. Petrography and diagenesis of silcrete from the Kalahari Basin and Cape coastal zone, southern Africa. Journal of Sedimentary Research 53, 895-909.

Summerfield, M.A., 1983b. Silcrete. In Goudie, A.S. & Pye, K. (eds.), Chemical Sediments and Geomorphology. Precipitates and Residua in the Near-surface Environment. Academic Press, London, pp. 59-91.

Szymański, W., Skiba, M. & Skiba, S., 2012. Origin of reversible cementation and brittleness of the fragipan horizon in Albeluvisols of the Carpathian Foothills, Poland. Catena 99, 66-74.

Thiry, M., 1978. Silicification des sédiments sablo-argileux de l'Yprésien du sudest du bassin de Paris. Genèse et évolution des dalles quarzitiques et silcrètes. Bulletin du Bureau de Recherches Gèologiques et Minières 1, 19-46.

Thiry, M., 1988. Les Grès lustrés de l'Éocène du Bassin de Paris: des silcrètes pédologiques. Bulletin d'Information des Géologues du Bassin de Paris 25, 15-24.

Thiry, M., 1992. Pedogenic silicifications: structures, micromorphology, mineralogy and their interpretation. Terra 10, 46-59.

Thiry, M., 1997. Continental silicifications: a review. In Paquet, H. & Clauer, N. (eds.), Soils and Sediments: Mineralogy and Geochemistry. Springer, Berlin, pp. 191-221.

Thiry, M., 1999. Diversity of continental silicification features: examples from the Cenozoic deposits in the Paris Basin and neighbouring basement. In Thiry, M. & Simon-Coinçon, R. (eds.), Palaeoweathering, Palaeosurfaces and Related Continental Deposits. International Association of Sedimentologists Special Publication 27. Blackwell Science, Oxford, pp. 87-128.

Thiry, M. & Ben Brahim, M., 1990. Silicification pédogénétiques dans les depôts hamadiens du piémont de Boudenin (Maroc). Geodynamica Acta 4, 237-251.

Thiry, M. & Millot, G., 1987. Mineralogical forms of silica and their sequence of formation in silcretes. Journal of Sedimentary Petrology 57, 343-352.

Thiry, M. & Milnes, A.R., 1991. Pedogenic and groundwater silcretes at Stuart Creek opal field, South Australia. Journal of Sedimentary Petrology 61, 11-127.

Thiry, M. & Ribet, I., 1999. Groundwater silicification in Paris Basin limestones: fabrics, mechanisms, and modeling. Journal of Sedimentary Research 69, 171-183.

Thiry, M. & Simon-Coinçon, R., 1996. Tertiary paleoweatherings and silcretes in the southern Paris Basin. Catena 26, 1-26.

Thiry, M., Milnes, A.R., Rayot, V. & Simon-Coinçon, R., 2006. Interpretation of palaeoweathering features and successive silicifications in the Tertiary regolith of inland Australia. Journal of the Geological Society 163, 723-736.

Thiry, M., Fernandes, P., Milnes, A. & Raynal, J.P., 2014. Driving forces for the weathering and alteration of silica in the regolith: implications for studies of prehistoric flint tools. Earth-Science Reviews 136, 141-154.

Thiry, M., Milnes, A.R. & Ben Brahim, M., 2015. Pleistocene cold climate groundwater silicification, Jbel Ghassoul region, Missour Basin, Morocco. Journal of the Geological Society 172, 125-137.

Thompson, C.H., Bridges, E.M. & Jenkins, D.A., 1994. An exploratory examination of some relict hardpans in the coastal lowlands of southern Queensland. In Ringrose-Voase, A. & Humphreys, G.S. (eds.), Soil Micromorphology: Studies in Management and Genesis. Developments in Soil Science, Volume 22. Elsevier, Amsterdam, pp. 233-245.

Tófalo, O.R. & Pazos, P.J., 2010. Paleoclimatic implications (Late Cretaceous-Paleogene) from micromorphology of calcretes, palustrine limestones and silcretes, southern Paraná Basin, Uruguay. Journal of South American Earth Sciences 29, 665-675.

Tshidibi, N.Y.B., 1985. Evolution du quartz au sein des cuirasses latéritiques et des sols ferrugineux. Geo-Eco-Trop 8, 93-110.

Ullyott, J.S. & Nash, D.J., 2006. Micromorphology and geochemistry of groundwater silcretes in the eastern South Downs, UK. Sedimentology 53, 387-412.

Ullyott, J.S. & Nash, D.J., 2016. Distinguishing pedogenic and non-pedogenic silcretes in the landscape and geological record. Proceedings of the Geologists' Association 127, 311-319.

Ullyott, J.S., Nash, D.J. Whiteman, C.A. & Mortimore, E.N., 2004. Distribution, petrology and mode development of silcretes (sarsens and puddingstones) on the eastern South Downs, UK. Earth Surface Processes and Landforms 29, 1509-1539.

Ullyott, J.S., Nash, D.J. & Huggett, J.M., 2015. Cap structures as diagnostic indicators of silcrete origin. Sedimentary Geology 325, 119-131.

Usai, M.R. & Dalrymple, J.B., 2003. Characteristics of silica-rich pedofeatures in a buried paleosol. Catena 54, 557-572.

Vaniman, D.T., Steve, J., Chipera, S.J. & Bish, D.L., 1994. Pedogenesis of siliceous calcretes at Yuca Mountain, Nevada. Geoderma 63, 1-17.

Verstraten, J.M. & Sevink, J., 1979. Clay soils on limestone in South Limburg, the Netherlands. 3. Soil formation. Geoderma 21, 281-295.

von Buch, M.W., 1967. Mikromorphologische Untersuchungen von Strukturelementen und Kieselsäurebildungen in älteren vulkanischen Böden der Collipulli-Serie, Frontera, Südchile. Geoderma 1, 249-276.

Watts, N.L., 1980. Quaternary pedogenic calcretes from the Kalahari (southern Africa): mineralogy, genesis and diagenesis. Sedimentology 27, 661-686.

Watts, S.H., 1978. A petrographic study of silcrete from inland Australia. Journal of Sedimentary Petrology 48, 987-994.

Webb, J.A. & Goulding, S.D., 1998. Geochemical mass-balance and oxygen-isotope constraints on silcrete formation and its paleoclimatic implications in Southern Australia. Journal of Sedimentary Research 68, 981-993.

Webb, J.A., Finlayson, B., Cochrane, G., Doelman, T. & Domanski, M., 2013. Silcrete quarries and artefact distribution in the Central Queenslkand Highlands, Eastern Australia. Archaeology in Oceania 48, 130-140.

Weisenborn, B.N., 2001. A Model of Fragipan Evolution in Michigan Soils. MSc Dissertation, Michigan State University, 428 p.

Weisenborn, B.N. & Schaetzl, R.J., 2005a. Range of fragipan expression in some Michigan soils. I. Morphological, micromorphological, and pedogenic characterization. Soil Science Society of America Journal 69, 168-177.

Weisenborn, B.N. & Schaetzl, R.J., 2005b. Range of fragipan expression in some Michigan soils. II. A model for fragipan evolution. Soil Science Society of America Journal 69, 178-187.

Williams, L.A., Parks, G.A. & Crerar, D.A., 1985. Silica diagenesis. I. Solubility controls. Journal of Sedimentary Petrology 55, 301-311.

Wopfner, H., 1978. Silcretes in northern South Australia and adjacent regions. In Langford-Smith, T. (ed.), Silcretes in Australia. University of New England Press, Armidale, pp. 93-141.

Wopfner, H., 1983. Environment of silcrete formation: a comparison of examples from Australia and the Cologne Embayment, West Germany. In Wilson, R.C.L (ed.), Residual Deposits: Surface Related Weathering Processes and Materials. Geological Society of London Special Publication 11, London, pp. 151-157.

Yarilova, E.A., 1956. Mineralogical study of a Sub-alpine Chernozem on andesite basalt [*in Russian*]. Kora Vyvetrivaniya 2, 45-60.

Zhang, M. & Moxon, T., 2014. Infrared absorption spectroscopy of SiO_2-moganite. Mineralogical Society of America 99, 671-680.

Chapter 7

Biogenic Siliceous Features

Danuta Kaczorek[1,5], Luc Vrydaghs[2], Yannick Devos[2], Ákos Pető[3], William R. Effland[4]

[1]WARSAW UNIVERSITY OF LIFE SCIENCES, WARSAW, POLAND;
[2]UNIVERSITÉ LIBRE DE BRUXELLES, BRUSSELS, BELGIUM;
[3]SZENT ISTVÁN UNIVERSITY, GÖDÖLLÓ, HUNGARY;
[4]NATURAL RESOURCES CONSERVATION SERVICE, U.S. DEPARTMENT OF AGRICULTURE, BELTSVILLE, MD, UNITED STATES;
[5]LEIBNIZ-CENTRE FOR AGRICULTURAL LANDSCAPE RESEARCH, MÜNCHEBERG, GERMANY

CHAPTER OUTLINE

1. Introduction

This chapter discusses the interpretation of micromorphological observations of biogenic siliceous features in soils, palaeosoils and regoliths. Forms of biogenic silica may be grouped according to the organisms associated with their formation: (i) phytogenic silica, defined as silica precipitates in roots, stems, branches, leafs and needles of plants, making a distinction between phytoliths (≥ 5 μm) and undefined remnants (<5 μm); (ii) microbial silica, including bacterial and fungal deposits; (iii) protozoic silica, comprising diatom frustules and testate amoeba shells; and (iv) zoogenic silica, such as sponge spicules (Sommer et al., 2006). Using micromorphology, selective dissolution techniques and other investigation procedures, biogenic siliceous features are studied to help to observe, record, interpret and conceptualise processes of soil genesis and to provide information on landscape evolution.

The microscopic study of biogenic silica has a fairly long tradition. The first microscopic descriptions of phytoliths were recorded around 1843 by Ehrenberg (1843, 1844, 1845, 1846); however, Ruprecht (1866) is credited with being the first to have noted the occurrence of opal phytoliths in soils (see Drees et al., 1989). Bobrov (2003) summarised the micropaleontological methods for studying biogenic silica in soils and palaeosoils with profile depth distribution data for phytoliths, diatoms and sponge spicules. The role of phytogenic siliceous features in geoarchaeological studies was more recently reviewed by Shillito (2011a) and Vrydaghs et al. (2016). Analysis of biogenic silica can provide insight into current and past conditions of pedogenesis and reveal evolutionary trends in individual soils and pedosediments or more widely in the whole soil cover (e.g., Golyeva, 2008).

2. Phytoliths

2.1 Nature and Origin

Opal phytoliths are plant microfossils (5-250 μm), characterised by high negative relief and optical isotropism. They are generally colourless, although they can grade to light tan, brown and black (Gutiérrez-Castorena & Effland, 2010) because of carbon inclusions (Stoops 2003) or the presence of a coating. They generally show a clear whitish colour under OIL or transmitted dark-field illumination (Stoops, 2003). Some opal phytoliths are autofluorescent under ultraviolet or blue light (FitzPatrick, 1984; Stoops, 2003), a feature that appears to be lacking for recently formed phytoliths (Altemüller & Van Vliet-Lanoë, 1990). Strong autofluorescence has been reported for phytoliths from Podzols and Andosols (Van Vliet-Lanoë, 1980).

Opal phytoliths are produced by polymerisation of monosilicic acid ($Si(OH)_4$) within plant tissues, taken up by the roots and transported by the vascular system. Within the plant tissue, silica precipitation can occur in three different loci: the cell lumen, the intercellular space, and the cell wall. Polymerisation results in the formation of a morphological identifiable opal cast. Opal phytoliths can accumulate in all kinds of plant organs (Rapp & Mulholland, 1992): leaves (Piperno, 1988, 2006), inflorescence bracts (Ball et al., 1999), fruits (Piperno, 1989), seeds (Lentfer, 2009; Eichhorn et al., 2010), culms (Blackman, 1968), truncks (Vrydaghs et al., 2001), woody tissue (Amos, 1952; Balan Menon, 1965; ter Welle, 1976a, 1976b; Vrydaghs et al., 1995) and roots (Tomlinson, 1961). Each plant organ is characterised by an association of different types of phytoliths, each with varying abundance. This multiplicity is made of phytoliths with specific morphologies and less specific or redundant morphologies (Rovner, 1971). In addition, their size varies according to their botanical origin (Wilding & Drees, 1974; Piperno, 2006).

Grasses are known to be an important plant group that can produce opal phytoliths (up to 10% silica by weight); however, opal phytoliths are also reported for Pteridophytes (Piperno, 1988, 2006), Gymnosperms (Klein & Geist, 1978; Sangster et al., 1997) and Angiosperms (Metcalfe & Chalk, 1950; Twiss et al., 1969; Brown, 1984; Bozarth, 1992; Kealhoffer & Piperno, 1998; Ball et al., 2006).

2.2   Occurrence in Soils

In soils and sediments, phytoliths can occur either within plant fragments (Fig. 1A) or within the groundmass (Fig. 1B).

Phytoliths have been observed for a variety of soils, such as Andosols (Stoops, 2003), Alfisols (Boettinger, 1994), Vertisols (Boettinger, 1994; see Kovda & Mermut, 2010, 2018), Podzols (Simons et al., 2000), Chernozems (Drees et al., 1989), Latosols (Kondo & Iwasa, 1981), Ultisols (Stoops et al., 1994), Histosols (Benvenuto et al., 2013) and Anthrosols (Devos et al., 2009). They are most commonly observed for A horizons and less commonly for B horizons. They are generally found in the groundmass and occasionally as infillings (Bullock et al. 1985; Stoops, 2003). In H horizons, phytoliths may constitute an important part of the coarse fraction. In soil profiles, their concentration commonly decreases with increasing depth within the upper 50 cm of the soil (Pető, 2010, 2013). Translocation into deeper horizons can result from bioturbation (Clarke, 2003), vertic behaviour (Boettinger, 1994; see also Kovda & Mermut, 2010, 2018) or transport by percolating water within the pore network (Stoops et al., 2001; Clarke, 2003; Sommer et al., 2006). Subsurface occurrences can also originate from burial of soil profiles (Golyeva, 2008). In alluvial soils, phytoliths may migrate laterally by water flow (Fishkis et al., 2010).

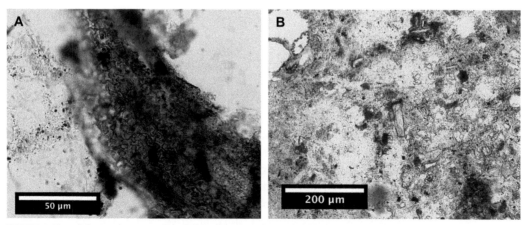

FIGURE 1 Phytoliths in plant remains and in the groundmass (PPL). (A) Articulated elongate dendritic phytoliths embedded within partly decomposed organic tissue (Site of Petite Rue des Bouchers, Brussels, Belgium). (B) Isolated elongate entire phytolith (Orthic-eutric Vertisol, Ecuador; Ghent University archive).

Records of phytoliths in soil thin sections appeared first in Russian literature (e.g., Parfenova & Yarilova, 1965). Over the last decades, several studies combining classical phytolith analysis with micromorphology were conducted (Macphail, 1981; Shahack-Gross et al., 2005; Barczi et al., 2006, 2009; Albert et al., 2008; French et al., 2009; Villagran et al., 2010; Pető & Barczi, 2011; Shillito & Ryan, 2013).

Several authors indicated the potential of phytolith analysis on thin sections to detail the depositional history of phytoliths (Osterrieth et al. 2009; Shillito 2011a, 2011b; Vrydaghs et al., 2016, 2017). Recently, a proposal for quantitative assessments of phytoliths based on the study of soil thin sections has been proposed (Devos et al., 2013a, 2013b; Vrydaghs et al., 2016). The occurrence and interpretation of phytoliths in archaeological soil materials and sediments is discussed in Vrydaghs et al. (2017).

Assuming silica is available in the environment, the formation of redundant morphotypes seems to be linked mainly to environmental conditions, whereas the formation of distinctive morphotypes would be genetically determined (Piperno et al., 2002; Piperno, 2006). As a consequence, the observation of redundant morphotypes in thin sections calls for further observations, as their occurrence should be correlated to more distinctive phytoliths. If this is not the case, such phytoliths could be intrusive.

The absence of phytoliths in thin sections should be considered with caution. As proposed by Stoops (2003), one should better state 'not recorded' rather than 'absent' when phytoliths are not observed. Besides a true absence of phytoliths, the quality of the thin sections, preservation conditions, the presence of coatings and the nature of the groundmass can also account for a lack of recorded phytoliths. This is particularly the case for clayey soils, without being restricted to that soil type. Contextual data should therefore be considered. The occurrence of other types of biogenic silica, particularly diatom remains, chrysophycean cysts and sponge spicules, could at least indicate that poor preservation is not an issue. In addition, the occurrence of plant tissue or organ residues points to the presence of plant remains, which may have released phytoliths into the soil.

2.3   Distribution Patterns

The basic paradigm of phytolith taphonomy is that phytoliths are released into the environment through the decay of plant remains, before or after the burial of those remains (e.g., Osterrieth et al., 2009; Madella & Lanceoletti, 2012). Besides natural decay of organic matter, human-induced processes, such as fire (Madella, 2003), dung deposition (Powers-Jones, 1994; Albert 2002) and plant processing (Harvey & Fuller, 2005), can also be at the origin of phytolith release. After being released, phytoliths can be transported, either before or after burial, by water (Vrydaghs, 2003), soil fauna or wind (Twiss et al., 1969; Fredlund & Tieszen, 1994; Osterrieth et al., 2009). This transport might be at the origin of altered distribution patterns as well as physical degradation of the phytoliths. Subsequently, recording the distribution patterns (*sensu* Stoops, 2003) of opal phytoliths in thin sections and their relative frequencies compose one of the major issues of phytolith analysis using soil thin sections.

Three major basic distribution patterns of opal phytoliths have been recorded: isolated (Figs. 1B and 2), clustered (Fig. 3A to D) and articulated (Fig. 3E and F) (Vrydaghs et al., 2016). Isolated phytoliths are totally disarticulated and well separated. They appear in different orientations in thin sections. The origins of these phytoliths cannot be established, but their occurrence points either to perturbation or to the occurrence of intrusive material. Clustered phytoliths are groups of disarticulated phytoliths in which not all phytoliths share the same orientation (Fig. 3A and B) or are necessarily of the same type (Fig. 3C and D). Because the phytoliths in a cluster may be intrusive, they may have a different depositional history and originate from a different plant tissue, organ, individual or species. Clusters are a clear signature of a certain degree of disturbance. Articulated phytoliths are those that appear to maintain the relative distribution they had within the plant tissues in which they were accumulated

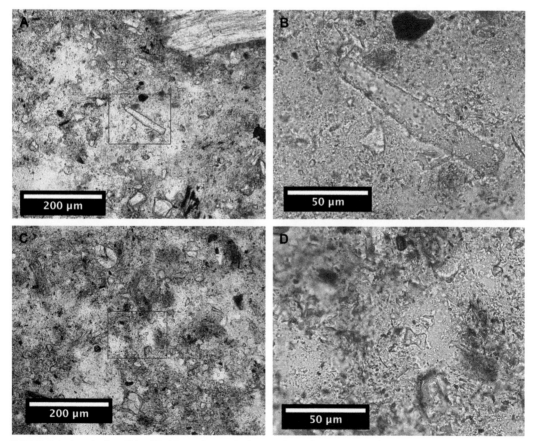

FIGURE 2 Isolated phytoliths (Orthic-eutric Vertisol, Ecuador; Ghent University archive) (PPL). (A, B) Isolated elongate entire phytolith (overview and detail). (C, D) Isolated bilobate phytolith (overview and detail).

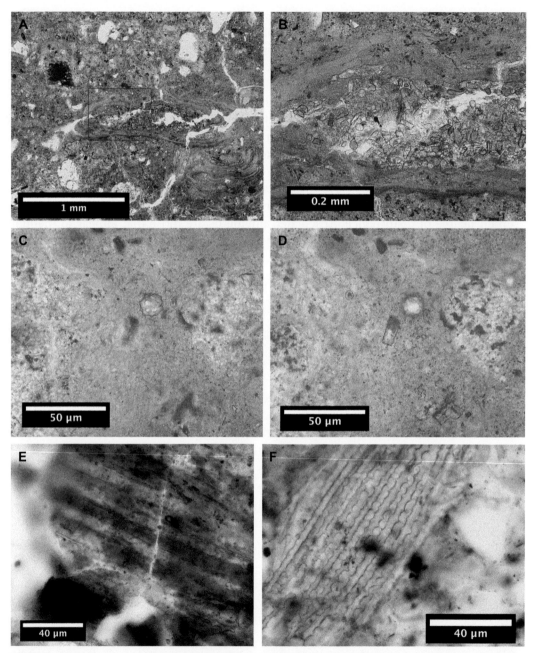

FIGURE 3 Clustered and articulated phytoliths (PPL). (A, B) Cluster of several phytoliths with different orientations, all more or less in focus in the same view (overview and detail) (Typic Haplustult, Peru; Ghent University archive). (C, D) Cluster of different types of phytoliths that are not in focus in the same view (Typic Haplustult, Peru; Ghent University archive). (E) Silica skeleton, composed of articulated phytoliths, showing some breakage and subsequent displacement (Site of the Rue des Bouchers, Brussels, Belgium). (F) Articulated phytoliths clearly showing the original anatomical distribution; the phytoliths are interlocked and show a wave pattern (Site of the Poor Clares, Brussels, Belgium).

(see Albert et al., 2008; Shillito, 2011b; Vrydaghs et al., 2016, 2017) (Fig. 3E and F). A distinction needs to be made here between complete sections of silicified epidermal tissue in the form of contiguous cells (silica skeletons; Rosen, 1992) (Fig. 3E) and groups of phytoliths preserving their original anatomical connection but without the phytoliths being bounded together (Cabanes et al., 2010) (Fig. 3F). This latter pattern indicates that plant material was buried and then decayed *in situ* with no or minimal postdecay disturbance. Silica skeletons, on the other hand, do not necessarily involve *in situ* decomposition of plant material and can therefore also be intrusive.

2.4 State of Preservation

Thin section analysis documents the state of preservation of opal phytoliths. Up to now, no standard terminology for features produced by physical breakdown and chemical corrosion has been proposed. Weathered phytoliths might appear as broken, rounded or chipped particles. Broken phytoliths might present clear cuts, which do not coincide with the normal morphology of the phytoliths. Jenkins (2009) proposed that surface pitting of isolated phytoliths and breakdown of silica skeletons are alteration features. Benayas (1963) described corrosion patterns on phytoliths in soil thin sections.

Terms that have been proposed to describe weathering of mineral grains (Stoops et al., 1979; Bullock et al., 1985; Stoops, 2003) cannot be easily applied to phytoliths. A major difficulty in describing corrosion patterns of phytoliths in relation to taphonomical processes is that some features that look like weathering patterns have been recorded for phytoliths extracted from fresh plant reference material. Another difficulty is that phytoliths deriving from the same botanical source can present different aspects despite having been preserved in the same conditions. For instance, distinctive grass phytoliths are reported to be more resistant than associated redundant phytoliths (Fredlund & Tieszen, 1997; Albert et al., 2006; Cabanes et al., 2011).

Opal phytoliths deposited through fire represent a special case. Because of the relatively high melting point of opal, most phytoliths are unaffected by heating up to 800°C (Runge 1998), although thin phytoliths might start melting at 600°C (Brochier, 2002). Siliceous glasses will start forming between 900 and 1000°C (Brochier, 2002), unless compounds which can lower the melting temperatures, such as sodium carbonates, are present (Canti, 2003). However, temperatures also appear to vary according to plant taxa and organ (Wu et al., 2012). In some cases, phytoliths present a blackish coating (Fig. 4A) resulting from the combustion of the organic tissues that surrounded them. Occurrences of millimetre- to centimetre-sized lumps of vesicular glassy slag derived from phytoliths by heating has also been observed (Canti, 2003; Devos et al., 2009). Molten phytoliths appear as deformed disarticulated particles (Fig. 4B), molten silica skeletons, droplets (Fig. 4C) or silica glass (Fig. 4D). They can remain whitish or turn blackish. Autofluorescence under UV and blue light has been reported by several authors (Gebhardt & Langohr, 1999; Devos et al., 2009).

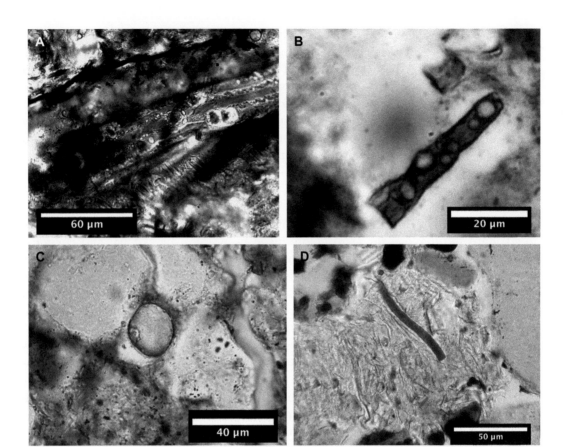

FIGURE 4 Phytoliths affected by heat (PPL). (A) Blackish silica skeleton composed of various types of grass phytoliths with blackish aspect due to carbon formed during the combustion of organic tissues in which the phytoliths occurred (Site of Achlum 12, the Netherlands). (B) Molten isolated elongate phytolith, with bumpy surface and blackish appearance (Site of the Poor Clares, Brussels, Belgium). (C) Silica droplet resulting from melting of phytoliths (Site of the Poor Clares, Brussels, Belgium). (D) Silica glass originating from a hearth structure (Site of Rue de Dinant, Brussels, Belgium).

2.5   Inventorying

The shape of opal phytoliths varies widely. Stoops (2003) was the first to elaborate a two-dimensional analytical and descriptive system of phytoliths in soil thin sections. He proposed a series of two-dimensional descriptors classifying phytoliths as angular (e.g., square, rectangular, polyhedral) or rounded (e.g., circular, oval). However, this system does not account for the huge morphological diversity of opal phytoliths that can be of taxonomical relevance. Based on modern plant reference material, different classification systems can be developed, such as monocot versus eudicot plants (Piperno, 1988) or inflorescence versus bark phytoliths (Albert et al., 2006). In relation to the analysis of soil thin sections, one of the most practical systems has been proposed by Runge (1999).

It consists of an open system comprising seven broad morphological categories. Each of these can be further subdivided into as much subcategories as needed. It makes the classification system flexible and as such easily applicable to almost all geographical and ecological zones (Mercader et al., 2000; Vrydaghs et al., 2004; Devos et al., 2009).

Phytoliths being plant microfossils, one concern of studies of phytoliths in soil profiles is to establish a correlation between vertical changes in phytolith inventory and changes in the corresponding vegetation cover through time (Lu et al., 2007; Golyeva, 2008; Evett & Bartolome, 2013). However, the palaeoenvironmental relevancy of each distribution pattern differs, the isolate pattern being a non-local or regional record whilst the articulate pattern composes a local one. Hence, the study of vertical changes needs to fully consider distribution patterns.

3. Diatoms

Diatoms are eukaryotic unicellular algae with an external skeleton of opal (frustules). The frustules, occurring exclusively as groundmass components, are transparent, colourless and optically isotropic, with high negative relief and distinct symmetrical ribbing (Fig. 5).

Diatoms are found on all continents, living in soils (Schuttler & Weaver, 1986; Van de Vijver & Beyens, 1998), peat (Kokfeld et al., 2009), lakes, rivers and marine environments (Clarke, 2003; Round et al., 2007). The frustules, with a size of 5-100 μm, exhibit a sieve pore-like surface morphology that is regularly spaced on the valve surface. They may have either a bilateral or a radial symmetry (Fig. 6). The latter occurs only in marine environments and, when found in soils, is thus inherited from the parent material (Clarke, 2003; Round et al., 2007). In acid conditions, the frustules are well preserved and

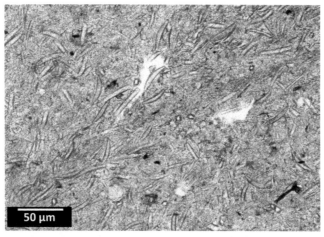

FIGURE 5 Diatoms with random distribution pattern (3C horizon, Anthrosol on dredge spoil island, Chinampas, Mexico City) (PPL).

FIGURE 6 Abundant diatom frustules, both with radial symmetry and with bilateral symmetry, in a paddy soil (Philippines; Ghent University archive) (PPL). *Image by G. Stoops.*

can be used as indicators of past environmental conditions. As they have the potential to reach high levels of productivity, diatoms exert an important control on the biogeo-chemical silica cycle in soil (Sommer et al., 2013; Puppe et al., 2016). The species composition of diatom assemblages can indicate the extent of soil hydromorphism, allowing a distinction between soils with periodical water flooding, submerged soil and alluvial deposits (Golyeva, 2000, 2001, 2008). The expansion and contraction of large lakes in low-relief continental environments can result in deposition of diatomaceous sediments over large areas and then expose them to pedogenic processes (Clarke, 2003). Diatoms, e.g., in excrements of herbivores, can be important for the interpretation of archaeological materials (Verleyen et al., 2017).

4. Sponge Spicules

Sponge spicules are cylindrical siliceous bodies with a small central channel (Bullock et al., 1985). The latter serves as the main diagnostic feature and distinguishes them from phytoliths (Drees et al., 1989). Siliceous sponges are colourless and isotropic (Fig. 7) and typically have a smooth surface. Spicules are generally found in the groundmass but can also occur as infillings (Stoops et al., 2001).

Sponge spicules are common components of marine, lacustrine (Harrison, 1988) and fluvial deposits (Chauvel et al., 1996), but they are also found in marshland (Volkmer-Ribeiro, 1992) and waterlogged soils (Smithson, 1959; Schwandes & Collins, 1994), as well as in other types of soils (e.g., Wilding & Drees, 1968). They have been found to date in the soils of all continents except Antarctica. In soils, most spicules are derived freshwater sponges, which produce much smaller spicules (10-500 μm) (Ricciardi & Reiswig, 1993)

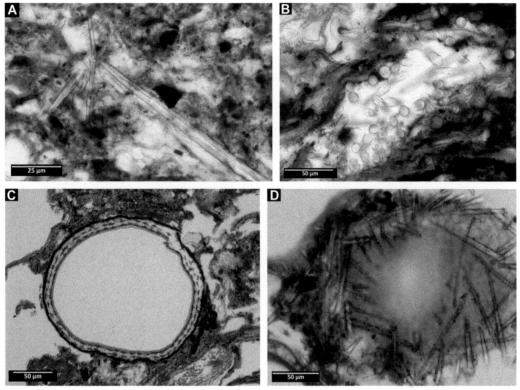

FIGURE 7 Sponge spicules (soil on lacustrine sediments, 163 cm depth, Uckermark, Germany) (PPL). (A) Spicules in organic matrix. (B) Spicules as infilling. (C) Spicules *in situ* in sponge remain. (D) Spicule accumulation.

than marine sponges (Clarke, 2003). Khangarot and Wilding (1973) observed sponge spicules in the 20-50 μm particle size fraction of terrace deposits, confirming the aquatic origin of the soil parent materials. Stoops et al. (2001) observed spicules in pedons within the spray zone of the Victoria Falls, Zimbabwe (Fig. 8), derived from sponges growing on wet trees and rocks. Spicule-rich sediments were reported by Chauvel et al. (1996) for rivers in Amazonia that have extremely low dissolved silica contents. Sponge spicules represent the least soluble form of biogenic silica. They are therefore very resistant in soil environments and can be transported over large distances by water and wind. Their abundance in a soil sample may indicate the occurrence, duration and intensity of floods, for example in soils on alluvial deposits (Golyeva, 2008). Their interpretation in archaeological contexts is discussed by Vrydaghs (2017).

5. Testate Amoebae

Testate amoebae (Protista) are eukaryotic unicellular organisms occurring abundantly in freshwater and estuarine environments, peatlands (Bobrov et al., 1999; Lamentowicz

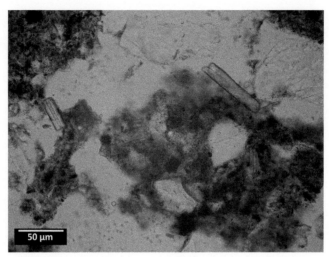

FIGURE 8 Sponge spicules in a soil within the spray zone of the Victoria Falls (Zimbabwe; Ghent University archive) (PPL). *Image by G. Stoops.*

et al., 2007), and soils from the tropics to polar areas (e.g., Foissner, 1987, 1999). They produce an organic shell (test) on which other components can be adhered (agglutinated). The agglutinated material can consist of mineral particles that are present in the environment (xenosomic taxa) or calcareous or siliceous platelets produced by the amoebae (idiosomic taxa) (Meisterfeld, 2002a, 2002b). Most xenosomic testate amoebae have larger and heavier shells (60 μm) than idiosomic individuals (30 μm) (Meisterfeld, 2002a, 2002b). Siliceous scales produced by idiosomic testate amoebae are formed from silica occurring in solution. The scales are very small (10 μm) and thin (1 μm) and probably dissolve readily in soil solutions (Aoki et al., 2007). Terrestrial testate amoebae occur mainly in organic layers, especially in Of and Oh horizons, and their abundance decreases downward towards the underlying mineral soil horizons (Kaczorek, 2009; Ehrmann et. al., 2012). They are one of the very first colonisers of newly exposed soil substrates (Wanner et al., 2008; Puppe et al., 2016). They can build up populations of some 100 million individuals per square metres within a few months, facilitating plant succession (Hodkinson et al., 2002; Wanner & Xylander, 2005).

In soil thin sections, a distinction can be made between testate amoeba shells composed of self-synthesised silica platelets (idiosomic) and those covered with extraneous material (xenosomic) (Fig. 9), but identification at the species level is in most cases impossible, because the shell opening (pseudostome) of the amoebae, important for determination, is not always visible. Best observation results for unstained samples were obtained using dark field or phase contrast microscopy (Ehrmann et al., 2012).

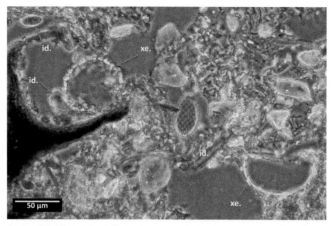

FIGURE 9 Xenosomic (xe.) and idiosomic (id.) testate amoebae *in situ* in soil material (Bildarchiv Boden, Otto Ehrmann, Creglingen, Germany) (phase contrast microscopy).

6. Radiolaria

Radiolaria are single-celled marine planktonic protozoa that secrete an opal skeleton composed of a number of architectural elements (radial spicules, internal bars, external spines) that are joined together to form regular symmetrical structures. The skeletons are usually smaller than 2 mm and commonly between 100 and 250 μm in diameter. Radiolaria can be solitary and colonial, the latter producing centimetre- to metre-sized aggregates (Flügel, 2004). Radiolaria are found exclusively in marine sediments, proving some information about the nature of soil parent materials (Clarke, 2003; Stoops, 2003).

7. Conclusions

Among biogenic siliceous features, mainly phytoliths have received attention in micro-morphological studies, mainly through analysis of archaeological materials. It has been shown that phytolith analysis using soil thin sections can yield valuable information about taphonomy and local to regional changes in vegetation. Development of diatom analysis using soil thin sections proves to be much more complex, as taxonomical identification of this biogenic opal is much more complex than for phytoliths.

References

Albert, R.M., 2002. Phytolith and spherulites study of herbivores dung from the African savannah as a tool for palaeoecological reconstruction. Pyrenae 33/34, 11-23.

Albert, R.M., Bamford, M.K. & Cabanes, D., 2006. Taphony of phytolith and macroplants in different soils from Olduvai Gorge (Tanzania) and the application to Plio-Pleistocene palaeoanthropological sample. Quaternary International 148, 78-94.

Albert, R.M., Shahack-Gross, R., Cabanes, D., Gilboa, A., Lev-Yadun, S., Portillo, M., Sharon, I., Boaretto, E. & Weiner, S., 2008. Phytolith-rich layers from the Late Bronze and Iron Ages at Tel Dor (Israel): mode of formation and archaeological significance. Journal of Archaeological Science 35, 57-75.

Altemüller, H.J. & Van Vliet-Lanoë, B., 1990. Soil thin section fluorescence microscopy. In Douglas, L.A. (ed.), Soil Micromorphology: A Basic and Applied Science. Developments in Soil Science, Volume 19. Elsevier, Amsterdam, pp. 565-579.

Amos, G.L., 1952. Silica in timbers. CSIRO Bulletin 267, 5-59.

Aoki, Y., Hoshino, M. & Matsubara, T., 2007. Silica and testate amoebae in a soil under pine-oak forest. Geoderma 142, 29-35.

Balan Menon, P.K., 1965. Guide to the distribution of silica in Malayan woods. Malayan Forester 28, 284-288.

Ball, T.B., Gardner, J.S. & Anderson, N., 1999. Identifying inflorescence phytoliths from selected species of wheat (*Triticum monococum, T. Dicoccon, T. Dicocoides,* and *T. Aestivum*) an barley (*Hordeum vulgare* and *H. Spontaneum*) (Gramineae). American Journal of Botany 86, 1615-1623.

Ball, T.B., Vrydaghs, L., Van Den Hauwe, I., Manwaring, J. & De Langhe, E., 2006. Differentiating banana phytoliths: wild and edible *Musa acuminata* and *Musa balbisiana.* Journal of Archaeological Sciences 33, 1228-1236.

Barczi, A., Tóth, T.M., Csanádi, A., Sümegi, P. & Czinkota I., 2006. Reconstruction of the paleo-environment and soil evolution of the Csípő-halom kurgan, Hungary. Quarternary International 156/157, 49-59.

Barczi, A., Golyeva, A.A. & Pető, Á., 2009. Palaeoenvironmental reconstruction of Hungarian kurgans on the basis of the examination of palaeosoils and phytolith analysis. Quaternary International 193, 49-60.

Benayas, J., 1963. Disolución parcial de sílice orgánica en suelos. Anales de Edafología y Agrobiología 22, 623-626.

Benvenuto, L.M., Fernández Honaine, M., Osterrieth, M., Coronato, A. & Rabassa, J., 2013. Silicophytoliths in Holocene peatlands and fossil peat layers from Tierra del Fuego, Argentina, southernmost South America. Quaternary International 287, 20-33.

Blackman, E., 1968. The pattern and sequence of opaline deposition in rye (*Secale cereale* L.). Annals of Botany 32, 207-218.

Bobrov, A.A., 2003. Micropaleontological methods for studying biogenic silica in soils. Eurasian Soil Science 36, 1307-1316.

Bobrov, A.A., Charman, D.J. & Warner, B.G., 1999. Ecology of testate amoebae (Protozoa: Rhizopoda) on peatlands in western Russia with special attention to niche separation in closely related taxa. Protist 150, 125-136.

Boettinger, J.L., 1994. Biogenic opal as an indicator of mixing in an Alfisol/Vertisol landscape. In Ringroase-Voase, A.J. & Humphreys, G.S. (eds.), Soil Micromorphology: Studies in Management and Genesis. Developments in Soil Science, Volume 22. Elsevier, Amsterdam, pp. 17-26.

Bozarth, S., 1992. Classification of opal phytoliths formed in selected dicotyledons native to the Great Plains. In Rapp, G. & Mulholland, S.C. (eds.), Phytolith Systematics. Emerging Issues. Springer Science, New York, pp. 193-214.

Brochier, J.E., 2002. Les sédiments anthropiques. Méthodes d'étude et perspectives. In Miskovsky, J.C. (ed.), Géologie de la Préhistoire: Méthodes, Techniques, Applications. Géopré Editions, Paris, pp. 453-477.

Brown, D.A., 1984. Prospects and limits of a phytolith key for grasses in the central United States. Journal of Archaeological Science 11, 345-368.

Bullock, P., Fedoroff, N., Jongerius, A., Stoops, G., Tursina, T. & Babel U., 1985. Handbook for Soil Thin Section Description. Waine Research Publications, Wolverhampton, 152 p.

Cabanes, D., Mallol, C., Expósitio, I. & Baena, J., 2010. Phytolith evidence for hearths and beds in the Late Mousterian occupations of Esquilleu cave (Cantabria, Spain). Journal of Archaeological Science 37, 2947-2957.

Cabanes, D., Wiener, S. & Shahack-Gross, R., 2011. Stability of phytoliths in the archaeological record: a dissolution study of modern and fossil phytoliths. Journal of Archaeological Science 38, 2480-2490.

Canti, M.G., 2003. Aspects of the chemical and microscopic characteristics of plant ashes found in archaeological soils. Catena 54, 339-361.

Chauvel, A., Walker, I. & Lucas, Y., 1996. Sedimentation and pedogenesis in a central Amazonian black water basin. Biogeochemistry 33, 77-95.

Clarke, J., 2003. The occurrence and significance of biogenic opal in the regolith. Earth-Science Reviews 60, 175-194.

Devos, Y., Vrydaghs, L., Degraeve, A. & Fechner, K., 2009. Palaeoenvironmental research on the site of rue the Dinant (Brussels): an interdisciplinary study of dark earth. Catena 78, 270-284.

Devos, Y., Nicosia, C., Vrydaghs, L. & Modrie, S., 2013a. Studying urban stratigraphy: Dark Earth and a microstratified sequence on the site of the Court of Hoogstraeten (Brussels, Belgium). Integrating archaeopedology and phytolith analysis. Quaternary International 315,147-166.

Devos, Y., Wouters, B., Vrydaghs, L., Tys, D., Bellens, T. & Schryvers, A., 2013b. A soil micromorphological study on the origins of the early medieval trading centre of Antwerp (Belgium). Quaternary International 315, 167-183.

Drees, L.R., Wilding, L.P., Smeck, N.E. & Senkayi, A.L., 1989. Silica in soils: quartz and disordered silica polymorphs. In Dixon, J.B. & Weed, S.B. (eds.), Minerals in Soil Environments. Second Edition, SSSA Book Series No. 1, Soil Science Society of America, Madison, pp. 913-974.

Ehrenberg, C.G., 1843. Über 2 neue Lager von Gebirgsmassen aus Infusorien als Meeres-Absatz in Nord-Amerika und eine Vergleichung derselben mit den organischen Kreide-Gebilden in Europa und Afrika. Bericht über die zur Bekanntmachung geeigneten Verhandlungen der Königlich Preussischen Akademie der Wissenschaften zu Berlin 1843, 57-97.

Ehrenberg, C.G., 1844. Mikroskopische Organismen, Phytholitharien. Bericht über die zur Bekanntmachung geeigneten Verhandlungen der Königlich Preussischen Akademie der Wissenschaften zu Berlin 1844, 66-359.

Ehrenberg, C.G., 1845. Über einen bedeutenden Infusorien haltenden vulkanischen Aschen-Tuff (Pyrobiolith) aus der Insel Ascension. Bericht über die zur Bekanntmachung geeigneten Verhandlungen der Königlich Preussischen Akademie der Wissenschaften zu Berlin, 1845, 140-142.

Ehrenberg, C.G., 1846. Zusätze zu den Mittheilungen über die vulkanischen Phytolitharien der Insel Ascension. Bericht über die zur Bekanntmachung geeigneten Verhandlungen der Königlich Preussischen Akademie der Wissenschaften zu Berlin, 1846, 191-202.

Ehrmann, O., Puppe, D., Wanner, M., Kaczorek, D. & Sommer, M., 2012. Testate amoebae in 31 mature forest ecosystems − densities and micro-distribution in soils. European Journal Protistology 48, 161-168.

Eichhorn, B., Neumann, K. & Garnier, A., 2010. Seed phytoliths in West African Commelinaceae and their potential for paleoecological studies. Palaeogeography, Palaeoclimatology, Palaeoecology 298, 300-310.

Evett, R.R. & Bartolome, J.W., 2013. Phytolith evidence for the extent and nature of prehistoric Californian grasslands. The Holocene 23, 1644-1649.

Fishkis, O., Ingwersen, J., Lamers, M., Denysenko, D. & Streck, T., 2010. Phytolith transport in soil: a laboratory study on intact soil cores. European Journal of Soil Science 61, 445-455.

FitzPatrick, E.A., 1984. Micromorphology of Soils. Chapman and Hall, London, 433 p.

Flügel, E., 2004. Fossils in thin section: it is not that difficult. In Flügel, E. (ed.), Microfacies of Carbonate Rocks. Springer-Verlag, Berlin, pp. 399-574.

Foissner, W., 1987. Soil protozoa: fundamental problems, ecological significance, adaptations in ciliates and testaceans, bioindicators, and guide to the literature. Progress in Protistology 2, 69-212.

Foissner, W., 1999. Soil protozoa as bioindicators: pros and cons, methods, diversity, representative examples. Agriculture, Ecosystems & Environment 74, 95-112.

Fredlund, G.G. & Tieszen, L.T., 1994. Modern phytolith assemblages from the North American Great Plains. Journal of Biogeography 21, 321-335.

Fredlund, G. & Tieszen, L.L., 1997. Phytolith and carbon evidence for late quaternary vegetation and climate change in the Southern Black Hills, South Dakota. Quaternary Research 47, 206-217.

French, C., Sulas, F. & Madella, M., 2009. New geoarchaeological investigations of the valley system in the Aksum area of northern Ethiopia. Catena 78, 218-233.

Gebhardt, A. & Langohr, R., 1999. Micromorphological study of construction materials and living floors in the medieval motte of Werken (West Flanders, Belgium). Geoarchaeology 14, 595-620.

Golyeva, A.A., 2000. Phytoliths and their Information Role in Natural and Archaeological Objects [in Russian]. Russian Academy of Science, Moscow, 200 p.

Golyeva, A.A., 2001. Biomorphic analysis as a part of soil morphological investigations. Catena 43, 217-230.

Golyeva, A.A., 2008. Microbiomorphic Analysis as Tool for Natural and Anthropogenic Landscape Investigation: Genesis, Geography, Interpretation [in Russian]. Russian Academy of Science, Moscow, 240 p.

Gutiérrez-Castorena, Ma. del C. & Effland, W.R., 2010. Pedogenic and biogenic siliceous features. In Stoops, G., Marcelino, V. & Mees, F. (eds.), Interpretation of Micromorphological Features of Soils and Regoliths. Elsevier, Amsterdam, pp. 471-496.

Harrison, F.W., 1988. Methods in Quaternary ecology. 4. Freshwater sponges. Geoscience Canada 15, 193-198.

Harvey, E.L. & Fuller, D., 2005. Investigating crop processing using phytolith analysis: the example of rice and millets. Journal of Archaeological Sciences 32, 732-752.

Hodkinson, I.D., Webb, N.R. & Coulson, S.J., 2002. Primary community assembly on land - the missing stages: why are the heterotropic organisms always there first? Journal of Ecology 90, 569-277.

Jenkins, E., 2009. Phytolith taphonomy: a comparison of dry ashing and acide extraction on the breakdown of conjoined phytolith formed in *Triticum durum*. Journal of Archaeological Sciences 36, 2402-2407.

Kaczorek, D., 2009. Identification of biogenic silica in soils with the application of microscope and microprobe analysis [in Polish]. Soil Science Annual 60, 42-49.

Kealhoffer, L. & Piperno, D.R., 1998. Opal phytoliths in Southeast Asian flora. Smithsonian Contribution to Botany 88, 1-39.

Khangarot, A.S. & Wilding, L.P., 1973. Biogenic opal in Wisconsin and Illinoian-age terraces of east-central Ohio. Journal of the Indian Society of Soil Science 21, 505-508.

Klein, R.L. & Geist, J.W., 1978. Biogenic silica in the Pinaceae. Soil Science 126, 145-156.

Kokfeld U., Struyf, E. & Randsalu, L., 2009. Diatoms in peat — dominant producers in a changing environment? Soil Biology & Biochemistry 41, 1764-1766.

Kondo, R. & Iwasa, Y., 1981. Biogenic opals of humic yellow latosol and yellow latosol in the Amazon region. Research Bulletin of Obihiro University 12, 231-239.

Kovda, I. & Mermut, A., 2010. Vertic features. In Stoops, G., Marcelino, V. & Mees, F. (eds.), Interpretation of Micromorphological Features of Soils and Regoliths. Elsevier, Amsterdam, pp. 109-127.

Kovda, I. & Mermut, A., 2018. Vertic features. In Stoops, G., Marcelino, V. & Mees, F. (eds.), Interpretation of Micromorphological Features of Soils and Regoliths. Second Edition. Elsevier, Amsterdam, pp. 605-632.

Lamentowicz, L., Gabka, M. & Lamentowicz, M., 2007. Species composition of testate amoebae (protists) and environmental parameters in a Sphagnum peatland. Polish Journal of Ecology 55, 749-759.

Lentfer, C.J., 2009. Tracing domestication and cultivation of bananas from phytoliths: an update from Papua New Guinea. Ethnobotany Research & Applications 7, 247-270.

Lu, H.Y., Wu, N.Q., Liu, K.B., Jiang, H. & Liu, T.S., 2007. Phytoliths as quantitative indicators for the reconstruction of past environmental conditions in China. II: palaeoenvironmental reconstruction in the Loess Plateau. Quaternary Science Reviews 26, 759-772.

Macphail, R.I., 1981. Soil and botanical studies of the 'Dark Earth'. In Jones, M. & Dimbleby, G.W. (eds.), The Environment of Man: the Iron Age to the Anglo-Saxon Period. British Archaeological Reports, British Series, Volume 87, pp. 309-331.

Madella, M., 2003. Investigating agriculture and environment in South Asia: present and future contributions from opal phytoliths. In Weber, S. & Belcher, W. (eds.), Indus Ethnobiology: New Perspectives from the Field. Lexington Books, Lanham, pp.199-249.

Madella, M. & Lancelotti, C., 2012. Taphonomy and phytoliths: a user manual. Quaternary International 275, 76-83.

Meisterfeld, R., 2002a. Order Arcellinida Kent, 1880. In Lee, J.J., Leedale, G.F. & Bradbury, P. (eds.), The Illustrated Guide to the Protozoa, Volume 2. Society of Protozoologists, Lawrence, pp. 827-860.

Meisterfeld, R., 2002b. Testate amoebae with filopodia. In Lee, J.J., Leedale, G.F. & Bradbury, P. (eds.), The Illustrated Guide to the Protozoa, Volume 2. Society of Protozoologists, Lawrence, pp. 1054-1084.

Mercader, J., Runge, F., Vrydaghs, L., Doutrelepont, H., Corneille, E. & Juan-Tresserras, J., 2000. Phytoliths from archaeological sites in the tropical forest of Ituri, Democratic Republic of Congo. Quaternary Research 54, 102-112.

Metcalfe, C.R. & Chalk, L., 1950. Anatomy of Dicotyledons. Clarendon Press, Oxford, 1500 p.

Osterrieth, M., Madella, M., Zurro, D. & Fernanda Alvarez, M., 2009. Taphonomical aspects of silica phytoliths in the loess sediments of the Argentinean Pampas. Quaternary International 193, 70-79.

Parfenova, E.I. & Yarilova, E.A., 1965. Mineralogical Investigations in Soil Science. Israel Program of Scientific Translations, Jerusalem, 178 p.

Pető, Á., 2010. Phytolith Profile Cadastre of the Most Significant and Abundant Soil Types of Hungary [*in Hungarian*]. PhD Dissertation, Szent István University, Gödöllő, 222 p.

Pető, Á., 2013. Studying modern soil profiles of different landscape zones in Hungary: an attempt to establish a soil-phytolith identification key. Quaternary International 287, 149-161.

Pető, Á. & Barczi, A., (eds.), 2011. Kurgan Studies. An Environmental and Archaeological Multiproxy Study of Burial Mounds in the Eurasian Steppe Zone. British Archaeological Reports, International Series, Volume 2238, 350 p.

Piperno, D.R., 1988. Phytolith Analysis: An Archaeological and Geological Perspective. Academic Press, San Diego, 280 p.

Piperno, D.R., 1989. The occurrence of phytoliths in the reproductive structures of selected tropical Angiosperms and their significance in tropical paleoecology, paleoethnobotany and systematics. Review of Palaeobotany and Palynology 61, 147-173.

Piperno, D.R., 2006. Phytoliths: A Comprehensive Guide for Archaeologists and Paleoecologists. AltaMira, Lanham, 238 p.

Piperno, D.R., Holst, I., Wessel-Beaver L. & Andres, T.C., 2002. Evidence for the control of phytolith formation in Cucurbita fruits by the hard rind (Hr) genetic locus: archaeological and ecological implications. Proceedings of the National Academy of Sciences 99, 10923-10928.

Powers-Jones, A.H., 1994. The use of phytolith analysis in the interpretation of archaeological deposits: an Outer Hebridean example. In Luff, R. & Rowley-Conwy, P. (eds.), Whither Environmental Archaeology? Oxbow Books, Oxford, p. 41-50.

Puppe, D., Höhn, A., Kaczorek, D., Wanner, M. & Sommer, M., 2016. As time goes by – spatiotemporal changes of biogenic Si pools in initial soils of an artificial catchment in NE Germany. Applied Soil Ecology 105, 9-16.

Rapp, G. & Mulholland, S.C. (eds.). 1992. Phytolith Systematics. Emerging Issues. Springer Science, New York, 350 p.

Ricciardi, A. & Reiswig, H.M., 1993. Freshwater sponges (Perifera, Spogillidae) of eastern Canada: taxonomy, distribution, and ecology. Canadian Journal of Zoology 71, 665-682.

Rosen, A.M., 1992. A Preliminary identification of silica skeletons from Near Eastern archaeological sites: an anatomical approach, In Rapp, G. & Mulholland, S.C. (eds.), Phytolith Systematics. Emerging Issues. Springer Science, New York, pp. 129-147.

Round, F.E., Crawford, R.M. & Mann, D.G., 2007. The Diatoms. Biology and Morphology of the Genera. Cambridge University Press, Cambridge, 758 p.

Rovner, I., 1971. Potential of opal phytoliths for use in paleoecological reconstruction. Quaternary Research 1, 343-359.

Runge, F., 1998. The effect of dry oxidation temperatures (500°C - 800°C) and of natural corrosion on opal phytoliths. In Abstracts of the Deuxième Congrès International de Recherches sur les Phytolithes, Aix-en-Provence, p. 73.

Runge, F., 1999. The opal phytolith inventory of soils in central Africa - quantities, shapes, classification and spectra. Review of Palaeobotany and Palynology 107, 23-53.

Ruprecht, F.J., 1866. Geobotanical Investigation of Chernozems [in Russian]. Proceedings of the Academy of Sciences, Volume 10, Annex n° 6. Academy of Sciences, St Petersburg, pp. 1-131.

Sangster, A.G., Williams, S.E. & Hodson, M.J., 1997. Silica deposition in the needles of the Gymnosperms. II Scanning electron microcopy and X-ray microanalysis. In Pinilla, A., Juan-Tresserras, J. &, Machado, M.J. (eds.), The State of the Art of Phytoliths in Soils and Plants. Centro de Ciencias Medioambientales, Monografia 4. CSIC, Madrid, pp. 135-146.

Schuttler, P.L. & Weaver, T., 1986. Concentrating soil diatoms for assemblage description. Soil Biology & Biochemistry 4, 389-394.

Schwandes, I.P. & Collins, M.E., 1994. Distribution and significance of freshwater sponge spicules in selected Florida soils. Transactions of the American Microscopical Society 113, 242-257.

Shahack-Gross, R., Albert, R.M., Gilboa, A., Nagar-Hilman, O., Sharon, I. & Weiner, S., 2005. Geoarchaeology in an urban context: the uses of space in a Phoenician monumental building at Tel Dor (Israel). Journal of Archaeological Science 32, 1417-1431.

Shillito, L.M., 2011a. Grains of truth or transparent blindfolds? A review of current debates in archaeological phytolith analysis. Vegetation History and Archaeobotany 22, 71-82.

Shillitto, L.M., 2011b. Taphonomic observations of archaeological wheat phytoliths from Neolithics Çatalhöyück, Turkey, and the use of conjoined phytolith size as an indicator of water availability. Archaeometry 53, 631-641.

Shillito, L.M. & Ryan, P., 2013. Surfaces and streets: phytoliths, micromorphology and changing use of space at Neolithic Çatalhöyük (Turkey). Antiquity 87, 684-700.

Simons, N.A., Hart, D.M. & Humphreys, G.S., 2000. Phytolith depth functions in a podzol and a texture contrast soil, Sydney. In Adams, J.A. & Metherell, A. (eds.), Soil 2000: New Horizons for a New Century. New Zealand Society of Soil Science, Lincoln, pp. 193-194.

Smithson, F., 1959. Opal sponge spicules in soils. Journal of Soil Science 10, 105-109.

Sommer, M., Kaczorek, D., Kuzyakov, Y. & Breuer, J., 2006. Silicon pools and fluxes in soils and land-scapes - a review. Journal of Plant Nutrition and Soil Science 169, 310-329.

Sommer, M., Jochheim, H., Höhn, A., Breuer, J., Zagorski, Z., Busse, J., Barkusky, D., Meier, K., Puppe, D., Wanner, M. & Kaczorek, D., 2013. Si cycling in a forest biogeosystem - the importance of transient state biogenic Si pools. Biogeosciences 10, 4991-5007.

Stoops, G., 2003. Guidelines for Analysis and Description of Soil and Regolith Thin Sections. Soil Science Society of America, Madison, 184 p.

Stoops, G., Altemüller, H.J., Bisdom, E.B.A., Delvigne, J., Dobrovolsky, V.V., FitzPatrick, E.A., Paneque, G. & Sleeman, J., 1979. Guidelines for the description of mineral alterations in soil micromorphology. Pedologie 29, 121-135.

Stoops, G., Marcelino, V., Zauyah, S. & Maas, A., 1994. Micromorphology of soils of the humid tropics. In Ringrose-Voase, A & Humphreys, G.S. (eds.), Soil Micromorphology: Studies in Management and Genesis. Developments in Soil Science, Volume 22. Elsevier, Amsterdam, pp. 1-15.

Stoops, G., Van Ranst, E. & Verbeek, K., 2001. Pedology of soils within the spray zone of the Victoria Falls (Zimbabwe). Catena 46, 63-83.

ter Welle, B.J.H., 1976a. On the occurrences of silica grains in the secondary xylem of the Chrysobalanaceae. IAWA Bulletin 2, 19-29.

ter Welle, B.J.H., 1976b. Silica gains in the woody plants of the Neotropics, especially Surinam. In Bass, P., Bolton, A.J. & Catling, D.M. (eds.), Wood Structure in Biological and Technological Research. Leiden Botanical Series, Volume 3. Leiden University Press, Leiden, pp. 107-142.

Tomlinson, P.B., 1961. Anatomy of the Monocotyledons. II. Palmae. Clarendon Press, Oxford, 453 p.

Twiss, P.C., Suess, E. & Smith, R.M., 1969. Morphological classification of grass phytoliths. Soil Science Society of America Proceedings 33, 109-115.

Van de Vijver B. & Beyens L., 1998. A preliminary study on the soil diatom assemblages from Ile de la Possession (Crozet, Subantarctica). European Journal of Soil Biology, 34, 133-141.

Van Vliet-Lanoë, B., 1980. Approche des conditions physico-chimiques favorisant l'autofluorescence des minéraux argileux. Pedologie 30, 369-390.

Verleyen, E., Sabbe, K., Vyverman, W. & Nicosia, C., 2017. Siliceous microfossils from single-celled organisms: diatoms and chrysophycean stomatocysts. In Nicosia, C. & Stoops, G. (eds.), Archaeological Soil and Sediment Micromorphology. John Wiley & Sons Ltd, Chichester, pp. 165-170.

Villagran, X.S., Balbl, A.L., Madella, M., Vila, A. & Estevez, J., 2010. Experimental micromorphology in Tierra del Fuego (Argentina): building a reference collection for the study of shells middens in cold climates. Journal of Archaeological Science 38, 588-604.

Volkmer-Ribeiro, C., 1992. The freshwater sponges in some peatbog ponds in Brazil. Amazoniana 12, 317-335.

Vrydaghs, L., 2003. Studies in Opal Phytoliths. Material and Identification Criteria. PhD Dissertation, Ghent University, 179 p.

Vrydaghs, L., 2017. Opal sponge spicules. In Nicosia, C. & Stoops, G. (eds.) Archaeological Soil and Sediment Micromorphology. John Wiley & Sons Ltd, Chichester, pp. 171-172.

Vrydaghs, L., Doutrelepont, H., Jansen, S., Robbrecht, E. & Beeckman, H., 1995. De la valeur systématique des phytolithes dans le bois des Naucleeae (Rubiaceae). Biologisch Jaarboek Dodonaea 63, 161-173.

Vrydaghs, L., Doutrelepont, H., Beeckman, H. & Haerinck, E., 2001. Identification of a morphotype association of *Phoenix dactylifera* L. lignified tissues origin at ed-Durr (1st AD), Umm Al-Qaiwain (U.A.E.). In Meunier, J.D. & Colin, F. (eds.), Phytoliths: Applications in Earth Science and Human History. Balkema, Lisse, pp. 239-250.

Vrydaghs, L., Cocquyt, C., Van de Vijver, T. & Goetghebeur, P., 2004. Phytolithic evidence for the introduction of *Schoenoplectus californicus* subsp. *tatora* at Easter Island. Rapa Nui Journal 18, 95-106.

Vrydaghs, L., Ball, T.B. & Devos, Y., 2016. Beyond redundancy and multiplicity. Integrating phytolith analysis and micromorphology to the study of Brussels Dark Earth. Journal of Archaeological Science 68, 79-88.

Vrydaghs, L., Devos, Y. & Pető, Á., 2017. Opal phytoliths. In Nicosia, C. & Stoops, G. (eds.), Archaeological Soil and Sediment Micromorphology. John Wiley & Sons Ltd, Chichester, pp. 155-163.

Wanner, M., Elmer, M., Kazda, M. & Xylander, W.E.R., 2008. Community assembly of terrestrial testate amoebae: how is the very first beginning characterized? Microbial Ecology 56, 43-54.

Wanner, M. & Xylander, W.E.R., 2005. Biodiversity development of terrestrial testate amoebae: is there any succession at all? Biology and Fertility of Soils 41, 428-438.

Wilding, L.P. & Drees, L.R., 1968. Distribution and implications of sponge spicules in surficial deposits in Ohio. Ohio Journal of Science 68, 92-99.

Wilding, L.P. & Drees, L.R., 1974. Contributions of forest opal and associated crystalline phases to the fine silt and clay fractions of soils. Clay Minerals 22, 295-306.

Wu, Y., Changsui, W. & Hill, D., 2012. The transformation of phytolith morphology as the result of their exposure to high temperatures. Microscopic Research and Technique 75, 852-855.

Chapter 8

Authigenic Silicate Minerals — Sepiolite-Palygorskite, Zeolites and Sodium Silicates

Florias Mees

ROYAL MUSEUM FOR CENTRAL AFRICA, TERVUREN, BELGIUM

CHAPTER OUTLINE

1. Introduction

This chapter deals with micromorphological features of silicate neoformation in soil environments, covering the three silicate mineral groups whose formation in soils has most widely been investigated using microscopic and submicroscopic techniques. The nature of crystalline and amorphous SiO_2 phases is discussed elsewhere (Gutiérrez-Castorena, 2018). Micromorphological features of the neoformation of clays other than sepiolite and palygorskite have mainly been described for weathering contexts with leaching of released elements, whose discussion is beyond the scope of this chapter.

Occurrences in present soil environments are covered by this chapter, as well as those attributed explicitly to past processes of pedogenesis or near-surface diagenesis. Many relevant petrographical studies of silicate neoformation in various other diagenetic settings are available, but they are not considered here. Some information about occurrences as groundmass components inherited from the parent material is included, because a distinction between inherited and pedogenic phases is often important. Microscopic and submicroscopic features receive nearly equal attention in this review, reflecting the nature of available micromorphological data.

2. Sepiolite and Palygorskite

Sepiolite $(Mg_4Si_6O_{15}(OH)_2 \cdot 6H_2O)$ and palygorskite $((Mg,Al)_2Si_4O_{10}(OH) \cdot 4H_2O)$ are common constituents of soils in arid and semiarid regions, where they occur as pedogenic or inherited phases (e.g., Singer, 1989, 2002). Palygorskite is often described as attapulgite, especially in older publications, but the latter is not the correct mineral name (see Bailey et al., 1971).

2.1   Thin Section Studies

Optical properties of sepiolite and palygorskite are very similar (e.g., Fleischer et al., 1984). Both are length-slow, with (nearly) straight extinction and with a birefringence of about 0.010, which is significantly lower than that of smectite and other 2:1 phyllosilicate minerals. Only the lower refractive indices of sepiolite are a possible criterion for distinction between sepiolite and palygorskite. Selective staining of sepiolite with methyl orange, leaving palygorskite unaffected, is an option (Mifsud et al., 1979), but this method has only rarely been applied. It has only been used to reveal sepiolite distribution patterns in hand specimens (e.g., Hay & Wiggins, 1980; Blank & Fosberg, 1991; Francis et al., 2013), which is complicated by the observation that magnesium smectites can become stained as well (Hay & Wiggins, 1980; Hay et al., 1986). Another approach is element mapping of uncovered thin sections, which can be translated into relative distribution patterns of sepiolite and palygorskite (Cuadros et al., 2016).

Sepiolite and palygorskite are generally difficult to recognise as groundmass components because of their low birefringence, the random orientation of elongated particles and the common masking effect of associated micritic carbonates. In some formations, they are only clearly observed in carbonate depletion hypocoatings (Fig. 1A and B; see also Mees & Van Ranst, 2011). Vanden Heuvel (1966) reports stipple-speckled and granostriated b-fabrics for sepiolite- and palygorskite-bearing samples, revealed after acid treatment of uncovered thin sections. Similar b-fabrics are recognised for noncalcareous materials (e.g., Francis et al., 2012, 2013) (Fig. 1C). The b-fabric of palygorskite-rich duricrusts is mainly mosaic-speckled (Stahr et al., 2000; Sauer et al., 2015). The same deposits also show granostriation, related by the authors to the pressure exerted during intergranular crystallisation and assumed to have resulted in

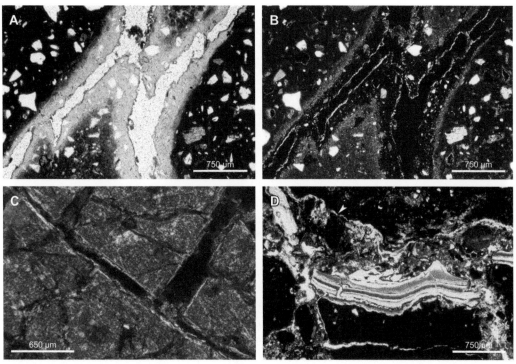

FIGURE 1 Palygorskite and sepiolite as groundmass components. (A) Carbonate depletion hypocoatings in palygorskite-rich lacustrine calcareous deposits of the Kalahari Group, with clearly orthic nature of the hypocoatings and gradual boundaries (Okamapu pan, Namibia) (PPL). (B) View of same field in XPL, illustrating the undifferentiated b-fabric of the decalcified material (statistical optical isotropy) and the occurrence of an overlying thin clay coating with high birefringence. (C) Non-calcareous sepiolite-dominated groundmass, with mosaic-speckled b-fabric, and with porostriation along the sides of the angular peds (Tugus pan, Namibia) (XPL). (D) Layered synsedimentary sepiolite layer in palustrine deposits, with sepiolite coatings (*arrows*) and sepiolite-rich infillings in part of the over- and underlying deposits (Witpan, Namibia) (PPL).

an increase of the spacing between the coarse grains of the groundmass (Stahr et al., 2000). Striated b-fabrics are more likely to develop by wetting and drying (Armenteros et al., 1995). Synsedimentary sepiolite-rich layers that are part of the parent material have been recognised by penecontemporaneous cracking and deformation (Hay et al., 1986) and by the presence of layering and unistrial b-fabrics (Mees, 2002) (Fig. 1D). Sepiolite aggregates (intraclasts) can be part of the groundmass, but the manner and degree of transport are not always certain (Khoury et al., 1982; Hay et al., 1986). Inherited palygorskite aggregates in a microcrystalline gypsum matrix have also been described (Porta & Herrero, 1990). Silicate dissolution features in horizons with palygorskite or sepiolite indicate that silica (and aluminium) is derived from a local source (Monger & Daugherty, 1991). Zones with higher interference colours in a palygorskite groundmass have been interpreted as recording transformation of palygorskite to smectite (Sauer et al., 2015).

FIGURE 2 Sepiolite as pedofeature. (A) Layered sepiolite coatings along the top and base of a horizontal planar void (top and base of coatings indicated by *arrows*), with a heterogeneous loose discontinuous infilling (Witpan, Namibia) (PPL). (B) View of same field in XPL, illustrating that layering is caused by variations in micritic carbonate content and degree of parallel alignment of sepiolite particles.

Pedofeatures include coatings with low, high or varying degrees of alignment of the clay particles parallel to the covered surface (Silva et al., 1965; Singer & Norrish, 1974; Brewer, 1976; Kapur et al., 1993; Mees, 1999; Sauer et al., 2015). Authigenic silicate formation within the pore system is evident for (layered) coatings with authigenic carbonates and that are formed in horizontal planar voids (Mees & Van Ranst, 2011) (Fig. 2). This can also be said for a rare example of palygorskite fibres that are formed perpendicular to the sides of detrital mineral grains (Jenkins, 1981), which has also been observed by SEM (e.g., Robins et al., 2012, 2014). Coatings surrounding individual sand grains (Achyuthan, 2003) are not necessarily formed *in situ*. In soils with sepiolite or palygorskite as part of the groundmass, coatings can be predominantly illuvial, based on criteria such as strong similarities between coating and micromass material (Mees & Van Ranst, 2011) (Fig. 3A).

The coatings can be part of compound pedofeatures, such as palygorskite coatings with associated iron oxide hypocoatings (Silva et al., 1965), non-palygorskitic illuvial clay (Kapur et al., 1993) or loose calcite infillings (Mees & Van Ranst, 2011), recording changes in soil conditions (Fig. 3B). This was already noted by Beattie and Haldane (1958), for palygorskite coatings with associated quartz-kaolinite and manganese oxides, in a study that apparently did not involve micromorphological observations. Coatings continuing along the sides of carbonate nodules document a difference in timing between carbonate enrichment and silicate neoformation (Watts, 1980). The formation of palygorskite coatings has also been reported to post-date the partial filling of channels by mechanically transported carbonate grains (Becq-Giraudon & Freytet, 1976; Freytet & Plaziat, 1982). Sepiolite coatings composed of relatively large crystals oriented perpendicular to the surface of fine-grained sepiolite aggregates have been described as recrystallisation features (Hay et al., 1986). Khoury et al. (1982) describe similar features as infillings of circumgranular desiccation cracks. The presence of mineral grains as inclusions in sepiolite-rich pendents has been interpreted as an indication for sepiolite

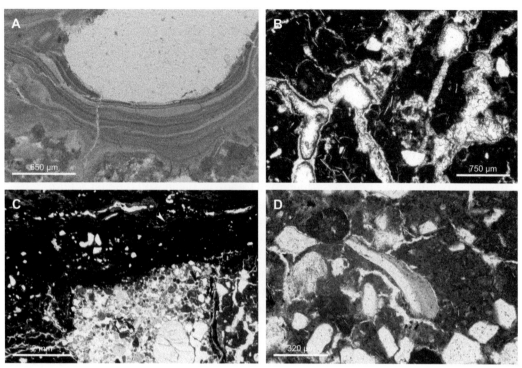

FIGURE 3 Sepiolite as pedofeature. (A) Thick layered sepiolite coating, composed of groundmass-derived material, with variations in relative amount of sepiolite and micritic calcite between layers (Toasis pan, Namibia; colour version of a figure in Mees & Van Ranst, 2011) (PPL). (B) Sepiolite coatings along the sides of voids, with juxtaposed carbonate coatings/infillings (Witpan, Namibia) (PPL). (C) Loose continuous infilling composed of sand grains and sepiolite aggregates, truncated by a layer (top indicated by *arrow*) that developed along the base of a horizontal planar void (see also Mees, 1999) (Witpan, Namibia) (PPL). (D) Fragment of a layered sepiolite coating, in deposits overlying palustrine sediments with sepiolite coatings/infillings (Witpan, Namibia) (PPL).

formation by pressure solution (Brock & Buck, 2005). Palygorskite and sepiolite can occur as part of infillings composed of palygorskite/sepiolite aggregates and mineral grains with palygorskite/sepiolite coatings (Vanden Heuvel, 1966; Kautz & Porada, 1976; Mees, 1999) (Fig. 3C). For the site studied by Vanden Heuvel (1966), all occurrences of fibrous clays in pores, including coatings along the sides of the pores, are assumed to be related to residual enrichment by carbonate dissolution of the palygorskite- or sepiolite-bearing groundmass. Another aspect of carbonate depletion is the possible super-position of porostriation in response to later stress (Mees & Van Ranst, 2011). Palygorskite along cracks in certain schists and granites has been described as a weathering product, progressively replacing the rock and being replaced in turn by calcite (Millot et al., 1977; Paquet, 1983; Wang et al., 1994).

The coatings can become fragmented, producing clustered to isolated fragments, including groups of aligned elongated fragments (Mees & Van Ranst, 2011). Fragments of coatings can be reworked along the soil surface, as illustrated by the occurrence of fragments of sepiolite coatings in deposits overlying the formation from which they are derived (Mees, 1999; Mees & Van Ranst, 2011) (Fig. 3D).

Possible nodules include aggregates composed of micritic carbonates and palygorskite (Alonso-Zarza & Silva, 2002) or sepiolite (Hay & Wiggins, 1980; Francis et al., 2013), both at least partly with concentric internal fabrics (see also Robins et al., 2015). For these concentric patterns, Francis et al. (2013) invoke a role of organic compounds, referring to the radial arrangement of enclosed elongated calcite crystals, whereas Robins et al. (2015) propose a model including pressure dissolution of detrital cores and deposition of illuvial or authigenic palygorskite/sepiolite in the created circumgranular space (see also Robins et al., 2012). Palygorskite nodules, surrounded by sparitic carbonate coatings, have been described for a Cenozoic palaeosoil (Freytet & Plaziat, 1982; Freytet & Verrecchia, 2002). Some nodular sepiolite occurrences have been interpreted as parts where sepiolite-rich formations have been protected from dolomitisation (Stoessell & Hay, 1978). Nahon et al. (1975) describe domains of neoformed palygorskite, formed at the expense of palygorskite-rich marls with a unistrial to crystallitic b-fabric.

Bustillo and García Romero (2003) describe relict fibrous textures in silcretes, inherited from palygorskite-bearing calcretes and possibly enhanced by sepiolite neoformation. Silicification of sepiolite deposits can be marked by the inheritance of a striated b-fabric and a granular microstructure (Bustillo & Alonso-Zarza, 2007). The lack of disturbance of the orientation of palygorskite particles around calcite aggregates in horizons with carbonate enrichment has been seen as an indication that a replacement reaction is involved (Nahon et al., 1975).

2.2 Transmission Electron Microscopy

TEM observations of dried suspensions have often been used to detect or confirm the presence of palygorskite or sepiolite, based on the typical fibrous morphology of these minerals (e.g., Yaalon, 1955; Barshad et al., 1956; Elgabaly, 1962; Elgabaly & Khadr, 1962; Al-Rawi & Sys, 1967; Altaie et al., 1969; Abtahi, 1977; Viani et al., 1983; Güzel & Wilson, 1985; Heidari et al., 2008; Fraser et al., 2016). It can therefore also be used to determine relative abundances of these minerals (e.g., Abtahi, 1980; Abtahi et al., 1980; Shadfan et al., 1985), although the good imaging characteristics of fibrous particles may lead to an overestimation of their concentration (Hillier & Pharande, 2008). Some caution is always needed in TEM identification of palygorskite and sepiolite, as various other clay minerals may have a fibrous morphology. Phyllosilicates that may be confused with palygorskite or sepiolite in TEM (or SEM) images include nontronite, hectorite and chrysotile, as well as halloysite and imogolite in some cases (see e.g., Bates, 1952, 1958; Beutelspacher & van der Marel, 1968; Grim, 1968; Gard, 1971; Sudo et al., 1981). More conclusive identification during TEM investigations is made possible by element analysis (EDS) and especially by electron diffraction analysis.

Some early studies suggested a difference in morphology between sepiolite and palygorskite fibres, for example with a greater width/thickness ratio for sepiolite particles

(Bates, 1958; Grim, 1968; Martin Vivaldi & Robertson, 1971), which would allow an easy distinction between these two related phases. However, this has not been confirmed by the many subsequent TEM studies of these minerals (see also García-Romero & Suárez, 2013). A relationship between fibre length and iron content has been suggested for both palygorskite and sepiolite, with shorter lengths for Fe-rich varieties (Weaver & Pollard, 1973; Zelazny & Calhoun, 1977), again with little subsequent confirmation.

Large well-developed or well-preserved palygorskite crystals are most likely to be inherited (Singer, 1971; Mashhady et al., 1980), although crystals of this type have also been interpreted as authigenic (Soong, 1992; Khademi & Mermut, 1998). Other studies found no difference in morphology between inherited and neoformed fibrous clays (Mackenzie et al., 1984). Vertical or lateral variations in size and shape have been used to assess variations in degree of weathering (Bigham et al., 1980; Shadfan & Dixon, 1984; Botha & Hughes, 1992) or transport (Botha & Hughes, 1992; Khademi & Mermut, 1998; Hojati et al., 2013), whereby both processes are considered to be responsible for the breakdown of the fibrous particles. Alteration during soil development has also been documented by TEM analyses for palaeosoils in loess deposits (Coudé-Gaussen et al., 1984).

TEM studies provide evidence of sepiolite or palygorskite formation by trans-formation of other clay minerals, by documenting the development of fibres along the edge of platy particles (Heysteck & Schmidt, 1953; Trauth, 1977; Vogt & Larqué, 1998; Kaplan et al., 2014). Other types of precursors include opaline material (Blank & Fosberg, 1991) and unidentified spherical aggregates (Martin de Vidales et al., 1987). A relation-ship between palygorskite formation and calcite dissolution has also been suggested (Bustillo & García Romero, 2003). TEM observations of replacement of palygorskite or sepiolite by another phase are reported by Vogt and Larqué (1998), who illustrate the replacement of sepiolite by opal.

2.3   Scanning Electron Microscopy

SEM analysis of soils with fibrous clays is partly used with the same purpose as TEM analysis of suspensions, but it has the added advantage of providing information about their mode of occurrence within the soil. Similar results were obtained by TEM analysis of replicas of undisturbed samples before SEM became widely available (Silva et al., 1965; Beattie, 1970; Alonso & Benayas, 1972; Singer & Norrish, 1974). TEM analysis of ultrathin sections is another alternative (Khademi & Mermut, 1999), but this has only rarely been used for the study of fibrous clays. In SEM studies of calcareous materials, the quality of observations is improved by acid treatment of the sample surface (e.g., Yaalon & Wieder, 1976; Blank & Fosberg, 1991). Another approach is the removal of the resin in thin sections by low-temperature ashing (Jenkins, 1981). SEM analysis can obviously also be done for dried suspensions, for which spray drying was advocated in an early study by Bohor and Hughes (1971) (see also Hughes & Bohor, 1970).

In most SEM studies, the fibrous morphology of sepiolite and palygorskite is illustrated, as in nearly all papers that are cited in this section. One exception is sepiolite coatings with a non-fibrous surface morphology in phreatic calcretes (Arakel & McConchie, 1982). Fibre development has been related to bacterial activity according to Folk and Rasbury (2007), referring to the presence of spherical bodies at the tip of fibres and to the occurrence of fibres that are composed entirely of a string of spheres.

For undisturbed samples, the delicate morphology of the fibres has been cited as an indication of a lack of transport (e.g., Schwaighofer, 1980; Blank & Fosberg, 1991). The presence of spherical aggregates of palygorskite particles has been interpreted in the same manner (Yaalon & Wieder, 1976). Similar spherical aggregates, composed of sepiolite, are described by Gehring et al. (1995), who relate their formation to conditions with alternating supersaturation and undersaturation. Palygorskite aggregates with broken and deformed fibres along the surface and with indications of compaction in the outer parts can be particles that have been subjected to eolian transport (Coudé-Gaussen & Blanc, 1985; Coudé-Gaussen, 1987, 1991). In these studies of eolian material, the protective effect of the felt-like surface layer of the aggregates is pointed out.

The presence of palygorskite or sepiolite fibres on the surface of pedogenic or authigenic minerals is a strong indication of silicate neoformation in soil environments or during diagenesis (Fig. 4). This was first proposed for palygorskite covering pedogenic gypsum (Eswaran & Barzanji, 1974; Abtahi et al., 1980) and subsequently for palygorskite or sepiolite on dolomite (e.g., Anand et al., 1997; Pimentel, 2002; Xie et al., 2013), calcite (e.g., Kapur et al., 1987; Khademi & Mermut, 1998; Kaplan et al., 2014) and halite (Hojati

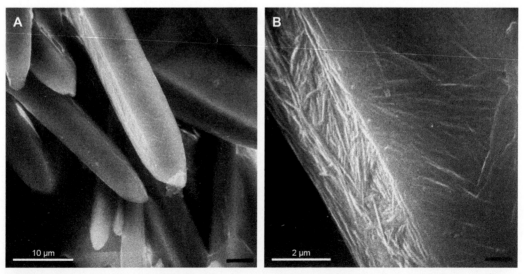

FIGURE 4 SEM images of palygorskite (gypsiferous soils, Iraq). (A) Palygorskite fibres covering the surface of authigenic celestine crystals. (B) View of same occurrence at higher magnification. *Images from Ghent University archive, sample provided by A.J. Barzanji.*

et al., 2010). An illuvial origin is nevertheless proposed for some occurrences of this type (Khademi & Mermut, 1999). Fibres that are oriented more or less perpendicular to a ped or grain surface are most unambiguously non-illuvial (Weaver & Beck, 1977; Monger & Daugherty, 1991; Singer et al., 1995; Dhir et al., 2004; Owliaie et al., 2006). Occurrences of fibres bridging the space between grains and aggregates suggest development from pore solutions that formed a meniscus in vadose conditions (e.g., Long et al., 1997; Colson et al., 1998; Kaplan et al., 2014), although palygorskite/sepiolite formation in a meniscus composed of microbial organic deposits has also been proposed (Miller & James, 2012). For water-deposited bridging accumulations, illuvial transport cannot be entirely excluded. This is also the case for films or coatings of delicately interwoven fibres on the surface of peds or grains, which are generally considered to be authigenic (Hassouba & Shaw, 1980; Watts, 1980; Verrecchia & Le Coustumer, 1996; Dhir et al., 2004). Beattie (1970), using replica-TEM, considered purity, discontinuity, a relatively large crystal size and juxtaposition on illuvial clay coatings as criteria for identifying palygorskite on ped surfaces as neoformed (see also Halitim et al., 1983a). The parallel alignment of palygorskite fibres is another criterion for authigenesis rather than illuviation (Nahon et al., 1975). In one palaeosoil study, the irregular SEM morphology of palygorskite coatings has been related to the non-planar form of the constituent crystals (Freytet & Plaziat, 1982).

Like TEM analyses, SEM studies have documented the transformation of other clay minerals to palygorskite or sepiolite (Bachman & Machette, 1977; Khoury et al., 1982; Pozo & Casas, 1999; Hillier & Pharande, 2008; Kadir & Eren, 2008; Xie et al., 2013; Kaplan et al., 2014). A similar relationship is deduced for sepiolite fibres emanating from opaline material in nodules (Blank & Fosberg, 1991). Indications for palygorskite formation by the coalescence of opal spheres have also been reported (Weaver & Beck, 1977).

SEM images have suggested a relationship between the formation of palygorskite or sepiolite and the dissolution of the material they surround, such as quartz (Halitim et al., 1983b; Colson et al., 1998; Francis et al., 2012), calcite (Courty et al., 1987) and dolomite (Kadir et al., 2010). A relationship with associated calcite formation and quartz dissolution has also been suggested (Halitim et al., 1983a). An alternative explanation for the formation of palygorskite or sepiolite surrounding mineral grains is that the latter have acted as nucleation sites (Morrás et al., 1982; Pérez-Rodríguez et al., 1990). Whether transformation or nucleation is involved cannot always be determined (Rodas et al., 1994). The sides of sand grains have also been considered to be sites where water preferentially circulates or resides (Sancho et al., 1992; Verrecchia & Le Coustumer, 1996). Epitaxial growth has also been suggested, for palygorskite on hornblende (Eswaran & Barzanji, 1974). A different type of relationship has been proposed for palygorskite covering gypsum crystals, whereby palygorskite formation is thought to have been favoured by increased magnesium/calcium ratios in soil solutions during gypsum formation (Khormali & Abtahi, 2003; Farpoor & Krouse, 2008).

Transformation of sepiolite or palygorskite to other materials has been documented in some studies. Reaction products include other clay minerals (Singer et al., 1995) and non-pseudomorphic fibrous opal (Bustillo, 1995; Bustillo & Bustillo, 2000). Opal formed by silicification of sepiolite deposits has been described as a mass of discrete pseudomorphic fibres that is subsequently transformed to aggregates of blades composed of joined fibres and finally to structureless opaline material (Bustillo & Alonso-Zarza, 2007).

3. Zeolites

Zeolites are quite rare in soil environments, occurring either as inherited components or as pedogenic minerals. The latter form nearly exclusively in saline alkaline soils, partly with volcanic parent materials (e.g., Ming & Boettinger, 2001).

Most zeolite minerals have a characteristic crystal morphology, which has been extensively documented for sedimentary and volcanic rocks (Mumpton & Ormsby, 1976; Gottardi & Galli, 1985). The morphology of zeolites in soil environments seems to be identical with that of zeolites formed in other settings (e.g., Ming & Mumpton, 1989) (Table 1; Fig. 5). Morphological studies of euhedral zeolite crystals, using thin sections or SEM observations, can therefore contribute to the identification of zeolite mineral species in soil samples (e.g., Allen & Jacob, 1983; Ping et al., 1988), although no identification of all phases appears to have been possible in some cases (e.g., Smale, 1968; Zarei et al., 1987, 1990). If a crystal morphology is recognised that is not clearly compatible with that of common soil zeolites (e.g., Brogowski et al., 1979, 1983; Verrecchia & Le Coustumer, 1996), the identification of those crystals as zeolites is doubtful. In addition to generally allowing mineral identification, micromorphological studies are also important for the recognition of associations and sequences (e.g., Renaut, 1993; Mees et al., 2005).

3.1 Thin Section Studies

Zeolites in soils can be groundmass components inherited from the parent material, occurring as single grains or as part of rock fragments (Stoops et al., 2001). Fracturing of

Table 1 Typical morphology of zeolite minerals that have been described as authigenic minerals in soils

Mineral	Formula[a]	Morphology	Selected references
Analcime	$Na(AlSi_2O_6) \cdot H_2O$	Equant polyhedral	Renaut (1993)
Chabazite	$Ca_2[Al_4Si_8O_{24}] \cdot 13H_2O$	Cubic	Dickinson & Grapes (1997)
Clinoptilolite	$K_6(Si_{30}Al_6)O_{72} \cdot 20H_2O$	Thick tabular	Ming & Dixon (1986)
Erionite	$Na_{10}[Si_{26}Al_{10}O_{72}] \cdot 30H_2O$	Elongated prismatic	Mees et al. (2005)
Mordenite	$(Na_2,Ca,K_2)_4(Al_8Si_{40})O_{96} \cdot 28H_2O$	Fibrous	Maglione & Tardy (1971)
Natrolite	$Na_2(Si_3Al_2)O_{10} \cdot 2H_2O$	Tabular	Mees et al. (2005)
Phillipsite	$K_6(Si_{10}Al_6)O_{32} \cdot 12H_2O$	Elongated prismatic	McFadden et al. (1987)

[a]Formula for the most common species if several exist.

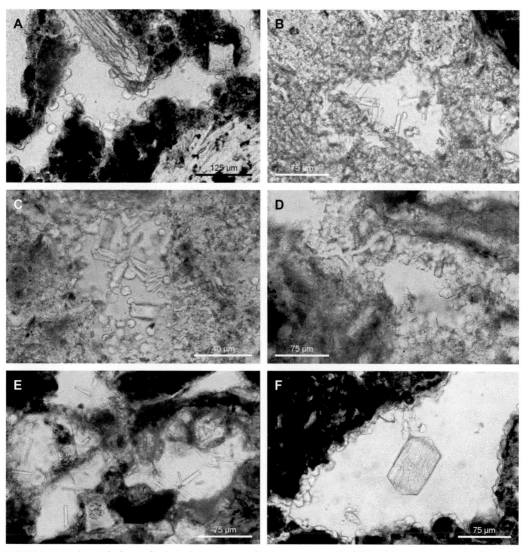

FIGURE 5 Typical morphology of selected common zeolite minerals in soils (Olduvai Basin, Tanzania; Beds I to III) (PPL). (A) Analcime (Bed III). (B) Phillipsite (Bed I). (C) Analcime and phillipsite (Bed II). (D) Chabazite (Bed I). (E) Erionite (Bed II). (F) Natrolite (Bed III).

zeolite grains in this study (laumontite) reveals susceptibility to physical breakdown. The fragmented crystal aggregates with an assumed zeolite composition illustrated by Sauer and Zöller (2006) should also be considered to be part of the groundmass. Grain mount studies of the sand fraction have also been used for zeolite-bearing soils, documenting the occurrence of zeolitised plagioclase and kaolinitised zeolites in laumontite-bearing parent materials (Taylor et al., 1990). They also show the presence of iron oxide coatings on altered zeolite grains (Bhattacharyya et al., 1999), whose significance is unclear. Reworked fragments of chabazite coatings, with aberrant crystal morphology, represent another type of zeolitic groundmass component (Alonso-Zarza et al., 2016).

Zeolite minerals are more commonly recognised in pores than as groundmass constituents (see Fig. 5). They can occur exclusively or preferentially in pores of specific types, providing information about the period of zeolite development, for example relative to the development of porosity by dissolution of feldspar grains (Dickinson & Grapes, 1997; Dickinson & Bleakley, 2005) or formation of planar voids (Mees et al., 2005). The recognition of zeolites in pores confirms that they are formed from solution rather than by direct transformation of clay minerals in the groundmass (Remy & Ferrell, 1989; Renaut, 1993); the latter has been suggested for some analcime occurrences observed in thin sections (e.g., Scott et al., 2008). Occurrences in pores that formed during subaerial exposure, particularly channels, generally indicate a pedogenic nature of the zeolites (Haesaerts et al., 1983; Renaut, 1993; Ashley & Driese, 2000; Yakimenko et al., 2004; Driese & Ashley, 2016). However, analcime coatings in root passages in Carboniferous lake margin deposits seem to have formed after burial rather than during soil development (Gall & Hyde, 1989). A non-pedogenic origin is also inferred for analcime infillings of mudcracks in Triassic lake sediments (Van Houten, 1962, 1964) and for laumontite coatings in Cretaceous palaeosoils (Howe & Francis, 2005). In one study, analcime coatings that formed in a soil environment are considered to be partly illuvial, composed of transported silt-sized crystals (Ashley & Driese, 2000).

Pores with zeolite coatings can contain other pedogenic materials, such as illuvial or authigenic clay coatings, carbonate coatings or infillings and iron/manganese oxide coatings or hypocoatings (Hay, 1978; Renaut, 1993; Ashley & Driese, 2000; Mees et al., 2005; Owen et al., 2008; Beverly et al., 2014; Alonso-Zarza et al., 2016) (Fig. 6). They record sequences of pedogenic processes, which may be related to changes in palaeoenvironmental conditions in some cases. The possibility that these processes took place during overlapping or multiple periods should always be considered (e.g., Dickinson & Grapes, 1997; Dickinson & Bleakley, 2005).

For settings other than soil environments, zeolite formation during alteration of volcanic material has been extensively documented. Studies of the alteration of this type of material in soil-related environments include the work of Boekschoten et al. (1983), describing the transformation of volcanic glass to microcrystalline offretite and the development of analcime veins with epistilbite spherulites. Le Gal et al. (1999) describe zeolites associated with a gel that is formed by alteration of volcanic glass. Hay (1980) describes occurrences of zeolites in eolian tuffs that were altered along the land surface, largely based on thin section observations. Reported features include the relatively low authigenic mineral content of cavities formed by dissolution of nepheline, indicating that this process also contributed to zeolite formation in other parts of the deposits (Hay, 1980). Other papers by the same author about zeolite occurrences at the same location are less explicitly based on thin section observations (Hay, 1963, 1964) or they concern lake-margin or alluvial environments with intermittent exposure rather than stable soil environments (e.g., Hay, 1970). One buried alteration profile was recognised, for which

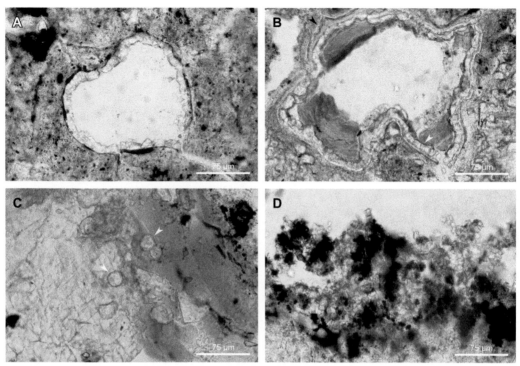

FIGURE 6 Zeolites as part of compound pedofeatures (Olduvai Basin, Tanzania; Beds I to IV) (PPL). (A) Analcime coating in a channel with a discontinuous Fe/Mn oxide hypocoating (Bed III). (B) Thin analcime coating (*black arrow*), followed by pale illuvial clay, followed by a thicker hypidiotopic analcime coating (*white arrow*), followed by orange brown illuvial clay (Bed III). (C) Analcime crystals (*arrows*) enclosed by sparitic calcite (left) and illuvial fine clay (right) (Bed IV). (D) Chabazite coating with superimposed iron/manganese oxide impregnation (Bed I).

analcime-chabazite coprecipitation and later formation of natrolite and dawsonite could be deduced (Hay, 1970). Zeolites in continental sediments with volcanic intercalations, deposited in intermittently dry marsh environments, were also studied by Teruggi (1964), documenting various types of analcime occurrences, including coatings, infillings, and partially replacive aggregates.

An entirely different context in which thin section and SEM analyses have been applied to the study of zeolites is the investigation of soils on iron production waste (Sauer & Burghardt, 2006).

3.2   Transmission Electron Microscopy

TEM has only rarely been used for the study of zeolites in soils. Gorbunov and Bobrovitskiy (1973) mention that it allows mineral identification based on external morphology, illustrated by an unconvincing image for alluvial soils. The identification of fibrous material as zeolites in alluvial soils by Brogowski et al. (1983) is equally doubtful. The use of TEM is also mentioned or implied by Nørnberg and Dalsgaard (1985) and

Nørnberg (1990), who used electron diffraction analysis to confirm the recognition of inherited clinoptilolite-heulandite. The same methods were used in a study of a clay fraction that includes inherited mordenite (Kirkman, 1976), and in a study that indicates the presence of smectite as an alteration product of clinoptilolite in the upper part of a soil profile (Ming & Boettinger, 2001).

3.3 Scanning Electron Microscopy

SEM studies of inherited zeolites that are part of the groundmass provide evidence for their stability in certain environments, based on the presence of euhedral shapes and unweathered surfaces (Ming & Dixon, 1986). Pitted surfaces and rounding are indicative of chemical weathering, with variations between horizons (Taylor et al., 1990) or climate zones (Bhattacharyya et al., 1999), although roundness can also be related to transport of the grains (Taylor et al., 1990; Bhattacharyya et al., 1999). Weathering of inherited zeolites may result in the appearance of alteration products such as halloysite (Taylor et al., 1990). A difference in resistance against weathering relative to associated components such as volcanic glass has been revealed by SEM studies (Ming & Dixon, 1986). Jacob and Allen (1990) found that clinoptilolite occurs in the interior of inherited altered glass shards in subsurface horizons, where it is protected from weathering.

SEM studies of pedogenic zeolites include investigations of fractions obtained by mineral separation, whereby the euhedral nature of zeolite crystals is indicative of formation from soil solutions and the presence of dissolution features records instability in specific horizons (Spiers et al., 1984). In SEM studies of undisturbed fragments, indications of authigenesis again include a lack of corrosion (Gibson et al., 1983) as well as zeolite occurrences in pores (e.g., McFadden et al., 1987; Dickinson & Rosen, 2003; Inozemtsev & Targulian, 2010) (Fig. 7). Other studies have shown that quartz grains in contact with zeolites are partly corroded (Dickinson & Grapes, 1997) or that detrital sand grains in the same profile show dissolution features (Gibson et al., 1983), documenting a relationship between zeolite formation and the dissolution of other silicates. A difference in timing with opal-CT development, taking place at a later stage than zeolite formation, has also been noted (Huggett et al., 2005; Pozo et al., 2006). Renaut (1993) describes analcime mixed with clay in root casts, developed either as precipitates around roots or as infillings of moulds at a later stage. SEM and optical observations of altered volcanic material from Lanzarote suggest a contribution of zeolites to soil aggregate stability (Zarei et al., 1987).

Buch and Rose (1996) illustrate the morphology of euhedral K-feldspar, assumed to have formed by alteration of analcime that was part of the parent material. This transformation has been documented in various studies of non-pedogenic authigenic silicates related to conditions with higher salinities and alkalinities than those that favour zeolite formation.

4. Sodium Silicates

Sodium silicate minerals can form by evaporation of highly alkaline interstitial brines in near-surface horizons. The most common sodium silicates are magadiite, kenyaite and

FIGURE 7 SEM image of clinoptilolite in a palaeosoil of the Omo Group, Ethiopia. *Image from Ghent University archive, sample provided by B. Van Vliet-Lanoë.*

kanemite. The morphology of individual crystals, recognised in SEM and TEM images, seems to be quite characteristic in soils or soil-related environments (Table 2). Spherulitic aggregates have been described for all three minerals (Rooney et al., 1969; Hay, 1970; Johan & Maglione, 1972; Maglione, 1976; Mees et al., 2005) (Fig. 8).

Kenyaite has been recognised in pores (Hay, 1970; Mees et al., 2005), associated with other authigenic silicate minerals that formed at an earlier stage (Mees et al., 2005), suggesting a late-stage pedogenic origin. Rooney et al. (1969) describe microscopic features of infillings of wide vertical cracks in playa deposits, composed of a fine-grained

Table 2   Morphology of sodium silicate minerals

Mineral	Composition	Morphology	Selected references
Magadiite	$Na_2Si_{14}O_{29} \cdot 11H_2O$	Plate-shaped with nearly square forms	McAtee et al. (1968); Maglione (1970)
Kanemite	$HNaSi_2O_5 \cdot 3H_2O$	Tabular parallel to (010), with {110} faces	Johan & Maglione (1972); Maglione (1976)
Kenyaite	$Na_2Si_{22}O_{41}(OH)_8 \cdot 6H_2O$	Lath-shaped crystals with pointed terminations	Icole & Perinet (1984)

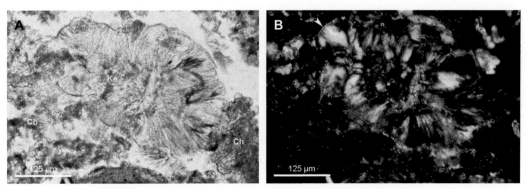

FIGURE 8 Sodium silicate occurrence (Olduvai Basin, Tanzania; Bed II). (A) Spherulitic kenyaite aggregate (Ke) in a pore with authigenic chabazite (Ch) (PPL). (B) View of same field in XPL, illustrating the presence of a thin (clay) coating with higher interference colours (*arrow*).

magadiite mass that is crossed by veins with a spherulitic fabric and which encloses possible magadiite pseudomorphs after unidentified mineral grains.

Icole et al. (1983) mention the occurrence of enclosed organic material in tubular magadiite concretions that seem to have formed around roots. Sebag et al. (2001) report various textural features for layers, concretions and mineralised plant remains from the same region. Fabric units include massive microcrystalline magadiite, partly grading to more coarse-grained material toward pores, and void-filling phases such as spherulites and palisade-textured coatings, whereby all coarse-grained material is tentatively identified as kenyaite (Sebag et al., 1999, 2001).

Magadiite is replaced by silhydrite ($Si_3O_6 \cdot H_2O$) by leaching, producing more or less pseudomorphic aggregates (Gude & Sheppard, 1972). Similar patterns are reported for the partial replacement of spherulitic kenyaite by chalcedony (Hay, 1970). Maglione and Servant (1973) mention SEM observations that indicate the transformation of magadiite and kenyaite to quartz. The transformation of magadiite beds to chert layers is well documented for the Lake Magadi area (e.g., Hay, 1968), which may in part be related to leaching in soil environments. Sheppard and Gude (1969) illustrate rhodesite coatings ($KHCaSi_8O_{19}.5H_2O$) in cavities within microcrystalline magadiite that are formed by the alteration of volcanic rocks, without indications for magadiite-rhodesite replacement.

5. Conclusions

Micromorphological techniques have been widely applied to the study of silicate neoformation, both in the field of soil science and for the study of diagenetic processes in sedimentary geology. Their use has greatly contributed to the recognition of silicate mineral authigenesis in earth-surface conditions. Micromorphology also provides information about the transformation of authigenic silicates and their inherited counterparts. Electron microscopy has been most commonly used, as a technique that is ideally suited for the study of submicroscopic particles. For palygorskite and sepiolite, this has led to an overrepresentation in the literature of SEM and TEM images that merely illustrate a fibrous particle morphology rather than textural features with genetic implications. Electron microscopy studies of zeolites and sodium silicates show a highly characteristic and invariable crystal morphology, which makes SEM and TEM important tools for their identification. Thin section studies of authigenic silicates and their inherited counterparts are rarer, but they invariably provide important information about the significance of their occurrence as groundmass components and pedofeatures.

References

Abtahi, A., 1977. Effect of a saline and alkaline ground water on soil genesis in semiarid southern Iran. Soil Science Society of America Journal 41, 583-588.

Abtahi, A., 1980. Soil genesis as affected by topography and time in highly calcareous parent materials under semiarid conditions of Iran. Soil Science Society of America Journal 44, 329-336.

Abtahi, A., Eswaran, H., Stoops, G. & Sys, C., 1980. Mineralogy of a soil sequence formed under the influence of saline and alkaline conditions in the Sarvestan Basin (Iran). Pedologie 30, 283-304.

Achyuthan, H., 2003. Petrologic analysis and geochemistry of the Late Neogene-Early Quaternary hardpan calcretes of Western Rajasthan, India. Quaternary International 106-107, 3-10.

Allen, B.L. & Jacob, J.S., 1983. Alteration of zeolitic tuffs during pedogenesis in Southwest Texas, USA. Sciences Géologiques, Mémoires 73, 7-17.

Alonso, J. & Benayas, J., 1972. Electron and petrographic microscopy applied to soil micromorphology. In Kowalinski, S. & Drozd, J. (eds.), Soil Micromorphology. Panstwowe Wydawnicto Naukowe, Warsawa, pp. 63-75.

Alonso-Zarza, A.M. & Silva, P.G., 2002. Quaternary laminar calcretes with bee nests: evidences of small-scale climatic fluctuations, Eastern Canary Islands, Spain. Palaeogeography, Palaeoclimatology, Palaeoecology 178, 119-135.

Alonso-Zarza, A.M., Bustamante, L., Huerta, P., Rodríguez-Berriguete, A. & Huertas, M.J., 2016. Chabazite and dolomite formation in a dolocrete profile: an example of a complex alkaline paragenesis in Lanzarote, Canary Islands. Sedimentary Geology 337, 1-11.

Al-Rawi, G.J. & Sys, C., 1967. A comparative study between Euphrates and Tigris sediments in the Mesopotamian Flood Plain. Pedologie 17, 187-211.

Altaie, F.H., Sys, C. & Stoops, G., 1969. Soil groups of Iraq. Their classification and characterization. Pedologie 19, 65-148.

Anand, R.R., Phang, C., Wildman, J.E. & Lintern, M.J., 1997. Genesis of some calcretes in the southern Yilgarn Craton, Western Australia: implications for mineral exploration. Australian Journal of Earth Sciences 44, 97-103.

Arakel, A.V. & McConchie, D., 1982. Classification and genesis of calcrete and gypsite lithofacies in paleodrainage systems of inland Australia and their relationship to carnotite mineralization. Journal of Sedimentary Petrology 52, 1149-1170.

Armenteros, I., Bustillo, M.A. & Blanco, J.A., 1995. Pedogenic and groundwater processes in a closed Miocene basin (northern Spain). Sedimentary Geology 99, 17-36.

Ashley, G.M. & Driese, S.G., 2000. Paleopedology and paleohydrology of a volcaniclastic paleosol interval: implications for Early Pleistocene stratigraphy and paleoclimate record, Olduvai Gorge, Tanzania. Journal of Sedimentary Research 70, 1065-1080.

Bachman, G.O. & Machette, M.N., 1977. Calcic Soils and Calcretes in the Southwestern United States. U. S. Geological Survey Open-File Report 77-794, 163 p.

Bailey, S.W., Brindley, G.W., Johns, W.D., Martin, R.D. & Ross, M., 1971. Summary of national and international recommendations on clay mineral nomenclature. 1969-70 CMS Nomenclature Committee. Clays and Clay Minerals 19, 129-132.

Barshad, I., Halevy, E., Gold, H.A. & Hagin, J., 1956. Clay minerals in some limestone soils from Israel. Soil Science 81, 423-437.

Bates, T.F., 1952. Electron microscopy as a method of identifying clays. Clays and Clay Minerals 1, 130-150.

Bates, T.F., 1958. Selected Electron Micrographs of Clays and Other Fine-Grained Minerals. Circular No. 51, Mineral Industries Experiment Station, Pennsylvania State University, 61 p.

Beattie, J.A., 1970. Peculiar features of soil development in parna deposits in the eastern Riverina, N.S.W. Australian Journal of Soil Research 8, 145-156.

Beattie, J.A. & Haldane, A.D., 1958. The occurrence of palygorskite and barytes in certain parna soils of the Murrumbidgee region, New South Wales. Australian Journal of Science 20, 274-275.

Becq-Giraudon, J.F. & Freytet, P., 1976. L'Oligocène du fossé de Saint-Maixent (Deux-Sèvres): observations paléontologiques et pétrographiques sur les calcaires 'lacustres' à attapulgite. Comptes Rendus de l'Académie des Sciences 282, 1943-1946.

Beutelspacher, H. & van der Marel, H.W., 1968. Atlas of Electron Microscopy of Clay Minerals and their Admixtures. A Picture Atlas. Elsevier, Amsterdam, 333 p.

Beverly, E.J., Ashley, G.M. & Driese, S.G., 2014. Reconstruction of a Pleistocene paleocatena using micromorphology and geochemistry of lake margin paleo-Vertisols, Olduvai Gorge, Tanzania. Quaternary International 322/323, 78-94.

Bhattacharyya, T., Pal, D.K. & Srivastava, P., 1999. Role of zeolites in persistence of high altitude ferruginous Alfisols of the humid tropical Western Ghats, India. Geoderma 90, 263-276.

Bigham, J.M., Jaynes, W.F. & Allen, B.L., 1980. Pedogenic degradation of sepiolite and palygorskite on the Texas High Plains. Soil Science Society of America Journal 44, 159-167.

Blank, R.R. & Fosberg, M.A., 1991. Duripans of the Owyhee Plateau region of Idaho: genesis of opal and sepiolite. Soil Science 152, 116-133.

Boekschoten, G.J., Buurman, P. & van Reeuwijk, L.P., 1983. Zeolites and palygorskite as weathering products of pillow lava in Curaçao. Geologie en Mijnbouw 62, 409-415.

Bohor, B.F. & Hughes, R.E., 1971. Scanning electron microscopy of clays and clay minerals. Clays and Clay Minerals 19, 49-54.

Botha, G.A. & Hughes, J.C., 1992. Pedogenic palygorskite and dolomite in a late Neogene sedimentary succesion, northwestern Transvaal, South Africa. Geoderma 53, 139-154.

Brewer, R., 1976. Fabric and Mineral Analysis of Soils. Robert E. Krieger Publishing Company, Huntington, New York, 482 p.

Brock, A.L. & Buck, B.J., 2005. A new formation process for calcic pendants from Pahranagat Valley, Nevada, USA, and implication for dating Quaternary landforms. Quaternary Research 63, 359-367.

Brogowski, Z., Dobrzanski, B. & Kocon, J., 1979. Morphology of natural zeolites occurring in soil as determined by electron microscopy. Bulletin de l'Académie Polonaise des Sciences, Série des Sciences de la Terre 27, 115-117.

Brogowski, Z., Dobrzanski, B., Kocon, J. & Zeniewska-Chlipalska, E., 1983. The possibility of zeolites occurrence in the soils of Poland. Zeszyty Problemowe Postepow Nauk Rolniczych 220, 489-494.

Buch, M.W. & Rose, D., 1996. Mineralogy and geochemistry of the sediments of the Etosha Pan Region in northern Namibia: a reconstruction of the depositional environment. Journal of African Earth Sciences 22, 355-378.

Bustillo, M.A., 1995. Una nueva ultraestructura de opalo CT en silcretas. Posible indicador de influencia bacteriana. Estudios Geológicos 51, 3-8.

Bustillo, M.A & Alonso-Zarza, A.M., 2007. Overlapping of pedogenesis and meteoric diagenesis in distal alluvial and shallow lacustrine deposits in the Madrid Miocene Basin, Spain. Sedimentary Geology 198, 255-271.

Bustillo, M.A. & Bustillo, M., 2000. Miocene silcretes in argillaceous playa deposits, Madrid Basin, Spain: petrological and geochemical features. Sedimentology 47, 1023-1037.

Bustillo, M.A. & García Romero, E., 2003. Arcillas fibrosas anómalas en encostramientos y sedimentos superficiales: características y génesis (Esquivias, Cuenca de Madrid). Boletin de la Sociedad Española de Cerámica y Vidrio 42, 289-297.

Colson, J., Cojan, I. & Thiry, M., 1998. A hydrogeological model for palygorskite formation in the Danian continental facies of the Provence Basin (France). Clay Minerals 33, 333-347.

Coudé-Gaussen, G., 1987. Observations au MEB de fibres de palygorskite transportée en grains par le vent. In Fedoroff, N., Bresson, L.M. & Courty, M.A. (eds.), Micromorphologie des Sols, Soil Micromorphology. AFES, Plaisir, pp. 199-205.

Coudé-Gaussen, G., 1991. Les Poussières Sahariennes. Editions John Libby Eurotext, Montrouge, 485 p.

Coudé-Gaussen, G. & Blanc, P., 1985. Présence de grains isolés de palygorskite dans les poussières actuelles et les sédiments récents d'origine désertique. Bulletin de la Société Géologique de France 8, 571-579.

Coudé-Gaussen, G., Le Coustumer, M.N. & Rognon, P., 1984. Paléosols d'âge Pleistocène supérieur dans les loess des Matmata (Sud Tunisien). Sciences Géologiques Bulletin 37. 359-386.

Courty, M.A., Dhir, P. & Raghavan, H., 1987. Microfabrics of calcium carbonate accumulations in arid soils of western India. In Fedoroff, N., Bresson, L.M. & Courty, M.A. (eds.), Micromorphologie des Sols, Soil Micromorphology. AFES, Plaisir, pp. 227-234.

Cuadros, J., Diaz-Hernandez, J.L., Sanchez-Navas, A., Garcia-Casco, A. & Yepes, J., 2016. Chemical and textural controls on the formation of sepiolite, palygorskite and dolomite in volcanic soils. Geoderma 271, 99-114.

Dhir, R.P., Tandon, S.K., Sareen, B.K., Ramesh, R., Rao, T.K.G., Kailath, A.J. & Sharma, N., 2004. Calcretes in the Thar desert: genesis, chronology and palaeoenvironment. Proceedings of the Indian Academy of Sciences (Earth and Planetary Sciences) 113, 473-515.

Dickinson, W. & Bleakley, N., 2005. Facies control on diagenesis in frozen sediments: the Sirius Group, Table Mountains and Mount Feather, Dry Valleys, Antarctica. Terra Antartica 12, 25-34.

Dickinson, W.W. & Grapes, R.H., 1997. Authigenic chabazite and implications for weathering in Sirius Group diamictite, Table Mountain, Dry Valleys, Antarctica. Journal of Sedimentary Research 67, 815-820.

Dickinson, W.W. & Rosen, M.R., 2003. Antarctic permafrost: an analogue for water and diagenetic minerals on Mars. Geology 31, 199-202.

Driese, S.G. & Ashley, G.M., 2016. Paleoenvironmental reconstruction of a paleosol catena, the Zinj archeological level, Olduvai Gorge, Tanzania. Quaternary Research 85, 133-146.

Elgabaly, M.M., 1962. The presence of attapulgite in some soils of the western desert of Egypt. Soil Science 93, 387-390.

Elgabaly, M.M. & Khadr, M., 1962. Clay mineral studies of some Egyptian desert and Nile alluvial soils. Journal of Soil Science 13, 333-342.

Eswaran, H. & Barzanji, A.F., 1974. Evidence for the neoformation of attapulgite in some soils of Iraq. Transactions of the 10th International Congress of Soil Science, Volume VII, Moscow, pp. 154-161.

Farpoor, M.H. & Krouse, H.R., 2008. Stable isotope geochemistry of sulfur bearing minerals and clay mineralogy of some soils and sediments in Loot Desert, central Iran. Geoderma 146, 283-290.

Fleischer, M., Wilcox, R.E. & Matzko, J.J., 1984. Microscopic Determination of the Nonopaque Minerals. U.S. Geological Survey Bulletin 1627, 453 p.

Folk, R.L. & Rasbury, E.T., 2007. Nanostructure of palygorskite/sepiolite in Texas caliche: possible bacterial origin. Carbonates and Evaporites 22, 113-122.

Francis, M.L., Fey, M.V., Ellis, F. & Poch, R.M., 2012. Petroduric and 'petrosepiolitic' horizons in soils of Namaqualand, South Africa. Spanish Journal of Soil Science 2, 8-25.

Francis, M.L., Ellis, F., Lambrechts, J.J.N. & Poch, R.M., 2013. A micromorphological view through a Namaqualand termitaria (Heuweltjie, a Mima-like mound). Catena 100, 57-73.

Fraser, M.B., Churchmanb, G.J., Chittleborough, D.J. & Rengasamy, P., 2016. Effect of plant growth on the occurrence and stability of palygorskite, sepiolite and saponite in salt-affected soils on limestone in South Australia. Applied Clay Science 124/125, 183-196.

Freytet, P. & Plaziat, J.C., 1982. Continental Carbonate Sedimentation and Pedogenesis – Late Cretaceous and Early Tertiary of Southern France. Contributions to Sedimentology, Volume 12. E. Schweizerbart'sche Verlagsbuchhandlung, Stuttgart, 213 p.

Freytet, P. & Verrecchia, E.P., 2002. Lacustrine and palustrine carbonate petrography: an overview. Journal of Paleolimnology 27, 221-237.

Gall, Q. & Hyde, R., 1989. Analcime in lake and lake-margin sediments of the Carboniferous Rocky Brook Formation, western Newfoundland, Canada. Sedimentology 36, 875-887.

García-Romero, E. & Suárez, M., 2013. Sepiolite-palygorskite: textural study and genetic considerations. Applied Clay Science 86, 129-144.

Gard, J.A. (ed.), 1971. The Electron-Optical Investigation of Clays. Mineralogical Society, London, 383 p.

Gehring, A.U., Keller, P., Frey, B. & Luster, J., 1995. The occurrence of spherical morphology as evidence for changing conditions during the genesis of a sepiolite deposit. Clay Minerals 30, 83-86.

Gibson, E.K., Wentworth, S.J. & McKay, D.S., 1983. Chemical weathering and diagenesis of a cold desert soil from Wright Valley, Antarctica: an analogue of Martian weathering processes. Journal of Geophysical Research 88 Supplement, A912-A928.

Gorbunov, N.I. & Bobrovitskiy, A.V., 1973. Distribution, genesis, structure, and properties of zeolite. Soviet Soil Science 5, 351-360.

Gottardi, G. & Galli, E., 1985. Natural Zeolites. Springer-Verlag, Berlin, 409 p.

Grim, R.E., 1968. Clay Mineralogy, 2nd Edition. McGraw-Hill, London, 596 p.

Gude, A.J. & Sheppard, R.A., 1972. Silhydrite, $3SiO_2.H_2O$, a new mineral from Trinity County, California. American Mineralogist 57, 1053-1065.

Güzel, N. & Wilson, J., 1985. High-magnesium clays from alluvial soils of the Acipayam plain, southern Turkey. In Konta, J. (ed.), Proceedings of the 5th Meeting of the European Clay Groups. Charles University, Prague, pp. 117-128.

Gutiérrez-Castorena, Ma. del C., 2018. Pedogenic siliceous features. In Stoops, G., Marcelino, V. & Mees, F. (eds.), Interpretation of Micromorphological Features of Soils and Regoliths. Second Edition. Elsevier, Amsterdam, pp. 127-155.

Haesaerts, P., Stoops, G. & VanVliet-Lanoë, B., 1983. Data on sediments and fossils soils. In de Heinzelin, J. (ed.), The Omo Group. Archives of the International Omo Research Expedition. Annales du Musée Royal de l'Afrique Centrale, Serie In-8 , n 85, pp. 149-186.

Halitim, A., Robert, M. & Berrier, J., 1983a. A microscopic study of quartz evolution in arid areas. In Bullock, P. & Murphy, C.P. (eds.), Soil Micromorphology. Volume 2. Soil Genesis. AB Academic Publishers, Berkhamsted, pp. 615-621.

Halitim, A., Robert, M. & Pédro, G., 1983b. Etude expérimentale de l'épigénie calcaire des silicates en milieu confiné — caractérisation des conditions de son développement et des modalités de sa mise en jeu. Sciences Géologiques, Mémoires 71, 63-73.

Hassouba, H. & Shaw, H.F., 1980. The occurrence of palygorskite in Quaternary sediments of the coastal plain of north-west Egypt. Clay Minerals 15, 77-83.

Hay, R.L., 1963. Zeolitic weathering in Olduvai Gorge, Tanganyika. Geological Society of America Bulletin 74, 1281-1286.

Hay, R.L., 1964. Phillipsite in lakes and soils. American Mineralogist 49, 1366-1387.

Hay, R.L., 1968. Chert and its sodium-silicate precursors in sodium-carbonate lakes of East Africa. Contributions to Mineralogy and Petrology 17, 255-274.

Hay, R.L., 1970. Silicate reactions in three lithofacies of a semi-arid basin, Olduvai Gorge, Tanzania. Mineralogical Society of America, Special Paper 3, 237-255.

Hay, R.L., 1978. Melilitite-carbonatite tuffs in the Laetolil Beds of Tanzania. Contributions to Mineralogy and Petrology 67, 357-367.

Hay, R.L., 1980. Zeolitic weathering of tuffs in Olduvai Gorge, Tanzania. In Rees, L.V.C. (ed.), Proceedings of the 5th International Conference on Zeolites. Heyden, London, pp. 155-163.

Hay, R.L. & Wiggins, B., 1980. Pellets, ooids, sepiolite and silica in three calcretes of the southwestern United States. Sedimentology 27, 559-576.

Hay R.L., Pexton R.E., Teague T.T. & Kyser T.K., 1986. Spring-related carbonate rocks, Mg clays and associated minerals in Pliocene deposits of the Amargosa Desert, Nevada and California. Geological Society of America Bulletin 97, 1488-1503.

Heidari, A., Mahmoodi, S. & Stoops, G., 2008. Palygorskite dominated Vertisols of southern Iran. In Kapur, S., Mermut, A. & Stoops, G. (eds.), New Trends in Soil Micromorphology. Springer-Verlag, Berlin, pp. 137-150.

Heystek, H. & Schmidt, E.R., 1953. The mineralogy of the attapulgite-montmorillonite deposit in the Springbok Flats, Transvaal. Transactions of the Geological Society of South Africa 56, 99-115.

Hillier, S. & Pharande, A.L., 2008. Contemporary pedogenic formation of palygorskite in irrigation-induced, saline-sodic, shrink-swell soils of Maharashtra, India. Clays and Clay Minerals 56, 531-548.

Hojati, S., Khademi, H. & Cano, A.F., 2010. Palygorskite formation under the influence of saline and alkaline groundwater in Central Iranian soils. Soil Science 175, 303-312.

Hojati, S., Khademi, H., Cano, A.F., Ayoubi, S. & Landi, A., 2013. Factors affecting the occurrence of palygorskite in central iranian soils developed on Tertiary sediments. Pedosphere 23, 359-371.

Howe, J. & Francis, J.E., 2005. Metamorphosed palaeosoils associated with Cretaceous fossil forests, Alexander Island, Antarctica. Journal of the Geological Society of London 162, 951-957.

Huggett, J.M., Gale, A.S. & Wray, D.S., 2005. Diagenetic clinoptilolite and opal-CT from the middle Eocene Wittering Formation, Isle of Wight, U.K. Journal of Sedimentary Research 75, 585-595.

Hughes, R.E. & Bohor, B.F., 1970. Random clay powders prepared by spray-drying. American Mineralogist 55, 1780-1786.

Icole, M. & Perinet, G., 1984. Les silicates sodiques et les milieux évaporitiques carbonatés bicarbonatés sodiques: une revue. Revue de Géographie Physique et de Géologie Dynamique 25, 167-176.

Icole, M., Durand, A., Perinet, G. & Lafont, R., 1983. Les silicates de sodium du Manga (Niger), marqueurs de paleoenvironnement? Palaeogeography, Palaeoclimatology, Palaeoecology 42, 273-284.

Inozemtsev, S.A. & Targulian, V.O., 2010. Upper Permian paleosols of the East European Platform: diagnostics of pedogenesis and paleogeographic reconstruction. Eurasian Soil Science 43, 127-140.

Jacob, J.S. & Allen, B.L., 1990. Persistence of a zeolite in tuffaceous soils of the Texas Trans-Pecos. Soil Science Society of America Journal 54, 549-554.

Jenkins, D.A., 1981. The transition from optical microscopic to scanning electron microscopic studies of soils. In Bisdom, E.B.A. (ed.), Submicroscopy of Soils and Weathered Rocks, Proceedings of the First Workshop of the International Working Group on Submicroscopy of Undisturbed Soil Materials. Pudoc, Wageningen, pp. 163-172.

Johan, Z. & Maglione, G.F., 1972. La kanemite, nouveau silicate de sodium hydraté de néoformation. Bulletin de la Societé Française de Minéralogie et de Cristallographie 95, 371-382.

Kadir, S. & Eren, M., 2008. The occurrence and genesis of clay minerals associated with Quaternary caliches in the Mersin area, southern Turkey. Clays and Clay Minerals 56, 244-258.

Kadir, S., Eren, M. & Atabey, E., 2010. Dolocretes and associated palygorskite occurrences in siliciclastic red mudstones of the Sariyer Formation (Middle Miocene), southeastern side of the Canakkale Strait, Turkey. Clays and Clay Minerals 58, 205-219.

Kaplan, M.Y., Eren, M., Kadir, S., Kapur, S. & Huggett, J., 2014. A microscopic approach to the pedogenic formation of palygorskite associated with Quaternary calcretes of the Adana area, southern Turkey. Turkish Journal of Earth Sciences 23, 559-574.

Kapur, S., Cavusgil, V.S. & FitzPatrick, E.A., 1987. Soil-calcrete (caliche) relationship on a Quaternary surface of the Cukurova region, Adana (Turkey). In Fedoroff, N., Bresson, L.M. & Courty, M.A. (eds.), Micromorphologie des Sols, Soil Micromorphology. AFES, Plaisir, pp. 597-603.

Kapur, S., Yaman, S., Gökçen, S.L. & Yetis, C., 1993. Soil stratigraphy and Quaternary caliche in the Misis area of the Adana basin, southern Turkey. Catena 20, 431-445.

Kautz, K. & Porada, H., 1976. Sepiolite formation in a pan of the Kalahari, South West Africa. Neues Jahrbuch für Mineralogie, Monashefte 1976, 545-559.

Khademi, H. & Mermut, A.R., 1998. Source of palygorskite in gypsiferous Aridisols and associated sediments from central Iran. Clay Minerals 33, 561-578.

Khademi, H. & Mermut, A.R., 1999. Submicroscopy and stable isotope geochemistry of carbonates and associated palygorskite in Iranian Aridisols. European Journal of Soil Science 50, 207-216.

Khormali, F. & Abtahi, A., 2003. Origin and distribution of clay minerals in calcareous arid and semi-arid soils of Fars Province, southern Iran. Clay Minerals 38, 511-527.

Khoury, H.N., Eberl, D.D. & Jones, B.F., 1982. Origin of magnesium clays from the Amargosa desert, Nevada. Clays and Clay Minerals 30, 327-336.

Kirkman, J.H., 1976. Clay mineralogy of Rotomahana sandy loam soils, North Island, New Zealand. New Zealand Journal of Geology and Geophysics 19, 35-41.

Le Gal, X., Crovisier, J.L., Gauthier-Lafaye, F., Honnorez, J. & Grambow, B., 1999. Altération météorique de verres volcaniques d'islande: changement du mécanisme à long terme. Comptes Rendus de l'Académie des Sciences 329, 175-181.

Long, D.G.F., McDonald, A.M., Yi Facheng, Li Houjei, Zheng Zili & Tian Xu, 1997. Palygorskite in palaeosoils from the Miocene Xiacaowan Formation of Jiangsu and Anhui Provinces, P.R. China. Sedimentary Geology 112, 281-295.

Mackenzie, R.C., Wilson, M.J. & Mashhady, A.S., 1984. Origin of palygorskite in some soils of the Arabian Peninisula. In Singer, A. & Galan, E. (eds.), Palygorskite – Sepiolite. Occurrences, Genesis and Uses. Developments in Sedimentology, Volume 37, Elsevier, Amsterdam, pp. 177-186.

Maglione, G., 1970. La magadiite, silicate sodique de néoformation des faciès évaporitiques du Kanem (littoral nord-est du Lac Tchad). Bulletin du Service de la Carte Géologique d'Alsace et de Lorraine 23, 177-189.

Maglione, G., 1976. Géochimie des Evaporites et Silicates Néoformés en Milieu Continental Confiné. Les Dépressions Interdunaires du Tchad. Travaux et Documents de l'ORSTOM n° 50. ORSTOM, Paris, 335 p.

Maglione, G. & Servant, M., 1973. Signification des silicates de sodium et des cherts néoformés dans les variations hydrologiques et climatiques holocènes du bassin tchadien. Comptes Rendus de l'Académie des Sciences 277, 1721-1724.

Maglione, G. & Tardy, Y., 1971. Néoformation pédogénétique d'une zéolite, la mordénite, associée aux carbonates de sodium dans une dépression interdunaire des bords du lac Tchad. Comptes Rendus de l'Académie des Sciences 272, 772-774.

Martin de Vidales, J.L., Jimenez Ballestra, R. & Guerra, A., 1987. Pedogenic significance of palygorskite in palaeosoils developed on terraces of the river Tajo (Spain). In Rodriguez Clemente, R. & Tardy, Y. (eds.), Geochemistry and Mineral Formation in the Earth Surface. CSIC, Madrid, pp. 535-548.

Martin Vivaldi, J.L. & Robertson, R.H., 1971. Palygorskite and sepiolite (the hormites). In Gard, J.A. (ed.), The Electron-Optical Investigation of Clays. Mineralogical Society, London, pp. 255-275.

Mashhady, A.S., Reda, M., Wilson, M.J. & Mackenzie, R.C., 1980. Clay and silt mineralogy of some soils from Qasim, Saudi Arabia. Journal of Soil Science 31, 101-115.

McAtee, J.L., House, R. & Eugster, H.P., 1968. Magadiite from Trinity County, California. American Mineralogist 53, 2061-2069.

McFadden, L.D., Wells, S.G. & Jercinovich, M.J., 1987. Influences of eolian and pedogenic processes on the origin and evolution of desert pavements. Geology 15, 504-508.

Mees, F., 1999. The unsuitability of calcite spherulites as indicators for subaerial exposure. Journal of Arid Environments 42, 149-154.

Mees, F., 2002. The nature of calcareous deposits along pan margins in eastern central Namibia. Earth Surface Processes and Landforms 27, 719-735.

Mees, F. & Van Ranst, E., 2011. Micromorphology of sepiolite occurrences in recent lacustrine deposits affected by soil development. Soil Research 49, 547-557.

Mees, F., Stoops, G., Van Ranst, E., Paepe, R. & Van Overloop, E., 2005. The nature of zeolite occurrences in deposits of the Olduvai basin, northern Tanzania. Clays and Clay Minerals 53, 659-673.

Mifsud, A., Huertas, F., Barahona, E., Linares, J. & Fornés, V., 1979. Test de couleur pour la sépiolite. Clay Minerals 14, 247-248.

Miller, C.R. & James, N.P., 2012. Autogenic microbial genesis of Middle Miocene palustrine ooids; Nullarbor Plain, Australia. Journal of Sedimentary Research 82, 633-647.

Millot, G., Nahon, D., Paquet, H., Ruellan, A. & Tardy, Y., 1977. L'épigénie calcaire des roches silicatées dans les encroûtements carbonatés en pays subaride Antiatlas, Maroc. Sciences Géologiques Bulletin 30, 129-152.

Ming, D.W. & Boettinger, J.L., 2001. Zeolites in soil environments. In Bish, D.L. & Ming, D.W. (eds.), Natural Zeolites: Occurrence, Properties, Applications. Reviews in Mineralogy & Geochemistry 47, Mineralogical Society of America, pp. 323-345.

Ming, D.W. & Dixon, J.B., 1986. Clinoptilolite in south Texas soil. Soil Science Society of America Journal 50, 1618-1622.

Ming, D.W. & Mumpton, F.A., 1989. Zeolites in soils. In Dixon, J.B. & Weed, S.B. (eds.), Minerals in Soil Environments. 2nd Edition. SSSA Book Series, No. 1. Soil Science Society of America, Madison, pp. 873-911.

Monger, H.C. & Daugherty, L.A., 1991. Neoformation of palygorskite in a southern New Mexico Aridisol. Soil Science Society of America Journal 55, 1646-1650.

Morrás, H.J.M., Robert, M. & Bocquier, G., 1982. Caractérisation minéralogique de certains sols salso-diques et planosoliques du 'Chaco Deprimido' (Argentine). Cahiers ORSTOM, Série Pédologie 19, 151-169.

Mumpton, F.A. & Ormsby, W.C., 1976. Morphology of zeolites in sedimentary rocks by scanning electron microscopy. Clays and Clay Minerals 24, 1-23.

Nahon, D., Paquet, H., Ruellan, A. & Millot, G., 1975. Encroûtements calcaires dans les altérations des marnes éocènes de la falaise de Thiès (Sénégal). Organisation morphologique et minéralogie. Sciences Géologiques Bulletin 28, 29-46.

Nørnberg, P., 1990. A potassium-rich zeolite in soil development on Danian chalk. Mineralogical Magazine 54, 91-94.

Nørnberg, P. & Dalsgaard, K., 1985. The origin of clay minerals in soils on Danian Chalk. In Konta, J. (ed.), Proceedings of the 5th Meeting of the European Clay Groups. Charles University, Prague, pp. 553-561.

Owen, R.A., Owen, R.B., Renaut, R.W., Scott, J.J., Jones, B. & Ashley, G.M., 2008. Mineralogy and origin of rhizoliths on the margins of saline, alkaline Lake Bogoria, Kenya Rift Valley. Sedimentary Geology 203, 143-163.

Owliaie, H.R., Abtahi, A. & Heck, R.J., 2006. Pedogenesis and clay mineralogical investigation of soils formed on gypsiferous and calcareous materials, on a transect, southwestern Iran. Geoderma 134, 62-81.

Paquet, H., 1983. Stability, instability and significance of attapulgite in the calcretes of mediterranean and tropical areas with marked dry season. Sciences Géologiques, Mémoires 72, 131-140.

Pérez-Rodríguez, J.L., Maqueda, C. & Morillo, E., 1990. Occurrence of palygorskite in soils of Ecija (Spain). Australian Journal of Soil Research 28, 117-128.

Pimentel, N.L.V., 2002. Pedogenic and early diagenetic processes in Palaeogene alluvial fan and lacustrine deposits from the Sado Basin (S Portugal). Sedimentary Geology 148, 123-138.

Ping, C.L., Shoji, S. & Ito, T., 1988. Properties and classification of three volcanic ash-derived pedons from Aleutian Islands and Alaska Peninsula, Alaska. Soil Science Society of America Journal 52, 455-462.

Porta, J. & Herrero, J., 1990. Micromorphology and genesis of soils enriched with gypsum. In Douglas, L. A. (ed.), Soil Micromorphology: A Basic and Applied Science. Developments in Soil Science, Volume 19. Elsevier, Amsterdam, pp. 321-339.

Pozo, M. & Casas, J., 1999. Origin of kerolite and associated Mg clays in palustrine-lacustrine environments. The Esquivias deposit (Neogene Madrid Basin, Spain). Clay Minerals 34, 395-418.

Pozo, M., Casas, J., Medina, J.A., Carretero, M.I. & Martín Rubí, J.A., 2006. Pedogenic origin of clinoptilolite and heulandite in magnesium-rich sedimentary deposits (Madrid Basin, Spain). In Bowman, R.S. & Delap, S.E. (eds.), Zeolite '06, Book of Abstracts, pp. 210-211.

Remy, R.R. & Ferrell, R.E., 1989. Distribution and origin of analcime in marginal lacustrine mudstones of the Green River Formation, south-central Uinta Basin, Utah. Clay and Clay Minerals 37, 419-432.

Renaut, R.W., 1993. Zeolitic diagenesis of late Quaternary fluviolacustrine sediments and associated calcrete formations in the Lake Bogoria Basin, Kenya Rift Valley. Sedimentology 40, 271-301.

Robins, C.R., Brock-Hon, A.L. & Buck, B.J., 2012. Conceptual mineral genesis models for calcic pendants and petrocalcic horizons, Nevada. Soil Science Society of America Journal 76, 1887-1903.

Robins, C.R., Buck, B.J., Spell, T.L., Soukup, D. & Steinberg, S., 2014. Testing the applicability of vacuum-encapsulated 40Ar/39Ar geochronology to pedogenic palygorskite and sepiolite. Quaternary Geochronology 20, 8-22.

Robins, C.R., Deurlington, A., Buck, B.J. & Brock-Hon, A.L., 2015. Micromorphology and formation of pedogenic ooids in calcic soils and petrocalcic horizons. Geoderma 251/252, 10-23.

Rodas, M., Luque, F.J., Mas, R. & Garzón, M.G., 1994. Calcretes, palycretes and silcretes in the palaeogene detrital sediments of the Duero and Tajo basins, Central Spain. Clay Minerals 29, 273-285.

Rooney, T.P., Jones, B.F. & Neal, J.T., 1969. Magadiite from Alkali Lake, Oregon. American Mineralogist 54, 1034-1043.

Sancho, C., Melendez, A., Signes, M. & Batisda, J., 1992. Chemical and mineralogical characteristics of Pleistocene caliche deposits from the central Ebro Basin, NE Spain. Clay Minerals 27, 293-308.

Sauer, D. & Burghardt, W., 2006. The occurrence and distribution of various forms of silica and zeolites in soils developed from wastes of iron production. Catena 65, 247-257.

Sauer, D. & Zöller, L., 2006. Mikromorphologie der Paläoböden der Profile Femés und Guatiza, Lanzarote. In Zöller, L. & von Suchodoletz, H. (eds.), Östliche Kanareninseln — Natur, Mensch, Umweltprobleme. Bayreuther Geographische Arbeiten 27, pp. 105-130.

Sauer, D., Stein, C., Glatzel, S., Kühn, J., Zarei, M. & Stahr, K., 2015. Duricrusts in soils of the Alentejo (southern Portugal) — types, distribution, genesis and time of their formation. Journal of Soils Sediments 15, 1437-1453.

Sebag, D., Verrecchia, E.P. & Durand, A., 1999. Biogeochemical cycle of silica in an apolyhaline interdunal Holocene lake (Chad, N'Guigmi region, Niger). Naturwissenschaften 86, 475-478.

Sebag, D., Verrecchia, E.P., Lee, S.J. & Durand, A., 2001. The natural hydrous sodium silicates from the northern bank of Lake Chad: occurrence, petrology and genesis. Sedimentary Geology 139, 15-31.

Schwaighofer, A., 1980. Pedogentischer Palygorskit im einem Lössprofil bei Stillfried an der March (Niederösterreich). Clay Minerals 15, 283-284.

Scott, J.J., Renaut, R.W. & Owen, R.B., 2008. Preservation and paleoenvironmental significance of a footprinted surface on the Sandai Plain, Lake Bogoria, Kenya Rift Valley. Ichnos 15, 208-231.

Shadfan, H. & Dixon, J.B., 1984. Occurrence of palygorskite in soils and rocks of the Jordan Valley. In Singer, A. & Galan, E. (eds.), Palygorskite — Sepiolite. Occurrences, Genesis and Uses. Developments in Sedimentology, Volume 37, Elsevier, Amsterdam, pp. 187-198.

Shadfan, H., Dixon, J.B. & Kippenberger, L.A., 1985. Palygorskite distribution in the Tertiary limestone and associated soil of northern Jordan. Soil Science 140, 206-212.

Sheppard, R.A. & Gude, A.J., 1969. Rhodesite from Trinity County, California. American Mineralogist 54, 251-255.

Silva, S.R., Spiers, V.M. & Gross, K.A., 1965. Soil Fabric Study with the Electron Microscope. A progress report. Australian Defence Scientific Service, Defence Standards Laboratories, Report 282, 13 p.

Singer, A., 1971. Clay minerals in the soils of the southern Golan Heights. Israel Journal of Earth-Sciences 20, 105-112.

Singer, A., 1989. Palygorskite and sepiolite group minerals. In Dixon, J.B. & Weed, S.B. (eds.), Minerals in Soil Environments. 2nd Edition. SSSA Book Series, No. 1, Soil Science Society of America, Madison, pp. 829-872.

Singer, A., 2002. Palygorskite and sepiolite. In Dixon, J.B. & Schulze, D.G. (eds), Soil Mineralogy with Environmental Applications. SSSA Book Series, No. 7, Soil Science Society of America Madison, pp. 555-583.

Singer, A. & Norrish, K., 1974. Pedogenic palygorskite occurrences in Australia. American Mineralogist 59, 508-517.

Singer, A., Kirsten, W. & Bühmann, C., 1995. Fibrous clay minerals in the soils of Namaqualand, South Africa: characteristics and formation. Geoderma 66, 43-70.

Smale, D., 1968. The occurrence of clinoptilolite in pan sediments in the Nata area, northern Botswana. Transactions of the Geological Society of South Africa 71, 147-152.

Soong, R., 1992. Palygorskite in northwest Nelson, South Island, New Zealand. New Zealand Journal of Geology and Geophysics 35, 325-330.

Spiers, G.A., Pawluk, S. & Dudas, M.J., 1984. Authigenic mineral formation by solodization. Canadian Journal of Soil Science 64, 515-532.

Stahr, K., Kühn, J., Trommler, J., Papenfuss, K.H., Zarei, M. & Singer, A., 2000. Palygorskite-cemented crusts (palycretes) in Southern Portugal. Australian Journal of Soil Research 38, 169-188.

Stoessell, R.K. & Hay, R.L., 1978. The geochemical origin of sepiolite and kerolite at Amboseli, Kenya. Contributions to Mineralogy and Petrology 65, 255-267.

Stoops, G., Van Ranst, E. & Verbeek, K., 2001. Pedology of soils within the spray zone of the Victoria Falls (Zimbabwe). Catena 46, 63-83.

Sudo, T., Shimoda, S., Yotsumoto, H. & Aita, S., 1981. Electron Micrographs of Clay Minerals. Developments in Sedimentology, Volume 31. Elsevier, Amsterdam, 203 p.

Taylor, K., Graham, R.C. & Ervin, J.O., 1990. Laumontite in soils of the San Gabriel Mountains, California. Soil Science Society of America Journal 54, 1483-1489.

Teruggi, M.E., 1964. A new and important occurrence of sedimentary analcime. Journal of Sedimentary Petrology 34, 761-767.

Trauth, N., 1977. Argiles Evaporitiques dans la Sédimentation Carbonatée Continentale et Epicontinentale Tertiaire. Bassins de Paris, de Mormoiron et de Salinelles (France), Jbel Ghassoul (Maroc). Sciences Géologiques, Mémoires 49, 195 p.

Vanden Heuvel, R.C., 1966. The occurrence of sepiolite and attapulgite in the calcareous zone of a soil near Las Cruces, New Mexico. In Bradley, W.F. & Bailey, S.W. (eds.), Proceedings of the 13th National Conference on Clays and Clay Minerals, Pergamon Press, Oxford, pp. 193-207.

Van Houten, F.B., 1962. Cyclic sedimentation and the origin of analcime-rich Upper Triassic Lockatong formation, west-central New Jersey and adjacent Pennsylvania. American Journal of Science 260, 561-576.

Van Houten, F.B., 1964. Cyclic lacustrine sedimentation, Upper Triassic Lockatong Formation, central New Jersey and adjacent Pennsylvania. In Merriam, D.F. (ed.), Symposium on Cyclic Sedimentation. Kansas Geological Survey Bulletin 169, pp. 497-531.

Verrecchia, E.P. & Le Coustumer, M.N., 1996. Occurrence and genesis of palygorskite and associated clay minerals in a Pleistocene calcrete complex, Sde Boqer, Negev Desert, Israel. Clay Minerals 31, 183-202.

Viani, B.E., Al-Mashhady, A.S. & Dixon, J.B., 1983. Mineralogy of Saudi Arabian soils: central alluvial basin. Soil Science Society of America Journal 47, 149-157.

Vogt, T. & Larqué, P., 1998. Transformations and neoformations of clay in the cryogenic environment: examples from Transbaikalia (Siberia) and Patagonia (Argentina). European Journal of Soil Science 49, 367-376.

Wang, Y., Nahon, D. & Merino, E., 1994. Dynamic model of the genesis of calcretes replacing silicate rocks in semi-arid regions. Geochimica et Cosmochimica Acta 58, 5131-5145.

Watts, N.L., 1980. Quaternary pedogenic calcretes from the Kalahari (southern Africa): mineralogy, genesis and diagenesis. Sedimentology 27, 661-686.

Weaver, C.E. & Beck, K.C., 1977. Miocene of the S.E. United States: a model for chemical sedimentation in a peri-marine environment. Sedimentary Geology 17, 1-234.

Weaver, C.E. & Pollard, L.D., 1973. The Chemistry of Clay Minerals. Developments in Sedimentology, Volume 15. Elsevier, Amsterdam, 213 p.

Xie, Q., Chen, T., Zhou, H., Xu, X., Xu, H., Ji, J., Lu, H. & Balsam, W., 2013. Mechanism of palygorskite formation in the Red Clay Formation on the Chinese Loess Plateau, northwest China. Geoderma 192, 39-49.

Yaalon, D.H., 1955. Clays and some non-carbonate minerals in limestones and associated soils of Israel. Bulletin of the Research Council of Israel 5B, 161-167.

Yaalon, D.H. & Wieder, M., 1976. Pedogenic palygorskite in some arid brown (Calciorthid) soils of Israel. Clay Minerals 11, 73-80.

Yakimenko, E., Inozemtsev, S. & Naugolnykh, S., 2004. Upper Permian paleosols (Salarevskian Formation) in the central part of the Russian Platform: paleoecology and paleoenvironment. Revista Mexicana de Ciencias Geológicas 21, 110-119.

Zarei, M., Stahr, K. & Jahn, R., 1987. Mikromorphologie der Verwitterung und Mineralneubilding aus jungen vulkanischen Aschen Lanzarotes. Mitteilungen der Deutschen Bodenkundlichen Gesellschaft 55, 1025-1030.

Zarei, M., Jahn, R. & Stahr, K., 1990. Zeolithbildung während der Basaltverwitterung in Lanzarote. Mitteilungen der Deutschen Bodenkundlichen Gesellschaft 62, 157-160.

Zelazny, L.W. & Calhoun, F.G., 1977. Palygorskite (attapulgite), sepiolite, talc, pyrophyllite, and zeolites. In Dixon, J.B. & Weed, S.B. (eds.), Minerals in Soil Environments. SSSA Book Series, No. 1, Soil Science Society of America, Madison, pp. 435-470.

Chapter 9

Calcium Carbonate Features

Nicolas Durand[1], H. Curtis Monger[2], Matthew G. Canti[3],
Eric P. Verrecchia[4]

[1]MINES PARIS-TECH, CENTRE DE GÉOSCIENCES, FONTAINEBLEAU, FRANCE;
[2]USDA-NATURAL RESOURCES CONSERVATION SERVICE, LINCOLN, NE, UNITED STATES;
[3]HISTORIC ENGLAND, EASTNEY, UNITED KINGDOM;
[4]UNIVERSITY OF LAUSANNE, LAUSANNE, SWITZERLAND

CHAPTER OUTLINE

1. Introduction 206
   1.1 Carbonate Mineralogy 207
   1.2 Microscopic Techniques 208
   1.3 Terminology 209
2. Groundmass 209
   2.1 Coarse Fraction 209
   2.2 Micromass 211
3. Pedofeatures 212
   3.1 Coatings 212
      3.1.1 General Aspects 212
      3.1.2 Pendants 212
      3.1.3 Laminar and Lamellar Crusts 215
   3.2 Hypo- and Quasicoatings 217
   3.3 Infillings 218
      3.3.1 General Aspects 218
      3.3.2 Needle-Fibre Calcite 218
      3.3.3 Coarse-Grained Cement 221
   3.4 Nodules 221
      3.4.1 General Aspects 221
      3.4.2 Pisoliths (Ooids) 224
   3.5 Root-Related Features 225
      3.5.1 Calcified Root Hairs 225
      3.5.2 Rhizoliths 226
      3.5.3 Microcodium 228

Interpretation of Micromorphological Features of Soils and Regoliths. https://doi.org/10.1016/B978-0-444-63522-8.00009-7

1. Introduction

The precipitation of calcium carbonate is common in soils and regoliths and occurs in a variety of climatic settings, ranging from arid to subhumid and from warm to cold (e.g., Courty et al., 1994). It produces a wide variety of forms, from isolated crystals and discrete elementary features to continuous carbonate-rich horizons. The accumulation of carbonate in soils and regoliths through time may lead to the formation of differentiated profiles with thick carbonate-enriched horizons, especially in arid to Mediterranean environments.

The horizon of prominent carbonate accumulation, in the form of fine-grained precipitates, has been named a 'K' horizon (Gile et al., 1965), which usually corresponds to the calcic (Bkk) or petrocalcic (Bkkm) horizons of Soil Taxonomy (Soil Survey Staff, 2014). Soils with calcic and petrocalcic horizons are grouped with the Calcisols in the World Reference Base (IUSS Working Group WRB, 2015). The term 'calcrete', widely used in the literature, has very different meanings depending on the chosen definition (see Wright, 2007; Alonso-Zarza & Wright, 2010). In its broader sense, calcrete designates near-surface, terrestrial accumulation of calcium carbonate, which occurs in a variety of forms, from powdery to highly indurated and from nodular to continuous. Aridisols, Vertisols, Mollisols and Alfisols (Soil Survey Staff, 1999) are the most typical soils containing indurated carbonate features (Wright & Tucker, 1991). In the classification of palaeosoils, calcretes are considered to be Aridisols (Retallack et al., 1993), Calcisols (Mack et al., 1993) or palaeo-Aridisols (Nettleton et al., 2000).

1.1 Carbonate Mineralogy

The precipitation and dissolution processes of calcium carbonate in soil are governed by the conventional equation:

$$Ca^{++} + 2HCO_3^- \leftrightarrow CaCO_3 + CO_2 + H_2O$$

However, this equation does not predict the mineralogy, morphology or crystal size of what will be formed. These parameters are closely linked to the presence of other ions and molecules in the environment and also depend on the biological activity involved in the precipitation of carbonate species.

The most common minerals of the ternary carbonate system $CaCO_3$-$MgCO_3$-$FeCO_3$, which may precipitate in soil or sediment, represent a complex series of solid solutions and ordering states. Calcite, aragonite and vaterite are the three polymorphs of $Ca(CO_3)$. Calcite (low-magnesium calcite) is the most abundant and the most thermodynamically stable mineral at surface conditions and forms the main part of the carbonate features discussed in this chapter. Aragonite is commonly found in shells of aquatic organisms, but it is rare in soil profiles because it gradually recrystallises into ordered calcite at earth-surface conditions. Aragonite has been observed in some calcrete profiles (Nahon et al., 1980; Watts, 1980; Milnes & Hutton, 1983) and as part of pendant coatings (Courty et al., 1994). Vaterite, which also mainly occurs in aquatic shells, has only rarely been identified in soils but is a common biogenic precipitate of soil microorganisms in liquid and solid growth media in laboratory studies (Lindemann et al., 2002; Braissant et al., 2003; Cacchio et al., 2003; Rodriguez-Navarro et al., 2007; Millo et al., 2012).

An important subgroup of the $CaCO_3$ minerals is the magnesian calcites. High magnesian calcite (HMC) contains between 11 and to 19 mol% $MgCO_3$, Mg ions being incorporated in a quite disordered way in the crystal lattice. Dolomite ($CaMg(CO_3)_2$) is a mineral whose mode of formation in surface environments is still poorly understood, since well-ordered stoichiometric dolomite has never been synthesised under earth-surface conditions, without biological and chemical mediation (Tribble et al., 1995; Sherman & Barak, 2000). In addition, only a few restricted sedimentary environments are considered as modern analogues for dolomite precipitation (see Last, 1990; Warren, 2000). In soil environments, high concentrations of dolomite and HMC have been observed in some calcretes and calcareous soils (Watts, 1980; Bui et al., 1990; El-Sayed et al., 1991; Milnes, 1992; Capo et al., 2000; Whipkey et al., 2002). Still, the vast majority of pedogenic carbonates consist of calcite, even those that formed in dolomitic deposits (Kraimer et al. 2005).

Siderite ($Fe(CO_3)$) occurs in hydromorphic soils (see Lindbo et al., 2010; Vepraskas et al., 2018), as xenotopic coatings on pore walls and as small pore infilling (Stoops, 1983). Iron can substitute for the cations in the crystal lattice of dolomite to give ankerite (up to 25 mol% $FeCO_3$). For sodium carbonates, the reader may refer to the chapter on salts and salts crusts (Mees & Tursina, 2018).

1.2   Microscopic Techniques

Carbonate minerals can easily be recognised in soil thin sections, as their birefringence is extreme resulting in upper-order creamy white interference colours. They are also characterised by high relief generally from strongly negative to strongly positive, depending upon their orientation. Details of the crystallographic and optical properties of the most common carbonate minerals are summarised in Table 1.

Most micromorphological analyses of thin sections are performed with a petrographic microscope. Because the optical properties of the various carbonate minerals are quite similar, additional techniques are needed for their identification; details about these techniques may be found in Stoops (2003).

In cathodoluminescence (CL) microscopy, calcite is either yellow, orange to red or dark blue to purple, depending mainly on the concentrations of activators (e.g., Mn^{2+}) and quenchers (e.g., Fe^{2+}). A red to purple colour is more typical of dolomite. Aragonite may have yellow-green colour. CL microscopy is useful to distinguish calcite of different origins and generations and to reveal patterns in features affected by recrystallisation (see Wright & Peeters, 1989; Budd et al., 2002; Bajnóczi et al., 2006; Khormali et al., 2006; Violette et al., 2010; Mintz et al., 2011; Michel et al., 2013).

Fluorescence microscopy can give results that are similar to those of CL microscopy, but it has only been rarely applied to carbonate studies (Dravis & Yurewicz, 1985; Dravis, 1991). Soil carbonate generally gives a bluish colour in ultraviolet-oblique incident light (UV-OIL) and a yellow colour in blue OIL. The original fabrics of carbonate nodules or lithic fragments obliterated by later recrystallisations may become visible by fluorescence microscopy. Fluorescence and CL studies proved to be useful in distinguishing between pedogenic and geogenic calcite (Khormali et al., 2006).

Staining is a common technique for the differentiation of carbonate minerals (Dickson, 1965). Typical products are alizarine-red for staining calcite (red colour), alizarine-red and

Table 1   Crystallographic and optical properties of the main carbonate minerals

Mineral	Formula	Crystal system	Refractive indices	Birefringence	Uniaxial/Biaxial
Ankerite	$CaFe(CO_3)_2$	Trigonal	n_e 1.536; n_o 1.741	0.205	U^-
Aragonite	$Ca(CO_3)$	Orthorhombic	n_x 1.530; n_y 1.680; n_z 1.685	0.155	B^-
Calcite	$Ca(CO_3)$	Trigonal	n_e 1.486; n_o 1.658	0.172	U^-
Dolomite	$CaMg(CO_3)_2$	Trigonal	n_e 1.501; n_o 1.680	0.179	U^-
Siderite	$Fe(CO_3)$	Trigonal	n_e 1.633; n_o 1.875	0.242	U^-
Vaterite	$Ca(CO_3)$	Hexagonal	n_e 1.550; n_o 1.650	0.100	U^+

Data from Fleischer et al. (1984).

potassium ferricyanide for Fe-bearing carbonates (purple to blue), Feigl solution for aragonite (black) and calcite titanium yellow for HMC and magnesite (yellow).

Selective extraction of components from thin sections is sometimes necessary to separate similar constituents and to make the masked b-fabric visible, for example a striated b-fabric masked by a calcitic crystallitic one. For the extraction procedure, weak HCl is used on uncovered thin sections (Wilding & Drees, 1990).

Moreover, submicroscopic techniques (SEM, TEM) are commonly used in the study of calcitic features, frequently in combination with thin section observations (James, 1972; Freytet, 1973; Freytet & Plaziat, 1982; Tucker & Wright, 1990; Kovda et al., 2003; Bajnóczi et al., 2006; Kuznetsova & Khokhlova, 2012).

1.3   Terminology

Geologists have long shown an interest in calcitic features in thin sections. A rich literature exists on diagenetic carbonate features in sedimentary petrology (e.g., Dunham, 1969a, 1969b; Walls et al., 1973; Harrison & Steinen, 1978; Adams, 1980; Freytet & Plaziat, 1982; Esteban & Klappa, 1983; James & Choquette, 1984; Harwood, 1988). A comprehensive overview of continental calcitic features was presented by Reeves (1976). In addition, a number of soil scientists who have studied microscopically discrete calcitic features in soils have proposed a classification of those features (Blokhuis et al., 1969; Sehgal & Stoops, 1972; Wieder & Yaalon, 1974; Bal, 1975a, 1975b; Ruellan et al., 1979; Verrecchia, 1987; Wright, 1990; Becze-Deák et al., 1997; Zamanian et al., 2016), including a classification developed for the discussion of carbon sequestration by pedogenic carbonates (Monger et al., 2015). Other accounts of the genesis and properties of petrocalcic horizon and calcrete profiles are given by Gile et al. (1966), Elloy and Thomas (1981), Braithwaite (1983), Goudie (1983), Esteban and Klappa (1983), Verrecchia and Freytet (1987), Wright and Tucker (1991) and Alonso-Zarza and Tanner (2010).

For general descriptive terms of pedofeatures, the reader is referred to Bullock et al. (1985) and Stoops (2003). Some of the terms and concepts used for pedogenic calcitic features are taken from sedimentary petrography. These terms include micrite (crystal size <4 µm in diameter, seen but not readily measured under the optical microscope), microsparite (5-20 µm) and sparite (>20 µm) (Folk, 1962; Chilingar et al., 1965). Isolated pedogenic crystals are considered as crystalline pedofeatures only when larger than 20 µm, while smaller crystals are part of the micromass.

2.   Groundmass

2.1   Coarse Fraction

Different groups of coarse particles can be distinguished in the groundmass: single mineral grains, compound mineral grains or rock fragments, inorganic residues of biological origin and anthropogenic materials.

Among mineral grains and rock fragments, limestone fragments, or calcite inherited from the parent material, should not be confused with pedogenic nodules or other types of carbonate pedofeatures. Different studies tried to use the morphology of calcite to discriminate between lithogenic (inherited from the parent rock) and pedogenic (formed in the soil itself) carbonates (e.g., Sehgal & Stoops, 1972; Tandon & Narayan, 1981; West et al., 1988; Kraimer & Monger, 2009). Sehgal and Stoops (1972) described sand-size inherited calcite crystals in soil profiles of the Indo-Gangetic alluvial plain, as the grains exhibit a rounded shape, probably linked to abrasion during transport, and a wavy extinction suggesting their origin from metamorphic rock.

Inorganic residues of biological origin include remains of bivalves, such as oyster (Ostreidae), mussel (e.g., *Mytilus edulis*) and clam (e.g., *Mercenaria mercenaria* and *Mya arenaria*). These organisms typically have substantial shells characterised by dense thicknesses of interwoven fibrous crystallites with high-order interference colours (Fig. 1A and B), which can be released as isolated needle-fibre crystals by weathering (Villagran & Poch, 2014). Gastropods also have thick shells in some cases, for example

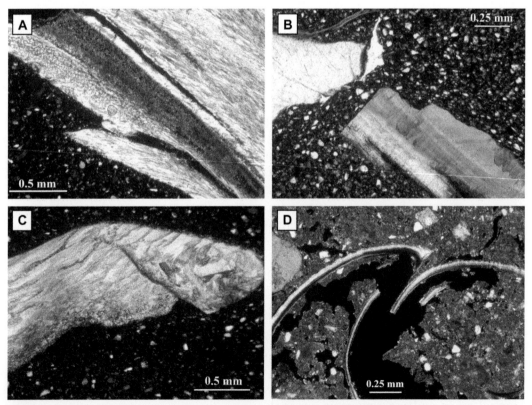

FIGURE 1 Mollusk shells (XPL). (A) Oyster. (B) Mussel. (C) Limpet. (D) Unidentified land snail. *A, B and C, modern material from UK beaches; D, material from archaeological layers at Solesbridge, Herts, UK.*

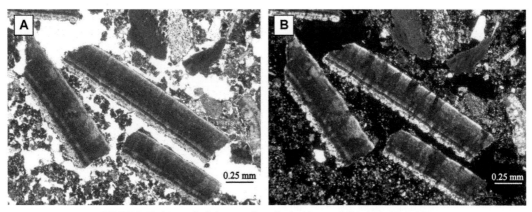

FIGURE 2 Birds' eggshell from archaeological layers. (A) In PPL. (B) In XPL.

cockle (*Cerastoderma edule*) and limpet (e.g., *Patella vulgata*, Fig. 1C), but this group also contains snails, producing thinner shells (Fig. 1D) and operculi, as well as arionid slugs, which produce granules (see Section 3.7.3).

Another group of calcareous coarse material of biological origin is represented by birds' eggshells. The shells consist of columnar calcite crystals (known as the 'palisade layer') arranged roughly at right angles to the eggshell surface (Keepax, 1981; Board, 1982). These form after an initial growth of aragonite crystals in a radial pattern producing a layer of crude 50-75 μm spherules known as mammillary knobs, which line the innermost part of the shell (Halstead, 1974). The palisade layer consists of crystals running the full thickness of the shell, mostly aligned perpendicular to the surface (Fig. 2). Consequently, in cross-polarised light, they all enter extinction within about 20° of the vertical position. Eggshell fragments can be rapidly spotted in thin sections by holding the whole slide against a dark background and viewing obliquely. The eggshell stands out as a dense white line.

2.2   Micromass

Calcareous soils generally have a high carbonate content of the micromass, occurring in the form of micritic or microsparitic crystals. The b-fabric is crystallitic (Bullock et al., 1985), due to the bright small punctuations caused by the optical anisotropy of the small crystals. This fabric has also been termed K-fabric (Gile et al., 1965), crystic plasmic fabric (Brewer, 1964; Sehgal & Stoops, 1972), calciasepic (Jongerius & Rutherford, 1979) and calcitic-crystallitic b-fabric (Stoops, 2003) (Fig. 3). When not inherited from sedimentary parent material, this fabric can develop in soils when fine calcite crystals gradually precipitate in the clayey micromass and when the pore space between soil framework grains is progressively filled by precipitated silt- and clay-size pedogenic carbonate crystals (Wieder & Yaalon, 1974; Monger et al., 1991b).

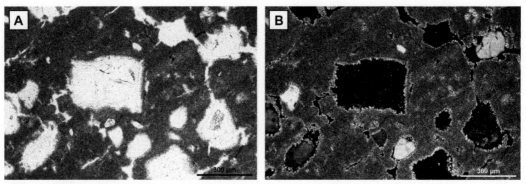

FIGURE 3 Calcite-rich groundmass with calcitic crystallitic b-fabric in calcareous soil (southern Tunisia). Note also calcitic coatings around quartz grains. (A) In PPL. (B) In XPL.

3. Pedofeatures

3.1 Coatings

3.1.1 General Aspects

Void coatings exhibit a wide variety of size, shape and basic distribution of calcite crystals (Fig. 4). Most frequently, carbonate coatings are composed of grey or brown micrite (sometimes laminated), microsparite and less frequently sparite. The crystals can be equant, elongated prismatic or needle-shaped (Verrecchia & Verrecchia, 1994; Kemp, 1995; Bajnóczi & Kovács-Kis, 2006; Khormali et al., 2006; Cailleau et al., 2009b).

Carbonate coatings appear to have two main origins: (i) unrelated to biological activity, formed by evaporation or mechanical translocation, the latter with admixtures of non-carbonate material (Siesser, 1973; Hay & Wiggins, 1980), and (ii) related to biogenic activity, e.g., coatings consisting of microbial tubules or certain needle-fibre calcite types (Knox, 1977; Calvet & Julia, 1983; Wright, 1986; see Section 3.5).

3.1.2 Pendants

At all latitudes, pendants are frequently observed under the face of gravelly stony coarse material. They consist of conical deposits that radiate downward from the bottom surfaces of pebbles into the soil (Fig. 5). Pendants are composed of alternating irregular laminae of pedogenic calcium carbonate, which can be identified at the microscopic level. In sedimentary geology literature, they are named 'vadose gravitational cements'.

The pendants range from limpid to dark brown depending on the content of impurities, such as organic matter or iron oxides. The surface of the carbonate coatings runs roughly parallel to the lower clast surface on which they form, but there is a wide variety of surface irregularities, the outermost layers often presenting a columnar morphology (Courty et al., 1994). Pendants range in thickness from about a millimetre to a centimetre. The thickness of most of the observed coatings is of the order of one to a few millimetres (Chadwick et al., 1989; Courty et al., 1994; Treadwell-Steitz & McFadden,

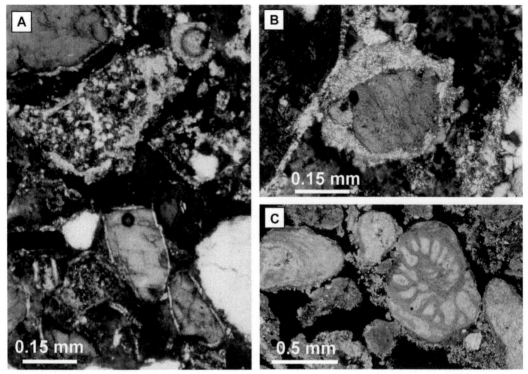

FIGURE 4 Carbonate coatings (XPL). (A) Chert fragment with carbonate coating (Btk horizon, Argic Petrocalcid). (B) Thick coating on quartz grain (Bk horizon, Typic Petroargid). (C) Foraminiferous limestone grains very thinly coated by pedogenic carbonate (Ck horizon, Typic Torriorthent formed in limestone alluvium). *All samples from Las Cruces, New Mexico, USA.*

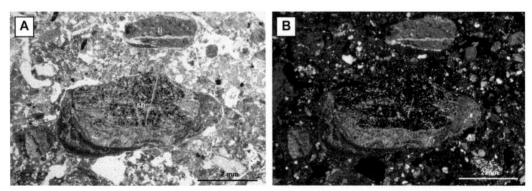

FIGURE 5 Pendants. (A) Calcite pendants below a shale pebble (sh) and a limestone fragment (li) in highly calcareous soils of Southern Iran (PPL). (B) The same as A in XPL. *Ghent University Archive; sample also studied by F. Khormali (see Khormali et al., 2006)*

2000; Khormali et al., 2006), but can be less than 0.5 mm for coatings in Late Pleistocene and Holocene desert soils (Wang et al., 1996) or as thick as 10-30 mm in older desert soils (Gile et al., 1966; Treadwell-Steitz & McFadden, 2000).

The calcitic pendants display a complex layering, which could result from sequential aggradation. Changes in crystal size, with alternating micritic and microsparitic laminae, have been recognised as the major factor causing the lamination (Chadwick et al., 1989), in addition to the common occurrence of dark rims that suggest organic and/or detrital enrichment. The lightest, limpid microlaminae of the coatings consist of relatively pure microsparitic calcite, loosely packed, with minor detrital content, whereas the dark ones are built up from micrite and include abundant organic and mineral impurities and traces of fungal hyphae (Courty et al., 1994). X-ray diffraction analyses have shown pendants to be composed mostly of calcite (Chadwick et al., 1989; Courty et al., 1994; Pustovoytov, 1998) even when they are formed on dolostone clasts (Kraimer et al., 2005). Chemical mapping has shown a substantial drop in magnesium concentration from dolostone to attached pendants. Chadwick et al. (1989) and Courty et al. (1994) also mention inclusions, such as quartz, feldspars, smectite, mica and kaolinite.

Ducloux et al. (1984) and Ducloux and Laouina (1989) showed that the genesis of the pendant is the result of an evolutionary mineralogical sequence involving different types of calcitic crystallisations: (i) randomly oriented, more or less complex, monocrystalline needles (1 μm width), (ii) heterogeneous polycrystalline needles of larger size (50 μm), and (iii) sparitic rhombohedra. This succession is accompanied by changes in chemical composition, in particular with magnesium rates. Ducloux and Laouina (1989) showed also that the genesis of the pendants seems to be the result of different processes according to the climatic setting, involving the same mineralogical sequence. Only the polarity of these crystallisations is different. In a semiarid climate, the monocrystalline needles first develop along the contact with the bearing pebbles, followed by successive crystallisations toward the outside, creating an evolutionary mineralogical sequence with a centrifugal differentiation (Ducloux & Laouina, 1989). In that case, pendants correspond to an absolute accumulation of carbonate. In temperate climate, the evolutional mineral sequence is inverted, the first crystallisations appearing in the lumen of the void and the subsequent crystallisations are formed closer to the pebble surface, representing centripetal differentiation (Ducloux et al., 1984). In this case, the pendant seems to be controlled by the progression of weathering during which, carbonate of the pebble is destabilised and not associated with a simple continuous carbonate accretion.

The fact that pendants do not always form like stalactites, in which the innermost laminae are the oldest and outermost laminae are the youngest, has been emphasised by Brock and Buck (2005). In some cases, void spaces at the clast-pendant contact, are sites of carbonate precipitation and are thus the location of the youngest laminae.

Depending on the crystal form and lamina morphology, it is possible to link the formation of the calcitic pendants to climatic changes and modifications of soil conditions through time (Dupuis et al., 1984; Chadwick et al., 1989; Courty et al., 1994). Light microlaminae, composed of pure calcium carbonate with well-formed and parallel-oriented crystals, seem to be indicative of drier periods that were less favourable for biological activity (hotter in warm climates and colder in cool climates). The dark microlaminae with high content of detrital particles, and composed of less well-developed and more randomly oriented calcite crystals with fungal hyphae, presumably mark wetter periods characterised by activation of soil biota and amelioration of climatic conditions. Therefore, the light laminae may be of physicochemical origin, and the dark laminae may have formed through mediation of organisms. Accretion of successive calcitic layers that compose the pendant represents a stratigraphic sequence useful for dating soils and landforms (Blank & Fosberg, 1990; Pustovoytov, 2003; Sharp et al., 2003) and reconstruct palaeoenvironmental conditions (Courty et al., 1994; Pustovoytov, 1998, 2002).

3.1.3 Laminar and Lamellar Crusts

Macroscopically, laminar and lamellar crusts are finely laminated dense horizons located either at the top of calcrete profiles (along the surface or beneath a thin soil cover), interbedded within sedimentary deposits, or at the top of any type of rock sub-strate, including igneous and metamorphic rocks (Fig. 6A to C). At the microscopic scale, two types of crust, with different origins, are distinguished – laminar and lamellar crusts (Freytet & Verrecchia, 1989; Freytet et al., 1997).

Laminar crusts are composed of a succession of micritic to microsparitic light-coloured bands alternating with darker brown laminae (Fig. 6D and E). The dark colour of the latter is due to the presence of Fe and Mn oxides, organic matter and/or clays. The total thickness of the crusts ranges from a few millimetres to a few tens of centimetres. Spherulites are a very common feature of this type of crust. They are fibro-radial low-Mg calcite aggregates, which are about 50 μm in diameter (Fig. 6F) (Verrecchia et al., 1995; Zhou & Chafetz, 2009). The presence of growth structures exhibiting a columnar habit (Verrecchia, 1994) is a characteristic feature of the crusts (Fig. 6E). Detrital silt-sized grains (20-40 μm), which are commonly well-sorted, are abundant (Huerta et al., 2015). It has been proposed that microorganisms, such as cyanobacteria (Vogt, 1984; Verrecchia et al., 1995), control the formation of the laminae, based on the presence of spherulites (Fig. 6E) and columnar growth structures (Rabenhorst et al., 1991; Verrecchia, 1994, 2016; Zhou & Chafetz, 2009). Spherulites have been grown experimentally and their formation seems to be related to cyanobacterial mats often requiring direct light exposure, meaning they have to form at the calcrete-atmosphere interface (Verrecchia et al., 1995). Bacteria (Braissant et al., 2003), fungi (Verrecchia & Verrecchia, 1994) and lichens (Klappa, 1979b) may also either contribute to or alter the original laminar features. However, Mees (1999) observed calcitic spherulites along subhorizontal cracks within Quaternary palustrine deposits, implying possible development in conditions with no or limited light. Several authors also suggested that the formation of laminar crusts corresponds to the last stages of horizon development when the hydraulic conductivity of the soil is sharply reduced by cemen-tation (Rabenhorst & Wilding, 1986; Brock & Buck, 2009).

Lamellar crusts have a less finely laminated fabric than laminar crusts, whereby the crust consists of layers that are a few millimetres to a few centimetres thick (Fig. 6G).

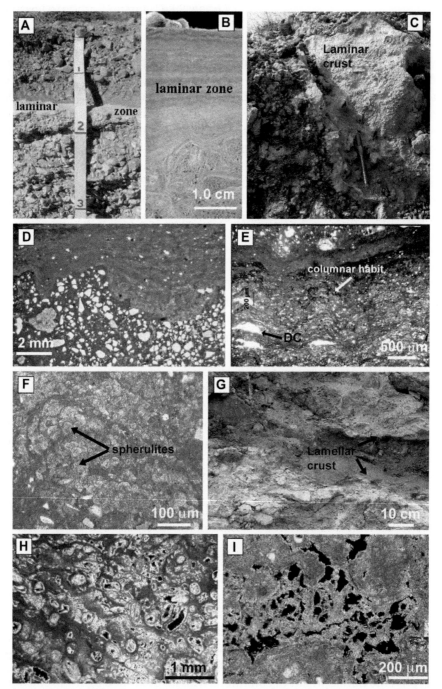

FIGURE 6 Laminar and lamellar crusts. (A) Laminar crust of a petrocalcic horizon in a loamy-skeletal Argic Petrocalcid; scale in feet. (B) Polished section of a laminar crust in the Bkkm horizon of a coarse-loamy Argic Petrocalcid. (C) Laminar crust (Karnataka, South India). (D) Thin section of a laminar crust (Las Cruces, New Mexico, USA) (PPL). (E) Laminar crust exhibiting columnar accretions, desiccation cracks (DC) and microsparitic layers trapping quartz, at the top of a calcrete profile developed on weathered gneiss (Tamil Nadu, South India) (PPL). (F) Close-up of D showing calcitic spherulites with microradial structure (XPL). (G) Lamellar crust that is part of a calcrete profile developed on weathered gneiss (Tamil Nadu, South India). (H) Spongelike fabric in a lamellar crust, with numerous pores related to a network of fine roots (Nari calcrete, Nazareth, Israel) (XPL). (I) Convoluted fabric, composed of needle-fibre calcite with associated micrite (Mediterranean carbonate soil, Israel) (XPL). *(D) Reprinted from Durand et al. (2006), with permission from Elsevier.*

Millimetric tubular pores and alveolar septal structures are common, forming a spongelike fabric (fenestral fabric; Wright et al., 1988; Fig. 6H) or convoluted fabric (Rabenhorst & Wilding, 1986; Verrecchia & Verrecchia, 1994; Bindschedler et al., 2012; Fig. 6I, see also Section 3.3.2). Crusts with spongelike fabric typically also show laminated micritic hypocoatings and sparitic or needle-fibre calcite infillings. The spongelike fabric is considered to represent densely interwoven rootlet horizons, which were calcified, possibly while the rootlets were alive, by micritic and microspar calcite (Wright et al., 1988; Mack & James, 1992; Wright, 1994; Alonso-Zarza, 1999; Alonso-Zarza & Silva, 2002). In some cases, thick lamellar crusts consist of centimetre-scale alternations of micrite laminae and others, rich in detrital sediments, ooids and clays or coated micritic grains (Sanz & Wright, 1994; Fedoroff et al., 1994).

3.2 Hypo- and Quasicoatings

Hypo- and quasicoatings, composed of micritic or microsparitic carbonates (Fig. 7), are formed from soil solutions percolating along the pores or fissures and penetrating into

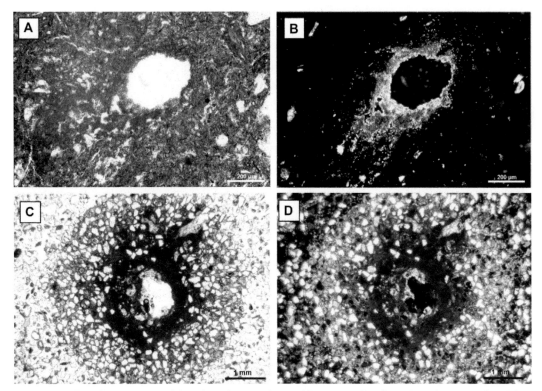

FIGURE 7 Calcite hypocoatings. (A) Calcite hypocoating, with an associated coating, in a sample with a non-calcareous groundmass (Ovikokorero pan, Namibia) (PPL). (B) The same as A in XPL. (C) Calcite hypocoating (Arenosol, Israel) (PPL). (D) The same as C in XPL. *(A) and (B) Images by F. Mees.*

the soil matrix (Sehgal & Stoops, 1972; Courty & Fedoroff, 1985; Kemp, 1995; Zamanian et al., 2016). They may also be considered as resulting from rapid precipitation of calcium carbonate due to root metabolism (Wieder & Yaalon, 1982; Zamanian et al., 2016). The degree of impregnation of the groundmass by calcite is variable. Where impregnation is weak, the fabric of the groundmass is relatively undisturbed and the distribution of the coarse fraction unchanged. Where impregnation is moderate to strong, the fabric of the host material is variously disturbed. Becze-Deák et al. (1997) consider that the amount of hypocoatings is not correlated with the texture of the material or with the total porosity. A quasicoating may form by partial dissolution of a hypocoating.

In present-day surface soils, calcium carbonate hypocoatings have been commonly described in arid and semiarid regions (Sehgal & Stoops, 1972; Courty, 1990; Monger et al., 1991b; Zamanian et al., 2016), in areas with fluctuating water tables (Sehgal & Stoops, 1972) and in soils on loess (Mermut et al., 1983; Kemp, 1995; Becze-Deák et al., 1997; Barta, 2011, 2014). In the latter, they are often described as pseudomycelium in the field and correspond to needle-fibre calcite in thin section (see Section 3.3.2).

3.3 Infillings

3.3.1 General Aspects

Carbonate infillings are frequent in soils mainly of xeric regimes. Both dense and loose infillings are abundant, occupying all types of voids, except packing voids by definition (Stoops, 2003), and exhibiting a wide variety in size, shape and basic distribution of the component crystals. The infilling of channels and planes, as well as packing voids, by calcite enlarges the original voids sometimes considerably (Monger & Daugherty, 1991b).

3.3.2 Needle-Fibre Calcite

Needle-fibre calcite refers to needles and fibres of low-magnesium calcite, which are a few micrometres wide and up to several hundred micrometres long. In older micromorphological literature, especially from Eastern Europe and Russia, it has often been designated as lublinite (e.g., Ulrich, 1938; Kowalinski et al., 1972; Bal, 1975a, 1975b). The needles fill the pores in which they occur in various proportions, creating a fine interlacing partial infilling (Fig. 8A). Different types of arrangements of these fibres have been described, such as random (e.g., James, 1972; Sehgal & Stoops, 1972; Knox, 1977; Phillips & Self, 1987), convoluted (Knox, 1977; Rabenhorst & Wilding, 1986; Jones & Ng, 1988; Verrecchia & Verrecchia, 1994) and tangential surface coatings (e.g., James, 1972; Calvet & Julia, 1983).

A specific optical characteristic of these needles is that the extinction angle and elongation of the needle do not correspond to that of calcite crystals whose longitudinal axis is parallel to the c-axis, representing the preferential growth axis of calcite (straight extinction, negative elongation). A relatively constant optical extinction angle of 40°-50° (Iwanoff, 1906), 32°-39° (Stoops, 1976), 45° (Vergès et al., 1982) and 30°-50° (Cailleau et al., 2009a, 2009b) is measured, with some exceptions (up to 60°; Richter et al., 2008).

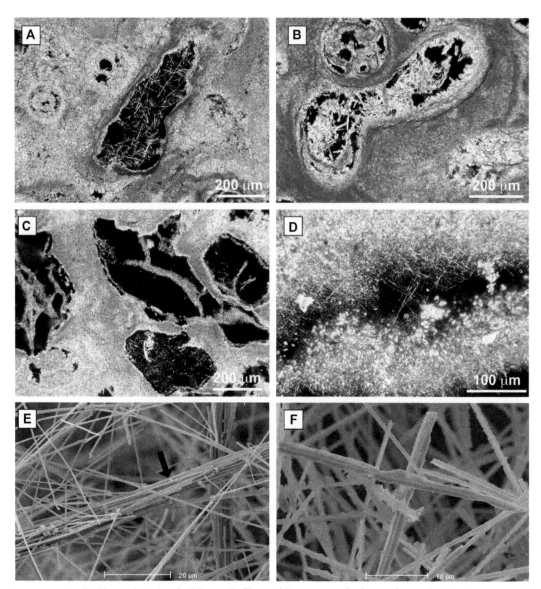

FIGURE 8 Needle-fibre calcite (NFC). (A) Needle-fibre calcite composed of smooth needles in a root channel (temperate Calcisol, Champagne, France) (XPL). (B) NFC composed of thick needles with syntactic/epitactic growths in a root channel (Bk2 horizon, Nari calcrete, Nazareth, Israel). (C) Convoluted/alveolar fabric in the pores with NFC crystals grouped in bundles forming an arcuate mesh of needles bridging pore walls (Calcisol, Champagne, France) (XPL). (D) Coatings of NFC layers on pore walls (Calcisol, Champagne, France) (XPL). (E) SEM image of smoothed NFC, including a bundle that probably resulted from biomineralisation in fungal hyphae (*arrow*) (Ferralsol, Cameroon). (F) SEM image of serrated-edged NFC, with syntactic and epitactic growths on smooth needles (Mediterranean Calcisol, France).

The use of scanning electron microscopy allowed the description of different sizes and morphological types, producing a better understanding of their nature and genesis (Fig. 8B to D). Three types of 'needles' are distinguished by their size: (i) microrods (<0.5 µm wide, <2 µm long); (ii) crystals (0.5-2 µm wide, up to 100 µm long), representing most commonly described type, and (iii) needles (2-20 µm wide, 30-1000 µm long).

A review on the morphology of needle-fibre calcite was compiled by Verrecchia and Verrecchia (1994) and updated by Cailleau et al. (2009b). In their classification, two groups are distinguished, monocrystalline rods and polycrystalline chains. The monocrystalline needles can be smooth single or paired rods (MA type) (Fig. 8E), serrated-edge paired rods (MB type) (Fig. 8F) or a transitional form from the smooth coupled rods to rods with serrated edges. The polycrystalline chains are composed of rhombohedra of calcite joined together in different ways; they correspond to the 'lenslike crystals' of Ducloux and Laouina (1989), the lublinite of Iwanoff (1906), the whisker crystals of Supko (1971), the 'calcite en échelon' of Stoops (1976) and the rhomb chains of Jones and Ng (1988). The microrods (M type) described by Verrecchia and Verrecchia (1994), which correspond to nanofibres as defined by Borsato et al. (2000) and Bindschedler et al. (2014), are not anymore considered to be a needle-fibre calcite type by Cailleau et al. (2009b).

Needle-fibre calcite occurs in a variety of vadose settings, in the voids of a large variety of soil types and palaeosoils, calcretes, calcarenites, Quaternary palustrine limestones and eolianites and it is present at all latitudes, from circumpolar to tropical areas (see e.g., Jones & Kahle, 1993; Becze-Deák et al., 1997; Cailleau et al., 2009b; Barta, 2011, 2014). Needle-fibre calcite can be found in soils of different ages. In old soils, needle-fibre calcite also exists but in channels that cross-cut older carbonate horizons. Such needle-fibre crystals can be used for stable isotope analysis and radiocarbon dating if the needles can be cleanly separated from the surrounding groundmass (Monger et al., 1998; Bajnóczi & Kovács-Kis, 2006; Millière et al., 2011a, 2011b).

The origin of needle-fibre calcite has been discussed for many years and is usually interpreted in two ways: (i) a pure physicochemical origin, and (ii) a direct or indirect biological origin. The monocrystalline forms are thought to originate from directly from biological processes, whereby elongated crystals are formed by fungal biomineralisation inside mycelia bundles and subsequently released upon decay of the organic matter of the fungal walls (Callot et al., 1985a, 1985b; Verrecchia & Verrecchia, 1994; Verrecchia & Dumont, 1996; Cailleau et al., 2009a, 2009b; Bindschedler et al., 2016). The polycrystalline chains seem to be crystals formed by physicochemical process, related to rapid evaporation and/or desiccation (Cailleau et al., 2009a, 2009b). This interpretation is based on the principle that under supersaturation, the development of polycrystalline calcite needles is possible such as rhomb chains (Jones & Ng, 1988), dendritic needle-shaped crystals and whisker crystals (James, 1972; Riche et al., 1982). Moreover, microrods (Verrecchia & Verrecchia, 1994), also known as nanofibres (Bindschedler et al., 2014, 2016), which are always associated with needle-fibre calcite, result from mineralisation of organic fibres during or after decay of fungal walls (Bindschedler et al., 2010, 2012, 2014, 2016).

Mineralogical and morphological transformations may affect needle-shaped calcite. Mineralised rods undergo diagenesis and their arrangements can evolve from a random mesh fabric to recrystallised micritic platelets, to microsparite (Fedoroff et al., 1994; Loisy et al., 1999). As mentioned above, some forms of needle-fibre calcite observed for soils can also be produced by physical weathering of shells (Villagran & Poch, 2014).

3.3.3   Coarse-Grained Cement

In contrast to the clay- and silt-sized crystals characteristic of pedogenic carbonate that formed in the vadose zone of soil, groundwater carbonates are characteristically sparitic (Fig. 9). The crystals engulf silicate grains, forming a coarse cement of interlocking crystals with a poikilotopic fabric. The micromorphology of the resulting deposit is characterised by phreatic cement (sparry, with palissadic or equant fabric) and the absence, in general, of obvious biogenic features. Profiles generally exhibit an increasing calcite crystal size and occurrence of euhedral crystals toward the profile base (Nash & Smith, 1998).

The relatively large size of groundwater carbonates suggests that they formed slowly from phreatic groundwater that continually supplied ions to growing crystals, in contrast with the smaller pedogenic carbonate crystals that formed rapidly from a drying vadose solution. Also, in contrast with most pedogenic carbonate horizons, the sedimentary structures such as bedding tend to be preserved when cemented by groundwater carbonate (Mack et al., 2000). It occurs in arid to semiarid climates, where evaporation and evapotranspiration rates are high. They may be associated with drainage channels (Nash & Smith, 1998), playas and lake deposits (Arakel, 1991).

3.4   Nodules

3.4.1   General Aspects

Carbonate nodules are common and occur in a wide variety of forms. Orthic nodules with gradual boundaries are generally formed *in situ* (Fig. 10A). In stable soils, they generally have a diffuse and irregular outline, whereas disorthic nodules indicate the effect of some pedoturbation (e.g., rotations; Wieder & Yaalon, 1974). Disorthic nodules are typical of Vertisols, shrinking and swelling of clay minerals leading to the displacement of the nodules (Driese et al., 2003; Mintz et al., 2011; Michel et al., 2013; see Kovda & Mermut, 2010, 2018). Certain nodules have a fabric that differs in composition from the soil in which they are incorporated and thus are allochthonous.

In general, micritic nodules are more common in fine- or medium-textured soils, while large sparite crystals predominate in coarse-textured soils (Wieder & Yaalon, 1974, 1982; Machette, 1985). Nodules formed by accumulations of microcrystalline calcite may exhibit a 'clotted' fabric, i.e., a flocculent structure which appears as coalescing irregular or rounded clusters (pellets) with a homogeneous micritic internal fabric (Tandon &

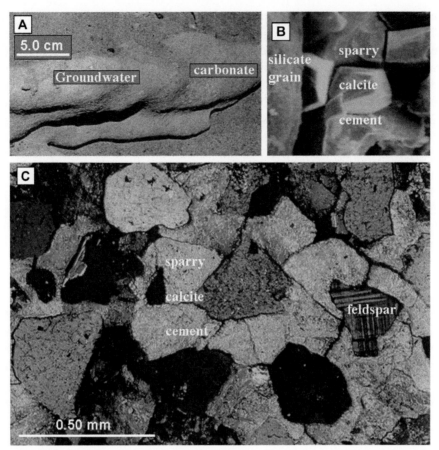

FIGURE 9 Groundwater carbonate. (A) Photograph of groundwater carbonate *in situ* (C2′ horizon, Argic Petrocalcid, New Mexico). (B) SEM image of sparry calcite cement bordering a detrital silicate grain (C horizon, Typic Petroargid, Las Cruces, New Mexico, USA). (C) Groundwater carbonates forming a cement composed of interlocking sparitic calcite crystals that engulf detrital grains, such as the feldspar grain (C2′ horizon, Argic Petrocalcid, New Mexico) (XPL). *(A) Image by L. Gile.*

Friend, 1989) or a 'mottled' fabric, i.e., patches of various colours or 'globular halos' (Wright & Tucker, 1991) reflecting the complex and highly irregular variability in crystal size and the amount of clay and Fe oxides present. According to the soil moisture regime, Mn oxide inclusions may occur (e.g., Blokhuis et al., 1969; Wieder & Yaalon, 1982). In some cases, especially in recrystallised nodules, quartz grains are corroded and sometimes surrounded by dissolution pores (see Section 5.1).

Many factors influence the final morphology of the nodules (Fig. 10B). Several episodes of precipitation and dissolution may affect the nodule and lead to a complex fabric (Courty et al., 1987; Durand et al., 2007) exhibiting cavities, circumgranular and shrinkage cracks (Fig. 10C and D) and recrystallised phases. Salt concentration, hydromorphism (Sehgal & Stoops, 1972), soil texture (Wieder & Yaalon, 1982), swell-shrink behaviour

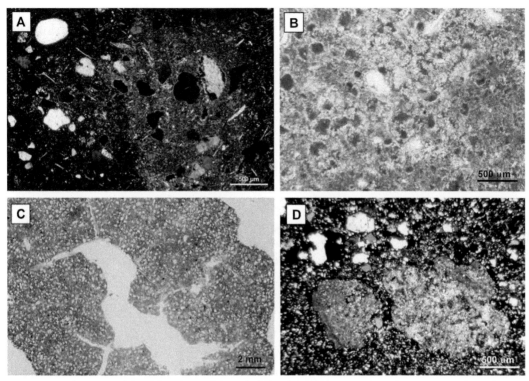

FIGURE 10 Carbonate nodules. (A) Orthic nodule with diffuse boundaries, in a non-calcareous groundmass (Owingi C pan, Namibia) (XPL). (B) Residuum of soil material in a carbonate nodule (Bca, Alfisol, Karnataka, South India) (PPL). (C) Septaric carbonate nodule, with quartz grains embedded in the carbonate groundmass (Vertisol, Shomeron, Israel) (PPL). (D) Orthic carbonate nodules in silty groundmass (Vertisol, Cameroon) (XPL). *(A) Image by F. Mees; (B) Reprinted from Durand et al. (2007), with permission from Elsevier.*

(Khadkikar et al., 2000), erosion (Miller et al., 2007) and conditions favouring rapid carbonate precipitation, such as frequent desiccation or CO_2 loss from the soil solution (Sobecki & Wilding, 1983), are also factors that affect the typology of the nodules.

The origins of pedogenic carbonate nodules are poorly understood (Zamanian et al., 2016). In early stages of calcite deposition, diffuse impregnative nodules of microcrystalline calcite are formed, by weak impregnation of the groundmass at specific sites. When calcium carbonate accumulation increases, larger nodules can form, mainly composed of microcrystalline calcite, but with a tendency to recrystallise to a larger grain size (Sehgal & Stoops, 1972; Wieder & Yaalon 1974; Courty & Fedoroff, 1985; Achyuthan & Rajaguru, 1997).

Wright and Tucker (1991) consider that the diffusion of carbonate to certain sites is a critical factor, followed by precipitation and displacive growth, as nodules frequently contain little groundmass material. Some nodules have been described as root calcification products, which stresses the role of plant respiration in the CO_3^{2-}-HCO_3^- equilibrium and the preferential flow along root channels in some soils (Zamanian et al., 2016).

3.4.2 Pisoliths (Ooids)

Pisoliths, also referred as ooids or vadoids in the literature (Peryt, 1983; Tucker & Wright, 1991; Alonso-Zarza & Wright, 2010), are variably concentric, accretionary structures 2 mm or larger, with or without a nucleus. They mainly consist of calcium carbonate but may also contain iron, manganese (Fig. 11), organic matter and clay minerals, such as palygorskite and sepiolite (Brock & Buck, 2005; Alonso-Zarza & Silva, 2002). When these layers are developed around a nucleus, they may be associated in different ways. According to Elloy and Thomas (1981), they may be concentric (the initial layers following all the unconformities of the nucleus), dissymmetric (the layers are absent or thin in one place and well developed on the opposite side) or polyphased (a series of layers resting unconformable on another set of layers; see also Nagtegaal, 1969; Siesser, 1973).

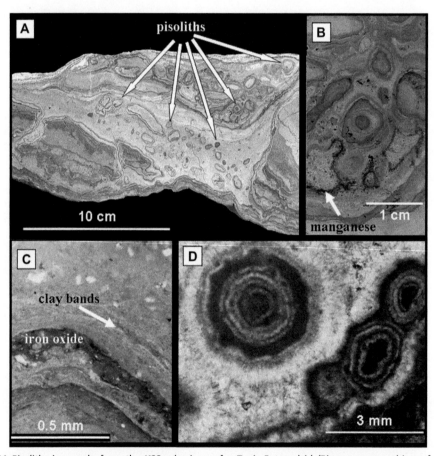

FIGURE 11 Pisoliths in sample from the K22m horizon of a Typic Petrocalcid (Rincon geomorphic surface, Dona Ana County, New Mexico, USA). (A) Polished section of highly developed Stage VI petrocalcic horizon (calcrete) containing pisoliths. (B) Magnified view of pisoliths in polished section. (C) Bands of iron oxide and clay in a laminar zone between pisoliths (XPL). (D) Pisoliths showing concentric rings of calcite (XPL). *(A) From Gile et al. (1966); reproduced with permission from the authors.*

Laminations in calcareous concretions are thinner over edges and corners of angular nuclei, thus increasing the sphericity of the particle. As yet, no satisfactory explanation has been given for this phenomenon, but the effect of surface tension due to opal gels has been suggested (Hay & Wiggins, 1980). Robins et al. (2015) propose a model of ooid formation that considers plastic behaviour of clay laminae and grain movement through calcite crystallisation pressure as factors. Features composed by aggregates and grains surrounded by calcite coatings that are thick compared with the diameter of the initial grain are more often classified as nucleic nodules.

Carbonate pisoliths are common in some marine limestones affected by subaerial diagenesis (vadose pisoliths of Dunham, 1969a; James 1972) and palustrine carbonates (Freytet & Verrecchia, 2002). They are also common in highly developed petrocalcic horizons or calcretes (stage V and VI of Machette, 1985), when the carbonated horizon is well indurated and subject to pedogenesis and alteration by chemical–mechanical weathering (Arakel, 1982).

In some cases, silicate grains are at the centre of pisoliths, suggesting that they served as nuclei around which carbonate layers precipitated. Concentric bands of clay in pisoliths may result from clay neoformation as Si, Al and other elements are released from silicate grains undergoing pressure solution (Monger & Daugherty, 1991a, 1991b). In many cases, however, there are no silicate grains in the centre of pisoliths, possibly because they were completely dissolved or because tangential sections through nucleic nodules often do not show the nucleus (Stoops, 2003).

The individualisation of the nucleus may be linked to the desiccation of preexisting carbonate deposits (Freytet & Plaziat, 1982) or to root activity (Alonso-Zarza et al., 1992a). Microorganisms associated with roots may control the formation of concentric layers (Calvet & Julia, 1983; Jones, 1987). Freytet and Plaziat (1965) have described pisolith occurrences linked to the activity of cyanobacteria in Upper Cretaceous and Eocene continental formation.

3.5   Root-Related Features

Roots play a key role in the dynamics of calcium carbonate in the terrestrial surface environments due to chemical exchanges with the soil (Verrecchia, 2011; Zamanian et al., 2016). Petrifaction or impregnation of root cells by calcite is also common and many carbonate features in soils and within calcretes are attributed to the former presence of plant roots and associated microorganisms. Carbonate is commonly found precipitated around living or decayed plant roots in various forms, as described below.

3.5.1   *Calcified Root Hairs*

Calcified root hairs are tubular, non-branching extensions of epidermal cells (Fig. 12). They are about 20 μm in diameter and are typically visible with a hand lens in the field. Some calcified rootlets can have euhedral calcite crystals that are larger than 50 μm, such as those precipitated on mesquite roots in the Chihuahuan Desert (Fig. 12B). Other

FIGURE 12 Calcified root hairs. (A) Photograph of soil containing calcified roots of a mesquite shrub (*Prosopis glandulosa*) in the C horizon of a Torripsamment (Tularosa Basin, New Mexico, USA). (B) SEM image of calcite crystals growing on the surface of a mesquite root. (C) Cross section of creosotebush (*Larrea tridentate*) containing calcified root hairs from the Ck horizon of a Torriorthent (Las Cruces, New Mexico, USA) (XPL). (D) SEM image of calcified root hairs of a creosotebush from the same horizon as C. *Images by H.C. Monger.*

shrub roots have much smaller crystal encrustations, such as those on creosote bush root hairs that sinuously surround soil particles (Fig. 12D).

These structures form by encrustation of root hair of the living plant by calcite crystals. A cast subsequently remains upon the decomposition of the organic tissue (Klappa, 1979a).

3.5.2 Rhizoliths

Rhizoliths are organosedimentary structures that have been described from many parts of the world and are produced by the activity and decay of plant roots (Klappa, 1980; Cohen, 1982; Mount & Cohen, 1984; Jones & Ng, 1988; Jones & Squair, 1989). They occur as straight to sinuous cylindrical structures formed of more or less well-cemented sediment, with lengths ranging from a few centimetres to several metres. They are also characterised by a circular cross-section, with diameters that range from 0.1 mm to about 20 cm (decreasing with depth), and by downward bifurcations.

The micromorphology and structure of rhizoliths depend on the position in which the calcification occurred in the rhizosphere, on the organisms involved and whether the plant was alive or dead when calcification occurred (Jaillard et al., 1991; Alonso-Zarza, 1999). The place of calcification and processes involved may correspond to (i) the cementation of soil material around the living roots, (ii) the filling of voids after the death of the roots, (iii) the calcification of root tissues, and (iv) a complex association of all these processes and the interaction of roots with other organisms, such as fungi.

Klappa (1980) identified five basic types of rhizolith: (i) root moulds, which are tubular voids that mark the positions of now-decayed roots; (ii) root casts, which are sediment-filled and/or cement-filled root moulds; (iii) root tubules, which are cemented cylinders around root moulds; (iv) rhizocretions *sensu stricto*, which are pedodiagenetic mineral accumulations that occur around plant roots; and (v) root petrifications, which are mineral impregnations or mineral replacements of organic matter that have partly or totally preserved the anatomical features of roots.

The rhizolith type that corresponds to the impregnation or replacement of root tissues is of particular interest, as the structure of the original cells of the root may be preserved. Several authors have described calcified root cells in sedimentary strata, palaeosoils or recent soils (e.g., Klappa, 1978; Ducloux & Butel, 1983; Jaillard, 1987; Jaillard et al., 1991; Becze-Deák et al., 1997; Alonso-Zarza et al., 1998; Alonso-Zarza, 1999; Verrecchia, 2011; Zamanian et al., 2016). Under the optical or electronic microscope, it is possible to distinguish between calcified parenchymatic cells (medulla or cortex), calcification of the whole cell or of the cell walls only, epidermal cells, calcitic vessel infillings or intercellular calcifications. The calcified cortical cells vary in size from 30 to 125 μm; they are more or less equigranular and are characterised by a subangular polyhedral shape, with smooth edges, sometimes elongated radially (cytomorphic calcite; see Herrero & Porta, 1987; Jaillard, 1987; Jaillard et al., 1991; Zamanian et al., 2016). They may be arranged in concentric layers around a central channel, the stele being rarely preserved.

A specific calcification-decalcification pedofeature associated with the impregnation of root tissues is frequently observed in Mediterranean and semiarid calcareous soils. It corresponds to channel infillings of coarse cytomorphic calcite, surrounded by a zone of non-calcareous material (Fig. 13). This combination has been described in detail by Porta and Herrero (1988), Herrero (1991) and Herrero et al., (1992), who proposed the following specific terminology. The complex feature as a whole is called a 'quera'. It consists of a channel (1-2 mm wide, <2 cm long), filled with equigranular sparitic crystals (the quesparite) (60-90 μm), surrounded by a calcite-depleted hypocoating and quasicoating (the quedecal) in a calcareous groundmass, with associated perpendicular microchannels (the quevoids) 35 μm wide and from 100 to 600 μm long. Bio- or pedoturbation can destroy the queras, incorporate the cytomorphic calcite in the groundmass and significantly increase the carbonate content in the soil (the cytomorphic sands of Jaillard, 1987).

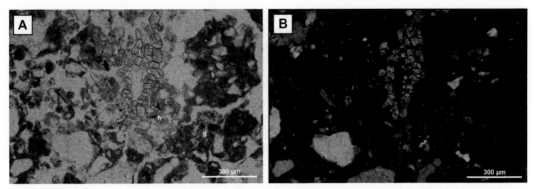

FIGURE 13 Quera composed of infilling with coarse phytomorphic calcite (i) and decarbonatation hypocoating (h), in a calcareous red soil (Anatolia, Turkey). (A) In PPL. (B) In XPL. Note calcitic crystallitic b-fabric of the groundmass (g). *Ghent University Archive; image by G. Stoops.*

3.5.3 Microcodium

Microcodium is a problematic calcitic feature of some calcretes and calcareous palaeosoils. It consists of a calcitic millimetric body composed of calcite prisms arranged around a central channel. Three types of *Microcodium* have been distinguished. The typical or classical *Microcodium* (type 1 and 2 of Bodergat, 1974; *Microcodium a* of Esteban, 1972, 1974) is composed of a single layer of radiating prismatic crystals in palisades around small nuclei or around an axis. This fabric shows a 'rosette' structure in transverse section and a 'corncob' structure in a longitudinal section. A second type is the lamellar monolaminar palissadic type, also having one layer of calcite prisms. Intermediate morphologies between this last type and the 'corncob'-type exist (Freytet & Plaziat, 1982). The third type corresponds to the *Microcodium* type III of Bodergat (1974) and the *Microcodium b* of Esteban (1972). It consists of several layers of isodiametric crystals arranged irregularly around a central void (for details about description and classification, see Klappa, 1978; Plaziat, 1984). *Microcodium* is differentiated from rhizoliths by its form and size. *Microcodium* typically has a 'rosette' structure in cross section and a 'corncob' structure in longitudinal section at the millimetre scale, which contrasts with rhizoliths that have various circular shapes in cross section and dendritic structure in longitudinal section at a millimetre to centimetre scale.

Microcodium has been described from various calcareous and hydromorphic palaeosoils, especially in alluvial cones, on flood plains, in lacustrine and paludal deposits (Freytet, 1972; Bodergat, 1974; Freytet & Plaziat, 1982), from the Cretaceous to the Pliocene (Klappa, 1978; Wright & Tucker, 1991; Wright et al., 1995).

The origin of *Microcodium* is still subject to controversies, including the possible relationship with calcified plant roots. Studies on recent roots (Jaillard, 1987; Jaillard & Callot, 1987; Jaillard et al., 1991; Morin, 1993) have described the intracellular calcification of living root cells, which closely resemble *Microcodium*. Examples from the sedimentary record (e.g., Alonso-Zarza et al., 1998) indicate that the *Microcodium* type composed of several layers of isodiametric calcite grains (*Microcodium b* of Esteban, 1972 and type III of Bodergat, 1974) seems to be formed by the calcification of root structures. The root origin of *Microcodium* has also been emphasised for the typical *Microcodium* type (Kosir, 2004). Other origins such as calcification of a mycorrhizae-cortical root cell association (Klappa, 1978) and the action of *Actinomycetes* (Freytet & Plaziat, 1982) have also been invoked. Kabanov et al. (2008) refer to a role of associated fungal substrate corrosion and bacterial carbonate precipitation.

3.6   Calcified Filaments

Calcified filaments are threadlike structures, often branching, that are encrusted with carbonate or oxalate crystals (Kahle, 1977; Klappa, 1979a; Vogt, 1984; Phillips et al., 1987; Verrecchia, 1990a, 1990b, 2011; Verrecchia et al., 1993). In thin section, they appear as thin fibres of calcite or oxalate that replicate the mycelial growth of fungal hyphae (Fig. 14A). With scanning electron microscopy, calcified filaments are revealed to be about 4 μm in diameter (the typical size of fungal hyphae) with hollow centres of 0.5-2 μm in diameter (Bruand & Duval, 1999). Their lengths are 100 μm or longer with multiple branches in three dimensions. SEM also reveals that the shape of individual carbonate crystals encrusting one hypha may be different from the shape of carbonate crystals encrusting a neighbouring hypha (Fig. 14B). Crystals are mainly equant calcite rhombs (1-2 μm in diameter), micron-sized plate-shaped crystals or needles (1-2 μm long, 0.1 μm wide), which may be perpendicular to the substrate (Klappa, 1979a; Phillips et al., 1987; Verrecchia et al., 1993; Bindschedler et al., 2016).

Calcified filaments may be associated with non-calcified filaments, characterised by a smooth, warty or hairy surface (Klappa, 1979a) or with partially calcified filaments. This

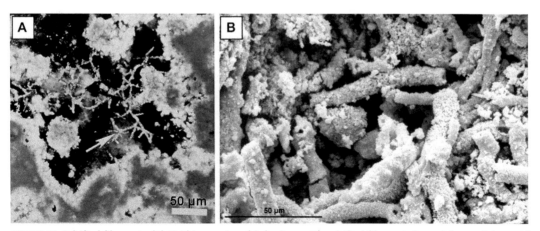

FIGURE 14 Calcified filaments. (A) Void in a petrocalcic horizon with calcified filaments (*arrow*) from a Petroargid (Las Cruces, New Mexico, USA) (XPL). (B) SEM image of calcified filaments in the Holocene loess from southern Pampa (Argentina). *Ghent University Archive; sample collected by M. Blanco.*

range reflects the various stages between primary filaments and the calcite- and/or oxalate-encrusted structures.

The filaments frequently occur on the walls of channels and voids where they may form dense mats, but they may also occur within the matrix where they are intimately associated with micrite. They have been recorded from Quaternary calcretes worldwide in different environments (e.g., Kahle, 1977; Klappa, 1979a; Scoppa & Pasos, 1981; Phillips et al., 1987; Jones & Ng, 1988; Verrecchia, 1990a; Monger et al., 1991b; Verrecchia et al., 1993; Monger & Adams, 1996; Loisy & Pascal, 1998; Bruand & Duval, 1999; Jiménez-Espinoza & Jiménez-Millán, 2003; Blanco & Stoops, 2007).

The origin of calcified filaments is biogenic, based on their size, shape and arrangement (Klappa, 1979a), as well as on laboratory and field experiments (Monger et al., 1991a; Bindschedler et al., 2014, 2016; Khormali et al., 2014). The alkaline mucilaginous sheath surrounding the hyphae provides an aqueous environment where Ca^{2+} combines with bicarbonate generated by respiration. With time, carbonate crystals are precipitated in the mucilaginous sheath and form a crust around the hyphae. When the fungus dies, a carbonate tube remains after the organic tissue decomposes (Klappa, 1979a; Phillips et al., 1987). Verrecchia (1990b) and Verrecchia et al. (1993) propose a chronological sequence of fungal filament mineralisation involving the precipitation and transformation of calcium oxalates. Fungi may play a key role in transformation of limestone or calcareous soil material by a redistribution of calcium carbonate (Verrecchia, 1990a; Bindschedler et al., 2016). Due to the activity of microorganisms (fungi and bacteria), a combination of solution and precipitation can gradually convert soft powdery limestone into a harder crust (Verrecchia & Dumont, 1996).

3.7 Biogenic Crystal Intergrowths

3.7.1 Earthworm Granules

Many species of earthworm produce calcium carbonate in the form of 0.1-2.5 mm large granules (Darwin, 1881). These granules are aggregates of individual calcite crystals arranged in a radial, parallel or random pattern depending partly on species (Fig. 15; Canti & Piearce, 2003).

A typical example consists of a roughly 0.5-2.5 mm spheroidal to ellipsoidal aggregate made up of 0.05-0.2 mm calcite crystals arranged approximately radially (Fig. 15A and B; see also Leiber & Maus, 1969; Weicek & Messenger, 1972; Preece et al., 1995; Becze-Deák et al., 1997). Amongst northern European earthworms, granules of this general morphology are produced by the surface feeding species such as *Lumbricus terrestris*, *Lumbricus rubellus* and *Lumbricus castaneus*. Soil-ingesting species (e.g., *Aporrectodea caliginosa* and *Aporrectodea icterica*) living predominantly in the A and upper B horizons, produce much smaller granules, often with irregular shapes (Canti & Piearce, 2003). In some species, the cementation of the granules is so weak that they easily fragment into individual crystals after entering the soil, and all species produce some single rhombohedral calcite crystals alongside their normal granule output.

FIGURE 15 Thin section and SEM photographs of earthworm granules showing radial, parallel and random crystal layout. (A) (B) of *Lumbricus terrestris* (archaeological layers at Solesbridge, Herts and at Kirton, Lincs, UK). (C) (D) of *Octolasion cyaneum*. (E) (F) of *Aporrectodea caliginosa*. *(C-F) Experimental cultures; from Canti & Piearce (2003), reproduced with permission from the authors.*

3.7.2 Fecal Spherulites

Calcium carbonate spherulites (Fig. 16) typically sized between 5 and 15 μm are produced in large numbers as a result of animal digestive processes (Brochier, 1983; Brochier et al., 1992; Canti, 1997, 1998a, 1999). The spherulites are made up of nanosized crystals (Fig. 16D) organised radially (Fig. 16B), producing an extinction cross (Fig. 16A) with negative elongation of the acicular crystals (Fig. 16C).

Although fecal spherulites are produced in the gut of many different animals, they are more abundant in herbivores such as cow, sheep, goat (Canti, 1999), llama and alpaca (Coil et al., 2003). The reasons for spherulite production are not known and many modern dung samples do not contain any. Among samples that do contain spherulites, there is a strong tendency for higher numbers to be produced by animals grazing plants from soils with pH higher than 6 (Canti, 1999). Preservation is partly dependent on pH, and soil or sediment matrix values above pH 7.5 offer the best conditions. Spherulites are valuable indicators of dung in studies of archaeological soils and sediments (Brochier, 1990, 1996; Coil et al., 2003).

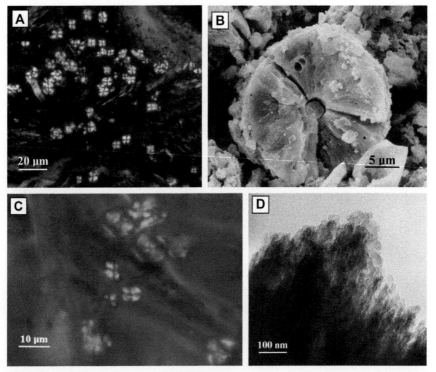

FIGURE 16 Fecal spherulites in modern sheep dung (Abergele, North Wales, UK). (A) Fecal spherulites showing the cross of extinction (XPL). (B) SEM image of a single specimen showing the radial structure. (C) Spherulites with gypsum (XPL, with retardation plate inserted) showing pseudouniaxial negative figure. (D) TEM image showing nanoscale crystallisation.

3.7.3   Remains of Mollusca

Operculi are mineralised seals produced by snails to block their shell openings for seasonal resting stages. They are roughly elliptical, but with one slightly recurved and pointed end. The structure is built up of concentric layers varying in thickness and starting from a central boss. Crystallisation consists of rather irregular growths on the inside of the operculum, but vertically arranged individual crystals of around 1-4 µm develop toward the outside (Bandel, 1990). A thin section shows mostly unstructured calcium carbonate, but some cuts near the edge yield subparallel layout of 50- to 150-µm-thick crystals rather like an eggshell (Fig. 17).

Arionid granules are single or multiple calcium carbonate crystals (Fig. 18) produced in the shield (beneath the skin) of slugs of the genus *Arion* and released only on death (Canti, 1998b). The single crystals are small, mostly elongated prismatic shapes with slightly rounded faces (Fig. 18D). In some cases, the crystals are twinned, cruciform or joined in threefold symmetry (Fig. 18D). Many have one or two outgrowths from the main prism which appear to develop into numerous lobate shapes (Fig. 18B). In crossed polarised light (Fig. 18A and C), *Arion* granules present very disparate morphologies. Constituent crystals range from 25 to 500 µm and appear to have no pattern of growth. Clusters of the smallest crystals are found next to the largest; small crystals appear to have been enveloped by larger ones; boundaries between crystals are sharp and frequently jagged due to intergrowth. Generally, granules smaller than 250 µm tend to be single crystals and the lobes on polycrystalline granules are clearly individual crystals as well (Canti, 1998b).

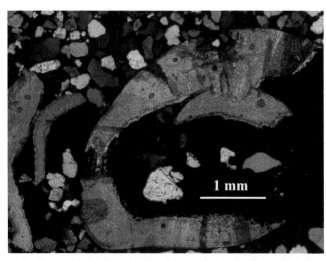

FIGURE 17 Snail operculi (set in sand, XPL) from archeological layers (Solesbridge, UK).

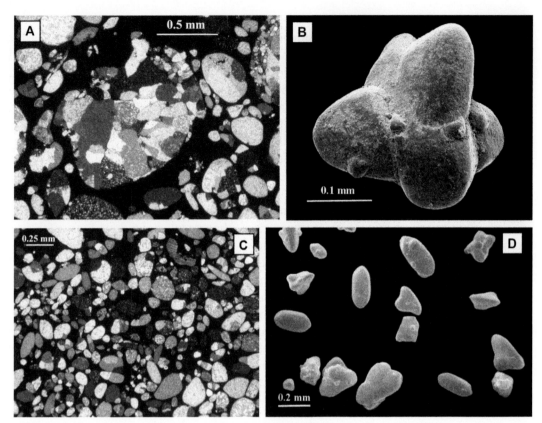

FIGURE 18 Polycrystalline (main objects in A and B) and monocrystalline (much of C and D) *Arion* granules from modern material (Thatcham, Berks, UK) viewed in thin section (XPL) and under SEM.

4. Recrystallisation and Dissolution

4.1 Recrystallisation

Soil carbonates, once precipitated, are very frequently subjected to recrystallisation, which increases in development with the age of the soil (Zamanian et al., 2016). It can be deduced from a range of microscopic characteristics (Fig. 19). For a complete review of recrystallisation features in sedimentary petrology, the reader is referred to other sources (e.g., Folk, 1962; Barthurst, 1975).

The distinctive features of carbonate recrystallisation in soils include the following:

- irregular distribution of crystal size, with gradual transition between zones of sparite crystals with distinct sizes that lack detectable boundaries and between micritic and sparry calcite domains (Ringrose et al., 2002). The crystals occur clustered in domains, which vary in a very irregular way over short distances with clusters of microsparite, pseudosparite and sparite;

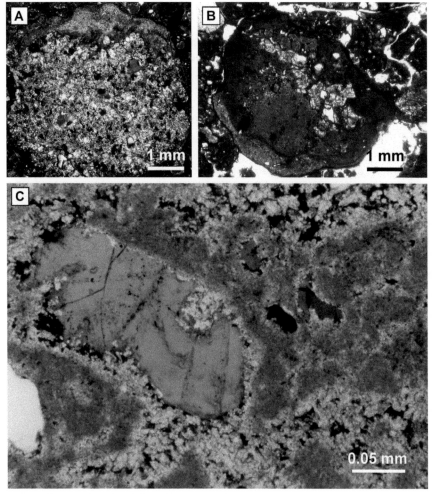

FIGURE 19 Recrystallisation features. (A) Carbonate nodule recrystallised (Alfisol, Karnataka, South India) (XPL). (B) Other example of the same feature (PPL). (C) Recrystallised K-fabric.

- microfabrics exhibiting 'mottled' or irregular crystal mosaics, where crystal size ranges from micrite to spar in patches with diffuse margins are thought by many authors to be the result of the replacement of finer crystals by coarse ones (Tandon & Narayan, 1981; Wieder & Yaalon, 1982; Tandon & Friend, 1989);
- starlike masses with a central zone of microsparite or pseudosparite, surrounded by sparite crystals. The sparite crystals are frequently elongated and radially arranged;
- sparitic crystals with a radial fibrous fabric, frequently with undulating extinction, forming coatings. Cytomorphic calcite may exhibit a fabric close to recrystallisation features, but the latter is characterised by irregular or curved intercrystalline

boundaries and very irregular crystal-size distribution (Tucker, 1991), whereas calcified root cells are characterised by the isodiametric morphology of crystals and specific arrangements (Jaillard et al., 1991; see Section 3.5);
- curved contacts between neighbouring sparitic carbonate crystals.

4.2 Dissolution

Because of their relatively high solubility, at least in comparison with silicate minerals, carbonates may dissolve from horizons in which they originally accumulated, as a reaction to changes in climate or local conditions toward a higher humidity or acidity. Some of the most common indications of dissolution and remobilisation of previously accumulated carbonates are reported below.

In some cases, the grains display well-rounded shapes, usually ellipsoidal, although spherical ones are not unknown, and show some large intergranular spaces. The most distinctive feature of this process is the presence of crystals with pronounced serration as a result of dissolution (e.g., James & Choquette, 1984).

The presence of mouldic voids (e.g., dissolved shells) is very diagnostic for dissolution processes. They are linked to dissolution in deep horizons where the lack of pedoturbation allows the preservation of crystal morphologies.

Another dissolution feature is the presence of clay-coating networks, without carbonate crystals (Alonso et al., 2004). They seem to have formed first by partial dissolution of carbonate grains forming an intergranular pore, followed by clay illuviation filling the pore, further dissolution of the carbonate grain and thus formation of the clay-coating networks.

Depletion hypocoatings can form by removal of part of the micromass in solution. Calcite depleted hypocoatings are frequently observed around roots in calcareous Mediterranean soils (Herrero & Porta, 1987) and in gypsic horizons (Barzanji & Stoops, 1974).

5. Associated features

5.1 Dissolution of Silicate Minerals

The replacement of silicate material by carbonate has been discussed by numerous authors (Reeves, 1970; Millot et al., 1977; Ruellan et al., 1979; Watts, 1980) and is referred to as epigenesis by, e.g., Nahon and Ruellan (1975) and Ruellan et al. (1979). This process is common in horizons with high calcite content.

Dissolution of silicates can be related to the well-expressed alkaline pH of these horizons, which causes a destabilisation of the quartz and other minerals, especially feldspars and micas. They are replaced by calcite due to the opposite trend in solubility of silica and carbonate as a function of pH. Very high pH is not required for the whole horizon, but it is sufficient that it is reached periodically and very locally. In a calcic horizon, the pH can be locally quite high and spatially variable (Callot et al., 1978). It requires a high concentration of carbonate and is favoured by the presence of coarse-grained carbonates that absorb less CO_2 on their surface. An intermediate phase would be the transformation of the primary silicates to palygorskite (Verrecchia & Le Coustumer, 1996), which in turn is transformed to calcite. Wang et al. (1994) suggested a dynamic model for calcrete genesis involving the pseudomorphic replacement of silicate parent material. A critical requirement is the presence of some amount of magnesium in the medium for the formation of transitory fibrous clay minerals and thus creation of unsaturated conditions with respect to SiO_2. The replacements are especially visible when illuvial clay coatings (exhibiting a strong orientation) are substituted (Alonso et al., 2004).

Floating grains (quartz, feldspars) are isolated grains embedded in a micritic matrix, surrounded by spar rims or by pore space. According to several authors (Nagtegaal, 1969; Reeves, 1976; Millot et al., 1977; Reheis, 1988; Paquet & Ruellan, 1993), the presence of feldspar and quartz grains floating in a carbonate matrix (Fig. 20D) is linked to the

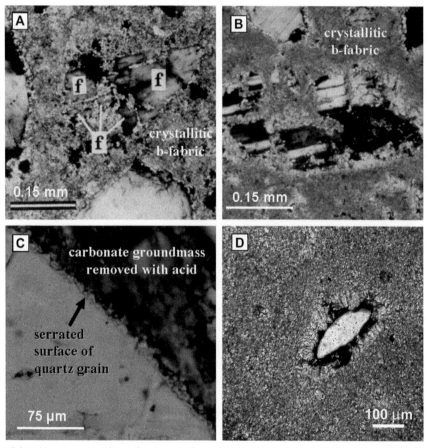

FIGURE 20 (A) Feldspar grain whose fragments (f) are in optical continuity (XPL). (B) Feldspar grain whose fragments have been moved out of optical alignment (XPL). (C) Serrated surface of quartz grain (XPL). (D) Quartz grains floating in a carbonate matrix and separated from the latter by pore (XPL). *(A-C) Sample from Bkkm-horizon, Petroargid, Las Cruces, USA; (D) Sample from massive calcrete, Tamil Nadu, South India.*

replacement of silicate minerals, the more or less large porous spaces between the grains and the matrix being attributed to the (congruent) dissolution of quartz and feldspar at very high pH values, in arid areas (see Section 4.1). However, some authors consider that the surrounding spar rims around the grains may have formed by the filling of a crack around the grain (circumgranular cracking) caused by dissolution or shrinkage of the micrite (Tandon & Friend, 1989).

Dissolution of silicates may also be related to a pressure effect. Displacement can occur by crystallisation pressure (Maliva & Siever, 1988), pushing preexisting grains and displacing them, when solutions carrying dissolved $CaCO_3$ penetrate along grain fractures or other planes of weakness. Displaced grain fragments often appear to retain grain outlines (Watts, 1978). Some indications of the process of displacement are (i) the presence of fragmented and separate mineral grains, showing important changes in position of the different fragments; (ii) the presence of expanded micaceous grains with separated and twisted flakes (this process has been called 'plasmasprengung' by Kerpen et al., 1960); and (iii) areas of original material compressed in zones of high density. When the K-fabric reaches an advanced stage and pedogenic carbonate crystals are densely crowded together, as in highly indurated petrocalcic horizons, silicate grains often become displaced and fragmented (Fig. 20A to C). This is accompanied by the formation of serrated edges on silicate grains (Fig. 20C). Because the shapes of the serrated edges match the shapes of impacting calcite crystals, a hypothesis has arisen that dissolution of silicate grains may result from pressure exerted by calcite precipitation: the force-of-crystallisation hypothesis (Maliva & Siever, 1988; Reheis, 1988). The ability of a growing crystal to exert a linear force on its surroundings is well documented in both laboratory and field studies (Wieder & Yaalon, 1974; Buczynski & Chafetz, 1987). Force of crystallisation can be strong enough to cause quartz grain breakages and separation by the growth of carbonate cement. It has been further hypothesised that as a consequence of force-of-crystallisation, the released silicon and aluminium reprecipitates as palygorskite and sepiolite clays (see Mees, 2010, 2018).

5.2   Calcium Oxalate Occurrences

The calcium oxalates whewellite ($CaC_2O_4 \cdot H_2O$) and weddellite ($CaC_2O_4 \cdot 2H_2O$) are organic minerals, occurring in various geological environments (e.g., Del Monte et al., 1987; Watchman, 1990; Verrecchia et al., 1993; Russ et al., 1996).

Whewellite crystals have an isodiametric or elongated morphology. The isodiametric crystals have the form of small oblong parallelepipeds, prisms and pseudorhombohedra (Frey-Wyssling, 1981) or tablets (Wadsten & Moberg, 1985; Jones & Wilson 1986). Weddellite crystals can be dipyramidal, elongated prismatic with pyramidal terminations or superimposed tablets (Verrecchia et al., 1993). Elongation of these crystals leads to needle shapes. The optical properties of the calcium oxalates are very similar to those of calcite.

5.2.1   *Lichens and Fungi*

Lichens are well known to produce calcium oxalates (Syers et al., 1967; Wadsten & Moberg, 1985) that can form crusts on rock surfaces (Del Monte et al., 1987; Russ et al., 1994; 1996). Russ et al. (1996) showed that lichens produced a calcium oxalate residue during dry periods on exposed limestone surfaces in open-air shelters (Lower Pecos region, southwest Texas) and that the oxalate may serve as an indicator of past xeric climate changes. In desert environments, fungi (Staley et al., 1982; Verrecchia, 2000) can also produce calcium oxalate when associated with various soils (Verrecchia, 1990a; Verrecchia et al., 1990b). In both cases, the oxalic acid can react with the available calcium in the environment or in tissues to form calcium oxalate salts (weddellite or whewellite). The importance of the precipitation of calcium oxalates by fungal filaments (Simkiss & Wilbur, 1989) in the redistribution of calcium in calcretes and limestone hardening by secondary calcite precipitation has been highlighted by Verrecchia (1990b) and Verrecchia and Dumont (1996). Those studies suggest that weddellite ($CaC_2O_4 \cdot 2H_2O$ or $CaC_2O_4 \cdot 3H_2O$) forms at the first stage of fungal filament mineralisation, followed by their evolution into more stable monohydrate crystals (whewellite, $CaC_2O_4 \cdot H_2O$) and then the transformation of calcium oxalate to calcium carbonate by bacteria, resulting in the formation of calcified filaments.

Recent experimental work on the oxalate-carbonate biogeochemical pathway has emphasised the important role of soil bacteria in the consumption of oxalate, resulting in carbonate precipitation (Verrecchia et al., 2006; Cailleau et al., 2011). Cailleau et al. (2005) showed that calcium carbonate accumulation can occur in orthox acidic equatorial soils during a dry period due to the presence of a large amount of oxalate in soil and of oxalotropic bacteria for oxalate oxidation into carbonate.

5.2.2   *Calcium Oxalate Phytoliths*

Most trees and herbaceous plants contain large amounts of 10-50 μm large calcium oxalate crystals either as whewellite or weddellite. These take the form of individual single or twinned prismatic crystals, druses (approximately radial aggregates) and raphides (needle-shaped crystals) according to species and location within the plant (Pobeguin, 1943; Arnott & Pautard, 1970; Franceschi & Horner, 1980; Horner & Wagner, 1995; Franceschi & Nakata, 2005).

Calcium oxalate can dominate the dry weight of some plants. Among herbaceous and tree species, quantities are usually smaller but still significant. Leaves of most common deciduous trees, for example, contain large amounts of prismatic oxalate crystals armouring the veins (Fig. 21), as well as druse crystals (Fig. 22) in the parenchyma.

Calcium oxalate crystals are found in the soil within plant litter and live plant tissue, but they are rapidly broken down by soil processes (Brochier, 1996). Pseudomorphs of calcium oxalate phytoliths formed of calcium carbonate are commonly found in anthropogenic soils as a result of burning plants (ash). When plants are burned, calcium oxalate loses its hydrocarbon portion and forms CaO, which recarbonates from

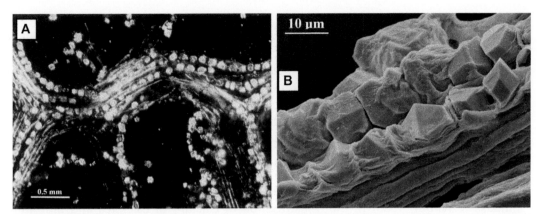

FIGURE 21 Prismatic calcium oxalate crystals in veins from leaves of oak (*Quercus robur*). (A) In XPL. (B) SEM image. *(B) Reprinted from Canti (2003), with permission from Elsevier.*

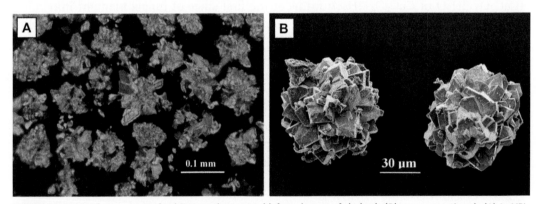

FIGURE 22 Druses (aggregates of calcium oxalate crystals) from leaves of rhubarb (*Rheum raponticum*). (A) In XPL. (B) SEM image. *(B) Reprinted from Canti (2003), with permission from Elsevier.*

atmospheric CO_2 to form $CaCO_3$. Despite the chemical change involved, this process does not cause the original calcium oxalate shapes to fragment, and perfect druse or prismatic crystal shapes (Fig. 23) are thus retained in the soil as pseudomorphs (Brochier, 1983, 2002; Canti, 2003; Cailleau et al., 2005).

6. Formations with Combinations of Calcareous Features

6.1 Palustrine Carbonates

Palustrine limestones result from episodic subaerial exposure and associated pedogenesis of lacustrine mud during the lake-level low-stands. Thus, palustrine carbonates are characterised by a range of sedimentary features of the primary lacustrine deposit and pedogenic features due to later transformations (Freytet, 1984). Reviews of these

FIGURE 23 Burnt druses from leaves of rhubarb (*Rheum raponticum*). (A) SEM image. (B) Close-up of (A) showing the surface texture caused by pseudomorphic replacement of oxalate by carbonate. (C) Calcium carbonate pseudomorphs of prismatic oxalate crystals (*arrow*) in a thin section of wood ash (PPL). (D) SEM image of a single burnt prismatic crystal from laurel (*Laurus nobilis*). *(A-C) Reprinted from Canti (2003), with permission from Elsevier.*

palustrine facies and microfabrics can be found in Freytet (1973), Freytet and Plaziat (1982), Freytet and Verrecchia (2002), Alonso-Zarza (2003) and Verrecchia (2007). Lacustrine limestones may be varved, laminated or homogeneous, with remains of charophytes, gastropods and ostracods and algae, sometimes bioturbated (burrows). Brecciated layers formed by mechanical reworking may be intercalated, as detrital beds (Freytet, 1984; Freytet & Verrecchia, 2002).

Pedogenic modifications are due to desiccation processes, root and soil organism activity, redistribution of carbonate and iron within the soil and sediments and physical and chemical phenomena such as dissolution and cementation and alternately phreatic and vadose environments. Typical palustrine features include mottling, channels, zones with a granular microstructure, desiccation cracks, and horizontal, planar and circumgranular cracks, brecciation and nodule development, rhizoliths, coated grains, dissolution cavities with complex vadose carbonate coatings and internal sediment infillings, laminar structure, *Microcodium* and early and late diagenetic crystals or crystal intergrowths (Freytet, 1973, 1984; Freytet & Plaziat, 1982; Alonso-Zarza et al., 1992b; Platt & Wright, 1992; Wright & Platt, 1995; Verrecchia, 2007).

6.2 Calcrete Profiles

The formation of thick calcrete profiles is the result of prolonged carbonate accumulation that goes through different stages of development. The concept of morphogenetic carbonate stages, whereby progressively older geomorphic surfaces have associated soils with progressively more developed carbonate stages, was introduced by Gile et al. (1966). Features of these soils include Stage I filaments and pebble coatings, Stage II nodules and interpebble fillings, Stage III calcic horizons with carbonate-plugged zones and Stage IV laminar layers overlying plugged horizons. Stage V calcretes, with cemented pisoliths and laminae, and Stage VI calcretes, with multiple generations of recemented breccia, pisoliths and laminae, were later introduced by Machette (1985). Recently, the features described by Gile et al. (1966) have been revised by Zamanian et al. (2016), who invoked a more important role of the groundwater during late stages of calcrete development. Itkin et al. (2016) consider the potential role of human activities in the formation of calcretes associated with anthropogenic soils and sediments.

The morphological classification system of Netterberg (1980) and Goudie (1983), modified by Wright and Tucker (1991), is also widely used. However, the formation of a calcrete profile is often more complex than a single sequence of stages for calcrete profile development. In some situations, one single stage may include several phases of erosion, soil formation and sedimentary processes. These interacting processes may be highlighted by the thin section analysis and the observation of elementary microfacies (Verrecchia, 1987, 2002; Verrecchia & Freytet, 1987). Indeed, calcrete profiles often include many of the carbonate features described in this chapter, with various degrees of development and in various combinations (Zhou & Chafetz, 2009; Pfeiffer et al., 2012; Kaplan et al., 2013; Khalaf & Al-Zamel, 2016).

Such massive terrestrial carbonates often contain as much carbon as Histosols, but as inorganic carbon rather than organic carbon (see Zamanian et al., 2016). Whether these massive carbonate horizons sequester atmospheric carbon dioxide depends on the source of calcium (Verrecchia et al., 2006; Cailleau et al., 2011), whose identification can be supported by thin section and SEM studies, for instance focusing on Ca-silicate weathering. If calcium is derived directly from silicates, carbonates sequester CO_2, in contrast to calcium derived from preexisting carbonates (Monger et al., 2015).

7. Conclusions

The precipitation and the accumulation of calcium carbonate in soils and regoliths are very complex phenomena. As they are linked to the interaction among the lithosphere, the biosphere and the atmosphere, pedogenic carbonate may be important proxies of palaeoenvironmental changes.

Even if controversies still exist on the genesis of carbonate features, the study of the morphological expression and hierarchical organisation of calcitic pedofeatures in thin sections allows climatic, geochemical and biological influences on the precipitation of carbonate in soils to be partially deciphered. With the contribution of submicroscopic techniques, progress has been made in the understanding of the relationships between the biological activity and the precipitation of carbonate, and many calcitic features seem to be linked to biological processes, in a direct or indirect manner.

For complex calcic and petrocalcic horizons, micromorphology also highlights the processes, which affect the primary accumulation of carbonate through time (Brasier, 2011). Carbonates may represent a highly active phase, undergoing intense transformation such as recrystallisation, dissolution and secondary precipitation. For these types of ancient carbonate horizons, future studies have to be performed to distinguish the active traits from inherited ones.

References

Achyuthan, H. & Rajaguru, S.N., 1997. Genesis of ferricretes and Calcic and petrocalcic of the Jayal gravel ridge: a micromorphological approach. In Wijayananda, N.P., Cooray P.G. & Mosley, P. (eds.), Geology in South Asia - II. Geological Survey and Mines Bureau, Sri Lanka, Professional Paper 7, Colombo, pp. 51-59.

Adams, A.E., 1980. Calcrete profiles in the Eyam limestone (Carboniferous) of Derbyshire: petrology and regional significance. Sedimentology 27, 651-660.

Alonso, P., Dorronsoro, C. & Egido, J.A., 2004. Carbonatation in palaeosols formed on terraces of the Tormes river basin (Salamanca, Spain). Geoderma 118, 261-276.

Alonso-Zarza, A.M., 1999. Initial stages of laminar calcrete formation by roots: examples from the Neogene of central Spain. Sedimentary Geology 126, 177-191.

Alonso-Zarza, A.M., 2003. Palaeoenvironmental significance of palustrine carbonates and calcretes in the geological record. Earth-Science Reviews 60, 261-298.

Alonso-Zarza, A.M. & Silva, P.G., 2002. Quaternary laminar calcretes with bee nests: evidences of small-scale climatic fluctuations, Eastern Canary Islands, Spain. Palaeogeography, Palaeoclimatology, Paleoecology 178, 119-135.

Alonso-Zarza, A.M. & Tanner, L.H. (eds.), 2010. Carbonates in Continental Settings: Facies, Environments, and Processes. Developments in Sedimentology, Volume 61. Elsevier, Amsterdam, 400 p.

Alonso-Zarza, A.M. & Wright, V.P., 2010. Calcretes. In Alonso-Zarza, A.M. & Tanner, L.H. (eds.), Carbonates in Continental Settings: Facies, Environments, and Processes. Developments in Sedimentology, Volume 61. Elsevier, Amsterdam., pp. 225-268.

Alonso-Zarza, A.M., Calvo, J.P. & García del Dura, M.A., 1992a. Palustrine sedimentation and associated features − grainification and pseudo-microkarst − in the Middle Miocene (Intermediate Unit) of the Madrid Basin, Spain. Sedimentary Geology 76, 43-61.

Alonso-Zarza, A.M., Wright, V.P., Calvo, J.P. & García del Dura, M.A., 1992b. Soil-landscape and climatic relationship in the middle Miocene of the Madrid Basin. Sedimentology 39, 17-35.

Alonso-Zarza, A.M., Sanz, M.E., Calvo, J.P. & Estévez, P., 1998. Calcified root cells in Miocene pedogenic carbonates of the Madrid Basin: evidences for the origin of *Microcodium* b. Sedimentary Geology 116, 81-97.

Arakel, A.V., 1982. Genesis of calcrete in Quaternary soil profiles, Hutt and Leeman Lagoons, Western Australia. Journal of Sedimentary Petrology 52, 109-125.

Arakel, A.V., 1991. Evolution of Quaternary duricrusts in Karinga Creek drainage system, central Australia groundwater discharge zone. Australian Journal of Earth Sciences 38, 333-347.

Arnott, H.J. & Pautard, F.G.E., 1970. Calcification in plants. In Schraer, H. (ed.), Biological Calcification: Cellular and Molecular Aspects. Appleton Century Crofts, New York, pp. 375-446.

Bajnóczi, B. & Kovács-Kis, V., 2006. Origin of pedogenic needle-fiber calcite revealed by micromorphology and stable isotope composition - a case study of a Quaternary paleosol from Hungary. Chemie der Erde 66, 203-212.

Bajnóczi, B., Horváth, Z., Demény, A. & Mindszenty, A., 2006. Stable isotope geochemistry of calcrete nodules and septarian concretions in a Quaternary 'red clay' paleovertisol from Hungary. Isotopes in Environmental and Health Studies 42, 335-350.

Bal, L., 1975a. Carbonate in soil: a theoretical consideration on, and proposal for its fabric analysis. 1. Crystic, calcic and fibrous plasmic fabric. Netherlands Journal of Agricultural Science 23, 18-35.

Bal, L., 1975b. Carbonate in soil: a theoretical consideration on, and proposal for its fabric analysis. 2. Crystal tubes, intercalary crystals, K fabric. Netherlands Journal of Agricultural Science 23, 163-176.

Bandel, K., 1990. Shell structure of the gastropoda excluding archaeogastropoda. In Carter, J.G. (ed.), Skeletal Biomineralisation: Patterns, Processes and Evolutionary Trends. Van Nostrand Rheinhold, New York, pp. 117-134.

Barta, G., 2011. Secondary carbonates in loess-paleosoil sequences: a general review. Central European Journal of Geosciences 3, 129-146.

Barta, G., 2014. Paleoenvironmental reconstruction based on the morphology and distribution of secondary carbonates of the loess-paleosol sequence at Sutto, Hungary. Quaternary International 319, 64-75.

Barthurst, R.G.C., 1975. Carbonate Sediments and their Diagenesis. 2nd Edition, Developments in Sedimentology, Volume 12. Elsevier, Amsterdam, 658 p.

Barzanji, A.F. & Stoops, G., 1974. Fabric and mineralogy of gypsum accumulations in some soils of Iraq. Transactions of the 10th International Congress of Soil Science, Volume VII, Moscow, pp. 271-277.

Becze-Deák, J., Langohr, R. & Verrecchia, E.P., 1997. Small scale secondary $CaCO_3$ accumulations in selected sections of the European loess belt. Morphological forms and potential for paleoenvironmental reconstruction. Geoderma 76, 221-252.

Bindschedler, S., Millière, L., Cailleau, G., Job, D. & Verrecchia, E.P., 2010. Calcitic nanofibres in soils and caves: a putative fungal contribution to carbonato-genesis. Geological Society of London, Special Publications 336, 225-238.

Bindschedler, S., Millière, L., Cailleau, G., Job, D. & Verrecchia, E.P., 2012. An ultrastructural approach to analogies between fungal structures and needle fibre calcite. Geomicrobiology Journal 29, 301-313.

Bindschedler, S., Cailleau, G., Braissant, O., Millière, L., Job, D. & Verrecchia, E.P., 2014. Unravelling the enigmatic origin of calcitic nanofibres in soils and caves: purely physicochemical or biogenic processes? Biogeosciences 11, 2809-2825.

Bindschedler, S., Cailleau, G. & Verrecchia, E.P., 2016. Role of fungi in the biomineralization of calcite. Minerals 6, 1-19.

Blanco, M. & Stoops, G., 2007. Genesis of pedons with discontinuous argillic horizons in the Holocene loess mantle of the southern Pampean landscape, Argentina. Journal of South American Earth Sciences 23, 30-45.

Blank, R.R. & Fosberg, M.A., 1990. Micromorphology and classification of secondary calcium carbonate accumulations that surround or occur on the undersides of coarse fragments in Idaho (USA). In Douglas, L.A. (ed.), Soil Micromorphology: A Basic and Applied Science. Developments in Soil Science, Volume 19. Elsevier, Amsterdam, pp. 341-346.

Blokhuis, W.A, Pape, T. & Slager, S., 1969. Morphology and distribution of pedogenic carbonate in some Vertisols of the Sudan. Geoderma 2, 173-200.

Board, R.G., 1982. Properties of avian egg shells and their adaptive value. Biological Reviews 57, 1-28.

Bodergat, A.M., 1974. Les Microcodiums: milieux et modes de developpement. Documents du Laboratoire de Géologie de la Faculté des Sciences de Lyon 62, pp. 137-235.

Borsato, A., Frisia, S., Jones, B. & Van Der Borg, K., 2000. Calcite moonmilk: crystal morphology and environment of formation in caves in the Italian Alps. Journal of Sedimentary Research 70, 1171-1182.

Braissant, O., Cailleau, G., Dupraz, C. & Verrecchia, E.P., 2003. Bacterially induced mineralization of calcium carbonate in terrestrial environments: the role of exopolysaccharides and amino acids. Journal of Sedimentary Research 73, 485-490.

Braithwaite, C.J.R., 1983. Calcrete and other soils in Quaternary limestones: structures, processes and applications. Journal of the Geological Society of London B 272, 1-32.

Brasier, A.T., 2011. Searching for travertines, calcretes and speleothems in deep time: processes, appearances, predictions and the impact of plants. Earth-Science Reviews 104, 213-239.

Brewer, R., 1964. Fabric and Mineral Analysis of Soils. John Wiley and Sons, New York, 470 p.

Brochier, J.E., 1983. Bergeries et feux de bois néolithiques dans le Midi de la France. Caractérisation et incidence sur le raisonnement sédimentologique. Quartär 33/34, 181-193.

Brochier, J.E., 1990. Des techniques géo-archéologiques au service de l'étude des paysages et de leur exploitation. Archéologie et Espaces: Xe Rencontres Internationales d'Archaeologie et d'Histoire d'Antibes. Editions APDCA, Juan-les-Pins, pp. 453-472.

Brochier, J.E., 1996. Feuilles ou fumiers? Observations sur le rôle des poussières sphérolitiques dans l'interprétation des dépôts archéologiques holocènes. Anthropozoologica 24, 19-30.

Brochier, J.E., 2002. Les sediments anthropiques. Méthodes d'étude et perspectives. In Miskovsky, J.C. (ed.), Géologie de la Préhistoire: Méthodes, Techniques, Applications. Géopré Editions, Paris, pp. 453-477.

Brochier, J.E., Villa, P. & Giacomarra, M., 1992. Shepherds and sediments: geo-ethnoarchaeology of pastoral sites. Journal of Anthropological Archaeology 11, 47-102.

Brock, A.L. & Buck, B.J., 2005. A new formation process for calcic pendants from Pahranagat Valley, Nevada, USA, and implication for dating Quaternary landforms. Quaternary Research 63, 359-367.

Brock, A.L. & Buck, B.J., 2009. Polygenetic development of the Mormon Mesa, NV petrocalcic horizons: geomorphic and paleoenvironmental interpretations. Catena 77, 65-75.

Bruand, A. & Duval, O., 1999. Calcified fungal filaments in the petrocalcic horizon of Eutrochrepts in Beauce, France. Soil Science Society of America Journal 63, 164-169.

Buczynski, C. & Chafetz, H.S., 1987. Silisiclastic grain breakage and displacement due to carbonate crystal-growth: an example from the Lueders Formation (Permian) of north-central Texas, USA. Sedimentology 34, 837-843.

Budd, D.A., Pack, S.M. & Fogel, M.L., 2002. The destruction of paleoclimatic isotopic signals in Pleistocene carbonate soil nodules of Western Australia. Palaeogeography, Palaeoclimatology, Palaeoecology 188, 249-273.

Bui, E.N., Loeppert, R.H. & Wilding, L.P., 1990. Carbonate phases in calcareous soils of the Western United States. Soil Science Society of America Journal 54, 39-45.

Bullock, P., Fedoroff, N., Jongerius A., Stoops, G., Tursina, T. & Babel, U., 1985. Handbook for Soil Thin Section Description. Waine Research Publications, Wolverhampton, 152 p.

Cacchio, P., Ercole, C., Cappuccio, G. & Lepidi, A., 2003. Calcium carbonate precipitation by bacterial strains isolated from a limestone cave and from a loamy soil. Geomicrobiology Journal 20, 85-98.

Cailleau, G., Braissant, O., Dupraz, C., Aragno, M. & Verrecchia, E.P., 2005. Biologically induced accumulations of $CaCO_3$ in orthox soils of Biga, Ivory Coast. Catena 59, 1-17.

Cailleau, G., Dadras, M., Abolhassani-Dadras, S., Braissant, O. & Verrecchia, E.P., 2009a. Evidence for an organic origin of pedogenic calcitic nanofibres. Journal of Crystal Growth 311, 2490-2495.

Cailleau, G., Verrecchia, E.P., Braissant, O. & Emmanuel, L., 2009b. The biogenic origin of needle fibre calcite. Sedimentology 56, 1858-1875.

Cailleau, G., Braissant, O. & Verrecchia, E.P., 2011. Turning sunlight into stone: the oxalate-carbonate pathway in a tropical tree ecosystem. Biogeosciences 8, 1755-1767.

Callot, G., Chamayou, H. & Dupuis, M., 1978. Variations of pH in solution of calcareous material as related to water dynamics — Data for analysis of a carbonate system. Annales Agronomiques 29, 37-57.

Callot, G., Mousain, D. & Plassard, C., 1985a. Concentrations de carbonate de calcium sur les parois des hyphes mycéliens. Agronomie 5, 143-150.

Callot, G., Guyon, A. & Mousain, D., 1985b. Inter-relations entre aiguilles de calcite et hyphes mycéliens. Agronomie 5, 209-216.

Calvet, F. & Julia, R., 1983. Pisoids in the caliche profiles of Tarragona northeast Spain. In Peryt, T.M. (ed.), Coated grains. Springer-Verlag, Berlin, pp. 456-473.

Canti, M.G., 1997. An investigation into microscopic calcareous spherulites from herbivore dungs. Journal of Archaeological Science 24, 219-231.

Canti, M.G., 1998a. The micromorphological identification of faecal spherulites from archaeological and modern materials. Journal of Archaeological Science 25, 435-444.

Canti, M.G., 1998b. Origin of calcium carbonate granules found in buried soils and Quaternary deposits. Boreas 27, 275-288.

Canti, M.G., 1999. The production and preservation of faecal spherulites: animals, environment and taphonomy. Journal of Archaeological Science 26, 251-258.

Canti, M.G., 2003. Aspects of the chemical and microscopic characteristics of plant ashes found in archaeological soils. Catena 54, 339-361.

Canti, M.G. & Piearce, T., 2003. Morphology and dynamics of calcium carbonate granules produced by different earthworm species. Pedobiologia 47, 511-521.

Capo, R.C., Whipkey, C.E., Blachere, J.R. & Chadwick, O.A., 2000. Pedogenic origin of dolomite in a basaltic weathering profile, Kohala peninsula, Hawaii. Geology 28, 271-274.

Chadwick, O.A., Sowers, J.M. & Amundson, R.G., 1989. Morphology of calcite crystals in cluster coatings from four soils in the Mojave desert region. Soil Science Society of America Journal 53, 219-221.

Chilingar, G.V., Bissell, H.J. & Wolf, K.H., 1965. Diagenesis of carbonate rocks. In Larsen, G. & Chilingar, G.V. (eds.), Diagenesis in Sediments. Elsevier, Amsterdam, pp. 179-322.

Cohen, A.S., 1982. Paleoenvironments of root casts from the Koobi Fora Formation Kenya. Journal of Sedimentary Petrology 52, 401-414.

Coil, J., Korstanje, M.A., Archer, S. & Hastorf, C.A., 2003. Laboratory goals and considerations for multiple microfossil extraction in archaeology. Journal of Archaeological Science 30, 991-1008.

Courty, M.A., 1990. Pedogenesis on Holocene calcareous parent materials under semi-arid conditions (Ghaggar Plain. NW India). In Douglas, L.A. (ed.), Soil Micromorphology: A Basic and Applied Science. Developments in Soil Science, Volume 19. Elsevier, Amsterdam, pp. 361-366.

Courty, M.A. & Fedoroff, N., 1985. Micromorphology of recent and buried soils in semi-arid region of Northwest India. Geoderma 35, 287-332.

Courty, M.A., Dhir, R.P. & Raghavan, H., 1987. Microfabrics of calcium carbonate accumulations in arid soils of western India. In Fedoroff, N., Bresson, L.M. & Courty, M.A. (eds.), Micromorphologie des Sols, Soil Micromorphology. AFES, Plaisir, pp. 227-234.

Courty, M.A., Marlin, C., Dever, L., Tremblay, P. & Vachier, P., 1994. The properties, genesis and environmental significance of calcitic pendants from the High Arctic (Spitsbergen). Geoderma 61, 71-102.

Darwin, C., 1881. The Formation of Vegetable Mould through the Action of Worms, with Observations on their Habits. John Murray, London, 298 p.

Del Monte, M., Sabbioni, C. & Zappia, G., 1987. The origin of calcium oxalates on historical buildings, monuments and natural outcrops. Science of the Total Environment 67, 17-39.

Dickson, J.A.D., 1965. A modified staining technique for carbonates in thin sections. Nature 205, 587.

Dravis, J.J., 1991. Carbonate petrography. Update on new techniques and applications. Journal of Sedimentary Petrology 61, 626-628.

Dravis, J.J. & Yurewicz, D.A., 1985. Enhanced carbonate petrography using fluorescence microscopy. Journal of Sedimentary Petrology 55, 795-804.

Driese, S.G., Jacobs, J.R. & Nordt, L.C., 2003. Comparison of modern and ancient Vertisols developed on limestone in terms of their geochemistry and parent material. Sedimentary Geology 157, 49-69.

Ducloux, J. & Butel, P., 1983. Micromorphology of calcretes in a slope deposit in the Poitevine plain, France. In Bullock, P. & Murphy, C.P. (eds.), Soil Micromorphology, Volume 2, Soil Genesis. AB Academic Publishers, Berkhamstead, pp. 637-646.

Ducloux, J. & Laouina, A., 1989. The pendent calcretes in semi-arid climates: an example located near Taforalt, NW Morocco. Catena 16, 237-249.

Ducloux, J., Butel, P. & Dupuis, T., 1984. Microsequence minéralogique des carbonates de calcium dans une accumulation carbonate sous galets calcaires, dans l'ouest de la France. Pedologie 34, 161-177.

Dunham, R.J., 1969a. Vadose pisolites in the Capitan Reef (Permian) of New Mexico and Texas. In Friedman, G. (ed.), Depositional Environments in Carbonates Rocks. Society of Economic Paleontologists and Mineralogists, Special Publication 14, pp. 182-191.

Dunham, R., 1969b. Early vadose silt in Townsend Mound (Reef), New Mexico. In Friedman, G. (ed.), Depositional Environments in Carbonate Rocks. Society of Economic Paleontologists and Mineralogists, Special Publication 14, pp. 139-181.

Dupuis, T., Ducloux, J., Butel, P. & Nahon, D., 1984. Etude par spectrographie infrarouge d'un encroûtement calcaire sous galet; mise en évidence et modélisation expérimentale d'une suite minérale évolutive à partir de carbonate de calcium amorphe. Clay Minerals 19, 605-614.

Durand, N., Gunnell, Y., Curmi, P. & Ahmad, S.M., 2006. Pathways of calcrete development on weathered silicate rocks in Tamil Nadu, India: mineralogy, chemistry and paleoenvironmental implications. Sedimentary Geology 192, 1-18.

Durand, N., Gunnell, Y., Curmi, P. & Ahmad, S.M., 2007. Pedogenic carbonates on Precambrian silicate rocks in South India: origin and paleoclimatic significance. Quaternary International 162/163, 35-49.

Elloy, R. & Thomas, G., 1981. Dynamique de la genèse des croûtes calcaires (calcrètes) développées sur séries rouges Pleistocènes en Algérie nord-occidentale. Bulletin des Centres de Recherche et Exploration - Production Elf-Aquitaine 5, 53-112.

El-Sayed, M.J., Fairchild, I.J. & Spiro, B., 1991. Kuwaiti dolocrete: petrology, geochemistry and groundwater origin. Sedimentary Geology 73, 59-75.

Esteban, M., 1972. Una nueva forma de prismas de *Microcodium elegans* Glück 1912 y su relación com el caliche del Eoceno Inferior, Marmellá, provincia de Tarragona (España). Revista del Instituto de Investigaciones Geológicas, Universidad de Barcelona 27, 65-81.

Esteban, M., 1974. Caliche textures and '*Microcodium*'. Bollettino della Societa Geologica Italiana 92, 105-125.

Esteban, M. & Klappa, C.F., 1983. Subaerial exposure environment. In Scholle, P.A., Debout, D.G. & Moore, C.H. (eds.), Carbonate Depositional Environment. American Association of Petroleum Geologists, Memoir 33, pp. 2-95.

Fedoroff, N., Courty, M.A., Lacroix, E. & Oleschko, K., 1994. Calcitic accretion on indurated volcanic materials (example of tepetates, Altiplano, Mexico). Transactions of 15th World Congress of Soil Science, Acapulco, Volume 6a, pp. 460-473.

Fleischer, M., Wilcox, R.E. & Matzko, J.J., 1984. Microscopic Determination of the Nonopaque Minerals. U.S. Geological Survey Bulletin 1627, 453 p.

Folk, R.L., 1962. Spectral subdivision of limestone types. In Ham, W.E. (ed.), Classification of Carbonate Rocks. American Association of Petroleum Geologists, Memoir 1, pp. 62-84.

Franceschi, V.R. & Horner, H.T., 1980. Calcium oxalate crystals in plants. Botanical Review 46, 361-427.

Franceschi, V.R. & Nakata, P.A., 2005. Calcium oxalate in plants: formation and function. Annual Review of Plant Biology 56, 41-71.

Frey-Wyssling, A., 1981. Crystallography of the two hydrates of crystalline calcium oxalate in plants. American Journal of Botany 68, 130-141.

Freytet, P., 1972. Réflexions sur la paléoécologie des *Microcodium* à partir d'observations faites dans le Crétacé Supérieur et l'Eocène Inférieur du Languedoc. Comptes Rendus du 97e Congrès National des Sociétés Savantes, Volume II, pp. 9-16.

Freytet, P., 1973. Caractères distinctifs et essai de classification des carbonates fluviatiles, lacustres et palustres. Comptes Rendus de l'Académie des Sciences 276, 1937-1940.

Freytet, P., 1984. Les sédiments lacustes carbonatés et leurs transformations par émersion et pédo-genèse. Importance de leur identification pour les reconstructions paléogéographiques. Bulletin des Centres de Recherche et Exploration-Production Elf-Aquitaine 8, 223-247.

Freytet, P. & Plaziat, J.C., 1965. Importance des constructions algaires dues à des Cyanophycées dans les formations continentales du Crétacé Supérieur et de l'Eocène du Languedoc. Bulletin de la Société Géologique de France 7, 679-694.

Freytet, P. & Plaziat, J.C., 1982. Continental Carbonate Sedimentation and Pedogenesis - Late Cretaceous and Early Tertiary of Southern France. Contributions to Sedimentology, Volume 12. E. Schweizerbart'sche Verlagsbuchhandlung, Stuttgart, 213 p.

Freytet, P. & Verrecchia, E.P., 1989. Les carbonates continentaux du pourtour méditerranéen: microfa-ciès et milieux de formation. Méditerranée 68, 5-28.

Freytet, P. & Verrecchia, E.P., 2002. Lacustrine and palustrine carbonate petrography: an overview. Journal of Paleolimnology 27, 221-237.

Freytet, P., Plaziat, J.C. & Verrecchia, E.P., 1997. A classification of rhizogenic (root-formed) calcretes, with examples from the Upper Jurassic-Lower Cretaceous of Spain and Upper Cretaceous of southern France - Discussion. Sedimentary Geology 110, 299-303.

Gile, L.H., Peterson, F.F. & Grossman, R.B., 1965. The K horizon: a master soil horizon of carbonate accumulation. Soil Science 99, 74-82.

Gile, L.H., Peterson, F.F. & Grossman, R.B., 1966. Morphological and genetic sequences of carbonate accumulation in desert soils. Soil Science 101, 347-360.

Gile, L.H., Grossman, R.B., Hawley, J.W., & Monger, H.C., 1996. Ancient soils of the Rincon Surface, northern Don Ana County. In Gile, L.H. & Ahrens, R.J. (eds.), Supplement to the Desert Project Soil Monograph. Soil Survey Investigations Report No. 44. USDA-Natural Resources Conservation Service, Lincoln, pp. 1-110.

Goudie, A.S., 1983. Calcrete. In Goudie, A.S. & Pye, K. (eds.), Chemical Sediments and Geomorphology. Academic Press, London, pp. 93-131.

Halstead, L.B., 1974. Vertebrate Hard Tissues. Wykeham Publications, London, 170 p.

Harrison, R.S. & Steinen, R.P., 1978. Subaerial crusts, caliche profiles, and breccia horizons: comparison of some Holocene and Mississippian exposure surfaces, Barbados and Kentucky. Geological Society of American Bulletin 89, 385-396.

Harwood, G., 1988. Microscopic techniques: II. Principles of sedimentary petrography. In Tucker, M. (ed.), Techniques in Sedimentology, Blackwell Scientific Publications, Oxford, pp. 108-173.

Hay, R.L. & Wiggins, B., 1980. Pellets, ooids, sepiolite and silica in three calcretes of the southwestern United States. Sedimentology 27, 559-576.

Herrero, J., 1991. Morfología y Génesis de Suelos Sobre Yesos. Monografías INIA 77, Instituto Nacional de Investigación y Tecnologia Agraria y Alimentaria, Madrid, 447 p.

Herrero, J. & Porta, J., 1987. Gypsiferous soils in the North-East of Spain. In Fedoroff, N., Bresson, L.M. & Courty, M.A. (eds.), Micromorphologie des Sols, Soil Micromorphology. AFES, Plaisir, pp. 186-192.

Herrero, J., Porta, J. & Fedoroff, N., 1992. Hypergypsic soil micromorphology and landscape relationships in north eastern Spain. Soil Science Society of America Journal 56, 1188-1194.

Horner, H.T. & Wagner, B.L., 1995. Calcium oxalate formation in higher plants. In Khan, S.R. (ed.), Calcium Oxalate in Biological Systems. CRC Press, Boca Raton, pp. 53-72.

Huerta, P., Rodríguez-Berriguete, A., Martín-García, R., Martín-Pérez, A., La Iglesia Fernández, A. & Alonso-Zarza, A.M., 2015. The role of climate and aeolian dust input in calcrete formation in volcanic islands (Lanzarote and Fuerteventura, Spain). Palaeogeography, Palaeoclimatology, Palaeoecology 417, 66-79.

Itkin, D., Goldfus, H. & Monger, H.C., 2016. Human induced calcretisation in anthropogenic soils and sediments: field observations and micromorphology in a Mediterranean climatic zone, Israel. Catena 146, 48-61.

Iwanoff, L.L., 1906. Ein Wasserhaltiges Calcium Carbonat Aussen Umgebungen von Nowo-Alexandria (guv. Lublin). Annalen der Geologie und Mineralogie der Russland 8, 23-25.

IUSS Working Group WRB, 2015. World Reference Base for Soil Resources 2014, Update 2015. World Soil Resources Reports No. 106. FAO, Rome, 192 p.

Jaillard, B., 1987. Les Structures Rhizomorphes Calcaires: Modèle de Réorganisation des Minéraux du Sol par les Racines. PhD Dissertation, Université des Sciences et Techniques du Languedoc, 220 p.

Jaillard, B. & Callot, G., 1987. Action des racines sur la ségrégation minéralogique des constituants minéraux du sol. In Fedoroff, N., Bresson, L.M. & Courty, M.A. (eds.), Micromorphologie des Sols, Soil Micromorphology. AFES, Plaisir, pp. 371-376.

Jaillard, B., Guyon, A. & Maurin, A.F., 1991. Structure and composition of calcified roots, and their identification in calcareous soils. Geoderma 50, 197-210.

James, N.P., 1972. Holocene and Pleistocene calcareous crust (caliche) profiles: criteria for subaerial exposure. Journal of Sedimentary Petrology 42, 817-836.

James, N.P. & Choquette, P.W., 1984. Diagenesis 9. Limestones – The meteoritic diagenetic environment. Geoscience Canada 11, 161-194.

Jiménez-Espinoza, R. & Jiménez-Millán, J., 2003. Calcrete development in Mediterranean colluvial carbonate systems from SE Spain. Journal of Arid Environments 53, 479-489.

Jones, B., 1987. The alteration of sparry calcite crystals in a vadose setting, Grand Cayman Island. Canadian Journal of Earth Sciences 24, 2292-2304.

Jones, B. & Kahle, C.F., 1993. Morphology, relationship, and origin of fibre and dendrite calcite crystals. Journal of Sedimentary Petrology 63, 1018-1031.

Jones, B. & Ng, K.C., 1988. The structure and diagenesis of rhizoliths from Cayman Brac, British West Indies. Journal of Sedimentary Petrology 58, 457-467.

Jones, B. & Squair, C.A., 1989. Formation of peloids in plant rootlets, Grand Cayman, British West Indies. Journal of Sedimentary Petrology 59, 1002-1007.

Jones, D. & Wilson, M.J., 1986. Biomineralization in crustose lichens. In Leadbeater, B.S.C. & Riding, R. (eds.), Biomineralization in Lower Plants and Animals. The Systematics Association, Special Volume 30. Clarendon Press, Oxford, pp. 91-105.

Jongerius, A. & Rutherford, G.K. (eds.), 1979. Glossary of Soil Micromorphology. Centre for Agricultural Publishing and Documentation, Wageningen, 138 p.

Kabanov, P., Anadón, P. & Krumbein, W.E., 2008. Microcodium: an extensive review and a proposed non-rhizogenic biologically induced origin for its formation. Sedimentary Geology 205, 79-99.

Kahle, C.F., 1977. Origin of subaerial Holocene calcareous crusts: role of algae, fungi and sparmicritisation. Sedimentology 24, 413-435.

Kaplan, M.Y., Eren, M., Kadir, S. & Kapur, S., 2013. Mineralogical, geochemical and isotopic characteristics of Quaternary calcretes in the Adana region, southern Turkey: implications on their origin. Catena 101, 164-177.

Keepax, C.A., 1981. Avian eggshell from archaeological sites. Journal of Archaeological Science 8, 315-335.

Kemp, R.A., 1995. Distribution and genesis of calcitic pedofeatures within a rapidly aggrading loess-paleosol sequence in China. Geoderma 65, 303-316.

Kerpen, W., Gewehr, H. & Scharpenseel, H.W., 1960. Zur Kentniss der ariden Irrigationsböden des Sudan. II Teil. Mikroskopische Untersuchungen. Pedologie 10, 303-323.

Khadkikar, A.S., Chamyal, L.S. & Ramesh, R., 2000. The character and genesis of calcrete in Late Quaternary alluvial deposits, Gujarat, western India, and its bearing on the interpretation of ancient climates. Palaeogeography, Palaeoclimatology, Palaeoecology 162, 239-261.

Khalaf, F.I. & Al-Zamel A., 2016. Petrography, micromorphology and genesis of Holocene pedogenic calcrete in Al-Jabal Al-Akhdar, Sultanate of Oman. Catena 147, 496-510.

Khormali, F., Abtahi, A. & Stoops, G., 2006. Micromorphology of calcitic features in highly calcareous soils of Fars Province, Southern Iran. Geoderma 132, 31-46.

Khormali, F., Monger, H.C. & Feng, Y., 2014. Experimental micropedology: a technique for investigating soil carbonate biogenesis along a desert-grassland-forest transect, New Mexico, USA. Spanish Journal of Soil Science 4, 1-18.

Klappa, C.F., 1978. Biolithogenesis of *Microcodium*: elucidation. Sedimentology 25, 489-522.

Klappa, C.F., 1979a. Calcified filaments in Quaternary calcretes: organo-mineral interactions in the subaerial vadose environment. Journal of Sedimentary Petrology 49, 955-968.

Klappa, C.F., 1979b. Lichen stromatolites: criterion for subaerial exposure and a mechanism for the formation of laminar calcretes (caliche). Journal of Sedimentary Petrology 49, 387-400.

Klappa, C.F., 1980. Rhizoliths in terrestrial carbonates: classification, recognition, genesis and significance. Sedimentology 27, 613-629.

Knox, G.J., 1977. Caliche profile formation, Saldanha Bay (South Africa). Sedimentology 24, 657-674.

Kosir, A., 2004. Microcodium revisited: root calcification product of terrestrial plants on carbonate-rich substrates. Journal of Sedimentary Research 74, 845-857.

Kovda, I. & Mermut, A., 2010. Vertic features. In Stoops, G., Marcelino, V. & Mees, F. (eds.), Interpretation of Micromorphological Features of Soils and Regoliths. Elsevier, Amsterdam, pp. 109-127.

Kovda, I. & Mermut, A., 2018. Vertic features. In Stoops, G., Marcelino, V. & Mees, F. (eds.), Interpretation of Micromorphological Features of Soils and Regoliths. Second Edition. Elsevier, Amsterdam, pp. 605-632.

Kovda, I.V., Wilding, L.P. & Drees, L.R., 2003. Micromorphological, submicroscopy and microprobe study of carbonate pedofeatures in a Vertisol gilgai soil complex, South Russia. Catena 54, 457-476.

Kowalinski, S., Pons, L.J. & Singer, S., 1972. Micromorphological comparison of three soils derived from loess in different climatic regions. Geoderma 7, 141-158.

Kraimer, R.A. & Monger, H.C., 2009. Carbon isotopic subsets of soil carbonate — a particle size comparison of limestone and igneous parent materials. Geoderma 150, 1-9.

Kraimer, R.A., Monger, H.C. & Steiner, R.L., 2005. Mineralogical distinctions of carbonates in desert soils. Soil Science Society of America Journal 69, 1773-1781.

Kuznetsova, A.M. & Khokhlova, O.S., 2012. Submicromorphology of pedogenic carbonate accumulations as a proxy of modern and paleoenvironmental conditions. Boletín de la Sociedad Geológica Mexicana 64, 199-205.

Last, W.M., 1990. Lacustrine dolomite — an overview of modern, Holocene, and Pleistocene occurrences. Earth-Science Reviews 27, 221-263.

Leiber, J. & Maus, H., 1969. Konkretionen organischen Ursprungs im Lösz. Jahresheft Geologisches Landesamt Baden-Württemberg 11, 299-308.

Lindbo, D.L., Stolt, M.H. & Vepraskas, M.J., 2010. Redoximorphic features. In Stoops, G., Marcelino, V. & Mees, F. (eds.), Interpretation of Micromorphological Features of Soils and Regoliths. Elsevier, Amsterdam, pp. 129-147.

Lindemann, W.C., Monger, H.C. & Kraimer, H.C., 2002. Carbon isotope fractionation by soil bacteria during pedogenic carbonate formation. American Society of Agronomy Abstract n° 171744. [CD ROM].

Loisy, C. & Pascal, A., 1998. Les encroûtements carbonatés ('calcrètes') en Champagne crayeuse: rôles de la diagenèse et des actions biologiques sous climat tempéré. Bulletin de la Société Géologique de France 169, 189-201.

Loisy, C., Verrecchia, E.P. & Dufour, P., 1999. Microbial origin for pedogenic micrite associated with a carbonate paleosol (Champagne, France). Sedimentary Geology 126, 193-204.

Machette, M.N., 1985. Calcic soils of southwestern United States. In Weide, D.L. (ed.), Soil and Quaternary Geology of the Southwestern United States. Geological Society of America Special Paper 203, pp. 1-21.

Mack, G.H. & James, W.C., 1992. Calcic paleosols of the Plio-Pleistocene Camp Rice and Palomas Formations, southern Rio Grande rift, USA. Sedimentary Geology 77, 89-109.

Mack, G.H., James, W.C. & Monger, H.C., 1993. Classification of paleosols. Geological Society of America Bulletin 105, 129-136.

Mack, G.H., Cole, D.R. & Trevino, L., 2000. The distribution and discrimination of shallow, authigenic carbonate in Pliocene-Pleistocene Palomas Basin, southern Rio Grande rift. Geological Society of America Bulletin 112, 643-656.

Maliva, R.G. & Siever, R., 1988. Diagenetic replacement controlled by force of crystallization. Geology 16, 688-691.

Mees, F., 1999. The unsuitability of calcite spherulites as indicators for subaerial exposure. Journal of Arid Environments 42, 149-154.

Mees, F., 2010. Authigenic silicate minerals — sepiolite-palygorskite, zeolites and sodium silicates. In Stoops, G., Marcelino, V. & Mees, F. (eds.), Interpretation of Micromorphological Features of Soils and Regoliths. Elsevier, Amsterdam, pp. 497-520.

Mees, F., 2018. Authigenic silicate minerals — sepiolite-palygorskite, zeolites and sodium silicates. In Stoops, G., Marcelino, V. & Mees, F. (eds.), Interpretation of Micromorphological Features of Soils and Regoliths. Second Edition. Elsevier, Amsterdam, pp. 177-203.

Mees, F. & Tursina, T.V., 2018. Salt Minerals in Saline Soils and Salt Crusts. In Stoops, G., Marcelino, V. & Mees, F. (eds.), Interpretation of Micromorphological Features of Soils and Regoliths. Second Edition. Elsevier, Amsterdam, pp. 289-321.

Mermut, A.R., Rostad, H.P.W. & St. Arnaud, R.J., 1983. Micromorphological studies of some loess soils in Western Saskatchewan, Canada. In Bullock, P. & Murphy, C.P. (eds.), Soil micromorphology. Volume 1, Techniques and Applications. AB Academic Publishers, Berkhamsted, pp. 309-315.

Michel, L.A., Driese, S.G., Nordt, L.C., Breecker, D.O., Labotka, D.M. & Dworkin, S.I., 2013. Stable-isotope geochemistry of vertisols formed on marine limestone and implications for deep-time paleoenvironmental reconstructions. Journal of Sedimentary Research 83, 300-308.

Miller, D.L., Mora, C.I. & Driese, S.G., 2007. Isotopic variability in large carbonate nodules in Vertisols: implications for climate and ecosystem assessments. Geoderma 142, 104-111.

Millière, L., Hasinger, O., Bindschedler, S., Cailleau, G., Spangenberg, J.E. & Verrecchia, E.P., 2011a. Stable carbon and oxygen isotope signatures of pedogenic needle fibre calcite. Geoderma 161, 74-87.

Millière, L., Spangenberg, J.E., Bindschedler, S., Cailleau, G. &, Verrecchia, E.P., 2011b. Reliability of stable carbon and oxygen isotope compositions of pedogenic needle fibre calcite as environmental indicators: examples from Western Europe. Isotopes in Environmental and Health Studies 47, 341-358.

Millo, C., Dupraz, S., Ader, M., Guyot, F., Thaler, C., Foy, E. & Ménez, B., 2012. Carbon isotope fractionation during calcium carbonate precipitation induced by ureolytic bacteria. Geochimica et Cosmochimica Acta 98, 107-124.

Millot, G., Nahon, D., Paquet, H., Ruellan, A. & Tardy, Y., 1977. L'épigénie calcaire des roches silicatées dans les encroûtements carbonatés en pays subaride. Antiatlas, Maroc. Sciences Géologiques Bulletin 30, 129-152.

Milnes, A.R., 1992. Calcretes. In Martini, I.P. & Chesworth, W. (eds.), Weathering, Soils and Paleosols. Developments in Earth Surface Processes, Volume 2. Elsevier, Amsterdam, pp. 309-347.

Milnes, A.R. & Hutton, J.T., 1983. Calcretes in Australia. In Soils: An Australian Viewpoint. Division of Soils, CSIRO. CSIRO, Melbourne and Academic Press, London, pp 119-162.

Mintz, J.S., Driese, S.G., Breecker, D.O. & Ludvigson, G.A., 2011. Influence of changing hydrology on pedogenic calcite precipitation in Vertisols, Dance Bayou, Brazoria County, Texas, USA: implications for estimating paleoatmospheric pCO_2. Journal of Sedimentary Research 81, 394-400.

Monger, H.C. & Adams, H.P., 1996. Micromorphology of calcite-silica deposits, Yucca Mountain, Nevada. Soil Science Society of America Journal 60, 519-530.

Monger, H.C. & Daugherty, L.A., 1991a. Neoformation of palygorskite in a southern New Mexico Aridisol. Soil Science Society of America Journal 55, 1646-1650.

Monger, H.C. & Daugherty, L.A., 1991b. Pressure solution: possible mechanism for silicate grain dissolution in a petrocalcic horizon. Soil Science Society of America Journal 55, 1625-1629.

Monger, H.C., Daugherty, L.A., Lindemann, W.C. & Liddell, C.M., 1991a. Microbial precipitation of pedogenic calcite. Geology 19, 997-1000.

Monger, H.C., Daugherty, L.A. & Gile, L.H., 1991b. A microscopic examination of pedogenic calcite in an Aridisol of southern New Mexico. In Nettleton, W.D. (ed.), Occurrence, Characteristics, and Genesis of Carbonate, Gypsum, and Silica Accumulations in Soils. Soil Science Society of America Journal Special Publication 26, Madison, pp. 37-60.

Monger, H.C., Cole, D.R., Gish, J.W. & Giordano, T.H., 1998. Stable carbon and oxygen isotopes in Quaternary soil carbonates as indicators of ecogeomorphic changes in the northern Chihuahuan Desert, USA. Geoderma 82, 137-172.

Monger, H.C., Kraimer, R.A., Khresat, S., Cole, D.R., Wang, X.J. & Wang, J.P., 2015. Sequestration of inorganic carbon in soil and groundwater. Geology 43, 375-378.

Morin, N., 1993. Les *Microcodium*: architecture, structure et composition. Comparaison avec les racines calcifies. PhD Dissertation, Université de Montpellier 2, 132 p.

Mount, J.F. & Cohen, A.S., 1984. Petrology and geochemistry of rhizoliths from Plio-Pleistocene fluvial and marginal lacustrine deposits, East Lake Turkana, Kenya. Journal of Sedimentary Petrology 54, 263-275.

Nagtegaal, P.J.C., 1969. Microtextures in recent and fossil caliches. Leidse Geologische Mededelingen 42, 131-142.

Nahon, D. & Ruellan, A., 1975. Les accumulations de calcaire sur les marnes éocènes de la Falaise de Thiès (Senegal). Mise en évidence de phénomènes d'épigénie. In Vogt, T. (ed.), Colloque Types de Croûtes Calcaires et leur Répartition Régionale. Université Louis Pasteur, Strasbourg, pp. 7-12.

Nahon, D., Ducloux, J., Butel, P., Augas, C. & Paquet, H., 1980. Néoformations d'aragonite, première étape d'une suite minéralogique évolutive dans les encroûtements calcaires. Comptes Rendus de l'Académie des Sciences de Paris 291, 725-727.

Nash, D.J. & Smith, R.F., 1998. Multiple calcrete profiles in the Tabernas Basin, southeast Spain: their origins and geomorphic implications. Earth Surface Processes and Landforms 23, 1009-1029.

Netterberg, F., 1980. Geology of southern African calcretes. I. Terminology, description, macrofacies and classification. Transactions of the Geological Society of South Africa 83, 255-283.

Nettleton, W.D., Olson, C.G. & Wysocki, D.A., 2000. Paleosol classification: problems and solutions. Catena 41, 61-92.

Paquet, H. & Ruellan, A., 1993. Epigénie et encroûtements calcaire (calcrètes). In Paquet, H. & Cauer, N. (eds.), Sédimentologie et Géochimie de la Surface. A la mémoire de Georges Millot. Les Colloques de l'Académie des Sciences et du Cadas. Institut de Géologie, Strasbourg, pp. 19-39.

Peryt, T.M. (ed.), 1983. Coated Grains. Springer-Verlag, Berlin, 655 p.

Pfeiffer, M., Aburto, F., Le Roux, J.P., Kemnitz, H., Sedov, S., Solleiro-Rebolledo, E. & Seguel, O., 2012. Development of a Pleistocene calcrete over a sequence of marine terraces at Tongoy (north-central Chile) and its paleoenvironmental implications. Catena 97, 104-118.

Phillips, S.E. & Self, P.G., 1987. Morphology, crystallography and origin of needle-fibre calcite in Quaternary pedogenic carbonates of South Australia. Australian Journal of Soil Research 25, 429-444.

Phillips, S.E., Milnes, A.R. & Foster, R.C., 1987. Calcified filaments: an example of biological influences in the formation of calcrete in South Australia. Australian Journal of Soil Research 25, 405-428.

Platt, N.H. & Wright, V.P., 1992. Palustrine carbonates at the Florida Everglades: towards an exposure index for the freshwater environment. Journal of Sedimentary Petrology 62, 1058-1071.

Plaziat, J.C., 1984. Le problème des *Microcodium*: une mise au point. In Le Domaine pyréneen de la fin du Crétacé à la fin de l'Eocène: stratigraphie, paléoenvironnements et évolution paléogéographique. PhD Dissertation, Université Paris-Sud, pp. 637-662.

Pobeguin, T., 1943. Les oxalates de calcium chez quelques angiospermes. Annales des Sciences Naturelles. Botanique et Biologie Végétale 4, 1-95.

Porta, J. & Herrero, J., 1988. Micromorfología de suelos con yeso. Anales de Edafología y Agrobiología 47, 179-197.

Preece, R.C., Kemp, R.A. & Hutchinson, J.N., 1995. A late-glacial colluvial sequence at Watcombe Bottom, Ventnor, Isle of Wight, England. Journal of Quaternary Science 10, 107-121.

Pustovoytov, K.E., 1998. Pedogenic carbonate cutans as a record of the Holocene history of relic tundra-steppes of the Upper Kolyma Valley (North-Eastern Asia). Catena 34, 185-195.

Pustovoytov, K.E., 2002. Pedogenic carbonate cutans on clasts in soils as a record of history of grassland ecosystems. Palaeogeography, Palaeoclimatology, Palaeoecology 177, 199-214.

Pustovoytov, K.E., 2003. Growth rates of pedogenic carbonate coatings on coarse clasts. Quaternary International 106/107, 131-141.

Rabenhorst, M.C. & Wilding, L.P., 1986. Pedogenesis on the Edwards Plateau, Texas: III. New model for the formation of petrocalcic horizons. Soil Science Society of America Journal 50, 693-699.

Rabenhorst, M.C., West, L.T. & Wilding, L.P., 1991. Genesis of calcic and petrocalcic horizons in soils over carbonate rocks. In Nettleton, W.D. (ed.), Occurrence, Characteristics, and Genesis of Carbonate, Gypsum, and Silica Accumulations in Soils. Soil Science Society of America Special Publication 26. SSSA, Madison, pp. 61-74.

Reeves, C.C., 1970. Origin, classification and geologic history of caliche on the southern high plains, Texas and eastern New Mexico. Journal of Geology 78, 352-362.

Reeves, C.C., 1976. Caliche: Origin, Classification, Morphology and Uses. Estacado Books, Lubbock, 233 p.

Reheis, M.C., 1988. Pedogenic replacement of aluminosilicate grains by $CaCO_3$ in Ustollic Haplargids, south-central Montana, U.S.A. Geoderma 41, 243-261.

Retallack, G.J., James, W.C., Mack, G.H. & Monger, H.C., 1993. Classification of paleosols: discussion and reply. Geological Society of America Bulletin 105, 1635-1637.

Riche, G., Rambaud, D. & Riera, M., 1982. Etude morphologique d'un encroûtement calcaire, Région d'Irecê, Bahia, Brésil. Cahiers Orstom, Série Pédologie 19, 257-270.

Richter, D.K., Immenhauser, A. & Neuser, R.D., 2008. Electron backscatter diffraction documents randomly orientated c-axes in moon-milk calcite fibres: evidence for biologically induced precipitation. Sedimentology 55, 487-497.

Ringrose, S., Kampunzu, A.B., Vink, B.W., Matheson, W. & Downey, W.S., 2002. Origin and palaeo-environments of calcareous sediments in the Moshaweng Dry Valley, Southeast Botswana. Earth Surface Processes and Landforms 27, 591-611.

Robins, C.R., Deurlingtona, A., Buck, B.J. & Brock-Hon, A.L., 2015. Micromorphology and formation of pedogenic ooids in calcic soils and petrocalcic horizons. Geoderma 251/252, 10-23.

Rodriguez-Navarro, C., Jimenez-Lopez, C., Rodriguez-Navarro, A., Gonzalez-Muñoz, M.T. & Rodriguez-Gallego, M., 2007. Bacterially mediated mineralization of vaterite. Geochimica et Cosmochimica Acta 71, 1197-1213.

Ruellan, A., Beaudet, G., Nahon, D., Paquet, H., Rognon, P. & Millot, G., 1979. Rôle des encroûtements calcaires dans le façonnement des glacis d'ablation des regions arides et semi-arides du Maroc. Comptes Rendus de l'Académie des Sciences 289, 619-622.

Russ, J., Palma, R.L. & Booker, J.L., 1994. Whewellite rock crusts in the Lower Pecos region of Texas. Texas Journal of Science 46, 165-172.

Russ, J., Palma, R.L., Lloyd, D.H., Boutton, T.W. & Coy, M.A., 1996. Origin of the whewellite-rich rock crust in the Lower Pecos region of southwest Texas and its significance to paleoclimate reconstructions. Quaternary Research 46, 27-36.

Sanz, M.E. & Wright, V.P., 1994. Modelo alternativo para el desarrollo de calcretas: un ejemplo del Plio-Cuaternario de la Cuenca de Madrid Basin, Spain. Sedimentology 42, 437-452.

Scoppa, C.O. & Pasos, M.S., 1981. Caracterización integral y análisis pedogenético de molisoles del borde sur de la pampa deprimida. Revista de Investigaciones Agropecuarias 16, 43-88.

Sehgal, J.L. & Stoops, G., 1972. Pedogenic calcic accumulation in arid and semiarid regions of the Indo-Gangetic alluvial plain of the erstwhile Punjab (India). Their morphology and origin. Geoderma 8, 59-72.

Siesser, W.G., 1973. Diagenetically formed ooids and intraclasts in South African calcretes. Sedimentology 20, 539-551.

Sharp, W.D., Ludwig, K.R., Chadwick, O.A., Amundson, R. & Glaser, L.L., 2003. Dating fluvial terraces by Th-230/U on pedogenic carbonate, Wind River Basin, Wyoming. Quaternary Research 59, 139-150.

Sherman, L.A. & Barak, P., 2000. Solubility and dissolution kinetics of dolomite in $Ca-Mg-HCO_3/CO_3$ solutions at 25°C and 0.1 Mpa carbon dioxide. Soil Science Society of America Journal 64, 1959-1968.

Simkiss, K. & Wilbur, K.M., 1989. Biomineralization - Cell Biology and Mineral Deposition. Academic Press, San Diego, 337 p.

Sobecki, T.M. & Wilding, L.P., 1983. Formation of calcic and argillic horizons in selected soils of the Texas Coast Prairie. Soil Science Society of America Journal 47, 707-715.

Soil Survey Staff, 1999. Soil Taxonomy. A Basic System of Soil Classification for Making and Interpreting Soil Surveys. 2nd Edition. USDA, Natural Resources Conservation Service, Agriculture Handbook Number 436. U.S. Government Printing Office, Washington, 871 p.

Soil Survey Staff, 2014. Keys to Soil Taxonomy. 12th Edition. USDA-Natural Resources Conservation Service, Washington, 360 p.

Staley, J.T., Palmer, F. & Adams, J.B., 1982. Microcolonial fungi: common inhabitants on desert rocks? Science 215, 1093-1094.

Stoops, G.J., 1976. On the nature of 'lublinite' from Hollanta (Turkey). American Mineralogist 61, 172.

Stoops, G., 1983. SEM and light microscopic observations of minerals in bog-ores of the Belgian Campine. Geoderma 30, 179-186.

Stoops, G., 2003. Guidelines for Analysis and Description of Soil and Regolith Thin Sections. Soil Science Society of America, Madison, 184 p.

Supko, P.R., 1971. 'Whisker' crystal cement in a Bahamian rock. In Bricker, O.P. (ed), Carbonate Cements. Johns Hopkins University Press, Studies in Geology 19, Baltimore, pp. 143-146.

Syers, J.K., Birnie A.C. & Mitchell, B.D., 1967. The calcium oxalate content of some lichens growing on limestone. The Lichenologist 3, 409-414.

Tandon, S.K. & Friend, P., 1989. Near surface shrinkage and carbonate replacement processes, Arran Cornstone Formation, Scotland. Sedimentology 36, 1113-1126.

Tandon, S.K. & Narayan, D., 1981. Calcrete conglomerate, case hardened conglomerate and cornstone-comparative account of pedogenic and non-pedogenic carbonates from the continental Siwalik Group, Punjab, India. Sedimentology 28, 353-367.

Treadwell-Steitz, C. & McFadden, L., 2000. Influence of parent material and grain size on carbonate coatings in gravelly soils, Palo Duro Wash, New Mexico. Geoderma 94, 1-22.

Tribble, J.S., Arvidson, R.S., Lane III, M. & Mackenzie, F.T., 1995. Crystal chemistry, and thermodynamic and kinetic properties of calcite, dolomite, apatite, and biogenic silica: applications to petrologic problems. Sedimentary Geology 95, 11-37.

Tucker, M.E., 1991. Sedimentary Petrology. Blackwell, Oxford, 259 p.

Tucker, M.E. & Wright, V.P., 1990. Carbonate Sedimentology. Blackwell Science Ltd, Oxford, 482 p.

Tucker, M.E. & Wright, V.P., 1991. Ooides and Pisoides, In Tucker, M.E. & Wright, P.V. (eds.), Carbonate Sedimentology. Blackwell Scientific Publications, Oxford, pp. 3-9.

Ulrich, F., 1938. Houby jako rozrusovatele a tvurci nerosbi a hornin. Veda Prirodni 19, 45-50.

Vepraskas, M.J., Lindbo, D.L. & Stolt, M.H., 2018. Redoximorphic features. In Stoops, G., Marcelino, V. & Mees, F. (eds.), Interpretation of Micromorphological Features of Soils and Regoliths. Second Edition. Elsevier, Amsterdam, pp. 425-445.

Vergès, V., Madon, M., Bruand, A. & Bocquier, G., 1982. Morphologie et crystallogenèse de monocristaux supergènes de calcite en aiguilles. Bulletin of Mineralogy 105, 351-356.

Verrecchia, E.P., 1987. Le contexte morpho-dynamique des croûtes calcaires: apport des analyses séquentielles à l'échelle microscopique. Zeitschrift für Geomorphologie 31, 179-193.

Verrecchia, E.P., 1990a. Incidence de l'activité fungique sur l'induration des profils de type calcrète pédologique. L'exemple du cycle oxalate-carbonate de calcium dans les encroûtements calcaires de Galilée (Israel). Comptes Rendus de l'Académie des Sciences 311, 1367-1374.

Verrecchia, E.P., 1990b. Litho-diagenetic implications of the calcium oxalate-carbonate biogeochemical cycle in semiarid calcretes, Nazareth, Israel. Geomicrobiology Journal 8, 87-99.

Verrecchia, E.P., 1994. L'origine biologique et supeficielle des croûtes zonaires. Bulletin de la Société Géologique de France 165, 583-592.

Verrecchia, E.P., 2000. Fungi and sediments. In Riding, R. & Awramik, S.M. (eds.), Microbial Sediments. Springer-Verlag, New York, pp. 68-75.

Verrecchia, E.P., 2002. Géodynamique du carbonate de calcium à la surface des continents. In Miskovsky, J.C. (ed.), Géologie de la Préhistoire: Méthodes, Techniques, Applications. Editions Géopré, Paris, pp. 233-258.

Verrecchia, E.P., 2007. Lacustrine and palustrine geochemical sediments. In Nash, D.J. & McLaren, S.J. (eds.), Geochemical Sediments and Landscapes. Blackwell, Oxford, 298-329.

Verrecchia, E.P., 2011. Pedogenic carbonates. In Reitner, J. & Thiel, V. (eds.), Encyclopedia of Geobiology. Springer Netherlands, Dordrecht, pp. 721-725.

Verrecchia, E.P., 2016. Vertical variation in porosity of nãri (calcrete) on chalk, Galilee, Israel: a new interpretation as a tribute to Dan H. Yaalon. Catena 146, 62-72.

Verrecchia, E.P. & Dumont, J.L., 1996. A biogeochemical model for chalk alteration by fungi in semiarid environments. Biogeochemistry 35, 447-470.

Verrecchia, E.P. & Freytet, P., 1987. Interférence pédogénèse sédimentation dans les croûtes calcaires — Proposition d'une nouvelle méthode d'étude: l'analyse séquentielle. In Fedoroff, N., Bresson, L.M. & Courty, M.A. (eds.), Micromorphologie des Sols, Soil Micromorphology. AFES, Plaisir, pp. 555-561.

Verrecchia, E.P. & Le Coustumer, M.N., 1996. Occurrence and genesis of palygorskite and associated clay minerals in a Pleistocene calcrete complex, Sde Boqer, Negev Desert, Israel. Clay Minerals 31, 183-202.

Verrecchia, E.P. & Verrecchia, K.E., 1994. Needle-fiber calcite: a critical review and a proposed classification. Journal of Sedimentary Research A64, 650-664.

Verrecchia, E.P., Ribier, J., Patillon, M. & Rolko, K.E., 1990a. Stromatolitic origin for desert laminar limecrusts. Naturwissenschaften 78, 505-507.

Verrecchia, E.P., Dumont, J.L. & Rolko, K.E., 1990b. Do fungi building limestones exist in semi-arid regions? Naturwissenschaften 77, 584-586.

Verrecchia, E.P., Dumont, J.L. & Verrecchia, K.E., 1993. Role of calcium oxalate biomineralization by fungi in the formation of calcretes: a case study from Nazareth, Israel. Journal of Sedimentary Petrology 63, 1000-1006.

Verrecchia, E.P., Freytet, P., Verrecchia, K.E. & Dumont, J.L., 1995. Spherulites in calcrete laminar crusts: biogenic CaCO$_3$ precipitation as a major contributor to crust formation. Journal of Sedimentary Research A65, 690-700.

Verrecchia, E.P., Braissant, O. & Cailleau, G., 2006. The oxalate-carbonate pathway in soil carbon storage: the role of fungi and oxalotrophic bacteria. In Gadd, G.M. (ed.), Fungi in Biogeochemical Cycles. Cambridge University Press, Cambridge, pp. 289-310.

Villagran, X.S. & Poch, R.M., 2014. A new form of needle-fiber calcite produced by physical weathering of shells. Geoderma 213, 173-177.

Violette, A., Riotte, J., Braun, J.J., Oliva, P., Marechal, J.C., Sekhar, M., Jeandel, C., Subramanian, S., Prunier, J., Barbiero, L. & Dupre, B., 2010. Formation and preservation of pedogenic carbonates in South India, links with paleomonsoon and pedological conditions: clues from Sr isotopes, U-Th series and REEs. Geochimica et Cosmochimica Acta 74, 7059-7085.

Vogt, T., 1984. Problèmes de genèse des croûtes calcaires Quaternaires. Bulletin des Centres de Recherche et Exploration-Production Elf-Aquitaine 8, 209-221.

Wadsten, T. & Moberg, R., 1985. Calcium oxalate hydrates on the surface of lichens. The Lichenologist 17, 239-245.

Walls, R.A., Harris, W.B. & Nunan, W.E., 1973. Calcareous crust (caliche) profiles and early subaerial exposure of Carboniferous carbonates, northeastern Kentucky. Sedimentology 22, 417-440.

Wang, Y., Nahon, D. & Merino, E., 1994. Dynamic model of the genesis of calcretes replacing silicate rocks in semi-arid regions. Geochimica et Cosmochimica Acta 58, 5131-5145.

Wang, Y., McDonald, E., Amundson, R., McFadden, L. & Chadwick, O., 1996. An isotopic study of soils in chronological sequences of alluvial deposits, Providence Mountains, California. Geological Society of America Bulletin 108, 379-391.

Warren, J., 2000. Dolomite: occurrence, evolution and economically important associations. Earth-Science Reviews 52, 1-81.

Watchman, A.L., 1990. A summary of occurrences of oxalate-rich crusts in Australia. Rock Art Research 7, 44-50.

Watts, N.L., 1978. Displacive calcite: evidence from recent and ancient calcretes. Geology 6, 699-703.

Watts, N.L., 1980. Quaternary pedogenic calcretes from the Kalahari (southern Africa): mineralogy, genesis and diagenesis. Sedimentology 27, 661-686.

Weicek, C.S. & Messenger, A.S., 1972. Calcite contributions by earthworms to forest soils in Northern Illinois. Soil Science Society of America Proceedings 36, 478-480.

West, L.T., Drees, L.R., Wilding, L.P. & Rabenhorst, M.C., 1988. Differentiation of pedogenic and lithogenic carbonate forms in Texas. Geoderma 43, 271-287.

Whipkey, C.E., Capo, R.C., Hsieh, J.C.C. & Chadwick, O.A., 2002. Development of magnesian carbonates in Quaternary soils on the Island of Hawaii. Journal of Sedimentary Research 72, 158-165.

Wieder, M. & Yaalon, D.H., 1974. Effect of matrix composition of carbonate nodule crystallisation. Geoderma 11, 95-121.

Wieder, M. & Yaalon, D.H., 1982. Micromorphological fabrics and developmental stages of carbonate nodular forms related to soil characteristics. Geoderma 28, 203-220.

Wilding, L.P. & Drees, L.R., 1990. Removal of carbonates from thin sections for microfabric interpretation. In Douglas, L.A. (ed.), Soil Micromorphology: A Basic and Applied Science. Developments in Soil Science, Volume 19. Elsevier, Amsterdam, pp. 613-620.

Wright, V.P., 1986. The role of fungal biomineralization in the formation of early Carboniferous soil fabrics. Sedimentology 33, 831-838.

Wright, V.P., 1990. A micromorphological classification of fossil and recent calcic and petrocalcic microstructures. In Douglas, L.A. (ed.), Soil Micromorphology: A Basic and Applied Science. Developments in Soil Science, Volume 19. Elsevier, Amsterdam, pp. 401-407.

Wright, V.P., 1994. Paleosols in shallow marine carbonate sequences. Earth-Science Reviews 35, 367-395.

Wright, V.P., 2007. Calcretes. In Nash, D. & McLaren, S. (eds.), Geochemical Sediments and Landscapes. Wiley-Blackwell, Oxford, pp. 10-45.

Wright, V.P. & Peeters, C., 1989. Origins of some early Carboniferous calcretes fabrics revealed by cathodoluminescence: implications for interpreting the sites of calcrete formation. Sedimentary Geology 65, 345-353.

Wright, V.P. & Platt, N.H., 1995. Seasonal wetland carbonate sequences and dynamic catenas: a reappraisal. Sedimentary Geology 99, 65-71.

Wright, V.P. & Tucker, M.E., 1991. Calcretes: an introduction. In Wright, V.P. & Tucker, M.E. (eds.), Calcretes. IAS Reprint Series, Volume 2. Blackwell, Oxford, pp. 1-22.

Wright, V.P., Platt, N.H. & Wimbledon, W.A., 1988. Biogenic laminar calcrete: evidence of calcified root-mat horizons in paleosols. Sedimentology 35, 603-620.

Wright, V.P., Platt, N.H., Marriott, S.B. & Beck, V.H., 1995. A classification of rhizogenic (root-formed) calcretes, with examples from Upper Jurassic-Lower Cretaceous of Spain and Upper Cretaceous of southern France. Sedimentary Geology 100, 143-158.

Zamanian, K., Pustovoytov, K. &, Kuzyakov, Y., 2016. Pedogenic carbonates: forms and formation processes. Earth-Science Reviews 157, 1-17.

Zhou, J. & Chafetz, H.S., 2009. Biogenic caliches in Texas: the role of organisms and effect of climate. Sedimentary Geology 222, 207-225.

Chapter 10

Gypsic Features

Rosa M. Poch[1], Octavio Artieda[2], Marina Lebedeva[3]

[1]UNIVERSITY OF LLEIDA, LLEIDA, SPAIN;
[2]EXTREMADURA UNIVERSITY, PLASENCIA, SPAIN;
[3]DOKUCHAEV SOIL SCIENCE INSTITUTE, MOSCOW, RUSSIA

CHAPTER OUTLINE

1. Introduction

Gypsic features in soils and sediments are composed of gypsum $(Ca(SO_4) \cdot 2H_2O)$, a mineral with rather low solubility (2.6 g/L), or they are at least related to gypsum formation. Gypsum is formed when Ca^{2+} and SO_4^{2-} concentrations in the soil are high. These ions can come from the dissolution of calcium sulphates contained in rocks in the

Interpretation of Micromorphological Features of Soils and Regoliths. https://doi.org/10.1016/B978-0-444-63522-8.00010-3

catchment, but they can also be separately derived from the alteration of limestone, carbonates of biological origin, sulphide-bearing formations. After the development of gypsum crystals in soils and sediments, they can be transported by wind, producing eolian gypsiferous or gypseous parent materials.

Bulk determination of total gypsum content, necessary for soil behaviour forecasting, may be insufficient for genetic interpretations (Artieda et al., 2006), for instance, in areas with gypsiferous parent materials where a distinction between pedogenic and lithogenic gypsum is important (Carter & Inskeep, 1988; Stoops & Poch, 1994). Micromorphological studies can help in understanding the distribution and genesis of gypsic features, including those that are not recognised in the field.

For information on the worldwide distribution of gypsiferous soils, and on their characteristics, the reader is referred to general reviews (e.g., Watson, 1983; FAO, 1990; Herrero & Boixadera, 2002; Herrero et al., 2009). Gypsum, both inherited and pedogenic, is most abundant in soils under arid or semiarid climates, because evaporative concentration is generally required for its formation and because it is easily leached in humid climates. Under extreme hot, dry conditions gypsum can transform to bassanite $(Ca(SO_4) \cdot 0.5H_2O)$ or anhydrite $(Ca(SO_4))$ (see Mees & Tursina, 2010, 2018). Gypsum also occurs in acid sulphate soils, whose micromorphological characteristics are discussed elsewhere (Mees & Stoops, 2010, 2018).

A summary of their characteristics is given in Table 1. Other habits have also been reported, such as prismatic pseudohexagonal, with {010}, {111}, {$\bar{1}$11} and {120}

Table 1 Main features of gypsum crystal morphologies found in soils

	Crystal forms	Most common morphology in cross section	Selected other names
Prismatic	{120} and {010}, and less well developed {$\bar{1}$11} and {$\bar{1}$03}, often flattened parallel to (010)	Lath, parallelogram	Tabular (Ortí, 1992), pseudohexagonal (Carenas et al., 1982), lath-shaped (Buck et al., 2006)
Tabular	{$\bar{1}$11} and {$\bar{1}$03}; {120} and {010} very reduced or absent	Lozenge with small prism faces, pseudohexagonal	Prismatic/hemi-bipyramidal (Cody & Cody, 1988), tabular pseudorhombohedral, prismatic pinacoidal (Jafarzadeh & Burnham, 1992), tabular hexagonal and pseudohexagonal (Buck et al., 2006)
Hemi-bipyramidal	{$\bar{1}$11} and {$\bar{1}$03}; elongated parallel to intersection between {$\bar{1}$11} faces	Lozenge, parallelogram	—
Lenticular	Curved crystal faces replacing {$\bar{1}$11} and {$\bar{1}$03} faces; flattened parallel to a plane between ($\bar{1}$01) and ($\bar{1}$03)	Lenticular	Discoid (Masson, 1955), spindle-shaped (Carter & Inskeep, 1988)
Acicular	Similar to prismatic, elongated parallel to c-axis	Acicular	—

Source: Mees (1999).

2. Crystal Morphology

2.1 Occurrences in Nature

In thin sections, gypsum is easily recognised by its first-order grey interference colours, well-expressed cleavage parallel to (010), low negative relief (n 1.520-1.529) and biaxial positive sign. Cleavage and optical sign are helpful to distinguish gypsum from quartz.

The crystal morphology of gypsum is highly variable and has been described using various terms. Descriptions are based on observations at different scales, using different techniques, and they partly involve interpretation of 3D shapes based on 2D cross sections. Common habits of gypsum in soils and sediments of arid and semiarid regions are prismatic, tabular, hemi-bipyramidal, lenticular and acicular. A summary of their characteristics is given in Table 1. Other habits have also been reported, such as prismatic pseudohexagonal, with {010}, {111}, {$\overline{1}$11} and {120} forms (Jafarzadeh & Burnham, 1992). The pseudohexagonal aspect of non-prismatic gypsum crystals in cross sections parallel to the plane of flattening is commonly indicated in the descriptive terms that are used (e.g., Buck & Van Hoesen, 2002).

Lenticular gypsum represents the most common morphology of gypsum in soils and has been described by many authors (e.g., Barzanji & Stoops, 1974; Watson, 1979; Stoops & Ilaiwi, 1981; Allen, 1985; Herrero & Porta, 1987; Pankova & Yamnova, 1987; Arricibita et al., 1988; Benayas et al., 1988; Carter & Inskeep, 1988; Halitim, 1988; Porta & Herrero, 1988; Eswaran & Zi-Tong, 1991; Herrero, 1991; Jafarzadeh & Burnham, 1992; Poch, 1992; Amit & Yaalon, 1996; Artieda, 1996, 2004; Chen, 1997; Poch et al., 1998; Buck & Van Hoesen, 2002; Buck et al., 2006). Some authors suggest that a lenticular morphology derives from a prismatic crystal habit, whereby the preferential development of the {$\overline{1}$03} face results in a lenticular shape (Mandado & Tena, 1989; Cody, 1991). Other authors consider gypsum lenses as hemi-bipyramidal or pseudohexagonal crystals with curved faces and rounded edges (Siesser & Rogers, 1976; Logan, 1987). Lenticular gypsum occurs as single crystals or as crystal intergrowths, as in the case of desert roses. They may also occur as massive crystallisations.

Acicular crystals are relatively rare. Some authors consider that the confining space is one of the controlling factors for the development of this crystal habit, indicating that when the crystal grows under confining pressure, as it does in the soil matrix, this would favour acicular or tabular habits (e.g., Eswaran & Zi-Tong, 1991). Watson (1985, 1989) mentions the presence of acicular gypsum in subsurface crusts, indicating an origin by displacive crystallisation. Likewise, Verboom et al. (2003) attribute the formation of fibrous gypsum pendants to downward displacive crystallisation, controlled by preferential flows through highly porous media or by hydration/dehydration processes. On the contrary, Jafarzadeh and Burnham (1992) consider that for acicular gypsum, the space to extend in length may be the key factor, rather than any type of confinement.

Fibrous gypsum occurs mainly in geological formations, as part of extensional veins (e.g., Testa & Lugli, 2000). Fibrous gypsum in soils is generally interpreted as inherited from parent material (Nettleton et al., 1982; Porta, 1998; Cantón et al., 2001). Nevertheless, fibrous gypsum has been described as pedogenic precipitates in some studies (Worrall, 1957; Labib, 1970; El-Sayed, 1993), and a fibrous texture is also recognised for pendants (Verboom et al., 2003; Lebedeva & Konyshkova, 2011; Lebedeva et al., 2014).

2.2   Experimental Studies

Numerous studies on gypsum crystal formation have been published. Only those that are directly relevant for soil environments are discussed here.

Cody and Shanks (1974) studied the influence of pH, temperature and growing media on gypsum crystallisation. They obtained elongated prismatic crystals growing in silica gel and tabular crystals in bentonite gel. Cody (1979) showed that lenticular gypsum is formed in the presence of organic matter (mangrove leaves), at high pH and in conditions favouring slow growth. Organic matter in solution promoted a lower yield, a longer induction period and fewer crystals. This is in agreement with Porta (1986), who found that gypsum crystals growing in free solutions, both with and without sand, had prismatic and tabular habits, while those growing in the presence of cereal straw were lenticular. Cody and Cody (1988) explained the relationship between the lenticular crystal morphology and the presence of organic material by the inhibition of growth due to high-molecular weight organic complexes under alkaline pH, which are fixed along {010} planes that include water molecules. In contrast, Jafarzadeh and Burnham (1992) produced lenticular crystals in sands without organic matter at pH 3.6. They observed that the first crystals to form in soils are small euhedral tabular or anhedral crystals, which subsequently develop a lenticular morphology in a wide range of conditions. In addition to lenticular crystal development, Cody and Cody (1988) also report the development of twinning in the presence of organic compounds. Harouaka et al. (2014) obtained small, twinned, flat tabular crystals at fast precipitation rates and both flat tabular crystals and longer needles at slower rates.

Kushnir (1980) considers the Ca:SO_4 ratio as the key factor for crystal morphology, whereby high ratios promote the development of lenticular crystals and low values favour prismatic forms. This has been related to bonds of Ca^{2+} ions, lodged in $\{\bar{1}11\}$ planes, with organic compounds, inhibiting the growth in this direction (Aref, 1998). High temperatures (60 °C) have also been found to promote lenticular habits, whereas lower temperatures (30 °C) promote prismatic habits, in the absence of organic matter (Cody, 1979; Cody & Cody, 1988).

A prismatic habit is promoted by rapid growth, because in those conditions the organic molecules do not have enough time to interact with the faces of the growing crystal (Cody, 1979). The growth of long prismatic crystals is also promoted by low pH, due to the adsorption of H^+ ions on {010} and {110} faces. It favours the growth perpendicular to {111} and $\{\bar{1}03\}$ faces, parallel to the c-axis (Cody, 1979). The occurrence of elongated prismatic crystals in a context of pyrite oxidation (see Mees & Stoops, 2010, 2018) is probably due to pH effects (Cody, 1979). Similarly, Barta et al. (1971) obtained acicular crystals in gels at pH < 7 and more tabular crystals at pH > 8.

Salts (NaCl) in solution promote the growth of larger and fewer crystals, due to the higher solubility of gypsum (Zen, 1965) which leads to less nucleation. Monovalent ions favour the growth of larger and better prismatic crystals (Cody, 1979; Cody & Cody, 1988).

3. Groundmass

Gypsum that is part of the parent material should clearly be described as a groundmass component. This is, for instance, the case for remnants of gypsum rock, eolian gypsum sands and gypsiferous salt lake deposits. Other gypsum occurrences in soils are generally pedofeatures, formed by pedogenic processes. Nevertheless, from the descriptive point of view there are some situations in which all gypsum in a thin section could be considered as groundmass, especially when it forms the base material of the soil. One reason can be the difficulty to distinguish between inherited and pedogenic gypsum using standard microscopic techniques, especially for gypsum with a very small crystal size (Herrero & Porta, 2000; see also Carter & Inskeep, 1988) for non-microcrystalline gypsum). Bullock et al. (1985) recommend considering pedogenic micritic calcite within the groundmass as part of the micromass, which could also be applied for microcrystalline gypsum. A second reason can be the large total volume of pedogenic gypsum in many soils, whereby a description as groundmass material is in line with the procedure suggested for continuous pedogenic features such as laterite crusts (Bullock et al., 1985). A third reason is that, regardless of its origin, gypsum in gypseous materials is the base material of the soil, which determines its physical and chemical properties and its macro- and mesomorphological appearance. In view of these considerations, gypsum in gypseous horizons is considered as groundmass material in this chapter. However, a different approach is recommended in studies related to soil genesis, because the concept of pedofeatures has a connotation of pedogenic origin and because the pedofeature category of crystals and crystal intergrowths should apply without considering abundance (Mees, pers. comm.).

Gypsum can appear in soil materials of very different properties and origins, and therefore there is not a specific set of characteristics of the non-gypsic groundmass. In calcareous soils, the b-fabric is normally crystallitic micritic, but in non-calcareous materials it can also be stipple-speckled (Moghiseh & Heidari, 2012) or striated (Laya et al., 1993; Scalenghe et al., 2016). Granostriated b-fabrics related to gypsum crystals are sometimes observed, indicating pressures due to displacive crystal growth within the matrix. It is best shown in clayey, non-calcareous matrices, being masked in other cases. In clay-rich soils showing vertic features, gypsum crystals can be oriented parallel to porostriation along planar voids (Scalenghe et al., 2016). Another possible general feature is layering, in soils developed on parent materials with synsedimentary gypsum beds (e.g., Poch, 1992; Mees et al., 2012). Displacive growth of lenticular gypsum can disturb the original orientation of sedimentary groundmass components (Basyoni & Aref, 2015), and it can shatter rock fragments in reg soils (Amit et al., 1993).

Gypseous horizons are considered to be the end points of the gypsification process, which implies progressive precipitation of gypsum in the horizon from a soil solution saturated with respect to gypsum. Herrero et al. (1992) describe gypsum enrichment of a soil as displacive crystallisation, in combination with biological activity, giving rise to an 'isles fabric', which is characterised by non-gypsic 'isles' surrounded by lenticular gypsum occurrences (see Fig. 1A), which can appear as dense infillings (Lebedeva & Konyshkova, 2010, 2011). This fabric can also occur in horizons with xenotopic gypsum (Stoops, pers. comm.) and microcrystalline gypsum (Aznar et al., 2013). Barzanji and Stoops (1974) propose progressive enrichment of the groundmass by gypsum crystals consisting of the precipitation of gypsum in channels. It forms pore infillings that are later disrupted and incorporated into the groundmass by biological activity, which in turn creates new pore space for gypsum to precipitate as new pore infillings. The end point is a groundmass formed almost exclusively of gypsum crystals with a distinct vermicular microstructure. Reheis (1987) and Buck and Van Hoesen (2002) propose a mechanism of gypsification in four stages, similar to the one of Gile et al. (1966) for calcareous soils. Toomanian et al. (2001) also find four stages of gypsification, from loose crystal intercalations, infillings and nodules, gypsum pendants under gravels to fine-grained hypergypsic horizons or coarse-crystalline petrogypsic horizons, which are both firm continuous layers. The micromorphological classification of gypsiferous materials by Stoops and Poch (1994), based on morphological observations, indicates stages of gypsification as well, from few infillings and coatings in pores, to massive accumulations, with fabrics from lenticular to microcrystalline or coarse-crystalline xenotopic. Despite these observations, some hypergypsic horizons are not necessarily

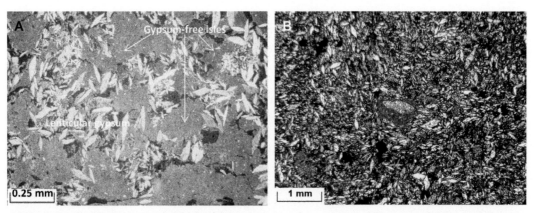

FIGURE 1 Powdery gypsic material. (A) Isles fabric, consisting of aggregates of fine-grained soil material masses (isles) surrounded by occurrences of lenticular gypsum (Bwy horizon, Gypsic Xerochrept, Pla d'Urgell, Catalonia, Spain) (PPL). (B) Hypergypsic material, mainly composed of silt-sized lenticular and tabular gypsum, with some calcareous nodules and coatings; some of the crystals have undergone dehydration during sample preparation (*gahza soil*, Uzbekistan) (XPL). *(A) Image by R.M. Poch. (B) Image by M. Lebedeva.*

the result of a progressive enrichment with gypsum, for instance, when soil forms on gypsum rock or on gypsum-rich eolian or lacustrine deposits.

The groundmass of massive accumulations of gypsum has several typologies depending on the degree of cementation and crystal size. Four main types are distinguished: powdery gypsic materials, microcrystalline gypsum, rupture-resistant gypsic materials and eolian gypsic materials.

3.1 Powdery Gypsic Materials

Powdery gypsic materials (commonly > 60% gypsum) are generally apedal, with a roughly coarse monic c/f-related distribution pattern, although sometimes a very coarse prismatic structure due to vertical fissures is recognised in the field. Gypsic powdery materials are classified as 'hypergypsic lenticular' or 'hologypsic lenticular' sub-formations according to the proposal of Stoops and Poch (1994). The coarse material in the groundmass is composed mainly of sand-sized lenticular gypsum crystals. It may also consist of quartz grains and coarse gypsum crystals with various habits (Aref, 2003). The fine material mainly appears as part of the non-gypsic fraction, which occurs as 'isles' separated by the gypsum ('isles fabric'; Herrero et al., 1992) (Fig. 1A). Gypsum infillings and coatings are irregularly distributed and strongly reworked by faunal activity, which reorganises gypsum crystals in passage features with crescent patterns, according to the gypsification process proposed by Barzanji and Stoops (1974). Powdery gypsum is well documented for the so-called *gazha* soils in Russia and Uzbekistan (Pankova & Yamnova, 1987; Verba, 1990; Verba & Yamnova, 1997) (Fig. 1B), attributed to groundwater precipitation (Yamnova & Pankova, 2013) but also to physical weathering of gypsum rocks (Goryachkin et al., 2003).

3.2 Microcrystalline Gypsic Materials

Massive accumulations of microcrystalline gypsum are described in the field as having a flourlike consistency (Neher & Bailey, 1976; Heinze & Fiedler, 1984; Herrero, 1987), owing to the silt size (<20 μm) of the gypsum crystals. They have a pinkish-white colour, and they are rupture-resistant when dry and friable when moist (Herrero et al., 2009). Due to their small size they can be best observed by SEM. In thin sections, they appear as faint yellow masses in plane-polarised light (Fig. 2B), with an almost undifferentiated b-fabric in cross-polarised light, showing only few clearly birefringent spots (Fig. 2A) (Porta & Herrero, 1988, 1990). SEM observations confirm a predominantly lenticular crystal habit (Fig. 2C). Similar materials have been described by Warren (1982), Pankova and Yamnova (1987), Watson (1988) and Amit and Yaalon (1996). Their microfabric corresponds to hypergypsic or hologypsic microcrystalline (Stoops & Poch, 1994).

Microcrystalline gypsic horizons are interpreted as being formed by *in situ* alteration of gypseous materials such as gypsum rock (Herrero, 1991; Artieda, 1996, 2013) or by transformation along the surface of pedogenic gypsic or petrogypsic horizons (Stoops & Ilaiwi, 1981). Massive accumulations can form by replacement of coarse-grained

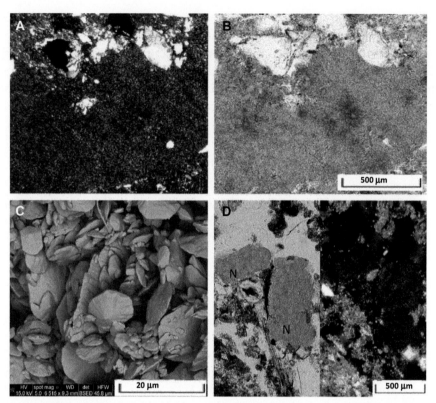

FIGURE 2 Microcrystalline gypsic material. (A) Microcrystalline gypsum (XPL). (B) Same field in PPL, illustrating the yellowish colour of microcrystalline gypsum (Quinto, Zaragoza, Spain). (C) SEM observations confirm a predominantly lenticular crystal habit. (D) Irregular microcrystalline gypsum nodules (N), with clearly lower birefringence than the fine-grained calcareous groundmass (Mediana de Aragón, Zaragoza, Spain) (PPL, XPL). *Images by O. Artieda.*

gypsum by fine-grained gypsum, proceeding from grain boundaries, similar to micritisation in calcrete development (Aref, 2003). Microcrystalline gypsum forms by quick dissolution/reprecipitation processes, which would hinder ion migration to gypsum nucleation sites, thus promoting precipitation of multiple small crystals. Microcrystalline horizons have also been identified in terrace or alluvial fan deposits (Herrero et al., 1996; Lebedeva et al., 2015) and also as layers between limestone strata at the top of calcareous platforms (Artieda, 1996). These horizons are commonly found at such surface positions, because microcrystalline textures are more likely to develop in zones of high water potential gradient.

3.3   Rupture-Resistant Gypsic Materials

Rupture resistance (Soil Survey Staff, 1993) seems to be a better suited concept than cementation when applied to gypsiferous soils, as discussed by Herrero and Porta (2000) and

Herrero (2004). Rupture-resistant behaviour occurs when gypsum crystals interlock, giving rise to a xenotopic fabric, typical of some petrogypsic horizons (El-Sayed, 1993; Stoops & Poch, 1994), a characteristic reflected in the WRB classification (IUSS Working Group WRB, 2015) and in the Russian system (*arzyk* horizon; Minashina & Khaknazarov, 1988). These horizons, composed of interlocking anhedral crystals (Fig. 3), can be extremely porous, with a crumb or sugarlike appearance in the field. They can also be either continuous or discontinuous, the latter with gypsum occurring in the form of pockets or pendants (Poch, 1992; Artieda, 1996; Toomanian et al., 2001; Hamdi-Aissa, 2002). Coarse-crystalline gypsic materials form the groundmass when gypsum precipitates at slow rates from shallow water bodies due to evaporation of sulphate-bearing waters (Yamnova & Pankova, 2013). Large lenticular gypsum crystals, with highly corroded boundaries, can be found 'floating' in a more fine-grained groundmass of microcrystalline or lenticular gypsum, resulting in a porphyrotopic fabric (Aref, 2003).

In the case of pendants, elongated anhedral crystals, perpendicular to the base of the gravel grains, interlock to form a dense fabric (Amit & Yaalon, 1996). Similar features have been described for compact horizons in palaeosoils, characterised by elongated vertically oriented gypsum crystals forming a spongy mass, sometimes with stony inclusions varying in size (Gubin, 1984; Verba & Yamnova, 1997). These features are

FIGURE 3 Rupture-resistant gypsic material. (A) Travertine gypsum (By horizon, Typic Haplogypsid, Quinto, Zaragoza, Spain) (XPL). (B) Travertine gypsum (Bym horizon, Petrogypsic Xerochrept, Algerri, Catalonia, Spain) (XPL). *(A) Image by O. Artieda. (B) Image by R.M. Poch.*

similar to those of *travertinic gypsum* (Artieda, 1996) and to those in soils with *gypsic threads* (Toomanian et al., 2001). Travertinic gypsum is highly porous (up > 50%) and made up of gravel of various types, including gypsum fragments with clear dissolution features. Gypsum between these gravels occurs as inequigranular hypidiotopic mosaics, together with patches of microcrystalline lenticular gypsum and with lenticular or anhedral gypsum along the sides of the gravel particles (Artieda, 1996). Differences between hypergypsic and petrogypsic horizons in the same profile are discussed by Herrero et al. (2009).

Other occurrences described by some authors as petrogypsic horizons have a microcrystalline gypsum fabric (e.g., Chen, 1997; Aref, 2003). This is not in disagreement with the hypothesis of interlocking of crystals as the cause of cementation, because the small crystal size does not allow direct observation of crystal boundaries.

3.4 Aeolian Gypsic Materials

Aeolian gypsarenites are commonly characterised by mixtures of gypsum crystals and rounded clay aggregates (Fig. 4; see also Mees et al., 2012). They are the result of reworking of loose gypsum deposits that can become disrupted by salt growth (Bowler, 1976, 1983; Watson, 1983; Warren, 1986; Khalaf et al., 2014). They may occur as laminated deposits with alternating layers of lenticular crystals (Magee, 1991) and also as uniform sand consisting of slightly rounded gypsum crystals. Warren (1982) ascribes the lenticular crystal habit to partial dissolution of wind-blown prismatic precursors. In these arid environments, gypsum behaves more as a clast than as a soluble material. Aeolian gypsum deposits with associated palaeosoils can be preserved in the rock record and used to interpret palaeoenvironments (Hamdi-Aissa et al., 2002; Lawton & Buck,

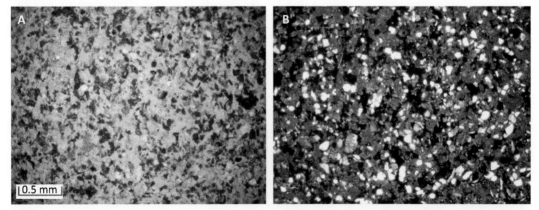

FIGURE 4 Aeolian gypsic material. Aeolian gypsarenite from White Sands, New Mexico, USA (A) PPL. (B) Same field in XPL. *Images by R.M. Poch.*

2006; Khalaf et al., 2014). Possible confusion with deposits formed by water sorting along the margins of sabkhas (Stoops & Ilaiwi, 1981) should be kept in mind.

Gypsiferous dust from terrestrial sources (Amit et al., 1993; El-Sayed, 1993; Buck & Van Hoesen, 2002; Buck et al., 2011; Boixadera et al., 2015), as well as marine aerosols and sulphur-rich volcanic ashes (Podwojewski, 1994), has been claimed as the origin of pedogenic gypsum in several environments.

4. Pedofeatures

Gypsic pedofeatures are morphologically classified as intrusive (Stoops, 2003) because the amount of incorporated groundmass material is generally very small. Gypsic pedofeatures include infillings, coatings, nodules, and crystals and crystal intergrowths. These types are commonly found together in the same horizon. Their shape in various 2D cross sections in a single sample generally allows identification of their 3D morphology. Various terms have been used for some gypsic pedofeatures, which complicate comparisons between studies.

All variants of gypsic pedofeatures may be found in soils affected by fluctuations of the groundwater table (Yamnova, 1990; Verba, 1996; Yamnova & Pankova, 2013) but also in relation to the depth of the wetting front in a leaching environment (Boixadera et al., 2015). The crystal morphology seems to be related to the distance from the water table, in the sense of increasing crystal size, from microcrystalline gypsum at the top to coarse-grained closer to the water table (Pankova & Yamnova, 1980, 1987; Minashina & Khaknazarov, 1988; Verba, 1990). In the zone of complete water saturation, the largest gypsum crystals are formed, without dissolution features and with indications of enlargement of crystals formed during an earlier growth stage. They occur as infillings in almost all large voids, resulting in dense complete infillings that plug all pores (Lim et al., 1975). In contrast, gypsum crystals are larger and more abundant in the upper part of some Vertisol profiles than in lower parts (Podwojewski, 1995).

4.1 Infillings and Coatings

Infillings and coatings are clearly pedogenic since they are accumulations in pores or around coarse grains or rock fragments. Commonly, the presence of gypsum-saturated groundwater, under a non-percolating soil water regime, creates the conditions for precipitation of gypsum in macropores and the formation of these pedofeatures.

Infillings are normally composed of lenticular crystals, which can eventually form mosaic intergrowths. Infillings composed of non-lenticular crystals have been observed by Mees (1999) and Mees and De Dapper (2005). The infillings may be dense complete (Fig. 5A), dense incomplete (Fig. 5B), loose continuous or loose discontinuous (Fig. 5C). Some of these types of infillings may occur in root channels (Aref, 2003). The constituent crystals of dense infillings can be perpendicular to the pore walls (Brock-Hon et al.,

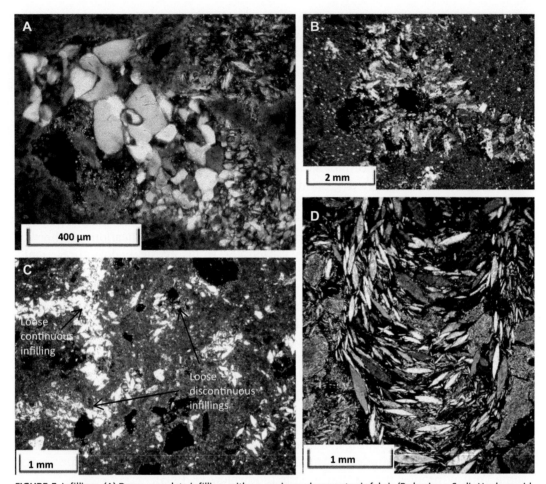

FIGURE 5 Infillings. (A) Dense complete infilling, with an equigranular xenotopic fabric (By horizon, Sodic Haplogypsid, Belchite, Zaragoza, Spain) (XPL). (B) Dense incomplete infilling, with an equigranular hypidiotopic fabric (Bky horizon, Typic Petrogypsid, Quinto, Zaragoza, Spain) (XPL). (C) Loose continuous and discontinuous infillings composed of lenticular gypsum (4Bwy horizon, Gypsic Xerochrept, Pla Urgell, Catalonia, Spain) (XPL). (D) Loose continuous infilling, composed of lenticular gypsum with crescent fabric (Cry horizon, Typic Haplogypsid, Quinto, Zaragoza, Spain) (XPL). *(A) Image by O. Artieda. (B) Image by O. Artieda. (C) Image by R.M. Poch. (D) Image by O. Artieda.*

2012). Structures with a crescent fabric, described earlier as a groundmass feature, are in fact infillings, formed by biological activity (Fig. 5D).

Miedema et al. (1974) use the term 'crystal tubes' for gypsum in channels, based on terminology of Brewer (1964). Caldwell (1976), Jacobson et al. (1988) and Magee (1991) use similar terms for the same features. They have also been described as 'vermiform gypsum' (e.g., Porta & Herrero, 1988; Herrero et al., 1992; Artieda, 1996; Poch et al., 1998; Porta, 1998). Other authors use 'pseudomycelia' (Stoops & Ilaiwi, 1981), which might in fact be best confined to calcium carbonate accumulations (Porta & Herrero, 1988).

Gypsum coatings are much more rare than infillings. Aznar et al. (2013) describe coatings of lenticular gypsum that are interpreted as recrystallisation, along pores, of gypsum that was part of the parent material. Coatings of hypidiotopic and xenotopic gypsum along pore walls have also been reported by Boixadera et al. (2015), Lebedeva and Konyushkova (2016), Pérez-Lambán et al. (2016) and Scalenghe et al. (2016). Dultz and Kühn (2005) attribute a biological origin to gypsum coatings associated with roots, where they show an increase in crystal size with distance from the root, together with a change in crystal morphology from equant to tabular and lenticular. In other cases the reverse trend is observed, as for gypsum rhyzoliths in Mediterranean loesslike

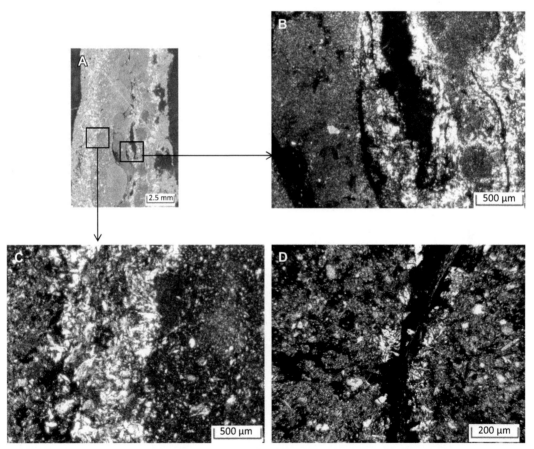

FIGURE 6 Infillings. (A) Gypsum rhyzolith consisting of a dense incomplete microcrystalline gypsum infilling with remains of the root epidermis, with associated hypo- and quasicoatings composed of lenticular gypsum (soil on calcareous loess, Ebro Valley, Spain) (XPL). (B) Detail of the dense incomplete microcrystalline gypsum infilling, with typic xenotopic gypsum coatings along the sides of the remaining voids (centre) (XPL). (C) Detail of the quasicoating composed of hypidiotopic lenticular gypsum (centre), showing also some microcrystalline gypsum infillings with variable crystal size outside this feature (XPL). (D). Gypsum along a pore with plant remains, suggesting a link between root and gypsum occurrences (Caspian lowlands, Russia) (XPL). *(C) Images by R.M. Poch. (D) Image by M. Lebedeva.*

materials (Boixadera, pers. comm.), corresponding to microcrystalline gypsum infillings with associated hypocoatings composed of coarser lenticular gypsum (Fig. 6A to C). Ryzoliths described by other authors (Khechai & Daoud, 2016) consist of impregnative hypocoatings along root channels in sandy materials. The formation of gypsum coatings around roots can be rather fast, as observed for Caspian lowland soils (Fig. 6D; Lebedeva & Konyushkova, 2010). Fast growth is also apparent for gypsum crystals precipitated in roots of nursery-grown pine seedlings (Carey et al., 2002). Coatings in pores other than those occupied by roots consist of lenticular gypsum, where no root action is assumed to influence the crystallisation process, although being pores they can act as a preferential precipitation site (Fig. 7A). Lebedeva et al. (2015) describe gypsum coatings covered by silty coatings around vesicles in topsoils of extremely arid, saline soils in Kazakhstan, but their genesis is not clear.

Thin gypsum crusts covering the soil surface can be considered as a type of coating. Aref et al. (2014) describe gypsum crusts in sebkhas that are composed of elongated prismatic gypsum crystals and swallowtail twins growing perpendicular to microbial laminae. A relationship between cyanobacteria and gypsum precipitation in saline soil crusts was suggested by Canfora et al. (2016).

Another type of coating consists of pedogenic gypsum along the sides of gypsum rock fragments. These fragments sometimes show dissolution features along the top and gypsum coatings along the base (Fig. 7B). Gypsum coatings without a fixed orientation have also been observed. Pendants are a relatively common type of coating, described in an earlier section. Khademi and Mermut (2003) and Lebedeva et al. (2015) describe large pendants, composed of randomly oriented lenticular and equant crystals that formed partly in channels and planar voids, produced by downward movement of water.

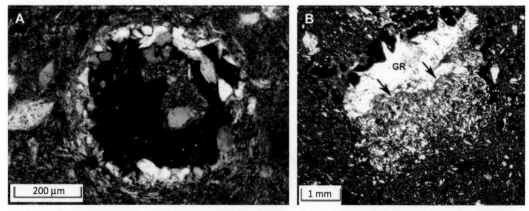

FIGURE 7 Coatings. (A) Typic coating with lenticular and anhedral gypsum (By (Y) horizon, Typic Calcigypsid, Quinto, Zaragoza, Spain) (XPL). (B) Coating consisting of lenticular gypsum, along the surface (*arrow*) of a gypsum rock fragment (Gr) with dissolution features along the top (Y horizon, Leptic Haplogypsid, Quinto, Zaragoza, Spain) (XPL). *Images by O. Artieda.*

4.2 Nodules

Gypsum nodules are probably formed by gypsum precipitation from a single point and can be polyphasic. Processes of dissolution-precipitation affecting a gypsic parent material can also produce nodules. The constituent crystals of nodules show a variety of fabrics, but a lenticular hypidiotopic fabric is most widespread (Fig. 8). Buck and Van Hoesen (2002) describe small round aggregates composed of small (0.5-3 mm) gypsum crystals of different habits, initially assumed to have been produced by actinomycetes. In a later study, it was found that these pedofeatures can form without any biological interaction (Buck et al., 2006; see also Verba & Yamnova, 1997; Yamnova & Golovanov, 2010a). Microcrystalline gypsum nodules (Fig. 2D), formed by recrystallisation, have been described for soils on gypsum-rich parent materials (Aznar et al., 2013). Nodules composed of fanlike or rosettelike aggregates of lenticular gypsum crystals have been described for gypseous palaeosoils (Pérez-Lambán et al., 2016), Technosols (Heidari & Asadi, 2015) and Vertisols (Podwojewski, 1994).

4.3 Crystals and Crystal Intergrowths

In the groundmass of gypseous soils it is common to find scattered gypsum crystals, with sizes between 20 μm and some millimetres with a variety of crystal morphologies. According to Mees et al. (2012), synsedimentary gypsum in ephemeral saline lake

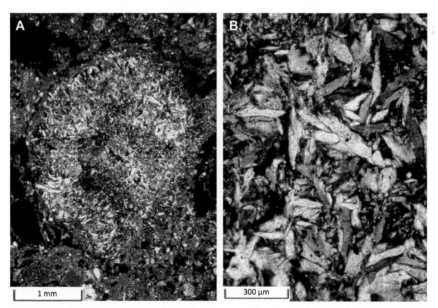

FIGURE 8 Gypsum nodule composed of lenticular gypsum (Mediana de Aragón, Zaragoza, Spain) (XPL). (A) Overview (XPL). (B) Detail of previous. *Images by O. Artieda.*

deposits can be distinguished from diagenetic gypsum by layered fabrics with variations in gypsum crystal morphology, size and/or abundance, often with horizontal alignment and with occasional occurrences of grading, without any indications of clastic reworking. Other features include reverse grading and vertical changes in crystal morphology. The same authors indicate that when the layered structure is missing, the occurrence of non-lenticular habits can be a criterion for the recognition of synsedimentary gypsum.

The most common crystal habit of gypsum in the groundmass of soils is lenticular, a habit that has also been recognised to develop subaqueously in salt lake basins (Mees et al., 2012). Occurrences that formed within the groundmass often contain groundmass-derived impurities, often as thin bands (dust lines) that are either parallel to the crystal faces or not, corresponding to earlier growth stages.

These isolated crystals form not only by *in situ* crystallisation from gypsum-saturated water but also by disruption of infillings and coatings by animal activity or other processes, incorporating them into the groundmass (Stoops et al., 1978). In some cases, the crystals grow at the expense of gypsum rock fragments, which act as nuclei. These overgrowths are generally in optical continuity with the fragment, which can be partly lined by impurities (Fig. 9A). Plate-shaped fragments can be bent by the pressure exerted between crystals that develop along its surface (Fig. 9B).

Zonation within gypsum crystals may be used as a record of previous growth stages. It can develop in dry or ephemeral saline lake environments, where seasonal variations in groundwater level promote discontinuous growth of the crystals. Mees (1999) and Mees et al. (2012) observed non-lenticular growth stages within lenticular gypsum, which show lateral extension in directions within the plane of flattening from tabular to lenticular habits. These are interpreted as a record of variations of environmental conditions during gypsum growth.

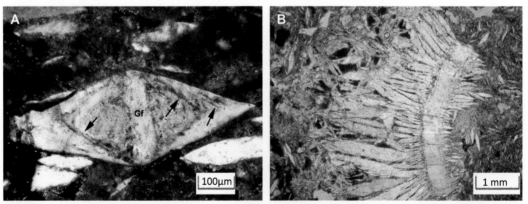

FIGURE 9 Gypsum overgrowths. (A) Lenticular crystal formed by precipitation around a gypsum fragment (Gr), lined by impurities (*arrows*) (Cry horizon, Leptic Haplogypsid, Belchite, Zaragoza, Spain) (XPL). (B) Lenticular crystals growing from and bending a gypsum rock fragment (Torre La Piedra, Huesca, Spain) (PPL). *(A) Image by O. Artieda. (B) Image by R.M. Poch.*

Very coarse crystals or crystal intergrowths (up to 50 cm) are described by various field terms, including desert roses. Their formation is related to conditions with an oscillating groundwater table, present-day or relict, whereby gypsum crystallises at or just above the water table (Watson, 1985; Samy & Metwally, 2012). Their occurrence has also been related to waterlogging in other studies (Tolchel'nikov, 1962; Stoops & Ilaiwi, 1981; Pankova & Yamnova, 1987). In extreme aridic desert soils of Mongolia starlike arrangements (roses) of lenticular crystals have been described as a result of exchange reactions (Lebedeva et al., 2009).

Depending on the texture of the host material, these coarse crystals may show a poikilotopic fabric, with important amounts of sand and host sediment inclusions (Masson, 1955; Tolchel'nikov, 1962; Watson, 1985; Aref, 2003; Fitzpatrick et al., 2014; Khalaf et al., 2014). Twinning, both swallow-tail and other types, has also been reported (Carenas et al., 1982; Aref et al., 2014; Basyoni & Aref, 2015) and might be most common in saline environments.

5. Other Features Common in Soils With Gypsum

5.1 Dissolution Features

Dissolution features can characterise gypsum in irrigated soils. Decomposition, preservation and development of gypsum crystals in such soils have been studied not only for virgin and reclaimed soils (Gagarina et al., 1988; Gracheva et al., 1988; Gubin & Kovda, 1988, Verba et al., 1995; Lebedeva & Konyshkova, 2011) but also in laboratory experiments (Ramazanov & Lim, 1981; Halitim & Robert, 1987; Verba & Deviatykh, 1992). Irrigation-induced dissolution of gypsum in the upper part of gypsic horizons can occur simultaneously with illuviation of carbonates or clay, so that gypsum crystals become coated and their further dissolution is hindered (Minashina, 1964, 1973). This phenomenon has been observed in various studies (Lim et al., 1975; Yamnova, 1995, 1996; Verba & Yamnova, 1997).

Under furrow irrigation, small gypsum crystals in the topsoil are dissolved, and when irrigation is long lasting the larger lenticular crystals are dissolved as well (Ramazanov & Lim, 1981; Verba et al., 1995; Verba & Yamnova, 1997). Corrosion of the crystals is revealed by SEM observations: the outer faces become fissured, and the crystals undergo gradual decay along cleavage planes (Fig. 10). The crystals also become rounded. In experiments with monoliths percolated by large volumes of water, the initially compact gypsic horizons became porous and friable, and secondary tabular crystals were formed in voids in all subsurface horizons (Verba & Deviatykh, 1992). Alternative dissolution and reprecipitation under contrasted rainfall regimes as in Mediterranean climates leads to neoformation of lenticular gypsum crystals on lithogenic gypsum (Bellanca & Neri, 1993; Laya et al., 1993; Pérez-Lambán et al., 2016).

In an SEM study of diagenesis of gypsum sands, Schenk and Fryberger (1988) found that gypsum dissolution in the capillary fringe and below the watertable creates

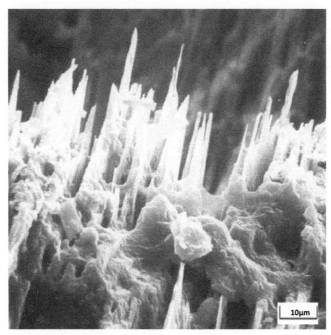

FIGURE 10 Dissolution features. SEM image of dissolution features of a gypsum crystal in a soil from Egypt (Ghent University archive).

crystallographically controlled elongate, prismatic etch pits on the surface of gypsum grains.

Void pseudomorphs after gypsum have been used as palaeoclimatic indicators of changes from arid to more humid conditions (Mestdagh et al., 1999) or of changes in drainage conditions (Junyent et al., 2012). Depending on the original mode of occurrence, they can be starlike (Mestdagh et al., 1999) or lenticular (Hamdi-Aissa, 2002). These mouldic voids have been described for the upper part of gypsic horizons (Verba & Yamnova, 1997; Gagarina, 2004) (Fig. 11A), as well as for surface horizons (Poch et al., 2004, 2009). When the groundmass is not calcitic, porostriation may be visible along the sides of these lenticular voids (Fig. 11B).

Microtomography has shown the possibility to study the gypsum leaching stages and evolution of porosity in 3D (Yamnova & Abrosimov, 2013).

5.2   Calcite and Other Pseudomorphs

Micritic or microsparitic calcite pseudomorphs after gypsum are quite frequent in soils from arid and semiarid areas (Feofarova, 1956; Yarilova, 1958; Carter & Inskeep, 1988; Gumuzzio & Casas, 1988; Sullivan, 1990; Stoops & Delvigne, 1990; Gerasimova et al., 1996; Verba & Yamnova, 1997; Mestdagh et al., 1999; Hamdi-Aissa, 2002; Artieda, 2004; Golovanov et al., 2005; Artieda et al., 2012; Yamnova, 2016) (Fig. 12). They are interpreted

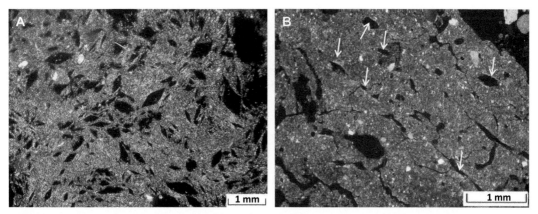

FIGURE 11 Dissolution features. (A) Moulds of lenticular and tabular gypsum crystals (Gypsic Solonchak, Uzbekistan) (XPL). (B) Layered surface horizon with porostriation along the sides of lenticular mouldic voids (*arrows*); image tilted to better show the b-fabric (A horizon, Orthithionic Fluvisol) (XPL). *(A) Image by M. Lebedeva. (B) From Poch et al.(2009), reproduced with permission of the publisher.*

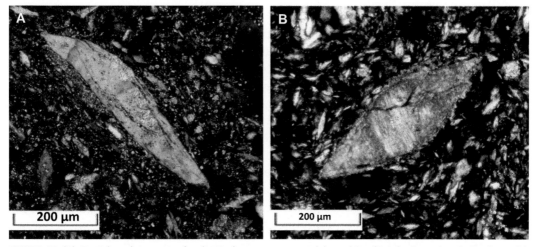

FIGURE 12 (A) Partial replacement of a lenticular gypsum crystal by calcite (By1 horizon, Leptic Haplogypsid, Mediana de Aragón, Zaragoza, Spain) (XPL). (B) Total replacement of a lenticular gypsum crystal by calcite (Ay horizon, Leptic Haplogypsid, Quinto, Zaragoza, Spain) (XPL). *Images by O. Artieda.*

as indicators of a more arid climate in the past, followed by more humid conditions. High biological activity (higher pCO_2) have been assumed to promote dissolution of gypsum and substitution by calcite. The process normally progresses from the sides of the crystal and along cleavage planes.

Sullivan (1990) considers two possible mechanisms for the formation of these pseudomorphs: (i) dissolution of gypsum followed by calcite precipitation in the void pseudomorph, and (ii) replacement of gypsum by calcite. The former could explain occurrences of pseudomorphs in the groundmass, with the voids acting as moulds, but not occurrences as loose infillings in pores. The second mechanism is supported by the occurrence of partially replaced crystals (Thompson et al., 1991; Stoops, 2003).

Replacement of gypsum by a more soluble mineral, such as halite, has also been proposed (Hussain & Warren, 1989). Pseudomorphs reported for pre-Quaternary deposits, including quartz pseudomorphs after lenticular gypsum (e.g., El Khoriby, 2005), are generally not pedogenic.

6. Discussion and Conclusions

The most widely used soil classification systems, Soil Taxonomy (Soil Survey Staff, 2014) and WRB (IUSS Working Group WRB, 2015), define gypsic and petrogypsic horizons on the basis of the gypsum content, and both systems require 'secondary visible gypsum' to be greater than 1% by volume. The WRB system also lists micromorphological characteristics of groundmass, micromass, microstructure and nodules for the petrogypsic horizon. This enumeration cannot be exhaustive because new forms are likely to be found as soil science advances.

When reading the definitions of gypsic and petrogypsic horizons of Soil Taxonomy (Soil Survey Staff, 1999), some researchers may get the erroneous idea that the mere increase in gypsum content leads to cementation. In WRB (IUSS Working Group WRB, 2015), the petrogypsic horizon does not require a higher gypsum content than the gypsic horizon, but it is described as having a compacted microstructure with only a few cavities, with a matrix composed of densely packed lenticular gypsum crystals mixed with a small amount of detrital material. However, depending on the host material and other factors, enrichment with gypsum may cause either disaggregation of the original material or an increase of its cohesion (Plet-Lajoux et al., 1971; Artieda, 1996, 2004). Indeed, although most hypergypsic horizons have a high rupture resistance when dry, the criteria for cementation are not fulfilled, because they are formed by individual crystals. Gypsum micromorphology may also be applied to soil mapping and soil classification, as different micromorphological features correspond to different soil types (Yamnova & Golovanov, 2010b; Golovanov & Yamnova 2013).

In studies of soil genesis, micromorphology is a powerful tool for determining a pedogenic origin for gypsum, despite the difficulties pointed out in this chapter. In addition, micromorphology can also help to understand and predict the behaviour of soils containing gypsum for agricultural, environmental and civil engineering land uses, as well as in geoarcheology.

References

Allen, B.L., 1985. Micromorphology of Aridisols. In Douglas, L.A. & Thompson, M.L. (eds.), Soil Micromorphology and Soil Classification. Soil Science Society of America Special Publication 15, SSSA, Madison, pp. 197-216.

Amit, R. & Yaalon, D.H., 1996. The micromorphology of gypsum and halite in Reg soils – the Negev Desert, Israel. Earth Surface Processes and Landforms 21, 1127-1143.

Amit, R., Gerson, R. & Yaalon, D.H., 1993. Stages and rate of the gravel shattering process by salts in desert Reg soils. Geoderma 57, 295-324.

Aref, M.A.M., 1998. Holocene stromatolites and microbial laminites associated with lenticular gypsum in a marine-dominated environment, Ras El Shetan area, Gulf of Aqaba, Egypt. Sedimentology 45, 245-262.

Aref, M.A.M., 2003. Classification and depositional environments of Quaternary pedogenic gypsum crusts (gypcrete) from east of the Fayum Depression, Egypt. Sedimentary Geology 155, 87-108.

Aref, M.A.M., Basyoni, M.H. & Bachmann, G.H., 2014. Microbial and physical sedimentary structures in modern evaporitic coastal environments of Saudi Arabia and Egypt. Facies 60, 371-388.

Arricibita, F.J., Íñiguez, J. & Val, R.M., 1988. Estudio de los gypsiorthids de Navarra. Anales de Edafología y Agrobiología 47, 199-220.

Artieda, O., 1996. Génesis y Distribución de Suelos en un Medio Semiárido. Quinto (Zaragoza). Ministerio de Agricultura, Pesca y Alimentación, Madrid, 222 p.

Artieda, O., 2004. Materiales Parentales y Geomorfología en la Génesis de Aridisoles en un Sector del Centro del Valle del Ebro. PhD Dissertation, Universidad de Zaragoza, 567 p.

Artieda, O., 2013. Morphology and micro-fabrics of weathering features on gyprock exposures in a semiarid environment (Ebro Tertiary Basin, NE Spain). Geomorphology 196, 198-210.

Artieda, O., Herrero, J. & Drohan, P.J., 2006. A refinement of the differential water loss method for gypsum determination in soils. Soil Science Society of America Journal 70, 1932-1935.

Artieda, O., Rodríguez-Ochoa, R. & Herrero, J., 2012. Calcite pseudomorphs after lenticular gypsum crystals in Aridisols of the central Ebro Valley, Spain. In Poch, R.M., Casamitjana, M. & Francis M.L. (eds.), Proceedings of the 14th International Working Meeting on Soil Micromorphology, Edicions de la Universitat de Lleida, pp. 59-63.

Aznar, J.M., Poch, R.M. & Badía, D., 2013. Soil catena along gypseous woodland in the middle Ebro Basin: soil properties and micromorphology relationships. Spanish Journal of Soil Science 3, 28-44.

Barta, C., Zemlicka, J. & Rene, V., 1971. Growth of $CaCO_3$ and $CaSO_4.2H_2O$ crystals in gels. Journal of Crystal Growth 10, 158-162.

Barzanji, A.F. & Stoops, G., 1974. Fabric and mineralogy of gypsum accumulations in some soils of Iraq. Transactions of the 10th International Congress of Soil Science, Volume VII, Moscow, pp. 271-277.

Basyoni, M.H. & Aref, M.A., 2015. Sediment characteristics and microfacies analysis of Jizan supratidal sabkha, Red Sea coast, Saudi Arabia. Arab Journal of Geosciences 8, 9973-9992.

Bellanca, A. & Neri, R., 1993. Dissolution and precipitation of gypsum and carbonate minerals in soils on evaporite deposits, central Sicily: isotope geochemistry and microfabric analysis. Geoderma 59, 263-277.

Benayas, J., Guerra, A., Batlle, J. & Gumuzzio, J., 1988. Micromorfología de algunos suelos con acumulaciones de yeso en la región central española (Toledo, Ciudad Real). Anales de Edafología y Agrobiología 47, 221-241.

Boixadera, J., Poch, R.M., Lowick, S.E. & Balasch, J.C., 2015. Loess and soils in the eastern Ebro Basin. Quaternary International 376, 114-133.

Bowler, J.M., 1976. Aridity in Australian soils: age, origin and expression in aeolian landforms. Earth Science Reviews 12, 179-310.

Bowler, J.M., 1983. Lunettes as indices of hydrologic change: a review of Australian evidence. Proceedings of the Royal Society of Victoria 95, 147-168.

Brewer, R., 1964. Fabric and Mineral Analysis of Soils. John Wiley and Sons, New York, 470 p.

Brock-Hon, A.L., Robins, C.R. & Buck, B.J., 2012. Micromorphological investigation of pedogenic barite in Mormon Mesa petrocalcic horizons, Nevada USA: implication for genesis. Geoderma 179/180, 1-8.

Buck, B.J. & Van Hoesen, J., 2002. Snowball morphology and SEM analysis of pedogenic gypsum, Southern New Mexico, USA. Journal of Arid Environments 51, 469-487.

Buck, B.J., Wolff, K., Merkler, D.J. & McMillan, N.J., 2006. Salt mineralogy of Las Vegas Wash, Nevada: morphology and subsurface evaporation. Soil Science Society of America Journal 70, 1639-1651.

Buck, B.J., King, J. & Etyemezian, V., 2011. Effects of salt mineralogy on dust emissions, Salton Sea, California. Soil Science Society of America Journal 75, 1971-1985.

Bullock, P., Fedoroff, N., Jongerius, A., Stoops, G., Tursina, T. & Babel, U., 1985. Handbook for Soil Thin Section Description. Waine Research Publications, Wolverhampton, 152 p.

Caldwell, R.H., 1976. Holocene Gypsum Deposits of the Bullara Sunkland, Carnarvon Basin, Western Australia. PhD Dissertation, University of Western Australia, 123 p.

Canfora, L., Vendramin, E., Vittori Antisari, L., Lo Papa, G., Dazzi, C., Benedetti, A., Iavazzo, P., Adamo, P., Jungblut, A.D. & Pinzari, F., 2016. Compartmentalization of gypsum and halite associated with cyanobacteria in saline soil crusts. FEMS Microbiology Ecology, 92, n° fiw080, 13 p.

Cantón, Y., Solé-Benet, A., Queralt, I. & Pini, R., 2001. Weathering of a gypsum-calcareous mudstone under semi-arid environment at Tabernas, SE Spain: laboratory and field-based experimental approaches. Catena 44, 111-132.

Carenas, B., Marfil, R. & De la Peña, J.A., 1982. Modes of formation and diagnostic features of recent gypsum in a continental environment, La Mancha, Spain. Estudios Geológicos 38, 345-359.

Carey, W.A., South, D.B. & Albrecht-Schmitt, T.E., 2002. Gypsum crystals on roots of nursery-grown pine seedlings. Communications in Soil Science and Plant Analysis 33, 1131-1137.

Carter, B.J. & Inskeep, W.P., 1988. Accumulation of pedogenic gypsum in western Oklahoma soils. Soil Science Society of America Journal 52, 1107-1113.

Chen, X.Y., 1997. Pedogenic gypcrete formation in arid central Australia. Geoderma 77, 39-61.

Cody, R.D., 1979. Lenticular gypsum: occurrences in nature, and experimental determinations of effects of soluble green plant material on its formation. Journal of Sedimentary Petrology 49, 1015-1028.

Cody, R.D., 1991. Organo-crystalline interactions in evaporite systems: the effects of crystallization inhibition. Journal of Sedimentary Petrology 61, 704-718.

Cody, R.D. & Cody, M., 1988. Gypsum nucleation and crystal morphology in analog saline terrestrial environments. Journal of Sedimentary Petrology 58, 247-255.

Cody, R.D. & Shanks, H.R., 1974. A comparison of calcium sulphate dihydrate grown in clay gels and in sodium silicate gels. Journal of Crystal Growth 23, 275-281.

Dultz, S. & Kühn, P., 2005. Occurrence, formation and micromorphology of gypsum in soils from the Central-German Chernozem region. Geoderma 129, 230-250.

El Khoriby, E.M., 2005. Origin of the gypsum-rich silica nodules, Moghra Formation, Northwest Qattara depression, Western Desert, Egypt. Sedimentary Geology 177, 41-55.

El-Sayed, M.I., 1993. Gypcrete of Kuwait: field investigation, petrography and genesis. Journal of Arid Environments 25, 199-209.

Eswaran, H. & Zi-Tong, G., 1991. Properties, genesis, classification, and distribution of soils with gypsum. In Nettleton, W.D. (ed.), Occurrence, Characteristics, and Genesis of Carbonate, Gypsum, and Silica Accumulations in Soils. Soil Science Society of America Special Publication 26, SSSA, Madison, pp. 89-119.

FAO, 1990. Management of Gypsiferous Soils. FAO Soils Bulletin 62. FAO, Rome, 81 p.

Feofarova, I.I., 1956. Micromorphological characteristics of takyrs. Takyrs in Western Turkmenia and the ways of their development for agricultural purposes [in Russian]. USSR Academy of Sciences Publication, pp. 351-381.

Fitzpatrick, R., Thomas, B., Merry, R., Marvanek, S. & Poch, R.M., 2014. Acid sulfate soils in Spencer Gulf. In Shepherd, S.A., Madigan, S.M., Gillanders, B.M., Murray-Jones, S. & Wiltshire, D.J. (eds.), Natural History of Spencer Gulf. Royal Society of South Australia, Adelaide, pp. 92-106.

Gagarina, E.I., 2004. Micromorphological Methods of Soil Studies [*in Russian*]. St. Petersburg University Publications, 155 p.

Gagarina, E.I., Panasenko, I.N. & Kulin N.I., 1988. Changes in microfabric of southern chernozem induced by drop irrigation [*in Russian*]. In Micromorphology of Human-modified Soils. Nauka, Moscow, pp. 98-106.

Gerasimova, M.I., Gubin, S.V. & Shoba, S.A., 1996. Soils of Russia and Adjacent Countries: Geography and Micromorphology. Moscow State University & Agricultural University Wageningen, 204 p.

Gile, L.H., Peterson, F.F. & Grossman, R.B., 1966. Morphological and genetic sequences of carbonate accumulation in desert soils. Soil Science 101, 347-360.

Golovanov, D.L. & Yamnova, I.A., 2013. Digital mapping of gypsic horizon morphotypes and soil salinity in an old alluvial piedmant plain of Uzbekistan. In Shahid, S.A., Taha, F.K. & Abdelfattah, M.A. (eds.), Developments in Soil Classification, Land Use Planning and Policy Implications: Innovative Thinking of Soil Inventory for Land Use Planning and Management of Land Resources. Springer, Dordrecht, pp. 257-267.

Golovanov, D.L., Lebedeva, M.P., Dorokhova, M.F., & Slobodkin, A.I., 2005. Micromorphological and microbiological characterization of elementary soil-forming processes in desert soils of Mongolia. Eurasian Soil Science 38, 1290-1300.

Goryachkin S.V., Spiridonova I.A., Sedov S.N. & Targulian, V.O., 2003. Boreal soils on hard gypsum rocks: morphology, properties, and genesis. Eurasian Soil Science 36, 691-703.

Gracheva, M.V., Tursina, T.V. & Korolyuk, T.V., 1988. Microstructure of non-irrigated and irrigated soils of chestnut alkali complexes in Stavropol region [*in Russia*]. In Dobrovol'skiy, G.V. (ed.), Micromorphology of Anthropogenic Soils. Nauka, Moscow, pp. 106-114.

Gubin, S.V., 1984. Micromorphological diagnostics of brown and grey-brown soils of northern Ust'-Urt [*in Russian*]. In Lobova, E.V. (ed.), Nature, Soils and Reclamation of the Ust'-Urt Desert, Institute of Soil Science and Agrochemistry, Pushchino, pp. 127-134.

Gubin, S.V. & Kovda, I.V., 1988. Changes in microfabric of dark-chestnut soils in the Terek river valley induced by prolonged irrigation [*in Russian*]. In Micromorphology of Human-modified Soils. Nauka, Moscow, 124-128.

Gumuzzio, J. & Casas, J., 1988. Accumulations of soluble salts and gypsum in soils of the Central Region, Spain. Cahiers ORSTOM, Série Pédologie 24, 215-226.

Halitim, A., 1988. Sols des Régions Arides d'Algérie. Office des Publications Universitaires, Alger, 384 p.

Halitim, A. & Robert, M., 1987. Interactions du gypse avec les autres constituants du sol. Analyse microscopique de sols gypseux en zone aride (Algérie) et études expérimentales. In Fedoroff, N., Bresson, L.M. & Courty, M.A. (eds.), Micromorphologie des Sols, Soil Micromorphology. AFES, Plaisir, pp. 179-185.

Hamdi-Aissa, B., 2002. Paleogeochemical interpretation of some gypsic microfabrics in hyperdesert soils. Proceedings of the 17th World Congress of Soil Sciences, Bangkok, Paper no. 1861, 9 p.

Hamdi-Aissa, B., Fedoroff, N., Halitim, A. & Valles, V., 2002. Short and long term soil-water dynamic in Chott soils in hyper-arid areas (Sahara of Algeria). Proceedings of the 17th World Congress of Soil Sciences, Bangkok, Paper no. 672, 8 p.

Harouaka, K., Eisenhauer, A. & Fantle, M.S., 2014. Experimental investigation of Ca isotopic fractionation during abiotic gypsum precipitation. Geochimica et Cosmochimica Acta 129, 157-176.

Heidari, A. & Asadi, P., 2015. Micromorphological characteristics of polluted soils in Tehran petroleum refinery. Journal of Agricultural Science and Technology 17, 1041-1055.

Heinze, M. & Fiedler, H.J., 1984. Physikalische Eigenshaften von Gipsböden und ihren Begleitbodenformen in Kyffhäuser-Gebirge. Hercynia 21, 190-203.

Herrero, J., 1987. Suelos sobre los Yesos Paleógenos Barbastro-Balaguer-Torà. PhD Dissertation, Universidad de Zaragoza, 468 p.

Herrero, J., 1991. Morfologia y Génesis de Suelos Sobre Yesos. Monografias INIA 77, Instituto Nacional de Investigación y Tecnologia Agraria y Alimentaria, Madrid, 447 p.

Herrero, J., 2004. Revisiting the definitions of gypsic and petrogypsic horizons in Soil Taxonomy and World Reference Base for Soil Resources. Geoderma 120, 1-5.

Herrero, J. & Boixadera, J., 2002. Gypsic soils. In Lal, R. (ed.), Encyclopedia of Soil Science. Marcel Dekker Inc., New York, pp. 635-639.

Herrero, J. & Porta, J., 1987. Gypsiferous soils in the North-East of Spain. In Fedoroff, N., Bresson, L.M. & Courty, M.A. (eds.), Micromorphologie des Sols, Soil Micromorphology. AFES, Plaisir, pp. 186-192.

Herrero, J. & Porta, J., 2000. The terminology and the concepts of gypsum-rich soils. Geoderma 96, 47-61.

Herrero, J., Porta, J. & Fedoroff, N., 1992. Hypergypsic soil micromorphology and landscape relationships in north eastern Spain. Soil Science Society of America Journal 56, 1188-1194.

Herrero, J., Poch, R.M., Porta, J. & Boixadera, J., 1996. Soils with Gypsum of the Central Catalan Depression. Excursion Guide. Edicions de la Universitat de Lleida, 87 p.

Herrero, J., Artieda, O. & Hudnall, W.H., 2009. Gypsum, a tricky material. Soil Science Society of America Journal 73, 1757-1763.

Hussain, M. & Warren, J.K., 1989. Nodular and enterolithic gypsum: the 'sabkha-tization' of Salt Flat playa, west Texas. Sedimentary Geology 64, 13-24.

IUSS Working Group WRB, 2015. World Reference Base for Soil Resources 2014, Update 2015. World Soil Resources Reports No. 106. FAO, Rome, 192 p.

Jacobson, G., Arakel, A.V. & Chen X.Y., 1988. The central Australian groundwater discharge zone: evolution of associated calcrete and gypcrete deposits. Australian Journal of Earth Sciences 35, 549-565.

Jafarzadeh, A.A. & Burnham, C.P., 1992. Gypsum crystals in soils. Journal of Soil Science 43, 409-420.

Junyent, E., Balasch, J.C. & Poch, R.M., 2012. The Iberian Fortress of Els Vilars. Mid-Meeting Excursion Guide, 14th International Working Meeting on Soil Micromorphology. Edicions de la Universitat de Lleida, 32 p.

Khademi, H. & Mermut, A.R., 2003. Micromorphology and classification of Argids and associated gypsiferous Aridisols from central Iran. Catena 54, 439-455.

Khalaf, F.I., Al-Zamel, A. & Gharib, I., 2014. Petrography and genesis of Quaternary coastal gypcrete in North Kuwait, Arabian Gulf. Geoderma 226/227, 223-230.

Khechai, S. & Daoud, Y., 2016. Characterization and origin of gypsum rhizoliths of Ziban oases soil-Algeria. World Applied Sciences Journal 34, 948-955.

Kushnir, J., 1980. The coprecipitation of strontium, magnesium, sodium, potassium and chloride ions with gypsum. An experimental study. Geochimica and Cosmochimica Acta 44, 1471-1482.

Labib, F., 1970. Contribution to the Mineralogical Characterisation of the Most Important Soil Parent Materials in the United Arab Republic (Egypt). PhD Dissertation, Ain Shams University, Cairo, 242 p.

Lawton, T., & Buck, B.J., 2006. Implications of diapir-derived detritus and gypsic paleosols in Lower Triassic strata near the Castle Valley salt wall, Paradox basin, Utah. Geology 34, 885-888.

Laya, H., Benayas, J. & Marfil, R., 1993. Yesos lenticulares de origin detritico en suelos de la España central. Cuaternario y Geomorfologia 7, 49-56.

Lebedeva, M.P. & Konyshkova, M.V., 2010. Changes in micromorphological features of semidesert soils in the southeast of European Russia upon the recent increase in climate moistening. In Proceedings of the 19th World Congress of Soil Science, Soil Solutions for a Changing World, Brisbane, pp. 12-15.

Lebedeva, M.P. & Konyshkova, M.V., 2011. Temporal changes in the microfabrics of virgin and reclaimed solonetzes at the Dzhanybek research station. Eurasian Soil Science 44, 753-765.

Lebedeva, M.P. & Konyushkova, M.V., 2016. Solonetzic soilscapes in the Northern Caspian lowland: local and spatial heterogeneity of pedofeatures and their changes in time. Byulleten Pochvennogo Instituta im. V.V. Dokuchaeva 86, 77-95.

Lebedeva (Verba), M.P., Golovanov, D.L. & Inozemtsev, S.A., 2009. Microfabrics of desert soils of Mongolia. Eurasian Soil Science 42, 1204-1217.

Lebedeva, M.P., Kutovaya, O.V., Sizemskaya, M.L. & Khokhlov, S.F., 2014. Micromorphological and microbiological diagnostics of initial pedogenesis on the bottom of an artificial mesodepression in the Northern Caspian semidesert. Eurasian Soil Science 47, 1123-1137.

Lebedeva, M.P., Gerasimova, M.I., Golovanov, D.L. & Yamnova, I.A., 2015. Extremely arid soils of the Ili Depression in Kazakhstan. Eurasian Soil Science 48, 11-26.

Lim, V.D., Shoba, S.A., Ramazanov, A.R. & Safonov V.F., 1975. Micromorphological studies of grey-meadow gypsiferous soils in Dzhizak steppe [*in Russian*]. In Engineering Measures to Control against the Salinity of Irrigated Soils. Tashkent 144, pp. 76-83.

Logan, B.W., 1987. The MacLeod Evaporite Basin, Western Australia. American Association of Petroleum Geologists Memoir 44, 140 p.

Magee, J.W., 1991. Late Quaternary lacustrine, groundwater, aeolian and pedogenic gypsum in the Prungle Lakes, Southeastern Australia. Palaeogeography, Palaeoclimatology, Palaeoecology 84, 3-42.

Mandado, J. & Tena, J.M., 1989. Génesis del yeso primario lenticular en sedimentos evaporíticos. Revista de la Academia de Ciencias, Zaragoza, 44, 199-214.

Masson, P.H., 1955. An occurrence of gypsum in Southwest Texas. Journal of Sedimentary Petrology 25, 72-77.

Mees, F., 1999. Distribution patterns of gypsum and kalistrontite in a dry lake basin of the southwestern Kalahari (Omongwa pan, Namibia). Earth Surface Processes and Landforms 24, 731-744.

Mees, F. & De Dapper, M., 2005. Vertical variations in bassanite distribution patterns in near-surface sediments, southern Egypt. Sedimentary Geology 181, 225-229.

Mees, F. & Stoops, G., 2010. Sulphidic and sulphuric materials. In Stoops, G., Marcelino, V. & Mees, F. (eds.), Interpretation of Micromorphological Features of Soils and Regoliths. Elsevier, Amsterdam, pp. 543-568.

Mees, F. & Stoops, G., 2018. Sulphidic and sulphuric materials. In Stoops, G., Marcelino, V. & Mees, F. (eds.), Interpretation of Micromorphological Features of Soils and Regoliths. Second Edition. Elsevier, Amsterdam, pp. 347-376.

Mees, F. & Tursina, T.V., 2010. Salt minerals in saline soils and salt crusts. In Stoops, G., Marcelino, V. & Mees, F. (eds.), Interpretation of Micromorphological Features of Soils and Regoliths. Elsevier, Amsterdam, pp. 441-469.

Mees, F. & Tursina, T.V., 2018. Salt minerals in saline soils and salt crusts. In Stoops, G., Marcelino, V. & Mees, F. (eds.), Interpretation of Micromorphological Features of Soils and Regoliths. Second Edition. Elsevier, Amsterdam, pp. 289-321.

Mees, F., Castañeda, C., Herrero, J. & Van Ranst, E., 2012. The nature and significance of variations in gypsum crystal morphology in dry lake basins. Journal of Sedimentary Research 82, 41-56.

Mestdagh, H., Haesaerts, P., Dodonov, A. & Hus, J., 1999. Pedosedimentary and climatic reconstruction of the last interglacial and early glacial loess-paleosol sequence in South Tadzhikistan. Catena 35, 197-218.

Miedema, R., Jongmans, A.G. & Slager, S., 1974. Micromorphological observations on pyrite and its oxidation products in four Holocene alluvial soils in the Netherlands. In Rutherford, G.K. (ed.), Soil Microscopy. The Limestone Press, Kingston, pp. 772-794.

Minashina, N.G., 1964. Changes in micromorphological and other properties of desertic (takyric) soils under irrigation [in Russian]. In Reports to the VIII International Congress of Soil Scientists. Nauka Publications, Moscow, pp. 371-379.

Minashina, N.G., 1973. Micromorphological properties of desertic (takyric) soils of Central Asia [in Russian]. In Micromorphology of soils and sediments. Nauka Publications, Moscow, pp. 45-83.

Minashina, N.G. & Khaknazarov, Y.R., 1988. Technological Properties and Reclamation of Gypsum-Containing Soils for Irrigation (Guidelines) [in Russian]. Dokuchaev Soil Science Institute, Moscow, 74 p.

Moghiseh, E. & Heidari, A., 2012. Polygenetic saline gypsiferous soils of the Bam region, Southeast Iran. Journal of Soil Science and Plant Nutrition 12, 729-746.

Neher, R.E. & Bailey, O.A., 1976. Soil Survey of White Sands Missile Range, New Mexico. U. S. Department of Agriculture, Soil Conservation Service, 66 p.

Nettleton, W.D., Nelson, R.E., Brasher, B.R. & Derr, P.S., 1982. Gypsiferous soils in the Western United States. In Kittrick, I.A., Fanning, D.S. & Hossner, L.R. (eds.), Acid Sulfate Weathering. Soil Science Society of America Special Publication 10, SSSA, Madison, pp. 147-168.

Ortí, F., 1992. Evaporitas marinas. In Arche, A. (ed.), Sedimentología. I. Colección Nuevas Tendencias, 11. CSIC, Madrid, pp. 89-177.

Pankova, Y.I. & Yamnova, I.A., 1980. Forms of salt accumulations in hydromorphic chloride and sulphate solonchaks in Mongolia [in Russian]. Pochvovedenie 2, 99-108.

Pankova, Y.I. & Yamnova, I.A, 1987. Forms of gypsic neoformations as a controlling factor affecting the meliorative properties of gypsiferous soils. Soviet Soil Science 19, 94-102.

Pérez-Lambán, F., Poch, R.M., Badia, D., Picazo, J.V., Peña-Monné, J.L. & Sampietro-Vattuone, M.M., 2016. Mid Holocene buried paleo-gypsisol in the semiarid Ebro Basin (NE Spain). Book of Abstracts, 15th International Conference on Soil Micromorphology. Universidad Nacional Autónoma de México, Mexico City, 58 p.

Plet-Lajoux, C., Monnier, G. & Pédro, G., 1971. Etude expérimentale sur la genèse et la mise en place des encroûtements gypseux. Comptes Rendus de l'Académie de Sciences 272, 3017-3020.

Poch, R.M., 1992. Fabric and Physical Properties of Soils with Gypsic and Hypergypsic Horizons of the Ebro Valley. PhD Dissertation, Ghent University, 285 p.

Poch, R.M., De Coster, W. & Stoops, G., 1998. Pore space characteristics as indicators of soil behaviour in gypsiferous soils. Geoderma 87, 87-109.

Poch, R.M., Fitzpatrick, R.W., Thomas, B.P., Merry, R.H., Self, P.G. & Raven, M.D., 2004. Contemporary and relict processes in a coastal acid sulfate soil sequence: microscopic features. SuperSoil 2004, 3rd Australian New Zealand Soils Conference, University of Sydney (CD-ROM), 8 p.

Poch, R.M., Thomas, B.P., Fitzpatrick, R.W. & Merry, R.H., 2009. Micromorphological evidence for mineral weathering pathways in a coastal acid sulfate soil sequence with Mediterranean-type climate, South Australia. Australian Journal of Soil Research 47, 403-422.

Podwojewski, P., 1994. Signification pédologique et paléoclimatique de la présence de gypse dans des vertisols de la côte ouest de l'île de Mulekula (Vanuatu). Comptes Rendus de l'Académie des Sciences, Série II, 319, 111-117.

Podwojewski, P., 1995. The occurrence and interpretation of carbonate and sulfate minerals in a sequence of Vertisols in New-Caledonia. Geoderma 65, 223-248.

Porta, J., 1986. Edafogénesis en Suelos Yesíferos en Medio Semiárido. Trabajo Original de Investigación para Acceder a Catedrático de Universidad, 136 p.

Porta, J., 1998. Methodologies for the analysis and characterization of gypsum in soils: a review. Geoderma 87, 31-46.

Porta, J. & Herrero, J., 1988. Micromorfología de suelos con yeso. Anales de Edafología y Agrobiología 47, 179-197.

Porta, J. & Herrero, J., 1990. Micromorphology and genesis of soils enriched with gypsum. In Douglas, L. A. (ed.), Soil Micromorphology: A Basic and Applied Science, Developments in Soil Science, Volume 19. Elsevier, Amsterdam, pp. 321-339.

Ramazanov, A.R. & Lim, V.D., 1981. Changes in micromorphological properties of gypsiferous soils under salt washing off and in vegetation experiments [*in Russian*]. Bulletin of the Dokuchaev Soil Institute 38, 57-58.

Reheis, M.C., 1987. Gypsic Soils on the Kane Alluvial Fans, Big Horn County, Wyoming. U.S. Geological Survey Bulletin 1590-C., 39 p.

Samy, Y. & Metwally, H.I., 2012. Gypsum crystal habits as an evidence for aridity and stagnation, northeast of the Nile delta coast, Egypt. Australian Journal of Basic and Applied Sciences 6, 442-450.

Scalenghe, R., Territo, C., Petit, S., Terribile, F. & Righi, D., 2016. The role of pedogenic overprinting in the obliteration of parent material in some polygenetic landscapes of Sicily (Italy). Geoderma Regional 7, 49-58.

Schenk, C.J. & Fryberger, S.G., 1988. Early diagenesis of eolian dune and interdune sands at White Sands, New Mexico. Sedimentary Geology 55, 109-120.

Siesser, W.G. & Rogers, J., 1976. Authigenic pyrite and gypsum in South West African continental slope sediments. Sedimentology 23, 567-577.

Soil Survey Staff, 1993. Soil Survey Manual. USDA Handbook Number 18. U.S. Government Printing Office, Washington, 437 p.

Soil Survey Staff, 1999. Soil Taxonomy. A Basic System of Soil Classification for Making and Interpreting Soil Surveys, 2nd Edition. USDA, Natural Resources Conservation Service, Agriculture Handbook Number 436. U.S. Government Printing Office, Washington, 871 p.

Soil Survey Staff, 2014. Keys to Soil Taxonomy. 12th Edition. USDA-Natural Resources Conservation Service, Washington, 360 p.

Stoops, G., 2003. Guidelines for Analysis and Description of Soil and Regolith Thin Sections. Soil Science Society of America, Madison, 184 p.

Stoops, G. & Delvigne, J., 1990. Morphology of mineral weathering and neoformation. II Neoformation. In Douglas, L.A. (ed.), Soil Micromorphology: A Basic and Applied Science. Developments in Soil Science, Volume 19. Elsevier, Amsterdam, pp. 483-492.

Stoops, G. & Ilaiwi, M., 1981. Gypsum in arid soils, morphology and genesis. In Proceedings of the Third International Soil Classification Workshop. Damascus, pp. 175-185.

Stoops, G. & Poch, R.M., 1994. Micromorphological classification of gypsiferous soil materials. In Ringrose-Voase, A.J. & Humphreys, G.S. (eds.), Soil Micromorphology: Studies in Management and Genesis. Developments in Soil Science, Volume 22. Elsevier, Amsterdam, pp. 327-332.

Stoops, G., Eswaran, H. & Abtahi, A., 1978. Scanning electron microscopy of authigenic sulphate minerals in soils. In Delgado, M. (ed.), Soil Micromorphology. University of Granada, pp. 1093-1113.

Sullivan, L.A., 1990. Micromorphology and genesis of some calcite pseudomorphs after lenticular gypsum. Australian Journal of Soil Research 28, 483-485.

Testa, G. & Lugli, S., 2000. Gypsum-anhydrite transformations in Messinian evaporites. Sedimentary Geology 130, 249-268.

Thompson, T.L., Hossner, L.R. & Wilding, L.P., 1991. Micromorphology of calcium carbonate in bauxite processing waste. Geoderma 48, 31-42.

Tolchel'nikov, Yu. S., 1962. On calcium sulfates and carbonates neoformations in sandy desert soils. Soviet Soil Science 6, 643-650.

Toomanian, N., Jalalian, A. & Eghbal, M.K., 2001. Genesis of gypsum enriched soils in north-west Isfahan, Iran. Geoderma 99, 199-224.

Verba, M.P., 1990. Macro-and microfabric of soils in Syr-Darya experimental station. In Conditions of development and properties of hard-to-reclaim soils of Dzhizak steppe [in Russian]. Dokuchaev Soil Institute, Moscow. pp. 48-55.

Verba, M.P., 1996. The forms of gypsum neoformation in the soils of Central Asian Deserts. In Poch, R.M. (ed.), Proceedings of the International Symposium of Soils With Gypsum. Edicions de la Universitat de Lleida, pp. 52.

Verba, M.P. & Deviatykh, V.A., 1992. Changes in micromorphology of gypsiferous meadow Sierozem during leaching (model experiment). Eurasian Soil Science 24, 12-22.

Verba, M.P. & Yamnova, I.A., 1997. Gypsum neoformations in nonirrigated and irrigated soils of the serozem zone. In Shoba, S., Gerasimova, M. & Miedema, R. (eds.), Soil Micromorphology: Studies on Soil Diversity, Diagnostics, Dynamics. Moscow, Wageningen, pp. 187-195.

Verba, M.P., Al-Kasiri, A.S., Goncharova, N.A. & Chizhikova, N.P., 1995. Impact of irrigation on micromorphological features of desert soils of Hadramout Vallet (Yemen). Eurasian Soil Science 27, 108-124.

Verboom, B., Grealish, G., Schoknecht, N. & Samira, O., 2003. Influence of gravel on the accumulation of pedogenic gypsum in Kuwait. Arid Land Research and Management 17, 71-84.

Warren, J.K., 1982. The hydrological setting, occurrence and significance of gypsum in late Quaternary salt lakes in South Australia. Sedimentology 29, 609-637.

Warren, J.K., 1986. Shallow-water evaporitic environments and their source rock potential. Journal of Sedimentary Petrology 56, 442-454.

Watson, A., 1979. Gypsum crusts in deserts. Journal of Arid Environments 2, 3-20.

Watson, A., 1983. Gypsum crusts. In Goudie, A.S. & Pye, K. (eds.), Chemical Sediments and Geomorphology: Precipitates and Residua in the Near-surface Environment. Academic Press, London, pp. 133-161.

Watson, A., 1985. Structure, chemistry and origins of gypsum crusts in southern Tunisia and the Central Namib Desert. Sedimentology 32, 855-875.

Watson, A., 1988. Desert gypsum crusts as palaeoenvironmental indicators: a micropetrographic study of crusts from southern Tunisia and the central Namib Desert. Journal of Arid Environments 15, 19-42.

Watson, A., 1989. Desert crusts and rock varnish. In Thomas, D.S.G. (ed.), Arid Zone Geomorphology. Belhaven Press, London, pp. 25-55.

Worrall, G.A., 1957. Features of some semi-arid soils in the district of Khartoum, Sudan. I. The high-level dark clays. Journal of Soil Science 8, 193-202.

Yamnova, I.A., 1990. Gypsiferous soils of Dzhizak steppe [in Russian]. In Conditions of development and properties of hard-to-reclaim soils of Dzhizak steppe. Dokuchaev Soil Institute, Moscow, pp. 13-20.

Yamnova, I.A., 1995. Salt minerals in soils of deserts in Mongolia. Abstracts of the International Conference on Asian Ecosystems and their Protection, Ulaanbaatar, pp. 65.

Yamnova, I.A., 1996. The micromorphological features and genesis of gypsum neoformations in gypsic solonchaks of Goby desert (Mongolia). In Poch, R.M. (ed.), Proceedings of the International Symposium of Soils With Gypsum. Edicions de la Universitat de Lleida, pp. 73.

Yamnova, I.A., 2016. Salt and gypsum pedofeatures as indicators of soil processes. Byulleten Pochvennogo Instituta im. V.V. Dokuchaeva 86, 96-102.

Yamnova, I.A. & Abrosimov, K.N., 2013. Experience with the X-ray microtomography for morpho-mineralogical analysis of gypsum and gypsum-bearing horizons of carbonate soil [*in Russian*]. In Proceedings of the All-Russian Conference 'Practical Microtomography', SSI Soil Science Institute VV Dokuchaev, Moscow, pp. 175-183.

Yamnova, I.A. & Golovanov, D.L., 2010a. Macro- and micromorphotypes of gypsic horizons in soils of arid regions. Arid Zone Research 27, 44-50.

Yamnova, I.A. & Golovanov, D.L., 2010b. Morphology and genesis of gypsum pedofeatures and their representation on detailed soil maps of arid regions. Eurasian Soil Science 43, 848-857.

Yamnova, I.A. & Pankova, E. I., 2013. Gypsic pedofeatures and elementary pedogenetic processes of their formation. Eurasian Soil Science 46, 1117-1129.

Yarilova, E.A., 1958. Mineralogical characteristics of solonetzes in the chernozemic zone [*in Russian*]. Proceedings of the Dokuchaev Soil Institute 53, 131-142.

Zen, E.A., 1965. Solubility measurements in the system $CaSO_4$-NaCl-H_2O at 35°, 50° and 70° and one atmosphere pressure. Journal of Petrology 6, 124-164.

Chapter 11

Salt Minerals in Saline Soils and Salt Crusts

Florias Mees[1], Tatiana V. Tursina[2]

[1]*ROYAL MUSEUM FOR CENTRAL AFRICA, TERVUREN, BELGIUM;*
[2]*DOKUCHAEV SOIL SCIENCE INSTITUTE, MOSCOW, RUSSIA*

CHAPTER OUTLINE

Interpretation of Micromorphological Features of Soils and Regoliths. https://doi.org/10.1016/B978-0-444-63522-8.00011-5

1. Introduction

In this chapter, the significance of micromorphological features of soluble salt occurrences in soil environments is discussed. It is largely confined to evaporite minerals in soils of arid regions, exclusive of gypsum (see Poch et al., 2018). Sulphate minerals that formed in a context of sulphide oxidation are treated elsewhere (Mees & Stoops, 2018). The morphology, type of aggregation and mode of occurrence are discussed for various mineral species, listed in Table 1. This table includes some information about optical properties as an aid for mineral identification in thin sections, but for more detailed information the reader is referred to other publications (e.g., Braitsch, 1971; Arnold, 1984; Fleischer et al., 1984).

The micromorphological study of salt minerals in soils can be complicated by a number of specific problems that are not encountered for most other soil materials. Firstly, salt minerals may crystallise during drying of moist samples (e.g., Hanna & Stoops, 1976; Kooistra, 1983), which is generally required for thin section preparation. Secondly, soils containing hydrated salts can become unstable during transport, storage or further handling (e.g., Driessen & Schoorl, 1973; Tursina & Yamnova, 1987; Vizcayno et al., 1995). Dehydration during SEM analysis, due to vacuum conditions or interaction with the electron beam, is one example of this problem (Shayan & Lancucki, 1984; Doner & Lynn, 1989), which can be avoided by using ESEM equipment (e.g., Rodriguez-Navarro & Doehne, 1999). Thirdly, partial or complete dissolution of salts can occur during thin section preparation, either during impregnation (Hanna & Stoops, 1976; Stoops et al., 1978; Kooistra, 1983; Courty et al., 1989; Stoops & Delvigne, 1990) or at a later stage during cutting or grinding (Eswaran et al., 1983; Allen, 1985). Despite these practical problems, micromorphological techniques have been successfully applied in various studies of salt-affected soils.

In this chapter, the three soluble salts for which most micromorphological data are available are first discussed (halite, thénardite, blödite), illustrating various interpretations that have been given to microscopic and submicroscopic features of salt occurrences in soils. This is followed by paragraphs about other highly soluble salts (chlorides, sulphates, carbonates, nitrates), about bassanite and anhydrite, and about poorly soluble sulphate minerals (celestine, kalistrontite, baryte). Finally, mineral assemblages are briefly discussed.

The first to describe the presence of salts (halite) in soil thin sections was probably Valérien Agafonoff (Agafonoff, 1935/1936), whose pioneering work on soil micromorphology deserves much wider recognition. Next came the work of Iraida Feofarova in

Table 1 Overview of minerals described in this paper, with information about their most common crystal habit in soil environments (if missing or limited, general information on crystal morphology between brackets) and selected optical properties (refractive index, birefringence, optical orientation)

Name	Composition	Morphology	Optical properties
Bischofite	$MgCl_2 \cdot 6H_2O$	– (Short prismatic)	n_x 1.494, n_z 1.528, Δn 0.034, X = b, $Y^\wedge c$ = 9°
Carnallite	$KMgCl_3 \cdot 6H_2O$	– (Tabular)	n_x 1.466, n_z 1.495, Δn 0.029, X = c, Z = a
Halite	$NaCl$	Cubic, elongated	n 1.544
Kainite	$KMg(SO_4)Cl \cdot 3H_2O$	Equant to short prismatic	n_x 1.494, n_z 1.516, Δn 0.022, Y = b, $Z^\wedge c$ = 13°
Sylvite	KCl	Cubic	n 1.490
Anhydrite	$Ca(SO_4)$	Elongated, tabular	n_x 1.570, n_z 1.614, Δn 0.044, X = a, Z = c
Aphthitalite	$K_3Na(SO_4)_2$	Equant (tabular)	n_o 1.487, n_e 1.492, Δn 0.005, Z = c
Baryte	$Ba(SO_4)$	Elongated, tabular	n_x 1.636, n_z 1.648, Δn 0.012, X = c, Y = b
Bassanite	$Ca(SO_4) \cdot 0.5H_2O$	Fibrous	n_x 1.559, n_z 1.584, Δn 0.025, Z // fibre axis
Blödite	$Na_2Mg(SO_4)_2 \cdot 4H_2O$	Equant, elongated, tabular	n_x 1.483, n_z 1.487, Δn 0.004, Y = b, $X^\wedge c$ = 37°
Burkeite	$Na_4(SO_4)(CO_3)$	Equant, lath-/plate-shaped	n_x 1.448, n_z 1.493, Δn 0.045, X = c, Z = b
Celestine	$Sr(SO_4)$	Elongated prismatic	n_x 1.622, n_z 1.631, Δn 0.009, Y = b, Z = c
Epsomite	$Mg(SO_4) \cdot 7H_2O$	Lath-shaped (prismatic)	n_x 1.433, n_z 1.461, Δn 0.028, X = b, Z = a
Eugsterite	$Na_4Ca(SO_4)_3 \cdot 2H_2O$	Acicular	n_x 1.492, n_z 1.496, Δn 0.004, Z = b, $Y^\wedge c$ = 27°
Glauberite	$Na_2Ca(SO_4)_2$	Tabular, acicular, equant	n_x 1.515, n_z 1.536, Δn 0.021, Z = b, $Y^\wedge a$ = 12°
Hexahydrite	$Mg(SO_4) \cdot 6H_2O$	Tabular pseudohexagonal	n_x 1.438, n_z 1.465, Δn 0.027, Y = b, $X^\wedge c$ = -25°
Kieserite	$Mg(SO_4) \cdot H_2O$	Tabular	n_x 1.520, n_z 1.584, Δn 0.064, Y = b, $Z^\wedge c$ = -77°
Kogarkoite	$Na_3(SO_4)F$	Pseudohexagonal	n_o 1.439, n_e 1.442, Δn 0.003
Kalistrontite	$K_2Sr(SO_4)_2$	Prismatic	n_o 1.549, n_e 1.569, Δn 0.020, X = c
Konyaite	$Na_2Mg(SO_4)_2 \cdot 5H_2O$	Plate-shaped	n_x 1.464, n_z 1.474, Δn 0.010, X = b, $Z^\wedge c$ = 70°
Mirabilite	$Na_2(SO_4) \cdot 10H_2O$	Variable	n_x 1.393, n_z 1.397, Δn 0.004, X = b, $Z^\wedge c$ = 30°
Omongwaite	$Na_2Ca_5(SO_4)_6 \cdot 3H_2O$	Elongated prismatic	n_x 1.556, n_z 1.567, Δn 0.011, $Z^\wedge c$ = 11°
Starkeyite	$Mg(SO_4) \cdot 4H_2O$	- (Fibrous)	n_x 1.460, n_z 1.474, Δn 0.014
Syngenite	$K_2Ca(SO_4)_2 \cdot H_2O$	Acicular to tabular	n_x 1.501, n_z 1.518, Δn 0.017, $X^\wedge c$ = 2°, Z = b
Thenardite	$Na_2(SO_4)$	Elongated or bipyramidal	n_x 1.469, n_z 1.484, Δn 0.015, X = c, Y = b
Gaylussite	$Na_2Ca(CO_3)_2 \cdot 5H_2O$	Elongated prismatic	n_x 1.445, n_z 1.523, Δn 0.078, X = b, $Z^\wedge c$ = 15°
Nahcolite	$NaH(CO_3)$	Lath-shaped	n_x 1.377, n_z 1.583, Δn 0.206, Y = b, $X^\wedge c$ = 27°
Natron	$Na_2(CO_3) \cdot 10H_2O$	- (Fibrous)	n_x 1.405, n_z 1.440, Δn 0.035, X = b, $Z^\wedge c$ = 41°
Pirssonite	$Na_2Ca(CO_3)_2 \cdot 2H_2O$	Short lath-shaped	n_x 1.504, n_z 1.573, Δn 0.069, X = a, Z = b
Trona	$Na_3(HCO_3)(CO_3) \cdot 2H_2O$	Lath-shaped, acicular	n_x 1.416, n_z 1.542, Δn 0.126, X = b, $Z^\wedge c$ = 83°
Darapskite	$Na_3(SO_4)(NO_3) \cdot H_2O$	– (Elongated prismatic)	n_x 1.391, n_z 1.486, Δn 0.095, X = b, $Z^\wedge c$ = 12°
Humberstonite	$K_3Na_7Mg_2(SO_4)_6(NO_3)_2 \cdot 6H_2O$	– (Platy)	n_o 1.474, n_e 1.426, Δn 0.038, X = c
Niter	$K(NO_3)$	– (Acicular)	n_x 1.332, n_z 1.504, Δn 0.172, X = c, Z = b
Nitratine	$Na(NO_3)$	– (Rhombohedral)	n_o 1.587, n_e 1.336, Δn 0.251, X = c

Russia, which included the study of salt minerals in density separates (Feofarova, 1958). Other early investigations, using thin section and SEM observations, were done in the 1970s and early 1980s, in Wageningen (e.g., Driessen, 1970; Vergouwen, 1981a), Bondy (Cheverry et al., 1972), Ghent (e.g., Stoops, 1974; Eswaran et al., 1980) and Moscow (e.g., Tursina et al., 1980; Shoba et al., 1983). The petrographical study of bassanite and anhydrite in soil-related environments became important around the time of the discovery of anhydrite in recent sabkha deposits (e.g., Kinsman, 1966). The earliest micromorphological descriptions of the poorly soluble sulphates that are covered by this review date from the 1970s (e.g., Lynn et al., 1971; Barzanji & Stoops, 1974), but they have only more recently been the subject of specific investigations.

2. Halite

Halite in soils occurs mainly as coatings or cements composed of anhedral crystals, which is observed both in thin sections (Tursina et al., 1980; Tursina & Yamnova, 1987; Shahid & Jenkins, 1994; Mees, 2003; Smith et al., 2004; Mees & Singer, 2006) and in SEM images (Driessen, 1970; Eswaran & Carrera, 1980; Eswaran et al., 1980, 1983; Vergouwen, 1981a; Eswaran, 1984; Gerasimova et al., 1996) (Fig. 1A). The general absence of well-developed crystal faces has been related to the strongly hygroscopic nature of halite, resulting in partial dissolution following the absorption of water (Eswaran & Carrera, 1980; Eswaran et al., 1980); this aspect of halite behaviour can be investigated using ESEM and low-temperature SEM (cf. Wierzchos et al., 2012). A lack of euhedral forms can be expected for occurrences of highly soluble salts unless they formed very recently (e.g., Buck et al., 2006), but xenotopic fabrics are clearly exceptionally common for halite in soils.

Euhedral to subhedral cubic crystals are largely confined to surface crusts or efflorescences (Eswaran et al., 1980; Vergouwen, 1981a; Herrero Isern et al., 1989; Mees & Stoops, 1991; Vizcayno et al., 1995; Joeckel & Ang Clement, 1999; Gore et al., 2000; Chernousenko et al., 2003; Mees & Singer, 2006; Buck et al., 2011) (Fig. 1B and C). These halite crystals generally formed from brines that covered the surface. However, cubic halite crystals have also been described as forming a meniscus cement in surface crusts (Pueyo Mur, 1978/1979, 1980). In salt lake studies, thick accumulations of cubic crystals, transformed to beds of interlocking equant crystals during diagenesis, have been interpreted as subarial precipitates, formed by capillary evaporation of groundwater (Bobst et al., 2001; Lowenstein et al., 2003).

Subsurface occurrences of euhedral cubic crystals are only rarely reported (Gibson et al., 1983; Amit & Yaalon, 1996; Joeckel & Ang Clement, 1999; Van Hoesen et al., 2001; Schiefelbein et al., 2002; Buck et al., 2006; Moghiseh & Heidari, 2012). They include occurrences of skeletal cubic crystals, interpreted to indicate a high degree of supersaturation (Buck et al., 2006; see also Eswaran & Drees, 2004). Another example are subhedral cubic crystals that formed within the groundmass by displacive and

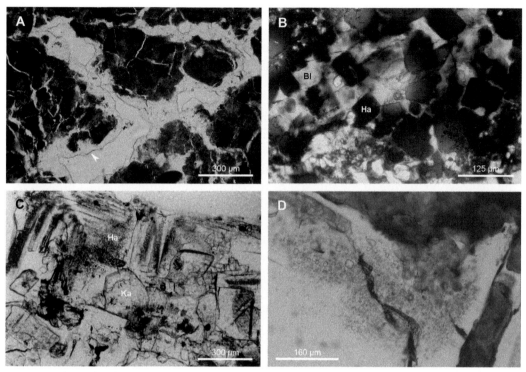

FIGURE 1 Halite. (A) Halite coating (*arrow*) composed of anhedral crystals (Otjomongwa pan, Namibia; F. Mees, unpublished data) (PPL). (B) Euhedral cubic halite crystals (Ha), enclosed by xenotopic blödite cement (Bl) (Timimoun, Algeria; F. Mees, unpublished data) (XPL). (C) Halite crystals with high concentration of fluid inclusions (Ha), with variations between bands parallel to the crystal faces, in a surface crust with associated kainite (non-cubic crystals; Ka) (southern Aral Sea basin, Uzbekistan; F. Mees, unpublished data) (PPL). (D) Detail of halite crystal with fluid inclusions, illustrating their cubic shape (southern Aral Sea basin, Uzbekistan; F. Mees, unpublished data) (PPL).

partly incorporative growth (Hussain &Warren, 1989; Amit & Yaalon, 1996). Occurrences of halite crystals in the groundmass are rare, because of the high solubility of the mineral.

A subordinate but common crystal habit of halite in soils is a fibrous form (see Kooistra, 1983). The crystals are perfectly straight or curved, elongated parallel to one of the crystallographic axes. Halite of this type mainly forms along or near the soil surface, where it develops as parallel fibres, perpendicular to the surface (Hanna & Stoops, 1976; Eswaran et al., 1980; Bullock et al., 1985). Random orientations (Joeckel & Ang Clement, 1999), subsurface occurrences (Hanna & Stoops, 1976; Yamnova, 2016) and fibrous cements in sedimentary gypsum crusts (Pueyo Mur, 1978/1979) have also been described. The development of fibrous forms of halite in soil environments has been related to fast drying (Eswaran et al., 1980), capillary evaporation (Joeckel & Ang Clement, 1999) and the continual localised upward movement of soil solutions (von Hodenberg & Miotke, 1983). Fibrous halite perpendicular to a surface is also common in efflorescences on building materials (e.g., Arnold & Kueng, 1985). In this context, it is interpreted to form in conditions with low water content of the covered material, low rates of water supply from the substrate and low evaporation rates (Arnold & Zehnder, 1985).

Shorter prismatic forms (Eswaran et al., 1980) seem to represent an intermediate stage in the development of halite fibres. Another related form is elongated crystals occurring as radial aggregates (Gore et al., 2000).

Features related to the dissolution of halite, other than xenotopic fabrics, include the presence of rounded cavities (Tursina et al., 1980; Chernousenko et al., 2003). Joeckel and Ang Clement (1999) describe more or less cubic hollows in the centre of crystal faces as dissolution features (see also Taher & Abdel-Motelib, 2015), related to the hygroscopic nature of the mineral or to contact with dilute solutions derived from the surface. Other possible examples are pitted surfaces (Gore et al., 2000) and grooves parallel to the longitudinal axis of fibrous crystals (Joeckel & Ang Clement, 1999). Halite casts include angular indentations along the base of silty intercalations in mudstone, recording a period of near-surface displacive halite growth followed by associated halite dissolution and deposition of coarse-grained sediments during flooding (Eriksson et al., 2005).

Other features displayed by halite crystals are the presence of fluid inclusions, which is indicative of fast growth (Fig. 1C and D), and a 'stair-step' pattern along the underside of a surface crust, which has been attributed to growth rather than dissolution (Joeckel & Ang Clement, 1999). A central tubular cavity was observed for elongated halite crystals by von Hodenberg and Miotke (1983), who suggest that it acted as a capillary during crystal growth.

Partial or complete replacement of lenticular gypsum crystals by halite has been described by Hussain and Warren (1989), without commenting on the processes that would explain this relationship between unrelated phases. An association of halite with remains of roots has also been reported (Tursina et al., 1980; Tursina & Yamnova, 1987; Young, 1987; Tovey & Dent, 2002), but the significance of this feature is not apparent. Canfora et al. (2016) imply that halite distribution in surface crusts can be determined by prior biogenic gypsum precipitation around cyanobacterial filaments.

The presence of xenotopic halite coatings has a sealing effect on the underlying soil material (Driessen, 1970; Vergouwen, 1981a; Vizcayno et al., 1995). This type of halite can also act as a cement that binds soil aggregates (Hanna & Stoops, 1976; Eswaran et al., 1980; Schroeder et al., 1984; Zhang & Wang, 1987; Shahid & Jenkins, 1994; Joeckel & Ang Clement, 1999), including vadoze meniscus cements between sand grains (McLaren, 2001). The type of aggregation may vary with salt mineral composition, whereby spheroidal shapes are recognised for halite-dominated materials and a vermiform morphology for thénardite-cemented aggregates (Tursina et al., 1980).

In contrast to cementation by halite, the development of elongated halite crystals may cause separation of soil aggregates (Smith et al., 2004). This includes the uplift of surface material, if deposition of eolian material following salt formation can be excluded (Hanna & Stoops, 1976). Crystal size has been mentioned as a factor for the degree of disruption of soil material, with large halite crystals having a stronger effect than smaller thénardite crystals (Tursina et al., 1980), although various other factors obviously also need to be considered (e.g., Rodriguez-Navarro & Doehne, 1999). A disruptive effect is implied for halite occurrences that are described as being displacive, formed within cracks in gravel-sized grains (Amit et al., 1993; Amit & Yaalon, 1996) or along cleavage planes and other structures in shales (Aref et al., 2002). In other studies, no disruptive effect of halite growth is observed, whereas a close association between halite crystals and altered quartz grain surfaces is interpreted to suggest that the main effect of the presence of salts is an increase of quartz dissolution rates (Young, 1987).

3. Thénardite and Mirabilite

Thénardite has been frequently observed during micromorphological investigations of saline soils. The most commonly reported crystal habit for this mineral is acicular or elongated prismatic (Driessen, 1970; Driessen & Schoorl, 1973; Stoops, 1974; Stoops et al., 1978; Pueyo Mur, 1978/1979; Eswaran & Carrera, 1980; Tursina et al., 1980; Tursina, 1981; Vergouwen, 1981a; Shoba et al., 1983; Bullock et al., 1985; Tursina & Yamnova, 1987; FitzPatrick, 1993; Shahid & Jenkins, 1994; Vizcayno et al., 1995; Gore et al., 2000; Mees, 2003; Buck et al., 2004; Hamdi-Aissa et al., 2004; Joeckel & Ang Clement, 2005; Van Ranst et al., 2008; Shahid, 2013) (Fig. 2A). The second-most common form is more or less equant bipyramidal, with lozenge-shaped cross-sections (Vergouwen, 1981a; Mees & Stoops, 1991; Shahid & Jenkins, 1994; Singer et al., 1999; Mees, 2003; Joeckel & Ang Clement, 2005; Van Ranst et al., 2008; Buck et al., 2011) (see Figs. 2D and 9A). Elongated crystals, with roughly lenticular (Mees & Stoops, 1991; Vizcayno et al., 1995), lath-shaped (Shahid & Jenkins, 1994) or irregular forms (FitzPatrick, 1993), have also been described. Plate-shaped crystal habits (Gibson et al., 1983; Joeckel & Ang Clement, 2005; Wentworth et al., 2005) seem to represent an aberrant morphology.

Stoops and Delvigne (1990) relate the development of acicular thénardite to crystal growth during fast evaporation in pores and the formation of lenticular crystals to slow evaporation within the groundmass. Similarly, Vizcayno et al. (1995) invoke a difference in speed of crystal growth for lenticular crystals and radial intergrowths of acicular crystals, without referring to the place where they form. Equant crystals with lozenge-shaped cross-sections do seem to represent the equilibrium form that also character-ises subaqueous precipitates in salt lakes (e.g., Mees, 1999b).

Mirabilite has been described by a few authors using SEM analysis, reporting equant, lath-shaped and tabular crystal habits (Eswaran & Carrera, 1980; Shahid & Jenkins, 1994; Buck et al., 2004, 2011; Bisson et al, 2015). However, these observations might in most cases pertain to thénardite that formed by dehydration of mirabilite, in view of the instability of the hydrated form, for example under vacuum conditions during SEM

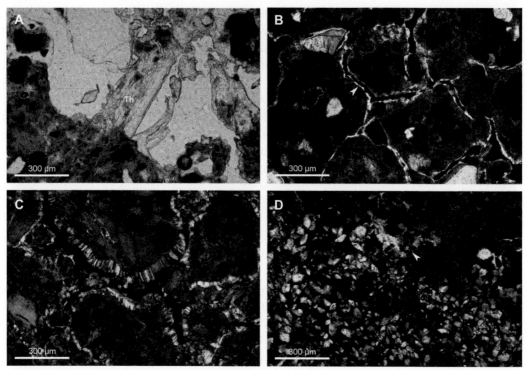

FIGURE 2 Thénardite (Otjomongwa pan, Namibia; F. Mees, unpublished data). (A) Elongated prismatic thénardite crystals (Th), perpendicular to the sides of a pore (PPL). (B) Thin thénardite coating with constant optical orientation (*arrow*) (XPL). (C) Thénardite coating composed of parallel elongated thénardite crystals, perpendicular to the sides of pores (XPL). (D) Loose continuous infilling composed of euhedral bipyramidal thénardite crystals, with associated xenotopic thénardite along the side of the pore (*arrow*) (XPL).

investigations or during impregnation for thin section preparation (e.g., Gumuzzio et al., 1982; Stoops & Delvigne, 1990; Vizcayno et al., 1995).

Thénardite formed by dehydration of mirabilite has been studied by several authors (Pueyo Mur, 1978/1979, 1980; Tursina, 1980, 1981; Tursina et al., 1980; Tursina & Yamnova, 1987; Vizcayno et al., 1995; Lebedeva-Verba et al., 2004). A difference in size and crystal habit between thénardite that formed directly from soil solutions (acicular) and thénardite formed following dehydration of mirabilite (tabular) has been reported (Tursina et al, 1980; Tursina, 2002). Pseudomorphic aggregates after large mirabilite crystals studied by Pueyo Mur (1978/1979) are composed of bipyramidal thénardite crystals that are larger in the interior than in outer parts of the same pseudomorph, and they also include groups of acicular thénardite crystals along voids in the interior. In the same study, the dehydration of fine-grained mirabilite was found to involve both pseudomorphic replacement and the growth of acicular thénardite crystals perpendicular to the surface of the affected grains.

Modes of occurrence of thénardite in soils include coatings (Tursina, 1981; Tursina & Yamnova, 1987; Mees, 2003; Mees & Singer, 2006), formed by evaporation of water contained in the peds or covering the ped surface as a thin film (Mees, 2003). The fabric of these coatings varies with depth, ranging from coatings with a constant optical orientation that represent large skeletal crystals (Fig. 2B) to coatings composed of parallel elongated crystals (Fig. 2C), recording strong vertical variations in nucleation behaviour within the same profile (Mees, 2003). Fibrous coatings also seem to have been observed during SEM studies (Gerasimova et al., 1996; Joeckel & Ang Clement, 2005). Radial aggregates of acicular crystals along the sides of pores (Stoops, 1974; Stoops et al., 1978; Eswaran & Carrera, 1980) represent a type of discontinuous coating. Another mode of occurrence is represented by loose continuous infillings composed of euhedral to subhedral crystals (Cheverry et al., 1972; Eswaran & Carrera, 1980; Tursina & Yamnova, 1987; Mees, 2003), formed in brine-filled macropores (Mees, 2003) (Fig. 2D). They can be associated with coatings or hypocoatings with a xenotopic fabric, formed by evaporation of solutions contained in the groundmass (Mees, 2003) (Fig. 2D).

Thénardite nodules include dense microcrystalline aggregates with a xenotopic fabric (Mees & Stoops, 1991) and aggregates of loosely packed crystals of thénardite and associated other minerals (Eswaran & Carrera, 1980). Some nodules are microcrystalline aggregates that are pseudomorphic after mirabilite crystals (e.g., Vizcayno et al., 1995).

Special features of thénardite occurrences in soils include an association with plant remains (Tursina et al., 1980; Tursina, 1981; Shoba et al., 1983), indications for local dissolution and reprecipitation (Gumuzzio & Casas, 1988) and the occurrence of granostriation related to the growth of mirabilite crystals (Tursina et al., 1980). Thénardite aggregates can be covered by clay, which hinders the recognition of the salts if no thin sections are used (Tursina, 1981). Driessen and Schoorl (1973) mention thénardite crystals in surface crusts showing damage caused by eolian processes or trampling.

4. Blödite

Blödite crystals in soils are typically elongated, with a rod-like, lath-shaped or tabular morphology (Driessen, 1970; Driessen & Schoorl, 1973; Pueyo Mur, 1978/1979; Eswaran & Carrera, 1980; Vergouwen, 1981a; Vizcayno et al., 1995; Singer et al., 1999; Van Hoesen et al., 2001; Buck et al., 2011). Equant polyhedral forms are rarer in soil environments (Feofarova, 1958; Gibson et al., 1983; von Hodenberg & Miotke, 1983; Wentworth et al., 2005; Mees & Singer, 2006). Those equant forms are characteristic of subaqueous precipitates (e.g., Mees, 1999b), which include surface crusts (Mees & Singer, 2006). Buck et al. (2006) recognise various forms in surface horizons, contributing to their view that crystal habits are most diverse at levels where conditions are most variable. As aggregate type, radial intergrowths have been reported in several studies (Pueyo Mur, 1980; Vizcayno et al., 1995; Mees & Singer, 2006; Mees et al., 2011) (Fig. 3A). A related type of

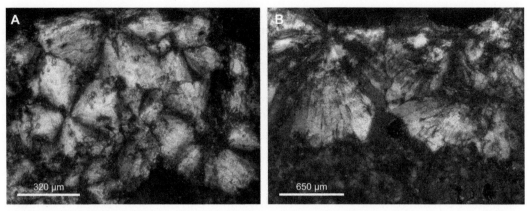

FIGURE 3 Blödite (Mediana, Spain; F. Mees, unpublished data). (A) Radial aggregates (XPL). (B) Fan-like aggregates with downward widening (XPL).

occurrence are fan-like aggregates with upward or downward widening (Mees et al., 2011; Rodríguez-Ochoa et al., 2012), recording poor nucleation followed by growth in a direction controlled by brine availability (Mees et al., 2011) (Fig. 3B). Other examples of aggregate types are groups of parallel crystals (Pueyo Mur, 1978/1979) and small round aggregates of randomly oriented crystals (Buck et al., 2006).

Blödite-dominated salt crusts are composed of overlapping or intergrown crystals, suggesting that they have a strong sealing effect on the underlying soil (Driessen, 1970; Driessen & Schoorl, 1973; Eswaran & Carrera, 1980). In contrast, Vergouwen (1981a) reports a high total porosity for blödite-rich crusts, composed of loosely stacked crystals. Driessen (1970) recognised the presence of blödite coatings in pores within a blödite-dominated crust, which further reduces its porosity. A second generation of blödite in pores was also recognised by Eswaran and Carrera (1980), who mention a difference in morphology between these crystals and those that form the bulk of the crust, relating this difference to the availability of space.

Another type of blödite occurrence in salt efflorescences is a xenotopic cement (Vizcayno et al., 1995) (see Fig. 1B). Eswaran and Carrera (1980) illustrate the occurrence of small etch pits as dissolution features. A highly irregular surface morphology of a blödite intergrowth has been attributed to repeated dissolution and reprecipitation (Van Hoesen et al., 2001). The development of subhedral forms and of foliated crystals or aggregates, as opposed to individual or radially intergrown prismatic crystals, has been related to rapid crystallisation (Vizcayno et al., 1995). Similar conditions have been invoked to explain the formation of small round aggregates in surface crusts (Buck et al., 2006).

Blödite crystals in contact with the groundmass may contain bands with sediment inclusions, recording successive growth stages (Vizcayno et al., 1995; Mees et al., 2011).

In the soil beneath a salt crust, blödite can occur as xenotopic coatings (Mees et al., 2011), but also as impregnative nodules and hypocoatings (Vizcayno et al., 1995). A related occurrence is represented by xenotopic blödite with high concentration of groundmass material as inclusion, along the base of thénardite coatings (Mees & Singer, 2006).

5. Other Highly Soluble Salt Minerals

5.1   Chlorides

Thin section and SEM observations for halides other than halite are rare and generally inconclusive. Sylvite is difficult to distinguish from (associated) halite on the basis of crystal morphology alone, but the significantly lower refractive index of sylvite allows a distinction between both minerals. Cubic forms (Yechieli & Ronen, 1997; Gore et al., 2000), as well as a lath-like morphology (Gore et al., 2000), have been reported for sylvite in salt efflorescences, based on SEM observations. Yechieli and Ronen (1997) illustrate occurrences of bischofite and carnallite that appear to be xenotopic, but they also mention a partly euhedral morphology as an indication for mineral authigenesis. A possible bischofite occurrence, with octahedral aspect of euhedral crystals, is also reported by Bisson et al. (2015). Wongpokhom et al. (2008) illustrate an assumed calcium chloride feature. Eswaran et al. (1980) tentatively identify lath-shaped crystals associated with halite as carnallite. Kainite has been identified as equant to short prismatic crystals associated with halite in the uppermost part of a salt crust (Mees & Singer, 2006) (see Fig. 1C) and possibly as pseudohexagonal crystals in a surface crust (Buck et al., 2006).

5.2   Sulphates

5.2.1   Aphthitalite
The morphology of two reported aphthitalite occurrences in salt efflorescences is poorly documented (Frenzel, 1963; Vergouwen, 1981a). Aphthitalite is also recognised as small euhedral equant crystals in the upper part of a thin halite crust, formed from brines that covered the surface (Fig. 4A).

5.2.2   Burkeite
The morphology of burkeite in soil environments ranges from equant and short prismatic (Mees & Stoops, 1991) to lath- and plate-shaped (Vergouwen, 1979, 1981a; Vizcayno et al., 1995). In a salt crust from Peru (Mees & Stoops, 1991), it occurs both as coatings and infillings and as pseudomorphs after thénardite crystals, indicating late-stage development. The coatings formed preferentially along the upper side of pores, which is related to the downward movement of solutions, in agreement with other micromorphological characteristics of the crust.

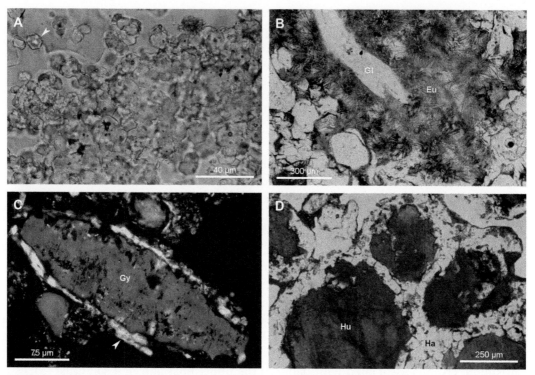

FIGURE 4 Highly soluble salt minerals other than halite, thénardite and blödite. (A) Cluster of small aphthitalite crystals, including isolated euhedral crystals (*arrow*), occurring in a halite-dominated surface crust (Timimoun, Algeria; F. Mees, unpublished data) (PPL). (B) Radial aggregates of eugsterite crystals (Eu), surrounding a subhedral glauberite crystal (Gl) (Timimoun, Algeria; F. Mees, unpublished data) (PPL). (C) Syngenite (*arrow*) surrounding a lenticular gypsum crystal (Gy) (Omongwa pan, Namibia; F. Mees, unpublished data) (XPL). (D) Humberstonite aggregates (Hu), surrounded by halite coatings (Ha) (Timimoun, Algeria; F. Mees, unpublished data) (PPL).

5.2.3 Eugsterite

Eugsterite crystals are invariably acicular (Vergouwen, 1981a, 1981b; Mees & Stoops, 1991; Buck et al., 2006; Mees & Singer, 2006). These crystals partly occur in the form of radial aggregates (Mees & Singer, 2006; Mees et al., 2011) (Fig. 4B). In SEM-EDS studies, eugsterite can be difficult to distinguish from acicular forms of compositionally related glauberite (Buck et al., 2011; Bisson et al., 2015).

5.2.4 Glauberite

Glauberite typically forms thick tabular crystals in salt lake deposits (e.g., Mees, 1999b; Mees et al., 2011). The same crystal habit is recognised for secondary glauberite appearing within moulds of larger synsedimentary glauberite crystals (Mees, 1998). Other occurrences reported in the same study are xenotopic, including glauberite intergrowths surrounding gypsum crystals that were partly transformed to bassanite and anhydrite (see Fig. 5A). In soils, glauberite crystals can have the same tabular shape as

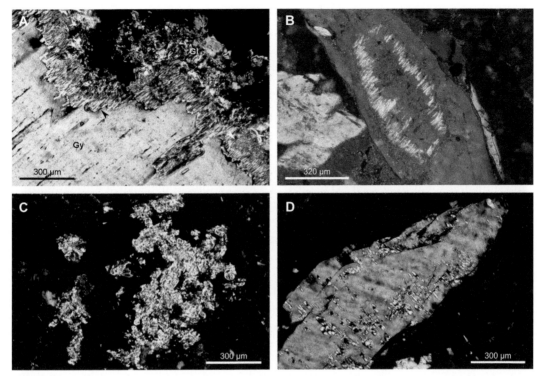

FIGURE 5 Bassanite, anhydrite, omongwaite. (A) Gypsum crystal (Gy), transformed to bassanite (*black arrow*) and anhydrite (*white arrow*) along the sides (topotactic replacement), surrounded by glauberite-bearing xenotopic material (Gl) (Taoudenni, Mali; colour version of a figure in Mees, 1998) (XPL). (B) Omongwaite inclusions in lenticular gypsum, occurring as bands of parallel crystals developed along opposite sides of the gypsum crystal (Omongwa pan, Namibia; colour image of crystal illustrated by Mees et al., 2008) (XPL). (C) Aggregate of bassanite crystals in a pore (El Adaima, Egypt; F. Mees, unpublished data) (XPL). (D) Lenticular gypsum crystals with enclosed, parallel-oriented, elongated prismatic bassanite crystals (El Adaima, Egypt; F. Mees, unpublished data) (XPL).

they do in salt lake deposits (see Fig. 4B), but they can also be plate-shaped (Shahid & Jenkins, 1994), equant (Schiefelbein et al., 2002) or acicular (Feofarova, 1958; Pueyo Mur, 1980; Tursina et al., 1980; Hamdi-Aissa et al., 2004; Shahid, 2013). Tursina and Yamnova (1987) report acicular to equant forms. Benayas et al. (1992) tentatively identify spherulites in pores as glauberite aggregates. Lebedeva-Verba et al. (2004) mention the occurrence of glauberite around thénardite crystals, but it is not clear whether the identification of these and other reported minerals (polyhalite, trona) was confirmed by other methods.

5.2.5 Magnesium Sulphates
Hexahydrite, epsomite, kieserite and starkeyite are the only magnesium sulphates for which micromorphological observations have been reported. Hexahydrite crystals

generally have a tabular pseudohexagonal form, as observed in SEM images (Stoops et al., 1978; Eswaran & Carrera, 1980; Vergouwen, 1981a; Van Hoesen et al., 2001; Hamdi-Aissa et al., 2004; Buck et al., 2006). In a subaqueously deposited blödite-dominated crust, hexahydrite occurs as partly fibrous aggregates that formed in part by transformation of blödite (Mees & Singer, 2006). Buck et al. (2011) report an occurrence of massive fine-grained hexahydrite affected by cracking due to dehydration. Ducloux et al. (1994) describe acicular and lath-shaped crystals, showing parallel alignment in some parts, identified as either hexahydrite or epsomite. Epsomite should be expected to become unstable after sampling or during SEM analysis. Epsomite crystals have been described as lath-shaped (Gore et al., 2000), prismatic (Lebedeva & Konyushkova, 2011) and anhedral (Shayan & Lancucki, 1984). Lebedeva and Konyushkova (2011) report a possible occurrence of tabular kieserite crystals. An efflorescence from Congo contains microcrystalline iron-rich starkeyite aggregates with a mosaic-like texture, but this mineral may have formed by dehydration after sampling (Van Tassel, 1958).

5.2.6 Kogarkoite
Kogarkoite has been reported for fluorine-rich environments, as tabular (Darragi et al., 1983) or bipyramidal (Van Ranst et al., 2008) pseudohexagonal crystals.

5.2.7 Konyaite
Konyaite occurs as elongated plate-shaped crystals with a pseudohexagonal form (Shayan & Lancucki, 1984). In another study, konyaite was observed to form rounded aggregates of plate-shaped crystals (Ducloux et al., 1994).

5.2.8 Syngenite
Syngenite has been reported to occur as acicular to tabular crystals in a salt efflorescence on a rock face (Frenzel, 1963). In a similar context, acicular syngenite forms the upper part of a salt crust on sandstone, whereby syngenite is assumed to have formed from solutions that ascended through the lower part of the crust, composed of gypsum (Schweigstillova et al., 2005). In soil environments, syngenite has been recognised as predominantly elongated crystals along the sides of gypsum crystals (Fig. 4C) and glauberite crystals (Timimoun, Algeria; Mees, unpublished data), formed by an interaction of those minerals with concentrated potassium-rich brines.

5.3 Sodium Carbonates

Trona generally has an acicular or lath-shaped crystal habit (Vergouwen, 1981a; Bullock et al., 1985; Mees & Stoops, 1991; FitzPatrick, 1993; Shahid & Jenkins, 1994; Vizcayno et al., 1995; Owen et al., 2008; Taher & Abdel-Motelib, 2015). This morphology, as well as the random to radial arrangement of the crystals, contributes to a high porosity of trona efflorescences (Vergouwen, 1981a). Skeletal trona crystals are assumed to have formed by rapid crystal growth (Datta et al., 2002; see also Jones & Renaut, 1996).

The morphology of nahcolite and natron has hardly been documented for soil environments (Shahid & Jenkins, 1994; Smith et al., 2004). Small pirssonite crystals have been recognised as possible pseudomorphic and non-pseudomorphic replacement products of calcium carbonates (Mees & Stoops, 1991). Pedogenic gaylussite has been described as occurring in the form of large elongated subhedral crystals with ground-mass material as inclusion (Mees & Stoops, 1990; see also Rodríguez-Ochoa et al., 2012). Gaylussite, together with trona, is also mentioned as a component observed for crusts from the Lake Chad area (Cheverry et al., 1972).

5.4   Nitrates

Little information is available on nitrate mineral occurrences in soil environments. Thin section observations of complex nitrate ores (e.g., Searl & Rankin, 1993; Pueyo et al., 1998; Collao et al., 2002), as well as descriptions of the morphology of specific minerals (e.g., Ericksen & Mrose, 1970; Mrose et al., 1970), are available for the Chilean nitrate deposits, but the pedogenic nature of these formations is not always apparent. The four nitrate minerals that have been reported for this type of setting are niter, nitratine, humberstonite and darapskite. Figure 4D shows an occurrence of microcrystalline humberstonite aggregates in a surface crust from a sabkha environment.

6. Bassanite and Anhydrite

The presence of bassanite and anhydrite is not uncommon at or near the land surface in arid regions. They include occurrences in near-surface settings that are not explicitly identified as soil environments but which show relevant micromorphological features. It should be kept in mind that bassanite in thin sections can be an artifact of heating during thin section preparation (Agafonoff, 1935/1936; Milton, 1942; Parfenova & Yarilova, 1965, 1977; Artiedo Cabello, 1996), which is a possibility that has been mentioned for bassanite occurrences in soil samples by several authors (Bullock et al, 1985; Stoops & Delvigne, 1990; FitzPatrick, 1993; Porta, 1998). The instability of gypsum or bassanite may also account for the lack of correspondence between XRD results and thin section or SEM observations (e.g., Miller et al., 1989).

Bassanite and anhydrite in soils mainly form by dehydration of gypsum. This process is recognised by the occurrence of these minerals along the sides of gypsum crystals, representing a pellicular alteration pattern (Fig. 5A). Bassanite-anhydrite successions away from the gypsum core suggest that bassanite is an intermediate phase in the transformation of gypsum to anhydrite (Moiola & Glover, 1965; Arakel, 1980; Mees, 1998) (Fig. 5A). The dehydration products of gypsum are generally composed of parallel fibres (Stoops et al., 1978; Simón et al., 1980; Eswaran et al., 1983; Smykatz-Kloss et al., 1985; Moghiseh & Heidari, 2012; Khalaf et al., 2014; Basyoni & Aref, 2015), parallel to the (010) cleavage direction of gypsum (Hunt et al., 1966; Stoops & Delvigne, 1990) (Fig. 5A), which is described as perpendicular to the long axis of lenticular crystals by some authors (Dronkert, 1977; Gunatilaka et al., 1980; Tursina & Yamnova, 1987). The fibre axis is in fact parallel to a specific crystallographic orientation within the gypsum, whereby the alignment of calcium and sulphate ions is preserved in the dehydration products (Moiola & Glover, 1965; Mees, 1998). This type of topotactic replacement is also recognised for bassanite developed around sand grains that are enclosed in a gypsum crust with a poikilotopic fabric (Mees & Stoops, 2003). Another type of topotactic replacement of gypsum is represented by omong-waite formation, through interaction of gypsum crystals with highly saline solutions, resulting in surface deposits that are preserved when covered by gypsum overgrowths (Mees et al., 2008) (Fig. 5B).

The possibility of replacement of gypsum by bassanite or anhydrite through a dissolution-reprecipitation process at a non-molecular scale is suggested for some occurrences by an apparent absence of fibrous textures (Arakel, 1980; Hamdi-Aissa, 2002; Hamdi-Aissa et al., 2004) and by the development of randomly oriented or radiating bassanite/anhydrite crystals (Aref et al., 1997; Aref, 1998; Mees & De Dapper, 2005). Dissolution-reprecipitation replacement is also inferred from associations between anhydrite occurrences and dissolution features along gypsum surfaces (Shearman, 1985; Gunatilaka, 1990).

The replacement of gypsum crystals by bassanite or anhydrite can result in the formation of perfectly pseudomorphic aggregates (Bush, 1973; Gunatilaka et al, 1980; Bullock et al., 1985; Stoops & Delvigne, 1990; FitzPatrick, 1993; Golovanov et al., 2005; Yamnova & Pankova, 2013), which may enclose small corroded gypsum remnants (Arakel, 1980). Pseudomorphism has also been suggested for lamellar bassanite, assumed to have formed at high salinities in frozen ground, by dehydration of cryogenic gypsum with prominent cleavage (Vogt & Larqué, 2002). SEM observations suggest that pseudomorphic anhydrite aggregates are susceptible to collapse (Gunatilaka, 1990). Similarly, the fine crystal size and loose consistency of anhydrite aggregates have been cited to explain the scarcity of pseudomorphs by Butler (1970), who also refers to the limited coherence of the sediment enclosing the original gypsum crystals. In the same study, thin section observations of samples with pseudomorphs after lenticular gypsum crystals reveal the presence of associated moulds of those crystals, with varying degrees of infilling by anhydrite, indicating a difference in timing between gypsum dissolution and anhydrite formation.

Further anhydrite growth following pseudomorphic replacement, requiring the addition of Ca^{2+} and SO_4^{2-} ions from other sources, results in the formation of aggregates with a more irregular outline and ultimately in the development of nodules that can no longer be recognised as pseudomorphic features (Bush, 1973; Shearman, 1978). These nodules are composed of cleavage fragments, formed by the growth of lath-shaped crystals in a confined space, which results in breakage (Shearman, 1978). The continued growth of the fine-grained anhydrite nodules requires conditions or processes favouring nucleation rather than enlargement of existing crystals (Shearman, 1978).

In contrast to these models relating bassanite and anhydrite formation to dehydration of gypsum, direct precipitation of these minerals from solution has also been proposed for some occurrences. It has been invoked for anhydrite nodules by Kinsman (1966), referring to the absence of pseudomorphic aggregates and of partially replaced gypsum crystals (see also Wilson et al., 2013), and to the difference between gypsum crystals and anhydrite nodules in amount of groundmass material occurring as inclusion. In the study by Kinsman (1966), randomly oriented single anhydrite crystals and small aggregates, occurring in the groundmass away from gypsum crystals, are interpreted as representing the initial stage of anhydrite nodule development. Lack of enclosed groundmass material and co-occurrence with unaltered gypsum are also invoked by Basyoni & Aref (2015) to argue an authigenic origin of anhydrite nodules, which show parallel alignment of plate-shaped crystals along the sides as peculiar feature. Formation of bassanite as direct precipitate has been proposed for radial aggregates and clusters of randomly oriented euhedral crystals, citing micromorphological evidence for a role of high salinity and the presence of organic compounds (Khalaf et al., 2014).

Most descriptions of anhydrite and bassanite concern occurrences as crystals or nodules within the groundmass. Both minerals have also been observed as aggregates in pores (Bullock et al., 1985; Tursina & Yamnova, 1987; Shahid & Jenkins, 1994; Mees & De Dapper, 2005), where they are probably derived from a gypsum precursor (Mees & De Dapper, 2005) (Fig. 5C).

Hydration of the dehydration products around gypsum crystals can result in the development of non-topotactic microcrystalline gypsum coatings (Mees & Singer, 2006). Topotactic replacement accounts for the presence of secondary gypsum along the sides of sand grains surrounded by bassanite, enclosed in poikilotopic gypsum with the same optical orientation as the secondary gypsum (Mees & Stoops, 2003). The presence of parallel-oriented bassanite/anhydrite inclusions in gypsum crystals, related to repeated dehydration and hydration (Dronkert, 1977; Mees & De Dapper, 2005), most likely involves epitaxial gypsum growth on the gypsum core (Fig. 5D). The development of gypsum crystals enclosing small, randomly oriented and distributed anhydrite crystals, interpreted as a feature formed by dehydration (Aref, 2003), may also have involved a rehydration stage. Bassanite along the contact between gypsum crystals can be protected from hydration by the development of gypsum overgrowths (Mees & De Dapper, 2005).

Features described as rehydration products of anhydrite include lath-shaped gypsum crystals, interpreted as having formed by hydration of lath-shaped fragments of complete pseudomorphs that formed by topotactic replacement of gypsum (Gunatilaka et al, 1980; Gunatilaka, 1990). These secondary gypsum crystals can show parallel alignment, which is deformed around remaining anhydrite crystals and nodules, representing a texture that has been related to liquefaction (Gunatilaka et al, 1980; Gunatilaka, 1990). Experimental studies indicate that textural features of hydration products have an impact on the geotechnical properties of soils with calcium sulphate minerals (Azam et al., 1998; Azam, 2003).

The microcrystalline nature of the dehydration products has been seen as an indication that their formation and rehydration contributes to the development of fine-grained gypcrete on gypsum deposits (Chen, 1997). Dehydration has also been assumed to have caused the development of a spongy microstructure in surface crusts (Hamdi-Aissa, 2002; Hamdi-Aissa et al., 2004), but the mechanism behind this change is not apparent.

Anhydrite nodules that formed in near-surface environments can be preserved in the rock record, but their recognition is complicated by possible confusion with nodules that formed in other conditions (e.g., Machel & Burton, 1991). The nodules are partly preserved as quartz nodules, characterised by a length-slow fibrous quartz composition and the occurrence of small calcium sulphate inclusions (e.g., Siedlecka, 1972; Chowns & Elkins, 1974; Tucker, 1976).

7. Poorly Soluble sulphate Minerals

7.1   Celestine

Celestine appears to be common in highly gypsiferous soils (Stoops et al., 1978; Eswaran & Zi-Tong, 1991), although relatively few published reports are available. The crystal morphology of celestine in soil environments is always elongated prismatic, with a lozenge-shaped cross-section and with pyramidal terminations composed of two or four crystal faces (e.g., Evans & Shearman, 1964; Barzanji & Stoops, 1974; Stoops et al., 1978; McFadden et al., 1987; Stoops, 2003) (Fig. 6A). Weathering can result in the development of parallel grooves on crystal faces (Stoops et al., 1978), and it can also account for the grooves and associated sawtooth patterns illustrated by Benayas et al. (1992).

Celestine crystals generally occur in groups, formed within the groundmass (e.g., Barzanji & Stoops, 1974; Bullock et al., 1985; Stoops & Delvigne, 1990; Rodríguez-Ochoa et al., 2012; Boixadera et al., 2015) or as infillings of pores (Stoops et al., 1978; Eswaran & Zi-Tong, 1991; Khormali, 2005) (Fig. 6B). The latter are most clearly authigenic, but the

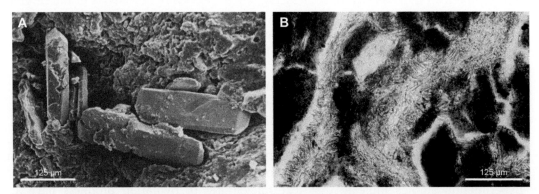

FIGURE 6 Celestine. (A) SEM image of celestine crystals in a gypsic horizon (Zebra Plain, Morocco; Image from Ghent University archive). (B) Celestine infillings of pores, with varying degree of intergrowth (El Adaima, Egypt; F. Mees, unpublished data) (PPL).

common euhedral forms, good sorting and clustering of crystals are also indications of authigenesis (Herrero, 1991). Fan-like or radial orientation patterns (e.g., McFadden et al., 1987; El-Sayed, 2000) are another feature that has been cited as an indication of authigenic celestine formation (Barzanji & Stoops, 1974). Evans and Shearman (1964) recognised sheaf-like and more simple radiating aggregates as two distinct types of intergrowths, each associated with a specific type of gypsum, formed at different stages in the diagenesis of sabkha deposits.

The preferential occurrence of celestine in decalcified parts of the groundmass in some gypsiferous soils suggests that celestine formation is related to calcite dissolution (Barzanji & Stoops, 1974). The authors refer to associated calcite dissolution and gypsum formation, with higher strontium concentrations in carbonates than in gypsum, resulting in a local excess of Sr, in a SO_4-rich environment. Herrero (1991) points out that this process is unlikely unless aragonite dissolution is involved and also found no systematic association between decalcification features and celestine occurrences in his study area. The possible role of aragonite was previously invoked by Evans and Shearman (1964), who refer to extensive dolomitisation of aragonite (and calcite) in their study area.

Hamdi-Aissa et al. (2004) describe celestine crystals occurring along the sides of gypsum crystals and as aggregates enclosed by gypsum, interpreted as an indication for co-precipitation of both minerals. Gypsum growth following celestine formation is illustrated by celestine crystals that are partly or completely enclosed by gypsum (Evans & Shearman, 1964; Barzanji & Stoops, 1974; Herrero, 1991; Stoops, 1998). Celestine occurrences within gypsum and associated anhydrite/bassanite shows that they are predictably unaffected by conditions resulting in dehydration and rehydration of calcium sulphate minerals (Hunt et al., 1966).

7.2 Kalistrontite

Kalistrontite occurs as microcrystalline nodules with a xenotopic fabric and as a thin discontinuous horizontal layer composed of euhedral to subhedral prismatic crystals, both developed along a lithological discontinuity below a dry lake bed (Mees, 1999a) (Fig. 7A). It encloses pedogenic lenticular gypsum crystals, demonstrating its pedogenic origin. The low solubility of kalistrontite is recorded by the presence of enclosed gypsum moulds, which are also present in the groundmass around the nodules (Fig. 7B).

7.3 Baryte

Baryte, which typically forms in hydromorphic conditions, displays two main types of crystal habit in soils, i.e., acicular or elongated prismatic (Lynn et al., 1971; Stoops & Zavaleta, 1978; Crum & Franzmeier, 1980; Fitzpatrick et al., 1992; Sullivan & Koppi, 1995; Wierzchos et al., 1995) (Fig. 8A) and tabular or plate-shaped with mainly hexagonal forms (Lynn et al., 1971; Stoops et al., 1978; Stoops & Zavaleta, 1978; Crum & Franzmeier, 1980; Carson et al., 1982; Dixon et al., 1982; Eswaran et al., 1983; Shoba et al., 1983;

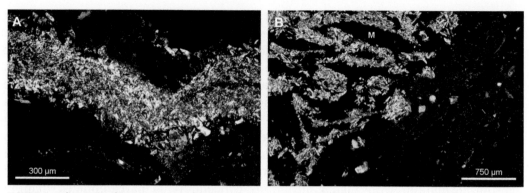

FIGURE 7 Kalistrontite (Omongwa pan, Namibia; F. Mees, unpublished data). (A) Horizontal kalistrontite layer, formed by evaporation of groundwater, partly separated from the calcareous groundmass in which it occurs (XPL). (B) Kalistrontite nodule with moulds of lenticular gypsum crystals (M) (XPL).

Rognon et al., 1987; Darmody et al., 1989; Sullivan & Koppi, 1993; Eswaran & Drees, 2004). Variations in crystal habit within baryte aggregates have been noted, with tabular forms in the centre and elongated prismatic crystals in the outer parts (Stoops et al., 1978; Stoops & Zavaleta, 1978). Aggregates with fan-like and radial arrangements of the elongated or plate-shaped crystals have also been described (Shoba et al., 1983; Sullivan & Koppi, 1995; McCarthy & Plint, 2003; Cumba & Imbellone, 2004; Retallack & Kirby, 2007). The factors determining variations in crystal habit and aggregation of baryte in soil environments are unknown. Irregular forms of crystals have been related to dissolution (Sullivan & Koppi, 1995).

Euhedral morphologies of baryte in soils are an indication of authigenesis (Lynn et al., 1971; Stoops & Zavaleta, 1978). Another criterion for authigenic mineral formation is the occurrence of baryte crystals in pores (Stoops & Zavaleta, 1978; Darmody et al., 1989) (Fig. 8B and C), which has also been observed in several other studies (Lynn et al., 1971; Crum & Franzmeier, 1980; Eswaran et al., 1983; Bullock et al., 1985; Rognon et al., 1987; Sullivan & Koppi, 1995; Brock-Hon et al., 2012). Baryte occurring as nodules enclosing groundmass material is also clearly pedogenic (Retallack & Kirby, 2007; see also Jennings et al., 2015). Preferential precipitation of baryte in parts of the groundmass that are rich in fibrous clay minerals has been attributed to local moisture conditions (Brock-Hon et al., 2012). Baryte occurrences in association with other pedogenic materials, such as carbonates (Sullivan & Koppi, 1995; Brock-Hon et al., 2012), illuvial clay (Sullivan & Koppi, 1995; Clarke et al., 2016) and iron oxides (Lee & Gilkes, 2005), record differences in timing of their development (Fig. 8B to D). The presence of baryte in pores in an interval with fragments of clay coatings indicates recent baryte formation, relative to a period of clay illuviation and pedoturbation (Stoops & Zavaleta, 1978). Illuvial clay covering baryte occurrences in some parts is seen as an indication that baryte authigenesis is not a very recent short-lived process (Darmody et al., 1989).

Baryte pseudomorphs after lenticular gypsum crystals are microcrystalline aggregates with admixed micritic carbonates (Sullivan & Koppi, 1993). The replaced gypsum crystals occurred in pores, which excludes pseudomorphism by infilling of moulds formed by dissolution (Sullivan & Koppi, 1993).

An association of baryte with silicified or carbonaceous root remains has been reported in a few studies (Sullivan & Koppi, 1995; McCarthy & Plint, 1998, 2003). In several more recent publications, microscopic features are cited to argue mediation of baryte precipitation by microorganisms (Brock-Hon et al., 2012; Jennings & Driese, 2014; Jennings et al., 2015). Features of baryte occurrences whose significance is not clear include associations with spherulitic aragonite aggregates (Podwojewski, 1995) and halloysite (Brewer et al., 1983). Other examples are baryte occurrences in the centre of quartz spherulites in silcretes (Smith, 2005), baryte infillings of planar voids in manganese oxide nodules (Blackburn et al., 1975), and baryte pseudomorphs after whewellite

FIGURE 8 Baryte (Witpan, Namibia; F. Mees, unpublished data). (A) SEM image of baryte crystals. (B) Baryte crystals, in pores with sepiolite coatings (*arrow*) (XPL). (C) Baryte in a channel lined by an opal-A hypocoating (Op) (XPL). (D) Baryte infilling, in a pore within a baryte-bearing sepiolite infilling, lined by a sepiolite coating and a calcite-depletion hypocoating (XPL).

($Ca(C_2O_4) \cdot H_2O$) crystals in plant remains (Smieja-Krol et al., 2014). McKee and Brown (1977) illustrate the SEM morphology of the interior of a baryte nodule, characterised by a high microporosity and a random orientation of euhedral crystals, but this nodule is not explicitly identified as either authigenic or inherited.

8. Mineral Assemblages

Spatial relationships between salt mineral species, in surface crusts and subsurface horizons, can record a sequence of stages with differences in brine composition, environmental conditions or mode of mineral formation related to degree of water saturation. This can result in the consecutive formation of different minerals, with or without involving the replacement of a pre-existing phase.

Non-replacive successive mineral formation produces coatings or infillings which cover mineral occurrences that developed at an earlier stage. During SEM investigations, this type of successive mineral formation is mainly recognised by the presence of small crystals covering the surface of larger sub/euhedral crystals (e.g., blödite on thénardite and hexahydrite; Vergouwen, 1981a). Examples observed in thin sections include infillings with a different salt mineral composition than the groundmass (Cheverry et al., 1972), juxtaposed coatings of different minerals (e.g., halite on thénardite; Mees, 2003), crystals enclosed by coatings along the sides of peds (e.g., eugsterite enclosed by halite; Mees & Singer, 2006) and crystals covering another mineral without signs of alteration (Fig. 9A). Successive formation can also result in complete cementation by pervasive development of infillings, most commonly composed of halite, for example halite cementation of part of a blödite-dominated crust (Vizcayno et al., 1995) (see also Fig. 1B). Sequences of mineral formation may include a dissolution stage, which can result in the presence of crystals inside dissolution cavities (e.g., kainite in halite; Mees & Singer, 2006). A special case of cumulative

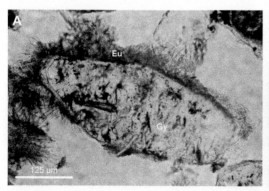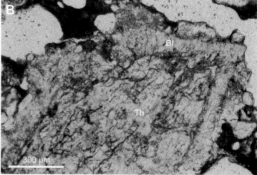

FIGURE 9 Mineral assemblages. (A) Acicular eugsterite crystals (Eu) partly covering the sides of a gypsum crystal (Gy) showing no indications of replacement (southern Aral Sea basin, Uzbekistan; F. Mees, unpublished data) (PPL). (B) Thénardite crystal (Th), transformed to blödite (Bl) along the sides by pseudomorphic replacement (Timimoun, Algeria; F. Mees, unpublished data) (PPL).

successive mineral formation is the development of layered surface crusts by subaqueous mineral formation (Mees & Stoops, 1991; Vizcayno et al., 1995; Mees & Singer, 2006).

Successive mineral formation involving replacement reactions has been recognised for various salt minerals. A common type of replacement reaction is the dehydration of hydrated minerals, as previously described for the Na-SO$_4$ and Ca-SO$_4$ systems.

A new mineral phase can also be produced by an interaction of a crystal with more concentrated solutions that involves the addition of elements. Examples of this type of replacement along the sides of a pre-existing phase include the transformation of gypsum to glauberite (Pueyo Mur, 1980), trona to burkeite (Vergouwen, 1981a), thénardite to eugsterite (Vergouwen, 1981a; Mees & Stoops, 1991), thénardite to blödite (Fig. 9B), gypsum to eugsterite (Rodríguez-Ochoa et al., 2012), and gypsum to syngenite (see Fig. 4C). Greater degrees of transformation have also been observed, resulting in complete or nearly complete pseudomorphic replacement, for example trona by burkeite (Vergouwen, 1981a) and thénardite by burkeite (Mees & Stoops, 1991). Replacement can take place during a period with subaqueous mineral formation (e.g., blödite by hexahydrite; Mees & Singer, 2006). It can also be followed by the development of overgrowths, which are in many cases composed of halite. This type of sequence is illustrated by halite enclosing thénardite crystals that are partly transformed to eugsterite (Mees & Stoops, 1991).

When paragenetic sequences are studied, the possibility of co-precipitation of certain minerals must be considered. Co-precipitation is recognised by an intimate association of two or more phases. Examples for which this process has been inferred include blödite-halite and halite-kainite associations (Vergouwen, 1981a; Mees & Singer, 2006).

Besides co-precipitation and replacive or non-replacive successive mineral formation, other events during the development of salt occurrences are recorded as well. In a salt crust from Peru, local redistribution of carbonates resulted in the development of trona-depleted thénardite-rich nodules, and salt distribution patterns in another crust are compatible with a period of leaching from the surface (Mees & Stoops, 1991). Cheverry et al. (1972) illustrate the presence of organic matter covering sodium carbonate crystals in a crust that was impregnated with organic compounds during a period of capillary evaporation, following the precipitation of trona and gaylussite from surface brines at an earlier stage.

9. Conclusions

Thin section studies of soil materials with soluble salts are still relatively few in number. SEM studies are more common but provide little information on some relevant aspects, such as the nature of internal features of crystals and aggregates. Many published micromorphological studies of salt occurrences are rather descriptive, with little attention for the significance of observed features.

The interpretation of variations in crystal habit and type of aggregation is hindered by a lack of experimental studies on possible controlling factors. Available results for gypsum growth, which has been intensively investigated, indicate that a great complexity should be expected. One consequence of the variable morphology of many salt minerals is that crystal shape can only be used with great care for mineral identification, which should be largely based on optical properties and supported by EDS analyses.

Micromorphology is more successful when it deals with modes of occurrence or with mineral assemblages, providing information that cannot be deduced from chemical data or XRD results. It allows a distinction between subaqueous crystal growth and mineral formation from subsurface brines, in various parts of a soil horizon (macropores, ped faces, groundmass). During the study of salt occurrences, the ephemeral nature of their presence, composition and fabric should be kept in mind.

References

Agafonoff, V., 1935/1936. Sols types de Tunisie. Annales du Service Botanique et Agronomique de Tunisie 12/13, 43-413.

Allen, B.L., 1985. Micromorphology of Aridisols. In Douglas, L.A. & Thompson, M.L. (eds.), Soil Micromorphology and Soil Classification. Soil Science Society of America Special Publication 15, SSSA, Madison, pp. 197-216.

Amit, R. & Yaalon, D.H., 1996. The micromorphology of gypsum and halite in Reg soils – the Negev Desert, Israel. Earth Surface Processes and Landforms 21, 1127-1143.

Amit, R., Gerson, R. & Yaalon, D.H., 1993. Stages and rate of the gravel shattering process by salts in desert reg soils. Geoderma 57, 295-324.

Arakel, A.V., 1980. Genesis and diagenesis of Holocene evaporitic sediments in Hutt and Leeman lagoons, Western Australia. Journal of Sedimentary Petrology 50, 1305-1326.

Aref, M.A.M., 1998. Holocene stromatolites and microbial laminites associated with lenticular gypsum in a marine-dominated environment, Ras El Shetan area, Gulf of Aqaba, Egypt. Sedimentology 45, 245-262.

Aref, M.A.M., 2003. Classification and depositional environments of Quaternary pedogenic gypsum crusts (gypcrete) from east of the Fayum Depression, Egypt. Sedimentary Geology 155, 87-108.

Aref, M.A.M., Attia, O.E.A. & Wali, A.M.A., 1997. Facies and depositional environment of the Holocene evaporites in the Ras Shukeir area, Gulf of Suez, Egypt. Sedimentary Geology 110, 123-145.

Aref, M.A.M., El-Khoriby, E. & Hamdan, M.A., 2002. The role of salt weathering in the origin of the Qattara depression, Western Desert, Egypt. Geomorphology 45, 181-195.

Arnold, A., 1984. Determination of mineral salts from monuments. Studies in Conservation 29, 129-138.

Arnold, A. & Kueng, A., 1985. Crystallization and habits of salt efflorescences on walls I. Methods of investigation and habits. In Félix, G. (ed.), Proceedings of the Fifth International Congress on Deterioration and Conservation of Stone. Presses Polytechniques Romandes, Lausanne, pp. 255-267.

Arnold, A. & Zehnder, K., 1985. Crystallization and habits of salt efflorescences on walls II. Conditions of crystallization. In Félix, G. (ed.), Proceedings of the Fifth International Congress on Deterioration and Conservation of Stone. Presses Polytechniques Romandes, Lausanne, pp. 269-277.

Artiedo Cabello, O., 1996. Génesis y Distribución de Suelos en un Medio Semiárido. Quinto (Zaragoza). Ministerio de Agricultura, Pesca y Alimentación, Madrid, 222 p.

Azam, S., 2003. Influence of mineralogy on swelling and consolidation of soils in eastern Saudi Arabia. Canadian Geotechnical Journal 40, 964-975.

Azam, S., Abduljauwad, S.N., Al-Shayea, N.A. & Al-Amoudi, O.S.B., 1998. Expansive characteristics of gypsiferous/anhydritic soil formations. Engineering Geology 51, 89-107.

Barzanji, A.F. & Stoops, G., 1974. Fabric and mineralogy of gypsum accumulations in some soils of Iraq. Transactions of the 10th International Congress of Soil Science, Moscow, Volume VII, pp. 271-277.

Basyoni, M.H. & Aref, M.A., 2015. Sediment characteristics and microfacies analysis of Jizan supratidal sabkha, Red Sea coast, Saudi Arabia. Arab Journal of Geosciences 8, 9973-9992.

Benayas, J., de la Cruz, M.T. & Rey Benayas, J.M., 1992. Caracteristicas micromorfologicas de suelos de humedales en zonas semiaridas (Cuenca del Duero). Suelo y Planta 2, 13-25.

Bisson, K.M., Welch, K.A., Welch, S.A., Sheets, J.M., Lyons, W.B., Levy, J.S. & Fountain, A.G., 2015. Patterns and processes of salt efflorescences in the McMurdo region, Antarctica. Arctic, Antarctic, and Alpine Research 47, 407-425.

Blackburn, G., Sleeman, J.R. & Cent, J.H., 1975. Gilgai in western Victoria. CSIRO, Land Resources Laboratories, Division of Soils, Biennial Report 1974-1975, pp. 32-33.

Bobst, A.L., Lowenstein, T.K., Jordan. T.E., Godfrey, L.V., Ku, T.L. & Luo, S., 2001. A 106 ka paleoclimate record from drill core of the Salar de Atacama, Northern Chile. Palaeogeography, Palaeoclimatology, Palaeoecology 173, 21-42.

Boixadera, J., Poch, R.M., Lowick, S.E. & Balasch, J.C., 2015. Loess and soils in the eastern Ebro Basin. Quaternary International 376, 114-133.

Braitsch, O., 1971. Salt Deposits. Their Origin and Composition. Springer-Verlag, Berlin, 297 p.

Brewer, R., Sleeman, J.R. & Foster, R.C., 1983. The fabric of Australian soils. In Soils: An Australian Viewpoint. CSIRO, Melbourne & Academic Press, London, pp. 439-476.

Brock-Hon, A.L., Robins, C.R. & Buck, B.J., 2012. Micromorphological investigation of pedogenic barite in Mormon Mesa petrocalcic horizons, Nevada USA: implication for genesis. Geoderma 179/180, 1-8.

Buck, B.J., Brock, A.L., Johnson, W.H. & Ulery, A.L., 2004. Corrosion of depleted uranium in an arid environment: soil-geomorphology, SEM/EDS, XRD, and electron microprobe analyses. Soil and Sediment Contamination 13, 545-561.

Buck, B.J., Wolff, K., Merkler, D.J. & McMillan, N.J., 2006. Salt mineralogy of Las Vegas Wash, Nevada: morphology and subsurface evaporation. Soil Science Society of America Journal 70, 1639-1651.

Buck, B.J., King, J. & Etyemezian, V., 2011. Effects of salt mineralogy on dust emissions, Salton Sea, California. Soil Science Society of America Journal 75, 1971-1985.

Bullock, P., Fedoroff, N., Jongerius A., Stoops, G., Tursina, T. & Babel, U., 1985. Handbook for Soil Thin Section Description. Waine Research Publications, Wolverhampton, 152 p.

Bush, P., 1973. Some aspects of the diagenetic history of the sabkha in Abu Dhabi, Persian Gulf. In Purser, B.H. (ed.), The Persian Gulf. Holocene Carbonate Sedimentation and Diagenesis in a Shallow Epicontinental Sea. Springer-Verlag, Berlin, pp. 395-407.

Butler, G.P., 1970. Holocene gypsum and anhydrite of the Abu Dhabi sabkha, Trucial Coast: an alternative explanation of origin. In Rau, J.L. & Dellwig, L.F. (eds.), Proceedings of the Third Symposium on Salt, Volume 1. Northern Ohio Geological Society, Cleveland, pp. 120-152.

Canfora, L., Vendramin, E., Vittori Antisari, L., Lo Papa, G., Dazzi, C., Benedetti, A., Iavazzo, P., Adamo, P., Jungblut, A.D. & Pinzari, F., 2016. Compartmentalization of gypsum and halite associated with cyanobacteria in saline soil crusts. FEMS Microbiology Ecology, 92, n° fiw080, 13 p.

Carson, C.D., Fanning, D.S. & Dixon, J.B., 1982. Alfisols and Ultisols with acid sulfate weathering features in Texas. In Kittrick, J.A., Fanning, D.S. & Hossner, L.R. (eds.), Acid Sulfate Weathering. Soil Science Society of America Special Publication 10, SSSA, Madison, pp. 127-146.

Chen, X.Y., 1997. Pedogenic gypcrete formation in arid central Australia. Geoderma 77, 39-61.

Chernousenko, G.I., Yamnova, I.A. & Skripnikova, M.I., 2003. Anthropogenic salinization of soils in Moscow. Eurasian Soil Science 36, 92-100.

Cheverry, C., Fromaget, M. & Bocquier, G., 1972. Quelques aspects micromorphologiques de la pédogenèse des sols de polders conquis sur le lac Tchad. Cahiers ORSTOM, Série Pédologie 10, 373-387.

Chowns, T.M. & Elkins, J.E., 1974. The origin of quartz geodes and cauliflower cherts through the silicification of anhydrite nodules. Journal of Sedimentary Petrology 44, 885-903.

Clarke, C.E., Majodina, T.O., du Plessis, A. & Andreoli, M.A.G., 2016. The use of X-ray tomography in defining the spatial distribution of barite in the fluvially derived palaeosols of Vaalputs, Northern Cape Province, South Africa. Geoderma 267, 48-57.

Collao, S., Arce, E. & Andía, A., 2002. Mineralogía, química e inclusiones fluidas en los depositos de nitratos de María Elena, IIa Región, Chile. Boletín de la Sociedad Chilena de Química 47, 181-190.

Courty, M.A., Goldberg, P. & Macphail, R., 1989. Soils and Micromorphology in Archaeology. Cambridge University Press, Cambridge, 344 p.

Crum, J.R. & Franzmeier, D.P., 1980. Soil properties and chemical composition of tree leaves in southern Indiana. Soil Science Society of America Journal 44, 1063-1069.

Cumba, A. & Imbellone, P., 2004. Micromorphology of palaeosoils at the continental border of the Buenos Aires province, Argentina. Revista Mexicana de Ciencias Geológicas 21, 18-29.

Darmody, R.G., Harding, S.D. & Hassett, J.J., 1989. Barite authigenesis in surficial soils of mid-continental United States. In Miles, D.L. (ed.), Water-Rock Interaction, WRI-6. Proceedings of the 6th International Symposium on Water-Rock Interaction, Malvern. A.A. Balkema, Rotterdam, pp. 183-186.

Darragi, F., Gueddari, M. & Fritz, B., 1983. Mise en évidence d'un fluoro-sulfate de sodium, la kogarkoite, dans les croûtes salines du Lac Natron en Tanzanie. Comptes Rendus de l'Académie des Sciences, Série II, 297, 141-144.

Datta, S., Thibault, Y., Fyfe, W.S., Powell, M.A., Hart, B.R., Martin, R.R. & Tripthy, S., 2002. Occurrence of trona in alkaline soils of the Indo-Gangetic Plains of Uttar Pradesh (U.P.), India. Episodes 24, 236-239.

Dixon, J.B., Hossner, L.R., Senkayi, A.L. & Egashira, K., 1982. Mineralogical properties of lignite overburden as they relate to mine spoil reclamation. In Kittrick, J.A., Fanning, D.S. & Hossner, L.R. (eds.), Acid Sulfate Weathering. Soil Science Society of America, Special Publication 10, SSSA, Madison, pp. 169-191.

Doner, H.E. & Lynn, W.C., 1989. Carbonate, halide, sulfate, and sulfide minerals. In Dixon, J.B. & Weed, S.B. (eds.), Minerals in Soil Environments. 2nd Edition. SSSA Book Series, No. 1, Soil Science Society of America, Madison, pp. 279-330.

Driessen, P.M., 1970. Soil Salinity and Alkalinity in the Great Konya Basin, Turkey. Pudoc, Wageningen, 99 p.

Driessen, P.M. & Schoorl, R., 1973. Mineralogy and morphology of salt efflorescences on saline soils in the Great Konya Basin, Turkey. Journal of Soil Science 24, 436-442.

Dronkert, H., 1977. A preliminary note on a recent sabkha deposit in S. Spain. Revista del Instituto de Investigaciones Geológicas Diputación Provincial Universidad de Barcelona 32, 153-166.

Ducloux, J., Guero, Y., Fallavier, P. & Valet, S., 1994. Mineralogy of salt efflorescences in paddy field soils of Kollo, southern Niger. Geoderma 64, 57-71.

El-Sayed, M.I., 2000. Karstic features associated with unconformity surfaces, a case study from the United Arab Emirates. Journal of Arid Environments 46, 295-312.

Ericksen, G.E. & Mrose, M.E., 1970. Mineralogical studies of the nitrate deposits of Chile. II. Darapskite, $Na(NO_3)(SO_4).H_2O$. American Mineralogist 55, 1500-1517.

Eriksson, K.A., Simpson, E.L., Master, S. & Henry, G., 2005. Neoarchaean (c. 2.58 Ga) halite casts: implications for palaeoceanic chemistry. Journal of the Geological Society of London 162, 789-799.

Eswaran, H., 1984. Scanning electron microscopy of salts. McGraw-Hill Yearbook of Science & Technology 1984, McGraw-Hill Book Company, New York, pp. 403-405.

Eswaran, H. & Carrera, M., 1980. Mineralogical zonation in salt crust. Proceedings of the International Symposium on Salt-affected Soils, Karnal, pp. 20-30.

Eswaran, H. & Drees, R., 2004. Soil under the Microscope: Evaluating Soils in Another Dimension. Soil Micromorphology Committee of the Soil Science Society of America, Madison. [CD ROM].

Eswaran, H. & Zi-Tong, G., 1991. Properties, genesis, classification, and distribution of soils with gypsum. In Nettleton, W.D. (ed.), Occurrence, Characteristics, and Genesis of Carbonate, Gypsum, and Silica Accumulations in Soils. Soil Science Society of America, Special Publication 26, SSSA, Madison, pp. 89-119.

Eswaran, H., Stoops, G. & Abtahi, A., 1980. SEM morphologies of halite (NaCl) in soils. Journal of Microscopy 120, 343-352.

Eswaran, H., Ilaiwi, M. & Osman, A., 1983. Mineralogy and micromorphology of Aridisols. In Beinroth, F. H. & Osman, A. (eds.), Proceedings of the Third International Soil Classification Workshop. ACSAD, Damascus, pp. 153-174.

Evans, G. & Shearman, D.J., 1964. Recent celestine from the sediments of the Trucial Coast of the Persian Gulf. Nature 202, 385-386.

Feofarova, I.I., 1958. Sulfates in saline soils [*in Russian*]. Trudy Pochvennogo Instituta V.V. Dokuchaeva 53.

FitzPatrick, E.A., 1993. Soil Microscopy and Micromorphology. John Wiley & Sons, Chichester, 304 p.

Fitzpatrick, R.W., Naidu, R. & Self, P.G., 1992. Iron deposits and microorganisms in saline sulfidic soils with altered soil water regimes in South Australia. In Skinner, H.C.W. & Fitzpatrick, R.W. (eds.), Biomineralization Processes of Iron and Manganese − Modern and Ancient Environments. Catena Supplement 21, 263-286.

Fleischer, M., Wilcox, R.E. & Matzko, J.J., 1984. Microscopic Determination of the Nonopaque Minerals. U.S. Geological Survey Bulletin 1627, 453 p.

Frenzel, B., 1963. Salzausblühungen im südpfälzischen Buntsandstein. Neues Jahrbuch für Mineralogie Abhandlungen 100, 130-144.

Gerasimova, M.I., Gubin, S.V. & Shoba, S.A., 1996. Soils of Russia and Adjacent Countries: Geography and Micromorphology. Moscow State University & Agricultural University Wageningen, 204 p.

Gibson, E.K., Wentworth, S.J. & McKay, D.S., 1983. Chemical weathering and diagenesis of a cold desert soil from Wright Valley, Antarctica: an analogue of Martian weathering processes. Journal of Geophysical Research 88 Supplement, A912-A928.

Golovanov, D.L., Lebedeva-Verba, M.P., Dorokhova, M.F. & Slobodkin, A.I., 2005. Micromorphological and microbiological characterization of elementary soil-forming processes in desert soils of Mongolia. Eurasian Soil Science 38, 1290-1300.

Gore, D.B., Nichols, G.T., Lehmann, C.E.R., Burgess, J.S., Baird, A.S. & Creagh, D.C., 2000. An Atlas of Surficial Salts of the Vestfold Hills, East Antarctica: Composition, Distribution and Origin. ANARE Report 143, Australian Antarctic Division, Kingston, 146 p.

Gumuzzio, J. & Casas, J., 1988. Accumulations of soluble salts and gypsum in soils of the Central Region, Spain. Cahiers ORSTOM, Série Pédologie 24, 215-226.

Gumuzzio, J., Battle, J. & Casas, J., 1982. Mineralogical composition of salt efflorescences in a Typic Salorthid, Spain. Geoderma 28, 38-51.

Gunatilaka, A., 1990. Anhydrite diagenesis in a vegetated sabkha, Al-Khiran, Kuwait, Arabian Gulf. Sedimentary Geology 69, 95-116.

Gunatilaka, A., Saleh, A. & Al-Temeemi, A., 1980. Plant-controlled supratidal anhydrite from Al-Khiran, Kuwait. Nature 288, 257-260.

Hamdi-Aissa, B., 2002. Paleogeochemical interpretation of some gypsic microfabrics in hyperdesert soils. Proceedings of the 17th World Congress of Soil Sciences, Bangkok, Paper no. 1861, 9 p.

Hamdi-Aissa, B., Valles, V., Aventurier, A. & Ribolzi, O., 2004. Soils and brine geochemistry and mineralogy of hyperarid desert playa, Ouargla Basin, Algerian Sahara. Arid Land Research and Management 18, 103-126.

Hanna, F.S. & Stoops, G.J., 1976. Contribution to the micromorphology of some saline soils of the North Nile Delta in Egypt. Pedologie 26, 55-73.

Herrero, J., 1991. Morfologia y Génesis de Suelos Sobre Yesos. Monografias INIA 77, Instituto Nacional de Investigación y Tecnologia Agraria y Alimentaria, Madrid, 447 p.

Herrero Isern, J., Rodriguez Ochoa, R. & Porta Casanellas, J., 1989. Colmatacion de Drenes en Suelos Afectados por Salinidad. Institución Fernando el Católico, Zaragoza, 133 p.

Hunt, C.B., Robinson, T.W., Bowles, W.A. & Washburn, A.L., 1966. Hydrologic Basin Death Valley California. U.S. Geological Survey Professional Paper 494-B, 138 p.

Hussain, M. & Warren, J.K., 1989. Nodular and enterolithic gypsum: the 'sabkha-tization' of Salt Flat playa, west Texas. Sedimentary Geology 64, 13-24.

Jennings, D.S. & Driese, S.G., 2014. Understanding barite and gypsum precipitation in upland acid-sulfate soils: an example from a Lufkin Series toposequence, south-central Texas, USA. Sedimentary Geology 299, 106-118.

Jennings, D.S., Driese, S.G. & Dworkin, S.I., 2015. Comparison of modern and ancient barite-bearing acid-sulphate soils using micromorphology, geochemistry and field relationships. Sedimentology 62, 1078-1099.

Joeckel, R.M. & Ang Clement, B., 1999. Surface features of the Salt Basin of Lancaster County, Nebraska, USA. Catena 34, 243-275.

Joeckel, R.M. & Ang Clement, B.J., 2005. Soils, surficial geology, and geomicrobiology of saline-sodic wetlands, North Platte River Valley, Nebraska, USA. Catena 61, 63-101.

Jones, B. & Renaut, R.W., 1996. Skeletal crystals of calcite and trona from hot-spring deposits in Kenya and New Zealand. Journal of Sedimentary Research 66, 265-274.

Khalaf, F.I., Al-Zamel, A. & Gharib, I., 2014. Petrography and genesis of Quaternary coastal gypcrete in North Kuwait, Arabian Gulf. Geoderma 226/227, 223-230.

Khormali, F., 2005. Occurrence of celestite in arid soils of southern Iran. Geophysical Research Abstracts 7, EGU05-A-00798, 2 p.

Kinsman, D.J.J., 1966. Gypsum and anhydrite of recent age, Trucial Coast, Persian Gulf. In Rau, J.L. (ed.), Proceedings of the Second Symposium on Salt. Northern Ohio Geological Society, Cleveland, pp. 302-326.

Kooistra, M.J., 1983. Light microscope and submicroscope observations of salts in marine alluvium. Geoderma 30, 149-160.

Lebedeva, M.P. & Konyushkova, M.V., 2011. Temporal changes in the microfabrics of virgin and reclaimed solonetzes at the Dzhanybek research station. Eurasian Soil Science 44, 753-765.

Lebedeva-Verba, M., Chizhikova, N.P. & Mochalova, E.F., 2004. Crystal chemistry and the fabric of salt accumulations in crust Solochaks of Uzbekistan. In Kapur, S., Akça, E., Montanarella, L., Ozturk, A. & Mermut, A. (eds.), Extended Abstracts of the 12th International Meeting on Soil Micromorphology, Adana, pp. 92-95.

Lee, S.Y. & Gilkes, R.J., 2005. Groundwater geochemistry and composition of hardpans in southwestern Australian regolith. Geoderma 126, 59-84.

Lowenstein, T.K., Hein, M.C., Bobst, A.L., Jordan, T.E., Ku, T.L. & Luo, S., 2003. An assessment of stratigraphic completeness in climate-sensitive closed-basin lake sediments: Salar de Atacama, Chile. Journal of Sedimentary Research 73, 91-104.

Lynn, W.C., Tu, H.Y. & Franzmeier, D.P., 1971. Authigenic barite in soils. Soil Science Society of America Proceedings 35, 160-161.

Machel, H.G. & Burton, E.A., 1991. Burial-diagenetic sabkha-like gypsum and anhydrite nodules. Journal of Sedimentary Petrology 61, 394-405.

McCarthy, P.J. & Plint, A.G., 1998. Recognition of interfluve sequence boundaries: integrating paleo-pedology and sequence stratigraphy. Geology 26, 387-390.

McCarthy, P.J. & Plint, A.G., 2003. Spatial variability of palaeosoils across Cretaceous interfluves in the Dunvegan Formation, NE British Columbia, Canada: palaeohydrological, palaeogeomorphological and stratigraphic implications. Sedimentology 50, 1187-1220.

McFadden, L.D., Wells, S.G. & Jercinovich, M.J., 1987. Influences of eolian and pedogenic processes on the origin and evolution of desert pavements. Geology 15, 504-508.

McKee, T.R. & Brown, J.L., 1977. Preparation of specimens for electron microscopic investigation. In Dixon, J.B. & Weed, S.B. (eds.), Minerals in Soil Environments. SSSA Book Series, No. 1, Soil Science Society of America, Madison, pp. 809-846.

McLaren, S.J., 2001. Effects of sea spray on vadose diagenesis of Late Quaternary aeolianites, Bermuda. Journal of Coastal Research 17, 228-240.

Mees, F., 1998. The alteration of glauberite in lacustrine deposits of the Taoudenni-Agorgott basin, northern Mali. Sedimentary Geology 117, 193-205.

Mees, F., 1999a. Distribution patterns of gypsum and kalistrontite in a dry lake basin of the southwestern Kalahari (Omongwa pan, Namibia). Earth Surface Processes and Landforms 24, 731-744.

Mees, F., 1999b. Textural features of Holocene perennial saline lake deposits of the Taoudenni-Agorgott basin, northern Mali. Sedimentary Geology 127, 65-84.

Mees, F., 2003. Salt mineral distribution patterns in soils of the Otjomongwa pan, Namibia. Catena 54, 425-437.

Mees, F. & De Dapper, M., 2005. Vertical variations in bassanite distribution patterns in near-surface sediments, southern Egypt. Sedimentary Geology 181, 225-229.

Mees, F. & Singer, A., 2006. Surface crusts on soils/sediments of the southern Aral Sea basin, Uzbekistan. Geoderma 136, 152-159.

Mees, F. & Stoops, G., 1990. Micromorphological study of a sediment core from the Malha crater lake, Sudan. In Douglas, L.A. (ed.), Soil Micromorphology: A Basic and Applied Science. Developments in Soil Science, Volume 19. Elsevier, Amsterdam, pp. 295-301.

Mees, F. & Stoops, G., 1991. Mineralogical study of salt efflorescences on soils of the Jequepeteque Valley, northern Peru. Geoderma 49, 255-272.

Mees, F. & Stoops, G., 2003. Circumgranular bassanite in a gypsum crust from eastern Algeria — a potential paleosurface indicator. Sedimentology 50, 1139-1145.

Mees, F. & Stoops, G., 2018. Sulphidic and sulphuric materials. In Stoops, G., Marcelino, V. & Mees, F. (eds.), Interpretation of Micromorphological Features of Soils and Regoliths. Second Edition. Elsevier, Amsterdam, pp. 347-376.

Mees, F., Hatert, F. & Rowe, R., 2008. Omongwaite, $Na_2Ca_5(SO_4)_6.3H_2O$, a new mineral from recent salt lake deposits, Namibia. Mineralogical Magazine 72, 1307-1318.

Mees, F., Castañeda, C., Herrero, J. & Van Ranst, E., 2011. Bloedite sedimentation in a seasonally dry saline lake (Salada Mediana, Spain). Sedimentary Geology 238, 106-115.

Miller, J.J., Pawluk, S. & Beke, G.J., 1989. Evaporite mineralogy, and soil solution and groundwater chemistry of a saline seep from southern Alberta. Canadian Journal of Soil Science 69, 273-286.

Milton, C., 1942. Note on the occurrence of calcium sulphate hemihydrate ($CaSO_4.1/2H_2O$) in thin sections of rocks. American Mineralogist 27, 517-518.

Moghiseh, E. & Heidari, A., 2012. Polygenetic saline gypsiferous soils of the Bam region, Southeast Iran. Journal of Soil Science and Plant Nutrition 12, 729-746.

Moiola, R.J. & Glover, E.D., 1965. Recent anhydrite from Clayton Playa, Nevada. American Mineralogist 50, 2063-2069.

Mrose, M.E., Fahey, J.J. & Ericksen, G.E., 1970. Mineralogical studies of the nitrate deposits of Chile. III. Humberstonite, $K_3Na_7Mg_2(SO_4)_6(NO_3)_2.6H_2O$, a new saline mineral. American Mineralogist 55, 1518-1533.

Owen, R.A., Owen, R.B., Renaut, R.W., Scott, J.J., Jones, B. & Ashley, G.M., 2008. Mineralogy and origin of rhizoliths on the margins of saline, alkaline Lake Bogoria, Kenya Rift Valley. Sedimentary Geology 203, 143-163.

Parfenova, E.I. & Yarilova, E.A., 1965. Mineralogical Investigations in Soil Science. Israel Program for Scientific Translations, Jerusalem, 178 p.

Parfenova, E.I. & Yarilova, E.A., 1977. A Manual of Micromorphological Research in Soil Science [in Russian]. Nauka, Moscow, 198 p.

Poch, R.M., Artieda, O. & Lebedeva, M., 2018. Gypsic features. In Stoops, G., Marcelino, V. & Mees, F. (eds.), Interpretation of Micromorphological Features of Soils and Regoliths. Second Edition. Elsevier, Amsterdam, pp. 259-287.

Podwojewski, P., 1995. The occurrence and interpretation of carbonate and sulfate minerals in a sequence of Vertisols in New Caledonia. Geoderma 65, 223-248.

Porta, J., 1998. Methodologies for the analysis and characterization of gypsum in soils: a review. Geoderma 87, 31-46.

Pueyo Mur, J.J., 1978/1979. La precipitación evaporitica actual en las lagunas saladas del area: Bujaraloz, Sástago, Caspe, Alcañiz y Calanda (provincias de Zaragoza y Teruel). Revista del Instituto de Investigaciones Geológicas, Diputación Provincial, Universidad de Barcelona 33, 5-56.

Pueyo Mur, J.J., 1980. Procesos diagenéticos observados en las lagunas tipo playa de la zona Bujaraloz-Alcañiz (provincias de Zaragoza y Teruel). Revista del Instituto de Investigaciones Geológicas, Diputación Provincial, Universidad de Barcelona 34, 195-207.

Pueyo, J.J., Chong, G. & Vega, M., 1998. Mineralogía y evolución de las salmueras madres en el yacimiento de nitratos Pedro de Valdivia, Antofagasta, Chile. Revista Geológica de Chile 25, 3-15.

Retallack, G.J. & Kirby, M.X., 2007. Middle Miocene global change and paleogeography of Panama. Palaios 22, 667-679.

Rodriguez-Navarro, C. & Doehne, E., 1999. Salt weathering: influence of evaporation rate, supersaturation and crystallization pattern. Earth Surface Processes and Landforms 24, 191-209.

Rodríguez-Ochoa, R., Olarieta, J.R. & Castañeda, C., 2012. Micromorphology of salt accumulations in soils of north Monegros, Spain: optical microscopy and SEM. In Poch, R.M., Casamitjana, M. & Francis, M.L. (eds.), Proceedings of the 14th International Working Meeting on Soil Micromorphology, Universitat de Lleida, Lleida, pp. 31-34.

Rognon, P., Coudé-Gaussen, G., Fedoroff, N. & Goldberg, P., 1987. Micromorphology of loess in the northern Negev (Israel). In Fedoroff, N., Bresson, L.M. & Courty, M.A. (eds.), Micromorphologie des Sols, Soil Micromorphology. AFES, Paris, pp. 631-638.

Schiefelbein, I.M., Buck, B.J., Lato, L. & Merkler, D., 2002. SEM analyses of pedogenic salt minerals in a toposequence, Death Valley, California, USA. Proceedings of the 17th World Congress of Soil Sciences, Bangkok, Paper no. 457, 9 p.

Schroeder, J.H., Kachholz, K.D. & Heuer, M., 1984. Eolian dust in the coastal desert of the Sudan: aggregates cemented by evaporites. Geo-Marine Letters 4, 139-144.

Schweigstillova, J., Simova, V. & Hradil, D., 2005. New investigations on the salt weathering of Cretaceous sandstones, Czech Republic. Ferrantia 44, 177-179.

Searl, A. & Rankin, S., 1993. A preliminary petrographic study of the Chilean nitrates. Geological Magazine 130, 319-333.

Shahid, S.A., 2013. Development in soil salinity assessment, modeling, mapping, and monitoring from regional to submicroscopic scales. In Shahid, S.A., Abdelfattah, M.A. & Taha, F.K. (eds.), Developments in Soil Salinity Assessment and Reclamation. Springer, Dordrecht, pp. 3-43.

Shahid, S.A. & Jenkins, D.A., 1994. Mineralogy and micromorphology of salt crusts from the Punjab, Pakistan. In Ringrose-Voase, A.J. & Humphries, G.S. (eds.), Soil Micromorphology: Studies in Management and Genesis. Developments in Soil Science, Volume 22. Elsevier, Amsterdam, pp. 799-810.

Shayan, A. & Lancucki, C.J., 1984. Konyaite in salt efflorescences from the Tertiary marine deposit near Geelong, Victoria, Australia. Soil Science Society of America Journal 48, 939-942.

Shearman, D.J., 1978. Evaporites of coastal sabkhas. In Dean, W.E. & Schreiber, B.C. (eds.), Marine Evaporites. SEPM Short Course Lecture Notes 4, pp. 6-42.

Shearman, D.J., 1985. Syndepositional and late diagenetic alteration of primary gypsum to anhydrite. In Schreiber, B.C. & Hamer, H. (eds.), Proceedings of the Sixth Symposium on Salt. Salt Institute, Alexandria, pp. 41-50.

Shoba, S.A., Tursina, T.V. & Yamnova, I.A., 1983. Scanning electron microscopy of salt neoformations in soils [*in Russian*]. Biologicheskie Nauki 3, 91-98.

Siedlecka, A., 1972. Length-slow chalcedony and relicts of sulphates − evidences of evaporitic environments in the Upper Carboniferous and Permian bends of Bear Island, Svalbard. Journal of Sedimentary Petrology 42, 812-813.

Simón, M., Aguilar, J. & Dorronsoro, C., 1980. Los suelos halomorfos de la Provincia de Granada. IV. Estudio mineralógico. Anales de Edafología y Agrobiología 39, 429-438.

Singer, A., Kirsten, W.F.A. & Bühmann, C., 1999. A proposed fog deposition mechanism for the formation of salt efflorescences in the Mpumalanga highveld, Republic of South Africa. Water, Air, and Soil Pollution 109, 313-325.

Smieja-Krol, B., Janeczek, J. & Wiedermann, J., 2014. Pseudomorphs of barite and biogenic ZnS after phyto-crystals of calcium oxalate (whewellite) in the peat layer of a poor fen. Environmental Science and Pollution Research 21, 7227-7233.

Smith, M.S., 2005. A glimpse at the geochemistry of alkaline salt-affected soils. In Roach, I.C. (ed.), Regolith 2005 − Ten Years of CRC LEME. CRC LEME, Canberra, pp. 289-293.

Smith, M.S., Kirste, D. & McPhail, D.C., 2004. Mineralogy of alkaline-saline soils on the western slopes of northern New South Wales. In Roach, I.C. (ed.), Regolith 2004. CRC LEME, Kensington, pp. 330-334.

Smykatz-Kloss, W., Istrate, G., Hötzl, H., Kössl, H. & Wohnlich, S., 1985. Vorkommen und Entstehung von bassanit, $CaSO_4.1/2H_2O$, im Gipskarstgebiet von Foum Tatahouine, Südtunesien. Chemie der Erde 44, 67-77.

Stoops, G., 1974. Optical and electron microscopy. A comparison of their principles and their use in micropedology. In Rutherford, G. (ed.), Soil Microscopy. The Limestone Press, Kingston, pp. 101-118.

Stoops, G., 1998. Minerals in soil and regolith thin sections. Natuurwetenschappelijk Tijdschrift 77, 3-11.

Stoops, G., 2003. Guidelines for Analysis and Description of Soil and Regolith Thin Sections. Soil Science Society of America, Madison, 184 p.

Stoops, G. & Delvigne, J., 1990. Morphology of mineral weathering and neoformation. II Neoformation. In Douglas, L.A. (ed.), Soil Micromorphology: A Basic and Applied Science. Developments in Soil Science, Volume 19. Elsevier, Amsterdam, pp. 483-492.

Stoops, G. & Zavaleta, A., 1978. Micromorphological evidence of barite neoformation in soils. Geoderma 20, 63-70.

Stoops, G., Eswaran, H. & Abtahi, A., 1978. Scanning electron microscopy of authigenic sulphate minerals in soils. In Delgado, M. (ed.), Soil Micromorphology. University of Granada, pp. 1093-1113.

Sullivan, L.A. & Koppi, A.J., 1993. Barite pseudomorphs after lenticular gypsum in a buried soil from central Australia. Australian Journal of Soil Research 31, 393-396.

Sullivan, L.A. & Koppi, A.J., 1995. Micromorphology of authigenic celestobarite in a duripan from central Australia. Geoderma 64, 357-361.

Taher, A.G. & Abdel-Motelib, A., 2015. New insights into microbially induced sedimentary structures in alkaline hypersaline El Beida Lake, Wadi El Natrun, Egypt. Geo-Marine Letters 35, 341-353.

Tovey, K. & Dent, D., 2002. Microstructure and microcosm chemistry of tidal soils. Proceedings of the 17th World Congress of Soil Sciences, Bangkok, Paper no. 892, 7 p.

Tucker, M.E., 1976. Replaced evaporites from the late Precambrian of Finnmark, Arctic Norway. Sedimentary Geology 16, 193-204.

Tursina, T., 1980. The microstructure and the origin of new salt formations of salt affected soils. Proceedings of the International Symposium on Salt-affected Soils, Karnal, pp. 35-43.

Tursina, T.V., 1981. Micromorphology of salt affected soils. Problems of Soil Science. Soviet Pedologists to the XIIth International Congress of Soil Science. Nauka, Moscow, pp. 177-182.

Tursina, T.V., 2002. Micromorphological methods in the analysis of pedogenetic problems. Eurasian Soil Science 35, 777-791.

Tursina, T.V. & Yamnova, I.A., 1987. Identification of salt minerals in soils. Soviet Soil Science 19, 97-110.

Tursina, T.V., Yamnova, I.A. & Shoba, S.A., 1980. Combined stage-by-stage morphological, mineralogical and chemical study of the composition and organization of saline soils. Soviet Soil Science 12, 81-94.

Van Hoesen, J.G., Buck, B.J., Merkler, D.J. & McMillan, N., 2001. An investigation of salt micromorphology and chemistry, Las Vegas Wash, Nevada. In Luke, B.A., Jacobson, E. & Werle, J. (eds.), Proceedings of the 36th Annual Symposium on Engineering Geology and Geotechnical Engineering, Idaho State University, Pocatello, p. 729-737.

Van Ranst, E., Shitumbanuma, V., Tembo, F. & Mees, F., 2008. Soil salinization and dental fluorosis as a result of water use in Zambia. Bulletin des Séances de l'Académie Royale des Sciences d'Outre-Mer 53, 431-446.

Van Tassel, R., 1958. Notes minéralogiques. XI. Jarosite, natrojarosite, beaverite, leonhardtite et hexahydrite du Congo belge. Bulletin de l'Institut Royal des Sciences Naturelles de Belgique 34, n° 44, 12 p.

Vergouwen, L., 1979. Two new occurrences and the Gibbs energy of burkeite. Mineralogical Magazine 43, 341-345.

Vergouwen, L., 1981a. Scanning electron microscopy applied on saline soils from the Konya Basin in Turkey and from Kenya. In Bisdom, E.BA. (ed.), Submicroscopy of Soils and Weathered Rocks, Proceedings of the First Workshop of the International Working Group on Submicroscopy of Undisturbed Soil Materials. Pudoc, Wageningen, pp. 237-248.

Vergouwen, L., 1981b. Eugsterite, a new salt mineral. American Mineralogist 66, 632-636.

Vizcayno, C., Garcia-Gonzalez, M.T., Gutierrez, M. & Rodriguez, R., 1995. Mineralogical, chemical and morphological features of salt accumulations in the Flumen-Monegros district, NE Spain. Geoderma 68, 193-210.

Vogt, T. & Larqué, P., 2002. Clays and secondary minerals as permafrost indicators: examples from the circum-Baikal region. Quaternary International 95/96, 175-187.

von Hodenberg, R. & Miotke, F.D., 1983. Einige besondere Salzkristallbildungen im Süd-Viktoria-Land der Antarktis und erste Ergebnisse der Untersuchung eines neuen Minerals, eines Na-Ca-Doppelsulfats. Kali und Steinsalz 8, 374-383.

Wentworth, S.J., Gibson, E.K., Velbel, M.A. & McKay, D.S., 2005. Antarctic Dry Valleys and indigenous weathering in Mars meteorites: implications for water and life on Mars. Icarus 174, 383-395.

Wierzchos, J., García-González, M.T. & Ascaso, C., 1995. Advantages of application of the backscattered electron scanning image in the determination of soil structure and soil constituents. International Agrophysics 9, 41-47.

Wierzchos, J., Davila, A.F., Sánchez-Almazo, I.M., Hajnos, M., Swieboda, R. & Ascaso, C., 2012. Novel water source for endolithic life in the hyperarid core of the Atacama Desert. Biogeosciences 9, 2275-2286.

Wilson, M.A., Shahid, S.A., Abdelfattah, M.A., Kelley, J.A. & Thomas, J.E., 2013. Anhydrite formation on the coastal sabkha of Abu Dhabi, United Arab Emirates. In Shahid, S.A., Taha, F.K. & Abdelfattah, M. A. (eds.), Developments in Soil Classification, Land Use Planning and Policy Implications. Springer, Dordrecht, pp. 175-201.

Wongpokhom, N., Kheoruenromne, I., Suddhiprakarn, A. & Gilkes, R.J., 2008. Micromorphological properties of salt affected soils in Northeast Thailand. Geoderma 144, 158-170.

Yamnova, I.A., 2016. Salt and gypsum pedofeatures as indicators of soil processes. Byulleten Pochvennogo Instituta im. V.V. Dokuchaeva 86, 96-102.

Yamnova, I.A. & Pankova, E.I., 2013. Gypsic pedofeatures and elementary pedogenetic processes of their formation. Eurasian Soil Science 46, 1117-1129.

Yechieli, Y. & Ronen, D., 1997. Early diagenesis of highly saline lake sediments after exposure. Chemical Geology 138, 93-106.

Young, A.R.M., 1987. Salt as an agent in the development of cavernous weathering. Geology 15, 962-966.

Zhang, C. & Wang, Z., 1987. Microscopical research on the crystallization of salts in saline soils [*in Chinese*]. Acta Pedologica Sinica 24, 281-285.

Chapter 12

Phosphatic Features

Panagiotis Karkanas[1], Paul Goldberg[2,3]

[1]*AMERICAN SCHOOL OF CLASSICAL STUDIES, ATHENS, GREECE;*
[2]*BOSTON UNIVERSITY, BOSTON, MA, UNITED STATES;*
[3]*UNIVERSITY OF WOLLONGONG, WOLLONGONG, NSW, AUSTRALIA*

CHAPTER OUTLINE

1. Introduction

Phosphate finds its way into the soil by a number of pathways. It can be inherited from phosphate-rich parent materials, such as phosphorites and pegmatites (Table 1). In soils on other rock types, phosphate can be inherited as a minor constituent. The amount of total soil phosphorus in normal soil environments is low, with an average value of only 0.05% (Lindsay et al., 1989). Hydroxylapatite, referred to as apatite throughout this chapter, is the major phosphate mineral in soils. Under conditions of diagenesis, other phosphate minerals are produced, such as crandallite, montgomeryite, variscite and vivianite (see Table 1).

Petrographic and micromorphological analysis of soils and sediments have contributed significantly to the understanding of the origin and evolution of phosphate in soils and sediments. This is for example the case for environments with bog ore formation (Stoops, 1983; Landuydt, 1990; Kaczorek & Sommer, 2003) and for lake basins (Stamatakis & Koukouzas, 2001; Kaczorek & Sommer, 2003; Lücke & Brauer, 2004; Fagel et al., 2005; Sapota et al., 2006; Gamzikov & Marmulev, 2007). In studies of weathering profiles developed on marine phosphorites, bauxites, laterites, and carbonaceous and coal deposits, the microscopic study of soil phosphates has revealed a suite of Ca and Al

Table 1 General characteristics and representative references on mode of occurrence of phosphate minerals in soils and archeological deposits

Mineral	Crystal habit	Aggregate type	Optical data	Representative references on mode of occurrence
Apatite (hydroxylapatite) $Ca_5(PO_4)_3(OH)$	Prismatic	Granular, nodular, crusts, massive, colloform	PPL: usually colourless, also yellowish, white, black, brown, greenish, pinkish, reddish, blue XPL: first order white or grey, length fast BLF: yellowish, white	Shahack-Gross et al. (2004), Macphail et al. (2004), Cullen (1988), Lindsay et al. (1989), Bergadà et al. (2013)
Ardealite $Ca_2(PO_3OH)(SO_4)\cdot4H_2O$	Plate-shaped	Powdery masses, crusts	PPL: light yellowish brown, colourless XPL: first order yellow	Shahack-Gross et al. (2004)
Brushite $Ca(PO_3OH)\cdot2H_2O$	Prismatic, needle-shaped	Powdery masses	PPL: pale yellow, white, colourless XPL: first order yellow	Shahack-Gross et al. (2004), Lindsay et al. (1989)
Crandallite $CaAl_3(PO_4)(PO_3OH)(OH)_6$	Prismatic, fibrous	Sheaflike, spherulitic, rosettes, fibroradial crusts, nodular, massive	PPL: yellow, white, grey, colourless XPL: first order grey	Karkanas et al. (2000), Flicoteaux & Lucas (1984), Dill et al. (1991)
Leucophosphite $KFe^{3+}_2(PO_4)_2(OH)\cdot2H_2O$	Prismatic	Massive, spherulitic	PPL: white, brown, pink, greenish XPL: second order yellowish green	Karkanas et al. (1999), Wilson & Bain (1976), Lindsay et al. (1989), Tiessen et al. (1996), Pereira et al. (2013)
Millisite $NaCaAl_6(PO_4)_4(OH)_9\cdot3H_2O$	Fibrous	Fibroradial crusts, spherulites	PPL: white, grey, greenish XPL: first order red	Flicoteaux & Lucas (1984)
Mitridatite $Ca_2Fe^{3+}_3O_2(PO_4)_3\cdot3H_2O$	—	Nodules, crusts, veinlets, coatings	PPL: red, green, brownish XPL: second or high order	Stamatakis & Koukouzas (2001)
Montgomeryite $Ca_4MgAl_4(PO_4)_6(OH)_4\cdot12H_2O$	Lath-shaped, plate-shaped	Massive, fibroradial	PPL: green, colourless; no to weak pleochrism (light orange brown to pale magenta pink) XPL: first order white BLF: greenish white	Goldberg & Nathan (1975)

Mineral	Crystal shape	Aggregate form	Optical properties	References
Newberyite $Mg(PO_3OH) \cdot 3H_2O$	Prismatic, needle-shaped, plate-shaped	Fibroradial	PPL: grey, brown, colourless XPL: first order red	Karkanas et al. (2002), Lindsay et al. (1989)
Strengite $Fe^{3+}(PO_4) \cdot 2H_2O$	Lath-shaped, fibrous	Fibroradial, botryoidal or spherical aggregates and crusts	PPL: colourless, red, violet XPL: second order yellow-green	Lindsay et al. (1989)
Struvite $(NH_4)Mg(PO_4) \cdot 6H_2O$	Plate-shaped	Crusts, botryoidal	PPL: white, yellowish white, brownish white XPL: first order grey	Cullen (1988), Lindsay et al. (1989), Pereira et al. (2013)
Taranakite $K_3Al_5(PO_3OH)_6(PO_4)_2 \cdot 18H_2O$	Plate-shaped	Nodular, massive	PPL: white, yellowish, brown, red or green XPL: first order grey BLF: light green	Karkanas et al. (1999), Simas et al. (2007), Lindsay et al. (1989)
Variscite $Al(PO_4) \cdot 2H_2O$	–	Massive, crusts, nodules, fibroradial	PPL: greenish, yellowish, colourless XPL: second order yellow-green	Karkanas et al. (1999), Flicoteaux & Lucas (1984), Lindsay et al. (1989)
Vivianite $(Fe^{2+})_3(PO_4)_2 \cdot 8H_2O$	Prismatic, lath-shaped	Radial aggregates, concretionary, crusts	PPL: colourless, greenish, becoming blue with oxidation, strong pleochroism (pale yellow to dark blue) XPL: third or higher order, length slow	Fagel et al. (2005), Landuydt (1990), Stoops (1983), Gebhardt & Langohr (1999), McGowan & Prangnell (2006), Lindsay et al. (1989), Dill & Techmer (2009)
Wavellite $Al_3(PO_4)_2(OH)_3 \cdot 5H_2O$	Prismatic	Spherical radial aggregates, crusts, massive	PPL: white, greenish, yellowish, blue, brown, weak pleochroism (greenish to yellowish) XPL: second order yellow	Flicoteaux & Lucas (1984), Dill et al. (1991)

XPL – highest interference colours for 30 μm thickness.

phosphate minerals and diagenetic pathways (Altschuler, 1973; Flicoteaux et al., 1977; Flicoteaux & Lucas, 1984; Nahon, 1991; Dill et al., 1991; Tiessen et al., 1996; Dill, 2001).

Although the mentioned occurrences unrelated to faunal activity are significant, animal inputs are considered to be the major source of soil phosphate (Altschuler, 1973; Flicoteaux & Lucas, 1984; Dill, 2001). The most obvious sources are guano (bird and bat guano), animal dung and coprolites. For example, extensive ornithogenic insular phosphate deposits are well known from throughout the world (Hutchinson, 1950; Wilson & Bain, 1976; Cullen, 1988; Gregory & Rodgers, 1989; Landis & Craw, 2003; Simas et al., 2006, 2007; Schaefer et al., 2008; Pereira et al., 2013; Villagran et al., 2013). Furthermore, anthropogenic inputs, like those of other vertebrates add P to the soil in the form of human waste, forming typical phosphatic anthropic horizons (Macphail et al., 1990, 2003; Schaefer et al., 2004). Micromorphological studies of archeological sediments and experimental strategies have demonstrated that phosphatic features can be the result of animal husbandry (e.g., stabling) and other agricultural and human activities (e.g., manuring, fertilising, waste deposition) (Bertran & Raynal, 1991; Macphail, 2000; Simpson et al., 2000; Schaefer et al., 2004; Goldberg & Macphail, 2006; Villagran et al., 2013). Even for stabling, pens and byres, however, the major input of phosphate is not directly the original vegetal remains but their reprocessed, phosphate-rich contributions derived from animals (e.g., horses, cattle, goats and sheep; see e.g., Macphail et al., 2004). Modern chemical fertilisers are by no means the main anthropogenic input of phosphorus in soils today. A vast literature exists on the reaction of phosphate fertilisers with soils (see e.g., Bhujbal & Mistry, 1986; Lindsay et al., 1989; Beauchemin et al., 2003). However, due to the low amounts of reaction products the use of petrography is limited, with some notable exceptions (Lehr & Brown, 1958; Qureshi et al., 1978; Qureshi & Jenkins, 1987).

Micromorphological studies in cave environments represent some of the best sources of information on the accumulation and alteration of soil phosphate related to guano and anthropogenic inputs (Goldberg & Nathan, 1975; Schiegl et al., 1996; Karkanas et al., 1999, 2000, 2002; Weiner et al., 2002; Shahack-Gross et al., 2004). The alteration sequence of the phosphate minerals has been used to reconstruct the palaeochemical environment in the sediment and assess the integrity of the archeological material.

In this chapter we summarise available information about micromorphological features of phosphates in soils, regoliths and altered terrestrial sediments. Many of the studies in sediments do not follow a uniform micromorphological terminology, but we have tried to report micromorphological features using the terminology of Bullock et al. (1985) and Stoops (2003).

2. Phosphate Occurrences as Groundmass Components

2.1 Coarse Fraction

The most prominent groundmass phosphate mineral is apatite (Table 1), which is the most common phosphate mineral found in igneous and sedimentary rocks. Discrete sand- or silt-size grains of apatite inherited from the parent rock are often found in soils. Apatite is very stable in calcareous soils but weathers easily in acidic soils and in areas of relatively intensive leaching (Walker & Syers, 1976; Retallack, 1990). In the initial stages of weathering, dissolution of apatite may lead to secondary phosphates often enriched in rare-earth elements; these mineralisations are formed along rims, edges and pits in primary apatite crystals (Banfield & Eggleton, 1989; Nedachi et al., 2005).

The second most frequent form of groundmass apatite is bone, which is common in the coarse fraction of archeological soils and sediments, and varies from whole bones to millimetre-size splinters. In thin sections, both spongy bone and cortical bone shows a ropy internal structure in cross-polarised light (XPL). Concentric layers of fibrous apatite surround voids called haversian canals, which typically appear as elliptical points in thin sections (Fig. 1A). Fish bone shows a mixture of lamellar and woven bone tissue, with clear lamellar organisation dominating in fish species with acellular bone (Fig. 1C and D) (Cohen et al., 2012). Interference colours of bone are low (first order grey) and commonly abnormal (dark brown/olive or blue) (Fig. 1B and D). However, the collagen and not the

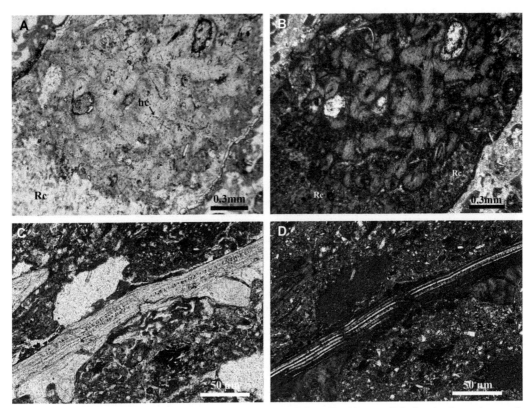

FIGURE 1 Recrystallisation of bone. (A, B) Recrystallisation to a light yellowish mass (Rc), showing haversian canals (hc) in PPL (A) and 'ropy' fine-grained mosaic fabric and abnormal blue and dark olive brown interference colours in CPL (B). (C, D) Fish spine bone with fine lamellar structure, in PPL (C) and in XPL (D).

bone mineral is responsible for the greater part of the birefringence of fresh bones. Hence, although apatite is optically uniaxial negative, the optical orientation observed for bone fragments (length fast or length slow) depends upon the preservation of the collagen (Watson, 1975). Recrystallisation of bone results in partial or total replacement of the ropy structure, with development of a mosaic-speckled b-fabric (with domains ca. 50 μm in diameter) (Fig. 1). In plane-polarised light (PPL), unaltered bone is typically yellow to yellowish brown, but upon heating it becomes locally blackish brown due to carbon accumulation (Hanson & Cain, 2007) and it loses its birefringence and histological structure. When heated above ca. 650 °C (calcination) the bone shows bright white low order interference colours. Cracking is another feature associated with burning. Duration and temperature of heating have an impact on the degree of calcination but the final composition and structure of bone are influenced by temperature alone (Snoeck et al., 2014). The presence of bone, particularly burnt bone, is common in archeological soils and sediments. It is usually a reflection of human inputs (e.g., hunting, waste accumulation), although carnivore denning (e.g., hyenas) also produces high concentrations of bone fragments. In some cases, intense physical and chemical alteration of bone can produce nodular apatitic features that can hardly be identified as bone (e.g., Morrás, 1983; Schaefer et al., 2004).

Apatite replacing plant ash (Fig. 2) is a common feature related to the alteration of anthropogenic combustion products by phosphate-rich solutions, most likely derived from decaying organic material such as guano. In this case, the commonly microlaminated calcitic ash is replaced by yellowish apatite with undifferentiated b-fabric. However, the general structure of the burnt layer, i.e., ashes overlying an organic/charcoal-rich or fire reddened substrate, can still be recognised. The use of micromorphological techniques is often the only way to prove that burning activities occurred in the past.

Leucophosphite and montgomeryite have been frequently found in severely altered anthropogenic combustion features (Schiegl et al., 1996; Karkanas et al., 1999, 2002). They usually occur as individual yellow, greenish or colourless spherulites, typically 100-500 μm in diameter (Fig. 3) and as radiating acicular grains or nodules (see also Section 3) (Fig. 4). The features that relate them to the original ash are basically the same as described above for apatite-altered ash remains. The presence of these Al phosphate minerals in severely altered anthropogenic sediment indicates that apatite is not stable in this environment. Thus if bone had been present in the past, it has now been dissolved, altering the archeological record and posing problems in its interpretation (Karkanas et al., 2000).

2.2 Micromass

In lateritic profiles formed in argilophosphatic sediments, Al-Ca phosphates (mainly crandallite, wavellite and millisite) occur in the groundmass as massive grey

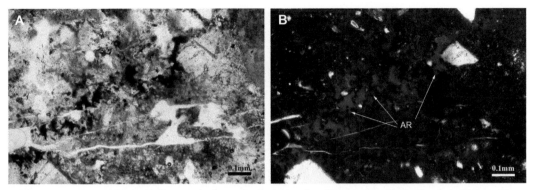

FIGURE 2 Apatite replacing ash in sample from the interior of a Middle Palaeolithic fireplace (Kebara Cave, Israel). (A) In PPL. (B) In XPL, showing the remains of ash rhombs (AR), which have a milky appearance when photographed with the substage auxiliary condensing lens (as used in conoscopic illumination; enhances details and contrast of minerals having a weak birefringence).

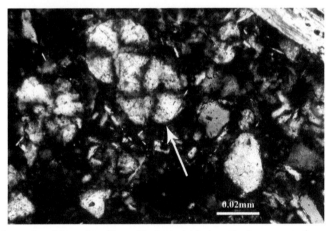

FIGURE 3 Scattered leucophosphite spherulites with typical black extinction crosses (*arrow*) in a sample from Theopetra Cave (Greece) (XPL).

accumulations of plate-shaped, needle-like and arrow-head crystals (Altschuler, 1973; Flicoteaux & Lucas, 1984; Nahon, 1991). They, thus form a groundmass with a crystallitic b-fabric. Detailed micromorphological studies have established a complex alteration pathway, with Al phosphates forming at the expense of apatite (Flicoteaux & Lucas, 1984). The phosphate mineral sequence is characterised by concentration of aluminium and leaching of calcium and sodium. The formation of a micromass with millisite or crandallite and then wavellite is due to leaching under conditions with permanent groundwater in a wet tropical climate (Flicoteaux & Lucas, 1984; Nahon, 1991). Undifferentiated assemblages of goethite, leucophosphite and strengite occur in the matrix and along pores in some ferruginous lateritic crusts (Tiessen et al., 1996). Phosphate minerals of the crandallite group have been also reported in other soils that

FIGURE 4 Greenish yellow radiating acicular montgomeryite (Mt) inside a grey cryptocrystalline matrix of non-stochiometric apatite in a sample from Grotte XVI (France) (PPL).

have undergone intensive leaching and weathering (Norrish, 1968; Adams et al., 1973; Dill et al., 1991). In weathered black shales, apatite forms macroscopically massive aggregates consisting of prismatic to acicular crystals, whereas crandallite appears in the form of fibrous rosettes and spherulites. In addition, Zn-rich turquoise $(CuAl_6(PO_4)_4(OH)_8 \cdot 4H_2O)$ occasionally constitutes the massive cryptocrystalline matrix (Dill et al., 1991).

In weathered shell middens, secondary calcium phosphate phases form at interfaces between the phosphatic clayey micromass and unaltered bone and shells. The formation of this phosphate phase, most likely apatite, is associated with development of a microgranular structure. An illuvial origin is suggested for its formation, probably derived from alteration of overlying phosphate-rich materials such as wood ashes (Corrêa et al., 2013).

In cultivated soils, concentration of phosphate phases, presumably apatite, is observed associated with ped faces, probably related to surface addition of P-rich fertilisers (Qureshi & Jenkins, 1987). However, these concentrations are detected only by electron probe microanalysis and are not distinguished from the homogeneous clay-rich groundmass by optical microscopy.

Materials with a phosphate-dominated micromass display both undifferentiated and crystallitic b-fabrics. An undifferentiated b-fabric, in the absence of significant amounts of amorphous organometallic complexes, is an indication of the presence of apatite (Fig. 5). One of the best ways to resolve such ambiguous cases is the use of fluorescence microscopy (Altemüller & Van Vliet-Lanoë, 1990). Apatite is autofluorescent under blue light (450-500 nm), usually displaying yellow, orange-yellow or whitish colours (Fig. 6)

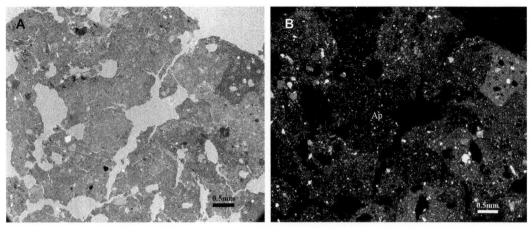

FIGURE 5 Phosphatised clayey and silty groundmass in a sample of Upper Palaeolithic deposits (Hohle Fels Cave, Germany). (A) In PPL. (B) In XPL, showing zones with undifferentiated b-fabric (Ap), corresponding to phosphate accumulations in the groundmass.

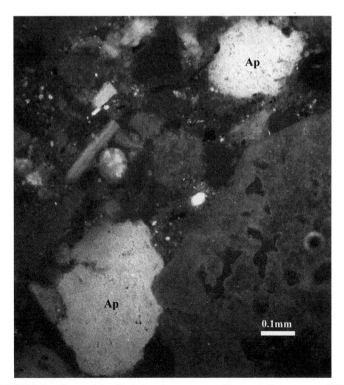

FIGURE 6 Apatite nodules (Ap) showing orange-yellow autofluorescence (BLF).

(Macphail et al., 2004; Goldberg & Macphail, 2006). Bright orange fine material in the form of concave meniscus-like bridges between coarse grains, in a chito-gefuric c/f-related distribution pattern, was observed for anthropogenic horizons with combustion features in sealing shelters. Based on UVL fluorescence characteristics, a phosphatic composition for the bright orange micromass was suggested. The phosphates could derive from seal remains (Villagran et al., 2013).

A number of phosphate minerals other than apatite are found both in materials from caves and soil environments. The most common are crandallite, leucophosphite, millisite, montgomeryite, strengite, taranakite, variscite, vivianite, and wavellite (Table 1). They often give a brownish, yellow, orange-yellow or grey tint to the groundmass. If confusion with other cryptocrystalline features (or amorphous or masked features) can be avoided, their presence is indicated by a groundmass with an undifferentiated b-fabric. Since all phosphate minerals are usually autofluorescent, microscopic study of fluorescence under blue light can be helpful to confirm their occurrence (Altemüller & Van Vliet-Lanoë, 1990) (Figs. 6 and 7). Nevertheless, aluminium complexes are also sometimes autofluorescent (Altemüller & Van Vliet-Lanoë, 1990; Goldberg & Macphail, 2006). In any case, it is not possible to unequivocally identify a specific phosphate mineral without the use of additional mineralogical identification techniques (e.g., XRD, FTIR, EDS, WDS). SEM studies accompanied by microprobe analysis, in particular, have proven to be invaluable in identifying some phosphate occurrences (Qureshi & Jenkins, 1987; Karkanas et al., 1999, 2000; Macphail & Goldberg, 2000, Thiry et al., 2006).

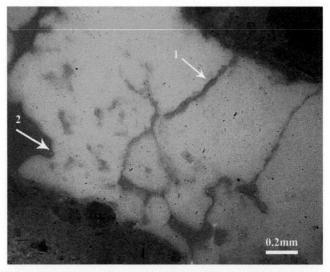

FIGURE 7 Fracture infilling composed of autofluorescent microcrystalline taranakite (BLF) in a sample from Theopetra Cave (Greece). Note straight and curved planar voids (1) and vughs with slightly undulating smooth sides (2).

2.3 Guano Deposits

Guano is a by-product of animal excretion. Its characteristics depend on the type of animal and its feeding habits. Guano deposition is believed to be the most important agent of phosphate enrichment of cave deposits, some insular sediment and orthogenic soils, producing a number of Al phosphate minerals as described below (Section 3) (Wilson & Bain, 1976; Karkanas et al, 2000, 2002; Shahack-Gross et al., 2004; Simas et al., 2007; Schaefer et al., 2008; Pereira et al., 2013; Bergadà et al., 2013).

Bird guano often has an undulating layered appearance (Hutchinson, 1950) and a spongy microstructure (Cullen, 1988) (Fig. 8). It consists of dark yellowish cryptocrystalline apatite (Fig. 9), with undifferentiated b-fabric. Microcrystalline struvite (Table 1) can be also a major constituent of fresh bird (gull) guano (Cullen, 1988; Pereira et al., 2013). It usually contains different types of inclusions, such as dissolved fine bone fragments, insect scales and plant remains, depending on the diet of the bird.

In contrast, fresh bat guano does not contain apatite. However, apatite is one of its first alteration products and occurs as groundmass material (Fig. 5), light yellow nodules (Fig. 9) or incomplete infillings of planar voids (Shahack-Gross et al., 2004; Bergadà et al., 2013). As in bird guano deposits, apatite is microcrystalline. Gypsum, chitin fragments and decayed amorphous fine plant residues are also associated with fresh bat guano. Zanin et al. (2005) observed tubular structures (10-20 µm in diameter) that they believe to be of microbial origin, interpreting these features as capsules of cyanobacterial filaments.

Ardealite, brushite, struvite and newberyite (Table 1) are other phosphate minerals that are related to guano (Shahack-Gross et al., 2004; Karkanas et al., 2002), but they are

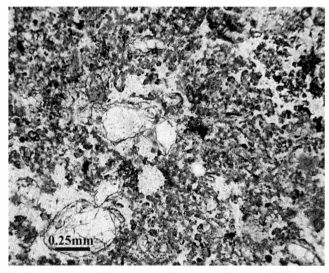

FIGURE 8 Slightly oxidised guano crust with a spongy microstructure in a sample from Cave 13B (Pinnacle Point, Mossel Bay, South Africa) (PPL).

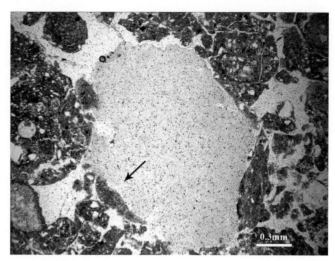

FIGURE 9 Yellow cryptocrystalline apatite nodule in an altered guano deposit from Kebara Cave (Israel) (PPL). Note the *curved* and mammillated external surface (*arrow*).

rarely preserved in the sediment or soil record. Newberyite, in particular, has an acicular crystal habit and occurs as rosette-like aggregates; it has been reported in altered cave sediments from archeological sites (Karkanas et al., 2002).

The presence of recognisable guano crusts (Fig. 8) and other guano deposits indicates stable surface exposure with lack of sedimentation. Black crusts consisting of dark reddish yellow cryptocrystalline apatite are often formed in cave environments and mark periods of stabilisation and exposure of the sediment on the surface. The black colour is due to the presence of manganese and iron oxides and amorphous compounds related to the decay and mineralisation of organic matter. Microcrystalline apatite crusts have also been reported in stabling floors (Macphail et al., 2004), interbedded with layered plant fragments and fecal calcitic spherulites (Canti, 1998).

3. Phosphate Occurrences as Pedofeatures

Crystalline and amorphous phosphate pedofeatures include nodules, coatings, hypo-coatings and infillings. Since most of these features are difficult to identify using optical microscopy, supplementary SEM observations and elemental microanalyses are typically needed (Morrás, 1978, 1987; Stoops, 1983; Karkanas et al., 1999; Simpson et al., 2000; Schaefer et al., 2004; Thiry et al., 2006; Simas et al., 2007) (see also Section 2.2). Amorphous and cryptocrystalline phosphate pedofeatures often exhibit features like a fine-banded, concentric or wavy internal fabric (Fig. 10) and mammillated boundaries (Figs. 7 and 9). Similar expressions are embodied by crandallite and leucophosphite spherulites (Fig. 3), radial aggregates of acicular montgomeryite (Fig. 4) and other non-equilibrium forms produced by rapid precipitation.

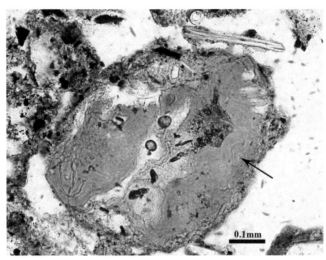

FIGURE 10 Orange cryptocrystalline crandallite nodule in a sample from Theopetra cave (Greece) (PPL). Note the convoluted wavy internal laminations (*arrow*).

Phosphate mineralogy can be used to infer the palaeochemical environment (Nriagu, 1976; Dill et al., 1991) and thus to assess the integrity of the archeological record and soil environment. The occurrence of Al-rich phosphates implies an acidic environment in which apatite was not stable. Under these conditions, bone apatite should have been dissolved (e.g., Berna et al., 2004), whereas other materials of interest, such as phytoliths and chert, are stable. This strategy would make possible to distinguish between environments where some of the primary constituents are absent because they were never incorporated in the archeological and soil record, as opposed to being absent or partially altered because they were subject to diagenesis (Weiner et al., 1993, 2002; Karkanas et al., 2000; Karkanas, 2010).

Excrement pedofeatures, such as carnivore coprolites, are often recognised in thin sections of samples from archeological sites (Horwitz & Goldberg, 1989; Fernández Rodríguez et al., 1995; Goldberg et al., 2001; Allen et al., 2002). Hyena coprolites generally contain rounded fragments that are composed of microcrystalline apatite, with a pale yellow colour, dusty limpidity and almost undifferentiated b-fabric (Fig. 11). In many cases, partially dissolved bone fragments can be observed along with plant remains and phytoliths (see also Macphail & Goldberg, 2018). In addition, smooth voids such as vesicles, probably resulting from gastric gas, and elongated pores, presumably from decomposed hair/fur, can be observed. Finally, depending on the mineralogy of local soils, inclusions of sand- and silt-sized quartz can also be found. In archeological sites, the occurrence of carnivore coprolites is an indication that the site was temporarily abandoned and that bone assemblages as excavated may to a certain extent have non-human origins. Other types of phosphate excrement pedofeatures are reported by

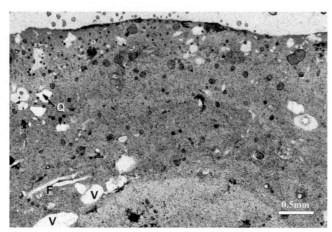

FIGURE 11 Hyena coprolite from Kebara Cave (Israel) (PPL). Yellowish brown groundmass mostly composed of apatite, with few quartz grains (Q), vesicles (V) and elongated voids (F).

Schaefer et al. (2004), who observed secondary complex Al-rich phosphates as infilling of channels, attributed to digestion of Ca phosphates by earthworms and other organisms.

Phosphate pedofeatures associated with guano occurrences include cryptocrystalline typic apatite nodules, found in insular arid soils associated with bird guano (Morrás, 1978); similar features are recognised in anthropic sediments associated with animal dung and human waste (Fig. 12). Gregory and Rodgers (1989) report microlaminated crusts of microcrystalline apatite with a crescent internal fabric, associated with insular bird guano. In permafrost-affected ornithogenic soils formed by accumulation of penguin guano, nodular phosphate-rich features and phosphatised organo-mineral

FIGURE 12 Apatite nodules (Ap) in a sample from dung-rich anthropic sediments in the Makri Neolithic settlement (Greece) (PPL).

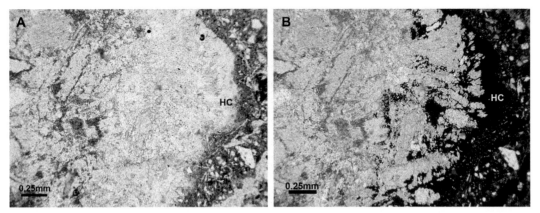

FIGURE 13 Cryptocrystalline apatite hypocoating (HC) on a limestone fragment from Theopetra Cave (Greece). (A) In PPL; apatite has a yellowish colour. (B) In XPL; apatite shows optically isotropic behaviour.

aggregates are also observed (Simas et al., 2006, 2007; Schaefer et al., 2008). These granules are usually covered by coatings consisting of yellowish illuvial phosphates, which also occur as infillings of internal cracks. Leucophosphite, minyulite ($KAl_2(PO_4)_2F \cdot 4H_2O$), taranakite, and metavariscite have been documented in these soils, together with several amorphous phosphate compounds (Simas et al., 2006, 2007; Schaefer et al., 2008; Pereira et al., 2013). Taranakite microcrystalline surface crusts together with leucophosphite have been observed on phosphatised basalts associated with bird guano (Landis & Craw, 2003). Wilson & Bain (1976) also report leucophosphite in the form of hypocoatings (reaction rims) around phyllitic rock fragments in an ornithogenic soil profile.

In limestone and along the sides of coarse calcareous fragments, apatite is often found as yellowish hypocoatings (reaction rims) (Fig. 13). Their presence in limestone and karstic terrains is a good indicator of guano-related phosphate enrichment in active and eroded cave environments, where phosphate-rich solutions derived from the decay of organic matter react with the calcareous host material. On the other hand, in settings with calcareous materials, irregular yellowish nodules with undifferentiated b-fabric and low to moderate relief, enclosed by calcite accumulations, are indications of calcite replacing apatite (Fig. 14). Such features indicate a reversal of the geochemical environment and may be related to changes to a more humid climate favouring massive calcite precipitation in the form of speleothems, or to local hydrological changes enhancing flow of calcium-saturated waters, or to past human activities that interrupted the accumulation of guano.

Additional phosphatic pedofeatures in cave environments include taranakite occurrences in the form of typic nodules and incomplete infillings of platy and lenticular voids, formed in sediments affected by freeze-thaw activity (Karkanas, 2001). These occurrences are brownish and cryptocrystalline and characterised by shrinkage planar and curved voids (Figs. 7 and 15). A Ca-Fe phosphate phase approaching that of

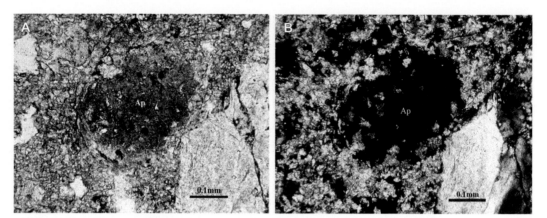

FIGURE 14 Sparitic calcite replacing a cryptocrystalline apatite pedofeature (Ap) in a sample from Cave 13B (Pinnacle Point, Mossel Bay, South Africa). (A) In PPL. (B) In XPL.

FIGURE 15 Brown cryptocrystalline taranakite (Tr) as incomplete infillings in voids separating lenticular aggregates produced by freeze-thaw activity in a sample from Theopetra Cave (Greece) (PPL).

calcioferrite ($Ca_4MgFe^{3+}_4(PO_4)_6(OH)_4\cdot12H_2O$) has been reported by Jenkins (1994) in archeological cave sediments. It was found in the form of orange-brown coatings and infillings of voids, where it formed by translocation of phosphate colloids derived from the dissolution of bone apatite. Crandallite has been observed in altered cave sediments mostly as coatings and hypocoatings but also as orange yellowish nodules with a banded, wavy or geodic internal fabric (Fig. 10). In clayey cave sediments banded

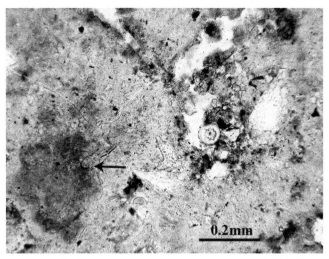

FIGURE 16 Vivianite nodule (arrow) with blue and brown colours within a yellow cryptocrystalline Fe-Ca phosphate groundmass in a sample from night soil deposits (Medieval St Julien, Tours, France) (PPL). *Image by R. Macphail.*

coatings of reddish yellow or colourless variscite or strengite have been also observed. In most cases, rhythmic precipitation is evident (Karkanas, unpublished data).

In water-logged sediments, iron-rich phosphates such as vivianite are found. These occurrences can be related to the presence of human waste (Fig. 16), animal waste and phosphate-rich iron slugs, and they are usually associated with crusts of manganese and iron (Bertran & Raynal, 1991; Gebhardt & Langohr, 1999; McGowan & Prangnell, 2006; Dill & Techmer, 2009). Vivianite is also found in lacustrine sediments, in the form of crystalline and alteromorphic nodules (Stamatakis & Koukouzas, 2001; Fagel et al., 2005; Sapota et al., 2006) or as layers (Lücke & Brauer, 2004). In addition, vivianite is frequently reported for bog ores (Fig. 17) (Stoops, 1983; Landuydt, 1990; Stoops & Delvigne, 1990; Kaczorek & Sommer, 2003; Gamzikov & Marmulev, 2007). It occurs as microcrystalline and coarser void infillings in former root channels and in decaying plant remains, as radial aggregates and nodules, as single crystals and in combination with siderite and iron oxides. Vivianite is predominant in the reduced zone of the lower horizons of bog ores, but its occurrence in compound coatings in the oxidised parts suggests temporary changes of redox potential, probably due to temporary higher water table levels (Stoops, 1983; Landuydt, 1990). Another relevant example is the formation of vivianite in lateritic iron crusts submitted to hydromorphic conditions (Lemos et al., 2007). It forms infillings and isolated crystals, and it replaces concentric iron nodules. The formation of vivianite is generally attributed to the activity of reducing bacteria (Dong et al., 2000; Lemos et al., 2007). Vivianite mineralisation has also been reported for water-logged ferricretes, in which it is associated with siderite or strengite. Cappings and crystal intergrowths, the latter consisting of acicular or plate-shaped crystals, are some of the forms that are found

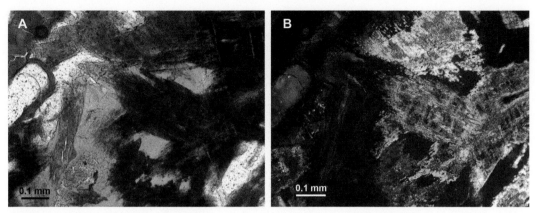

FIGURE 17 Lath-shaped crystals of weathered vivianite from a bog ore (region of Antwerp, Belgium) (Ghent University Archive). (A) In PPL showing the characteristic blue colour of oxidised vivianite crystals altering at the edges to a yellowish orange mass. (B) In XPL.

in these environments. The origin of vivianite in ferricretes is attributed to parent material composition and to weathering of phosphate-rich iron slags (Dill & Techmer, 2009). Vivianite is easily identified in the field and in thin sections by its blue oxidation products (Fig. 17A). Weathering also transforms it to an optically isotropic, yellowish substance (Stoops & Delvigne 1990) (Fig. 17A and B) that in some cases seems to be santabarbaraite ($Fe^{3+}_3(PO_4)_2(OH)_3 \cdot 5H_2O$) (Pratesi et al., 2003). In addition, vivianite can react with Ca-rich groundwater in conditions of increasing pH and Eh to produce mitridatite usually in the form of infillings (Stamatakis & Koukouzas, 2001).

Below lateritic profiles formed in argilophosphatic sediments, a second stage of alteration follows the development of the phosphate groundmass described above (Section 2.1). Laminated or compound layered coatings of cryptocrystalline Al-Ca phosphates, kaolinite and goethite are observed in voids (Flicoteaux & Lucas, 1984; Nahon, 1991). Macroscopic banded and pseudobrecciated structures resulting from pedological differentiation have been reported in the same soil environments. In thin sections, these structures have the form of nodular compound pedofeatures containing crystalline phosphate fragments (Nahon, 1991). Parron and Nahon (1980) have also reported the formation of alteromorphic nodules of crandallite after original apatite nodules in lateritic weathering of glauconitic quartzites. In weathered carbonaceous schists, crandallite and wavellite occur as layered coatings surrounding chert fragments, whereas variscite forms nodules within bedrock fractures (Dill et al., 1991). Most of the above features occur close to the soil surface and probably result from superficial weathering in conditions with non-permanent groundwater and with a drier climate than during the formation of the phosphate groundmass (Flicoteaux & Lucas, 1984; Dill et al., 1991; Dill, 2001).

In other environments, non-stochiometric apatite infillings have been reported in samples from a solodised Solonetz (Morrás, 1987). Ca-Fe phosphate coatings and infillings in an anthropic sediment have been interpreted as the result of fish bone decomposition and recrystallisation, based on micromorphological and instrumental microchemical analyses (Simpson et al., 2000). Additional crystal textural studies using microfocus synchrotron X-ray scattering has confirmed that these materials are of fish bone origin (Adderley et al., 2004). Groundwater cementation at a site with archeological deposits resulted in the development of juxtaposed convolute laminated (hypo)coatings and infillings of microcrystalline apatite (Thiry et al., 2006) and other phosphate minerals, observed in voids and around grains.

4. Conclusions

The identification of phosphate minerals under the microscope is not an easy task, mainly because they are mostly present in cryptocrystalline form, both as pedofeatures and as groundmass components. However, phosphates are typically autofluorescent in blue light and this can be a useful tool for their identification. SEM studies provide additional information on their submicroscopic structure, but their specific mineralogy can only be determined by instrumental mineralogical analysis.

Most of the information regarding the microscopy of phosphates comes from altered phosphorites and phosphate-rich laterites, cave environments and unique soil environments such as bog ores and ornithogenic soils. In most cases apatite is the common phosphate mineral, which with advancing diagenesis and decreasing pH is replaced first by Ca-Al phosphates and then by Al phosphates. In decalcified clay-rich sediments and soils, Al phosphates are formed directly. The development of a relatively coarse-crystalline phosphate groundmass is favoured by permanent groundwater, whereas more fine-grained features are formed by superficial alteration processes. Fe phosphates, such as vivianite and its alteration products, are particularly valuable to monitor redox changes in waterlogged sediments and soils.

The micromorphology of phosphates in normal soil environments is not well known, mainly because they constitute a minor, almost undetected, component. Most of the features seem to be derived from anthropogenic and biogenic inputs. Future studies should focus on this poorly known soil environments providing a more complete picture of the micromorphology of phosphates.

References

Adams, J.A., Howarth, D.T. & Campbell, A.S., 1973. Plumbogummite minerals in a strongly weathered New Zealand soil. Journal of Soil Science 24, 224-232.

Adderley, W.P., Alberts, I.L., Simpson, I.A. & Wess, T.J., 2004. Calcium-iron-phosphate features in archaeological sediments: characterization through microfocus synchrotron X-ray scattering analyses. Journal of Archaeological Science 31, 1215-1224.

Allen, S.D.A., Matthew, A.J., Bell, M.G., Hollins, P., Marks, S. & Mortimore, J.L., 2002. Infrared spectroscopy of the mineralogy of coprolites from Brean Down: evidence of past human activities and animal husbandry. Spectrochimica Acta A 58, 959-965.

Altemüller, H.J. & Van Vliet-Lanoë, B., 1990. Soil thin section fluorescence microscopy. In Douglas, L.A. (ed.), Soil Micromorphology: A Basic and Applied Science. Developments in Soil Science, Volume 19. Elsevier, Amsterdam, pp. 565-579.

Altschuler, Z.S., 1973. The weathering of phosphate deposits. Geochemical and environmental aspects. In Griffith, E.J., Beeton, A., Spencer, J.M. & Mitchel, D.T. (eds.), Environmental Phosphorus Handbook. Wiley, New York, pp. 33-96.

Banfield, J. F. & Eggleton, R., 1989. Apatite replacement and Rare Earth mobilization, fractionation and fixation during weathering. Clays and Clay Minerals 37, 113-127.

Beauchemin, S., Chou, J., Beauchemin, M., Simard, R.R. & Sayers, D.E., 2003. Speciation of phosphorus in phosphorus-enriched agricultural soils using X-Ray Absorption Near-Edge Structure Spectroscopy and chemical fractionation. Journal of Environmental Quality 32, 1809-1819.

Bergadà, M.M., Villaverde, V. & Román, D., 2013. Microstratigraphy of the Magdalenian sequence at Cendres Cave (Teulada-Moraira, Alicante, Spain): formation and diagenesis. Quaternary International 315, 56-75.

Berna, F., Matthews, A. & Weiner, S., 2004. Solubilities of bone mineral from archaeological sites: the recrystallization window. Journal of Archaeological Science 31, 867-882.

Bertran, P. & Raynal, J-P., 1991. Apport de la micromorphologie à l'étude archéologique du village médiéval de Saint-Victor de Massiac (Cantal, France). Revue Archéologique du Centre de la France 30, 137-150.

Bhujbal, B.M. & Mistry, K.B., 1986. Reaction products of ammonium nitrate phosphate fertilizers of varying water-soluble phosphorus content in different Indian soils. Fertilizer Research 10, 59-71.

Bullock, P., Fedoroff, N., Jongerius, A., Stoops, G., Tursina, T. & Babel, U., 1985. Handbook for Soil Thin Section Description. Waine Research Publications, Wolverhampton, 152 p.

Canti, M.G., 1998. The micromorphological identification of faecal spherulites from archaeological and modern materials. Journal of Archaeological Science 25, 435-444.

Cohen, L., Dean, M., Shipov, A., Atkins, A., Monsonego-Ornan, E. & Shahar, R., 2012. Comparison of structural, architectural and mechanical aspects of cellular and acellular bone in two teleost fish. Journal of Experimental Biology 215, 1983-1993.

Corrêa, G.R., Schaefer, C.E. & Gilkes, R.J., 2013. Phosphate location and reaction in an archaeoanthrosol on shell-mound in the Lakes Region, Rio de Janeiro State, Brazil. Quaternary International 315, 16-23.

Cullen, D.J., 1988. Mineralogy of nitrogenous guano on the Bounty Islands, SW Pacific Ocean. Sedimentology 93, 421-428.

Dill, H.G., 2001. The geology of aluminium phosphates and sulphates of the alunite group minerals: a review. Earth-Science Reviews 53, 35-93.

Dill, H.G. & Techmer A., 2009. The geogene and anthropogenetic impact on the formation of *per descensum* vivianite-goethite-siderite mineralization in Mesozoic and Cenozoic siliciclastic sediments in SE Germany. Sedimentary Geology 217, 95-111.

Dill, H.G., Klaus Busch, K. & Blum, N., 1991. Chemistry and origin of vein-like phosphate mineralization, Nuba Mountains (Sudan). Ore Geology Reviews 6, 9-24.

Dong, H., Fredrickson, J.K., Kennedy, D.W., Zachara, J.M., Kukkadapu, R.K. & Onstott, T.C., 2000. Mineral transformation associated with the microbial reduction of magnetite. Chemical Geology 169, 299-318.

Fagel, N., Alleman, L., Granina, L., Hatert, F., Thamo-Bozso, E., Cloots, R. & André, L., 2005. Vivianite formation and distribution in Lake Baikal sediments. Global and Planetary Change 46, 315-336.

Fernández Rodríguez, C., Ramil, P. & Martínez, A., 1995. Characterization and depositional evolution of hyaena (*Crocuta crocuta*) coprolites from La Valiña cave (northwest Spain). Journal of Archaeological Science 22, 597-607.

Flicoteaux, R. & Lucas, J., 1984. Weathering of phosphate minerals. In Nriagu, J.O. & Moore, P.B. (eds.), Phosphate Minerals. Springer-Verlag, Berlin, pp. 292-317.

Flicoteaux, R., Nahon, D. & Paquet, H., 1977. Genèse des phosphates alumineux a partir des sédiments argilo-phosphatés du Tertiaire de Lam-Lam (Sénégal). Suite minéralogique, permanences et changements de structures. Sciences Géologiques, Bulletin 30, 153-174.

Gamzikov, G.P. & Marmulev, A.N., 2007. Agrochemical assessment of bog phosphates in Western Siberia. Eurasian Soil Science 40, 986-992.

Gebhardt, A. & Langohr, R., 1999. Micromorphological study of construction materials and living floors in the medieval motte of Werken (West Flandres, Belgium). Geoarchaeology 14, 595-620.

Goldberg, P. & Macphail, R.I., 2006. Practical and Theoretical Geoarchaeology. Blackwell Publishing, Oxford, 455 p.

Goldberg, P. & Nathan, Y., 1975. The phosphate mineralogy of et-Tabun cave, Mount Carmel, Israel. Mineralogical Magazine 40, 253-258.

Goldberg, P., Weiner, S., Bar-Yosef, O., Xu, Q. & Liu, J., 2001. Site formation processes at Zhoukoudian, China. Journal of Human Evolution 41, 483-530.

Gregory, M.R. & Rodgers, K.A., 1989. Phosphate minerals from the Bounty Islands, South Pacific Ocean. Neues Jahrbuch für Mineralogie, Abhandlungen 160, 117-131.

Hanson, M. & Cain, C.R., 2007. Examining histology to identify burned bone. Journal of Archaeological Science 34, 1902-1913.

Horwitz, L.K. & Goldberg, P., 1989. A study of Pleistocene and Holocene hyena coprolites. Journal of Archaeological Science 16, 71-94.

Hutchinson, G.E., 1950. Survey of contemporary knowledge of biogeochemistry. 3. The biogeochemistry of vertebrate excretion. Bulletin of the American Museum of Natural History 96, 1-554.

Jenkins, D.A., 1994. Interpretation of interglacial cave sediments from a hominid site in North Wales: translocation of Ca-Fe phosphates. In Ringrose-Voase, A.J. & Humphreys, G.S. (eds.), Soil Micromorphology: Studies in Management and Genesis. Developments in Soil Science, Volume 22. Elsevier, Amsterdam, pp. 293-305.

Kaczorek, D. & Sommer, M., 2003. Micromorphology, chemistry, and mineralogy of bog iron ores from Poland. Catena 54, 393-402.

Karkanas, P., 2001. Site formation processes in Theopetra cave: a record of climatic change during the Late Pleistocene and early Holocene in Thessaly, Greece. Geoarchaeology 16, 373-399.

Karkanas, P., 2010. Preservation of anthropogenic materials under different geochemical processes: a mineralogical approach. Quaternary International 210, 63-69.

Karkanas, P., Kyparissi-Apostolika, N., Bar-Yosef, O. & Weiner, S., 1999. Mineral assemblages in Theopetra, Greece: a framework for understanding diagenesis in a prehistoric cave. Journal of Archaeological Science 26, 1171-1180.

Karkanas, P., Bar-Yosef, O., Goldberg, P. & Weiner, S., 2000. Diagenesis in prehistoric caves: the use of minerals that form in situ to assess the completeness of the archaeological record. Journal of Archaeological Science 27, 915-929.

Karkanas, P., Rigaud, J.P., Simek, J.F., Albert, R.A. & Weiner, S. 2002. Ash, bones and guano: a study of the minerals and phytoliths in the sediment of Grotte XVI, Dordogne, France. Journal of Archaeological Science 29, 721-732.

Landis, C.A. & Craw, D., 2003. Phosphate minerals formed by reaction of bird guano with basalt at Cooks Head Rock and Green Island, Otago, New Zealand. Journal of the Royal Society of New Zealand 33, 487-495.

Landuydt, C.J., 1990. Micromorphology of iron minerals from bog ores of the Belgian Campine area. In Douglas, L.A. (ed.), Soil Micromorphology: A Basic and Applied Science. Developments in Soil Science, Volume 19. Elsevier, Amsterdam, pp. 289-294.

Lehr, J.R. & Brown, W.E., 1958. Calcium phosphate fertilizers: II. A petrographic study of their alteration in soils. Soil Science Society of America Proceedings 22, 29-32.

Lemos, V.P., da Costa, M.L., Lemos, R.L. & de Faria, M.S.G., 2007. Vivianite and siderite in lateritic iron crusts: an example of bioreduction. Quimica Nova 30, 36-40.

Lindsay, W.L., Vlek, P.L.G. & Chien, S.H., 1989. Phosphate minerals. In Dixon, B. J & Weed, S.B (eds.), Minerals in soil environments, Second Edition. Soil Science Society of America Book Series 1, SSSA, Madison, pp. 1089-1130.

Lücke, A. & Brauer, A., 2004. Biochemical and micro-facial fingerprints of ecosystem response to rapid Late Glacial climatic changes in varved sediments of Meerfelder Maar (Germany). Palaeogeography, Palaeoclimatology, Palaeoecology 211, 139-155.

Macphail, R.I., 2000. Soils and microstratigraphy: a soil micromorphological and micro-chemical approach. In Lawson, A.J. (ed.), Potterne 1982-5: Animal Husbandry in Later Prehistoric Wiltshire. Wessex Archaeology Report Vol. 17, Wessex Archaeology, Salisbury, pp. 47-71.

Macphail, R.I. & Goldberg, P., 2000. Geoarchaeological investigations of sediments from Gorham's and Vanguard Caves, Gibraltar: microstratigraphical (soil micromorphological and chemical) signatures. In Stringer, C.B., Barton, R.N.E. & Finlayson, C. (eds.), Neanderthals on the Edge. Oxbow, Oxford, pp. 183-200.

Macphail, R.I. & Goldberg, P., 2018. Archaeological materials. In Stoops, G., Marcelino, V. & Mees, F. (eds.), Interpretation of Micromorphological Features of Soils and Regoliths. Second Edition. Elsevier, Amsterdam, pp. 779-819.

Macphail, R.I., Courty, M.A. & Gebhardt, A., 1990. Soil micromorphological evidence of early agriculture in north-west Europe. World Archaeology 22, 53-69.

Macphail, R.I., Galinié, H. & Verhaeghe, F., 2003. A future for Dark Earth? Antiquity 296, 349-358.

Macphail, R.I., Cruise, G.M., Allen, J.R.M., Linderholm, J. & Reynolds, P., 2004. Archaeological soil and pollen analysis of experimental floor deposits; with special references to Butser Ancient Farm, Hampshire, UK. Journal of Archaeological Science 31, 175-191.

McGowan, G. & Prangnell, J., 2006. The significance of vivianite in archaeological settings. Geoarchaeology 21, 93-111.

Morrás, H.J.M., 1978. Phosphatic nodules from a soil profile of Santa Fé island, Galapagos. In Delgado, M. (ed.), Soil Micromorphology. University of Granada, pp. 1007-1018.

Morrás, H., 1983. Submicroscopic characterization of phosphatic and sequioxidic nodules of some soils of the 'Chao Deprimido'(Argentina): preliminary results. Geoderma 30, 187-194.

Morrás, H., 1987. Identification microanalytique de traits phosphatés et calcitiques de solonetz solodisés (Santa Fé, Argentine). In Fedoroff, N., Bresson, L.M. & Courty, M.A. (eds.), Micromorphologie des Sols, Soil Micromorphology. AFES, Plaisir, pp. 235-241.

Nahon, D.B., 1991. Introduction to the Petrology of Soils and Chemical Weathering. John Wiley & Sons, New York, 313 p.

Nedachi, Y., Nedachi, M., Bennett, G. & Ohmoto, H., 2005. Geochemistry and mineralogy of the 2.45 Ga Pronto palaeosols, Ontario, Canada. Chemical Geology 214, 21-44.

Norrish, K., 1968. Some phosphate minerals of soils. Transactions of the 9th International Congress of Soil Science, Volume II, Adelaide, pp. 713-723.

Nriagu, O.J., 1976. Phosphate-clay mineral relations in soils and sediments. Canadian Journal of Earth Sciences 13, 717-736.

Parron, C. & Nahon, D., 1980. Red bed genesis by lateritic weathering of glauconitic sediments. Journal of the Geological Society of London 137, 689-693.

Pereira, T.T.C., Schaefer, C.E.G.R., Ker, J.C., Almeida, C.C., Almeida, I.C.C. & Pereira, A.B., 2013. Genesis, mineralogy and ecological significance of ornithogenic soils from a semi-desert polar landscape at Hope Bay, Antarctic Peninsula. Geoderma 209/210, 98-109.

Pratesi, G., Cipriani, C., Guili, G. & Birch, W.D., 2003. Santabarbaraite: a new amorphous phosphate mineral. European Journal of Mineralogy 15, 185-192.

Qureshi, R.H. & Jenkins, D.A., 1987. Concentration of phosphorus and sulphur at soil ped surfaces. Journal of Soil Science 38, 255-265.

Qureshi, R.H., Jenkins, D.A. & Davies, R.I., 1978. Electron probe microanalytical studies of phosphorous distribution within soils fabric. Soil Science Society of American Journal 42. 698-703.

Retallack, G.J., 1990. Soils of the Past. Unwin Hyman, Boston, 520 p.

Sapota, T., Aldahan, A. & Al-Aasm, I.S., 2006. Sedimentary facies and climate control on formation of vivianite and siderite microconcretions in sediments of Lake Baikal, Siberia. Journal of Paleolimnology 36, 245-257.

Schaefer, C.E.G.R., Lima, H.N., Gilkes, R.J. & Mello, J.W.V., 2004. Micromorphology and electron microprobe analysis of phosphorus and potassium forms of an Indian Black Earth (IBE) Anthrosol from Western Amazonia. Australian Journal of Soil Research 42, 401-409.

Schaefer, C.E.G.R., Simas, F.N.B., Gilkes, R.J., Mathison, C., da Costa, L.M. & Albuquerque, M.A., 2008. Micromorphology and microchemistry of selected Cryosols from maritime Antarctica. Geoderma 144, 104-115.

Schiegl, S., Goldberg, P., Bar-Yosef, O. & Weiner, S., 1996. Ash deposits in Hayonim and Kebara caves, Israel: macroscopic, microscopic and mineralogical observations, and their archaeological implications. Journal of Archaeological Science 23, 763-781.

Shahack-Gross, R., Berna, F., Karkanas, P. & Weiner, S., 2004. Bat guano and preservation of archaeological remains in cave sites. Journal of Archaeological Science 31, 1259-1272.

Simas, F.N.B., Schaefer, C.E.G.R., Melo, V.F., Guerra, M.B.B., Saunders, M. & Gilkes, R.J., 2006. Clay-sized minerals in permafrost-affected soils (Cryosols) from King George Island, Antarctica. Clays and Clay Minerals 54, 721-736.

Simas, F.N.B., Schaefer C.E.G.R., Melo, V.F., Albuquerque-Filho, M.R., Michel, R.F.M., Pereira, V.V., Gomes, M.R.M. & da Costa, L.M., 2007. Ornithogenic cryosols from maritime Antarctica: phosphatization as a soil forming process. Geoderma 138, 191-203.

Simpson, I.A., Perdikaris, S., Cook, G., Campbell, J. & Teesdaly, W.J., 2000. Cultural sediment analyses and transitions in early fishing activity at Langenesværet, Vesterålen, Northern Norway. Geoarchaeology 15, 743-763.

Snoeck, C., Lee-Thorp, J.A. & Schulting, R.J., 2014. From bone to ash: compositional and structural changes in burned modern and archaeological bone. Palaeogeography, Palaeoclimatology, Palaeoecology 416, 55-68.

Stamatakis, M.G. & Koukouzas, N.K., 2001. The occurrence of phosphate minerals in lacustrine clayey diatomite deposits, Thessaly, Central Greece. Sedimentary Geology 139, 33-47.

Stoops, G., 1983. SEM and light microscopic observations of minerals in bog ores of the Belgian Campine. Geoderma 30, 179-186.

Stoops, G., 2003. Guidelines for Analysis and Description of Soil and Regolith Thin Sections. Soil Science Society of America, Madison, 184 p.

Stoops, G. & Delvigne, J., 1990. Morphology of mineral weathering and neoformation. II Neoformations. In Douglas, L.A. (ed.), Soil Micromorphology: A Basic and Applied Science. Developments in Soil Science, Volume 19. Elsevier, Amsterdam, pp. 483-492.

Thiry, M., Galbois, J. & Schmitt, J.M., 2006. Unusual phosphate concentrations related to groundwater flow in continental environment. Journal of Sedimentary Research 76, 86-870.

Tiessen, H., Lo Monaco, S., Ramirez, A., Santos, M.C.D. & Shang, C., 1996. Phosphate minerals in a lateritic crust from Venezuela. Biogeochemistry 34, 1-17.

Villagran, X.S., Schaefer, C.E.G.R. & Ligouis, B., 2013. Living in the cold: geoarchaeology of sealing sites from Byers Peninsula (Livingston Island, Antarctica). Quaternary International 315, 184-199.

Walker, T.W. & Syers, J.K., 1976. The fate of phosphorus during pedogenesis. Geoderma 15, 1-19.

Watson, J.P.N., 1975. Domestication and bone structure in sheep and goats. Journal of Archaeological Science 2, 375-383.

Weiner, S., Goldberg, P. & Bar-Yosef, O., 1993. Bone preservation in Kebara Cave, Israel using on-site Fourier Transform Infrared spectrometry. Journal of Archaeological Science 20, 613-627.

Weiner, S., Goldberg, P. & Bar-Yosef, O., 2002. Three-dimensional distribution of minerals in the sediments of Hayonim cave, Israel: diagenetic processes and archaeological implications. Journal of Archaeological Science 29, 1289-1308.

Wilson, M.J. & Bain, D.C., 1976. Occurrence of leucophosphite in a soil from Elephant Island, British Antarctic Territory. American Mineralogist 61, 1027-1028.

Zanin, Y.N., Tsykin, R.A. & Dar'in, A.V., 2005. Phosphorites of the Arckehologicheskaya Cave (Khakassia, East Siberia). Lithology and Mineral Resources 40, 48-55.

Chapter 13

Sulphidic and Sulphuric Materials

Florias Mees[1], Georges Stoops[2]

[1]*ROYAL MUSEUM FOR CENTRAL AFRICA, TERVUREN, BELGIUM;*
[2]*GHENT UNIVERSITY, GHENT, BELGIUM*

CHAPTER OUTLINE

1. Introduction

Sulphidic sediments commonly occur in environments with reducing conditions, decaying organic matter and sufficient availability of iron and sulphur. These conditions are met in a wide range of coastal lowlands. They can also develop in certain continental environments, where sulphide-bearing geological formations or mining-related materials are exposed to weathering.

The development of aerobic conditions, typically following natural emergence or artificial drainage, leads to oxidation of the sulphides. This results in acidification unless sufficient carbonate minerals are present to neutralise the sulphuric acid that is produced. In this manner, acid sulphate soils (cat clays) are formed, derived from sulphidic potential acid sulphate soils. As a result of widespread land reclamation in coastal lowlands, these types of soils occupy a considerable area of important agricultural regions worldwide, especially in the tropics. The high acidity and associated release of toxic metals, particularly aluminium, leads to serious problems in their agricultural use and management. Acid weathering in soils and other near-surface formations can also result in infrastructure damage, off-site surface water pollution and ecosystem deterioration.

In the World Reference Base for Soil Resources, sulphidic materials and thionic horizons are recognised, with acid sulphate soils appearing in several soil groups, classified with Thionic as suffix qualifier (IUSS Working Group WRB, 2015). In Soil Taxonomy, the equivalent concepts are sulphidic materials and sulphuric horizons, and the soils mainly fit the requirements of Sulfaquents (potential acid sulphate soils) and Sulfaquepts (acid sulphate soils) (Soil Survey Staff, 1999, 2014).

2. Sulphidic Materials

It is commonly accepted that the following conditions are needed for pyrite formation in soils and sediments: (i) a source of iron (e.g., iron oxides, iron compounds absorbed at the surface of clay minerals); (ii) a source of sulphate ions (e.g., brackish or saline water); (iii) reducing conditions; (iv) the presence of sulphate-reducing bacteria; and (v) a source of energy for bacterial activity (organic matter). These requirements all contribute to various micromorphological characteristics of sulphidic materials. Pyrite either forms from solution or from amorphous or crystalline precursors, but no micromorphological evidence seems to be available for this issue.

Based on thin section observations, Pons (1964) distinguished three types of pyrite occurrences in acid sulphate soils: (i) primary pyrite, occurring as particles scattered in marine sediments without vegetation; (ii) secondary pyrite, formed in sediments with a vegetation cover before soil development, with pyrite associated with plant remains; and (iii) tertiary pyrite, formed when soils are inundated by brackish water rich in sulphate. The morphologies of secondary and tertiary pyrite are similar, and destruction of pyrite-hosting plant remains or pedoturbation can result in a dispersion of pyrite that renders a distinction between the three types impossible (Pons, 1964). Overall, the terminology and use of this classification are problematic. Van Dam and Pons (1973) propose a more genetically neutral classification, distinguishing pyrite occurrences (i) within the groundmass, (ii) associated with organic material, (iii) in pores, and (iv) as infillings of diatom or foraminifer remains.

2.1   Pyrite

Pyrite (FeS_2) is by far the most common sulphide mineral in unconsolidated sediments and sulphidic soil materials. In thin sections, pyrite is opaque (Fig. 1A), with a characteristic brass-like metallic lustre in oblique incident light (OIL) (Fig. 1B). Associated organic material is often also opaque, but it shows a dark brown colour and/or a cellular fabric where the section is thinner and it remains black in OIL (Fig. 1B). In polished thin sections or blocks, studied with reflected light, pyrite is characterised by a strong light-yellow reflectance, using microscopic methods for sulphide mineral identification in ore deposits (e.g., Craig & Vaughan, 1995). In SEM studies, pyrite crystals always have an equant shape, with cubic, octahedral or dodecahedral habits (see below). In back-scattered electron mode (BSE), they appear as bright particles, because of the high effective atomic number (e.g., Rabenhorst & Fanning, 1989). SEM imaging of uncovered thin sections shows a lower density of small pyrite crystals than thin section observations with transmitted or reflected light, because crystals below the surface of the specimen are not revealed (e.g., Rabenhorst & Haering, 1990; Bush & Sullivan, 1997a).

2.1.1   Pyrite Framboids

In soils and recent sediments, pyrite occurs mainly as framboids and much less commonly as isolated crystals. Framboids are more or less round aggregates of small crystals, which are typically well sorted (Fig. 2A and B). They are illustrated in many published reports dealing with soil-related occurrences, mainly by SEM images of undisturbed samples (Miedema et al., 1974; Arora et al., 1978; Moormann & Eswaran, 1978; Paramananthan et al., 1978; Pugh et al., 1981; Dent, 1986; Wada & Seisuwan, 1988; Rabenhorst & Fanning, 1989; Rabenhorst & Haering, 1990; Rabenhorst, 1990; Willett et al., 1992; Bush & Sullivan, 1997a, 1999; Fanning et al., 2002; Osterrieth et al., 2016). Framboids are also observed in thin sections, using incident light (van Dam & Pons, 1973; Eswaran & Joseph, 1974; Bullock et al., 1985; Dent, 1986; Stoops, 2003), transmitted

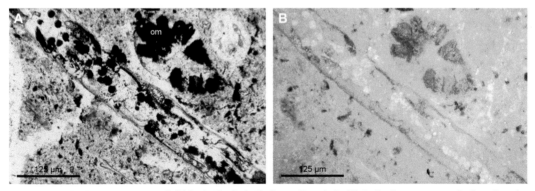

FIGURE 1 Pyrite in a mangrove soil (Thionic Fluvisol, Pitahaya, Ecuador). (A) Pyrite in a root remain (crossing the image from top left to bottom right), next to an occurrence of opaque organic material (om) (PPL). (B) Idem in OIL, illustrating the bright lustre of pyrite and the dull dark appearance of the organic matter.

FIGURE 2 Pyrite. (A) Pyrite framboids in a pore, with associated brownish organic material (Haarlemmermeer, the Netherlands; thin section of A. Jongerius) (PPL). (B) Cluster of pyrite framboids in an acid sulphate soil (The Netherlands) (Ghent University archive). (C) Pyrite crystals as part of a framboid in an acid sulphate soil (Malaysia) (Ghent University archive; see also Eswaran & Joseph, 1974).

light (Rabenhorst & Haering, 1990; Rabenhorst, 1990; Rabenhorst & James, 1992) or BSE imaging (Rabenhorst & Haering, 1990; Fitzpatrick et al., 1992, 1993, 1996, 2008a; Poch et al., 2004, 2009). The size of framboids ranges from about 5 to 50 μm and can vary considerably within a horizon (Arora et al., 1978). The size of individual crystals within framboids ranges from about 0.5 to 8 μm. The ratio of crystal size to framboid diameter shows a considerable range (e.g., 1:6 to 1:40; Bush & Sullivan, 1999).

Pyrite crystals within the framboids are nearly exclusively octahedral (bipyramidal) (e.g., Eswaran & Joseph, 1974; Miedema et al., 1974; Rabenhorst & Haering, 1990; Bush & Sullivan, 1999; Fanning et al., 2002; Bush et al., 2004) (Fig. 2C). Several published SEM images show crystals with a combination of octahedral and hexahedral (cubic) forms (Arora et al., 1978; Pugh et al., 1981; Postma, 1982; Dent, 1986; Willett et al., 1992). Pentagonal dodecahedral (pyritohedral) crystals, which are common in geological formations, are sometimes mentioned for soils and recent sediments (Eswaran & Joseph, 1974; Rabenhorst & Fanning, 1989; Bush & Sullivan, 1997b, 1999; Dharmasri et al., 2004; Skwarnecki & Fitzpatrick, 2008), but no convincing illustrations seem to have been published. In some cases, these accounts might refer to combinations of the octahedral and hexahedral forms. Cubic forms, another common habit of pyrite in geological materials, are only rarely mentioned for framboids (Bush & Sullivan, 1999). In field experiments, promoting fast growth of pyrite framboids, crystal shapes that are less well developed than in natural settings are produced (Rabenhorst, 1990).

The organisation of the crystals in framboids is generally random, although regular patterns have been documented for non-soil-related materials (e.g., Rickard, 1970). Loose packing results in framboids with a rough surface, whereas tighter packing produces a smooth surface (Bush & Sullivan, 1999). Surface roughness can also be related to crystal habit, whereby framboids composed of octahedral crystals have a greater surface roughness than those composed of dodecahedral or rounded crystals (Bush & Sullivan, 1997a, 1999). Crystals that protrude from the framboids could represent continued growth of the aggregate by addition of crystals along the surface (Wada & Seisuwan, 1988).

Pyrite framboids occur both within the groundmass and in association with plant remains, commonly within the same horizon (e.g., Eswaran & Joseph, 1974; Moormann & Eswaran, 1978; Paramananthan et al., 1978; Willett et al., 1992; Bush & Sullivan, 1999). The framboids can be clustered (e.g., Eswaran & Joseph, 1974; Miedema et al., 1974; Poch et al., 2004, 2009) (see Fig. 2B), although this grouping is susceptible to being destroyed by pedoturbation in the upper part of the soil profile (Eswaran & Joseph, 1974). Occasionally, these clusters form coherent aggregates, termed 'polyframboids' (e.g., Arora et al., 1978; Dixon et al., 1982; Postma, 1982). Clustered framboids with interstitial amorphous iron sulphides in pores were described by Kooistra (1978, 1981).

The nature and formation of framboids is discussed in various papers dealing with settings that are largely unrelated to soil environments, whose review is beyond the scope of this chapter. With regard to the genesis of framboids, processes suggested in these papers include pseudomorphism after organic aggregates (e.g., Kalliokoski & Cathles, 1969; Rickard, 1970), development as infillings of spherical voids (Rickard, 1970), replacement of a greigite ($Fe^{2+}Fe^{3+}_2S_4$) or iron monosulphide precursor (e.g., Sweeney & Kaplan, 1973; Sawlowicz, 1993; Wilkin & Barnes, 1997), and transformation of sulphur aggregates (Kribek, 1975). In a soil-related on-site experiment, framboids formed only in macropores, not in the clayey groundmass, which is explained by the greater availability of free sulphides, the more aerobic conditions compared with those within the groundmass, and the possibility of growth in non-restricting open space (Rabenhorst, 1990). Similarly, Wada and Seisuwan (1988) suggest that framboids only form at levels where oxygen-bearing waters can penetrate. Pyrite framboid morphology may be an indicator of redox conditions (Wilkin et al., 1996), but this is not confirmed by a study of pyrite in salt marsh sediments (Roychoudhury et al., 2003). Prakongkep et al. (2012) describe an occurrence of sulphur aggregates as possible precursors of pyrite framboids.

Framboids can be coated with a thin film of organic material (e.g., Miedema et al., 1974; Wada & Seisuwan, 1988; Willett et al., 1992; Bush & Sullivan, 1997a, 1999), possibly derived from fungi (Bush & Sullivan, 1997a, 1999). Coatings composed of a mixture of clay and organic material have been recognised for macropores without root remains (Bush & Sullivan, 1997a, 1999). Both types of coatings probably retard the rate of oxidation (Bush & Sullivan, 1997a, 1999). As fine-grained aggregates, framboids are anyway more susceptible to fast oxidation than single crystals of a similar size (e.g., Pugh et al., 1981). Framboids in roots, discussed further in Section 2.3, can have a coating of iron monosulphides, which could increase the rate of oxidation (Bush & Sullivan, 1997a, 1999). Another type of degradation of framboids is dispersion of the constituent crystals, as a result of sediment transport (van Dam & Pons, 1973), wave action (Fitzpatrick et al., 1993, 2008a) or bioturbation (Fitzpatrick et al., 2008a). Coherence of undisturbed framboids may require cementation by iron sulphides (Arora et al., 1978).

2.1.2   Other Aspects of Pyrite Occurrences

Pyrite in soils and recent sediments partly occur as isolated crystals in the groundmass (Eswaran, 1967; Slager et al., 1970; van Dam & Pons, 1973; Eswaran & Joseph, 1974; Miedema et al., 1974; Fanning & Rabenhorst, 1990; Willett et al., 1992; Bush & Sullivan, 1997a, 1999). The shape of these isolated crystals has been described as spherical (Slager et al., 1970), cubic (Eswaran & Joseph, 1974) and octahedral (e.g., Miedema et al., 1974). The octahedral form is occasionally combined with small faces of the hexahedral form (van Breemen, 1976; Rabenhorst & Fanning, 1989; Shamshuddin et al., 2004; Burton et al., 2006; López-Buendía et al., 2007). In some profiles, isolated crystals in the groundmass are more common in deeper parts than at other depths (Pons, 1970; Eswaran & Joseph, 1974). This type of pyrite has been considered to be synsedimentary (van Dam & Pons, 1973; Eswaran & Joseph, 1974). Isolated pyrite crystals can also be derived from occurrences associated with plant remains, following the decomposition of organic material (Wada & Seisuwan, 1988). In other studies, the occurrence of isolated crystals rather than framboids has been related to relatively fast pyrite formation (e.g., Luther et al., 1982).

Another type of non-framboidal pyrite is represented by clusters of crystals associated with plant remains, with a greater crystal size than pyrite in framboids (Paramananthan et al., 1978; Willett et al., 1992; Roychoudhury et al., 2003). Bush and Sullivan (1997a, 1999) also mention non-framboidal clusters in the groundmass and pyrite occurrences as coatings or infillings of macropores. Infillings of pores without associated organic material have been interpreted as synsedimentary precipitates (van Dam & Pons, 1973). However, infillings of this type can also develop later as diagenetic features (Rabenhorst, 1990), unrelated to the presence of organic material (Rabenhorst & James, 1992).

Associations between pyrite and plant remains, particularly roots, have commonly been reported. The pyrite crystals or framboids sometimes occupy the central hollow of the remains of roots or reeds (Slager et al., 1970; Miedema et al., 1974; Dent, 1986; Rabenhorst & Haering, 1990; Rabenhorst & James, 1992; FitzPatrick, 1993). They can also form within the cells of the plant remains (van Dam & Pons, 1973; Paramananthan et al., 1978; Eswaran & Shoba, 1983; Wada & Seisuwan, 1988), where the growth of pyrite can disrupt the tissue structure (van Dam & Pons, 1973). Pyrite occurrences in cells can be connected by filaments composed of pyrite, suggesting a relationship with fungal hyphae (Wada & Seisuwan, 1988). In some cases, pyrite occurs in larger voids within plant remains (Bouma et al., 1990; Fitzpatrick et al., 1993). Associations between organic material and pyrite without a restriction to specific parts of the plant remains have also been illustrated (Rabenhorst & Haering, 1990; Rabenhorst & James, 1992). A related feature is pyrite pseudomorphs after organic material (Kooistra, 1978), which is a fairly common diagenetic feature in the geological record. Wada and Seisuwan (1988) observed that the amount of pyrite associated with plant remains increases with increasing degree of decomposition, suggesting a relationship between decomposition and pyrite formation, but exceptions to this general rule indicate that both processes are not completely interdependent.

Eswaran (1967) observed the development of pyrite hypocoatings around plant remains in an initial stage, followed by impregnation of the remains and subsequently by decomposition of all organic matter, leaving pyrite occurrences as coatings or infillings of channels. Rabenhorst and Haering (1990) observed a preferential occurrence of pyrite in root channels and vughs within a mineral horizon, explained by a genetic relationship between pyrite formation and the presence of organic material, rather than by the slow diffusion of sulphates into the groundmass that would also explain the observed pyrite distribution pattern. In the overlying organic layer of the same soil profile, pyrite is closely associated with plant fragments but randomly distributed. Other examples of pyrite occurrences in organic layers have been described for continental environments (Stoops, 1983; Jakobsen, 1988; Gudmundsson & FitzPatrick, 2004). The presence of pyrite inside shells (van Dam & Pons, 1973; Miedema et al., 1974; Fanning & Rabenhorst, 1990; Bush & Sullivan, 1997a, 1999) has also been related to the decay of organic matter derived from the dying organism (Fanning & Rabenhorst, 1990).

Pyrite occurrences in palaeosoils (e.g., Moormann & Eswaran, 1978; McSweeney & Fastovsky, 1987; Stoops, 1992; Joeckel, 1995; Wright et al., 1997) do not necessarily represent buried sulphidic soil materials, as their formation can be related to mineralisation during flooding or at a later stage.

2.2   Sulphides Other Than Pyrite

Marcasite (FeS_2) is opaque in transmitted light and has an appearance similar to that of pyrite in OIL. In reflected light microscopy, the strong anisotropy of marcasite can allow a distinction with pyrite. In a rare study of marcasite in recent sediments, a platy crystal habit has been observed (Bush et al., 2004). In this study, the marcasite crystals are observed to cover the surface of pyrite framboids associated with organic material, and a marcasite-pyrite association is considered to be a proxy for freshwater to brackish depositional swamp environments. Pugh et al. (1981) illustrate spherical marcasite aggregates, with platy crystals (see Dixon et al., 1982), in lignite overburden, which represents an atypical setting.

Greigite ($Fe^{2+}Fe^{3+}_2S_4$) is unstable under ambient conditions and might require fast drying for its detection in micromorphological studies (Bush & Sullivan, 1997a, 1997b). It occurs as tabular to platy crystals (Bush & Sullivan, 1997a, 1997b; Burton et al., 2011), but subhedral cubic forms have also been reported for a possible greigite occurrence (Skwarnecki & Fitzpatrick, 2008). Burton et al. (2011) cite micromorphological evidence against a role of greigite as pyrite precursor.

Mackinawite (($Fe,Ni)_{1+x}S$, $x = 0$-0.07) forms small convoluted plate-shaped crystals (Burton et al., 2006; see also Herbert et al., 1998). Unidentified iron monosulphides were described by Rabenhorst (1990), as hypocoatings surrounding iron-rich soil fragments placed in tidal marsh sediments. These features were only observed in thin sections stored in anaerobic conditions and are rapidly oxidised in contact with air. Boman et al. (2008) illustrate iron monosulphide occurrences with a composition that is intermediate between that of mackinawite and greigite.

Amorphous iron sulphides with variable iron/sulphur ratios have also been reported, apparently confined to the upper part of pyrite-bearing intervals (Kooistra, 1981). The former presence of iron sulphides that are more reactive than pyrite has been suggested for jarosite-bearing materials containing unaltered pyrite framboids (Plumlee et al., 2016). In potential acid sulphate soil areas where the geological substrate is characterised by base metal mineralisation, sulphides such as chalcopyrite ($CuFeS_2$), galena (PbS) and sphalerite (ZnS) can form as authigenic precipitates (Skwarnecki & Fitzpatrick, 2008).

2.3   General Fabric of Sulphidic Layers

Little attention has been given in the literature to the general fabric characteristics of sulphidic layers. These features are partly unrelated to the sulphidic nature of the soil, but their characterisation can contribute to understanding soil development.

Mineral horizons with sulphidic materials are generally apedal, with a channel or vugh microstructure (Rabenhorst & Haering, 1990), compatible with a lack of soil structure development in a permanently wet environment. Planar voids do occur in the upper part of acid sulphate soils (Miedema et al., 1974). Changes in submicroscopic structure during drainage of tidal soils, particularly the collapse of micropores, have been illustrated by Tovey and Dent (2002). Organic layers in sulphidic soils typically have a high porosity, as deposits composed of a mixture of intact organ residues and amorphous fine organic material (Rabenhorst & Haering, 1990).

The groundmass commonly has a porphyric c/f-related distribution pattern, typical of poorly aerated soils. The composition of the coarse fraction is a reflection of the marine or alluvial origin of the sediments, which include components such as glauconite (Miedema et al., 1974) and diatom remains (e.g., Paramananthan et al., 1978). The occurrence of layering (e.g., Miedema et al., 1974) or a parallel orientation of plant remains (Wada & Seisuwan, 1988) is also compatible with a sedimentary origin. Organ and tissue residues are common constituents of the groundmass, and fresh or decayed roots can be abundant. The micromass can show speckled or striated b-fabrics (Paramananthan et al., 1978), unless masking by fine organic matter results in an undifferentiated b-fabric. The recognition of striated b-fabrics is in agreement with SEM observations of zones with oriented clay along roots and microped faces in dewatered tidal soils (Tovey & Dent, 2002). Unistrial b-fabrics are typical of undisturbed sediments in the lower horizons of soil profiles (Spaargaren et al., 1981). The absence of preferred orientations in dredged sulphidic materials has been related to a high degree of clay dispersion (Fanning & Rabenhorst, 1990).

3. Sulphuric Materials

Oxidation of pyrite results in the release of Fe^{2+}, SO_4^{2-} and H^+ ions. If no buffering occurs, most commonly by a reaction with carbonate minerals, conditions with low pH will develop. This strong acidity leads to alteration of other soil constituents such as clay minerals, resulting in the liberation of additional ion species. These reactions can ultimately result in the formation of various authigenic minerals, whose micromorphological features are discussed below.

Oxidation of sulphides first occurs along pores, resulting in preferential alteration of pyrite along ped faces (Bronswijk & Groenenberg, 1988) and determining the distribution of secondary minerals (e.g., Fanning & Rabenhorst, 1990). The preferential occurrence of pyrite along pores is another factor that determines these patterns (e.g., Bush & Sullivan, 1999). In some studies, only dissolution of pyrite is observed, without associated mineral formation (Eswaran, 1967). Corrosion of remaining pyrite crystals is recognised by the presence of pervasively pitted crystal faces (e.g., Shamshuddin et al., 2004, 2014). The groundmass around weathered pyrite occurrences can be affected as well, resulting in alteration of clay particles and contributing to the development of circumgranular voids around framboids embedded within a clay groundmass (Willett et al., 1992).

3.1   Jarosite

In thin sections, jarosite ($KFe^{3+}_3(SO_4)_2(OH)_6$) is yellowish in plane-polarised light and orange brown in cross-polarised light (e.g., van Dam & Pons, 1973; Eswaran & Joseph, 1974). Fine-grained jarosite aggregates can be dark, but such occurrences can be recognised by a yellowish to greenish colour in reflected light (Dent, 1986; FitzPatrick, 1993).

In SEM images, jarosite crystals are typically equant, with a pseudocubic habit of euhedral crystals (e.g., Cheverry et al., 1972; van Breemen & Harmsen, 1975; van Breemen, 1976; Wagner et al., 1982; Eswaran & Shoba, 1983; Shamshuddin et al., 1986; Shamshuddin & Auxtero, 1991), which in TEM images is compatible with square forms (Shamshuddin & Auxtero, 1991; Simas et al., 2006) and hexagonal shapes (Prakongkep et al., 2012). The pseudocubic shape has led some authors to suggest pseudomorphism after pyrite crystals (Carson et al., 1982; Carson & Dixon, 1983; Doner & Lynn, 1989). This is partly based on the recognition of octahedral forms for jarosite crystals (Carson et al., 1982; Carson & Dixon, 1983), although the presence of triangular crystal faces is compatible with the development of the {0001} form in combination with the rhombohedral {0112} form (see e.g., Gasharova et al., 2005). Aberrant jarosite crystal morphologies are represented by tabular (Eswaran & Joseph, 1974), elongated (Paramananthan et al., 1978) and plate-shaped crystals (Shamshuddin et al., 2004, 2014; Plumlee et al., 2016).

Jarosite in acid sulphate soils occurs as infillings (Slager et al., 1970; van Dam & Pons, 1973; Eswaran & Joseph, 1974; Moormann & Eswaran, 1978; FitzPatrick, 1993), as coatings (Dent, 1986; Fanning & Rabenhorst, 1990; Poch et al., 2004), and as hypo- or quasicoatings (Slager et al., 1970; van Dam & Pons, 1973; Miedema et al., 1974; Spaargaren et al., 1981; Bouma et al., 1990; Poch et al., 2004, 2009) (Fig. 3A).

When the transition from reduced to oxidising conditions is fast, as it can be along pores, jarosite will form near the pyrite from which most of its constituents are derived during oxidation (van Dam & Pons, 1973). This can result in development of occurrences that are pseudomorphic after pyrite aggregates, which may have some remaining pyrite

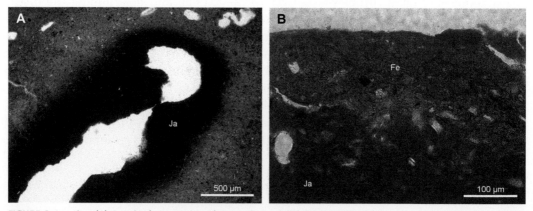

FIGURE 3 Jarosite. (A) Jarosite hypocoating along a channel (Ja) (Histic Sulfaquept, Sedu series, Malaysia; sample studied by Paramananthan et al., 1978) (PPL). (B) Jarosite hypocoating (Ja), grading to an iron oxide hypocoating toward the pore walls (Fe) (Haarlemmermeer, The Netherlands; thin section of A. Jongerius) (PPL).

in their centres (Miedema et al., 1974; Bouma et al., 1990). Local transformation of pyrite has also been suggested for dense complete jarosite infillings of channels (FitzPatrick, 1993). Other possible results of local processes are the development of a honeycomb structure, apparently by jarosite formation around the constituent crystals of pyrite framboids (Wallace et al., 2005), and the development of thin jarosite coatings around pyrite framboids (Shand et al., 2016).

When oxidation of sulphidic materials is slower, ions released by pyrite alteration are able to migrate toward ped faces, where jarosite infillings, coatings or hypocoatings will ultimately develop (van Dam & Pons, 1973; Miedema et al., 1974), possibly with an intermediate stage of iron oxide formation (Miedema et al., 1974).

Once jarosite has formed, it can be affected by dissolution, as suggested for fine-grained hypidiotopic aggregates in intertidal settings (Keene et al., 2010; Johnston et al., 2011), which might also apply to floodplain occurrences (Fall et al., 2014). A more commonly reported type of jarosite alteration is transformation to iron oxides through interaction with percolating water (van Dam & Pons, 1973; Eswaran & Joseph, 1974; Miedema et al., 1974; Dent, 1986; Fanning & Rabenhorst, 1990). This process can produce iron oxide features that are pseudomorphic after jarosite aggregates (Miedema et al., 1974). More commonly, jarosite coatings/hypocoatings (Miedema et al., 1974; Dent, 1986; Fanning & Rabenhorst, 1990) or infillings (van Dam & Pons, 1973) are only partly transformed to iron oxides, along the sides of the pores. The alteration of jarosite hypocoatings can result in development of iron oxide hypocoatings or of associated iron oxide hypocoatings and jarosite quasicoatings (Miedema et al., 1974) (Fig. 3B). Complete transformation of a jarosite hypocoating, followed by partial leaching, has been suggested for pores with iron oxide quasicoatings (Dent, 1986). An association between jarosite and derived iron oxides is also observed at the submicroscopic scale (Simas et al., 2006).

In addition to susceptibility to chemical alteration, jarosite (and iron oxide) hypo-coatings can become separated from ped faces, because of a difference in shrink-swell behaviour relative to the non-cemented ped interiors (Fanning & Rabenhorst, 1990). This can result in development of infillings composed of fragments of jarosite hypocoatings, together with sand grains and plant fragments (Miedema et al., 1974). Illuviation has also been invoked for the development of layered jarosite coatings (Fanning & Rabenhorst, 1990), implying transportation of jarosite as individual grains or small aggregates. Another type of translocation is mechanical disturbance by ploughing, resulting in the presence of nodules composed of jarosite and iron oxides in the Ap horizon, derived from coatings and hypocoatings in the B horizon (Spaargaren et al., 1981).

Varying conditions during and after the formation of jarosite result in complex patterns for compound pedofeatures, composed of jarosite, iron oxides, groundmass material and organic matter (Slager et al., 1970; Miedema et al., 1974; Fanning & Rabenhorst, 1990).

Jarosite in palaeosoils occurs as infillings of channels (Buurman, 1975; PiPujol & Buurman, 1994) and other pores (Moormann & Eswaran, 1978). Other examples are jarosite-dominated rhizoliths (McSweeney & Fastovsky, 1987; Kraus & Hasiotis, 2006), round aggregates associated with organic material in pores (Kraus, 1998), and impregnative features around channels (Kraus, 1998). These occurrences are mainly interpreted as an indication that the soils were pyrite-bearing, whereby jarosite formed following exposure of the deposits at a much later stage, unrelated to the period of palaeosoil development (Buurman, 1975; PiPujol & Buurman, 1994; Kraus, 1998; Kraus & Hasiotis, 2006).

3.2 Iron Oxides

Similar to patterns of jarosite formation, the distribution of iron oxides is partly determined by the speed of establishment of oxidising conditions, with iron oxide precipitation near pyrite occurrences when oxidation is fast and iron oxide hypo- or quasicoating development along pores when oxidation is slower (van Dam & Pons, 1973; Miedema et al., 1974). In this way, fast oxidation can result in pseudomorphism after pyrite aggregates (Slager et al., 1970; Miedema et al., 1974) (see also Tovey, 1986; Wierzchos et al., 1995).

Iron oxides in acid sulphate soils occur in the form of nodules (Slager et al., 1970; Paramananthan et al., 1978), coatings (Slager et al., 1970; Miedema et al., 1974), and hypo- or quasicoatings (Slager et al., 1970; van Dam & Pons, 1973; Paramananthan et al., 1978; Spaargaren et al., 1981) (Fig. 4). Occurrences of iron oxides in pores are partly associated with plant remains (Eswaran & Joseph, 1974; Miedema et al., 1974). Slager et al. (1970) mention an increase in the amount of nodules relative to the amount of hypo- and quasicoatings with decreasing depth, as well as a greater abundance and sharpness of the nodules in older soils, suggesting that the nodules are derived from fragmented hypocoatings. Nodules can also form by impregnation of the groundmass,

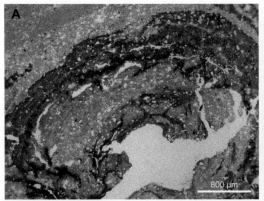
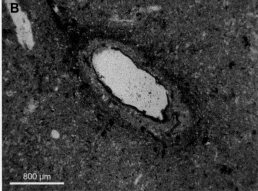

FIGURE 4 Iron oxides. (A) Thick iron oxide hypocoating, layered and with high iron oxide content in the outer part (Typic Tropaquept, Briah series, Malaysia; sample studied by Paramananthan et al., 1978) (PPL). (B) Channel with iron oxide quasicoating and thin iron oxide coating (Sulfic Tropaquept, Kangkong series, Malaysia; sample studied by Paramananthan et al., 1978) (PPL).

possibly confined to parts with a low degree of compaction (Paramananthan et al., 1978). A uniform size distribution suggests continuous growth without initiation of the development of additional nodules (Paramananthan et al., 1978).

The amorphous iron oxides that are initially formed can at a later stage be transformed to goethite or hematite by dehydration (Slager et al., 1970; van Dam & Pons, 1973; Miedema et al., 1974). This alteration is recognised by a change in colour (Slager et al., 1970) or by the appearance of relatively coarse-crystalline iron oxides (van Dam & Pons, 1973; Miedema et al., 1974). Hematite coatings illustrated by Spaargaren et al. (1981) seem to occur along the sides of pores with (amorphous) iron oxide quasicoatings. Well-developed crystalline iron oxides and amorphous iron compounds occurring at the same site are also described by Eswaran and Joseph (1974), without commenting on a possible relationship between both phases. Another possible type of iron compound in acid sulphate soils are complexes of iron and organic matter, described to occur as hypocoatings and as coatings with cracked surfaces (Paramananthan et al., 1978). Microscopic features of akaganeite $((Fe^{3+},Ni^{2+})_8(OH,O)_{16}Cl_{1.25} \cdot nH_2O)$, originally described as an Fe oxide, have also been documented for acid sulphate soils (Bibi et al., 2011).

3.3   Gypsum

Gypsum crystals $(Ca(SO_4) \cdot 2H_2O)$ in acid sulphate soils are typically elongated prismatic (Slager et al., 1970; van Dam & Pons, 1973; Stoops et al., 1978; Moormann & Eswaran, 1978; Paramananthan et al., 1978; Wagner et al., 1982; Eswaran, 1984; Eswaran & Shoba, 1983; Eswaran & Zi-Tong, 1991) (Fig. 5A). More equant prismatic forms (Miedema et al., 1974; FitzPatrick, 1984, 1993; Bullock et al., 1985) and roughly tabular crystals (Aslan &

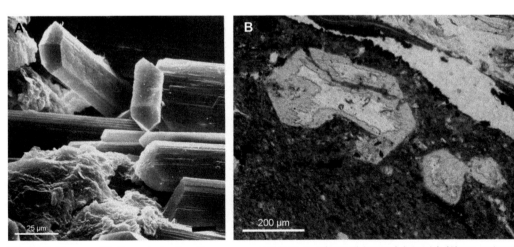

FIGURE 5 Gypsum. (A) Elongated prismatic gypsum crystals in a palaeosoil interval (Nigeria) (Ghent University archive; see also Moormann & Eswaran, 1978) (B) Tabular gypsum crystals (partly dehydrated to bassanite during thin section preparation) (Haarlemmermeer, the Netherlands) (thin section of A. Jongerius) (PPL).

Autin, 1996; Sukitprapanona et al., 2014) (Fig. 5B) have also been described. This difference in morphology between gypsum in acid soils and lenticular pedogenic gypsum in soils of arid regions (see Poch et al., 2010, 2018) is related to a difference in soil water chemistry whose nature is poorly understood. An effect of high aluminium concentrations in acid environments has been suggested (Moormann & Eswaran, 1978), but the absence or inactivity of habit-modifying compounds that cause the development of lenticular forms in other settings is a more likely cause. For an exceptional occurrence of predominantly lenticular gypsum in an acid sulphate soil, a stage with less acidic conditions in a calcareous soil is inferred (Poch et al., 2004, 2009).

The elongated gypsum crystals partly occur as radial or fan-like aggregates (Stoops et al., 1978; Moormann & Eswaran, 1978; Eswaran, 1984). Some clusters are closely associated with soil components that act as a source of sulphur or calcium, such as pyrite aggregates (Moormann & Eswaran, 1978) or shell fragments (Poch et al., 2004). A different interpretation of a gypsum-pyrite association is given by Tsatskin and Nadel (2003), who refer to the establishment of reducing conditions with pyrite formation, following a period with gypsum formation.

Gypsum occurrences in acid sulphate soils include infillings (Slager et al., 1970; van Dam & Pons, 1973; Miedema et al., 1974; Bouma et al., 1990; FitzPatrick, 1993) and clustered or isolated crystals in the groundmass (FitzPatrick, 1984, 1993; Ritsema & Groenenberg, 1993). Gypsum formation in pores that is restricted to the upper parts of soil profiles has been related to an interaction between jarosite and calcium-bearing water derived from the surface (van Dam & Pons, 1973). FitzPatrick (1993) illustrates gypsum associated with a jarosite infilling, both assumed to be derived from local oxidation of pore-filling pyrite.

Special features related to gypsum occurrences include granostriation around crystals (Poch et al., 2004) and infillings (van Dam & Pons, 1973), related to crystallisation pressure during displacive growth. These features are also present around moulds of lenticular crystals that remain after dissolution of the gypsum (Poch et al., 2004, 2009). Gypsum crystals incorporated in jarosite hypocoatings record that the latter formed at a later stage, when the availability of calcium was lower (Poch et al., 2004, 2009).

3.4   Silica

Pedofeatures composed of silica include hypo- and quasicoatings (Slager et al., 1970; van Dam & Pons, 1973) and nodules (Slager et al., 1970). Uniform impregnation of the groundmass has also been suggested, related to alteration of clay (van Dam & Pons, 1973). Buurman et al. (1973) describe silicified plant remains, characterised by the presence of thin silica coatings on cell walls (see also van Breemen, 1982). Bush and Sullivan (1997a) recognise aggregates of mixed microcrystalline silica and iron sulphides, whose formation is apparently unrelated to oxidation of pyrite. Joeckel (1995) found a close association of pyrite, authigenic silica and plant remains in hydromorphic palaeosoils.

3.5   Other Minerals

Reports of other authigenic minerals in acid sulphate soils are rare. Montoroi (1995) illustrates features of salt efflorescences with lath-shaped halotrichite, acicular alunogen, rounded rozenite crystals and plate-shaped tamarugite and soda-alum, commenting on the relationship between type of aggregation and porosity (see Table 1 for chemical formulas). Fitzpatrick et al. (2008b, 2010) document occurrences of bladed sideronatrite $(Na_2Fe^{3+}(SO_4)_2(OH)\cdot 3H_2O)$, xenotopic botryogen $(MgFe^{3+}(SO_4)_2(OH)\cdot 7H_2O)$ and needle- to lath-shaped redingtonite $(Fe^{2+}Cr_2(SO_4)_4\cdot 22H_2O)$, which were in part interpreted as having formed after sampling (Fitzpatrick et al., 2009). Gouleau et al. (1982) describe authigenic aluminium hydroxides, occurring as short prismatic crystals, isolated and as part of small aggregates (see also Marius, 1985), and implied amorphous deposits are described as coatings on halite (NaCl) by Fitzpatrick et al. (2008c). Micromorphological features of a pre-Quaternary example are illustrated for a Jurassic swamp deposit containing sodium-rich alunite that formed before compaction (Goldbery, 1978). Alunite is also abundant in certain Cenozoic silcretes, as cubic crystals and associated moulds of such crystals, whose present distribution is not clearly related to the swamp environment that is inferred for deposition of part of the parent material (Meyer & Pena dos Reis, 1985).

4. Sulphide Oxidation Products in Contexts Other Than Acid Sulphate Soils

Processes similar to those in acid sulphate soils on aerated pyrite-bearing sediments take place in various other environments where sulphide minerals are oxidised. Using studies cited in Table 1 as examples, these contexts include soils developed on sulphide-bearing shales (e.g., Mermut et al., 1985; Mermut & Arshad, 1987), exposures of sulphide-bearing rocks (e.g., Tien, 1968; Clayton, 1980; Parafiniuk, 1991; Flohr et al., 1995a; Lauf, 1997; Hammarstrom & Smith, 2002; Hammarstrom et al., 2003a, 2005a, 2005b; Bowell & Parshley, 2005; Joeckel et al., 2005, 2007, 2011; Gomes & Favas, 2006; Eskola & Peuraniemi, 2008), mining waste deposits (e.g., Harris et al., 2003; Alvarez-Valero et al., 2008; Sracek et al., 2010), soils affected by mine tailings spills (e.g., Dorronsoro et al., 2002; Simón et al., 2002), alluvial material in mining areas (e.g., Buckby et al., 2003), and banks of rivers affected by acid mine drainage (e.g., Romero et al., 2006).

In these settings, oxidation of sulphides is more commonly expressed by the presence of sulphates other than jarosite and gypsum, in comparison with typical acid sulphate soils. Below, an overview of the morphology and mode of occurrence of the various relevant sulphate minerals is given, together with a summary of published interpretations of micromorphological features. Iron oxides are not considered here. Subaqueous settings, cave environments and experimental studies are also excluded. A more directly relevant context is mineral formation during storage of sulphide-bearing samples (e.g., Wiese et al., 1987).

4.1 Jarosite

The morphology reported for jarosite is predominantly pseudocubic (Dixon et al., 1982; Flohr et al., 1995a, 1995b; Lin, 1997; Hochella et al., 1999; Charpentier et al., 2001; Simón et al., 2002; Sucha et al., 2002; Jamieson et al., 2005b; Kostova & Zdravkov, 2007; Park & Kim, 2016). Exceptions are hexagonal prismatic crystals (Gillott, 1980), plate-shaped crystals (Krasilnikov & Shoba, 1997; Civeira et al., 2016), acicular crystals (Boulet & Larocque, 1998; Courtin-Nomade et al., 2003), thick tabular to plate-shaped crystals (Frost et al., 2007), as well as possible combinations of forms producing triangular crystal faces (Darmody et al., 2007). Some studies have suggested a relationship between crystal size and crystallisation rates (Frost et al., 2007) and between both crystal size and morphological uniformity and water availability (Miller, 2011).

Jarosite partly forms directly on sulphides in ore deposits (e.g., Haubrich & Tichomirowa, 2002). Partial or complete pseudomorphic replacement of pyrite by jarosite has also been reported (Furbish, 1963). Nickel (1984) describes pseudomorphs after large pyrite crystals, characterised by a boxwork structure with septa composed of fine-grained jarosite that formed along cracks. These pseudomorphs also include sulphur crystals that formed at a later stage, as well as pyrite remnants. Another example of this type of replacement is jarosite aggregates that are pseudomorphic after glauconite grains (Briggs, 1951). A related feature is replacement of potassium feldspar by jarosite and kaolinite (Mascaro et al., 2001).

In alluvium derived from mining areas, euhedral jarosite crystals have been considered to be detrital, with good preservation during transport due to flow density (Bustillo et al., 2010). Alluvium in similar settings can also contain jarosite as part of detrital coarse grains (Hudson-Edwards et al., 1999). The same deposits also contain iron-bearing sulphates developed around reworked pyrite grains after deposition. Similar processes take place in waste piles, for example resulting in cross-linear alteration of pyrrhotite (Fe_7S_8) to jarosite and sulphur (Hammarstrom et al., 2003b). Away from the inherited sulphides, jarosite occurrences include infillings or nodules described by Strawn et al. (2002), who used micro-XRF and micro-EXAFS analyses to confirm mineral identification. Possible jarosite infillings in pyrite-bearing volcanic rock fragments are illustrated by Schaefer et al. (2008).

Occurrences in mine passages include silica-cemented jarosite aggregates (Jamieson et al., 2005b). Features with a similar macroscopic appearance but composed of jarosite and schwertmannite ($Fe^{3+}_{16}O_{16}(OH)_{9.6}(SO_4)_{3.2}\cdot10H_2O$) have been described for an identical setting (Gammons, 2006).

Mermut et al. (1985) and Mermut and Arshad (1987) describe natrojarosite occurrences ($NaFe^{3+}_3(SO_4)_2(OH)_6$) that are very similar to those of jarosite in acid sulphate soils, for example with regard to crystal morphology, association with pores, presence of pyrite remnants and alteration to iron oxides. Gieré et al. (2003) describe a hydroniumjarosite occurrence ($(H_3O)Fe^{3+}_3(SO_4)_2(OH)_6$) whose development predates jarosite formation. These minerals partly occur as spherical hydroniumjarosite nodules and as coatings composed of small jarosite oolites. Textural evidence identifies anglesite

Table 1 Overview of the morphology of common sulphate minerals, other than jarosite and gypsum, formed in a context of sulphide oxidation

Mineral	Formula	Morphology	References
Aluminite	$Al_2(SO_4)(OH)_4 \cdot 7H_2O$	Lath-shaped	Joeckel et al. (2011)
Alunite	$KAl_3(SO_4)_2(OH)_6$	Pseudocubic	Flohr et al. (1995b)
Alunogen	$Al_2(SO_4)_3(H_2O)_{12} \cdot 5H_2O$	Plate-shaped	Gomes & Favas (2006)
Apjohnite	$Mn^{2+}Al_2(SO_4)_4 \cdot 22H_2O$	Fibrous	Lauf (1997), Onac et al. (2003), Carmona et al. (2009)
Chalcanthite	$Cu(SO_4) \cdot 5H_2O$	Equant to short prismatic	Harris et al. (2003), Sracek et al. (2010)
Copiapite	$Fe^{2+}Fe^{3+}_4(SO_4)_6(OH)_2 \cdot 20H_2O$	Plate-shaped	Flohr et al (1995a), Lauf (1997), Buckby et al. (2003), Hammarstrom & Smith (2002), Hammarstrom et al. (2005a, b), Joeckel et al. (2005), Gomes & Favas (2006), Alvarez-Valero et al. (2008), Carmona et al. (2009)
		Acicular	Bowell & Parshley (2005)
Coquimbite	$Fe^{3+}_2(SO_4)_3 \cdot 9H_2O$	Plate-shaped	Buckby et al. (2003), Joeckel et al. (2005)
		Equant	Romero et al. (2006)
Dietrichite	$ZnAl_2(SO_4)_4 \cdot 22H_2O$	Acicular	Onac et al. (2003), Hammarstrom et al. (2005a)
Felsőbányaite	$Al_4(SO_4)(OH)_{10} \cdot 4H_2O$	Plate-shaped	Tien (1968), Clayton (1980)
Ferricopiapite	$Fe^{3+}_{0.67}Fe^{3+}_4(SO_4)_6(OH)_2 \cdot 20H_2O$	Lath-shaped	Equeenuddin et al. (2010)
Fibroferrite	$Fe^{3+}(SO_4)(OH) \cdot 5H_2O$	Acicular	Parafiniuk (1991), Hammarstrom et al. (2005a)
Halotrichite	$Fe^{2+}Al_2(SO_4)_4 \cdot 22H_2O$	Acicular	Sclar (1961), Flohr et al (1995a), Lauf (1997), Buckby et al. (2003), Hammarstrom & Smith (2002), Hammarstrom et al. (2005a, b), Gomes & Favas (2006), Frost et al. (2007), Joeckel et al. (2007)
Hinsdaleite	$PbAl_3(SO_4)(PO_4)(OH)_6$	Lenticular	Flohr et al (1995b)
Magnesiocopiapite	$MgFe^{3+}_4(SO_4)_6(OH)_2 \cdot 20H_2O$	Plate-shaped	Equeenuddin et al. (2010)
Melanterite	$Fe(SO_4) \cdot 7H_2O$	Lath-shaped, curved	Sclar (1961)
Osarizawaite	$Pb(Al_2Cu^{2+})(SO_4)_2(OH)_6$	Equant	Hammarstrom et al. (2005a)
		Equant	Harris et al. (2003)
Pickeringite	$MgAl_2(SO_4)_4 \cdot 22H_2O$	Acicular	Parafiniuk (1991), Hammarstrom et al. (2003a, 2005a), Onac et al. (2003), Gomes & Favas (2006), Romero et al. (2006), Eskola & Peuraniemi (2008)
Rozenite	$Fe^{2+}(SO_4) \cdot 4H_2O$	Equant to thick tabular	Buckby et al. (2003)
Slavikite	$(H_3O)_3Mg_6Fe_{15}(SO_4)_{21}(OH)_{18} \cdot 98H_2O$	Hexagonal plate-shaped	Parafiniuk (1991), Flohr et al (1995a), Lauf (1997), Hammarstrom et al. (2003a), Joeckel et al. (2007)
Szomolnokite	$Fe(SO_4) \cdot H_2O$	Equant	Buckby et al. (2003)
Tamarugite	$NaAl(SO_4)_2 \cdot 6H_2O$	Plate-shaped	Harris et al. (2003), Fernández-López et al. (2014)

($Pb(SO_4)$) formation as an intermediate step (Farkas et al., 2009) in alteration of galena to plumbojarosite ($Pb_{0.5}Fe^{3+}_3(SO_4)_2(OH)_6$), characterised by a roughly tabular morphology (Farkas et al., 2009).

Pronounced compositional zoning with varying K-Na-H_3O proportions has been interpreted to record a strong influence of variations in local conditions during crystal growth (Jamieson et al., 2005b). Papike et al. (2007) discuss similar features in terms of a miscibility gap between end-members with different solubilities, implying that phases with intermediate compositions will not form at low temperatures.

4.2 Gypsum

Gypsum crystals formed in acid environments are almost invariably elongated prismatic (Dixon et al., 1982; Fitzpatrick et al., 1992, 1996; Flohr et al., 1995a, 1995b; Lauf, 1997; Boulet & Larocque, 1998; Puura et al., 1999; Charpentier et al., 2001; Darmody et al., 2002; Dorronsoro et al., 2002; Simón et al., 2002; Aguilar et al., 2003; Buckby et al., 2003; Joeckel et al., 2005; Gomes & Favas, 2006; Darmody et al., 2007; Kostova & Zdravkov, 2007; Eskola & Peuraniemi, 2008). Lath- and plate-shaped crystals have also been observed (Gillott, 1980).

In some studies, occurrences of gypsum near pyrite aggregates (Charpentier et al., 2001; Kostova & Zdravkov, 2007) or limestone fragments (Puura et al., 1999) are mentioned. In soils affected by a tailings spill, gypsum crystals occur in the outer part of coatings around carbonate grains, with an inner part composed of aluminium compounds whose formation is related to an increase in pH during alteration of the grain they surround (Aguilar et al., 2003). Other aspects of gypsum occurrences are widening of planar voids by crystal growth (Gillott, 1980), rounding of crystals by partial dissolution (Joeckel et al., 2005) and development of Cu-rich reaction rims (Mees et al., 2013).

4.3 Other Sulphate Minerals

An overview of the morphology of common sulphate minerals in environments with sulphide oxidation, other than jarosite and gypsum, is given in Table 1. Occurrences whose mineralogical composition could not be identified precisely are not included in this table. Various less common sulphate minerals formed by sulphide alteration are illustrated by Szakáll et al. (1997).

Harris et al. (2003) mention a difference in degree of preservation between crystals of poorly soluble and highly soluble salt minerals, illustrated by subhedral osarizawaite and euhedral chalcanthite and tamarugite. Joeckel et al. (2005) recognise two size classes for plate-shape crystals, possibly related to copiapite-coquimbite transformations. An association of small and larger plate-shaped iron sulphate crystals is also recognised by Jamieson et al. (2005a), who identify the smaller crystals as aluminium-rich copiapite that formed on coarser magnesium-rich copiapite after sampling (see also Jambor et al., 2000, p. 318-319). In another study, a difference in iron content between (iron-poor) acicular copiapite and associated lozenge-shaped tabular copiapite crystals is suggested (Dill et al., 2002).

Several studies document the patterns of alteration of sulphides to secondary sulphate minerals (e.g., Jamieson et al., 1995; Boulet & Larocque, 1998; Shaw et al., 1998; Bowell & Parshley, 2005; Koski et al., 2008). SEM studies have also been used to deduce sequences of mineral formation. Paragenetic sequences recognised by Buckby et al. (2003) are copiapite-halotrichite, coquimbite-szomolnokite and rozenite-rhomboclase/szomolnokite-gypsum. Hammarstrom et al. (2005a) describe copiapite formation on pyrite (see also Joeckel et al., 2007), together with rozenite, followed by the formation of halotrichite on rozenite. In the same study, dietrichite aggregates enclosing fine-grained zinc sulphates are observed, but the significance of this association is not apparent. Frost et al. (2007) interpret the absence of halotrichite in certain jarosite-bearing samples as an indication of the conversion of halotrichite to jarosite. Flohr et al. (1995b) refer to micromorphological features suggesting late-stage development of hinsdaleite and alunite, and Valente & Gomes (2009) mention aluminium sulphate deposition following melanterite formation as a result of iron depletion. Ring-shaped osarizawaite crystals have been interpreted as remnants of tabular hexagonal crystals of the alunite-jarosite group with compositional zoning (Nickel, 1980). Alteration of halotrichite to hydrated iron sulphates occurs preferentially at the ends of halotrichite bundles (Joeckel et al., 2007). Another aspect of sulphate mineral formation for which microscopic evidence has been cited is the possible role of microorganisms (e.g., Laskou et al., 2010).

A rare thin section study records the occurrence of felsőbányaite (basaluminite) in residual clays derived from limestone along the contact with pyrite-bearing shale (Milton et al., 1955).

5. Conclusions

Micromorphological studies of sulphidic materials and sulphuric horizons have well documented the mode of occurrence of the various authigenic minerals they contain, showing a great similarity between profiles worldwide. Micropedology has considerably contributed to understanding the main processes involved in the formation and behaviour of sulphidic and sulphuric soil materials. Further progress in understanding the development of these materials might mainly be achieved by concentrating on occurrences with aberrant mineralogical compositions or textural characteristics. For acid environments other than acid sulphate soils, the potential of micromorphology to understand paragenetic sequences is largely unexplored but bears considerable promise.

References

Aguilar, J., Dorronsoro, C., Bellver, R., Fernández, E., Fernández, J., García, I., Iriarte, A., Martin, F., Ortiz, I. & Simón, M., 2003. Contaminación de los Suelos tras el Vertido Tóxico de Aznalcollar. Departamento de Edafología y Química Agrícola, Universidad de Granada, 184 p.

Alvarez-Valero, A.M., Pérez-López, R., Matos, J., Capitán, M.A., Nieto, J.M., Sáez, R., Delgado, J. & Caraballo, M., 2008. Potential environmental impact at São Domingos mining district (Iberian Pyrite Belt, SW Iberian Peninsula): evidence from a chemical and mineralogical characterization. Environmental Geology 55, 1797-1809.

Arora, H.S., Dixon, J.B., Hossner, L.R., 1978. Pyrite morphology in lignitic coal and associated strata of east Texas. Soil Science 125, 151-159.

Aslan, A. & Autin, W.J., 1996. Depositional and pedogenic influences on the environmental geology of Holocene Mississippi River floodplain deposits near Ferriday, Louisiana. Engineering Geology 45, 417-432.

Bibi, I., Singh, B. & Silvester, E., 2011. Akagenéite (β-FeOOH) precipitation in inland acid sulfate soils of south-western New South Wales. Geochimica et Cosmochimica Acta 75, 6429-6438.

Boman, A., Åström, M. & Fröjdö, S., 2008. Sulfur dynamics in boreal acid sulfate soils rich in metastable iron sulfide − The role of artificial drainage. Chemical Geology 255, 68-77.

Boulet, M.P. & Larocque, A.C.L., 1998. A comparative mineralogical and geochemical study of sulfide mine tailings at two sites in New Mexico, USA. Environmental Geology 33, 130-142.

Bouma, J., Fox, C.A. & Miedema, R., 1990. Micromorphology of hydromorphic soils: applications for soil genesis and land evaluation. In Douglas, L.A. (ed.), Soil Micromorphology: A Basic and Applied Science. Developments in Soil Science, Volume 19. Elsevier, Amsterdam, pp. 257-278.

Bowell, R.J. & Parshley, J.V., 2005. Control of pit-lake water chemistry by secondary minerals, Summer Camp pit, Getchell mine, Nevada. Chemical Geology 215, 373-385.

Briggs, L.I., 1951. Jarosite from the California Tertiary. American Mineralogist 36, 902-906.

Bronswijk, J.J.B. & Groenenberg, J.E., 1988. A simulation model for acid sulphate soils. I. Basic principles. In Dost, H. (ed.), Selected Papers of the Dakar Symposium on Acid Sulphate Soils. Publication 44, International Institute for Land Reclamation and Improvement, Wageningen, pp. 341-355.

Buckby, T., Black, S., Coleman, M.L. & Hodson, M.E., 2003. Fe-sulphate-rich evaporative mineral precipitates from the Río Tinto, southwest Spain. Mineralogical Magazine 67, 263-278.

Bullock, P., Fedoroff, N., Jongerius, A., Stoops, G., Tursina, T. & Babel, U., 1985. Handbook for Soil Thin Section Description. Waine Research Publications, Wolverhampton, 152 p.

Burton, E.D., Bush, R.T. & Sullivan, L.A., 2006. Sedimentary iron geochemistry in acidic waterways associated with coastal lowland acid sulfate soils. Geochimica et Cosmochimica Acta 70, 5455-5468.

Burton, E.D., Bush, R.T., Johnston, S.G., Sullivan, L.A. & Keene, A.F., 2011. Sulfur biogeochemical cycling and novel Fe-S mineralization pathways in a tidally re-flooded wetland. Geochimica et Cosmochimica Acta 75, 3434-3451.

Bush, R.T. & Sullivan, L.A., 1997a. Sulfide micromorphology in some eastern Australian Holocene sediments. In Shoba, S., Gerasimova, M. & Miedema, R. (eds.), Proceedings of the 10th International Working Meeting on Soil Micromorphology, Moscow, pp. 290-303.

Bush, R.T. & Sullivan, L.A., 1997b. Morphology and behaviour of greigite from a Holocene sediment in Eastern Australia. Australian Journal of Soil Research 35, 853-861.

Bush, R.T. & Sullivan, L.A., 1999. Pyrite micromorphology in three Australian Holocene sediments. Australian Journal of Soil Research 37, 637-653.

Bush, R.T., McGrath, R. & Sullivan, L.A., 2004. Occurrence of marcasite in an organic-rich Holocene estuarine mud. Australian Journal of Soil Research 42, 617-621.

Bustillo, M.A., Aparicio, A. & García, R., 2010. Surface saline deposits and their substrates in a polluted arid valley (Murcia, Spain). Environmental Earth Sciences 60, 1215-1225.

Buurman, P., 1975. Possibilities of palaeopedology. Sedimentology 22, 289-298.

Buurman, P.N., van Breemen, N. & Henstra, S., 1973. Recent silicification of plant remains in acid sulphate soils. Neues Jahrbuch für Mineralogie Monatshefte 1973, 117-124.

Carmona, D.M., Faz Cano, Á. & Arocena, J.M., 2009. Cadmium, copper, lead, and zinc in secondary sulfate minerals in soils of mined areas in Southeast Spain. Geoderma 150, 150-157.

Carson, C.D. & Dixon, J.B., 1983. Mineralogy and acidity of an inland acid sulfate soil of Texas. Soil Science Society of America Journal 47, 828-833.

Carson, C.D., Fanning, D.S. & Dixon, J.B., 1982. Alfisols and Ultisols with acid sulfate weathering features in Texas. In Kittrick, J.A., Fanning, D.S. & Hosner, L.R. (eds.), Acid Sulfate Weathering. Soil Science Society of America, Special Publication 10, SSSA, Madison, pp. 127-146.

Charpentier, D., Cathelineau, M., Mosser-Ruck, R. & Bruno, G., 2001. Evolution minéralogique des argilites en zone sous-saturée oxydée: exemple des parois du tunnel de Tournemire (Aveyron, France). Comptes Rendus de l'Académie des Sciences 332, 601-607.

Cheverry, C., Fromaget, M. & Bocquier, G., 1972. Quelques aspects micromorphologiques de la pédogenèse des sols de polders conquis sur le lac Tchad. Cahiers ORSTOM, Série Pédologie 10, 373-387.

Civeira, M., Oliveira, M.L.S., Hower, J.C., Agudelo-Castañeda, D.M., Taffarel, S.R., Ramos, C.G., Kautzmann, R.M. & Silva, L.F.O., 2016. Modification, adsorption, and geochemistry processes on altered minerals and amorphous phases on the nanometer scale: examples from copper mining refuse, Touro, Spain. Environmental Science and Pollution Research 23, 6535-6545.

Clayton, T., 1980. Hydrobasaluminite and basaluminite from Chickerell, Dorset. Mineralogical Magazine 43, 931-937.

Courtin-Nomade, A., Bril, H., Neel, C. & Lenain, J.F., 2003. Arsenic in iron cements developed within tailings of a former metalliferous mine − Enguialès, Aveyron, France. Applied Geochemistry 18, 395-408.

Craig, J.R. & Vaughan, D.J., 1995. Ore Microscopy and Ore Petrography. 2nd Edition. John Wiley and Sons, New York, 434 p.

Darmody, R.G., Campbell, S.W., Dixon, J.C. & Thorn, C.E., 2002. Enigmatic efflorescence in Kärkevagge, Swedish Lapland: the key to chemical weathering? Geografiska Annaler 84A, 187-192.

Darmody, R.G., Thorn, C.E. & Dixon, J.C., 2007. Pyrite-enhanced chemical weathering in Kärkevagge, Swedish Lapland. Geological Society of America Bulletin 119, 1477-1485.

Dent, D., 1986. Acid Sulphate Soils: A Baseline for Research and Development. Publication 39, International Institute for Land Reclamation and Improvement, Wageningen, 204 p.

Dharmasri, L.C., Hudnall, W.H. & Ferrell, R.E., 2004. Pyrite formation in Louisiana coastal marshes: scanning electron microscopy and X-ray diffraction evidence. Soil Science 169, 624-631.

Dill, H.G., Pöllmann, H., Bosecker, K., Hahn, L. & Mwiya, S., 2002. Supergene mineralization in mining residues of the Matchless cupreous pyrite deposit (Namibia) − a clue to the origin of modern and fossil duricrusts in semiarid climates. Journal of Geochemical Exploration 75, 43-70.

Dixon, J.B., Hosner, L.R., Senkayi, A.L. & Egashira, K., 1982. Mineralogical properties of lignite overburden as they relate to mine spoil reclamation. In Kittrick, J.A., Fanning, D.S. & Hosner, L.R. (eds.), Acid Sulfate Weathering. Soil Science Society of America, Special Publication 10, SSSA, Madison, pp. 169-191.

Doner, H.E. & Lynn, W.C., 1989. Carbonate, halide, sulfate, and sulfide minerals. In Dixon, J.B. & Weed, S. B. (eds.), Minerals in Soil Environments. 2nd Edition. SSSA Book Series, No. 1, Soil Science Society of America, Madison, pp. 279-330.

Dorronsoro, C., Martin, F., Ortiz, I., García, I., Simón, M., Fernández, E., Aguilar, J. & Fernández, J., 2002. Migration of trace elements from pyrite tailings in carbonate soils. Journal of Environmental Quality 31, 829-835.

Equeenuddin, S.M., Tripathy, S., Sahoo, P.K. & Panigrahi, M.K., 2010. Hydrogeochemical characteristics of acid mine drainage and water pollution at Makum Coal field, India. Journal of Geochemical Exploration 105, 75-82.

Eskola, T. & Peuraniemi, V., 2008. Secondary sulphate precipitates in the gravel pit at Kumpuselkä esker, northern Finland. Mineralogical Magazine 72, 415-417.

Eswaran, H., 1967. Micromorphological study of a 'cat-clay' soil. Pedologie 17, 259-265.

Eswaran, H., 1984. Scanning electron microscopy of salts. McGraw-Hill Yearbook of Science & Technology 1984, McGraw-Hill Book Company, New York, pp. 403-405.

Eswaran, H. & Joseph, K.T., 1974. Pyrite and its oxidation products in an acid sulphate soil. MARDI Research Bulletin 2, 50-57.

Eswaran, H. & Shoba, S.A., 1983. Scanning electron microscopy in soil research. In Bullock, P. & Murphy, C.P. (eds.), Soil Micromorphology, Volume 1, Techniques and Applications. AB Academic Publishers, Berkhamsted, pp. 19-51.

Eswaran, H. & Zi-Tong, G., 1991. Properties, genesis, classification, and distribution of soils with gypsum. In Nettleton, W.D. (ed.), Occurrence, Characteristics, and Genesis of Carbonate, Gypsum, and Silica Accumulations in Soils. Soil Science Society of America, Special Publication 26, SSSA, Madison, pp. 89-119.

Fall, A.C.A.L., Montoroi, J.P. & Stahr, K., 2014. Coastal acid sulfate soils in the Saloum River basin, Senegal. Soil Research 52, 671-684.

Fanning, D.S. & Rabenhorst, M.C., 1990. Micromorphology of acid sulphate soils in Baltimore Harbor dredged materials. In Douglas, L.A. (ed.), Micromorphology: A Basic and Applied Science. Elsevier, Amsterdam, pp. 279-288.

Fanning, D.S., Rabenhorst, M.C., Burch, S.N., Islam, K.R. & Tangren, S.A., 2002. Sulfides and sulfates. In Dixon, J.B. & Schulze, D.G. (eds.), Soil Mineralogy with Environmental Applications. SSSA Book Series, No. 7. Soil Science Society of America, Madison, pp. 229-260.

Farkas, I.M., Weiszburg, T.G., Pekker, P. & Kuzmann, E., 2009. A half-century of environmental mineral formation on a pyrite-bearing waste dump in the Mátra Mountains, Hungary. Canadian Mineralogist 47, 509-524.

Fernández-López, C., Faz Cano, A., Arocena, J.M. & Alcolea, A., 2014. Elemental and mineral composition of salts from selected natural and mine-affected areas in the Poopó and Uru-Uru lakes (Bolivia). Journal of Great Lakes Research 40, 841-850.

FitzPatrick, E.A., 1984. Micromorphology of Soils. Chapman and Hall, London, 433 p.

FitzPatrick, E.A., 1993. Soil Microscopy and Micromorphology. John Wiley & Sons, Chichester, 304 p.

Fitzpatrick, R.W., Naidu, R. & Self, P.G., 1992. Iron deposits and microorganisms in saline sulfidic soils with altered soil water regimes in South Australia. In Skinner, H.G.W. & Fitzpatrick, R.W. (eds.), Biomineralization Processes of Iron and Manganese – Modern and Ancient Environments. Catena Supplement 21, 263-286.

Fitzpatrick, R.W., Hudnall, W.H., Self, P.G. & Naidu, R., 1993. Origin and properties of inland and tidal saline acid sulphate soils in South Australia. In Dent, D.L. & van Mensvoort, M.E.F. (eds.), Selected Papers of the Ho Chi Minh City Symposium on Acid Sulphate Soils. Publication 53, International Institute for Land Reclamation and Improvement, Wageningen, pp. 71-80.

Fitzpatrick, R.W., Fritsch, E. & Self, P.G., 1996. Interpretation of soil features produced by ancient and modern processes in degraded landscapes: V. Development of saline sulfidic features in non-tidal seepage areas. Geoderma 69, 1-29.

Fitzpatrick, R.W., Thomas, B.P., Merry, R.H. & Marvanek, S., 2008a. Acid sulfate soils in Barker Inlet and Gulf St. Vincent Priority Region. CSIRO Land and Water Science Report 35/08, 21 p.

Fitzpatrick, R.W., Shand, P., Merry, R.H., Thomas, B., Marvanek, S., Creeper, N., Thomas, M., Raven M. D., Simpson, S.L., McClure, S. & Jayalath, N., 2008b. Acid sulfate soils in the Coorong, Lake Alexandrina and Lake Albert: properties, distribution, genesis, risks and management of sub-aqueous, waterlogged and drained soil environments. CSIRO Land and Water Science Report 52/08, 163 p.

Fitzpatrick, R., Degens, B., Baker, A., Raven, M., Shand, P., Smith, M., Rogers, S. & George, R., 2008c. Avon Basin, WA Wheatbelt: acid sulfate soils and salt efflorescences in open drains and receiving environments. In Fitzpatrick, R. & Shand, P. (eds.), Inland Acid Sulfate Soil Systems across Australia. CRC LEME Open File Report No. 249, pp. 189-204.

Fitzpatrick, R.W., Shand, P. & Merry, R.H., 2009. Acid sulfate soils. In Jenning, J.T. (ed.), Natural History of the Riverland and Murraylands. Royal Society of South Australia, Adelaide, pp. 65-111.

Fitzpatrick, R., Shand, P., Raven, M. & McClure, S., 2010. Occurrence and environmental significance of sideronatrite and other mineral precipitates in acid sulfate soils. In Gilkes, R. & Prakongkep, N. (eds.), Proceedings of the 19th World Congress of Soil Science. Australian Society of Soil Science, Warragul, pp. 1770-1773.

Flohr, M.J., Dillenberg, R.G. & Plumlee, G.S., 1995a. Characterization of secondary minerals formed as the result of weathering of the Anakeesta Formation, Alum Cave, Great Smoky Mountains National Park, Tennessee. U.S. Geological Survey Open File Report 95-477, 22 p.

Flohr, M.J., Dillenberg, R.G., Nord, G.L. & Plumlee, G.S., 1995b. Secondary mineralogy of altered rocks, Summitville Mine, Colorado. U.S. Geological Survey Open File Report 95-808, 27 p.

Frost, R.L., Weier, M., Martinez-Frias, J., Rull, F. & Reddy, B.J., 2007. Sulphate efflorescent minerals from El Jaroso Ravine, Sierra Almagrera − an SEM and Raman spectroscopic study. Spectrochimica Acta A 66, 177-183.

Furbish, W.J., 1963. Geologic implications of jarosite, pseudomorphic after pyrite. American Mineralogist 48, 703-706.

Gammons, C.H., 2006. Geochemistry of perched water in an abandoned underground mine, Butte, Montana. Mine Water and the Environment 25, 114-123.

Gasharova, B., Göttlicher, J. & Becker, U., 2005. Dissolution at the surface of jarosite: an in situ AFM study. Chemical Geology 215, 499-516.

Gieré, R., Sidenko, N.V. & Lazareva, E.V., 2003. The role of secondary minerals in controlling the migration of arsenic and metals from high-sulfide wastes (Berikul gold mine, Siberia). Applied Geochemistry 18, 1347-1359.

Gillott, J.E., 1980. Use of scanning electron microscope and Fourier methods in characterization of microfabric and texture of sediments. Journal of Microscopy 120, 261-277.

Goldbery, R., 1978. Early diagenetic, nonhydrothermal Na-alunite in Jurassic flint clays, Makhtesh Ramon, Israel. Geological Society of America Bulletin 89, 687-698.

Gomes, M.E.P. & Favas, P.J.C., 2006. Mineralogical controls on mine drainage of the abandoned Ervedosa tin mine in north-eastern Portugal. Applied Geochemistry 21, 1322-1334.

Gouleau, D., Kalck, Y., Marius, C. & Lucas, J., 1982. Cristaux d'hydroxydes d'aluminium néoformés dans les sédiments actuels des mangroves du Sénégal (Sine Saloum − Casamance). Mémoire de la Société Géologique de France 144, 147-154.

Gudmundsson, T. & FitzPatrick, E.A., 2004. Micromorphology of an Icelandic Histosol. In Oskarsson, H. & Arnalds, O. (eds.), Volcanic Soil Resources in Europe. Abstracts. RALA Report 214. Agricultural Research Institute, Reykjavik, pp. 79-80.

Hammarstrom, J.M. & Smith, K.S., 2002. Geochemical and mineralogic characterization of solids and their effects on waters in metal-mining environments. In Seal, R.R. & Foley, N.K. (eds.), Progress on Geoenvironmental Models for Selected Mineral Deposit Types. U.S. Geological Survey Open File Report 02-195, pp. 8-54.

Hammarstrom, J.M., Seal, R.R., Meier, A.L. & Jackson, J.C., 2003a. Weathering of sulfidic shale and copper mine waste: secondary minerals and metal cycling in Great Smoky Mountains National Park, Tennessee, and North Carolina, USA. Environmental Geology 45, 35-57.

Hammarstrom, J.M., Piatak, N.M., Seal, R.R., Briggs, P.H., Meier, A.L. & Muzik, T.L., 2003b. Geochemical Characteristics of TP3 Mine Wastes at the Elizabeth Copper Mine Superfund Site, Orange Co., Vermont. U.S. Geological Survey Open File Report 03-431, 40 p.

Hammarstrom, J.M., Seal, R.R., Meier, A.L. & Kornfeld, J.M., 2005a. Secondary sulfate minerals associated with acid drainage in the eastern US: recycling of metals and acidity in surficial environments. Chemical Geology 215, 407-431.

Hammarstrom, J.M., Brady, K. & Cravotta, C.A., 2005b. Acid-rock drainage at Skytop, Centre County, Pennsylvania, 2004. U.S. Geological Survey Open File Report 05-1148, 45 p.

Harris, D.L., Lottermoser, B.G. & Duchesne, J., 2003. Ephemeral acid mine drainage at Montalbion silver mine, North Queensland. Australian Journal of Earth Sciences 50, 797-809.

Haubrich, F. & Tichomirowa, M., 2002. Sulfur and oxygen isotope geochemistry of acid mine drainage – the polymetallic sulfide deposit 'Himmelfahrt Fundgrube' in Freiberg (Germany). Isotopes in Environmental and Health Studies 38, 121-138.

Herbert, R.B., Benner, S.G., Pratt, A.R. & Blowes, D.W., 1998. Surface chemistry and morphology of poorly crystalline iron sulfides precipitated in media containing sulfate-reducing bacteria. Chemical Geology 144, 87-97.

Hochella, M.F., Moore, J.N., Golla, U. & Putnis, A., 1999. A TEM study of samples from acid mine drainage systems: metal-mineral association with implications for transport. Geochimica et Cosmochimica Acta 63, 3395-3406.

Hudson-Edwards, K.A., Schell, C. & Macklin, M.G., 1999. Mineralogy and geochemistry of alluvium contaminated by metal mining in the Rio Tinto area, southwest Spain. Applied Geochemistry 14, 1015-1030.

IUSS Working Group WRB, 2015. World Reference Base for Soil Resources 2014, Update 2015. World Soil Resources Reports No. 106. FAO, Rome, 192 p.

Jakobsen, B.H., 1988. Accumulation of pyrite and Fe-rich carbonate and phosphate minerals in a lowland moor area. Journal of Soil Science 39, 447-455.

Jambor, J.L., Nordstom, D.K. & Alpers, C.N., 2000. Metal-sulfate salts from sulfide mineral oxidation. In Alpers, C.N., Jambor, J.L. & Nordstrom, D.K. (eds.), Sulfate Minerals. Crystallography, Geochemistry, and Environmental Significance. Reviews in Mineralogy & Geochemistry 40, Mineralogical Society of America, pp. 303-350.

Jamieson, H.E., Shaw, S.C. & Clark, A.H., 1995. Mineralogical factors controlling metal release from tailings at Geco, Manitouwadge, Ontario. Proceedings of the Sudbury '95 Conference on Mining and the Environment. CANMET, Natural Resources Canada, Ottawa, Volume 1, pp. 405-413.

Jamieson, H.E., Robinson, C., Alpers, C.N., McCleskey, R.B., Nordstrom, D.K. & Peterson, R.C., 2005a. Major and trace element composition of copiapite-group minerals and coexisting water from the Richmond mine, Iron Mountain, California. Chemical Geology 215, 387-405.

Jamieson, H.E., Robinson, C., Alpers, C.N., Nordstrom, D.K., Poustovetov, A. & Lowers, H.A., 2005b. The composition of coexisting jarosite-group minerals and water from the Richmond Mine, Iron Mountain, California. Canadian Mineralogist 43, 1225-1242.

Joeckel, R.M., 1995. Paleosols below the Ames marine unit (Upper Pennsylvanian, Conemaugh Group) in the Appalachian Basin, USA: variability on an ancient depositional landscape. Journal of Sedimentary Research 65, 393-407.

Joeckel, R.M., Ang Clement, B.J. & VanFleet Bates, L.R., 2005. Sulfate-mineral crusts from pyrite weathering and acid rock drainage in the Dakota Formation and Graneros Shale, Jefferson County, Nebraska. Chemical Geology 215, 433-452.

Joeckel, R.M., Wally, K.D., Fischbein, S.A. & Hanson, P.R., 2007. Sulfate mineral paragenesis in Pennsylvanian rocks and the occurrence of slavikite in Nebraska. Great Plains Research 17, 17-33.

Joeckel, R.M., Wally, K.D., Ang Clement, B.J., Hanson, P.R., Dillon, J.S. & Wilson, S.K., 2011. Secondary minerals from extrapedogenic *per latus* acidic weathering environments at geomorphic edges, Eastern Nebraska, USA. Catena 85, 253-266.

Johnston, S.G., Keene, A.F., Bush, R.T., Burton, E.D., Sullivan, L.A., Isaacson, L., McElnea, A.E., Ahern, C.R., Smith, C.D. & Powell, B., 2011. Iron geochemical zonation in a tidally inundated acid sulfate soil wetland. Chemical Geology 280, 257-270.

Kalliokoski, J. & Cathles, L., 1969. Morphology, mode of formation, and diagenetic changes in framboids. Bulletin of the Geological Society of Finland 41, 125-133.

Keene, A., Johnston, S., Bush, R., Sullivan, L. & Burton, E., 2010. Reductive dissolution of natural jarosite in a tidally inundated acid sulfate soil: geochemical implications. In Gilkes, R. & Prakongkep, N. (eds.), Proceedings of the 19th World Congress of Soil Science. Australian Society of Soil Science, Warragul, pp. 5712-5715.

Kooistra, M.J., 1978. Soil Development in Recent Marine Sediments of the Intertidal Zone in the Oosterschelde – The Netherlands. A Soil Micromorphological Approach. Soil Survey Paper 14, Nertherlands Soil Survey Institute, Wageningen, 183 p.

Kooistra, M.J., 1981. The determination of iron, manganese, sulphur and phosphorus in thin sections of recent marine intertidal sediments in the south-west of the Netherlands by SEM-EDXRA. In Bisdom, E.BA. (ed.), Submicroscopy of Soils and Weathered Rocks, Proceedings of the First Workshop of the International Working Group on Submicroscopy of Undisturbed Soil Materials. Pudoc, Wageningen, pp. 217-236.

Koski, R.A., Munk, L., Foster, A.L., Shanks, W.C. & Stillings, L.L., 2008. Sulfide oxidation and distribution of metals near abandoned copper mines in coastal environments, Prince William Sound, Alaska, USA. Applied Geochemistry 23, 227-254.

Kostova, I. & Zdravkov, A., 2007. Organic petrology, mineralogy and depositional environment of the Kipra lignite seam, Maritza-West basin, Bulgaria. International Journal of Coal Geology 71, 527-541.

Krasilnikov, P.V. & Shoba, S.A., 1997. Morphological diagnostics of sulfides oxidation in soils of Karelia. In Shoba, S., Gerasimova, M. & Miedema, R. (eds.), Proceedings of the 10th International Working Meeting on Soil Micromorphology, Moscow, pp. 114-119.

Kraus, M.J., 1998. Development of potential acid sulfate paleosols in Paleocene floodplains, Bighorn Basin, Wyoming, USA. Palaeogeography, Palaeoclimatology, Palaeoecology 144, 203-224.

Kraus, M.J. & Hasiotis, S.T., 2006. Significance of different modes of rhizolith preservation to interpreting paleoenvironmental and paleohydrologic settings: examples from Paleogene paleosols, Bighorn Basin, Wyoming, U.S.A. Journal of Sedimentary Research 76, 633-646.

Kribek, B., 1975. The origin of framboidal pyrite as a surface effect of sulphur grains. Mineralium Deposita 10, 389-396.

Laskou, M., Economou-Eliopoulos, M. & Mitsis, I., 2010. Bauxite ore as an energy source for bacteria driving iron-leaching and bio-mineralization. Hellenic Journal of Geosciences 45, 163-174.

Lauf, R.J., 1997. Secondary sulfate minerals from Alum Cave Bluff: microscopy and microanalysis. Oak Ridge National Laboratory, Report ORNL/TM-13471, 44 p.

Lin, Z., 1997. Mobilization and retention of heavy metals in mill-tailings from Garpenberg sulfide mines, Sweden. Science of the Total Environment 198, 13-31.

López-Buendía, A.M., Whateley, M.K.G., Bastida, J. & Urquiola, M.M., 2007. Origins of mineral matter in peat marsh and peat bog deposits, Spain. International Journal of Coal Geology 71, 246-262.

Luther, G.W., Giblin, A., Howarth, R.W. & Ryans, R.A., 1982. Pyrite and oxidized iron mineral phases formed from pyrite oxidation in salt marsh and estuarine sediments. Geochimica et Cosmochimica Acta 46, 2665-2669.

Marius, C., 1985. Mangroves du Sénégal et de la Gambie. Ecologie, Pédologie, Géochimie, Mise en Valeur et Aménagement. Travaux et Documents de l'ORSTOM n° 193, ORSTOM, Paris, 357 pp.

Mascaro, I., Benvenuti, B., Corsini, F., Costagliola, P., Lattanzi, P., Parrini, P. & Tanelli, G., 2001. Mine wastes at the polymetallic deposit of Fenice Capanne (southern Tuscany, Italy). Mineralogy, geochemistry, and environmental impact. Environmental Geology 41, 417-429.

McSweeney, K. & Fastovsky, D.E., 1987. Micromorphological and SEM analysis of Cretaceous-Paleogene Petrosols from eastern Montana and western North Dakota. Geoderma 40, 49-63.

Mees, F., Masalehdani, M.N.N., De Putter, T., D'Hollander, C., Van Biezen, E., Mujinya, B.B., Potdevin, J.L. & Van Ranst, E., 2013. Concentrations and forms of heavy metals around two ore processing sites in Katanga, Democratic Republic of Congo. Journal of African Earth Sciences 77, 22-30.

Mermut, A.R. & Arshad, M.A., 1987. Significance of sulfide oxidation in soil salinization in southeastern Saskatchewan. Soil Science Society of America Journal 51, 247-251.

Mermut, A.R., Curtin, D. & Rostad, H.P.W., 1985. Micromorphological and submicroscopical features related to pyrite oxidation in an inland marine shale from east central Saskatchewan. Soil Science Society of America Journal 49, 256-261.

Meyer, R. & Pena dos Reis, R.B., 1985. Paleosols and alunite silcretes in continental Cenozoic of western Portugal. Journal of Sedimentary Petrology 55, 76-85.

Miedema, R., Jongmans, A.G. & Slager, S., 1974. Micromorphological observations on pyrite and its oxidation products in four Holocene alluvial soils in the Netherlands. In Rutherford, G.K. (ed.), Soil Microscopy. The Limestone Press, Kingston, pp. 772-794.

Miller, K., 2011. Jarosite morphology as indicator of water saturation levels on Mars. Abstracts of the Meeting on the Importance of Solar System Sample Return Missions to the Future of Planetary Science, n° 5015, 1 p.

Milton, C., Conant, L.C. & Swanson, V.E., 1955. Sub-Chattanooga residuum in Tennessee and Kentucky. Geological Society of America Bulletin 66, 805-810.

Montoroi, J.P., 1995. Mise en évidence d'une séquence de précipitation des sels dans les sols sulfatés acides d'une vallée aménagée de Basse-Casamance (Sénégal). Comptes Rendus de l'Académie des Sciences 320, 395-402.

Moormann, F.R. & Eswaran, H., 1978. A study of a paleosol from east Nigeria. Pedologie 28, 251-270.

Nickel, E.H., 1980. Ring crystals of osarizawaite from Whim Creek, Western Australia. American Mineralogist 65, 1287-1290.

Nickel, E.H., 1984. An unusual pyrite-sulphur-jarosite assemblage from Arkaroola, South Australia. Mineralogical Magazine 48, 139-142.

Onac, B.P., Veres, D.S., Kearns, J., Chirienco, M., Minut, A. & Breban, R., 2003. Secondary sulfates found in an old adit from Rosia Montana, Romania. Studia Universitatis Babes-Bolyai Geologia 48, 29-44.

Osterrieth, M., Borrelli, N., Alvarez, M.F., Nóbrega, G.N., Machado, W. & Ferreira, T.O., 2016. Iron biogeochemistry in Holocene palaeo and actual salt marshes in coastal areas of the Pampean Plain, Argentina. Environmental Earth Sciences 75, n° 672, 12 p.

Papike, J.J., Burger, P.V., Karner, J.M., Shearer, C.K. & Lueth, V.W., 2007. Terrestrial analogs of martian jarosites: major, minor element systematics and Na-K zoning in selected samples. American Mineralogist 92, 444-447.

Parafiniuk, J., 1991. Fibroferrite, slavikite and pickeringite from the oxidation zone of pyrite-bearing schists in Wiesciszowice (Lower Silesia). Mineralogia Polonica 22, 3-15.

Paramananthan, S., Sooryanarayana, V., Syed Sofi, S.O. & Eswaran, H., 1978. Micromorphology of some soils developed on marine clays in Peninsular Malaysia. In Delgado, M. (ed.), Soil Micromorphology. University of Granada, pp. 589-609.

Park, S. & Kim, Y., 2016. Mineralogical changes and distribution of heavy metals caused by the weathering of hydrothermally altered, pyrite-rich andesite. Environmental Earth Sciences 75, n° 1125, 16 p.

PiPujol, M.D. & Buurman, P., 1994. The distinction between ground-water gley and surface-water gley phenomena in Tertiary paleosols of the Ebro basin, NE Spain. Palaeogeography, Palaeoclimatology, Palaeoecology 110, 103-113.

Plumlee, G.S., Benzel, W.M., Hoefen, T.M., Hageman, P.L., Morman, S.A., Reilly, T.J., Adams, M., Berry, C. J., Fischer, J.M. & Fisher, I., 2016. Environmental implications of the use of sulfidic back-bay sediments for dune reconstruction – lessons learned post Hurricane Sandy. Marine Pollution Bulletin 107, 459-471.

Poch, R.M., Fitzpatrick, R.W., Thomas, B.P., Merry, R.H., Self, P.G. & Raven, M.D., 2004. Contemporary and relict processes in a coastal acid sulfate soil sequence: microscopic features. SuperSoil 2004, 3rd Australian New Zealand Soils Conference, University of Sydney (CD-ROM), 8 p.

Poch, R.M., Thomas, B.P., Fitzpatrick, R.W. & Merry, R.H., 2009. Micromorphological evidence for mineral weathering pathways in a coastal acid sulfate soil sequence with Mediterranean-type climate, South Australia. Australian Journal of Soil Research 47, 403-422.

Poch, R.M., Artieda, O., Herrero, J. & Lebedeva-Verba, M., 2010. Gypsic features. In Stoops, G., Marcelino, V. & Mees, F. (eds.), Interpretation of Micromorphological Features of Soils and Regoliths. Elsevier, Amsterdam, pp. 195-216.

Poch, R.M., Artieda, O. & Lebedeva, M., 2018. Gypsic features. In Stoops, G., Marcelino, V. & Mees, F. (eds.), Interpretation of Micromorphological Features of Soils and Regoliths. Second Edition. Elsevier, Amsterdam, pp. 259-287.

Pons, L.J., 1964. A quantitative microscopical method of pyrite determination in soils. In Jongerius, A. (ed.), Soil Micromorphology. Elsevier, Amsterdam, pp. 401-410.

Pons, L.J., 1970. Acid sulphate soils (soils with cat clay phenomena) and the prediction of their origin from pyrites muds. In Field to Laboratory, Fysisch Geografisch en Bodemkundig Laboratorium, Amsterdam, Publication 16, pp. 93-107.

Postma, D., 1982. Pyrite and siderite formation in brackish and freshwater swamp sediments. American Journal of Science 282, 1151-1183.

Prakongkep, N., Gilkes, B., Singh, B. & Wong, S., 2012. Pyrite and other sulphur minerals in giant aquic spodosols, Western Australia. Geoderma 181/182, 78-90.

Pugh, C.E., Hossner, L.R. & Dixon, J.B., 1981. Pyrite and marcasite surface area as influenced by morphology and particle diameter. Soil Science Society of America Journal 45, 979-982.

Puura, E., Neretnieks, I. & Kirsimäe, K., 1999. Atmospheric oxidation of the pyritic waste rock in Maardu, Estonia. 1. Field study and modelling. Environmental Geology 39, 1-19.

Rabenhorst, M.C., 1990. Micromorphology of induced iron sulfide formation in a Chesapeake Bay (USA) tidal marsh. In Douglas, L.A. (ed.), Micromorphology: A Basic and Applied Science. Developments in Soil Science, Volume 19. Elsevier, Amsterdam, pp. 303-310.

Rabenhorst, M.C. & Fanning, D.S., 1989. Pyrite and trace metals in glauconitic parent materials of Maryland. Soil Science Society of America Journal 53, 1791-1797.

Rabenhorst, M.C. & Haering, K.C., 1990. Soil micromorphology of a Chesapeake Bay tidal marsh: implications for sulfur accumulation. Soil Science 147, 329-347.

Rabenhorst, M.C. & James, B.R., 1992. Iron sulfidization in tidal marsh soils. In Skinner, H.G.W. & Fitzpatrick, R.W. (eds.), Biomineralization Processes of Iron and Manganese – Modern and Ancient Environments. Catena Supplement 21, 203-217.

Rickard, D.T., 1970. The origin of framboids. Lithos 3, 269-293.

Ritsema, C.J. & Groenenberg, J.E., 1993. Pyrite oxidation, carbonate weathering, and gypsum formation in a drained potential acid sulfate soil. Soil Science Society of America Journal 57, 968-976.

Romero A., González, I. & Galán, E., 2006. The role of efflorescent sulfates in the storage of trace elements in stream waters polluted by acid mine-drainage: the case of Peña del Hierro, southwestern Spain. Canadian Mineralogist 44, 1431-1446.

Roychoudhury, A.N., Kostka, J.E. & Van Cappellen, P., 2003. Pyritization: a palaeoenvironmental and redox proxy reevaluated. Estuarine, Coastal and Shelf Science 57, 1183-1193.

Sawlowicz, Z., 1993. Pyrite framboids and their development: a new conceptual mechanism. Geologische Rundschau 82, 148-156.

Schaefer, C.E.G.R., Simas, F.N.B., Gilkes, R.J., Mathison, C., da Costa, L.M. & Albuquerque, M.A., 2008. Micromorphology and microchemistry of selected Cryosols from maritime Antarctica. Geoderma 144, 104-115.

Sclar, C.B., 1961. Decomposition of pyritized carbonaceous shale to halotrichite and melanterite. American Mineralogist 46, 754-756.

Shamshuddin, J. & Auxtero, E.A., 1991. Soil solution compositions and mineralogy of some active acid sulfate soils in Malaysia as affected by laboratory incubation with time. Soil Science 152, 365-376.

Shamshuddin, J., Paramananthan, S. & Mokhtar, N., 1986. Mineralogy and surface charge properties of two acid sulfate soils from Peninsular Malaysia. Pertanika 9,167-176.

Shamshuddin, J., Muhrizal, S., Fauziah, I. & Van Ranst, E., 2004. Laboratory study of pyrite oxidation in acid sulfate soils. Communications in Soil Science and Plant Analysis 35, 117-129.

Shamshuddin, J., Elisa Azura, A., Shazana, M.A.R.S., Fauziah, C.I., Panhwar, Q.A. & Naher, U.A., 2014. Properties and management of acid sulfate soils in Southeast Asia for sustainable cultivation of rice, oil palm, and cocoa. Advances in Agronomy 124, 91-142.

Shand, P., Gotch, T., Love, A., Ravena, M., Priestley, S. & Grocke, S., 2016. Extreme environments in the critical zone: linking acidification hazard of acid sulfate soils in mound spring discharge zones to groundwater evolution and mantle degassing. Science of the Total Environment 568, 1238-1252.

Shaw, S.C., Groat, L.A., Jambor, J.L., Blowes, D.W., Hanton-Fong, C.J. & Stuparyk, R.A., 1998. Mineralogical study of base metal tailings with various sulfide contents, oxidized in laboratory columns and field lysimeters. Environmental Geology 33, 209-217.

Simas, F.N.B., Schaefer, C.E.G.R., Melo, V.F., Guerra, M.B.B., Saunders, M. & Gilkes, R.J., 2006. Clay-sized minerals in permafrost-affected soils (Cryosols) from King George Island, Antarctica. Clays and Clay Minerals 54, 721-736.

Simón, M., Dorronsoro, C., Ortiz, I., Martin, F. & Aguilar, J., 2002. Pollution of carbonate soils in a Mediterranean climate due to a tailings spill. European Journal of Soil Science 53, 321-330.

Skwarnecki, M.S. & Fitzpatrick, R.W., 2008. Regional geochemical dispersion in materials associated with acid sulfate soils in relation to base-metal mineralisation of the Kanmantoo Group, Mt Torrens-Strathalbyn region, eastern Mt Lofty Ranges, South Australia. CRC LEME Open File Report No. 205, 45 pp.

Slager, S., Jongmans, A.G. & Pons, L.J., 1970. Micromorphology of some tropical alluvial clay soils. Journal of Soil Science 21, 233-241.

Soil Survey Staff, 1999. Soil Taxonomy. A Basic System of Soil Classification for Making and Interpreting Soil Surveys, 2nd Edition. USDA, Natural Resources Conservation Service, Agriculture Handbook Number 436. U.S. Government Printing Office, Washington, 871 p.

Soil Survey Staff, 2014. Keys to Soil Taxonomy, 12th Edition. USDA, Natural Resources Conservation Service, Washington, 360 p.

Spaargaren, O.C., Creutzberg, D., van Reeuwijk, L.P. & van Diepen, C.A., 1981. Thionic Fluvisol (Sulfic Tropaquept). Central Plain Region, Thailand. Soil Monolith Paper no. 1, International Soil Museum, Wageningen, 48 p.

Sracek, O., Veselovsky, F., Kríbek, B., Malec, J. & Jehlicka, J., 2010. Geochemistry, mineralogy and environmental impact of precipitated efflorescent salts at the Kabwe Cu-Co chemical leaching plant in Zambia. Applied Geochemistry 25, 1815-1824.

Stoops, G., 1983. SEM and light microscopic observations of minerals in bog-ores of the Belgian Campine. Geoderma 30, 179-186.

Stoops, G., 1992. Micromorphological study of pre-Cretaceous weathering in the Brabant Massif (Belgium). In Schmitt, J.M. & Gall, Q. (eds.), Mineralogical and Geochemical Records of Paleoweathering. ENSMP, Mémoires des Sciences de la Terre 18, 69-84.

Stoops, G., 2003. Guidelines for Analysis and Description of Soil and Regolith Thin Sections. Soil Science Society of America, Madison, 184 p.

Stoops, G., Eswaran, H. & Abtahi, A., 1978. Scanning electron microscopy of authigenic sulphate minerals in soils. In Delgado, M. (ed.), Soil Micromorphology. University of Granada, pp. 1093-1113.

Strawn, D., Doner, H., Zavarin, M. & McHugo, S., 2002. Microscale investigation into the geochemistry of arsenic, selenium, and iron in soil developed in pyritic shale materials. Geoderma 108, 237-257.

Sucha, V., Dubikova, M., Cambier, P., Elsass, F. & Pernes, M., 2002. Effect of acid mine drainage on the mineralogy of a dystric cambisol. Geoderma 110, 151-167.

Sukitprapanona, T., Suddhiprakarn, A., Kheoruenromne, I., Anusontpornperm, S. & Gilkes, R.J., 2014. Mineralogy and geochemistry of some Thai acid sulfate soils. Abstracts of the 23rd Australian Clay Minerals Conference, Perth, pp. 11-15.

Sweeney, R.E. & Kaplan, I.R., 1973. Pyrite framboid formation: laboratory synthesis and marine sediments. Economic Geology 68, 618-634.

Szakáll, S., Foldvári, M., Papp, G., Kovács-Pálffy, P. & Kovács, A., 1997. Secondary sulphate minerals from Hungary. Acta Mineralogica-Petrographica 38, Supplement, 7-63.

Tien, P., 1968. Hydrobasaluminite and basaluminite in Cabaniss Formation (Middle Pennsylvanian), southeastern Kansas. American Mineralogist 53, 722-732.

Tovey, N.K., 1986. Microfabric, chemical and mineralogical studies of soils: techniques. Geotechnical Engineering 17, 131-166.

Tovey, K. & Dent, D., 2002. Microstructure and microcosm chemistry of tidal soils. Proceedings of the 17th World Congress of Soil Sciences, Bangkok, Paper no. 892, 7 p.

Tsatskin, A. & Nadel, D., 2003. Formation processes at the Ohalo II submerged prehistoric campsite, Israel, inferred from soil micromorphology and magnetic susceptibility studies. Geoarchaeology 18, 409-432.

Valente, T.M. & Gomes, C.L., 2009. Occurrence, properties and pollution potential of environmental minerals in acid mine drainage. Science of the Total Environment 407, 1135-1152.

van Breemen, N., 1976. Genesis and Solution Chemistry of Acid Sulfate Soils in Thailand. Agricultural Research Reports no. 848, Pudoc, Wageningen, 263 p.

van Breemen, N., 1982. Genesis, morphology, and classification of acid sulphate soils in coastal plains. In Kittrick, J.A., Fanning, D.S. & Hosner, L.R. (eds.), Acid Sulphate Weathering. Soil Science Society of America Special Publication 10, SSSA, Madison, pp. 95-108.

van Breemen, N. & Harmsen, K., 1975. Translocation of iron in acid sulfate soils: I. Soil morphology, and the chemistry and mineralogy of iron in a chronosequence of acid sulfate soils. Soil Science Society of America Proceedings 39, 1140-1148.

van Dam, D. & Pons, L.J., 1973. Micropedological observations on pyrite and its pedological reaction products. In Drost, H. (ed.), Acid Sulphate Soils. II. Research Papers. Publication 18, International Institute for Land Reclamation and Improvement, Wageningen, pp. 169-196.

Wada, H. & Seisuwan, B., 1988. The process of pyrite formation in mangrove soils. In Dost, H. (ed.), Selected Papers of the Dakar Symposium on Acid Sulphate Soils. Publication 44, International Institute for Land Reclamation and Improvement, Wageningen, pp. 24-37.

Wagner, D.P., Fanning, D.S., Foss, J.E., Patterson, M.S. & Snow, P.A., 1982. Morphological and mineralogical features related to sulfide oxidation under natural and disturbed land surfaces in Maryland. In Kittrick, J.A., Fanning, D.S. & Hosner, L.R. (eds.), Acid Sulphate Weathering. Soil Science Society of America Special Publication 10, SSSA, Madison, pp. 109-125.

Wallace, L., Welch, S.A., Kirste, D., Beavis, S. & McPhail, D.C., 2005. Characteristics of inland acid sulfate soils of the lower Murray floodplains, South Australia. In Roach, I.C. (ed.), Regolith 2005 – Ten Years of CRC LEME. CRC LEME, Kensington, pp. 326-328.

Wiese, R.G., Powell, M.A. & Fyfe, W.S., 1987. Spontaneous formation of hydrated iron sulfates on laboratory samples of pyrite- and marcasite-bearing coals. Chemical Geology 63, 29-38.

Wierzchos, J., García-González, M.T. & Ascaso, C., 1995. Advantages of application of the backscattered electron scanning image in the determination of soil structure and soil constituents. International Agrophysics 9, 41-47.

Wilkin, R.T. & Barnes, H.L., 1997. Formation processes of framboidal pyrite. Geochimica et Cosmochimica Acta 61, 323-339.

Wilkin, R.T., Barnes, H.L. & Brantley, S.L., 1996. The size distribution of framboidal pyrite: an indicator of redox conditions. Geochimica et Cosmochimica Acta 60, 3897-3912.

Willett, I.R., Crockford, R.H. & Milnes, A.R., 1992. Transformation of iron, manganese and aluminium during oxidation of a sulfidic material from an acid sulfate soil. In Skinner, H.G.W. & Fitzpatrick, R.W. (eds.), Biomineralization Processes of Iron and Manganese – Modern and Ancient Environments. Catena Supplement 21, 287-302.

Wright, V.P., Vanstone, S.D. & Marshall, J.D., 1997. Contrasting flooding histories of Mississippian carbonate platforms revealed by marine alteration effects in palaeosoils. Sedimentology 44, 825-842.

Chapter 14

Textural Pedofeatures and Related Horizons

Peter Kühn[1], José Aguilar[2], Rienk Miedema[3], Maria Bronnikova[4]

[1]*EBERHARD KARLS UNIVERSITÄT TÜBINGEN, TÜBINGEN, GERMANY;*
[2]*UNIVERSITY OF GRANADA, GRANADA, SPAIN;*
[3]*WAGENINGEN UNIVERSITY, WAGENINGEN, THE NETHERLANDS (RETIRED);*
[4]*INSTITUTE OF GEOGRAPHY, RUSSIAN ACADEMY OF SCIENCES, MOSCOW, RUSSIA*

CHAPTER OUTLINE

Interpretation of Micromorphological Features of Soils and Regoliths. https://doi.org/10.1016/B978-0-444-63522-8.00014-0

1. Introduction

Textural pedofeatures are characterised by a difference in grain size with the adjacent groundmass and comprise coatings, hypocoatings, infillings and intercalations. Because the most prominent textural pedofeatures are clay coatings, this chapter focuses mainly on this type of textural feature, its development and occurrence. Information about other textural features, such as coarse-grained coatings and depletion hypocoatings, is reviewed as well.

Clay illuviation was one of the first pedogenic processes to be recognised in micro-pedology (Kubiëna, 1943; Frei & Cline, 1949), and the resulting features may first have been documented by Agafonoff (1936, p. 73). This process is associated with the removal of clay from upper horizons and its redeposition as clay coatings and infillings in deeper horizons. The fundamental micromorphological characteristics of clay coatings and their occurrence were first described in detail by Minashina (1958) and Stephen (1960). Much earlier, in 1915, key micromorphological features of clay coatings and infillings had already been described by Polynov, who considered these features to have formed by mineral authigenesis (Polynov, 1915). Next was Kubiëna (1937), who described clay coatings, without using this term, as 'a layer of fine particles with varnish-like lustre covering voids' and used the expression 'fluidal structure' (Fluidalgefüge) to describe this feature. The terms 'optically oriented clay' (e.g., Brewer & Blackmore, 1956) and 'clay films' (e.g., Miedema et al., 1999) have also been used synonymously. Several other terms have been used in micromorphological publications to describe clay coatings (Table 1).

Clay coating formation due to neoformation of clay - first proposed by Polynov (1915) - was a widely accepted explanation amongst early Russian micromorphologists (e.g., Parfenova & Yarilova, 1957). This presumed a complete decay of minerals in the groundmass, setting free silicium, aluminium or iron oxides, which moved slowly down the profile. They were supposed to recombine in the course of this movement, forming new authigenic clay minerals (Parfenova et al., 1964). These neoformed clay minerals were named polynite, after B.B. Polynov.

Table 1 Terms used in literature to indicate coatings of well-oriented clay

Term	Author
Birefringent streaks (*'doppelbrechende Schlieren'*)	Kubiëna (1953)
Flow plasma (*'Fließplasma'*)	Kubiëna (1956)
Polynite	Parfenova & Yarilova (1957)
Clay skins	Buol & Hole (1959, 1961)
Argillans and ferriargillans	Brewer (1964)
Transparent collomorphic clay	Parfenova & Yarilova (1965)
Clay flow accumulations	Dobrovol'ski (1991)

2. Clay Coatings

2.1 General Characteristics

Clay coatings (Fig. 1) are typically characterised by a continuous orientation of clay particles parallel to the surface with which they are associated. Since the thickness of a thin section is about 30 μm, the light passes through many clay particles. If these are randomly oriented, the effects of all individual particles, in terms of birefringence, will compensate one another, resulting in optically isotropic behaviour.

If clay particles are oriented parallel to one another, addition of the effects of the individual particles will take place and the domain will act as a single crystal with distinctive optical properties (e.g., Fernández et al., 2006). Therefore, clay coatings display parallel extinction and positive elongation, relative to the covered surface.

FIGURE 1 Limpid illuvial clay coatings with characteristic extinction bands (XPL). Note microlaminations in the clay coating (right).

If the coating is curved, extinction lines sweep through it when the stage is rotated (e.g., FitzPatrick, 1993).

Interference colours depend on the mineralogical composition: first-order grey indicates 1:1 or 2:1:1 phyllosilicates, whereas higher interference colours indicate 2:1 phyllosilicates. Interference colours may be modified or masked by various organic and inorganic compounds. Pure clay coatings are greyish, but in most cases the presence of iron oxides gives them a yellowish or reddish colour. Polynov (1915), Parfenova and Yarilova (1957) and Stephen (1960) mentioned distinct pleochroism of some clay coatings, e.g., from dark brown to yellowish brown. Nevertheless no attention has been paid to this phenomenon and the descriptions of clay coatings usually have no data on pleochroism, although it appears to be common in palaeosoils (Kühn, unpublished data).

Other common characteristics of clay coatings in thin sections include (i) a marked textural contrast with the adjacent groundmass, in both PPL and XPL (this may be less clear in some clayey soils), with sharp boundaries, (ii) their location on the surface of peds and other macropore walls, (iii) the presence of microlaminations, and (iv) the presence of organic particles reflected by darker colours and a more speckled appearance. The third and fourth of these characteristics are common but not ubiquitous.

2.2 Confusion With Micromorphologically Similar Features

In general, illuvial clay coatings can be easily identified by the aforementioned characteristics (Section 2.1). However, they may occasionally be confused with other features, such as those described below.

Authigenic clay coatings − Coatings formed by clay neoformation can in some cases be distinguished based on the orientation of the clay particles, because it is occasionally perpendicular to the pore walls in authigenic clay coatings and always parallel to the pore walls for illuvial clay coatings (Bullock & Thompson, 1985; Mulyanto & Stoops, 2003; Stoops, 2003; Kühn & Pietsch, 2013; see also Mees, 2010, 2018). Neoformation of clay is often observed in soils with a high percentage of volcanic fragments (Jongmans et al., 1994). Optically isotropic thin clay coatings and infillings in buried Andosols were described as authigenic precipitates by Sedov et al. (2003), who interpret their optical behaviour as indicative of their amorphous nature or of poor orientation of halloysite-dominant clay. Isotropic clay coatings partly recrystallised to birefringent clay coatings were described in Oxisols on andesitic volcanic material in Indonesia (Buurman & Jongmans, 1987). However, isotropic coatings, as well as optically anisotropic features along pore walls, can also be artifacts of resin polymerisation during sample impregnation for thin section preparation (Morrás, 1983a).

Pressure faces − In the field it is difficult to distinguish between intrusive clay coatings and pressure faces (slickensides), which are formed by orientation of clay particles due to repeated swelling and shrinking in clayey materials. However, in thin sections pressure faces show characteristics different from clay coatings. In PPL, pressure faces have diffuse boundaries and no occurrences of lamination are visible. In XPL, the presence of pressure faces is indicated by porostriation along the sides of peds (see also Kovda & Mermut, 2010, 2018) and extinction lines, which are typical of clay coatings, are absent. Stress-deformed clay coatings are somewhat more difficult to distinguish from pressure faces than undisturbed coatings in clayey soils. Nevertheless, clay coatings usually contain more fine clay than the groundmass, whereas pressure faces have the same texture as the groundmass.

Goethite – Goethite surrounding pores and peds may display an apparent orientation and interference colours similar to those of illuvial clay and may even be laminated (Bullock & Thompson, 1985; Sauer et al., 2013). At high magnification a distinction is possible, if goethite crystals are identifiable (see also Stoops & Marcelino, 2010, 2018).

Straw – In surface horizons, radial sections through straw may look similar to channel coatings: egg-yellow colour, limpid, grey interference colours masked by the colour of the material (Stoops, 2003). Careful examination of the morphology, however, can avoid mistakes.

Altered rock fragments and mineral grains – Aggregates consisting of strongly parallel-oriented fine clay, formed by weathering of phyllosilicate grains (often biotite) or shales (Fig. 2) (e.g., Mermut & Pape, 1973) may be confused with fragments of illuvial clay coatings. Fragments of (unaltered) clayey parent materials, e.g., in colluvial soils or fragments of surface crusts, may also have the appearance of fragments of clay coatings (Fig. 2). Opaline and/or allophanic bodies with micromorphological characteristics similar to those of fragments of clay coatings are common in some Andosols (see Sedov et al., 2010; Stoops et al., 2018).

Phosphate coatings and infillings – These pedofeatures occur in waterlogged environments and in soils enriched in phosphates by bird or bat droppings or through various types of human activity. Their occurrence and nature have been described by Goldberg and Macphail (2006) and Karkanas and Goldberg (2010, 2018). Amorphous and cryptocrystalline phosphate pedofeatures commonly have colours similar to those typical

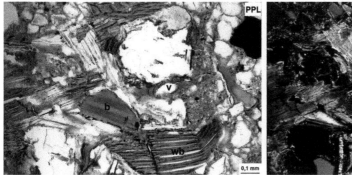
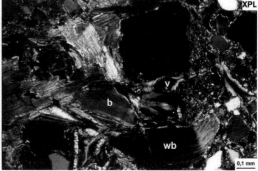

FIGURE 2 Coating of well-oriented, yellow-brown clay around a channel (v), weathered (wb) and fresh biotite (b) grains in cross-sections parallel and perpendicular to the *c*-axis. Grain b could be confused with a fragment of a clay coating.

for clay coatings or infillings (colourless, yellow, brown, reddish), with first-order white to second-order yellow interference colours. However, they can be easily distinguished from clay coatings by their characteristic green colour under blue light fluorescence. In addition, phosphate coatings have often a radiating fanlike internal orientation.

2.3 Formation

2.3.1 General Aspects

Based on evidence from detailed micromorphological studies from a wide variety of soils and experimental studies, both in laboratory conditions and in the field, the formation of clay coatings in B horizons is related with clay eluviation/illuviation processes and generally attributed to the vertical translocation of fine clay suspended in percolating soil water (Altemüller, 1956; Brewer & Haldane, 1957; Karpachevskiy, 1960; Buol & Hole, 1961; Brewer & Sleeman, 1970; Bullock et al., 1974; Gagarina & Tsyplenkov, 1974; De Coninck et al., 1976; Dobrovol'ski et al., 1976; Howitt & Pawluk, 1985; Theocharopoulos & Dalrymple, 1987; Ponomarenko et al., 1988; Stephan, 2000). This is supported by non-micromorphological evidence, such as the observation that clay coatings and the clay fraction of the eluvial horizon commonly have similar mineralogical compositions, whereas the mineralogical compositions of clay coatings and the clay fraction of the peds in the illuvial horizon can be different (Khalifa & Buol, 1968; Ranney & Beatty, 1969).

In other studies, clay coatings in Retisols and Luvisols have been demonstrated to include a large relative amount of minerals with unstable lattice, so that their quantitative mineralogical composition is intermediate between that of the clay fraction of E and Bt horizons (Grossman et al., 1964; Miller et al., 1971; Targulian et al., 1974b; Sokolova et al., 1988). In other cases, the quantitative mineralogical composition of clay coatings (fraction <1 μm) reveals some variability but is generally closer to the mineralogical composition of the same fraction in the groundmass of the related Bt horizon than to that of the overlying A and E horizons (Beke & Zwarich, 1971; Ogleznev, 1971; Targulian et al., 1974b; Rusanova, 1978; Birina, 1980; Targulian & Tselischeva, 1983; Bronnikova & Targulian, 2005). The quantitative mineralogical composition of the 10-50 μm and 50-250 μm fractions of silt-clay coatings in Retisols is closer to the composition of the same fractions in the eluvial horizons than in groundmass of the Bt horizons (Bronnikova & Targulian, 2005). The authors hypothesise alteration of less stable minerals in the upper horizons caused by acid hydrolysis, accompanied by clay eluvation, followed by eluvation of other components of the fine earth fraction of the eluvial horizon, implying that clay illuviation started earlier than vertical silt-clay and sand-clay translocation in the studied profiles. This conclusion is supported by micromorphological evidence obtained by semiquantitative analysis of layer sequences within compound coatings (Bronnikova & Targulian, 2005).

However, differences between the clay mineral associations in clay coatings and in peds of Bt horizons can be also related to the difficulty in obtaining pure samples of clay coatings (Beke & Zwarich, 1971).

Although clay coatings exclusively composed of coarse clay have been reported (e.g., Eswaran, 1979), and although clay coatings commonly contain a more or less considerable admixture of silt and even sand particles (Fedoroff, 1972; Targulian et al., 1974b), the eluviation/illuviation process usually involves fine clay particles, smaller than 0.2 µm (Grossman et al., 1964; Kovenya, 1972). This process can be subdivided into the following three stages (Blume, 1964; Duchaufour, 1982; Stephan, 2000; Miedema, 2002), favoured by alternating wet and dry conditions (Gombeer & D'Hoore, 1971):

Dispersion – First, clay dispersion is required, because most clay particles are aggregated in small clusters up to 250 µm (Edwards & Bremner, 1967). The main factors affecting clay dispersion and thus its mobilisation are the type of clay minerals, the particle size, the presence or absence of cementing agents (e.g., iron oxides, carbonates, humus), pH, the types of cations present, the electrolyte concentration in the soil solution and its content in dissolved organic matter (e.g., Atanasescu et al., 1974; Duchaufour, 1982; Dorronsoro & Aguilar, 1988; van den Broek, 1989).

Downward transport – Water, carrying clay particles in suspension, percolates through conducting pores and infiltrates into deeper, drier horizons as laminar flow along the pore walls (Theocharopoulos & Dalrymple, 1987). Experiments with radioactive tracers showed more difficult particle (<0.5 µm) movement in materials with a large specific surface (Melnikov & Kovenya, 1974). Clay particles are transported without undergoing any major chemical or physical transformation. However, the presence and concentrations of certain cations in solution influence the movement of kaolinite and montmorillonite differently, whereas some ion species have no effect on the mobility of clay (Hallsworth, 1963). In calcareous soils, high Ca^{2+} concentrations may inhibit clay illuviation (e.g., Elliot & Drohan, 2009), but this is not always the case (e.g., Pal et al., 2003). Dispersion of clay in non-calcareous A horizons of calcareous soils can be sufficient to develop illuvial clay coatings in lower parts of the profile (Poch et al., 2013).

Because of the higher suction by micropores compared to macropores, water is absorbed into unsaturated materials. Since the clay particles cannot enter the micropores, a thin film of suspension coats the macropore walls (Bullock & Thompson, 1985; Dorronsoro & Aguilar, 1988; Fernández et al., 2006). The greater the depth reached by the suspension, the higher its concentration in clay (Gombeer & D'Hoore, 1971). The translocation of clay suspensions can reach several metres (e.g., Teruggi & Andreis, 1971; Goss et al., 1973; Targulian et al., 1974a; Johnson et al., 2003; Moretti & Morrás, 2013). The main factors involved are the amount and the velocity of percolating water, the concentration of suspended clay and the soil porosity.

Deposition – Clay deposition occurs when (i) the infiltration of the suspension stops because the water supply has stopped, particularly if percolating water is hindered by active roots at the depth of their maximum concentration (Runge, 1973), (ii) a level with low macroporosity or entrapped air is reached, or (iii) a higher electrolyte concentration causes flocculation (Dorronsoro & Aguilar, 1988). Also a decrease in water velocity may result in the deposition of suspended clay (Duchaufour, 1982). Upon suction of the water into fine voids, clay particles are filtered by the pore walls and a clay coating is formed with clay particles oriented parallel to the walls (Bullock & Thompson, 1985). In horizontal channels clay coatings have usually the same thickness at the upper and lower side (Fig. 3) (Van Ranst et al., 1980; Dorronsoro & Aguilar, 1988).

During the next wet period, this process may be repeated, resulting in an increasingly thicker clay coating. Besides suction, evaporation is also an important process. This is illustrated by the formation of clay coatings in sandy materials, which form by slow evaporation causing a gradual reduction of the water menisci and the retention of clay suspensions on the surface of sand grains by adsorption and capillary forces (Van Ranst et al., 1980; Sullivan, 1994). Eventually, very stable, concave (Fig. 4) and sometimes convex clay bridges between sand grains are formed.

Experiments with drainage of clay suspensions (Sullivan, 1994) through columns of coarse quartz sand show that clay coatings can also develop from flow of suspensions within wet soil material. This suggests that dry soil material is not required for the formation of coatings of well-oriented clay. The thickness of clay coatings formed on impermeable or wet sandy material is a direct function of the clay concentration of the suspension, which implies that the higher the clay concentration, the thicker the clay coatings in this type of material.

Thick clay coatings with sharp boundaries, continuous orientation and sharp extinction bands have been also experimentally obtained by repeated addition of the same clay suspension to Bt blocks after 40 application/drying cycles (Akamigbo & Dalrymple, 1985) or 100 cycles applied on synthetic sponges (Theocharopoulos & Dalrymple, 1987).

It should be stressed that not all clay coatings point to a process of clay translocation from one horizon into another, i.e., that the presence of clay coatings does not automatically imply the (former) occurrence of an overlying eluvial horizon. Vertical or lateral translocations within the same horizon are probably more common than presumed (Laves & Thiere, 1970; Mermut & Pape, 1971; Laves, 1972; Smith & Wilding, 1972; Targulian et al., 1974a; Kovalev, 1975; Baize, 1989; van den Broek, 1989) but are

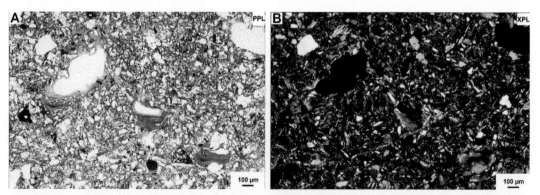

FIGURE 3 Dusty yellow-brown clay and silt-clay coatings (thicker along the lower side of the channels) and coatings of well-oriented, limpid, yellow-brown clay with nearly the same thickness. Note that both types of coatings occur also as part of layered textural coatings, indicating different phases of clay illuviation.

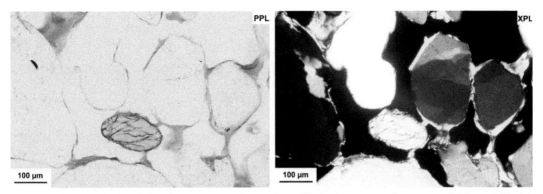

FIGURE 4 Gefuric c/f-related distribution pattern in sandy argillic horizon. Note the concave bridges of illuvial yellow-brown clay with extinction bands.

morphologically not distinguishable from clay coatings resulting from accumulation of clay derived from A and E horizons.

Snow meltwater can be responsible for the formation of clay coatings and layered clay coatings as described for surficial sediments of the Transantarctic Mountains (van der Meer et al., 1993; van der Meer & Menzies, 2011), showing that clay illuviation can occur even in areas with low temperatures, which complicates the interpretation of clay coatings in terms of palaeoclimatic conditions.

2.3.2 Specific Types of Clay Coatings

Independent of the parent material, the occurrence and characteristics of clay coatings may indicate different (in many cases at least three) phases of clay illuviation leading to the formation of Bt horizons (e.g., Thiere & Laves, 1968; Weir et al., 1971; Fedoroff, 1972; Eimberck-Roux, 1977; Bullock & Murphy, 1979; Felix-Henningsen, 1979; Fedoroff & Courty, 1987; Tursina, 1989; Payton, 1992, 1993; Van Vliet-Lanoë et al., 1992; Rogaar et al., 1993; Kemp et al., 1998; Kühn, 2003b; Bronnikova & Targulian, 2005). Different types of clay coatings, with different limpidity and degree of sorting, may develop depending on the grain size and type of compounds involved and on the phase of the clay translocation process.

Very dusty, greyish-black clay coatings (commonly with a small amount of silt), moderately to well sorted, usually with weak parallel orientation of the clay particles (diffuse or absent extinction pattern, low brightness of interference colours), sometimes microlaminated, are preferentially found covering the lower part of horizontal channels (Fig. 5) (Kühn, 2003a) or as the inner layer of layered coatings.

The dark colour is mainly caused by finely dispersed organic matter. They mostly correspond to the youngest phase of clay illuviation, typically in E and upper Bt horizons of loamy material (Gagarina & Tsyplenkov, 1974; Kühn, 2003a), below plough layers (see Section 4.3) or in colluvial materials (Kwaad & Mücher, 1977, 1979; Bolt et al., 1980). These very dusty clay coatings also occur in Luvic Phaeozems of deciduous forests

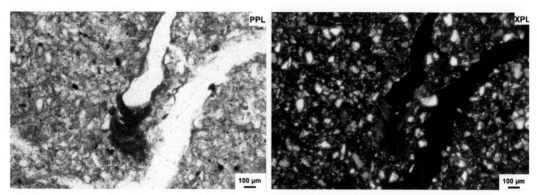

FIGURE 5 Very dusty clay coatings with moderately oriented clay particles on the lower side of a channel.

(Urusevskaya et al., 1987; Gerasimova et al., 1996), as well as in natric horizons of Solonetzes (Morrás, 1978; Gerasimova et al., 1996; Lebedeva-Verba & Gerasimova, 2009). The occurrence of dark clay coatings rich in organic matter can be related to degradation of mollic horizons (Alexandrovskiy & Birina, 1987; Loyko et al., 2015). Similar coatings were also described for soils in arid environments, including Gypsisols as well as Solonchaks assumed to have formed by degradation of Solonetz that developed in milder conditions (Lebedeva et al., 2015). They have also been reported for Holocene palaeosoils considered to have developed in a slightly redoximorphic environment in mountain valleys (Moretti et al., 2012).

Dusty greyish-black clay coatings are also commonly associated with deforestation (Thompson et al., 1990) or other anthropogenic changes (e.g., Slager & Van de Wetering, 1977; Macphail, 1986; Kühn, 2003a) and have been used as indicators of early cultivation (e.g., Macphail et al., 1987, 1990; Usai, 2001). Their absence can obviously not be taken for the absence of human activities, as very dusty clay coatings have been recognised for horizons influenced by well-documented land use (Carter & Davidson, 1998; Adderley et al., 2010, 2018). A particular case of combined clay and organic carbon illuviation is contamination of clay by coal-mining products (Solntseva & Rubilina, 1987). Admixtures of coal dust (<250 μm) were discovered in clay coatings at depths of up to 1.7 m 30 years after its surface deposition. Clay coatings enriched with dispersed organic matter or fine charcoal particles have also been reported for soil profiles with Bt horizons that became buried by habitation deposits of medieval settlements (Murasheva et al., 2012).

Dusty clay coatings (see Fig. 6), moderately to well sorted, with moderate to good parallel orientation of the clay particles, containing quartz and small opaque particles of different origins, such as Fe and Ti oxides (Smolikova, 1968; Chartres, 1987) and heavy mineral grains (Felix-Henningsen, 1979), with a particle size between 1 and 5 μm, are frequent in E and upper Bt horizons. The speckled appearance in PPL is caused by the admixture of opaque particles, but sharp extinction lines are visible in XPL. These coatings have been also described as clay coatings with granulation (*Granulierung*; Kubiëna, 1986) or as grainy coatings (Chartres, 1987) and should not be confused with

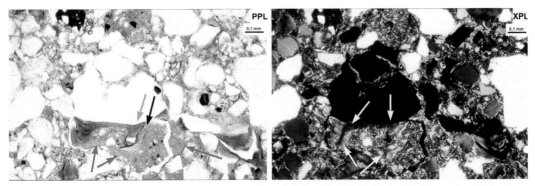

FIGURE 6 Layered coating suggesting three different phases of clay illuviation: (i) silt and clay coating (*red arrows* in PPL) with poor orientation of the clay, showing no extinction bands in XPL, (ii) limpid, yellow-brown clay coating (*black arrow* in PPL), and (iii) dusty brown clay coating with microlamination (*grey arrow* in PPL). The sharp extinction bands (*white arrows* in XPL) indicate good orientation of the clay particles in (ii) and (iii).

clay coatings having a speckled b-fabric (granular extinction pattern; FitzPatrick, 1993) resulting from reorientation of clay (see Section 2.4.1). Dusty clay coatings do not clearly correspond to a specific phase of the clay illuviation process and their dusty appearance has been attributed to ageing of former clay coatings (e.g., Bronger, 1969/1970). However, this is questionable, since these dusty coatings have been found also in young soils (Kühn, 2003a).

Limpid clay coatings (Figs. 1, 3 and 6), with continuous orientation, sharp extinction bands, high intensity of interference colours (with colours depending on mineralogical composition and iron oxide masking), good sorting, no cracks, sharp boundaries and usually completely covering the pore walls with a constant thickness (mostly 50-100 μm) are associated with a final phase of clay illuviation (e.g., Rogaar et al., 1993; Miedema et al., 1999; Kühn, 2003a) and are frequent in Bt and often in C horizons (Stephan, 2000). These limpid coatings are lacking small contrasting particles. Slow transport of clay explains the good sorting.

Non-gravitational migration and slow evaporation lead to the formation of these coatings, since fine clay suspensions are very stable and can be held by capillary water at low matrix suctions (Gombeer & D'Hoore, 1971). This is not only the case of ascended water transport (Brewer & Haldane, 1957) but also of water migrations from the groundmass to the evaporating surfaces of peds and channel walls (Gombeer & D'Hoore, 1971).

The above types of coatings may occur independently or as part of layered textural coatings (Fig. 6) (see Section 3.2).

2.4 Destruction and Alteration

Clay coatings do not remain undisturbed in the soil. Chemical and mechanical processes may result in their alteration or fragmentation.

2.4.1 Mechanical Processes

The most important processes leading to mechanical fragmentation or deformation of clay coatings are described below. The former process results in the presence of fragments of clay coatings, termed 'papules' by Brewer (1964). However, this term is not restricted to clay-coating fragments but also applies to fragments of weathered shale, surface crusts and weathered biotite.

Reorientation – The original continuous orientation of clay in clay coatings can be lost by internal stress that produces a stipple-speckled b-fabric (cf. McCarthy & Plint, 1998) (flecked orientation; Brewer, 1976). The reorganisation of the clay results in randomly oriented clay domains within the coatings (Fig. 6). FitzPatrick (1993) attributes the reorganisation to ageing of the clay coating and proposes four stages, from weak to complete, whereby a stipple-speckled b-fabric is reached at the final stage.

Bioturbation – This process fragments the coatings, generally without deforming them (Mermut & Jongerius, 1980). A microstructure dominated by channels and passage features suggests that clay-coating fragmentation was caused by bioturbation (Fig. 7) (see also Kooistra & Pulleman, 2010, 2018). The degree of reworking can be expressed quantitatively (see Section 5). Features related to bioturbation have been found to be more abundant in soils lacking clay-illuvial features than in associated soils with Bt horizons, suggesting that bioturbation can destroy illuvial horizons or prevent their formation (Blanco & Stoops, 2007).

Colluvial transport – If colluvial processes involve erosion of argic horizons or clayey parent materials, this can induce fragmentation of clay coatings and lead to the presence of a large amount of fragments of clay coatings or of clayey aggregates in the soil (Fig. 8)

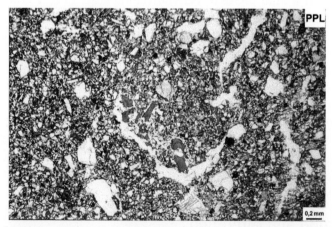

FIGURE 7 Limpid, reddish-brown fragments of clay coatings within dense complete infilling (passage feature) in Ahb horizon of a buried Luvisol.

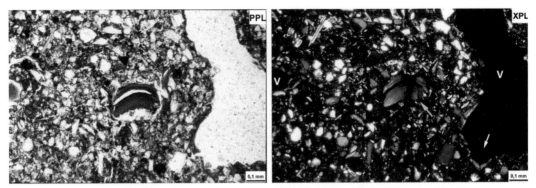

FIGURE 8 Fragment of coating of well-oriented, microlaminated, reddish brown clay (*turquoise arrow*, XPL) in colluvial deposit and thin coatings of well-oriented, yellow-brown clay (*white arrow*, XPL) around voids (v) indicating a later period of clay illuviation.

(Mücher, 1974; see also Mücher et al., 2010, 2018). The fragments very often have angular shapes, although they can also be subrounded.

They can be derived from clay coatings exposed and dried at the soil surface, which are rather stable. When fragments of Bt horizons are not completely distorted and still contain complete clay coatings, care has to be taken with interpretation, avoiding confusion with occurrences of coatings that formed *in situ*. The recognition of other features pointing to colluvial processes (see Mücher et al., 2010, 2018) can confirm the presence of fragments of Bt horizons.

Argilloturbation − Differential movements in the soil, caused by shrinking and swelling of clays, can destroy the coatings and infillings by shear in materials with a porphyric c/f-related distribution pattern (Fig. 9) (Castellet & FitzPatrick, 1974). The continuous orientation in the coatings is thus lost and replaced by a striated orientation. This situation does not always develop (Altemüller & Bailly, 1976), but it is often recognised in well-developed hydromorphic Bt horizons ('dynamic B horizons'; Fedoroff, 1968) and in clayey argic horizons (see Section 4.2.3). Fragmented clay coatings are also common in Vertisols and sometimes show a kinkband fabric (Morrás et al., 1993; Stoops, 2003; see Kovda & Mermut, 2010, 2018). Kinkband fabrics have also been observed for nonfragmented deformed coatings in clayey soils (Stoops, 2013).

Cryoturbation − Frost action can be responsible for fragmentation of clay coatings and infillings (Van Vliet-Lanoë, 1985; Bouza et al., 2005) (see also Van Vliet-Lanoë, 2010; Van Vliet-Lanoë & Fox, 2018). Its role is again confirmed by the recognition of other freeze-thaw features (Fig. 10) (e.g., Kühn, 2003b).

Crystallisation of pedogenic minerals − The growth of, for example, calcite crystals (Fig. 9) or gypsum crystals (Fig. 11) can bring about the fragmentation and/or disorganisation of clay coatings (e.g., Gile & Grossman, 1968; Bouza & del Valle, 1998; Khademi & Mermut, 2003; Dultz & Kühn, 2005).

FIGURE 9 Fragmentation of clay coating (mechanical) and disorganisation of clay coatings as a result of micritic carbonate precipitation (chemical).

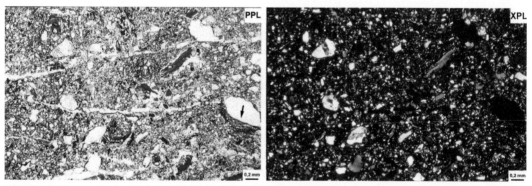

FIGURE 10 Fragments of reddish-brown clay coatings (*turquoise arrows* in PPL) within frost-induced lenticular aggregates and a dusty silt and clay coating (*black arrow* in PPL) in a channel in a Btwb horizon. Note abundant vertical passage features.

2.4.2   Alteration and Secondary Illuviation

Chemical alteration of clay coatings is limited, because these features are relatively stable. Decomposition by ferrolysis in seasonally wet acid soils has been documented (Brinkman et al., 1973). These authors observed that the original limpid coatings of well-oriented clay gradually become seemingly optically isotropic and greyish, with a characteristic grainy limpidity. They exhibit a bright bluish-white 'opalescent' lustre when viewed with oblique incident light. As these patterns may cross the original microlaminations, it is clear that they postdate deposition. In extreme cases, only scattered

FIGURE 11 Growth of gypsum crystals around a root leading to fragmentation of clay coatings in a Bt/E horizon.

fine black grains between bare grains of silt and sand are visible, pointing to an almost complete destruction of the clay. XRD analyses indicate a high percentage of fine quartz in the altered coatings, and microchemical analyses point to a high Ti content (Beke & Zwarich, 1971; Chartres, 1987).

Some studies (Morrás, 1979; Van Ranst & De Coninck, 2002; Boixadera et al., 2003; Boivin et al., 2004; Van Ranst et al., 2011) question the validity of ferrolysis, offering alternative mechanisms to explain the degradation of clay in clay coatings. However, the authors do not suggest a solution for the occurrence of fine quartz particles across clay coatings as described in detail by Brinkman et al. (1973). Breakdown of coarser quartz grains and illuviation of such fine quartz can happen but should be reflected by the presence of microlamina in the coatings. Moreover, the process of ferrolysis resulting in the degradation of clay in clay coatings (Brinkman et al., 1973) has been also documented for clays in the groundmass (e.g., Miedema et al., 1987).

Chemical alteration of clay coatings may also lead to a phenomenon described as secondary illuviation (De Coninck & Herbillon, 1969; Fedoroff, 1972; Ducloux, 1976; Eimberck-Roux, 1977). When a large amount of channels are filled with illuvial clay, redoximorphic conditions may develop, primarily in Bt/E transition zones and in the upper part of Bt horizons. This induces the reduction and removal of iron oxides from the clay in the groundmass and in the coatings. This Fe-free clay is more easily dispersed and translocated than clay containing iron oxides (Jamagne, 1972; De Coninck et al., 1976). Deposited in lower parts of glossic tongues and Bt horizons, this clay forms thick white or yellowish white clay coatings with continuous orientation or silt-clay coatings with weak orientation of the clay (Fedoroff, 1972; Jamagne & Jeanson, 1978; Dorronsoro & Aguilar, 1988), which macroscopically correspond to thick greyish clay coatings (Jamagne, 1972). In soils with redoximorphic features, these coatings can occur together with clay depletion phenomena (Jamagne et al., 1987) (see also Lindbo et al., 2010; Vepraskas et al., 2018). Another possible indication of redistribution of illuvial clay is the occurrence of coatings with high microporosity, without any signs of chemical weathering (Morrás, 1983b).

For tropical soils, Nahon (1991) distinguishes two main types of transformation: (i) those involving redistribution of constituents within the coating, such as iron oxide migration (e.g., Stoops, 1967; Bocquier & Nalovic, 1972; Boulangé et al., 1975), and (ii) those involving a change in chemical or mineralogical composition. The latter can be 'subtractive', referring to the occurrence of leaching and residual enrichment (e.g., kaolinite alteration with gibbsite formation; Boulangé et al., 1975; Mpiana, 1980), or 'additive', referring to processes involving take-up of elements from the soil solution (e.g., Al-phosphate impregnation of kaolinite coatings; Flicoteaux et al., 1977).

2.5 Related Features

Infillings of illuvial clay must be interpreted cautiously, because tangential sections of coatings in channels may look like infillings (Stoops, 2003). Clay infillings frequently occur in the lower part of Bt horizons or, unrelated to depth, in Btb horizons (e.g., Sedov et al., 2001). Clay infillings were found in takyric Solonchaks at a depth of 40 cm (Gerasimova et al., 1996) and in Bt horizons of Ultisols between 2 and 4 m depth (Moretti & Morrás, 2013).

Clay cappings with a minor admixture of silt grains have been attributed to meltwater processes during thawing of permafrost (see Van Vliet-Lanoë, 2010; Van Vliet-Lanoë & Fox, 2018) and have mainly been reported for E horizons (Kühn, 2003a).

Clay intercalations have been reported for Takyr soils of the Near East, below and at the base of argic horizons of the Niger delta, and were attributed to waterlogging (Fedoroff & Courty, 1987). Sauer et al. (2013) explained the formation of clay intercalations in Stagnic Albeluvisols from Norway as a result of penetration, under pressure, of clay-carrying and clay-mobilising solutions from albeluvic tongues into the groundmass of the Btg horizon. Intercalations are not always clearly distinguishable (Stoops, 2003) and still little information exists about intercalations since the term was introduced by Bullock et al. (1985). Fragmentation of intercalations by cracking, and deformation by calcite nodule growth, was reported by Badía et al. (2013).

3. Textural Pedofeatures Other Than Clay Coatings

3.1 Coarse-Grained Coatings

Coarse-grained textural pedofeatures, in the form of silty coatings on ped surfaces, occur predominantly in Retisols, Luvisols and Solonetzes. Although some occurrences have been attributed to authigenic quartz formation (e.g., Parfenova, 1956; Brewer, 1964), they typically form by depletion or illuviation.

Development of sandy-silty coatings and infillings as residual *in situ* accumulation of coarse-grained material due to removal of clay-sized material along ped surfaces was first suggested by Frei and Cline (1949) and further elaborated in many later publications (e.g., Arnold, 1963; Grossman et al., 1964; Curmi, 1987; Szymański et al., 2011). Depletion in fine material can involve interaction with acid solutions and eluviation of weathering products along preferential flow paths (Grossman et al., 1964; Nettleton et al., 1994). Redox conditions can play a role, whereby clay becomes more mobile when iron and manganese have been removed (e.g., Vepraskas et al., 1994, 2018; Lindbo et al., 2010).

Coarse-grained coatings and infillings can also form as illuvial features, which can be recognised based on various characteristics, such as the occurrence of underlying illuvial clay coatings, sharp lower boundaries and dissimilarities in distribution and orientation patterns of coarse material between the groundmass and the coating or infilling (Targulian et al., 1974a; Nettleton et al., 1994; Bronnikova & Targulian, 2005). The possibility of vertical translocation of silt- and sand-sized particles was demonstrated by non-micromorphological laboratory and field experiments (Wright & Foss, 1968; Locke, 1986; Fishkis et al., 2010).

Illuvial silt and very fine sand coatings have been reported for the lower part of clay-eluvial horizons (McKeague et al., 1973a, 1973b; von Zezschwitz, 1979/80; Gerasimova et al., 1996; Bronnikova et al., 2000), the upper part of Bt horizons (Targulian et al., 1974a, 1974b; Targulian & Vishnevskaya, 1975; Morrás, 1983b), in fissures in Bt horizons (Targulian, 1974a, 1974b; Targulian & Vishnevskaya, 1975; Gerasimova et al., 1996; Spiridonova et al., 1999) and horizons underlying Bt horizons (Nettleton et al., 1994). In a toposequence study, abundant silt coatings have been recognised to form by vertical illuviation, whereas clay was mainly transported laterally, resulting in partial clogging of pores downslope (Lahmar & Bresson, 1987). The detachment of coarse particles is favoured by rapid wetting of dry soils, drainage of saturated soil materials and thawing of frozen soils. It occurs when the shear stress of water flowing in channels, cracks or fissures exceeds forces retaining particles in the groundmass. Low organic carbon, Fe, Ca and Mg content, low aggregate stability, pore size discontinuities and high silt content favour the formation of silt and very fine sand coatings (Nettleton et al., 1994). In Retisols (Albeluvisols, Alfisols), coarse-grained illuvial and depletion features can occur within the same profile or even within the same horizon (Targulian et al., 1974a; Bullock &Thompson, 1985; Bronnikova & Targulian, 2005; Szymánski et al., 2011).

The common occurrence of silt coatings in frost-affected soils (see Van Vliet-Lanoë, 2010; Van Vliet-Lanöe & Fox, 2018) has led to their use as indicators of downward translocation of coarse material in soils formerly affected by frost action (Fedoroff & Goldberg, 1982, Kemp, 1985; Rose et al., 2000).

Silt coatings and infillings have also been attributed to a rearrangement of silt-sized particles in vesicles of vesicular layers of desert soils (Sullivan & Koppi, 1991; Lebedeva et al., 2015). Silt coatings in duripans within argic horizons of subtropical soils in semiarid regions were considered to be relict features (Nettleton et al., 1989).

Silty clay and clayey silt coatings as well as unsorted coatings are widely spread in argic horizons of boreal and subboreal soils of Eurasia and North America (Fedoroff, 1974; Targulian et al., 1974a; Targulian & Tselischeva, 1983; Bullock & Thompson, 1985; Gerasimova et al., 1996). Silty clay and clayey silt coatings contain variable proportions of clay and silt. Unsorted coatings are a mixture of clay, silt and sand particles, with a grain size distribution that is different from that of the surrounding groundmass.

Silty clay, clayey silt and unsorted coatings that are enriched in organic matter, mostly in the form of silt-sized particles, have been regarded as evidence of agricultural practices (see Section 4.3). The term 'agricutans' was introduced by Jongerius (1970) for this type of coatings. However, not all occurrences can be linked to former agricultural land use (e.g., Usai, 1998, 2001), although they do characterise contemporary ploughed soils (e.g., Eimberck-Roux, 1977).

Weakly oriented coarse clay and silt covering the larger conducting pores have been reported in soils flooded with water containing fine material in suspension (the so-called flood coatings). These coatings are assumed to form within one to 2 years and to be quickly destroyed (Brammer, 1971). Thompson (1983) pointed out the possibility that coatings with the same properties as mentioned above may form in well-drained soils without any flooding, possibly as a result of cultivation.

3.2 Layered Coatings

Layered coatings consist of layers with materials of different size classes, including material coarser than clay (Stoops, 2003). Brewer (1964), describing them as a type of 'compound cutans', attributed their layered fabric to alternating conditions during their formation. This is supported by experiments reported by Theocharopoulos and Dalrymple (1987; see Section 2.3.1), who produced layered coatings by imposing variations in suspension composition (clay, silt, Fe oxides organic matter). This implies that variations in source material and/or soil conditions are required, whereby illuvial material can be derived from different parts of the eluvial horizons or it can record different stages of weathering or pedogenesis. A sequence of different layers can in principle be interpreted as time-sequence of specific migration and accumulation conditions.

The occurrence of layered coatings often increases with increasing depth in Bt horizons (Chartres, 1987), with possible differences between intrapedal and interpedal voids (Bronnikova &Targulian, 2005). Clear sharp boundaries of the layers indicate that the illuviation process was either active over several events or had variable intensity (Kühn, 2003a). Gradual merging boundaries could indicate development by chemical alteration rather than deposition (Chartres, 1987).

The formation of silt-clay coatings, without layering (Figs. 4 and 5), is explained by turbulent flow (Miedema, 2002) or secondary illuviation (see Section 2.4.2). They are mostly found in Bt horizons (e.g., Kühn et al., 2006a; Nikorych et al., 2012). Formation of layered coatings (Fig. 6) with clay and silt layers probably occurred in different phases: silt particles were translocated by rapid water movement during spring melt or heavy rain, whereas purely fine clay translocation took place during slow percolation (Dasog et al., 1987). On the other hand, mineralogical and microprobe analysis of coatings with alternating clay and silt layers showed in some cases that the silt layers have the same composition as the horizon in which they occur, whereas the composition of the clay layers is different from the surrounding groundmass but similar to that of the overlying horizons (Curmi, 1987).

The detailed description of the different layers and their order of appearance, as well as the relative distribution of various types of non-layered coatings, may provide relevant information about the different phases of material translocation and the relative age of the coatings (Jamagne & Jeanson, 1978; Remmelzwaal, 1979; Jamagne et al., 1987; Payton, 1993; Fedoroff, 1997; Bronnikova & Targulian, 2005).

3.3   Infillings

Abundant clay-silt infillings with variable organic matter content were described for A and B horizons of Mollisols from the humid Lowlands in Argentina (Pazos & Stoops, 1987). Those infillings, mainly formed by bioturbation, consist of material of under- and/or overlying horizons and are a characteristic feature of transitional horizons of Greyzems (Miedema et al., 1999), many Chernozems (Gerasimova et al., 1996) and Nitisols (Creutzberg & Sombroek, 1987).

Silt and sand infillings were observed (i) in Bt horizons (Spiridonova et al., 1999) or below Bt horizons (Nettleton et al., 1994), (ii) along surfaces of dry lakes and depressions of arid areas, (iii) in cracks and fissures (Fedoroff & Courty, 1987), and (iv) in vertical and horizontal fissures of takyric Solonchaks (Gerasimova et al., 1996). Similar processes as discussed in Section 2.3.1 for coarse-grained coatings may lead to the formation of those infillings. Infillings with reverse grading (i.e., increasing coarseness from bottom to the top) have been attributed to percolation of meltwater (Van Vliet & Langohr, 1981). Silt and sand infillings in the lower part of loess deposits have been related to erosional/ depositional processes coupled with synsedimentary loess deposition (Rognon et al., 1987).

3.4   Silt Cappings

Silt cappings, consisting of mainly silty material accumulated on top of coarse fragments or peds, were described for buried palaeosoils (e.g., Ransom et al., 1987; Fedoroff et al., 1990) and as the result of periglacial processes in Fragipan horizons (e.g., Nettleton et al., 1968; FitzPatrick, 1976; Van Vliet & Langohr, 1981). These pedofeatures were also reported for well-drained coarse-grained Arctic soils, where they were attributed to silt mobilisation by direct rainfall, creating water-saturated conditions (Locke, 1986).

4. Horizons Related to Occurrences of Textural Pedofeatures

4.1 General Aspects

Textural pedofeatures have served as a diagnostic criterion for clay illuviation in various soil classification systems for many years (e.g., Soil Survey Staff, 1960, 2014; Avery, 1980; Soil Classification Working Group, 1998; Shishov et al., 2004; IUSS Working Group WRB, 2006, 2015). The presence of clay coatings is currently one of the requirements for the classification of diagnostic illuvial horizons such as the argic horizon (IUSS Working Group WRB, 2015) and the argillic horizon (Soil Survey Staff, 2014).

The requirement that argillic/argic horizons must have illuvial clay coatings was questioned by McKeague (1983) particularly for strongly weathered soils of tropical and subtropical regions, which have B horizons with higher clay content than the material from which they are derived but no illuvial clay coatings. Those horizons were described as argillic/argic horizons without clay coatings (Nettleton et al., 1969) and the increase in clay and the absence of clay coatings attributed to either (i) neoformation of clay in the Bt horizon, or (ii) destruction of clay coatings and incorporation of the clay into the groundmass by reworking processes (Bronger & Bruhn, 1989, 1990). In contrast to these examples of tropical and subtropical soils with clay-rich horizons without clay coatings, illuvial clay coatings are in fact widespread in soils of this type in those regions. They were documented in argillic/argic horizons of tropical Acrisols/Ultisols (e.g., Fedoroff & Eswaran, 1985; Stoops, 1989), in kandic horizons of subtropical Ultisols (Moretti & Morrás, 2013) and also beneath the ferralic/oxic horizon where they have a dark colour caused by iron staining (e.g., Nettleton et al., 1987; see also Marcelino et al., 2010, 2018). Brownish-red and partially weathered fragments of clay coatings have been also observed within oxic horizons in Brazil (Muggler & Buurman, 2000).

Apart from the generally accepted vertical clay illuviation, increased clay content in the B horizon has been attributed to other causes such as biological activity, textural discontinuity of the parent material, clay formation by *in situ* weathering of silicates, residual enrichment by dissolution of carbonates, chemical destruction of clay in the surface horizon and loss of clay by lateral superficial water flow (e.g., Smeck & Wilding, 1980; Baize, 1995; Soil Survey Staff, 2014; IUSS Working Group WRB, 2015). These processes, however, do not lead to the formation of clay coatings in subhorizons. The possibility of clay accumulation and formation of clay coatings due to lateral influx of clayey suspensions has to be considered (Kovalev, 1975; Legros, 1976).

In the last versions of Soil Taxonomy and WRB, the presence of illuvial clay is one of the diagnostic features, either required or optional, for agric, argillic, glossic, kandic and natric horizons (Soil Survey Staff, 2014) and for argic, natric and nitic horizons (IUSS Working Group WRB, 2015). Diagnostic features include clay coatings along the sides of peds or channels, and clay bridges between mineral grains.

In the WRB system (IUSS Working Group WRB, 2015), evidence of the illuvial nature of clay in argic and natric horizons is not a requirement. Soils with argic or natric horizons for which such evidence does exist are assigned with the cutanic qualifier.

Besides the discussed diagnostic horizons, the so-called beta-horizon also contains clay coatings as described in a number of micromorphological studies (see Section 4.6).

4.2 Argic Horizons

The micromorphological expression of argic horizons (IUSS Working Group WRB, 2015) depends to a large extent on the texture and the porosity of the soil. This affects not only the particle size that can be transported but also influences the nature and size of the pores through which the illuviated particles move and the stability of the illuvial deposits formed. Different types of illuvial clay coatings and other textural pedofeatures occur depending on the parent material.

A classification of argic horizons on basis of the type and the micromorphological characteristics of the textural pedofeatures present was proposed by Fedoroff (1974) but rarely used.

4.2.1   Sandy Argic Horizons

In sandy materials, illuvial clay is usually concentrated in more or less thick bands, separated by zones without or with less clay, at a macroscopic scale (Miedema, 1987; Kemp & McIntosh, 1989). These bands can be found up to a depth of several metres (Rawling, 2000). Macroscopically similar bands can be of sedimentary origin, but these are easily distinguishable based on micromorphological properties, such as lack of orientation of clay around sand grains and the presence of close-porphyric c/f-related distribution patterns. Illuvial lamellae are also unrelated to bedding planes and may cut across the original stratification of the sediment (Dijkerman et al., 1967).

Soil porosity is made up of fairly continuous simple packing voids through which clay particles can pass. Illuvial clay occurs as concave bridges between grains ('meniscus-like clay bridges' after Altemüller, 1962) and as coatings on pebbles and grains (Dijkerman et al., 1967; Gile & Grossman, 1968; West et al., 1987) also in calcareous material (Wieder & Yaalon, 1978). These coatings and bridges are usually characterised by strong parallel orientation and by high intensity of the interference colours (Dijkerman et al., 1967; Bullock & Thompson, 1985). Concave bridges of oriented clay build up a concave gefuric c/f-related distribution pattern (see Fig. 4). Clay coatings cover and bridge the individual sand grains and give rise to a chitonic or chito-gefuric c/f-related distribution pattern (Fig. 12), which evolves to a close porphyric distribution when all packing pores are filled.

In a single horizon different degrees of infilling may occur. As long as the porphyric stage is not reached, stress deformation of coatings and infillings is absent, and their disturbance, mainly caused by bioturbation, yields fragments in which the continuous orientation of the clay is preserved. Bioturbation creates additional porosity which may become infilled by illuvial clay, eventually leading to an open porphyric c/f-related

FIGURE 12 Chito-gefuric c/f-related distribution pattern in a sandy argillic horizon. Note that nearly all grains are coated and connected by concave bridges of yellow-brown oriented clay.

distribution pattern. Once the close porphyric stage is passed, all fine material is composed of illuvial clay, which is susceptible to argilloturbation (see Section 2.4.1).

Whereas clay coatings on peds can be explained by suction and filtering of the clay suspension on the pore surface, this is not the case for sandy argic horizons. Besides the processes discussed in Section 2.3, *in situ* clay redistribution or weathering may lead to the formation of thin coatings composed of unsorted coarser clay without continuous orientation. Similar features have been also attributed to the formation of sericite along the weathering surface of feldspar grains (Arocena et al., 1992) and to the mechanical infiltration of materials external to the soil, such as muddy waters from overland flow (Buurman et al., 1998).

It has been demonstrated in experiments with sand columns that clay coatings on sand grains form in percolations using chloritic clay but not in those with illitic or smectitic clays (Matlack et al., 1989).

4.2.2   Loamy Argic Horizons

In soils with a medium texture, clay coatings occur mainly along conducting pores, i.e., in interpedal fissures, common in soils containing smectite, and in transpedal channels (Fedoroff & Eswaran, 1985). The illuvial clay may gradually fill the pores as infillings (Fig. 13; e.g., Szymański et al., 2011). Fragmentation of clay coatings is common in loamy argic horizons, which are often subjected to pedoturbation (see Section 2.4.1).

4.2.3   Clayey Argic Horizons

In argic horizons with clayey texture of arid and Mediterranean climates, it may be difficult, if not impossible, to recognise clay coatings (Nettleton et al., 1969, Osman & Eswaran, 1974; Reynders, 1974; Verheye & Stoops, 1974; Scoppa, 1978; Imbellone & Giménez, 1990; Morrás et al., 1993). This is due to (i) the small contrast between illuvial clay and clayey groundmass, and (ii) argilloturbation (see Section 2.4.1) related to the presence of swelling clays. The latter results in rapid incorporation of the coatings in the

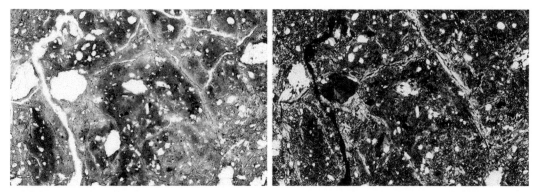

FIGURE 13 Abundant illuvial clay coatings in planes and channels of a loamy argic horizon. Note that some are deformed by shear.

groundmass or it can lead to deformation which renders the coatings therefore difficult to distinguish from the striated groundmass and slickensides.

Nevertheless, clay coatings between large hard particles, such as rock fragments, are protected and may remain intact. They are also more evident in the deepest horizons, because structural dynamics decrease with increasing depth and the coatings are more stable, remaining on the surface of the aggregates.

4.3   Agric Horizons

Agric horizons are subsurface horizons formed by illuviation of mechanically dispersed fine material in soils that have been under cultivation for a long time (Soil Survey Staff, 2014). They frequently occur immediately below Ap horizons.

In thin sections, agric horizons typically show thick coatings and infilling of weakly oriented, dark brown, coarse, dusty clay on ped surfaces and in channels. These dusty coatings may also contain silt-sized mineral grains and charcoal fragments, mixed with considerable amounts of organic material. They have also been called agricutans (Jongerius, 1970; see Section 2.3.2) and may occur juxtaposed on coatings of well-oriented fine clay (Jongerius, 1970; Kwaad & Mücher, 1979).

In Bt horizons of cultivated red soils in the Mediterranean region, these coatings are difficult to identify because of similarities with the adjacent groundmass, but they are easily detectable in C horizons (Fedoroff, 1997).

4.4   Natric Horizons

Natric horizons (Soil Survey Staff, 2014; IUSS Working Group WRB, 2015) are characterised by a well-developed angular blocky microstructure. The groundmass is mostly very rich in clay and has a well-developed striated b-fabric (e.g., Gerasimova et al., 1996; Oliveira et al., 2004a, 2004b). Natric horizons usually have thick clay coatings (Alexander & Nettleton, 1977; Lebedeva-Verba & Gerasimova, 2009), mostly with a yellowish grey colour, a dotted or speckled appearance and moderately continuous orientation. These coatings are generally more massive and less yellow than clay coatings in argic horizons. According to Fedoroff and Courty (1986), clay coatings in natric horizons are thicker than those in argic horizons and are never laminated. However, Gerasimova et al. (1996) describes microlaminations with finely dispersed organic matter. Fragmentation of clay coatings as a result of argilloturbation, not to be confused with neoformation of clay from micas, have been also documented in natric horizons (Gerasimova et al., 1996; Oliveira et al., 2004a, 2004b).

4.5 Nitic Horizons

The presence of clay coatings is often observed in nitic horizons, but it is not a requirement for its classification. This is the diagnostic horizon of Nitisols (IUSS Working Group WRB, 2015), which cover large areas in tropical and subtropical regions.

Rather thick (up to 50 μm), microlaminated, well-oriented clay coatings as well as very thin (2-3 μm) coatings of strongly oriented clay around grains and peds have been reported in nitic horizons (Creutzberg & Sombroek, 1987). These thin clay coatings (also called leptocoatings) are barely detectable with plane polarised light and are suggested to form by clay translocation within the same horizon (Creutzberg & Sombroek, 1987). Clear red or yellowish clay coatings with strong clay orientation were also described as characteristic non-relic features in B and BC horizons of Nitisols in the lower Congo (Stoops, 1968), in Brazil (Oliveira et al., 2004b; Cooper & Vidal-Torrado, 2005) and in SW Ethiopia (De Wispelaere et al., 2015).

4.6 Beta-Horizons

The beta-horizon is a zone of clay accumulation, beneath a Bt horizon, just above the contact between a calcareous substrate and the overlying decalcified soil material. It is generally separated from the Bt horizon by a Bw or C horizon and is frequently situated relatively deep in the profile (often >2 m) and therefore not always recognised during soil surveys.

Beta-horizons have a number of typical characteristics in thin sections (Bartelli & Odell, 1960; Ducloux, 1970; Mathieu & Stoops, 1974) such as:
- high porosity, mainly composed of channels and smooth vughs, reflecting intense biological activity; planar voids associated with moderately developed blocky microstructure were also documented for more shallow beta-horizons;
- coarse mineral grains with the same composition as the coarse material of the overlying layers and with no evidence of important *in situ* weathering;
- abundant thick coatings of strongly oriented fine clay around grains or on pore walls, frequently disturbed by faunal activity and/or physical processes, mostly cryoturbation; these coatings have the same composition as the illuvial clay coatings of the overlying argic horizon;

- coatings of coarse, dark, dotted clay, often covering the limpid clay coatings; these dark coatings are rich in organic matter and the organic matter content of beta-horizons is often higher than that of the overlying subsurface horizons.

There are several hypotheses concerning the nature of clay-dominated intervals above calcareous substrates: surface alteration of the limestone bedrock, subsurface limestone dissolution and accumulation of illuvial clay. The latter can be derived from overlying clayey sediments or loess (e.g., Mathieu & Stoops, 1974; Dewolf, 1976; Thiry & Trauth, 1976; Catt, 1986; Van Ranst et al., 2014). Illuvial beta-horizons can be confused in the field with buried *in situ* weathering layers of chalk. However, based on the above-mentioned characteristics, it is possible to distinguish between them (Ducloux, 1973; Mathieu & Stoops, 1974).

4.7   Argic Horizons in Palaeosoils

In palaeosoils with an argic horizon, the corresponding eluvial horizons are often missing in the soil profile. Generally, the Bt horizon is polygenetic and thus the result of more than one pedogenic process. In addition to different phases of clay illuviation, they can be affected by other pedogenic processes, such as water-logging, calcification, freezing and thawing, and bioturbation. Micromorphological investigations (e.g., Bronger, 1969/1970; Velichko & Morozova, 1970; Bullock & Murphy, 1979; Aguilar et al., 1983; Kemp, 1985; Dorronsoro & Aguilar, 1988; Bronger & Heinkele, 1989; Günster & Skowronek, 2001; Kühn, 2003c; Kühn et al., 2006b; Kadereit et al., 2010; Murasheva et al., 2012; Sycheva & Sedov, 2012; Alexandrovskiy et al., 2014) have allowed the identification of these processes and their chronology in buried Bt horizons. Micromorphological features reflecting a certain soil forming process can be related to environments in which those processes took place (Bronger et al., 1994). Clay coatings in intrapedal channels may point to an older phase of clay illuviation (Kemp, 1998). Also, joining of two or more stratigraphically different horizons must be taken into consideration (Stephan, 2000; Kemp, 2001). Conversion of buried A horizons into B horizons has been suggested for intervals in which excremental aggregates are covered by clay coatings (Kemp & Zárate, 2000; Zárate et al., 2002).

In European palaeosoils and pedosequences, some Bt horizons are characterised by a high content of illuvial clay, up to 30%, compared to less than 8% for those formed since the last glaciation (Bullock & Thompson, 1985). Most illuvial clay can be present as fragmented clay coatings and sometimes has a different colour and pleochroism than clay coatings in Holocene argillic horizons (Bullock, 1985; Fedoroff, 1997). In the Mediterranean region, red clay coatings are thicker and more abundant in buried Bt horizons than in surface soils, in which fragments of red clay coatings frequently occur (Ortiz et al., 2002). Furthermore, thick, undisturbed compound clay coatings composed of layers of fragments of clay coatings alternating with limpid, yellow-brown clay coatings (Fig. 14) have been observed (Kühn et al., 2006b).

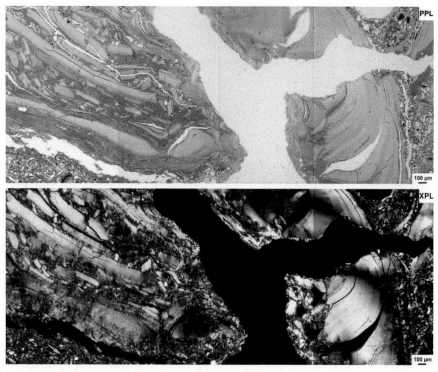

FIGURE 14 Compound clay coatings composed of layers with accumulation of fragments of clay coatings alternating with coatings of well-oriented, limpid, yellow-brown clay (thickness >2000 μm) in the 4Btg horizon of a middle Pleistocene palaeosoil, Italy. Note ferruginous hypocoating on fragment of limpid, orange yellow-brown clay coating (*turquoise arrow*, lower left).

Relevant palaeoenvironmental information can be derived from assemblages of feature sets and their relationships in Bt horizons of palaeosoils (see Fedoroff et al., 2010, 2018). Thin section observations allow also a better evaluation of geochemical data used as proxies for palaeoclimate. For example, the occurrence of fragments of clay coatings in buried mollic horizons explains higher weathering indexes than expected, because the horizons developed on redeposited material derived from soils with argic horizons (Kühn et al., 2013).

4.8 Eluvial Horizons

By definition, illuvial features are generally absent in clay-eluvial horizons. These horizons may contain some illuvial clay-silt cappings, silt coatings and relatively thin clay coatings. However, clay coatings occur sometimes in the lower part of eluvial horizons (Thompson, 1987; Bronnikova et al., 2000), where they can be residual features of degrading argic horizons (Thompson, 1987; Imbellone et al., 2016) and natric horizons (Morrás, 1983b). A stipple-speckled b-fabric is occasionally present and has been interpreted as indicating clay neoformation in this setting (Kühn, 2003a). Gorniak (1991) proposed new formation of clay coatings in E horizons of Luvisols in middle Europe, based on TEM studies. In tropical and subtropical regions, the E horizon generally shows no evidence of clay illuviation features (Gerasimova, 2003). However, examples studied in Chad (Bocquier, 1973) and Burkina Faso (Boulet, 1974) do show the presence of illuvial clay coatings accompanied by iron oxides in the E horizon, where they are juxtaposed on depletion features.

Sandy E horizons with coarse monic c/f-related distribution pattern contain mineral grains or rock fragments without coatings, but local occurrences of thin rims of oriented fine material have been reported (Kühn, 2003a).

In frost-affected soils, including soils affected by frost action in the past, the lower part of E horizons and E/Bt horizons are commonly observed to contain silt-clay cappings on coarse grains and Bt streaks (Kühn, 2003b). These cappings are attributed to freezing and thawing (see Van Vliet-Lanöe, 2010; Van Vliet-Lanoë & Fox, 2018). Bt streaks ('Schmitze' after Murawski, 1998) differ in composition from the surrounding material and have b-fabrics and other characteristics similar to those of the underlying Bt or Btb horizons. The formation of Bt streaks is attributed to uprooting (e.g., Schaetzl et al., 1990), to bioturbation (e.g., Gabet et al., 2003), to the further development of E into Bt horizons (Bullock, 1968; Bullock et al., 1974) or to lateral, vertical and/or horizontal transport (Pawluk & Bal, 1985). Despite micromorphological investigations, fine material cappings have not been described in E horizons from loamy to clayey sediments in temperate regions (e.g., Stephan, 1981, 1993; Miedema, 1987) or in tropical/subtropical environments (Nettleton et al., 1989).

The combined occurrence of lenticular microstructure and undisturbed, mostly limpid clay coatings in eluvial horizons indicates former ice segregation processes followed by clay illuviation; otherwise the clay coatings would have been fragmented or destroyed (Van Vliet-Lanoë, 1988; Kühn, 2003a). The presence of fragments of limpid clay coatings within lenticular plates in E/Bt horizons implies that clay illuviation has preceded ice segregation (Kühn, 2003b). Three different micromorphological types (eluvial platy, eluvial platy-differentiated and eluvial simple apedal nodular) were proposed by Gerasimova (2003) for loamy E horizons, each with specific micromorphological properties.

With increasing depth in transitional horizons that show tonguing and/or an irregular lower boundary, Bt streaks (from mm to several cm in diameter) occur with small *in situ* limpid clay coatings and, less commonly, fragments of clay coatings (Fig. 15) (Miedema et al., 1999; Kühn, 2003b; Kühn et al., 2006a). The transitional horizons can also include silt and sand infillings that sometimes exhibit reverse grading indicating former periglacial influence (see Van Vliet-Lanoë, 2010; Van Vliet-Lanoë & Fox, 2018).

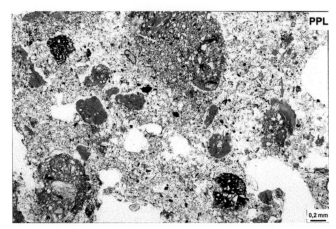

FIGURE 15 Fragments of yellow-brown clay coatings (*red arrows*) in Bt streaks of a Bt/E horizon.

Degradation features such as reorientation phenomena of clay coatings (FitzPatrick, 1993) and incorporation of fragments of clay coatings into the groundmass are often found in E, Bt/E and E/Bt horizons (Bullock, 1968; Payton, 1993; Gerasimova et al., 1996; Bronnikova et al., 2000). These features may indicate a destruction of the upper part of Bt horizons (Bullock, 1968) or changes in the palaeoenvironment, also in the case of Bt formation in layered parent material (Kühn et al., 2006a).

5. Quantification of Clay Illuviation

The quantitative micromorphological criteria proposed for the classification of argillic horizons was first 10% or more of oriented clays in thin sections (Soil Survey Staff, 1960). The limit has been lowered in later issues of Soil Taxonomy and is now 1% (Soil Survey Staff, 2014).

McKeague et al. (1981) evaluated criteria for argillic horizons in Canadian soils using semiquantitative counts of illuvial clay and found that only 32 of the 54 horizons designated Bt in the field had at least 1% apparently illuvial clay. This could be due to the fact that (i) Bt horizons were identified in the field by increased clay content rather than the presence of illuvial clay coatings, (ii) pressure surfaces were confused with illuvial clay coatings, or (iii) as discussed below, that the microscopic estimation was incorrect.

Quantitative studies of clay illuviation generally involve volume determinations for the clay coatings, using thin sections (Brewer, 1968). Another approach is the estimation of gains and losses in clay during soil formation and evolution, on the basis of stable mineral contents in the different horizons of the profile (Smeck & Wilding, 1980) or on the basis of the current clay contents and the thickness of the horizons under the hypothesis of a constant soil volume (Van Wambeke, 1972, 1974).

Point counting has been the most frequently used technique to quantify illuvial clay in thin sections (e.g., Brewer, 1968; Eswaran, 1968; Miedema & Slager, 1972; McKeague et al., 1978, 1980; McKeague, 1983; Murphy & Kemp, 1984). However, the reproducibility of results has been questioned because different trained operators obtained different results on the same samples, with a coefficient of variance of up to 50% (McKeague et al., 1980, Murphy, 1983, Miedema, 2002). This could be explained by difficulties in identifying illuvial clay, neoformed clay resulting from weathering (e.g., Mermut & Pape, 1973; Magaldi & Tallini, 2000) and porostriated b-fabrics or stress deformed illuvial clay coatings. Repeated counting by the same operator using one set of interpretative rules leads to much more consistent counting results (McKeague et al., 1980). It should also be taken into account that the amount of micromorphologically detected illuvial clay is usually not in linear relationship with the gravimetric clay content (Schlichting & Blume, 1961; Murphy & Kemp, 1984).

With the development of image analysis techniques in the last decades, new methods became available for the quantification of micromorphological features. Some attempts have been made to quantify clay coatings in thin sections using multilayer digital image processing (Terribile & FitzPatrick, 1992, 1995).

Based on counts of illuvial features in thin sections, some indices have been developed (Table 2), such as the degree of illuviation, the degree of reworking and the profile clay illuviation index (Miedema & Slager, 1972). The degree of clay illuviation corresponds to the area (volume) occupied by all illuvial features (both *in situ* and reworked) and expressed as a percentage of the total volume of the thin section. The degree of reworking is the ratio between the volume of fragmented and deformed illuviation features and that of all illuvial features. The profile illuviation index is the sum of the products of the percentage of illuvial clay in each horizon and the thickness of the horizon in centimetres.

Table 2   Degree of illuviation, degree of reworking and profile illuviation index

Illuviated clay		Reworked illuvial features		Profile illuviation index	
%[a]	Degree of illuviation	%[b]	Degree of reworking	% cm[c]	
<0.3	Negligible	<30	Weak	<50	Very low
0.3–1.0	Weak	30–70	Moderate	50–300	Low
1.0–4.0	Moderate	>70	Strong	300–700	Moderately high
4.0–7.0	Strong			>700	Very high
>7.0	Very strong				

[a]Illuvial features expressed as % of total thin section area.
[b]Fragmented and deformed illuvial features expressed as % of total volume of illuvial features.
[c]Sum of the product of the % of illuviated clay in each horizon and the horizon thickness in cm.
After Miedema & Slager, 1972.

These indices have been used in various micromorphometric studies of argillic horizons in Late Glacial deposits in Europe (Miedema et al., 1978, 1983, 1999; Schwan et al., 1982; Miedema, 1987; Feijtel et al., 1988; Jongmans et al., 1991) and South America (Pazos & Stoops, 1987). The highest degree of illuviation values recorded for subhorizons of the argillic horizon were 6.2%-19.9%. These values are consistent with the typical range of illuvial clay (1%-20%) in thin sections of argillic horizons as mentioned by Murphy and Kemp (1987). The degree of reworking ranged from more than 70% in well-drained soils to less than 25% in imperfectly drained soils, indicating that biological activity is strongly related with the drainage conditions. In a study of zonal Mollisols, a greater abundance and thickness of illuvial clay coatings was measured for soils with calcite-cemented subsurface horizons than for soils lacking such interval (Pazos, 1984).

To evaluate the degree of soil development, some indices were developed that include the degree of clay illuviation (Harden, 1982, Ferrari & Magaldi, 1983; Dorronsoro, 1994). The Micromorphological Soil Development Index (MISODI; Magaldi & Tallini, 2000) was defined on basis of the size and the frequency of occurrence of both illuvial and authigenic clay coatings in thin sections. The study of a chronosequence of relicts of palaeosoils has shown a good relationship between the proposed relative age of the palaeosoils and the degree of weathering and pedogenesis in the B and B/C horizons as indicated by the MISODI (Magaldi & Tallini, 2000). Khormali et al. (2003) suggested a modified MISODI that they named MISECA (Micromorphological Index of Soil Evolution in highly Calcareous Arid to Semiarid Conditions). The authors tested MISECA in soils from Iran and found that it could be used for relative dating of argillic horizons when soil-forming factors have been uniform during the entire period of soil development.

6. Conclusions

The correct recognition and description of clay coatings is the key to a precise interpretation of genesis and properties of the horizons and soils under consideration. Illuvial clay coatings are often difficult to recognise in the field, but in most cases they are easily distinguishable under the microscope. So far, micromorphology is the only reliable method to validate clay illuviation, and it also provides additional information about the conditions of formation of the various textural illuvial features.

Many field and laboratory experiments were carried out with the aim of investigating the genesis and formation of clay coatings. Nevertheless, it is hardly possible to infer from micromorphological data alone if the illuvial clay is a result of (i) vertical translocation from the overlying horizon, (ii) lateral translocation, or (iii) translocation processes within the same horizon. The combination of micromorphological studies with laser ablation ICP-MS, FTIR, XRF and SEM along the horizons may reveal more information about the source of the clay within specific horizons.

Acknowledgements

We are grateful to Georges Stoops, for giving us his lecture notes as the starting point for this chapter. We thank Maria Gerasimova for additional information on the history of polynite. Many thanks also to Juan Carlos Loaiza and Hector Morrás for their valuable remarks and for providing references for French and Spanish publications that were not covered in the first edition.

References

Adderley, W.P., Wilson, C.A., Simpson, I.A. & Davidson, D.A., 2010. Anthropogenic Features. In Stoops, G., Marcelino, V. & Mees, F. (eds.), Interpretation of Micromorphological Features of Soils and Regoliths. Elsevier, Amsterdam, pp. 569-588.

Adderley, W.P., Wilson, C.A., Simpson, I.A. & Davidson, D.A., 2018. Anthropogenic features. In Stoops, G., Marcelino, V. & Mees, F. (eds.), Interpretation of Micromorphological Features of Soils and Regoliths. Second Edition. Elsevier, Amsterdam, pp. 753-777.

Agafonoff, V., 1936. Les Sols de France au Point de Vue Pédologique. Dunod, Paris, 154 p.

Aguilar, J., Guardiola, J.L., Barahona, E., Dorronsoro, C. & Santos, F., 1983. Clay illuviation in calcareous soils. In Bullock, P. & Murphy, C.P. (eds.), Soil micromorphology, Volume 2, Soil Genesis. AB Academic Publishers, Berkhamsted, pp. 541-550.

Akamigbo, F.O.R. & Dalrymple, J.B., 1985. Experimental simulation of the results of clay translocation in the B horizons of soils; the formation of intrapedal cutans. Journal of Soil Science 36, 401-409.

Alexander, E.B. & Nettleton, W.D., 1977. Post-Mazama Natrargids in Dixie Valley, Nevada. Soil Science Society of America Journal 41, 1210-1212.

Alexandrovskiy, A.L. & Birina, A.G., 1987. Evolution of grey forest soils in the foothills of North Caucasus [*in Russian*]. Pochvovedenie 8, 28-39.

Alexandrovskiy, A.L., Sedov, S.N. & Shishkov, V.A., 2014. The development of deep soil processes in ancient kurgans of the North Caucasus. Catena 112, 65-71.

Altemüller, H.J., 1956. Mikroskopische Untersuchungen einiger Löß-Bodentypen mit Hilfe von Dünnschliffen. Zeitschrift für Pflanzenernährung und Bodenkunde 72, 152-167.

Altemüller, H.J., 1962. Beitrag zur mikromorphologischen Differenzierung von durchschlämmter Parabraunerde, Podsol-Braunerde und Humus-Podsol. Zeitschrift für Pflanzenernährung und Bodenkunde 98, 247-258.

Altemüller, H.J. & Bailly, F., 1976. Mikromorphologische Untersuchungen an einer nord-westdeutschen Parabraunerde-Pseudogley-Sequenz aus Löss. Geoderma 16, 327-343.

Arnold, R.W., 1963. Silans in Some Brunizem Soils. PhD Dissertation, Iowa State University, 236 p.

Arocena, J.M., Pawluk, S. & Dudas, M.J., 1992. Genesis of selected sandy soils in Alberta, Canada as revealed by microfabric, leachate- and soil composition. Geoderma 54, 65-90.

Atanasescu, R., Florea, N. & Morgenstern, S., 1974. Consideratii referitoare la migrarea argilei in soluri. Analele Institutului de Cercetari pentru Pedologie si Agrochimie 42, 177-184.

Avery, B.W., 1980. Soil Classification in England and Wales (Higher Categories). Volume 8. Rothamsted Experimental Station, Harpenden, 172 p.

Badía, D., Poch. R., Martí, C. & García-Gozález, M., 2013. Paleoclimatic implications of micromorphic features of a polygenetic soil in the Monegros Desert (NE- Spain). Spanish Journal of Soil Science 3, 95-115.

Baize, D., 1989. Planosols in the 'Champagne Humide' region, France. A multi-approach study. Pedologie 39, 119-151.

Baize, D., 1995. Les sols argileux appauvris en argile sous climat temperé humide. Planosols texturaux, Pélosols Differenciés et autres solumns. Étude et Gestion des Sols 2, 227-240.

Bartelli, L.J. & Odell, R.T., 1960. Laboratory studies and genesis of a clay-enriched horizon in the lowest part of the solum of some Brunizem and Grey-Brown Podzolic Soils in Illinois. Soil Science Society of America Proceedings 23, 390-395.

Beke, G.J. & Zwarich, M.A., 1971. Chemical and mineralogical characteristics of cutans from B horizons of three Manitoba Soils. Canadian Journal of Soil Science 51, 221-228.

Birina, A.G. 1980. Mineralogical and Chemical Compositions of Structural Elements in Sod-Podzolic and Gley Soils [in Russian]. PhD Dissertation, Moscow State University, 236 p.

Blanco, M. & Stoops, G., 2007. Genesis of pedons with discontinuous argillic horizons in the Holocene loess mantle of the southern Pampean landscape, Argentina. Journal of South America Earth Sciences 23, 30-45.

Blume, H.P., 1964. Zum Mechanismus der Tonverlagerung. Transactions of the 8th International Congress of Soil Science, Volume V, Bucharest, pp. 715-722.

Bocquier, G., 1973. Genèse et Evolution de Deux Toposéquences de Sols Tropicaux du Tchad. Interpretation Biogéodynamique. Mémoires de l'ORSTOM n° 62, 350 p.

Bocquier, G. & Nalovic, L., 1972. Utilisation de la microscopie électronique en pédologie. Cahiers ORSTOM, Série Pédologie, 10, 411-434.

Boivin, P., Saejiew, A., Grunberger, O. & Arunin, S., 2004. Formation of soils with contrasting textures by translocation of clays rather than ferrolysis in flooded rice fields in Northeast Thailand. European Journal of Soil Science 55, 713-724.

Boixadera, J., Poch, R.M., García-González, M.T. & Vizcayno, C., 2003. Hydromorphic and clay-related processes in soils from the Llanos de Moxos (northern Bolivia). Catena 54, 403-424.

Bolt, A.J.J., Mücher, H.J., Sevink, J. & Verstraten, J.M., 1980. A study on loess-derived colluvia in southern Limbourg (The Netherlands). Netherlands Journal of Agricultural Science 28, 110-126.

Boulangé, B., Paquet., H. & Bocquier, G., 1975. Le rôle de íargile dans la migration et íaccumulation de íalumine de certaines bauxites tropicales. Comptes Rendus de l'Académie des Sciences, Série D, 280, 2183-2186.

Boulet, R., 1974. Toposéquences de Sols Tropicaux en Haute-Volta. Équilibre et Déséquilibre Pédobioclimatique. Mémoires de l'ORSTOM n° 85, 272 p.

Bouza, P. & del Valle, H., 1998. Propiedades micromorfológicas del suelo superficial y subsuperficial en un ambiente pedemontano árido de Patagonia, Argentina. Ciencia del Suelo 16, 30-38.

Bouza, P., Simon, M., Aguilar, J., Rostagno, M. & del Valle, H., 2005. Genesis of some selected soils in the Valdés Peninsula, NE Patagonia, Argentina. In Faz Cano, A., Ortiz, R. & Mermut, A. (eds.), Genesis, Classification and Cartography of Soils. Advances in GeoEcology 36, Catena Verlag, Reiskirchen, pp. 1-12.

Brammer, H., 1971. Coatings in seasonal flooded soils. Geoderma 6, 5-16.

Brewer, R., 1964. Fabric and Mineral Analysis of Soils. John Wiley and Sons, New York, 470 p.

Brewer, R., 1976. Fabric and Mineral Analysis of Soils. Robert E. Krieger Publishing Company, Huntington, New York, 482 p.

Brewer, R., 1968. Clay illuviation as a factor in particle-size differentiation in soil profiles. Transactions of the 9th International Congress of Soil Science, Volume IV, Adelaide, pp. 489-499.

Brewer, R. & Blackmore, A.V., 1956. The effects of entrapped air and optically oriented clay on aggregate breakdown and soil consistence. Australian Journal of Applied Science 7, 59-68.

Brewer, R. & Haldane, A.D., 1957. Preliminary experiments in the development of clay orientations in soils. Soil Science 84, 301-309.

Brewer, R. & Sleeman, J.R., 1970. Some trends in pedology. Earth Science Reviews 6, 297-335.

Brinkman, R., Jongmans, A.G., Miedema, R. & Maaskant, P., 1973. Clay decomposition in seasonally wet, acid soils: micromorphological, chemical and mineralogical evidence from individual argillans. Geoderma 10, 259-270.

Bronger, A., 1969/70. Zur Mikromorphogenese und zum Tonmineralbestand quartärer Lößböden in Südbaden. Geoderma 3, 281-320.

Bronger, A. & Bruhn, N., 1989. Relict and recent features in tropical Alfisols from South India. Catena Supplement 16, 107-128.

Bronger, A. & Bruhn, N., 1990. Clay illuviation in semiarid-tropical (SAT) Alfisols? A first approach to a new concept. In Douglas, L.A. (eds.), Soil Micromorphology: A Basic and Applied Science. Developments in Soil Science, Volume 19. Elsevier, Amsterdam, pp. 175-181.

Bronger, A. & Heinkele, Th., 1989. Micromorphology and genesis of paleosols in the Luochuan loess section, China: pedostratigraphic and environmental implications. Geoderma 45, 123-143.

Bronger, A., Bruhn-Lobin, N. & Heinkele, T., 1994. Micromorphology of paleosols — genetic and paleoenvironmental deductions: case studies from central China, south India, NW Morocco and the Great Plains of the USA. In Ringrose-Voase, A.J. & Humphreys, G.S. (eds.), Soil Micromorphology: Studies in Management and Genesis. Developments in Soil Science, Volume 22. Elsevier, Amsterdam, pp. 187-206.

Bronnikova, M.A. & Targulian, V.O., 2005. Assemblage of Cutans in Texturally Differentiated Soils (Case Study for Albeluvisols of the East European Plain) [*in Russian*]. PBMC Akademkniga, Moscow, 197 p.

Bronnikova, M.A., Sedov, S.N. & Targulian, V.O., 2000. Clay, iron-clay, and humus-clay coatings in the eluvial part of soddy-podzolic soils profile. Eurasian Soil Science 33, 577-584.

Bullock, P., 1968. The Zone of Degradation at the Eluvial-illuvial Interface of some New York Soils. PhD Dissertation, Cornell University, Ithaca, 187 p.

Bullock, P., 1985. The role of micromorphology in the study of Quaternary soil processes. In Boardman, J. (ed.), Soils and Quaternary Landscape Evolution. Wiley, Chichester, pp. 117-157.

Bullock, P. & Murphy, C.P., 1979. Evolution of a paleo-argillic brown earth (paleudalf) from Oxfordshire, England. Geoderma 22, 225-252.

Bullock, P. & Thompson, M.L., 1985. Micromorphology of Alfisols. In Douglas, L.A. & Thompson, M.L. (eds.). Soil Micromorphology and Soil Classification. Soil Science Society of America Special Publication 15, SSSA, Madison, pp. 17-48.

Bullock, P., Milford, M.H. & Cline, M.G., 1974. Degradation of argillic horizons in Udalf soils of New York State. Soil Science Society America Proceedings 38, 621-628.

Bullock, P., Fedoroff, N., Jongerius, A., Stoops, G., Tursina, T. & Babel, U., 1985. Handbook for Soil Thin Section Description. Waine Research Publications, Wolverhampton, 152 p.

Buol, S.W. & Hole, F.D., 1959. Some characteristics of clay skins on peds in the B horizon of a gray-brown podzolic soil. Soil Science Society of America Journal 23, 239-241.

Buol, S.W. & Hole, F.D., 1961. Clay skin genesis in Wisconsin soils. Soil Science Society of America Proceedings 25, 377-379.

Buurman, P. & Jongmans, A.G., 1987. Amorphous clay coatings in a lowland Oxisol and other andesitic soils of West Java, Indonesia. Pemberitaan Penelitian Tanah dan Pupuk 7, 31-40.

Buurman, P., Jongmans, A.G. & PiPujol, M.D., 1998. Clay illuviation and mechanical clay infiltration — Is there a difference? Quaternary International 51, 66-69.

Canada Soil Survey Committee, 1978. The Canadian System of Soil Classification. Canada Department of Agriculture, Ottawa, Publication n° 1646, 164 p.

Carter, S.P. & Davidson, D.A., 1998. An evaluation of the contribution of soil micromorphology to the study of ancient arable agriculture. Geoarchaeology 13, 535-547.

Castellet, J.T. & FitzPatrick, E.A., 1974. Clay cutans and matrix disruption in some soils form central Spain. In Rutherford, G.K. (ed.), Soil Microscopy. The Limestone Press, Kingston, pp. 632-641.

Chartres, C.J., 1987. The composition and formation of grainy void coatings in some soils with textural contrast in Southeastern Australia. Geoderma 39, 209-233.

Catt, J.A., 1986. The nature, origin and geomorphological significance of Clay-with-Flints. In Sieveking, G. & Hart, M.B. (eds.), The Scientific Study of Flint and Chert. Cambridge University Press, Cambridge, pp. 151-159.

Cooper, M. & Vidal-Torrado, P., 2005. Caracterização morfológica, micromorfológica e fisico-hídrica de solos com horizonte B nítico. Revista Brasileira de Ciência do Solo 29, 581-595.

Creutzberg, D. & Sombroek, W.G., 1987. Micromorphological characteristics of Nitosols. In Fedoroff, N., Bresson, L.M. & Courty, M.A. (eds.), Micromorphologie des Sols, Soil Micromorphology. AFES, Plaisir, pp. 151-155.

Curmi, P., 1987. Sur la significacion des revetements complexes argileux et limoneux dans les sols lessivés acides. In Fedoroff, N., Bresson, L.M. & Courty, M.A. (eds.), Micromorphologie des Sols, Soil Micromorphology. AFES, Plaisir, pp. 251-254.

Dasog, G.S., Mermut, A.R. & Acton, D.F., 1987. Micromorphology and submicroscopy of illuviated mineral particles in Boreal Clay Soils of Saskatchewan, Canada. Geoderma 40, 193-208.

De Coninck, F. & Herbillon, A., 1969. Étude minéralogique et chimique des fractions argileuses dans les Alfisols et les Spodosols de la Campine (Belgique). Pedologie 19, 159-272.

De Coninck, F., Favrot, J.C., Tavernier, R. & Jamagne, M., 1976. Dégradation dans les sols lessivés hydromorphes sur matériaux argilo-sableux. Exemples des sols de la nappe détritique Bourbonnaise (France). Pedologie 26, 105-151.

De Wispelaere, L., Marcelino, V., Regassa, A., De Grave, E., Dumon, M., Mees, F. & Van Ranst, E., 2015. Revisiting nitic horizon properties of Nitisols in SW Ethiopia. Geoderma 243/244, 69-79.

Dewolf, Y., 1976. A propos des argiles à silex - essai de typologie. Revue de Géomorphologie Dynamique 25, 113-138.

Dijkerman, J.C., Cline, M.G. & Olson, G.W., 1967. Properties and genesis of textural subsoil lamellae. Soil Science 1041, 7-15.

Dobrovol'ski, G.V. (ed.), 1991. A Methodological Manual of Soil Micromorphology. International Training Centre for Post-Graduate Soil Scientists, Publication Series 3, Ghent, 63 p.

Dobrovol'ski, G.V., Nikitin, E.D. & Fedorov, K.N. 1976. Experimental study of oriented clays formation in soil [in Russian]. Pochvovedenie 4, 140-143.

Dorronsoro, C., 1994. Micromorphological index for the evaluation of soil evolution in central Spain. Geoderma 61, 237-250.

Dorronsoro, C. & Aguilar, J., 1988. El proceso de iluviación de arcilla. Annales de Edafologia y Agrobiologia 47, 311-350.

Duchaufour, P., 1982. Pedology. Pedogenesis and Classification. G. Allen & Unwin, London, 448 p.

Ducloux, J., 1970. L'horizon Beta des sols lessivés sur substratum calcaire de la plaine poitevine. Bulletin de l' Association Française pour l'Etude du Sol 3, 15-25.

Ducloux, J., 1973. Essai de quantification au niveau micromorphologique. Application aux sols d'une toposéquence sur substratum calcaire de la plaine vendéenne. Science du Sol 2, 81-89.

Ducloux, J., 1976. Essai de caractérisation semi-quantitative des revêtements argileux des sols lessivés glossiques du bocage vendéen meridional à Iaide du microscope électronique à balayage. Science du Sol 1, 23-36.

Dultz, S. & Kühn, P., 2005. Occurrence, formation, and micromorphology of gypsum in soils from the Central-German Chernozem region. Geoderma 129, 230-250.

Edwards, A.P. & Bremner, J.M., 1967. Microaggregates in soils. Journal of Soil Science 18, 64-73.

Eimberck-Roux, M., 1977. Les sols lessivés glossiques à pseudogley de l'Argonne méridionale: caractérisation micromorphologique et minéralogique. Science du Sol 15, 81-92.

Elliott, P. & Drohan, P., 2009. Clay accumulation and argillic-horizon development as influenced by aeolian deposition vs. local parent material on quartzite and limestone-derived alluvial fans. Geoderma 151, 98-108.

Eswaran, H., 1968. Point counting analysis as applied to soil micromorphology. Pedologie 18, 238-252.

Eswaran, H., 1979. Micromorphology of Alfisols and Ultisols with low activity clays. In Beinroth, F.H & Panichapong, S. (eds.), Proceedings of the Second International Soil Classification Workshop, Part II; Land Development Department, Bangkok, pp. 53-76.

Fedoroff, N., 1968. Genèse et morphologie de sols à horizon B textural en France atlantique. Science du Sol 1, 29-65.

Fedoroff, N., 1972. The clay illuviation. (A micromorphological study). In Kowalinski, S. & Drozd, J. (eds.), Soil Micromorphology. Panstwowe Wydawnicto Naukowe, Warsawa, pp. 195-207.

Fedoroff, N., 1974. Classification of accumulations of translocated particles. In Rutherford, G.K. (ed.), Soil Microscopy. The Limestone Press, Kingston, pp. 695-713.

Fedoroff, N., 1997. Clay illuviation in Red Mediterranean soils. Catena 28, 171-198.

Fedoroff, N. & Courty, M.A., 1986. Micromorphology of natric horizons. In Transactions of the 13th Congress International Society Soil Science, Volume 5, pp. 1551-1552.

Fedoroff, N. & Courty, M.A., 1987. Morphology and distribution of textural features in arid and semiarid regions. In Fedoroff, N., Bresson, L.M. & Courty, M.A. (eds.), Micromorphologie des Sols, Soil Micromorphology. AFES, Plaisir, pp. 213-219.

Fedoroff, N. & Eswaran, H., 1985. Micromorphology of Ultisols. In Douglas, L.A. & Thompson, M.L. (eds.), Soil Micromorphology and Soil Classification. Soil Science Society of America Special Publication 15, SSSA, Madison, pp. 145-164.

Fedoroff, N. & Goldberg, P., 1982. Comparative micromorphology of two late Pleistocene paleosols (in the Paris Basin). Catena 9, 227-251.

Fedoroff, N., Courty, M.A. & Thompson, M.L., 1990. Micromorphological evidence of paleoenvironmental change in pleistocene and holocene paleosols. In Douglas, L.A. (ed.), Soil Micromorphology: A Basic and Applied Science. Developments in Soil Science, Volume 19. Elsevier, Amsterdam, pp. 653-665.

Fedoroff, N., Courty, M.A., & Zhengtang Guo, 2010. Palaeosols and relict soils. In Stoops, G., Marcelino, V. & Mees, F. (eds.), Interpretation of Micromorphological Features of Soils and Regoliths, Elsevier, Amsterdam, pp. 623-662.

Fedoroff, N., Courty, M.A., & Zhengtang Guo, 2018. Palaeosols and relict soils, a conceptual approach. In Stoops, G., Marcelino, V. & Mees, F. (eds.), Interpretation of Micromorphological Features of Soils and Regoliths. Second Edition. Elsevier, Amsterdam, pp. 821-862.

Feijtel, T.C.J., Jongmans, A.G., van Breemen, N. & Miedema, R., 1988. Genesis of two Planosols in the Massif Central, France. Geoderma 43, 249-269.

Felix-Henningsen, P., 1979. Merkmale, Genese und Stratigraphie Fossiler und Reliktischer Bodenbildungen in Saalezeitlichen Geschiebelehmen Schleswig-Holsteins und Süd-Dänemarks. PhD Dissertation, University of Kiel, 218 p.

Fernández, E., Dorronsoro-Fernández, C., Aguilar, J., Dorronsoro, B., Stoops, G. & Dorronsorro Díaz, C., 2006. IlluviaSol. Clay eluviation/illuviation processes in soils. Internet Resource.

Ferrari, G. & Magaldi, D., 1983. Significato et applicazione della paleopedologia nella stratigrafia del Quaternio. Bollettino del Museo Civico di Storia Naturale di Verona 10, 315-340.

Fishkis, O., Ingwersen, J., Lamers, M., Denysenko, D. & Streck, T., 2010. Phytolith transport in soil: a field study using fluorescent labeling, Geoderma 157, 27-36.

FitzPatrick, E.A., 1976. Cryons and Isons. Proceedings of the North of England Soils Discussion Group 11, 31-43.

FitzPatrick, E.A., 1993. Soil Microscopy and Micromorphology. John Wiley & Sons, Chichester, 304 p.

Flicoteaux, R., Nahon, D. & Paquet., H., 1977. Genèse des phosphates alumineux à partir des sédiments argilo-phosphatés du Tertiaire de Lam-Lam (Sénegal). Suite minéralogique, permanences et changements de structures. Sciences Géologiques Bulletin 30, 153-174.

Frei, E. & Cline, M.G., 1949. Profile studies of normal soils of New York: II. Micromorphological studies of the gray brown podzolic - brown podzolic soil sequence. Soil Science 68, 333-344.

Gabet, E.J., Reichmann, O.J. & Seabloom, E.W., 2003. The effects of bioturbation on soil processes and sediment transport. Annual Review of Earth and Planetary Sciences 31, 249-273.

Gagarina, E.I. & Tsyplenkov, V.P., 1974. Use of micromorphological method in modelling the modern process of soil formation [in Russian]. Pochvovedenie 4, 20-27.

Gerasimova, M., 2003. Higher levels of description − approaches to the micromorphological characterisation of Russian soils. Catena 54, 319-337.

Gerasimova, M.I., Gubin, S.V. & Shoba, S.A., 1996. Soils of Russia and Adjacent Countries: Geography and Micromorphology. Moscow State University & Agricultural University Wageningen, 204 p.

Gile, L.H. & Grossman, R.B., 1968. Morphology of the argillic horizon in desert soils of southern New Mexico. Soil Science 106, 6-15.

Goldberg, P. & Macphail, R.I., 2006. Practical and Theoretical Geoarchaeology. Blackwell, Oxford, 472 p.

Gombeer, R. & D'Hoore, J., 1971. Induced migration of clay and other moderately mobile constituents. Pedologie 21, 311-342.

Gorniak, A., 1991. Pedogenic perstructions of the grain surface in loess luvisols in the light of TEM investigations. Polish Journal of Soil Science 24, 99-107.

Goss, D.W., Smith, S.J. & Stewart, B.A., 1973. Movement of added clay through calcareous materials. Geoderma 9, 97-103.

Grossman, R.B., Odell, R.T. & Beavers, A.H., 1964. Surfaces of peds from B horizons of Illinois soils. Soil Science Society of America Proceedings 28, 792-798.

Günster, N. & Skowronek, A., 2001. Sediment-soil sequences in the Granada Basin as evidence for long- and short-term climatic changes during the Pliocene and Quaternary in the Western Mediterranean. Quaternary International 78, 17-32.

Hallsworth, E.G., 1963. An examination of some factors affecting the movement of clay in an artificial soil. Journal of Soil Science 14, 360-371.

Harden, J.W., 1982. A quantitative index of soil development from field: examples from a chronosequence in central California. Geoderma 28, 51-98.

Howitt, R.W. & Pawluk, S., 1985. The genesis of a Gray Luvisol within the Boreal Forest region: I. Static pedology. Canadian Journal of Soil Science 65, 1-8.

Imbellone, P.A. & Giménez, J.E., 1990. Propiedades físicas, mineralógicas y micromorfológicas de suelos con características vérticas del partido de La Plata (provincia de Buenos Aires). Ciencia del Suelo 8, 231-236.

Imbellone, P.A., Beilinson, E. & Aguilera, E.Y., 2016. Micromorfología de suelos. In Pereyra, F.X. & Torres Duggan, M. (eds.), Suelos y Geología Argentina. Una Visión Integradora Desde Diferentes Campos Disciplinarios. Asociación Geológica Argentina, Asociación Argentina de la Ciencia del Suelo, UNDAV Ediciones, Buenos Aires, pp. 159-183.

IUSS Working Group WRB, 2006. World Reference Base for Soil Resources. 2nd Edition. World Soil Resources Reports No. 103. FAO, Rome, 128 p.

IUSS Working Group WRB, 2015. World Reference Base for Soil Resources 2014, update 2015. World Soil Resources Reports No. 106. FAO, Rome, 192 p.

Jamagne, M., 1972. Some micromorphological-aspects of soils developed in loess deposits of Northern France. In Kowalinski, S. & Drozd, J. (eds.), Soil Micromorphology. Panstwowe Wydawnicto Naukowe, Warsawa, pp. 559-582.

Jamagne, M. & Jeanson, C., 1978. Illuviations primaire et secondaire dans les sols lessivés sur matériaux limoneux. Micromorphologie et microanalyse élémentaire. In Delgado, M. (ed.), Soil Micromorphology. University of Granada, pp. 935-965.

Jamagne, M., Jeanson, C. & Eimberck, M., 1987. Données sur la composition des argilanes en régions tempérées et continentales. In Fedoroff, N., Bresson, L.M. & Courty, M.A. (eds.), Micromorphologie des Sols, Soil Micromorphology. AFES, Plaisir, pp. 279-285.

Johnson, D.L., Johnson, D.N. & Moore, D.M., 2003. Deep and actively forming illuvial clay in the regolith and on bedrock. In Dominguez, E., Mas, G. & Cravero, F. (eds.), 2001, A Clay Odyssey. Elsevier, Amsterdam, pp. 205-210.

Jongerius, A., 1970. Some morphological aspects of regrouping phenomena in Dutch soils. Geoderma 4, 311-331.

Jongmans, A.G., Feijtel, T.C.J., Miedema, R., van Breemen, N. & Veldkamp, A., 1991. Soil formation in a Quaternary terrace sequence of the Allier, Limagne, France. Macro- and micromorphology, particle size distribution, chemistry. Geoderma 49, 215-239.

Jongmans, A.G., Van Oort, F., Buurman, P. & Jaunet, A.M., 1994. Micromorphology and submicroscopy of isotropic and anisotropic Al/Si coatings in a Quaternary Allier terrace, (France). In Ringrose-Voase, A.J. & Humphreys, G.S. (eds.), Soil Micromorphology: Studies in Management and Genesis. Developments in Soil Science, Volume 22, Elsevier, Amsterdam, pp. 285-291.

Kadereit, A., Kühn, P. & Wagner, G., 2010. Holocene relief and soil changes in loess-covered areas of south-western Germany – the pedosedimentary archives of Bretten-Bauerbach (Kraichgau). Quaternary International 222, 96-119.

Karkanas, P. & Goldberg, P., 2010. Phosphatic features. In Stoops, G., Marcelino, V. & Mees, F. (eds.), Interpretation of Micromorphological Features of Soils and Regoliths. Elsevier, Amsterdam, pp. 521-541.

Karkanas, P. & Goldberg, P., 2018. Phosphatic features. In Stoops, G., Marcelino, V. & Mees, F. (eds.), Interpretation of Micromorphological Features of Soils and Regoliths. Second Edition. Elsevier, Amsterdam, pp. 323-346.

Karpachevskiy, L.O., 1960. Micromorphological study of leaching and podzolisation of soils in a forest. Soviet Soil Science 5, 493-500.

Kemp, R.A., 1985. Soil Micromorphology and the Quaternary. Quaternary Research Association Technical Guide, Volume 2. Cambridge, 80 p.

Kemp, R.A., 1998. Role of micromorphology in paleopedological research. Quaternary International 51/52, 133-141.

Kemp, R.A., 2001. Pedogenic modification of loess: significance for palaeoclimatic reconstructions. Earth-Science Reviews 54, 145-156.

Kemp, R.A. & McIntosh, P.D., 1989. Genesis of a texturally banded soil in Southland, New Zealand. Geoderma 45, 65-81.

Kemp, R.A. & Zárate, M.A., 2000. Pliocene pedosedimentary cycles in the southern Pampas, Argentina. Sedimentology 47, 3-14.

Kemp, R.A., McDaniel, P.A. & Busacca, A.J., 1998. Genesis and relationship of macromorphology and micromorphology to contemporary hydrological conditions of a welded Argixeroll from the Palouse in Idaho. Geoderma 83, 309-329.

Khademi, H. & Mermut, A., 2003. Micromorphology and classification of Argids and associated gypsiferous Aridisols from central Iran. Catena 54, 439-455.

Khalifa, E.M. & Buol, S.W., 1968. Studies of clay skins in Cecil (Typic Hapludult) soil: I. Composition and genesis. Soil Science Society of America Proceedings 32, 857-861.

Khormali, F., Abtahi, A., Mahmoodi, S. & Stoops, G., 2003. Argillic horizon development in calcareous soils of arid and semiarid regions of southern Iran. Catena 53, 273-301.

Kooistra, M.J. & Pulleman, M.M., 2010. Features related to faunal activity. In Stoops, G., Marcelino, V. & Mees, F. (eds.), Interpretation of Micromorphological Features of Soils and Regoliths. Elsevier, Amsterdam, pp. 397-418.

Kooistra, M.J. & Pulleman, M.M., 2018. Features related to faunal activity. In Stoops, G., Marcelino, V. & Mees, F. (eds.), Interpretation of Micromorphological Features of Soils and Regoliths. Second Edition. Elsevier, Amsterdam, pp. 447-469.

Kovalev, R.V. (ed.), 1975. Soil Ecology and Soil Resources of Kemerovo Region [in Russian]. Nauka, Novosibirsk, 299 p.

Kovenya, S.V., 1972. Studies of Fine-Grained Particles Translocation in Sod-Podzolic Soil Applying Radioactive Tracers [in Russian]. Extended abstract of Dissertation, Leningrad, 14 p.

Kovda, I. & Mermut, A., 2010. Vertic features. In Stoops, G., Marcelino, V. & Mees, F. (eds.), Interpretation of Micromorphological Features of Soils and Regoliths. Elsevier, Amsterdam, pp. 109-127.

Kovda, I. & Mermut, A., 2018. Vertic features. In Stoops, G., Marcelino, V. & Mees, F. (eds.), Interpretation of Micromorphological Features of Soils and Regoliths. Second Edition. Elsevier, Amsterdam, pp. 663-689.

Kubiëna, W.L., 1937. Beiträge zur Kenntnis des Gefüges kohärenter Bodenmassen. Zeitschrift für Pflanzenernährung und Bodenkunde 2, 1-23.

Kubiëna, W.L., 1943. Gefügeuntersuchungen an tropischen und subtropischen Rotlehmen. Beiträge zur Kolonialforschung 3, 48-58.

Kubiëna, W.L., 1953. Bestimmungsbuch und Systematik der Böden Europas. Enke, Stuttgart. 392 p.

Kubiëna, W.L., 1956. Zur Mikromorfologie, Systematik und Entwicklung der rezenten und fossilen Lössboden. Eiszeitalter und Gegenwart 7, 102-112.

Kubiëna, W.L., 1986. Grundzüge der Geopedologie und der Formenwandel der Böden. Österreichischer Agrarverlag, Wien. 128 p.

Kühn, P., 2003a. Spätglaziale und holozäne Lessivégenese auf jungweichselzeitlichen Sedimenten Deutschlands. Greifswalder Geographische Arbeiten 28, Greifswald. 167 p.

Kühn, P., 2003b. Micromorphology and Late Glacial/Holocene Genesis of Luvisols in Mecklenburg-Vorpommern (NE-Germany). Catena 54, 537-555.

Kühn, P., 2003c. Besonderheiten pedogenetischer Prozesse in fluvialen und kolluvialen Sedimenten im Mamertal bei Mersch (Luxemburg). Bulletin de la Societé Préhistorique Luxembourgeoise 23/24, 21-30.

Kühn, P. & Pietsch, D., 2013. Soil micromorphogenesis and Early Holocene palaeoclimate at the desert margin of Southern Arabia. Spanish Journal of Soil Science 3, 59-77.

Kühn, P., Billwitz, K., Bauriegel, A., Kühn, D. & Eckelmann, W., 2006a. Distribution and genesis of Fahlerden (Albeluvisols) in Germany. Journal of Plant Nutrition and Soil Science 169, 420-433.

Kühn, P., Terhorst, B. & Ottner, F., 2006b. Micromorphology of Middle Pleistocene palaeosols in northern Italy. Quaternary International 156/157, 156-166.

Kühn, P., Techmer, A. & Weidenfeller, M., 2013. Lower to middle Weichselian pedogenesis and palaeoclimate in Central Europe using combined micromorphology and geochemistry: the loess-paleosol sequence of Alsheim (Mainz Basin, Germany). Quaternary Science Reviews 75, 43-58.

Kwaad, F.J.P.M. & Mücher, H.J., 1977. The evolution of soils and slope deposits in the Luxembourg Ardennes near Wiltz. Geoderma 17, 1-37.

Kwaad, F.J.P.M. & Mücher, H.J., 1979. The formation and evolution of colluvium on arable land in northern Luxembourg. Geoderma 22,173-192.

Lahmar, R. & Bresson, L., 1987. Genèse et fonctionnement des sols fersiallitiques sur micaschiste du Massif de Thenia. Algérie. In Fedoroff, N., Bresson, L.M. & Courty, M.A. (eds.), Micromorphologie des Sols, Soil Micromorphology. AFES, Plaisir, pp. 171-177.

Laves, D., 1972. Beitrag zur Mikromorphologie und Mikromorphogenese von Fahlerden (Lessivés). In Kowalinski, S. & Drozd, J. (eds.), Soil Micromorphology. Panstwowe Wydawnicto Naukowe, Warsawa, pp. 323-335.

Laves, D. & Thiere, J., 1970. Mikromorphologische, chemische und mineralogische Untersuchungen zur Entstehung körnungsdifferenzierter Böden im Jungmoränengebiet der DDR. Albrecht-Thaer-Archiv 14, 691-699.

Lebedeva-Verba, M.P. & Gerasimova, M.I., 2009. Macro- and micromorphological features of genetic horizons in a solonetzic soil complex at the Dzhanybek Research Station. Eurasian Soil Science 42, 239-252.

Lebedeva, M.P., Gerasimova, M.I., Golovanov, D.L. & Yamnova I.A., 2015. Extremely arid soils of the Ili Depression in Kazakhstan. Eurasian Soil Science 48, 11-26.

Legros, J.P., 1976. Migrations latérales et accumulations litées dans les arènes du massif cristallin et crystallophylien du Pilat (Ardèche, Loire, Haute-Loire). Science du Sol 3, 205-220.

Lindbo, D.L., Stolt, M.H. & Vespraskas, M.J., 2010. Redoximorphic features. In Stoops, G., Marcelino, V. & Mees, F. (eds.), Interpretation of Micromorphological Features of Soils and Regoliths. Elsevier, Amsterdam, pp 129-147.

Locke, W.W., 1986. Fine particle translocation in soils developed on glacial deposits, Southern Baffin Island, NWT, Canada. Arctic and Alpine Research 18, 33-43.

Loyko, S.V., Geras'ko, L.I., Kulizhskii, S.P., Amelin, I.I., Istigechev, G.I., 2015. Soil cover patterns in the northern part of the area of Aspen-Fir Taiga in the Southeast of Western Siberia. Eurasian Soil Science 48, 359-372.

Macphail, R.I., 1986. Paleosols in archaeology: their role in understanding Flandrian pedogenesis. In Wright, V.P. (ed.), Paleosols: their recognition and interpretation. Princeton University Press, Princeton, pp. 263-290.

Macphail, R., Romans, J.C.C. & Robertson, L., 1987. The application of micromorphology to the understanding of holocene soil development in the British Isles; with special reference to early cultivation. In Fedoroff, N., Bresson, L.M. & Courty, M.A. (eds.), Micromorphologie des Sols, Soil Micromorphology. AFES, Plaisir, pp. 647-656.

Macphail, R.I., Courty, M.A. & Gebhardt, A., 1990. Soil micromorphological evidence of early agriculture in north-west Europe. World Archaeology 22, 53-69.

Magaldi, D. & Tallini, M., 2000. A micromorphological index of soil development for the Quaternary geology research. Catena 41, 261-276.

Marcelino, V., Stoops, G. & Schaefer, C.E.G.R., 2010. Oxic and related materials. In Stoops, G., Marcelino, V. & Mees, F. (eds.), Interpretation of Micromorphological Features of Soils and Regoliths. Elsevier, Amsterdam, pp. 305-327.

Marcelino, V., Schaefer, C.E.G.R. & Stoops, G. 2018. Oxic and related materials. In Stoops, G., Marcelino, V. & Mees, F. (eds.), Interpretation of Micromorphological Features of Soils and Regoliths. Second Edition. Elsevier, Amsterdam, pp. 663-689.

Mathieu, C. & Stoops, G., 1974. Genesis of deep clay-rich horizons at the contact with the chalky substratum, in the north of France. In Rutherford, G.K. (ed.), Soil Microscopy. The Limestone Press, Kingston, pp. 455-480.

Matlack, K.S., Houseknecht, D.W. & Applin, K.R., 1989. Emplacement of clay into sand by infiltration. Journal of Sedimentary Petrology 59, 77-87.

McCarthy, P.J. & Plint, A.G., 1998. Recognition of interfluve sequence boundaries; integrating paleo-pedology and sequence stratigraphy. Geology 26, 387-390.

McKeague, J.A., 1983. Clay skins and argillic horizons. In Bullock, P. & Murphy, C.P. (eds.), Soil micro-morphology, Volume 2, Soil Genesis. AB Academic Publishers, Berkhamsted, pp. 367-387.

McKeague, J.A., MacDougall, J.I. & Miles, N.M., 1973a. Micromorphological, physical, chemical and mineralogical properties of a catena of soils from Prince Edward Island in relation to their classification and genesis. Canadian Journal of Soil Science 53, 281-295.

McKeague, J.A., Miles, N.M., Peters, T.W. & Hoffman, D.W., 1973b. A comparison of luvisolic soils from three regions in Canada. Geoderma 7, 49-69.

McKeague, J.A., Guertin, R.K., Page, F. & Valentine, K.W.G., 1978. Micromorpholgical evidence of illuvial clay in horizons designated Bt in the field. Canadian Journal of Soil Science 58, 179-186.

McKeague, J.A., Guertin, R.K., Valentine, K.W., Belisle, J., Bourbeau, G.A., Michalyna, W., Hopkins, L., Howell, L., Page, F. & Bresson, L.M., 1980. Variability of estimates of illuvial clay in soils by micro-morphology. Soil Science 129, 386-388.

McKeague, J.A., Wang, C., Ross, G.J., Acton, C.J., Smith, R.E., Anderson, D.W., Pettapiece, W.W. & Lord, T.M., 1981. Evaluation of criteria for argillic horizons (Bt) of soils in Canada. Geoderma 25, 63-74.

Mees, F., 2010. Authigenic silicate minerals – sepiolite-palygorskite, zeolites and sodium silicates. In Stoops, G., Marcelino, V. & Mees, F. (eds.), Interpretation of Micromorphological Features of Soils and Regoliths. Elsevier, Amsterdam, pp. 497-520.

Mees, F., 2018. Authigenic silicate minerals – sepiolite-palygorskite, zeolites and sodium silicates. In Stoops, G., Marcelino, V. & Mees, F. (eds.), Interpretation of Micromorphological Features of Soils and Regoliths. Second Edition. Elsevier, Amsterdam, pp. 177-203.

Melnikov, M.K. & Kovenya, S.V., 1974. Simulation studies of lessivage [in Russian]. Transactions of the 10th International Congress of Soil Science, Volume VI, Moscow, pp. 601-608.

Mermut, A. & Jongerius, A., 1980. A micromorphological analysis of regrouping phenomena in some Turkish soils. Geoderma 24, 159-175.

Mermut, A. & Pape, Th., 1971. Micromorphology of two soils from Turkey, with special reference to in-situ formation of clay cutans. Geoderma 5, 271-281.

Mermut, A. & Pape, Th., 1973. Mikromorphologie von in situ gebildeten Tonhäutchen in Böden. Leitz-Mitteilungen für Wissenschaft und Technik 5, 243-246.

Miedema, R., 1987. Soil formation, Microstructure and Physical Behaviour of Late Weichselian and Holocene Rhine Deposits in the Netherlands. PhD Dissertation, Agricultural University Wageningen, 340 p.

Miedema, R., 2002. Alfisols. In Lal, R. (ed.), Encyclopedia of Soil Science. Marcel Dekker Inc., New York, pp. 45-49.

Miedema, R. & Slager, S., 1972. Micromorphological quantification of clay illuviation. Journal of Soil Science 23, 309-314.

Miedema, R., van Engelen, E. & Pape, Th., 1978. Micromorphology of a toposequence of Late Pleistocene fluviatile soils in the eastern part of the Netherlands. In Delgado, M. (ed.), Soil Micromorphology. University of Granada, pp. 469-501.

Miedema, R., Slager, S., Jongmans, A.G. & Pape, Th., 1983. Amount, characteristics and significance of clay illuviation features in Late Weichselian Meuse terraces. In Bullock, P. & Murphy, C.P. (eds.), Soil micromorphology, Volume 2, Soil Genesis. AB Academic Publishers, Berkhamsted, pp. 519-529.

Miedema, R., Jongmans, A.G. & Brinkman, R., 1987. The micromorphology of a typical catena from Sierra Leone, West Africa. In Fedoroff, N., Bresson, L.M. & Courty, M.A. (eds.), Micromorphologie des Sols, Soil Micromorphology. AFES, Plaisir, pp. 137-145.

Miedema, R., Koulechova, I.N. & Gerasimova, M.I., 1999. Soil formation in Greyzems in Moscow district: micromorphology, chemistry, clay mineralogy and particle size distribution. Catena 34, 315-347.

Miller, F.R., Wilding, L.P. & Holowaychuk, N., 1971. Canfield silt loam, a Fragiudalf. II. Micromorphology, physical and chemical properties. Soil Science Society of America Proceedings 35, 324-331.

Minashina, N.G., 1958. Optically oriented clays in soils. Soviet Soil Science 4, 424-430.

Moretti, L. & Morrás, H., 2013. New microscopic evidences of the autochthony of the ferrallitic pedological mantle in the Misiones Province, Argentina. Latin American Journal of Sedimentology and Basin Analysis 20, 129-142.

Moretti, L., Morrás, H., Sanabria, J. & Argüello, G., 2012. Mineralogía y micromorfología de paleosuelos en la Pampilla de Los Gigantes, Córdoba. In Gianelli, V., Hernández, K., Reussi, N. & Studdert, G. (eds.), XIX Congreso Latinoamericano de la Ciencia del Suelo, Mar del Plata (CDROM).

Morrás, H., 1978. Contribution à la Connaissance Pédologique des 'Bajos Submeridionales', Province de Santa Fe, Argentine. Influence de ÍEnvironnement sur la Formation et ÍEvolution des Sols Halomorphes. PhD Dissertation, Université de Paris 7, 184 p.

Morrás, H., 1979. Quelques élements de discussion sur les mécanismes de pédogenèse des planosols et d'autres sols apparentés. Science du Sol 1, 57-66.

Morrás, H., 1983a. Characteristics and composition of some cutans and glaebules of probable artificial origin. In Bullock, P. & Murphy, C.P. (eds.), Soil micromorphology, Volume 1, Techniques and applications. AB Academic Publishers, Berkhamsted, pp. 253-263.

Morrás, H., 1983b. Some properties of degraded argillans from A2 horizons of Solodic Planosols. In Bullock, P. & Murphy, C.P. (eds.), Soil micromorphology, Volume 2, Soil Genesis. AB Academic Publishers, Berkhamsted, pp. 575-581.

Morrás, H., Bayarski, A., Benayas, J. & Vesco, C., 1993. Algunas características genéticas y litológicas de una toposecuencia de suelos vérticos de la Provincia de Entre Ríos, Argentina. In Gallardo, J. (ed.), El Estudio del Suelo y de su Degradación en Relación con la Desertificación. Sociedad Española de la Ciencia del Suelo, Salamanca, pp. 1054-1061.

Mpiana, K., 1980. Contribution à l'Etude des Profils Bauxitiques de Cote d'Ivoire et du Cameroun. Relations entre les Microstructures et la Minéralogie. Dissertation, Université Aix-Marseille 3, 187 p.

Mücher, H.J., 1974. Micromorphology of slope deposits: the necessity of a classification. In Rutherford, G.K. (ed.), Soil Microscopy. The Limestone Press, Kingston, pp. 553-566.

Mücher, H., van Steijn, H. & Kwaad, F., 2010. Colluvial and mass wasting deposits. In Stoops, G., Marcelino, V. & Mees, F. (eds.), Interpretation of Micromorphological Features of Soils and Regoliths. Elsevier, Amsterdam, pp. 37-48.

Mücher, H., van Steijn, H. & Kwaad, F., 2018. Colluvial and mass wasting deposits. In Stoops, G., Marcelino, V. & Mees, F. (eds.), Interpretation of Micromorphological Features of Soils and Regoliths. Second Edition. Elsevier, Amsterdam, pp. 21-36.

Muggler, C.C. & Buurman, P., 2000. Erosion, sedimentation and pedogenesis in a polygenetic oxisol sequence in Minas Gerais, Brazil. Catena 41, 3-17.

Mulyanto, B. & Stoops, G., 2003. Mineral neoformation in pore spaces during alteration and weathering of andesitic rocks in humid tropical Indonesia. Catena 54, 385-391.

Murasheva, V., Bronnikova, M., Panin, A., Pushkina, T., Adamiec, T. & Sheremetskaya, E., 2012. Geoarchaeology of the upper Dnieper River valley at Gnezdovo: field excursion. In Geoarchaeological Issues of the Upper Dnieper - Western Dvina River Region (Western Russia): Fieldtrip Guide. Universum, Moscow-Smolensk, pp. 20-48.

Murawski, H., 1998. Geologisches Wörterbuch. 10th Edition, Enke, Stuttgart, pp. 276.

Murphy, C.P., 1983. Point counting pores and illuvial clay in thin sections. Geoderma 31, 133-150.

Murphy, C.P. & Kemp, R.A., 1984. The over-estimation of clay and the under-estimation of pores in soil thin sections. Journal of Soil Science 35, 481-495.

Murphy, C.P. & Kemp, R.A., 1987. Micromorphology and the argillic horizon – a reappraisal. In Fedoroff, N., Bresson, L.M. & Courty, M.A. (eds.), Micromorphologie des Sols, Soil Micromorphology. AFES, Plaisir, pp. 257-261.

Nahon, D.B., 1991. Introduction to the Petrology of Soils and Chemical Weathering. John Wiley & Sons, New York, 313 p.

Nettleton, W.D., McCracken, R.J. & Daniels, R.B., 1968. Two North Carolina coastal plain catenas: II Micromorphology, composition and fragipan genesis. Soil Science Society of America Proceedings 32, 582-587.

Nettleton, W.D., Flach, K.W. & Brasher, B.R., 1969. Argillic horizons without clay skins. Soil Science Society of America Journal 33, 121-125.

Nettleton, W.D., Eswaran, H., Holzhey, C.S. & Nelson, R.E., 1987. Micromorphological evidence of clay translocation in poorly dispersible soils. Geoderma 40, 37-48.

Nettleton, W.D., Gamble, E.E., Allen, B.L., Borst, G. & Peterson, F.F., 1989. Relict soils of subtropical regions of the United States. Catena Supplement 16, 59-63.

Nettleton, W.D., Brasher, B.R., Baumer, O.W. & Darmody, R.G., 1994. Silt flow in soils. In Ringrose-Voase, A.J. & Humphreys, G.S. (eds.), Soil Micromorphology: Studies in Management and Genesis. Developments in Soil Science, Volume 22. Elsevier, Amsterdam, pp. 361-371.

Nikorych, V., Szymánski, W., Skiba, S. & Kryzhanivskiy, O., 2012. Features of cutans complex in Albeluvisols of the Ukrainian Precarpathians. Gruntoznavsto 13, 40-51.

Ogleznev, A.K., 1971. Diagnostic significance of clay crusts and soil clay in evaluation of podzol- and gley formation [in Russian]. Pochvovedenie 12, 12-22.

Oliveira, L.B., Ribeiro, M.R., Ferraz, F.B., Ferreira, M.G.V.X. & Mermut, A.R., 2004a. Mineralogia, micromorfologia e gênese de solos planossolicos do Sertão do Araripe, Estado de Pernambuco. Revista Brasileira de Ciência do Solo 28, 665-678.

Oliveira, L.B., Ferreira, M.G.V.X. & Marques, A.F., 2004b. Characterization and classification of two soils derived from basic rocks in Pernambuco State Coast, northeast Brazil. Scientia Agricola 61, 615-625.

Ortiz, I., Simón, M., Dorronsoro, C., Martín, F. & García, I., 2002. Soil evolution over the Quaternary period in a Mediterranean climate (SE Spain). Catena 48, 131-148.

Osman, A. & Eswaran, H., 1974. Clay translocation and vertic properties of some red Mediterranean soils. In Rutherford, G.K. (ed.), Soil Microscopy. The Limestone Press, Kingston, pp. 846-857.

Pal, D.K., Srivastava, P. & Bhattacharyya, T., 2003. Clay illuviation in calcareous soils of the semiarid part of the Indo-Gangetic Plains, India. Geoderma 115, 177-192.

Parfenova, E.I., 1956. A study of the minerals of podzolic soils with regard to their formation process [in Russian]. Kora Vyvetrivaniya 2, 31-44.

Parfenova, E.I. & Yarilova, E.A., 1957. Newly formed clay minerals in soils. Pochvovedenie 9, 37-48.

Parfenova, E.I. & Yarilova, E.A., 1965. Mineralogical Investigations in Soil Science. Israel Program for Scientific Translations, Jerusalem, 178 p.

Parfenova E.I., Mochalova, E.F. & Titova, N.A., 1964. Micromorphology and chemism of humus-clay new-formations in grey forest soils. In Jongerius, A. (ed.), Soil Micromorphology. Elsevier, Amsterdam, pp. 201-212.

Pawluk, S. & Bal, L., 1985. Micromorphology of selected mollic epipedons. In Douglas, L.A. & Thompson, M.L. (eds.), Soil Micromorphology and Soil Classification. Soil Science Society of America Special Publication 15, SSSA, Madison, pp. 63-83.

Payton, R.W., 1992. Fragipan formation in argillic brown earths (Fragiudalfs) of the Milfield Plain, north-east England: I. Evidence for a periglacial stage of development. Journal of Soil Science 43, 621-644.

Payton, R.W., 1993. Fragipan formation in argillic brown earths (Fragiudalfs) of the Milfield Plain, north-east England: II. Post Devensian developmental processes and the origin of Fragipan consistence. Journal of Soil Science 44, 703-723.

Pazos, M.S., 1984. Relación arcilla iluvial/arcilla total en Molisoles del sudeste de la Provincia de Buenos Aires. Ciencia del Suelo 2, 131-136.

Pazos, M.S. & Stoops, G., 1987. Micromorphological aspects of soil formation in mollisols from Argentina. In Fedoroff, N., Bresson, L.M. & Courty, M.A. (eds.), Micromorphologie des Sols, Soil Micromorphology. AFES, Plaisir, pp. 263-270.

Poch, R., Simó, I. & Boixadera, J., 2013. Benchmark soils on alluvial, fluvial, and fluvio-glacial formations of the upper-Segre valley. Spanish Journal of Soil Science 3, 78-94.

Polynov, B.B., 1915. Secondary minerals in ortsteinigenic soil horizons [*in Russian*]. Izvestija Pochvennoj Komissii 2, 135-138.

Ponomarenko, S.V., Targulian, V.O. & Shoba, S.A., 1988. Initial stages of soil formation in the forest zone on loamy sediments [*in Russian*]. In Dobrovolski, G.V. (ed.), Micromorphology of Anthropogenically Transformed Soils, Moscow, pp. 167-183.

Ranney, R.W. & Beatty, M.T., 1969. Clay translocation and albic tongue formation in two Glossoboralfs of West Central Wisconsin. Soil Science Society of America Proceedings 33, 768-775.

Ransom, M.D., Smeck, N.E. & Bigham, J.M., 1987. Micromorphology of seasonally wet soils on the Illinoian till plain, U.S.A. Geoderma 40, 83-99.

Rawling, J.E., 2000. A review of lamellae. Geomorphology 35, 1-9.

Remmelzwaal, A., 1979. Translocation and transformation of clay in Alfisols in early, middle and late Pleistocene coastal sands of southern Italy. Catena 6, 379-398.

Reynders, J.J., 1974. A study of a soil type with a textural B-horizon and expanding clays. Transactions of the 10th International Congress of Soil Science, Volume VII, Moscow, pp 218-227.

Rogaar, H., Lothammer, H., Van der Plas, L., Jongmans, A.G. & Bor, J., 1993. Phaeozem and Luvisol development in relation to relief and climate in Southwestern Rheinhessen, Germany. Mainzer Geowissenschaftliche Mitteilungen 22, 227-246.

Rognon, P., Coudé-Gaussen, G., Fedoroff, N. & Goldberg, P., 1987. Micromorphology of loess in the northern Negev (Israel). In Fedoroff, N., Bresson, L.M. & Courty, M.A. (eds.), Micromorphologie des Sols, Soil Micromorphology. AFES, Plaisir, pp. 631-638.

Rose, J., Lee, J.A., Kemp, R.A. & Harding, P.A., 2000. Palaeoclimate, sedimentation and soil development during the Last Glacial Stage (Devensian), Heathrow Airport, London UK. Quaternary Science Reviews 19, 827-847.

Runge, E.C.A., 1973. Soil development sequences and energy models. Soil Science 115, 183-193.

Rusanova, G.V., 1978. Micromorphology of Taiga Soils. Nauka, Leningrad, 152 p.

Sauer, D., Schülli-Maurer, I., Sperstad, R. & Sørensen, R., 2013. Micromorphological characteristics reflecting soil-forming processes during Albeluvisol development in S Norway. Spanish Journal of Soil Science 3, 38-58.

Schaetzl, R.J., Burns, S.F., Johnson, D.L. & Small, T.W., 1990. Tree uprooting: a review of types and patterns of soil disturbance. Physical Geography 11, 277-291.

Schlichting, E. & Blume, H.P., 1961. Das typische Bodenprofil auf jungpleistozänem Geschiebemergel in der westbaltischen Klimaprovinz und seine grundsätzliche Deutung. Zeitschrift für Pflanzenernährung und Bodenkunde 95, 193-208.

Schwan, J., Miedema, R. & Cleveringa, P., 1982. Pedogenic and sedimentary characteristics of a Late Glacial-Holocene solifluction deposit at Hjerupsgyde, Funen, Denmark. Catena 9, 109-138.

Scoppa, C., 1978. Micropedología de series de suelos característicos del noreste bonaerense. Revista de Investigaciones Agropecuarias, INTA, Serie 3, Clima y Suelo 14, 37-69.

Sedov, S., Solleiro-Rebolledo, E., Gama-Castro, J.E., Vallejo-Gómez, E. & González-Velázquez, A., 2001. Buried palaeosols of the Nevado de Toluca: an alternative record of Late Quaternary environmental change in central Mexico. Journal of Quaternary Science 16, 375-389.

Sedov, S.N., Solleiro-Rebolledo, E. & Gama-Castro, J.E., 2003. Andosol to Luvisol evolution in Central Mexico: timing, mechanisms and environmental setting. Catena 54, 495-513.

Sedov, S., Stoops, G. & Shoba, S., 2010. Regoliths and soils on volcanic ash. In Stoops, G., Marcelino, V. & Mees, F. (eds.), Interpretation of Micromorphological Features of Soils and Regoliths. Elsevier, Amsterdam, pp. 275-303.

Shishov, L.L., Tonkonogov, V.D., Lebedeva, I.I. & Gerasimova, M.I. (eds.), 2004. Classification and Diagnostic of Soils of Russia [in Russian]. Oykumena Publishing House, Smolensk, 342 p.

Slager, S. & Van de Wetering, H.T.J., 1977. Soil formation in archaeological pits and adjacent soils in southern Germany. Journal of Archaeological Science 4, 259-267.

Smeck, N.E. & Wilding, L.P., 1980. Quantitative evaluation of pedon formation in calcareous glacial deposits in Ohio. Geoderma 24, 1-16.

Smith, H. & Wilding, P., 1972. Genesis of Argillic horizons on Ochraqualfs derived from fine textured till deposits of Northwestern Ohio and Southeastern Michigan. Soil Science Society of America Proceedings 36, 808-815.

Smolikova, L., 1968. Mikromorphologie und Mikromorphometrie der Pleistozänen Bodenkomplexe. Rozpravy Československe Akademie Věd, Řada Matematických a Přírodních Věd 78, 75 p.

Soil Classification Working Group, 1998. The Canadian System of Soil Classification. Agriculture and Agri-Food Canada, Publication n° 1646 (revised). NRC Research Press, Ottawa, 187 p.

Soil Survey Staff, 1960. Soil Classification. A Comprehensive System. 7th Approximation. USDA, Government Printing Office, Washington, 265 p.

Soil Survey Staff, 2014. Keys to Soil Taxonomy. 12th Edition. USDA-Natural Resources Conservation Service, Washington, 360 p.

Sokolova, T.A., Shoba, S.A., Bgantsov, V.N. & Urusevskaya, I.S., 1988. Profile and intra-horizon differentiation of clay material in sod-podzolic soils on moraine. Soviet Soil Science 20, 39-49.

Solntseva, N.P. & Rubilina, N.E., 1987. Morphological analysis of soils transformed by coal-mining [in Russian]. Pochvovedenie 2, 105-118.

Spiridonova, I.A., Sedov, S.N., Bronnikova, M.A. & Targulian, V.O., 1999. Arrangement, composition, and genesis of bleached components of loamy Soddy-Podzolic soils. Eurasian Soil Science 32, 507-513.

Stephan, S., 1981. Zur Mikromorphologie der Tonverlagerung im Alluvium des Niederrheins. Sonderveröffentlichung des Geologischen Institutes der Universität Köln 41, 243-247.

Stephan, S., 1993. Mikromorphologie und Genese von Böden auf den Niederterrassen des Niederrheingebietes und der Kölner Bucht. Eiszeitalter und Gegenwart 43, 67-86.

Stephan, S., 2000. Bt-Horizonte als Interglazial-Zeiger in den humiden Mittelbreiten: Bildung, Mikromorphologie, Kriterien. Eiszeitalter und Gegenwart 50, 95-106.

Stephen, I., 1960. Clay orientation in soils. Science Progress 48, 322-331.

Stoops, G., 1967. Le profil d'altération au Bas Congo (Kinshasa). Sa description et sa genèse. Pedologie 17, 60-105.

Stoops, G., 1968. Micromorphology of some characteristic soils of the Lower Congo (Kinshasa). Pedologie 18, 110-149.

Stoops, G., 1989. Relict properties in soil of humid tropical regions with special reference to central Africa. In Bronger, A. & Catt, J.A. (eds.), Paleopedology. Nature and Application of Paleosols. Catena Supplement 16, pp. 95-106.

Stoops, G., 2003. Guidelines for Analysis and Description of Soil and Regolith Thin Sections. Soil Science Society of America, Madison, 184 p.

Stoops, G., 2013. A micromorphological evaluation of pedogenesis on Isla Santa Cruz (Galápagos). Spanish Journal of Soil Science 3, 14-37.

Stoops, G. & Marcelino, V., 2010. Lateritic and bauxitic materials. In Stoops, G., Marcelino, V. & Mees, F. (eds.), Interpretation of Micromorphological Features of Soils and Regoliths. Elsevier, Amsterdam, pp. 329-350.

Stoops, G. & Marcelino, V., 2018. Lateritic and bauxitic materials. In Stoops, G., Marcelino, V. & Mees, F. (eds.), Interpretation of Micromorphological Features of Soils and Regoliths. Second Edition. Elsevier, Amsterdam, pp. 691-720.

Stoops, G., Sedov, S. & Shoba, S., 2018. Regoliths and soils on volcanic ash. In Stoops, G., Marcelino, V. & Mees, F. (eds.), Interpretation of Micromorphological Features of Soils and Regoliths. Second Edition. Elsevier, Amsterdam, pp. 721-751.

Sullivan, L.A., 1994. Clay coating formation on impermeable materials: deposition by suspension retention. In Ringrose-Voase, A.J. & Humphreys, G.S. (eds.), Soil Micromorphology: Studies in Management and Genesis. Developments in Soil Science, Volume 22. Elsevier, Amsterdam, pp. 373-380.

Sullivan, L.A. & Koppi, A.J., 1991. Morphology and genesis of silt and clay coatings in the vesicular layer of a desert loam soil. Australian Journal of Soil Research 29, 579-586.

Sycheva, S. & Sedov, S., 2012. Paleopedogenesis during the Mikulino interglacial (MIS 5E) in the East-European Plain: buried toposequence of the key-section 'Alexandrov Quarry'. Boletin de la Sociedad Geologica Mexicana 64, 189-197.

Szymański, W., Skiba, M. & Skiba, S., 2011. Fragipan horizon degradation and bleached tongues formation in Albeluvisols of the Carpathian Foothills, Poland. Geoderma 167/168, 340-350.

Targulian, V.O. & Tselischeva, L.K., 1983. Net of fissures in the sod-podzolic soils and partluvation in the profile (macro- and micromorphological study) [*in Russian*]. In Targulian, V.O. (ed.), Micromorphological Diagnostic of Soils and Soil Forming Processes. Nauka, Moscow, pp. 33-68.

Targulian, V.O. & Vishnevskaya, I.V., 1975. Translocation of clay and silt particles in the profile of sod-podzolic soil [*in Russian*]. In Fridland, V.N. (ed.) Geochemical and Soil Aspects in Landscape Studies. MSU Publishing House, Moscow, pp. 26-42.

Targulian, V.O., Birina, A.G., Kulikov A.V., Sokolova, T.A. & Tselischeva, L.K., 1974a. Arrangement, Composition and Genesis of Sod-pale-podzolic Soil Derived from Mantle Loam. Morphological Investigation. Nauka, Moscow, 47 p.

Targulian, V.O., Sokolova, T.A., Birina, A.G., Kulikov, A.V. & Tselischeva, L.K., 1974b. Arrangement, Composition and Genesis of Sod-pale-podzolic Soil Derived from Mantle Loam. Analytical Investigation. Nauka, Moscow. 109 p.

Terribile, F. & FitzPatrick, E.A., 1992. The application of multilayer digital image processing techniques to the description of soil thin sections. Geoderma 55, 159-174.

Terribile, F. & FitzPatrick, E.A., 1995. The application of some image-analysis techniques to recognition of some micromorphological features. European Journal of Soil Science 46, 29-45.

Teruggi, M.E. & Andreis, R.R., 1971. Micromorphological recognition of paleosolic features in sediments and sedimentary rocks. In Yaalon, D.H. (ed.), Paleopedology - Origin, Nature and Dating of Paleosols. International Soil Science Society and Israel University Press, Jerusalem, pp.161-172.

Theocharopoulos, S.P. & Dalrymple, J.B., 1987. Experimental construction of illuviation cutans (channel argillans) with differing morphological and optical properties. In Fedoroff, N., Bresson, L.M. & Courty, M.A. (eds.), Micromorphologie des Sols, Soil Micromorphology. AFES, Plaisir, pp. 245-250.

Thiere, J. & Laves, D., 1968. Untersuchung zur Entstehung der Fahlerden, Braunerden und Staugleye im norddeutschen Jungmoränengebiet. Albrecht-Thaer-Archiv 12, 659-677.

Thiry, M. & Trauth, N., 1976. Evolution historique de la notion d'argile à silex. Bulletin d'Information des Géologues du Bassin de Paris 13, 41-48.

Thompson, M.L., 1987. Micromorphology of four argialbolls in Iowa. In Fedoroff, N., Bresson, L. M. & Courty, M.A. (eds.), Micromorphologie des Sols, Soil Micromorphology. AFES, Plaisir, pp. 271-277.

Thompson, M.L., Fedoroff, N. & Fournier, B., 1990. Morphological features related to agriculture and faunal activity in three loess-derived soils in France. Geoderma 46, 329-349.

Thompson, T.R.E., 1983. Translocation of fine earth in some soils from an area in mid Wales. In Bullock, P. & Murphy, C.P. (eds.), Soil micromorphology, Volume 2, Soil Genesis. AB Academic Publishers, Berkhamsted, pp. 531-539.

Tursina, T.V., 1989. Genesis and lithologic homogeneity of texturally differentiated soils. Soviet Soil Science 21, 25-39.

Urusevskaya, I.S., Sokolova, T.A., Shoba, S.A., Bagnovets, O.S. & Kujbysheva, I.P., 1987. Morphogenetic and genetic features of light-grey forest soil over mantle loams [in Russian]. Pochvovedenie 4, 5-16.

Usai, M.R., 1998. Textural Pedofeatures and Pre-Hadrian's Wall Ploughed Paleosols at Stanwix, Carlisle, Cumbria, UK. Reports from the Environmental Archaeology Unit, York 99/24, 22 p.

Usai, M.R., 2001. Textural features and Pre-Hadrian's Wall ploughed Paleosols at Stanwix, Carlisle, Cumbria, U.K. Journal of Archaeological Science 28, 541-553.

van den Broek, T.M.W., 1989. Clay Dispersion and Pedogenesis of Soils With an Abrupt Contrast in Texture. PhD Dissertation, Universiteit van Amsterdam, 109 p.

van der Meer, J.J.M. & Menzies, J., 2011. The micromorphology of unconsolidated sediments. Sedimentary Geology 238, 213-232.

van der Meer, J.J.M., Mücher, H.J. & Höfle, H.Ch., 1993. Micromorphological observations on till samples from the Shackleton Range and North Victoria Land, Antarctica. Polarforschung 62, 57-65.

Van Ranst, E. & De Coninck, F., 2002. Evaluation of ferrolysis in soil formation. European Journal of Soil Science 53, 513-520.

Van Ranst, E., Righi, D., De Coninck, F., Robin, A.M. & Jamagne, M., 1980. Morphology, composition and genesis of argillans and organans in soils. Journal of Microscopy 120, 353-361.

Van Ranst, E., Dumon, M., Tolossa, A.R., Cornelis, J.T., Stoops, G., Vandenberghe, R.E. & Deckers, J., 2011. Revisiting ferrolysis processes in the formation of Planosols for rationalizing the soils with stagnic properties in WRB. Geoderma 163, 265-274.

Van Ranst, E., Mees, F., Bock, L. & Langohr, R., 2014. Development of a clay-rich interval above a limestone substrate in the Condroz region of southern Belgium. Catena 121, 204-213.

Van Vliet, B. & Langohr, R., 1981. Correlation between fragipans and permafrost with special reference to silty Weichselian deposits in Belgium and northern France. Catena 8, 137-154.

Van Vliet-Lanoë, B., 1985. Frost effects in soils. In Boardman, J. (ed.), Soils and Quaternary Landscape Evolution. Wiley Publishers, London, pp. 117-158.

Van Vliet-Lanoë, B., 1988. Le Rôle de la Glace de Ségrégation Dans les Formations Superficielles de l'Europe de l'Ouest. Processus et Héritages. PhD Dissertation, Université de Paris 1, 854 p.

Van Vliet-Lanoë, B., 2010. Frost action. In Stoops, G., Marcelino, V. & Mees, F. (eds.), Interpretation of Micromorphological Features of Soils and Regoliths. Elsevier, Amsterdam, pp. 81-108.

Van Vliet-Lanoë, B. & Fox, C., 2018. Frost action. In Stoops, G., Marcelino, V. & Mees, F. (eds.), Interpretation of Micromorphological Features of Soils and Regoliths. Second Edition. Elsevier, Amsterdam, pp. 575-603.

Van Vliet-Lanoë, B., Fagnart, J.P., Langohr, R. & Munaut, A., 1992. Importance de la succession des phases écologiques anciennes et actuelles dans la différenciation des sols lessivés de la couverture loessique d'Europe occidentale: argumentation stratigraphique et archéologique. Science du Sol 30, 75-93.

Van Wambeke, A.W., 1972. Mathematical expression of eluviation illuviation processes and the computation of the effects of clay migration in homogeneous soil parent materials. Journal of Soil Science 23, 325-332.

Van Wambeke, A.W., 1974. Horizon thickness relationships in eluviation illuviation processes. Transactions of the 10th International Congress of Soil Science, Volume VI, Moscow, pp. 212-218.

Velichko, A.A. & Morozova, T.D., 1970. Peculiarities of Middle and Lower Pleistocene fossil soils in the Russian Plain. Palaeogeography, Palaeoclimatology, Palaeoecology 8, 221-236.

Verheye, W. & Stoops, G., 1974. Micromorphological evidences for the identification of an argillic horizon in Terra Rossa soils. In Rutherford, G.K. (ed.), Soil Microscopy. The Limestone Press, Kingston, pp. 817-831.

Vepraskas, M.J., Wilding, L.P. & Drees, L.R., 1994. Aquic conditions for Soil Taxonomy: concepts, soil morphology and micromorphology. In Ringrose-Voase, A.J. & Humphreys, G.S. (eds.), Soil Micromorphology: Studies in Management and Genesis. Developments in Soil Science, Volume 22. Elsevier, Amsterdam, pp. 117-131.

Vepraskas, M.J., Lindbo, D.L. & Stolt, M.H., 2018. Redoximorphic features. In Stoops, G., Marcelino, V. & Mees, F. (eds.), Interpretation of Micromorphological Features of Soils and Regoliths. Second Edition. Elsevier, Amsterdam, pp. 425-445.

von Zezschwitz, E., 1979/80. Reliktisches und jungholozänes Tonfließplasma in bronzezeitlichen Grabhügeln auf der Paderborner Hochfläche. Fundberichte aus Hessen 19/20, 423-447.

Weir, A.H., Catt, J. & Madgett, P.A., 1971. Postglacial soil formation in the loess of Pegwell Bay, Kent (England). Geoderma 5, 131-149.

West, L.T., Wilding, L.P. & Calhoun, F.G., 1987. Argillic horizons in sandy soils of the Sahel, West Africa. In Fedoroff, N., Bresson, L.M. & Courty, M.A. (eds.), Micromorphologie des Sols, Soil Micromorphology. AFES, Plaisir, pp. 221-225.

Wieder, M. & Yaalon, D.H., 1978. Grain cutans resulting from clay illuviation in calcareous soil material. In Delgado, M. (ed.), Soil Micromorphology. University of Granada, pp. 1133-1158.

Wright, W.R. & Foss, J.E., 1968. Movement of silt-sized particles in sand columns. Soil Science Society of America Proceedings 32, 446-448.

Zárate, M.A., Kemp, R.A. & Blasi, A.M., 2002. Identification and differentiation of Pleistocene paleosols in the northern Pampas of Buenos Aires, Argentina. Journal of South America Earth Sciences 15, 303-313.

Chapter 15

Redoximorphic Features

Michael J. Vepraskas[1], David L. Lindbo[2], Mark H. Stolt[3]

[1]NORTH CAROLINA STATE UNIVERSITY, RALEIGH, NC, UNITED STATES;
[2]UNITED STATES DEPARTMENT OF AGRICULTURE, WASHINGTON, DC, UNITED STATES;
[3]UNIVERSITY OF RHODE ISLAND, KINGSTON, RI, UNITED STATES

CHAPTER OUTLINE

1. Introduction

Redoximorphic features form in water-saturated and reduced soils. They can be identified by their distinctive colour patterns and are usually visible to the naked eye. Many terms have been used to describe redoximorphic features, including mottles, red mottles, grey mottles, gley spots, iron concretions, iron nodules, plinthite, iron stone, gley mottles and gley, to name just a few (e.g., Schlichting & Schwertmann,

1973; Stoops & Eswaran, 1985; Bouma et al., 1990; Veneman et al., 1998). The term 'mottle' is used by some authors to describe any feature that has a colour different from that of the surrounding soil matrix (e.g., Schoeneberger et al., 2012). Redoximorphic features are a specific subset of mottles. Soil scientists define redoximorphic features as those formed by the reduction, translocation and oxidation of iron and manganese compounds in the soil after water saturation and desaturation, respectively (Soil Science Society of America, 2001; Vepraskas, 2015). Schoeneberger et al. (2012) provide a flow chart to determine whether or not a feature is redoximorphic. This flow chart, although designed for field descriptions, can also be used in thin section studies.

The terminology used to describe redoximorphic features in thin sections is derived from sedimentary petrology and soil micromorphology (Brewer, 1976; Blatt, 1982; Bullock et al., 1985; FitzPatrick, 1993). Although these studies focus on viewing the features with the microscope, it is important to note that these features are also visible in hand samples. Many soil scientists unfamiliar with micromorphology are familiar with redoximorphic features and their interpretation.

The relevance of these features has increased over the last 20 to 30 years from being used primarily to understand soil genesis and palaeoenvironments (Veneman et al., 1976; Bullock et al., 1985; Vepraskas et al., 1994; Kemp, 1999; Vepraskas, 2015) to being used in the regulation of land-use issues related to waterlogged soils and wetlands. For example, these features are used to determine the depth to seasonal water saturation for onsite wastewater-system design in many jurisdictions within the United States (Lindbo et al., 2004). They are also used in the identification and delineation of hydric soils found in wetlands in the United States (USDA-NRCS, 2010). The formation of these features is explained in a monograph by Vepraskas (2015).

Although this book is focused on micromorphological interpretations, there is considerable overlap between redoximorphic features seen in thin section and those observed in the field. Soil micromorphological investigations must be considered with regard to the whole profile (Wilding & Drees, 1983). Accordingly, features in both hand sample and thin section will be discussed. Most detailed investigations begin by examining the soil with the naked eye, progress to hand lenses and eventually to microscopic evaluation. Since redoximorphic features are related to environmental conditions, some measurements of environmental conditions (e.g., water table levels and redox potentials) are critical to the interpretation of modern redoximorphic features. Once current environmental conditions related to redoximorphic feature formation are established, redoximorphic features in palaeosoils can be used to elucidate its palaeoenvironmental context.

2. General Genetic Aspects

Redoximorphic features form from the reduction, translocation and oxidation of Fe and Mn. An understanding of Fe and Mn chemistry is essential in the interpretation of redoximorphic features. In viewing a soil with the naked eye or a hand lens, the colours that are seen are due to oxide coatings (or lack of coatings) on mineral grains and/or impregnation of the groundmass. This is evident when the soil is viewed in thin section. In general, red, brown and yellow colours are due to Fe oxide coatings, dark purplish black colours are due to Mn oxides, and dull black colours are due to organic matter. More specifically in PPL, goethite and lepidocrocite are yellow, hematite is red and Mn oxides are black. If soil particles are uncoated they will often appear grey, white or the colour of the uncoated mineral grain.

Fe and Mn oxides are removed from particles if they are chemically reduced in the soil. This occurs under certain specific conditions: (i) saturation with stagnant, oxygen-depleted (anaerobic) water; (ii) presence of sufficient organic matter and microorganisms; and (iii) soil temperatures above biological zero (5°C) to allow biological activity (Vepraskas, 2015; Vepraskas et al., 2016).

As microorganisms decompose organic matter, electrons are transferred from the organic material to another compound or element that is called an electron acceptor (Megonigal et al., 1993). The electron acceptors are reduced in a well-defined order (Stumm & Morgan, 1981). Oxygen (O_2) is reduced first. Once the oxygen has been depleted (anaerobic conditions), nitrate (NO_3^-) in solution is reduced to nitrogen gas (N_2). These first two reactions leave no morphological signature in the soil. The reduction of Mn (Mn^{4+} or Mn^{3+} to Mn^{2+}) occurs next followed by reduction of Fe (Fe^{3+} to Fe^{2+}). The reduced forms of these elements being highly soluble, the Fe/Mn oxides dissolve off the soil particles. As the oxide coatings are removed, the underlying mineral grain colour becomes visible, the soil typically becomes grey and redox depletion features develop. If anaerobic conditions persist, sulphate (SO_4^{2-}) is next reduced to hydrogen sulphide (H_2S). Finally, carbon dioxide (CO_2) is reduced to methane (CH_4).

A result of Fe and Mn reduction is the destabilisation of clay particles because these oxides act as cementing agents. If the oxide coating on clay particles dissolves through reduction, clay particles are more easily dispersed in subsequent water saturation events (Brinkman, 1970; Brinkman et al., 1973; Morrás, 1983) and can be translocated. This process has been used in part to describe the degradation of clay (Aurousseau, 1990) and argillic or fragipan horizons (Brinkman et al., 1973; Bullock et al., 1974; Ransom et al., 1987; Lindbo et al., 2000) and the formation of E horizons (Tselishcheva, 1972; Morrás, 1983). If, on the other hand, the soil has sufficient buffering capacity, this process may not occur (Van Ranst & De Coninck, 2002; Boixadera et al., 2003).

As reduced Fe and Mn in solution move into oxidising environments, Fe will oxidise and precipitate from the soil solution before Mn (i.e., at a lower Eh) (e.g., Hseu & Chen, 1996; Kyuma, 2004). Even within a predominantly anaerobic soil, aerobic conditions occur wherever there are air-filled channels, entrapped air in intrapedal voids or aerobic water within peds which lacked sufficient organic matter to cause reducing conditions to occur. In these areas, the reduced elements oxidise and accumulate on the soil particles or ped surfaces (Callame & Dupuis, 1972; Vepraskas et al., 1974; Evans & Franzmeier, 1986; Stolt et al., 1994; Vepraskas & Vaughn, 2016).

3. Nature and Identification

The specific terminology of redoximorphic features depends on whether one is viewing a thin section under the microscope or a hand sample in the field (Table 1). In thin sections, redoximorphic features can be categorised into three groups: intrusive redox pedofeatures, impregnative redox pedofeatures and depletion redox pedofeatures (Figs. 1 to 4) (Stoops, 2003). Hand samples fall into three distinct groups: redox depletions, redox concentrations and reduced matrices (Vepraskas, 2015).

3.1   Intrusive Redox Pedofeatures

Intrusive redox pedofeatures are zones where oxidised Fe and/or Mn have accumulated as coatings or infillings in a void or as coatings on a grain or aggregate, due to changes in the oxidation state of the element (Fig. 2). These features were called mangans and ferrans in older literature (Brewer, 1976).

The identification of minerals present is done in part by colour (in reflected and transmitted plain polarised light). However, colour alone is not enough to identify Fe and Mn oxide minerals as a small amount of Mn can greatly darken a feature (e.g., Rhoton et al., 1993). Therefore, other means such as X-ray diffraction and SEM-EDS analyses may be necessary to confirm visual observations.

3.2   Impregnative Redox Pedofeatures

Impregnative redox pedofeatures are zones where oxidised Fe and/or Mn have accumulated in the matrix as nodules or as Fe/Mn oxide hypo- or quasicoatings along voids

Table 1   Terms commonly used to designate redoximorphic features in hand samples in the field and corresponding terms for thin sections

Redoximorphic features	
In hand samples in the field[a]	**In thin sections**[b]
Redox depletions:	Depletion redox pedofeatures:
Fe/Mn depletions	Fe/Mn oxide depletion hypocoatings
Clay depletions	Clay and Fe oxide depletion hypocoatings
Depleted matrix	Fe oxide-depleted groundmass
Redox concentrations:	Impregnative redox pedofeatures:
Fe/Mn masses	Fe/Mn oxide quasicoatings
Fe/Mn pore linings	and impregnative nodules
Fe/Mn nodules	Fe/Mn oxide hypocoatings
Fe/Mn concretions	Fe/Mn oxide impregnative nodules
	Fe/Mn oxide impregnative nodules
Fe/Mn pore linings	Intrusive redox pedofeatures:
	Fe/Mn oxide coatings and infillings

[a]Vepraskas (2015).
[b]Stoops (2003).

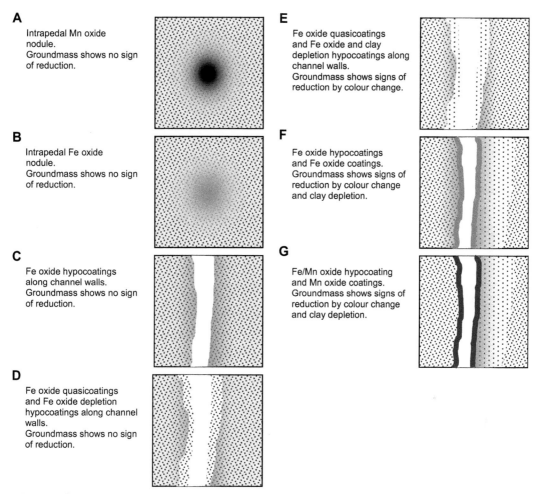

A
Intrapedal Mn oxide nodule. Groundmass shows no sign of reduction.

B
Intrapedal Fe oxide nodule. Groundmass shows no sign of reduction.

C
Fe oxide hypocoatings along channel walls. Groundmass shows no sign of reduction.

D
Fe oxide quasicoatings and Fe oxide depletion hypocoatings along channel walls. Groundmass shows no sign of reduction.

E
Fe oxide quasicoatings and Fe oxide and clay depletion hypocoatings along channel walls. Groundmass shows signs of reduction by colour change.

F
Fe oxide hypocoatings and Fe oxide coatings. Groundmass shows signs of reduction by colour change and clay depletion.

G
Fe/Mn oxide hypocoating and Mn oxide coatings. Groundmass shows signs of reduction by colour change and clay depletion.

FIGURE 1 Schematic overview of typical redoximorphic pedofeatures. In general, the degree of water saturation and reduction increases from A to G. Depth increases from A to G except in paddy soils where F will be closer to the surface and Mn oxide features will occur deeper in the profile.

or coarse mineral grains (Stoops, 2003) (Table 1; Fig. 3A to C), due to changes in the oxidation state of the element. In older literature (before the mid-1980s), they have been referred to as reddish or rusty mottles, neoferrans and quasiferrans, and neomangans (Brewer, 1976). Their colour in PPL is dependent on Fe and Mn oxide mineralogy and on the amount of oxides present (Rhoton et al., 1993). If the pedofeatures are dominated by Fe oxides, they have a higher chroma (i.e., appear more reddish, yellowish or brownish) than the adjacent groundmass. Typical Fe oxide minerals in these concentrations are goethite, ferrihydrite and lepidocrocite (Schwertmann & Taylor, 1989). If Mn oxides are present in the features, they have a lower value and chroma (appear darker or black) than the groundmass. Commonly hypocoatings will appear denser on the side proximal

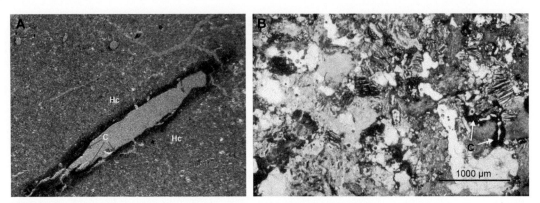

FIGURE 2 Intrusive redox pedofeature. (A) Fe oxide coatings (c) and Fe oxide hypocoatings (Hc), along a void in an Fe oxide-depleted groundmass as indicated by its grey colour (partially XPL). (B) Fe oxide coatings (C) along a void; the yellowish-brown groundmass shows no sign of reduction (PPL).

to the natural surface and will have a diffuse boundary on their distal side. They may appear denser (darker, opaque) throughout when the oxides completely coat smaller grains. An Fe-rich mineral grain affected by weathering may be surrounded by an Fe oxide hypocoating (Fig. 3C), but this would not be interpreted as an Fe redox pedofeature as key processes for their formation were not involved (reduction-oxidation related to water saturation and reduction).

Nodules (Fig. 3D and E) have been described in the literature using a number of terms including concretions, masses, glaebules and mottles (Brewer, 1976; Vepraskas et al., 1994; D'Amore et al., 2004; Vepraskas, 2015). In thin section descriptions, these terms, particularly 'concretions', have not been recommended for use (Bullock et al., 1985). These pedofeatures are orthic, disorthic or anorthic and can be further described based on their internal fabric, external morphology and boundaries (Stoops, 2003). Any coatings related to the nodule should be described for correct interpretation of these features (Tucker et al., 1994; D'Amore et al., 2004). For example a nodule coated by clay may no longer be actively forming.

3.3 Depletion Redox Pedofeatures

Depletion redox pedofeatures are zones where Fe/Mn oxides and/or clay have been removed, due to the reduction of Fe/Mn and subsequent mobilisation of these elements (Fe oxide depletions or Mn oxide depletions) (Fig. 4A and B), or are associated with the loss of clay (clay depletions) (Fig. 4C) (Schoeneberger et al., 2012; Vepraskas, 2015). All are identified in part by a grey colouration having a low chroma (less than 2, with a value of 4 or more).

Fe/Mn oxide depletions are the most common of the redox depletion pedofeatures. Such features have been referred to as albans, neoalbans, eluvial bodies, grey or gley mottles, depletions, wetness mottles (Daniels et al., 1968; Brewer, 1976; Veneman et al.,

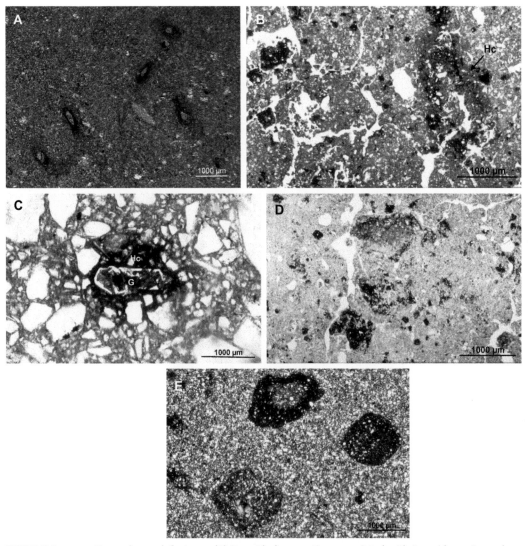

FIGURE 3 Impregnative redox pedofeatures. (A) Fe oxide hypocoatings associated with Fe oxide coatings along channels, in an Fe oxide-depleted groundmass as indicated by its grey colour (partially XPL). (B) Fe oxide hypo-coating (Hc) and yellowish-brown groundmass with no signs of reduction (PPL). (C) Fe oxide hypocoating (Hc) whose formation is related to oxidation of a mineral grain (G); this is not a redox feature because it is not due to reduction of Fe in water-saturated conditions. (D) Irregular Fe/Mn oxide nodules (PPL). (E) Fe/Mn oxide nodules with regular outline (XPL).

1976; Payton, 1993a, 1993b; James et al., 1995). These features have a lower Fe or Mn oxide content than the matrix as their colour suggests, yet their fabric is similar to that of the matrix. These features most often occur adjacent to voids and are described as depletion hypocoatings (Stoops, 2003). They often occur in association with Fe or Mn oxide quasicoatings (Veneman et al., 1976; Bouma et al., 1990; Vepraskas et al., 1994;

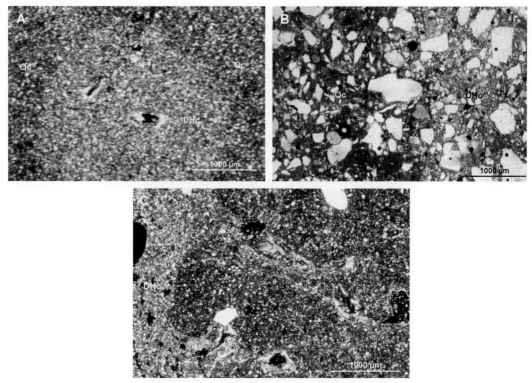

FIGURE 4 Depletion redox pedofeatures. (A) Fe oxide depletion hypocoating (DHc) in part of a sample with channels containing illuvial clay coatings; an Fe oxide quasicoating (Qc) occupies the remainder of the image (XPL). (B) Fe oxide depletion hypocoating (DHc) on right side of the image (lighter in colour) and Fe oxide quasicoating (Qc) on the left side (dark in colour) (PPL). (C) Clay and Fe oxide depletion hypocoating (DHc), identified by the grey colour; illuvial clay coatings in channels are evident in the right side of the image but absent in the left side (XPL).

Vepraskas, 2015). Fe/Mn oxide depletion hypocoatings have a porphyric c/f-related distribution pattern and an undifferentiated, stipple-speckled or unistrial b-fabric (Stoops & Eswaran, 1985; Hseu & Chen, 1999; Hseu et al., 1999). James et al. (1995) indicated that the fabric of the depletion feature is similar to that of the unaffected matrix and/or associated Fe oxide quasicoatings.

Clay depletions have been referred to as albic neoskeletans (Vepraskas & Wilding, 1983). These are zones where, besides Fe/Mn oxides, clay is removed as well. Clay depletions have a low chroma and are often related to pores in which case they are described as clay depletion hypocoatings. Unlike Fe oxide depletions, clay depletion features have a fabric that is different from that of the groundmass, because clay has been stripped from the zone (Figs. 1 and 4C). Clay depletion hypocoatings appear to have a better expressed stippled speckled b-fabric as compared to the groundmass (Brinkman et al., 1973; Bouma et al., 1990; Nettleton et al., 1994). Differentiation between clay depletion hypocoatings and coatings of illuvial coarse material can be difficult and may require investigation of thin sections of the full profile. One possible way to differentiate them is to look closely at any remaining clay in or adjacent to the clay depletion hypocoating. If this clay is highly oriented, the clay, and by default the coarser material, has moved by leaching and illuviation (Van Ranst & De Coninck, 2002; Boixadera et al., 2003). If the size and mineralogy of the coarse grains in the pedofeature are similar to that of the adjacent groundmass, then it is likely that the pedofeature is formed by clay depletion rather than as an illuvial accumulation of coarse material (Lindbo & Veneman, 1993; Payton, 1993b; Ajmone-Marsan et al., 1994; Boixadera et al., 2003). The presence of layering within a coarse-textured coating indicates that the feature is illuvial. A final approach that can assist in identification of clay depletion is to look deeper in the profile for the presence of eluviated clay (Vepraskas, 2015).

3.4   Quantification

Some land-use assessments are based on estimates of the percentage of redoximorphic features found in a given soil horizon (USDA-NRCS, 2010). The percentage of redoximorphic features in wetland soils, for example, has been related to the frequency and duration of saturation (He et al., 2003), which are expensive and time-consuming to measure. Estimates of percentages should be made on a 'representative elementary area' (REA) which is the smallest area that can be used to get a consistent estimate of a given feature, such as total porosity (VandenBygaart & Protz, 1999). Estimates of areal percentage for a feature in a sample should remain constant for cross-sectional areas larger than the REA. Estimates made on areas smaller than the REA will change and will not represent field conditions.

REA increases with the size of the features being described (VandenBygaart & Protz, 1999). O'Donnell et al. (2011) estimated that the REA needed to describe redoximorphic features is approximately 50 cm^2 for impregnative redox pedofeatures ('redox concentrations') in horizons for which field estimates of redox feature abundance are lower than 2%. This REA will probably vary among soil types. Where the thin section area is less than the REA, observations should be made on multiple thin sections until the REA is equalled or exceeded. REA can be best determined using digital imaging equipment (O'Donnell et al., 2010). Alternatively, abundance estimates can be made in the field on areas larger than the REA, and those values can be used in conjunction with thin section analysis where needed.

4. Significance

The type and location of redox pedofeatures within the soil profile are used to interpret the degree of water saturation (Table 2) (Veneman et al., 1976; Vepraskas & Bouma, 1976; Richardson & Hole, 1979; Stoops & Eswaran, 1985; Bouma et al., 1990; PiPujol & Buurman, 1994; Vepraskas et al., 1994; Hseu & Chen, 2001; Vepraskas, 2015). The pattern

Table 2 Relationship between the occurrence of specific redoximorphic pedofeatures and the depth and duration of water saturation in paddy soils

Depth	Saturation	Intrusive pedofeatures	Impregnative and depletion pedofeatures	Groundmass
Surface	Long	Fe oxide coatings	Fe oxide hypocoatings (less than Fe oxide coatings)	Fe oxide-depleted
Saturated interface	Long	Fe oxide coatings (more than at surface)	Fe oxide hypocoatings	Fe oxide-depleted
Unsaturated interface	Intermittent	Fe oxide coatings (less than hypocoatings) Mn oxide coatings	Fe oxide hypocoatings Fe oxide nodules Fe oxide depletion hypocoatings	
Below unsaturated interface	Short to none	Mn oxide coatings	Mn oxide nodules Fe/Mn oxide nodules Fe oxide nodules	
Deep in profile	None		Fe oxide nodules	

Hseu & Chen (1999) and Kyuma (2004).

of redox pedofeatures varies depending on the movement of water and air through the soil, the duration of water saturation and anaerobic conditions, and the presence or absence of organic matter.

4.1 Duration of Water Saturation

Short duration of water saturation or rapid fluctuation of water table levels results in intrapedal Fe/Mn oxide nodules and hypocoatings (Schwertmann & Fanning, 1976; Richardson & Hole, 1979; Arshad & St. Arnaud, 1980). At microsites where organic matter is present, the dissolved oxygen will be consumed rapidly by organic matter decomposing microorganisms when soil temperatures are above 5°C. This eventually leads first to the reduction of Mn and then to the reduction of Fe. If reduced Mn and Fe diffuse into the ped interior where carbon is not present and some dissolved oxygen remains in the water, they will oxidise to form an intrapedal impregnative nodule (Bouma et al., 1990; Tucker et al., 1994). If the soil contains enough Mn, then Mn nodules occur higher in the profile (closer to the surface) than Fe oxide nodules due to the difference in redox potential required for precipitation from solution (Veneman et al., 1976; Stoops & Eswaran, 1985).

If only Mn nodules are present the duration of water saturation has been estimated to be 2 to 3 days (Veneman et al., 1976), but the actual duration may vary among soils depending on the soil organic C content and soil temperature. If the duration of water saturation is long enough to reduce Fe and Mn, then hypocoatings made of these oxides can be observed. Nodules may still be present but the amount of Mn in the pedofeatures tends to be lower (Veneman et al., 1976; Stoops & Eswaran, 1985). This duration of water saturation is often not long enough to produce redox depletion hypocoatings (Richardson & Hole, 1979; Murphy, 1984; Bouma et al., 1990). As duration of water saturation increases and the soil becomes reduced for longer periods, goethite and lepidocrocite predominate in the redoximorphic features (Arocena et al., 1994; PiPujol & Buurman, 1997). In more oxidising and warmer environments, the Fe in redox pedo-features tends to be dominated by hematite and goethite (Arocena et al., 1994).

The formation of concentric nodules occurs over multiple wet/dry cycles or fluctu-ations in the water table (Morrás, 1983; Bouma et al., 1990; D'Amore et al., 2004). The Fe and Mn oxides present in the layers of the concentric nodules can be used to determine changes in the environment in which each layer formed (Menning & Shoba, 1980; Landuydt, 1990; White & Dixon, 1996; Liu et al., 2002).

As the duration of water saturation increases to a few weeks, the type and pattern of redox pedofeature changes from only impregnative to both impregnative and depletion-related pedofeatures (Veneman et al., 1976; Stoops & Eswaran, 1985; Bouma et al., 1990; He et al., 2003). The increased duration of water saturation results in longer periods of reduction in areas adjacent to a source of organic carbon. In these zones, redox depletion hypocoatings are formed. The reduced Fe and Mn may diffuse into the ped where they oxidise to form Fe/Mn redox quasicoatings.

Intrapedal nodules become smoother and less abundant as the duration of water saturation increases (D'Amore et al., 2004). Manganese remains reduced longer than Fe and may be leached deeper into the profile (Stoops & Eswaran, 1985; Kyuma, 2004). Clay depletion hypocoatings can also occur. They are most often associated with root channels or large voids containing organic material such as decaying roots (Payton, 1993b; Vepraskas et al., 1994). High concentrations of base cations in the soil may limit the formation of clay depletion hypocoatings (Boixadera et al., 2003).

The patterns of features associated with short and medium duration of water satu-ration have been related mostly to surface water gley (Murphy, 1984; PiPujol & Buurman, 1994) but also to episaturated horizons and the zone of greatest water table fluctuation in endosaturated soils (Vepraskas et al., 1994; Boixadera et al., 2003).

Under long duration (many weeks to months) of water saturation and reduction, Fe oxide depletion pedofeatures occupy the entire groundmass (Bouma et al., 1990; Vepraskas, 2015). The Fe oxide-depleted groundmass appears greyer or lighter in colour as compared to the groundmass of soils profiles or horizons where water saturation occurs for a short to medium duration. Several researchers noted that the fine material in an Fe oxide-depleted pedofeature and/or groundmass has a more speckled appear-ance (grainy limpidity), which they explained by 'ferrolysis' leading to the destruction of some clay minerals (Brinkman, 1970, 1977; Brinkman et al., 1973; Brammer & Brinkman, 1977; Murphy, 1984; Payton 1993a, 1993b; Hseu et al., 1999; Zhang & Gong, 2003; Kyuma, 2004). However, this process has been questioned by Van Ranst and De Coninck (2002).

If conditions are such that SO_4^{2-} is reduced in the presence of reduced Fe, pyrite may form (Slager et al., 1970; Stoops & Eswaran, 1985; see also Mees & Stoops, 2010, 2018) indicating extremely reduced conditions. If the pore water is saline then barite may form, and if carbonates are present siderite may result (Stoops & Eswaran, 1985; PiPujol & Buurman, 1997; McMillan & Schwertmann, 1998).

When these saturated horizons are drained, the macropores will fill with air first while the pores in the soil matrix remain filled with water. Reduced Fe in the soil matrix adjacent to the macropores diffuses toward areas of lower concentrations of dissolved Fe that are found along the air-filled pores, where it oxidises and precipitates to form Fe oxide coatings and hypocoatings (Callame & Dupuis, 1972; Veneman et al., 1976; Evans & Franzmeier, 1986; Bouma et al., 1990; Lindbo & Veneman, 1993; Stolt et al., 1994). The resulting pattern is one of grey ped interiors with Fe oxide hypocoatings and coatings along voids and ped faces (Vepraskas et al., 1974; Veneman et al., 1976; Vepraskas & Bouma, 1976; Bouma et al., 1990; Kemp et al., 1998). This pattern representing long duration of water saturation has been related to ground water gley (Murphy, 1984; Stoops & Eswaran, 1985) as well as to endosaturation (Vepraskas et al., 1994; Vepraskas, 2015).

Assuming that the organic matter content is sufficient to promote Fe reduction and that the groundmass contains at least 3% free iron, Vepraskas and Guertal (1992) developed a nomograph to quantitatively assess the time required for the formation of a 2 mm thick Fe oxide depletion hypocoating, relating the length of time during which Fe^{2+} is in solution (i.e., the amount of time the soil is saturated) to the concentration of Fe^{2+} in solution. Thus, data on carbon content, Fe content and duration of water saturation are necessary to fully estimate the time required for a feature to form. Furthermore, as the amount of depletion (size and percentage) increases, both the duration and the frequency of water saturation increase (He et al., 2003; Morgan & Stolt, 2006; Severson et al., 2008).

4.2   Special Cases

4.2.1   Stratified Soils

Stratified soils, specifically where a layer with fine texture overlies a layer with coarse texture, exhibit redoximorphic patterns that do not completely agree with the patterns discussed in the previous section. In these soils, the fine-textured layer may have a high water content with the pores in the soil matrix being filled with water and reduced, but the macropores along ped faces and channels may remain air-filled. This leads to the formation of an (intrapedal) Fe-depleted groundmass and Fe depletion hypocoatings along ped faces or around channels (Vepraskas et al., 1974; Veneman et al., 1976; Clothier et al., 1978; Morgan & Stolt, 2006). The underlying coarser material may be devoid of redoximorphic features (Clothier et al., 1978). If this pattern occurred in a groundwater gley (reduced groundmass), the underlying horizon also would show redoximorphic features (Vepraskas, 2015).

4.2.2   Paddy Soils

Anthrosaturation (human controlled) occurs as soil is intentionally flooded and ponded for agronomic reasons. Most often this occurs in the cultivation of rice (paddy soils). The long duration of water saturation and reduction, coupled with the agronomic practices, result in a distinctive pattern of redoximorphic features (Table 2) (Hseu & Chen, 1996, 1999; Zhang & Gong, 2003; Kyuma, 2004). In these soils, which are both saturated and reduced (Fe^{2+} present in pore water) and where actively respiring plant roots bring oxygen into the rooting zone, iron coatings and hypocoatings are formed along channel walls associated with living roots (oxidised rhizospheres; USDA-NRCS, 2010) (Chen et al., 1980; Xu & Zhu, 1983), by diffusion of reduced Fe toward the actively respiring root where dissolved O_2 is available. If reduced Mn is present, it will oxidise in a less-reducing zone, often closer to the living root or on the proximal edge of the coating. Translocation of reduced Mn deeper into the soil results in the occurrence of Mn nodules and hypo-coatings at the interface between the saturated zone of the paddy soil and the underlying unsaturated soil (Kyuma, 2004). Nodules are also more common as the depth to the saturated/unsaturated soil interface increases and where the water saturation has the greatest fluctuations (Hseu & Chen, 2001; Kyuma, 2004). The shape of nodules seems to be affected by cultivation, as they can be broken and disrupted physically (Hseu & Chen, 2001). The Fe oxide-depleted groundmass of paddy soils is similar in appearance to the groundmass in groundwater gley soils (Hseu et al., 1999; Zhang & Gong, 2003; Kyuma, 2004).

4.2.3   Bog Iron

'Bog iron' is a large-scale redoximorphic feature observed in areas where groundwater rich in reduced Fe moves to a more oxidising environment (Stoops, 1983; Landuydt, 1990; Curmi et al., 1994a, 1994b; Kaczorek & Sommer, 2003). Bog iron is highly porous and composed of different intervals, which have formed due to contrasting microenvironmental conditions, such as changes in redox status (Stoops, 1983; Landuydt, 1990). In some alluvial environments, the upper interval can be a layer composed predominately of clay and diatoms (Stoops, 1983). The next interval contains optically isotropic Fe oxides and zones of banded radiating acicular goethite needles on void walls and ferrihydrite nodules (Landuydt, 1990). It has been assumed that the upper intervals are periodically oxidised zones. The next layer is dominated by minerals such as vivianite and siderite and assumed to be predominantly reduced. Siderite is often observed as coatings on root channels as well as discrete crystals and spherulites (Stoops, 1983; Landuydt, 1990). Vivianite also occurs as coatings on voids and may be associated with decaying roots in channels (see Karkanas & Goldberg, 2010, 2018). The lowest interval may occasionally contain discrete pyrite framboids. Landuydt (1990) observed that within each major zone some microzonation (both vertical and horizontal) of ferric-ferrous mineralogy can be observed, representing fluctuating redox conditions.

4.2.4   Dark Parent Material

In soils derived from dark parent materials (value 4 or less, chroma 2 or less), redoximorphic features may be difficult if not impossible to recognise in the field. Fe oxide depletion pedofeatures (hypocoatings) were observed in thin sections of soils on dark carboniferous glacial till (Stolt et al., 2001). These observations coupled with water table data confirm both water saturation and reduction at the site. Similarly, soils derived from high chroma Fe-rich parent materials may not show the typical low chroma depletion features (chroma 2 or less). However, thin section analysis revealed depletion hypo-coatings where Fe had indeed been reduced and translocated (Elles & Rabenhorst, 1994).

4.3   Relict Versus Contemporary Features

Relict redoximorphic features are those that have stopped forming in the soil horizon in which they now occur or in some cases were inherited from the parent material. They are of interest because their presence is good evidence that the soil's hydrology has changed from the time the features first formed. In most cases, soils with relict redoximorphic features should be drier and more oxidised than when the features initially developed. The redoximorphic features most likely to be relict are Fe/Mn nodules. When these features are hard, they are very resistant to being broken down and mixed into the groundmass. Most other redoximorphic features are friable and therefore more likely to be destroyed.

Separating relict features from contemporary ones that are still forming in present time is often problematic. Combining micromorphological observations with data on current hydrology, such as records of water table fluctuation, can aid in determining whether features are contemporary or not (Tucker et al., 1994; Greenberg & Wilding, 1998; D'Amore et al., 2004). Conversely, relict features can be used as a less expensive but also a less reliable way to assess current hydrology.

A further question that can be addressed by micromorphological investigation is whether the feature is actively forming or degrading (Vepraskas & Wilding, 1983; Kemp et al., 1998).

The nature of the boundaries of redoximorphic features, nodules in particular, have been the focus of several investigations that have studied relict redoximorphic features. The presence of a diffuse boundary of a nodule is often considered to be indicative of contemporary development (Suddhiprakarn & Kheoruenromne, 1994; Tucker et al., 1994; Costantini and Priori, 2007). A sharp boundary suggests the feature is no longer forming (Tucker et al., 1994). This does not always mean the nodule did not form *in situ* (Stolt et al., 1994; Lindbo et al., 2000; D'Amore et al., 2004). Stolt et al. (1994) found that certain nodules having abrupt boundaries may form in place if there are isolated areas of contrasting particle size within a horizon.

The roundness or sphericity of the nodules may also be used. Rounded orthic nodules have been considered to reflect current hydrology, especially if drainage is restricted (Schwertmann & Fanning, 1976; Richardson & Hole, 1979; Elles & Rabenhorst, 1994; Elles et al., 1996). However, McCarthy et al. (1998) consider rounded nodules as being transported and not formed *in situ*. Nodules that are degrading will often have sand or silt grains protruding from their boundaries and will have irregular digitate or mammillate shapes (Tucker et al., 1994). They may also have Fe oxide depletion hypocoatings along their sides (Mücher & Coventry, 1994). Tucker et al. (1994) indicated that the more pronounced the degradation of the nodules, the greater the degree of water saturation.

The occurrence of Fe oxide hypocoatings and nodules in the same horizon (Greenberg & Wilding, 1998) and the presence of mycelia or fungal hyphae around an Fe oxide nodule (Zauyah & Bisdom, 1983) have also been used to suggest that the feature is contemporary.

If the redoximorphic feature is related to current structural voids, it is likely to be pedogenic in origin and formed *in situ* but not necessarily currently forming (Tucker et al., 1994; Wagner & Rabenhorst, 1994; Greenberg & Wilding, 1998; Lindbo et al., 2000). If the redoximorphic features are coated with illuvial clay but still related to the structural voids then they are most likely not contemporary (Vepraskas & Wilding, 1983; Tucker et al., 1994). On the other hand, if there are associated clay depletion hypocoatings, as well as nodules with irregular boundaries or features with diffuse boundaries, the redoximorphic features are more likely to be contemporary.

All types of features (contemporary, relict, *in situ* or inherited) may be found in the same horizon (Tucker et al., 1994; Busacca & Cremaschi, 1998; Muggler et al., 2001). In this case, the presence of contemporary features suggests some features are related to current hydrology. In the end, the deduced sequence of formation of the features may be used to understand the stages of pedogenesis in the profile (Bullock et al., 1974; Richardson & Hole, 1979; Lindbo et al., 2000; D'Amore et al., 2004).

Redoximorphic features in palaeosoils may help determine the climatic conditions during soil formation. However, these features are not completely static or stable. Several investigations have examined diagenetic changes in mineralogy associated with palaeosoils (McCarthy et al., 1998; Kemp, 1999; Reolid et al., 2008). Still other investigations have used changes in mineralogy to look at how the reducing environment and therefore drainage changed after the formation of a given feature (McCarthy & Plint, 1998; Choi, 2005). In both of these studies a more reducing environment was determined based on the presence of spherulitic siderite. Despite changes in mineralogy, the type of features appears to remain unchanged, so interpretation based on the morphology of a feature rather than on its mineralogical composition is valid (McCarthy et al., 1998).

5. Conclusions

Redoximorphic features are common in most soils where water saturation occurs and their presence is used extensively in making land use decisions. Although redoximorphic features are visible in the field with both the naked eye and a hand lens, micromorphological analysis can further enhance our understanding of how these features form and how to interpret them correctly. These interpretations are best made when supportive information such as water table and climatic data are available (Bouma et al., 1990; Vepraskas, 2015), but they can to a large extent be based on extensive micromorphological data correlating redoximorphic pedofeatures to environmental conditions.

References

Ajmone-Marsan, F., Pagliai M. & Pini, R., 1994. Identification and properties of Fragipan soils in the Piemonte region of Italy. Soil Science Society of America Journal 58, 891-900.

Arocena, J.M., Pawluk, S. & Dudas, M.J., 1994. Iron oxides in iron-rich nodules of sandy soils from Alberta (Canada). In Ringrose-Voase, A.J. & Humphreys, G.S. (eds.), Soil Micromorphology: Studies in Management and Genesis. Developments in Soil Science, Volume 22. Elsevier, Amsterdam, pp. 83-97.

Arshad, M.A. & St. Arnaud, R.J., 1980. Occurrence and characteristics of ferromanganiferous nodules in some Saskatchewan soils. Canadian Journal of Soil Science 60, 685-695.

Aurousseau, P., 1990. A microscopic and mineralogical study of clay degradation in acid and reducing conditions. In Douglas, L.A. (ed.), Soil Micromorphology: A Basic and Applied Science. Developments in Soil Science, Volume 19. Elsevier, Amsterdam, pp. 289-294.

Blatt, H., 1982. Sedimentary Petrology. Freeman W.H. & Company, New York, 564 p.

Boixadera, J., Poch, R.M., García-González, M.T. & Vizcayno, C., 2003. Hydromorphic and clay-related processes in soils from the Llanos de Moxos (northern Bolivia). Catena 54, 403-424.

Bouma, J., Fox, C.A. & Miedema, R., 1990. Micromorphology of hydromorphic soils: applications for soil genesis and land evaluation. In Douglas, L.A. (ed.), Soil Micromorphology: A Basic and Applied Science. Developments in Soil Science, Volume 19. Elsevier, Amsterdam, pp. 257-278.

Brammer, H. & Brinkman, R., 1977. Surface-water gley soils in Bangladesh: environment, landforms and soil morphology. Geoderma 17, 91-109.

Brewer, R., 1976. Fabric and Mineral Analysis of Soils. Robert E. Krieger Publishing Company, Huntington, New York, 482 p.

Brinkman, R., 1970. Ferrolysis: a hydromorphic soil forming process. Geoderma 3, 199-206.

Brinkman, R., 1977. Surface-water gley soils in Bangladesh: genesis. Geoderma 17, 111-144.

Brinkman, R., Jongmans, A.G., Miedema, R. & Maaskant, P., 1973. Clay decomposition in seasonally wet, acid soils: micromorphological, chemical and mineralogical evidence from individual argillans. Geoderma 10, 259-270.

Bullock, P., Milford, M.H. & Cline, M.G., 1974. Degradation of argillic horizons in Udalf soils of New York. Soil Science Society of America Proceedings 38, 621-628.

Bullock, P., Fedoroff, N., Jongerius, A., Stoops, G., Tursina, T. & Babel, U., 1985. Handbook for Soil Thin Section Description. Waine Research Publications, Wolverhampton, 152 p.

Busacca, A. & Cremaschi, M., 1998. The role of time versus climate in the formation of deep soils of the Apennine fringe of the Po Valley, Italy. Quaternary International 51/52, 95-107.

Callame, B. & Dupuis, J., 1972. Iron transformations and salt-marsh soil ferruginous deposits of the intertidal zone of the Point D'Arcay (Vendee). Science du Sol 2, 33-60.

Chen, C.C., Dixon, J.B. & Turner, F.T., 1980. Iron coatings on rice roots: morphology and models of development. Soil Science Society of America Journal 44, 1113-1119.

Choi, K., 2005. Pedogenesis of late Quaternary deposits, northern Kyonggi Bay, Korea: implications for relative sea-level change and regional stratigraphic correlation. Palaeogeography, Palaeoclimatology, Palaeoecology 220, 387-404.

Clothier, B.E., Polluk, J.A. & Scotter, D.R., 1978. Mottling in soil profiles containing a coarse-textured horizon. Soil Science Society of America Journal 42, 761-763.

Costantini, E.A.C. & Priori, S., 2007. Pedogenesis of plinthite during early Pliocene in the Mediterranean environment: case study of a buried paleosol at Podere Renieri, central Italy. Catena 71, 425-443.

Curmi, P., Soulier, A. & Trolard, F., 1994a. Forms of iron oxides in acid hydromorphic soil environments. Morphology and characterization by selective dissolution. In Ringrose-Voase, A.J. & Humphreys, G.S. (eds.), Soil Micromorphology: Studies in Management and Genesis. Developments in Soil Science, Volume 22. Elsevier, Amsterdam, pp. 141-148.

Curmi, P., Widiatmaka, Pellerin, J. & Ruellan, A., 1994b. Saprolite influence on formation of well-drained hydromorphic horizons in an acid soil system as determined by structural analysis. In Ringrose-Voase, A.J. & Humphreys, G.S. (eds.), Soil Micromorphology: Studies in Management and Genesis. Developments in Soil Science, Volume 22. Elsevier, Amsterdam, pp. 133-141.

D'Amore, D.V., Stewart, S.R. & Huddleston, J.H., 2004. Saturation, reduction, and the formation of iron-manganese concretions in the Jackson-Frazier Wetland, Oregon. Soil Science Society of America Journal 68, 1012-1022.

Daniels, R.B., Gamble, E.E. & Bartelli, L.J., 1968. Eluvial bodies in B-horizons of some Ultisols. Soil Science 106, 200-206.

Elles, M.P. & Rabenhorst, M.C., 1994. Micromorphological interpretations of redox processes in soils derived from Triassic red bed parent materials. In Ringrose-Voase, A.J. & Humphreys, G.S. (eds.), Soil Micromorphology: Studies in Management and Genesis. Developments in Soil Science, Volume 22. Elsevier, Amsterdam, pp. 171-178.

Elles, M.P., Rabenhorst, M.C. & James, B.R., 1996. Redoximorphic features in soils of the Triassic Culpepper Basin. Soil Science 161, 58-69.

Evans, C.V. & Franzmeier, D.P., 1986. Saturation, aeration, and color patterns in a toposequence of soils in north-central Indiana. Soil Science Society of America Journal 50, 975-980.

FitzPatrick, E.A., 1993. Soil Microscopy and Micromorphology. John Wiley & Sons, Chichester, 304 p.

Greenberg, W.A. & Wilding, L.P., 1998. Evidence of contemporary and relict redoximorphic features of an Alfisol in East-Central Texas. In Rabenhorst, M.C. Bell, J.C. & McDaniel, P.A. (eds.), Quantifying Soil Hydromorphology. SSSA Special Publication 54, SSSA, Madison, pp. 227-246.

He, X., Vepraskas, M.J., Lindbo, D.L., & Skaggs, R.W., 2003. A method to predict soil saturation frequency and duration from soil color. Soil Science Society of America Journal 57, 961-969.

Hseu, Z.Y. & Chen, Z.S., 1996. Saturation, reduction, and redox morphology of seasonally flooded Alfisols in Taiwan. Soil Science Society of America Journal 60, 941-949.

Hseu, Z.Y. & Chen, Z.S., 1999. Micromorphology of redoximorphic features of subtropical anthraquic Ultisols. Food Science and Agricultural Chemistry 1, 194-202.

Hseu, Z.Y. & Chen, Z.S., 2001. Quantifying soil hydromorphology of a rice-growing Ultisol Toposequence in Taiwan. Soil Science Society of America Journal 65, 270-278.

Hseu, Z.Y., Chen, Z.S. & Wu, Z.D., 1999. Characterization of placic horizons in two subalpine forest Inceptisols. Soil Science Society of America Journal 63, 941-947.

James, H.R., Ransom, M.D. & Miles, R.J., 1995. Fragipan genesis in poly genetic soils on the Springfield Plateau of Missouri. Soil Science Society of America Journal 59, 151-160.

Kaczorek, D. & Sommer, M., 2003. Micromorphology, chemistry, and mineralogy of bog iron ores from Poland. Catena 54, 393-402.

Karkanas, P. & Goldberg, P., 2010. Phosphatic features. In Stoops, G., Marcelino, V. & Mees, F. (eds.), Interpretation of Micromorphological Features of Soils and Regoliths. Elsevier, Amsterdam, pp. 521-541.

Karkanas, P. & Goldberg, P., 2018. Phosphatic features. In Stoops, G., Marcelino, V. & Mees, F. (eds.), Interpretation of Micromorphological Features of Soils and Regoliths. Second Edition. Elsevier, Amsterdam, pp. 323-346.

Kemp, R.A., 1999. Micromorphology of loess-paleosol sequences: a record of paleoenvironmental change. Catena. 35, 179-196.

Kemp, R.A., McDaniel, P.A. & Busacca, A.J., 1998. Genesis and relationship of macromorphology and micromorphology to contemporary hydrological conditions of a welded Argixeroll from the Palouse in Idaho. Geoderma 83, 309-329.

Kyuma, K., 2004. Paddy Soil Science. Kyoto University Press and Trans Pacific Press. Melbourne, 280 p.

Landuydt, C.J., 1990. Micromorphology of iron minerals from bog ore of the Belgian Campine Area. In Douglas, L.A. (eds.), Soil Micromorphology: A Basic and Applied Science. Developments in Soil Science, Volume 19. Elsevier, Amsterdam, pp. 289-294.

Lindbo, D.L. & Veneman, P.L.M., 1993. Micromorphology of selected Massachusetts fragipan soils. Soil Science Society of America Journal 57, 437-442.

Lindbo, D.L., Rhoton, F.E., Bigham, J.M., Smeck, N.E., Hudnall, W.H. & Tyler, D.D., 2000. Fragipan degradation and nodule formation in Glossic Fragiudalfs of the Lower Mississippi River Valley. Soil Science Society of America Journal 64, 1713-1722.

Lindbo, D.L., Vepraskas, M.J., He, X. & Severson, E., 2004. A comparison of soil wetness by morphological and modeling methods. In Mankin, K. (ed.), On-Site Wastewater Treatment. Proceedings of the Tenth National Symposium on Individual and Small Community Sewage Systems. ASAE, St. Joseph, pp. 52-58.

Liu, F., Colombo, C., Adamo, P., He, J.Z. & Violante A., 2002. Trace elements in manganese-iron nodules from a Chinese Alfisol. Soil Science Society of America Journal 66, 661-670.

McCarthy, P.J. & Plint, A.G., 1998. Recognition of interfluve sequence boundaries: integrating paleo-pedology and sequence stratigraphy. Geology 26, 387-390.

McCarthy, P.J., Martini, I.P. & Leckie D.A., 1998. Use of micromorphology for palaeoenvironmental interpretation of complex alluvial palaeosols: an example from the Mill Creek Formation (Albian), southwestern Alberta, Canada. Palaeogeography, Palaeoclimatology, Palaeoecology 143, 87-110.

McMillan, S.G. & Schwertmann, U., 1998. Morphological and genetic relations between siderite, calcite and goethite in a Low Moor Peat from southern Germany. European Journal of Soil Science 49, 283-293.

Mees, F. & Stoops, G., 2010. Sulphidic and sulphuric materials. In Stoops, G., Marcelino, V. & Mees, F. (eds.), Interpretation of Micromorphological Features of Soils and Regoliths. Elsevier, Amsterdam, pp. 543-568.

Mees, F. & Stoops, G., 2018. Sulphidic and sulphuric materials. In Stoops, G., Marcelino, V. & Mees, F. (eds.), Interpretation of Micromorphological Features of Soils and Regoliths. Second Edition. Elsevier, Amsterdam, pp. 347-376.

Megonigal, J.P., Patrick, W.H. & Faulkner, S.P., 1993. Wetland identification in seasonally flooded forest soils: soil morphology and redox dynamics. Soil Science Society of America Journal 57, 140-149.

Menning, P. & Shoba, S.A., 1980. Makro- und mikromorphologische Untersuchungen an festen Fe-Mn-Kronkretionen vernasster Boden. Archiv Für Acker- und Pflanzenbau und Bodenkunde 24, 625-637.

Morgan, C.P. & Stolt, M.H., 2006. Soil morphology-water table cumulative duration relationships in southern New England. Soil Science Society of America Journal 70, 816-824.

Morrás, H., 1983. Some properties of degraded argillans from A2 horizons of solodic planosols. In Bullock, P. & Murphy, C.P. (eds), Soil Micromorphology, Volume 2, Soil Genesis. AB Academic Publishers, Berkhamsted, pp. 575-581.

Mücher, H.J. & Coventry, R.J., 1994. Soil and landscape processes evident in hydromorphic grey earth (Plintusalf) in semiarid tropical Australia. In Ringrose-Voase, A.J. & Humphreys, G.S. (eds.), Soil Micromorphology: Studies in Management and Genesis. Developments in Soil Science, Volume 22. Elsevier, Amsterdam, pp. 221-232.

Muggler, C.C., van Loef, J.J., Buurman, P. & van Doesburg, J.D.J., 2001. Mineralogical and (sub) microscopic aspects of iron oxides in polygenetic Oxisols from Minas Gerais, Brazil. Geoderma 100, 147-171.

Murphy, C.P., 1984. The morphology and genesis of eight surface – water gley soils developed in till in England and Wales. Journal of Soil Science 35, 251-272.

Nettleton, W.D., Brasher, B.R., Baumer, O.W. & Darmody, R.G., 1994. Silt flows in soil. In Ringrose-Voase, A.J. & Humphreys, G.S. (eds.), Soil Micromorphology: Studies in Management and Genesis. Developments in Soil Science, Volume 22. Elsevier, Amsterdam, pp. 361-371.

O'Donnell, T.K., Goyne, K.W., Miles, R.J., Baffaut, C., Anderson, S.H. & Sudduth, K.A., 2010. Identification and quantification of soil redoximorphic features by digital image processing. Geoderma 157, 86-96.

O'Donnell, T.K., Goyne, K.W., Miles, R.J., Baffaut, C., Anderson, S.H. & Sudduth, K.A., 2011. Determination of representative elementary areas for soil redoximorphic features identified by digital image processing. Geoderma 161, 138-146.

Payton, R.W., 1993a. Fragipan formation in argillic brown earths (Fragiudalfs) of the Milfield Plain, north-east England: II. Post Devensian developmental processes and the origin of Fragipan consistence. Journal of Soil Science 44, 703-723.

Payton, R.W., 1993b. Fragipan formation in argillic brown earths (Fragiudalfs) of the Milfield Plain, north-east England: III. Micromorphology, SEM and EDXRA studies of Fragipan degradation and the development of Glossic features. Journal of Soil Science 44, 725-739.

PiPujol, M.D. & Buurman, P., 1994. The distinction between ground-water gley and surface-water gley phenomena in Tertiary paleosols of the Ebro Basin, NE Spain. Palaeogeography, Palaeoclimatology, Palaeoecology 110, 103-113.

PiPujol, M.D. & Buurman, P., 1997. Dynamics of iron and calcium carbonate redistribution and palaeohydrology in middle Eocene alluvial paleosols of the southeast Ebro Basin margin (Catalonia, northeast Spain). Palaeogeography, Palaeoclimatology, Palaeoecology 134, 87-107.

Ransom, M.D., Smeck, N.E. & Bigham, J.M., 1987. Micromorphology of seasonally wet soils on the Illinoian till plain, U.S.A. Geoderma 40, 83-99.

Richardson, J.L. & Hole, F.D., 1979. Mottling and iron distribution in a Glossoboralf-Haplaquoll hydrosequence on a glacial moraine in northwestern Wisconsin. Soil Science Society of America Journal 43, 552-558.

Reolid, M., Abad, I. & Martín-Gracía, J.M., 2008. Palaeoenvironmental implications of ferruginous deposits related to a Middle-Upper Jurassic discontinuity (Prebetic Zone, Betic Cordillera, Southern Spain). Sedimentary Geology 203, 1-16.

Rhoton, F.E., Bigham, J.M. & Schulze, D.G., 1993. Properties of iron-manganese nodules from a sequence of eroded Fragipan soils. Soil Science Society of America Journal 57, 1386-1392.

Schlichting, E. & Schwertmann, U., 1973. Pseudogley and Gley –Genesis and Use of Hydromorphic Soils. Transactions of Commission V and VI of the International Soil Science Society. Verlag Chemie, Weinheim, 771 p.

Schoeneberger, P.J., Wysocki, D.A., Benham, E.C. & Soil Survey Staff, 2012. Field Book for Describing and Sampling Soils. Version 3.0. USDA-NRCS, National Soil Survey Center, Lincoln, 228 p.

Schwertmann, U. & Fanning, D.S., 1976. Iron-manganese concretions in a hydrosequence of soil in loess in Bavaria. Soil Science Society of America Journal 40, 731-738.

Schwertmann, U. & Taylor, R.M., 1989. Iron oxides. In Dixon, J.B. & Weed, S.B. (eds.), Minerals in Soil Environments. SSSA Book Series, No. 1, Soil Science Society of America, Madison, pp. 379-438.

Severson, E.D., Lindbo, D.L. & Vepraskas, M.J., 2008. Hydropedology of a coarse-loamy catena in the lower Coastal Plain. Catena 73, 189-196.

Slager, S., Jongmans, A.G. & Pons, L.J., 1970. Micromorphology of some tropical alluvial clay soils. Journal of Soil Science 21, 233-241.

Soil Science Society of America, 2001. Glossary of Soil Science Terms. Soil Science Society of America, Madison, 44 p.

Stolt, M.H., Lesinski, B.C. & Wright, W., 2001. Micromorphology and seasonally saturated soils in carboniferous glacial till. Soil Science 166, 406-414.

Stolt, M.H., Ogg, C.M. & Baker, J.C., 1994. Strongly contrasting redoximorphic patterns in Virginia Valley and Ridge paleosols. Soil Science Society of America Journal 58, 477-484.

Stoops, G., 1983. SEM and light microscopic observations of minerals in bog-ores of the Belgian Campine. Geoderma 30, 179-186.

Stoops, G., 2003. Guidelines for Analysis and Description of Soil and Regolith Thin Sections. Soil Science Society of America, Madison, 184 p.

Stoops, G. & Eswaran, H., 1985. Morphological characteristics of wet soils. In Wetland Soils: Characterization, Classification and Utilization. IRRI, Los Baños, pp. 177-189.

Stumm, W. & Morgan, J.J., 1981. Aquatic Chemistry: An Introduction Emphasizing Chemical Equilibria in Natural Waters. John Wiley & Sons, New York, 780 p.

Suddhiprakarn, A. & Kheoruenromne, I., 1994. Fabric features in laterite and plinthite layers in Ultisols in northeast Thailand. In Ringrose-Voase, A.J. & Humphreys, G.S. (eds.), Soil Micromorphology: Studies in Management and Genesis. Developments in Soil Science, Volume 22. Elsevier, Amsterdam, pp. 51-64.

Tselishcheva, L.K., 1972. Micromorphological characteristics of Pseudopodzolic soil in the Transcarpathian region. In Kowalinski, S. & Drozd, J. (eds.), Soil Micromorphology. Panstwowe Wydawnicto Naukowe, Warsawa, pp. 313-321.

Tucker, R.J., Drees, L.R. & Wilding, L.P., 1994. Signposts old and new: active and inactive redoximorphic features; and seasonal wetness in two Alfisols of the gulf coast region of Texas, U.S.A. In Ringrose-Voase, A.J. & Humphreys, G.S. (eds.), Soil Micromorphology: Studies in Management and Genesis. Developments in Soil Science, Volume 22. Elsevier, Amsterdam, pp. 99-106.

USDA-NRCS, 2010. Field Indicators of Hydric Soils in the United States, Version 7.0. In Vasilas, L.M., Hurt, G.W., & Noble, C.V. (eds.), USDA-NRCS in cooperation with the National Technical Committee for Hydric Soils, Fort Worth, 44 p.

VandenBygaart, A.J. & Protz, R., 1999. The representative elementary area (REA) in studies of quantitative soil micromorphology. Geoderma 89, 333-346.

Van Ranst, E. & De Coninck, F., 2002. Evaluation of ferrolysis in soil formation. European Journal of Soil Science 53, 513-520.

Veneman, P.L.M., Vepraskas, M.J. & Bouma, J., 1976. The physical significance of soil mottling in a Wisconsin toposequence. Geoderma 15, 103-118.

Veneman, P.L.M., Spokas, L.A. & Lindbo, D.L., 1998. Soil moisture and redoximorphic features: a historical perspective. In Rabenhorst, M.C., Bell, J.C. & McDaniel, P.A. (eds.), Quantifying Soil Hydromorphology. SSSA Special Publication 54, SSSA, Madison, pp. 1-23.

Vepraskas, M.J., 2015. Redoximorphic Features for Identifying Aquic Conditions. Technical Bulletin 301, North Carolina Agricultural Research Service, Raleigh, 33 p.

Vepraskas, M.J. & Bouma, J., 1976. Model experiments on mottle formation simulating filed conditions. Geoderma 15, 217-230.

Vepraskas, M.J. & Guertal, W.R., 1992. Morphological indicators of soil wetness. In Kimble, J.M. (ed.), Proceedings of the 8th International Correlation Meeting on Wetland Soils. USDA Soil Conservation Service, Washington, pp. 307-312.

Vepraskas, M.J. & Vaughn, K.L., 2016. Morphological features of hydric and reduced soils. In Vepraskas, M.J. & Craft, C.B., (eds.), Wetland Soils: Genesis, Hydrology, Landscapes and Classification. Lewis Publisher, Boca Raton, pp. 189-217.

Vepraskas, M.J. & Wilding, L.P., 1983. Albic neoskeletans in argillic horizons as indicators of seasonal saturation. Soil Science Society of America Journal 47, 1202-1208.

Vepraskas, M.J., Baker, F.G. & Bouma, J., 1974. Soil mottling and drainage in a Mollic Hapludalf as related to suitability for septic tank construction. Soil Science of America Proceedings 38, 497-501.

Vepraskas, M.J., Wilding, L.P. & Drees, L.R., 1994. Aquic conditions for Soil Taxonomy: concepts, soil morphology and micromorphology. In Ringrose-Voase, A.J. & Humphreys, G.S. (eds.), Soil Micromorphology: Studies in Management and Genesis. Developments in Soil Science, Volume 22. Elsevier, Amsterdam, pp. 117-131.

Vepraskas, M.J., Polizzotto, M. & Faulkner, S.P., 2016. Redox chemistry of hydric soils. In Vepraskas, M.J. & Craft, C.B. (eds.), Wetland Soils: Genesis, Hydrology, Landscapes and Classification. Lewis Publisher, Boca Raton, pp. 105-132.

Wagner, D.P. & Rabenhorst, M.C., 1994. Micromorphology of an Aquic Paleudult developed in clayey, kaolinitic, coastal plain sediments in Maryland, USA. In Ringrose-Voase, A.J. & Humphreys, G.S. (eds.), Soil Micromorphology: Studies in Management and Genesis. Developments in Soil Science, Volume 22. Elsevier, Amsterdam, pp. 161-170.

Wilding, L.P. & Drees, L.R., 1983. Spatial variability and pedology. In Wilding, L.P., Smeck, N.E. & Hall, G. F. (eds.), Pedogenesis and Soil Taxonomy: 1. Concepts and Interactions. Elsevier, New York, pp. 83-116.

White, G.N. & Dixon, J.B., 1996. Iron and manganese distribution in nodules from a young Texas Vertisol. Soil Science Society of America Journal 60, 1254-1262.

Xu, Q. & Zhu, H., 1983. The characteristics of spotted horizons in paddy soils. Acta Pedologica Sinica 20, 53-59.

Zauyah, S. & Bisdom, E.B.A., 1983. SEM-EDXRA investigations of tubular features and iron nodules in lateritic soils from Malaysia. Geoderma 30, 219-232.

Zhang, G.L. & Gong, Z.T., 2003. Pedogenic evolution of paddy soils in different soil landscapes. Geoderma 115, 15-29.

Chapter 16

Features Related to Faunal Activity

Maja J. Kooistra[1], Mirjam M. Pulleman[2]

[1]*INTERNATIONAL SOIL REFERENCE AND INFORMATION CENTRE AND KOOISTRA MICROMORPHOLOGICAL SERVICES, RHENEN, THE NETHERLANDS (RETIRED);*
[2]*WAGENINGEN UNIVERSITY, WAGENINGEN, THE NETHERLANDS*

CHAPTER OUTLINE

1. Introduction

1.1 Soil Fauna and Micromorphology

Soil animals pass one or more active phases of their life cycle in the soil for different reasons, such as protection, feeding and reproduction. These organisms play an important role in transporting and altering various soil components, including organic matter decomposition and soil structure formation, and they thereby influence physical and chemical processes in soils. Faunal features are found in all types of soils, and the nature of entire soil horizons can be determined by faunal activity. There is a surprising lack of knowledge, however, on the precise functional roles played by many animals occurring in soils (Davidson et al., 2002).

Micromorphological techniques enable us (i) to distinguish between features originating from different groups of organisms and their modifications, (ii) to reconstruct the conditions under which these features were formed and the chronological order in which they developed, and (iii) to quantify the effect of different groups of soil fauna on soil (micro)structure development and organic matter dynamics.

This chapter is focused on micromorphological features resulting from soil fauna activity. Features related to other soil biota, such as microorganisms and plants, are described elsewhere in this book (e.g., Ismail-Meyer et al., 2018). Interactions between soil fauna and these other soil organisms, although frequently observed in thin sections, are beyond the scope of this chapter. Also excluded are large features resulting from the activities of soil megafauna (e.g., mice, moles), of large flying insects (e.g., wasps, bumblebees) and birds, whose study does not require a micromorphological approach.

The effects of soil fauna on soils and sediments can be multifold and complex. One group alone, e.g., earthworms, termites or mollusks (Fig. 1), can produce over 50 different types of features. In addition, many modifications of soil faunal features occur through interaction with other soil processes. This chapter aims to provide a basic

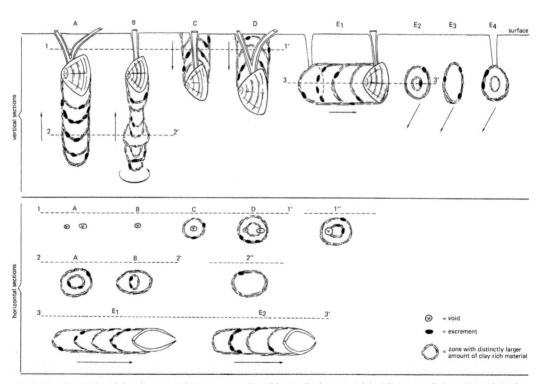

FIGURE 1 Example of the diversity of features produced by mollusks in a tidal sediment. *After Kooistra (1981).*

outline of the diversity of soil fauna, their activity in soils and the manner in which these affect micromorphological features in general terms. To interpret soil faunal features in thin sections, a basic understanding of the behaviour of the most important species is essential.

Until the first half of the 20th century, the knowledge of soil animals was virtually restricted to a few prominent species, such as earthworms and termites. Kubiëna (1938, 1948) described the important role of soil fauna in the decomposition of organic matter and mixing of organic and mineral soil constituents, based on micromorphological observations of excrements and other features. Starting in 1955, with a keynote publication by Kubiëna on the impact of animal activity in soils, much attention has been paid to the micromorphological aspects of soil fauna and their role in the development of humus forms and soils (e.g., Hartmann, 1965; Zachariae, 1965; Bal, 1970; Babel, 1971, 1975; De Coninck & Righi, 1983; Douglas & Thompson, 1985; Pawluk, 1987). Micromorphological studies dealing with faunal activity have been published in journals and handbooks covering a range of disciplines, including soil science, biology, forestry, chemistry and agricultural management. A selection of key publications is given in Table 1.

Since the late 1980s, the interest for sustainable land use has grown steadily and micromorphological studies of soil fauna, in combination with soil physical and chemical analyses, have been focused mostly on the effects of land use or agricultural management on the activity of soil animals and their role in soil structure development (Kooistra et al., 1985; Thompson et al., 1990; Boersma & Kooistra, 1994; Jongmans et al., 2001; Davidson et al., 2002, 2004; Pulleman et al., 2003, 2005b). Kooistra et al. (1985) studied the role of different groups of soil biota and their effects on soil physical

Table 1 List of key publications with relevant micromorphological information on features produced by specific soil organisms

Soil fauna	References
Meso- and macrofauna	Kubiëna (1955), Jongerius & Schelling (1960), Bal (1973, 1982), Pawluk (1985, 1987), Rusek (1985), Kooistra & Brussaard (1995), Phillips & FitzPatrick (1999), Davidson et al. (2002)
Earthworms	Jongerius (1957), Slager (1966), Shaw & Pawluk (1986), Kretzschmar (1987), Joschko et al. (1989), Babel et al. (1992), Babel & Kretzschmar (1994), Jongmans et al. (2001), Pulleman et al. (2005b)
Termites	Lee & Wood (1971), Sleeman & Brewer (1972), Miedema & Van Vuure (1977), Mando & Miedema (1997), Schaefer (2001), Sleeman & Brewer (1972), Jungerius et al. (1999)
Collemboles	Zachariae (1963)
Enchytraeids	Zachariae (1964), Babel (1968), Dawod & FitzPatrick (1993)
Millipedes, sawbugs, and diplopods	Striganova (1967, 1971)
Snails and mollusks	Kooistra (1978, 1981)

properties under selected management systems. They quantified the different types of void systems by image analysis, assigned them to soil fauna groups and related the results to measured soil physical properties. The data were then used to calculate soil-water regimes and land qualities in simulation models. The results also illustrated the contribution of soil fauna under different management systems to crop response (Kooistra & Boersma, 1994). An integrated approach using micromorphology and physical and chemical soil analyses has been also used by Pulleman et al. (2005b) to determine the effect of earthworms on macro- and microaggregate formation and carbon sequestration under permanent grassland and conventional arable land. This study revealed that earthworm-induced microaggregates played an important role in the stabilisation of soil carbon in permanent pasture soils but not in arable land under conventional management.

1.2   Diversity of the Impact of Soil Fauna on Soils

Awareness and understanding of the diversity and impact of soil fauna is crucial when identifying and interpreting micromorphological features related to soil faunal activities. Figure 2 gives an overview of the different kinds of soil organisms encountered in soils and their classification according to body diameter. Body width, rather than length, determines their impact on soil structure (Swift et al., 1979). All groups feed in essence on soil organic matter, either as primary consumers of live or dead plant materials or as

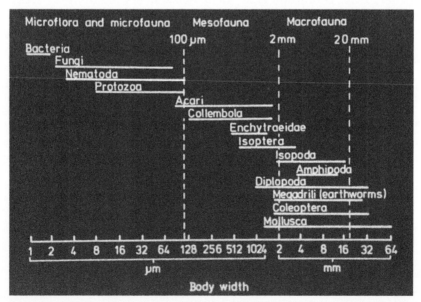

FIGURE 2 Main taxonomic groups of soil organisms according to body width. *After Swift et al. (1979).*

secondary consumers feeding on other fauna, microorganisms or their products (Brussaard & Juma, 1996). The whole set of organisms present in a certain soil environment forms a complex community in which each of the various species has its own habitat and role in the soil foodweb. Based on the kind of organic substrate they consume, several trophic groups are distinguished, e.g., detrivores, fungivores, bacterivores, phytophages, predators and omnivores (De Ruiter et al., 1993). Within one group of soil organisms, different trophic levels may be represented, for example beetle mites, mould mites and predatory mites.

Based on the impact of soil fauna on soil formation and soil structure development, a distinction is made between structure-following and structure-forming soil fauna (Kooistra et al., 1989). Structure-followers depend on the existing soil structure without being able to modify it, while structure-formers actively form or modify soil pores. Microfauna (<100 μm) rarely has a direct impact on the soil structure, but its presence attracts larger organisms (e.g., predators) that do have such an impact. Mesofauna (0.1-2 mm) can be structure-following or structure-forming. Macrofauna (>2 mm) is always structure-forming.

Soil fauna plays a major role in the decomposition of living or dead organic material of plant or animal origin. Many species also consume mineral soil material for different reasons: to enable passage through the soil, to take up the organic matter from organomineral complexes, to activate the intestines or to detoxify poisonous compounds. Different groups of soil fauna may have a very diverse diet or selectively consume specific types of tissues or particle sizes of organic and/or mineral material (Rusek, 1985; Kooistra et al., 1990; Schaefer, 2001). The ingested materials are fragmented, partly decomposed, dispersed and mixed, whereas microbes and mucus are added during passage through the intestinal tract. Ingested materials may vary throughout the year and depend on parent materials and past soil development. Consequently, the impact of specific groups of soil fauna on soils is not always well defined. Close examination of soil thin sections, combined with sampling and identification of soil animals in the field and studies of their behaviour under laboratory conditions, however, often leads to recognition of the effects of specific groups of soil fauna on soils.

Some soil animals are very mobile and transport or rearrange organic and/or inorganic soil constituents while moving from one place to another. Their impact ranges from displacement of soil materials while creating a temporary channel, as some beetles do (Brussaard, 1985), to the creation of a complex system of plastered channels and chambers, including new organomineral structures, as is common to termites (Wielemaker, 1984; Abe, 1990; Mando & Miedema, 1997; Jungerius et al., 1999; Schaefer, 2001). Consequently, soils that contain an active soil fauna population are often characterised by an intimate relationship between faunal activity and composition on the one hand and soil horizon formation on the other hand (Davidson et al., 2002).

Whole soil horizons can be bioturbated (Fig. 3) or completely depleted in fine-grained particles (Kooistra et al., 1990; Jungerius et al., 1999; Schaefer, 2001).

Interactions between different soil organisms within the same environment frequently cause modifications of faunal features. Excreta produced by one type of soil fauna can be an attractive substrate to other organisms (Fig. 4) (e.g., Pulleman et al., 2003; Davidson et al., 2004; Kooistra et al., 2006). Also faunal voids can be modified

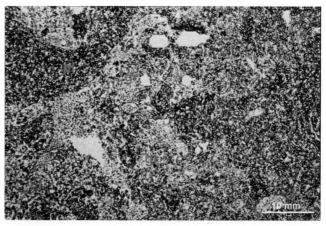

FIGURE 3 Completely bioturbated oxic horizon with granular microstructure (gr) and plastered channels (pl) by generations of different kinds of termites (Mozambique) (PPL). *Image by M.J. Kooistra.*

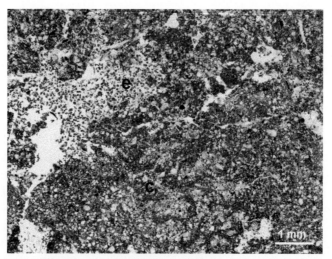

FIGURE 4 Infilling with excrements of enchytraeids (e) which feed on earthworm casts (c) in polder soil under permanent pasture (The Netherlands) (PPL). *Image by M.M. Pulleman.*

by roots or by soil organisms other than those that created the voids (Kooistra & Brussaard, 1995).

As described below, three main categories of fauna-induced features can be distinguished in thin sections: (i) voids, (ii) excrements and other faunal products, and (iii) coatings and infillings. Besides these features, thin sections can show cross sections through whole organisms, such as worms or nematodes, and resistant remnants of soil animals, such as bones, teeth, nails, skeleton parts of centipedes or isopods, wing-sheaths of beetles, snails or mollusk shells. Remnants of soil fauna and their features do not necessarily occur at the location where they were initially deposited. Resistant remnants can accumulate in sedimentary layers (Kooistra, 1978) or may contribute substantially to prehistoric waste deposits (Van Heeringen & Theunissen, 2001).

2. Faunal Voids

Faunal voids are formed by consumption or removal of soil material, by exerting pressure on the soil material or by combinations of these. Three types of voids can be distinguished: (i) channels, (ii) chambers, and (iii) modified voids. Formerly, faunal voids were classified according to their diameter into micro-, meso- and macropores (Jongerius, 1957). Slager (1966) used the term biopores for all voids that are formed by organisms. The terms channels and chambers were introduced by Brewer (1964) and are still in use (e.g., Stoops, 2003). Kooistra (1982a) distinguished between primary biological voids and modified biological voids, the latter being existing voids that are subsequently modified by biological activity.

2.1 Channels

Channels (Fig. 5) are the most common faunal voids. They can be straight, curved or convoluted, with or without branching. Cross sections through a channel can be circular and have constant diameter (e.g., dung beetle channels) (Brussaard, 1985) or be non-circular and of variable diameter (e.g., enchytraeids channels) (Babel, 1968). Some functional groups of soil fauna, such as anecic earthworms which feed at the soil surface, produce one permanent, deep vertical channel (Fig. 6B). Other functional groups, such as endogeic earthworms, produce many more or less regular channel systems confined to the upper soil layers (Kooistra, 1982b; Joschko et al., 1993). Certain termites can construct tunnel systems with oval cross sections (Kooistra, 1982a; Abe, 1990) or produce channels with flat floors and arched roofs (Brewer, 1964; Abe, 1990), where the termites can pass each other. Regular primary faunal channels with smooth walls can be distinguished from primary root channels by their less perfectly round shapes and by the presence of irregular compressed zones in the soil material along the pore walls (Fig. 6) (Kooistra, 1982b; Kooistra & Brussaard, 1995). Channel types and diameters can be used to identify certain species or species groups (Fig. 7), when the species groups present in

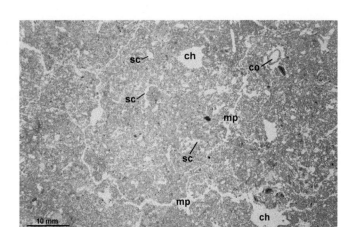

FIGURE 5 Different types of faunal voids in an alluvial plain soil (B horizon, 65 cm depth, Uttar Pradesh, India) with a subangular blocky microstructure and features recording strong earthworm activity: large irregular anecic earthworm channels (ch); modified planar voids (mp) and small channels inside the subangular blocky peds produced by endogeic earthworms and mesofauna (sc); excremental cocoon in a faunal channel (co) (PPL). *Image by M.J. Kooistra.*

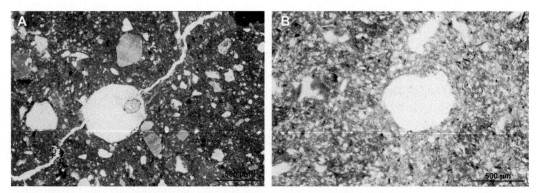

FIGURE 6 Root channel versus faunal channel. (A) Nearly circular cross section of a wheat root and equally compressed zones around it (PPL). (B) Irregular cross section of an anecic earthworm channel in old alluvial plain soil (west Haryana, India) (PPL). *Images by M.J. Kooistra.*

the soil differ substantially in body width and type of channel system they produce. However, care has to be taken because juveniles of larger species can produce channels with smaller diameters than adults of smaller species (Kooistra & Brussaard, 1995).

2.2 Chambers

Chambers are enlargements in void systems produced by soil fauna for different purposes. Generally they are round or oval, but irregularly shaped chambers also occur.

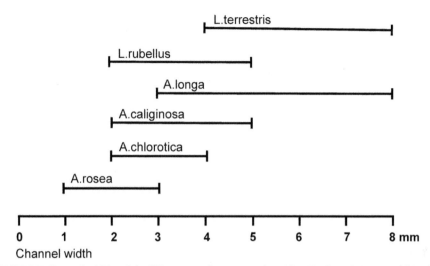

FIGURE 7 Channel widths of six different earthworm species. *After Kooistra & Brussaard (1995).*

Non-social species (e.g., dung beetles, diplopods, earthworms) produce enlargements for retreat, reproduction or deposition of reproduction-related materials such as eggs, seeds, capsules or nutrition for larvae (see Fig. 5) (Kooistra, 1982b). Chambers are often associated with channel systems. However, they can also be constructed along existing vertical planar voids (Fig. 8) (Kooistra, 1978).

The most complex and variable channel and chamber systems, however, are built by the social soil fauna such as ants and termites. The large galleries or chamber systems are used for food production (e.g., fungal cones), temperature regulation in the nest and bypassing. Special chambers are built for the queen, the eggs and the early stages of new generations. Several termite and ant species construct not only subterranean

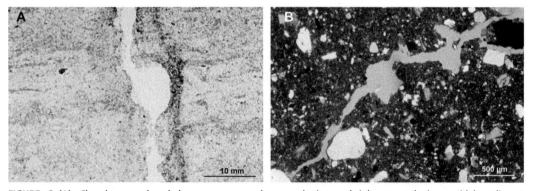

FIGURE 8 (A) Chamber produced by a common shore crab in a shrinkage crack in a tidal sediment (The Netherlands) (PPL). (B) Planar void in cambic horizon with local, irregular modifications by mesofauna and endogeic earthworms (PPL). *Images by M.J. Kooistra.*

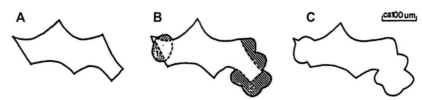

FIGURE 9 Formation of a complex modified void (C) starting from a void between clusters of earthworm excrements (A). A plant root finds its way through this void (B1) and mesofauna, e.g., enchytraeids or mites, and feeds on the organomineral earthworm excrements (B2), resulting in a considerable modification of the original morphology. *After Kooistra & Brussaard (1995).*

nests and galleries but also large aboveground mounds. To study the large and more complex channel systems, chambers, galleries, nests and mounds formed by soil fauna, researchers have often used macromorphological techniques rather than micromorphology.

2.3   Modified Voids

Smaller mesofauna, such as Acarina, Collembola and enchytraeids, do not generally produce their own void systems but use existing voids and locally enlarge them, which results in irregularly shaped, modified voids (Fig. 8B) (Kooistra & Brussaard, 1995). These organisms often feed on bacteria, algae and fungi that live on the void walls, thereby consuming also fine-grained soil material from the walls. Macrofauna such as earthworms can also enlarge existing planar voids (see Fig. 5). Interactions in time and space between different soil organisms can lead to the formation of complex modified voids (Fig. 9). Ultimately soil horizons can become depleted in fine-grained soil material leading to secondary packing voids between coarse-grained soil particles (Fig. 10A).

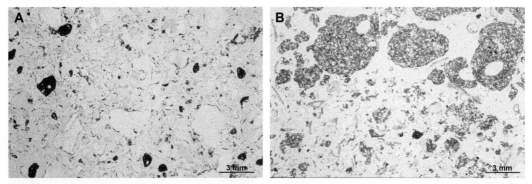

FIGURE 10 Impact of the activities of a tropical earthworm species, member of the Eudrilidae (e.g., *Hyperodryllus* spp.), on a natural bush fallow soil. (A) B horizon consisting of loosely packed single coarse particles (>180 μm) after virtually all the fine material of an argillic horizon has been consumed by the earthworms (PPL). (B) Clay-rich excrements, often with a central excretion channel, deposited by the earthworms in the surface horizon (PPL). *Images by M.J. Kooistra.*

3. Pedofeatures

3.1   Excrements

Only the soil fauna that consume solid organic and mineral materials produce recognisable excrements. Excrements can be described according to shape, size, composition, smoothness of the surface, surface features, internal fabric and organisation (Bal, 1973). They vary largely in shape, ranging from individual units with specific shape to shapeless compact masses. Mites, enchytraeids, diplopods and isopods usually produce individual units (Fig. 11A to C), whereas earthworm excreta (Figs. 10B, 11D and 12) are often more or less coalescent (Fig. 11D). A summary of the main shapes of intact excrements and the corresponding soil mesofauna species is given by Bullock et al. (1985) and Stoops (2003). The size of excrements may vary considerably even for one species, because of differences between juveniles and adults. Clearly shaped excrements have also been called droppings (Kubiëna, 1938), fecal pellets (Brewer, 1964) and modexi (Bal, 1973).

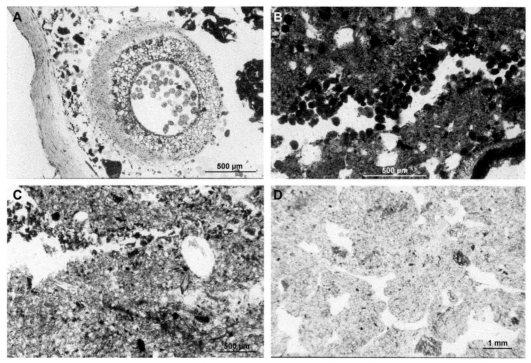

FIGURE 11 Common occurrences of excrements in arable land. (A) Smooth, ellipsoidal organic excrements of mites in a decomposing root (PPL). (B) Smooth, ellipsoidal, organomineral, mite excrements in a modified planar void near a wheat root of which the interior part is consumed by mites (PPL). (C) Local accumulation of rough, ellipsoidal, organomineral, enchytraeid excrements in a tillage void near a nearly completely consumed wheat root (PPL). (D) Coalesced smooth, ellipsoidal, organomineral excrements of an endogeic earthworm species (*Aporrectodea caliginosa*) (PPL). *Images by M.J. Kooistra.*

Excrements are composed of organic material or mixtures of organic and mineral materials. The organic material may contain recognisable cell structures (Fig. 12A) and/ or consist of more decomposed material, and it can include fungal hyphae and pollen. The mineral material is generally fine-grained, whereby the upper grain-size limit may be used for identification of the species (Kooistra et al., 1990). Excrements of earthworms (Jongmans et al., 2001) and of several mollusk species (Kooistra, 1978) may comprise a coating of clay-sized material around a core which includes coarse mineral grains (Fig. 12B). Excrements with an organic coating, also called 'peritrophic membrane' (Bal, 1970, 1973) are generally produced by epigeic earthworms.

For a number of invertebrate species the characteristic excreta are known. An overview of the excreta of Acarina, Collembola, Enchytraeidae, Diptera, Coleoptera and larvae of other insects, diplopods, isopods and lumbricids can be found in Rusek (1985). Figure 11 illustrates some types of excrements occurring in cultivated soils.

Excreta can frequently not be identified without doubt and assignment of excrement types to specific soil fauna needs to be made with care (Davidson et al., 2002). Many taxonomically different organisms produce very similar excrements, and excrements of the same species may have a different composition and morphology depending on the type of substrate consumed (Babel & Vogel, 1989; FitzPatrick, 1993). Consequently, not only shape and size but also location in the soil profile and environmental conditions in relation to the occurrence of soil faunal species need to be considered when attempting to relate excrements to specific groups or species of organisms.

Ageing and associated microbiological, physical and chemical processes lead to gradual morphological changes in the excrements (Fig. 13A), which eventually are no longer recognisable as excrements (Fig. 13B) (e.g., Davidson et al., 2002; Pulleman et al., 2005b). Excrements can also be modified by human interference, e.g., through ploughing (Spek et al., 2003), changes in land use or land drainage. Several types of ageing of excrements have been described (Bullock et al., 1985; Stoops, 2003).

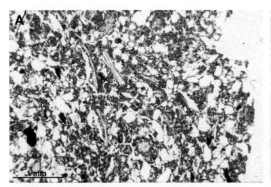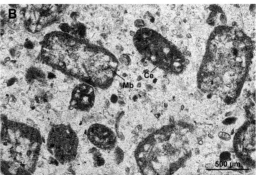

FIGURE 12 (A) Part of an organomineral excrement of a tropical Eudrilidae earthworm (Nigeria), composed of recognisable tissue residues, fine mineral material and coarse mineral particles (<180 μm) (PPL). (B) Organomineral excrements of mollusks in tidal sediments (The Netherlands): the excrements of *Macoma balthica* are cylindrical, have a clay coating around and contain coarse mineral grains (Mb); the excrements of *Cardium edule* are ellipsoidal and essentially composed of fine mineral material (Ce) (PPL). *Images by M.J. Kooistra.*

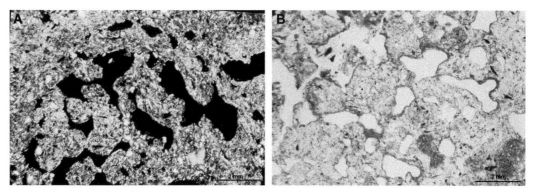

FIGURE 13 Ageing of earthworm excrements in an alluvial plain (North India). (A) Deformed and moderately coalesced excrements in a faunal void (XPL). (B) A later stage in which excrements of the same kind are more deformed and strongly coalesced. Note the presence of partial thin dark brown coatings of fine material along voids due to incidental flooding (PPL). *Images by M.J. Kooistra.*

3.2   Coatings

Faunal coatings are found on the walls of channels that have been used for long periods (Jongerius, 1957) or of channels crossing unstable soils or sediments or regularly flooded soils (Kooistra, 1978, 1982b). In older literature they are described as cutans (Brewer, 1960). The material used for the coatings can be derived from fine-grained soil fragments transported by termites (Fig. 14A) from excrements (Fig. 14B) or from locally accumulated wet or dry displaced soil material. It can also consist of a mixture of materials from different origins. Coatings composed of excreta are often darker than the surrounding groundmass due to the admixture of organic material. In unstable materials, such as soils in deserts, regularly flooded areas and intertidal flats, multilayered coatings composed of infillings material fixed to the walls by excreta are common. In wet areas,

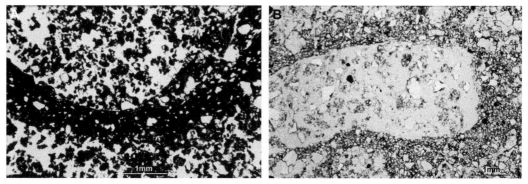

FIGURE 14 (A) Coating of fine soil material on channel walls (plasters) made by termites to stabilise channels in loosely packed soil fragments and infilling with groundmass material (PPL). (B) Coating composed of fine-grained excreta made by an earthworm, stabilising the channel and infilling with soil material from the surface (PPL). *Images by M.J. Kooistra.*

several species of worms and pelecypods produce these multilayered coatings (Kooistra, 1978, 1981). Social soil fauna such as ants and termites build large complexes of sub-terranean nests or mounds with radiating plastered channels to the surrounding areas. These animals use aggregates of fine-grained soil material mixed with saliva to plaster their channels, chambers and galleries for stabilisation (see Fig. 3) (Miedema & Van Vuure, 1977; Kauffman, 1987).

3.3   Infillings

The large variety of infillings by soil fauna can be split up in three groups: (i) infillings with shaped units, (ii) infillings with a sharp external boundary, and (iii) infillings with a gradual external boundary. In older literature, channel infillings are termed pedotubules (Brewer & Sleeman, 1963) and crotovinas (Jongerius & Rutherford, 1979). Nearly all morphological types of infillings described in the micromorphological handbooks (Bullock et al., 1985; Stoops, 2003) can be produced by soil fauna as the result of their wide range of activities.

Any type of faunal void, with or without coating, can be filled to different degrees with shaped units. Such shaped units can be excrements but also small soil aggregates or single coarse mineral grains that were transported by soil animals such as termites (Fig. 14A) or they can be derived from the soil surface and have accumulated into the channel by different processes (Fig. 14B). Excrements can have different compositions and be present as accumulations of individual units or as clusters of more or less coalesced material (Fig. 15A).

Faunal voids with sharp external boundaries can be filled with soil material, as in the case of dung beetles (Fig. 15B) and cockchafers, or with shapeless excreta of a rather homogeneous composition produced by certain earthworm species (Fig. 15C). In some cases, the voids are filled with soil material as well as excrements deposited by the animal during passage. Such infillings can be formed by mollusks, beetle larvae and earthworms, and they often show a crescent fabric (Fig. 15D; see also Fig. 1).

Infillings with a gradual external boundary (Fig. 16) are formed during passage of soil animals through wet, plastic soils. The material flows back into the track immediately after passage, whereby elongated particles may become oriented. Excrements are commonly encountered in these tracks. Passage features are generally produced by snails, mollusks and worms (Kooistra, 1978).

3.4   Other Pedofeatures

Earthworms produce with their excreta cocoon-shaped features with a spherical to slightly ellipsoidal internal chamber (Fig. 17). These features, found in chambers, channel systems or planar voids, are used during diapause or to survive during unfavourable conditions.

Several biogenic carbonate products can be formed by the soil fauna (see Durand et al., 2018). Biogenic calcite nodules (e.g., Ponomareva, 1948; Bal, 1977) are common

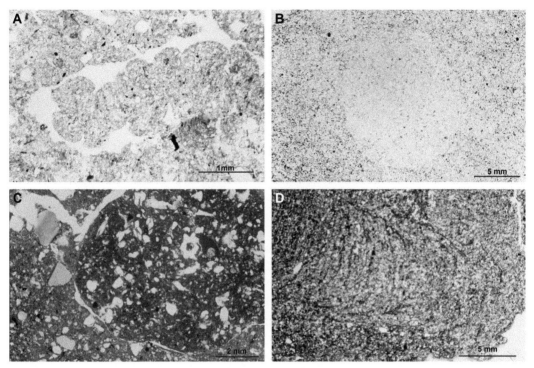

FIGURE 15 Faunal infillings with sharp external boundaries. (A) Infilling with coalesced earthworm excrements (PPL). (B) Channel infilling produced by dung beetle (PPL). (C) Infilling produced by earthworm (*Lumbricus terrestris*) with shapeless excreta of a rather homogeneous composition (PPL). (D) Infilling with bow-like structure probably formed by an earthworm (India) (PPL). *Image by M.J. Kooistra.*

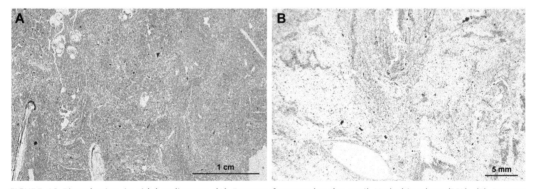

FIGURE 16 Bioturbation in tidal sediments. (A) Passage features by the snail *Hydrobia ulvae* (PPL). (B) Passage features by the mollusk *Macoma balthica* (PPL). *Images by M.J. Kooistra.*

features in clayey alluvial plains of humid, temperate and tropical climates (Kooistra, 1982b). They are produced by certain earthworm species to regulate the pH in the intestines (Piearce, 1972). The excreted calcite spheroids consist of sparitic calcite crystals, frequently arranged in a radial pattern (Jongmans et al., 2001), with sizes up to

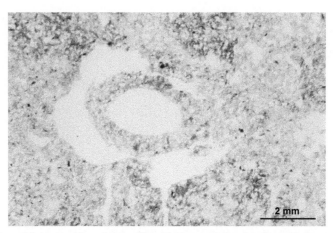

FIGURE 17 Cocoon-shaped feature with a slightly ellipsoidal internal chamber formed of excreta by an earthworm in faunal void in Ganges alluvial plain soil (India) (PPL). *Image by M.J. Kooistra.*

several millimeters (see Durand et al., 2018). Biogenic calcite nodules observed in thin sections can be used as an indicator of current or past earthworm activity (Jongmans et al., 2001).

4. Features Related to Faunal Impact on Profile Development

Organic residues in O and A horizons are the primary source of food for soil fauna near the surface. In B and C horizons and less vegetated areas, the primary organic substrate for soil fauna is provided by soil organic matter incorporated into the soil by other organisms, roots, fungi and microorganisms, such as bacteria and algae. The decomposition of soil organic matter in O and A horizons is closely related to the presence of specific groups of soil micro-, meso- and macrofauna, which in turn depends on environmental conditions such as moisture availability, drainage and pH (Fig. 18). Clear stratifications into L, F and H layers are found under acid, cool and wet conditions where the activity of structure-forming soil fauna is limited. Under conditions of limited aeration and low pH, organic matter decomposition in the soil is slowed down and dominated by bacteria, microfauna and fungi, resulting in the formation of peat, anmoor or a mor humus type. Under drier and less acidic conditions, moder and mull humus forms are developed under the impact of meso- and macrofauna. Moder humus largely consists of accumulations of distinctly shaped excreta of mesofauna and is formed under moderately acidic conditions, mainly in sandy soils. Mull humus develops under neutral to basic soil conditions when earthworms dominate the decomposition

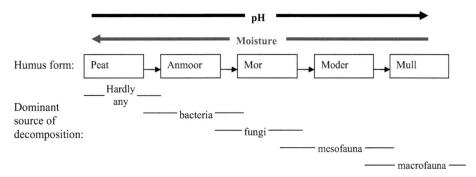

FIGURE 18 Main groups of soil organisms responsible for the decomposition of organic matter under varying moisture and pH conditions.

process, including the incorporation and intimate mixing of relatively fresh litter into the mineral soil matrix. Under conditions of moderate acidity and humidity, development of either a moder or a mull humus form is largely determined by the nutrient status and palatability of the leave litter, which in turn determines the composition of the soil fauna community (Pulleman et al., 2005a). For detailed description of micromorphological features related to soil organic matter decomposition, the reader is referred to Ismail-Meyer et al., 2018.

In soil horizons where organic inputs are limited, the abundance and diversity of soil fauna species is low. However, when these species are very mobile, they can have a large impact on soil structure. Their activity results in intensively bioturbated soil materials showing different kinds of superimposed infillings, with or without coatings. It can also produce granular textures consisting of organomineral excrements, with or without admixed coarse-grained groundmass material.

Consequently, the nature of a number of diagnostic A horizons, such as the mollic, umbric, antropic and plaggen epipedons, is determined by soil fauna activity. This is also the case for certain B horizons, such as oxic horizons with a granular microstructure (see Fig. 3) formed by termite activity (e.g., Kauffman, 1987; Jungerius et al., 1999; Schaefer, 2001) and former argillic horizons with a coarse granular microstructure (see Fig. 11) produced by tropical earthworms (Kooistra et al., 1990), as well as some spodic and cambic horizons (see Fig. 5) (e.g., Phillips & FitzPatrick, 1999).

Transitions between mineral soil and parent material and/or weathered rock may be strongly affected by bioturbation (see Stoops & Schaefer, 2018). In extreme climates with high temperatures and/or distinct drought periods, the greatest faunal impact is found in the deeper B and C horizons (Barzanji & Stoops, 1974; Kooistra, 1982a). The impact of soil fauna is not necessarily directly related to the size of the populations or number of species. A few mollusks in a sandy tidal deposit can erase completely the original lamination and incorporate, by feeding on the suspension and excreting in the sediment, 5%–8% fine-grained organomineral material that could not have been deposited during sedimentation due to the high velocities of the tidal currents (Kooistra, 1978).

In agricultural soils, as long as the climate, soil type and crop management practices are favourable for the development of a substantial soil fauna population, high soil fauna activity can lead to the formation of a strongly biogenic soil structure, dominated by excrements and channels. This is the case of finely textured soils in (semi) humid climates under pasture (see Fig. 4), or to a lesser extent, under reduced or zero tillage systems. However, under the same climate and soil conditions, a compact physicogenic soil structure can be formed when management practices are not favourable for structure-forming soil fauna, as for example in the case of conventional tillage systems with low organic inputs or with high pesticide use (Kooistra et al., 1989; Jongmans et al., 2001, 2003; Pulleman et al., 2003, 2005b). Moreover, crop residues incorporated into tillage-induced aggregates can often not be reached by soil fauna (Fig. 19).

Important factors that determine soil fauna composition and activity, i.e., food sources, moisture, pH, temperature and soil disturbance, can change over time, for example due to sedimentation or erosion, drainage of waterlogged soils, changes in land use and amelioration or deep tillage of agricultural soils. Complex features resulting from sudden disturbances or gradually changing environmental conditions and populations are commonly encountered in thin sections (e.g., Hammond & Collins, 1983; Dawod & FitzPatrick, 1993; Spek et al., 2003; Kooistra et al., 2006) and may help in the reconstruction of past conditions and their chronology. On the other hand, human-induced changes in a landscape, such as drainage or amelioration, can lead to increased activity of soil fauna whereby all information in soil archives can become completely erased (Van Heeringen & Theunissen, 2001).

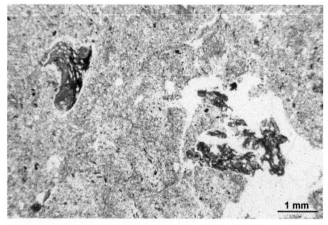

FIGURE 19 Ap horizon of arable land: crop residues at the edge of tillage voids (right) are consumed by enchytraeids, but those inside compact soil fragments (left) resulting from tillage remain untouched (PPL). *Image by M.J. Kooistra.*

5. Conclusions

The decomposition and redistribution of soil organic matter, the excretion of fecal pellets that contribute to stable soil aggregation and the formation of continuous voids by soil fauna can have important implications for soil functions such as susceptibility to erosion, water infiltration, aeration, root growth, leaching of agrochemicals and carbon and nutrient dynamics. Moreover, organisms that alter the soil environment through burrowing and other structure-forming activities also modify the habitat of other biota. Important applications of micromorphological studies in relation to soil fauna are studies related to soil formation and humus development, landscape reconstruction, archaeology and faunal impact on soil functions as affected by soil management.

Regarding future research, micromorphological studies of soil faunal features can offer a valuable contribution as part of an integrated approach aimed at elucidating the role of soil fauna diversity and their functions in soils. Currently, the growing interest in understanding soil biodiversity and the functional role of soil fauna in soils and ecosystems (Davidson & Grieve, 2006; Brussaard et al. 2007) calls for a more integrated approach to the study of soil fauna and its function in soils. The combination of micromorphological methods with other techniques such as field and laboratory manipulations of soil fauna, isolation of fauna-induced soil structural elements and novel physical, chemical or biological soil analyses provides a promising approach for future studies of the role of soil fauna.

Acknowledgement

We thank Wouter Bomer (ISRIC) for the preparation of the figures.

References

Abe, T., 1990. Distribution and abundance of subterranean fungus-growing termites (Isoptera) in the grassland of Kajiado, Kenya. In Veeresh, G.K., Rajagopal, D. & Viraktamath, C.A. (eds.), Advances in Management and Conservation of Soil Fauna. Oxford & IBH Publishing, New Delhi, pp. 111-121.

Babel, U., 1968. Enchytraeen-Losungsgefüge in Löss. Geoderma 2, 57-63.

Babel, U., 1971. Gliederung und beschreibung des Humusprofils in Mitteleuropäicher Wäldern. Geoderma 5, 297-324.

Babel, U., 1975. Micromorphology of soil organic matter. In Gieseking, J.E. (ed.), Soil Components. Volume 1: Organic Components. Springer-Verlag, New York, pp. 369-473.

Babel, U. & Kretzschmar, A., 1994. Micromorphological observations of casts and burrow walls of the Gippsland giant earthworm (*Megascolides australis*, Mc Coy 1878). In Ringrose-Voase, A.J. & Humphreys, G.S. (eds.), Soil Micromorphology: Studies in Management and Genesis. Developments in Soil Science, Volume 22. Elsevier, Amsterdam, pp. 451-457.

Babel, U., Ehrmann, O. & Krebs, M., 1992. Relationships between earthworms and some plant species in a meadow. Soil Biology and Biochemistry 24, 1477-1481.

Babel, U. & Vogel, H.J., 1989. Zur Beurteilung der Enchyträen-und Collembolen-Aktivität mit Hilfe von Bodendünnschliffen. Pedobiologia 33, 167-172.

Bal, L., 1970. Morphological investigation in two moder-humus profiles and the role of the soil fauna in their genesis. Geoderma 4, 5-36.

Bal, L., 1973. Micromorphological Analysis of Soils - Lower Levels in the Organization of Organic Soil Materials. Soil Survey Papers 6. Soil Survey Institute, Wageningen, 174 p.

Bal, L., 1977. The formation of carbonate nodules and intercalary crystals in the soil by the earthworm *Lumbricus rubellus*. Pedobiologia 17, 102-106.

Bal, L., 1982. Zoological Ripening of Soils. Agricultural Research Report n° 850. Centre for Agricultural Publishing and Documentation, Wageningen, 365 p.

Barzanji, A.F. & Stoops, G., 1974. Fabric and mineralogy of gypsum accumulations in some soils of Iraq. Transactions of the 10th International Congress of Soil Science, Moscow, Volume VII, pp. 271-277.

Boersma, O.H. & Kooistra, M.J., 1994. Differences in soil structure of silt loam Typic Fluvaquents under various agricultural management practices. Agriculture, Ecosystems & Environment 51, 21-42.

Brewer, R., 1960. Cutans: their definition, recognition and classification. Journal of Soil Science 11, 280-292.

Brewer, R., 1964. Fabric and Mineral Analysis of Soils. John Wiley and Sons, New York, 470 p.

Brewer, R. & Sleeman, J.R., 1963. Pedotubules: their definition, classification and interpretation. Journal of Soil Science 15, 66-78.

Brussaard, L., 1985. A Pedological Study of the Dung Beetle Typhaeus Typhoeus (Coleoptera, Geotrupidae). Agricultural University Wageningen, 168 p.

Brussaard, L. & Juma, N.G., 1996. Organisms and humus in soils. In Piccolo, A. (ed.), Humic Substances in Terrestrial Ecosystems. Elsevier, Amsterdam, pp. 329-359.

Brussaard, L., Pulleman, M.M., Ouedraogo, E., Mando, A. & Six, J., 2007. Soil fauna and soil function in the fabric of the food web. Pedobiologia 50, 447-462.

Bullock, P., Fedoroff, N., Jongerius, A., Stoops, G., Tursina, T. & Babel, U., 1985. Handbook for Soil Thin Section Description. Waine Research Publications, Wolverhampton, 152 p.

Davidson, D.A. & Grieve, I.C., 2006. Relationships between biodiversity and soil structure and function: evidence from laboratory and field experiments. Applied Soil Ecology 33, 176-185.

Davidson, D.A., Bruneau, P.M.C., Grieve, I.C. & Young, I.M., 2002. Impacts of fauna on an upland grassland soil as determined by micromorphological analysis. Applied Soil Ecology 20, 133-143.

Davidson, D.A., Bruneau, P.M.C., Grieve, I.C. & Wilson, C.A., 2004. Micromorphological assessment of the effect of liming on faunal excrement in an upland grassland soil. Applied Soil Ecology 26, 169-177.

Dawod, V. & FitzPatrick, E.A., 1993. Some population sizes and effects of Enchytraeidae (Oligochaeta) on soil structure in a selection of Scottish soils. Geoderma 56, 173-178.

De Coninck, F. & Righi D., 1983. Podzolisation and the spodic horizon. In Bullock, P. & Murphy, C.P. (eds.), Soil Micromorphology, Volume 2. Soil Genesis. AB Academic Publishers, Berkhamsted, pp. 389-417.

De Ruiter, P.C., Moore, J.C., Zwart, K.B., Bouwman, L.A., Hassink, J., Bloem, J., De Vos, J.A., Marinissen, J. C.Y., Didden, W.A.M., Lebbink, G. & Brussaard, L., 1993. Simulation of nitrogen mineralization in the below-ground food webs of two winter wheat fields. Journal of Applied Ecology 30, 95-106.

Douglas, L.A. & Thompson, M.L. (eds.), 1985. Soil Micromorphology and Soil Classification. Soil Science Society of America Special Publication 15, SSSA, Madison, 216 p.

Durand, N., Monger, H.C. & Canti, M.G., 2018. Calcium carbonate features. In Stoops, G., Marcelino, V. & Mees, F. (eds.), Interpretation of Micromorphological Features of Soils and Regoliths. Second Edition. Elsevier, Amsterdam, pp. 205-258.

FitzPatrick, E.A., 1993. Soil Microscopy and Micromorphology. John Wiley & Sons, Chichester, UK, 304 p.

Hammond, R.F. & Collins, J.F., 1983. Microfabrics of a sphagnofibrist and related changes resulting from soil amelioration. In Bullock, P. & Murphy, C.P. (eds.), Soil Micromorphology, Volume 2, Soil Genesis. AB Academic Publishers, Berkhamsted, pp. 689-697.

Hartmann, F., 1965. Waldhumusdiagnose auf Biomorphologischer Grundlage. Springer-Verlag, Wien, New York. 88 p.

Ismail-Meyer, K., Stolt, M. & Lindbo, D., 2018. Soil organic matter. In Stoops, G., Marcelino, V. & Mees, F. (eds.), Interpretation of Micromorphological Features of Soils and Regoliths. Second Edition. Elsevier, Amsterdam, pp. 471-512.

Jongerius, A., 1957. Morfologische Onderzoekingen over de Bodemstructuur. Mededelingen van de Stichting voor Bodemkartering, Bodemkundige Studies, 2, 93 p.

Jongerius, A. & Schelling, J., 1960. Micromorphology of organic matter formed under the influence of soil organisms, especially soil fauna. Transactions of the 7th International Congress of Soil Science, Volume III, Madison, pp. 702-710.

Jongerius, A. & Rutherford, G.K., 1979. Glossary of Soil Micromorphology. Centre for Agricultural Publishing and Documentation, Wageningen, 138 p.

Jongmans, A.G., Pulleman, M.M. & Marinissen, J.C.Y., 2001. Soil structure and earthworm activity in a marine silt loam under pasture versus arable land. Biology and Fertility of Soils 33, 279-285.

Jongmans, A.G., Pulleman, M.M., Balabane, M., van Oort, F. & Marinissen, J.C.Y., 2003. Soil structure and characteristics of organic matter in two orchards differing in earthworm activity. Applied Soil Ecology 24, 219-232.

Joschko, M., Diestel, H. & Larink, O., 1989. Assessment of earthworm burrowing efficiency in compacted soil by a combination of morphological and soil physical measurements. Biology and Fertility of Soils 8, 191-196.

Joschko, M., Müller, P.C., Kotzke, K., Döhring, W. & Larink, O., 1993. Earthworm burrow system development assessed by means of X-ray computed tomography. Geoderma 56, 209-221.

Jungerius, P.D., van den Ancker, J.A.M. & Mücher, H.J., 1999. The contribution of termites to the microgranular structure of soils on the Uasin Gishu Plateau, Kenya. Catena 34, 349-363.

Kauffman, J.H., 1987. Comparative Classification of Some Deep, Well-Drained Red Clay Soils of Mozambique. Technical Paper no. 16, International Soil Reference and Information Centre, Wageningen, 61 p.

Kooistra, M.J., 1978. Soil Development in Recent Marine Sediments of the Intertidal Zone in the Oosterschelde - The Netherlands. A Soil Micromorphological Approach. Soil Survey Paper 14, Nertherlands Soil Survey Institute, Wageningen, 183 p.

Kooistra, M.J., 1981. The interpretation and classification of features produced by Pelecypods (Mollusca) in marine intertidal deposits in The Netherlands. Geoderma 26, 83-94.

Kooistra, M.J., 1982a. Micromorphology. In Murthy, R.S., Hirekerur, L.R., Deshpande, S.B. & Venkata Rao, B.V. (eds.), Benchmark Soils of India. Morphology, Characteristics and Classification for Resource Management. National Bureau of Soil Survey and Land Use Planning (ICAR), Nagpur, pp. 71-89.

Kooistra, M.J., 1982b. Micromorphological Analysis and Characterization of 70 Benchmark Soils of India. A Basic Reference Set. Netherlands Soil Survey Institute, Wageningen, 788 p.

Kooistra, M.J. & Boersma, O.H., 1994. Subsoil compaction in Dutch marine sandy loams: loosening practises and effects. Soil and Tillage Research 29, 237-249.

Kooistra, M.J. & Brussaard, L., 1995. A micromorphological approach to the study of soil structure-soil biota interactions. In Edwards, C.A., Abe, T. & Striganova, B.R. (eds.), Structure and Function of Soil Communities. Kyoto University Press, Kyoto, pp. 55-69.

Kooistra, M.J., Bouma, J., Boersma, O.H. & Jager, A., 1985. Soil-structure differences and associated physical properties of some loamy Typic Fluvaquents in The Netherlands. Geoderma 36, 215-228.

Kooistra, M.J., Lebbink, G. & Brussaard, L., 1989. The Dutch programme on soil ecology of arable farming systems. II. Geogenesis, agricultural history, field site characteristics and present farming systems at the Lovinkhoeve experimental farm. Agriculture, Ecosystems & Environment 27, 361-387.

Kooistra, M.J., Juo, A.S. & Schoonderbeek, D., 1990. Soil degradation in cultivated Alfisols under different management systems in Southern Nigeria. In Douglas, L.A. (ed.), Soil Micromorphology: A Basic and Applied Science. Developments in Soil Science, Volume 19. Elsevier, Amsterdam, pp. 61-68.

Kooistra, M.J., Kooistra, L.I., Van Rijn, P. & Sass-Klaassen, U., 2006. Woodlands of the past − The excavation of wetland woods at Zwolle-Stadshagen (The Netherlands): reconstruction of the wetland wood in its environmental context. Netherlands Journal of Geosciences 85, 37-60.

Kretzschmar, A., 1987. Caractérisation microscopique de l'activité des lombriciens endogés. In Fedoroff, N., Bresson, L.M. & Courty, M.A. (eds.), Micromorphologie des Sols, Soil Micromorphology. AFES, Plaisir, pp. 325-330.

Kubiëna, W.L., 1938. Micropedology. Collegiate Press, Ames, 242 p.

Kubiëna, W.L., 1948. Entwicklungslehre des Bodens. Springer-Verlag, Wien, 215 p.

Kubiëna, W.L., 1955. Animal activity in soils as a decisive factor in establishment of humus forms. In Kevan, D. (ed.), Soil Zoology. Butterworths, London, pp. 73-82.

Lee, K.E. & Wood, T.G, 1971. Termites and Soils. Academic Press, London, New York, 251 p.

Mando, A. & Miedema, R., 1997. Termite-induced change in soil structure after mulching degraded (crusted) soil in the Sahel. Applied Soil Ecology 6, 241-249.

Miedema, R. & Van Vuure, W., 1977. The morphological, physical and chemical properties of two mounds of Macrotermes bellicosus (Smeathman) compared with surrounding soils in Sierra Leone. Journal of Soil Science 28, 112-124.

Pawluk, S., 1985. Soil micromorphology and soil fauna: problems and importance. Quaestiones Entomologicae 21, 473-496.

Pawluk, S., 1987. Faunal micromorphological features in moder humus of some western Canadian soils. Geoderma 40, 3-16.

Phillips, D.H. & FitzPatrick, E.A., 1999. Biological influences on the morphology and micromorphology of selected Podzols (Spodosols) and Cambisols (Inceptisols) from the eastern United States and north-east Scotland. Geoderma 90, 327-364.

Piearce, T.G., 1972. The calcium relations of selected Lumbricidae. Journal of Animal Ecology 41, 167-188.

Ponomareva, S.I., 1948. The rate of formation of calcite in the soil by earthworms. Report of the Academy of Science of the USSR 61, 505-507.

Pulleman, M., Jongmans, A., Marinissen, J. & Bouma, J., 2003. Effects of organic versus conventional arable farming on soil structure and organic matter dynamics in a marine loam in the Netherlands. Soil Use and Management 19, 157-165.

Pulleman, M.M., Kooistra, M.J., Hommel, P.W.F.M. & De Waal, R.J., 2005a. Strooiselafbraak onder Verschillende Loofboomsoorten op de Stuwwal bij Doorwerth. Micromorfologisch Onderzoek van de Humusprofielen. Alterra Report 1052, Wageningen, 82 p.

Pulleman, M.M., Six, J., Uyl, A., Marinissen, J.C.Y. & Jongmans, A.G., 2005b. Earthworms and management affect organic matter incorporation and microaggregate formation in agricultural soils. Applied Soil Ecology 29, 1-15.

Rusek, J., 1985. Soil microstructures − contributions on specific soil organisms. Quaestiones Entomologicae 21, 497-514.

Schaefer, C.E.R., 2001. Brazilian latosols and their B horizon microstructure as long-term biotic constructs. Australian Journal of Soil Research 39, 909-926.

Shaw, C. & Pawluk, S., 1986. The development of soil structure by *Octolasium tyrtaeum, Aporrectodea turgida* and *Lumbricus terrestris* in patent materials belonging to different texture classes. Pedobiologia 29, 296-302.

Slager, S., 1966. Morphological Studies of Some Cultivated Soils. PhD Dissertation, Agricultural University Wageningen, 111 p.

Sleeman, J.R. & Brewer, R., 1972. Micro-structures of some Australian termite nests. Pedobiologia 12, 347-373.

Spek, T., Groenman-van Waateringe, W., Kooistra, M. & Bakker, L., 2003. Formation and land-use history of Celtic fields in north-west Europe – an interdisciplinary case study at Zeijen, The Netherlands. European Journal of Archaeology 6, 141-173.

Stoops, G., 2003. Guidelines for Analysis and Description of Soil and Regolith Thin Sections. Soil Science Society of America, Madison, 184 p.

Stoops, G. & Schaefer, C.E.G.R., 2018. Pedoplasmation: formation of soil material. In Stoops, G., Marcelino, V. & Mees, F. (eds.), Interpretation of Micromorphological Features of Soils and Regoliths. Second Edition. Elsevier, Amsterdam, pp. 205-258.

Striganova, B.R., 1967. Ueber die Zersetzung von überwinterter Laubstreu duch Tausendfüssler und Landasseln. Pedobiologia 7, 125-134.

Striganova, B.R., 1971. Significance of diplopod activity in leaf litter decomposition. Annales de Zoologie et d'Ecologie Animale 71, 409-415.

Swift, M.J., Heal, O.W. & Anderson, J.M., 1979. Decomposition in Terrestrial Ecosystems. Studies in Ecology, Volume 5. Blackwell Scientific Publications, 372 p.

Thompson, M.L., Fedoroff, N. & Fournier, B., 1990. Morphological features related to agriculture and faunal activity in three loess-derived soils in France. Geoderma 46, 329-349.

Van Heeringen, R.M. & Theunissen, E.M., 2001. Kwaliteitsbepalend onderzoek ten behoeve van duurzaam behoud van Neolitische terreinen in West-Friesland en de Kop van Noord-Holland. Nederlandse Archeologische Rapporten 21. Deel 1 Waardestelling. Rijksdienst voor het Oudheidkundig Bodemonderzoek, Amersfoort, 295 p.

Wielemaker, W.G., 1984. Soil Formation by Termites, a Study in the Kisii Area, Kenya. PhD Dissertation, Agricultural University Wageningen, 132 p.

Zachariae, G., 1963. Was leisten Collembolen für den Waldhumus? In Doeksen J. & van der Drift, J. (eds.), Soil Organisms: Proceedings of the Colloquium on Soil fauna, Soil Microflora and their Relationships, Oosterbeek, North-Holland. Amsterdam, pp. 109-124.

Zachariae, G., 1964. Welche Bedeutung haben Enchytraeen im Waldboden? In Jongerius, A. (ed.), Soil Micromorphology. Elsevier, Amsterdam, pp. 57-68.

Zachariae, G., 1965. Spuren Tierischer Tätigkeit im Boden des Buchenwaldes. Frostwissenschaftliche Forschungen 20. Parey, Hamburg, 68 p.

Chapter 17

Soil Organic Matter

Kristin Ismail-Meyer[1], Mark H. Stolt[2], David L. Lindbo[3]

[1]*UNIVERSITY OF BASEL, BASEL, SWITZERLAND;*
[2]*UNIVERSITY OF RHODE ISLAND, KINGSTON, RI, UNITED STATES;*
[3]*UNITED STATES DEPARTMENT OF AGRICULTURE, WASHINGTON, DC, UNITED STATES*

1. Introduction

Organic matter is universal to all soils. The form and distribution of soil organic matter is dependent on a number of processes that may act independently or in concert. Most of these processes are related to the initial decomposition of accumulated plant residues by a suite of soil fauna and to the continued decomposition, transport and accumulation of the by-products. Descriptions of the forms, distribution and genesis of organic components in soils can range from quite simple to extremely complex. Micromorphological approaches to descriptions are comparably varied. For example, in one of the seminal works focused on concepts and terminology in micromorphology, Brewer (1964) essentially ignored the soil organic matter component, whereas others have proposed detailed approaches to describe its form and distribution in soils (Kubiëna, 1953; Barratt, 1969; Bal, 1973; De Coninck et al., 1974; Babel, 1975; Fox, 1984; Pawluk, 1987). Reviewing each of these approaches is beyond the scope of this chapter. Instead, we will use the descriptive system outlined by Bullock et al. (1985) and Stoops (2003) as our context and relate the various features and horizons or layers described in other works to this system to assist the reader in understanding the source and genesis of soil organic matter observed in thin sections using optical microscopy, at scales of tens of micrometres and greater. The descriptive system of Bullock et al. (1985) and Stoops (2003) for organic matter was primarily based on an extensive review by Babel (1975). Although Babel's work is more than four decades old, readers are encouraged to also consider his text in their studies.

Soil organic matter includes all organic matter derived from living organisms in soils. Examples range from a newly fallen leaf, roots, animal residues and droppings to carbon-rich gel, as dopplerite or organic pigment (Stoops, 2003). The term humus refers to materials containing 60-80% of degraded soil organic matter. Humus is rich in lignin (see Section 2.1) and forms amorphous colloids (Osman, 2013; Eash et al., 2016). Of all the features, horizons and constituents of soils examined under the microscope, soil organic matter is arguably the most dynamic. In environments where microbes, fungi and macroinvertebrates are very active (moist, aerobic conditions during the warmest of seasons), soil organic matter features such as fine roots may decompose in a matter of days (Hendrick & Pregitzer, 1997; Arnone et al., 2000; Tingey et al., 2000). In such cases, the thin section represents only a snapshot of soil conditions at the moment of sampling. Conversely, under highly anaerobic, waterlogged conditions (e.g., in peat; Aber et al., 2012), or in subsurface horizons such as spodic horizons and buried surface horizons rich in soil organic matter (Blazejewski et al., 2005), decomposition of organic matter may be very slow and the thin section represents stable soil conditions relative to this component.

Several difficulties are encountered when examining thin sections of soil material rich in organic matter. Because soil organic matter is much more susceptible to shrinkage upon drying than mineral components, and because it typically holds a considerable amount of water, thin sections of these materials often contain artefacts of the production process. Thus, care should be taken during thin section preparation to ensure that the soil organic matter features observed are representative of the soil materials and not artificial. Reviews of various approaches to remove water and prepare thin sections of soil organic matter can be found in a number of publications (e.g., FitzPatrick & Gudmundsson, 1978; Fox, 1985; Murphy, 1986; Fox & Parent, 1993; Lopez-Buendia, 1998; De Vleeschouwer et al., 2008; Kooistra, 2015). Because organic matter is so dynamic at the soil surface, sampling of organic material in surface horizons should include not only the central part of the A horizon but also its top and the litter layer.

Determination of specific organic remains in thin sections is possible (see Section 2) but can be challenging due to the state of preservation, as well as due to the random orientation of cross sections through the remains (radial, transversal, tangential or intermediate). Other difficulties encountered when studying soil organic matter and related features in thin sections include distinguishing between dark organic features and other components that are equally dark, typically rich in Fe or Mn oxides. Depending on the material and the objectives of the study, these problems can be overcome by studying the thin sections using fluorescence microscopy or by chemical treatment of the samples (Stoops, 2003). Some of these methods can also be used to assist in the identification of the various plant cells and tissues (e.g., Babel, 1972b; Stoops, 2003). Available treatments for distinguishing organic material from other compounds include bleaching with sodium hypochlorite (see Babel, 1964a, 1964b, 1975). An example of a common treatment in fluorescence microscopy is staining with fluorochromes (Tippkötter, 1990), such as acridine orange. Cell walls and roots in samples treated with acridine orange can be distinguished by a greenish colour under blue light excitation, whereas other materials such as clay appear orange to yellow (Altemüller & Van Vliet-Lanoë, 1990).

2. Organic Materials

2.1 Fresh Materials

Litter and roots from plants that are currently growing at a given site are the primary sources of the soil organic matter in surface and near-surface horizons. To identify specific plant residues in thin sections, an understanding of the anatomy of plants in general, as well as that of the particular plant species growing at the site, is necessary. As an aid to identify the components of higher plants under the microscope, several handbooks and atlases in the field of botany, plant anatomy and plant physiology are available (e.g., Bracegirdle & Miles, 1971; Esau, 1977; Schweingruber, 1982; Gifford & Foster, 1989; Schoch et al., 2004; Botanical Society of America, 2008; Upton et al., 2011). Especially in waterlogged, anaerobic environments, where plant remains are mainly well preserved, it is often possible to differentiate between wood, wood bark, twigs (and/or roots), deciduous leaves, remains of true grasses (Poaceae), seeds/fruits and mosses (see Ismail-Meyer, 2017).

The higher vascular plants consist of different types of tissues such as the epidermis (single-cell surface layer), the vascular tissues or vessels for conduction (phloem and xylem) and the ground tissue. The latter can be divided into parenchyma (thin-walled cells, filler tissue, food storage cells), sclerenchyma (lignin-rich tissue, as fibres and stone cells, phlobaphene-containing tissues) and collenchyma (irregularly thickened cells in leaves and stems) (Babel, 1975; FitzPatrick, 1993; Gifford & Foster, 1989; Upton et al., 2011) (see Figs. 1 to 3; Table 1).

Wood, bark and twigs from conifers (gymnosperms), deciduous trees and shrubs (angiosperms) can often be distinguished under the microscope (Fig. 1A to E). In temperate zones, annual rings are built in the sapwood and heartwood (xylem). Conifer wood shows very distinct year rings, uniseriate pith rays and small vessels (tracheids) (Fig. 1A and B). Deciduous wood has uni- or multiseriate pith rays and bigger vessels (trachea), which are arranged ring-porous or diffuse-porous (Fig. 1C and D). Well-preserved wood shows first-order interference colours and a blue colour under fluorescent light because of preserved cellulose in the cell walls. The outer bark or cork consists of sclerenchymatic cells, which show a characteristic parallel arrangement and are often squashed and deformed. They contain tannins (phlobaphenes), which give them a dark brown to reddish colour (Fig. 1E and F). Gymnosperm bark may contain pale sclerenchymatic stone cells and fibres (Gifford & Foster, 1989). Good wood/twig sections can be determined by the use of wood atlases (e.g., Schweingruber, 1982; Schoch et al., 2004).

Deciduous leaves show a characteristic cross section (Fig. 2A and B) with the upper epidermis, underlain by a palisade layer, followed by an interval consisting of loose parenchyma tissue (mesophyll) and that also comprises leaf veins (ribs), followed by the lower epidermis with stomata. In thin sections, leaves appear as elongated, yellowish to dark brown structures; the leaf veins with their complex form and the palisade layer are often the most striking characteristics (Babel, 1964b; Upton et al., 2011). Conifer needles consist of a loose parenchyma tissue (mesophyll) enclosing one or two central veins and are covered by an epidermis with stomata. The shape in cross section may indicate to which tree species the needles belong: fir needles (*Abies*) exhibit a 'coat hanger' shape with two resin ducts (Fig. 2C and D); spruce needles (*Picea*) are four-sided to diamond-shaped, whereas pine needles (*Pinus*) have a semicircular shape and show two resin ducts towards the flat side (Schoch et al., 1988; Gifford & Foster, 1989; Ismail-Meyer, 2017). The epidermis, the stomata and part of the veins (xylem) may show interference colours (Babel, 1975; FitzPatrick, 1993; Tian et al., 1997).

Leaves of true grasses (Poaceae) consist of an epidermis (the upper side containing stomata) and spongy parenchyma with embedded veins, in the form of three pores arranged in a manner producing a skull-like appearance. Under the microscope the leaves are usually elongated and show distinct shapes in cross section, such as flat, folded, boat-shaped, or with rolled-in edges (Fig. 2E and F). Poaceae stems have a round cross section, and the characteristic leaf veins are usually arranged around the outer rim

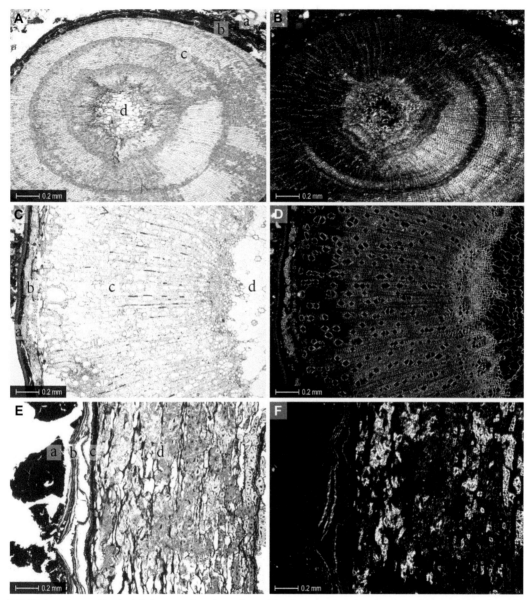

FIGURE 1 Twigs and bark. (A) Cross section through a white fir twig, showing the main characteristics of conifer wood: dark outer bark (periderm) (a); dense, slightly decayed phloem (b); xylem (c) consisting of small vessels (tracheids) and showing three year rings; pith (d); uniseriate pith rays (>) (Zug-Riedmatt, Switzerland) (PPL). (B) Idem in XPL, showing the first-order interference colours of the cellulose, indicating a good preservation. (C) Cross section through a deciduous porous twig consisting of outer bark (periderm) (a), phloem (b), xylem (c) with large and small vessels (tracheae and tracheids) and the (slightly darker) uniseriate pith rays (>) and the pith (d) (Pfyn-Hinterried, Switzerland) (PPL). (D) Idem in XPL, illustrating the interference colours of well-preserved cellulose and a diffuse year ring border (>). (E) Cross section through the outer bark or cork (periderm) of a tree, probably a conifer, consisting of several layers: a somewhat degraded, black, gelified outer layer (a); a striped layer (epidermis) (b); a dense dark brown layer (c); a brown, lignin-rich layer, showing shrinkage and degradation, containing beige sclerenchymatic stone cells (d) (Zug-Riedmatt, Switzerland) (PPL). (F) Idem in XPL, illustrating the interference colours of the sclerenchymatic stone cells. *Images by K. Ismail-Meyer.*

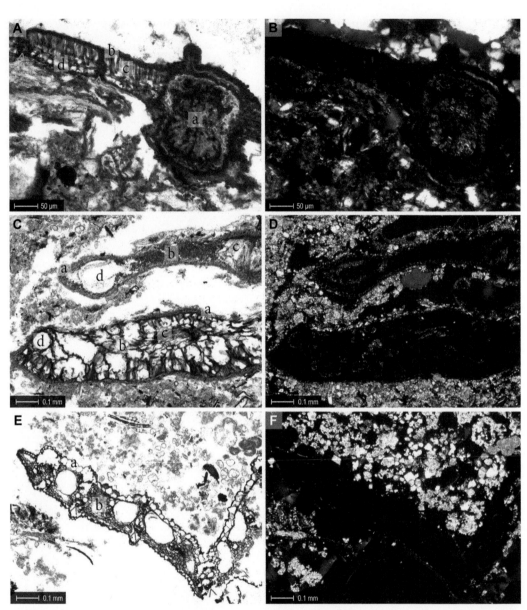

FIGURE 2 Leaves and needles. (A) Cross section through a foliage leave, showing midrib xylem (a) and leaf veins (>), epidermis (b), palisade layer (c) and spongy mesophyll (d) (Cham-Hagendorn, Switzerland) (PPL). (B) Idem in XPL, showing low-intensity interference colours of the midrib xylem. (C) Cross sections through two white fir needles, showing epidermis (a), mesophyll (b) (with poor preservation in the upper needle), veins (vascular bundle) (c) and resin ducts (d) (limnic calcareous deposit, Zug-Riedmatt, Switzerland) (PPL). (D) Idem in XPL, showing low-intensity interference colours of the epidermis and veins. (E) Cross section through a well-preserved folded leaf of a true grass (Poaceae), in calcitic wood ashes, showing epidermis (a), brown mesophyll (b) and several leaf veins (>; 'skulls') (Zug-Riedmatt, Switzerland) (PPL). (F) Idem in XPL, showing low-intensity interference colours of the leave. *Images by K. Ismail-Meyer.*

(Gifford & Foster, 1989; Ismail-Meyer, 2017). Grasses usually contain phytoliths, which may be hidden by the organic tissues (see Vrydaghs et al., 2017; Kaczorek et al., 2018).

Plant seeds are often small (<1 mm) and show a complex anatomy, with several layers and differing tissues. The central, nutritive tissue of the seeds (endosperm) is degraded very quickly; thus seeds are often empty (Gifford & Foster, 1989; Kenward & Hall, 2000). In thin sections, colour and interference colours can be distinct (see Fig. 3). Seeds such as cereals, rice and maize may contain starch grains, which look similar to calcite spherulites (see Canti & Brochier, 2017a). Usually, cereal grains are preserved in a charred state, with burst starch easily recognised by its popcorn-like appearance (Ismail-Meyer, 2017).

Roots show many different patterns, but usually they consist of an epidermis, a cortex containing reddish-brown tannins and a central cylinder with conductive vessels (Fig. 4A and B). The central cylinder is often completely decomposed. Woody roots may look very similar to twigs. Roots are often better preserved than organic matter from other sources in the same sample (Babel, 1975; FitzPatrick, 1990; Hather, 1993; Blazejewski et al., 2005).

Mosses, as non-vascular plants, show a simple structure, consisting of stalks with a thickened epidermis, a parenchyma rim and a central duct. The wing-like leaves may still be in connection to the stalks and consist of one to three identical cell layers. In thin sections, leaf mosses have a yellowish to brown colour (Fig. 4C and D), whereas turf mosses are almost transparent (Fig. 4E and F). Mosses do not contain lignin, but the cellulose in the cell walls shows interference colours (Collins & Kuehl, 2001; Mauquoy & Van Geel, 2007; De Vleeschouwer et al., 2008; Müller & Frahm, 2013; Ismail-Meyer, 2017).

Aquatic plants, such as algae, generally have no supporting tissues and contain no lignin or cellulose. Algal tissues are therefore usually not preserved, even in waterlogged environments (Collins & Kuehl, 2001; Blume et al., 2016), but their seeds (Fig. 3C to F) and oogonia can occur as remnants (e.g., Nymphaceae, Charales). Algal remains can be abundant in limnic carbonate environments, in the form of specific biogenic calcium carbonate precipitates (e.g., Freytet & Verrecchia, 2002).

In some soils, simple plants such as algae play an important role in soil organic matter dynamics (Lund, 1962; MacEntee et al., 1972; Hunt et al., 1979). Dodge and Shubert (1996) present an overview of the most common algal types and associated soil settings. Additions of carbon from lichens, fungi, mosses and algae are particularly important in desert and arctic environments where soil crusting occurs (Cameron & Devaney, 1970; Verrecchia, 2000; Asta et al., 2001; Hu et al., 2003; Williams et al., 2018) and in subaqueous environments where remains of marine or limnic fauna accumulate (Kubiëna, 1953).

In depositional environments such as floodplains, alluvial fans and lake margins, as well as in settings with anthropogenic accumulation such as agricultural fields where manure has been added, the origin and initial degree of decomposition of the plant remains is often unknown. Identifying the pathways leading to decomposition of soil organic matter *in situ* in these cases can be quite difficult (Macphail et al., 2003). However, recent studies of opal phytoliths using thin sections provide important information about degradation and *in situ* preservation of organic material

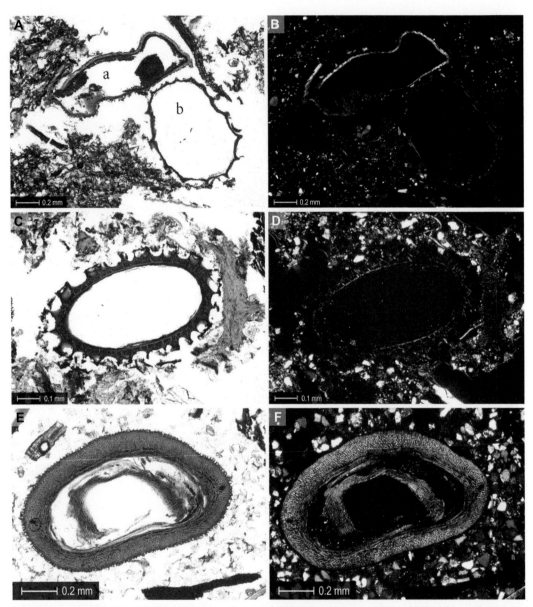

FIGURE 3 Seeds. (A) Cross sections through a flax seed (a) and a poppy seed (b), both with well-preserved epidermis and degraded internal part (endosperm), in groundmass with high organic matter content (Zug-Riedmatt, Switzerland) (PPL). (B) Idem in XPL, illustrating the characteristic yellow colour of the flax seed and dark-red colour of the poppy seed. (C) Cross section through a water lily seed (Nymphaceae), with characteristic shape, in alluvial sediments containing orange, gelified organic matter (Zug-Riedmatt, Switzerland) (PPL). (D) Idem in XPL, illustrating optical anisotropism of the sclerenchymatic tissue. (E) Water nut seed (*Trapa natans*) (alluvial sandy sediments, Steinhausen-Chollerpark, Switzerland) (PPL). (F) Idem in XPL, illustrating optical anisotropism of the sclerenchymatic tissue. *Images by K. Ismail-Meyer.*

Table 1 Plant tissue types

Tissue type	Description	Figures
Dermal tissues		
Epidermis	Outermost single layer of cells, covering several plant parts, composed of parenchyma and sclerenchyma cells. Found on leaves, flowers, fruits, seeds, roots and stems.	2A-F, 3A-B, 4A-D
Periderm (bark)	Outermost layer of stems and roots of woody plants (trees) corresponding to the epidermis in non-woody plants, composed of multilayered, dead, sclerenchymatic cells and phlobaphene-impregnated tissues.	1A-F
Vascular tissues/bundles		
Xylem (wood)	Dead tissue of thick-walled, tubular cells, containing cellulose and impregnated with lignin, which form a duct for water and mineral transport in roots, stems and leaves. Composed of thick trachea (xylem vessels) and tracheids in angiosperms, but only of tracheids in conifers and ferns. The main part of wood consists of xylem with embedded pith rays; the xylem may show annual rings.	1A-D, 2A-B, 4A, 6A
Phloem	Sieve tubes of living tissue, composed of thin-walled tubular cells, for transport of food and nutrients in roots, stems and leaves.	1A-D
Ground tissues		
Parenchyma	Large, thin-walled cells in green leaves, roots and stems.	2A-F, 4C-F
Sclerenchyma	Thick-walled dead cells, rich in lignin and containing phlobaphene, in stems, leaf veins, hard outer layer of seeds and nuts. Subdivided into sclereids (stone cells) and fibres (bast).	1E-F, 3C-F
Collenchyma	Irregularly thick-walled living cells, thickened by cellulose at the corners, which provide structure and support in marginal parts and petioles of leaves and stems. Absent in monocots and roots.	–

Source: Babel (1975), Gifford & Foster (1989), FitzPatrick (1993), Schoch et al. (2004), Upton et al. (2011).

(Vrydaghs et al., 2017). On the other hand, organic matter accumulations related to waste deposition or *in situ* combustion may reveal significant information regarding former vegetation, land use and human occupation (Kooistra & Kooistra, 2003; Ismail-Meyer et al., 2013).

If horizons rich in organic matter of natural or anthropogenic origin are buried under anaerobic condition, preservation of macrobotanical remains and pollen may as well help in identifying the use of plants and soil that characterised the former land surface (Bernabo & Webb, 1977; Brugam, 1978; Nelson, 1984; Winkler, 1985; Parshall et al., 2003; Kühn et al., 2013). Information about pollen analysis in thin sections can be found in publications by Allison et al. (1986), Van Mourik (1999, 2003) and Kooistra and Kooistra (2003). The study of insect remains and diatoms has not yet been a specific subject in micromorphological research, but it may provide useful environmental insights in the future.

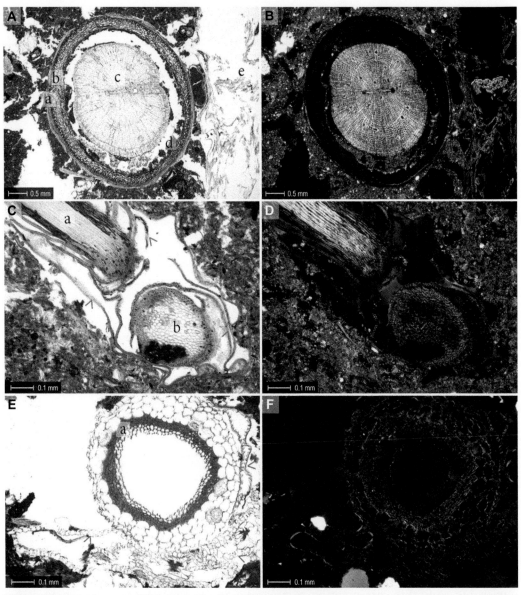

FIGURE 4 Roots and mosses. (A) Cross section through a young, well-preserved woody root, showing epidermis (a), cortex (b), central cylinder (xylem) (c) and degraded dark brown phloem (d) (the beige material in the lower part of the root section is an artefact); the image also shows a longitudinal section through a non-woody root (e) (drained, decayed peat deposit, Pfyn-Hinterried, Switzerland) (PPL). (B) Idem in XPL, showing first-order interference colours of the cellulose of the central cylinder and the epidermis. (C) Stem of a leaf moss in longitudinal cross section (a) and in transversal cross section (b), showing epidermis (denser rim of the stem), parenchymatic tissue, a cluster of iron oxide aggregates and thin moss leaves connected to the stem (>) (limnic calcareous deposit, Zug-Riedmatt, Switzerland) (PPL). (D) Idem in XPL, showing interference colours of the cellulose in the moss tissue. (E) Cross section through the stem of sphagnum peat moss, showing a dense, brown band (a) and large parenchyma cells; note the leaves around the stem (>) (Lake Luokesa, Lithuania) (PPL). (F) Idem in XPL, showing interference colours of the cellulose in the moss tissue. *Images by K. Ismail-Meyer.*

2.2   Partially and Fully Decomposed Materials or Humus

Plant decomposition occurs in aerobic and anaerobic environments and is analogous to mineral weathering, whereby organic components go through a series of transformations representing a fundamental pedogenic process (Simonson, 1959; Fanning & Fanning, 1989; Buol et al., 2011). Plants consist of not only different tissues (see Section 2.1; Table 1) but also diverse molecular components, i.e., approximately 15-40% cellulose, 10-43% hemicellulose, 25-40% lignin, 1-10% waxes and lipids and 1-15% proteins (Osman, 2013). Which organic remains or plant parts decay depends not only on tissue type and chemical composition but also on moisture and oxygen content as well as temperature and pH of the environment (Eash et al., 2016). Easily degradable substances (sugars, proteins, amino acids, lipids, fats, oil, waxes) are quickly lost. Tissues made up of cellulose, such as collenchyma, phloem and parenchyma, also decay rapidly. Lignin-rich issues such as sclerenchyma (xylem, vascular bundles, pith rays) and phlobaphene-containing tissues such as cork (outer bark) are usually better preserved (Babel, 1965; Bal, 1973; Gifford & Foster, 1989; Tian et al., 1997; Osman, 2013; Strawn et al., 2015; Eash et al., 2016). Lignin, one of the most resistant organic compounds, decays only at the final stage of aerobic decomposition and is therefore the main constituent of humus, consisting of approximately 45% lignin, 35% amino acids, 4% cellulose, 7% hemicellulose and 9% fats, waxes, resins and other substances (Osman, 2013; Blume et al., 2016).

The decay of organic matter can be divided into alteration and decomposition (see Figs. 5 and 6). Alteration includes fungal and bacterial degradation, which can be seen in thin sections as discolouration and deformation of cells, browning, blackening, gelification, colonisation by fungi and dislocation of cells (Fig. 6A to E). Altered remains may keep their shape for a long time, which allows their identification (Babel, 1997; Stoops, 2003; Kooistra, 2015). Decomposition of organic matter, mainly due to fragmentation caused by meso- and macrofauna (Fig. 6F), can produce organ residues and organic fine material. Further decomposition can be caused by physical processes such as desiccation, wetting and drying, freezing and thawing, resulting in cell deformation, development of openings in tissues (Fig. 6A) and formation of organic fine material (Blazejewski et al., 2005; Strawn et al., 2015). Humic substances (fulvic acid, humic acid and humin) arise from microbial decomposition (Osman, 2013; Strawn et al., 2015; Eash et al., 2016). As a result of this process, gelified organic matter and dopplerite may be formed (Fig. 6G and H). Dopplerite, or waterhard, is pale-yellowish brown to dark brown, translucent material, commonly crossed by cracks upon drying. It forms during early diagenesis of plant tissues in an anaerobic, sub-aquatic or waterlogged environment, as in peat and hydric soils (see Section 5) (Lewis, 1882; Babel, 1975; FitzPatrick, 1993; Koopman, 1988; Sebag et al., 2006; Feller et al., 2010; Ismail-Meyer et al., 2013; Ismail-Meyer, 2014). Further investigation regarding the formation of gelified organic matter and dopplerite is clearly needed. Complete decomposition of organic matter leads to breakdown into inorganic compounds, such

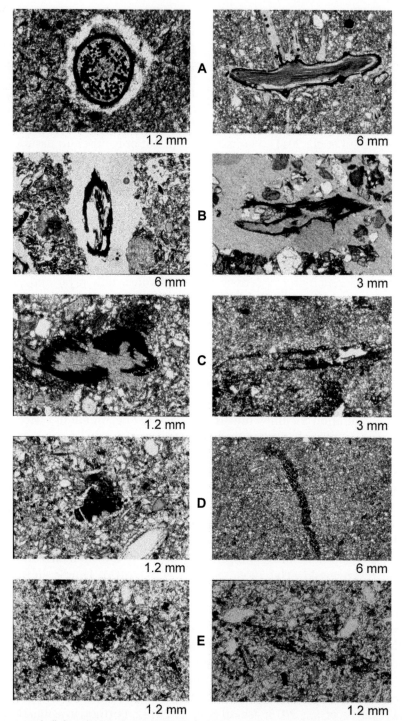

FIGURE 5 Transversal (left column) and longitudinal (right column) cross sections of the different root decomposition classes, according to Blazejewski et al. (2005), on a scale of zero (minimal to no decomposition) to four (nearly completely decomposed) (PPL). Adjacent pairs do not represent the same root but roots with similar degrees of decomposition. (A) Class 0 root, inner portion and sheath are complete. (B) Class 1 root, inner portion and sheath are incomplete. (C) Class 2 root, organic matter is dispersing into surrounding soil, and tissue fragments are still present. (D) Class 3 root, visible tissue remains are absent, and root shape is discernible. (E) Class 4 root, visible tissue remains are absent, and original root shape is indefinite. Frame widths are indicated at bottom right corner of photographs.

as carbon dioxide, water, phosphates, sulphates and cations (Osman, 2013; Strawn et al., 2015; Eash et al., 2016).

Different environments favour specific degradation patterns. Under aerobic, neutral or basic conditions, macro- and mesofaunal decay may be strong. In contrast, at low pH, fungal activity can be dominant, whereas under anaerobic, waterlogged conditions, bacterial activity results in gelification of organic matter and decomposition of cellulose. The dominant degrading agent and decomposition environment can in some cases be identified (see Blanchette, 2000; Kim & Singh, 2000; Schwarze, 2007; Nilsson & Björdal, 2008).

Turnover times, i.e., the average time that soil organic matter resides in a soil from its entry to complete decay, range from days to more than several centuries, with most decomposition occurring shortly after deposition and mostly within less than 10 years (Arnone et al., 2000; Kenward & Hall, 2000; Tingey et al., 2000; Blazejewski et al., 2005; Lützow et al., 2006; Osman, 2013; Strawn et al., 2015). For example, Babel (1964b) observed that beech leaves in the litter horizon of a mull (see Section 4.3) had lost the main part of their phloem and collenchyma after six months.

2.3 Charcoal and Related Materials

Black carbon results from the incomplete combustion of vegetation and fossil fuels (Goldberg, 1985; Jones et al., 1997; Masiello, 2004; Forbes et al., 2006). The two primary components of black carbon are charcoal and ash. The source of most charcoal fragments is organic materials burned at the soil surface (e.g., during forest fires) or charred organic materials burned off-site and applied to the soil together with ash. The charcoal can be incorporated into the soil through sedimentation, bioturbation or tillage. Fine charcoal fragments (<50 μm; soot) may be deposited by water or wind. The black colour, opaque nature and sharp boundaries help discern charcoal from other organic materials (Babel, 1964c; Bullock et al., 1985). Large pieces (>30 μm) of charcoal typically exhibit cell structure (Fig. 7A and B). Calcitic wood ashes (Fig. 7C and D) may be identified because of their rhomboidal crystal shape and pseudomorphism after plant cells (Canti & Brochier, 2017b). The ashes are easily blown away by wind or dissolved. Wood ashes may be present in caves, in open-air sites or in waterlogged anthropogenic accumulations (Schiegl et al., 1996; Ismail-Meyer et al., 2013; Canti & Brochier, 2017b). Another remnant of burnt plant material can be vesicular glass derived from phytoliths that were present (see Kaczorek et al., 2018).

3. Soil Organic Matter Horizons and Profiles

Soil organic matter accumulates when microbial decomposition occurs at rates that are slower than those of the addition of organic matter (Simonson, 1959; Collins & Kuehl, 2001).

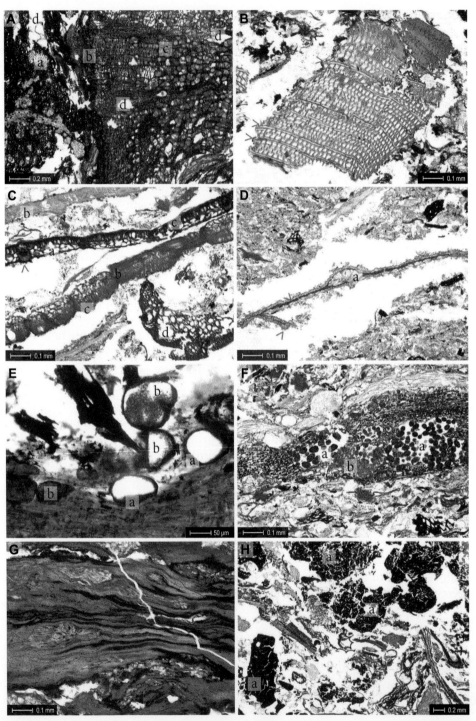

FIGURE 6 Altered and decomposed organic remains. (A) Deciduous wood from a peat layer, showing dark-brown outer bark with beige sclereids (a), dense, decayed phloem (b) and xylem (c), with openings in the latter tissue (d) (Sennhausen-Chollerpark, Switzerland) (PPL). (B) Conifer wood, with clearly visible pith rays (>) and with strong gelification resulting in complete loss of the cell structure (top right), due to microbial alteration in waterlogged

Organic soils belong to the soil order Histosols and include the suborders Folists, Fibrists, Hemists and Saprists (Soil Survey Staff, 2014). Folists are not permanently wet and form in upland areas and forests (moder, mull, mor and anmoor; see Section 4). Fibrists, Hemists and Saprists form under permanent wet conditions (hydric soils, peat; see Section 5). Soils with an organic surface layer that is 20-40 cm thick belong to mineral soils with a histic or folistic epipedon (Collins & Kuehl, 2001; Soil Survey Staff, 2014; for correlation with the European classification of humus forms, see Zanella et al., 2011). Histosols often occur in transitional zones between dry upland areas and lacustrine or marine environments, but they are also present in grassland, permafrost regions, anthropogenic deposits, floodplains and alluvial fans (dy, gyttja, sapropel; see Section 5.5) (Aber et al., 2012; Kooistra, 2015).

In Histosols, various stages of decomposition or preservation of soil organic matter are often clearly identifiable within L, F and H horizons, as used in the Canadian Soil Classification (Soil Classification Working Group, 1998), and in the underlying A horizon. The horizon symbols refer to litter (L), fragmentation or fermentation (F) and humus (H), corresponding to fibric (Oi), hemic (Oe) and sapric (Oa) horizons in Soil Taxonomy (see Buol et al., 2011; Osman, 2013). On the basis of micromorphological observations, Babel (1972a) further subdivided these horizons according to degree of decomposition, types of decomposers, degree of mixing and relative amounts of fine material, residues and roots. Similar observations were made by Ponge (1999) and described as a series of Oi, Oe, Oa and A horizons that follow the Soil Taxonomy system for horizon designation (Soil Survey Staff, 2014). Micromorphological examination and quantifications by Ponge (1999) showed clear boundary distinctness between Oi and Oe horizons and between Oe and Oa horizons. The boundary between the Oa and upper A horizons, however, was not objectively distinguishable under the microscope.

The thickness and macro- and micromorphological characteristics of the various organic horizons and mineral horizons rich in organic matter depend on a number of factors, such as frequency and duration of water saturation, pH, nutrient status, plant community and soil fauna (Jongerius & Schelling, 1960; Kooistra, 1991; Phillips & FitzPatrick, 1999; Ponge, 1999; Davidson et al., 2002). For example, in boreal forests, thick surface horizons that are rich in organic matter develop because of slow litter

---

conditions (Zug-Riedmatt, Switzerland) (PPL). (C) Leaves with different states of preservation: (a) well-preserved leaf, with clear palisade tissue (c) and leaf veins (>); (b) leaves that are partly gelified due to microbial activity in waterlogged conditions, with hardly recognisable palisade tissue (c) and leaf veins (>); the plant remain marked (d) is a grass stem fragment (Arbon-Bleiche 3, Switzerland) (PPL). (D) Partly altered, strongly shrunken white fir needle, with preserved central vein (a) and epidermis (>), but no remaining mesophyll (limnic calcareous deposit, Zug-Riedmatt, Switzerland) (PPL). (E) Fungal sclerotia, showing a dark-brown mantle (a) and some containing spores (b), recording fungal attack of wood in an aerated environment (Zug-Riedmatt, Switzerland) (PPL). (F) Bark remain showing signs of mesofaunal attack by at least two species, in the form of probable mite excrements (a) and *Diptera* excrements (b) (Zug-Riedmatt, Switzerland) (PPL). (G) Gelified organic matter in peat, probably resulting from alteration of a woody root (Sennhausen-Chollerpark, Switzerland) (PPL). (H) Black dopplerite (a), with characteristic cracking, formed by *in situ* alteration of organic material in waterlogged conditions (anthropogenic sediment, Zug-Riedmatt, Switzerland) (PPL). *Images by K. Ismail-Meyer.*

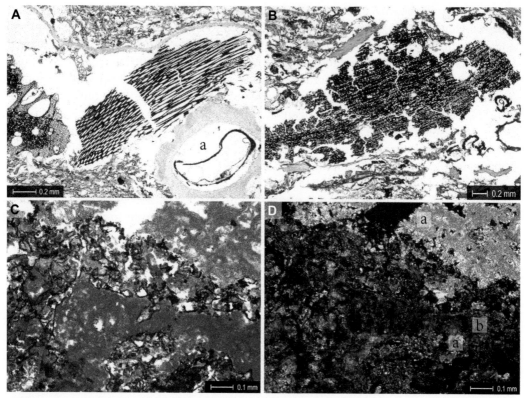

FIGURE 7 Charcoal and ashes. (A) Charcoal fragments with clearly visible cell structure, from conifer wood (right) and deciduous wood (left); the feature marked (a) is a blackberry seed (Zug-Riedmatt, Switzerland) (PPL). (B) Undisturbed fragmented charcoal (Zug-Riedmatt, Switzerland) (PPL). (C) Slightly degraded calcareous wood ashes, with probable iron/manganese oxide impregnations (black rims) (Zug-Riedmatt, Switzerland) (PPL). (D) Idem in XPL, showing clear distinction between well-preserved ashes (a), decalcified ashes (b) and phosphatised ashes (>). *Images by K. Ismail-Meyer.*

decomposition, whereas tropical forests do not include such surface intervals because the large amounts of litter that are produced are quickly decomposed, within approximately one year (Osman, 2013). Although the majority of soil organic matter is contained in surface or near-surface organic layers and A horizons, certain B horizons can be dominated by illuvial organic matter. For example, organic matter in spodic horizons typically occurs as coatings or infillings (see Van Ranst et al., 2018). In addition to spodic horizons, buried O and A horizons may occur well below the soil surface in palaeosoils, e.g., interglacial soils covered by glacial tills or palaeosoils in loess, or in streamside riparian zones that may contain significant amounts of organic matter (Blazejewski et al., 2005; Blume et al., 2016).

Soil scientists have long recognised that, under most conditions, organic matter will become distributed throughout the soil profile in an organised and predictable pattern. The majority of the literature points to Müller (1887) as the first to recognise these patterns and to start introducing the various distribution names such as mull and mor. Kubiëna (1953) described 34 different profile types dominated by horizons rich in soil organic matter. The main types include moder, mull, mull-like moder, gyttja, dy, anmoor and raw, raw-soil and tangel humus. Most of these types are discussed in the following sections. Since the work of Kubiëna (1953), these terms have been used, amended and redefined by numerous authors studying soils under the microscope (see Green et al., 1993). Discussions of the use and misuse of the terms in micromorphological studies related to soil morphology and genesis can be found in various publications (e.g., Jongerius & Schelling, 1960; Barratt, 1964, 1969; Bal, 1970, 1973; Bullock, 1974; Babel, 1975; Ponge, 1999; Ampe & Langohr, 2003). General problems are the use of multiple terms to describe the same horizon or distribution (e.g., mor vs. raw humus) and the use of the same terms to describe both a single horizon rich in soil organic matter and a group of horizons that are dominated by soil organic matter (humus profile; Bal, 1970). Babel (1975) suggests that the terms represent the entire profile but that the profile is typified by a particular horizon (e.g., the Ah horizon for mull). Resolving the inconsistencies in the use of soil organic matter distribution terminology is beyond the scope of this chapter.

4. Forest Soils

4.1 Moder

Moder, one of the most common terrestrial humus forms, is intermediate between mor and mull (see Sections 4.2 and 4.3). It is characterised by the presence of mineral-deficient coprogenic constituents (Jongerius & Rutherford, 1979) that are mixed with tissue residues and fine organic materials (Fig. 8). Moder humus is found in organic surface layers (L, F, H horizons), in A horizons and in Bh horizons of Spodosols. Moder profiles consist of L, F, H and A horizons. Kubiëna (1953) indicates that moders can be present in both deciduous and coniferous forests, with a range of soil fauna being responsible for decomposition. Micromorphological descriptions of moders can be found in numerous publications (e.g., Dalrymple, 1958; Jongerius & Schelling, 1960; Barratt, 1964; Babel, 1972a; Pawluk, 1987; Fox & Parent, 1993; Dijkstra, 1998; Ponge, 1999; Frouz & Novakova, 2005; Mori et al., 2009; Zaiets & Poch, 2016). Micromorphological descriptions by Bal (1970) of two moder profiles developed under red oak and Douglas fir provide the foundation for the following synthesis of the morphology and genesis of a typical moder. These descriptions illustrate the formation of surface and near-surface horizons that are rich in soil organic matter, and at the same time they indicate the complexity of the morphology, distribution and genesis of soil organic matter in soils.

Under deciduous vegetation, such as oak, L horizons are relatively thin, consisting primarily of relatively unaltered leaves (up to 75% in a temperate forest; Osman, 2013). Some leaves may show the first signs of faunal decomposition. Leaves in the upper F horizon are much denser than those in the overlying L horizon, and in places they are

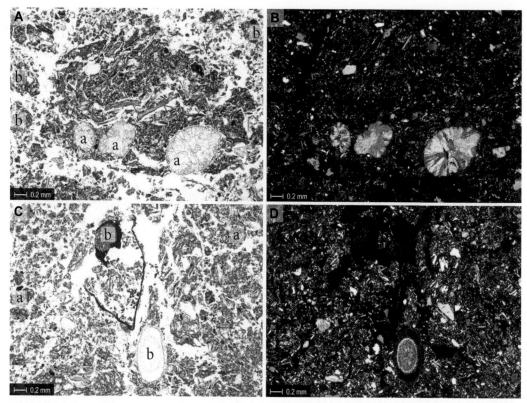

FIGURE 8 Moder profile (alpine valley bottom, 684 m a.s.l.; Gamsen, Switzerland). (A) Mesofaunal excrement, containing mainly organic fine material, some brown tissue residues (>) and three spheroidal calcite aggregates (a), probably from slugs; the features marked (b) are smaller excrements, possibly from *Diptera* larvae (H horizon) (PPL). (B) Idem in XPL. (C) Mineral soil material with crumb microstructure, some mesofaunal excrements (a) and several root residues (b) (A horizon, 4-5 cm depth) (PPL). (D) Idem in XPL, illustrating differences in preservation state of the roots. *Images by K. Ismail-Meyer.*

covered with dark excrements as a result of the activity of fauna such as earthworms and true fly larvae (Diptera) and bacteria. The size and shape of the excrements, in the F horizon and below, provide clues to the predominant type of faunal activity (see Kooistra & Pulleman, 2018). Leaf residues of the upper F horizon are primarily epidermis and lignin-rich veins (and stalks). The lower F horizon often contains much excrements and generally few small leaf residues. Small organisms such as Diptera larvae are likely the most important producers of excrement. The process leading to the integration of soil organic matter into excrements has been termed 'vermoderung' (Bal, 1970).

Under coniferous trees, such as Douglas fir, L horizons of moder profiles are typically thicker than those under deciduous trees because needles are more resistant to decomposition than leaves of deciduous plants (Bal, 1970; Ampe & Langohr, 2003). The only indications of needle decomposition are the few areas of dark excrements resulting from the activity of primary decomposers such as potworms (Enchytraeidae), mites

(Phtiracaridae) and crane fly larvae (Tipulidae). The upper F horizon of coniferous moder profiles is composed of needle organ residues, and their alteration is a product of the activity of fauna such as fungus gnats (Mycetophilidae), mites and crane fly larvae. The predominance of particular soil fauna in specific horizons rich in organic matter has been observed in numerous studies (e.g., Dinc et al., 1976; Phillips & FitzPatrick, 1999; Ponge, 1999; Davidson et al., 2002). For example, mite larvae excrements are often found within decomposing needles and roots, especially in the Fm horizon (Oe2 horizon). The H horizon is primarily composed of pellets (also observed in the overlying Fm horizon) with minor amounts of needle tissues. The pellets are the result of 'ageing', which refers to the decomposition, collapse and coalescence of excrements as a result of physical, chemical and microbial processes (Bal, 1970; Jongerius & Rutherford, 1979; Phillips & FitzPatrick, 1999; Davidson, 2002; Stoops, 2003). The combination of pellets and needle organ residues provides a very loose and spongy character to the F and underlying H horizons that is not observed under deciduous forest (red oak). The absence of earthworms under coniferous forest (Douglas fir moder) was also noted by Bal (1970) as contributing to differences in organic matter micromorphology and distribution between soils formed under coniferous versus deciduous vegetation.

Roots are apparent in the lower F and H horizons and in the mineral soil horizons of moder profiles. Decomposition of the roots follows a number of pathways that depend on the type of soil fauna. The morphology of final stages of root decomposition observed by Bal (1970) is similar to that of the root traces and irregular nodules described by Blazejewski et al. (2005). Over time, roots break down and root-derived carbon disperses into the surrounding soil to form root traces. Eventually, the carbon associated with root traces can become partially consumed by soil microbes or mixed with the surrounding mineral material by pedoturbation.

Excrements, with a colour that grades to black with depth, are abundant in moder H horizons, which consist almost entirely of a dense layer of excrements. The excrements show a range of sizes (<0.3-2.0 mm), with the larger excrements generally associated with larger potworms and some smaller epigeic earthworms. The denser fabric and darker colours in the lower H horizon are attributed to ageing of the excrements. In A horizons of moders, small black excrements are commonly found. In sandy soils, aggregates of this type are dispersed between sand grains in an enaulic c/f-related distribution pattern (Ampe & Langohr, 2003; Macphail et al., 2003). Bal (1970) suggested that the pellet distribution is a result of illuviation of the smallest pellets from the H horizon. Evidence of illuviation is given by the layered arrangement of the aggregates between the sand grains. Davidson et al. (2002) questioned this interpretation and suggested that the arrangement of the pellets could be the result of bioturbation (Davidson et al., 1999). Similar pellet-dominated fabrics have been observed for spodic horizons (see Van Ranst et al., 2018) and plaggen soils (see Adderley et al., 2018).

Under certain conditions, variants of the moder profiles described in the previous paragraphs can form. Babel (1975) provided descriptions of rendzina, alpine pitch and mull-like moders. Rendzina moders are composed primarily of very small, black, loosely packed droppings consisting mostly of calcium humates (Jongerius & Rutherford, 1979). This humus form is found predominately in shallow soils that develop on chalk or limestone (Kubiëna, 1953). Alpine pitch moders also form on limestone but at high elevations and under grasslands (Kubiëna, 1953). They are typically blackish to reddish brown. The H horizon is dominant, with a loose upper part, having an abundance of roots, and a dense lower part, showing cracking upon desiccation. Large plant residues are absent, but moderate quantities of very small residues are observed. Babel (1975) suggested that excrements were those of potworms or springtails (Collembola). Mull-like moders are dominated by Ah horizons, whereas F horizons are rare and H horizons are absent. Mixing is accomplished through a variety of soil fauna that occur in abundance. This humus form develops in grasslands and is differentiated from mulls by the small amount of clay that precludes formation of a clay-soil organic matter complex that is associated with mulls (Szabo et al., 1964; Jongerius & Rutherford, 1979).

4.2 Mor

Mor or raw humus (Kubiëna, 1953) is a terrestrial organic matter form consisting predominantly of well-preserved, though often fragmented, plant remains from forests, heaths and alpine ecosystems (twigs, branches, leaves, cones, grasses), with few excrements (Jongerius & Rutherford, 1979; Fox & Tarnocai, 2011). Generally, vegetation that creates acidic soil conditions, such as conifers, favours the formation of mor, in cool and humid environments, forming thick accumulations of folic material. The tissue residues that remain under the low pH conditions are selectively resistant to the primary decomposing organisms in the system, which are mostly fungi (white, brown and soft rot), soil mesofauna (i.e., Acarina species) and microorganisms (bacteria). Earthworms are usually absent below a pH of 4.8 (Meyer, 1964; Babel, 1975; Beyer, 1996; Fox & Tarnocai, 2011; Osman, 2013). As a result of microbial decomposition, organic matter in these L, F and H horizons is dark brown to black, shows loss of internal cell structures and breakdown of the leaf epidermis and can contain fungal mantles and hyphae. While in the F layer recognisable leaves make up to 90% of the organic fraction, they have almost completely disappeared in the H horizon, where only cork remains and root bark are still well preserved (Babel, 1965; Fox & Tarnocai, 2011). Because of the acidic environment, dissolved humic substances may form a spodic horizon (Beyer, 1996).

4.3 Mull

Jongerius & Rutherford (1979) defined a mull, from a micromorphological perspective, as a terrestrial humus form in which soil organic matter is mainly fine material in association with mineral components (mainly clay). Plant residues are few or absent; signs of earthworm activity can be strong. Mulls (Fig. 9) are usually found in nutrient-rich broadleaf forests, steppes, grassland and arable soils (Beyer, 1996; Kooistra, 2015). Thick mulls, typically thicker than 25 cm, are classified as mollic or umbric epipedons in

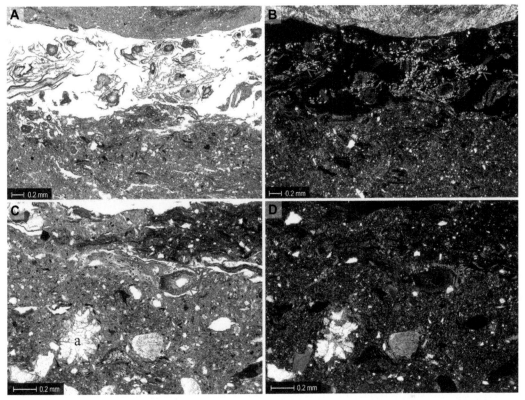

FIGURE 9 Mull profile (buried beneath slope wash deposit composed of marl, Niederdorf, Switzerland). (A) Porous L horizon with well-preserved mosses, overlain by the marl at the top and underlain by an F horizon, which is mainly composed of fine organic matter and which includes some preserved moss stems in the upper part (>) (PPL). (B) Idem in XPL, showing birefringence of well-preserved moss remains in the L horizon, and in the same layer, abundant small calcium oxalate aggregates (>), derived from burnt deciduous leaves. (C) Detail of the F horizon, showing impregnative dark organic fine material (top right), two moss stems (>) and a partially dissolved spheroidal calcite aggregate (a), probably from an earthworm (PPL). (H) Idem in XPL. *Images by K. Ismail-Meyer.*

Soil Taxonomy (Soil Survey Staff, 2014). More information about mollic epipedons and mulls can be found in the review by Gerasimova and Lebedeva (2018).

4.4 Anmoor

Anmoor profiles are typically ponded (semiterrestrial), often with black or dark grey horizons containing strongly decomposed soil organic matter, primarily in the form of aquatic animal excrements (Jongerius & Rutherford, 1979). In conditions with moderate water content, they have an earthy structure, which becomes muddy with persistent water saturation (Babel, 1975). Micromorphological descriptions of anmoors can be found in a number of publications (e.g., Kubiëna, 1953; Jongerius & Schelling, 1960; Jongerius, 1961; Menge, 1965; Loustau & Toutain, 1987). Diatoms are often present in anmoors under permanent water. Excrements are generally well decomposed and difficult to identify, especially where surface water saturation conditions vary throughout the year. Most of the soil organic matter is colloidal, with any plant residues that are present being fine. Kubiëna (1953) provided micromorphological descriptions of several anmoor subtypes, including eutrophic, dystrophic, peat, pitch and pitch peat anmoors. Dystrophic anmoors contain considerable amounts of plant tissue and much less coprogenic material than eutrophic anmoors. Peat anmoors have a low mineral content because they form within the uppermost layer of submerged peat. Pitch anmoors are a eutrophic form that in the wet state is primarily massive but may crack into angular blocks upon drying.

5. Wetland Soils or Hydric Histosols

Wetland soils with high organic matter content can have characteristic organic surface horizons, classified according to degree of degradation as fibric Oi horizon (peat), hemic Oe horizon (mucky peat) and sapric Oa horizon (muck) (IUSS Working Group WRB, 2015; Soil Survey Staff, 2014). Wetland soils develop where land remains waterlogged or inundated for most of the time, favouring hydrophytic vegetation. They can be divided into marine wetlands (e.g., coastal, estuarine, tidal) and inland wetlands (e.g., riverine, lacustrine, palustrine), the latter including bogs, swamps, marshes, fens and peatlands.

In depressions and sloping areas that are water-saturated up to the soil surface, considerable soil organic matter will accumulate, because of slower and less efficient decomposition than in aerobic environments, with 5 to 40 times lower rates of degradation. Characteristics of wetland soils include reducing conditions, occurrence of dissolved organic matter and the presence of anaerobic bacteria that cause reduction of iron, manganese and sulphur (Kroetsch et al., 2011; Aber et al., 2012). Few micromorphological studies have focused just on peat (Bullock, 1974; Cohen & Spackman, 1992; Mooney et al., 2000; De Vleeschouwer et al., 2008). Early studies were reviewed by Babel (1975) and Lee (1983). Most of the discussion by Babel (1975) concerns the work of Kubiëna (1943), Gracanin (1962), Grosse-Brauckmann (1963, 1964), Grosse-Brauckmann and Puffe (1964) and Krause (1964), reported in terms of primary peats (continuous saturation) and secondary peats (affected by drainage or cultivation; see also Section 5.4). Dinc et al. (1976) reviewed the various approaches to classify peat based on micromorphology. Fox (1985) presented a review of peat micromorphology in relation to the US soil classification system. Other approaches to classify peat, based on decomposition, were suggested by Von Post (1924) and Malterer et al. (1992). Detailed overviews of characteristics of wetlands, including hydrology, chemistry and peat decomposition, can be found in various publications (e.g., Richardson & Vepraskas, 2001; Mitsch & Gosselink, 2007; Keddy, 2010; Aber et al., 2012).

5.1 Fibric Horizons

Most fibric horizons (Fig. 10A to F), designated as Oi horizons or peat, are dominated by coarse plant residues (as many as 70% larger than 2 mm), which are light yellowish to reddish brown or dark brown and appear to have undergone little decomposition. Fibric horizons are common in raised bogs of boreal forest zones (Manoch, 1970; Dinc et al., 1976; Levesque & Dinel, 1982; Fox, 1985; Collins & Kuehl, 2001). Residues originate from herbaceous materials such as sedges and grasses, wood, moss (*Sphagnum*) (Fig. 10C and D) and seeds. They may be relatively large and are easily recognised under the microscope. Some fibric horizons may also contain minor amounts of materials that appear much darker than the yellow tissues and are similar to hemic or sapric material in degree of decomposition (Fig. 10A and B) (Lee & Manoch, 1974; Dinc et al., 1976).

5.2   Hemic Horizons

In general, hemic materials (Fig. 10E to H; Fig. 11A and B), assigned to the Oe horizon or mucky peat, are dominated by coarse plant residues, but a large portion of the groundmass consists of barely distinguishable plant residues and organic fine material (Manoch, 1970; Dinc et al., 1976; Levesque & Dinel, 1982; Fox, 1984, 1985; Mooney et al., 2000; Collins & Kuehl, 2001). They represent an intermediate degree of decomposition between fibric and sapric horizons. Coarse residues are often quite decomposed, and hemic horizons are thus generally darker than fibric horizons. The fabric is generally loosely packed and porous, with denser, poorly preserved excrements dispersed among the coarse tissue residues. Dissolved organic matter can be present and can react with other organic components or impregnate tissues (Kibblewhite et al., 2015). Dopplerite (see Section 2.2) may be present as pore infillings or thin layers (Babel, 1975).

5.3   Sapric Horizons

Most sapric materials (Fig. 11A to D), corresponding to the Oa horizon or muck, are dominated by black or dark brown organic fine material when viewed in incident light (Manoch, 1970; Lee & Manoch, 1974; Dinc et al., 1976; Levesque & Dinel, 1982; Fox, 1985; Collins & Kuehl, 2001). The few organ residues that are observed are dark brown and quite decomposed, which can make it difficult to identify the species from which they are derived (Fig. 11A and B). Sapric horizons are much denser than hemic or fibric horizons. A range of pore types is observed, including vughs, channels and packing voids (Dinc et al., 1976; Fox, 1985). Earthworm and mite excrements are often present (Fox, 1985). Sapric materials are common in many Histosols that have been drained and cultivated (Collins & Kuehl, 2001).

5.4   Peat Ripening

Although the transformation of soil organic matter deposits into peat (paludification) is considered an important process in the formation of Histosols (Buol et al., 2011), some authors do not consider peat to undergo pedogenesis until these water-saturated deposits are artificially drained or become dry during periods of major drought. When such changes in water saturation conditions occur, the pores fill with air and decomposition

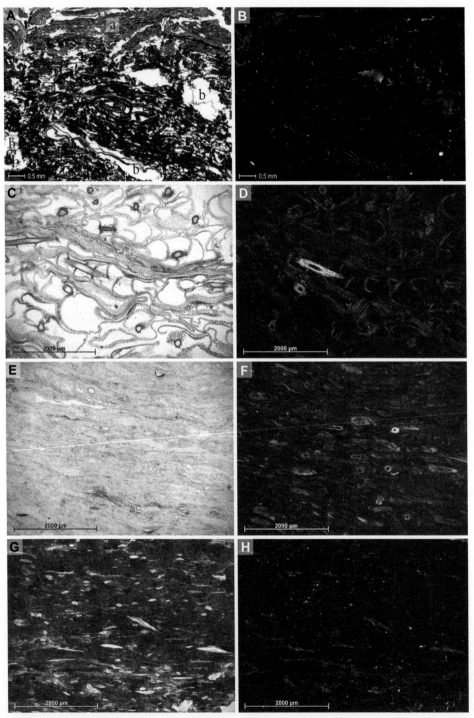

FIGURE 10 Fibric and hemic horizons. (A) Fibric horizon rich in bark (a), overlying a porous hemic horizon composed of altered, dark-brown organic matter, disturbed by root growth (b) (buried riparian peat profile, 160 cm depth, Steinhausen-Chollerpark, Switzerland) (PPL). (B) Idem in XPL showing that only few bark remains in

rates increase, resulting in subsidence of the peat ('ripening'; Pons, 1960). Babel (1975) and Bouma et al. (1990) reviewed micromorphological studies focused on peat that has undergone ripening; the peat is transformed to a much denser material having prismatic, granular, blocky (Fig. 11E and F) or platy microstructure (Kuiper & Slager, 1963; Lee & Manoch, 1974; Fox, 1984, 1985). The increase in faunal activity as a result of ripening greatly reduces the volume of plant residues and increases the relative amount of organic fine material, which can be easily removed by water and/or wind (Babel, 1975; Dinc et al., 1976; Hammond & Collins, 1983; Mitsch & Gosselink, 2007). Extreme ripening may result in formation of mulls or moders from previously water-saturated peats (Jongerius & Pons, 1962).

5.5   Dy, Gyttja and Sapropel

Non-peaty layers with high organic matter content that developed in low-energy sub-aqueous environments include dy, gyttja and sapropel. Such deposits can be found in floodplains, fans and deltas of rivers and streams. Organic matter accumulates at the sediment-water interface from suspension or when its specific weight exceeds that of water. Degradation of the organic matter occurs under anaerobic conditions. Micromorphological studies focused on dy, gyttja and sapropel are very limited (Kubiëna, 1953; Babel, 1975).

Dys (or peat mud), occurring in subaquatic Fluvisols (IUSS Working Group WRB, 2015; Blume et al., 2016), are black to brown humus layers deposited in acidic waters that are poor in nutrients and have high concentrations of soluble organic compounds. Such organic deposits have a gel-like form indicative of an abundance of amorphous organic matter. Small amounts of residues from plants associated with acidic environments, such as *Sphagnum* and other mosses, can be found. Those forms of dy having a dark glassy appearance and conchoidal fracture correspond to dopplerite (see Fig. 6H).

Gyttja (or eutrophic mud) forms in subaqueous environments rich in nutrients. It is defined as fine-textured, plastic, often gelatinous material when wet but becomes hard when dry, as the material shrinks and cracks along horizontal planes (Kroetsch et al., 2011). Gyttja (Fig. 11G and H) consists largely of droppings, diatoms and residues of aquatic animals, accumulated as loosely stacked particles and aggregates (Babel, 1975; Jongerius & Rutherford, 1979). Plant residues are commonly derived from algae, but remains of rooted plants of the shore vegetation are also often observed. In thin sections, the organic materials in gyttja are yellow (Kubiëna, 1953). Gyttja layers are referred to as

---

the fibric horizon are still birefringent. (C) Fibric horizon with well-preserved sphagnum moss remains (56-64 cm depth, Nieuw-Dordrecht, The Netherlands) (PPL). (D) Idem in XPL, showing the birefringence of well-preserved moss remains. (E) Fibric to hemic horizon with some, still recognisable sphagnum moss remains (>) (grassy peat profile, 12-20 cm depth, Nieuw-Dordrecht) (PPL). (F) Idem in XPL, showing the birefringence of moss remains. (G) Hemic horizon, with no clearly recognisable organic remains (reed peat underneath a medieval house floor, 69-76 cm depth, Peizermade, The Netherlands) (PPL). (H) Idem in XPL, showing that only a few tissue remains are birefringent. *Images (A) and (B) by K. Ismail-Meyer; images (C) to (H) by H. Huisman.*

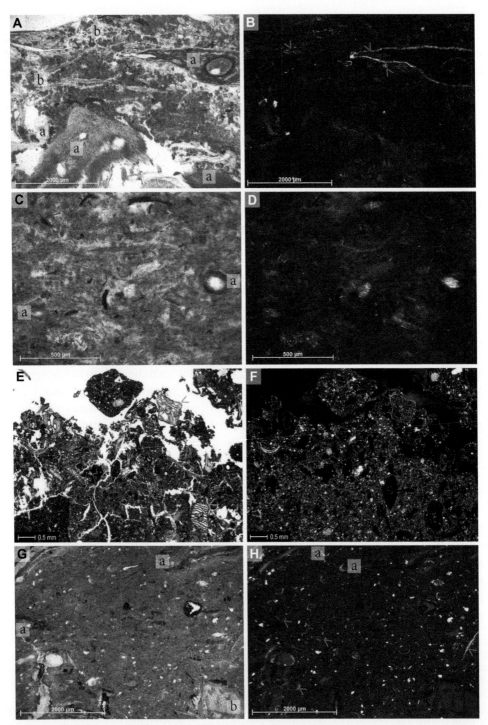

FIGURE 11 Hemic and sapric horizons, gyttja and degraded peat. (A) Hemic to sapric horizon with wood and/or root remains (a), embedded in fine organic material, partly with a fine granular microstructure produced by mesofaunal activity (b) (wood peat profile, 269-277 cm depth, Weesp, The Netherlands) (PPL). (B) Idem in XPL,

sedimentary peat, coprogenous earth or limnic materials in Soil Taxonomy (see Fox, 1985; Soil Survey Staff, 2014). Few studies have examined gyttja under the microscope (e.g., Uggla et al., 1972; Fox & Tarnocai, 1990). Fox & Tarnocai (1990) reported that the composition of sedimentary peat from a Pacific temperate wetland was dominated by amorphous organic matter, besides recognisable diatoms and plant tissues. At high magnification, the amorphous organic matter was identified as being composed largely of diatoms frustule fragments and organic remains with cell structure.

Sapropels are subaqueous layers formed at the bottom of nutrient-rich waters under anaerobic conditions. They can develop from gyttja or accumulate on top of the latter. Sapropels contain various amounts of more or less recognisable organic debris, and they are often highly enriched in sulphides, occurring as Fe-monosulphides or pyrite. Colours of sapropel horizons in the field are typically black, changing to grey upon drying. In marine settings, sea grasses such as *Zostera* spp. may provide considerable amounts of plant residues and tissues to surface and near-surface sapropel horizons (Babel, 1975; Bradley & Stolt, 2006; Blume et al., 2016). In marine intertidal zones, organic layers may also contain shell fragments, algae and remains of higher plants. Pseudomorphs after organic matter, composed of carbonates, iron sulphides, iron oxides and manganese oxides, are commonly encountered (Kooistra, 1978).

5.6   Anthropogenic Histosols

Deposits with high organic matter content that accumulate through anthropogenic activities may develop in waterlogged environments. Those sediments usually are not comparable with Anthrosols, which are characterised by long-term human impact and are described as human-altered or human-transported soils, mainly formed by irrigation or cultivation (Osman, 2013; Kooistra, 2015; see also Adderley et al., 2018). We prefer to allocate them to Histosols and not organic Anthrosols, as their genesis is much more comparable with that of natural waterlogged soils, such as peat soils (see below Section 5). Anthropogenic Histosols are found not only, for example, in Neolithic and Bronze Age lakeside settlements of Europe but also in urban areas near rivers or lakes (Kenward & Hall, 2000; Ismail-Meyer et al., 2013; Ismail-Meyer, 2014). Such accumulations may contain different types of organic remains, such as wood chips, charcoal, ash, bones, leather, woven fabrics, different coprolites, moss, grass, needles, leaves, seeds and shells

---

with probable roots showing interference colours (>). (C) Sapric horizon rich in fine organic material, with few moss stems (a) and fungal hyphae (>) (weakly degraded peat, 38-46 cm depth, Nieuw-Dordrecht, The Netherlands) (PPL). (D) Idem in XPL, showing low-intensity interference colours of moss stems. (E) The uppermost part of degraded peat, with subangular microstructure, some dense dark aggregates and a groundmass rich in organic fine material; the features marked (>) are spruce needles of recent origin (degraded peat, affected by ripening following drainage and peat cutting, Pfyn-Hinterried, Switzerland) (PPL). (F) Idem in XPL, showing that only the epidermis and central veins of the spruce needles display interference colours. (G) Calcareous gyttja containing altered leaves (a), a probable root (b), shells and moss stems (anthropogenic deposit, Jelsum, The Netherlands) (PPL). (H) Idem in XPL, clearly showing the presence of gastropod and ostracod shells (>), as well as some degraded moss stems (a). *Images (A) to (D), (G) and (H) by H. Huisman; images (E) and (F) by K. Ismail-Meyer.*

(Fig. 12). The origin of such accumulations can be diverse, including house constructions (wood, bark, grass or peat sods), animal husbandry (twigs, leaves and grasses in connection with herbivore and omnivore coprolites), and harvesting, tillage and hunting activities (seeds, fruits, bones, possibly shells) (e.g., Guttmann et al., 2003; Huisman et al., 2009; Verrill & Tipping, 2010; Ismail-Meyer et al., 2013; Ismail-Meyer, 2014; Huisman & Milek, 2017; Brönnimann et al., 2017a, 2017b). In terms of formation and degradation processes, these accumulations can be compared with natural wetland soils, with the difference that accretion of organic matter was not natural but anthropogenic. The organic layers of anthropogenic Histosols can correspond to fibric, hemic or sapric horizons (Fig. 12; see also Section 5). The sediments may be reworked by running water, contain fluvial or limnic intercalations or can be overprinted by ripening during phases of lower water table (Fig. 12E and F). Because of erosion and breaks in sedimentation, contrasting layers can accumulate in succession, forming complex stratigraphic sequences. Thick anthropogenic layers can be produced in settings with high sedimentation rates (up to several centimetres per year) and short periods of settlement. They therefore represent brief events, not comparable with a normal peat soil formation with accumulation rates of 0.01-3.8 mm per year (Kroetsch et al., 2011; Ismail-Meyer et al., 2013).

6. Organic Pedofeatures in Mineral Horizons

Organic pedofeatures commonly observed in thin sections of mineral horizons include excrements, infillings, coatings and nodules. Illuvial horizons rich in soil organic matter, occurring as pedofeatures, include spodic (e.g., De Coninck & McKeague, 1985; Bardy et al., 2008; Coelho et al., 2012), placic (Hseu et al., 1999) and humilluvic materials (van Heuveln & De Bakker, 1972). Detailed information about the micromorphology of spodic and placic materials is provided in the reviews by Wilson and Righi (2010) and Van Ranst et al. (2018). Illuvial organic material, sometimes referred to as illuvial humus, accumulates at the base of some drained and cultivated acidic organic soils (Soil Survey Staff, 2014). These illuvial horizons generally occur near the contact with underlying sandy materials. Such illuvial organic materials have only been identified in a few soils; thus little is known regarding their nature (Fox, 1985).

Excrements can be dispersed in mineral horizons or occur in groups, typically within pores or associated with plant remains (see Kooistra & Pulleman, 2018). The source of soil organic matter that was ingested and the type of soil fauna will primarily determine whether the excrements are composed of organic fine material or contain plant residues, the latter potentially including some that can be identified by tissue type. Excrements containing recognisable plant residues, produced by soil macrofauna such as earthworms, are often consumed again by soil mesofauna, such as potworms. Description and discussion of the genesis of excrements is presented by Kooistra & Pulleman (2018).

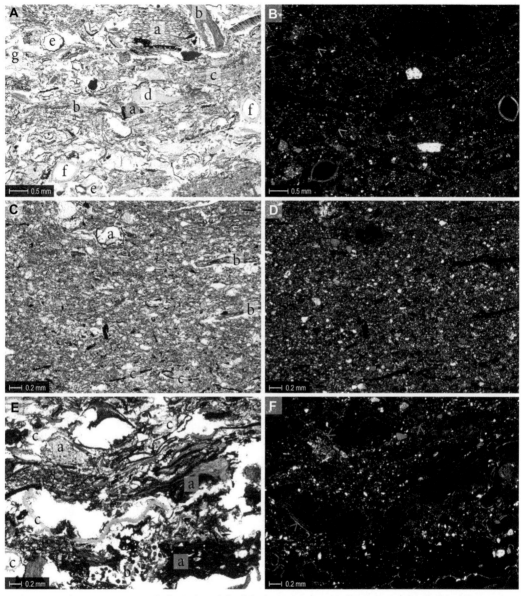

FIGURE 12 Anthropogenic Histosols (buried Neolithic lakeshore site, Zug-Riedmatt, Switzerland). (A) Anthropogenic layer with fibric and hemic intercalations, containing rounded and angular charcoal fragments (a), wood/bark remains (b), a white fir needle (c), a bone fragment (d) (probably from an amphibian), numerous poppy seeds (e), strawberry seeds (f) and moss remains (g) (PPL). (B) Idem in XPL, illustrating the characteristic dark-red colour of the poppy seeds (>), the birefringence of the strawberry seeds and the regular distribution of fine sand grains in the hemic parts rich in organic fine material. (C) Hemic horizon rich in organic fine material and containing few fibric remains, such as poppy seeds (a), white fir needles (b) and altered Poaceae remains (c) (PPL). (D) Idem in XPL, illustrating the abundance and distribution of calcareous silt and quartz fine sand in the groundmass. (E) Hemic to sapric layer mainly composed of strongly degraded and gelified wood and bark remains (a), excrements (b) and remains of mistletoe bark (c) (PPL). (F) Idem in XPL, showing that most wood and bark remains have lost their birefringence but not the mistletoe remains (>). *Images by K. Ismail-Meyer.*

Soils can contain infillings that are composed of a variety of soil organic matter basic components, including plant residues and organic fine materials (Kooistra et al., 2006), although such infillings are a less common type of soil organic matter pedofeatures than excrements. In agricultural soils that are well populated with earthworms, these infillings are common in unplowed near-surface horizons (Ligthart, 1997). In contrast, Blazejewski et al. (2005) found that infillings were the least common OM-rich pedofeatures found in subsurface horizons of hydric riparian (streamside) soils and the few that were observed were directly below A horizons or buried surface horizons.

Coatings composed of soil organic matter occur in several forms. One of the most common types of coating is associated with coarse grains, in a chitonic c/f-related distribution pattern, as in spodic horizons. Detailed descriptions of these coatings and their genesis are provided elsewhere (see Van Ranst et al., 2018). Other types of organic matter coatings include coatings in channels, resulting from earthworm activity. Worms moving through the soils create channels and leave exudates along the channel walls that may be rich in organic carbon (Kretzschmar, 1987; Blackwell et al., 1990; Binet & Curmi, 1992; Babel & Kretzschmar, 1994). Ligthart (1997) found that earthworm channels at shallow depths (<22 cm) lacked coatings, in contrast to channels at greater depths in the same profile. However, earthworm-related coatings have also been observed at deeper levels in soils (Kretzschmar, 1987). Coatings rich in organic matter are also often observed along ped surfaces of gleyed horizons (Bg or Btg). Some illuvial clay coatings are rich in organic matter pigment or amorphous fine material (see Kühn et al., 2018).

Irregular impregnative (non-cemented) nodules rich in organic matter have been described in some studies. They are sometimes clearly associated with incipient or incomplete spodic horizon formation (Macphail, 1983; Blazejewski et al., 2005), but in other cases the apparent genetic process cannot be identified. These nodules are often found in the same horizons as other organic matter features; thus a number of different pathways for their formation may exist. Based on the observation that these nodules, together with roots, were the most abundant features or components rich in organic matter, Blazejewski et al. (2005) suggested that they formed primarily by transformation of roots. This transformation involves a series of decomposition steps, from a fibrous root into a root trace (decomposed root that still maintains the orientation and dimensions of the original root), into a barely recognisable soil organic matter aggregate and finally into an irregular impregnative nodule (see Fig. 5). Other processes involved in the formation of these features may include pedoturbation (Hole, 1961; Johnson et al., 1987; Peacock & Fant, 2002) and illuviation (Fanning & Fanning, 1989).

7. Organic Sedimentary Features

In environments where deposition is a major process, such as floodplains, footslopes and alluvial fans, layered sediments indicative of water deposition are often encountered at various depths in the profile. These layers, referred to as lenses here, contain a range of soil organic materials from various sources and range in thickness from millimetres to a few centimetres (e.g., Kooistra, 1978; Kooistra et al., 1989, 2006; Macphail et al., 2003; Blazejewski et al., 2005). Samples can include several parallel layers of this type. Some of these layers are composed almost entirely of plant residues, whereas most are dominated by organic fine materials or pigments. In sandy materials, organic matter is primarily associated with fine organomineral aggregates that occur between coarse grains (enaulic c/f-related distribution pattern). In fine-textured soils, organic matter is so intimately associated with the silt and clay particles that differentiating between amorphous, fine organic materials and pigments is difficult. Boundaries between lenses rich in organic matter and those strata containing much less carbon are typically prominent. Their limited thickness, abrupt boundaries, abundance of plant residues and parallel orientation and distribution relative to the soil surface suggest that the lenses formed during short periods of land surface stability.

8. Applications and Conclusions

Some of the most important issues that soil scientists face are related to the effects of land-use change, global warming, pollution and invasive species on the soil environment. Because soil organic matter is the most dynamic of all soil components, effects of changing soil environment are often recorded first in the quality and distribution of organic matter. Thin sections provide a window to directly view these changes in soils at mesoscopic and microscopic scales.

Understanding changes relative to anthropogenic activities and the effects of cultivation, erosion, sedimentation, ground raising, compaction, waste deposition and fire may require integrative approaches using micromorphology, macrobotanical analyses, palynology and palynofacies analyses, possibly supplemented by molluscs, insect and diatom analyses, to interpret the form and distribution of soil organic matter, as well as to understand natural and anthropogenic soil formation processes (Kooistra & Kooistra, 2003; Sebag et al., 2006; Kühn et al., 2013). These plant and animal remains can be useful, for instance, in establishing the succession of changes in natural communities and understanding land-use changes. Wood fibres, cereal pollen and grains representing allochthonous organic matter can be used to identify the depth of anthropogenic activity.

Another important marker for anthropogenic activity is black carbon, especially charcoal (Masiello, 2004). Numerous studies have used micromorphology and the occurrence and distribution of charcoal to understand anthropogenic occupation and activities (e.g., Courty et al., 1989; Macphail et al., 1990; Davidson, 2002; Stephens et al., 2005; Macphail & Crowther, 2007). Fine black carbon (soot) in the atmosphere may be responsible for climate change in certain regions (McConnell et al., 2007; Ramanathan & Carmichael, 2008), and it can also be a significant sink of carbon in the global cycle (Kuhlbusch, 1998; Dai et al., 2005). Future micromorphological research may assist in understanding deposition rates of black carbon relative to global change and warming, and carbon turnover and storage.

Land-use change, global warming and atmospheric deposition of N may contribute to increased soil organic matter turnover rates such that carbon stocks in soils decrease and consequently CO_2 concentrations in the atmosphere increase. Records of soil organic matter quality, its spatial distribution and its position relative to mineral components may assist in understanding soil organic carbon and CO_2 responses to these changes. For example, soil horizons with lignin- or phlobaphene-rich materials, resistant to decomposition, will have slower turnover rates and thus will not be affected as much by environmental change. Turnover rates of soil organic matter may also be slowed by occlusion within aggregates and by intercalation or interaction with the mineral fraction or metals (Lützow et al., 2006). Understanding the spatial distribution of the organic matter in the soil and the interactions with mineral components may assist in understanding which soils will be more prone to higher turnover rates than others.

The distribution of soil organic matter in profiles such as moders and anmoors is the result of soil, plant, mesofauna and microfauna interactions. Soil micromorphology provides a tool to observe these interrelationships and to understand ecosystem processes (Ponge, 2003).

Acknowledgements

Many thanks to Hans Huisman for the images he provided, and to Philippe Rentzel, David Brönnimann, Michel Guélat, Urs Leuzinger, Christine Pümpin and Reto Jagher for useful discussions.

References

Aber, J.S., Pavri, F. & Ward Aber, S., 2012. Wetland Environments: A Global Prespective. John Wiley & Sons, Chichester, 421 p.

Adderley, W.P., Wilson, C.A., Simpson, I.A. & Davidson, D.A., 2018. Anthropogenic features. In Stoops, G., Marcelino, V. & Mees, F. (eds.), Interpretation of Micromorphological Features of Soils and Regoliths. Second Edition. Elsevier, Amsterdam, pp. 753-777.

Allison, T.D., Moeller, R.E. & Davis, M.B., 1986. Pollen in laminated sediments provides evidence for a mid-Holocene forest pathogen outbreak. Ecology 67, 1101-1105.

Altemüller, H.J. & Van Vliet-Lanoë, B., 1990. Soil thin section fluorescence microscopy. In Douglas, L.A. (ed.), Soil Micromorphology: A Basic and Applied Science. Developments in Soil Science, Volume 19. Elsevier, Amsterdam, pp. 565-579.

Ampe, C. & Langohr, R., 2003. Morphological characterization of humus forms in recent coastal dune ecosystems in Belgium and northern France. Catena 54, 363-383.

Arnone, J.A., Zaller, J.G., Spehn, E., Hirschel, G., Niklaus, P. & Korner, C., 2000. Dynamics of root systems in native grasslands: effects of elevated atmospheric CO_2. New Phytologist 147, 73-85.

Asta, J., Orry, F., Toutain, F., Souchier, B. & Villemin, G., 2001. Micromorphological and ultrastructural investigations of the lichen-soil interface. Soil Biology and Biochemistry 33, 323-337.

Babel, U., 1964a. Chemische Reaktionen an Bodendünnschliffen. Leitz-Mitteilungen für Wissenschaft und Technik 3, 12-14.

Babel, U., 1964b. Dünnschliffuntersuchungen über den Abbau lignifizierter Gewebe im Boden. In Jongerius, A. (ed.), Soil micromorphology. Elsevier, Wageningen, pp. 15-22.

Babel, U., 1964c. Opake organische Gemengteile in Auflagehumusformen. In Scheffer-Festschrift 1964. Institut für Bodenkunde, Göttingen, pp. S.95-S.110.

Babel, U., 1965. Humuschemische Untersuchung eines Buchen-Rohhumus mittels mikroskopischer Methoden. Mitteilungen des Vereins für Forstliche Standortskunde und Forstpflanzenzüchtung 15, 33-38.

Babel, U., 1972a. Moderprofile in Wäldern: Morphologie und Umsetzungsprozesse. Verlag Eugen Ulmer, Stuttgart, 120 p.

Babel, U., 1972b. Fluoreszenzmikroskopie in der Humusmikromorphologie. In Kowalinski, S. & Drozd, J. (eds.), Soil Micromorphology. Panstwowe Wydawnicto Naukowe, Warsawa, pp. 111-127.

Babel, U., 1975. Micromorphology of soil organic matter. In Gieseking, J.E. (ed.), Soil Components. Volume 1: Organic Components. Springer-Verlag, New York, pp. 369-473.

Babel, U., 1997. Zur mikromorphologischen Untersuchung der organischen Substanz des Bodens. In Babel, U., Fischer, W.R., Kaupenjohann, M., Roth, K. & Stahr, K. (eds), Mikromorphologische Methoden in der Bodenkunde. Ergebnisse eines Workshops der Deutschen Bodenkundlichen Gesellschaft (DBG), Kommission VII, 9-11. Oktober 1995 an der Universität Hohenheim. Hohenheimer Bodenkundliche Hefte 40, Universität Hohenheim, Stuttgart, pp. 7-14.

Babel, U. & Kretzschmar, A., 1994. Micromorphological observations of casts and burrow walls of the Gippsland giant earthworm (*Megascolides australis*, McCoy 1878). In Ringrose-Voase, A.J. & Humphreys, G.S. (eds.), Soil Micromorphology: Studies in Management and Genesis. Developments in Soil Science, Volume 22. Elsevier, Amsterdam, pp. 451-457.

Bal, L., 1970. Morphological investigation in two moder-humus profiles and the role of the soil fauna in their genesis. Geoderma 4, 5-36.

Bal, L., 1973. Micromorphological Analysis of Soils. Lower Levels in the Organization of Organic Soil Materials. Soil Survey Papers No. 6. Netherlands Soil Survey Institute, Wageningen, 174 p.

Bardy, M., Frisch, E., Derenne, S., Allard, T., do Nascimento, N.R. & Bueno, G.T., 2008. Micromorphology and spectroscopic characteristics of organic matter in waterlogged podzols of the upper Amazon basin. Geoderma 145, 222-230.

Barratt, B.C., 1964. A classification of humus forms and micro-fabrics of temperate grasslands. Journal of Soil Science 15, 342-356.

Barratt, B.C., 1969. A revised classification and nomenclature of microscopic soil materials with particular reference to organic components. Geoderma 2, 257-271.

Bernabo, J.C. & Webb, T., 1977. Changing patterns in the Holocene pollen record of northeastern North America: a mapped summary. Quaternary Research 8, 64-96.

Beyer, L., 1996. Humusformen und -typen. In Blume, H.P. (ed.), Handbuch der Bodenkunde. Ecomed, Landsberg am Lech, pp. 1-20.

Binet, F. & Curmi, P., 1992. Structural effects of Lumbricus terrestris (Oligochaeta: Lumbricidae) on the soil-organic matter system: micromorphological observations and autoradiographs. Soil Biology and Biochemistry 24, 1519-1523.

Blackwell, P.S., Green, T.W. & Mason, W.K., 1990. Responses of biopore channels from roots to compression by vertical stresses. Soil Science Society of America Journal 54, 1088-1091.

Blanchette, R.A., 2000. A review of microbial deterioration found in archaeological wood from different environments. International Biodeterioration & Biodegradation 46, 189-204.

Blazejewski, G.A., Stolt, M.H., Gold, A.J. & Groffman, P.M., 2005. Macro- and micromorphology of subsurface carbon in riparian zone soils. Soil Science Society of America Journal 69, 1320-1329.

Blume, H.P., Brümmer, G.W., Fleige, H., Horn, R., Kandeler, E., Kögel-Knabner, I., Kretzschmar, R., Stahr, K. & Wilke, B.M., 2016. Scheffer/Schachtschabel Soil Science. Springer, Heidelberg, 618 p.

Botanical Society of America, 2008. On-Line Image Collection. On-line database.

Bouma, J., Fox, C.A. & Miedema, R., 1990. Micromorphology of hydromorphic soils: applications for soil genesis and land evaluation. In Douglas, L.A. (ed.), Soil Micromorphology: A Basic and Applied Science. Developments in Soil Science, Volume 19, Elsevier, Amsterdam, pp. 257-278.

Bracegirdle, B. & Miles, P.H., 1971. An Atlas of Plant Structure. Heinemann, London, 123 p.

Bradley, M.P. & Stolt, M.H., 2006. Landscape-level seagrass-sediment relationships in a coastal lagoon. Aquatic Botany 84, 121-128.

Brewer, R., 1964. Fabric and Mineral Analysis of Soils. John Wiley and Sons, New York, 470 p.

Brönnimann, D., Pümpin, C., Ismail-Meyer, K., Rentzel, P. & Égüez, N., 2017a. Excrements of omnivores and carnivores. In Nicosia, C. & Stoops, G. (eds.), Archaeological Soil and Sediment Micromorphology. John Wiley & Sons Ltd, Chichester, pp. 67-81.

Brönnimann, D., Ismail-Meyer, K., Rentzel, P., Pümpin, C. & Lisá, L., 2017b. Excrements of herbivores. In Nicosia, C. & Stoops, G. (eds.), Archaeological Soil and Sediment Micromorphology. John Wiley & Sons Ltd, Chichester, pp. 55-65.

Brugam, R.B., 1978. Pollen indicators of land-use change in southern Connecticut. Quaternary Research 9, 349-362.

Bullock, P., 1974. The micromorphology of soil organic matter. A synthesis of recent research. In Rutherford, G.K. (ed.), Soil Microscopy. Limestone Press, Kingston, pp. 49-66.

Bullock, P., Fedoroff, N., Jongerius A., Stoops, G., Tursina, T. & Babel, U., 1985. Handbook for Soil Thin Section Description. Waine Research Publications, Wolverhampton, 152 p.

Buol, S.W., Southard, R.J., Graham, R.C. & McDaniel, P.A., 2011. Soil Genesis and Classification. Sixth Edition. John Wiley & Sons, Chichester, 543 p.

Cameron, R.E. & Devaney, J.R., 1970. Antarctic soil algal crusts: scanning electron and optical microscope study. Transactions of the American Microscopical Society 89, 264-273.

Canti, M.C. & Brochier, J.É., 2017a. Faecal spherulites. In Nicosia, C. & Stoops, G. (eds.), Archaeological Soil and Sediment Micromorphology. John Wiley & Sons Ltd, Chichester, pp. 51-54.

Canti, M.C. & Brochier, J.É., 2017b. Plant ash. In Nicosia, C. & Stoops, G. (eds.), Archaeological Soil and Sediment Micromorphology. John Wiley & Sons Ltd, Chichester, pp. 147-154.

Coelho, M.R., Martins, V.M., Otero Pérez, X.L., Vázquez, F.M., Gomes, F.H., Cooper, M. & Vidal-Torrado, P., 2012. Micromorfologia de horizontes espódicos nas Restingas do Estado de São Paulo. Revista Brasileira de Ciência do Solo 36, 1380-1394.

Cohen, A.D. & Spackman, W., 1992. Methods in peat petrology and their application to reconstrucition of paleoenvironments. Geological Society of America Bulletin 83, 129-142.

Collins, M.E. & Kuehl, R.J., 2001. Organic matter accumulation and organic soils. In Richardson, J.L. & Vepraskas, M.J. (eds.), Wetland Soils: Genesis, Hydrology, Landscapes, and Classification. CRC Press, Boca Raton, pp. 137-162.

Courty, M.A., Goldberg, P. & Macphail, R.I., 1989. Soils and Micromorphology in Archaeology. Cambridge Manuals in Archaeology. Cambridge University Press, Cambridge, 344 p.

Dai, X., Boutton, T.W., Glaser, B., Ansley, R.J. & Zech, W., 2005. Black carbon in a temperate mixed-grass savanna. Soil Biology and Biochemistry 37, 1879-1881.

Dalrymple, J.B., 1958. The application of soil micromorphology to fossil soil and other deposits from archaeological sites. Journal of Soil Science 9, 199-227.

Davidson, D.A., 2002. Bioturbation in old arable soils: quantitative evidence from soil micromorphology. Journal of Archaeological Science 29, 1247-1253.

Davidson, D.A., Carter, S., Boag, B., Long, D., Tipping, R. & Tyler, A., 1999. Analysis of pollen in soils: processes of incorporation and redistribution of pollen in five soil profile types. Soil Biology and Biochemistry 31, 643-653.

Davidson, D.A., Bruneau, P.M.C., Grieve, I.C. & Young, I.M., 2002. Impacts of fauna on an upland grassland soil as determined by micromorphological analysis. Applied Soil Ecology 20, 133-143.

De Coninck, F. & McKeague, J.A., 1985. Micromorphology of Spodosols. In Douglas, L.A. & Thompson, M.L. (eds.), Soil Micromorphology and Soil Classification. Soil Science Society of America Special Publication 15, SSSA, Madison, pp. 121-144.

De Coninck, F., Righi, D. Maucorps, J. & Robin, A.M., 1974. Origin and micromorphological nomenclature of organic matter in sandy Spodosols. In Rutherford, G.K. (ed.), Soil Microscopy. The Limestone Press, Kingston, pp. 263-280.

De Vleeschouwer, F., Van Vliet-Lanoë, B., Fagel, N., Richter, T. & Boës, X., 2008. Development and application of high-resolution petrography on resin-impregnated Holocene peat columns to detect and analyse tephras, cryptotephras, and other materials. Quaternary International 178, 54-67.

Dijkstra, E.F., 1998. A micromorphological study on the development of humus profiles in heavy metal polluted and non-polluted forest soils under Scots pine. Geoderma 82, 241-358.

Dinc, U., Miedema, R., Bal, L. & Pons, L.J., 1976. Morphological and physico-chemical aspects of three soils developed in the Netherlands and their classification. Netherlands Journal of Agricultural Science 24, 247-265.

Dodge, J.D. & Shubert, L.E., 1996. Algae in terrestrial and aquatic ecosytems. In Hall, G.S. (ed.), Methods for the Examination of Organismal Diversity in Soils and Sediments. CAB International, New York, pp. 67-78.

Eash, N.S., Saurer, T.J., O'Dell, D. & Odoi, E., 2016. Soil Science Simplified. Sixth Edition, John Wiley & Sons, Chichester, 260 p.

Esau, K., 1977. Anatomy of Seed Plants, Second Edition. Wiley, New York, 576 pp.

Fanning, D.S. & Fanning, M.C.B., 1989. Soil: Morphology, Genesis and Classification. John Wiley and Sons Inc., New York, 395 p.

Feller, C., Brossard, M., Chen, Y., Landa, E.R. & Trichet, J., 2010. Selected pioneering works on humus in soils and sediments during the 20th century: a retrospective look from the International Humic Substances Society view. Physics and Chemistry of the Earth 35, 903-912.

FitzPatrick, E.A., 1990. Roots in thin section of soils. In Douglas, L. A. (ed.) Soil Micromorphology: A Basic and Applied Science. Developments in Soil Science 19. Elsevier, Amsterdam, pp. 9-23.

FitzPatrick, E.A., 1993. Soil Microscopy and Micromorphology. John Wiley & Sons, Chichester, 304 p.

FitzPatrick, E.A. & Gudmundsson, T., 1978. The impregnation of wet peat for the production of thin sections. Journal of Soil Science 29, 585-587.

Forbes, M.S., Raison, R.J. & Skjemstad, J.O., 2006. Formation, transformation and transport of black carbon (charcoal) in terrestrial and aquatic ecosystems. Science of the Total Environment 370, 190-206.

Fox, C.A., 1984. A morphometric system for describing the micromorphology of organic soils and organic layers. Canadian Journal of Soil Science 64, 495-503.

Fox, C.A., 1985. Micromorphological characterization of Histisols. In Douglas, L.A. & Thompson, M.L. (eds.), Soil Micromorphology and Soil Classification. Soil Science Society of America Special Publication 15. SSSA, Madison, pp. 85-104.

Fox, C.A. & Parent, L.E., 1993. Micromorphological methodology for organic soils. In Carter, M.R. (ed.), Soil Sampling and Methods of Analysis. Canadian Society of Soil Science, Lewis Publishers, Ann Arbor, pp. 473-485.

Fox, C.A. & Tarnocai, C., 1990. The micromorphology of a sedimentary peat deposit from the Pacific temperate wetland region of Canada. In Douglas, L.A. (ed.), Soil Micromorphology: A Basic and Applied Science. Developments in Soil Science, Volume 19. Elsevier, Amsterdam, pp. 311-319.

Fox, C.A. & Tarnocai, C., 2011. Organic soils of Canada: Part 2. Upland organic soils. Canadian Journal of Soil Science 91, 823-842.

Freytet, P. & Verrecchia, E.P., 2002. Lacustrine and palustrine carbonate petrography: an overview. Journal of Paleolimnology 27, 221-237.

Frouz, J. & Novakova, A., 2005. Development of soil microbial properties in topsoil layer during spontaneous succession in heaps after brown coal mining in relation to humus microstructure development. Geoderma 129, 54-64.

Gerasimova, M. & Lebedeva, M., 2018. Organo-mineral surface horizons. In Stoops, G., Marcelino, V. & Mees, F. (eds.), Interpretation of Micromorphological Features of Soils and Regoliths. Second Edition. Elsevier, Amsterdam, pp. 513-538.

Gifford, E.M. & Foster, A.S., 1989. Morphology and Evolution of Vascular Plants. Third Edition. Freeman, New York, 626 p.

Goldberg, E.D., 1985. Black Carbon in the Environment: Properties and Distribution. John Wiley and Sons, New York, 198 p.

Gracanin, Z., 1962. Zur Genese, Morphologie und Mikromorhplogie der Hangtorfbildung auf Kalksteinen in Kroatien. Zeitschrift für Pflanzenernahrung und Bodenkunde 98, 264-272.

Green, R.N., Trowbridge, R.L. & Kliny, K., 1993. Towards a Taxonomic Classification of Humus Forms. Monograph 29, Supplement to Forest Science, Volume 39, 49 p.

Grosse-Brauckmann, G., 1963. Zur Artenzusammensetzung von Torfen (Einige Befunde und Uberlegungen zur Frage der Zersetzlichkeit und Erhaltungsfahigkeit von Pflanzenresten.) Berichte der Deutschen Botaischen Gesellschaft 76, 22-37.

Grosse-Brauckmann, G., 1964. Einige wenig beachtete Pflanzenreste in nordwestdeutschen Torfen und die Art ihres Vorkommes. Geologisches Jahrbuch 81, 621-644.

Grosse-Brauckmann, G. & Puffe, D., 1964. Untersuchungen an Torf-Dunnschnitten aus einem Moorprofil vom teufelsmoor bei Bremen. In Jongerius, A. (ed.), Soil Micromorphology. Elsevier, Amsterdam, pp. 83-93.

Guttmann, E.B.A., Simpson, I.A. & Dockrill, S.J., 2003. Joined-up archaeology at Old Scatness, Shetland: thin section analysis of the site and hinterland. Environmental Archaeology 8, 17-31.

Hammond, R.F., & Collins, J.F., 1983. Microfabrics of a Sphagnofibrist and related changes resulting from soil amelioration. In Bullock, P. & Murphy, C.P. (eds.), Soil Micromorphology. Volume 2. Soil Genesis. AB Academic Publishers, Berkhamsted, pp. 689-697.

Hather, J., 1993. An Archaeobotanical Guide to Root and Tuber Identification. Volume 1, Europe and South West Asia. Oxbow Monograph 28, Oxbow Books, Oxford, 154 p.

Hendrick, R.L. & Pregitzer, K.S., 1997. The relationship between fine root demography and the soil environment in northern hardwoods forests. Ecoscience 4, 99-105.

Hole, F.D., 1961. A classification of pedoturbations and some other processes and factors of soil formation in relation to isotropism and anisotropism. Soil Science 91, 375-377.

Hseu, Z.Y., Chen, Z.S. & Wu, Z.D., 1999. Characterization of placic horizons in two subalpine forest Inceptisols. Soil Science Society of America Journal 63, 941-947.

Hu, C., Zhang, D., Huang, Z. & Liu, Y., 2003. The vertical microdistribution of cyanobacteria and green algae within desert crusts and the development of the algal crusts. Plant and Soil 257, 97-111.

Huisman, D.J., Jongmans, A.G. & Raemaekers, D.C.M., 2009. Investigating Neolithic land use in Swifterbant (NL) using micromorphological techniques. Catena 78, 185-197.

Huisman, D.J. & Milek, K.B., 2017. Turf as construction material. In Nicosia, C. & Stoops, G. (eds.), Archaeological Soil and Sediment Micromorphology. John Wiley & Sons Ltd, Chichester, pp. 113-119.

Hunt, M.E., Floyd, G.L. & Stout, B.B., 1979. Soil algae in field and forest environments. Ecology 60, 362-375.

Ismail-Meyer, K., 2014. The potential of micromorphology for interpreting sedimentationprocesses in wetland sites: a case study of a Late Bronze-early Iron Age lakeshore settlement at Lake Luokesa (Lithuania). Vegetation History and Archaeobotany 23, 367-382.

Ismail-Meyer, K., 2017. Plant remains. In Nicosia, C. & Stoops, G. (eds.), Archaeological Soil and Sediment Micromorphology. John Wiley & Sons Ltd, Chichester, pp. 121-135.

Ismail-Meyer, K., Rentzel, P., & Wiemann, P., 2013. Neolithic lakeshore settlements in Switzerland: new insights on site formation processes from micromorphology. Geoarchaeology 28, 317-339.

IUSS Working Group WRB, 2015. World Reference Base for Soil Resources 2014, Update 2015. World Soil Resources Reports No. 106. FAO, Rome, 192 p.

Johnson, D.L., Watson-Stegner, D., Johnson, D.N. & Schaetzl, R.J., 1987. Proisotropic and proanisotropic processes of pedoturbation. Soil Science 143, 278-292.

Jones, T.P., Chaloner, W.G. & Kuhlbusch, T.A.J., 1997. Proposed bio-geological and chemical based terminology for fire-altered plant matter. In Clark, J.S., Cachier, H., Goldammer, J.G. & Stocks, B.J. (eds.). Sediment Records of Biomass Burning and Global Change. Springer-Verlag, Berlin, pp. 9-22.

Jongerius, A., 1961. De micromorfologie van de organische stof. In Bodemkunde. Ministerie van Landbouw en Visserij, The Hague, pp. 43-58.

Jongerius, A. & Pons, L.J., 1962. Soil genesis in organic soils. Auger and Spade 12, 156-168.

Jongerius, A & Rutherford, G.K. (eds.), 1979. Glossary of Soil Micromorphology. Centre for Agricultural Publishing and Documentation, Wageningen, 138 p.

Jongerius, A. & Schelling, J., 1960. Micromorphology of organic matter formed under the influence of soil organisms, especially soil fauna. Transactions of the 7th International Congress of Soil Science, Volume III, Madison, pp. 702-710.

Kaczorek, D., Vrydaghs, L., Devos, Y., Pető, Á. & Effland, W.R., 2018. Biogenic siliceous features. In Stoops, G., Marcelino, V. & Mees, F. (eds.), Interpretation of Micromorphological Features of Soils and Regoliths. Second Edition. Elsevier, Amsterdam, pp. 157-176.

Kenward, H. & Hall, A., 2000. Decay of delicate organic remains in shallow urban deposits: are we at a watershed? Antiquity 74, 519-525.

Keddy, P.A., 2010. Wetland Ecology: Principles and Conservation. Second Edition. Cambridge University Press, New York, 497 p.

Kibblewhite, M., Tóth, G. & Hermann, T., 2015. Predicting the preservation of cultural artefacts and buried materials in soil. Science of the Total Environment 528, 249-263.

Kim, Y.S. & Singh, A.P., 2000. Micromorphological characteristics of wood biodegradation in wet environments: a review. IAWA Journal 21, 135-155.

Kooistra, M.J., 1978. Soil Development in Recent Marine Sediments of the Intertidal Zone in the Oosterschelde – The Netherlands. A Soil Micromorphological Approach. Soil Survey Paper 14, Nertherlands Soil Survey Institute, Wageningen, 183 p.

Kooistra, M.J., 2015. Descripción de los componentes orgánicos del suelo. In Loaiza, J.C., Stoops, G., Poch, R. & Casamitjana, M. (eds.), Manual de Micromorfología de Suelos y Técnicas Complementarias. Fondo Editorial Pascual Bravo, Medellin, pp. 261-292.

Kooistra, M.J., 1991. A micromorphological approach to the interactions between soil structure and soil biota. Agriculture, Ecosystems and Environment 34, 315-328.

Kooistra, M.J. & Kooistra, L.I., 2003. Integrated research in archaeology using soil micromorphology and palynology. Catena 54, 603-618.

Kooistra, M.J. & Pulleman, M.M., 2018. Features related to faunal activity. In Stoops, G., Marcelino, V. & Mees, F. (eds.), Interpretation of Micromorphological Features of Soils and Regoliths. Second Edition. Elsevier, Amsterdam, pp. 447-469.

Kooistra, M.J., Lebbink, G. & Brussaard, L., 1989. The Dutch programme on soil ecology of arable farming systems. II. Geogenesis, agricultural history, field site characteristics and present farming systems at the Lovinkhoeve experimental farm. Agriculture, Ecosystems and Environment 27, 361-387.

Kooistra, M.J., Kooistra, L.I., van Rijn, P. & Sass-Klaassen, U., 2006. Woodlands of the past - The excavation of wetland woods at Zwolle-Stadshagen (The Netherlands). Reconstruction of the wetland wood in its environmental context. Netherlands Journal of Geosciences 85, 37-60.

Koopman, G.J., 1988. 'Waterhard': a hard brown layer in sand below peat, The Netherlands. Geoderma 42, 147-157.

Krause, W., 1964. Zur Technik der mikromorphologischen Untersuchung nicht getrockneter Unterwasserboden. In Jongerius, A. (ed.), Soil Micromorphology. Amsterdam, Elsevier, pp. 361-370.

Kretzschmar, A., 1987. Caractérisation microscopique de l'activité de lombriciens endogés. In Fedoroff, N., Bresson, L.M. & Courty, M.A. (eds.), Micromorphologie des Sols, Soil Micromorphology. AFES, Plaisir, pp. 325-330.

Kroetsch, D.J., Geng, X., Chang, S.X. & Saurette D.D., 2011. Organic soils of Canada: Part 1. Wetland organic soils. Canadian Journal of Soil Science 91, 807-822.

Kubiëna, W.L., 1943. Die mikroskopische Humusuntersuchung. Zeitschrift für Weltforstwirtschaft 10, 387-410.

Kubiëna, W.L., 1953. The Soils of Europe. Thomas Murby & Co., London, 317 p.

Kuhlbusch, T.A.J., 1998. Black carbon and the carbon cycle. Science 280, 1903-1904.

Kühn, M., Maier, U., Herbig, C., Ismail-Meyer, K., Le Bailly, M. & Wick, L., 2013. Methods for the examination of cattle, sheep and goat dung in prehistoric wetland settlements with examples of the sites Alleshausen-Täschenwiesen and Alleshausen-Grundwiesen (around cal 2900 BC) at Lake Federsee, south-west Germany. Environmental Archaeology 18, 43-57.

Kühn, P., Aguilar, J., Miedema, R. & Bronnikova, M., 2018. Textural Pedofeatures and Related Horizons. In Stoops, G., Marcelino, V. & Mees, F. (eds.), Interpretation of Micromorphological Features of Soils and Regoliths. Second Edition. Elsevier, Amsterdam, pp. 377-423.

Kuiper, F. & Slager, S., 1963. The occurrence of distinct prismatic and platy structures in inorganic soil profiles. Netherlands Journal of Agricultural Science 11, 418-421.

Lee, G.B., 1983. The micromorphology of peat. In Bullock, P. & Murphy, C.P. (eds.), Soil Micromorphology. Volume 2. Soil Genesis. AB Academic Publishers, Berkhamsted, pp. 485-501.

Lee, G.B. & Manoch B., 1974. Macromorphology and micromorphology of a Wisconsin Saprist. In Stelly, M. & Dinauer, R.C. (eds.), Histosols: Their Characteristics, Classification and Use. Soil Science Society of America Special Publication 6, Madison, pp. 47-62.

Levesque, M.P. & Dinel, H., 1982. Some morphological and chemical aspects of peats applied to the characterization of Histosols. Soil Science 133, 324-332.

Lewis, H., 1882. On a new substance resembling dopplerite from a peat bog at Scranton. Proceedings of the American Philosophical Society 20, 112-117.

Ligthart, T.N., 1997. Thin section analysis of earthworm burrow disintegration in a permanent pasture. Geoderma 75, 135-148.

Lopez-Buendia, A.M., 1998. A new method for the preparation of peat samples for petrographic study by transmitted and reflected light microscopy. Journal of Sedimentary Research 68, 214-217.

Loustau, D. & Toutain, F., 1987. Micromorphologie et fonctionnement de quelques humus des formes hydromull et anmoor de l'est de la France. In Fedoroff, N., Bresson, L.M. & Courty, M.A. (eds.), Micromorphologie des Sols, Soil Micromorphology. AFES, Plaisir, pp. 385-389.

Lund, J.W.G., 1962. Soil algae. In Lewin, R.D. (ed.), Physiology and Biochemistry of Algae. Academic Press, New York, pp. 759-766.

Lützow, M. v., Kögel-Knabner, I., Ekschmitt, K., Matzner, E., Guggenberger, G., Marschner, B. & Flessa, H., 2006. Stabilization of organic matter in temperate soils: mechanisms and their relevance under different soil conditions – a review. European Journal of Soil Science 57, 426-445.

MacEntee, F.J., Schreckenberg, G. & Bold, H.C., 1972. Some observations on the distribution of edaphic algae. Soil Science 114, 196-219.

Macphail, R.I., 1983. The micromorphology of dark earth from Gloucester, London and Norfolk: an analysis of urban anthropogenic deposits from the late Roman to early Medieval periods. In Bullock, P. & Murphy, C.P. (eds.), Soil Micromorphology. Volume 1. Techniques and Applications. AB Academic Publishers, Berkhamsted, pp. 245-252.

Macphail, R.I. & Crowther, J., 2007. Soil micromorphology, chemistry and magnetic susceptibility studies at Huizui (Yiluo region, Henan Province, northern China), with special focus on a typical Yangshao floor sequence. Bulletin of the Indo-Pacific Prehistory Association 27, 93-113.

Macphail, R.I., Courty, M.A. & Gebhardt, A., 1990. Soil micromorphological evidence of early agriculture in north-west Europe. World Archaeology 22, 53-69.

Macphail, R.I., Crowther, J., Acott, T.G., Bell, M.G. & Cruise, J.M., 2003. The experimental earthwork at Wareham, Dorset after 33 years: changes to the buried LFH and Ah horizons. Journal of Archaeological Science 30, 77-93.

Malterer, T.J., Verry, E. S. & Erjavec, J., 1992. Fiber content and degree of decomposition in peats: review of national methods. Soil Science Society of America Journal 56, 1200-1211.

Manoch, B., 1970. Micromorphology of a Saprist. MSc Dissertation, University of Wisconsin, Madison, 98 p.

Masiello, C.A., 2004. New directions in black carbon organic geochemistry. Marine Chemistry 92, 201-213.

Mauquoy, D. & Van Geel, B., 2007. Mire and peat macros. In Elias, S.A. (ed.), Encyclopedia of Quaternary Science. Elsevier, Amsterdam, pp. 2324-2330.

McConnell, J.R., Edwards, R., Kok, G.L., Flanner, M.G., Zender, C.S., Saltzman, E.S., Banta, J.R., Pasteris, D.R., Carter, M.M. & Kahl, J.D.W., 2007. 20th-century industrial black carbon emissions altered arctic climate forcing. Science 317, 1381-1384.

Menge, L., 1965. Untersuchungen uber das Gefuge und die Wasserbindungsintensitat einiger Anmoor und Moorboden. Dissertation, Landwirtschaftliche Hochschule Stuttgart-Hohenheim, 126 p.

Meyer, F.H., 1964. The role of the fungus Cenococum Granifrome (Sow.) Ferd. et Winge in the formation of a mor. In Jongerius, A. (ed.), Soil Micromorphology. Elsevier, Amsterdam, pp. 23-31.

Mitsch, W.J. & Gosselink, J.G., 2007. Wetlands. Fifth Edition. John Wiley & Sons, Chichester, 456 p.

Mooney, S.J., Holden, N.M., Ward, S.M. & Collins, J.F., 2000. The micromorphology of selected Irish milled peats. Plant and Soil 222, 15-23.

Mori, K., Bernier, N., Kosaki, T. & Ponge, J.F., 2009. Tree influence on soil biological activity: what can be inferred from the optical examination of humus profiles? European Journal of Soil Biology 45, 290-300.

Müller, P.E. 1887. Recherches sur les formes naturelles de l'humus et leur influence sur la végétation et le sol. Annales de la Science Agronomique Française et Etrangère 6, 85-423.

Müller, R.D. & Frahm, J.P., 2013. Moose unter dem Mikroskop. Archive of Bryology, Special Volume 13. 40 p.

Murphy, C.P., 1986. Thin Section Preparation of Soils and Sediments, AB Academic Publishers, Berkhamsted, 149 p.

Nelson, S., 1984. Upland and wetland vegetational changes in southeastern Massachusetts: a 12,000 year record. Northeastern Geology 6, 181-191.

Nilsson, T. & Björdal, C., 2008. Culturing wood-degrading erosion bacteria. International Biodeterioration & Biodegradation 61, 3-10.

Osman, K.T., 2013. Soils: Principles, Properties and Management. Springer, Dordrecht, 271 p.

Parshall, T., Foster, D.R., Faison, E., MacDonald, D. & Hansen, B.C.S., 2003. Long-term history of vegetation and fire in pitch-pine oak forests on Cape Cod Massachusetts. Ecology 84, 736-748.

Pawluk, S., 1987. Faunal micromorphological features in moder humus of some western Canadian soils. Geoderma 40, 3-16.

Peacock, E. & Fant, D.W., 2002. Biomantle formation and artifact translocation in upland sandy soils: an example from the Holly Springs National Forest, North-Central Mississippi, U.S.A. Geoarchaeology 17, 91-114.

Phillips, D.H. & FitzPatrick, E.A., 1999. Biological influences on the morphology and micromorphology of selected Podzols (Spodosols) and Cambisols (Inceptisols) from the eastern United States and northeast Scotland. Geoderma 90, 327-364.

Ponge, J.F., 1999. Horizons and humus forms in beech forests of the Belgian Ardennes. Soil Science Society of America Journal 63, 1888-1901.

Ponge, J.F., 2003. Humus forms in terrestrial ecosystems: a framework to biodiversity. Soil Biology and Biochemistry 35, 935-945.

Pons, L.J., 1960. Soil genesis and classification of reclaimed peat soils in connection with initial soil formation. Transactions of the 7th International Congress of Soil Science, Volume IV, Madison, pp. 205-211.

Ramanathan, V. & Carmichael, G., 2008. Global and regional climate changes due to black carbon. Nature Geoscience 1, 221-227.

Richardson, J.L. & Vepraskas, M.J., 2001. Wetland Soils: Genesis, Hydrology, Landscapes, and Classification. CRC Press, Boca Raton, 417 p.

Schiegl, S., Goldberg, P., Bar-Yosef, O. & Weiner, S., 1996. Ash deposits in Hayonim and Kebara Caves, Israel: macroscopic, microscopic and mineralogical observations, and their archaeological implications. Journal of Archaeological Science 23, 763-781.

Schoch, W.H., Pawlik, B. & Schweingruber, F.H., 1988. Botanical Macro-Remains: An Atlas for the Determination of Frequently Encountered and Ecologically Important Plant Remains. Paul Haupt Publishers, Bern, 227 p.

Schoch, W.H., Heller, I., Schweingruber, F.H. & Kienast, F., 2004. Wood Anatomy of Central European species. On-line database.

Schwarze, F.W.M.R., 2007. Wood decay under the microscope. Fungal Biology Reviews 21, 133-170.

Schweingruber, F.H., 1982. Microscopic Wood Anatomy, Structural Variability of Stems and Twigs in Recent and Sufossil Woods from Central Europe. Fluck-Wirth, Teufen, 226 p.

Sebag, D., Di Giovanni, C., Ogier, S., Mesnage, V., Laggoun-Défarge, F. & Durand, A., 2006. Inventory of sedimentary organic matter in modern wetland (Marais Vernier, Normandy, France) as source-indicative tools to study Holocene alluvial deposits (Lower Seine Valley, France). International Journal of Coal Geology 67, 1-16.

Simonson, R.W., 1959. Outline of a generalized theory of soil genesis. Soil Science Society of America Proceedings 23, 152-156.

Soil Classification Working Group, 1998. The Canadian System of Soil Classification. Agriculture and Agri-Food Canada, Publication no. 1646 (revised). NRC Research Press, Ottawa, 187 p.

Soil Survey Staff, 2014. Keys to Soil Taxonomy. 12th Edition. USDA-Natural Resources Conservation Service, Washington, 360 p.

Stephens, M., Rose, J., Gilbertson, D. & Canti, M.G., 2005. Micromorphology of cave sediments in the humid tropics: Niah Cave, Sarawak. Asian Perspectives 44, 42-55.

Stoops, G., 2003. Guidelines for Analysis and Description of Soil and Regolith Thin Sections. Soil Science Society of America, Madison, 184 p.

Strawn, D.G., Bohn, H.L. & O'Connor, G.A., 2015. Soil Chemistry. Fourth Edition. John Wiley & Sons, Chichester, 375 p.

Szabo, I., Marton, M. & Partai, G., 1964. Micro-milieu studies in the A-horizon of a mull-like rendzina. In Jongerius, A. (ed.) Soil Micromorphology, Elsevier, Amsterdam, pp. 33-45.

Tian, X., Takeda, H. & Tatsuo, A., 1997. Application of a rapid thin section method for observations on decomposing litter in mor humus form in a subalpine coniferous forest. Ecological Research 12, 289-300.

Tingey, D.T., Phillips, D.L. & Johnson, M.G., 2000. Elevated CO_2 and conifer roots: effects on growth, life span and turnover. New Phytologist 147, 87-103.

Tippkötter, R., 1990. Staining of soil microorganisms and related materials with fluorochromes. In Douglas, L.A. (ed.), Soil Micromorphology: A Basic and Applied Science. Developments in Soil Science, Volume 19. Elsevier, Amsterdam, pp. 605-612.

Uggla, H., Rog, Z. & Woclawek, T., 1972. Micromorphology of a gyttja muck soil of Jawty Male. In Kowalinski, S. & Drozd, J. (eds.), Soil Micromorphology. Panstwowe Wydawnicto Naukowe, Warsawa, pp. 481-489.

Upton, R., Graff, A., Joliffe, G., Länger, R. & Williamson, E., 2011. American Herbal Pharmacopoeia: Botanical Pharmacognosy – Microscopic Characterization of Botanical Medicines. American Herbal Pharmacopoeia, CRC Press, Boca Raton, 734 p.

van Heuveln, B. & De Bakker, H., 1972. Soil-forming processes in Dutch peat soils with special reference to humus-illuviation. In Proceedings of the 4th International Peat Congress, Helsinki, pp. 289-297.

Van Mourik, J.M., 1999. The use of micromorphology in soil pollen analysis. The interpretation of the pollen content of slope deposits in Galicia, Spain. Catena 35, 239-258.

Van Mourik, J.M., 2003. Life cycle of pollen grains in mormoder humus farms of young acid forest soils: a micromorphological approach. Catena 54, 651-664.

Van Ranst, E., Wilson, M.A. & Righi, D., 2018. Spodic materials. In Stoops, G., Marcelino, V. & Mees, F. (eds.), Interpretation of Micromorphological Features of Soils and Regoliths. Second Edition. Elsevier, Amsterdam, pp. 633-662.

Verrecchia, E., 2000. Fungi and sediments. In Riding, R.E. & Awramik, S.M. (eds.), Microbial Sediments. Springer, Berlin, pp. 68-75.

Verrill, L. & Tipping, R., 2010. A palynological and geoarchaeological investigation into Bronze Age farming at Belderg Beg, Co. Mayo, Ireland. Journal of Archaeological Science 37, 1214-1225.

Vrydaghs, L., Devos, Y. & Pető, Á., 2017. Opal phytoliths. In Nicosia, C. & Stoops, G. (eds.), Archaeological Soil and Sediment Micromorphology. John Wiley & Sons Ltd, Chichester, pp. 155-163.

Von Post, L., 1924. Das genetische System der organogenen Bildungen Schwedens. Comité International de Pédologie, IV Commission, Communication 22, 287-304.

Williams, A.J., Pagliai, M. & Stoops, G., 2018. Physical and biological surface crusts and seals. In Stoops, G., Marcelino, V. & Mees, F. (eds.), Interpretation of Micromorphological Features of Soils and Regoliths. Second Edition. Elsevier, Amsterdam, pp. 539-574.

Wilson, M.A. & Righi, D., 2010. Spodic materials. In Stoops, G., Marcelino, V. & Mees, F. (eds.), Interpretation of Micromorphological Features of Soils and Regoliths. Elsevier, Amsterdam, pp. 251-273.

Winkler, M.G., 1985. A 12,000-year history of vegetation and climate for Cape Cod, Massachusetts. Quaternary Research 23, 301-312.

Zaiets, O. & Poch, R.M., 2016. Micromorphology of organic matter and humus in Mediterranean mountain soils. Geoderma 272, 83-92.

Zanella, A., Jabiol, B., Ponge, J.F., Sartori, G., De Waal, R., Van Delft, B., Graefe, U., Cools, N., Katzensteiner, K., Hager, H. & Englisch, M., 2011. A European morpho-functional classification of humus forms. Geoderma 164, 138-145.

Chapter 18

Organo-mineral Surface Horizons

Maria Gerasimova[1,2], Marina Lebedeva[2]

[1]*MOSCOW LOMONOSOV UNIVERSITY, MOSCOW, RUSSIA;*
[2]*DOKUCHAEV SOIL SCIENCE INSTITUTE, MOSCOW, RUSSIA*

CHAPTER OUTLINE

1. Introduction

Three concepts and associated terms are used in soil science to indicate the organo-mineral surface horizons: (i) topsoil, the most general term used by the World Reference Base for Soil Resources (IUSS Working Group WRB, 2006, 2015; see also Spaargaren, 1998); (ii) epipedon, used in Soil Taxonomy (Soil Survey Staff, 1999) for diagnostic organo-mineral horizons complementary to subsoil horizons; and (iii) humus forms introduced by Kubiëna (1938) and widely used by micromorphologists. Kubiëna (1953) gives clear and comprehensive definitions of 17 humus forms, which were later subdivided into narrower categories, mostly by pedologists in France (e.g., Duchaufour, 1959, 1960; Baize & Girard, 1995; Jabiol et al., 1995) and Russia (Chertov, 1966).

Micromorphologists followed Kubiëna's ideas, either extending them over broader sets of soils (Jongerius & Marsman, 1971; Kowalinski & Kollender-Szych, 1972; Parfenova & Yarilova, 1977; Romashkevich & Gerasimova, 1982) or proposing alternative systems that refer more to the properties of organic matter in the surface horizons than to the horizons themselves (Barratt, 1964; Bal, 1970). At present, organic components are described in micromorphology strictly according to their properties (Bullock et al., 1985; Stoops, 2003), whose combinations serve as a basis for characterising organic soil horizons (Stolt & Lindbo, 2010; Ismail-Meyer et al., 2018).

This chapter summarises the micromorphological characteristics of organo-mineral surface horizons of natural soils, weakly or not subjected to human influence. Topsoils with prominent anthropogenic features are referred to in another chapter (Adderley et al., 2018). Most of the horizons considered here occur as part of soils on unconsolidated parent materials such as loess and clays, in which manifestations of pedogenesis are easily observed.

Special attention is given to mollic diagnostic horizon with typical mull humus of steppe, forest-steppe and prairie soils, and its variants under more humid or more arid climatic conditions, or in particular geomorphological settings. Also considered are umbric horizons, which mainly form in rather humid and cool climate under deciduous forests (e.g., southern taiga and mountainous forest). In addition to the organo-mineral surface horizons mentioned above, intervals with takyric and yermic properties (IUSS Working Group WRB, 2015), corresponding roughly to the former takyric and yermic diagnostic horizons (IUSS Working Group WRB, 2006), are also discussed. Both properties are specific of desert areas and are essentially determined by the mineral solid phase. In contrast to other topsoils, organic constituents and biological activity are of minor importance. Takyric properties occur in periodically flooded soils of arid regions and comprise a surface crust above a platy structured layer. Yermic properties usually comprise residual surface accumulations of rock fragments, the so-called desert pavement, embedded in a loamy vesicular crust and covered by a thin layer of aeolian sand or silt. Comprehensive descriptions of the micromorphology of aridic soils were given by Allen (1985), Lebedeva-Verba and Gerasimova (2009) and Lebedeva et al. (2013). The micromorphology of desert topsoils has been reported in several publications by Russian authors (e.g., Feofarova, 1956; Romashkevich & Gerasimova, 1982; Gubin, 1984; Golovanov et al., 2005; Gerasimova & Lebedeva, 2008; Lebedeva et al., 2015).

The organo-mineral soils discussed in this chapter were chosen to illustrate the idea of soil diversity in terms of micromorphological properties, with emphasis on zonal soils of the temperate climate. Less attention was paid to forest soils to avoid overlapping with the description of humus forms (see Ismail-Meyer et al., 2018). However, overlap with the chapters on faunal activity (Kooistra & Pulleman, 2018), organic matter (Ismail-Meyer et al., 2018) and surface crusts (Williams et al., 2018) are sometimes inevitable. WRB terminology (IUSS Working Group WRB, 2015) will be used throughout this chapter; if correlation with WRB terms is ambiguous, soil names according to the

Russian classification systems (Cooperative Research Group of Chinese Soil Taxonomy, 2001; Shishov et al., 2004) are also reported for Russian examples.

To cover the diversity of surface horizons within the above-mentioned groups, we use the concept of orthotype, which applies to a horizon with the most typical and well-developed micromorphological characteristics, may it be a real horizon in a specific soil profile or a generalised set of the most conspicuous and recurring properties. Orthotypes reflect the typical results of pedogenic processes in certain environments. They more or less correspond to the concepts of higher levels of organisation (Stoops, 1994), or micromorphotypes and central images (Gerasimova, 2003). Deviations of micromorphological properties from those inherent to the orthotype indicate changes in pedogenic processes that produced horizons with similar basic properties but with some different and/or additional features (variants). In WRB, qualifiers are used to record such minor modifications of the orthotype. In the Russian soil classification system, they correspond to 'diagnostic properties', identifying the subtype level, which are soil types with additional or less expressed features than in the main type.

2. Organo-mineral Surface Horizons of Soils of Temperate Climates

2.1 General Features

Particular combinations of soil-forming factors, including loess-like parent material, easily decomposable litter, high biological activity, favourable soil moisture and temperature regimes, explain the micromorphological properties that are inherent to the majority of organo-mineral surface horizons in soils of temperate climates. These properties have been described for a broad set of soils, such as soddy-podzolic soils (Albeluvisols, Retisols), forest-steppe and steppe soils (Phaeozems and Chernozems), and are summarised below.

The microstructure is most commonly granular and crumb; in addition, elements of platy or lenticular microstructure have been observed for frost-affected soils. The aggregates being moderately to highly separated, the total void space is large in the majority of these topsoils. Packing voids are dominant, vughs may occur, and channels and chambers are less abundant. The main processes responsible for microstructure development are faunal activity (see Kooistra & Pulleman, 2010, 2018), root growth, coagulation, binding by organic substances (see Stolt & Lindbo, 2010; Ismail-Meyer et al., 2018) and ice crystal formation due to seasonal frost (see Van Vliet-Lanoë, 2010; Van Vliet-Lanoë & Fox, 2018).

The groundmass is mostly homogeneous. The c/f-related distribution pattern varies, although porphyric types are the most common. The micromass has a speckled or dotted limpidity and generally an undifferentiated or stipple-speckled b-fabric owing to the masking effect of organic compounds. Compared with other horizons, organo-mineral surface horizons are the most strongly affected by present-day pedogenesis and the first to reflect changes in environmental conditions. Being the main habitat of biota in the soil, organo-mineral surface horizons contain plant residues in various stages of alteration. The type of alteration and the properties of the organic fine material reflect the intensity and conditions of humus accumulation.

Except for excrements and passage features, few pedofeatures are observed. Earthworm activity is of primary interest as being responsible for microstructure development. The contribution of earthworms is well known and universally accepted since the time of Darwin (e.g., Kubiëna, 1938; Jeanson, 1964; Bal, 1973; FitzPatrick, 1984, 2006; Kretzschmar, 1987; Jabiol et al., 1995; Brown et al., 2003; Kooistra & Pulleman, 2010, 2018). Other mesofauna groups are common, such as oribatid mites and potworms in surface horizons of forest soils with moder or mull-moder humus forms (Zachariae, 1964; Bal, 1973; Babel & Vogel, 1989). In addition, organo-mineral surface horizons may contain loose infillings of channels and packing voids with extraneous material, mostly in the form of non-aggregated fine silt that is different from the enclosing groundmass. These infillings probably result from deposition, in large open voids, of fine material that was loosened and transported by surface water during rainstorms or that was added to the surface by wind.

Relic features preserved from earlier periods of pedogenesis, such as nodules, coatings and crystals, are uncommon in surface horizons, unless they are either transported (e.g., by colluvial processes) or part of a truncated soil.

2.2   Mollic and Chernic Horizons

Mollic horizons are part of the profile of a broad spectrum of soils, although they are most typical for chernozems. In the most recent version of the WRB system (IUSS Working Group WRB, 2015), the chernic horizon was introduced, in which the most prominent characteristics of mollic are clearly manifested. It is identified by higher organic matter content and lower chroma than the mollic horizon and by a strongly developed granular structure. In the majority of Russian chernozems, under both forest-steppe and steppe vegetation, the uppermost part of the mollic horizon may be identified as chernic.

The humus form in mollic and chernic horizons is mull, eumull (Baize & Girard, 1995), 'well-developed terrestrial mull' (Kubiëna, 1938, 1970) or calcic mull (Chertov, 1966; Romashkevich & Gerasimova, 1982). The major micromorphological features have been comprehensively described for North American Chernozems (Pawluk & Bal, 1985; Sanborn & Pawluk, 1989; Eswaran & Drees, 2004; Kovda et al., 2004). There are also many publications on Russian and Ukrainian Chernozems (e.g., Tselishcheva, 1966; Yarilova, 1972, 1974; Yarilova & Bystritskaya, 1976; Parfenova & Yarilova, 1977; Poliakov, 1981; Yarilova et al., 1983; Bystritskaya & Gerasimova, 1988; Khitrov et al., 2013). A review of the micromorphological properties of mollic horizons was presented by FitzPatrick (1984).

2.2.1 Orthotype

The upper part of typical mollic horizons, often corresponding to a chernic horizon, has a well-developed, highly separated complex crumb microstructure (Fig. 1A) and abundant channels. The peds are mostly organo-mineral excrements of various sizes, generally between 0.1 and 0.5 mm but occasionally up to 1 mm. They are mostly arranged in clusters, for which a three-level hierarchy can be recognised. Channel infillings composed of earthworm excrements (Fig. 1B), either separate or coalesced, are common, and a vermicular microstructure is often dominant. According to Poliakov (1981), the upper 0.5 m of chernozem profiles is at least five times fully reworked by earthworms. In the lower part of the topsoil, the aggregates are less separated, and the microstructure is a result of ageing and coalescence of earthworm excrements (Fig. 1C). Associated voids are mostly vughs, channels and chambers. Passage features are common. The described crumb, vermicular and spongy microstructures basically correspond to the 'spongy fabric' of Poliakov (1981) and Pazos and Stoops (1987) and to the 'mullgranic' and 'mullgranoidic' microstructures of Pawluk and Bal (1985).

The groundmass is homogeneous, generally with a close porphyric c/f-related distribution pattern, as the majority of Chernozems occur on loess or loess-like materials. The micromass is dark brown and dotted, because of fine organic material occurring as punctuations, with almost always undifferentiated b-fabric, which confirms the close association between organic and mineral fine material, one of the important mull humus characteristics. Few fresh plant residues, consisting of light yellowish parenchymatic root tissues with distinct interference colours, occur in channels.

Earthworm excrements are mostly spheroidal and mammilated, coalescing by ageing, porous to very porous and composed of the same material as the groundmass with minor admixtures of tissue fragments. Some excrements containing material from lower horizons can be identified as aggregates with a different colour and b-fabric than the groundmass (Fig. 1D and E). The degree of distinctness of their boundaries and contrast with the enclosing material indicate the stage of their assimilation into the groundmass of the surface horizon. This process can play an important role in maintaining the calcium carbonate equilibrium and provides the stability of the microstructure of a typical mollic or chernic horizon.

2.2.2 Variants

Variants have either additional or modified micromorphological features, which reflect different environments (climate, drainage conditions, parent material) and which are complementary to the main properties of the orthotype. The first three variants that are considered in this section (greyzemic, hydromorphic, lenticular/platy) basically correspond to surface horizons of Chernozems and Phaeozems. The next series of variants represent horizons in other Reference Soil Groups.

Greyzemic variant – Circular or semicircular arrangement of silt grains in the groundmass (Fig. 2) was observed for chernic horizons of Greyzemic Luvic Chernozems (Yarilova et al., 1983) and Greyzemic Luvic Phaeozems (Miedema et al., 1999). This

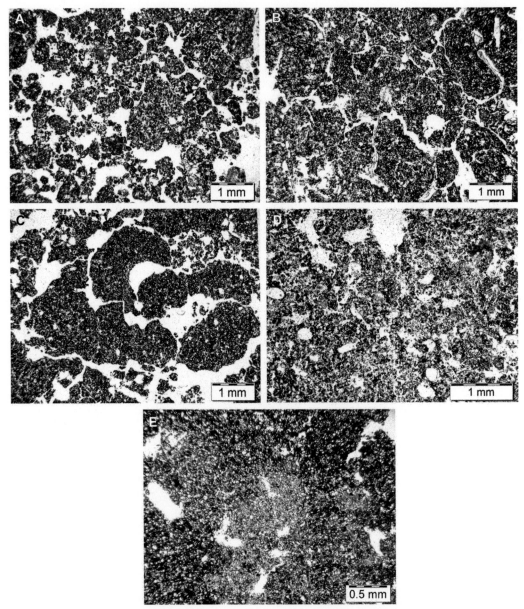

FIGURE 1 Orthotype of a mollic/chernic horizon. (A) Well-developed, highly separated crumb microstructure and infilling with earthworm excrements (upper right corner) (Haplic Chernozem, mown steppe, Kamennaya Step Reserve, European Russia) (PPL). (B) Disintegrating and partly coalescing coarse earthworm excrements composed of organo-mineral material similar to that of the groundmass in a pore (Haplic Chernozem, pasture, Kursk Biosphere Reserve, European Russia) (PPL). (C) Moderately developed and moderately separated aggregates resulting from ageing and coalescence of excrements in the lower part of a mollic horizon (Haplic Chernozem, forest belt, Kamennaya Step Reserve, European Russia) (PPL). (D) Channel microstructure and abundant excrements incorporated into the groundmass, recognised by a lighter colour (Luvic Chernozem, cropland, Tula region, European Russia) (PPL). (E) Brown earthworm excrements incorporated in the groundmass in the lower part of a mollic horizon; the excrements are lighter than the groundmass because they contain material derived from carbonate-rich horizons (Luvic Chernozem, ploughland, Tula region, European Russia) (PPL).

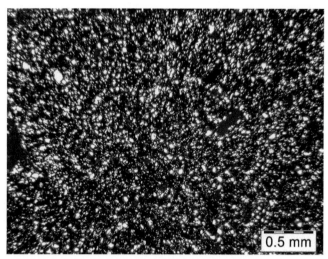

FIGURE 2 Greyzemic variant of the mollic horizon, showing a circular or semicircular arrangement of coarse silt and fine sand grains in the groundmass (Greyzemic Luvic Chernozem, cropland, Tula region, European Russia) (XPL).

feature is interpreted as having formed by the removal of fine material from the periphery of former crumbs or granules, which next became hardly recognisable through coalescence or disturbance. This variant corresponds to leached mollic material below the chernic horizon, mostly in forest-steppe soils. In terms of soil classification, this feature identifies Chernozems as Greyzemic and/or Luvic in the current WRB system, or as a podzolised subtype of clay-illuvial chernozem in the current Russian system.

Hydromorphic variant – Dark typic (orthic) nodules composed of iron oxides with admixture of organic matter (Fig. 3) have been encountered in organo-mineral surface horizons of Phaeozems of the Far Eastern plains (Gerasimova et al., 1996). Iron-depletion hypocoatings and light brown limpid non-laminated clay coatings around nodules and aggregates, described for Gleyic Phaeozems of the Amur River terraces (Gyninova & Shoba, 1988), suggest present-day mobilisation of iron and weak clay illuviation. Similar pedofeatures were reported for some Chernozems of European Russia with active hydromorphism (Yarilova et al., 1983), and for Luvic Phaeozems of small depressions in the Caspian semidesert (Verba et al., 2006). These pedofeatures of the hydromorphic variant reflect short periods of waterlogging.

Lenticular/platy variant – Crumbs arranged into wavy plates that form lenticular or platy microstructures (Fig. 4), typical of soils subjected to freeze-thaw cycles, have been reported for North American mollic horizons (Mermut & St. Arnaud, 1981; Sanborn & Pawluk, 1989). Pawluk and Bal (1985) described 'banded' cryic microstructures for A1-horizons of soils in the forest-grassland transition zone and interpreted them as degraded mollic horizons. In Russia, similar features were recorded for forest-steppe grey soils (Mollic Luvisols) (Gerasimova et al., 1996; Miedema et al., 1999). Another

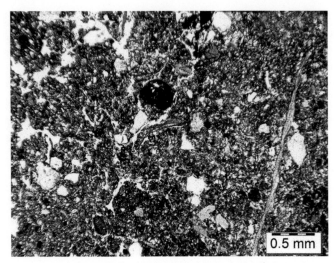

FIGURE 3 Hydromorphic variant of the mollic horizon, showing iron oxide depletion hypocoatings, impregnative iron oxide nodules (with admixture of organic matter) and clay coatings (Stagnic Phaeozem, meadow, Amur River old terraces, Russian Far East) (PPL).

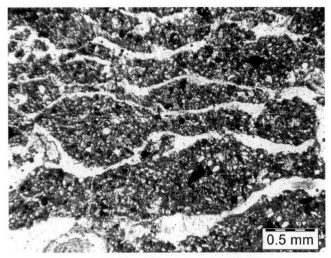

FIGURE 4 Lenticular/platy variant of the mollic horizon, with characteristic lenticular microstructure (Luvic Phaeozem, meadow, Dzhanybek Research station; Caspian Lowland, Kazakhstan) (PPL).

type of platy microstructure with welded round aggregates, forming partially accommodated plates, may form in depressions, as a result of the strong lateral runoff in spring that leads to the transport and deposition of granular or crumb aggregates. This is the case for the mollic horizon of Luvic Phaeozems in large 'steppe saucers' in the Caspian Lowland (Verba et al., 2006).

Melanic variant − A rather compact and very dark micromass is typical of mollic horizons of Phaeozems of the Argentinean Pampa (Pazos & Stoops, 1987). These horizons are characterised by intense biological activity. Compared with the orthotype, they contain less excremental pedofeatures and much more abundant organic pigment and have weakly developed microstructure (unpublished work by the authors). Dark coatings on coarse grains and abundant dark opaque amorphous aggregates (50-100 μm) (Fig. 5) have been observed as well. The latter are probably remnants of volcanic glass, which is known to be abundant in Pampean soils (Moscatelli & Pazos, 2000). The differences with the orthotype can be mostly attributed to the admixture of volcanic material, which can require an 'andic' qualifier.

Kastanozem variant − The mollic horizon of Kastanozems (Chestnut soils) of dry steppes (Medvedev, 1983; Gracheva et al., 1988) shows a more compact arrangement of the granular aggregates, with moderate to weak ped separation and few to common secondary peds, a total void space (vughs and channels) of less than 20% by volume, a lighter grey-brown colour of the micromass (lower content of organic matter content) and more diverse and abundant plant residues, owing to a shorter period available for decomposition. Traces of faunal activity are less abundant (Fig. 6). The Kastanozem variant is not very close to the orthotype, but it meets the requirements for mollic, which confirms its diagnostic significance for Kastanozems.

Serozem variant − This carbonate-enriched variant of the mollic horizon occurs in Calcisols on loess (serozems). It was first described by Minashina (1966) as strongly reworked by soil fauna and having preserved loess-inherited properties such as high pedality and high porosity, dominated by vesicles. The organic carbon content is 1.5-3.5% and the organic matter is bound to the mineral soil constituents, which also

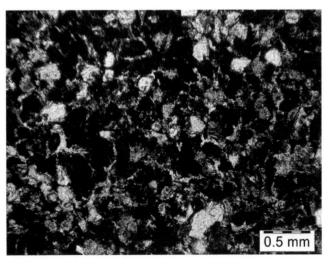

FIGURE 5 Melanic variant of the mollic horizon, showing abundant dark opaque aggregates and coatings (Luvic Phaeozem, cropland, Pampa Ondulada, Buenos Aires province, Argentina) (PPL).

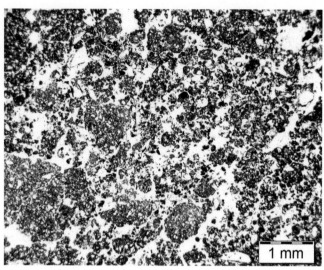

FIGURE 6 Kastanozem variant of the mollic horizon, having a granular microstructure with moderate to weak ped separation, and a micromass lighter in colour than the orthotype, due to its lower organic matter content (Kastanozem, Transvolga area, Sarpa depression, European Russia) (PPL).

justifies qualifying this topsoil as calcic mull. Aggregates are weakly to moderately separated, mostly rounded and not accommodated. They coalesce to create a sponge-like microstructure with various types of voids (packing voids, vesicles, channels, chambers, vughs) (Fig. 7). The micromass is greyish pale brown with a weak stipple-speckled and/or calcitic crystallitic b-fabric. It differs from the mollic horizons of

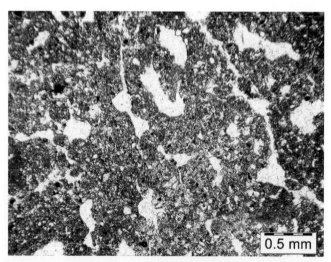

FIGURE 7 Serozem variant of the mollic horizon, with mostly coalesced aggregates producing a sponge-like microstructure, and with carbonate enrichment of the micromass (Calcisol, Turkestanskiy Range piedmont plain, Uzbekistan) (PPL).

Chernozems by colour, b-fabric and a greater abundance of channels and other passage features. The similarity in microstructure between this variant and the orthotype can be explained by the prominent contribution of soil fauna, whose excrements in serozems, although less numerous than in chernozems, are better preserved owing to the semiarid climate.

Rendzic variant − This limestone-derived variant essentially corresponds to the 'rendzina fabric' of Kubiëna (1938). The microstructure is subangular blocky rather than crumb, granular or vermicular. The dark brown coarse aggregates (0.2-0.5 mm) are weakly accommodated and have undulating to rough boundaries. The proportion of aggregates related to faunal origin strongly varies. The high porosity, resulting in low moisture content due to the excessive aeration, together with the great abundance of rock fragments seems to be the limiting factor for soil animal activity, recorded by the low abundance of excrement pedofeatures. Tissue residues are strongly altered, with deformed dark brown lignified walls only partly preserved (Kowalinski et al., 1972). The b-fabric is generally undifferentiated and locally weakly developed calcitic-crystallitic. Occurrences of the latter b-fabric type correspond to small zones presumably adjacent to dissolving fragments of the calcareous parent rock (Fig. 8), which are an essential component of the groundmass (Wilding et al., 1997).

Plowed variant − This variant is common among mollic topsoils, and its properties vary more in relation to field management practices than to natural factors. Under continuous low-input farming (e.g., limited fertilisation, plowing irrespective of soil moisture status, use of heavy machines), the topsoil hardly qualifies as mollic or chernic. Extreme cases of undue management may result in strong compaction, plow pan formation, sealing and development of vertic features. Some of these features may

FIGURE 8 Rendzic variant of the mollic horizon, with subangular blocky microstructure and calcitic crystallitic b-fabric (Rendzic Leptosol, northwestern foothills of the Caucasus) (XPL).

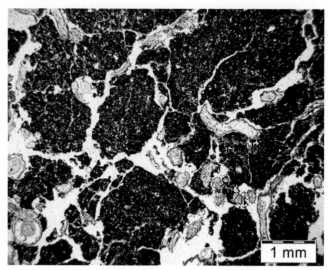

FIGURE 9 Plowed variant of the mollic horizon, having a rather compact groundmass crossed by planes, with abundant weakly decomposed root residues (Haplic Chernozem, Rostov oblast, Russia) (PPL).

disappear after several years of no-till practices or abandonment. The effect of traditional farming on the microstructure of chernozems is usually rather modest. The main effects are a decrease in porosity and a change in aggregate shape. The resulting microstructure is dominated by rounded, elongated or irregular coarse aggregates with uneven boundaries that are better accommodated. Many of them are former coprolites and may be identified as part of a rather compact soil material or as part of channel infillings. The volume of packing voids is smaller and few fine planes are present (Fig. 9). Preservation of microstructure is considered to be a prerequisite for the restoration of initial properties of mollic and chernic horizons following favourable changes in land use or climate (Bystritskaya & Gerasimova, 1988).

2.3   Umbric Horizon

Morphologically, the umbric horizon has much in common with the mollic horizon: it is dark, rich in organic matter, rather thick, and it has a strong to medium crumb structure. It is set apart from the mollic horizon by lower base saturation (IUSS Working Group WRB, 2015), and it typically also has a lower pH. Unlike the mollic horizon, it has a limited geographical occurrence, confined to deciduous forests of regions with a cool climate.

The umbric horizon is diagnostic for the Umbrisols in WRB, and umbric is a qualifier in other Reference Soil Groups. Only few micromorphological descriptions of the umbric horizon are available, and only its orthotype is discussed here. It is based on studies, by the authors, of Umbrisols, or Umbric Leptosols of the Altay Mountains in Southern Siberia and Khentey Mountains in Mongolia.

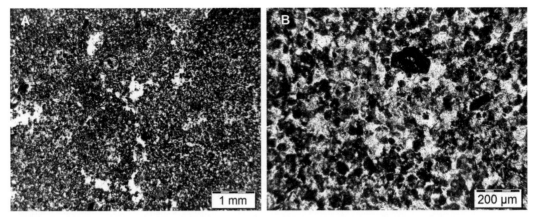

FIGURE 10 Umbric horizon (Haplic Umbrisol (Gelic), northern periphery of Khentey Mountains, Mongolia). (A) Moderately developed fine granular microstructure, with homogeneous dark colour and few vughs (PPL). (B) Detail showing many fine organic aggregates, partly arranged in a circular pattern (PPL).

The crumb microstructure of umbric horizon is moderately developed with weakly separated peds. In some coarser aggregates (>1 mm), finer aggregates (<0.5 mm) that are welded together are discernible (Fig. 10A). The outlines of the welded aggregates are indistinct and rugose. They consist of organo-mineral material and contain numerous fine voids. Packing voids between the coarse aggregates are few, and some associated vughs are present. Traces of faunal activity are limited to intact and coalesced earthworm excrements, occurring as infillings. Few residues of yellowish brown parenchymatic root tissues are found in the groundmass. These residues seem to be fragmented, possibly by frost. The groundmass is homogeneous, generally with a close porphyric c/f-related distribution pattern. The micromass is dusky brown to black with many fine organic punctuations, which display a trend to ring-like patterns. The b-fabric is undifferentiated (Fig. 10B).

The described micromorphological features of the umbric horizon confirm that, despite some common properties, it differs from the mollic orthotype by lower biological activity, more weakly developed microstructure and evidence of frost action.

3. Organo-mineral Surface Horizons of Desert Soils

Organo-mineral surface horizons of desert soils have a distinct set of common properties. The horizons are extremely shallow, being only a few centimetres thick. A very thin layer (0.5-1 cm) of wind-blown sand is commonly present above a more compact material, recording the contribution of aeolian processes (Lobova, 1967; Dregne, 1976; Souirji, 1991). Moreover, even in sandy desert soils (Arenosols) of Turkmenia, the thin surface layer of fine sand is continuously moved by wind, thus protecting the tender roots of ephemeral grasses from direct irradiation (Lobova, 1967).

The recent change in taxonomic position of intervals with takyric and yermic properties is of minor importance with regard to their micromorphological characteristics, as reviewed by Lebedeva-Verba and Gerasimova (2010), because the orthotypes remain distinct. In the definitions of both diagnostic properties (IUSS Working Group WRB, 2015), aridic properties are now an obligatory criterion, implying low organic matter content, evidence of aeolian activity, light colour and high base saturation (IUSS Working Group WRB, 2015).

Takyric properties occur in aridic, periodically flooded soils and comprise the presence of a surface crust above the layer with platy structure. Yermic properties usually comprise the occurrence of surface accumulations of rock fragments, the so-called desert pavement, embedded in a loamy vesicular crust and covered by a thin layer of aeolian sand or silt. A comprehensive description of the micromorphology of aridic soils was given by Allen (1985). The micromorphology of desert topsoils has been reported in several publications by Russian authors (e.g., Feofarova, 1956; Romashkevich & Gerasimova, 1982; Gubin, 1984; Golovanov et al., 2005; Gerasimova & Lebedeva, 2008; Lebedeva (Verba) et al., 2009; Lebedeva et al., 2015). In recent years, the interest of micromorphologists in desert soils has shifted towards studies of vesicular crusts and mechanisms of their formation (Lebedeva-Verba & Kutovaya, 2013; Lebedeva et al., 2013; Turk & Graham, 2014).

3.1 Intervals With Takyric Properties

Intervals with takyric properties (typically 2-6 cm thick) have a clear and limited set of micromorphological characteristics, which were first described by Feofarova (1956) for takyrs of southwestern Turkmenia. Subsequent research allowed specifying the micromorphological properties of takyrs with respect to the stages of their evolution (Romashkevich & Gerasimova, 1982; Gerasimova, 2003). These stages were described for southwestern Turkmenistan by Lavrov et al. (1976) and named after the associated successive plant communities: abiotic, algal and lichen stages, followed by degraded takyr with some higher plants. The orthotype may be regarded as representing the lichen stage of takyr development in Lavrov's evolution sequence, whereas two variants represent the early and final stages.

3.1.1 Orthotype

The takyric orthotype is characterised by a homogeneous, apedal, salt-free, clayey groundmass, with few to many vesicles and few planes. The c/f-related distribution pattern is mostly fine monic, with coarse mineral grains generally confined to infillings of planar voids. The micromass is yellowish brown to light brown with mainly calcitic crystallitic b-fabric (Fig. 11) and locally stipple-speckled b-fabric. Very few limpid, non-laminated clay coatings or fragments of clay coatings have been reported. Their occurrence has been described in terms of 'epi-illuvial argillification' for Chinese takyr soils (Cao Chenggeng, 1990). Organic constituents are rare or absent and may be

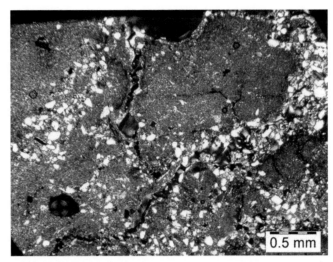

FIGURE 11 Interval with takyric properties (orthotype), showing a dense micromass with calcitic crystallitic b-fabric and fissures filled with aeolian sand grains (Endosalic Regosol (Loamic, Takyric), Misrian foothill plain, western Kopet-Dag Range, Turkmenistan) (XPL).

fragments of lichens blown into fissures. The presence of a thin discontinuous surface layer of wind-transported fine sand (less than 0.5 cm thick) has been mentioned for a number of takyrs (Fedoroff & Courty, 1987). Such deposits can become buried, resulting in the presence of sand layers.

3.1.2   Variants

The *algal variant* represents an earlier stage than the orthotype, starting after deposition of clayey material in depressions or playas. It survives a few weeks of waterlogging in spring, allowing development of algae and abundant zooplankton (Bazilevich & Rodin, 1956). Morphologically, algal deposits are identified by a grayish blue or greenish thin (<0.5 cm) discontinuous films with curling edges (Fig. 12A). Such films, both on the surface and embedded in the mineral groundmass, are observed in thin sections at high magnification as consisting of blue-green algae (cyanobacteria), occurring as algal filaments with high birefringence, mixed with micritic calcite aggregates (Fig. 12B). Besides algae, other manifestations of microbiota include small (5 μm), rounded or oval, dark brown, semitransparent biogenic bodies and flakes. The c/f-related distribution pattern is fine monic, but silt and fine sand grains occur in thin bands, with mica plates displaying a subparallel orientation. The micromass is yellowish brown to light brown with mainly crystallitic b-fabric (Fig. 12B) and locally stipple-speckled b-fabric. The microstructure is massive with few planes and closed equant pores in the compact groundmass. These testify to periodical and strong effects of living organisms on the takyr fabric (Lebedeva-Verba & Gerasimova, 2010). The filamentous cyanobacteria are assumed to maintain the stability of the crust on the soil surface, contributing to better cohesion of

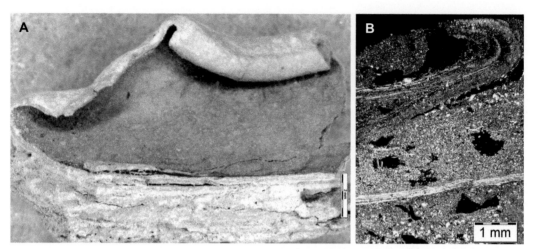

FIGURE 12 Interval with takyric properties (algal variant) (Endosalic Regosol (Loamic, Takyric), Kopet-Dag foothill plain, Kara-Chil area, Turkmenistan). (A) Thin algal surface crust (macroscopic image). (B) Filaments of blue-green algae with high birefringence, in soil material with crystallitic b-fabric, few vughs, and scattered aeolian coarse mineral grains (XPL).

the mineral mass and to dust accumulation (Pagliai & Stoops, 2010; Williams et al., 2018).

The *degraded variant*, or takyr of the final stage, is characterised by gradual incorporation of aeolian sand infillings into the groundmass. The boundaries of the infillings become gradually less distinct, and in older takyrs the former infillings are recognised only as linear features composed of aligned sand grains along former planes or clusters of sand grains filling former vesicles. The initial fine monic c/f-related distribution pattern becomes open porphyric (Fig. 13), some vesicles are transformed to vughs and a more open pore system is formed. Rare fragments of clay coatings with indistinct boundaries are impregnated by fine-grained calcite. The input of aeolian sand enhances the transformation of takyrs, which is also controlled by the changes in plant communities during alternating flooding stages and dry periods. When the amount of sand reaches a certain level, vesicular voids are transformed to vughs, and some of them become no longer closed. As a result, the surface horizon acquires certain water permeability, and water no longer stagnates on their surface in spring. These changes in moisture regime contribute to evolution of the pedon to another type of desert soil with residual takyric features (Lobova, 1967). Presumably, they gradually acquire yermic properties.

Comparing the orthotype and variants, it may be concluded that the micromorphological features of intervals with takyric properties are very conspicuous and that differences between them permit to decipher the history of these soils. The microfabric of each variant is indicative of its hydrological regime, which strongly depends on the shape and continuity of voids, including vesicles, which are further discussed in the next section.

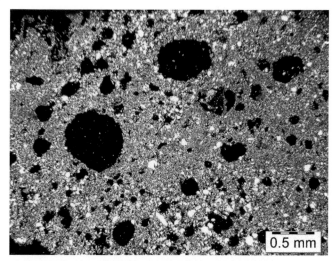

FIGURE 13 Interval with takyric properties (degraded variant), showing incorporation of aeolian sand to silt grains into the groundmass, as well as many vesicles (Endosalic Regosol (Loamic, Takyric), Misrian foothill plain, western Kopet-Dag area, Turkmenistan) (XPL).

3.2   Intervals With Yermic Properties

The most recent definition of yermic properties in WRB (IUSS Working Group WRB, 2015) includes aridic properties as requirement (see Section 3.1) as well as the required presence of desert pavement (identified by desert varnish and presence of ventifacts) and/or a vesicular layer (either associated with the desert pavement or overlain by a thin platy layer). The components of the associated desert pavement are too large for standard thin section studies and are not discussed in this chapter.

The fine material of the intervals with yermic properties usually comprises two thin sublayers (microlayers of Boiffin and Bresson, 1987) that are a few centimetres thick: the vesicular sublayer and the platy sublayer (IUSS Working Group WRB, 2006, 2015), with corresponding vesicular and platy microstructures (Fig. 14; see also Pagliai & Stoops, 2010; Williams et al., 2018). Both sublayers show prominent features of aeolian accumulation. In most cases, the vesicular sublayer is located on the top of the platy sublayer, but the opposite arrangement has also been recorded (Lebedeva (Verba) et al., 2009). The platy layer can also be absent in soils with desert pavement, as reported for Antarctic 'dry valleys' (Bockheim, 2010).

A layer with vesicular microstructure has been described for many desert soils (Springer, 1958; Evenari et al., 1974; McFadden et al., 1998; Romashkevich & Gerasimova, 1982; Figueira & Stoops, 1983; Gong Zitong & Gu Guoan, 1991; Gubin, 1984; Allen, 1985; Souirji, 1991; Anderson et al., 2002; Golovanov et al., 2005; Gerasimova & Lebedeva, 2008; Lebedeva & Shishkov, 2016). The vesicular sublayer is an important diagnostic element of intervals with yermic properties, corresponding to vesicular crusts, and it has been assumed to be genetically related to the desert pavement (see Pagliai & Stoops, 2010;

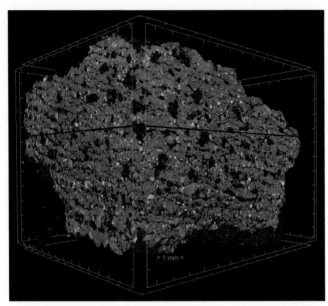

FIGURE 14 Interval with yermic properties (Endosalic Regosol (Loamic, Nudiyermic), Trans-Altay Gobi, Mongolia). Three-dimensional CT-rendering of a crust with distinct stratification: upper part with abundant large vesicles and lower part dominated by narrow subparallel planar voids. *Image by K. Abrosimov.*

Williams et al., 2018). The micromass of this sublayer is calcareous, with embedded fine sand-sized and silt-sized grains (open porphyric c/f-related distribution pattern) and calcitic crystallitic b-fabric (Fig. 15). Its most prominent feature is the presence of round, closed pores. This type of porosity can result in low water permeability, which can cause water stagnation for short periods after episodic rainstorms (Gorbunov & Bekarevich. 1955; Evenari et al., 1974). Investigating the hydrological function of vesicular micro-structures was one of the reasons to apply new methods to the study of pore patterns, including X-ray computed tomography (Turk & Graham, 2011, 2014; Lebedeva et al., 2016). As illustrated by such studies, vesicular pores prevail in the upper part of the microhorizon, and their number decreases downward, sometimes merging to form interconnected voids, with equant voids becoming substituted by vughs. This trend may be attributed to deformation mechanisms caused by swelling-shrinking or by freeze-thaw cycles (Springer, 1958; see Van Vliet-Lanoë & Fox, 2018). The latter can apply to aridic soils with yermic properties of Central Asia, for which low temperatures have been reported (Pankova, 1992; Golovanov et al., 2005; Lebedeva (Verba) et al., 2009; Lebedeva et al., 2015; Pankova & Lebedeva, 2015).

Vesicular microstructures are considered to be rather stable owing to microlaminated clay coatings along pore walls (Sullivan & Koppi, 1991), or to stress-deformed clay around the vesicles (porostriated b-fabric). In hyperaridic soils of the Gobi desert, hypocoatings and coatings with partially disordered fabric were observed along the sides of vesicular voids (Fig. 16) (unpublished work by the authors).

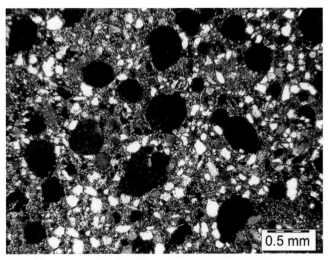

FIGURE 15 Interval with yermic properties showing typical vesicular crust with numerous vesicles, single-spaced porphyric c/f-related distribution pattern and crystallitic b-fabric (Haplic Regosol (Loamic, Endosalic, Nudiyermic), Baer Hummocks, Caspian Lowland, Russia) (XPL).

Vesicular pores have been described for substrates that are quite different in terms of particle size distribution and carbonate content. In these different materials, the origin of vesicles may be different, while remaining related to trapping of a gas phase. Thus, for aridic soils of Ili Depression (Eastern Kazakhstan), with a highly calcareous silty groundmass, vesicle formation may be related to shifts in carbonate-bicarbonate equilibrium in response to moisture and temperature fluctuations during summer rains, promoting carbon dioxide emission (Lebedeva et al., 2015). In an extremely aridic soil of the Gobi Desert (Mongolia), with a sandy clay groundmass, aeolian material has been found as loose discontinuous infillings of vesicular pores, suggesting another way of vesicle development, presumably involving liberation of air from a strongly desiccated material caused by short intense summer rains.

In subboreal deserts, the vesicular layer is mostly underlain by a platy layer. The latter has a complex microstructure composed of aeolian rounded aggregates displaying a more or less strongly developed banded arrangement, producing a platy or lenticular microstructure, with subparallel voids (Fig. 17). These rounded aggregates vary in micromass properties, c/f-related distribution pattern and degree of separation. Their micromass has a calcitic crystallitic b-fabric, and calcareous nodules or fragments of nodules are frequently included in the groundmass. Frost action has probably contributed to the platy rearrangement of the rounded aggregates.

In the majority of intervals with yermic properties, plant residues are scarce. Few traces of faunal activity are encountered as rather dense yellowish limpid spheroidal organo-mineral excrements. Pedozoologists have reported an abundance of Isopoda in aridic soils (Striganova, 1996).

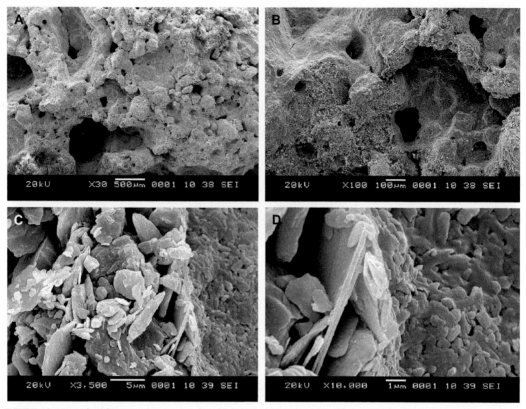

FIGURE 16 Interval with yermic properties (Haplic Gypsisol (Loamic, Yermic), Trans-Altay Gobi, Mongolia). SEM images of a vesicular crust. (A) and (B) Overview at low magnification, showing vesicles of varying sizes, lined by clay coatings. (C) Appearance at higher magnification, showing weakly compact micromass near a vesicle (right). (D) Detail of previous field, showing the contact between the micromass (with subparallel mica plates) and the clay coating that lines the vesicle. *Images by S. Inozemtsev.*

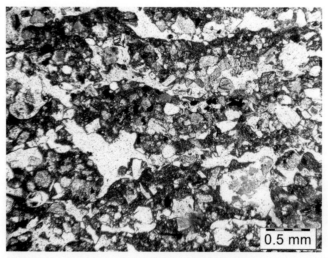

FIGURE 17 Interval with yermic properties, with lenticular microstructure and large vughs (Skeletic Gypsisol (Loamic, Yermic), Ili River basin, Eastern Kazakhstan) (PPL).

4. Conclusions

Organo-mineral surface horizons are more sensitive to environmental conditions and changes than other parts of soil profiles. This is reflected in thin sections by the presence of characteristic microstructures, organic matter types and features related to faunal activity.

The orthotype of the mollic horizon, coinciding with the recently named 'chernic horizon' (IUSS Working Group WRB, 2015), corresponds to steppe chernozemic mull and has been the subject of many micromorphological studies. Variants of this orthotype occur in a broad spectrum of soils, mostly under grasslands of boreal or subtropical steppes, meadows and prairies. The micromorphological characteristics of mollic horizons are essentially determined by their high organic matter content and strong faunal activity. Other organo-mineral surface horizons in temperate climate having diagnostic importance in WRB (umbric, melanic) are also dominated by organic matter features, but they have been far less investigated.

In contrast, the role of organic matter is insignificant in desert topsoils such as those with takyric and yermic properties. Such intervals have a distinct set of micromorphological properties that characterise the orthotypes. Their variants are related to the evolutionary sequence of takyric soils. For intervals with yermic properties, most attention has been given to layers with vesicular microstructure.

The relative scarcity of micromorphological information about organo-mineral surface horizons of desert regions is partly due to incomplete and difficult sampling of surface features for thin section preparation for undisturbed material.

The use of micromorphological data contributes to a better understanding of the origin, properties and evolution of the described surface horizons. Responses of soils to changes in climate, environment and land use may be efficiently unravelled by micromorphological studies.

References

Adderley, W.P., Wilson, C.A., Simpson, I.A. & Davidson, D.A., 2018. Anthropogenic features. In Stoops, G., Marcelino, V. & Mees, F. (eds.), Interpretation of Micromorphological Features of Soils and Regoliths. Second Edition. Elsevier, Amsterdam, pp. 753-777.

Allen, B.L., 1985. Micromorphology of Aridisols. In Douglas, L.A. & Thompson, M.L. (eds.), Soil Micromorphology and Soil Classification. Soil Science Society of America Special Publication 15, SSSA, Madison, pp. 197-216.

Anderson, K., Wells, S. & Graham, R., 2002. Pedogenesis of vesicular horizons, Cima volcanic field, Mojave Desert, California. Soil Science Society of America Journal 66, 878-887.

Babel, U. & Vogel, H.J., 1989. Zur Beurteilung der Enchyträen- und Collembolen-Aktivität mit Hilfe von Bodendünnschliffen. Pedobiologia 33, 167-172.

Baize, D. & Girard, M.C., 1995. Référentiel Pédologique. INRA Editions, Paris, 332 p.

Bal, L., 1970. Morphological investigation in two moder-humus profiles and the role of the soil fauna in their genesis. Geoderma 4, 5-36.

Bal, L., 1973. Micromorphological Analysis of Soils – Lower Levels in the Organization of Organic Soil Materials. Soil Survey Papers 6. Soil Survey Institute, Wageningen, 174 p.

Barratt, B., 1964. A classification of humus forms and microfabrics of temperate grasslands. Journal of Soil Science 15, 342-356.

Bazilevich, N.I. & Rodin, L.E., 1956. On the role of vegetation in the formation and evolution of takyrs on the Meshed-Messerian deltaic alluvial plain [in Russian]. In Takyrs of Southwestern Turkmenia. AN SSSR Publisher, Moscow, pp. 222-277.

Bockheim, J.G., 2010. Evolution of desert pavements and the vesicular layer in soils of the Transantarctic Mountains. Geomorphology 118, 433-443.

Boiffin, J. & Bresson, L.M., 1987. Dynamique de formation des croûtes superficielles: apport de l'analyse microscopique. In Fedoroff, N., Bresson, L.M. & Courty, M.A. (eds.), Micromorphologie des Sols, Soil Micromorphology. AFES, Plaisir, pp. 393-399.

Brown, G.G., Feller, Ch., Blanchart, E., Deleporte, P. & Chernyanskii, S.S., 2003. With Darwin, earthworms turn intelligent and become human friends: the 7th International Symposium on Earthworm Ecology, Cardiff, Wales, 2002. Pedobiologia 47, 924-933.

Bullock, P., Fedoroff, N., Jongerius A., Stoops, G., Tursina, T. & Babel, U., 1985. Handbook for Soil Thin Section Description. Waine Research Publications, Wolverhampton, 152 p.

Bystritskaya, T.L. & Gerasimova, M.I., 1988. On the annual cycle of current Chernozem formation [in Russian]. Pochvovedenie 6, 5-16.

Cao Chenggeng, 1990. Micromorphological study on the argillification in Aridisols of China. In Douglas, L.A. (ed.), Soil Micromorphology: A Basic and Applied Science. Developments in Soil Science, Volume 19. Elsevier, Amsterdam, pp. 347-354.

Chertov, O.G., 1966. Characterization of humus profile types in podzolic soils of Leningrad Oblast [in Russian]. Pochvovedenie 3, 26-37.

Cooperative Research Group of Chinese Soil Taxonomy, 2001. Chinese Soil Taxonomy. Science Press, Beijing, 202 pp.

Dregne, H.E., 1976. Soils of Arid Regions. Elsevier, Amsterdam, 237 p.

Duchaufour, Ph., 1959. La Dynamique du Sol Forestier en Climat Atlantique. Les Presses Universitaires Laval, Québec, 72 p.

Duchaufour, Ph., 1960. Précis de Pédologie. Masson & Cie, Paris, 437 p.

Evenari, J., Yaalon, D.H. & Gutterman, Y., 1974. Note on soils with vesicular structures in desert. Zeitschrift fur Geomorphologie 18, 162-172.

Eswaran, H. & Drees, R., 2004. Soil under the Microscope: Evaluating Soils in Another Dimension. Soil Micromorphology Committee of the Soil Science Society of America, Madison. [CD ROM].

Fedoroff, N. & Courty, M.A., 1987. Morphology and distribution of textural features in arid and semiarid regions In Fedoroff, N., Bresson, L.M. & Courty, M.A. (eds.), Micromorphologie des Sols, Soil Micromorphology. AFES, Plaisir, pp. 213-219.

Feofarova, I.I., 1956. Micromorphological Characteristics [in Russian]. In Takyrs of Southwestern Turkmenia. Collection of Papers. AN SSSR Publisher, Moscow, pp. 361-380.

Figueira, H. & Stoops, G., 1983. Application of micromorphometric techniques to the experimental study of vesicular layer formation. Pedologie 33, 77-89.

FitzPatrick, E.A., 1984. Micromorphology of Soils. Chapman and Hall, London, 433 p.

FitzPatrick, E.A., 2006. Soil Microscopy and Micromorphology, UK. [CD ROM].

Gerasimova, M.I., 2003. Higher levels of description – approaches to the micromorphological characterisation of Russian soils. Catena 54, 319-339.

Gerasimova, M. & Lebedeva, M., 2008. Contribution of micromorphology to classification of aridic soils. In Kapur, S., Mermut, A. & Stoops, G. (eds.), New Trends in Soil Micromorphology. Springer, Berlin, pp. 151-162.

Gerasimova, M.I., Gubin, S.V. & Shoba, S.A., 1996. Soils of Russia and Adjacent Countries: Geography and Micromorphology. Moscow State University & Agricultural University Wageningen, 204 p.

Golovanov, D.L., Lebedeva-Verba, M.P., Dorokhova, M.F. & Slobodkin, A.I., 2005. Micromorphological and microbiological characterization of elementary soil-forming processes in desert soils of Mongolia. Eurasian Soil Science 38, 1290-1300.

Gong Zitong & Gu Guoan, 1991. Altocryic aridisols in China. In Kimble, J.M. (ed.), Proceedings of the 6th International Soil Correlation Meeting (ISCOM), Characterization, Classification and Utilization of Cold Aridisols and Vertisols. USDA, Soil Conservation Service, National Soil Survey Center, Lincoln, pp. 54-60.

Gorbunov, N.I. & Bekarevich, N.E., 1955, Soil Crusting under Irrigation [*in Russian*]. AN SSSR Publisher, Moscow, 70 p.

Gracheva, M.V., Tursina, T.V. & Korolyuk, T.V., 1988. Microstructure of non-irrigated and irrigated soils of chestnut alkali complexes in Stavropol region [*in Russian*]. In Dobrovol'skiy, G.V. (ed.), Micromorphology of Anthropogenic Soils. Nauka, Moscow, pp. 106-114.

Gubin, S.V., 1984. Micromorphological diagnostics of brown and grey-brown soils of northern Ust'-Urt [*in Russian*]. In Lobova, E.V. (ed.), Nature, Soils and Reclamation of the Ust'-Urt Desert. Institute of Soil Science and Agrochemistry, Pushchino, pp. 127-134.

Gyninova, A. & Shoba, S.A., 1988. Micromorphology and structural state of meadow soils of Amur basin as related to drainage [*in Russian*]. In Dobrovol'skiy, G.V. (ed.), Micromorphology of Anthropogenic Soils. Nauka, Moscow, pp. 150-167.

Ismail-Meyer, K., Stolt, M. & Lindbo, D., 2018. Soil organic matter. In Stoops, G., Marcelino, V. & Mees, F. (eds.), Interpretation of Micromorphological Features of Soils and Regoliths. Second Edition. Elsevier, Amsterdam, pp. 471-512.

IUSS Working Group WRB, 2006. World Reference Base for Soil Resources. 2nd Edition. World Soil Resources Reports No. 103. FAO, Rome, 128 p.

IUSS Working Group WRB, 2015. World Reference Base for Soil Resources 2014, Update 2015. World Soil Resources Reports No. 106. FAO, Rome, 192 p.

Jabiol, B., Brêthes, A., Ponge, J.F., Toutain, F. & Brun, J.J., 1995. L'Humus sous toutes ses Formes. Ecole Nationale du Génie Rural, des Eaux et des Forêts, Nancy, 63 p.

Jeanson, C., 1964. Micromorphologie et pédozoologie expérimentale: contribution à l'étude sur plaques minces de grandes dimensions d'un sol artificiel structuré par les Lombricides. In Jongerius, A. (ed.), Soil Micromorphology. Elsevier, Amsterdam, pp. 47-55.

Jongerius, A. & Marsman, R., 1971. Humeuze laagjes in stuifzand. Boor en Spade 17, 7-22.

Khitrov, N.,Gerasimova, M., Bronnikova M. & Zazovskaya, E. 2013. Chernozem of Kursk Biosphere Station of the Institute of Geography, RAS. In Kovda, I.V. (ed.), Guide for Field Excursions, 12th International Symposium and Field Seminar on Paleopedology. Institute of Geography, RAS, Moscow, pp. 46-55.

Kooistra, M.J. & Pulleman, M.M., 2010. Features related to faunal activity. In Stoops, G., Marcelino, V. & Mees, F. (eds.), Interpretation of Micromorphological Features of Soils and Regoliths. Elsevier, Amsterdam, pp. 397-418.

Kooistra, M.J. & Pulleman, M.M., 2018. Features related to faunal activity. In Stoops, G., Marcelino, V. & Mees, F. (eds.), Interpretation of Micromorphological Features of Soils and Regoliths. Second Edition. Elsevier, Amsterdam, pp. 447-469.

Kovda, I.V., Brevic, E.C., Fenton, T.E. & Gerasimova, M., 2004. The impact of white pine (*Pinus strobus*) on a Mollisol after seven decades of soil development. Journal of the Iowa Academy of Science 111, 58-66.

Kowalinski, S. & Kollender-Szych, A., 1972. Micromorphological and physico-chemical investigations on the decomposition rate of organic matter in some muck soils. In Kowalinski, S. & Drozd, J. (eds.), Soil Micromorphology. Panstwowe Wydawnicto Naukowe, Warsawa, pp. 143-155.

Kowalinski, S., Licznar, S. & Licznar, M., 1972. Micromorphological properties of Rendzinas and soils on limestones developed out of Triassic carbonate-calcareous formations. In Kowalinski, S. & Drozd, J. (eds.), Soil Micromorphology. Panstwowe Wydawnicto Naukowe, Warsawa, pp. 455-479.

Kretzschmar, A., 1987. Caractérisation microscopique de l'activité des lombriciens endogés. In Fedoroff, N., Bresson, L.M. & Courty, M.A. (eds.), Micromorphologie des Sols, Soil Micromorphology. AFES, Plaisir, pp. 325-330.

Kubiëna, W.L., 1938. Micropedology. Collegiate Press, Ames, 242 p.

Kubiëna, W.L., 1953. The Soils of Europe. Thomas Murby & Co., London, 317 p.

Kubiëna, W.L., 1970. Micromorphological Features of Soil Geography. Rutgers University Press, New Brunswick, 256 p.

Lavrov, A.P., Larin, E.V. & Sanin, S.A., 1976. Regionalization of Takyrs in Turkmenia for Agricultural Purposes [*in Russian*]. Ylim, Ashkhabad, 170 p.

Lebedeva, M. & Kutovaya, O., 2013. Fabric of topsoil horizons in aridic soils of Central Asia. Spanish Journal of Soil Science 3, 148-168.

Lebedeva, M.P. & Shishkov, V.A., 2016. A comparative analysis of the microfabrics of surface horizons and desert varnish in extremely arid soils of the Mojave (USA) and Trans-Altai Gobi (Mongolia) deserts. Eurasian Soil Science 49, 163-179.

Lebedeva-Verba, M.P. & Gerasimova, M.I., 2009. Micromophology of diagnostic horizons in aridic soils (complementary to the new classification system of soils of Russia). Eurasian Soil Science 13, 1427-1434.

Lebedeva-Verba, M.P. & Gerasimova, M.I, 2010. Micromorphology of takyrs and the desert 'papyrus'of Southwestern Turkmenia. Eurasian Soil Science 43, 1220-1229.

Lebedeva (Verba), M.P., Golovanov, D.L. & Inozemtsev, S.A., 2009. Microfabrics of desert soils of Mongolia. Eurasian Soil Science 42, 1204-1217.

Lebedeva, M., Gerasimova, M. & Golovanov, D., 2013. Classification of the topsoil fabrics in arid soils of Central Asia. In Shahid, S.A., Adelfatth, M.A. & Taha, F.K. (eds.), Developments in Soil Classification, Land Use Planning and Policy Implications. Springer, Dordrecht, pp. 243-256.

Lebedeva, M.P., Gerasimova M.I., Golovanov, D.L. & Yamnova, I.A., 2015. Extremely arid soils of the Ili Depression in Kazakhstan. Eurasian Soil Science 48, 11-26.

Lebedeva, M.P., Golovanov, D.L. & Abrosimov, K.N., 2016. Micromorphological diagnostics of pedogenic, eolian, and colluvial processes from data on the fabrics of crusty horizons in differently aged extremely aridic soils of Mongolia. Quaternary International 418, 75-83.

Lobova, E.V., 1967. Soils of the Desert Zone of the USSR. Israel Programme for Scientific Translations, Jerusalem, 405 p.

McFadden, L.D., McDonald, E.V., Wells, S.G., Anderson, K., Quade, J. & Forman, S.L., 1998. The vesicular layer and carbonate collars of desert soils and pavements, formation, age and relation to climate change. Geomorphology 24, 101-145.

Medvedev, V.V., 1983. Comparative analysis of hydrological properties and microfabric of typical Chernozems and dark Chestnut soils in Ukraine [*in Russian*]. In Targulyan, V.O. (ed.), Micromorphological Diagnostics of Soils and Soil-Forming Processes. Nauka, Moscow, pp. 139-153.

Mermut, A.R. & St. Arnaud, R.J., 1981. Microband fabric in seasonally frozen soils. Soil Science Society of America Journal 45, 578-586.

Miedema, R., Koulechova, I.N. & Gerasimova, M.I., 1999. Soil formation in Greyzems in Moscow district: micromorphology, chemistry, clay mineralogy and particle size distribution. Catena 34, 315-347.

Minashina, N.G., 1966. Micromorphology of Loess, Serozems, Hei-Loo-Too and some problems of their paleogenesis [*in Russian*]. In Zonn, S.V. (ed.), Micromorphological Methods in the Study of Soil Genesis. Nauka, Moscow, pp. 76-93.

Moscatelli, G. & Pazos, M.S., 2000. Soils of Argentina: nature and use. In Kheoruenromne, I. & Theerawong, S. (eds.), Proceedings of the International Symposium on Soil Science: Accomplishments and Changing Paradigm Towards the 21st Century, Bangkok, pp. 81-92.

Pagliai, M. & Stoops, G., 2010. Physical and biological surface crusts and seals. In Stoops, G., Marcelino, V. & Mees, F. (eds.), Interpretation of Micromorphological Features of Soils and Regoliths. Elsevier, Amsterdam, pp. 419-440.

Pankova, E.I., 1992. Genesis of Salinity in Desert Soils [*in Russian*]. Izd-vo VASKhNIL, Moscow, 136 p.

Pankova, E.I. & Lebedeva, M.P., 2015. Similarity and difference in soil properties of boreal deserts in Central and Middle Asia. Byulleten Pochvennogo Instituta im. V.V. Dokuchaeva 81, 45-59.

Parfenova, E.I. & Yarilova, E.A., 1977. A Manual of Micromorphological Research in Soil Science [*in Russian*]. Nauka, Moscow, 198 p.

Pawluk, S. & Bal, L., 1985. Micromorphology of selected mollic epipedons. In Douglas, L.A. & Thompson, M.L. (eds.), Soil Micromorphology and Soil Classification. Soil Science Society of America Special Publication 15, SSSA, Madison, pp. 63-83.

Pazos, M.S. & Stoops, G., 1987. Micromorphological aspects of soil formation in mollisols from Argentina. In Fedoroff, N., Bresson, L.M. & Courty, M.A. (eds.), Micromorphologie des Sols, Soil Micromorphology. AFES, Plaisir, pp. 263-270.

Poliakov, A., 1981. Micromorphology of Chernozems of the forest-steppe Transvolga Province [*in Russian*]. Byulleten Pochvennogo Instituta im. V.V. Dokuchaeva 28, 49-50.

Romashkevich, A.I. & Gerasimova, M.I., 1982. Micromorphology and Diagnostics of Pedogenesis [*in Russian*]. Nauka, Moscow, 125 p.

Sanborn, P. & Pawluk, S., 1989. Microstructure diversity in Ah horizons of black chernozemic soils, Alberta and British Columbia (Canada). Geoderma 45, 221-240.

Shishov, L.L., Tonkonogov, V.D., Lebedeva, I.I. & Gerasimova, M.I. (eds.), 2004. Classification and Diagnostics of Soils of Russia [in *Russian*]. Oykumena Publishing House, Smolensk, 342 p.

Soil Survey Staff, 1999. Soil Taxonomy. A Basic System of Soil Classification for Making and Interpreting Soil Surveys, 2nd Edition. USDA, Natural Resources Conservation Service, Agriculture Handbook Number 436. U.S. Government Printing Office, Washington, 871 p.

Souirji, A., 1991. Classification of aridic soils, past and present: proposal of a diagnostic desert epipedon. In Kimble, J.M. (ed.), Characterization, Classification and Utilization of Cold Aridisols and Vertisols. Proceedings of the International Soil Correlation Meeting (ISCOM). USDA, Soil Conservation Service, National Soil Survey Center, Lincoln, pp. 175-185.

Spaargaren, O., 1998. Topsoil Characterization for Sustainable Land Management. Draft. FAO, Rome, 71p.

Springer, M.E., 1958. Desert pavement and vesicular layer of some soils of the desert of Lahontan Basin, Nevada. Soil Science Society of America Proceedings 22, 63-66.

Stolt, M.H. & Lindbo, D.L., 2010. Soil organic matter. In Stoops, G., Marcelino, V. & Mees, F. (eds.), Interpretation of Micromorphological Features of Soils and Regoliths. Elsevier, Amsterdam, pp. 369-396.

Stoops, G., 1994. Soil thin sections description: higher levels of classification of microfabrics as a tool for interpretation. In Ringrose-Voase, A. & Humphreys, G.S. (eds.), Soil Micromorphology: Studies in Management and Genesis. Developments in Soil Science 22, Elsevier, Amsterdam. pp. 317-325.

Stoops, G., 2003. Guidelines for Analysis and Description of Soil and Regolith Thin Sections. Soil Science Society of America, Madison, 184 p.

Striganova, B.R., 1996. Adaptive strategies of animals for populating the soil layer. Eurasian Soil Science 6, 643-650.

Sullivan, L.A. & Koppi, A.J., 1991. Morphology and genesis of silt and clay coatings in the vesicular layer of a desert loam soil. Australian Journal of Soil Research 29, 579-586.

Tselishcheva, L.K., 1966. Micromorphology of virgin chernozems and meadow-chernozemic soils of Streletskaya steppe [*in Russian*]. In Zonn, S.V. (ed.), Micromorphological Methods in the Study of Soil Genesis. Nauka, Moscow, pp. 5-16.

Turk, J.K. & Graham, R.C., 2011. Distribution and properties of vesicular horizons in the Western United States. Soil Science Society of America Journal 74, 1449-1461.

Turk, J.K. & Graham, R.C., 2014. A proposed master V horizon for the designation of near-surface horizons with vesicular porosity. Soil Science Society of America Journal 78, 868-880.

Van Vliet-Lanoë, B., 2010. Frost action. In Stoops, G., Marcelino, V. & Mees, F. (eds.), Interpretation of Micromorphological Features of Soils and Regoliths. Elsevier, Amsterdam, pp. 81-108.

Van Vliet-Lanoë, B. & Fox, C., 2018. Frost action. In Stoops, G., Marcelino, V. & Mees, F. (eds.), Interpretation of Micromorphological Features of Soils and Regoliths. Second Edition. Elsevier, Amsterdam, pp. 575-603.

Verba, M.P., Kulakova, N.I. & Yamnova, I.A., 2006. Genesis and properties of dark-coloured chernozem-like soils of mesodepressions under fallow in the northern Caspian region. Eurasian Soil Science 9, 990-1001.

Wilding, L.P., Drees, L.R. & Woodruffs, C.M., 1997. Mineralogy and microfabrics of limestone soils on stepped landscapes in Central Texas. In Shoba, S., Gerasimova, M. & Miedema, R. (eds.), Soil Micromorphology: Studies on Soil Diversity, Diagnostics, Dynamics. Moscow, Wageningen, pp. 204-219.

Williams, A.J., Pagliai, M. & Stoops, G., 2018. Physical and biological surface crusts and seals. In Stoops, G., Marcelino, V. & Mees, F. (eds.), Interpretation of Micromorphological Features of Soils and Regoliths. Second Edition. Elsevier, Amsterdam, pp. 539-574.

Yarilova, E.A., 1972. Comparative characteristics of fabric components in some chernozems of the USSR. In Kowalinski, S. & Drozd, J. (eds.), Soil Micromorphology. Panstwowe Wydawnicto Naukowe, Warsawa, pp. 357-369.

Yarilova, E.A., 1974. Micromorphology of Chernozems [*in Russian*]. In Fridland, V.M. & Lebedeva, I.I. (eds.), Chernozems of the USSR. Kolos, Moscow, pp. 156-163.

Yarilova, E.A. & Bystritskaya, T.L., 1976. Morphological and micromorphological structure of the soils of Khomutovskaya steppe in the Azov Sea region. Soviet Soil Science 8, 268-278.

Yarilova, E.A., Samoilova, E.M., Poliakov, A.N. & Makeeva, V.I., 1983. Micromorphology of Chernozems of the Russian Plain [*in Russian*]. In Targulyan, V.O. (ed.), Micromorphological Diagnostics of Soils and Soil-Forming Processes. Nauka, Moscow, pp. 130-139.

Zachariae, G., 1964. Welche Bedeutung haben Enchytraeen im Waldböden? In Jongerius, A. (ed.), Soil Micromorphology. Elsevier, Amsterdam, pp. 57-68.

Chapter 19

Physical and Biological Surface Crusts and Seals

Amanda J. Williams[1], Marcello Pagliai[2], Georges Stoops[3]

[1]SWCA ENVIRONMENTAL CONSULTANTS, LAS VEGAS, NV, UNITED STATES;
[2]CENTRO DI RICERCA PER L'AGROBIOLOGIA E LA PEDOLOGIA, FLORENCE, ITALY;
[3]GHENT UNIVERSITY, GHENT, BELGIUM

CHAPTER OUTLINE

1. Introduction

Soil crusts are classified into three morphological types. Physical soil crusts are modifications of topsoil caused by physical perturbation, such as raindrop impact or sedimentation, resulting in development of a compacted surface layer with reduced porosity; some authors prefer to use the term *seal* if no drying or hardening has taken place (Remley & Bradford, 1989). Biological soil crusts (BSCs) are living communities of lichens, cyanobacteria, algae or mosses growing on and near the soil surface and binding soil particles. Chemical crusts are salt crusts, which are discussed elsewhere (Mees & Tursina, 2018). These three main crust types are not mutually exclusive, as many forms of biological soil crusts will also exhibit some physical or chemical crusting (Cantón et al., 2003; Malam Issa et al., 2011; Williams et al., 2012). Although the word crust may also indicate hard layers of lateritic, calcareous and siliceous material, as well as horizons described as 'weathering crusts' by Russian authors, these terms have meanings that are different from those discussed in this chapter. Physical compaction by trampling (Rentzel et al., 2017) or tillage also forms physical soil crusts, but this is not covered here.

Micromorphological studies of vertically oriented thin sections of soil crusts, as well as SEM studies, have contributed considerably to our understanding of crusts and have enhanced field-plot to hand-lens scale observations of surface processes (see Valentin & Bresson, 1992; West et al., 1992; Bresson & Valentin, 1994; Slattery & Bryan, 1994; Pagliai & Stoops, 2010; Williams et al., 2012; Felde et al., 2014). Endpoint and sequential laboratory and field experiments combined with micromorphology have been important for understanding crust and seal formation (e.g., Southard et al., 1988; Luk et al., 1990; Poesen et al., 1990; see also Table 1), particularly the genesis and destruction of ephemeral structures (Luk et al., 1990; Mermut et al., 1995). Morphometric quantification of pore shape, size, continuity and distribution allows identification of microstructural changes associated with crust formation that correspond to changes in soil functions, which can support land management interpretations (Pagliai, 1994).

The aim of this chapter is to provide an overview of the micromorphological expressions of soil crust formation and to reconcile differences in terminology. Table 1 provides an overview of the most relevant papers dealing with the micromorphological characterisation of physical and biological crusts.

2. Physical Soil Crusts

Physical soil crusts form in a variety of climatic conditions but are widespread on soils of arid and semiarid regions, in both natural and cultivated systems (see also Gerasimova & Lebedeva, 2018). In temperate areas, surface crusts most commonly develop on unstable loamy soils with low organic matter content (Mücher & De Ploey, 1977), especially when cultivated (Pagliai et al., 1983a; Norton & Schroeder, 1987; Pagliai, 1987; Poesen & Nearing, 1993). In recent decades, soil degradation resulting from intensive land use has

Table 1   Overview of micromorphological studies of soil crusts

References	Material[a] and method[b]	Location	Crust or microlayer type
Kubiëna (1938)	n, optical	USA	Sedimentary
Duley (1939)	f, l, optical	USA	Disruptional
McIntyre (1958)	f, l, M	Australia	Disruptional, skin seal, washed-in, washed-out
Springer (1958)	n, l, M	USA	Vesicular[c]
Tackett & Pearson (1965)	l, optical	USA	Skin seal
Epstein & Grant (1967)	l, M	USA	Disruptional
Evans & Buol (1968)	n, optical	USA	Sedimentary, vesicular
Volk & Geyger (1970)	n, optical	Morocco, Spain	Vesicular
Ahmad & Roblin (1971)	l, optical	Trinidad	Coalescing, disruptional
Miller (1971)	l, s, M	USA	Vesicular
Evenari et al. (1974)	n, l, optical	Israel	Vesicular
Bishay & Stoops (1975)	c, optical	Egypt	Sedimentary
Falayi & Bouma (1975)	f, c, optical	USA	Disruptional, sedimentary
Ferry & Olsen (1975)	l, optical	USA	Disruptional
Labib et al. (1975)	n, optical	Egypt	Sedimentary, vesicular
Bunting (1977)	n, optical	Canada	Vesicular
Mücher & De Ploey (1977)	n, l, optical	Belgium	Sedimentary, structural
Farres (1978)	l, s, optical	UK	Disruptional, washed-in, washed-out
Campbell (1979)	f, l, SEM+	USA	Biological (cyanobacteria)
Pagliai & La Marca (1979)	n, l, optical	Italy	Sedimentary, vesicular
Chen et al. (1980)	l, s, n, SEM	Israel	Disruptional, sedimentary, skin seal, washed-out
Mücher et al. (1981)	l, optical	Belgium	Structural, sedimentary
Figueira & Stoops (1983)	n, l, q, optical	Argentina	Vesicular
Pagliai et al. (1983a)	c, q, SEM+	Italy	Disruptional, sedimentary, skin seal vesicular
Pagliai et al. (1983b)	c, q, optical	Italy	Sedimentary
Figueira (1984)	n, l, q, optical	Argentina	Vesicular
Onofiok & Singer (1984)	l, s, SEM	USA	Disruptional, skin seal, washed-in, washed-out
Tarchitzky et al. (1984)	l, s, n, SEM	Israel	Disruptional, skin seal
Collins et al. (1986)	l, optical	Ireland, Iraq	Skin seal, vesicular, washed-in, washed-out
Courty (1986)	n, c, optical	India	Coalescing, sedimentary, vesicular
Kooistra & Siderius (1986)	n, c, optical	Burkina Faso	Sedimentary, vesicular
Norton et al. (1986)	c, f, optical	USA	Disruptional, sedimentary, washed-out
Boiffin & Bresson (1987)	c, optical	France	Disruptional, sedimentary

Continued

Table 1 Overview of micromorphological studies of soil crusts — cont'd

References	Material[a] and method[b]	Location	Crust or microlayer type
Escadafal & Fedoroff (1987)	n, optical	Tunisia	Sedimentary, skin seal, vesicular
Fedoroff & Courty (1987)	n, optical	India, Niger	Sedimentary
Hall (1987)	c, optical	UK	Sedimentary
McFadden et al. (1987)	n, SEM	USA	Vesicular
Norton (1987)	l, optical	USA	Disruptional, skin seal, washed-in, washed-out
Norton & Schroeder (1987)	c, l, f, optical	USA	Disruptional, washed-out
Pagliai (1987)	c, q, optical	Italy	Sedimentary
Smillie et al. (1987)	c, l, SEM	Iraq, Ireland	Skin seal, washed-in, washed-out
Valentin & Ruiz Figueroa (1987)	c, f, optical	Ivory Coast	Disruptional, sedimentary, skin seal, vesicular
Arshad & Mermut (1988)	c, SEM+	Canada	Disruptional, sedimentary, skin seal
Mücher et al. (1988)	n, optical	Australia	Biological, sedimentary, vesicular, washed-in, washed-out
Southard et al. (1988)	l, SEM+	USA	Sedimentary
Chartres & Mücher (1989)	n, l, optical	Australia	Biological
Pagliai et al. (1989)	c, f, q, optical	Italy	Disruptional, sedimentary, vesicular
Remley & Bradford (1989)	l, SEM+	USA	Disruptional, skin seal
Bresson & Boiffin (1990)	c, f, s, optical	France	Coalescing, sedimentary, vesicular
Bresson & Valentin (1990)	c, g, f, s, optical	France, western Africa	Coalescing, disruptional, sedimentary, vesicular
Greene et al. (1990)	n, f, optical	Australia	Biological (lichens and others), vesicular
Le Bissonnais et al. (1990)	l, s, optical	France	Coalescing
Le Souder et al. (1990)	l, f, SEM	France	Coalescing, sedimentary, vesicular
Luk et al. (1990)	l, s, optical	China	Disruptional, sedimentary, skin seal, vesicular, washed-in, washed-out
Moore & Singer (1990)	l, s, M	USA	Disruptional, sedimentary
Poesen et al. (1990)	l, optical	Belgium	Vesicular washed-in, washed-out
West et al. (1990)	c, l, q, optical	USA	Disruptional, washed-out
Danin & Ganor (1991)	n, SEM	Israel	Biological (mosses)
Moss (1991a, 1991b)	l, optical	Australia	Disruptional, sedimentary, skin seal, vesicular, washed-in
Radcliffe et al. (1991)	c, f, optical	USA	Disruptional, sedimentary, washed-out
Sullivan & Koppi (1991)	n, optical	Australia	Vesicular
Valentin (1991)	n, SEM+	Niger	Coalescing
Casenave & Valentin (1992)	f, M	Western Africa	Coalescing, disruptional, erosional, sedimentary

Table 1   Overview of micromorphological studies of soil crusts — cont'd

References	Material[a] and method[b]	Location	Crust or microlayer type
Slattery & Bryan (1992)	l, s, optical	Canada	Disruptional
Tanaka & Kyuma (1992)	l, q, optical	Japan	N/A
Valentin & Bresson (1992)	n,c, SEM+ (literature review)	France, western Africa	Disruptional, sedimentary, skin seal, vesicular, washed-in, sieving, washed-out, biological
West et al. (1992)	SEM+ (literature review)	Worldwide review	Disruptional, skin seal, washed-in, washed-out, sedimentary
Belnap & Gardner (1993)	n, SEM	USA	Biological (cyanobacteria)
Bouza et al. (1993)	n, SEM+	Argentina	Vesicular
Le Bissonnais & Bruand (1993)	f, s, SEM	France	Disruptional, coalescence
Bresson & Valentin (1994)	SEM+ (literature review)	Worldwide review	Disruptional, washed-in, washed-out, sedimentary, coalescing, biological, vesicular, skin seal
Carnicelli et al. (1994)	c, q, optical	Zimbabwe	Disruptional, vesicular, sedimentary
Chiang et al. (1994)	c, s, optical	USA	Disruptional, washed-out, washed-in, skin seal
Eldridge & Greene (1994)	n, SEM	USA	Biological (lichens, cyanobacteria, mosses)
Greene & Ringrose-Voase (1994)	n, f, optical	Australia	Disruptional, vesicular
Humphreys (1994)	n, SEM+, X-ray radiography	Australia	Sedimentary
Kwaad & Mücher (1994)	c, s, q	The Netherlands	Coalescing, sedimentary
Pagliai (1994)	c, q (literature review)	Italy	Disruptional, sedimentary, vesicular
Singer et al. (1994)	l, SEM	South Africa	Washed-in
Slattery & Bryan (1994)	l, optical	Canada	Disruptional, sedimentary
Bielders & Baveye (1995)	l, SEM+	USA	Disruptional, washed-out, washed-in
Mermut et al. (1995)	l, s, optical	Belgium, Canada, China, USA	Disruptional, sedimentary, vesicular, washed-out
Reichert & Norton (1995)	l, q optical	Australia, Brazil, Puerto Rico, USA	Disruptional, skin seal, sedimentary
Sveistrup et al. (1995)	f, optical	Norway	N/A
Borselli et al. (1996)	f, q, optical	Zimbabwe	Structural
Eghbal et al. (1996)	c, f, s, optical	Iran	Disruptional, sedimentary
Garcia-Pichel & Belnap (1996)	n, SEM	USA	Biological (cyanobacteria)
Le Bissonnais (1996)	Literature review	France	Structural, sedimentary
McKenna Neuman et al. (1996)	l, SEM	Canada	Biological (algae, cyanobacteria)
Nicolau et al. (1996)	n, conceptual diagram	Spain	Disruptional, sieving
Hallaire et al. (1997)	n, q, optical	Burkina Faso	Sedimentary

Continued

Table 1   Overview of micromorphological studies of soil crusts — cont'd

References	Material[a] and method[b]	Location	Crust or microlayer type
Panini et al. (1997)	c, f, q, optical	Italy	Disruptional, sedimentary
Pérez (1997)	n, M	Venezuela	Biological, vesicular
Danin et al. (1998)	n, s, SEM, optical, XRD	Israel	Biological (mosses, lichens), vesicular
Eldridge (1998)	n, l, optical	Australia	Biological (lichens, bryophytes)
McFadden et al. (1998)	n, optical	USA	Vesicular
Bajracharya & Lal (1999)	c, l, s, optical	India	Disruptional
Clegg et al. (1999)	l, optical	Belgium	Coalescing, sedimentary
Malam Issa et al. (1999)	n, SEM+	Niger	Biological (cyanobacteria, chlorophytes), disruptional, sedimentary, sieving, vesicular
Puigdefábregas et al. (1999)	n, optical	Spain	Disruptional, sieving, washed-in, washed-out
Eldridge et al. (2000)	n, M	Israel	Biological (cyanobacteria, mosses, green algae, lichen), disruptional, sedimentary, vesicular
Usón & Poch (2000)	c, f, q, SEM	Spain	Disruptional, coalescing, sedimentary
Bresson et al. (2001)	f, c, s, optical	France	Coalescing, skin seal, vesicular
Malam Issa et al. (2001)	n, SEM+	Niger	Biological (cyanobacteria)
Noffke et al. (2001)	marine, SEM+		Marine microbial-induced sedimentary structures
Anderson et al. (2002)	n, optical	USA	Vesicular
Rousseva et al. (2002)	c, f, q, optical	Italy	Disruptional, sedimentary
Solé-Benet et al. (2002)	f, optical	Spain	Disruptional, sieving, washed-in, washed-out
Wakindiki & Ben-Hur (2002)	l, SEM	Israel, Kenya	Disruptional, skin seal, washed-in
Braissant et al. (2003)	l, SEM	Switzerland	Biological (bacteria)
Cantón et al. (2003)	n, SEM+	Spain	Biological (lichens), sedimentary, structural, vesicular
Hu et al. (2003)	n, SEM+	China	Biological (cyanobacteria, algae, moss)
Bresson & Moran (2004)	l, q, optical	Australia	Structural
Csotonyi & Addicott (2004)	n, l, morphological descriptions	Canada	Biological (mosses)
Fox et al. (2004)	l, q, SEM+	Canada	Structural, vesicular
Hoppert et al. (2004)	n, SEM+	Germany	Biological (algae, mosses, lichens)
Pagliai (2004)	Literature review, optical	Italy	Sedimentary
Pagliai et al. (2004)	c, f, q, optical	Italy	Sedimentary
Souza-Egipsy et al. (2004)	n, SEM	Spain	Biological (mosses, lichens, cyanobacteria)
Cousin et al. (2005)	l, s, q, SEM+	France	Structural
Trindade et al. (2005)	n, optical	Brazil	Biological
Belnap (2006)	n, SEM	USA	Biological (cyanobacteria)
Marsh et al. (2006)	n, optical	Canada	Biological (mosses, lichens, cyanobacteria)
Mees & Singer (2006)	n, SEM+	Uzbekistan	Vesicular

Table 1 Overview of micromorphological studies of soil crusts — cont'd

References	Material[a] and method[b]	Location	Crust or microlayer type
Zhang et al. (2006)	n, SEM+	China	Biological (cyanobacteria)
Blanco & Stoops (2007)	n, SEM+	Argentina	Paleosol crust
Escudero et al. (2007)	n, none	Spain	Biological (lichens)
Malam Issa et al. (2007)	l, SEM	South Africa	Biological (cyanobacteria)
Thomas & Dougill (2007)	n, SEM+	Africa	Biological (cyanobacteria)
Contreras et al. (2008)	n, optical	Spain	Disruptional, sieving, washed-in, washed-out
Ries & Hirt (2008)	n, f, optical	Spain	Disruptional, sedimentary, vesicular
Beraldi-Campesi et al. (2009)	n, M	USA	Biological (cyanobacteria)
Chen et al. (2009)	n, SEM+	China	Biological (mosses, lichens, cyanobacteria)
Fox et al. (2009)	g, q, optical	South Africa	Biological (algae, bryophytes, lichens, cyanobacteria), structural, vesicular
Malam Issa et al. (2009)	n, SEM+	Niger	Biological (cyanobacteria), vesicular
Yonovitz & Drohan (2009)	n, f, q SEM	USA	Vesicular
Fischer et al. (2010)	n, SEM+	Germany	Biological (cyanobacteria, algae, mosses)
Buck et al. (2011)	n, SEM	USA	Biological (cyanobacteria), vesicular,
Malam Issa et al. (2011)	n, optical	Niger	Biological (cyanobacteria)
Miralles-Mellado et al. (2011)	n, q, optical	Spain	Biological (lichens, cyanobacteria, mosses), structural
Williams (2011)	n, SEM+	USA	Biological (cyanobacteria, mosses, lichens), vesicular
Dietze et al. (2012)	n, f, optical	Mexico, USA	Vesicular
Fischer et al. (2012)	n, SEM	Germany, Israel	Biological (cyanobacteria, algae, mosses)
Lan et al. (2012)	n, optical	China	Biological (cyanobacteria, algae, lichens, mosses)
Williams et al. (2012)	n, SEM+	USA	Biological (mosses, lichens, cyanobacteria), vesicular
Badorreck et al. (2013)	f, q, XCT, neutron radiography	Germany	Sedimentary, structural, vesicular
Drahorad & Felix-Henningsen (2013)	n, optical	Israel	Biological (cyanobacteria)
Simpson et al. (2013)	n, optical	USA	Biological (modern and ancient)
Felde et al. (2014)	n, optical and XCMT, mercury intrusion porosimity	Israel	Biological (mosses, lichens, cyanobacteria)
Turk & Graham (2014)	n, XCT	USA	Vesicular
Zhao et al. (2015)	l, optical	USA	Structural

[a]n, natural soil; c, cultivated soil; l, laboratory experiments; f, field experiments; g, pasture soil; s, sequential study; q, quantitative study.

[b]optical, optical microscopy; SEM+, optical and scanning electron microscopy; SEM, only scanning electron microscopy; M, morphological descriptions or images; XRD, X-ray diffraction; XCT, X-ray computed tomography; XCMT, X-ray computed micro-tomography.

[c]Any report of vesicles is denoted as 'vesicular'.

caused physical surface crusts to develop on most soil types (Valentin & Janeau, 1989; Pagliai, 2004).

Soil seals and physical soil crusts have a strong negative influence on seedling emergence and water infiltration rates (Volk & Geyger, 1970; Sombroek, 1986; West et al., 1992; Bresson & Valentin, 1994; Pagliai, 2008). The decrease in water infiltration causes problems for irrigation, whereas increases in surface runoff lead to significant soil erosion (Moore & Singer, 1990). Several studies suggest that no-till systems (e.g., Pagliai et al., 1983a, 1989; Pagliai, 1987) and the use of soil conditioners and amendments may reduce crust formation by improving soil structure and aggregate stability, thereby increasing infiltration (Pagliai & La Marca, 1979; Pagliai et al., 1983b; Le Souder et al., 1990; Reichert & Norton, 1995). In such studies, soil micromorphology has enhanced our understanding of crust characteristics and management.

A first description and illustration of physical soil crusts in thin section, dealing with a sedimentary crust, was published by Kubiëna (1938, p. 196-197). The first broader micromorphological study of physical soil crusts was made in 1939 by Duley, which was followed only much later by other significant micromorphological studies, in the 1970s. More recently, research has expanded to include detailed conceptual models of crust development (e.g., Bresson & Valentin, 1990; Malam Issa et al., 1999; Ries & Hirt, 2008).

The three main morphological types of physical soil crusts, occurring individually or in combination, develop through both natural and anthropogenic processes. Structural crusts (Section 2.1) form through *in situ* modification of the soil surface by raindrop impact and by wetting and drying cycles. Vesicular surface horizons (Section 2.2) form when air is entrapped by advancing wetting fronts to form vesicular pores. Sedimentary crusts (Section 2.3) form through deposition of transported material. Although most crusts are strictly pedofeatures, some vesicular surface horizons and sedimentary crusts reach sufficient thickness to be considered soil horizons. Figure 1 illustrates the pathways that form common physical soil crust types.

Systematic descriptions of physical soil crusts are rare in the literature, and terminology varies greatly (Bresson & Valentin, 1994). The micromorphological data provided here, and outlined in Table 1, are based on the study and interpretation of published photographs and schematic drawings, with terminology drawn primarily from the reviews by West et al. (1992) and Bresson and Valentin (1994).

The morphology and characteristics of physical soil crusts vary as a function of soil and climatic conditions. Crusts commonly range from less than 1 mm up to 5 cm in thickness and are more compact, hard and brittle than the underlying soil and also have a different porosity, in terms of void size, number and arrangement (Valentin & Ruiz Figueroa, 1987; Bresson & Boiffin, 1990; West et al., 1992; Bresson & Valentin, 1994; Pagliai, 2008). Particle size, soil wetting, surface water flow and runoff, soil moisture, organic matter content, pore water chemistry and rainfall kinetic energy largely control the morphology, strength, structure, bulk density, infiltration rate and permeability of

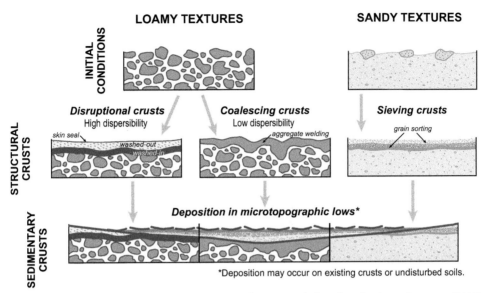

FIGURE 1 Conceptual model, adapted after the work of West et al. (1992), Valentin and Bresson (1992) and Anderson et al. (2002), illustrating the primary pathways that form physical soil crusts under various textural and environmental conditions.

physical soil crusts (Tackett & Pearson, 1965; Gupta & Larson, 1979; Valentin, 1986; Arshad & Mermut, 1988; Remley & Bradford, 1989; Greene & Ringrose-Voase, 1994).

2.1   Structural Crusts

Structural crusts are categorised into four types: (i) disruptional crusts, which form when aggregates are disrupted or disintegrate upon raindrop impact (Figs. 2 and 3) or when aggregates slake because of compression of entrapped air (Le Bissonnais, 1996), resulting in particle rearrangement (West et al., 1992); (ii) coalescing crusts (or aggradational crusts), which develop when well-separated aggregates of moist soil material coalesce under raindrop impact, whereby the aggregates are welded to form a single apedal layer (Bresson & Boiffin, 1990; Kwaad & Mücher, 1994; Bresson & Moran, 2004); (iii) erosional crusts, which are thin layers of non-oriented fine-grained material whose formation is related to erosion (Bresson & Valentin, 1994); and (iv) sieving crusts, which are composed of a coarse-grained surface layer overlying a fine-grained layer that results from winnowing or sieving processes initiated by raindrop impact (Bresson & Valentin, 1994). The mechanisms that form structural crusts, as well as differences in morphology, are strongly tied to soil surface conditions and rainfall characteristics (Bresson & Moran, 2004), with disruptional crusts forming on dry soils in summer and coalescing crusts forming on wet soils during winter (Le Bissonnais & Bruand, 1993).

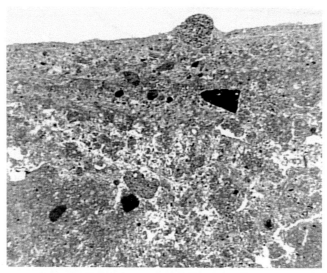

FIGURE 2 Structural crust on the surface layer of a conventionally ploughed loamy soil, subjected to the impact of raindrops (Italy). Soil aggregates, visible in the lower part, gradually disintegrate towards the surface, with loss of porosity (PPL, frame width 3 cm).

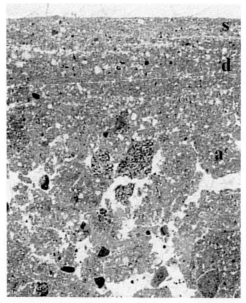

FIGURE 3 Disruptional layer (d) showing gradual collapse of the original aggregates (a), covered by a sedimentary layer (s) with narrow elongated pores parallel to the soil surface, in a conventionally ploughed loamy soil (Italy) (PPL, frame width 3 cm).

2.1.1 Disruptional Crusts

Disruptional crusts consist of four microlayers, which occur individually or in combination (Figs. 1 to 3): the disruptional microlayer, the skin seal, the washed-out microlayer, and the washed-in microlayer (McIntyre, 1958; Arshad & Mermut, 1988; West et al., 1992; Bresson & Valentin, 1994; Chiang et al., 1994). A model for the development of disruptional seals and crusts was initially proposed by McIntyre (1958), and the genesis and stages of formation of these crusts have been further discussed by other authors (e.g., Farres, 1978; Bresson & Boiffin, 1990; Chen et al., 1980; Tarchitzky et al., 1984; Collins et al., 1986; Luk et al., 1990; West et al., 1992; Bresson & Valentin, 1994; Chiang et al., 1994; Mermut et al., 1995; Bajracharya & Lal, 1999). In the model of McIntyre (1958), slaking drives the formation of the microlayers of disruptional crusts, which occur on bare, dry soils (Bresson & Valentin, 1994). This model of development involves (i) slaking, or disintegration of cohesive aggregates, resulting from compression of entrapped air or dispersion of clay (disruptional microlayer) (Bresson &Valentin, 1994); (ii) formation of a surface layer of coarse material with most of the fine material removed (washed-out layer); (iii) accumulation of fine material beneath that surface layer (washed-in layer), plugging available voids and hence reducing water movement; and (iv) development of a thin layer (c. 100 μm) of fine particles along the soil surface (skin seal; McIntyre, 1958), attributed to aggregate breakdown and compaction by raindrop impact (Chen et al., 1980; Tarchitzky et al., 1984) or to settling of fine particles from suspension at the end of the rainfall event (Pagliai et al., 1983b; Onofiok & Singer, 1984; Norton, 1987; Arshad & Mermut, 1988).

2.1.1.1 Disruptional Microlayer

The disruptional microlayer forms through slaking, due to compression of entrapped air or dispersion of clay, producing a dense layer at the soil surface. This microlayer is 100 μm to 6 mm thick (Arshad & Mermut, 1988) and shows no internal layering (Bresson & Valentin, 1994). The disruptional layer has a smaller aggregate size than the underlying soil, and it has a porosity that is 30 to 60% lower (Chiang et al., 1994).

In strongly aggregated soils, the disruptional microlayer forms slowly and is initially recognised by the presence of newly formed small aggregates that bind the original larger aggregates together. In weakly aggregated soils, detached micromass material quickly fills interpedal voids (Cousin et al., 2005). In early stages of formation, particles are oriented parallel to the surface (Luk et al., 1990). In sandy soil materials, coatings of fine particles on coarse grains are quickly removed, leaving uncoated grains (Chen et al., 1980; Tarchitzky et al., 1984). For granite-derived soils, Moss (1991a, 1991b) observed the formation of a silt layer on the soil surface, overlying a compacted layer. Other authors describe small depressions resulting from raindrop impact (e.g., Slattery & Bryan, 1994; Zhao et al., 2015), or small protrusions (Slattery & Bryan, 1992).

The c/f-related distribution pattern and the b-fabric of the disruptional microlayer depend on the soil material affected. The m1 microhorizon of Bresson and Boiffin (1990) is a discontinuous disruptional microlayer showing moderate to weak ped separation, whereas the overlying m2 microhorizon corresponds to a more continuous, apedal microlayer, and the m3 microhorizon is a local sedimentary surface.

The susceptibility of soils to slaking and formation of the disruptional microlayer are determined by soil texture, organic matter content, mineralogical composition of the clay fraction, cation saturation, and Fe and Al hydroxide content (see Le Bissonnais, 1996; Amézketa, 1999). Medium-textured soils with <20% clay are susceptible to slaking, and swelling clays such as smectite are more prone to crust formation than kaolinitic materials (e.g., Mermut et al., 1995). Clay dispersion is favoured by high exchangeable Na contents, whereas Fe and Al hydroxides have a stabilising effect (see Le Bissonnais, 1996).

2.1.1.2   Skin Seal

Skin seals, formed by raindrop-induced compaction or fine particle deposition (Chen et al., 1980; Onofiok & Singer, 1984), are very thin surface coatings composed of fine material (clay, silt, fine sand), with strial b-fabric. Skin seals vary in thickness from less than 10 μm (Luk et al., 1990) or 50 μm (Arshad & Mermut, 1988) to 500 μm (Chiang et al., 1994). In studies by Pagliai et al. (1983a), the skin seal is only a few micrometres thick and can directly overlie a relatively pure silt layer. McIntyre (1958) described the skin seal as having no visible pores, but SEM analysis shows many small (<5 μm wide) parallel planar voids (Arshad & Mermut, 1988). Seals are commonly local features, being confined to interridge depressions adjacent to microtopographic ridges. However, the extent of skin seals may increase or become more continuous as a rainfall event continues (Bresson et al., 2001).

2.1.1.3   Washed-Out Microlayer

The washed-out layer (Onofiok & Singer, 1984) is a thin, micromass-depleted hypocoating along the soil surface. This layer forms only after prolonged rainfall (Luk et al., 1990). It ranges from less than 1 to 5 mm in thickness, depending on surface topography, duration of rain impact and aggregate stability (Chiang et al., 1994), and it commonly displays a coarse monic c/f-related distribution pattern. The fine material that is removed in suspension from the washed-out layer is commonly thought to be translocated into the underlying washed-in microlayer (McIntyre, 1958; Poesen, 1981; Onofiok & Singer, 1984; Luk et al., 1990), but lateral evacuation of fine material (West et al., 1990, 1992; Radcliffe et al., 1991; Mermut et al., 1995; Clegg et al., 1999) and local surface deposition of coarse material (West et al., 1990) have also been suggested. Vertical translocation is observed when the grain size or sorting is different from that of the underlying soil. Sieving crusts (see Section 2.1.4) can be considered to be a type of washed-out layer (Bresson & Valentin, 1994).

2.1.1.4   Washed-In Microlayer

The washed-in microlayer lies below the washed-out layer. It is a thin layer in which fine material, in the form of particles or aggregates, has been washed in and accumulates, plugging the voids. Clay illuviation has been suggested as a mechanism for this process, but Bresson and Valentin (1994) contend that evidence for clay translocation has not been clearly demonstrated.

The morphology of the washed-in layer varies with soil texture. In sandy material, the washed-in layer starts as a discontinuous band of clay (Bielders & Baveye, 1995), which is morphologically similar to a sieving crust (see Section 2.1.4). In loess, the deposits of translocated clay have a speckled orientation pattern (Luk et al., 1990) and occur at a depth of 1 mm. Removal of the overlying washed-out layer can transform the washed-in layer to a surface seal.

2.1.2 Coalescing Crusts

Coalescing crusts, also called aggradational crusts, form by gradual welding of aggregates upon wetting of moist soil material (Bresson & Boiffin, 1990; Kwaad & Mücher, 1994; Clegg et al., 1999). Under moist soil conditions, little slaking occurs (Le Bissonnais et al., 1990; Kwaad & Mücher, 1994). As crust development proceeds, smaller aggregates disappear first. Eventually, packing voids are transformed to vughs, and planar voids may appear. The newly developed microstructure of the affected surface interval, with a diffuse lower boundary, is much denser than the original structure (Bresson & Boiffin, 1990). A special form of coalescing crusts, known as a 'swelling crust' (Valentin, 1991) forms because of swelling of clay. A similar process, known as slumping, occurs when overburden pressure causes structural collapse that welds soil aggregates and results in the formation of hard-setting seedbeds (Bresson & Moran, 2004).

2.1.3 Erosional Crusts

Erosional crusts, which occur primarily on arid rangelands, are thin fine-grained layers with poor grain orientation. These crusts form on the highest points of irregular soil surfaces and extend across lower parts (Valentin, 1986). Such crusts are thought to form by selective removal of coarse grains in conditions with strong raindrop impact (Chen et al., 1980; Valentin, 1991). They can also form by removal, through erosion, of the coarse-grained surface layer of disruptional or sieving crusts, exposing the underlying fine-grained layer (Valentin, 1986). Morphological similarities with biological soil crusts (e.g., filament knobs and filament sheets) suggest that some occurrences may be formed by biological processes (Williams et al., 2012).

2.1.4 Sieving Crusts

Sieving crusts, which are most commonly observed for sandy soils of arid and semiarid tropical regions, form when rainfall causes vertical sorting of soil material, resulting in the development of a succession of three layers (Valentin & Bresson, 1992): (i) an upper layer of loose coarse sand or fine gravel, (ii) a middle fine-grained layer of densely packed grains, with vesicles, and (iii) a lower layer with a high concentration of densely packed fine particles, resulting in reduced porosity (Valentin & Bresson, 1992; Nicolau et al., 1996; Puigdefábregas et al., 1999; Solé-Benet et al., 2002; Contreras et al., 2008). The depth of the fine layer seems to be controlled by mechanical energy (raindrop momentum) rather than by physicochemical properties, as no such sorting occurs when the material is protected from drop impact (Bielders & Baveye, 1995). Desert pavements, which are common in arid regions and form in combination with vesicular horizons (Section 2.2), are considered to be a special case of sieving crust (Springer, 1958).

2.2 Vesicular Surface Horizons

Vesicular surface horizons are characterised by dense concentrations of vesicular pores (Figs. 4 and 5). They were first discussed by Springer (1958), who proposed the horizon symbol A_V. Soils with vesicular surface horizons have also been called 'Schaumböden'

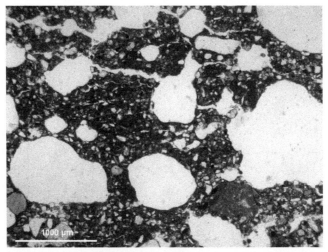

FIGURE 4 Vesicular surface horizon in calcite-rich material (Negev, Israel). Horizontal voids parallel to the surface interconnect some of the vesicles (PPL). *Ghent University archive.*

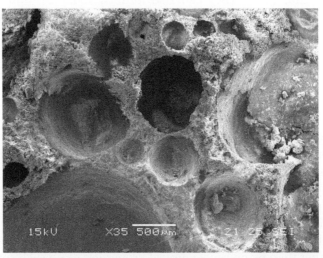

FIGURE 5 SEM image of a vesicular surface horizon underneath desert pavement (Death Valley, California, USA). Note clay coating on vesicle walls. *Image by B. Buck.*

(foam soils) (Volk & Geyger, 1970). Although vesicular surface horizons are most common in arid and semiarid climates (Turk & Graham, 2011), they are also observed within humid regions, alpine areas and arctic zones (Volk & Geyger, 1970; Badorreck et al., 2013; see also Van Vliet-Lanoë & Fox, 2018). They are commonly observed in association with biological soil crusts (Williams et al., 2012), salt crusts (Mees & Singer, 2006), or in soils with furrow irrigation (Miller, 1971). Vesicular horizons may be up to several centimetres thick below desert pavements (Turk & Graham, 2011). Vesicular surface horizons are distinct from vesicular layers that display similar micromorphological characteristics but occur in deeper parts of soil profiles, as observed for paddy soils (G. Stoops, unpublished data).

The mechanisms forming vesicular surface horizons have been investigated experimentally. Evenari et al. (1974) suggest that vesicle formation is the result of expanding air trapped by surface crusts or desert pavements (reg soils). However, studies with simulated desert pavements indicate that vesicles do not form directly under rocks (Poesen et al., 1990; Dietze et al., 2012). A comprehensive experimental study by Dietze et al. (2012) identified three mechanisms that lead to increasing gas pressures within the soil matrix, resulting in vesicular pore formation: (i) downward migration of the wetting front, (ii) surface sealing by fine materials, and (iii) water pooling above the surface seal. Pore development appears to be influenced by the number of wetting cycles, as artificial vesicular crusts may be obtained after 5-10 simulated wetting and drying cycles (Figueira & Stoops, 1983; Figueira, 1984) and the general appearance of the natural vesicular surface horizon is reached after 20-30 cycles (Figueira & Stoops, 1983). These studies and others (Miller, 1971) have demonstrated that with increasing number of cycles, the number of pores decreases but vesicle size increases. Following the disturbance of vesicular surface crusts, the vesicular pore structure can be reestablished in natural soils within one year (Yonovitz & Drohan, 2009).

Vesicles are by definition the dominant void type (Figs. 4 and 5) in vesicular surface horizons. The vesicles are commonly arranged in parallel horizontal bands (Brewer, 1964). Vesicle diameters range from 25 μm to 8 mm, with predominantly small pores (25-500 μm) in natural soils (Volk & Geyger, 1970; Escadafal & Fedoroff, 1987). Pore density ranges from 300 to 500 vesicles/cm^2 in vertical cross-sections, with a pore volume of up to 40% (Volk & Geyger, 1970), to more than 600 vesicles/cm^2 and a total pore volume of 31% (Figueira & Stoops, 1983). For vesicular horizons below desert pavements, porosities of up to 50% have been reported (Escadafal & Fedoroff, 1987).

Vesicle formation is dynamic, with vesicles being destroyed and reformed, depending on water saturation (Luk et al., 1990) and root activity (Dietze et al., 2012). Turk and Graham (2014) proposed a conceptual model of pore evolution in vesicular horizons, starting with equant vesicles, which evolve to non-equant vesicles, which in turn can merge into individual vughs or connected vughs, which will finally collapse to form planar voids. These observations are supported by other studies demonstrating that larger vesicles commonly occur in the upper 1-2 cm of the surface horizon, and smaller pores occur in the lower 3-5 cm (Bouza et al., 1993), where they grade to planar voids

(Anderson et al., 2002) or vughs (Volk & Geyger, 1970; Figueira, 1984; Collins et al., 1986). Similar phenomena occur within sedimentary crusts, where vesicles (0.2-3.0 mm) collapse upon drying, giving rise to vughs (Kooistra & Siderius, 1986).

Vesicular surface horizons form in silt-rich material, and while clay contents can be very low, high sand contents are common (Volk & Geyger, 1970). The c/f-related distribution pattern of vesicular surface horizons is porphyric (Figueira & Stoops, 1983; Valentin & Ruiz Figueroa, 1987; Luk et al., 1990; Bouza et al., 1993). A porostriated b-fabric was described by Bouza et al. (1993). In arid soils, a calcitic crystallitic b-fabric can occur, and strong calcite impregnation may affect part of the horizon (Collins et al., 1986). A thin, dense, often laminar crust of fine material with a strial or calcitic crystallitic b-fabric commonly occurs at the surface (e.g., Anderson et al., 2002). Vesicle walls are commonly coated with parallel-oriented, speckled clay (Volk & Geyger, 1970; Figueira & Stoops, 1983; Fedoroff & Courty, 1987), which may be most apparent at the base of the vesicle (Bouza et al., 1993). Textural coatings covering the sides of vesicles and interconnecting fissures are interpreted to have formed between air bubble and void wall in wet conditions (Sullivan & Koppi, 1991). In some cases, grain size of the groundmass increases from the vesicle wall to the midpoint between neighbouring vesicles (Collins et al., 1986). In addition to clay coatings, thin calcite coatings have also been reported (Evenari et al., 1974; Collins et al., 1986).

2.3 Sedimentary Crusts

Sedimentary crusts (Figs. 6 to 10) form by lateral transport and subsequent deposition of fine particles along the soil surface. Sedimentary crusts have also been called

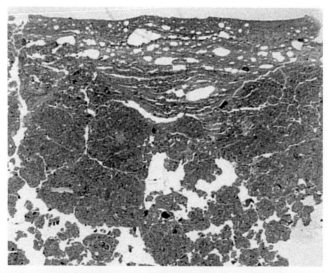

FIGURE 6 Sedimentary crust containing elongated vesicles superposed on disruptional crust, in a conventionally ploughed loam soil (Italy) (PPL, frame width 3 cm).

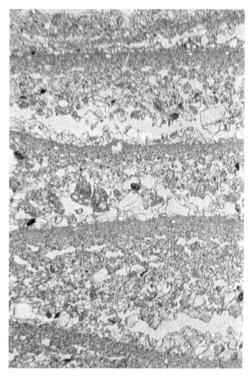

FIGURE 7 Complex sedimentary crust formed by several sedimentary microlayers, each showing typical particle size sorting, in a cultivated sandy loam soil (Italy) (PPL, frame width 3 mm).

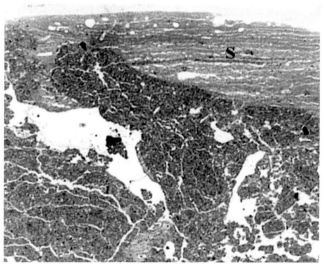

FIGURE 8 Thin laminated sedimentary crust (s) with platy microstructure, in a conventionally ploughed loamy soil (Italy) (PPL, frame width 3 cm).

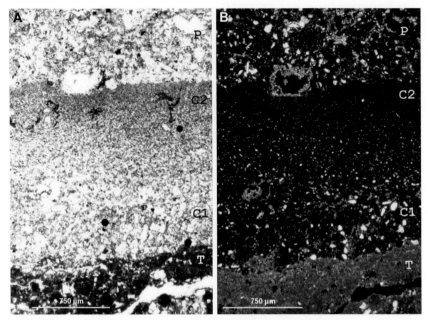

FIGURE 9 Buried sedimentary crust along the surface of an eroded calcic horizon (Tosca, T), covered by Pampa loess (P). The crust has a coarse texture at the bottom (C1), gradually becoming finer towards the top (C2), where it mainly consists of phytoliths (optically isotropic) (Argentina). (A) In PPL. (B) In XPL. *Images from Ghent University archive; sample collected by S. Pazos.*

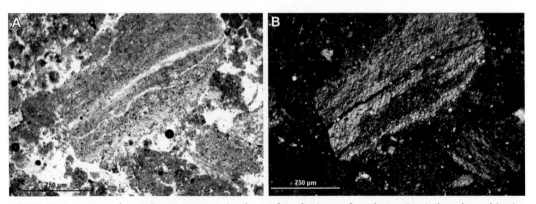

FIGURE 10 Fragment of a sedimentary crust in the surface horizon of a clayey Gypsisol under cultivation (Ecuador). (A) In PPL. (B) In XPL, showing unistrial b-fabric.

depositional crusts, but the former term will be used to avoid confusion, as suggested by Arshad and Mermut (1988) and West et al. (1992), because washed-in layers have also been termed 'depositional crusts'. Sedimentary crusts form in a variety of environments, where they can be local features (e.g., in furrows and rill channels) or widespread deposits (e.g., base of sheet erosion-affected slopes, irrigation basins) (Evans & Buol, 1968; Bishay & Stoops, 1975). In cultivated soils, sedimentary seals are very common and may occur in microlows or overlie disruptional crusts (Panini et al., 1997) and commonly consist of small stable aggregates (Falayi & Bouma, 1975). A special type of anthropogenic sedimentary crust is formed when liquid organic manure slurry is spread on the soil surface (Sveistrup et al., 1995).

Sedimentary crusts form when the fine soil particles, derived from disintegration of soil aggregates, are transported and deposited at a certain distance from the source to form a crust with a different composition and structure than the underlying soil (Chen et al., 1980) (Fig. 6). Deposition may occur on a previously formed crust, producing a layer with a different texture (Pagliai & La Marca, 1979). Salts can play an important role in depositional crust formation in clayey soils, as clays may be dispersed in moderately saline conditions, followed by deposition upon drying (Labib et al., 1975; Tayel et al., 1975). The thickness of these sedimentary crusts generally ranges from a few millimetres to a few centimetres, with the thickest crusts being composed of several layers (e.g., Kooistra & Siderius, 1986; Boiffin & Bresson, 1987). These layers can be sorted or laminar (Bresson & Valentin, 1994) (Figs. 7 and 8). Drain fillings often have the same micromorphological characteristics as sedimentary seals (Hall, 1987), including a banded distribution of the groundmass (Herrero Isern et al., 1989).

Sedimentary crusts on sandy soils consist of finer-grained layers (silt or clay) that include few pores that are larger than a few micrometres, alternating with coarse-grained layers with greater total porosity. Isolated larger voids may occur randomly within the fine-grained layers. With the exception of packing voids and surface-parallel planar voids, few voids occur within the coarse-textured layers, but vesicles can be present in association with surface crusting (Pagliai et al., 1989) or concentrate below clay seals (Courty, 1986).

Sedimentary crusts are characterised by a clear banded distribution parallel to the surface, with alternating clayey and more coarse-grained layers. In many crusts, with variable textures, some layers show a clear fining-upward textural gradient (Southard et al., 1988) (Fig. 9). The thickness of individual laminae ranges from 3 to 9 mm (Valentin & Ruiz Figueroa, 1987), and layers may be irregular because of small-scale erosion (Fig. 10). Clay particles and coarser plate-shaped grains are commonly oriented parallel to the surface (Evans & Buol, 1968). Parallel orientation of fine platy particles, with negligible porosity, is also observed in SEM studies (Arshad & Mermut, 1988). Laboratory experiments (Southard et al., 1988) indicate that the strongest parallel orientation occurs during sedimentation in distilled water, whereas phosphogypsum addition provokes clay flocculation, resulting in a more random orientation and a speckled b-fabric. Material suspended in pure water can penetrate up to 5 mm in the underlying layer, where it forms deposits with strong parallel orientation (Southard et al., 1988). In the same study, material suspended in water with a high electrolyte content penetrates to less than 0.5 mm. Below sedimentary crusts on sandy materials, textural coatings may form in horizontal planar voids (Fedoroff & Courty, 1987).

Poorly sorted laminae form by rainsplash and overland flow (Clegg et al., 1999), whereas well-sorted laminae result from afterflow (Mücher & De Ploey, 1977; Mücher et al., 1981). Sedimentary deposits in microlows clearly show wedging of the composing layers along the walls of the depression (Kooistra & Siderius, 1986). Multiple layers of foreset laminae of coarse silt to fine sand are thought to be overwash deposits (Kooistra & Siderius, 1986), which display an angle of 60 to 70 degrees and may be several centimetres thick.

The c/f-related distribution pattern of sedimentary crusts ranges from fine monic in the clayey parts, over porphyric to coarse monic in sandy parts, where it can also locally be enaulic or gefuric (Hall, 1987). In cultivated soils, tissue residues and charcoal fragments are common. In the fine-grained upper part of each layer, the commonly observed strial b-fabric (e.g., Southard et al., 1988) may be masked by a calcitic crystallitic b-fabric (Bishay & Stoops, 1975).

3. Biological Soil Crusts

BSCs are complexes of lichens, cyanobacteria, algae or mosses that bind and enmesh soil grains. BSCs are commonly observed in arid and semiarid regions, from hot sub-equatorial zones to the cold deserts of polar zones (Belnap et al., 2001a). They occur where soils are exposed or unvegetated and relatively stable and undisturbed. BSCs are thought to have a net positive impact on surface soils, by acting as a surface membrane that regulates the flow of water, gases and solutes across soils (Belnap et al., 2003), reducing erosion (McKenna Neuman et al., 1996; Miralles-Mellado et al., 2011; Rodríguez-Caballero et al., 2014), modifying water and energy balances (Belnap, 2006; Thomas et al., 2011; Rodríguez-Caballero et al., 2012; Couradeau et al., 2016), improving soil fertility (Kleiner & Harper, 1977; Evans & Belnap, 1999; Delgado-Baquerizo et al., 2016) and influencing plant community establishment (Li et al., 2005; Escudero et al., 2007; Hernandez & Sandquist, 2011; Lan et al., 2013). BSCs are extremely fragile and sensitive to disturbances such as off-road vehicles, hiking and grazing (Belnap, 1998) and are estimated to take decades to thousands of years to recover from disruption (Belnap & Warren, 2002; Williams et al., 2012).

Some of the earliest descriptions of BSCs were published in the 1960s and 1970s (e.g., Cameron & Blank, 1966; Friedmann & Galun, 1974). The primary focus of most early BSC studies was the biological taxonomy of component organisms and their ecological roles. Only recently have their biosedimentary structures been subject to micromorphological analysis (e.g., Campbell, 1979; Greene et al., 1990; Danin & Ganor, 1991; Belnap, 2001; Malam Issa et al., 2001; Hoppert et al., 2004; Miralles-Mellado et al., 2011; Lan et al., 2012; Williams et al., 2012; Felde et al., 2014).

Several soil crust classification systems have been proposed for BSCs, either characterising the morphology of the component organisms (e.g., Eldridge & Greene, 1994) and/or emphasising changes in soil morphological features during crust development. Belnap et al. (2001a, 2001b) developed a classification system based on climate-driven differences in crust surface microtopography, including flat and rugose crusts of hot deserts and rolling and pinnacled crusts of cool deserts. Other authors have proposed successional models that characterise changes in crust morphology and microtopography with time (e.g., Danin et al., 1998; Thomas & Dougill, 2007; Lan et al., 2012) or level of development (Belnap et al., 2008). The following discussion is based on the work of Williams et al. (2012), which integrates conclusions from these previous morphological studies, extended based on new observations.

3.1 Macromorphological Features

Several studies have identified two dominant zones within most BSC types, which are visible to the naked eye (Malam Issa et al., 2009; Miralles-Mellado et al., 2011; Fischer et al., 2012; Williams et al., 2012; Felde et al., 2014): (i) a highly cohesive surface layer (0.5 to 20 mm thick) with great abundance of biological components, capped by biological structures ('bio-rich zone'; Williams et al., 2012), and (ii) a lower non-cohesive layer composed of poorly consolidated and pedologically unstructured grains, with lower abundance of organic material ('bio-poor zone'; Williams et al., 2012). A planar void that is visible with a hand lens commonly separates both zones (Williams et al., 2012; Felde et al., 2014) (Fig. 11).

In most BSC successional models, crust formation commences as a layer dominated by cyanobacteria. If conditions are favourable, these develop into crusts composed of lichens and short mosses (short-moss-lichen crusts), eventually succeeded by pinnacled crusts consisting of lichens and tall mosses (tall-moss-lichen crusts). The latter may include vesicular pores (within the pinnacles) and vesicular horizons (beneath the pinnacles) (e.g., Williams et al., 2012). In the field, cyanobacteria-dominated crusts are characterised by a relatively smooth surface, pale colour, recognisable filamentous cyanobacteria. Short-moss-lichen crusts are recognisably dominated by lichens and mosses, with ≤2 cm of vertical relief and a rolling surface morphology. Tall-moss-lichen pinnacled crusts form small mounds (up to 5 cm high), which are commonly separated by polygonal surface cracks. The upper interval with high concentration of biological

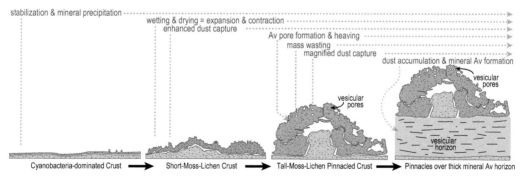

FIGURE 11 Conceptual model of biological soil crust development and succession, adapted from Williams et al. (2012). Processes (dark blue) proceed through time concurrent with changes in crust species composition and morphological features, including evolution of surface layer with great abundance of biological components (brown) and lower layer with lower abundance of organic material (yellow).

components is up to 1.5 mm thick in cyanobacteria-dominated crusts, 10 mm thick in short-moss-lichen crusts, and 20 mm thick in tall-moss-lichen crusts.

3.2   Micromorphological Features

Figure 11 illustrates the genetic processes that drive the formation of the three BSC types (Williams et al., 2012). In thin section, BSC components are mineral grains (predominantly clay to medium sand), biotic structures (filaments, lichens, plant roots and mosses) and intricate voids. As the component organisms change and formative processes become more complex, organic and mineral material accumulates, micro-topographic relief is enhanced, and increasingly intricate biosedimentary structures are formed (Williams et al., 2012). Precipitation and stabilisation of authigenic minerals dominate at the stage of cyanobacteria-dominated crust formation and continue throughout the succession. Expansion-contraction (caused by wetting-drying) and dust capture processes become more pronounced through the formation of short-moss-lichen crusts. The mentioned processes become more intense during formation of the tall-moss-lichen crusts, during which mass-wasting and vesicular pore development also take place.

3.2.1   Stabilisation and Authigenic Mineral Precipitation Features

Filamentous cyanobacteria, such as *Microcoleus*, play a primary role in stabilisation by weaving around fine sands and by trapping clay, silt and very fine sand grains that stick to their mucilaginous sheaths (Campbell, 1979; Belnap & Gardner, 1993; Zhang et al., 2006; Thomas & Dougill, 2007). Authigenic minerals (primarily carbonates) are precipitated as the sheath material of the cyanobacteria is mineralised (Campbell, 1979; Braissant et al., 2003; Souza-Egipsy et al., 2004), which results in cementation of the crust and further stabilises the soil surface (Williams et al., 2012).

Stabilisation processes and authigenic mineral precipitation occur in all crust types and form two main types of features. One type is filament sheets (up to 300 µm thick), which are horizontal surface structures consisting of compacted clays to very fine sands, enmeshed in likely cyanobacterial filaments, with variations in grain size between successive layers (Thomas & Dougill, 2007; Williams et al., 2012; Felde et al., 2014). Another type is filament knobs, which are surface protrusions with convex summits, consisting of clay to sand, and up to 2 mm wide and 4 mm tall, which form when the distribution of cyanobacteria is not uniform (Williams et al., 2012). The tallest knobs contain poorly developed to well-developed vesicular pores. The shortest knobs contain planar voids, aligned parallel to the knob summits. These voids are 1.5 mm wide and 0.25 mm thick and occur approximately 1 mm below the surface. Sharp protrusions formed by silt- and clay-coated filaments are commonly observed along the top of the knobs.

FIGURE 12 Moss crust on a sandy soil (inner Mongolia), mainly composed of mosses (m), which trap transported sand grains (PPL). *Thin section provided by Zhou Hu.*

3.2.2   Dust Capture and Shrink-Swell Features

Dust capture potential and heaving processes increase throughout the BSC succession, as component organisms change and surface roughness increases. Cyanobacteria-dominated crusts, with smooth surfaces, capture only relatively small amounts of dust, by trapping particles on mucilaginous filament sheaths (Campbell, 1979; Belnap & Gardner, 1993; Chen et al., 2009; Lan et al., 2012), whereas establishment of mosses (Fig. 12) and lichens increases dust capture potential by forming short-moss-lichen crusts with rolling surface topography and eventually tall-moss-lichen crusts with a more irregular surfaces and greater microtopographic relief (Williams et al., 2012) (Fig. 13).

Wetting-drying cycles cause dust capture to fluctuate through time and, in combination with shrinking-swelling of organic tissues, result in differential heaving that causes development of surface irregularities. Cracks become filled with dust during dry stages and close again by groundmass expansion during wet stages (Williams, 2011), eventually forming expansion-contraction features similar to those of Vertisols (Williams et al., 2012). Various unique biosedimentary features can also develop, including organic sheets that become partially or completely detached from the soil surface by curling upwards ('upturned features', 'rafts'; Williams et al., 2012), commonly with some attached mineral grains (Wynn-Williams, 1990; Williams et al., 2012). Other features are protrusions composed mainly of clay and silt that developed within or beneath lichen rhizine structures ('pedestals', 'towers'; Williams et al., 2012) (Fig. 14). In cool

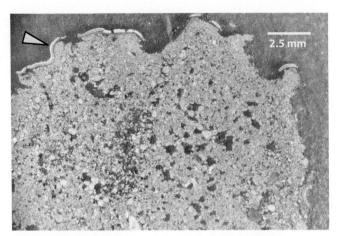

FIGURE 13 Tall-moss-lichen pinnacled crust with vesicular pores and discontinuously covered by lichen squamules (*arrow*) (surface biological (lichen) crust, Nevada) (PPL). *Collection of A.J. Williams.*

FIGURE 14 Surface protusion covered by a lichen squamule (l) and cemented by micritic carbonates. The structure contains a long, thick rhizine (r). Authigenic mineral precipitates are embedded within exudates emanating from lichen squamules (*arrow*) (surface biological (lichen) pinnacled crust, Nevada, USA) (XPL). *Collection of A.J. Williams, archive of the University of Nevada, Las Vegas.*

environments, considerable surface roughness may also develop when hoof prints are made by large mammals (Csotonyi & Addicott, 2004) or when the soil is affected by frost heave (Belnap, 2001; Marsh et al., 2006).

3.2.3 Destabilization and Mass-Wasting Features

Although several features of developing tall-moss-lichen crusts increase their stability and reduce erosion, sustained buildup leads to microscale instability and mass-wasting (Williams et al., 2012). Processes include toppling, collapse, creep, sloughing, differential erosion and dissolution of authigenic minerals. The resulting features include narrow columnar protrusions, overhanging curved extensions of surface-parallel layers and bridge-like structures extending along the top of large voids ('hoodoos', 'curved features', 'biosediment bridges'; Williams et al., 2012).

3.2.4 Voids

BSC formation leads to the development of a variety of void morphologies, including irregular planar voids (up to 5 mm wide), large vughs (to 24 by 36 mm) that can be partly lined by compacted clays to very fine sands containing organic remains and horizontal to vertical planar voids (0.5 to 1 mm wide) that are parallel to biotic structures (Williams et al., 2012).

Vesicles (0.1-0.8 mm) develop when a combination of organic matter accumulation, physical crusting and authigenic mineral precipitation seals the surface of the crust, favouring entrapment of air by the wetting front to form vesicles (Williams et al., 2012). A variety of studies have reported vesicles in association with BSCs (e.g., Danin et al., 1998; Marsh et al., 2006; Miralles-Mellado et al., 2011; Felde et al., 2014).

3.3 Implications of Biological Soil Crusts for Moisture Conditions

Micromorphology has been used to study the impact of BSC properties on soil moisture behaviour because the surface and internal structure of crusts control infiltration and runoff (Miralles-Mellado et al., 2011; Felde et al., 2014). The complex biotic structures and intricate void systems, with the greatest complexity being observed for tall-moss-lichen crusts, influence infiltration rates as well as moisture retention (Williams et al., 2012).

Other studies applied experimental micromorphological approaches to explore water infiltration properties of BSCs. Miralles-Mellado et al. (2011) combined infiltration measurements and micromorphological analyses to demonstrate that physical soil crusts and incipient BSCs had the fewest pores and lowest meso-macroporosity and that these crusts also had the lowest infiltration coefficients and greatest erosion potential. BSCs with lichens and cyanobacteria displayed the greatest meso- to macroporosity, but infiltration rates remained low for these crusts because they are partially detached from the underlying soil. In another study, Felde et al. (2014) used micromorphological observations, as well as mercury intrusion porosimetry and X-ray computed tomography, to demonstrate that porosity and pore size increase from cyanobacterial to lichen and short-moss crusts and that pore morphology changed from tortuous to straight, facilitating gas and solute transport. In early-stage cyanobacteria crusts, infiltration rates are also limited by pore systems composed of vesicles and other discontinuous voids, which do not conduct water as readily as interconnected pore systems (Felde et al., 2014).

4. Conclusions

This literature review has revealed some inconsistency in terminology and opportunities for continued research on formation of soil crusts. Historically, multiple terms have been used to describe the same common features of physical soil crusts, and competing theories have been proposed for their formation. Consistency and consensus within the scientific community would facilitate discussions of crust development. Micromorphological research of biological soil crusts having been comparatively limited, inconsistency in terminology has been less of an issue.

Because land management interpretations depend on an understanding of how crusts form and evolve, development of clear conceptual models for both physical and biological crusts would benefit scientists and policy-makers in the future. In interpreting the results of micromorphological studies, it is important to separate results obtained for natural and experimental crusts so that researchers can revise existing models more decisively. The usefulness of micromorphological studies would also be enhanced by more systematic quantification of features and functions, allowing greater applicability in land management and in ecological impact assessments.

Acknowledgements

We kindly thank Danielle Desruisseaux for assistance with technical editing. We are also grateful to Albert Solé-Benet and Vincent Felde whose careful reviews and thoughtful comments greatly improved this chapter.

References

Ahmad, N. & Roblin, A.J., 1971. Crusting of River Estate soil, Trinidad, and its effect of gaseous diffusion, percolation, and seedling emergence. Journal of Soil Science 22, 23-31.

Amézketa, E., 1999. Soil aggregate stability: a review. Journal of Sustainable Agriculture 14, 83-151.

Anderson, K., Wells, S. & Graham, R., 2002. Pedogenesis of vesicular horizons, Cima Volcanic Field, Mojave Desert, California. Soil Science Society of America Journal 66, 878-887.

Arshad, M.A. & Mermut, A.R., 1988. Micromorphological and physico-chemical characteristics of soil crust types in Northwestern Alberta, Canada. Soil Science Society of America Journal 52, 724-729.

Badorreck, A., Gerke, H.H. & Hüttl, R.F., 2013. Morphology of physical soil crusts and infiltration patterns in an artificial catchment. Soil and Tillage Research 129, 1-8.

Bajracharya, R.M. & Lal, R., 1999. Land use effects on soil crusting and hydraulic response of surface crusts on a tropical Alfisol. Hydrological Processes 13, 59-72.

Belnap, J., 1998. Environmental auditing choosing indicators of natural resource condition: a case study in Arches National Park, Utah, USA. Environmental Management 22, 635-642.

Belnap, J., 2001. Comparative structure of physical and biological soil crusts. In Belnap, J. & Lange, O.L. (eds.), Biological Soil Crusts: Structure, Function, and Management. Ecological Studies (Analysis and Synthesis), Volume 150. Springer, Berlin, pp. 177-191.

Belnap, J., 2006. The potential roles of biological soil crusts in dryland hydrologic cycles. Hydrological Processes 20, 3159-3178.

Belnap, J. & Gardner, J.S., 1993. Soil microstructure in soils of the Colorado Plateau: the role of the cyanobacterium Microcoleus vaginatus. Great Basin Naturalist 53, 40-47.

Belnap, J. & Warren, S.D., 2002. Patton's tracks in the Mojave Desert, USA: an ecological legacy. Arid Land Research and Management 16, 245-258.

Belnap, J., Büdel, B. & Lange, O. L., 2001a. Biological soil crusts: characteristics and distribution. In Belnap, J. & Lange, O.L. (eds.), Biological Soil Crusts: Structure, Function, and Management. Ecological Studies (Analysis and Synthesis), Volume 150. Springer, Berlin, pp. 3-30.

Belnap, J., Kaltnecker, J.H., Rosentreter, R., Williams, J., Leonard, S. & Eldridge, D., 2001b. Biological Soil Crusts: Ecology and Management. Technical Reference 1930-2-2001, Department of the Interior, Denver, 110 p.

Belnap, J., Hawkes, C.V. & Firestone, M.K., 2003. Boundaries in miniature: two examples from soil. BioScience 53, 739-749.

Belnap, J., Phillips, S.L., Witwicki, D.L. & Miller, M.E., 2008. Visually assessing the level of development and soil surface stability of cyanobacterially dominated biological soil crusts. Journal of Arid Environments 72, 1257-1264.

Beraldi-Campesi, H., Hartnett, H.E., Anbar, A., Gordon, G.W. & Garcia-Pichel, F., 2009. Effect of biological soil crusts on soil elemental concentrations: implications for biogeochemistry and as traceable biosignatures of ancient life on land. Geobiology 7, 348-359.

Bielders, C.L. & Baveye, P., 1995. Vertical particle segregation in structural crusts: experimental observations and the role of shear strain. Geoderma 67, 247-261.

Bishay, B.G. & Stoops, G., 1975. Micromorphology of irrigation crusts formed on a calcareous soil of the mechanized farm, north-west Egypt. Pedologie 25, 143.

Blanco, M. & Stoops, G., 2007. Genesis of pedons with discontinuous argillic horizons in the Holocene loess mantle of the southern Pampean landscape, Argentina. Journal of South American Earth Sciences 23, 30-45.

Boiffin, J. & Bresson, L.M., 1987. Dynamique de formation des croûtes superficielles : apport de l'analyse microscopique. In Fedoroff, N., Bresson, L.M. & Courty, M.A. (eds.), Micromorphologie des Sols, Soil Micromorphology. AFES, Plaisir, pp. 393-400.

Borselli, L., Carnicelli, S., Ferrari, G.A., Pagliai, M. & Lucamante, G., 1996. Effects of gypsum on hydrological, mechanical and porosity properties of a kaolinitic crusting soil. Soil Technology 9, 39-54.

Bouza, P., Del Valle, H. & Imbellone, P.A., 1993. Micromorphological, physical, and chemical characteristics of soil crust types of the central Patagonia region, Argentina. Arid Soil Research and Rehabilitation 7, 355-368.

Braissant, O., Cailleau, G., Dupraz, C. & Verrecchia, E.P., 2003. Bacterially induced mineralization of calcium carbonate in terrestrial environments: the roles of exopolysaccharides and amino acids. Journal of Sedimentary Research 73, 485-490.

Bresson, L.M. & Boiffin, J., 1990. Morphological characterisation of soil crust development on an experimental field. Geoderma 47, 301-325.

Bresson, L.M. & Moran, C.J., 2004. Micromorphological study of slumping in a hardsetting seedbed under various wetting conditions. Geoderma 118, 277-288.

Bresson, L.M. & Valentin, C., 1990. Comparative micromorphological study of soil crusting in temperate and arid environments. Transactions of the 14th International Congress of Soil Science, Volume VII, Kyoto, 238-243.

Bresson, L.M. & Valentin, C., 1994. Soil surface crust formation: contribution of micromorphology. In Ringrose-Voase, A.J. & Humphreys, G.S. (eds.), Soil Micromorphology: Studies in Management and Genesis. Developments in Soil Science, Volume 22. Elsevier, Amsterdam, pp. 737-762.

Bresson, L.M., Koch, C., Le Bissonnais, Y., Barriuso, E. & Lecomte, V., 2001. Soil surface structure stabilization by municipal waste compost application. Soil Science Society of America Journal 65, 1804-1811.

Brewer, R., 1964. Fabric and Mineral Analysis of Soils. John Wiley and Sons, New York, 470 p.

Buck, B.J., King, J. & Etyemezian, V., 2011. Effects of salt mineralogy on dust emissions, Salton Sea, California. Soil Science Society of America Journal 75, 1958-1972.

Bunting, B.T., 1977. The occurrence of vesicular structures in artic and subartic soils. Zeitschrift für Geomorphologie 21, 87-95.

Cameron, R.E. & Blank, G.B., 1966. Desert Algae: Soil Crusts and Diaphanous Substrata as Algal Habitats. Technical Report 32-971, Jet Propulsion Laboratory, California Institute of Technology, Pasadena, 41 p.

Campbell, S.E., 1979. Soil stabilization by a prokaryotic desert crust: implications for Precambrian land biota. Origins of Life 9, 335-348.

Cantón, Y., Solé-Benet, A. & Lázaro, R., 2003. Soil-geomorphology relations in gypsiferous materials of the Tabernas Desert (Almería, SE Spain). Geoderma 115, 193-222.

Carnicelli, S., Ferrari, G.A. & Pagliai, M., 1994. Pore space degradation in Zimbabwean crusting soils. In Ringrose-Voase, A.J. & Humphreys, G.S. (eds.), Soil Micromorphology: Studies in Management and Genesis. Developments in Soil Science, Volume 22. Elsevier, Amsterdam, pp. 677-686.

Casenave, A. & Valentin, C., 1992. A runoff capability classification system based on surface features criteria in semiarid areas of West Africa. Journal of Hydrology 130, 231-249.

Chartres, C.J. & Mücher, H.J., 1989. The effect of fire on the surface properties and seed germination in two shallow monoliths from a rangeland soil subjected to simulated raindrop impact and water erosion. Earth Surface Processes and Landforms 14, 407-417.

Chen, R., Zhang, Y., Li, Y., Wei, W., Zhang, J. & Wu, N., 2009. The variation of morphological features and mineralogical components of biological soil crusts in the Gurbantunggut Desert of Northwestern China. Environmental Geology 57, 1135-1143.

Chen, Y., Tarchitzky, J., Brouwer, J., Morin, J. & Banin, A., 1980. Scanning electron microscope observation on soil crusts and their formation. Soil Science 130, 49-55.

Chiang, S.C., West, L.T. & Radcliffe, D.E., 1994. Morphological properties of surface seals in Georgia soils. Soil Science Society of America Journal 58, 901-910.

Clegg, Z., Farres, P.J. & Poesen, J.W., 1999. Soil surface drip point features: an integrated approach using analytical photogrammetry and soil micromorphology. Catena 35, 303-316.

Collins, J.F., Smillie, G.W. & Hussain, S.M., 1986. Laboratory studies of crust development in Irish and Iraqi soils. III. Micromorphological observations of artificially-formed crusts. Soil and Tillage Research 6, 337-350.

Contreras, S., Cantón, Y. & Solé-Benet, A., 2008. Sieving crusts and macrofaunal activity control soil water repellency in semiarid environments: evidences from SE Spain. Geoderma 145, 252-258.

Couradeau, E., Karaoz, U., Lim, H.C., da Rocha, U.N., Northen, T., Brodie, E. & Garcia-Pichel, F., 2016. Bacteria increase arid-land soil surface temperature through the production of sunscreens. Nature Communications 7, Article 10373.

Courty, M.A., 1986. Morphology and genesis of soil surface crusts in semi-arid conditions (Hissar region, Northwest India). In Callebaut, F., Gabriels, D. & De Boodt, M., (eds.), Assessment of Soil Surface Sealing and Crusting. Flanders Research Centre for Soil Erosion and Conservation, Ghent, pp. 32-39.

Cousin, J., Malam Issa, O. & Le Bissonais, Y., 2005. Microgeometrical characterisation and percolation threshold evolution of a soil crust under rainfall. Catena 62, 173-188.

Csotonyi, J.T. & Addicott, J.F., 2004. Influence of trampling-induced microtopography on growth of the soil crust bryophyte Ceratodon purpureus in Jasper National Park. Canadian Journal of Botany 82, 1382-1392.

Danin, A. & Ganor, E., 1991. Trapping of airborne dust by mosses in the Negev Desert, Israel. Earth Surface Processes and Landforms 16, 153-162.

Danin, A., Dor, I., Sandler, A. & Amit, R., 1998. Desert crust morphology and its relations to microbiotic succession at Mt. Sedom, Israel. Journal of Arid Environments 38, 161-174.

Delgado-Baquerizo, M., Maestre, F.T., Eldridge, D.J., Bowker, M.A., Ochoa, V., Gozalo, B., Berdugo, M., Val, J. & Singh, B.K., 2016. Biocrust-forming mosses mitigate the negative impacts of increasing aridity on ecosystem multifunctionality in drylands. New Phytologist 209, 1540-1552.

Dietze, M., Bartel, S., Lindner, M. & Kleber, A., 2012. Formation mechanisms and control factors of vesicular soil structure. Catena 99, 83-96.

Drahorad, S.L. & Felix-Henningsen, P., 2013. Application of an electronic micropenetrometer to assess mechanical stability of biological soil crusts. Journal of Plant Nutrition and Soil Science 176, 904-909.

Duley, F.L., 1939. Surface factors affecting the rate of intake of water by soils. Soil Science Society of America Proceedings 3, 60-64.

Eghbal, M.K., Hajabbasi, M.A. & Golsefidi, H.T., 1996. Mechanism of crust formation on a soil in central Iran. Plant and Soil 180, 67-73.

Eldridge, D.J., 1998. Trampling of microphytic crusts on calcareous soils and its impact on erosion under rain-impacted flow. Catena 33, 221-239.

Eldridge, D.J. & Greene, R.S.B., 1994. Microbiotic soil crusts. A review of their roles in soil and ecological processes in the rangelands of Australia. Australian Journal of Soil Research 32, 389-415.

Eldridge, D.J., Zaady, E. & Shachak, M., 2000. Infiltration through three contrasting biological soil crusts in patterned landscapes in the Negev, Israel. Catena 40, 323-336.

Epstein, E. & Grant, W.J., 1967. Soil losses and crust formation as related to some physical properties. Soil Science Society of America Journal Proceedings, 31, 547-550.

Escadafal, R. & Fedoroff, N., 1987. Apport de la micromorphologie à une étude multi-scalaire de la surface des sols en région aride (Tunisie méridionale). In Fedoroff, N., Bresson, L.M. & Courty, M.A. (eds.), Micromorphologie des Sols, Soil Micromorphology. AFES, Plaisir, pp. 409-414.

Escudero, A., Martínez, I., de la Cruz, A., Otálora, M.A.G. & Maestre, F.T., 2007. Soil lichens have species-specific effects on the seedling emergence of three gypsophile plant species. Journal of Arid Environments 70, 18-28.

Evans, R.D. & Belnap, J., 1999. Long-term consequences of disturbance on nitrogen dynamics in an arid ecosystem. Ecology 80, 150-160.

Evans, D.D. & Buol, S.W., 1968. Micromorphological study of soil crusts. Soil Science Society of America Proceedings 32, 19-22.

Evenari, M., Yaalon, D.H. & Gutterman, Y., 1974. Note on soils with vesicular structure in deserts. Zeitschrift für Geomorphologie 18, 162-172.

Falayi, O. & Bouma, J., 1975. Relationships between the hydraulic conductance of surface crusts and soil management in a Typic Hapludalf. Soil Science Society of America Journal 39, 957-963.

Farres, P.J., 1978. The role of time and aggregate size in the crusting process. Earth Surface Processes and Landforms 3, 243-254.

Fedoroff, N. & Courty, M.A., 1987. Morphology and distribution of textural features in arid and semid-arid regions. In Fedoroff, N., Bresson, L.M. & Courty, M.A. (eds.), Micromorphologie des Sols, Soil Micromorphology. AFES, Plaisir, pp. 213-220.

Felde, V.J.M.N.L., Peth, S., Uteau-Puschmann, D., Drahorad, S. & Felix-Henningsen, P., 2014. Soil microstructure as an under-explored feature of biological soil crust hydrological properties: case study from the NW Negev Desert. Biodiversity and Conservation 23, 1687-1708.

Ferry, D.M. & Olsen, R.A., 1975. Orientation of clay particles as it relates to crusting soil. Soil Science 120, 367-375.

Figueira, H.L., 1984. Horizonte vesicular: morfologia y genesis en un aridisol del norte de la Patagonia. Ciencia del Suelo 2, 121-129.

Figueira, H. & Stoops, G., 1983. Application of micromorphometric techniques to the experimental study of vesicular layer formation. Pedologie 33, 77-89.

Fischer, T., Veste, M., Wiehe, W. & Lange, P., 2010. Water repellency and pore clogging at early successional stages of microbiotic crusts on inland dunes, Brandenburg, NE Germany. Catena 80, 47-52.

Fischer, T., Yair, A. & Veste, M., 2012. Microstructure and hydraulic properties of biological soil crusts on sand dunes: a comparison between arid and temperate climates. Biogeosciences Discussions 9, 12711.

Fox, D.M, Bryan, R.B. & Fox, C.A., 2004. Changes in pore characteristics with depth for structural crusts. Geoderma 120, 109-120.

Fox, S.J.C., Mills, A.J. & Poch, R.M., 2009. Micromorphology of surface crusts in the Knersvlakte, South Africa. Journal of Mountain Science 6, 189-196.

Friedmann, E.I. & Galun, M., 1974. Desert algae, lichens, and fungi. In Brown, G.W. (ed.), Desert Biology. Academic Press, New York, pp. 165-212.

Garcia-Pichel, F. & Belnap, J., 1996. Microenvironments and microscale productivity of cyanobacterial desert crusts. Journal of Phycology 32, 774-782.

Gerasimova, M. & Lebedeva, M., 2018. Organo-mineral surface horizons. In Stoops, G., Marcelino, V. & Mees, F. (eds.), Interpretation of Micromorphological Features of Soils and Regoliths. Second Edition. Elsevier, Amsterdam, pp. 513-538.

Greene, R.S.B. & Ringrose-Voase, A.J., 1994. Micromorphological and hydraulic properties of surface crusts formed on a red earth soil in the semi-arid rangelands of eastern Australia. In Ringrose-Voase, A.J. & Humphreys, G.S. (eds.), Soil Micromorphology: Studies in Management and Genesis. Developments in Soil Science, Volume 22. Elsevier, Amsterdam, pp. 763-776.

Greene, R.S.B., Chartres, C.J. & Hodgkinson, K.C., 1990. The effects of fire on the soil in a degraded semiarid woodland. I. Cryptogam cover and physical and micromorphological properties. Soil Research 28, 755-777.

Gupta, S.C. & Larson, W.E., 1979. A model for predicting packing density of soil using particle-size distribution. Soil Science Society of America Journal 43, 758-764.

Hall, N.W., 1987. An application of micromorphology to evaluating the distribution and significance of soil erosion by water In Fedoroff, N., Bresson, L.M. & Courty, M.A. (eds.), Micromorphologie des Sols, Soil Micromorphology. AFES, Plaisir, pp. 437-444.

Hallaire V., Degoumois, Y., Guenat, G. & Curmi, P., 1997. Coupling image analysis and hydrodynamic measurements to quantify pore space for a functional typology of surface crusts. In Shoba, S., Gerasimova, M. & Miedema, R. (eds.), Soil Micromorphology: Studies on Soil Diversity, Diagnostics, Dynamics. Moscow, Wageningen, pp. 390-398.

Hernandez, R.R. & Sandquist, D.R., 2011. Disturbance of biological soil crust increases emergence of exotic vascular plants in California sage scrub. Plant Ecology 212, 1709-1721.

Herrero Isern, J., Rodriguez Ochoa, R. & Porta Casanellas, J., 1989. Colmatacion de Drenes en Suelos Afectados por Salinidad. Institución Fernando el Católico, Zaragoza, 133 p.

Hoppert, M., Reimer, R., Kemmling, A., Schröder, A., Günzl, B. & Heinken, T., 2004. Structure and reactivity of a biological soil crust from a xeric sandy soil in Central Europe. Geomicrobiology Journal 21, 183-191.

Hu, C., Zhang, D., Huang, Z. & Liu, Y., 2003. The vertical microdistribution of cyanobacteria and green algae within desert crusts and the development of the algal crusts. Plant and Soil 257, 97-111.

Humphreys, G.S., 1994. Bowl-structures: a composite depositional soil crust. In Ringrose-Voase, A.J. & Humphreys, G.S. (eds.), Soil Micromorphology: Studies in Management and Genesis. Developments in Soil Science, Volume 22. Elsevier, Amsterdam, pp. 787-798.

Kleiner, E.F. & Harper, K.T., 1977. Soil properties in relation to cryptogamic groundcover in Canyonlands National Park. Journal of Range Management 30, 202-205.

Kooistra, M.J. & Siderius, W., 1986. Micromorphological aspects of crust formation in a savannah climate under rainfed subsistence agriculture. In Callebaut, F., Gabriels D. & De Boodt, M. (eds.), Assessment of Soil Surface Sealing and Crusting. Flanders Research Centre for Soil Erosion and Conservation, Ghent, pp. 9-17.

Kubiëna, W.L., 1938. Micropedology. Collegiate Press, Ames, 242 p.

Kwaad, F.J.P.M. & Mücher, H.J., 1994. Degradation of soil structure by welding – a micromorphological study. Catena 23, 253-268.

Labib, F., Tayel, M.Y. & Elrashidi, M.A., 1975. Crust formation in clayey and calcareous soils. 1. Micropedological study. Egyptian Journal of Soil Science Special Issue, pp. 19-27.

Lan, S., Wu, L., Zhang, D. & Hu, C., 2012. Successional stages of biological soil crusts and their microstructure variability in Shapotou region (China). Environmental Earth Sciences 65, 77-88.

Lan, S., Zhang, Q., Wu, L., Liu, Y., Zhang, D. & Hu, C., 2013. Artificially accelerating the reversal of desertification: cyanobacterial inoculation facilitates the succession of vegetation communities. Environmental Science & Technology 48, 307-315.

Le Bissonnais, Y., 1996. Aggregate stability and assessment of soil crustability and erodibility: I. Theory and methodology. European Journal of Soil Science 47, 425-437.

Le Bissonnais, Y. & Bruand, A., 1993. Crust micromorphology and runoff generation on silty soil materials during different seasons. Catena Supplement 24, 1-16.

Le Bissonnais, Y., Bruand, A. & Jamagne, M., 1990. Etude expérimentale sous pluie simultanée de la formation des croûtes superficielles. Apport à la notion d'érodibilité des sols. Cahiers ORSTOM, série Pédologie 25, 31-40.

Le Souder, C., Le Bissonnais, Y., Robert, M. & Bresson, L.M., 1990. Prevention of crust formation with a mineral conditioner. In Douglas, L.A. (ed.), Soil Micromorphology: A Basic and Applied Science. Developments in Soil Science, Volume 19. Elsevier, Amsterdam, pp. 81-88.

Li, X.R., Jia, X.H., Long, L.Q. & Zerbe, S., 2005. Effects of biological soil crusts on seed bank, germination and establishment of two annual plant species in the Tengger Desert (N. China). Plant and Soil 277, 375-385.

Luk, S.H., Dubbin, W.E. & Mermut, A.R., 1990. Fabric and mineral analysis of surface crust development under simulated rainfall on loess in China. Catena Supplement 17, 29-40.

Malam Issa, O.M., Trichet, J., Défarge, C., Couté, A. & Valentin, C., 1999. Morphology and microstructure of microbiotic soil crusts on a tiger bush sequence (Niger, Sahel). Catena 37, 175-196.

Malam Issa, O., Le Bissonnais, Y., Défarge, C. & Trichet, J., 2001. Role of a cyanobacterial cover on structural stability of sandy soils in the Sahelian part of western Niger. Geoderma 101, 15-30.

Malam Issa, O.M., Défarge, C., Le Bissonnais, Y., Marin, B., Duval, O., Bruand, A. & Annerman, M., 2007. Effects of the inoculation of cyanobacteria on the microstructure and the structural stability of a tropical soil. Plant and Soil 290, 209-219.

Malam Issa, O.M., Défarge, C., Trichet, J., Valentin, C. & Rajot, J.L., 2009. Microbiotic soil crusts in the Sahel of western Niger and their influence on soil porosity and water dynamics. Catena 77, 48-55.

Malam Issa, O., Valentin, C., Rajot, J.L., Cerdan, O., Desprats, J.F. & Bouchet, T., 2011. Runoff generation fostered by physical and biological crusts in semi-arid sandy soils. Geoderma 167, 22-29.

Marsh, J., Nouvet, S., Sanborn, P. & Coxson, D., 2006. Composition and function of biological soil crust communities along topographic gradients in grasslands of central British Columbia (Chilcotin) and southwestern Yukon (Kluane). Canadian Journal of Botany 84, 717-736.

McFadden, L.D., Wells, S.G. & Jercinovich, M.J., 1987. Influences of eolian and pedogenic processes on the origin and evolution of desert pavements. Geology 15, 504-508.

McFadden, L.D., McDonald, E.V., Wells, S.G., Anderson, K., Quade, J. & Forman, S.L., 1998. The vesicular layer and carbonate collars of desert soils and pavements: formation, age and relation to climate change. Geomorphology 24, 101-145.

McIntyre, D.S., 1958. Soil splash and the formation of surface crusts by raindrop impact. Soil Science 85, 261-266.

McKenna Neuman, C., Maxwell, C.D. & Boulton, J.W., 1996. Wind transport of sand surfaces crusted with photoautotrophic microorganisms. Catena 27, 229-247.

Mees, F. & Singer, A., 2006. Surface crusts on soils/sediments of the southern Aral Sea basin, Uzbekistan. Geoderma 136, 152-159.

Mees, F. & Tursina, T., 2018. Salt minerals in saline soils and salt crusts. In Stoops, G., Marcelino, V. & Mees, F. (eds) Interpretation of Micromorphological Features of Soils and Regoliths. Second Edition, Elsevier, Amsterdam. pp. 289-321.

Mermut, A.R., Luk, S.H., Römkens, M.J.M. & Poesen, J.W.A., 1995. Micromorphological and mineralogical components of surface sealing in loess soils from different geographic regions. Geoderma 66, 71-84.

Miller, D.E., 1971. Formation of vesicular structure in soil. Soil Science Society of America Proceedings 35, 635-637.

Miralles-Mellado, I., Cantón, Y. & Solé-Benet, A., 2011. Two-dimensional porosity of crusted silty soils: indicators of soil quality in semiarid rangelands? Soil Science Society of America Journal 75, 1330-1342.

Moore, D.C. & Singer, M.J., 1990. Crust formation effects on soil erosion processes. Soil Science Society of America Journal 54, 1117-1123.

Moss, A.J., 1991a. Rain-impact soil crust. I. Formation on a granite-derived soil. Australian Journal of Soil Research 29, 271-289.

Moss, A.J., 1991b. Rain-impact soil crust. II. Some effects of surface-shape, drop-size and soil variation. Australian Journal of Soil Research 29, 291-309.

Mücher, H.J. & De Ploey, J., 1977. Experimental and micromorphological investigation of erosion and redeposition of loess by water. Earth Surface Processes and Landforms 2, 117-124.

Mücher, H.J., De Ploey, J. & Savat, J., 1981. Response of loess materials to simulated translocations by water: micromorphological observations. Earth Surface Processes and Landforms 6, 331-336.

Mücher, H.J., Chartres, C.J., Tongway, D.J. & Greene, R.S.B., 1988. Micromorphology and significance of the surface crusts of soils in rangelands near Cobar, Australia. Geoderma 42, 227-244.

Nicolau, J.M., Solé-Benet, A., Puigdefábregas, J. & Gutiérrez, L., 1996. Effects of soil and vegetation on runoff along a catena in semi-arid Spain. Geomorphology 14, 297-309.

Noffke, N., Gerdes, G., Klenke, T. & Krumbein, W.E., 2001. Microbially induced sedimentary structures – a new category within the classification of primary sedimentary structures. Journal of Sedimentary Research 71, 649-646.

Norton, L.D., 1987. Micromorphological study of surface seals developed under simulated rainfall. Geoderma 40, 127-140.

Norton, L.D. & Schroeder, S.L., 1987. The effects of various cultivation methods on soil loss: a micromorphological approach. In Fedoroff, N., Bresson, L.M. & Courty, M.A. (eds.), Micromorphologie des Sols, Soil Micromorphology. AFES, Plaisir, pp. 431-436.

Norton, L.D., Schroeder, S.L. & Moldenhauer, W.C., 1986. Differences in surface crusting and soil loss as affected by tillage methods. In Callebaut, F., Gabriels, D. & De Boodt, M. (eds.), Assessment of Soil Surface Sealing and Crusting. Flanders Research Centre for Soil Erosion and Conservation, Ghent, pp. 64-71.

Onofiok, O. & Singer, M.J., 1984. Scanning electron microscope studies of surface crusts formed by simulated rainfall. Soil Science Society of America Journal 48, 1137-1143.

Pagliai, M., 1987. Effects of different management practices on soil structure and surface crusting. In Fedoroff, N., Bresson, L.M. & Courty, M.A. (eds.), Micromorphologie des Sols, Soil Micromorphology. AFES, Plaisir, pp. 415-422.

Pagliai, M., 1994. Micromorphology and soil management. In Ringrose-Voase, A.J. & Humphreys, G.S. (eds.), Soil Micromorphology: Studies in Management and Genesis. Developments in Soil Science, Volume 22. Elsevier, Amsterdam, pp. 623-640.

Pagliai, M., 2004. Soil degradation and land use. In Werner, D. (ed.), Biological Resources and Migration. Springer, Berlin, pp. 273-280.

Pagliai, M., 2008. Crust, crusting. In Chesworth, W. (ed.), Encyclopedia of Soil Science. Springer, Dordrecht, pp. 171-178.

Pagliai, M. & La Marca, M., 1979. Micromorphological study of soil crusts. Agrochimica 23, 16-25.

Pagliai, M. & Stoops, G., 2010. Physical and biological surface crusts. In Stoops, G., Marcelino, V. & Mees, F. (eds.), Interpretation of Micromorphological Features of Soils and Regoliths. Elsevier, Amsterdam. pp. 419-440.

Pagliai, M., La Marca, M. & Lucamante, G., 1983a. Micromorphometric and micromorphological investigations of a clay loam soil in viticulture under zero and conventional tillage. Journal of Soil Science 34, 391-403.

Pagliai, M., Bisdom, E.B.A. & Ledin, S., 1983b. Changes in surface structure (crusting) after application of sewage sludges and pig slurry to cultivated agricultural soils in northern Italy. Geoderma 30, 35-53.

Pagliai, M., Pezzarossa, B., Mazzoncini, M. & Bonari, E., 1989. Effects of tillage on porosity and microstructure of a loam soil. Soil Technology 2, 345-358.

Pagliai, M., Vignozzi, N. & Pellegrini, S., 2004. Soil Structure and the effect of management practices. Soil and Tillage Research 79, 131-143.

Panini, T., Torri, D., Pellegrini, S., Pagliai, M. & Salvador Sanchis, M.P., 1997. A theoretical approach to soil porosity and sealing development using simulated rainstorms. Catena 31, 199-219.

Pérez, F.L., 1997. Microbiotic crusts in the high equatorial Andes, and their influence on Paramo soils. Catena 31, 173-198.

Poesen, J., 1981. Rainwash experiments on the erodibility of loose sediments. Earth Surface Processes and Landforms 6, 258-307.

Poesen, J.W.A. & Nearing, M.A., 1993. Soil Surface Sealing and Crusting. Catena Supplement 24, 139 p.

Poesen, J., Ingelmo-Sanchez, F. & Mücher, H., 1990. The hydrological response of soil surfaces to rainfall as affected by cover and position of rock fragments in the toplayer. Earth Surface Processes and Landforms 15, 653-671.

Puigdefábregas, J., Solé-Benet, A., Gutiérrez, L., Del Barrio, G. & Boer, M., 1999. Scales and processes of water and sediment redistribution in drylands: results from the Rambla Honda field site in Southeast Spain. Earth-Science Reviews 48, 39-70.

Radcliffe, D.E., West, L.T., Hubbard, R.K. & Asmussen, L.E., 1991. Surface sealing in coastal plains loamy sands. Soil Science Society of America Journal 55, 223-227.

Reichert, J.M. & Norton, L.D., 1995. Surface seal micromorphology as affected by fluidized-bed combustion bottom-ash. Soil Technology 7, 303-317.

Remley, P.A. & Bradford, J.M., 1989. Relationship of soil crust morphology to inter-rill erosion parameters. Soil Science Society of America Journal 53, 1215-1221.

Rentzel, P., Nicosia, C., Gebhardt, A., Brönnimann, D., Pümpin, C. & Ismail-Meyer, K., 2017. Trampling, poaching and the effect of traffic. In Nicosia, C. & Stoops, G. (eds.) Archaeological Soil and Sediment Micromorphology. John Wiley & Sons Ltd., pp 281-297.

Ries, J.B. & Hirt, U., 2008. Permanence of soil surface crusts on abandoned farmland in the Central Ebro Basin/Spain. Catena 72, 282-296.

Rodríguez-Caballero, E., Cantón, Y., Chamizo, S., Afana, A. & Solé-Benet, A., 2012. Effects of biological soil crusts on surface roughness and implications for runoff and erosion. Geomorphology 145, 81-89.

Rodríguez-Caballero, E., Cantón, Y., Lazaro, R. & Solé-Benet, A., 2014. Cross-scale interactions between surface components and rainfall properties. Non-linearities in the hydrological and erosive behavior of semiarid catchments. Journal of Hydrology 517, 815-825.

Rousseva, S., Torri, D. & Pagliai, M., 2002. Effect of rain on the macroporosity at the soil surface. European Journal of Soil Science 53, 83-94.

Simpson, E.L., Heness, E., Bumby, A., Eriksson, P.G., Eriksson, K.A., Hilbert-Wolf, H.L, Linnevelt, S., Malenda, H. F, Modungwa, T. & Okafor, O.J., 2013. Evidence for 2.0 Ga continental microbial mats in a paleodesert setting. Precambrian Research 237, 36-50.

Singer, A., Kirsten, W.F.A. & Buhmann, C., 1994. Clay dispersivity and crusting of soils determined by Buchner funnel extractions. Australian Journal of Soil Research 32, 465-470.

Slattery, M.C. & Bryan, R.B., 1992. Laboratory experiments on surface seal development and its effect on interrill erosion processes. Journal of Soil Science 43, 517-529.

Slattery, M.C. & Bryan, R.B., 1994. Surface seal development under simulated rainfall on an actively eroding surface. Catena 22, 17-34.

Smillie, G.W., Collins, J.F. & Hussain, S.M., 1987. A microscopic study of phosphoric acid treatment on artificially-formed soil crusts. In Fedoroff, N., Bresson, L.M. & Courty, M.A. (eds.), Micromorphologie des Sols, Soil Micromorphology. AFES, Plaisir, pp 423-430.

Solé-Benet, A., Pini, R., & Raffaelli, M., 2002. Hydrological consequences of soil surface type and condition in colluvial mica-schist soils after agricultural abandonment. In Rubio, J.L., Morgan, R.P.C., Asins, S. & Andreu, V. (eds.), Man and soil at the third millennium. Proceedings of the Third International Congress of the European Society for Soil Conservation. Geoforma Ediciones, Logroño, pp. 523-533.

Sombroek, W.G., 1986. Introduction to the subject. In Callebaut, F., Gabriels, D. & De Boodt, M. (eds.), Assessment of Soil Surface Sealing and Crusting. Flanders Research Centre for Soil Erosion and Conservation, Ghent, pp. 1-7.

Southard, R.J., Shainberg, I. & Singer, M.J., 1988. Influence of electrolyte concentration on the micromorphology of artificial depositional crust. Soil Science 145, 278-288.

Souza-Egipsy, V., Wierzchos, J., Sancho, C., Belmonte, A. & Ascaso, C., 2004. Role of biological soil crust cover in bioweathering and protection of sandstones in a semi-arid landscape (Torrollones de Gabarda, Huesca, Spain). Earth Surface Processes and Landforms 29, 1651-1661.

Springer, M.E., 1958. Desert pavement and vesicular layer of some soils of the desert of Lahontan Basin, Nevada. Soil Science Society of America Proceedings 22, 63-66.

Sullivan, L.A. & Koppi, A.J., 1991. Morphology and genesis of silt and clay coatings in the vesicular layer of a desert loam soil. Australian Journal of Soil Research 29, 579-586.

Sveistrup, T., Marcelino, V. & Stoops, G., 1995. Effects of slurry application on the microstructure of the surface layers of soils from northern Norway. Norwegian Journal of Agricultural Sciences 9, 1-13.

Tackett, J.L. & Pearson, R.W., 1965. Some characteristics of soil crusts formed by simulated rainfall. Soil Science 99, 407-413.

Tanaka, U. & Kyuma, K., 1992. Quantification of morphological characteristics of crusts by thin section-imager analysing method (TS-IA method). Soil Science and Plant Nutrition 38, 369-373.

Tarchitzky, J., Banin, A., Morin, J. & Chen, Y., 1984. Nature, formation and effects of soil crusts formed by water drop impact. Geoderma 33, 135-155.

Tayel, M.Y., Elrashidi, M.A. & Labib, F., 1975. Crust formation in clayey and calcareous soils. II. Some physical properties of crusted soils. Egyptian Journal of Soil Science Special Issue, pp. 167-174.

Thomas, A.D. & Dougill, A.J., 2007. Spatial and temporal distribution of cyanobacterial soil crusts in the Kalahari: implications for soil surface properties. Geomorphology 85, 17-29.

Thomas, A.D., Hoon, S.R. & Dougill, A.J., 2011. Soil respiration at five sites along the Kalahari Transect: effects of temperature, precipitation pulses and biological soil crust cover. Geoderma 167, 284-294.

Trindade, E. de S., Schaefer, C.E.G.R., Abrahão, W.A.P., Ribeiro Jr., E.S., Oliveira, D.M.F. & Teixeira, P.C., 2005. Crostas biológicas de saprólitos da região do Quadrilátero Ferrífero, MG: ciclagem biogeoquímica e micromorfológica. Geonomos 13, 37-45.

Turk, J.K. & Graham, R.C., 2011. Distribution and properties of vesicular horizons in the western United States. Soil Science Society of America Journal 75, 1450-1461.

Turk, J.K. & Graham, R.C., 2014. Analysis of vesicular porosity in soils using high resolution X-ray computed tomography. Soil Science Society of America Journal 78, 868-880.

Usón, A. & Poch, R.M., 2000. Effect of tillage and management practices on soil crust morphology under a Mediterranean environment. Soil and Tillage Research 54, 191-196.

Valentin, C., 1986. Effects of soil moisture and kinetic energy on the mechanical resistance of surface crust. In Callebaut, F., Gabriels, D. & De Boodt, M. (eds.), Assessment of Soil Surface Sealing and Crusting. Flanders Research Centre for Soil Erosion and Conservation, Ghent, pp. 367-379.

Valentin, C., 1991. Surface crusting in two alluvial soils of northern Niger. Geoderma 48, 201-222.

Valentin, C. & Bresson, L.M., 1992. Morphology, genesis and classification of surface crusts in loamy and sandy soils. Geoderma 55, 225-245.

Valentin, C. & Janeau, J.L., 1989. Les risques de dégradation structurale de la surface des sols en savane humide. Cahiers ORSTOM, Série Pédologie 25, 41-52.

Valentin, C. & Ruiz Figueroa, J.F., 1987. Effect of kinetic energy and water application rate on the development of crusts in a fine sandy loam soil using sprinkling irrigation and rainfall simulation. In Fedoroff, N., Bresson, L.M. & Courty, M.A. (eds.), Micromorphologie des Sols, Soil Micromorphology. AFES, Plaisir, pp. 401-408.

Van Vliet-Lanoë, B. & Fox, C., 2018. Frost action. In Stoops, G., Marcelino, V. & Mees, F. (eds.), Interpretation of Micromorphological Features of Soils and Regoliths. Second Edition. Elsevier, Amsterdam, pp. 575-603.

Volk, O.H. & Geyger, E., 1970. 'Schaumboden' als Ursache der Vegetationslosigkeit in ariden Gebieten. Zeitschrift für Geomorphologie 14, 79-95.

Wakindiki, I.I.C. & Ben-Hur, M. 2002. Soil mineralogy and texture effects on crust micromorphology, infiltration, and erosion. Soil Science Society of America Proceedings 66, 897-905.

West, L.T., Bradford, J.M. & Norton, L.O., 1990. Crust morphology and infiltrability in surface soils from the Southeast and Midwest U.S. In Douglas, L.A. (ed.), Soil Micromorphology: A Basic and Applied Science. Developments in Soil Science, Volume 19. Elsevier, Amsterdam, pp. 107-114.

West, L.T., Chiang, S.C. & Norton, L.D., 1992. The morphology of surface crusts. In Summer, M.E. & Stewart, B.A. (eds.), Soil Crusting, Chemical and Physical Processes. Lewis Publishers, Boca Raton, pp. 73-92.

Williams, A.J., 2011. Co-Development of Biological Soil crusts, Soil-Geomorphology, and Landscape Biogeochemistry in the Mojave Desert, Nevada, U.S.A.: Implications for Ecological Management. PhD Dissertation. University of Nevada, 386 p.

Williams, A.J., Buck, B.J. & Beyene, M.A., 2012. Biological soil crusts in the Mojave Desert USA: micro-morphology and pedogenesis. Soil Science Society of America Journal 76, 1685-1695.

Wynn-Williams, D.D., 1990. Microbial colonization processes in Antarctic fellfield soils – an experimental overview. Proceedings of the NIPR Symposium on Polar Biology 3, 164-178.

Yonovitz, M. & Drohan, P.J., 2009. Pore morphology characteristics of vesicular horizons in undisturbed and disturbed arid soils; implications for arid land management. Soil Use and Management 25, 293-302.

Zhang, Y.M., Wang, H.L., Wang, X.Q., Yang, W.K. & Zhang, D.Y., 2006. The microstructure of microbiotic crust and its influence on wind erosion for a sandy soil surface in the Gurbantunggut Desert of Northwestern China. Geoderma 132, 441-449.

Zhao, R., Zhang, Q., Tjugito, H. & Cheng, X., 2015. Granular impact cratering by liquid drops: under-standing raindrop imprints through an analogy to asteroid strikes. Proceedings of the National Academy of Sciences 112, 342-347.

Chapter 20

Frost Action

Brigitte Van Vliet-Lanoë[1], Catherine A. Fox[2]

[1]*CNRS GEOSCIENCES OCÉAN, UNIVERSITY OF BREST, PLOUZANÉ, FRANCE;*
[2]*AGRICULTURE AND AGRI-FOOD CANADA, HARROW, ON, CANADA*

CHAPTER OUTLINE

1. Introduction

Frost activity is present in many regions of the world, from high latitudes to tropical arid zones, where it may penetrate the soil to a depth of 5-15 cm in winter. It is common in active to subactive Cryosols (IUSS Working Group WRB, 2015) from both high altitudes and high latitudes. However, frost-induced microfabrics are not specific of present-day cold environments. Features formed during the extension of the cold domain in the Quaternary are preserved in warmer soil climates as, for instance, in the C1 to E horizons of loess soils from the temperate zone. Specific horizons such as some periglacial fragipans (Soil Survey Staff, 2014) or foliated lodgement tills of formerly glaciated regions (subglacial permafrost) are related to the specific thermal regime of permafrost and frost diagenesis of sediments (Van Vliet-Lanoë & Langohr, 1981).

The effects of frost on soil have been widely described since the first observations of Kokkonen (1927), Taber (1929), Beskow (1935), Sharp (1942), McMillan and Mitchell (1953), FitzPatrick (1956) and Frese and Czeratzki (1957). These authors showed that ice segregation is an important pedogenetic agent, influencing soils by its action on structure development, consolidation, deformation and particle translocation. Frost features are related to the segregation of ice in the soil, usually in the form of lenses, and to the heaving and associated strains resulting from both freezing and thawing.

Frost is a temperature-driven desiccation which can open shrinking fissures in the still unfrozen substratum (Van Vliet-Lanoë, 1985). As the ground freezes, the segregation of ice develops by migration of pore solution to the freezing front. In soil, frost dynamics are controlled by the local drainage conditions and the grain-size distribution (Kaplar, 1974), which specifically affects the ability of the soil to retain capillary and adsorbed water. The latter is mainly related to the content in colloidal organic matter and clay. An average pore diameter of 50 μm, for intergranular porosity, is usually the most favourable for ice nucleation (Shumskii, 1964; Kaplar, 1974) explaining its common development in silty and loamy to sandy loam soils. The greater and steadier the water supply, the greater the amount of ice segregated. Soils with these textures at the capillary fringe of the water table or in moisture retention conditions close to field capacity have thus optimal conditions for ice lens development (Van Vliet-Lanoë, 1988a). A large capability to develop ice lenses, also called frost susceptibility, may lead to frost heave of the soil. In spring, the inversion of the thermal gradient, with the temperature of the frozen soil lower than that of the air, promotes the infiltration of melting water in the still frozen substratum. This can result in supplementary heave and stress in the soil (Parmuzina, 1978; Van Vliet-Lanoë, 1985) just before thaw settlement or thaw consolidation, i.e., volume loss by ice melt. Other associated processes are the development of desiccation crack networks and the lateral sliding of soil aggregates along melting ice lenses. Fragmentation of mineral grains is induced by physical weathering related to frost dehydration, hydraulic pressure development and ice crystallisation (McDowall, 1960), in association with chemical weathering and soil faunal activity. On the other hand, the translocation of particles during spring melt or the precipitation of gel-like clays or organic matter can influence or create frost-susceptibility gradients. As fissures left by ice lenses drastically reduce the shearing resistance of the thawed soil (Coutard et al., 1988), deformations can occur simultaneously with differential heave and solifluction on slopes.

Pioneering work on the micromorphology of (sub)Arctic and Antarctic Cryosols, where frost action is an active process, was undertaken by Kubiëna (1953, 1970, 1971) in the Antarctic and Arctic, FitzPatrick (1956) in Svalbard, Dumanski and St. Arnaud (1966) on the boreal soils of Canada and Morozova (1965) in central Yakoutia. They were followed by Parfenova and Yarilova (1967), Benedict (1969), Fedorova and Yarilova (1972), Bunting and Fedoroff (1974), Brewer and Pawluk (1975), Pawluk and Brewer (1975), Fox and Protz (1981), Mermut and St. Arnaud (1981), Tursina (1985) and Van Vliet-Lanoë (1988b). Other studies were performed on mountain soils of Europe (Ontañon, 1978, Harris & Ellis, 1980; Harris, 1987, 1990; Van Vliet-Lanoë, 1987a; Harris & Cook, 1988). More recent publications report research on soils of the Antarctic (Schaefer et al., 2008, Villagran et al., 2013) and Siberia (Kovda & Lebedeva, 2013; Bronnikova et al., 2014). The study of soil thin sections has also often been used to provide information on the classification of these soils (e.g., Tursina, 1994; Tarnocai & Bockheim, 2011). Micromorphological analysis is usually performed on unfrozen material from field samples collected in summer, but some attempts have been made to study the micromorphological characteristics of still frozen soils, in natural or in experimental conditions (Dumanski, 1964; Pissart, 1970; Van Vliet-Lanoë et al., 1984; Coutard & Mücher, 1985; Coutard et al., 1988; Van Vliet-Lanoë, 1988a; Van Vliet-Lanoë & Dupas, 1991; Gubin & Gulyaeva, 1997; White & Fox, 1997; Szymański et al., 2015).

Micromorphological studies of fossil Cryosols are much more common, especially of those on loess from the Russian Plain (Morozova, 1965; Mücher & Morozova, 1983; Tursina, 1985; Gubin, 1994), from Europe (Van Vliet-Lanoë, 1976, 1988b; Miedema, 1987; Cremaschi & Van Vliet-Lanoë, 1990; Huijzer, 1993; Mestdagh, 2005) and from North America (Mermut et al., 1983; Tarnocai & Valentine, 1989; Todisco & Bhiry, 2008). Micromorphological evidence of past frost activity has been also obtained from other materials such as tills and slope deposits (FitzPatrick, 1956, 1976; Romans et al., 1966, 1980; Menzies & Maltman, 1992) and in caves (Pissart et al., 1988, Cremaschi & Van Vliet-Lanoë, 1990; Laafar et al.,1995).

A few important papers (Van Vliet-Lanoë, 1976, 1985; Mücher & Morozova, 1983; Harris, 1985; Van Vliet-Lanoë et al., 2004) review the use of soil micromorphology for the study of the effects of frost on specific soils and sediments. In the present review, the micromorphological features resulting from the effects of freezing and thawing processes on soils, in both present-day and past environments, are summarised and systematically discussed.

2. Microstructure

The *sine qua non* for ice segregation is the availability of water. Ice segregation and related microstructures can therefore not develop in very cold but dry regions of the globe (Campbell & Claridge, 1987). Most microstructures discussed further are indicative of ice segregation process in permafrost and non-permafrost cold areas as well as in subtropical arid soils. They reflect frost activity at various depths, in different materials or in specific drainage conditions. They may represent very short to very long periods, at the scale of days to millions of years. They are generally well preserved when they were formed at very low temperatures because of intense compaction and ultradesiccation (Van Vliet-Lanoë, 1985). Transitions between the different microstructures described can be observed in a soil, and the same soil may show more than one microstructure type.

2.1 Platy, Angular Blocky and Lenticular Microstructure

Evidence of freezing in soils is given by the characteristic lenticular and platy microstructures parallel to the surface, but it may also appear as short prismatic to blocky micro- to mesostructures. They vary with depth usually from lenticular close to the soil surface (macroscopically foliated) to platy or angular blocky in deeper horizons, depending on the grain-size distribution and the initial consolidation status of the sediment (Tsytovich, 1975; FitzPatrick, 1976). Lenticular and platy microstructures may also be vertical, when they develop in association with ice wedge formation and thermal cracking (Mol et al., 1993). The thickness of the peds ranges from <200 μm to 2 cm, according to the rate of frost penetration, which is controlled by the thermal gradient at the soil surface and the water supply at the freezing front. The voids that delimit the peds are similar to planar voids resulting from desiccation but have smooth and unaccommodating walls (Fig. 1A and B). If the development of ice lenses is limited, these pores may be confused with desiccation features. Other pores such as wide star-shaped vughs are often observed associated with angular blocky microstructures (Van Vliet-Lanoë, 1976) (Fig. 1C and D). Also vertical planar voids can be observed, which correspond to vertical ice blades formed in conditions of rapid freezing (Pissart, 1969). Large pores (0.5-5 mm) often develop at the base of large rock fragments, mineral grains or compact aggregates (Fig. 2). This type of pore is probably related to locally increased retraction and accentuated suction, due to the high thermal conductivity of the fragment or aggregate, which results in a greater water supply for ice lens formation at the base of the coarse grains. The fact that a coarse grain creates a hydraulic barrier enhances this process (Van Vliet-Lanoë & Dupas, 1991).

McMillan and Mitchel (1953), Dumanski (1964) and Dumanski and St. Arnaud (1966), who provided some of the first micromorphological descriptions of frost-affected soils, referred to the platy and lenticular microstructures with uniformly distributed fine and coarse material as 'isoband fabric'. This fabric often occurs in frost-affected soils with silty textures or in deep horizons as a result of frost desiccation retraction (Van Vliet-Lanoë, 1985) (Fig. 1A and B).

As a consequence of the gradual decrease in thermal deficit with depth (Van Vliet-Lanoë, 1976, 1985), the size of the aggregates formed by ice lensing usually increases progressively with depth (Shumskii, 1964) in seasonally frozen soils, both macroscopically and at microscopic level. However, in permafrost soils, an abrupt change in aggregate size from coarse to fine occurs in the lower part of the active layer, at the contact with the permafrost table (Van Vliet-Lanoë, 1985). After the melting of ice or thaw consolidation, the interpedal fissures can still be open, leading to platy or foliated macroscopic soil structures. This has a direct impact on the hydraulic conductivity and further translocation of particles and also favours the lateral movement of soil water.

Frost-generated aggregates have various shapes depending on the pattern of ice accumulation in the soil (e.g., Shumskii, 1964). Ice lenses usually form if freezing is progressive and the sediments are moist but not saturated. The resulting peds are lenticular and have a tile-like arrangement in silt-rich homogenous materials (Fig. 1A), and they are curved upward in a clay-rich substratum susceptible to stronger retraction at desiccation (Fig. 1B). Rapid cooling in saturated clays or silty clays (e.g., marine or lacustrine sediments, fresh colluvium) causes rapid shrinking (Van Vliet-Lanoë et al.,

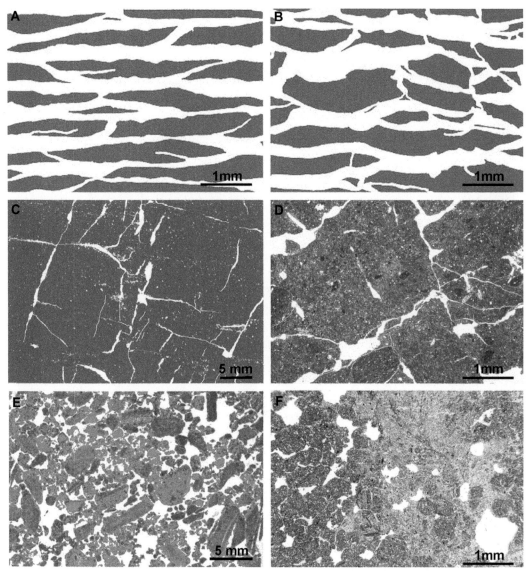

FIGURE 1 Frost-induced microstructures. (A) Lenticular with tile-like arrangement in silt-rich materials (experiment, Caen, France) (PPL, processed image). (B) Lenticular with aggregates curved upward in consolidated clays (experiment, Caen, France) (PPL, processed image). (C) Angular blocky in unconsolidated clays (experiment, Caen, France) (PPL, processed image). (D) Angular blocky with wide interaggregate star-shaped vughs in clayey silty materials (palsa, Ungava, Canada) (PPL). (E) Granular microstructure at the surface of a clayey mud boil (sorted circle, W Svalbard, Norway) (PPL). (F) Granular aggregates at the boundary between a former Bt horizon and an E horizon (Turbic Cryosol, French Alps) (PPL). *Images by B. Van Vliet-Lanoë.*

1984, 1992) and results in vertical (Fig. 3) or perpendicular (reticulate; Fig. 1C and D) ice growth directions, responsible for the development prismatic peds and angular blocky microstructures (Tsytovich, 1975; FitzPatrick, 1976), common in minerogenic palsas (mounds formed by the local accumulation of segregated ice in wetland sediments) and

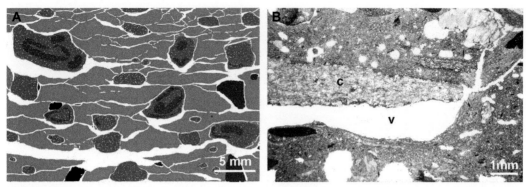

FIGURE 2 Pores associated with compact fragments. (A) Large pores open by enhanced ice lensing and frost-induced moisture retraction (cryosuction) underneath clayey aggregates (dark grey) and rock fragments (black) (experiment, Caen, France) (PPL, processed image). (B) Large pore (v) originating from ice lensing at the base of a coarse rock fragment (c) (onset of frost jacking) in a solifluction lobe (Southern French Alps) (PPL). *Images by B. Van Vliet-Lanoë.*

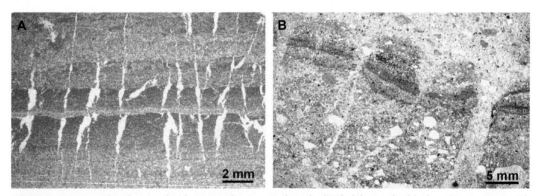

FIGURE 3 Features related to ice development along vertical cracks (PPL). (A) Angular blocky microstructure with prismatic peds caused by the development of vertically bladed ice in colluvial deposits (Bay of Mont St. Michel, France). (B) Same feature as in A, but the fissures opened by ice blade development were later filled with finer material (Port Racine archeological site, France). *Images by B. Van Vliet-Lanoë.*

in some fragipan horizons. Vertical ice blades may invade the non-frozen part of the sediment as in colluvial or tidal deposits (Fig. 3).

In stratified sediments, discontinuities affect the shape and size of the aggregates. If stratification is oblique to the frost penetration front, the sedimentary layers are obliquely cut by ice lenses. Obliquely oriented rock fragments in sediments, related to initial fluvial or slope deposition or soil stretching, and the disruption of thinly laminated sediments can enhance the development of frost-induced microstructures (Van Vliet-Lanoë, 1985).

During thaw consolidation, regular settling of the aggregates is observed when the ice content is moderate (Fig. 4A). When ice is abundant, like near the permafrost table, random lateral displacements of the peds (Fig. 4B) or oriented lateral displacement

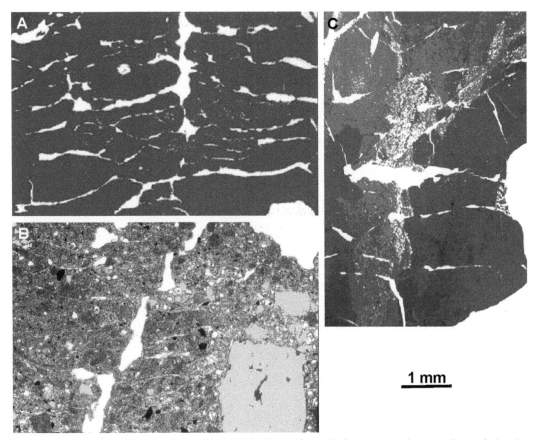

FIGURE 4 Aggregate settling after ice melting. (A) Settling without displacement, in the active layer of silty clay materials (Ungava, Canada) (UVF, fluorescent resin, processed image). (B) Settling with random displacement in marine clay above the permafrost table (Ungava, Canada) (PPL). (C) Settling with lateral displacement, above the permafrost table (Ny Alesund, Svalbard, Norway) (PPL). *Images by B. Van Vliet-Lanoë.*

(Fig. 4C) on slopes, can occur because of sliding along the top of melting ice lenses that promote reduced shearing resistance.

When repetitive freeze-thaw cycles affect a soil, the lenticular or platy microstructure is affected by internal microerosion, particle translocation and some plastic deformation. Fine particles can accumulate as cappings at the upper face of the platy or lenticular aggregates, giving rise to the so-called banded fabrics (Figs. 5 and 6), already mentioned in early descriptions of active boreal soils and slope deposits (e.g., Dumanski, 1964; Romans et al., 1966; Fedorova & Yarilova, 1972; FitzPatrick, 1976). While cappings are formed, sorting and accumulation of coarse grains may occur *in situ*, together with vertical jacking of grains and an inverted grading of grain sizes (Fig. 5) (Van Vliet-Lanoë, 1976). The size of the mineral grains determines whether they are propelled or not on the thin water film persisting at the front of growing ice crystals. Fine particles migrate ahead of the growing ice lens, whereas coarse grains are rapidly incorporated in the ice lens.

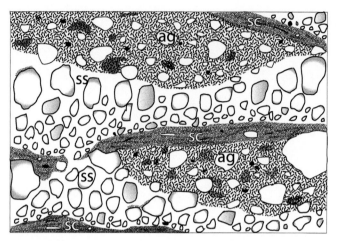

FIGURE 5 Basic characteristics of well-developed 'banded fabric': sorted coarse sand grains with inverted grading (ss), layered silt cappings with some enclosed fine sand grains (sc), lenticular aggregate (3-10 mm width) (ag). *After Van Vliet-Lanoë (1976); based on material from Piekray archeological site, Poland.*

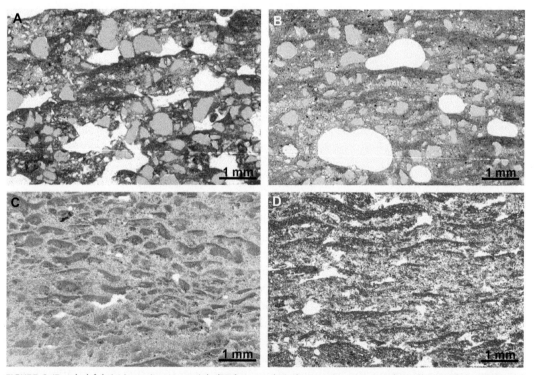

FIGURE 6 'Banded fabric' in various materials (PPL); note that the cappings survive the collapse of the platy and lenticular microstructures: (A) In sandy clay loam, with vertical alignment of the rock fragments (Southern French Alps). (B) In sandy loam, with the presence of vesicles (Kwadehuksletta, NW Svalbard, Norway). (C) In residual dolomitic silt (Southern French Alps). (D) In loamy fine sands (Heitatievat, Finnish Lapland). *Images by B. Van Vliet-Lanoë.*

The larger the specific surface of the grain and the thicker the water film, the longer the migration of the grains in front of the growing ice crystal will be (Van Vliet-Lanoë, 1982), resulting in vertical grading. Vertical translocation of fine particles during thaw periods further enhances the sorting of particles (Corte, 1966; Rowell & Dillon, 1972; Fox & Protz, 1981; Frenot et al., 1995). Along slopes, this sorting process is less prominent because of shearing or rotational movements (Van Vliet-Lanoë, 1982). Accumulation of sorted grains between the peds is also rather limited in materials rich in components with swell-shrink behaviour (e.g., smectite, organic matter). Accumulation of clean coarse grains has been experimentally obtained after 5-10 freezing cycles, and distinct capping along the upper face of the platy aggregate after 18-25 cycles (Coutard & Mücher, 1985), 30 cycles (Dumanski, 1964) and 100 cycles (Pawluk, 1988).

Dumanski and St. Arnaud (1966) defined three types of banded fabrics on the basis of the thickness of the 'bands' of fine material and the shape of the aggregates, which would reflect different degrees of soil development. Stoops (2003) refers to these banded fabrics as an example of complex pedofeatures related to lenticular or platy microstructures in which most peds have cappings of silt or coarse clay and are separated by more or less graded layers of clean coarse grains. The leached coarse grains correspond to the 'whitish powdering' of Russian authors (e.g., Morozova, 1965) or to 'podzol flour' (e.g., Dumanski & St. Arnaud, 1966). In present-day surface soils of Western and Eastern Europe, they are sporadically observed and often confused with a chemical or hydro-morphic degradation (e.g., Hardy et al., 1999).

In active cold environments, several frost cycles may occur each year, accelerating the process. Abundance of snowmelt water and low clay content can favour the expression of frost/thaw effects at depth. A distinctive layer on the cap or coating is often produced by each freeze-thaw cycle. In climatic zones with only one effective cycle of frost each year, one can use the development and thickness of the cappings to establish a relative chronology and to determine the rate of capping formation in a specific context, such as after an abrupt change of drainage conditions (Van Vliet-Lanoë, 1985).

The presence of platy/lenticular microstructures and/or banded fabrics with a different regular spacing in a same thin section (Fig. 7A) indicates that several generations of repetitive ice lensing occurred as a consequence of burial and/or erosion of the soil profile or a change in environmental conditions and pedogenesis (Van Vliet-Lanoë, 1976; Mestdagh, 2005). Micromorphological features indicating clay translocation or other pedogenetic processes may separate the two generations of cryogenic fabrics (Fig. 7B).

Freeze-thaw banded fabrics may be a transient feature in the history of the soil and may be further disturbed by permafrost, frost creep or mud flow, or be included in larger capped platy aggregates by burial and change in the freeze-thaw regime. Soils rich in swelling clays and organic colloids are highly susceptible to frost, but the cryogenic microstructures are poorly preserved after thaw because of rehydration swelling (Van Vliet-Lanoë, 1985).

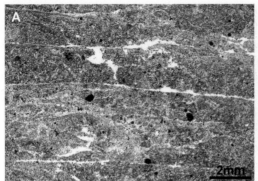
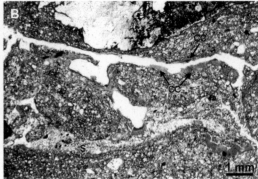

FIGURE 7 Frost features indicating two successive generations of ice lensing (PPL). (A) In a soil on loess (Rocourt, Belgium); the platy microstructure with open planar voids corresponds to the youngest stage. (B) In a fragipan soil on loess (Sonian Forest, Belgium), the oldest planar voids are filled with coarse mineral material (s), the most recent are open and were later covered by thin coatings of clay and iron oxides (cc). *(A) Image by H. Mestdagh; (B) Image by B. Van Vliet-Lanoë.*

2.2   Granular Microstructure

Granular microstructure (Fig. 1E and F) is very common in frost-affected soils with a very fine silt to clayey texture and/or rich in amorphous colloids (e.g., Crampton, 1977). It usually occurs in the upper part of the soil (Fig. 1E), which is most susceptible to multidirectional frost and frequent freeze-thaw cycles. At the macroscopic level this corresponds to so-called frost mulching (e.g., Sveistrup et al., 2005). The size of granular peds ranges from 1 to 8 mm. These aggregates were originally small angular blocky and became more and more rounded with age because of the cumulative effect of repeated frost-induced plastic deformations. Granular peds can also form through disruption of preexisting clay coatings or infillings produced by melt water infiltration (Van Vliet-Lanoë, 1988a). In this case, they are not microstructure elements and should be described as fragmented or deformed pedofeatures (Stoops, 2003).

Individual granules sometimes show stress features along the sides (Fig. 9B), often only visible in XPL and thus corresponding to poro- or granostriated b-fabrics (see Section 3.1). These features are formed by alignment of clay particles at the surface of the aggregates under the pressure exerted by the growth of ice crystals. Their presence seems to increase structure stability (Van Vliet-Lanoë, 1976; Smith et al., 1991). Ultradesiccation and colloid flocculation have a similar effect and make the granular units resistant to deformation by cryoturbation, mass wasting and collapse on thawing (Van Vliet-Lanoë, 1988a; Fox, 1994; Ostroumov, 1998, 2004). Other features associated with granular aggregates, such as cappings of fine material on granules (Figs. 9B and 15B) and occurrences of interpedal coarse grains around weakly developed granules, have been often documented in patterned grounds. These distinctive soil fabrics have been called 'ovoid' when the granules have associated stress features along the sides (poro- or granostriated b-fabric) or when the peds are covered by a capping (Morozova, 1965;

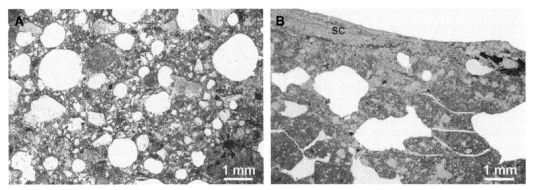

FIGURE 8 Vesicles (PPL). (A) Vesicles in the silty surface horizon of a mud boil (sorted circle; Boniface Lake, Nunavut, Canada). (B) Mammillated voids (deformed vesicles) in the silty clay surface horizon of a mud boil (*sorted circle*), with presence of a surface splash crust (sc) (Vars, French Alps). *Images by B. Van Vliet-Lanoë.*

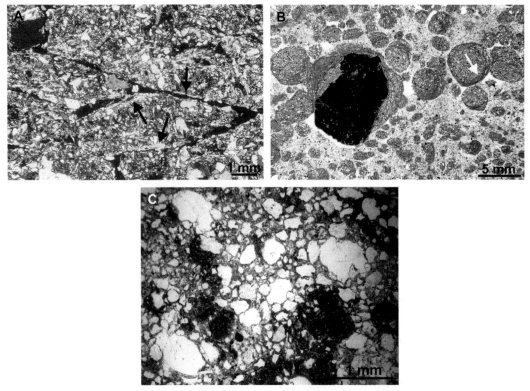

FIGURE 9 Stress features. (A) Porostriated b-fabric (arrows) along the sides of lenticular peds in loess (Harmignies, Belgium) (CPL). (B) Stress feature (*arrow*) associated with granular aggregates ('conglomeric fabric') in loamy sand, with stratified capping around the coarse grain (Boniface lake, Nunavut, Canada) (PPL). (C) Silt and sand-sized grains forming roughly circular patterns in the groundmass ('orbiculic fabric') of the surface horizon of an Orthic Turbic Cryosol (Mackenzie Mountains, Canada) (CPL). *(A) and (B) Images by B. Van Vliet-Lanoë; (C) Sampled by C. Tarnocai, image by C. Fox.*

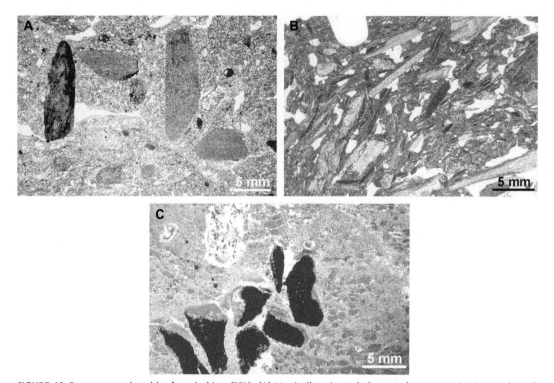

FIGURE 10 Features produced by frost jacking (PPL). (A) Vertically oriented elongated coarse grains in a paleosoil (Belgian Ardennes). (B) Oblique alignment of elongated coarse grains, resulting from frost jacking associated with frost creep (slow solifluction) (La Mortice, French Alps). (C) Rock fragments with silt cappings displaced to the side or bottom of the fragment (hummocks, Le Vallonet, French Alps). *Images by B. Van Vliet-Lanoë.*

Bunting & Fedoroff, 1974; Pawluk, 1988; Kovda & Lebedeva, 2013). Fabrics have been described in terms of 'orbiculic' textures (Fox & Protz, 1981) and 'microcircles' (Konishchev et al., 1973; Konishchev & Rogov, 1977) when coarse particle segregation has occurred between the peds, producing a two-dimensional pattern composed of rough circular bands of silt grains around granular units of more fine-grained material (Fig. 9C). Microcircles or orbiculic features seem to be often associated with very old, cryoturbated and often thixotrophic permafrost soils, subject to extremely frequent freeze-thaw cycles.

Rotation of soil aggregates and consequent formation of downturned cappings along the surface (Fig. 12C) is sometimes observed. It probably results from differential frost heave within a horizon at the microtopographical scale or from lateral displacement of the aggregates by solifluction on a slope (e.g., Aurousseau, 1976; Van Vliet-Lanoë, 1976, 1982; Harris & Ellis, 1980). Rotation events can be separated by periods of stability and non-frost pedogenesis, generating complex coatings on the grains or aggregates, often observed in downslope positions (Van Vliet-Lanoë, 1988a; Smith et al., 1991). This can produce a 'conglomeric fabric', which refers to a compound pattern consisting of heterometric granular aggregates, mineral grains and organic compounds (e.g., plant fragments, amorphous gels), enclosed by dense fine mineral material (e.g., Fox & Protz, 1981; Fox, 1983) (Figs. 9B and 15B). This fabric usually records a change in cryogenic dynamics, related to burial by colluvial or eolian sedimentation or to cryoturbation (differential frost heave).

A granular microstructure can be transitory, whereby granular aggregates are later included in larger capped platy aggregates, by burial or by changes in the freeze-thaw regime. Granular aggregates are often found in contemporary soils affected by past permafrost. Their persistence with time attests for their increased resistance. In Cryosols and palaeosoils, distinct granular aggregates often occur within the active layer, at sites with patterned ground features such as sorted circles and hummocks in the field. At greater depths, they are often found at the interface between two materials with different frost susceptibilities (Fig. 16B). They also form at depth in highly mobile solifluction lobes, in the case of temporary oversaturation in melt water (Van Vliet-Lanoë, 1976, 1982, 1995; Fox & Protz, 1981; Fox, 1983; Smith et al., 1991; Van Vliet-Lanoë et al., 2004).

Cryogenic granular peds may resemble aggregates reworked by rill wash in cold conditions, for example, accumulating at the foot slope of a fault, but they are clearly formed in the soil without any external input of material. Also, a cryogenic granular microstructure may sometimes be confused with granular or crumb microstructures produced by biological activity (Van Vliet-Lanoë et al., 1984) or salt mulching (Shein et al., 2013).

2.3 Microstructure Stability

Platy aggregates created solely by frost desiccation and without any compaction are often very unstable on moistening during thaw periods. This is usually the case in the A horizon and at levels near the permafrost table, where the propagation of frost is fast. In contrast, platy aggregates formed at depth by slow ice lensing under natural conditions are usually extremely firm and resistant, despite not being cemented with amorphous or other materials. This is related to ultradesiccation, causing vapour diffusion from the adsorbed water film at the surface of mineral grains and allowing the growth of ice lenses at very low temperature. The mechanical action of the enlarging ice lenses may lead to frost heave of freshly frozen material and to the development of platy aggregates. At very low temperatures (below -10°C), ice lens growth and ultradesiccation by frost contribute to compaction of these aggregates and the consequent increase in aggregate stability (e.g., Van Vliet-Lanoë et al., 1984; Dagesse, 2013).

In surface horizons rich in clay, irregular ice lensing induces fragmentation of the soil aggregates, known by farmers as frost mulching. Because soil particles and fine aggregates are detached, the susceptibility of the soil surface to splash erosion and to deflation increases. At depth, the stability of the aggregates created by frost increases when thawing is slow, the coldest winter temperatures are low and the concentration of organic or mineral colloids is high. These conditions can lead to the development of overconsolidated horizons (Van Vliet-Lanoë, 1988a, 1998).

Vesicles are commonly observed in frost-affected soils. Some authors (Van Vliet-Lanoë et al., 1984; Van Vliet-Lanoë, 1985, 1988a) attributed them to the expulsion of the air that was present between the ice crystals and inside peds in the frozen soil and that became trapped when the microstructure collapsed on sudden moistening during thawing. Other authors (FitzPatrick, 1956; Bunting, 1977; Harris & Ellis, 1980; Harris, 1983) explained vesicles as the result of air expelled from soil water during freezing. Vesicles are usually perfectly round in silt-rich or silt-and-sand-rich materials (Fig. 8A), but in clayey or organic soils they are often mammillated (Fig. 8B). Vesicles often occur in the upper 5 cm of the soil but may also occur along cracks or at depth when the presence of abundant rock fragments affects the thermal conductivity of the soil resulting in a relatively fast thawing.

The partial or total collapse of the microstructure of the soil surface under high precipitation and recurrent frost action enhances particle translocation processes. A direct consequence is the development at depth of B horizons with accumulations of fine material (FitzPatrick, 1956, 1976, 1993; Romans & Robertson, 1975; Forman & Miller, 1984; Van Vliet-Lanoë, 1985; Locke, 1986), which lead to differences in frost susceptibility within the profile. Differential frost susceptibility, due to variations in the clay and silt content, also exists between upslope and the downslope soils (Van Vliet-Lanoë, 1987b, 1995). The pedofeatures associated with these processes are further discussed in Section 4.

Needle ice formation, in combination with subsequent summer rainfall, also promotes the collapse of the microstructure by separating the peds and releasing particles. A splash crust (Fig. 8B) may form on the soil surface, and in deeper horizons layered or cross-layered infillings of silt and coarse clay can develop (see Section 4; Fig. 13B). One can confuse these structural splash crusts with the structure collapse occurring by liquefaction at the surface of patterned grounds. This process, or the percolation of surface runoff after snowmelt, is also responsible for the formation of the so-called mud cutans (FitzPatrick, 1956, 1997; Kubiëna, 1970; Wilkinson & Bunting, 1975), formed by massive infiltration of unsorted silt and coarse clay (see Section 4; Fig. 13A). The preservation of the platy microstructure during summer in cryoturbated grounds shows that liquefaction is accidental and that most of the plastic deformation of layers at macro- and microscopic scales occurs just before the frost consolidation (Van Vliet-Lanoë, 1998).

3. Groundmass

3.1 Micromass

Striated b-fabrics, in particular poro- and granostriated (Fig. 9A), have been commonly documented in present or past frost-affected soils and associated with alternating freeze-thaw (e.g., Morozova, 1965; Miedema, 1987) and other cryogenic processes (e.g., Konishchev et al., 1973, Konishchev & Rogov, 1977; Pécsi & Morozova, 1987; Tarnocai & Smith, 1989).

These b-fabrics correspond to the stress features (see Section 2.2) around cryogenic aggregates as described by Van Vliet-Lanoë (1976, 1982, 1985), who explains them as the result of the pressure exerted by growing ice crystals, inducing reorientation of fine material in the groundmass along pore surfaces or around aggregates. This interpretation has also been followed by Bunting (1983) and Fox (1983). However, striated b-fabrics alone cannot be considered diagnostic for frost activity because they can be produced by any type of stress, not necessarily exerted by the growing of ice crystals (Brewer & Pawluk, 1975). For instance, stress can be related to wetting and drying of clay-rich materials (Jamagne, 1972; Brewer, 1976).

3.2   Coarse Material

Vertically oriented grains (Fig. 10A and C) were already observed in early studies of frost-affected soils at both macroscopic and microscopic scales (Kaplar, 1965). A micro-structure dominated by them is often called a 'succitic microfabric' (Fox & Protz, 1981). The vertical position results from pressure exerted by growing ice lenses on the coarse grains (e.g., rock fragments, bone fragments, fragments of clay coatings, charcoal fragments, nodules) in a frost-susceptible groundmass, which leads to rotation and gradual vertical alignment of the coarse grains, creating a system in which ice growth pressure is minimised. This process was suggested in an early study by Beskow (1935) and later experimentally confirmed by Kaplar (1965) and Pissart (1969). The presence of large open pores (see Section 2.1; Fig. 2) at the base of coarse fragments is a good indicator that this process can take place (Van Vliet-Lanoë, 1976; Fox & Protz, 1981). The fact that the coarse grain sets up a hydraulic barrier enhances the process (Van Vliet-Lanoë & Dupas, 1991). Nevertheless, frost jacking is always in competition with other displacement processes, such as solifluction and gravity-controlled processes, which may result in alignments with other (oblique) predominant directions (Fig. 10B) (Harris & Ellis, 1980; Harris, 1990).

Frost-shattered rock fragments (Fig. 11A) and coarse mineral grains in the groundmass result from high moisture retention in fissures with subsequent ice growth. Fine particles are produced and loosened, especially if the rock is fractured by geological processes or weathered. This is particularly visible for soils on Arctic or alpine dolomitic limestones, in which the released silt-sized dolomite crystals form the micromass of many patterned grounds. Bones are also commonly fragmented in shards by frost (e.g., Todisco & Monchot, 2008). Frost shattering can also affect pedofeatures (Fig. 11B) such as nodules and coatings (Van Vliet-Lanoë & Langohr, 1983). Shattered grains can be further affected by other processes, such as dissolution, bioturbation and particle translocation (Van Vliet-Lanoë et al., 2004).

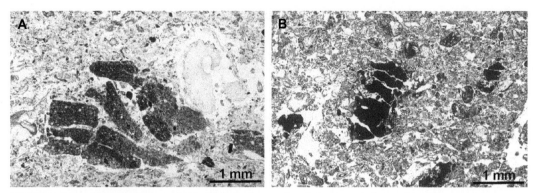

FIGURE 11 Frost shattering features (PPL). (A) Frost-shattered rock fragment in loess with silty groundmass material within the fissures (Spitsbergen, Norway). (B) Frost-shattered nodule in silty material (Southern Alps). *Images by B. Van Vliet-Lanoë.*

4. Pedofeatures

The most important pedofeatures associated with present or past alternating freeze and thaw are cappings of silt and coarse clay on coarse grains or peds, which are part of banded fabrics, as discussed in Section 2.1 (Fig. 12). They are usually called silt cappings and also correspond to the 'silt droplets' described by Romans and Robertson (1975). The thickness of these cappings varies between <20 μm and >1000 μm on large grains, depending on various factors such as the rate of infiltration during ice melting in spring, the susceptibility of fine material to translocation, the size and continuity of the interpedal porosity and the presence of rocks fragments. Cappings are generally finely layered and may contain inclusions of coarse mineral grains, charcoal, pollen grains or excrements. They can also bridge several generations of older cappings or form link cappings that either overlie or are juxtaposed to those cappings. Silt and coarse clay cappings can be intact, or they can be fragmented by pedoturbation or more recent frost. They are rather resistant to collapse or dispersion because they were formed by flocculation, ultradesiccation and mechanical compaction, in contrast with cappings induced by simple washing in aridic soils or tills. They often survive the complete collapse of the original platy microstructure, remaining as undulating accumulations of fine material that are not visibly associated with the present pore system (Fig. 6). As mentioned in Section 2.2, silt cappings that initially formed along the upper surface of an aggregate or coarse grain may appear at its side or bottom because of frost- or thaw-induced rotation (Figs. 10C and 12C).

Organic residues, such as mosses or leaves fragmented by ice lensing into coarse silt-sized particles, can be translocated and form cappings on grains or aggregates (Van Vliet-Lanoë, 1987a; Van Vliet-Lanoë & Sëppällä 2002; Todisco & Bhiry, 2008).

The partial or total superficial collapse of the soil at thaw promotes particle translocation from the surface horizons. This can occur as diffuse enrichment in the fine material of the groundmass (Schmertmann & Taylor, 1965; Forman & Miller, 1984;

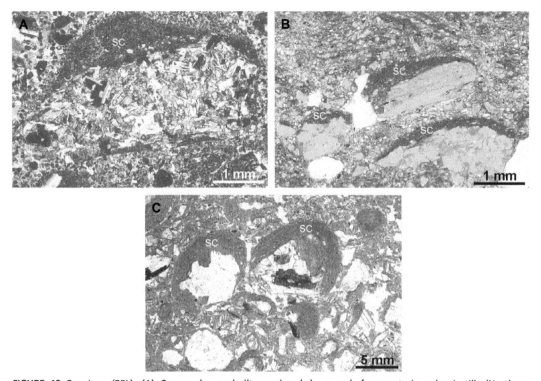

FIGURE 12 Cappings (PPL). (A) Coarse clay and silt capping (sc) on rock fragments in volcanic tills (Northern Iceland). (B) Coarse clay and silt cappings (sc) in solifluction deposits (frost creep; Spitzbergen, Norway). (C) Downturned cappings (sc) on coarse rock fragments (solifluction, French Massif Central). *Images by B. Van Vliet-Lanoë.*

Locke, 1986; Van Vliet-Lanoë, 1987b, 1988a) or as individual pedofeatures such as clay coatings and infillings. Thin coatings of illuviated fine silt or coarse clay ('mud cutans'; FitzPatrick, 1956, 1997; Kubiëna, 1970; Wilkinson & Bunting, 1975) around peds in subsurface horizons (Fig. 13A) and layered coatings and layered or cross-layered infillings of illuviated fine and coarse clay (Fig. 13B and C) have been documented in deeper horizons of freeze-thaw soils and attributed to liquefaction at the surface and/or fast percolation of surface runoff after snowmelt (Fedorova & Yarilova, 1972; McKeague et al., 1973; Van Vliet-Lanoë, 1988a; see also Kühn et al., 2010, 2018). Particle translocation is also enhanced by fast melting (liquefaction) and increased porosity (frost mulching) along the soil surface, as well as by rainfall (Frenot et al., 1995) or snowmelt (Wilkinson & Bunting, 1975). At depth, the accumulation of clay and silt leads to an increase in aggregate stability and consequently to the improvement of drainage and oxidation conditions, favouring the formation and preservation of coatings of clay and associated iron hydroxides. In stratified deposits, such as decalcified loess, illuviation below gentle slopes result, under repeated freeze-thaw, in the formation of a microbanded Bt horizon (Van Vliet-Lanoë, 1988a).

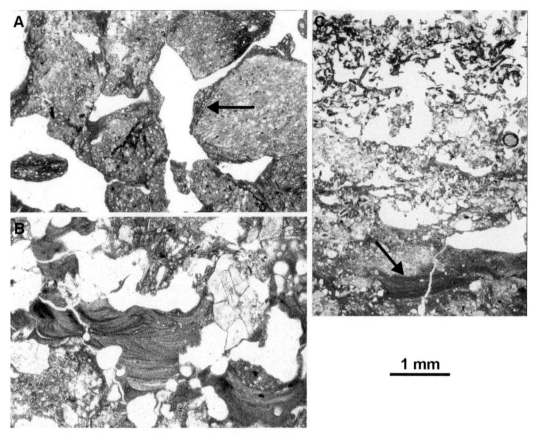

FIGURE 13 Illuvial features (PPL). (A) Silt and coarse clay coatings ('mud cutans') on angular blocky aggregates, showing some incorporation of illuvial material into the groundmass (Spitsbergen, Norway). (B) Cross-layered infillings of silt and clay reflecting rapid particle translocation during seasonal thaw, in saprolite (early Glacial, Normandy, France). (C) Layered infillings of silt and coarse clay below a cryptogamic crust (Boniface Lake, Nunavut, Canada). *Images by B. Van Vliet-Lanoë.*

Illuvial features are preserved only in well-drained soils with limited water availability and thus limited ice segregation, and if they are formed below the lower boundary of active ice segregation (the frost limit or desiccated layer). However, fragments of illuvial clay coatings integrated in the groundmass (Fig. 14) are often observed in Bt horizons (e.g., Brewer, 1976; Van Vliet-Lanoë & Langohr, 1983). Both the fragmentation and the integration in the groundmass result from mechanical stress due to differential frost heave between the groundmass and the clay-rich pedofeatures. Identical processes are responsible for the 'conglomeric' fabric integrating rounded fragments in the groundmass as observed for some spodic horizons (Fox & Protz, 1981; Jakobsen, 1989; see Fig. 15B).

Microinjection features are infillings produced by slow upward injection of still unfrozen material along desiccation or thermal contraction fissures, which occurs at the

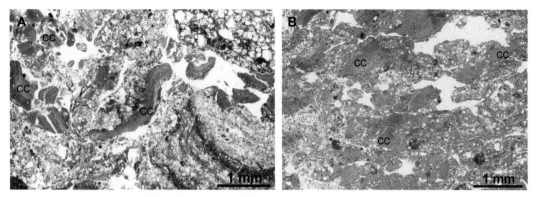

FIGURE 14 Fragmented illuvial features (PPL). (A) Coatings of clay and associated iron oxides (cc) fragmented by experimental cryoturbation. (B) Fragments of clay coatings (cc) incorporated into a silty clay groundmass by differential frost heave (Ungava Bay, Canada). *Images by B. Van Vliet-Lanoë.*

onset of soil freezing. The injected material may show deformed platy/lenticular or granular aggregates (Van Vliet-Lanoë et al., 1984). Their presence is related to lateral stress due to differential freezing rates and differential frost heave capabilities of the two materials. Microinjections are observed in all types of elongated voids in both modern and past environments (Van Vliet-Lanoë et al., 1984) subjected to repeated freezing and thawing. They have also been produced experimentally (Coutard & Mücher, 1985).

Increased salt concentration in the soil solution lowers the ice nucleation temperature, and liquid droplets may remain in the pores of the frozen soil enabling the crystallisation of salts independently from ice formation (Sweet, 1974; Ostroumov et al., 2001; Ostroumov, 2004). These authigenic silicates, carbonates and salt minerals (see also Mees, 2010, 2018; Mees & Tursina, 2010, 2018) occur as infillings, coatings or nodules. Although related to frost-induced desiccation, these features are not direct cryogenic pedofeatures (Sweet, 1974).

5. Implications for Pedostratigraphy

The micromorphological features discussed in this chapter are indicative of ice segregation, affecting soils in areas with and without permafrost, including aridic soils with available surface humidity (dew). They have no specific meaning in terms of environmental conditions, except the occurrence of frost activity at certain depths and in specific drainage conditions, but happening during very short (days) to very long periods (million years in permafrost). Specific horizons directly related to a permafrost thermal regime are, for instance, periglacial fragipans or foliated lodgement tills of formerly glaciated regions.

The impact of frost is often better expressed for cool periods than for very cold stages because water supply is essential for frost activity, whereby rainfall was higher during interglacial and interstadial periods than during glacial or stadial periods. Frost

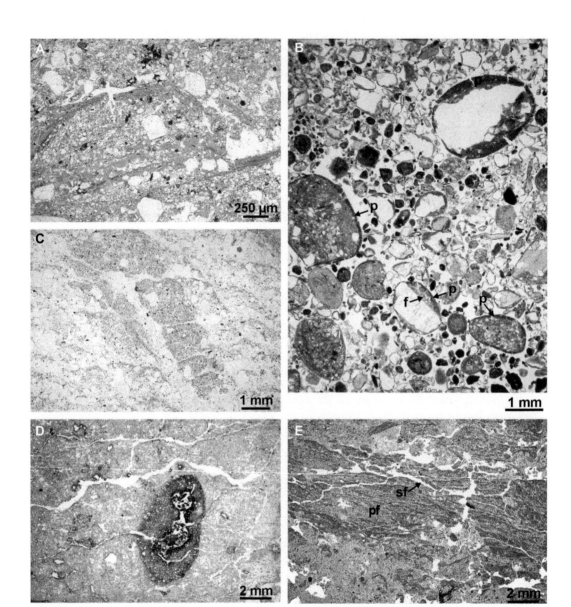

FIGURE 15 Compound features (PPL). (A) Two generations of illuvial clay on a former lenticular aggregate and separated by renewed frost activity (Piekary, Poland). (B) Silt capping (f) (frost activity) on coarse mineral grain, coated with a thin coating rich in organic material (p) ('conglomeric' fabric with podsolisation) in a peaty hummock (Finnish Lapland). (C) Frost-generated banded fabric developed in a channel infilling produced by beetles (Piekary, Poland). (D) Planar voids, associated with a platy/lenticular microstructure, cutting through an iron hydroxide nodule that formed at an earlier stage (Le Clypot, Belgium). (E) Material with a banded fabric (pf) (primary frost) incorporated in a till, subsequently affected as a whole by the development of a frost-generated platy/lenticular microstructure (sf) (secondary frost) (Mt. Jacques Cartier, Canada). *Images (A) to (D) by B. Van Vliet-Lanoë; image (E) by C. Tsao, Laval University.*

diagenesis affecting pedofeatures, usually at the end of an interglacial or after soil burial, has often been documented (Morozova, 1965; Van Vliet-Lanoë, 1976, 1988a; Mermut et al., 1983; Mücher & Morozova, 1983; Tursina, 1985; Miedema, 1987; Bertran, 1989; Cremaschi & Van Vliet-Lanoë, 1990; Huijzer, 1993; Mestdagh, 2005; Boixadera et al., 2008; Rellini et al., 2014; Bertran et al., 2015). Disturbance is mainly associated with deep seasonal frost and generally with an oligotrophic soil status with weak dispersion of clays (Van Vliet-Lanoë, 1998), except where organomineral complexes are present (Ugolini et al., 1977).

The development of cryogenic fabrics affects other soil features, for which the formation is related to physicochemical and biological processes. The expression of frost activity depends on whether frost action occurred before, during or after the other processes. If frost disturbance took place before temperate pedogenesis, the classical micromorphological features of temperate climates, such as clay coatings, appear superimposed on former frost-induced features (Van Vliet-Lanoë, 1998; Mestdagh, 2005; Van Vliet-Lanoë & Morawski, 2008) such as lenticular cryogenic aggregates (Figs. 7B and 15A). In the case of alternating periods with more and less cold climatic conditions, the changes of pedogenic processes associated with them are reflected in thin sections by the presence of compound pedofeatures (Jakobsen, 1989; Van Vliet-Lanoë & Seppälä, 2002; Mestdagh, 2005; Van Vliet-Lanoë & Morawski, 2008; Pawelec & Ludwikowska-Kedzia, 2016). Some examples are compound coatings on lenticular aggregates (Fig. 15A), coarse mineral grains with thin coatings rich in organic matter and silt cappings (Fig. 15B), banded fabric development on former biogenic infillings (Fig. 15C), zones with a banded fabric integrated in the groundmass or iron hydroxide nodules cut by planar voids associated with a recently developed platy microstructure (Fig. 15D and E).

On the other hand, the enrichment or depletion of the groundmass in clay or organic matter, decalcification and porosity changes resulting from pedo-, bio- and cryoturbation induce local changes in water retention capacity, resulting in zones with different frost susceptibility occurring side by side. Deformation of aggregates produced by frost heave at the contact between materials with different frost susceptibilities (Fig. 16) has been commonly observed in thin sections (e.g., Van Vliet-Lanoë, 1998; Van Vliet-Lanoë et al., 2004).

6. Summary and Conclusions

Ground freezing is a pedogenetic agent among other soil-forming processes. It results in temperature-driven soil desiccation associated with the main process of ice segregation. The impact of this process depends on various soil characteristics (e.g., grain-size distribution, clay and organic matter content, soil moisture content, drainage) and on the temperature regime. The segregation of ice lenses results in the development of distinctive soil features, the most typical being platy and lenticular structures at various scales. In conditions of alternating freeze and thaw, the characteristic lenticular and platy microstructures occur together with more or less developed silt and coarse clay

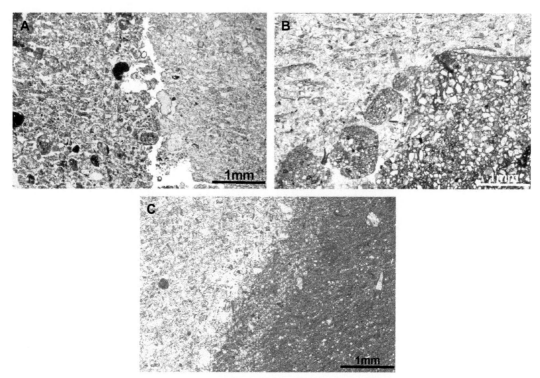

FIGURE 16 Deformation of aggregates induced by differential frost heave at the contact between materials with different frost susceptibility (PPL). (A) Contact between silty clay rich in organic matter (left, highly frost susceptible) and silty clay (right, less frost susceptible) (Gåsebu, Svalbard, Norway). (B) Contact between sandy material (left) and clayey sand (right, more frost susceptible) (Boniface Lake, Nunavut, Canada). (C) Vertical contact between sand (left) and dolomitic silts (right, more frost susceptible); note the lenticular aggregates in the silt bending toward the less frost-susceptible sand (Gåsebu, Svalbard, Norway). *Images by B. Van Vliet-Lanoë.*

cappings on lenticular aggregates and sorting of coarse grains, giving rise to different types of banded fabrics. Other micromorphological features related to ice segregation in soils are silt cappings on coarse grains, vertically oriented coarse grains, microinjection features, grano- and porostriated b-fabrics and granular microstructures. Freezing and thawing and solifluction processes further lead to compaction, displacement, rotation and deformation of the cryogenic features.

Cryogenic processes also have an effect on pore shape and size. The resulting change in porosity, composed of planes, vughs and vesicles, in turn affects the hydraulic conductivity of the soil after ice melt and thaw consolidation. Freezing combined with ice and snowmelt leads to the collapse of aggregates during thaw and hence to the translocation of fine particles and their accumulation in deeper horizons. Layered infillings of silt and coarse clay, indicating rapid translocation of fine material, are common. However, these illuvial pedofeatures are only preserved below the depth of seasonal frost penetration or in parts of the active layer that are dry in winter. In moist parts, they are rapidly incorporated to the soil by the frost-induced mechanical stress in winter.

Frost-induced microstructures are very typical and resistant. Their presence in palaeosoils and in soils from warmer regions thus constitutes a valuable proxy for past climatic conditions.

References

Aurousseau, P., 1976. Morphologie et Genese des Sols Sur Granite du Morvan. PhD Dissertation, Université de Rennes 1, 152 p.

Benedict, J.B., 1969. Microfabric of patterned ground. Arctic and Alpine Research 1, 45-46.

Bertran, P., 1989. L'Evolution de la Couverture Superficielle Depuis le Dernier Interglaciaire: Etude Micromorphologique de Quelques Profils Types du Sud de la France. PhD Dissertation, Université de Bordeaux 1, 210 p.

Bertran, P., Beauval, C., Boulogne, S., Brenet, M., Costamagno, S., Feuillet, T., Laroulandie, V., Lenoble, A., Malaurent, P. & Mallye, J.B., 2015. Experimental archaeology in a mid-latitude periglacial context: insight into site formation and taphonomic processes. Journal of Archaeological Science 57, 283-301.

Beskow, G., 1935. Soil freezing and frost heaving with special application to roads and railroads [*in Swedish*]. The Swedish Geological Society, Series C, No. 375. Translated by J.O. Osterberg, Technological Institute, Northwestern University, Evanston, 1947, 145 p.

Boixadera, J., Antúnez, M. & Poch, R.M., 2008. Soil evolution along a toposequence on glacial and periglacial materials in the Pyrenees Range. In Kapur, S., Mermut, A. & Stoops, G. (eds.), New Trends in Soil Micromorphology. Springer-Verlag, Berlin, pp 39-66.

Brewer, R., 1976. Fabric and Mineral Analysis of Soils. Robert E. Krieger Publishing Company, Huntington, 482 p.

Brewer, R. & Pawluk, S., 1975. Investigations of some soils developed in hummocks of the Canadian Sub-Arctic and Southern-Arctic regions. 1. Morphology and micromorphology. Canadian Journal of Soil Science 55, 301-319.

Bronnikova, M., Panin, A., Uspenskaya, O.G., Fuzeina, Y. & Turova, I., 2014. Late Pleistocene-Holocene environmental changes in ultra-continental subarid permafrost-affected landscapes of the Terekhol' Basin, South Siberia. Catena 112, 99-111.

Bunting, B.T., 1977. The occurrence of vesicular structures in artic and subartic soils. Zeitschrift für Geomorphologie 21, 87-95.

Bunting, B.T., 1983. High artic soils through the microscope: prospect and retrospect. Annals of the Association of American Geographers 73, 609-616.

Bunting, B.T. & Fedoroff, N., 1974. Micromorphological aspects of soil development in the Canadian High Arctic. In Rutherford, G.K. (ed.), Soil Microscopy. The Limestone Press, Kingston, pp. 350-365.

Campbell, I.B. & Claridge, G.G.C., 1987. Antarctica: Soils, Weathering Processes and Environment. Developments in Soil Science, Volume 6. Elsevier, Amsterdam, 368 p.

Corte, A.E., 1966. Particle sorting by repeated freezing and thawing. Biuletyn Peryglacjalny 15, 175-240.

Coutard, J.P. & Mücher, H., 1985. Deformation of laminated silt loam due to repeated freezing and thawing. Earth Surface Processes and Landforms 10, 309-319.

Coutard, J.P., Van Vliet-Lanoë, B. & Auzet, A.V., 1988. Frost heaving and frost creep on an experimental slope: results from soil structures and sorted stripes. Zeitschrift für Geomorphologie, Supplementband 71, 13-23.

Crampton, C.B., 1977. A study of the dynamics of hummocky microrelief in the Canadian North. Canadian Journal of Earth Sciences 14, 639-649.

Cremaschi, M. & Van Vliet-Lanoë, B., 1990. Traces of frost activity and ice segregation in Pleistocene loess deposits and till of Northern Italy: deep seasonal freezing or permafrost? Quaternary International 5, 39-48.

Dagesse, D.F., 2013. Freezing cycle effects on water stability of soil aggregates Canadian Journal of Earth Sciences 93, 473-483.

Dumanski, J., 1964. A Micropedological Study of Eluviated Horizons. MSc Dissertation, University of Saskatchewan, 124 p.

Dumanski, J. & St. Arnaud, R.J., 1966. A micropedological study of eluvial soil horizons. Canadian Journal of Earth Sciences 46, 287-292.

Fedorova, N.N. & Yarilova, E.A., 1972. Morphology and genesis of prolonged seasonaly frozen soils in Western Siberia. Geoderma 7, 1-13.

FitzPatrick, E.A., 1956. An indurated soil horizon formed by permafrost. Journal of Soil Science 7, 248-254.

FitzPatrick, E.A., 1976. Cryons and Isons. Proceedings of the North of England Soils Discussion Group 11, 31-43.

FitzPatrick, E.A., 1993. Soil Microscopy and Micromorphology. John Wiley & Sons, Chichester, 304 p.

FitzPatrick, E.A., 1997. Arctic soils and permafrost. In Woodin, S.J. & Marquiss, M. (eds.) Ecology of Arctic environments, British Ecological Society, Special Publication No. 13, Blackwell, Oxford, pp 1-39.

Forman, S. & Miller, G.H., 1984. Time-dependent soil morphologies and pedogenetic processes on raised beaches, Bröggerhalvöya, Spitsbergen, Svalbard. Arctic and Alpine Research 16, 381-394.

Fox, C.A., 1983. Micromorphology of an Orthic Turbic Cryosol, a permafrost soil. In Bullock, P. & Murphy, C.P. (eds.), Soil Micromorphology, Volume 2, Soil Genesis. AB Academic Publishers. Berkhamsted, pp. 699-705.

Fox, C.A., 1994. Micromorphology of permafrost-affected soils. In Kimble, J.M. & Ahrens R.J. (eds.), Proceedings of the Meeting on the Classification, Correlation and Management of Permafrost-affected Soils, Alaska. USDA, Washington, pp. 51-62.

Fox, C.A. & Protz, R., 1981. Definition of fabric distributions to characterize the rearrangement of soil particles in the Turbic Cryosols. Canadian Journal of Soil Science 61, 29-34.

Frenot, Y., Van Vliet-Lanoë, B. & Gloaguen, J.C., 1995. Particle translocation, and initial soil development on a glacier foreland, Kerguelen Islands, Subantarctic. Arctic and Alpine Research 27, 107-115.

Frese, H. & Czeratzki, W., 1957. Strukturbildung im Ackerboden. Die Umschau in Wissenschaft und Technik 16, 495-498.

Gubin, S.V., 1994. Relic features in recent tundra soil profiles and tundra soil classification. In Kimble, J. M. & Ahrens, R.J. (eds.), Proceedings of the Meeting on the Classification, Correlation and Management of Permafrost-affected Soils, Alaska. USDA, Washington, pp. 63-65.

Gubin, S.V. & Gulyaeva, L.A., 1997. Dynamic of soil micromorphology fabric. In Shoba, S., Gerasimova, M. & Miedema, R. (eds.), 10th International Working Meeting on Soil Micromorphology. Moscow-Wageningen, pp. 23-27.

Hardy, M., Jamagne, M., Elsass, F., Robert, M. & Chesneau, D., 1999. Mineralogical development of the silt fractions of a Podzoluvisol on loess in the Paris Basin (France). European Journal of Soil Science 50, 443-456.

Harris, C., 1983. Vesicles in thin sections of periglacial soils from north and south Norway. Proceedings of the Fourth International Conference on Permafrost, Fairbanks, pp. 445-449.

Harris, C., 1985. Geomorphological applications of soil micromorphology with particular reference to periglacial sediments and processes. In Richards, K.S., Arnett, R.R. & Ellis, S. (eds.), Geomorphology and Soils. George Allen and Unwin, London, pp. 219-232.

Harris, C., 1987. Mechanisms of mass movement in periglacial environments. In Anderson, M. &. Richards, K.S. (eds.), Slope Stability. John Wiley & Sons, Chichester, pp. 531-559.

Harris, C., 1990. Micromorphology and microfabrics of sorted circles, Front Range, Colorado USA. Proceedings of the 5th Canadian Permafrost Conference. Collection Nordicana 54, 89-94.

Harris, C. & Cook, J.D., 1988. Micromorphology and microfabrics of sorted circles, Jotunheimen, Southern Norway. Proceedings of the Fifth International Conference on Permafrost, Trondheim, Norway, TAPIR Publishers, Volume 1, pp. 776-783.

Harris, C. & Ellis, S., 1980. Micromorphology of soils in soliflucted materials, Okstindan, Northern Norway. Geoderma 23, 11-29.

Huijzer, A.S., 1993. Cryogenic Microfabrics and Macrostructures: Interrelations, Processes, and Paleoenvironmental Significance. PhD Dissertation, Vrije Universiteit Amsterdam, 245 p.

IUSS Working Group WRB, 2015. World reference base for soil resources 2014, update 2015. World Soil Resources Reports No. 106. FAO, Rome, 192 p.

Jakobsen, B.H., 1989. Evidence for translocations into the B horizon of a subarctic Podzol in Greenland. Geoderma 45, 3-17.

Jamagne, M., 1972. Some micromorphological aspects of soils developed in loess deposits of Northern France. In Kowalinski, S. & Drozd, J. (eds.), Soil Micromorphology. Panstwowe Wydawnicto Naukowe, Warsawa, pp. 559-582.

Kaplar, C.W., 1965. Stone migration by freezing of soil. Science 149, 1520-1521.

Kaplar, C.W., 1974. Phenomena and mechanism of frost heaving. Highway Research Record 304, 1-13.

Kokkonen, P., 1927. Beobachtungen über die Struktur des Bodenfrostes. Acta Forestalia Fennica 30, 1-55.

Konishchev, V.N. & Rogov, V.V., 1977. Micromorphology of cryogenic soils and sediments. Pochvovedenie 2, 119-125.

Konishchev, V.N., Faustova, M.A. & Rogov, V.V., 1973. Cryogenic processes as reflected in ground microstructure. Biuletyn Peryglacjalny 22, 213-219.

Kovda, I. & Lebedeva, M., 2013. Modern and relict features in clayey cryogenic soils: morphological and micromorphological identification. Spanish Journal of Soil Science 3, 130-147.

Kubiëna, W.L., 1953. The Soils of Europe. Thomas Murby, London, 317 p.

Kubiëna, W.L., 1970. Micromorphological Features of Soil Geography. Rutgers University Press, New Brunswick, 256 p.

Kubiëna, W.L., 1971. Micromorphology of polygenetic soils and paleosoils in Polar regions. Anales de Edafologia y Agrobiologia 30, 845-856.

Kühn, P., Aguilar, J. & Miedema, R., 2010. Textural features and related horizons. In Stoops, G., Marcelino, V. & Mees, F. (eds.), Interpretation of Micromorphological Features of Soils and Regoliths. Elsevier, Amsterdam, pp. 217-250.

Kühn, P., Aguilar, J., Miedema, R. & Bronnikova, M., 2018. Textural Pedofeatures and Related Horizons. In Stoops, G., Marcelino, V. & Mees, F. (eds.), Interpretation of Micromorphological Features of Soils and Regoliths. Second Edition. Elsevier, Amsterdam, pp. 377-423.

Laafar, S., Rousseau, L. & de Lumley, H., 1995. Evidence of palaeostructures related to freezing through soil micromorphology in Lazaret Cave, Nice (France). Comptes Rendus de l'Académie des Sciences, Série II, 321, 209-214.

Locke, W.W., 1986. Fine particle translocation in soils developed on glacial deposits, Southern Baffin Island, NWT, Canada. Arctic and Alpine Research 18, 33-43.

McDowall, I.C., 1960. Particle size reduction of clay minerals by freezing and thawing. New Zealand Journal of Geology and Geophysics 3, 337-343.

McKeague, J.A., MacDougall, J.I. & Miles, N.M., 1973. Micromorphological, physical, chemical and mineralogical properties of a catena of soils from Prince Edwards Island in relation to their classification and genesis. Canadian Journal of Soil Science 53, 281-295.

McMillan, N.J. & Mitchell, J., 1953. A microscopic study of platy and concretionary structures in certain Saskatchewan soils. study of platy and concretionary structures in certain Saskatchewan soils. Canadian Journal of Agriculture Science 33, 178-183.

Mees, F., 2010. Authigenic silicate minerals − sepiolite-palygorskite, zeolites and sodium silicates. In Stoops, G., Marcelino, V. & Mees, F. (eds.), Interpretation of Micromorphological Features of Soils and Regoliths. Elsevier, Amsterdam, pp. 497-520.

Mees, F., 2018. Authigenic silicate minerals − sepiolite-palygorskite, zeolites and sodium silicates. In Stoops, G., Marcelino, V. & Mees, F. (eds.), Interpretation of Micromorphological Features of Soils and Regoliths. Second Edition. Elsevier, Amsterdam pp. 177-203.

Mees, F. & Tursina, T.V., 2010. Salt minerals in saline soils and salt crusts. In Stoops, G., Marcelino, V. & Mees, F. (eds.), Interpretation of Micromorphological Features of Soils and Regoliths. Elsevier, Amsterdam, pp. 441-469.

Mees, F. & Tursina, T.V., 2018. Salt minerals in saline soils and salt crusts. In Stoops, G., Marcelino, V. & Mees, F. (eds.), Interpretation of Micromorphological Features of Soils and Regoliths. Second Edition. Elsevier, Amsterdam, pp. 289-321.

Menzies, J. & Maltman, A.J., 1992. Microstructures in diamictons − evidence of subglacial bed conditions. Geomorphology 6, 27-40.

Mermut, A.R. & St. Arnaud, R.J., 1981. Microband fabric in seasonally frozen soils. Soil Science Society of America Journal 45, 578-586.

Mermut, A.R., Rostad, H.P.W. & St. Arnaud, R.J., 1983. Micromorphological studies of some loess soils in Western Saskatchewan, Canada. In Bullock, P. & Murphy, C.P. (eds.), Soil micromorphology, Volume 1, Techniques and Applications. AB Academic Publishers, Berkhamsted, pp. 309-315.

Mestdagh, H., 2005. Environmental Reconstruction of the Last Interglacial and Early Glacial Based on Soil Characteristics of Pedocomplexes on Loess at Selected Sites From the Atlantic Coast to Central Asia. PhD Dissertation, Ghent University, 396 p.

Miedema, R., 1987. Soil Formation, Microstructure and Physical Behaviour of Late Weichselian and Holocene Rhine Deposits in the Netherlands. PhD Dissertation, Agricultural University Wageningen, 340 p.

Mol, J., Vandenberghe, J., Kasse, K. & Stel, H., 1993. Periglacial microjointing and faulting in Weichselian fluvio-aeolian deposits. Journal of Quaternary Science 8, 15-30.

Morozova, T.D., 1965. Micromorphological characteristics of pale yellow permafrost soils of Central Yakoutia in relation to cryogenesis. Soviet Soil Science 7, 1333-1342.

Mücher, H.J. & Morozova, T.D., 1983. The application of soil micromorphology in Quaternary Geology and Geomorphology. In Bullock, P. & Murphy, C.P. (eds.), Soil Micromorphology, Volume 1, Techniques and Applications. AB Academic Publishers, Berkhamsted, pp. 151-194.

Ontañon, J.M., 1978. Relation between two cryoturbation microforms at Sierra del Guadarrama (Spain). In Delgado, M. (ed.), Soil Micromorphology. University of Granada, Spain, pp. 581-588.

Ostroumov, V., 1998. Impact of freezing on distribution of ions in soils. Eurasian Soil Science, 31, 555-560.

Ostroumov, V., 2004. Physico-chemical processes in cryogenic soils. In Kimble, J. (ed.), Cryosols: Permafrost-Affected Soils. Springer, Berlin, pp. 353-368.

Ostroumov, V., Hoover, R., Ostroumova, N., Van Vliet-Lanoë, B., Siegert, Ch. & Sorokovikov, V., 2001. Redistribution of soluble components during ice segregation in freezing ground. Cold Regions Science and Technology 32, 175-182.

Parfenova, E.I. & Yarilova, I., 1967. Humus microforms in the soils of the USSR. Geoderma 1, 197-207.

Parmuzina, O.Y., 1978. Cryogenic structure and certain ice segregation phenomena in the active layer [*in Russian*]. In Popov, A.I. (ed.), Problems of Cryolithology. Moscow University Press, Moscow, pp. 141-164.

Pawelec, H. & Ludwikowska-Kedzia, M., 2016. Macro- and micromorphologic interpretation of relict periglacial slope deposits from the Holy Cross Mountains, Poland. Permafrost and Periglacial Processes 27, 229-247.

Pawluk, S., 1988. Freeze-thaw effects on granular structure reorganisation for soil materials of varying texture and moisture content. Canadian Journal of Soil Science 7, 1333-1342.

Pawluk, S. & Brewer, R., 1975. Micromorphological and analytical characteristics of some soils from Devon and King Christian islands, N.W.T. Canadian Journal of Soil Science 55, 349-361.

Pécsi, M. & Morozova, O., 1987. Micromorphological investigation of paleosol enclosures in the loess profile at Paks Hungary. In Fedoroff, N., Bresson, L.M., Courty, M.A. (eds.), Micromorphologie des Sols, Soil Micromorphology. AFES, Plaisir, pp. 619-624.

Pissart, A., 1969. Le mécanisme périglaciaire dressant les pierres dans le sol. Résultats d'expériences. Comptes Rendus de l'Académie des Sciences 268, 3015-3017.

Pissart, A., 1970. Les phénomènes physiques essentiels liés au gel, les structures périglaciaires qui en résultent et leur signification climatique. Annales de la Société Géologique de Belgique 93, 7-49.

Pissart A., Van Vliet-Lanoë, B., Ek, C. & Juvigné, E., 1988. Traces of ice in caves: evidence of former permafrost. Proceedings of the Fifth International Conference on Permafrost, Trondheim, TAPIR Publishers, Volume 1, pp. 840-845.

Rellini, I., Trombino, L., Rossi, P.M. & Firpo, M., 2014. Frost activity and ice segregation in a palaeosol in the Ligurian Alps (Beigua massif, Italy): evidence of past permafrost? Geografia Fisica e Dinamica Quaternaria 7, 29-42.

Romans, J.C.C. & Robertson, L., 1975. Soil and archaeology in Scotland. In Evans, J.G., Limbrey, S. & Cleere, H. (eds.), The Effect of Man on the Landscape: the Highland Zone. Council of British Archaeology, Research Report 11, pp. 37-39.

Romans, J.C.C., Stevens, J. & Robertson, L., 1966. Alpine soils in northeast Scotland. Journal of Soil Science 17, 184-199.

Romans, J.C.C., Robertson, L. & Dent, D.L., 1980. The micromorphology of young soils from South-East Iceland. Geografiska Analer 62A, 93-103.

Rowell, D. & Dillon, P., 1972. Migration and aggregation of Na-Ca clays by freezing of dispersed and flocculated suspensions. Journal of Soil Science 23, 442-447.

Schaefer, C.E.G.R., Simas, F.N.B., Gilkes, R.J., Mathison, C., da Costa, L.M. & Albuquerque, M.A., 2008. Micromorphology and microchemistry of selected Cryosols from maritime Antarctica. Geoderma 144, 104-115.

Schmertmann, J.H. & Taylor, R.S., 1965. Quantitative data from a patterned ground site over permafrost. CRREL Research Report 96, 76 p.

Sharp, R.P., 1942. Soil structures in the St Elias Range, Yukon Territory. Journal of Geomorphology 5, 274-301.

Shein, E.V., Kharitonova, G.V., Milanovskii, E.Yu., Dembovetskii, A.V., Fedotovac, A.V., Konovalovad, N.S., Sirotskii, S.E. & Pervova, N.E., 2013. Aggregate formation in salt-affected soils of the Baer mounds. Eurasian Soil Science 46, 401-412.

Shumskii, P.A., 1964. Principles of Structural Glaciology (translated by D. Krauss). Dover, New York. 497 p.

Smith, C.A.S., Fox, C.A. & Hargrave, A.E., 1991. Development of soil structure in some Turbic Cryosols in the Canadian Low Arctic. Canadian Journal of Earth Sciences 71, 11-29.

Soil Survey Staff, 2014. Keys to Soil Taxonomy. 12th Edition. USDA-Natural Resources Conservation Service, Washington, 360 p.

Stoops, G., 2003. Guidelines for Analysis and Description of Soil and Regolith Thin Sections. Soil Science Society of America, Madison, 184 p.

Sveistrup, T.E., Haralsen, T.K., Langohr, R., Marcelino, V. & Kvaerner, J., 2005. Impact of land use and seasonal freezing on morphological and physical properties of silty Norwegian soils. Soil and Tillage Research 81, 39-56.

Sweet, K., 1974. Calcrete crusts in an arctic permafrost environment. American Journal of Science 274, 1059-1063.

Szymański, W., Skiba, M., Wojtuń, B. & Drewnik, M., 2015. Soil properties, micromorphology, and mineralogy of Cryosols from sorted and unsorted patterned grounds in the Hornsund area, SW Spitsbergen. Geoderma 253/254, 1-11.

Taber, S., 1929. Frost heaving. Journal of Geology 37, 428-461.

Tarnocai, C. & Bockheim, J., 2011. Cryosolic soils of Canada: genesis, distribution, and classification. Canadian Journal of Soil Science 91, 749-762.

Tarnocai, C. & Smith, C.A.S., 1989. Micromorphology and development of some central Yukon paleosols, Canada. Geoderma 45, 145-162.

Tarnocai, C. & Valentine, K.W.G., 1989. Relict soil properties of the Arctic and Subarctic regions of Canada. Catena Supplement 16, 9-39.

Todisco, D. & Bhiry, N., 2008. Palaeoeskimo site burial by solifluction: periglacial geoarchaeology of the Tayara site (KbFk-7), Qikirtaq Island, Nunavik (Canada). Geoarchaeology 23, 177-211.

Todisco, D. & Monchot, H., 2008. Bone weathering in a periglacial environment: the Tayara site (KbFk-7), Qikirtaq Island, Nunavik (Canada). Arctic 61, 87-101.

Tsytovich, N.A., 1975. The Mechanism of Frozen Ground. Scripta Books Company, Washinghton, 152 p.

Tursina, T.V., 1985. Micromorphological diagnostics of cryogenic features in soils. Problems of soil cryogenesis. Nauka, Moscow, pp. 32-34.

Tursina, T.V., 1994. The peculiar features of microstructure and genesis of Cryosols of the permafrost regions of Russia. In Kimble, J.M. & Ahrens, R.J. (eds.), Proceedings of the Meeting on Classification, Correlation, and Management of Permafrost-affected Soils. USDA, Washington, pp. 155-159.

Ugolini, F.C., Dawson, H. & Zachara, J.M., 1977. Direct evidence of particle migration in the soil solution of a podzol. Science 188, 603-605.

Van Vliet-Lanoë, B., 1976. Traces de segrégations de glace associées aux sols et phénomènes périglaciaires fossiles. Biuletyn Peryglacjalny 26, 41-54.

Van Vliet-Lanoë, B., 1982. Structures et microstructures associées à la formation de glace de ségrégation. Leurs conséquences. In French, H.M. (ed.), Proceedings of the 4th Canadian Permafrost Conference, Calgary. NRC, Ottawa, pp. 116-122.

Van Vliet-Lanoë, B., 1985. Frost effects in soils. In Boardman, J. (ed.), Soils and Quaternary Landscape Evolution. Wiley Publishers, London, pp. 117-158.

Van Vliet-Lanoë B., 1987a. Interaction entre l'activité biologique et la glace de ségrégation en lentilles (Exemples en milieux arctiques et alpins). In Fedoroff, N., Bresson, L.M., Courty, M.A. (eds.), Micromorphologie des Sols, Soil Micromorphology. AFES, Plaisir, pp. 337-344.

Van Vliet-Lanoë, B., 1987b. Cryoreptation, gélifluxion et coulées boueuses: une dynamique de solifluxion continue en relation avec le drainage et la stabilité de l'agrégation cryogénique. In Pécsi, M. & French, H.M. (eds.), Loess and Periglacial Phenomena. Akadémiaì Kiadó, Budapest, pp. 203-226.

Van Vliet-Lanoë, B., 1988a. Le Rôle de la Glace de Ségrégation dans les Formations Superficielles de l'Europe de l'Ouest. Processus et Héritages. PhD Dissertation, Université de Paris 1, 854 p.

Van Vliet-Lanoë, B., 1988b. The significance of cryoturbation phenomena in environmental reconstruction. Journal of Quaternary Science 3, 85-96.

Van Vliet-Lanoë, B., 1995. Solifluxion et transferts illuviaux sur les versants périglaciaires lités. Etat de la question. Géomorphologie, Processus et Environnement 2, 85-113.

Van Vliet-Lanoë, B., 1998. Frost and soils: implications for paleosols, paleoclimates, and stratigraphy. Catena 34, 157-183.

Van Vliet-Lanoë, B. & Dupas, A., 1991. Development of soil fabric by freeze-thaw cycles. Its effect on frost heave. In Yu & Wang (eds.), Ground Freezing. Balkema, Rotterdam, pp. 189-195.

Van Vliet-Lanoë, B. & Langohr, R., 1981. Correlation between fragipans and permafrost with special reference to silty Weichselian deposits in Belgium and Northern France. Catena 8, 137-154.

Van Vliet-Lanoë, B. & Langohr, R., 1983. Evidence of disturbance by frost of pore ferriargilans in silty soils of Belgium and Northern France. In Bullock, P. & Murphy, C.P. (eds.), Soil Micromorphology, Volume 2, Soil Genesis. AB Academic Publishers, Berkhamsted, pp. 511-518.

Van Vliet-Lanoë, B. & Morawski, W., 2008. 4.1. Piekary IIa: 1971 excavations: stratigraphy and paleoenvironmental interpretation. In Sitlivy, V., Zieba, A. & Sobczyk, K. (eds.), Middle and early Upper Palaeolithic of the Krakow Region, Piekary IIa, Musées Royaux d'Art et d'Histoire, Bruxelles, pp. 19-36.

Van Vliet-Lanoë, B. & Seppälä, M., 2002. Stratigraphy, age and formation of peaty earth hummocks (pounus), Finnish Lapland. Holocene 12, 187-199.

Van Vliet-Lanoë, B., Coutard, J.P. & Pissart, A., 1984. Structures caused by repeated freezing and thawing in various loamy sediments. A comparison of active, fossil and experimental data. Earth Surface Processes and Landforms 9, 553-565.

Van Vliet-Lanoë, B., Helluin, M., Péllerin, J. & Valadas, B., 1992. Soil erosion in Western Europe: from the last interglacial to the present. In Bates, M. & Boardman, J. (ed.), Past and Present Erosion. Oxbow Monograph 22, Oxbow Books, Oxford, pp. 101-114.

Van Vliet-Lanoë, B., Fox, C.A. & Gubin, S., 2004. Micromorphology of Cryosols. In Kimble, J.M. (ed.), Cryosols: Permafrost-Affected Soils. Springer, Berlin, pp. 365-390.

Villagran, X.S., Schaefer, C.E.G.R. & Ligouis, B., 2013. Living in the cold: geoarchaeology of sealing sites from Byers Peninsula (Livingston Island, Antarctica). Quaternary International 315, 184-199.

White, T.L. & Fox, C.A., 1997. Comparison of cryogenic features dominant in permafrost affected soils with those produced experimentally. In Shoba, S., Gerasimova, M. & Miedema, R. (eds.), Soil micromorphology: Studies on Soil Diversity, Diagnostics, Dynamics. Moscow, Wageningen, pp. 436-444.

Wilkinson, A. & Bunting, B.T., 1975. Overland transport by rill water in a periglacial environment. Geografiska Annaler 57A, 105-116.

Chapter 21

Vertic Features

Irina Kovda[1,2], Ahmet R. Mermut[3]

[1]*DOKUCHAEV SOIL SCIENCE INSTITUTE, MOSCOW, RUSSIA;*
[2]*INSTITUTE OF GEOGRAPHY, RUSSIAN ACADEMY OF SCIENCES, MOSCOW, RUSSIA;*
[3]*HARRAN UNIVERSITY, SANLIURFA, TURKEY*

CHAPTER OUTLINE

1. Introduction

Vertic features result from shrink-swell processes, pedoturbation (or churning) and lateral shearing due to the alternating water regime in clayey materials. They are recognised for Vertisols and for various vertic soils that do not meet the requirements for this soil group or order. Cracking and pedoturbation are now believed not to be the most important but concomitant processes in the formation of vertic features by lateral shearing.

The World Reference Base for Soil Resources (IUSS Working Group WRB, 2015) defines Vertisols as churning heavy clay soils with high proportion of swelling clays, deep cracks and a vertic horizon, the latter being a clayey subsurface horizon that, as a result of shrinking and swelling, has slickensides and wedge-shaped peds. In Soil Taxonomy (Soil Survey Staff, 2014), similar criteria are used, with mention of slickensides, pedality, clay content and cracks in the soil order definition.

Vertic features occur worldwide in swelling clay soils, from boreal to tropical environments, with alternating wet and dry periods. They occur under a variety of moisture and temperature regimes, landforms, ecosystems and crops, which influence and transform their macromorphological and micromorphological, physical and chemical attributes but retain the diagnostic features.

The major morphological markers of Vertisols are slickensides and wedge-shaped peds (e.g., Mermut et al., 1996a). Other relevant characteristics are a clayey texture, high density when dry, granular structure in surface horizons, blocky structure at subsurface levels, deep wide cracks, monotonous colour coupled with weak horizonation or discontinuous cyclic sequences interrupted by diapiric intrusions (or 'chimneys') of underlying material, gilgai, and pedogenic carbonate, iron-manganese oxide and gypsum segregations. Except for the slickensides, wedge-shaped aggregates and clayey texture, the mentioned characteristics may all not be present in Vertisols.

Vertic behaviour results in the development of specific micromorphological features that are observed in thin sections. Jongerius and Bonfils (1964) were the first to describe in detail the micromorphology of a Vertisol. Micromorphological investigations were subsequently performed worldwide as part of studies on Vertisol genesis and commonly presented at international meetings on Vertisols or soil micromorphology. Selected examples of papers describing vertic features in tropical, subtropical and boreal Vertisols are listed in Table 1. Since the 1990's, studies focused mainly on classification issues and on the understanding of stress phenomena.

A number of papers present an overview of the micromorphological characteristics of Vertisols and vertic features and their pedogenic interpretation (Nettleton & Sleeman, 1985; Mermut et al., 1988, 1996b; Blokhuis et al., 1990). Illustrations of vertic features in thin sections can also be found in the G.D. Smith Memorial slide collection (Eswaran et al., 1999).

The purpose of this paper is to review and summarise the available micromorphological information about vertic features, in Vertisols and in the vertic intergrades of other soil groups. The specificity of these soils is largely related to shrink-swell, churning and shearing processes, which strongly influence all aspects of the microfabric.

2. Microstructure

The main processes responsible for the formation of aggregates in soils with vertic properties are desiccation, shear failure and biological activity (Wilding & Hallmark, 1984).

Table 1 Selected micromorphological studies of vertic features in Vertisols and vertic palaeosoils

Region	Publication
America	Mermut & St. Arnaud (1983), Stephan et al. (1983), Stephan & De Petre (1986), Yerima et al. (1987), Dasog et al. (1991), Nordt et al. (2004), Tabor & Montanez (2004), Driese et al. (2008, 2013), Díaz-Ortega et al. (2011), Dzenowski & Hembree (2012), González-Arqueros et al. (2013), Sánchez-Pérez et al. (2013)
Africa	Fedoroff & Fies (1968), De Vos & Virgo (1969), Blokhuis et al. (1970), Labib & Stoops (1970), Buursink (1971), Rodriguez Hernandez et al. (1979), Fedorov et al. (1991), Abdel Rahman et al. (1992), Acquaye et al. (1992), Blokhuis (1993), Beverly et al. (2014)
Europe	Krupenikov et al. (1966), Yarilova et al. (1969), Ghitulescu (1971), Kabakchiev & Boneva (1972), Kabakchiev & Galeva (1973), Bellinfante et al. (1974), de Olmedo Pujol & Pérez (1975), Tul'panov & Makeeva (1984), Bystritskaya et al. (1988), Fedorov & Soliyanik (1991), Kovda et al. (1992), Tsatskin et al. (1998, 2008), Martins & Pfefferkorn (2008)
Asia	Kooistra (1982a, 1982b), Rao et al. (1986), Dasog et al. (1988), Dabbakula et al. (1992), Sui Yaobing & Cao Shenggeng (1992), Zhang et al. (1993), Qui et al. (1994), Balpande et al. (1997), Kapur et al. (1997), Tsatskin & Ronen (1999), Heidari et al. (2005, 2008), Wieder et al. (2008), Pal et al. (2009), Srivastava et al. (2010), Rajamuddin et al. (2013), Shorkunov (2013)
Australia	Sleeman & Brewer (1984)

At subsurface levels, a moderately to strongly developed angular blocky microstructure (Fig. 1) with prismatic peds is typical for vertic horizons and corresponds to the wedge-shaped, angular blocky, prismatic-blocky or prismatic pedality observed in the field. A hierarchy of microstructures with up to three hierarchical levels of peds is commonly observed for vertic horizons (Sui Yaobing & Cao Shenggeng, 1992). Subangular blocky (Figs. 1 and 2), massive and complex microstructures also occur.

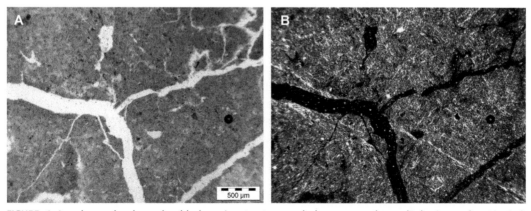

FIGURE 1 Angular and subangular blocky microstructure and dense groundmass (B horizon of a Vertisol, Ecuador). (A) In PPL. (B) Same field in XPL, showing various striated b-fabrics. *Ghent University archive.*

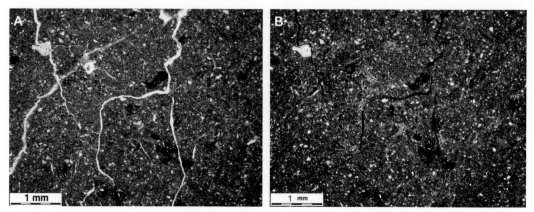

FIGURE 2 Subangular blocky microstructure, dense groundmass with open porphyric c/f-related distribution pattern and illuvial clay coatings (2Bkss2b horizon of a vertic palaeosoil, north-western Caucasus, Russia). Note that some clay coatings are related with present pores, whereas others have been incorporated into the groundmass. (A) In PPL. (B) Same field in XPL. *Images by I. Kovda.*

The surface horizon (1-5 cm) of Vertisols may have a fine granular microstructure produced by repeated self-mulching, i.e., cracking and swelling (Nettleton et al., 1983; Dasog et al., 1991; Sui Yaobing & Cao Shenggeng, 1992; Mermut et al., 1996b). Mermut et al. (1988) noted the importance of rapid desiccation in the formation of a granular structure. Experiments have shown that in the course of successive wet and dry cycles the structure of the surface of a Vertisol gradually changes from massive to complex crumb, blocky and platy and that the size of the peds progressively decreases (Hussein & Adey, 1998). Development of a granular microstructure may contribute to the appearance of other vertic features. For instance, granular peds falling down cracks may lead to heterogeneity of the groundmass at the initial stage, which is homogenised in mature Vertisols (Kovda et al., 2016). Rounded (spheroidal) microaggregates were also described for vertic horizons of Turkish and Israeli Vertisols formed on different parent materials. Their development and preservation are linked to high vermiculite and palygorskite contents, high intrinsic structural stability and high hydraulic conductivity (Kapur et al., 1997) or to high calcium contents (Tsatskin et al., 2008).

Vertic features have been described for soils with swell-shrink behaviour in strongly continental environments with permafrost (Kovda & Lebedeva, 2013; Kovda et al., 2014, 2017). Specific for these Vertisols is the transformation of cryogenic granular and platy structures at the macroscopic level and especially at the microscopic scale, whereby development of planar voids within granular peds results in the formation of finer wedge-shaped aggregates (Fig. 3).

The degree of pedality and ped separation is usually strong in dry field conditions but changes to weak in wet soils. Intraaggregate microporosity is usually weakly developed, but planar voids separating peds are common. Planar voids (Fig. 4) dominate in the vertic horizon at a depth between 60 and 160 cm (Blokhuis et al., 1990; Balpande et al.,

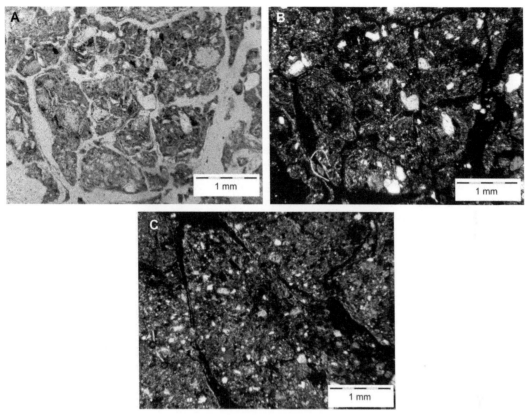

FIGURE 3 Angular blocky and wedge-shaped aggregates, resulting from transformation of cryogenic granular aggregates by shrink-swell processes (Bt,i,k,g horizon of a Gleyic Vertisol Glossic Gelistagnic, TransBaikal region, Buryatia, Russia). (A) In PPL. (B) Same field in XPL. (C) Detail of wedge-shaped aggregate (XPL). *Images by M. Lebedeva.*

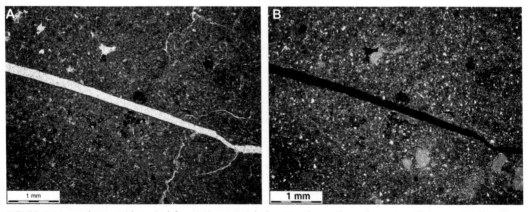

FIGURE 4 Large planar void, typical for vertic materials (vertic palaeosoil, north-western Caucasus, Russia). (A) In PPL. (B) Same field in XPL. *Images by I. Kovda.*

1997). Bui and Mermut (1989) found that the dominant orientation of planar voids is subhorizontal and oblique. It has been suggested that the frequency and orientation of planar voids could be used to distinguish between Vertisols and vertic subgroups (Mermut et al., 1988, 1996a). Comparative micromorphological studies of boreal and tropical Vertisols concluded that vughs were more common than planar voids in the boreal soils (Dasog et al., 1991), suggesting that swell-shrink and shearing processes are more active in tropical environments.

The characteristics and distribution of pores in Vertisols have been studied using image analysis (Puentes et al., 1992; Velde et al., 1996; Cabidoche & Guillaume, 1998, Moreau et al., 1999; Baer et al., 2009), but most of these studies were focused on the evaluation of image analysis techniques rather than on the characterisation of Vertisol porosity. However, image analysis has shown promising results in the assessment of compaction and subsequent structural regeneration of Vertisols (McKenzie et al., 1992; Pillai-McGarry, 1992).

Thin section preparation procedures may lead to structural changes in the clayey material. Freeze drying and critical point drying were compared, using SEM control, to find out which method of dehydration leads to more significant alteration of the microstructure (Bruand & Tessier, 1987; Tessier, 1987). It was observed that freeze drying gives rise to numerous microscopic cracks (1 μm wide). The critical point drying method modifies the microstructure less significantly, but it leads to a decrease in specimen volume.

3. Groundmass

Vertic horizons are mainly composed of clay with few coarse constituents. They usually have a dense groundmass with a double-spaced or open porphyric c/f-related distribution pattern. Fine monic c/f-related distribution patterns have been occasionally observed (Heidari et al., 2008).

Material from the surface horizons falls down the open cracks and is incorporated in deep horizons of the subsoil. In thin sections, this process can be recognised by a heterogeneous groundmass with incorporated dark aggregates (Fig. 5) or with materials of contrasting colour and composition occurring side by side (Fig. 6) (Mermut et al., 1996a; Kovda et al., 1999). Bioturbation can also explain the incorporation of abundant dark grains of surface-derived organic matter in the groundmass of deep subsurface horizons (Stephan & De Petre, 1986). Lateral shearing may lead to the lateral heterogeneity of the groundmass. These processes explain the complex morphology and soil cover pattern of some Vertisols (Wilding et al., 1990), especially in more wet conditions, and the strongly contrasting colours at the surface and in depth.

3.1 Coarse Fraction

Coarse material generally occurs in small amounts and rarely show relevant micromorphological features. Coarse grains can become fragmented by swell-shrink behaviour

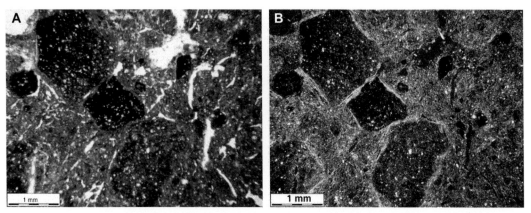

FIGURE 5 Heterogeneous groundmass (Bkss horizon of a Vertisol, northern Caucasus, Russia). Dark aggregates of material originated from the surface horizon are incorporated in the groundmass (A) In PPL. (B) Same field in XPL. Note the strongly developed granostriated b-fabric. *Images by I. Kovda.*

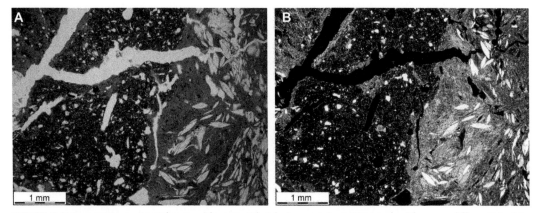

FIGURE 6 Heterogeneous groundmass with material rich in organic matter and with an undifferentiated to weakly developed stipple-speckled b-fabric occurring next to clayey material with random striated b-fabric and coarse gypsum grains (BCkssy horizon of a Vertisol, northern Caucasus, Russia). (A) In PPL. (B) Same field in XPL, showing a well-developed porostriated b-fabric in the left upper corner. *Images by I. Kovda.*

of the soil, showing an increase in degree of fragmentation with decreasing depth (Rodriguez Hernandez et al., 1979). Phytoliths in deep subsurface horizons can be an indication of the incorporation of surface-derived material through the vertical cracks (Fig. 7) (Boettinger, 1994; Kovda et al., 1999).

Yaalon and Kalmar (1978) observed that some Vertisols exhibit, toward the surface, a coarsening of the sand fraction, although the high clay content remains unchanged with depth. Their explanation is that repeated cycles of wetting and drying resulted in textural differentiation by uplifting of the sand grains, which is most effective on grains entirely

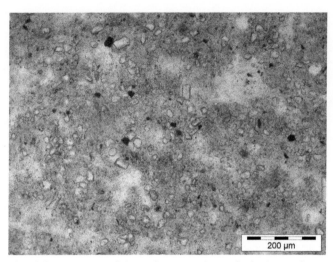

FIGURE 7 Numerous phytoliths incorporated into the groundmass (C horizon of an orthic-eutric Vertisol, Gigantones, Ecuador) (PPL). *Ghent University archive.*

embedded in the clayey groundmass. This process should be better expressed in older Vertisols formed along stable land surfaces with high content of swelling clays and relatively small amount of sand grains.

3.2 Micromass

3.2.1 Colour

A dark colour of the micromass is typical for many Vertisols and other soils with vertic properties. It has mainly been attributed to strong complexation between clay and organic matter (Singh, 1956). Bornand et al. (1984) considered the localisation of organic matter in relation to smectite layers and particles (quasicrystals), whereby a small part of the organic matter was found to have an external position and the main part has intraparticle localisation. Decomposition of organ and tissue residues may be responsible for the dark colour of the micromass (Fedorov & Soliyanik, 1991; Fedorov et al., 1991).

The colour of the fine material is strongly influenced by climate, parent material and carbonate content. Vertisols and vertic soils in arid and semiarid environments generally have a red-brown colour. In temperate or wet environments, they are often darker, which is generally attributed to high biological activity and high organic matter content (Dudal, 1965; Nordt et al., 2004). However, dark Vertisols can also be characterised by low biological activity and rare biological features (Fedorov & Soliyanik, 1991; Fedorov et al., 1991). Vertisols derived from basalt are usually dark. More than half of all grey, brown and red clayey Vertisols in Australia are dominated by illite and kaolinite (Norrish & Pickering, 1983).

3.2.2   b-Fabric

Shrink and swell processes in soils result in microshearing, which leads to reorientation of the individual clay plates into planar zones with face-to-face alignment of clay domains (Wilding & Tessier, 1988). These zones correspond to streaks with striated b-fabrics, characteristic of vertic materials (McCormack & Wilding, 1974; Mermut et al, 1996a). Mono-, parallel, cross-, poro- and granostriated b-fabrics (Figs. 1B and 6B) occur together with speckled b-fabrics (Fig. 6B). Strial b-fabrics can characterise Vertisols at depth, where these soils developed on sedimentary parent materials (e.g., Mees, 2001). The local occurrence of materials with this b-fabric in BC and C horizons can be used to estimate the amount of sediment already converted to soil material (Dasog et al., 1987).

SEM studies have shown strong orientation of clay particles in vertic horizons and, on the other hand, a decrease in the size of oriented clay domains due to their breakdown by high pressures under swelling (Tessier et al., 1992). Field et al. (1997), using SEM, observed that structural units of a Vertisol were typically composed of small spherical and elongated aggregates (10-50 μm diameter). These aggregates are clearly visible in SEM images, but in thin sections they appear as a pattern of small strongly anisotropic and weakly anisotropic areas, which correspond to cross-sections through oriented clay features enclosing the aggregates.

Calcitic crystallitic b-fabrics are predominant in calcareous vertic horizons from arid and semiarid environments (e.g., Blokhuis et al., 1970; Bellinfante et al., 1974; Kalbande et al., 1992; Pal et al., 2001). In these soils, the presence of calcite in the groundmass masks the occurrence of striation (Wilding, 1985; Kalbande, 1988; Mermut et al., 1996b; Aydemir et al., 2004; Heidari et al., 2005, 2008). This is demonstrated by the better recognition of striated b-fabrics in decalcified thin sections (Wilding & Drees, 1990). The expression of b-fabrics can also be obscured by the dark colour of the micromass related to a high organic matter content (Yerima et al., 1987; Eswaran et al., 1988; Kovda et al., 1992) or by a poor orientation resulting from a small size of the clay particles (Eswaran et al., 1988).

Vertic horizons with distinct slickensides show well-developed, striated b-fabrics (Fig. 8) (Mermut et al., 1996a). In SEM images (Fig. 9A), slickensides are recognised as bright, smooth surfaces with microgrooves approximately 0.3-20 μm wide, which are believed to be caused by the movement of coarse grains (Morgun Nobles et al., 2004). The thickness of the oriented clay along the slickenside surface is about 3-4 μm. A decrease in shrink-swell and shearing intensity is reflected by weakening of the expression of striated b-fabrics in thin sections and by lowering of the brightness of surfaces in SEM images.

SEM observations permitted to distinguish two types of coatings associated with peds in vertic palaeosoils (Shorkunov, 2013): (i) unidirectional 'shear coatings', which reflect both pressure and shearing processes (Fig. 9A) and which are similar to the slickensides described by Morgun Nobles et al. (2004), and (ii) 'pressure coatings', formed by repeated strong shrinking and swelling without shearing (Fig. 9B)

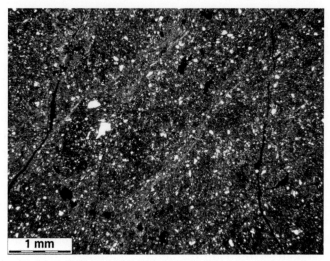

FIGURE 8 Parallel-striated b-fabric (2Bkss2b horizon of a vertic palaeosoil, north-western Caucasus, Russia). Note the striations corresponding to former slickensides, crossing present-day planar voids (XPL). *Image by I. Kovda.*

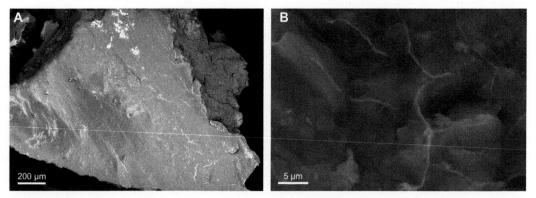

FIGURE 9 SEM images of ped faces (vertic palaeosoil, north-western Caucasus, Russia). (A) Continuous unidirectional 'shear coating'. (B) 'Pressure coating' along the surface of blocky ped, showing a microridge fractal pattern. *Images by I. Shorkunov.*

(see also Stoops & Mees, 2018). Features similar to 'pressure coatings' have also been observed for modern Vertisols (Kovda, unpublished data).

The degree of development of striated b-fabrics is affected by several factors. It seems to be greater in areas with high rainfall than in dry areas and to increase with increasing depth (Blokhuis et al., 1970). The same study indicates that the development of granostriated b-fabrics is influenced by the intensity of the churning process and by the size of the coarse grains.

A change in b-fabrics was also observed for Indian Vertisols formed along a climosequence extending from humid tropical to arid dry climates (Pal et al., 2009).

Despite the similar high shrink-swell potential, porostriated, parallel-striated and granostriated b-fabrics become less common and mosaic-speckled, stipple-speckled and crystallitic b-fabrics become better expressed in Vertisols of the drier climatic environments.

Attempts have been made to correlate b-fabrics with the shrink-swell potential of a soil measured by COLE values, whereby it was found that higher COLE values corresponded to better developed striated b-fabrics (Yerima et al., 1987). In turn, the shrink-swell potential of a soil is generally negatively correlated with the abundance of organic matter, carbonates, gypsum, iron-manganese oxides and low-activity clays and with electrolyte concentration (Mermut et al., 1991). Kalbande et al. (1992) found that weak swelling of clays, limited by chloritisation of smectite interlayers, was insufficient for the formation of a porostriated b-fabric and only resulted in either mosaic-speckled or granostriated b-fabrics. Based on variations in b-fabric, Kalbande (1988) suggested that shrink-swell processes are less pronounced in the presence of micritic calcite or coarse grains and in conditions with short dry periods.

Changes in b-fabric through wet/dry cycles were studied by Hussein and Adey (1998). They found that these changes were less well defined and more difficult to interpret than microstructure changes developed in the same conditions. The development of b-fabric generally decreased as the soil went through successive wet/dry cycles. b-Fabrics were better developed in vertical thin sections than in horizontal thin sections, suggesting that more clay orientation took place parallel to the wetting front. In this study, the b-fabrics were generally mosaic- to stipple-speckled with granostriation around large mineral grains and calcium carbonate nodules, and with weakly developed or absent porostriation.

The degree of development of vertic features may decrease in mature soils. Deep black soils on alluvium derived from weathered basalt, with typical properties of Vertisols except for the presence of slickensides, show only weakly developed porostriated b-fabrics (Paranjape et al., 1997). In these soils, shrink and swell processes were operative, but their extent was not great enough to reorganise particles and produce b-fabrics typical for vertic horizons. Clayey smectite-dominated soils may need at least 550 years of shrinking and swelling before slickensides and striated b-fabrics can form (Paranjape et al., 1997). Other studies found that in smectite-dominated soils in very cold environments these features need less time (several decades) to form and become visible (Kovda & Lebedeva, 2013; Kovda et al., 2014, 2017).

Drying of unconfined soils does not produce significant reorganisation of clay domains because this requires a rather strong force. However, some methods of thin section preparation may create artificial stress that could enhance the development of shear-related b-fabrics (Blokhuis et al., 1990). This happens when moist clayey soil is dried quickly and/or when the resin is excessively heated during hardening (60-80° C). Shear-related b-fabrics can also occur along the edges of a thin section, as a result of friction between the material and the sampling tool.

3.3   Pedofeatures

In comparison with the groundmass features discussed above, the type, distribution, orientation and abundance of pedofeatures in thin sections of vertic materials are much more variable, in function of the temperature and moisture regimes. They therefore provide useful information about environmental conditions.

3.3.1   Fe and Mn Oxide Pedofeatures

Fe and Mn oxide concentrations are very common in vertic horizons and Vertisols as a whole (Sleeman & Brewer, 1984; Blokhuis et al., 1990). The high water retention properties and low permeability of the dense clayey vertic materials are responsible for humid or water-saturated environments and consequently for the reduction and mobilisation of Fe and Mn, resulting in the formation of Fe and Mn oxide pedofeatures (see also Lindbo et al., 2010; Vepraskas et al., 2018).

The most common Fe and/or Mn oxide features are impregnative and intrusive nodules (Fig. 10) and poorly crystalline coatings and hypocoatings (Rodriguez Hernandez et al., 1979; Sleeman & Brewer, 1984; Blokhuis et al., 1990; Kovda et al., 1992; Nordt et al., 2004; Heidari et al., 2008). Concentric nodules (Fig. 10C) are typical for vertic horizons and are believed to reflect moisture regimes with repeated seasonal wet/dry cycles (Brewer, 1964; Van Ranst et al., 2011). Fe/Mn oxides followed by zeolite crystals in the pore system of palaeo-Vertisols were attributed to a transition from wetter to drier conditions during the evolution of these soils (Beverly et al., 2014). Image analysis and mapping with SEM/EDS analysis of rounded nodules in a young Vertisol from Texas have shown clear banding for Fe and Mn. Irregularly shaped nodules have patchy patterns for Mn with homogeneous Fe distribution. No banding for Al, Si, K, Ca or Ba was observed for any of the studied nodules (White & Dixon, 1996). Typic, aggregate or dendritic nodules (Fig. 10A, B and D), partly with gradual boundaries, have also often been described for vertic materials (Labib & Stoops, 1970; Buursink, 1971; Sleeman & Brewer, 1984). An upward increase in degree of fragmentation and dispersion of these pedofeatures has been reported (Rodriguez Hernandez et al., 1979).

The occurrence of abundant Fe and Mn oxide pedofeatures of various morphologies has been interpreted as evidence for prevailing wet conditions in the past (Blokhuis et al., 1969; Kabakchiev & Galeva, 1973). It indicates periodic anaerobic conditions and may occur in deep horizons of soils in both low and high parts of a gilgai landscape (Hallsworth & Beckmann, 1969). Backscattered electron images of thin sections show a large variability of size and shape of nodules within the clay matrix of Vertisols (Chittamart et al., 2010). SEM-EDS analysis of abundant nodules (smaller than 6 µm) in Vertisols from Argentina revealed their complicated structure and composition, with a combination of organic substances, sesquioxides and other absorbed compounds (Stephan et al., 1983). In addition, Fe/Mn oxide and magnetite nodules of bacterial origin were described for Vertisols (Chotte et al., 1994), emphasising the importance of investigating the distribution of soil microorganisms within soil pores and the relationships between microorganisms and mineral grains.

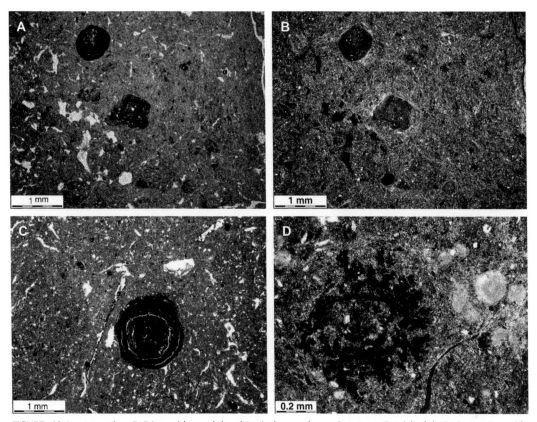

FIGURE 10 Impregnative Fe/Mn oxide nodules (Vertisols, northern Caucasus, Russia). (A) Typic Fe/Mn oxide nodules (Bkss horizon) (PPL). (B) Idem in XPL. Note granostriation around nodules. (C) Concentric Fe/Mn oxide nodule (Bss horizon) (PPL). (D) Dendritic Fe/Mn oxide nodule (Bk horizon). Note also the calcite nodules. (XPL). *Images by I. Kovda.*

3.3.2 Carbonate Pedofeatures

Vertic horizons may be free of carbonates or may have soft (powdery) or hard (concretionary) carbonate segregations. In thin sections, hard carbonate nodules are generally micritic or microsparitic and usually have sharp boundaries. Calcite nodules can be orthic, disorthic or anorthic (Wieder & Yaalon, 1974). Acicular calcite, referred to as lublinite, may also occur inside voids (Dabbakula et al., 1992).

Hard nodules also often contain Fe and Mn oxide impregnations (Figs. 11 and 12), including dendritic Mn oxide occurrences (Blokhuis et al., 1969; Labib & Stoops, 1970; Kooistra, 1982a). Their sharp boundary has been attributed to churning processes, i.e., argilloturbation (Blokhuis et al., 1969; Wieder & Yaalon, 1974). It has also been considered an indication that nodules were no longer actively growing (Kovda et al., 2003; Nordt et al., 2004).

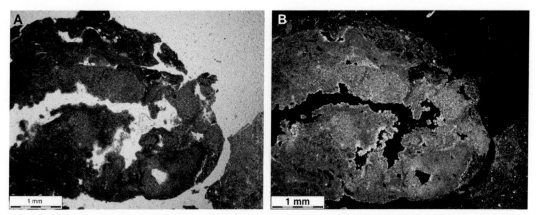

FIGURE 11 Geodic disorthic complex microsparitic carbonate nodule with Fe and Mn oxide impregnations (Bkss horizon of a Vertisol, northern Caucasus, Russia). (A) In PPL. (B) Same field in XPL. *Images by I. Kovda.*

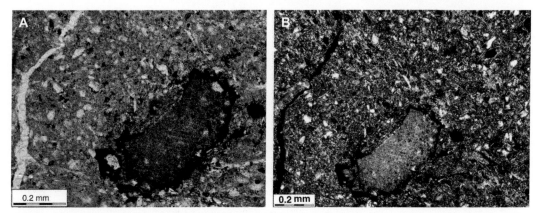

FIGURE 12 Typic disorthic calcite nodule with external and internal Fe oxide hypocoating (3Bkss1b horizon of a vertic palaeosoil, north-western Caucasus, Russia). (A) In PPL. (B) Same feature in XPL. *Images by I. Kovda.*

White and black carbonate nodules, with and without Fe/Mn oxide coatings, have been described for Indian and Sudanese Vertisols (Blokhuis et al., 1969; Rajan et al., 1972; Mermut & Dasog, 1986; Srivastava et al., 2002). Microsparitic or sparitic carbonate nodules with Fe/Mn oxide coatings and sharp boundaries have been interpreted as having formed elsewhere, followed by transportation and redeposition (Pal et al., 1999, 2001). Fe/Mn oxide coatings on hard carbonate nodules reflect a wetter environment and are often a relict feature (Mermut & Dasog, 1986). These coatings can protect carbonate nodules against recrystallisation, preventing rejuvenation of their radiocarbon age and changes in stable isotope geochemistry (Kovda, 2004).

Shrinking/swelling, lateral shearing and vertical churning have an impact on the morphology and location of carbonate pedofeatures. Many hard carbonate nodules are disorthic, whereas micritic soft carbonate masses may be elongated parallel to the shear direction (planar voids or slickensides) (Podwojewski, 1995). Carbonate nodules may have a clustered or semibanded distribution parallel to planar voids or slickensides. Some bands were reported to be oriented at an angle of 45 degrees with the horizontal, indicating soil displacement through shear failure (Mermut & Dasog, 1986). Blokhuis et al. (1969) explained the peculiarities of hard carbonate nodules in Vertisols by their formation in the substratum and further upward transport, during which they acquired a subangular to rounded shape and also became harder. The Fe/Mn oxide impregnations and coatings on those carbonate nodules were also inherited from the substratum.

Micritic/microsparitic calcite coatings and hypocoatings in voids have been documented as well in horizons with vertic properties (Blokhuis et al., 1990; Kovda et al., 2003; Heidari et al., 2005).

Other carbonate nodules such as porous spherulitic aragonite accumulations and dolomitic or magnesitic micritic nodules, as well as barite aggregates, were described by Podwojewski (1995) for magnesium-rich Vertisols.

3.3.3 Gypsum Pedofeatures

Gypsum pedofeatures often occur in deep vertic materials in arid environments (Dudal, 1965; Sehgal & Bhattacharjee, 1988; Ahmad, 1996; Heidari et al., 2008). Compared with Fe/Mn oxide and carbonate pedofeatures, gypsum pedofeatures are less common. Various gypsum features have been described (Barzanji & Stoops, 1974; Podwojewski & Arnold, 1994; Podwojewski, 1995; see also Poch et al., 2010, 2018) with no indications about the relationship between those features and vertic characteristics. Gypsum can be inherited from the parent material or formed by early translocation processes of soluble constituents of the parent material by groundwater or soil solutions and even from ocean source (Barzanji & Stoops, 1974; Parfenova & Yarilova, 1977; Podwojewski & Arnold, 1994; Coulombe et al., 1996a).

3.3.4 Textural Pedofeatures

Textural pedofeatures, in particular illuvial clay coatings, are uncommon in Vertisols and horizons with vertic properties (Blokhuis, 1993). Nevertheless, they have sometimes been described (e.g., Osman & Eswaran, 1974; Verheye & Stoops, 1974; Mermut & Jongerius, 1980; Yerima et al., 1987; Dasog et al., 1991; Gunal & Ransom, 2006a, 2006b). Osman and Eswaran (1974) combined data on the percentage of clay coatings in thin sections with total clay contents and incipient flocculation ratios, concluding that clay translocation is an active process in Vertisols but that the evidence of clay translocation is continuously being destroyed by vertic processes. Textural pedofeatures are often fragmented and/or deformed and incorporated in the groundmass by shrink-swell and shearing processes (Fig. 2) (Nettleton & Sleeman, 1985; Yerima et al., 1987; see also Kühn et al., 2010, 2018) and if present they may show a kink-band fabric instead of a continuous orientation (Stoops, 2003). They are more commonly found in deep horizons, where they are not disturbed by shrink-swell activity (Wilding & Drees, 1990).

Textural pedofeatures in Vertisols can be related to a former wetter environment or to a decrease in the activity of vertic processes, with the rate of illuviation exceeding the rate of destruction by churning processes (Nettleton & Sleeman, 1985). The development of clay coatings can also be due to ferrolysis in wet Vertisols that become very acid because of redox changes under alternating seasonal reduction and oxidation, resulting in clay disintegration in the upper part of the soil (Wilding, personal communication).

Illuvial clay coatings were observed in thin sections of smectite-rich horizons of buried palaeosoils combined with porostriated and granostriated b-fabrics, but they were absent in parts with cross-striated b-fabrics (Gunal & Ransom, 2006a). A more recent review of tropical Indian Vertisols suggests that the presence of coatings of impure and weakly oriented clay in some Bss horizons results from long-term clay illuviation combined with low shrink-swell activity (Pal et al., 2012).

Deformed illuvial clay coatings may be very similar in appearance to porostriated b-fabrics originating exclusively from vertic soil behaviour, and the distinction between both types of features may require close investigation (e.g., Gunal & Ransom, 2006b).

Fe-rich clay coatings were observed for Canadian Vertisols and interpreted as resulting from weak pedoturbation in a subhumid boreal climate (Dasog et al., 1991). Also, minor clay illuviation and leaching of carbonates and/or gypsum were described for Australian grey and brown clay soils (Hallsworth & Beckmann, 1969).

Pedofeatures composed of silt, in the form of nodules and intercalations, were occasionally found in some Australian cracking clay soils, generally in the lower part of the profiles (Sleeman & Brewer, 1984). Infillings composed of sorted sand were also described for a vertic palaeosoil in India (Srivastava et al., 2010).

3.3.5 Excrements
Some authors consider the absence of excrement pedofeatures to be a typical feature of Vertisols, due to low biological activity (Fedorov & Soliyanik, 1991; Fedorov et al., 1991).

4. Degradation of Vertic Features in Cultivated Soils

Well-illustrated examples of how structure and porosity of Vertisols are affected by land use were presented by Coulombe et al. (1996b), showing that continuous and intensive cultivation results in the formation of a massive structure, an increase in bulk density and a lowering of continuity and connectivity of macropores. McGarry (1987, 1989) reported that structural degradation may extend to a depth of 65 cm and result in the development of blocky, massive or platy microstructures in various parts of the profile. A comparison of the effects of wet and dry cultivation on soil structure and b-fabric using thin sections shows that cultivation of wet soils leads to better developed striated b-fabrics, mostly at a depth of 5-20 cm (McGarry, 1987, 1989). This is related to the greater incidence, at this depth, of shearing of soil in a plastic state by tillage tools and tractor wheels. Other studies suggest that management practices, in general, did not lead to significant modifications, except for a change from a complex striated b-fabric in a

Vertisol under dry farming to a mostly parallel-striated b-fabric in an irrigated Vertisol (Kalbande et al., 1986).

Waterlogging is a major risk in irrigated Vertisols and results in changes of the type, location and abundance of pedofeatures (calcite, gypsum and Fe/Mn oxide occurrences; clay coatings; depletion pedofeatures). Dissolution of calcite and gypsum and their accumulation in deep horizons as calcite nodules and gypsum infillings were described for wet Vertisols (Kovda et al., 2003). Newly formed pedofeatures reflect a more humid soil environment by the larger size of the crystals and the presence of illuvial clay coatings and iron oxide nodules. Fe oxides occur as impregnative typic nodules and hypocoatings instead of concentric nodules. In irrigated soils, the total porosity and the degree of development of b-fabrics increase, microaggregation and intraaggregate porosity decrease, and the colour of the micromass in the surface horizon changes to lighter brown (Dostovalova & Tursina, 1988).

The reclamation of sodic Vertisols by adding $CaCO_3$ results in a change in aggregation, an increase in clay translocation and the dispersion of fine organic material (Bystritskaya et al., 1988). Experiments with Br and Brilliant Blue FCF dyes have shown that slickensides do not act as significant barriers for the penetration of solutes and chemicals into the groundmass (Morgun Nobles et al., 2004). Solutes penetrate into the soil both vertically along the cracks and laterally through the slickenside surfaces.

5. Vertic Features in Palaeosoils

Diagnostic vertic features are usually well preserved and easily identifiable in palaeosoils that are up to Archaean-Proterozoic and Cambrian in age (Retallack, 2008; Bandopadhyay et al., 2010) even in case of assumed weak and/or short duration of vertic pedogenesis (Tovar et al., 2014). Vertic features have been recognised and studied in palaeosoils in several places worldwide (Matviishina, 1982; Gray & Nickelsen, 1989; Tsatskin & Chizhikova, 1990; Driese & Foreman, 1992; Joeckel, 1994, 1995; Tsatskin et al., 1998; Driese et al., 2003, 2008; Tabor & Montanez, 2004; Kovda et al., 2008; Martins & Pfefferkorn, 2008; Wieder et al., 2008; Díaz-Ortega et al., 2011; Beverly et al., 2014) and they have also been studied in archaeological contexts (Driese et al., 2013; Sánchez-Pérez et al., 2013).

In thin sections of vertic palaeosoils, typical vertic features such as an open porphyric c/f-related distribution pattern, blocky or massive microstructures and striated b-fabrics are commonly observed (Driese et al., 2003; Kovda et al., 2008). Voids related to non-active slickensides sometimes show only weakly developed porostriated b-fabrics (Fig. 4) and can be filled with carbonates, gypsum and silt-sized quartz grains and have Fe/Mn oxide coatings. Typic and complex carbonate nodules with distinct sharp boundaries (disorthic) have been reported (Caudill et al., 1996; Dzenowski & Hembree, 2012). Clay coatings (Fig. 2) and fragments of clay coating, as well as a variety of Fe/Mn oxide nodules, carbonate nodules and other impregnative pedofeatures, have been described for vertic palaeosoils (Tsatskin & Ronen, 1999; Kovda, 2004; Díaz-Ortega et al., 2011;

Dzenowski & Hembree, 2012). Concentric nodules with laminae composed of pure Mn oxides without any admixture of Fe compounds have been suggested to be associated with environments with high biological activity (Tsatskin & Ronen, 1999). Fragments of clay coatings were interpreted as illuvial pedofeatures fragmented by vertic processes (Díaz-Ortega et al., 2011). Several phases of carbonate precipitation and later recrystallisation have been documented in vertic palaeosoils of various ages (Caudill et al., 1996; Kovda, 2004). The common occurrence of Vertic palaeosoils, with a variety of pedofeatures that could not have developed simultaneously, suggests their polygenetic formation with the soil record of several evolutionary stages (Pal et al., 2001, Kovda et al., 2003; Sánchez-Pérez et al., 2013; Solis-Castillo et al., 2015).

Vertic palaeosoils can provide additional information for environmental reconstructions because they undergo little burial compaction (less than 10%) due to their high original density (Blodgett, 1985; Caudill et al., 1996, 1997). This fact was used to correlate the depth to the pedogenic carbonate horizon in Holocene Vertisols with the mean annual precipitation for estimating palaeoprecipitation values (Caudill et al., 1996; Nordt et al., 2006). The difficulties of this method are depth corrections for erosion and possible microrelief and/or subsurface cycling in the past. The wetter soil environments in the microlows and drier conditions in the microhighs complicate the estimation of mean annual precipitation for the area. Micromorphology could be helpful for identification of erosion and correlation with palaeomicrorelief.

Striated b-fabrics can not be used for the reconstruction of tropical or subtropical environments (with 4-8 dry months) or monsoonal climates, as done by some authors (Tsatskin & Chizhikova, 1990; Driese & Foreman, 1992; Beverly et al., 2014, 2015), because cold boreal Holocene Vertisols of continental climate zones may show the same features. Other micromorphological features of vertic materials can help the identification of various stages of palaeopedogenesis (see also Fedoroff et al., 2010, 2018).

Clayey materials such as palaeo-Vertisols may be one of the best substrates to preserve archaeological materials, but gilgai and/or subsurface cycling common to Vertisols should be taken into account during palaeoenvironmental reconstructions (Driese et al., 2013).

6. Conclusions

Research on the micromorphology of vertic features has been carried out since the 1960s. Vertic features, resulting from shrink-swell processes, vertical mixing and lateral shearing, due to alternating wetting and drying of clayey materials, are very prominent and rather easy to identify in thin sections.

Vertic materials show a characteristic combination of micromorphological features. The most relevant are striated b-fabrics, an open porphyric c/f-related distribution pattern, a blocky microstructure, a heterogeneous groundmass with incorporated dark aggregates, Fe/Mn oxide pedofeatures and, in more arid regions, carbonate and gypsum pedofeatures. However, classifying Vertisols and vertic subgroups of other soils based exclusively on their micromorphological characteristics is impossible (Blokhuis et al.,

1991; Mermut et al., 1991) because other clayey soils may be micromorphologically similar.

The micromorphological study of vertic features in modern soils and palaeosoils can provide reliable information about the genesis and evolution of the soil and on long-term and short-term environmental changes and trends. Good indicators for the recognition of soils with vertic behaviour include the presence of surface-derived granular aggregates preserved in deep cracks or as part of a heterogeneous groundmass, as well as the occurrence of striated b-fabrics and abundant complex nodules. Nevertheless, more micromorphological studies are required for a better understanding of the preservation and diagenetic transformations of vertic features in palaeosoils.

References

Abdel Rahman, S.I., Hanna, F.S., Rabie, F.H. & Stoops, G., 1992. Impact of the soil environment on the micromorphological and mineralogical characteristics of some minerals in Vertisols of Egypt. Abstracts of 9th International Working Meeting on Soil Micromorphology, Townsville, pp. 59.

Acquaye, K., Dowuona, G., Mermut, A. & St. Arnaud, R., 1992. Micromorphology and mineralogy of cracking soils from the Accra Plains of Ghana. Soil Science Society of America Journal 56, 193-201.

Ahmad, N., 1996. Management of Vertisols in rainfed conditions. In Ahmad, N. & Mermut, A. (eds.), Vertisols and Technologies for their Management. Developments in Soil Science, Volume 24. Elsevier, Amsterdam, pp. 363-428.

Aydemir, S., Drees, L.P., Hallmark, C.T. & Cullu, M.A., 2004. Micromorphological characteristics of vertisols and vertic-like soils in the Harran Plain of Southeastern Turkey. In Kapur, S., Akça, E., Montanarella, L., Ozturk, A. & Mermut, A. (eds.), Extended abstracts of 12th International Working Meeting on Soil Micromorphology, Adana, pp. 124-126.

Baer, J.U., Kent, T.F. & Anderson, S.H., 2009. Image analysis and fractal geometry to characterize soil desiccation cracks. Geoderma 154, 153-163.

Balpande, S.S., Deshpande, S.B. & Pal, D.K., 1997. Plasmic fabric of Vertisols of the Purna Valley of India in relation to their cracking. Journal of the Indian Society of Soil Science 45, 553-562.

Bandopadhyay, P.C., Eriksson, P.G. & Roberts, R.J., 2010. A vertic paleosol at the Archean-Proterozoic contact from the Singhbhum-Orissa craton, eastern India. Precambrian Research 177, 277-290.

Barzanji, A.F. & Stoops, G., 1974. Fabric and mineralogy of gypsum accumulations in some soils of Iraq. Transactions of the 10th International Congress of Soil Science, Moscow, Volume VII, pp. 271-277.

Bellinfante, N., Paneque, G., Olmedo, J. & Baños, C., 1974. Micromorphological study of Vertisols in southern Spain. In Rutherford, G.K. (ed.), Soil Microscopy. The Limestone Press, Kingston, pp. 296-305.

Beverly, E.J., Ashley, G.M. & Driese, S.G., 2014. Reconstruction of a Pleistocene paleocatena using micromorphology and geochemistry of lake margin paleo-Vertisols, Olduvai Gorge, Tanzania. Quaternary International 322/323, 78-94.

Beverly, E.J., Driese, S.G., Peppe, D.J., Arellano, L.N., Blegen, N., Faith, J.T. & Tryon, C.A., 2015. Reconstruction of a semi-arid late Pleistocene paleocatena from the Lake Victoria region, Kenya. Quaternary Research 84, 368-381.

Blodgett, R.H., 1985. Paleovertisols as indicators of climate. American Association of Petroleum Geologists Bulletin 69, 239.

Blokhuis, W.A., 1993. Vertisols in the Central Clay Plain of the Sudan. PhD Dissertation, Agricultural University Wageningen, 418 p.

Blokhuis, W.A, Pape, T. & Slager, S., 1969. Morphology and distribution of pedogenic carbonate in some Vertisols of the Sudan. Geoderma 2, 173-200.

Blokhuis, W.A., Slager, S. & Van Schagen, R.H., 1970. Plasmic fabric of two Sudan Vertisols. Geoderma 4, 127-137.

Blokhuis, W.A., Kooistra, M.J. & Wilding, L.P., 1990. Micromorphology of cracking clayey soils (Vertisols). In Douglas, L.A. (ed.), Micromorphology: A Basic and Applied Science. Developments in Soil Science, Volume 19. Elsevier, Amsterdam, pp. 123-148.

Blokhuis, W.A., Wilding, L.P. & Kooistra, M.J., 1991. Classification of vertic intergrades: macromorphological and micromorphological aspects. In Kimble, J.M. (ed.), Characterization, Classification and Utilization of Cold Aridisols and Vertisols. USDA, Soil Conservation Service, National Soil Survey Center, Lincoln, pp. 1-7.

Boettinger, J.L., 1994. Biogenic opal as an indicator of mixing in an Alfisol/Vertisol landscape. In Ringroase-Voase, A.J. & Humphreys, G.S. (eds.), Soil Micromorphology: Studies in Management and Genesis. Developments in Soil Science, Volume 22. Elsevier, Amsterdam, pp. 17-26.

Bornand, M., Dejou, J., Robert, M. & Roger, L., 1984. Composition minéralogique de la phase argileuse des Terres Noires de Limagne (Puy-de-Dôme). Le problème des liaisons argiles-matière organique. Agronomie 4, 47-62.

Brewer, R., 1964. Fabric and Mineral Analysis of Soils. John Wiley and Sons, New York, 470 p.

Bruand, A. & Tessier, D.,1987. Etude de l'organisation d'un matériau argileux en microscopie: modifications intervenant lors de la déshydratation. In Fedoroff, N., Bresson, L.M. & Courty, M.A. (eds.), Micromorphologie des Sols, Soil Micromorphology. AFES, Plaisir, pp. 31-35.

Bui, E.N. & Mermut, A.R., 1989. Orientation of planar voids in Vertisols and soils with vertic properties. Soil Science Society of America Journal 52, 171-178.

Buursink, J., 1971. Soils of Central Sudan. PhD Dissertation, University of Utrecht, 238 p.

Bystritskaya, T.L., Gubin, S.V., Tul'panov, V.I. & Skripnichenko, I.I., 1988. The properties of solonetzic swell-shrink Chernozems of Stavropol region as affected by chemical amelioration [in Russian]. Pochvovedenie 11, 108-118.

Cabidoche, Y.M. & Guillaume, P., 1998. A casting method for the three-dimensional analysis of the intraprism structural pores in Vertisols. European Journal of Soil Science 49, 187-196.

Caudill, M.R., Driese, S.G. & Mora, C.I., 1996. Preservation of a paleo-Vertisol and an estimate of Late Mississippian paleoprecipitation. Journal of Sedimentary Research 66, 58-70.

Caudill, M.R., Driese, S.G. & Mora, C.I., 1997. Physical compaction of vertic palaeosols: implications for burial diagenesis and palaeoprecipitation estimates. Sedimentology 44, 673-685.

Chittamart, N., Suddhiprakarn, A., Kheoruenromne, I. & Gilkes, R.J., 2010. The pedo-geochemistry of Vertisols under tropical savanna climate. Geoderma 159, 304-316.

Chotte, J.L., Villemin, G., Guilloré, P. & Jocteur Monrozier, L., 1994. Morphological aspects of microorganism habitats in a vertisol. In Ringrose-Voase, A.J. & Humphreys, G.S. (eds.), Soil Micromorphology: Studies in Management and Genesis. Developments in Soil Science, Volume 22. Elsevier, Amsterdam, pp. 395-404.

Coulombe, C.E., Dixon, J.B. & Wilding, L.P., 1996a. Mineralogy and chemistry of Vertisols. In Ahmad, N. & Mermut, A. (eds.), Vertisols and Technologies for their Management. Developments in Soil Science, Volume 24. Elsevier, Amsterdam, pp. 115-200.

Coulombe, C.E., Wilding, L.P. & Dixon, J.B., 1996b. Overview of Vertisols: characteristics and impacts on society. Advances in Agronomy 57, 289-375.

Dabbakula, M., Moncharoen, P., Yoothong, K., Vijarnson, P., Moncharoen, L. & Eswaran, H., 1992. Microvariability in Vertisols. Abstracts of 9th International Working Meeting on Soil Micromorphology, Townsville, pp. 60.

Dasog, G.S., Acton, D.F. & Mermut, A.R., 1987. Genesis and classification of clay soils with vertic properties in Saskatchewan. Soil Science Society of America Journal 51, 1243-1250.

Dasog, G.S., Mermut, A.R. & Acton, D.F., 1988. Micromorphology of some Vertisols in India. In Hirekerur, L.R., Pal, D.K., Sehgal, J.L. & Deshpande, S.B. (eds.), Transactions of the International Workshop on Swell-Shrink Soils. Oxford and IBH Publishing Co., New Delhi, pp. 147-149.

Dasog, G.S., Mermut, A.R. & Acton, D.F., 1991. Comparative micromorphology of swelling clay soils from Boreal and Tropical regions. Agropedology 1, 83-90.

de Olmedo Pujol, J. & Pérez, J.M., 1975. Estudio micromorfológico de una catena de Vertisoles en la provincia de Sevilla (Sud de España). Annales de Edafologia y Agrobiologia 34, 745-776.

De Vos, T.N.C. & Virgo, K.J., 1969. Soil structure in Vertisols of the Blue Nile clay plains, Sudan. Journal of Soil Science 20, 189-206.

Díaz-Ortega, J., Solleiro-Rebolledo, E. & Sedov, S., 2011. Spatial arrangement of soil mantle in Glacis de Buenovista, Mexico as a product and record of landscape evolution. Geomorphology 135, 248-261.

Dostovalova, E.V. & Tursina, T.V., 1988. Morphogenetic peculiarities of compact Chernozems of Stavropol region and their changes due to irrigation [*in Russian*]. In Mikromorphologia Antropogenno Izmenennikh Pochv. Nauka, Moscow pp. 114-123.

Driese, S.G. & Foreman, J.L., 1992. Paleopedology and paleoclimatic implications of Late Ordovician vertic paleosols, Juniata Formation, southern Appalachians. Journal of Sedimentary Petrology 62, 71-83.

Driese, S.G., Jacobs, J.R. & Nordt, L.C., 2003. Comparison of modern and ancient Vertisols developed on limestone in terms of their geochemistry and parent material. Sedimentary Geology 157, 49-69.

Driese, S.G., Li, Z.H. & McKay, L.D., 2008. Evidence for multiple, episodic, mid-Holocene Hypsithermal recorded in two soil profiles along an alluvial floodplain catena, southeastern Tennessee, USA. Quaternary Research 69, 276-291.

Driese, S.G., Nordt, L.C., Waters, M.R. & Keene, J.L., 2013. Analysis of site formation history and potential disturbance of stratigraphic context in Vertisols at the Debra L. Friedkin archaeological site in Central Texas, USA. Geoarchaeology 28, 221-248.

Dudal, R., 1965. Dark Clay Soils of Tropical and Subtropical Regions. FAO Agricultural Development Paper No 83. FAO, Rome, 161 p.

Dzenowski, N.D. & Hembree, D.I., 2012. Examining local climate variability in the Late Pennsylvanian through paleosols: an example from the Lower Conemaugh group of Southeastern Ohio, USA. Geosciences 2, 260-276.

Eswaran, H., Kimble, J. & Cook, T., 1988. Properties, genesis and classification of Vertisols. In Hirekerur, L.R., Pal, D.K., Sehgal, J.L. & Deshpande, S.B. (eds.), Transactions of the International Workshop on Swell-Shrink Soils. Oxford and IBH Publishing Co., New Delhi, pp. 1-22.

Eswaran, H., Beinroth, F.H., Reich, P.F. & Quandt, L.A., 1999. G.D. Smith Memorial Slide Collection. Vertisols: Their Properties, Classification, Distribution and Management. USDA, National Resources Conservation Service, Washington (CDROM).

Fedoroff, N. & Fies, J.C., 1968. Les Vertisols de Sud-Est de la Beauce. Bulletin de l'Association Française pour l'Etude du Sol 1, 19-31.

Fedoroff, N., Courty, M.A., & Zhengtang Guo, 2010. Palaeosoils and relict soils. In Stoops, G., Marcelino, V. & Mees, F. (eds.), Interpretation of Micromorphological Features of Soils and Regoliths. Elsevier, Amsterdam, pp. 623-662.

Fedoroff, N., Courty, M.A., & Zhengtang Guo, 2018. Palaeosoils and relict soils, a conceptual approach. In Stoops, G., Marcelino, V. & Mees, F. (eds.), Interpretation of Micromorphological Features of Soils and Regoliths. Second Edition. Elsevier, Amsterdam, pp. 821-862.

Fedorov, K.N. & Soliyanik, G.M., 1991. Micromorphology of compact Chernozems [*in Russian*]. Vestnik Moskovskogo Universiteta, Series 17, 2, 62-67.

Fedorov, K.N., Samoilova, E.M. & Fiza Salama Aly, 1991. Micromorphology of alluvial compact soils of Nile's delta [*in Russian*]. Vestnik Moskovskogo Universiteta, Series 17, 1, 19-25.

Field, D.J., Koppi, A.J. & Drees, L.R., 1997. The characterisation of microaggregates in Vertisols using scanning electron microscopy (SEM) and thin section. In Shoba, S., Gerasimova, M. & Miedema, R. (eds.), Soil Micromorphology: Studies on Soil Diversity, Diagnostics, Dynamics. Moscow, Wageningen, pp. 80-86.

Ghitulescu, N., 1971. Etude micromorphologique de quelques sols de la plaine de Cilnistea (Roumanie). Pedologie 21, 131-151.

Gonzáles-Arqueros, M.L., Vázquez-Selem, L., Gama Castro, J.E., McClung de Tapia, E. & Sedov, S., 2013. History of pedogenesis and geomorphic processes in the Valley of Teotihuacan, Mexico: micromorphological evidence from a soil catena. Spanish Journal of Soil Science 3, 201-216.

Gray, M.B. & Nickelsen, R.P., 1989. Pedogenic slickensides as indicators of strain and deformation processes in redbed sequences of Appalachian Foreland. Geology 17, 72-75.

Gunal, H. & Ransom, M.D., 2006a. Genesis and micromorphology of loess-derived soils from central Kansas. Catena 65, 222-236.

Gunal, H. & Ransom, M.D., 2006b. Clay illuviation and calcium carbonate accumulation along a precipitation gradient in Kansas. Catena 68, 59-69.

Hallsworth, E.G. & Beckmann, G.G., 1969. Gilgai in the Quaternary. Soil Science 107, 409-420.

Heidari, A., Mahmoodi, Sh., Stoops, G. & Mees, F., 2005, Micromorphological characteristics of Vertisols in Iran, including nonsmectitic soils. Arid Land Research and Management 19, 29-46.

Heidari, A., Mahmoodi, Sh., Stoops, G. & Mees, F., 2008. Micromorphological characteristics of palygorskite-dominated Vertisols of Iran. In Kapur, S., Akça, E., Montanarella, L., Ozturk, A. & Mermut, A. (eds.), Extended Abstracts of 12th International Meeting on Soil Micromorphology, Adana, pp. 96-99.

Hussein, J. & Adey, M.A., 1998. Changes in microstructure, voids and b-fabric of surface samples of a Vertisol caused by wet/dry cycles. Geoderma 85, 63-82.

IUSS Working Group WRB, 2015. World Reference Base for Soil Resources 2014, Update 2015. World Soil Resources Reports No. 106. FAO, Rome, 192 p.

Joeckel, R.M., 1994. Virgilian (Upper Pensylvanian) Paleosols in the upper Lawrence Formation (Douglas Group) and in the Snyderville shale member (Oread Formation, Shawnee Group) of the northern mid-continent, USA: pedologic contrasts in a cyclothem sequence. Journal of Sedimentary Research 64, 853-866.

Joeckel, R.M., 1995. Paleosols below the Ames marine unit (Upper Pensylvanian, Conemaugh group) in the Appalachian Basin, USA: variability on an ancient depositional landscape. Journal of Sedimentary Research 65, 393-407.

Jongerius, A. & Bonfils, G., 1964. Micromorfologia de un suelo negro grumosolico de la provincia de Entre Rios. Revista de Investigaciones Agropecuarias, INTA, Serie 3, Clima y Suelo 1, 33-53.

Kabakchiev, I. & Boneva, K., 1972. Investigation of iron-manganese concretions in Smonitzas [*in Bulgarian*]. Pochvoznanie i Agrokhimiya 7, 3-12.

Kabakchiev, I. & Galeva, V., 1973. Comparative micromorphological investigation of cernozemsmonitzas and chernozems [*in Bulgarian*]. Pochvoznanie i Agrokhimiya 8, 11-24.

Kalbande, A.R., 1988. Micromorphology of benchmark Vertisols in India. In Hirekerur, L.R., Pal, D.K., Sehgal, J.L. & Deshpande, S.B. (eds.), Transactions of the International Workshop on Swell-Shrink Soils. Oxford and IBH Publishing Co., New Delhi, pp. 150-152.

Kalbande, A.R., Kooistra, M.J. & Deshpande, S.B., 1986. Micromorphology of a Typic Chromustert under two management practices. Transactions of the 13th International Congress of Soil Science, Volume IV, Hamburg, pp. 1557.

Kalbande, A.R., Pal, D.K. & Deshpande, S.B., 1992. B-fabric of some benchmark Vertisols of India in relation to their mineralogy. Journal of Soil Science 43, 375-385.

Kapur, S., Karaman, C., Akca, E., Aydin, M., Dinc, U., FitzPatrick, E.A., Pagliai, M., Kalmar, D. & Mermut, A.R., 1997. Similarities and differences of the spheroidal microctructure in Vertisols from Turkey and Israel. Catena 28, 297-311.

Kooistra, M.J., 1982a. Micromorphology. In Murthy, R.S., Hirekerur, L.R., Deshpande, S.B. & Venkata Rao, B.V. (eds.), Benchmark Soils of India. Morphology, Characteristics and Classification for Resource Management. National Bureau of Soil Survey and Land Use Planning, Nagpur, pp. 71-89.

Kooistra, M.J., 1982b. Micromorphological Analysis and Characterization of 70 Benchmark Soils of India. A Basic Reference Set. Netherlands Soil Survey Institute, Wageningen, 788 p.

Kovda, I.V., 2004. Carbonate neoformations in soils: old and new problems of studying [*in Russian*]. In Glazovsky, N.F. (ed.), Soil, Biogeochemical Cycles and Biosphere. KMK Scientific Press Ltd., Moscow, pp. 115-136.

Kovda, I. & Lebedeva M., 2013. Modern and relict features in clayey cryogenic soils: morphological and micromorphological identification. Spanish Journal of Soil Science 3, 130-147.

Kovda, I.V., Morgun, E.G. & Yarilova, E.A., 1992. Micromorphological evidence of the gilgai soils poly-genesis [*in Russian*]. In Soil genesis, Geography and Evolution. Dokuchaev's Soil Institute, Moscow, pp. 115-125.

Kovda, I.V., Ermolaev, A.M., Gol'eva, A.A. & Morgun, E.G., 1999. Reconstruction of gilgai landscape elements according to botanical and biomorph analysis. Biology Bulletin 26, 297-306.

Kovda, I.V., Wilding, L.P. & Drees, L.R., 2003. Micromorphology, submicroscopy and microprobe study of carbonate pedofeatures in a Vertisol gilgai soil complex, South Russia. Catena 54, 457-476.

Kovda, I., Mora, C.I. & Wilding, L.P., 2008. PaleoVertisols of the Northwestern Caucasus: (micro) morphological, physical, chemical and isotopic constraints on early to late Pleistocene climate. Journal of Plant Nutrition and Soil Science 171, 498-508.

Kovda I., Goryachkin, S., Lebedeva, M., Chizhikova, N., Kulikov, A. & Badmayev, N., 2014. Shrinking-swelling soils in cryogenic environment. Abstracts of the 20th World Congress of Soil Science, Jeju, n° O37-3, 1 p.

Kovda, I.V., Lebedeva, M.P., Morgun, E.G., 2016. Central image of vertosol: evolution of concepts of their morphology and genesis. Bulleten Pochvennogo Instituta im. V.V. Dokuchaeva 86, 134-142.

Kovda, I., Goryachkin, S., Lebedeva, M., Chizhikova, N, Kulikov, A & Badmaev, N., 2017. Vertic soils and Vertisols in cryogenic environments of southern Siberia, Russia. Geoderma 288, 184-195.

Krupenikov, I.A., Sinkevitch, Z.A. & Grati, V.P., 1966. Micromorphological investigation of Moldavian soils [*in Russian*]. In Voprosi Issledovania i Ispol'zovania Pochv Moldavii. Sbornik 4. Kishinev, pp. 53-60.

Kühn, P., Aguilar, J. & Miedema, R., 2010. Textural features and related horizons. In Stoops, G., Marcelino, V. & Mees, F. (eds.), Interpretation of Micromorphological Features of Soils and Regoliths. Elsevier, Amsterdam, pp. 217-250.

Kühn, P., Aguilar, J., Miedema, R. & Bronnikova, M., 2018. Textural Pedofeatures and Related Horizons. In Stoops, G., Marcelino, V. & Mees, F. (eds.), Interpretation of Micromorphological Features of Soils and Regoliths. Second Edition. Elsevier, Amsterdam, pp. 377-423.

Labib, F. & Stoops, G., 1970. Micromorphological contribution to the knowledge of some alluvial soils in the U.A.R. (Egypt). Pedologie 20, 108-126.

Lindbo, D.L., Stolt, M.H. & Vepraskas, M.J., 2010. Redoximorphic features. In Stoops, G., Marcelino, V. & Mees, F. (eds.), Interpretation of Micromorphological Features of Soils and Regoliths. Elsevier, Amsterdam, pp. 129-147.

Martins, U.P. & Pfefferkorn, H.W., 2008. Genetic interpretation of a lower Triassic paleosol complex based on soil micromorphology. Palaeogeography, Palaeoclimatology, Palaeoecology 64, 1-14.

Matviishina, J., 1982. Micromorphology of Pleistocene Soils of Ukraine [in Russian]. Naukova Dumka, Kiev, 144 p.

McCormack, D.E. & Wilding, L.P., 1974. Proposed origin of lattisepic fabric. In Rutherford, G. (ed.), Soil Microscopy. The Limestone Press, Kingston, pp. 761-771.

McGarry, D., 1987. The effect of soil water content during land preparation on aspects of soil physical condition and cotton growth. Soil and Tillage Research 9, 287-302.

McGarry, D., 1989. The effect of wet cultivation on the structure and fabric of a Vertisol. Journal of Soil Science 40, 199-207.

McKenzie, D.C., Koppi, A.J., Moran, C.J. & McBratney, A.B., 1992. A role for image analysis when assessing compaction in Vertisols. Abstracts 9th International Working Meeting on Soil Micromorphology, Townsville, pp. 144.

Mees, F., 2001. An occurrence of lacustrine Mg-smectite in a pan of the southwestern Kalahari, Namibia. Clay Minerals 36, 547-556.

Mermut, A.R. & Dasog, G.S., 1986. Nature and micromorphology of carbonate glaebules in some Vertisols of India. Soil Science Society of America Journal 50, 382-391.

Mermut, A. & Jongerius, A., 1980. A micromorphological analysis of regrouping phenomena in some Turkish soils. Geoderma 24, 159-175.

Mermut, A.R. & St. Arnaud, R.J., 1983. Micromorphology of some Chernozem soils with grumic properties in Saskatchewan, Canada. Soil Science Society of America Journal 47, 536-541.

Mermut, A.R., Sehgal, J.L. & Stoops, G., 1988. Micromorphology of swell-shrink soils. Hirekerur, L.R., Pal, D.K., Sehgal, J.L. & Deshpande, S.B. (eds.), Transactions of the International Workshop on Swell-Shrink Soils. Oxford and IBH Publishing Co., New Delhi, pp.127-144.

Mermut, A.R., Acton, D.F. & Tarnocai, C., 1991. A review of recent research on swelling clay soils in Canada. In Kimble, J.M. (ed.), Characterization, Classification and Utilization of Cold Aridisols and Vertisols. USDA, Soil Conservation Service, National Soil Survey Center, Lincoln, pp. 112-121.

Mermut, A.R., Dasog., G.S. & Dowuona, G.N., 1996a. Soil morphology. In Ahmad, N. & Mermut, A. (eds.), Vertisols and Technologies for their Management. Developments in Soil Science, Volume 24. Elsevier, Amsterdam, pp. 89-114.

Mermut, A.R., Padmanabham, E., Eswaran, H. & Dasog, G.S., 1996b. Pedogenesis. In Ahmad, N. & Mermut, A. (eds.), Vertisols and Technologies for their Management. Developments in Soil Science, Volume 24. Elsevier, Amsterdam, pp. 43-61.

Moreau, E., Velde, B. & Terribile, F., 1999. Comparison of 2D and 3D images of fractures in a Vertisol. Geoderma 92, 55-72.

Morgun Nobles, M., Wilding, L.P. & McInnes, K.J., 2004. Submicroscopic measurements of tracer distribution related to surface features of soil aggregates. Geoderma 123, 84-97.

Nettleton, W.D. & Sleeman, J.R., 1985. Micromorphology of Vertisols. In Douglas, L.A. & Thompson, M.L. (eds.), Soil Micromorphology and Soil Classification. Soil Science Society of America Special Publication 15, SSSA, Madison, pp. 65-196.

Nettleton, W.D., Peterson, F.F. & Borst, G., 1983. Micromorphological evidence of turbation in Vertisols and soils in vertic subgroups. In Bullock, P. & Murphy, C.P. (eds.), Soil Micromorphology, Volume 2, Soil Genesis. AB Academic Publishers, Berkhamsted, pp. 441-458.

Nordt, L.C., Wilding, L.P., Lynn, W.C. & Crawford, C.C., 2004. Vertisol genesis in a humid climate of the coastal plain of Texas, U.S.A. Geoderma 122, 83-102.

Nordt, L., Oroz, M., Driese, S. & Tubbs, J., 2006. Vertisol carbonate properties in relation to mean annual precipitation: implications for paleoprecipitation estimates. Journal of Geology 114, 501-510.

Norrish, K. & Pickering, J.G., 1983. Clay minerals. In Soils: An Australian Viewpoint. CSIRO, Melbourne, Academic Press London, pp. 281-308.

Osman, A. & Eswaran, H., 1974. Clay translocation and vertic properties of some red mediterranean soils. In Rutherford, G.K. (ed.), Soil Microscopy, The Limestone Press, Kingston, pp. 846-857.

Pal, D.K., Dasog, G.S., Vadivelu, S., Ahuja, R.L. & Bhattacharyya, T., 1999. Secondary calcium carbonate in soils of arid and semi-arid regions of India. In Lal, R., Kimble, J.M., Eswaran, H. & Stewart, B.A. (eds.), Global Change and Pedogenic Carbonates. Lewis Publishers, Boca Raton, pp. 149-185.

Pal, D.K., Balpande, S.S. & Srivastava, P., 2001. Polygenetic Vertisols of the Purna valley of central India. Catena 43, 231-249.

Pal, D.K., Bhattacharyya, T., Chandran, P., Ray., S.K., Satyavati, P.L.A., Durge, S.L., Raja, P. & Maurya, U. K., 2009. Vertisols (cracking clay soils) in a climosequence of Peninsular India: evidence for Holocene climate changes. Quaternary International 209, 6-21.

Pal, D.K., Wani, S.P. & Sahrawat, K.L., 2012. Vertisols of tropical Indian environments: pedology and edaphology. Geoderma 189/190, 28-49.

Paranjape, M.V., Pal, D.K. & Deshpande, S.B., 1997. Genesis of non-vertic deep Black soils in a basaltic landform of Maharashtra. Journal of the Indian Society of Soil Science 45, 174-180.

Parfenova, E.I. & Yarilova, E.A., 1977. A Manual of Micromorphological Research in Soil Science [*in Russian*]. Nauka, Moscow, 198 p.

Pillai-McGarry, U.P., 1992. Image analysis of structure regeneration of a compacted Vertisol. Abstracts of 9th International Working Meeting on Soil Micromorphology, Townsville, pp. 145.

Poch, R.M., Artieda, O., Herrero, J. & Lebedeva-Verba, M., 2010. Gypsic features. In Stoops, G., Marcelino, V. & Mees, F. (eds.), Interpretation of Micromorphological Features of Soils and Regoliths. Elsevier, Amsterdam. pp. 195-216.

Poch, R.M., Artieda, O. & Lebedeva, M., 2018. Gypsic features. In Stoops, G., Marcelino, V. & Mees, F. (eds.), Interpretation of Micromorphological Features of Soils and Regoliths. Second Edition. Elsevier, Amsterdam, pp. 259-287.

Podwojewski, P., 1995. The occurrence and interpretation of carbonate and sulfate minerals in a sequence of Vertisols in New Caledonia. Geoderma 65, 223-248.

Podwojewski, P. & Arnold, M., 1994. The origin of gypsum in Vertisols in New Caledonia determined by isotopic composition of sulphur. Geoderma 63, 179-195.

Puentes, R., Wilding, L.P. & Drees, L.R., 1992. Microspatial variability and sampling concepts in soil porosity studies of Vertisols. Geoderma 53, 373-385.

Qui, R.L., Xiong, D.X., Huang, R.C., 1994. Swell-shrink properties and influencing factors of Vertisols [*in Chinese*]. Journal of Nanjing Agricultural University 17, 71-77.

Rajamuddin, U.A., Lopulisa, Ch., Husni, H. & Nathan, M., 2013. Mineralogy and micromorphology characteristic of Vertisol lying on limestone parent rocks at Jeneponto District of South Sulawesi Province, Indonesia. International Journal of Agricultural Systems 1, 92-97.

Rajan, S.V.G., Murthy, R.S., Kalbande, A.R. & Venugopal, K.R., 1972. Micromorphology and chemistry of carbonate concretions in black clayey soils. Indian Journal of Agricultural Science 42, 1020-1023.

Rao, Y.S., Mohan, N.G.R. & Rao, A.E.V., 1986. Micromorphology of some soils in Cuddapah Basin, India. Journal of the Indian Society of Soil Science 34, 225-229.

Retallack, G.J., 2008. Cambrian paleosols and landscapes of South Australia. Australian Journal of Earth Sciences 55, 1083-1106.

Rodriguez Hernandez, C.M., Fernandez-Caldas, E., Fedoroff, N. & Quantin, P., 1979. Les Vertisols des Iles Canaries occidentales. Etude physico-chimique, minéralogique et micromorphologique. Pedologie 29, 71-107.

Sánchez-Pérez, S., Solleiro-Rebolledo, E., Sedov, S., McClung de Tapia, E., Golyeva, A., Prado, B. & Ibarra-Morales, E., 2013. The black San Pablo paleosol of the Teotihuancan Valley, Mexico: pedogenesis, fertility, and use in ancient agricultural and urban systems. Geoarchaeology 28, 249-267.

Sehgal, J.L. & Bhattacharjee, J.C., 1988. Typic Vertisols of India and Iraq — their characterization and classification. Pedologie 38, 67-95.

Shorkunov, I.G., 2013. Types and genesis of stress cutans in buried early Pleistocene paleosols in North-West Ciscaucasia [in Russian]. Science Prospects 10, 37-46.

Singh, S., 1956. The formation of dark-coloured clay-organic complexes in Black Soils. Journal of Soil Science 7, 43-58.

Sleeman, J.R. & Brewer, R., 1984. Micromorphology of some Australian cracking clay soils. In McGarity, J. W., Hoult, E.H. & So, H.B. (eds.), The Properties and Utilization of Cracking Clay Soils. Reviews in Rural Sciences 5, University of New England, Armidale, pp. 73-82.

Soil Survey Staff, 2014. Keys to Soil Taxonomy. 12th Edition. USDA-Natural Resources Conservation Service, Washington, 360 p.

Solis-Castillo, B., Golyeva, A., Sedov, S., Solleiro-Rebolledo, E. & Lopez-Rivera, S., 2015. Phytoliths, stable carbon isotopes and micromorphology of a buried alluvial soil in Southern Mexico: a polychromous record of environmental change during Middle Holocene. Quaternary International 365, 150-158.

Srivastava, P., Bhattacharyya, T. & Pal, D.K., 2002. Significance of the formation of calcium carbonate minerals in the pedogenesis and management of cracking clay soils (Vertisols) of India. Clays and Clay Minerals 50, 111-126.

Srivastava, P., Rajak, M.K., Sinha, R., Pal, D.K. & Bhattacharyya, T., 2010. A high-resolution micromorphological record of the Late Quaternary paleosols from Ganga-Yamuna interfluve: stratigraphic and paleoclimatic implications. Quaternary International 227, 127-142.

Stephan, S. & De Petre, A.A., 1986. Micromorphology of Vertisols from Argentina. Transactions of the 13th International Congress of Soil Science, Volume IV, Hamburg, pp.1566-1567.

Stephan, S., Berrier, J., De Petre, A.A., Jeanson, C., Kooistra, M.J., Scharpenseel, H.W. & Schiffmann, H., 1983. Characterization of in situ organic matter constituents in Vertisols from Argentina, using submicroscopic and cytochemical methods — first report. Geoderma 30, 21-34.

Stoops, G., 2003. Guidelines for Analysis and Description of Soil and Regolith Thin Sections. Soil Science Society of America, Madison, 184 p.

Stoops, G. & Mees, F., 2018. Groundmass composition and fabric. In Stoops, G., Marcelino, V. & Mees, F. (eds.), Interpretation of Micromorphological Features of Soils and Regoliths. Second Edition. Elsevier, Amsterdam, pp. 73-125.

Sui Yaobing & Cao Shenggeng, 1992. Micromorphological study on some Vertisols in China [in Chinese]. Acta Pedologica Sinica 29, 18-25.

Tabor, N.J. & Montanez, I.P., 2004. Morphology and distribution of fossil soils in the Permo-Pennsylvanian Wichita and Bowie Groups, north-central Texas, USA: implications for western equatorial Pangean palaeoclimate during icehouse-greenhouse transition. Sedimentology 51, 851-884.

Tessier, D., 1987. Validité des techniques de déshydratation pour l'étude de la micro-organisation des sols — apport des matériaux argileux purs. In Fedoroff, N., Bresson, L.M. & Courty, M.A. (eds.), Micromorphologie des sols, Soil micromorphology. AFES, Plaisir, pp. 23-29.

Tessier, D., Bouzigues, B., Favrot, J.C. & Valles, V., 1992. Influence du microrelief sur l'évolution texturale des argiles dans les sols lessivés de la vallée de la Garonne. Différenciation des structures vertiques ou prismatiques. Comptes Rendus de l'Académie des Sciences 315, 1027-1032.

Tovar, R.E., Sedov, S., Montellano-Ballesteros, M., Solleiro, E. & Benammo, M., 2014. Paleosols, bones, phytoliths, and $\delta^{13}C$ signatures of humus and teeth in the alluvial sequence of Axamilpa, Puebla: inferences for landscape evolution and megafauna paleoecology during MIS 3-2 in Southern Mexico. Catena 112, 25-37.

Tsatskin, A.I. & Chizhikova, N.P., 1990. Soil formation in the Pleistocene within the upper Don basin as estimated by micromorphological and mineralogical data [*in Russian*]. Pochvovedenie 12, 94-106.

Tsatskin, A. & Ronen., A., 1999. Micromorphology of a Mousterian paleosol in aeolianites at the site Habonim, Israel. Catena 34, 365-384.

Tsatskin, A., Heller, F., Hailwood, E.A., Gendler, T.S., Hus, J., Montgomery, P., Sartori, M. & Virina, E.I., 1998. Pedosedimentary division, rock magnetism and chronology of the loess/palaeosol sequence at Roxolany (Ukraine). Palaeogeography, Palaeoclimatology, Palaeoecology 143, 111-133.

Tsatskin, A., Gendler, T.S. & Heller, F., 2008. Improved paleopedological reconstruction of vertic paleosols at Novaya Etulia, Moldova via integration of soil micromorphology and environmental magnetism. In Kapur, S., Mermut, A. & Stoops, G. (eds.), New Trends in Soil Micromorphology. Springer-Verlag, Berlin, pp. 91-110.

Tul'panov, V.I. & Makeeva, V.I., 1984. Physico-chemical properties and micromorphology of alkaline compact soils of Central Precaucasus [*in Russian*]. In Puti Povisheniya Plodorodia Pochv Stavropolia, Stavropol, pp. 54-59.

Van Ranst, E., Dumon, M., Tolossa, A.R., Cornelis, J.T., Stoops, G., Vanderberghe, R.E. & Deckers, J., 2011. Revisiting ferrolysis processes in the formation of Planosols for rationalizing the soils with stagnic properties in WRB. Geoderma 163, 265-274.

Velde, B., Moreau, E. & Terribile, F., 1996. Pore networks in an Italian Vertisol: quantitative characterization by two dimensional image analysis. Geoderma 72, 271-285.

Vepraskas, M.J., Lindbo, D.L. & Stolt, M.H., 2018. Redoximorphic features. In Stoops, G., Marcelino, V. & Mees, F. (eds.), Interpretation of Micromorphological Features of Soils and Regoliths. Second Edition. Elsevier, Amsterdam, pp. 425-445.

Verheye, W. & Stoops, G., 1974. Micromorphological evidence for the identification of an argillic horizon in Terra Rossa soils. In Rutherford, G.K. (ed.), Soil Microscopy. The Limestone Press, Kingston, pp. 817-831.

White, G.N. & Dixon, J.B., 1996. Iron and manganese distribution in nodules from a young Texas Vertisol. Soil Science Society of America Journal 60, 1254-1262.

Wieder, M. & Yaalon, D.H., 1974. Effect of matrix composition on carbonate nodule crystallization. Geoderma 11, 95-121.

Wieder, M., Gvirtzman, G., Porat, N. & Dassa, M., 2008. Paleosols of the southern coastal plain of Israel. Journal of Plant Nutrition and Soil Science 171, 533-541.

Wilding, L.P., 1985. Genesis of Vertisols. In Proceedings of the 5th International Soil Classification Workshop. Soil Survey Administration, Khartoum, pp. 47-62.

Wilding, L.P. & Drees, L.R., 1990. Removal of carbonates from thin sections for microfabric interpretations. In Douglas, L.A. (ed.), Soil Micromorphology: A Basic and Applied Science. Developments in Soil Science, Volume 19. Elsevier, Amsterdam, pp. 613-620.

Wilding, L.P. & Hallmark, C.T., 1984. Development of structural and microfabric properties in shrinking and swelling clays. In Bouma, J. & Raats, P.A.C. (eds.), Proceedings of the ISSS Symposium on Water and Solute Movement in Heavy Clay Soils. ILRI Publication 37, Wageningen, pp. 1-18.

Wilding, L.P. & Tessier, D., 1988. Genesis of Vertisols: shrink-swell phenomena. In Wilding, L.P. & Puentes, R. (eds.), Vertisols: their Distribution, Properties, Classification, and Management. Texas A & M University Printing Services, College Station, pp. 55-82.

Wilding, L.P., Williams, D., Miller, W., Cook, T. & Eswaran, H., 1990. Close interval spatial variability of Vertisols: a case study in Texas. In Kimble, J.M. (ed.), Proceedings of the 6th International Soil Correlation Meeting (ISCOM), Characterization, Classification and Utilization of Cold Aridisols and Vertisols. USDA, Soil Conservation Service, National Soil Survey Center, Lincoln, pp. 232-247.

Yaalon, D.H. & Kalmar, D., 1978. Dynamics of cracking and swelling clay soils: displacement of skeletal grains, optimum depth of slickensides, and rate of intra-pedonic turbation. Earth Surface Processes 3, 31-42.

Yarilova, E.A., Poliakov, A.N. & Kiziakov, Y.E., 1969. A comparative micromorphological study of Stavropol and Moldavian compacted Chernozems [in Russian]. Pochvovedenie 7, 103-112.

Yerima, B.P.K., Wilding, L.P., Calhoun, F.G. & Hallmark, C.T., 1987. Volcanic ash-influenced Vertisols and associated Mollisols of El Salvador: physical, chemical and morphological properties. Soil Science Society of America Journal 51, 699-708.

Zhang, M., Liu, L.W. & Gong, Z.T., 1993. Age and some genetic characteristics of Vertisols in China. Pedosphere 3, 81-88.

Chapter 22

Spodic Materials

Eric Van Ranst[1], Michael A. Wilson[2], Dominique Righi[3,†]

[1]*GHENT UNIVERSITY, GHENT, BELGIUM;*
[2]*USDA-NRCS, LINCOLN, NE, UNITED STATES;*
[3]*UNIVERSITÉ DE POITIERS, POITIERS, FRANCE (DECEASED)*

CHAPTER OUTLINE

1. Introduction

Spodic materials are mineral soil materials generally associated with vertical and/or lateral translocation of organically complexed (cheluviated) forms of Fe and Al. Spodic features in soils are expressed in specific horizons: (i) eluvial horizons (E, albic), predominantly light-coloured intervals depleted of clay, base cations, Fe and Al, and mainly composed of residual quartz; (ii) illuvial spodic horizons (Bhs, Bs, Bhm),

†Deceased on March 7, 2010.

Interpretation of Micromorphological Features of Soils and Regoliths. https://doi.org/10.1016/B978-0-444-63522-8.00022-X

reddish brown to brownish black depositional zones that have accumulated organic matter, Al, Fe and/or Mn; (iii) ortstein, composed of cemented spodic materials rich in organic matter and Al; and (iv) placic horizons, reddish brown cemented layers enriched in Fe with or without Mn.

The formation of spodic materials is principally controlled by specific biodegradation processes of the organic material, primarily facilitated by fungi. These processes favour the production of complexing organic acids that are preserved for a certain time and that are subjected to eluviation/illuviation. These specific biodegradation processes may be induced by cool and humid climatic conditions, seasonal subsurface groundwater, acid-producing vegetation (e.g., pines, subalpine fir, heath), quartz-rich and base-cation-depleted parent materials or a combination of two or more of these factors (De Coninck, 1983; McKeague et al., 1983; Jongmans et al., 1997; Lundström et al., 2000a). Formation, mobilisation and deposition of organo-metallic complexes are major soil-forming processes for occurrences of spodic features but are not requirements. For example, placic horizons are often associated with podzolisation but may develop in soils by redoximorphic and hydromorphic processes and consequently are not specific spodic features. Also, other processes besides organo-metallic cheluviation have been recognised in soils with spodic features. For instance, the translocation and illuviation of clay with and without organic materials has often been mentioned (Glinka, 1924; Fridland, 1958; De Coninck & Laruelle, 1960, 1964; De Coninck & Maucorps, 1972; Guillet et al., 1975; Condron & Rabenhorst, 1994; Li et al., 1998; Hseu et al., 2004) and the mobilisation of non-crystalline or nanocrystalline inorganic components (proto-imogolite allophane) has also been documented (Farmer et al., 1980, 1984, 1985; Kodama & Wang, 1989).

Soil Taxonomy (Soil Survey Staff, 1999, 2014) and the World Reference Base (WRB) (IUSS Working Group WRB, 2015) recognise the spodic diagnostic horizon, composed predominantly of spodic materials, at their highest categorical levels, as the Spodosol order and Podzols Reference Soil Group, respectively; in this chapter the WRB term Podzols will be used for these soils. The albic diagnostic horizon appears in Soil Taxonomy, whereas the albic horizon is replaced by albic diagnostic materials in the 2015 WRB system. Classification uses the spodic subsoil horizon rather than the more variable and transient albic horizon or materials. Placic features are used in Soil Taxonomy at the great group level, whereas placic is a supplementary qualifier of the Podzols Reference Soil Group in the WRB. In the latter system, ortsteinic, as well as albic, is the principal qualifier of the Podzols Reference Soil Group. Spodic horizons can be found in soils other than Spodosols or Podzols. Spodic horizons can occur below thick (>2 m) zones of well-expressed albic materials (giant podzols; e.g., Dubroeucq & Volkoff, 1988; Thompson et al., 1996; Van Herreweghe et al., 2003; D'Amico et al., 2015) in soils classified as Entisols in Soil Taxonomy (Soil Survey Staff, 2014) and Regosols in WRB (IUSS Working Group WRB, 2015).

Soils with spodic materials occur mainly in temperate and boreal regions of the northern hemisphere, but also in specific areas in the humid tropics, the Arctic and in temperate mountainous regions of the southern hemisphere.

Micromorphological studies have been carried out for soils with spodic features, developed in a variety of parent materials in temporal and boreal climates of northern latitudes (e.g., De Coninck & Laruelle, 1960; Polteva & Sokolva, 1967; De Coninck & Righi, 1969; Righi, 1975, 1977; De Coninck, 1980; McKeague et al., 1983; Pagé, 1987a; Jakobsen, 1989; Sanborn & Lavkulich, 1989; Pagé & Guillet, 1991; Jersak et al., 1995; Gerasimova et al., 1996; Taina et al., 1997; Melkerud et al., 2000; Buurman et al., 2005; Kristiansen et al., 2010; Wallinga et al., 2013; Waroszewski et al., 2013), as well as in temperate to subtropical climates along the east coast of the United States (Rabenhorst & Hill, 1994; Condron & Rabenhorst, 1994; Phillips & FitzPatrick, 1999). Podzols are also found in the southern hemisphere (e.g., Bullock & Clayden, 1980; Farmer et al., 1985; Milnes & Farmer, 1987; Jamet et al., 1990; Liu et al., 1997; Sullivan, 1997; Van Ranst et al., 1997; Horbe et al., 2004; Hseu et al., 2004; Liu & Chen, 2004), but to a lesser degree due to warmer winters and drier climate (Lundström et al., 2000a).

In tropical areas, soils with spodic features were studied in coastal areas of Sarawak (Andriesse, 1970), in African lowlands with a thick quartz sand substrate (e.g., Schwartz et al., 1986; Schwartz, 1988) and at higher elevations in Rwanda (Van Ranst et al., 1997) and South Africa (Hawker et al., 1992). Under tropical rain forest, soils with thick albic and spodic horizons occur in progressive transition with Oxisols and are the result of an evolution of Oxisols (Ferralsols) into white sands. These Oxisol-Spodosol sequences were studied along slopes (Lucas et al., 1987; Bravard & Righi, 1989, 1990; Schaefer et al., 2002), in plateau areas where the Oxisol (Ferralsol) profile has been truncated and the lateritic crust removed (Horbe et al., 2004) and in tropical lowlands (Altemüller & Klinge, 1964; Dubroeucq & Volkoff, 1988, 1998; Thomas et al., 1999; Bardy et al., 2008).

Podzols on coarse volcanic material are found in the North Island of New Zealand (De Coninck & Rijkse, 1986), and Podzols on tephra-derived parent materials with characteristics of Andisols were studied in Japan and the northwestern United States (Shoji & Ito, 1990; Dahlgren & Ugolini, 1991). These soils with spodic features occur at higher elevation regions where cooler, moister conditions prevail with subalpine fir vegetation, while Andisols develop in the lower, drier climatic zones (McDaniel et al., 1993).

Micromorphology has been established as a fundamental tool to characterise spodic materials and to assist in determining the nature and sequence of pedogenic processes. Early morphological studies helped define the properties and genesis of albic and spodic horizons by illustrating the nature of the intact arrangement of illuvial materials (Kubiëna, 1938, 1953; Flach, 1960). De Coninck and co-workers (De Coninck et al., 1974; De Coninck, 1980; De Coninck & Righi, 1983; De Coninck & McKeague, 1985) used micromorphology to help elucidate the chemical processes of podzolisation. McKeague and his colleagues also made major contributions toward understanding of spodic features, studying the pedogenesis of soils in Canada, especially of ortstein and Fe/Mn pans (McKeague et al., 1968; Brewer et al., 1973; McKeague & Sprout, 1975; Wang et al., 1978; McKeague & Wang, 1980; McKeague & Kodama, 1981).

Non-destructive microanalysis techniques, including EDS and WDS analyses, have enabled examination of the composition of constituents of cementing materials and the spatial distribution of elements in spodic horizons and ortstein (Lee et al, 1988; Schaefer et al., 2002; Horbe et al., 2004; Kaczorek et al., 2004) and in placic horizons (Righi et al, 1982; Breuning-Madsen et al., 2000; Jien et al., 2010). Micromorphological studies in combination with other types of chemical analyses have provided new results on the constitution and origin of organic matter in spodic materials, resulting in new concepts of Spodosol or Podzol formation (Buurman et al., 2005, 2008; Buurman & Jongmans, 2005).

2. Spodic Horizons and Ortstein

Spodic horizons are typically black (Bh horizons) or dark reddish brown to brown (Bs or Bhs horizons), reflecting deposition of various types of organic materials, organo-metallic complexes and Fe oxides. Darker spodic materials, richer in organic matter, typically accumulate above redder materials, rich in Fe and Al oxides (De Coninck et al., 1974; Bullock & Clayden, 1980). Spodic horizons have been intensively studied over a long period and different theories for their formation have been proposed. Conceptually, organic (fulvic) acids with relatively high molecular weight (HMW) that are leached from the O horizon form complexes with Al and Fe. These initially unsaturated organo-metallic complexes migrate through the soil and precipitate because of continuing addition of metals during the downward migration, until a certain carbon/metal ratio has been reached, causing the complex to precipitate and thus creating a spodic B horizon (McKeague et al., 1971; De Coninck, 1980). An alternative depositional process is microbial decomposition of organic ligands, mainly in the form of acids with low molecular weight (LMW), during downward migration, releasing ionic Al and Fe to the soil solution and resulting in the formation of Al-Si-OH and Fe-OH solid phases (Lundström, 1993; Lundström et al., 1995, 2000a). It has been hypothesised that metal-LMW acid complexes are transported by hyphae from the E horizon to the O layer, where Fe and Al are partly released and then complexed by HMW acids (Lundström et al., 2000b). Other theories of spodic horizon formation include eluviation and precipitation, or *in situ* formation, of proto-imogolite sols in Bs horizons followed by adsorption of mobile humus (Farmer et al., 1980; Anderson et al., 1982; Ugolini & Dahlgren, 1987), rather than involving formation of allophane-related phases through alteration of illuvial organo-mineral complexes (Farmer, 1982; Farmer et al., 1984, 1985). Besides a possible combination of various processes of spodic material deposition, the origin of the organic fraction is currently also considered to be varied, with dissolved organic matter (predominantly in poorly-drained Podzols) and decomposed root-derived material (predominantly in well-drained Podzols) as possible sources (Buurman & Jongmans, 2005; Buurman et al., 2013). Organic components are stabilised and the resistance to decomposition or biodegradation is increased by complexation with metals as well as by the presence of nanocrystalline minerals (Phillips & FitzPatrick, 1999; Buurman & Jongmans, 2005).

Ortstein (Bhsm, Bsm) is a black to dark brown, cemented layer rich in spodic materials (Wang et al., 1978; Freeland & Evans, 1993), defined as a diagnostic subsurface horizon in Soil Taxonomy (Soil Survey Staff, 1999, 2014). Ortstein is generally composed of translocated complex organo-mineral materials principally associated with Al (McKeague & Wang, 1980; McHardy & Robertson, 1983; Lee et al., 1988; Phillips & FitzPatrick, 1999). Other cementing agents such as Fe and Mn oxides and proto-imogolite allophane have been identified (Polteva & Sokolova, 1967; McKeague & Kodama, 1981; Farmer et al., 1984, 1985; Ross et al., 1989; Thompson et al., 1996; Kaczorek, et al., 2004). The colour of the horizon is generally linked to the relative amounts of Fe and Mn oxides and organic carbon. Gibbsite can form from degradation of organic ligands of the Al-organic-matter complex in ortstein horizons of sandy hydromorphic Podzols (Righi & De Coninck, 1977). The formation of ortstein is related to the eluviation/illuviation of spodic materials and can occur in both well-drained and poorly-drained soils (McKeague & Wang, 1980). Vertical eluviation/illuviation is most commonly enhanced by lateral translocation of organic and/or mineral compounds, induced by groundwater dynamics (Karavayeva, 1968; De Coninck et al., 1991). Ortstein can occur as prominent subhorizontal bands, resulting from lateral movement of soil water containing dissolved organic matter, which tends to follow stratification in the parent material (Buurman et al., 2013). Ortstein formation can begin as nodules (Polteva & Sokolova, 1967), subsequently forming a continuous accumulation of organo-mineral coatings that cement mineral grains and result in decreased biological activity (De Coninck et al., 1974, 1986; Van Ranst et al., 1997). The degree of hardness or cementation is related to the distribution of the illuvial organo-metallic complexes around coarse mineral grains and the infilling of pore space (McKeague & Wang, 1980; McHardy & Robertson, 1983; Rabenhorst & Hill, 1994). This is in turn influenced by texture, as coarser materials and lower void content require less number of agents for cementation (McKeague & Sprout, 1975; Righi, 1987).

2.1 Microstructure and c/f-Related Distribution Patterns

Spodic horizons are generally characterised by a basic microstructure, most accurately described in terms of their c/f-related distribution patterns. The typical gefuric, chitonic and enaulic c/f-related distribution patterns (according to Stoops, 2003) (see further) correspond to bridged grain, pellicular grain and intergrain microaggregate microstructures (according to Bullock et al., 1985).

The c/f-related distribution patterns in spodic horizons vary widely and reflect both the size of the coarse mineral components and the composition of the micromass (De Coninck & McKeague, 1985). The c/f-related distribution pattern may range from gefuric, chitonic (Fig. 1) or enaulic to porphyric (Fig. 2). Gradual differences in c/f-related distribution pattern, from gefuric to porphyric, can be viewed as an evolutionary

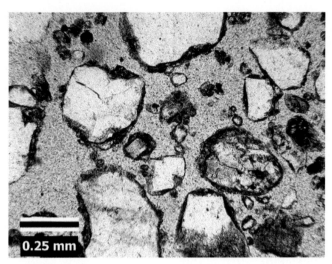

FIGURE 1 Chitonic c/f-related distribution pattern with simple packing voids in a spodic horizon (Bs horizon, Spodosol, Wisconsin, USA) (PPL). *Image by M. Wilson.*

FIGURE 2 Chito-porphyric c/f-related distribution pattern, with coarse material principally composed of subrounded and subangular quartz grains and with brownish-yellow to black coatings and infillings of polymorphic fine organic material in a spodic horizon (Bs3 horizon, Spodosol, W Oregon, USA) (PPL). *Image by M. Wilson.*

sequence of increasing accumulation of illuvial spodic materials (Pawluk, 1983). In an Oxisol-Spodosol (Ferralsol-Podzol) landscape, Bravard and Righi (1990) found that the c/f-related distribution patterns evolved from porphyric in the Oxisol (Ferralsol) on the plateau to enaulic, and finally to chitonic in the Spodosol (Podzol) on the footslope. Elemental and mineralogical similarities suggested a genetic link between the soils although they were classified in different and contrasted orders.

Individual thin sections may exhibit more than one distribution pattern, and intergrades of these fabrics are possible (e.g., chito-gefuric). The differences in c/f-related distribution patterns as well as in composition of the micromass result in different grades of consistency, ranging from friable to cemented. Spodic horizons with continuous or patchy cementation have chitonic and/or gefuric to porphyric c/f-related distribution patterns (see Fig. 7) and common vughs and planes (Rabenhorst & Hill, 1994; Phillips & FitzPatrick, 1999; Kaczorek et al., 2004).

Friable, sandy-textured spodic horizons have been described as having chito-gefuric (Schaefer et al., 2002) as well as enaulic c/f-related distribution patterns (Rabenhorst & Hill, 1994). The gefuric c/f-related distribution pattern may grade to enaulic as aggregates develop with increasing amounts of monomorphic organic material (see Section 2.3) (Salem Avad et al., 1982). Abundant complex packing voids are common between coarse sand grains, especially in the upper part of spodic horizons (Phillips & FitzPatrick, 1999). Aggregates (Fig. 3) are irregular to ovoid-shaped. Well-developed cappings and link cappings on the upper surfaces of the larger grains or rock fragments have been frequently observed for spodic horizons. The stratification of these cappings, in the form of silt-clay segregation, is presumably due to frost action and gelifluction (Van Vliet-Lanoë, 1998), emphasising the polygenetic character of some of these horizons (Phillips & FitzPatrick, 1999). Oriented fine sand/silt cappings and cryoturbation features (e.g., wedge-cast-like figures, platy microstructures) created during cold periods have been observed for relict Podzols in Alpine regions (D'Amico et al., 2015).

Silty or loamy spodic horizons commonly have a crumb microstructure (Fig. 4) or an equal to coarse enaulic c/f-related distribution pattern, with large (up to several millimetres in diameter), irregularly shaped, porous aggregates, composed of small, more or less coalescent granules. Some loamy spodic horizons have a subangular blocky microstructure and chitonic to porphyric c/f-related distribution patterns (Hseu et al., 2004).

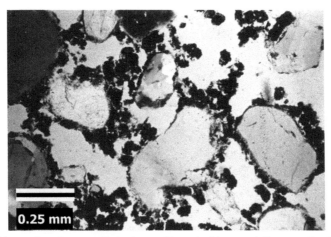

0.25 mm

FIGURE 3 Polymorphic coatings in a spodic horizon with a double-spaced fine enaulic c/f-related distribution pattern (sandy hydromorphic Spodosol, Landes, SW France) (partially XPL). *Image by D. Righi.*

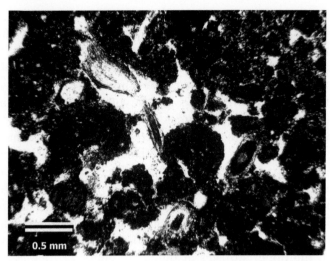

FIGURE 4 Crumb microstructure in a spodic horizon mainly consisting of aggregates of polymorphic amorphous organic fine material and with common root residues (Bhs horizon, Spodosol, Iron Co., Michigan, USA) (PPL). *Image by M. Wilson.*

2.2   Coarse Material

Micromorphological data on coarse material are rather scarce for spodic horizons, as they are predominantly formed in quartz-dominated parent materials, whose main mineral component contributes little to soil development.

The evolution of glauconite and its impact on the formation of spodic materials have been studied for Podzols of northern Belgium (De Coninck & Laruelle, 1960) and northern France (De Coninck & Maucorps, 1972). Glauconite is strongly weathered during podzolisation, undergoing transformation to swelling clay minerals as well as partial to complete dissolution.

Most published data refer to comparative studies of weathering of minerals in albic materials and spodic horizons (e.g., Fedoroff et al., 1977). Microprobe and SEM analyses have shown that feldspar grains in the albic materials exhibit a wide variety of dissolution features (prismatic etching pits), with Ca-feldspar being more severely etched than other feldspar types (Cruickshank et al., 1990).

Melkerud et al. (2000) noted, through thin section point-counting, an increase in weathering of plagioclase, alkali feldspar and biotite toward the surface of Podzols from northern Europe. Biotite is present in spodic B horizons as both fresh mineral grains and as exfoliated altered grains stained by Fe oxides, but it is absent in the overlying albic materials. The weathering rate of minerals is generally lower in spodic horizons than in E horizons because mineral grains are protected by Fe or Al oxides and organo-metallic complexes (McKeague et al., 1983).

The opposite trend was observed for quartz grains from a tropical Spodosol, as the result of the seasonal influence of groundwater and the greater water-holding capacity in the spodic horizon than in the E horizon (Marcelino & Stoops, 1996).

Jongmans et al. (1997) were the first to describe tunnel-like features within feldspar and hornblende grains of E horizons of Podzols. These features are different from (coalesced) etch pits and are assumed to be created by fungal hyphae (see also van Breemen et al., 2000; Hoffland et al., 2002). In feldspar, they were found to be concentrated in the uppermost centimetres of the E horizon (Hoffland et al., 2002), and they were observed more frequently for older Podzols (Hoffland et al., 2002; Smits et al., 2005), with differences between plagioclase and K-feldspar (Smits et al., 2005). Fungal weathering, without tunnel development, has also been studied for phyllosilicate minerals, using Atomic Force Microscopy (AFM) (McMaster, 2012). Mineral horizons of boreal forest Podzols are intensively colonised by mycorrhizal mycelia, which transfer protons and organic metabolites to mineral surfaces, resulting in mineral dissolution (Rosling et al., 2009).

2.3 Fine Organic Material

The fine organic material in spodic horizons and ortstein results from either *in situ* transformation of plant remains or illuviation. Two main types of secondary organic fine material may be observed, i.e., polymorphic and monomorphic material (De Coninck et al., 1974).

2.3.1 Polymorphic Material

This material is organised as porous aggregates consisting of accumulations of biologically and chemically decomposed fine organic material, limited amounts of clay- and silt-sized mineral grains and coarse organic elements with a recognisable cell or tissue structure. The material is very dark brown to black and occurs primarily in the A horizon and the upper part of B horizons where soil microorganisms are active (e.g., De Coninck et al., 1974, 1986; Buurman et al., 2005). This microbial activity is most intense near roots (Pagé, 1987b), with high fine root turnover rates (Sanborn & Lavkulich, 1989). Roots are mechanically and biochemically fragmented and transformed by soil mesofauna and microbial activity. These transformations result in multiple forms of degraded plant material (hence the name), including excrements (De Coninck & Righi, 1969; Eswaran et al., 1972; De Coninck et al., 1974, 1986; Righi & De Coninck, 1974; Robin & De Coninck, 1978; De Coninck, 1983).

Crumbs (Fig. 4) or granules of polymorphic material impart a loose and friable consistency to soils, also called 'pellety' (Eswaran et al., 1972; Bullock & Clayden, 1980; McHardy & Robertson, 1983; Gerasimova et al., 1996; Van Ranst et al., 1997; Li et al., 1998). These aggregates (pellets) may coalesce, evolving to a vughy or spongy mass of interlocking dense to very dense microaggregates in the lower B horizon (Phillips &

FitzPatrick, 1999). These transformations are generally regarded as biological, but physicochemical processes may influence polymorphic forms as well (Bruckert & Selino, 1978; Phillips & FitzPatrick, 1999). Buurman et al. (2005) categorised the arrangement of polymorphic aggregates as single, welded, strongly welded and flowed. They considered an increase in welding as a sign of abundant microbial and chemical decay of organic components.

Polymorphic materials typically occur in the presence of roots and can also form coatings without cracks on coarse sand grains (Robin & De Coninck, 1978). Faunal activity is the process that incorporates and mixes plant residues and silt-sized materials (Fig. 5). Elemental microprobe analysis shows that the bulk of structural aggregates are enriched with Al, suggesting that illuvial Al-organic-matter complexes are a substantial part of the material constituting polymorphic organic matter (Righi, 1977; Righi & De Coninck, 1977; Robin, 1979; Van Ranst et al., 1980).

Depending on their degree of decomposition, plant tissue residues in aggregates of polymorphic organic matter as well as aliphatic and poorly condensed organic compounds (fulvic acids) exhibit a more or less strong primary fluorescence when viewed with blue or ultraviolet (UV) light (Van Vliet et al., 1983; Altemüller & Van Vliet-Lanoë, 1990).

FIGURE 5 SEM image of an aggregate of polymorphic organic material in a spodic horizon; note the remnants of plant tissue (*arrow*) (sandy hydromorphic Spodosol, Landes, SW France). *Image by D. Righi.*

2.3.2 *Monomorphic Material*

This constituent is the primary spodic material of cemented spodic horizons or ortstein, resulting from the processes of mobilisation, transport and precipitation of illuvial organic matter (De Coninck et al., 1974; Righi & De Coninck, 1974; Righi, 1977; De Coninck & Righi, 1983; De Coninck & McKeague, 1985). The deposit is regarded as a precipitate of dissolved organic compounds that were water-transported (Buurman et al., 2005). Monomorphic material is composed of uniform colloidal-sized materials without coarse organic elements or microvoids (De Coninck, 1980). SEM examination reveals randomly stacked, small bodies of material (Van Ranst et al., 1980). It forms coatings on mineral grains (Fig. 6) and covers pre-existing organic features (polymorphic aggregates, plant remains) (De Coninck et al., 1974). The illuvial nature of this amorphous material is inferred from its presence as coatings in vertical channels and on coarse grains, and as deposits that bridge those grains (De Coninck et al., 1974, 1986; Robin & De Coninck, 1978; Van Ranst et al., 1980, 1997; Buurman et al., 2005). The material has a limpid aspect and its colour can range from black to reddish or yellowish brown. It is initially a gel but undergoes a transition from a liquid or dispersed phase by dehydration and solidification (De Coninck & McKeague, 1985). It becomes increasingly hydrophobic and typically exhibits polygonal cracking when dry, which is evident in both air- and oven-dried material (Fig. 7). Lack of a cracking pattern may be due to continuously moist conditions (Sanborn & Lavkulich, 1989). Monomorphic material may exhibit angular cracking, with detachment from the coarse grain surface (Van Ranst et al., 1980; De Coninck et al., 1986; Lee et al., 1988; McSweeney & FitzPatrick, 1990; FitzPatrick, 1993; Van Herreweghe et al., 2003). Except for areas that contain Al-organic-matter complexes, it is generally black under UV light (Jamet et al., 1990; Altemüller & Van Vliet-Lanoë, 1990; Thompson

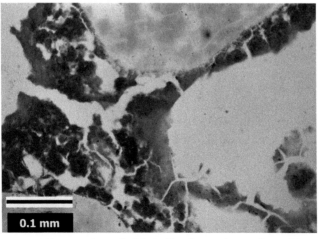

0.1 mm

FIGURE 6 Cracked monomorphic organic material (uniform) covering polymorphic organic matter (more heterogeneous aspect) in a cemented spodic horizon, suggesting two stages of development (sandy hydromorphic Spodosol, Landes, SW France) (PPL). *Image by D. Righi.*

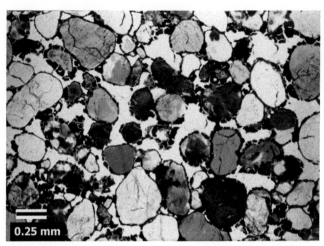

FIGURE 7 Cracked monomorphic coatings in a cemented spodic horizon (sandy hydromorphic Spodosol, Landes, SW France) (partially XPL). *Image by D. Righi.*

et al., 1996). In comparison with polymorphic organic matter, monomorphic materials are mainly composed of compounds with more aromatic and condensed molecular structures (humic acids), which strongly absorb blue and UV light without provoking fluorescence.

Monomorphic material is abundant in the lower part of the spodic B horizon, which is also characterised by the absence of recognisable fresh plant remains and roots. It is the most common form of organic material in hydromorphic Podzols (Buurman & Jongmans, 2005; Buurman et al., 2008; Kristiansen et al., 2010). However, Coelho et al. (2012) observed polymorphic organic matter and/or plant residues in different decomposition stages as most important soil features in spodic horizons of poorly-drained soils of a coastal tropical forest area. In these soils, *in situ* root decomposition by mesofauna and microorganisms is an important mechanism of organic matter accumulation in the deeper B horizons.

Elemental microprobe analysis shows that monomorphic material is composed of organo-Al or Al/Fe compounds with little Si (Fig. 8) (Righi, 1975; Van Ranst et al., 1980; De Coninck & McKeague, 1985; Thompson et al., 1996; Phillips & FitzPatrick, 1999). The lack of silica implies that clay minerals are absent.

2.3.3 Occurrence

Soil hydrology is an important factor in organic matter dynamics (Buurman & Jongmans, 2005). Microbial and mesofaunal activity increase the formation of polymorphic organic material in well-aerated horizons. Micromorphological evidence shows that the principal form of secondary organic materials is polymorphic in the A horizon and well-drained B horizons, where root activity is common (De Coninck,

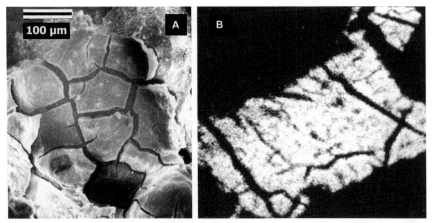

FIGURE 8 (A) SEM image of a thick monomorphic coating with polygonal cracking in ortstein (sandy hydromorphic Spodosol, Landes, SW France). (B) Element map obtained by EDS analysis, showing the distribution of Al in a monomorphic coating of the same sample. *Images by D. Righi.*

1980). In spodic B horizons, the bulk of the polymorphic materials are likely formed from *in situ* decomposition of roots that have followed the nutrient source into lower horizons (De Coninck, 1980; Van Ranst et al., 1997). Monomorphic materials dominate in lower B horizons or those horizons influenced by groundwater. Translocation of the dissolved organic compounds that constitute monomorphic materials can occur both vertically and laterally (Buurman & Jongmans, 2005; Buurman et al., 2005, 2008).

In general, polymorphic and monomorphic materials can form simultaneously in the soil (De Coninck et al., 1974; Brandon, et al., 1977; Buurman & Jongmans, 2005; Kristiansen et al., 2010). However, a preponderance of monomorphic organic material generally indicates a more advanced stage of podzolisation, with an increase in the amount of illuvial Al/Fe-complexes. Active processes of plant fragmentation by mesofaunal activity and biodegradation of plant residues create a polymorphic structure within a spodic horizon. If illuvial organo-metallic complexes are subsequently deposited, progressive cementation of the horizon will result in an increasingly higher amount of monomorphic forms. In turn, this accumulation results in the reduction of root development and faunal activity (De Coninck, 1980). Pervasive accumulation of monomorphic material in a horizon eventually leads to formation of the ortstein layer, especially in coarse-textured material (Righi, 1987).

The fine material of cemented horizons or nodules is brown or reddish brown to black with a generally undifferentiated b-fabric. It is commonly composed of monomorphic organic material enriched in Al, although it may contain non-crystalline material rich in Fe or Mn and complex polymorphic organo-mineral materials (Karavayeva, 1968; Miles et al., 1979; McKeague & Wang, 1980; Van Ranst et al., 1980; McHardy & Robertson, 1983;

Arocena & Pawluk, 1991; Rabenhorst & Hill, 1994). Nodules in a Canadian Spodosol or Podzol were identified to also contain polymorphic organic matter where active decomposition of organic materials by fungi prevented the dominance of monomorphic forms (Arocena & Pawluk, 1991).

Clay illuviation has been often recognised in soils with spodic features (e.g., De Coninck & Maucorps, 1972; Alekseyev, 1983; Liu et al., 1997; Li et al., 1998; Liu & Chen., 2004). Monomorphic organic material can occur as coatings that cover illuvial clay or polymorphic organic material (Fig. 6) (McKeague & Wang, 1980; Rabenhorst & Hill, 1994; Hseu et al., 2004; Kaczorek et al., 2004). This process is generally thought to occur sequentially in pedogenesis rather than simultaneously with spodic horizon development (e.g., Guillet et al., 1975; Condron & Rabenhorst, 1994).

In a region with associated Oxisols (Ferralsols) and Spodosols (Podzols), Schaefer et al. (2002) identified buried A horizons that subsequently have undergone organo-mineral illuviation and are currently interpreted as Bh horizons. In this study, micromorphological examination of organo-Fe/Al deposits in a Bh horizon provided evidence of active podzolisation processes taking place in the lower landscape positions.

2.4   Nanocrystalline Aluminosilicates

Nanocrystalline aluminosilicates (e.g., proto-imogolite allophane) have been identified in the fine material of spodic materials by means of selective dissolution, TEM observations and infrared spectroscopy (Farmer et al., 1980; McKeague & Kodama, 1981; Childs et al., 1983; Gustafsson et al., 1995; Lundström et al., 2000b). These compounds, as well as Al-polymers and Al-organic-matter complexes, are fluorescent under UV illumination (e.g., Altemüller & Van Vliet-Lanoë, 1990; Thompson et al., 1996).

The migration of aluminosilicates in Bs horizons is suggested by root pseudomorphs in which the cell wall structure is preserved in the allophanic material and by allophanic coatings covering sand grains (Farmer et al., 1984, 1985). Results presented by Milnes and Farmer (1987) suggest that allophane and organic matter are deposited separately and that gibbsite formed from Al-rich gels before the main allophane deposition phase in humus iron Podzols. Several generations of allophanic deposits are recognised, with evidence of dissolution of earlier phases by percolating waters. These observations support the views of Anderson et al. (1982) and Farmer (1982) that allophane accumulation in spodic B horizons is not controlled by illuviation and precipitation of organic matter. Kodama and Wang (1989), using selective dissolution and differential X-ray diffraction (XRD), identified opaline silica, proto-imogolite allophane and ferrihydrite in Canadian Podzols. Proto-imogolite allophane has been also cited as a contributor to pan formation in Australian Podzols and its occurrence was attributed to a previous soil-forming regime (Thompson et al., 1994, 1996). Interlayering of nanocrystalline aluminosilicates and organic materials, documented in the fine material of ortstein and other hardened spodic materials, suggests that multiple episodes of deposition are responsible for the formation of these occurrences (Farmer et al., 1984).

3. Related Horizons

3.1 Surface Horizons

O and A horizons are important sources of organic materials, including organic acids, and of soluble inorganic elements. Fabrics of O horizons of Podzols under forest vegetation are not unique to that order, but are typical of mor (raw humus) or moder (organic materials with varying degrees of decomposition partly mixed with inorganic components) (see also Stolt & Lindbo, 2010; Ismail-Meyer et al., 2018), with little incorporation of mineral material along the forest floor (De Coninck & McKeague, 1985; Sanborn & Lavkulich, 1989). The composition of O horizons depends on the type of material; the degree of degradation by fungi, bacteria and soil fauna; and the degree of decomposition and rearrangement (De Coninck et al., 1974; Phillips & FitzPatrick, 1999; Buurman et al., 2005).

Microscopic organic matter occurrences in O horizons of soils with spodic features range from recognisable fresh organ tissue and cell residues of plants to unrecognisable organic fine materials (including punctuations and organic pigments) (De Coninck et al., 1974; Van Ranst et al., 1997; Stoops, 2003; Hseu et al., 2004). Woody detritus and roots, often reduced to a melanised ring enclosing excrements, typically comprise the bulk of the O horizons (De Coninck & McKeague, 1985; Sanborn & Lavkulich, 1989). Fungal material, in the form of hyphae, sclerotia and mycorrhizal sheaths, is quite common. Fungi often penetrate into excrements, causing partial disintegration (De Coninck et al., 1974), and they also cause biochemical alteration of plant tissue. Excrements occur in various shapes and sizes, fresh or more or less transformed, and they increase in abundance with depth in the O horizon. The accumulation of organic material indicates microbial activity and consumption of plant remains by earthworms, enchytraeids, beetle larvae or mites (Phillips & FitzPatrick, 1999) (see also Kooistra & Pulleman, 2010, 2018). The organic O horizons may lie directly over albic soil materials, as in many boreal Podzols (e.g., Melkerud et al., 2000; Schaefer et al., 2002), or they may grade into an A horizon (e.g., Bravard & Righi, 1989; Van Ranst et al., 1997; Phillips & FitzPatrick, 1999).

A horizons generally have a dark grey to black colour and a 'salt-and-pepper' appearance in the field (Bullock & Clayden, 1980), due to a mixture of discrete, light-coloured mineral grains (generally quartz) and dark small organic particles. In many cases, excrements originally in O horizons enter the underlying mineral soil as channel infillings, composed of aggregates. In thin sections, black organic materials are generally highly decomposed, consisting of vegetal remains and amorphous fine organic materials. Plant remains are generally strongly fragmented but recognisable. The microstructure is spongy or granular, with a high total porosity (e.g., Van Ranst et al., 1997; Hseu et al., 2004). Fine and coarse granules may form from a combination of clay or silt particles with root remains and/or excrements (Phillips & FitzPatrick, 1999). Part of the plant remains and excrements are transformed into polymorphic organic matter (Robin & De Coninck, 1978; De Coninck, 1983). The c/f-related distribution pattern is coarse monic to enaulic and the soil is highly porous. It generally consists of loosely packed, mostly uncoated sand-sized grains, organic residues with various degrees of decomposition and loose aggregates of organic fine material (e.g., De Coninck & McKeague, 1985). The c/f-related distribution pattern is determined by the relative silt, clay and humus content, components that form aggregates of fine material between the sand-sized coarse grains (Phillips & FitzPatrick, 1999).

3.2   Eluvial Horizons and Albic Materials

Albic materials and eluvial (E) horizons are typically light grey, reflecting the absence of clay, Fe oxides and organic constituents (e.g., Melkerud et al., 2000). Iron depletion may result from reduction and dissolution of Fe oxides if water saturation occurs, as redoximorphic processes may be involved in the formation or expression of albic materials (Van Ranst et al., 1997; Hseu et al., 1999; Buurman et al., 2005; Wu & Chen, 2005). A second pale eluvial horizon (possibly albic materials) may occur beneath the spodic horizon, as the result of groundwater moving laterally or perching on an argillic, fragipan or placic horizon (Bullock & Clayden, 1980).

The thickness of eluvial horizons increases during soil development, extending into the upper B horizon. Formation of these horizons results from mineral dissolution and chelation of Fe and Al by complexing agents such as short-chain aliphatic and aromatic acids, followed by organo-metallic eluviation and/or microbial decay of the remaining organic matter at the boundary between the E and B horizons leaving only the most resistant compounds (Melkerud et al., 2000; Lundström et al., 2000b; Buurman & Jongmans, 2005; Buurman et al., 2013). Buurman et al. (2005) regarded microbial decomposition of organic matter, following removal of metals, as the principal mechanism in E horizon formation (rather than eluviation of organic materials) and indicated that the loss of Al increased the palatability of the remaining organic materials for microbes. E horizons are considered to be, in part, derived from former spodic B horizons by eluviation of organic matter, as witnessed by the presence of remnants of monomorphic coatings (Buurman et al., 2008). Thin Fe oxide coatings (<10 μm) have been observed in the lower part of E horizons and in vertical tongues of E horizon material (Buurman et al., 2008).

Enrichment in quartz of the eluvial horizon also occurs via congruent destruction of clay minerals by hydrolysis (Bravard & Righi, 1989; Horbe et al., 2004). This enrichment has been evaluated by calculation of quartz/feldspar ratios with depth (Salem Avad et al., 1982). Eluviation of clay, alone or in combination with organo-metallic complexes, have been documented, although this process may precede podzolisation (Guillet et al., 1975; Li et al., 1998).

Eluvial horizons from sandy soils typically have a basic microstructure, with a coarse monic c/f-related distribution pattern (Fig. 9). However, patterns can range to gefuric and possibly chitonic (e.g., Dubroeucq & Volkoff, 1998; Phillips and FitzPatrick, 1999), and the microstructure can also be porphyric (e.g., De Coninck & McKeague, 1985; Dubroeucq & Volkoff, 1998; Phillips & FitzPatrick, 1999). Sandy loam to loamy E horizons show more compact monic and gefuric c/f-related distribution patterns with closely packed bare grains and minor amounts of fine organic material (e.g., De Coninck & McKeague, 1985). Porphyric c/f-related distribution patterns (Fig. 10) have been reported for clay-rich E horizons of Podzols in Taiwan (Li et al., 1998; Hseu et al., 1999, 2004).

The coarse material of eluvial horizons is composed of uncoated grains, with a silt to coarse sand size, depending on the texture of the horizon. Phytoliths can be abundant in E horizons (Taina et al., 1997). Organic material and plant remains in various stages of decomposition are found in limited amounts (Salem Avad et al., 1982; De Coninck & McKeague, 1985; Li et al., 1998; Hseu et al., 2004).

Micromass material is generally present in limited amounts or absent. A weakly micaceous crystallitic b-fabric in an albic E horizon due to presence of sericite has been described (Van Ranst et al., 1997). Some small aggregates rich in organic matter (polymorphic material) might still be present (Kristiansen et al., 2010).

SEM examination of sand-sized grains (Dahlgren & Ugolini, 1991; Farmer et al., 1985; Skiba, 2007) illustrates the absence of organic surface coatings and the intense

FIGURE 9 Coarse monic and locally weakly developed enaulic and chitonic c/f-related distribution patterns in an E horizon (sandy hydromorphic Spodosol, Landes, SW France) (partially XPL). *Image by D. Righi.*

FIGURE 10 Single-spaced porphyric c/f-related distribution pattern and coarse material consisting mainly of quartz with lesser amounts of feldspar in an E horizon (Spodosol, NW Washington, USA) (PPL). *Image by M. Wilson.*

etching of mineral grains such as plagioclase and biotite. An increase in mineral weathering in eluvial horizons has been documented for Podzols from northern Europe (Melkerud et al., 2000) and northwest Russia (Salem Avad et al., 1982). Platy minerals (such as muscovite, biotite or their weathering products) have been observed in thin sections to be strongly cracked parallel or perpendicular to the basal planes of the mineral (Skiba, 2007). Jongmans et al. (1997) and van Breemen et al. (2000) report the presence of tubular micropores (3-10 μm) in feldspar and hornblende grains of albic materials, thought to be created by organic acids exuded by fungal hyphal tips. The observation by Phillips and FitzPatrick (1999) of fungi within biotite grains in E horizons is further evidence of biological weathering in these horizons. They also observed sclerotia within fractures in quartz grains. Quartz grains in albic horizons also often appear corroded (Dubroeucq & Volkoff, 1998; Schaefer et al., 2002; Horbe et al., 2004). Deep and abundant dissolution pits and cracks have been observed in quartz sand grains from tropical Podzols (Righi et al., 1990; Marcelino & Stoops, 1996). Quartz weathering induces grain fragmentation and results in abundant fine sand- and silt-sized quartz grains, in areas where the water table is close to the surface (Bravard & Righi, 1990; Dubroeucq & Volkoff, 1998).

3.3 Placic Horizons

Placic horizons generally have porphyric c/f-related distribution patterns with coarse mineral grains embedded in a micromass in which colour depends on the relative amount of Fe and Mn oxides and organic matter (Fig. 11) (De Coninck & McKeague, 1985; Hseu et al., 1999). It may be reddish brown and cracked, consisting of complex organo-mineral

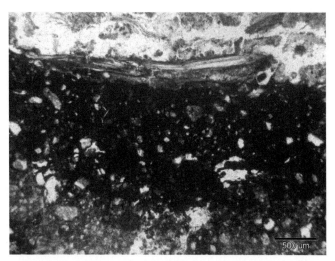

FIGURE 11 Double-spaced to open porphyric c/f-related distribution pattern in a placic horizon; note the root growing horizontally on top of the cemented horizon. *(Spodosol, Ireland) (Sample from M.J. Conry, see also Conry et al., 1996; Ghent University Archive).*

material associated principally with iron; pale to reddish brown with little cracking, composed of Fe oxides (De Coninck & Righi, 1983); or black, principally composed of Mn oxides (Brewer et al., 1973). The composition of this material has been investigated by microanalysis, including element mapping (Brewer et al., 1973; Righi et al., 1982; Breuning-Madsen et al., 2000; Jien et al., 2010). SEM and differential XRD analysis have shown that the fine material is composed of a variety of Fe or Fe/Mn oxides (e.g., poorly crystalline goethite, ferrihydrite, lepidocrocite) (Eswaran, 1971; Guillet et al., 1976; Campbell & Schwertmann, 1984). The lepidocrocite likely results from saturated, reducing conditions created during pan formation, and goethite remains poorly crystalline as a result of the carbon-rich environment (Campbell & Schwertmann, 1984; Clayden et al., 1990). Lepidocrocite is well preserved in periodically waterlogged Podzols and in placic horizons (Campbell & Schwertmann, 1984) because its stability is strongly enhanced by the presence of complexing organic molecules and dissolved Al (Schwertmann & Taylor, 1989; Chiang et al., 1999). Microscopic examination and electron microprobe analysis have show that Fe and Mn concentrations occur as separate zones or layers and that Mn concentration increases relative to Fe content with increasing depth (Brewer et al., 1973; D'Amico et al., 2015). SEM examination by Eswaran (1971) revealed the presence of fine, discontinuous voids not visible under the petrographic microscope.

Placic horizons, superimposed on horizons with spodic-like characteristics, were observed in soils derived from volcanic materials (Pinheiro et al., 2004). The placic horizons were described as generally having chitonic c/f-related distribution patterns that grade to close-porphyric. These horizons can have a lamellar microstructure at the top, composed of a layer of yellow to red or dark red globular material (Fe oxide) with an undifferentiated b-fabric, superimposed on a dark brown layer (organo-metallic complexes) with undifferentiated b-fabric.

The placic horizon is found in a wide range of climatic conditions (temperate to tropical) and landscapes, but always under udic or perudic moisture regimes (McKeague et al., 1983; Hseu et al., 1999; Pinheiro et al., 2004; Schawe et al., 2007; Wu & Chen, 2005; Jien et al., 2010). The formation process of these horizons is different from podzolisation (Clayden et al., 1990; Conry et al., 1996). Iron is reduced and mobilised in association with wet, acid soil conditions with high organic matter concentrations (Clayden et al., 1990; Lapen & Wang, 1999; Wu & Chen, 2005). Placic horizons form on the crystallisation of Fe/Mn oxides from dissolved divalent ions, triggered by changes in redox potential (Bockheim, 2011); thus periodical waterlogging is essential. These horizons are common in soils with spodic features that contain sufficient iron (De Coninck, et al., 1974; Phillips & FitzPatrick, 1999). The zone of saturation/reduction may represent a temporary condition within a seasonally water-saturated surface horizon with high organic matter content, or it may develop in organic and mineral horizons overlying a slowly permeable subsoil such as created by the placic layer (Righi et al., 1982; Hseu et al., 1999; Pinheiro et al., 2004). Placic deposition is generally at discontinuities where soil textural or structural changes affect permeability and hydraulic conductivity (Pinheiro et al., 2004). Therefore these horizons may be present in different positions within a soil profile (Conry et al., 1996). Hseu et al. (1999) found that best developed placic horizons occur in flat landscapes. In steeper landforms, placic horizons were bifurcated and thicker, suggesting lateral translocation. Well-developed placic horizons between E horizons and underlying ortstein have been reported for Podzols in the Italian Alps (D'Amico et al., 2015).

3.4   Other Horizons

Fragipans and duric horizons can form under spodic horizons (McKeague & Sprout, 1975; De Coninck & McKeague, 1985). Both fragipans (Weisenborn & Schaetzl, 2005) and duric horizons (McKeague & Sprout, 1975; McKeague & Protz, 1980) are generally characterised by a close porphyric c/f-related distribution pattern. Micromorphological examination shows few voids and coarse grains bridged or coated by thin, grey or brown silt. McKeague and Protz (1980) reported clay coatings on voids and coarse grains that vary from dark reddish brown to yellowish brown, which appear to cement the grains. Electron microprobe analysis indicates that the cement is composed of organo-Al or organo-Al/Fe compounds.

4. Conclusions

Fine organic and inorganic constituents composing spodic materials accumulate as a result of eluviation/illuviation processes and *in situ* transformations. Translocation of constituents is generally regarded to occur by organo-complexation of metals, but translocation of inorganic compounds is known to occur as well. These processes, in part governed by the properties of the soil in which they form, result in a variety of materials and distribution patterns in thin sections.

One of the earliest applications of microscopic examination of undisturbed soil materials was for elucidation of the nature of spodic processes. This technique is important to determine the relative development of spodic horizons, especially the type, amount and distribution of amorphous fine organic materials, the degree of cementation and the sequence of the soil-forming processes. Development of spodic features has a complexity that makes searching for changes in genesis over time feasible. In addition, small changes in volume during pedogenesis, common in these soils, increase the length of time that features remain recognisable, adding to the identifiable genetic complexity. The different types of spodic materials vary considerably over short distances, both vertically and laterally.

Information on the nature of components derived from the micromorphological study of spodic materials is increased when optical microscopy is used in conjunction with other analytical techniques such as selective dissolution, electron microscopy with microanalysis, XRD and infrared analysis. Micromorphology in combination with other analytical tools will continue to be important in future research on spodic features. Buurman and Jongmans (2005) illustrated that spodic processes differ by climate and soil hydrology, but further research is needed to increase the understanding of site-specific spodic features, especially under different weathering regimes and hydromorphic conditions. One issue is the identification of soil conditions that determine the formation and stability of the different types of amorphous fine organic materials. Also, a better understanding of the role of specific soil organisms in the transformation of organic forms is needed (Phillips & FitzPatrick, 1999).

Landscape studies focused on soil hydrologic conditions influencing vertical as well as lateral translocation of spodic materials will provide a more complete understanding of the processes involved. These studies should include the compounding influence of water saturation related to fluctuating or perched water tables on ortstein and placic horizon chemistry. Chemical and physical association of organic and inorganic forms involved in monomorphic materials and in the nature of cementation of ortstein, placic and associated horizons is still not completely resolved. Also, further study of the development of albic horizons via humic acid eluviation and/or degradation is warranted. Questions still remain regarding the process of clay mineral loss from eluvial horizons during podzolisation in terms of congruent or incongruent dissolution, subsequent mineral formation or translocation of clay minerals. The complexity of these questions is further enhanced by the presence of pedogenic features developed during earlier stages with different climatic conditions on which the modern soil is imprinted.

References

Alekseyev, V.V., 1983. Mineralogical analysis for the determination of podzolization, lessivage, and argillation. Soviet Soil Science 15, 21-28.

Altemüller, H.J. & Klinge, H., 1964. Mikromorphologische Untersuchungen über die Entwicklung von Podsolen im Amazonabecken. In Jongerius, A. (ed.), Soil Micromorphology. Elsevier, Amsterdam, pp. 295-305.

Altemüller, H.J. & Van Vliet-Lanoë, B., 1990. Soil thin section fluorescence microscopy. In Douglas, L.A. (ed.), Soil Micromorphology: A Basic and Applied Science. Developments in Soil Science, Volume 19. Elsevier, Amsterdam, pp. 565-579.

Anderson, H.A., Berrow, M.L., Farmer, V.C., Hepburn, A., Russell, J.D. & Walker, A.D., 1982. A reassessment of podzol formation processes. Journal of Soil Science 33, 125-136.

Andriesse, J.P., 1970. The development of the podzol morphology in the tropical lowlands of Sarawak (Malaysia). Geoderma 3, 261-279.

Arocena, J.M. & Pawluk, S., 1991. The nature and origin of nodules in podzolic soils from Alberta. Canadian Journal of Soil Science 71, 411-426.

Bardy, M., Fritsch, E., Derenne, S., Allard, T., do Nascimento, N.R. & Bueno, G.T., 2008. Micromorphology and spectroscopic characteristics of organic matter in waterlogged podzols of the upper Amazon basin. Geoderma 145, 222-230.

Bockheim, J., 2011. Distribution and genesis of ortstein and placic horizons in soils of the USA: a review. Soil Science Society of America Journal 75, 994-1005.

Brandon, C.E., Boul, S.W., Gamble, E.E. & Pope, R.A., 1977. Spodic horizon brittleness in Leon (Aeric Haplaquod) soils. Soil Science Society of America Journal 41, 951-954.

Bravard, S. & Righi, D., 1989. Geochemical differences in an Oxisol-Spodosol toposequence of Amazonia, Brazil. Geoderma 44, 29-42.

Bravard, S. & Righi, D., 1990. Micromorphology of an Oxisol-Spodosol catena in Amazonia (Brazil). In Douglas, L.A. (ed.), Soil Micromorphology: A Basic and Applied Science. Developments in Soil Science, Volume 19. Elsevier, Amsterdam, pp. 169-174.

Breuning-Madsen, H., Rønsbo, J. & Holst, M.K., 2000. Comparison of the composition of iron pans in Danish burial mounds with bog iron and spodic material. Catena 39, 1-9.

Brewer, R., Protz, R. & McKeague, J.A., 1973. Microscopy and electron microprobe analysis of some iron-manganese pans from Newfoundland. Canadian Journal of Soil Science 53, 349-361.

Bruckert, S. & Selino, D., 1978. Mise en évidence de l'origine biologique ou chimique des structures microagrégées foissonnantes des sols bruns ocreux. Pedologie 27, 46-59.

Bullock, P. & Clayden, B., 1980. The morphological properties of Spodosols. In Theng, B.K.G. (ed.), Soils with Variable Charge. New Zealand Society of Soil Science, Lower Hutt, pp. 45-65.

Bullock, P., Fedoroff, N., Jongerius, A., Stoops, G., Tursina, T. & Babel, U., 1985. Handbook for Soil Thin Section Description. Waine Research Publications, Wolverhampton, 152 p.

Buurman, P. & Jongmans, A.G., 2005. Podzolization and soil organic matter dynamics. Geoderma 125, 71-83.

Buurman, P., Van Bergen, P.F., Jongmans, A.G., Meijer, E.L., Duran, B. & Van Lagen, B., 2005. Spatial and temporal variation in podzol organic matter studied by pyrolysis-gas chromatography/mass spectrometry and micromorphology. European Journal of Soil Science 56, 253-270.

Buurman, P., Jongmans, A.G. & Nierop, K.G.J., 2008. Comparison of Michigan and Dutch podzolized soils: organic matter characterization by micromorphology and pyrolysis-GC/MS. Soil Science Society of America Journal 72, 1344-1356.

Buurman, P., Vidal-Torrado, P. & Martins, V.M., 2013. The Podzol hydrosequence of Itaguare (São Paulo, Brazil). 1. Geomorphology and interpretation of profile morphology. Soil Science Society of America Journal 77, 1294-1306.

Campbell, A.S. & Schwertmann, U., 1984. Iron oxide mineralogy of placic horizons. Journal of Soil Science 35, 569-582.

Childs, C.W., Parfitt, R.L. & Lee, R., 1983. Movement of aluminium as an inorganic complex in some podzolised soils, New Zealand. Geoderma 29, 139-155.

Chiang, H.C., Wang, M.K., Houng, K.H., White, N. & Dixon, J., 1999. Mineralogy of B horizons in alpine forest soils of Taiwan. Soil Science 164, 111-122.

Clayden, B., Daly, B.K., Lee, R. & Mew, G., 1990. The nature, occurrence, and genesis of placic horizons. In Kimble, J.M. & Yeck, R.D. (eds.), Proceedings of the 5th International Soil Correlation Meeting. Characterization, Classification, and Utilization of Spodosols. USDA, Soil Conservation Service, Lincoln, pp. 88-104.

Coelho, M.R., Martins, V.M., Otero Pérez, X.L., Vázquez, F.M., Gomes, F.H., Cooper, M. & Vidal-Torrado, P., 2012. Micromorfologia de horizontes espódicos nas Restingas do Estado de São Paulo. Revista Brasileira de Ciência do Solo 36, 1380-1394.

Condron, M.A. & Rabenhorst, M.C., 1994. Micromorphology of spodic horizons in a Psamment-Aquod toposequence on the Atlantic coastal plain of Maryland, U.S.A. In Ringrose-Voase, A.J. & Humphreys, G.S. (eds.), Soil Micromorphology: Studies in Management and Genesis. Developments in Soil Science, Volume 22. Elsevier, Amsterdam, pp. 179-186.

Conry, M.J., De Coninck, F. & Stoops, G., 1996. The properties, genesis and significance of a man-made iron pan podzol near Castletownbere, Ireland. European Journal of Soil Science 47, 279-284.

Cruickshank, C.L., Evans, L.J. & Spiers, G.A., 1990. Chemical and morphological features of mineral grains in some spodosolic soils. In Douglas, L.A. (ed.), Soil Micromorphology: A Basic and Applied Science. Developments in Soil Science, Volume 19. Elsevier, Amsterdam, pp. 557-564.

Dahlgren, R.A. & Ugolini, F.C., 1991. Distribution and characterization of short-range-order minerals in Spodosols from the Washington Cascades. Geoderma 48, 391-413.

D'Amico, M.E., Catoni, M., Terrible, F., Zanini, E. & Bonifacio, E., 2015. Contrasting environmental memories in relict soils on different parent rocks in the south-western Italian Alps. Quaternary International 418, 61-74.

De Coninck, F., 1980. Major mechanisms in formation of spodic horizons. Geoderma 24, 101-128.

De Coninck, F., 1983. Genesis of Podzols. Academiae Analecta 45, 65 p.

De Coninck, F. & Laruelle, J., 1960. L'influence de la composition minéralogique sur la formation des podzols. Transactions of the 7th International Congress of Soil Science, Volume IV, Madison, pp. 157-164.

De Coninck, F. & Laruelle, J., 1964. Soil development in sandy materials of the Belgian Campine. In Jongerius, A. (ed.), Soil Micromorphology. Elsevier, Amsterdam, pp. 169-188.

De Coninck, F. & Maucorps, J., 1972. Evolution of glauconite and kaolinite in Spodosols of northern France. In Serratosa, J.M. (ed.), Kaolin Symposium. International Clay Conference, Madrid, pp. 125-142.

De Coninck, F. & McKeague, J.A., 1985. Micromorphology of Spodosols. In Douglas, L.A. & Thompson, M.L. (eds.), Soil Micromorphology and Soil Classification. Soil Science Society of America Special Publication 15, SSSA, Madison, pp. 121-144.

De Coninck, F. & Righi, D., 1969. Aspects micromorphologiques de la podzolization en Forêt de Rambouillet. Science du Sol 1969, 55-77.

De Coninck, F. & Righi, D., 1983. Podzolisation and the spodic horizon. In Bullock, P. & Murphy, C.P. (eds.), Soil Micromorphology, Volume 2, Soil Genesis. AB Academic Publishers, Berkhamsted, pp. 389-417.

De Coninck, F. & Rijkse, W.C., 1986. Micromorphology of three podzols in coarse volcanic material (N. Island, New Zealand). Transactions of the 13th Congress of the International Society of Soil Science, Hamburg, pp. 1549-1550.

De Coninck, F., Righi, D., Maucorps, J. & Robin, A.M., 1974. Origin and micromorphological nomenclature of organic matter in sandy Spodosols. In Rutherford, G.K. (ed.), Soil Microscopy. The Limestone Press, Kingston, pp. 263-280.

De Coninck, F., Langohr, R., Embrechts, J. & Van Ranst, E., 1986. The Belgian soil classification system under the microscope. Pedologie 36, 235-261.

De Coninck, F., Bock, C., Grégoire, F., Maucorps, J. & Wicherek, S., 1991. Les transferts latéraux de solutions du sol dans un système Lande sur podzol-tourbière acide. Bulletin d'Ecologie 22, 343-362.

Dubroeucq, D. & Volkoff, B., 1988. Evolution des couvertures pédologiques sableuses à podzols géants d'Amazonie, bassin du haut rio Negro. Cahiers ORSTOM, Série Pédologie 24, 191-214.

Dubroeucq, D. & Volkoff, B., 1998. From Oxisols to Spodosols and Histosols: evolution of the soil mantles in the Rio Negro basin (Amazonia). Catena 32, 245-280.

Eswaran, H., 1971. Electron scanning studies of the fabric of fracture surfaces. Soil Science Society of America Proceedings 35, 787-790.

Eswaran, H., De Coninck, F. & Conry, M.J., 1972. A comparative micromorphological study of light and medium textured podzols. In Kowalinski, S. & Drozd, J. (eds.), Soil Micromorphology. Panstwowe Wydawnicto Naukowe, Warsawa, pp. 269-285.

Farmer, V.C., 1982. Significance of the presence of allophane and imogolite in podzol Bs horizons for podsolization mechanisms. A review. Soil Science and Plant Nutrition 28, 571-578.

Farmer, V.C., Russell, J.D. & Berrow, M.L., 1980. Imogolite and proto-imogolite allophane in spodic horizons: evidence for a mobile aluminium silicate complex in podzol formation. Journal of Soil Science 31, 673-684.

Farmer, V.C., Fraser, A.R., Robertson, L. & Sleeman, J.R., 1984. Proto-imogolite allophane in Podzol concretions in Australia: possible relationship to aluminous ferrallitic (lateritic) cementation. Journal of Soil Science 35, 333-340.

Farmer, V.C., McHardy, W.J, Robertson, L., Walker, A. & Wilson, M.J., 1985. Micromorphology and submicroscopy of allophane and imogolite in a Podzol Bs horizon: evidence for translocation and origin. Journal of Soil Science 36, 87-95.

Fedoroff, N., de Kimpe, C.R. & Bourbeau, G., 1977. L'altération des minéraux primaires en milieu podzolique en France Atlantique et au Québec. Catena 4, 29-40.

FitzPatrick, E.A., 1993. Soil Microscopy and Micromorphology. John Wiley & Sons, Chichester, 304 p.

Flach, K.W., 1960. Sols Bruns Acides in the Northeastern United States. Genesis, Morphology, and Relationship to Associated Soils. PhD Dissertation, Cornell University, Ithaca, 210 p.

Freeland, J.A. & Evans, C.V., 1993. Genesis and profile development of Success soils, northern New Hampshire. Soil Science Society of America Proceedings 57, 183-191.

Fridland, V.M., 1958. Podzolization and illimerization. Soviet Soil Science 1, 24-32.

Gerasimova, M.I., Gubin, S.V. & Shoba, S.A., 1996. Soils of Russia and Adjacent Countries: Geography and Micromorphology. Moscow State University & Agricultural University Wageningen, 204 p.

Glinka, K.D., 1924. Die Degradation und der podsolige Prozess. Internationale Mitteilungen für Bodenkunde 14, 40-49.

Guillet, B., Rouiller, J. & Souchier, B., 1975. Podzolization and clay migration in Spodosols of eastern France. Geoderma 14, 223-245.

Guillet, B., Rouiller, J. & Souchier, B., 1976. Accumulations de fer (lepidocrocite) superposées à des encroutements ferri-manganiques dans des sols hydromorphes vosgiens. Bulletin de la Sociéte Géologique de France 18, 55-58.

Gustafsson, J.P., Bhattacharya, P., Bain, D.C., Fraser, A.R. & McHardy, W.J., 1995. Podzolisation mechanisms and the synthesis of imogolite in northern Scandinavia. Geoderma 66, 167-184.

Hawker, L.C., van Rooyen, T.H. & Fitzpatrick, R.W., 1992. A slope sequence of Podzols in the southern Cape, South Africa. 1. Physical and micromorphological properties. South African Journal of Plant and Soil 9, 94-102.

Hoffland, E., Giesler, R., Jongmans, A.G. & van Breemen, N., 2002. Increasing feldspar tunneling by fungi across a north Sweden podzol chronosequence. Ecosystems 5, 11-22.

Horbe, A.M.C., Horbe, M.A. & Suguio, K., 2004. Tropical Spodosols in northeastern Amazonas State, Brazil. Geoderma 119, 55-68.

Hseu, Z.Y., Chen, Z.S. & Wu, Z.D., 1999. Characterization of placic horizons in two subalpine forest Inceptisols. Soil Science Society of America Journal 63, 941-947.

Hseu, Z.Y., Tsai, C.C., Lin, C.W. & Chen, Z.S., 2004. Transitional soil characteristics of Ultisols and Spodosols in the subalpine forest of Taiwan. Soil Science 169, 457-467.

Ismail-Meyer, K., Stolt, M. & Lindbo, D., 2018. Soil organic matter. In Stoops, G., Marcelino, V. & Mees, F. (eds.), Interpretation of Micromorphological Features of Soils and Regoliths. Second Edition. Elsevier, Amsterdam, pp. 471-512.

IUSS Working Group WRB, 2015. World Reference Base for Soil Resources 2014, Update 2015. World Soil Resources Reports No. 106. FAO, Rome, 192 p.

Jakobsen, B.H., 1989. Evidence for translocations into the B horizons of a subartic Podzol in Greenland. Geoderma 45, 3-17.

Jamet, R., Toutain, F., Guillet, B. & Rambaud, D., 1990. Forms and origin of alumina in the A2 horizon of the tropical podzols of Tahiti (French Polynesia). In Douglas, L.A. (ed.), Soil Micromorphology: A Basic and Applied Science. Developments in Soil Science, Volume 19. Elsevier, Amsterdam, pp. 213-218.

Jersak, J., Amundson, R. & Brimhall, G., 1995. A mass balance analysis of podzolization: examples from the northeastern United States. Geoderma 66, 15-42.

Jien, S.H., Hseu, Z.Y., Iizuka, Y., Chen, T.H. & Chiu, C.Y., 2010. Geochemical characterization of placic horizons in subtropical montane forest soils, northeastern Taiwan. European Journal of Soil Science 61, 319-332.

Jongmans, A.G., van Breemen, N., Lundström, U., van Hees, P.A.W., Finlay, R.D., Srinivasan, M., Unestam, T., Giesler, R., Melkerud, P.A. & Olsson, M., 1997. Rock-eating fungi. Nature 389, 682-683.

Kaczorek, D., Sommer, M., Andruschkewitsch, I., Oktaba, L., Czerwinski, Z. & Stahr, K., 2004. A comparative micromorphological and chemical study of 'Raseneisenstein' (bog iron ore) and 'Ortstein'. Geoderma 121, 83-94.

Karavayeva, N.A., 1968. Formation and evolution of cemented ortstein horizons in the Taiga zone. Transactions of the 9th International Congress of Soil Science, Volume 4, pp. 451-458.

Kodama, H. & Wang, C., 1989. Distribution and characterization of noncrystalline inorganic components in Spodosols and Spodosol-like soils. Soil Science Society of America Proceedings 53, 526-534.

Kooistra, M.J. & Pulleman, M.M., 2010. Features related to faunal activity. In Stoops, G., Marcelino, V. & Mees, F. (eds.), Interpretation of Micromorphological Features of Soils and Regoliths. Elsevier, Amsterdam. pp. 397-418.

Kooistra, M.J. & Pulleman, M.M., 2018. Features related to faunal activity. In Stoops, G., Marcelino, V. & Mees, F. (eds.), Interpretation of Micromorphological Features of Soils and Regoliths. Second Edition. Elsevier, Amsterdam, pp. 447-469.

Kristiansen, S.M., Dalsgaard, K., Thomsen, I.K., Knicker, H., Laubel, A., Schneider, D. & Odgaard, B.V., 2010. Genesis of spodic material underneath peat bogs in a Danish wetland. Soil Science Society of America Journal 74, 1284-1292.

Kubiëna, W.L., 1938. Micropedology. Collegiate Press, Ames, 242 p.

Kubiëna, W.L., 1953. The Soils of Europe. Thomas Murby & Co., London, 317 p.

Lapen, D.R. & Wang, C., 1999. Placic and ortstein horizon genesis and peatland development, southeastern Newfoundland. Soil Science Society of America Journal 63, 1472-1482.

Lee, F.Y., Yuan, T.L. & Carlisle, V.W., 1988. Nature of cementing materials in ortstein horizons of selected Florida Spodosols. I. Constituents of cementing materials. Soil Science Society of America Journal 52, 1411-1418.

Li, S.Y., Chen, Z.S. & Liu, J.C.,1998. Subalpine loamy Spodosols in Taiwan: characteristics, micromorphology and genesis. Soil Science Society of America Journal 62, 710-716.

Liu, J.C. & Chen, Z.S., 2004. Soil characteristics and clay mineralogy of two subalpine forest Spodosols with clay accumulation in Taiwan. Soil Science 169, 66-80.

Liu, J.C., Li, S.Y. & Chen, Z.S., 1997. Microscopic characterization and podzolization of Spodosols in central Taiwan. In Shoba, S., Gerasimova, M. & Miedema, R. (eds.), Soil Micromorphology: Studies on Soil Diversity, Diagnostics, Dynamics. Moscow, Wageningen, pp. 120-128.

Lucas, Y., Boulet, R., Chauvel, A. & Veillon, L., 1987. Systèmes sols ferrallitiques—podzols en région amazonienne. In Righi, D. & Chauvel, A. (eds.), Podzols et Podzolisation. AFES-INRA, Plaisir-Grignon, pp. 53-65.

Lundström, U.S., 1993. The role of organic acids in soil solution chemistry in a podzolized soil. Journal of Soil Science 44, 121-133.

Lundström, U.S., Van Breemen, N. & Jongmans, A.G., 1995. Evidence for microbial decomposition of organic acids during podzolization. European Journal of Soil Science 46, 489-496.

Lundström, U.S., van Breemen, N. & Bain, D., 2000a. The podzolization process: a review. Geoderma 94, 91-107.

Lundström, U.S., van Breemen, N., Bain, D., van Hees, P.A.W., Giesler, R., Gustafsson, J.P., Ilvesniemi, H., Karltun, E., Melkerud, P.A., Olsson, M., Riise, G., Wahlberg, O., Bergelin, A., Bishop, K., Finlay, R., Jongmans, A.G., Magnusson, T., Mannerkoski, H., Nordgren, A., Nyberg, L., Starr, M. & Tau Strand, L., 2000b. Advances in understanding the podzolization process resulting from a multidisciplinary study of three coniferous forest soils in the Nordic Countries. Geoderma 94, 335-353.

Marcelino, V. & Stoops, G., 1996. A weathering score for sandy soil materials based on the intensity of etching of quartz grains. European Journal of Soil Science 47, 7-12.

McDaniel, P.A., Fosberg, M.A. & Falen, A.L., 1993. Expression of andic and spodic properties in tephra-influenced soils of northern Idaho, USA. Geoderma 58, 79-94.

McHardy, W.J. & Robertson, L., 1983. An optical scanning microscopic and microanalytical study of cementation in some podzols. Geoderma 30, 160-171.

McKeague, J.A. & Kodama, H., 1981. Imogolite in cemented horizons of some British Columbia soils. Geoderma 25, 189-197.

McKeague, J.A. & Protz, R., 1980. Cement of duric horizons, micromorphology, and energy dispersive analysis. Canadian Journal of Soil Science 60, 45-52.

McKeague, J.A. & Sprout, P.B., 1975. Cemented subsoils (duric horizons) in some soils of British Columbia. Canadian Journal of Soil Science 55, 189-203.

McKeague, J.A. & Wang, C., 1980. Micromorphology and energy dispersive analysis of ortstein horizons of Podzolic soils from New Brunswick and Nova Scotia, Canada. Canadian Journal of Soil Science 60, 9-21.

McKeague, J.A., Damman, A.W.H. & Heringa, P.K., 1968. Iron-manganese and other pans in some soils of Newfoundland. Canadian Journal of Soil Science 48, 243-253.

McKeague, J.A., Brydon, J.E. & Miles, N.M., 1971. Differentiation of forms of extractable iron and aluminium in soils. Soil Science Society of America Proceedings 35, 33-38.

McKeague, J.A., De Coninck, F. & Franzmeier, D.P., 1983. Spodosols. In Wilding, L.P., Smeck, N.E. & Hall, G.F. (eds.), Pedogenesis and Soil Taxonomy. II. The Soil Orders. Developments in Soil Science, Volume 11B. Elsevier, Amsterdam, pp. 217-252.

McMaster, T.J., 2012. Atomic Force Microscopy of the fungi-mineral interface: applications in mineral dissolution, weathering and biogeochemistry. Current Opinion in Biotechnology 23, 562-569.

McSweeney, K. & FitzPatrick, E.A., 1990. Microscopic characterization of the spodic horizon. In Kimble, J. M. & Yeck, R.D. (eds.), Proceedings of the Fifth International Soil Correlation Meeting (ISCOM V): Characterization, Classification, and Utilization of Spodosols. USDA, Soil Conservation Service, Washington, pp. 211-220.

Melkerud, P.A, Bain, D.C., Jongmans, A.G. & Tarvainen, T., 2000. Chemical, mineralogical and morphological characterization of three podzols developed on glacial deposits in Northern Europe. Geoderma 94, 123-146.

Miles, N.M., Wang, C. & McKeague, J.A., 1979. Chemical and clay mineralogical properties of ortstein soils from the Maritime Provinces. Canadian Journal of Soil Science 59, 287-299.

Milnes, A.R. & Farmer, V.C., 1987. Micromorphological and analytical studies of the fine matrix of an Australian humus iron podzol. Journal of Soil Science 38, 593-605.

Pagé, F., 1987a. Etude Géochimique et Micromorphologique des Sols Podzolisés du Québec. Genèse, Fonctionnement et Classification. PhD Dissertation, Université Pierre et Marie Curie, Paris, 341 p.

Pagé, F., 1987b. Role des racines sur la différenciation des organisations micromorphologiques des horizons B podzoliques du Québec. In Fedoroff, N., Bresson, L.M. & Courty, M.A. (eds.), Micromorphologie des Sols, Soil Micromorphology. AFES, Plaisir, pp. 377-384.

Pagé, F. & Guillet, B., 1991. Formation of loose and cemented B-horizons in Podzolic soils - evaluation of biological actions from micromorphological features, C/N values and C-14 datings. Canadian Journal of Soil Science 71, 485-494.

Pawluk, S., 1983. Fabric sequences as related to genetic processes in two Alberta soils. Geoderma 30, 233-242.

Phillips, D.H. & FitzPatrick, E.A., 1999. Biological influences on the morphology and micromorphology of selected Podzols (Spodosols) and Cambisols (Inceptisols) from the eastern United States and north-east Scotland. Geoderma 90, 327-364.

Pinheiro, J., Tejedor Salguero, M. & Rodriguez, A., 2004. Genesis of placic horizons in Andisols from Terceira Island Azores − Portugal. Catena 56, 85-94.

Polteva, R.N. & Sokolova, T.A., 1967. Investigations of concretions in a strongly podzolized soil. Soviet Soil Science 7, 884-893.

Rabenhorst, M.C. & Hill, R.L., 1994. Strength characteristics of spodic horizons in soils of the Atlantic Coastal Plain, U.S.A. In Ringrose-Voase, A. & Humphreys, G.S. (eds.), Soil Micromorphology: Studies in Management and Genesis. Developments in Soil Science, Volume 22. Elsevier, Amsterdam, pp. 855-864.

Righi, D., 1975. Etude au microscope électronique à balayage de champ et au microanalyseur à sonde électronique des revêtements et agrégats organiques d'horizons B spodiques. Science du Sol 4, 315-321.

Righi, D., 1977. Genèse et Evolution des Podzols et des Sols Hydromorphes des Landes du Médoc. PhD Dissertation, Université de Poitiers, 144 p.

Righi, D., 1987. Microstructures des horizons B des sols podzolisés: influence de la texture et de la minéralogie de la roche-mère. In Righi, D. & Chauvel, A. (eds.), Podzols et Podzolisation. AFES-INRA, Plaisir-Grignon, pp. 107-116.

Righi, D. & De Coninck, F., 1974. Micromorphological aspects of Humods and Haplaquods of the Landes du Médoc, France. In Rutherford, G.K. (ed.), Soil Microscopy. The Limestone Press, Kingston, pp. 567-588.

Righi, D. & De Coninck, F., 1977. Mineralogical evolution in hydromorphic sandy soils and podzols in 'Landes du Médoc', France. Geoderma 19, 339-359.

Righi, D., Van Ranst, E., De Coninck, F. & Guillet, B., 1982. Microprobe study of a Placohumod in the Antwerp Campine (North Belgium). Pedologie 32, 117-134.

Righi, D., Bravard, S., Legros, J.P. & Falipou, P., 1990. Changes in the particle size distribution of quartzose sands affected by lateritization: an approach by computer simulation. Soil Science 149, 361-366.

Robin, A.M., 1979. Genèse et Evolution des Sols Podzoliques sur Affleurements Sableux du Bassin Parisien. PhD Dissertation, Université de Nancy 1, 173 p.

Robin, A.M. & De Coninck, F., 1978. Micromorphological aspects of some podzols in the Paris basin (France). In Delgado M. (ed.), Soil Micromorphology. Deparment of Edaphology, University of Granada, pp. l019-1050.

Rosling, A., Roose, T., Herrmann, A.M., Davidson, F.A., Finlay, R.D. & Gadd, G.M., 2009. Approaches to modelling mineral weathering by fungi. Fungal Biology Reviews 23, 138-144.

Ross, C.W., Mew, G. & Childs, C.W., 1989. Deep cementation in late Quaternary sands near Westport, New Zealand. Australian Journal of Soil Research 27, 275-288.

Salem Avad, M.Z., Gagarina, E.I. & Kulin, N.I., 1982. Mineralogy and micromorphology of sandy podzolic soils. Soviet Soil Science 14, 88-96.

Sanborn, P. & Lavkulich, L.M., 1989. Ferro-humic podzols of coastal British-Columbia. 2. Micromorphology and genesis. Soil Science Society of America Journal 53, 517-526.

Schaefer, C.E.R., Ker, J.C., Gilkes, R.J., Campos, J.C., da Costa, L.M. & Saadi, A., 2002. Pedogenesis on the uplands of the Diamantina Plateau, Minas Gerais, Brazil: a chemical and micropedological study. Geoderma 107, 243-269.

Schawe, M., Glatzel, S. & Gerold, G., 2007. Soil development along an altitudinal transect in a Bolivian tropical montane rainforest: podsolization vs. hydromorphy. Catena 69, 83-90.

Schwartz, D., 1988. Some podzols on Batéké sands and their origins, People's Republic of Congo. Geoderma 43, 229-247.

Schwartz, D., Guillet, B., Villemin, G. & Toutain, F., 1986. Les alios humiques des podzols tropicaux du Congo; constituants, micro- et ultra-structure. Pedologie 36, 179-198.

Schwertmann, U. & Taylor, R.M., 1989. Iron oxides. In Dixon, J.B. & Weed, S.B. (eds.) Minerals in Soil Environments. SSSA Book Series, No. 1, Soil Science Society of America, Madison, pp. 379-438.

Shoji, S. & Ito, I., 1990. Classification of tephra-derived Spodosols. Soil Science 150, 799-811.

Skiba, M., 2007. Clay mineral formation during podzolization in an alpine environment of the Tatra Mountains, Poland. Clays and Clay Minerals 55, 618-634.

Smits, M.M., Hoffland, E., Jongmans, A.G. & van Breemen, N., 2005. Contribution of mineral tunneling to total feldspar weathering. Geoderma 125, 59-69.

Soil Survey Staff, 1999. Soil Taxonomy. A Basic System of Soil Classification for Making and Interpreting Soil Surveys. 2nd Edition. USDA, Natural Resources Conservation Service, Agriculture Handbook Number 436. U.S. Government Printing Office, Washington, 871 p.

Soil Survey Staff, 2014. Keys to Soil Taxonomy. 12th Edition. USDA, Natural Resources Conservation Service, Washington, 360 p.

Stolt, M.H. & Lindbo, D.L., 2010. Soil organic matter. In Stoops, G., Marcelino, V. & Mees, F. (eds.), Interpretation of Micromorphological Features of Soils and Regoliths. Elsevier, Amsterdam. pp. 369-396.

Stoops, G., 2003. Guidelines for Analysis and Description of Soil and Regolith Thin Sections. Soil Science Society of America, Madison, 184 p.

Sullivan, L.A., 1997. Micromorphology of podzol reformation post sand mining in eastern Australia. In Shoba, S., Gerasimova, M. & Miedema, R. (eds.), Soil Micromorphology: Studies on Soil Diversity, Diagnostics, Dynamics. Moscow, Wageningen, pp. 282-288.

Taina, I., Opris, M., Balaceanu, V., Taina, S., Berde, S. & Craciun, C., 1997. Micromorphological aspects of podzolization in some soils from Bucegi mountains — Romania. In Shoba, S., Gerasimova, M. & Miedema, R. (eds.), Soil Micromorphology: Studies on Soil Diversity, Diagnostics, Dynamics. Moscow, Wageningen, pp. 167-177.

Thomas, M., Thorp, M. & McAlister, J., 1999. Equatorial weathering, landform development and the formation of white sands in north western Kalimantan, Indonesia. Catena 36, 205-232.

Thompson, C.H., Bridges, E.M., & Jenkins, D.A., 1994. Relict hardpans in coastal lowlands of Queensland. In Ringrose-Voase, A. & Humphreys, G.S. (eds.), Soil Micromorphology: Studies in Management and Genesis. Developments in Soil Science, Volume 22. Elsevier, Amsterdam, pp. 233-245.

Thompson, C.H., Bridges, E.M. & Jenkins, D.A., 1996. Pans in humus podzols (Humods and Aquods) in coastal southern Queensland. Australian Journal of Soil Research 34, 161-182.

Ugolini, R.C. & Dahlgren, R.A., 1987. The mechanism of podzolization as revealed through soil solution studies. In Righi, D. & Chauvel, A. (eds.), Podzols et Podzolisation. AFES-INRA, Plaisir-Grignon, pp. 195-203.

van Breemen, N., Lundström, U.S. & Jongmans, A.G., 2000. Do plants drive podzolization via rock-eating mycorrhizal fungi? Geoderma 94, 161-169.

Van Herreweghe, S., Deckers, S., De Coninck, F., Merckx, R. & Gullentops, F., 2003. The paleosol in the Kerkom sands near Pellenberg (Belgium) revisited. Netherlands Journal of Geosciences 82, 149-159.

Van Ranst, E., Righi, D., De Coninck, F., Robin, A.M. & Jamagne, M., 1980. Morphology, composition and genesis of argillans and organans in soils. Journal of Microscopy 120, 353-361.

Van Ranst, E., Stoops, G., Gallez, A. & Vandenberghe, R.E., 1997. Properties, some criteria of classification and genesis of upland forest Podzols in Rwanda. Geoderma 76, 263-283.

Van Vliet, B., Faivre, P., Andreux, F., Robin, A.M. & Portal, J.M., 1983. Behaviour of some organic components in blue and ultraviolet light: application to the micromorphology of Podzols, Andosols and Planosols. In Bullock, P. & Murphy, C.P. (eds.), Soil Micromorphology, Volume 1, Techniques and Applications. AB Academic Publishers, Berkhamsted, pp. 91-99.

Van Vliet-Lanoë, B., 1998. Frost and soils: implications for paleosols, paleoclimates and stratigraphy. Catena 34, 157-183.

Wallinga, J., van Mourik, J.M. & Schilder, M.L.M., 2013. Identifying and dating buried micropodzols in Subatlantic polycyclic drift sands. Quaternary International 306, 60-70.

Wang, C., Beke, G.J., & McKeague, J.A., 1978. Site characteristics, morphology and physical properties of selected ortstein soils from the Maritime Provinces. Canadian Journal of Soil Science 58, 405-420.

Waroszewski, J., Kalinski, K., Malkiewicz, M., Mazurek, R., Kozlowski, G. & Kabala, C., 2013. Pleistocene-Holocene cover-beds on granite regolith as parent material for Podzols — an example from the Sudeten Mountains. Catena 104, 161-173.

Weisenborn, B.N. & Schaetzl, R.J., 2005. Range of fragipan expression in some Michigan soils: I. Morphological, micromorphological, and pedogenic characterization. Soil Science Society of America Journal 69, 168-177.

Wu, S.P. & Chen, Z.S., 2005. Characteristics and genesis of Inceptisols with placic horizons in the sub-alpine forest soils of Taiwan. Geoderma 125, 331-341.

Chapter 23

Oxic and Related Materials

Vera Marcelino[1], Carlos E.G.R. Schaefer[2], Georges Stoops[1]

[1]*GHENT UNIVERSITY, GHENT, BELGIUM;*
[2]*FEDERAL UNIVERSITY OF VIÇOSA, VIÇOSA, MINAS GERAIS, BRAZIL*

CHAPTER OUTLINE

1. Introduction

The oxic horizon (Soil Taxonomy; Soil Survey Staff, 2014) is a strongly weathered subsurface horizon with a clay fraction dominated by low-activity clays and iron and aluminium oxides, and with silt and sand fractions mainly composed of quartz and

Interpretation of Micromorphological Features of Soils and Regoliths. https://doi.org/10.1016/B978-0-444-63522-8.00023-1

including little or no weatherable minerals. The oxic horizon essentially corresponds to the ferralic horizon of the World Reference Base for soil resources (WRB Classification; IUSS Working Group WRB, 2015). Oxic and ferralic horizons are diagnostic horizons of Oxisols and Ferralsols, respectively. Throughout this manuscript, only Soil Taxonomy terminology will be used.

Oxisols were generally named red soils, red loams, red earths, ferralitic soils, lateritic soils and latosols in early publications. The name Latosol (Brazilian Soil Classification System; Embrapa, 1999) is still often used in publications (e.g., Schaefer et al., 2004; Reatto et al., 2009).

These soils are widespread in humid tropical and subtropical regions, where they are mostly found on old, stable land surfaces or in sediments derived from them; they can also form on easily weatherable materials such as ultrabasic magmatic rocks (e.g., Van Wambeke et al., 1983). They are occasionally found in present-day aridic soil moisture regimes (e.g., north-eastern Brazil), where they developed in sedimentary deposits or saprolites, previously weathered under more humid conditions (Beinroth et al., 1996). Even in tropical and subtropical regions, the characteristics of oxic materials are not necessarily the result of current pedological processes and climates, but may be inherited from parent materials derived directly or indirectly from deeply weathered saprolites (Stoops, 1989; Stoops et al., 1994; Buol et al., 2003).

The micromorphology of oxic materials was first briefly studied by Kubiëna (1948). The earliest more detailed studies, on soils from D.R. Congo, were done by De Craene and Laruelle (1955, 1956a, 1956b), Laruelle (1956) and De Craene (1967). Other important early contributions were published by pedologists of Ghent University on strongly weathered soils from Central Africa, Central and South America and Southeast Asia (e.g., Stoops, 1964, 1968; Eswaran, 1972, 1979; Eswaran & Sys, 1976), by French pedologists on soils from Central and West Africa (e.g., Beaudou, 1972; Chauvel, 1972; Beaudou et al., 1977; Muller, 1977a, 1977b; Chauvel et al., 1978), and by French and Brazilian soil scientists on soils from Brazil (Bennema et al., 1970; Lepsch & Buol, 1974; Pédro et al., 1976; Chauvel et al., 1983). General reviews of the micromorphological characteristics of Oxisols have been published in the past by Buol and Eswaran (1978, 1999), Eswaran (1979), Stoops (1983) and Stoops and Buol (1985).

This review concerns the interpretation of the micromorphological characteristics of oxic materials in a broad sense. It covers not only oxic horizons but also related materials, such as horizons with clay content increase but with physical, chemical and mineralogical characteristics similar to those of oxic horizons (e.g., kandic horizons of certain Ultisols), and to horizons that do not fit the definition of oxic because of their very low clay content (e.g., ochric horizons of certain Entisols).

The main problems encountered when reviewing the existing literature are the imprecise definition of the materials under study, often classified using regional soil classification systems, and the lack of a clear and uniform terminology in the description of micromorphological features.

2. Microstructure

2.1   General Features

A well-developed granular microstructure is typical for oxic materials (Figs. 1A and B, and 2A and B). The granular aggregates can be well separated or partly to completely welded, the latter giving rise to more massive microstructures (Fig. 3A). Their diameter varies between 10 and 1000 μm, with various size ranges reported by different authors, e.g., 50-400 μm (Benayas & Pinto Ricardo, 1973), 30-500 μm (Chauvel, 1977), 100-500 μm (Chauvel & Pédro, 1978), 50-1000 μm (Eschenbrenner, 1986), 10-100 μm (Santos et al., 1989), 30-40 μm (Tandy et al., 1990), 100-1000 μm (Schaefer, 2001; Cooper et al., 2005) and 50-200 μm (Reatto et al., 2009).

The granular aggregates are responsible for the friable consistency of these soils in the field. They correspond to the so-called pseudosilts and pseudosands (Ahn, 1970), as

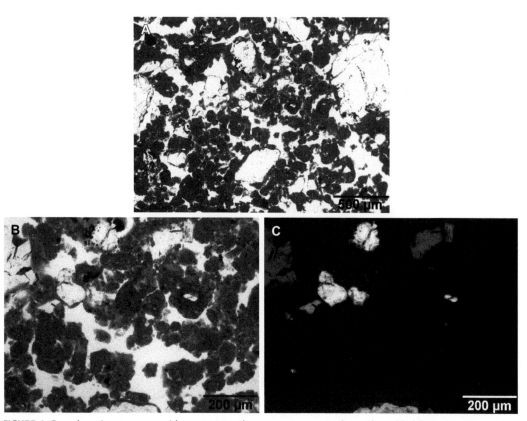

FIGURE 1 Granular microstructure with aggregates that are not or are only partly welded (Oxisol, Kasese region, D.R. Congo). (A) Overview (PPL). (B) Detail (PPL). (C) Same field in XPL, illustrating undifferentiated b-fabric typical for oxic materials.

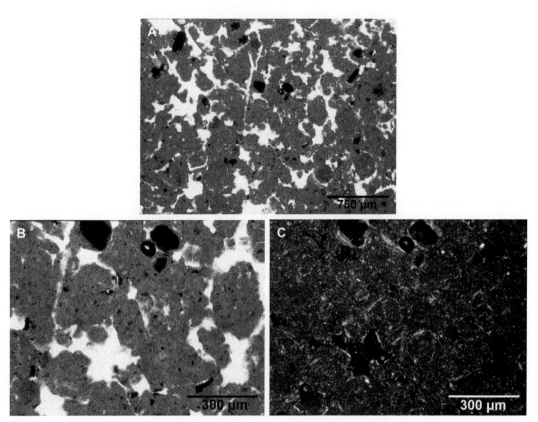

FIGURE 2 Granular microstructure with coalesced granular aggregates (Oxisol, Indonesia). (A) Overview (PPL). (B) Detail (PPL). (C) Same field in CPL, illustrating circular striated, porostriated and granostriated b-fabrics.

documented by the micromorphological study of these fractions after different dispersion treatments (Pédro et al., 1976; Guedez & Langohr, 1978; Embrechts & Stoops, 1987).

SEM studies of Oxisols from Hawaii (Tsuji et al., 1975) and TEM examination of Oxisol samples from Brazil (Bui et al., 1989; Santos et al., 1989; Bartoli et al., 1992), French Guiana (Tandy et al., 1990) and Thailand (Tawornpruek et al., 2005) revealed the presence of submicroscopic intra-aggregate pores. Different sizes were reported for the diameter of these pores: average of 40 nm (Santos et al., 1989; Bui et al., 1989), smaller than 100 nm (Tandy et al., 1990), 7-40 nm (Bartoli et al., 1992), 10-50 nm (Tawornpruek et al., 2005). Early TEM studies by Cambier and Prost (1981) reported the existence of several levels of organisation in a granular aggregate: (i) kaolinite crystallites (100 nm) formed by regular stacking of several layers; (ii) particles originated by face-to-face juxtaposition of kaolinite crystallites; (iii) domains composed by particles with same orientation and bound by iron compounds; and (iv) microaggregates composed of non-oriented domains. Soil aggregate hierarchy of a Brazilian Oxisol was also studied by Vrdoljak and Sposito (2002), who observed that organic components were an important

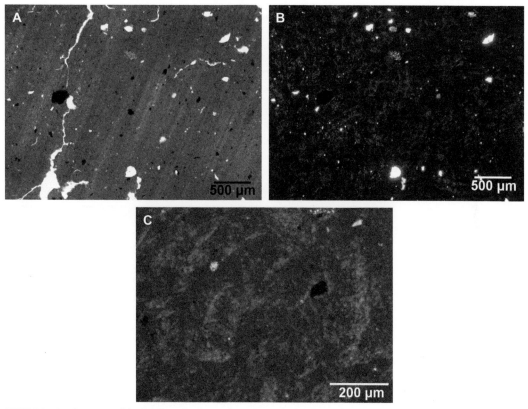

FIGURE 3 Massive to weakly developed subangular microstructure (Oxisol, Malaka, Malaysia). (A) Overview (PPL). Oblique parallel streaks are artefacts. (B) Same field in XPL, showing abundant passage features. (C) Detail, illustrating crescent striated b-fabric (XPL).

factor at all levels of the structural organisation. Other ultramicroscopic studies of Oxisol aggregates from Brazil and Thailand showed that the clay-size mineral particles consist of platy kaolinite crystals and almost equant iron oxide crystals (Hart et al., 2003) or gibbsite crystals (Bartoli et al., 1992), arranged in an open 'card house' manner (Tawornpruek et al., 2005). Humified hyphae and decomposed organic matter have been observed within the aggregates (Santos et al., 1989; Vrdoljak & Sposito, 2002), suggesting that fungal activity could play a relevant role in the formation of the aggregates.

The granular microstructure and the internal structure of the granular aggregates strongly affect the water retention behaviour of the soil (Sharma & Uehara, 1968, Tsuji et al., 1975; Uehara & Keng, 1975; Muller, 1983; Taworpruek et al., 2005), and they are thus important for soil management. They also have an effect on the results of soil analysis, for instance by the release of cations when soil samples are crushed (Moura Filho & Buol, 1976).

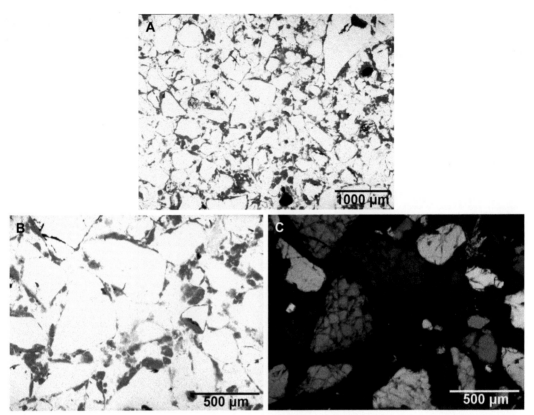

FIGURE 4 Basic microstructure with close enaulic and locally chitonic c/f-related distribution patterns in sandy oxic materials (Entisol, D.R. Congo). (A) Overview (PPL). (B) Detail (PPL). (C) Same field in XPL, illustrating undifferentiated b-fabric typical for oxic materials and showing fissuring of the quartz grains.

Sandy oxic materials have a basic microstructure with close enaulic and locally chitonic c/f-related distribution patterns, the granular aggregates being similar to those observed for clayey oxic materials (Fig. 4A and B).

In addition to intergranular porosity, channels are common and usually partially or completely filled with excrements or groundmass material (see Section 4.1).

2.2   Origin of Granular Aggregates

Granular aggregates have since long been recognised in thin sections of oxic materials (e.g., De Craene & Laruelle, 1955; Fölster, 1964; Condado, 1969; Bennema et al., 1970; Lepsch & Buol, 1974). Although various attempts have been made to explain their origin, no consensus has been reached between the different authors.

Some authors (e.g., Muller, 1977a, 1983; Bitom & Volkoff, 1991) suggest an essentially physical origin and explain the formation of granular aggregates by mechanical fracturing of a compact soil material, due to alternating wetting and drying. The originally speckled b-fabric of the compact soil would first evolve to a reticulate b-fabric, the streaks gradually becoming discoloured and acquiring a progressively better visibility of the interference colours of the micromass. Next, small fissures develop within these streaks, eventually leading to formation of individual small peds, which become more and more rounded. However, features related to these processes are not recognised for most soils studied by the present authors. The inferred mechanisms also do not apply to sandy oxic materials, which have clearly similar granular aggregates associated with enaulic c/f-related distribution patterns (Schaefer et al., 2004; Marcelino, unpublished data). Besides aggregates formed by fracturing, Muller (1977a) distinguished other less abundant genetic types of granular aggregates: zoogenetic, ferritic, relict and complex. The first three are in fact pedofeatures (see Section 4), namely excrements and inherited nodules.

According to other authors, physico-chemical interactions between kaolinite and iron oxides, in conditions with extreme leaching rates, can be at the origin of granular aggregate occurrences (e.g., Beaudou, 1972; Pédro et al., 1976; Chauvel, 1977; Chauvel et al., 1978; Chauvel & Pédro, 1978; Buol & Eswaran, 1978; Eswaran & Daud, 1980; Chauvel et al., 1983; Pédro, 1987; Santos et al., 1989; Bitom & Volkoff, 1991). A secondary role can be played by soil mesofauna, which could contribute to the separation of the pre-formed granular aggregates (Chauvel, 1972, 1977; Chauvel et al., 1976; Muller, 1977a; Stoops, 1983). Multiple processes, including faunal activity, were also invoked by Vidal-Torrado et al. (1999) for the formation of microaggregates in an Oxisol from Brazil.

Many studies attribute the granular microstructure mainly to biological activity (e.g., Benayas & Pinto Ricardo, 1973; Verheye & Stoops, 1975), more specifically long-term termite activity (Eschenbrenner, 1986; Trapnell & Webster, 1986; Stoops, 1991; Jungerius et al., 1999; Nunes et al., 2000; Schaefer, 2001; Reatto et al., 2009; Sarcinelli et al., 2009). Micromorphological evidence in favour of a role of termites includes the observation that granular peds in oxic materials are similar in size, shape and internal structure to pellets made by termites to build nests and mounds, both in nature and in laboratory experiments (Eschenbrenner, 1986; Jungerius et al., 1999; Cosarinsky et al., 2005). The presence of charcoal and well-sorted small quartz grains (<100 μm) within the granular aggregates, whereas larger quartz grains are observed in non-granular parts, further supports this hypothesis (Schaefer, 2001). Impregnation of soil material with saliva makes the granular aggregates very stable, much more stable than faecal pellets, which are only found as infillings (Jungerius et al., 1999). It has also been observed that Oxisols in regions with less intense termite activity (e.g., in East Sumatra) have a less developed granular microstructure than Oxisols with high termite activity (e.g., in Brazil) (Buurman, pers. comm.).

An experimental study on the effects of wetting and drying cycles has shown that physical, chemical and mineralogical factors alone could not explain the genesis of granular aggregates (Viana et al., 2004). Also, three different morphological types have been recognised for Oxisols from Brazil, each of them related to one or more genetic processes (Cooper et al., 2005): (i) oval granular aggregates with well-sorted quartz grains, produced by termites or ants; (ii) oval granular aggregates without well-sorted quartz grains, with either a biological or a geochemical origin; and (iii) polyhedral aggregates, formed by cracking of soil material due to alternating wet and dry conditions.

2.3 Degree of Development and Degradation of the Granular Microstructure

The formation of a granular structure in oxic materials does not seem to depend on the underlying lithology, but its degree of development and its stability seems to be related to both the composition of the parent material and climatic conditions, mostly because these factors determine the type and amount of aluminium and iron oxides present. Some authors observed that the granular microstructure was more developed in Ustic soil moisture regimes than in Udic soil moisture regimes (Buol & Eswaran, 1978; Utami et al., 1997; Tawornpruek et al, 2005; Simas et al., 2005). Well-drained red soils have also been reported to have better developed granular microstructure than less well-drained yellow soils (Beaudou et al., 1987; Muggler & Buurman, 1997; Skorupa et al., 2016), the colour being determined by the type of iron oxides present, in particular the relative proportion of hematite and goethite (Camacho et al., 1990). High gibbsite and iron oxide contents of the soil or the parent material have been associated with better developed and more stable granular microstructures (Cardoso de Lima & Eswaran, 1987; Dick & Schwertmann, 1996; Baert & Van Ranst, 1997a; Ferreira et al., 1999; Schaefer, 2001; Schaefer et al., 2002, 2004). However, Reatto et al. (2009) compared Oxisols with different mineralogical compositions and observed that this factor has no effect on microstructure development.

The age of the soil may play a role in the degree of development of the granular structure. For instance, Lepsch and Buol (1974), studying an Ultisol-Oxisol toposequence in Brazil, and West et al. (1997), studying a sequence of strongly weathered soils on volcanic deposits in the Philippines, observed that older soils had better developed granular microstructure in the oxic horizon than the younger soils. Degradation of a stable granular microstructure is usually linked to local changes in soil climate, which mainly resulted in the solubilisation and redistribution of iron and aluminium oxides and in increased clay dispersion and eluviation.

Collapse of the granular microstructure is one of the processes involved in present-day pedogenesis that leads to the formation of epipedons at the expense of the underlying oxic horizon, besides discolouration of the fine material, clay eluviation and accumulation of organic matter (Lepsch & Buol, 1974; Chauvel, 1977; Muller, 1977b, 1983; Muggler & Buurman, 1997). Similarly, changes in the relationship between clays and iron oxides, followed by changes of the c/f-related distribution patterns and the collapse of the granular microstructure, may explain the transformation of 'red soils' typical for humid climates into 'beige soils' in drier climates with alternating very dry and very wet periods (Chauvel et al., 1976; Chauvel, 1977; Chauvel & Pédro, 1978). In the same way, redistribution of iron oxides may lead to loss of aggregate stability and collapse of the granular microstructure in poorly drained soils (Stoops, 1989).

Transformation of a typical granular microstructure in oxic (ferralic) horizons to a more massive subangular blocky microstructure in argillic or kandic (argic or nitic) horizons has been often reported (e.g., Lepsch & Buol, 1974, Fedoroff & Eswaran, 1985; Vidal-Torrado et al., 1999; Hallaire et al., 1994; Cooper et al., 2010; Padmanabhan et al., 2012) and attributed to several processes, including wetting and drying, biological activity and clay illuviation. The model proposed by Cooper et al. (2010), for Brazilian soils with alternating wetting and drying cycles, in a region with a pronounced dry season, assumes that coalescence of granular aggregates occurred during a dry period in the past, followed by development of a blocky microstructure through fissuration of the coalesced material when the climate became more humid.

Aggregates in surface horizons of Oxisols in slope positions along an Oxisol-Spodosol toposequence had irregular and rough surfaces and UV-fluorescent hypocoatings, indicative of clay alteration to amorphous compounds, and they were more porous than those of Oxisols in plateau positions (Bravard & Righi, 1990). These observations suggest that the granular structure of surface horizons of Oxisols in slope positions is unstable in present-day conditions.

Compaction and degradation of the granular microstructure of Oxisols as a consequence of cultivation and land use changes has often been documented (Chauvel et al., 1991; Stoner et al., 1991; Curmi et al., 1994; Hartmann et al., 1994; Barros et al., 2001; Balbino et al., 2002, Silva et al., 2015). In cultivated soils, the main processes involved in the formation of a massive groundmass are the destruction of large aggregates and mechanical welding of small aggregates into larger, blocky clods (Fig. 5). Moreover, micromorphological evidence of pore clogging by fine-grained material produced by tillage has been reported (Stoner et al., 1991). Land use changes also affect soil structure by inducing changes in soil macrofauna composition (see Kooistra & Pulleman, 2018).

3. Groundmass

The groundmass of oxic materials is characterised by a marked homogeneity, which results from continuous bioturbation (Flach et al., 1968; Stoops, 1983, 1991; see also Stoops & Schaefer, 2018).

3.1 c/f-Related Distribution Patterns

The limit between coarse and fine material is generally set at 5 μm, to consider the numerous small opaque grains as part of the fine material. The c/f-related distribution pattern is generally open to close porphyric. In surface horizons and in sandy materials, it becomes enaulic (Fig. 4).

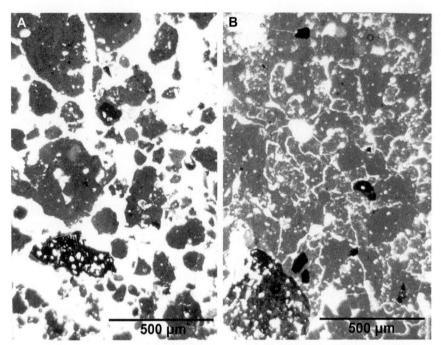

FIGURE 5 Impact of tillage on the soil microstructure (A horizon, Oxisol, Brazil). (A) Loose granular microstructure in soil with no tillage (PPL). (B) Compact microstructure with small aggregates welded into larger, blocky clods as a consequence of conventional disk-ploughing during 5 years.

3.2 Coarse Mineral Material

3.2.1 General Features

The coarse fraction predominantly consists of sand-sized quartz grains, with other resistant minerals, such as zircon, tourmaline and rutile, as the most common accessory components. Opaque minerals, such as ilmenite and magnetite, are common in soils formed on basic rocks (Buol & Eswaran, 1978; Rodriguez-Rodriguez et al., 1980; Tejedor Salguero et al., 1984/85). In soils on metamorphic rocks, sillimanite and kyanite may also be present (Stoops & Buol, 1985). Weatherable minerals are occasionally present when covered by iron oxide or gibbsite coatings (Stoops, 1968; Verheye & Stoops, 1975; Buol & Eswaran, 1978) and thus not available for interactions with soil solutions (Baert & Van Ranst, 1997b). Stable iron and aluminium oxide pseudomorphs are sometimes recognised, especially near the saprolite (see Zauyah et al., 2018). Examples include goethite pseudomorphs after garnet (Embrechts & Stoops, 1982), gibbsite after feldspar (Eswaran & Wong Chaw Bin, 1978), and various minerals after olivine, augite and biotite (Eswaran et al., 1979; Eswaran, 1990). Inherited iron oxide nodules (see Section 4.2) are also part of the coarse material.

In most oxic materials, the silt fraction is poorly represented, except for the abundant opaque grains. The silt is mostly composed of quartz, and it sometimes includes gibbsite (Eswaran et al., 1977), phytoliths and inherited iron oxide nodules (Stoops, 1983). The relative amount of silt-sized quartz grains has been reported to increase with increasing depth, which was explained by dissolution of small grains in the upper part of the profile (Chauvel, 1977). Fragmentation of sand-sized quartz grains during sample preparation is responsible for most of the silt fraction obtained in textural analyses. Quartz grains in oxic materials are generally brittle, and it is commonly observed in thin sections (more than for other materials) that the grains are strongly fractured and pieces are lost during polishing (see Fig. 6) (Stoops & Buol, 1985). Others authors found indications that partial dissolution of quartz is the most important process of silt formation in deep saprolites (Muggler et al., 1997).

In surface horizons, the coarse material is generally more abundant than in the oxic horizon. The presence of surface horizons with a coarser texture has often been reported for tropical soils and has been attributed to one or more of the following processes acting simultaneously or sequentially on surface and subsurface horizons: selective removal of clay from the surface, earthworm and termite activity, clay destruction in the epipedon, and clay eluviation/illuviation (e.g., Nye, 1955; Roose, 1970; Moorman, 1979; Buurman & Sukardi, 1980; Andrade et al., 1997).

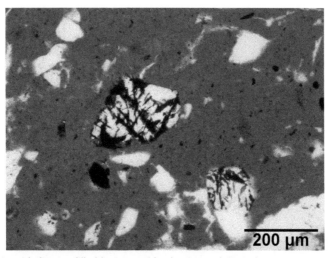

FIGURE 6 Quartz grain with fissures filled by iron oxides (runiquartz) (Oxisol, Kasese region, D.R. Congo) (PPL). A piece of the runiquartz grain (right, bottom) has been broken and lost during thin section preparation.

3.2.2 Weathering of Quartz

Fissuring of quartz grains (see Fig. 4C) was already observed in some early studies (Beaudou, 1972; Claisse, 1972). The fissures are often filled with Fe and Al oxides (Stoops, 1968; Claisse, 1972), which is a feature described as 'runiquartz' (Eswaran et al., 1975) (Fig. 6). According to the latter study, runiquartz is formed by infusion of amorphous iron- or aluminium-rich fine material into the cracks of quartz grains, which is then transformed to goethite, hematite or gibbsite. In some soils the same material may form coatings around the runiquartz grains (Eswaran et al., 1975, Padmanabhan & Mermut, 1996). Runiquartz is very common in laterites (see Stoops & Marcelino, 2018) and in most Oxisols, in which the material filling the cracks is usually different from the enclosing fine material. The latter has been interpreted as an indication that runiquartz grains are inherited, probably from a former lateritic material (Stoops, 1989; Muggler & Buurman, 1997). Runiquartz is very stable and persists even in soils subjected to strong weathering conditions, such as soils along slopes and poorly drained soils (Eswaran, 1979; Stoops, 1989). However, the infusion of fine material may also lead to physical breakdown of the quartz grains (see Fig. 6) (Eswaran & Stoops, 1979).

Thin sections show various signs of dissolution of quartz grains, such as corroded surfaces (Beaudou, 1972) and 'floating' grains in voids with a slightly larger size (Eswaran & Stoops, 1979; Muller et al., 1980/81; Muggler et al., 2007). In SEM images, marks of intense chemical weathering along the surface of quartz grains, such as shallow triangular features, deep pyramidal pits and other dissolution features (Fig. 7), are unequivocally recognised and have often been described and studied (e.g., Leneuf, 1973; Eswaran et al., 1979; Eswaran & Stoops, 1979; Flageollet, 1980/81; Fritsch, 1988). Although etch pits in

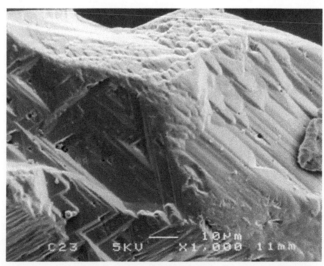

FIGURE 7 SEM image of a highly weathered quartz grain in sandy oxic material (Entisol, D.R. Congo), illustrating different types of dissolution figures such as the deep pyramidal etch pits on one crystal face (left) and the triangular plates with tile-like arrangement on another face (right).

the surfaces of quartz grains from oxic materials are often inherited from previous environments, their frequency and size reflect present-day soil-moisture relationships and can been used to assess the relative degree of weathering of tropical soils developed on the same parent material (Marcelino & Stoops, 1996; Marcelino et al., 1999).

3.3   Fine Mineral Material

The micromass in well-drained soils is characteristically very homogeneous. The colour is related to the amount, type and degree of crystallisation of the iron oxides present, which in turn depend on the parent material, degree of alteration, climate and internal drainage (Stoops, 1968; Bennema et al., 1970). When yellowish, the micromass can be limpid or speckled, and when reddish, it is generally speckled and frequently cloudy (Bennema et al., 1970). The cloudiness of the fine material has been related to incipient flocculation of iron oxides. However, a micromass free of iron oxides, as in mottled clays, has also been described as cloudy, whereby cloudiness is attributed to the presence of very fine-grained gibbsite (e.g., Dedecker & Stoops, 1993). Silt-sized red speckles, probably hematite aggregates, are often observed. They are inherited from plinthitized saprolitic material (see Zauyah et al., 2018) and incorporated into the soil by bio-turbation (Stoops, 1968; Schmidt-Lorenz, 1980; Stoops, 1989; Muggler & Buurman, 1997; Muggler et al., 1999). Very small gibbsite or goethite crystals may also be present.

The b-fabric in oxic materials is weakly expressed, most commonly undifferentiated (Figs. 1C and 4C) or weakly stipple-speckled. This is attributed to the high concentration and masking effect of iron oxides (e.g., Eswaran & Daud, 1980), the low birefringence of kaolinite (Stoops, 1983) and the random arrangement of clay particles (Stoops, 1968; Eswaran, 1972; Stoops & Buol, 1985).

Weakly striated b-fabrics are sometimes observed, if a strong light source is used and the section is sufficiently thin. The b-fabric is porostriated in the case of individual granular aggregates and circular striated in the case of welded granular aggregates (Fig. 2C). They are the result of clay particles being parallel orientated along the border of the aggregates (Buol & Eswaran, 1978; Santos et al., 1989). This, as mentioned above, could be due to pressures exerted by termites on the surface of pellets during regurgitation and when plastering channel walls (Eschenbrenner, 1986). Buol and Eswaran (1978) suggest that this specific orientation of the clay is attained during transport. The feature has also been related to the presence of easily dispersible clay around the granular aggregates (Embrechts & Stoops, 1986, 1987).

The undifferentiated b-fabric of soils developed on volcanic material becomes generally stipple-speckled to weakly striated in a first stage and finally, in the Oxisol, changes again to undifferentiated, with cloudy limpidity, typical for oxic materials (Rodriguez Rodriguez et al., 1988; West et al., 1997; see also Stoops et al., 2018).

Tubular features (50-100 μm) with crescent striated b-fabrics (Fig. 3B and C), not visible in plain light (Fig. 3A), have been reported for oxic materials from Central Africa (Stoops, 1968; Beaudou et al., 1977), Brazil (Stoops, 1991; Muggler & Buurman, 2000;

Schaefer et al., 2004) and Southeast Asia (Stoops & Buol, 1985). They are passage features, recording intense biological activity.

A clear relationship seems to exist between the degree of development of the b-fabric and other characteristics of Oxisols. Early studies related weakly developed b-fabrics to strong aggregate stability (Cagauan & Uehara, 1965). Speckled b-fabrics are common in yellow oxic materials with poorly developed granular microstructures, but they are rare or absent in materials with better developed granular microstructures (Beaudou et al., 1977). Well-developed b-fabrics have also been related to high concentrations of easily dispersible clays (Embrechts & Stoops, 1986, 1987), and weakly developed b-fabrics in clay-rich Oxisols have been related to low plasticity (Zainol & Stoops, 1986).

3.4 Organic Material

Root residues and other organ residues in different states of decomposition are frequent in surface horizons. Charcoal fragments, probably originated from bush fires, are sometimes observed in thin sections (unpublished observation by the authors).

4. Pedofeatures

Pedofeatures, except those related to faunal activity, are very restricted, because of the physico-chemical stability of the material and continuous bioturbation. Many of them are relict features formed in previous environments and thus no longer reflect present-day pedogenesis.

4.1 Channel Infillings and Coatings

Clay coatings that are not laminated and commonly have an admixture of very fine sand and silt are sometimes observed in channels, which may be filled with granular aggregates and more or less sorted quartz grains (Fig. 8A and B). In very clayey materials, these coatings or fragments of coatings are difficult to distinguish from the groundmass, but in sandy materials they are very clear (Fig. 8B) (Marcelino, unpublished data). These textural coatings are plasters that are applied to the walls of channels by termites (e.g., Kooistra, 1982; Kauffman, 1987; Miedema et al., 1994; see also Kooistra & Pulleman, 2018).

Channels and chambers often contain loose continuous or discontinuous infillings composed of groundmass material (Fig. 8A and B) or excrements (e.g., Chauvel, 1977, Stoops, 1983, Kauffman, 1987). Dense complete channel infillings with bow-like (crescent striated) internal fabric have also been observed for oxic materials, as well as in present-day termite mounds, and interpreted as passage features (e.g., Stoops, 1964, 1968; Lee & Wood, 1971; Sleeman & Brewer, 1972; Mermut et al., 1984).

4.2 Iron Oxide Pedofeatures

More or less rounded iron oxide nodules with sharp boundaries and various types of internal fabric, sometimes containing other pedofeatures such as iron oxide coatings and

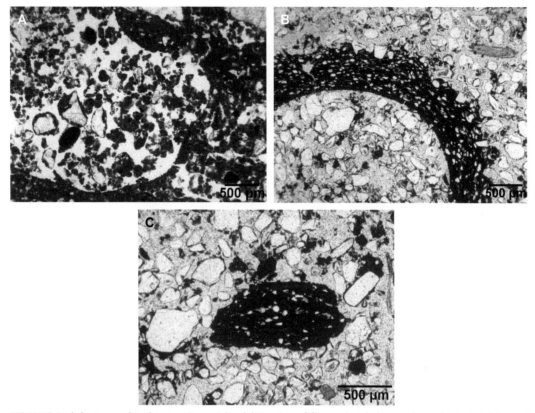

FIGURE 8 Pedofeatures related to termite activity. (A) Coating of fine material along a channel (plaster), followed by an infilling composed of groundmass material, in a clayey soil (Oxisol, Kasese region, D.R. Congo) (PPL). (B) The same feature in sandy oxic materials (Entisol, Bateke plateau, D.R. Congo); the thick coating is composed of fine material and perfectly sorted, very fine sand grains. (C) Fragment of a plaster with perfectly sorted quartz grains (Entisol, Bateke plateau, D.R. Congo) (PPL).

fragments of clay coatings, are common (Fig. 9). These nodules are inherited laterite nodules (e.g., Stoops, 1968, 1983; Buol & Eswaran, 1978; Faure, 1987; see also Stoops & Marcelino, 2018), unrelated to present-day pedogenesis and thus to be considered as part of the coarse material.

Newly formed impregnative, irregular and rather diffuse iron oxide nodules, coatings and hypocoatings, as well as spherulitic goethite coatings (e.g., Frei, 1964; Eswaran, 1972; Stoops, 1983), occur in moderately to poorly drained oxic materials (see Vepraskas et al., 2018).

Skorupa et al. (2016) report the presence of relict iron oxide nodules with signs of dissolution, due to the more humid current environment, in Oxisols formed on the same parent material but with different internal drainage intensity. However, newly formed impregnative nodules were only present in the Oxisol with slower internal drainage.

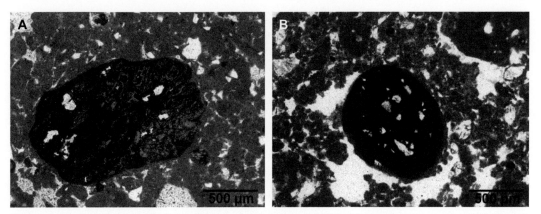

FIGURE 9 Inherited nodules with internal fabrics similar to lateritic material (Oxisol, Kasese region, D.R. Congo). (A) Nodule containing coatings and fragments of coatings of iron oxides and clay and some subrounded quartz grains (PPL). (B) Iron oxide nodule containing subangular quartz grains (PPL).

4.3   Gibbsite Coatings and Nodules

Gibbsite may be present as a pedogenic mineral in Oxisols. Its formation in highly weathered soils has been attributed to direct precipitation from soil solutions, in the form of gibbsite coatings along the sides of voids and grains and as coarse-grained gibbsite nodules (Eswaran et al. 1977, 1979; Stoops & Delvigne, 1990; Zeese et al., 1994; Muggler et al., 2007). The coatings generally consist of silt-sized crystals oriented perpendicularly to the surface. Gibbsitic pedofeatures are more frequent in oxic materials developed on acid rocks than on basic or ultrabasic rocks (Buol & Eswaran, 1978) and seem to be rather unstable (Stoops & Buol, 1985).

Small gibbsite crystals (fine silt to clay) that formed by replacement of kaolinite have commonly been described in thin sections of saprolites (see Zauyah et al., 2018) and C horizons of Oxisols. They occur as disseminated crystals or as nodules and have been related to microenvironments with *in situ* desilicification (Stoops & Delvigne, 1990; Stoops et al., 1990; Dedecker & Stoops, 1993; Zeese et al., 1994; Muggler et al., 2007).

Both coarse- and fine-grained inherited gibbsite nodules (e.g., bauxite nodules) can be found in Oxisols. Various gibbsite forms occurring in Oxisols and lateritic materials have been studied in detail by Eswaran et al. (1977).

4.4   Clay Coatings and Infillings

Illuviation of clay is not, in general, an active process in oxic horizons, and thus clay coatings are usually not present in large amounts. However, common clay coatings have been reported in early publications, for instance, for soils from Angola (Benayas & Pinto Ricardo, 1973; Benayas & Réfega, 1974), Brazil (Falci & Mendes, 1973) and Uganda (Pidgeon, 1976), and for poorly drained oxic materials from Central Africa (Beaudou, 1972; Beaudou & Chatelin, 1974). The horizons in which they occur would most probably be classified not as oxic but as kandic (argic or nitic) and, depending on the depth of their occurrence and on variations in clay content within the profile, the soils are either Oxisols or Ultisols. Referring to strong similarities in micromorphological characteristics between oxic and kandic horizons, and in clay composition between textural pedofeatures and the groundmass, Padmanabhan et al. (2012) suggested that the presence of moved clay in kandic horizons, in contrast to argillic horizons, may not result in significant changes of soil properties. Clay coatings and fragments of clay coatings have been observed in small amounts for Oxisols transitional to Ultisols (Lepsch & Buol, 1974; Eswaran & Sys, 1976; Fedoroff & Eswaran, 1985), where the presence of clay coatings is generally attributed to soil formation on less weathered parent materials. Infillings with oriented clay in interaggregate spaces have been reported by Ibraimo et al. (2004) for polycyclic Oxisols from Brazil, suggesting that clay illuviation is associated with the present-day arid climate, whereas the granular microstructure developed during an earlier period with a more equable, wetter climate. Illuvial clay features may be present in deep horizons of Oxisols (Chauvel, 1977; Stoops & Buol, 1985; Nettleton et al., 1987; Eswaran, 1990) and seem to be less frequent in soils on basic volcanic rocks than in those on other rocks (Stoops & Buol, 1985). Thin clay coatings around quartz grains were often noticed in surface horizons and are attributed to clay translocation after first heavy rains (Stoops, 1968).

Clay coatings and infillings in oxic materials are usually composed of limpid yellowish red clay, generally without layering (Fig. 10A and B). They show low interference colours typical for kaolinite, which often is masked by iron oxides (Fig. 10C).

Fragments of clay coatings have been observed in small amounts (Stoops, 1968; Beaudou, 1972; Eswaran et al., 1979; Beaudou et al., 1987; Eswaran, 1990; Muggler & Buurman, 2000). They are mostly rounded or subrounded and sometimes occur inside iron oxide nodules, suggesting that they are transported relict features.

Optically isotropic clay coatings with anisotropic zones were described in Oxisols on volcanic ash from West Java (Buurman & Jongmans, 1987). They originate from precipitation of allophane from recent ash fall, which recrystallised to halloysite under the influence of desiccation.

4.5 Siliceous Pedofeatures

Although neoformation of siliceous features is difficult to explain for environments producing oxic materials, Eswaran (1972) mentions the presence, in Oxisols on basalt, of quartz nodules and coatings, crystallised from soil solutions saturated with silica derived from weathering of primary minerals. Also, triangular plates with tile-like arrangement and parallel elongated prismatic crystals with pyramidal terminations have been observed in SEM images of the surface of quartz grains and interpreted as quartz neoformations (e.g., Krinsley & Doornkamp, 1973; Eswaran & Stoops, 1979). Other authors explain similar surface textures as dissolution features, related to differences in dissolution rates between crystallographic orientations, hypothesis supported by the occurrence

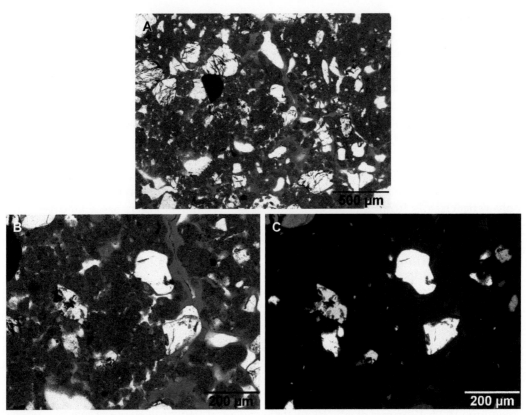

FIGURE 10 Clay coatings and infillings (Oxisol transitional to Ultisol, Kasese region, D.R. Congo). (A) Overview (PPL). (B) Detail, showing very limpid red illuvial clay (PPL). (C) Same field in XPL, illustrating the weak birefringence of the illuvial clay.

of triangular etch pits on other sides of the same grain (e.g., Leneuf, 1973; Muller et al., 1980/81; Fritsch, 1988) (see Fig. 7).

5. Conclusions

Micromorphological techniques have been often used in the study of oxic and related materials. Wide attention has been given to the study of the granular microstructure that is typical for these materials, but no consensus has been reached on its origin. Submicroscopic techniques were mainly used to study mineral weathering, mainly of quartz, and the formation of kaolinite, iron oxides and gibbsite occurrences.

Besides its support to soil classification revisions and the study of palaeosoils, the systematic use of micromorphological techniques in the study of well-defined oxic and related materials could help to further clarify various aspects, such as the role of soil animals in soil formation, and the preservation or degradation of features under changed environmental conditions.

References

Ahn, P.M., 1970. West African Soils. Oxford University Press, Oxford, 332 p.

Andrade, H., Schaefer, C.E.G.R., Demat-tê, J.L.I. & Andrade, F.V., 1997. Pedogeomorfologia e micro-pedologia de uma sequência Latossolo - Areia Quartzosa Hidromórfica sobre rochas cristalinas do estado do Amazonas. Geonomos 5, 55-66.

Baert, G. & Van Ranst, E., 1997a. Comparative micromorphological study of representative weathering profiles on different parent materials in the Lower Zaire. In Shoba, S., Gerasimova, M. & Miedema, R. (eds.), Soil Micromorphology: Studies on Soil Diversity, Diagnostics, Dynamics. Moscow, Wageningen, pp. 28-40.

Baert, G. & Van Ranst, E., 1997b. Total reserve in bases as an alternative for weatherable mineral content in soil classification: a micromorphological investigation. In Shoba, S., Gerasimova, M. & Miedema, R. (eds.), Soil Micromorphology: Studies on Soil Diversity, Diagnostics, Dynamics. Moscow, Wageningen, pp. 41-51.

Balbino, L.C., Bruand, A., Brossard, M., Grimaldi, M., Hajnos, M. & Guimarães, M.F., 2002. Changes in porosity and microaggregation in clayey Ferralsols of the Brazilian Cerrado on clearing for pasture. European Journal of Soil Science 53, 219-230.

Barros, E., Curmi, P., Hallaire, V., Chauvel, A. & Lavelle, P., 2001. The role of macrofauna in the trans-formation and reversibility of soil structure of an Oxisol in the process of forest to pasture conversion. Geoderma 100, 193-213.

Bartoli, F., Philippy, R., & Burtin, G., 1992. Influence of organic matter on aggregation in Oxisols rich in gibbsite or in goethite. I. Structures: the fractal approach. Geoderma 54, 231-257.

Beaudou, A.G., 1972. Expression micromorphologique de la micro-agrégation et de l'illuviation dans certains horizons de sols ferrallitiques centrafricains et dans les sols hydromorphes associés. Cahiers ORSTOM, Série Pédologie 10, 357-371.

Beaudou, A.G. & Chatelin, Y., 1974. Les mouvements d'argile dans certains sols ferrallitiques centrafricains. Transactions of the 10th International Congress of Soil Science, Volume VII, Moscow, 247-255.

Beaudou, A.G., Chatelin, Y., Collinet, J., Martin, D. & Sala, G.H., 1977. Notes sur la micromorphologie de certains sols ferrallitiques jaunes de régions équatoriales d'Afrique. Cahiers ORSTOM, Série Pédologie 15, 361-379.

Beaudou, A.G., Fromaget, M. & Guichard, E., 1987. Analyse des organisations macro et microstructurales de sols ferrallitiques centrafricains issus de roches basiques. In Fedoroff, N., Bresson, L.M. & Courty, M.A. (eds.), Micromorphologie des Sols, Soil Micromorphology. AFES, Plaisir, pp. 119-124.

Beinroth, F.H., Eswaran, H., Palmieri, F. & Reich, P.F., 1996. Properties, Classification and Management of Oxisols. World Soil Ressources, USDA, Natural Resources Conservation Services. Washington, 33 p.

Benayas, J. & Pinto Ricardo, R., 1973. Aspectos micromorfológicos de solos de uma topossequência ocorrendo em zona tropical húmida (Superfície da Quibala – Angola). Anais do Instituto Superior de Agronomia 34, 181-203.

Benayas, J. & Réfega, A., 1974. Aplicación de la micromorfologia a la clasificación de algunos suelos ferralíticos de Angola. Anales de Edafología y Agrobiología 33, 283-294.

Bennema, J., Jongerius, A. & Lemos, R., 1970. Micromorphology of some oxic and argillic horizons in South Brazil in relation to weathering sequences. Geoderma 4, 333-355.

Bitom, D. & Volkoff, B., 1991. Mise en évidence de deux modes de microstructuration dans une couverture de sols ferralitiques rouges du Sud-Cameroun. Science du Sol 29, 289-300.

Bravard, S. & Righi, D., 1990. Micromorphology of an Oxisol-Spodosl catena in Amazonia (Brazil). In Douglas, L.A. (ed.), Soil Micromorphology: A Basic and Applied Science. Developments in Soil Science, Volume 19. Elsevier, Amsterdam, pp. 169-174.

Bui, E.N., Mermut, A.R. & Santos, M.C.D., 1989. Microscopic and ultramicroscopic porosity of an Oxisol as determined by image analysis and water retention. Soil Science Society of America Journal 53, 661-665.

Buol, S.W. & Eswaran, H., 1978. The micromorphology of Oxisols. In Delgado, M. (ed.), Soil Micromorphology. University of Granada, pp. 325-347.

Buol, S.W. & Eswaran, H., 1999. Oxisols. Advances in Agronomy 68,151-195.

Buol, S.W., Southard, R.J., Graham, R.C. & McDaniel, P.A., 2003. Soil Genesis and Classification. Fifth Edition. Iowa State Press, Ames, 512 p.

Buurman, P. & Jongmans, A.G., 1987. Amorphous clay coatings in a lowland Oxisol and other andesitic soils of West Java, Indonesia. Pemberitaan Penelitian Tanah dan Pupuk 7, 31-40.

Buurman, P. & Sukardi, 1980. Brown Soils, Latosols or Podzolics. In Buurman, P. (ed.), Red Soils in Indonesia. Centre for Agricultural Publishing and Documentation, Wageningen, pp. 93-104.

Cagauan, B. & Uehara, G., 1965. Soil anisotropy and its relation to aggregate stability. Soil Science Society of America Proceedings 29, 198-200.

Camacho, E., Robert, M. & Jaunet, A.M., 1990. Mineralogy and structural organization of a red to yellow soil sequence in Cuba - relationships with soil properties. In Douglas, L.A. (ed.), Soil Micromorphology: A Basic and Applied Science. Developments in Soil Science, Volume 19. Elsevier, Amsterdam, pp. 183-190.

Cambier, P. & Prost, R., 1981. Etude des associations argile-oxyde: organisation des constituants d'un matériau ferrallitique. Agronomie 1, 713-722.

Cardoso de Lima, P. & Eswaran, H., 1987. The microfabric of soils belonging to the Acri-great groups of Oxisols. In Fedoroff, N., Bresson, L.M. & Courty, M.A. (eds.), Micromorphologie des Sols, Soil Micromorphology. AFES, Plaisir, pp. 145-150.

Chauvel, A., 1972. Observation micromorphologique de la partie supérieur des sols rouges ferrallitiques de Casamance (Sénégal). Cahiers ORSTOM, Série Pédologie 10, 343-356.

Chauvel, A., 1977. Recherches sur la Transformation des Sols Ferrallitiques dans la Zone Tropicale à Saisons Contrastées. Évolution et Réorganisation des Sols Rouges de Moyenne Casamance (Sénégal). Travaux et Documents de l'ORSTOM 62, 532 p.

Chauvel, A. & Pédro, G., 1978. Genèse de sols beiges (ferrugineux tropicaux lessivés) par transformation des sols rouges (ferralitiques) de Casamance (Sénégal). Cahiers ORSTOM, Série Pédologie 16, 231-249.

Chauvel, A., Pédro, G. & Tessier, D., 1976. Rôle du fer dans l'organisation de matériaux kaolinitiques. Etudes experimentales. Science du Sol 2, 101-113.

Chauvel, A., Bocquier, G. & Pédro, G., 1978. La stabilité et la transformation de la microstructure des sols rouges ferralitiques de Casamance (Senegal). Analyse microscopique et données experimentals. In Delgado, M. (ed.), Soil Micromorphology. University of Granada, pp. 779-813.

Chauvel, A., Soubies, F. & Melfi, A., 1983. Ferralitic soils from Brazil: formation and evolution of structure. In Nahon, D. & Noack, Y. (eds.), Pétrologie des Altérations et des Sols. Sciences Géologiques Mémoire 72, 37-46.

Chauvel, A., Grimaldi, M. & Tessier, D., 1991. Changes in soil pore-space distribution following deforestation and revegetation: an example from the Central Amazon Basin, Brazil. Forest Ecology and Management 38, 259-271.

Claisse, G., 1972. Etude sur la solubilization du quartz en voie d'altération. Cahiers ORSTOM, Série Pédologie 10, 97-122.

Condado, J.L., 1969. Micropedologia de Alguns dos Mais Representativos Solos de Angola. Memória da Junta de Investigação do Ultramar 59, 142 p.

Cooper, M., Vidal-Torrado, P. & Chaplot, V., 2005. Origin of microaggregates in soils with ferralic horizons. Scientia Agricola 62, 256-263.

Cooper, M., Vidal-Torrado, P. & Grimaldi, M., 2010. Soil structure transformations from ferralic to nitic horizons on a toposequence in southeastern Brazil. Revista Brasileira de Ciência do Solo 34, 1685-1699.

Cosarinsky, M.I., Bellosi, E.S. & Genise, J.F., 2005. Micromorphology of modern epigean termite nests and possible termite ichnofossils: a comparative analysis (Isoptera). Sociobiology 45, 745-778.

Curmi, P., Kertzman, F.F. & Queiroz Neto, J.P., 1994. Degradation of structure and hydraulic properties in an Oxisol under cultivation (Brazil). In Ringrose-Voase, A.J. & Humphreys, G.S. (eds.), Soil Micromorphology: Studies in Management and Genesis, Developments in Soil Science, Volume 22. Elsevier, Amsterdam, pp. 569-579.

De Craene, A., 1967. Diagnose des Latosols à Lehm et à Erde ainsi que de leurs latérites. In Rapport Annuel du Département de Géologie, Minéralogie, Paléontologie du Musée Royal de l'Afrique Centrale, Année 1966, pp. 112-121.

De Craene, A. & Laruelle, J., 1955. Genèse et altération des Latosols équatoriaux et tropicaux humides. Bulletin Agricole du Congo Belge 46, 1113-1243.

De Craene, A. & Laruelle, J., 1956a. Les Latosols à «scoriacé atypique» dans le nord-est du Congo-belge. Leur genèse. Transactions of the 6th International Congress of Soil Science, Volume E, Paris, 359-366.

De Craene, A. & Laruelle, J., 1956b. Les Latosols à «scoriacé atypique» dans le nord-est du Congo-belge. Leur altération. Transactions of the 6th International Congress of Soil Science, Volume E, Paris, 533-540.

Dedecker, D. & Stoops, G., 1993. Micromorphological and mineralogical study of a polygenetic laterite profile (Trombetas area, Brazil). Pedologie 43, 335-356.

Dick, D.P. & Schwertmann, U., 1996. Microaggregates from Oxisols and Inceptisols: dispersion through selective dissolutions and physico-chemical treatments. Geoderma 74, 49-63.

Embrapa, 1999. Sistema Brasileiro de Classificação de Solos. Empresa Brasileira de Pesquisa Agropecuária, Rio de Janeiro, 412 p.

Embrechts, J. & Stoops, G., 1982. Microscopical aspects of garnet weathering in a humid tropical environment. Journal of Soil Science 33, 535-545.

Embrechts, J. & Stoops, G., 1986. Relations between microscopical features and analytical characteristics of a soil catena in a humid tropical climate. Pedologie 36, 315-328.

Embrechts, J. & Stoops, G., 1987. Microscopic identification and quantitative determination of microstructure and potentially mobile clay in a soil catena in a humid tropical soil environment. In Fedoroff, N., Bresson, L.M. & Courty, M.A. (eds.), Micromorphologie des Sols, Soil Micromorphology. AFES, Plaisir, pp. 157-164.

Eschenbrenner, V., 1986. Contribution des termites à la micro-agrégation des sols tropicaux. Cahiers ORSTOM, Série Pédologie 22, 397-408.

Eswaran, H., 1972. Micromorphological indicators of pedogenesis in some tropical soils derived from basalts from Nicaragua. Geoderma 7, 15-31.

Eswaran, H., 1979. Micromorphology of Oxisols. In Beinroth, F.H. & Paramananthan, S. (eds.), Proceedings of the Second International Soil Classification Workshop, Part I: Malaysia. Soil Survey Division, Land Development Department, Bangkok, pp. 61-72.

Eswaran, H., 1990. Soils with ferralic attributes. Transactions of the 14th International Congress of Soil Science, Volume V, Kyoto, 65-70.

Eswaran, H. & Daud, N., 1980. A scanning electron microscopy evaluation of the fabric and mineralogy of some soils from Malaysia. Soil Science Society of America Journal 44, 855-861.

Eswaran, H. & Stoops, G., 1979. Surface textures of quartz in tropical soils. Soil Science Society of America Journal 43, 420-424.

Eswaran, H. & Sys, C., 1976. Micromorphological and mineralogical properties of the Quin Hill toposequence. Pedologie 26, 280-291.

Eswaran, H. & Wong Chaw Bin, 1978. A study of a deep weathering profile on granite in Peninsular Malaysia. I. Physico-chemical and micromorphological properties. Soil Science Society of America Journal 42, 144-149.

Eswaran, H., Sys, C. & Sousa, E.C., 1975. Plasma infusion. A pedological process of significance in the humid tropics. Anales de Edafología y Agrobiología 34, 665-674.

Eswaran, H., Stoops, G. & Sys, C., 1977. The micromorphology of gibbsite forms in soils. Journal of Soil Science 28, 136-143.

Eswaran, H., Van Wambeke, A. & Beinroth, F.H., 1979. A study of some highly weathered soils of Puerto Rico. Micromorphological properties. Pedologie 28, 139-162.

Falci, S.C. & Mendes, A.C.T., 1973. Identificação de cutans em perfis de latossol roxo e terra roxa estruturada. Anais da Escola Superior de Agricultura Luiz de Queiroz 30, 49-70.

Faure, P., 1987. Les héritages ferrallitiques dans les sols jaunes du Nord-Togo. Aspects micromorphologiques des élements figures. In Fedoroff, N., Bresson, L.M. & Courty, M.A. (eds.), Micromorphologie des Sols, Soil Micromorphology. AFES, Plaisir, pp. 111-118.

Fedoroff, N. & Eswaran, H., 1985. Micromorphology of Ultisols. In Douglas, L.A. & Thompson, M.L. (eds.), Soil Micromorphology and Soil Classification. Soil Science Society of America Special Publication 15, SSSA, Madison, pp. 145-164.

Ferreira, M.M., Fernandes, B. & Curi, N., 1999. Mineralogia da fração argila e estrutura de Latossolos da região sudeste do Brasil. Revista Brasileira de Ciência do Solo 23, 507-514.

Flach, K.W., Cady, J.G. & Nettleton, W.D., 1968. Pedogenic alteration of highly weathered parent materials. Transactions of the 9th International Congress of Soil Science, Volume IV, Adelaide, pp. 343-351.

Flageollet, J., 1980/1981. Aspects morphoscopiques et exoscopiques des quartz dans quelques sols ferralitiques de la région de Cechi (Cote d'Ivoire). Cahiers ORSTOM, Série Pédologie 18, 111-121.

Fölster, H., 1964. Die Pedi-sedimenten des Südsudanischen Pediplane. Herkunft und Bodenbildung. Pedologie 14, 64-84.

Frei, E., 1964. Eisenoxydkonkretionen und Schlierenbildungen in einigen Tropenböden Ecuadors. In Jongerius, A. (ed.). Soil Micromorphology. Elsevier, Amsterdam, pp. 291-294.

Fritsch, E., 1988. Morphologie des quartz d'une couverture ferralitique dégradée par hydromorphie. Cahiers ORSTOM, Série Pédologie 24, 3-15.

Guedez, J.E. & Langohr, R., 1978. Some characteristics of pseudosilts in a soil toposequence of the Llanos Orientales (Venezuela). Pedologie 28, 118-131.

Hallaire, V., Castro, S.S. & Curmi, P., 1994. Image analysis of the main horizons of an Oxisol/Ultisol toposequence in São Paulo State (Marilia, S.P., Brazil). Transactions of the 15th International Congress of Soil Science, Volume 6b, Acapulco, 94-95.

Hart, R.D., Wiriyakitnateekul, W. & Gilkes, R.J., 2003. Properties of soil kaolins from Thailand. Soil Science Society of America Journal 44, 855-861.

Hartmann, C., Tessier, D. & Pédro, G., 1994. Changes in sandy Oxisols microfabric after mechanical up-rooting of an oil palm plantation. In Ringrose-Voase, A.J. & Humphreys, G.S. (eds.), Soil Micromorphology: Studies in Management and Genesis. Developments in Soil Science, Volume 22. Elsevier, Amsterdam, pp. 687-695.

Ibraimo, M.M., Schaefer, C.E.G.R., Ker, J.C., Lani, J.L., Rolim-Neto, F.C., Albuquerque, M.A. & Miranda, V. J., 2004. Gênese e micromorfologia de solos sob vegetação xeromórfica (caatinga) na região dos Lagos (RJ). Revista Brasileira de Ciência do Solo 28, 695-712.

IUSS Working Group WRB, 2015. World Reference Base for Soil Resources 2014, Update 2015. World Soil Resources Reports No. 106. FAO, Rome, 192 p.

Jungerius, P.D., van den Ancker, J.A.M. & Mücher, H.J., 1999. The contribution of termites to the microgranular structure of soils on the Uasin Gishu Plateau, Kenya. Catena 34, 349-363.

Kauffman, J.H., 1987. Comparative Classification of Some Deep, Well-Drained Red Clay Soils of Mozambique. Technical Paper no. 16, International Soil Reference and Information Centre, Wageningen, 61 p.

Kooistra, M.J., 1982. Micromorphological Analysis and Characterization of 70 Benchmark Soils of India. A Basic Reference Set. Netherlands Soil Survey Institute, Wageningen, 788 p.

Kooistra, M.J. & Pulleman, M.M., 2018. Features related to faunal activity. In Stoops, G., Marcelino, V. & Mees, F. (eds.), Interpretation of Micromorphological Features of Soils and Regoliths. Second Edition. Elsevier, Amsterdam, pp. 447-469.

Krinsley, D.H. & Doornkamp, J.C., 1973. Atlas of Quartz Sand Surface Textures. Cambridge University Press, Cambridge, 91 p.

Kubiëna, W.L., 1948. Entwicklungslehre des Bodens. Springer-Verlag, Wien, 215 p.

Laruelle, J., 1956. Quelques aspects de la microstructure des sols du nord-est du Congo Belge. Pedologie 6, 38-58.

Lee, K.E. & Wood, T.G., 1971. Termites and Soils. Academic Press, London, 251 p.

Leneuf, N., 1973. Observations stéréoscopiques sur les figures de corrosion du quartz dans certaines formations superficielles. Cahiers ORSTOM, Série Pédologie 11, 43-51.

Lepsch, I.F. & Buol, S.W., 1974. Investigations in an Oxisol-Ultisol toposequence in S.Paulo State, Brazil. Soil Science Society of America Proceedings 38, 491-496.

Marcelino, V. & Stoops, G., 1996. A weathering score for sandy soil materials based on the intensity of etching of quartz grains. European Journal of Soil Science 47, 7-12.

Marcelino, V., Mussche, G. & Stoops, G., 1999. Surface morphology of quartz grains from tropical soils and its significance for assessing soil weathering. European Journal of Soil Science 50, 1-8.

Mermut, A.R., Arshad, M.A. & St. Arnaud, R.J., 1984. Micropedological study of termite mounds of three species of Macrotermes in Kenya. Soil Science Society of America Journal 48, 613-620.

Miedema, R., Brouwer, J., Geiger, S.C. & Vandenbeldt, R.J., 1994. Variability in the growth of Faidherbia albida near Niamey, Niger, Africa: micromorphological aspects of termite activity. In Ringrose-Voase, A.J. & Humphreys, G.S. (eds.), Soil Micromorphology: Studies in Management and Genesis. Developments in Soil Science, Volume 22. Elsevier, Amsterdam, pp. 411-419.

Moorman, F.R., 1979. Taxonomic problems of low activity clay Alfisols and Ultisols. In Beinroth, F.H & Panichapong, S. (eds.), Proceedings of the Second International Soil Classification Workshop, Part II. Land Development Department, Bangkok, pp. 53-76.

Moura Filho, W. & Buol, S.W., 1976. Studies of a Latosol Roxo (Eutrustox) in Brazil: micromorphology effect on ion release. Experientiae 21, 161-177.

Muggler, C.C. & Buurman, P., 1997. Micromorphological aspects of polygenetic soils developed on phyllitic rocks in Minas Gerais, Brazil. In Shoba, S., Gerasimova, M. & Miedema, R. (eds.), Soil Micromorphology: Studies on Soil Diversity, Diagnostics, Dynamics. Moscow, Wageningen, pp. 129-138.

Muggler, C.C. & Buurman, P., 2000. Erosion, sedimentation and pedogenesis in a polygenetic oxisol sequence in Minas Gerais, Brazil. Catena 41, 3-17.

Muggler, C.C., Pape, T. & Buurman, P., 1997. Laser grain-size determination in soil genetic studies: 2. Clay content, clay formation, and aggregation in some Brazilian Oxisols. Soil Science 162, 219-228.

Muggler, C.C., van Griethuysen, C., Buurman, P. & Pape, T., 1999. Aggregation, organic matter, and iron oxide morphology in Oxisols from Minas Gerais, Brazil. Soil Science 164, 759-770.

Muggler, C.C., Buurman, P. & van Doesburg, J.D.J., 2007. Weathering trends and parent material characteristics of polygenetic oxisols from Minas Gerais, Brazil. 1. Mineralogy. Geoderma 138, 39-48.

Muller, J.P., 1977a. Microstructuration des stutichrons rouges ferralitiques, à l'amont des modelés convexes (Centre-Cameroun). Aspects morphologiques. Cahiers ORSTOM, Série Pédologie 15, 239-258.

Muller, J.P., 1977b. La microlyse plasmique et la différenciation des épipédons dans les sols ferrallitiques rouges du Centre-Cameroun. Cahiers ORSTOM, Série Pédologie 15, 345-359.

Muller, J.P., 1983. Micro-organization of loose ferralitic materials in the Cameroons. In Bullock, P. & Murphy, C.P. (eds.), Soil Micromorphology. Volume 2. Soil Genesis. AB Academic Publishers, Berkhamsted, pp. 655-666.

Muller, D., Bocquier, G., Nahon, D. & Paquet, H., 1980/1981. Analyse des différenciations minéralogiques et structurales d'un sol ferrallitique à horizons nodulaires du Congo. Cahiers ORSTOM, Série Pédologie 18, 87-109.

Nettleton, W.D., Eswaran, H., Holzhey, C.S., & Nelson, R.E., 1987. Micromorphological evidence of clay translocation in poorly dispersible soils. Geoderma 40, 37-48.

Nunes, W.A.G.A., Schaefer, C.E.G.R., Ker, J.C. & Fernandes Filho, E.I., 2000. Caracterização micropedológica de alguns solos da Zona da Mata Mineira. Revista Brasileira de Ciência do Solo 24, 103-115.

Nye, P.H., 1955. Some soil-forming processes in the humid tropics. IV. The action of the soil fauna. Journal of Soil Science 6, 73-83.

Padmanabhan, E. & Mermut, A.R., 1996. Submicroscopic structure of Fe-coatings on quartz grains in tropical environments. Clays and Clay Minerals 44, 801-810.

Padmanabhan, E., Eswaran, H. & Mermut, A.R., 2012. Classifying soils at the ultimate stage of weathering in the tropics. Journal of Plant Nutrition and Soil Science 175, 86-93.

Pédro, G., 1987. Géochimie, minéralogie et organisation des sols. Aspects coordonnés des problèmes pédogénétiques. Cahiers ORSTOM, Série Pédologie 23, 169-186.

Pédro, G., Chauvel, A. & Melfi, A.J., 1976. Recherches sur la constitution et la genèse des Terre Roxa Estructurada du Brésil. Annales Agronomiques 27, 265-294.

Pidgeon, J.D., 1976. Contemporary pedogenic processes in a ferrallitic soil in Uganda. I. Identification. Geoderma 15, 425-436.

Reatto, A., Bruand, A., de Souza Martins, E., Muller, F., da Silva, E.M., de Carvalho Jr., O.A., Brossard, M. & Richard, G., 2009. Development and origin of the microgranular structure in latosols of the Brazilian Central Plateau: significance of texture, mineralogy, and biological activity. Catena 76, 122-134.

Rodríguez-Rodríguez, A., Fedoroff, N., Tejedor Salguero, M.L. & Fernández-Caldas, E., 1980. Suelos fersialíticos sobre cenizas volcánicas. III Características micromorfológicas. Interpretación y clasificación. Anales de Edafología y Agrobiología 39, 37-49.

Rodríguez-Rodríguez, A., Jimenez Mendoza, C.C. & Tejedor Salguero, M.L., 1988. Micromorfología de los suelos ferralíticos en las Islas Canarias. Anales de Edafología y Agrobiología 47, 409-430.

Roose, E.J., 1970. Importance relative de l'érosion du drainage oblique et vertical dans la pédogenèse actuelle d'un sol ferralitique de moyenne Côte d'Ivoire. Deux années de mesure sur parcelle expérimentale. Cahiers ORSTOM, Série Pédologie 8, 469-482.

Santos, M.C.D., Mermut, A.R. & Ribeiro, M.R., 1989. Submicroscopy of clay microaggregates in an Oxisol from Pernambuco, Brazil. Soil Science Society of America Journal 53, 1895-1901.

Sarcinelli, T.S., Schaefer, C.E.G.R., Lynch, L.de S., Arato, H.D., Viana, J.H.M., Filho, M.R.de A. & Gonçalves, T.T., 2009. Chemical, physical and micromorphological properties of termite mounds and adjacent soils along a toposequence in Zona da Mata, Minas Gerais State, Brazil. Catena 76, 107-113.

Schaefer, C.E.R., 2001. Brazilian latosols and their B horizon microstructure as long-term biotic constructs. Australian Journal of Soil Research 39, 909-926.

Schaefer, C.E.R., Ker, J.C., Gilkes, R.J., Campos, J.C., da Costa, L.M. & Saadi, A., 2002. Pedogenesis on the uplands of the Diamantina Plateau, Minas Gerais, Brazil: a chemical and micropedological study. Geoderma 107, 243-269.

Schaefer, C.E.G.R., Gilkes, R.J. & Fernandes, R.B.A., 2004. EDS/SEM study of microaggregates of Brazilian Latosols, in relation to P adsorption and clay fraction attributes. Geoderma 123, 69-81.

Schmidt-Lorenz, R., 1980. Soil reddening through hematite from plinthitized saprolite. In Joseph, K.T. (ed.), Conference on Classification and Management of Tropical Soils. Malaysian Society of Soil Science, Kuala Lumpur, pp. 101-106.

Sharma, M.L. & Uehara, G., 1968. Influence of soil structure on water relations in low humic latosols: I Water retention. Soil Science Society of America Proceedings 32, 765-770.

Silva, L.F.S.D., Marinho, M.D.A., Matsura, E.E., Cooper, M. & Ralisch, R., 2015. Morphological and micromorphological changes in the structure of a Rhodic Hapludox as a result of agricultural management. Revista Brasileira de Ciência do Solo 391, 205-221.

Simas, F.N.B., Schaefer, C.E.G.R., Fernandes Filho, E.I., Chagas, A.C. & Brandão, P.C., 2005. Chemistry, mineralogy and micropedology of highland soils on crystalline rocks of Serra da Mantiqueira, southeastern Brazil. Geoderma 125, 187-201.

Sleeman, J.R. & Brewer, R., 1972. Micro-structures of some Australian termite nests. Pedobiologia 12, 347-373.

Skorupa, A.L.A., Tassinari, D., Silva, S.H.G., Poggere, G.C., Zinn, Y.L. & Curi, N., 2016. Xanthic-and Rhodic-Acrudoxes under cerrado vegetation: differential internal drainage and covarying micromorphological properties. Ciência e Agrotecnologia 40, 443-453.

Soil Survey Staff, 2014. Keys to Soil Taxonomy. 12th Edition. USDA-Natural Resources Conservation Service, Washington, 360 p.

Stoner, E.R., Freitas, E., Macedo, J., Mendes, R.C.A., Cardoso, I.M., Amabile, R.F., Bryant, R.B. & Douglas, J.L., 1991. Physical Constraints to Root Growth in Savanna Oxisols. TropSoils Bulletin 91-01, North Carolina State University, Raleigh, 28 p.

Stoops, G., 1964. Application of some pedological methods to the analysis of termite mounds. In Bouillon, A. (ed.), Etudes sur les Termites Africains. Université de Léopoldville, pp. 379-398.

Stoops, G., 1968. Micromorphology of some characteristic soils of the Lower Congo (Kinshasa). Pedologie 18, 110-149.

Stoops, G., 1983. Micromorphology of the oxic horizon. In Bullock, P. & Murphy, C.P. (eds.), Soil Micromorphology. Volume 2. Soil Genesis. AB Academic Publishers, Berkhamsted, pp. 419-440.

Stoops, G., 1989. Relict properties in soils of humid tropical regions with special reference to Central Africa. In Bronger, A. & Catt, J.A. (eds.), Paleopedology. Nature and Application of Paleosols. Catena Supplement 16, pp. 95-106.

Stoops, G., 1991. The influence of the fauna on soil formation in the tropics. Micropedological aspects. Bulletin des Séances de l'Académie Royale des Sciences d'Outre-Mer 36, 461-469.

Stoops, G. & Buol, S.W., 1985. Micromophology of Oxisols. In Douglas, L.A. & Thompson, M.L. (eds.), Soil Micromorphology and Soil Classification. Soil Science Society of America Special Publication 15, SSSA, Madison, pp. 105-119.

Stoops, G. & Delvigne, J., 1990. Morphology of mineral weathering and neoformation. II Neoformations. In Douglas, L.A. (ed.), Soil Micromorphology: A Basic and Applied Science. Developments in Soil Science, Volume 19. Elsevier, Amsterdam, pp. 483-492.

Stoops, G. & Marcelino, V., 2018. Lateritic and bauxitic materials. In Stoops, G., Marcelino, V. & Mees, F. (eds.), Interpretation of Micromorphological Features of Soils and Regoliths. Second Edition. Elsevier, Amsterdam, pp. 691-720.

Stoops, G. & Schaefer, C.E.G.R., 2018. Pedoplasmation: formation of soil material. In Stoops, G., Marcelino, V. & Mees, F. (eds.), Interpretation of Micromorphological Features of Soils and Regoliths. Second Edition. Elsevier, Amsterdam, pp. 59-71.

Stoops, G., Sedov, S. & Shoba, S., 2018. Regoliths and soils on volcanic ash. In Stoops, G., Marcelino, V. & Mees, F. (eds.), Interpretation of Micromorphological Features of Soils and Regoliths. Second Edition. Elsevier, Amsterdam, pp. 721-751.

Stoops, G., Shi Guang Chun & Zauyah, S., 1990. Combined micromorphological and mineralogical study of a laterite profile on graphite sericite phyllite from Malacca (Malaysia). Bulletin de la Société Belge de Géologie 99, 79-92.

Stoops, G., Marcelino, V., Zauyah, S. & Maas, A., 1994. Micromorphology of soils of the humid tropics. In Ringrose-Voase, A.J., Humphreys, G.S. (eds.), Soil Micromorphology: Studies in Management and Genesis. Developments in Soil Science, Volume 22. Elsevier, Amsterdam. pp. 1-15.

Tandy, J.C., Grimaldi, M., Grimaldi, C. & Tessier, D., 1990. Mineralogical and textural changes in french Guyana Oxisols and their relation with microaggregation. In Douglas, L.A. (ed.), Soil Micromorphology: A Basic and Applied Science. Developments in Soil Science, Volume 19. Elsevier, Amsterdam, pp. 191-198.

Tawornpruek, S., Kheoruenromne, I., Suddhiprakarn, A. & Gilkes, R.J., 2005. Microstructure and water retention of Oxisols in Thailand. Australian Journal of Soil Research 43, 973-986.

Tejedor Salguero, M.L., Garcia-Lopez, L. & Fernández-Caldas, E, 1984/1985. Les sols ferrallitiques des Iles Canaries (Espagne). Cahiers ORSTOM, Série Pédologie 21, 109-116.

Trapnell, C.G. & Webster, R., 1986. Microaggregates in red earths and related soils in East and Central Africa, their classification and occurrence. Journal of Soil Science 37, 109-123.

Tsuji, G.Y., Watanabe, R.T. & Sakai, W.S., 1975. Influence of soil microstructure on water characteristics of selected Hawaiian soils. Soil Science Society of America Proceedings 39, 28-33.

Uehara, G. & Keng, J., 1975. Management implications of soil mineralogy in Latin America. Soil management in tropical America. Publication of the Soil Science Department. North Carolina State University, Raleigh, pp. 351-363.

Utami, S.R., Mulyanto, B., Stoops, G., Van Ranst, E. & Baert, G., 1997. Mineralogical and micromorphological characterization of a soil sequence in Pleihari, South Kalimantan (Indonesia). In Shoba, S., Gerasimova, M. & Miedema, R. (eds.), Soil Micromorphology: Studies on Soil Diversity, Diagnostics, Dynamics. Moscow, Wageningen, pp. 326-334.

Van Wambeke, A., Eswaran, H., Herbillon, A.J. & Comerma, J., 1983. Oxisols. In Wilding, L.P., Smeck, N.E. & Hall, G.F. (eds.), Pedogenesis and Soil Taxonomy. II The Soil Orders. Developments in Soil Science, Volume 11b. Elsevier, Amsterdam, pp. 325-354.

Vepraskas, M.J., Lindbo, D.L. & Stolt, M.H., 2018. Redoximorphic features. In Stoops, G., Marcelino, V. & Mees, F. (eds.), Interpretation of Micromorphological Features of Soils and Regoliths. Second Edition. Elsevier, Amsterdam, pp. 425-445.

Verheye, W. & Stoops, G., 1975. Nature and evolution of soils developed on the granite complex in the subhumid tropics (Ivory Coast). II. Micromorphology and mineralogy. Pedologie 25, 40-55.

Viana, J.H.M., Fernandes Filho, E.I. & Schaefer, C.E.G.R., 2004. Efeitos de ciclos de umedecimento e secagem na reorganização da estrutura microgranular de Latossolos. Revista Brasileira de Ciência do Solo 28, 11-19.

Vidal-Torrado, P., Lepsch, I.F., Castro, S.S. & Cooper, M., 1999. Pedogênese em uma seqüência Latossolo-Podzólico na borda de um platô na depressão periférica Paulista. Revista Brasileira de Ciência do Solo 23, 909-921.

Vrdoljak, G. & Sposito, G., 2002. Soil aggregate hierarchy in a Brazilian Oxisol. Developments in Soil Science 28, 197-217.

West, L.T., Lawrence, K.S., Dayot, A.A., Tomas, L.M. & Yeck, R.D., 1997. Micromorphology and soil development as indicators of ash age on Mindanao, the Philippines. In Shoba, S., Gerasimova, M. & Miedema, R. (eds.), Soil Micromorphology: Studies on Soil Diversity, Diagnostics, Dynamics. Moscow, Wageningen, pp. 335-344.

Zainol, E. & Stoops, G., 1986. Relationship between plasticity and selected physico-chemical and micromorphological properties of some inland soils from Malaysia. Pedologie 36, 263-275.

Zauyah, S., Schaefer, C.E.G.R. & Simas, F.N.B., 2018. Saprolites. In Stoops, G., Marcelino, V. & Mees, F. (eds.), Interpretation of Micromorphological Features of Soils and Regoliths. Second Edition. Elsevier, Amsterdam, pp. 37-57.

Zeese, R., Schwertmann, U., Tietz, G.F. & Jux, U., 1994. Mineralogy and stratigraphy of three deep laterite profiles of the Jos plateau (Central Nigeria). Catena 21, 195-214.

Chapter 24

Lateritic and Bauxitic Materials

Georges Stoops, Vera Marcelino

GHENT UNIVERSITY, GHENT, BELGIUM

CHAPTER OUTLINE

1. Introduction

1.1 Historical Background

Laterites were first described by Buchanan (1807) in southern India. They have subsequently been the subject of numerous studies, which is partly related to their importance in various disciplines. Examples include the fields of economic geology (e.g., source of elements such as aluminium, manganese and nickel, and use as building materials), agronomy (e.g., restriction of root development, scarcity of essential nutrients) and geomorphology (e.g., palaeoenvironmental change). Laterites and their residues cover about 33% of the continents (Tardy, 1993). Their formation is related primarily to a climate with high humidity and high mean annual temperature, a geomorphological environment with limited runoff and without aggressive erosion, and relative tectonic stability with minimal crustal deformation and only gradual regional uplift (Widdowson, 2007).

Although a large volume of literature is available on laterites and bauxites, micromorphological and petrographical studies are rather limited in number and often ambiguous. Up to now, no systematic review has been published on the nature and significance of the micromorphological features observed. The oldest thin section descriptions known to us are those by Lacroix (1913) and Harrison (1933), who studied laterites in Guinea and British Guiana. Important early studies by soil scientists include those by Humbert (1948) on lateritic weathering in New Guinea, Kubiëna (1954) on laterites from Guinea, the less well-known work of De Craene and Laruelle (1955) on laterites and associated soils from Congo and the book by Alexander and Cady (1962), which reports detailed microscopic studies of various lateritic materials from Central Africa. Another relevant early contribution is the micromorphological study on the evolution of soft lateritic material by Schmidt-Lorenz (1964). Research by French scientists has also been important, as summarised in the works of Nahon (1991) and Tardy (1993). For bauxites, attempts to classify microfabrics have been made by Valeton (1972, 1973), Bárdossy and Nicolas (1973), Mindszenty (1978, 1983), Bárdossy (1982), Boulangé (1984) and Aleva (1987), partly concentrating on karst bauxites and using a largely sedimentological approach.

Early thin section studies by Lelong (1967) concluded that alterites or laterites cannot be fully characterised if not all fractions, from the clay fraction to the largest components, are taken into consideration. This implies that the routine procedures in soil science laboratories, discarding the fraction larger than 2 mm, are not suitable for the study of lateritic materials.

This chapter deals with fabrics of lateritic materials recognised in thin sections for specific parts of the lateritic weathering mantle. Only weathering profiles and their erosion products, directly relevant for soil and regolith studies, are included in this review. Studies done in a context of mining, such as the exploitation of karst bauxites, are not considered. The preparation of this review was complicated by the lack of published detailed descriptions, and by the manifold and sometimes contradictory terms for field and microscopic characteristics, often influenced by the different genetic approaches of the various schools.

1.2   Definition of Laterite

Although lateritic and bauxitic materials have been widely studied, no generally accepted definition of laterite exists. Many definitions have been proposed since the first description of laterites by Buchanan (1807). Excellent reviews of the old definitions are given by Sivarajasingham et al. (1962) and Maignien (1966). The different backgrounds of the authors formulating these definitions lead sometimes to conflicting situations. Also the continent where the author has been mostly active influences his ideas about the nature and genesis of lateritic materials. As a result, similar materials were often indicated by different names, or the same name was used for different features.

A commonly used definition is the one proposed by Schellmann (1982, 1983) for IGCP Project 124 on Laterisation Processes and later adopted by EUROLAT and CORLAT. This definition, however, does not explicitly consider the role of lateral transport, absolute enrichment or the common polygenetic nature of laterite profiles (e.g., Fölster et al., 1971; Bourman & Ollier, 2002). Some authors limit the use of the term 'laterite' to those formed by residual enrichment (Schmidt-Lorenz, 1974b; Aleva, 1986). Other authors use the term 'laterite soil' or 'lateritic soil' for red tropical soils (e.g., Varghese & Byju, 1993; Fritsch et al., 2002). Additional terms found in literature for irreversibly hardened iron or aluminium oxide accumulations include petroplinthite, introduced by Sys (1968) and ferricrete, commonly used by French authors (e.g., Tardy, 1993), the latter being a general term that is not restricted to materials derived from processes of lateritisation *sensu strictum.*

Laterites have been classified according to position in the landscape, morphology, chemistry and mineralogy. A ternary mineralogical diagram with an iron-axis (hematite, goethite), an aluminium-axis (gibbsite, boehmite, diaspore) and a clay-axis (mainly kaolinite) has been commonly used (Bárdossy & Aleva, 1990; Aleva, 1994). This diagram has ferrite, bauxite and kaolinite as end-members and ferruginous laterite, bauxitic laterite and kaolinitic bauxite as intermediate phases.

In this chapter, the term laterite is used in a broad sense, as in the definition by Charman (1988). It refers to a highly weathered natural material with high concentration of hydrated oxides of iron or aluminium (so-called sesquioxides), as a consequence of residual accumulation and/or absolute enrichment, by components transported in solution or as detrital material. The distinction between these processes is not always clear and both may be involved in the formation of a laterite. Since the aim of this chapter is not to discuss genesis and classification in detail, only two groups will be considered: the residual or *in situ* laterites and the detrital laterites, resulting from accumulation of eroded laterite fragments. The accumulated hydrated oxides may be present as the only constituent of an unhardened soil or a hardened layer. They can also be the main constituent of concretionary nodules in a soil matrix or of a cemented matrix enclosing other materials. Laterites are thus the product of secondary physico-chemical processes and not of the normal primary sedimentary, metamorphic or igneous processes and they can be considered as metasomatic rocks. One exception is the deposition of reworked lateritic material in valleys.

The proposed definition corresponds essentially to that of the plinthic and petroplinthic diagnostic horizons of the World Reference Base for Soil Resources (IUSS Working Group WRB, 2015). In Soil Taxonomy (Soil Survey Staff, 2014), much of the materials that have been called laterite are included in categories of soils with plinthite, which is a diagnostic soil characteristic that refers to the iron-rich and humus-poor parts of a layer or horizon, which irreversibly harden to ironstone on exposure to repeated wetting and drying. The absence or low concentration of organic matter in ironstone differentiates it from other iron-rich horizons such as ortstein and placic horizons (Soil Survey Staff, 2014).

1.3   The Standard Laterite Profile

Not only has the definition of laterite been controversial, but also the typical sequence of a 'normal' laterite profile and its subdivision. Many laterite profiles are the result of truncation and subsequent deposition during successive cycles of pedimentation (e.g., Muggler & Buurman, 2000; Achyuthan & Fedoroff, 2008; see also Fedoroff et al., 2010, 2018) and therefore difficult to compare. A complete *in situ* profile consists in principle of four layers (Aleva, 1986; Bárdossy & Aleva, 1990; Nahon, 1991), from bottom to top: (i) the parent rock, (ii) a saprolite layer (see Zauyah et al., 2010, 2018), (iii) the lateritic layer, and (iv) a gravel-free or gravel-poor cover layer, often with oxic characteristics (see Marcelino et al., 2010, 2018).

Within the lateritic layer several sublayers can be distinguished, from bottom to top: a soft laterite (mottled clay zone), a hard laterite (laterite *sensu strictum*, for some authors) and a gravel-rich layer (Duchaufour, 1960). Soft laterite was called 'argile tachetée' or 'argile bigarée', by French authors, and hard laterite was named 'carapace' or 'cuirasse' (e.g., Maignien, 1966, Tardy, 1992). The absolute and relative thickness of these sublayers is highly variable.

The soft laterite corresponds to a zone with plinthite (Soil Survey Staff, 2014) and often gradually extends into the hard laterite above and the saprolite beneath. In the lower part, residual saprolite fragments are often present. In some laterites, the mottled zone (so called mottled clay) does not extend to the saprolite and an intermediate zone exists that is called the pallid zone, which has mainly been documented for Australian soils (McCrea et al., 1990). The mottled clay can be either saprolitic material or material that underwent pedoplasmation (see Stoops & Schaefer, 2010, 2018; Zauyah et al., 2010, 2018).

The hard laterite can be nodular or vermicular, depending on the nature of the underlying mottled clay interval. Nodular laterites consist of closely packed hard nodules, derived from hardening of the red mottles followed by loss of white or grey interstitial material of mottled clay. In vermicular laterites, a hard iron-rich body is intersected by numerous tubes filled with soft, generally white, kaolinitic clay and commonly coated by hard, dark ferruginous material, which can be the result of termite activity (de Barros Machado, 1983).

The gravel-rich layer is often allochthonous, which is most commonly documented for occurrences in Africa and India. In this case, it consists mainly of transported iron-rich nodules, mixed with other coarse components. They are often considered as pediment gravel (Fölster et al., 1971).

2. Ferritic Laterites

2.1   Soft Laterite or Mottled Clay Zone

2.1.1   General Characteristics

The microstructure of the mottled clay is mainly massive (Stoops, 1968; Stoops et al., 1990), although channels and vughs can be present in the upper part. The groundmass is homogeneous, with porphyric c/f-related distribution patterns and a grey micromass in the upper part, whereas in the lower part a saprolitic fabric may be present. The micromass usually has an undifferentiated b-fabric in the lower part (Achyuthan & Fedoroff, 2008). In the upper part, where the effects of pedogenic processes are better expressed, the b-fabric is more or less well-developed striated with low interference colours, compatible with a kaolinitic composition. The red spots or streaks, which correspond to plinthite, can occur either as impregnative nodules or as massive impregnative zones interrupted by linear or tubular (vermiform) zones of grey kaolinite. Toward the upper part of the mottled clay zone, the reddish areas become more strongly impregnated and the colour of the groundmass outside these areas changes to greyish yellow (Stoops et al., 1990).

2.1.2   Hematite Nodules and Coatings

Diffuse reddish impregnative clusters or streaks correspond to concentrations of small (0.5-1 µm) hematite crystals (Rosolen et al., 2002) or aggregates (Schmidt-Lorenz, 1980). They often display a specific cellular fabric, referred to as plinthitic cellular fabric throughout this chapter, which was first reported for 'groundwater laterites' from the DR Congo (De Craene & Laruelle, 1955). It was also described in detail and discussed by Schmidt-Lorenz (1964), as part of a study of the genesis and evolution of soft laterites from Sri Lanka and from the type locality Kerala in southern India. His ideas were soon confirmed for African laterites by Stoops (1968) and Réfega (1972). Starting from a homogeneous yellowish groundmass, soft laterite is formed by processes of leaching of silica, resulting in relative accumulation of iron. First, shrinkage of the kaolinitic groundmass leads to opening of fissures and to development of an angular blocky microstructure. The iron oxides dispersed in the micromass migrate to the walls of the blocky aggregates where they form hypocoatings. These hypocoatings, which are dark reddish to opaque in transmitted light and red in oblique incident light, consist of small (1-10 µm) hematite aggregates called 'laterite droplets' (Hamilton, 1964) or 'plinthitic hematite' (Schmidt-Lorenz, 1980). Further retraction gives rise to secondary fissures where similar hypocoatings gradually form. This process eventually results in the development of a cellular fabric (Figs. 1A and 2). The cells, generally 70-100 µm in diameter, have a white interior that is mainly composed of kaolinite (Stoops, 1970). Similar hematite hypocoatings have been observed below a hard laterite formed on saprolite, where it is interpreted as a gley feature (Muggler & Buurman, 2000). The hematite was reported to be essentially low in aluminium, which is in agreement with values reported by other authors (Zeese et al., 1994). The sequence of processes

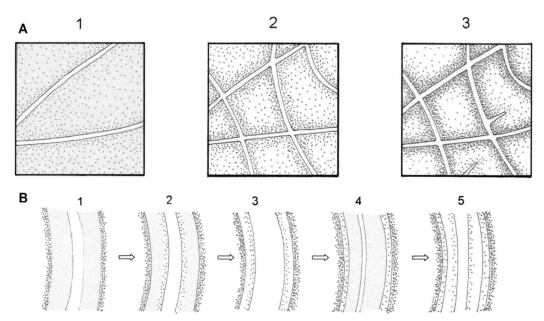

FIGURE 1 (A) Schematic representation of the formation of plinthitic cellular fabric (red points represent hematite grains or aggregates): (1) weakly impregnated hematitic hypocoatings start to develop along planar voids in the homogeneous yellowish mass of kaolinite and fine plinthitic hematite grains; (2) the number of planar voids and the degree of impregnation of the hematitic hypocoatings increases; (3) strongly impregnated hematitic hypocoatings are formed at the expense of nearly purely kaolinitic groundmass. (B) Schematic representation of the evolution of ferruginous clay coatings along the planar voids between plinthitic cellular fabric units (red points represent hematite grains or aggregates, yellow zones refer to illuvial ferrugineous clay coatings): (1) yellowish ferruginous kaolinitic clay coating; (2) development of hematitic hypocoating within the clay coating; (3) transformation of the yellowish coating into a thin white kaolinitic coating with hematitic hypocoating; (4) deposition of a new yellowish clay coating; (5) transformation of this coating to another thin white kaolinitic coating with hematitic hypocoating. *Based on Stoops (1968).*

described above results in relative accumulation of iron and aluminium. Absolute accumulation may occur by deposition of illuvial ferruginous clay in the planar voids formed by shrinkage (see Section 2.1.3).

Based on the study of a large number of laterites in thin section, Schmidt-Lorenz (1974b) concluded that the presence of small round hematite aggregates (plinthitic hematite), typical of laterites, is indicative of ferralitic palaeoweathering in soils from Europe. In African soils, the presence of plinthitic hematite is an indication that the soils developed on laterite-derived colluvial sediments (Stoops, 1989). Good SEM images of plinthitic hematite are difficult to obtain because of its close intermixture with kaolinite (Stoops, 1970; Magaldi & Bidini, 1991). It is not clear whether plinthitic hematite results exclusively from internal redistribution of iron (residual accumulation) or from an input from external sources (absolute accumulation). The formation of plinthitic cellular fabrics is best expressed in clayey materials. Based on thermodynamic considerations,

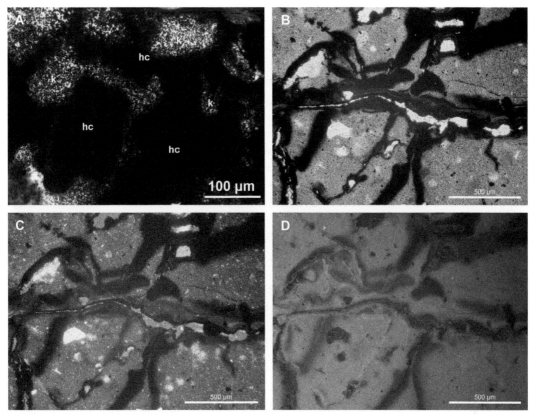

FIGURE 2 (A) Plinthitic cellular fabric in a laterite nodule (Lower Congo, DR Congo): small hematite crystals or aggregates (plinthitic hematite) are concentrated in hypocoatings (hc) along fissures surrounding cells whose core is composed of pale kaolinite (k) (PPL). (B) Strongly developed plinthitic cellular fabric with an illuvial clay coating developed at a later stage, in laterite (Southern India). (PPL). (C) In XPL, illustrating the faint interference colours of the clay coating. (D) In OIL, illustrating the yellowish colour of the clay coatings.

this has been related to the preferential formation of hematite as iron oxide mineral when porosity is low (Didier et al., 1983).

Impregnation of the hematite nodules and the surrounding groundmass with diffuse goethite was observed in a polygenetic profile in southern Sudan in which a pseudogley developed, due to truncation and changes in environmental conditions (Fölster et al., 1971).

The formation of red hematitic mottles has been explained by the build-up of a high iron oxide concentration as a result of two concurrent processes: precipitation of iron oxides from solution and dissolution of kaolinite, seemingly by a replacement reaction (Ambrosi et al., 1986).

Red mottles associated with the redistribution of iron as a result of hydromorphism can be distinguished from primary ferruginous mottles formed in saprolite derived from coarse-grained rocks by *in situ* weathering of ferriferous minerals (Fölster, 1971; Fölster et al., 1971), because they do not display a plinthitic cellular fabric.

Optical and submicroscopic studies of the nodular horizon of laterites on gneiss from Cameroon (Muller & Bocquier, 1987) revealed that the change from ferruginous nodules (with partly preserved lithic fabric) to the interstitial red groundmass was characterised by a gradual loss of lithic fabric and the formation of successive generations of kaolinite, each with smaller particle size (SEM), lower crystallinity and higher degree of iron substitution. This is accompanied by a decrease of the degree of orientation of the particles with respect to the original gneiss fabric. Simultaneously the content of hematite with low Al-substitution decreased with a corresponding increase in amount of goethite with high Al-substitution.

Optical and SEM observations of a transitional pallid zone have shown that most pedogenic clay particles (kaolinite, halloysite) are randomly oriented, giving rise to a very porous internal fabric (McCrea & Gilkes, 1987).

2.1.3 Other Coatings, Infillings, Nodules

Coatings and infillings of strongly oriented yellowish clay are often deposited in planar voids and channels (Fig. 3), especially in the upper part of the mottled clay zone (e.g., Magaldi & Bidini, 1991). Inside these coatings, redistribution of iron takes place, whereby hematite hypocoatings form near the wall of the void, eventually producing a plinthitic cellular fabric (see Section 2.1.2.; Figs. 1B and 3). This process can be repeated in successive clay illuviation lamellae (Schmidt-Lorenz, 1964; Stoops, 1968; Réfega, 1972; Miedema et al., 1987).

Iron oxide depletion hypocoatings lined by hypocoatings with plinthitic hematite have been described for the mottled clay interval of Brazilian laterites (Muggler & Buurman, 2000; Rosolen et al., 2002).

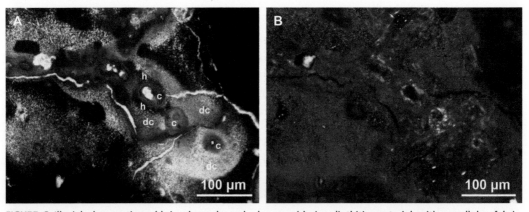

FIGURE 3 Illuvial clay coatings (c) in channels and planar voids in plinthitic material with a cellular fabric, surrounded by older degrading clay coatings (dc) (paler, with low interference colours) and hematite hypocoatings (h) formed at the expense of the degrading clay coating, in laterite nodule (Lower Congo, DR Congo). (A) In PPL. (B) In XPL.

Fragments of the underlying saprolite, impregnated with gibbsite, often occur, especially in the lower part of the mottled clay zone (e.g., Alexander & Cady, 1962; Stoops et al., 1990). Passage features and infillings with excrements or soil aggregates are also common and point to high biological activity in this zone, possibly by termites (Stoops et al., 1990; Horváth et al., 2000).

2.2   Hard Laterite

2.2.1   General Characteristics

The hard laterite in the middle of the profile can be nodular or vermicular. The nodular type is often matrix-supported at the bottom near the soft clay layer, and it evolves to purely nodule-supported in the upper part. In the lowest part of the layer, the nodules are generally large and uncoated, whereas toward the top of the layer they become smaller and have thick kaolinite-goethite coatings (see Section 2.2.3) (e.g., Muller et al., 1980/1981). The vermicular types have a reddish groundmass and are intersected by abundant channels. The latter are generally filled with white kaolinitic clay at the bottom of the layer, more yellowish clay higher in the layer and soil material with oxic characteristics near the top. The channels in the upper part of the layer often show kaolinite-goethite coatings (see Section 2.2.3.). Both types of hard laterite can also display other pedofeatures, such as illuvial clay coatings and infillings, infillings with excrements, coatings and infillings of pure goethite.

In autochthonous laterites, the nodules are typically orthic or disorthic. They have a same type of groundmass as the surrounding loose material, and a continuity exists between the coarse material inside and outside the nodules (Muller et al., 1980/1981). The nodules and the loose interstitial material are often subjected to later iron oxide impregnation, affecting mainly the interstitial material and giving rise to complex fabrics. When vertical or lateral input of iron is considerable, continuous hard layers form by cementation of the nodules.

A specific type of autochthonous nodular laterite was described by Lambiv Dzemua et al. (2011) for the top of mottled saprolite on serpentinite. It consists of packed fragments of iron oxide coatings and hypocoatings, which remain after total dissolution and leaching of the kaolinitic groundmass, producing a locally highly porous deposit. Although not often reported, this type of nodular laterite can be expected to be common on mafic rocks with low aluminium content.

The infillings in vermicular laterite and the material between the nodules in nodular laterites have similar characteristics. They often display a granular microstructure, a yellowish or reddish micromass, an undifferentiated b-fabric and signs of intense bioturbation, which are all features that are commonly recognised for oxic materials (see Marcelino et al., 2010, 2018). In other cases, the groundmass has characteristics of less strongly developed tropical soils, with abundant illuvial coatings of limpid fine clay and striated b-fabrics (Muller et al., 1980/1981; Costantini & Priori, 2007). Phytoliths are sometimes observed, indicating that the material is derived from overlying soils, which in many cases have been subsequently removed by erosion.

The evolution of soft matrix-supported mottles to hard nodule-supported zones has been attributed to purely relative accumulation of iron and/or aluminium by loss of alkali and earth alkali elements and silica (Nahon, 1986; Widdowson, 2007), which, however, has never been confirmed by thin section studies. On the contrary, inputs of dissolved or colloidal iron and aluminium and/or of mechanically transported ferruginous clay are clearly expressed by the presence of iron- and aluminium-rich clay coatings in voids formed by shrinkage of the original volume by loss of the more soluble elements. These coatings are subjected to the same processes as the groundmass (formation of hematite hypocoatings, destruction of kaolinite; see Section 2.1.2; Figs. 1B and 3) (Stoops, 1968; Réfega, 1972). In some cases, desilicification of kaolinite gives rise to gibbsite formation (Boulangé, 1983), which means that relative accumulation of aluminium takes place within materials formed as a result of absolute accumulation of iron, aluminium and silica in the form of clay.

2.2.2 Nodules

Authigenic nodules generally show a fabric similar to that of the underlying mottled clay (Stoops et al., 1990), but they have a denser concentration of plinthitic hematite, as well as goethite impregnations and coatings along voids (Fölster, 1971). They also have better expressed cellular fabrics than nodules in the mottled clay. Due to leaching of oxides and weathering of kaolinite, the intensity of the interference colours of the micromass inside the cells decreases and the micromass material is gradually removed. The result is an empty boxwork in iron-rich laterites, or a boxwork with fine-grained gibbsite inside the cells in aluminium-bearing laterites. SEM observations showing the corrosion of quartz grains (Eswaran & Raghu Mohan, 1973; Stoops et al., 1990) also point to strong leaching of silica. The groundmass of the nodules shows either a saprolitic fabric or a soil fabric, depending on the degree of pedoturbation before iron or aluminium enrichment took place.

2.2.3 Kaolinite-Goethite Coatings

The walls of channels and other voids, as well as the sides of nodules (mainly in the upper part of the hard laterite), are generally covered by kaolinite-goethite coatings (Figs. 4 to 6). These coatings are a few hundred micrometres to several millimetres thick (up to 1 cm), laminated, orange and brownish yellow in PPL and yellow in OIL (Table 1). In reflected light with polarisers at an angle of 60°, they are greyish blue, which has been interpreted as an indication for the presence of hydrated goethite (Alexandre, 2002). The thickness of individual laminae is not constant, ranging from 50 to 100 μm, and they do not always follow the irregularities of the covered surface (Figs. 4 to 6). The coatings do not show extinction lines, and in most cases they do not go into extinction at all but display a speckled extinction pattern (Fig. 4B). Occasionally, they contain thin layers (few μm) of pure goethite with palisade fabric. These coatings are very common in the upper part of the hard laterite, but they are also abundant on most pediment gravels (see Section 2.3), and they import great rigidity to the laterites (Alexander & Cady, 1962).

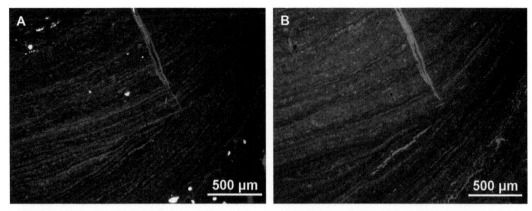

FIGURE 4 Detail of a thick kaolinite-goethite coating intersected by a fissure with a layered coating of pure goethite, in laterite (upper terrace of the Niger River, Burkina Faso). (A) The layers composing the coating are irregular, with diffuse boundaries, and not entirely parallel (PPL). (B) The same field in XPL, illustrating the speckled extinction pattern of the coating.

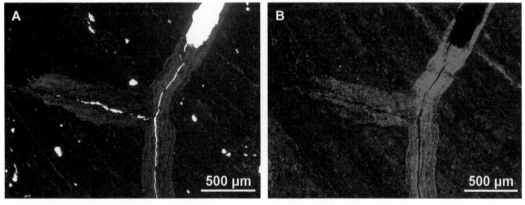

FIGURE 5 Layered coating of pure goethite along planar voids crossing a thick kaolinite-goethite coating, in laterite (upper terrace of the Niger river, Burkina Faso). (A) In PPL. (B) The same field in XPL, showing a parallel orientation of the elongated goethite crystals (perpendicular to the void walls).

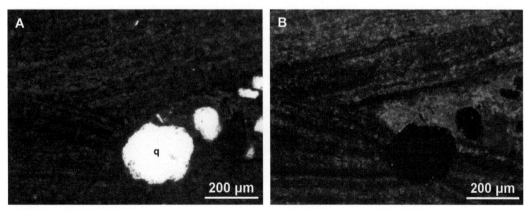

FIGURE 6 Detail of a thick kaolinite-goethite coating, in laterite (upper terrace of the Niger River, Burkina Faso), showing diffuse layering, and layers terminating against an enclosed quartz grain (q). (A) In PPL. (B) In XPL.

Table 1 Optical properties of various fabric units in laterites

Feature	Mineralogical composition	PPL	XPL	OIL
Dark red nodules	Hematite	Opaque	Intense red	Grey with bluish lustre
Light reddish nodules	Hematite, kaolinite	Dark red	Light intense red	Brownish, slightly purple
Yellow nodules	Goethite, kaolinite	Brown to dark brown	Orange brown to dark brown	Yellow; dark greyish yellow, sometimes with dark streaks
Hard lamellar light to dark brown coatings	Goethite, kaolinite	Light and dark orange-brown layers	Orange and dark brown layers, without complete extinction; speckled extinction	Dark and greyish yellow layers; weak reflection
Shiny black iron oxide coatings	Goethite	Orange	Bright orange to light green	Bluish grey to black
Clay coatings	Kaolinite, some goethite	Orange to light brown	Orange with extinction lines	Alternating greyish yellow to orange brown layers
Gibbsite nodules, infillings and coatings	Gibbsite	Colourless	First-order interference colours	Colourless or white

Tangential cross-sections through these thick coatings can produce concentric patterns similar to concentric nodules (see Stoops, 2003).

The origin of the kaolinite-goethite coatings is not yet well understood. Their internal and external fabric and their optical characteristics do not indicate an illuvial nature. The parallelism of the bands in the coatings is often very imperfect and irregular and, in addition, coarse quartz grains frequently interrupt the continuity of the bands (Fig. 5), which is incompatible with an illuvial origin. According to Horváth et al. (2000), this type of coatings would indicate vadose conditions and precipitation from solution. However, their formation has been also attributed to termite activity (de Barros Machado, 1983, 1987), based on structural similarities with termite galleries and on the common occurrence of organic particles that are also found in termite constructions (plant splitters, chitin fragments). Based on observations for samples from several regions, the coatings appear to be built by regular packing, layer by layer, of termite pellets, which are either fecal or buccal, later enriched in iron oxides by circulating groundwater (de Barros Machado, 1983, 1987). The piling of pellets could explain the speckled b-fabric. The influence of termite activity on the formation of laterites was earlier put forward by Yakushev (1968), based on similarities in fabric.

Similar coatings in outcrops of hard laterites, covering its surface and occurring along the sides of vertical cracks, have a layered fabric with local cross-bedding, attributed to iron oxide precipitation from surface water, partly controlled by microbial activity under temporarily arid conditions (Alexandre & Tshidibi, 1985). The color of these coatings, which can be covered by a siliceous deposit similar to desert varnish, has been related to the age of the surface (Alexandre, 2002).

According to the French school (see Nahon et al., 1977; Nahon, 1991), the formation of clay-goethite coatings around hematite-kaolinite nodules is the result of internal reorganisation: centripetal destruction of the kaolinite-hematite mass of the nodule, by desilicification and by obliteration of microporosity, would lead to loosening of thin shells of aluminium-rich hematite from the nodule surface and their transformation to aluminium-rich goethite layers; the size of the hematitic core is gradually reduced at the expense of the growing cortex, finally producing a concentric nodule. However, this theory does not explain the commonly observed difference in coarse material content between the nodules and the surrounding coatings. According to McFarlane (1983), this is the result of the growth of coatings around the nodule, pushing quartz grains toward the internodular space. Delvigne (1998) describes the coating as an accretion cortex formed by recrystallisation of iron oxides and disappearance of secondary voids. The formation of concentric nodules (pisolitisation) has been also associated with the degradation of laterites at the expense of the underlying continuous laterite layer with a lithic fabric (Boulangé, 1984).

2.2.4   Goethite Coatings and Infillings
Planar voids in kaolinite-goethite coatings and in nodules are often filled with pure goethite. Macroscopically, these features are recognised as very dark brown to black shiny surfaces. In thin sections, pure goethite coatings are layered, with platy crystals oriented perpendicular to the surface (Fig. 5). SEM observations have also documented this fabric, and they show that the individual crystals are fine lenticular and often split (Stoops, 1970; Eswaran & Raghu Mohan, 1973; Suddhiprakarn & Kheoruenromne, 1994) (Fig. 7). Acicular goethite has also been reported as the main constituent of layered goethite coatings (Eswaran et al., 1981; Zauyah & Bisdom, 1983; Stubendorff, 1986). Goethite in these coatings and infillings seems to have a relatively high degree of aluminium substitution (Stoops et al., 1990; Tardy, 1993), corroborated by analyses for similar coatings with a palisadic internal fabric (Zeese et al., 1994). Hematite and goethite coatings covering hard laterite surfaces are mentioned by Alexandre (2002), but it is not clear to what extent they are associated with the kaolinite-goethite coatings described above.

Goethite coatings and infillings have been considered to be a diagnostic feature of hardened laterite (Rutherford, 1972).

2.2.5   Manganese Features
Micromorphological descriptions of manganese features in laterites are rare; but a number of petrographic studies have been published (see Varentsov, 2013). This scarcity of data may be due in part to the opaque nature of manganese oxides, requiring reflected light microscopy combined with SEM-EDS to be differentiated. Manganese features are common on parent materials containing manganese minerals, ranging from carbonates (rhodochrosite) to silicates (e.g., braunite, spessartine garnet, ottrelite). Manganese laterites are economically important secondary ores, both as source of manganese and, particularly on serpentinite deposits, as source of metals such as cobalt and nickel.

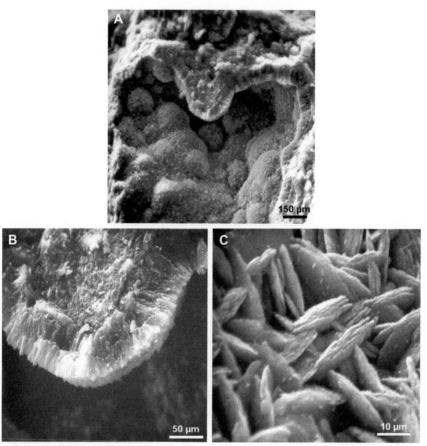

FIGURE 7 SEM images of a goethite coating around a vugh, in a nodule from a hard laterite (Lower Congo, DR Congo). (A) Overview. (B) Detail showing that the coating consists of two layers of crystals oriented perpendicular to the surface. (C) Detail of the surface of the coating, showing split lenticular goethite crystals.

In saprolites on metasedimentary manganese deposits rich in rhodochrosite ($Mn(CO_3)$) and spessartine ($Mn^{2+}_3Al_2(SiO_4)_3$), the first weathering product is manganite ($Mn^{3+}O(OH)$), forming well-crystallised infillings in fissures and, in some cases, producing pseudomorphs after primary minerals (Nahon et al., 1983). Near the edge of the voids clear needle-shaped crystals are visible with SEM. For similar deposits, Grandin and Perseil (1977, 1983) describe the formation of lithiophorite ((Al,Li)(Mn^{4+},Mn^{3+})$_2O_2(OH)_2$) pseudomorphs after spessartine in the lower part of the profile. In the upper part of the same profile, a manganese-rich zone is formed, generally preserving a lithic fabric, containing predominantly lithiophorite, besides other manganese oxides such as cryptomelane ($K(Mn^{4+}_7Mn^{3+})O_{16}$), nsutite ($Mn^{2+}_xMn^{4+}_{1-x}O_{2-2x}(OH)_{2x}$) and pyrolusite ($MnO_2$), depending on the original mineralogical composition of the deposits. For another region, a sequence of cryptomelane over nsutite and ramsdallite

(MnO_2) to pyrolusite, during pseudomorphic replacement of primary Mn minerals and kaolinite, was described by Beauvais et al. (1987). Rod- or needle-shaped crystals of cryptomelane are recognised with SEM. Gradual replacement of kaolinite by lithiophorite, with progressive development of lithiophorite pseudomorphs after kaolinite booklets, was also described by Nahon et al. (1989).

For the top of the hard laterite on ultrabasic rocks in Cameroon, Yongue-Fouateu et al. (2006) describe the presence of zones with lithic fabrics, dominated by cryptomelane and lithiophorite evolving to minerals of the asbolane group ($Mn^{4+}(O,OH)_2 \cdot (Co,Ni,Mg,Ca)_x(OH)_{2x} \cdot nH_2O$). Manganese coatings composed of globular aggregates and acicular crystals have been reported for the higher parts of *in situ* hard laterites in Bangalore (Eswaran & Raghu Mohan, 1973). Reflected light microscopy and microprobe studies by Lambiv Dzemua et al. (2013) document *in situ* formation of Co-Ni-rich lithiophorite in the lower part of a laterite on serpentinite in Cameroon. The lithiophorite encloses magnetite grains in the basal part of the profile, just above the saprolite, and it occurs as coatings and hypocoatings of voids with coarse-crystalline gibbsite infillings in the overlying layer, where it records a more recent stage of absolute aluminium accumulation.

2.2.6   Other Features

Desilicification resulting in separation of fragments of fractured quartz grains, with dissolution progressing along the fractures, has been commonly reported to occur from the saprolite onward (Chauvel et al., 1983; Nahon, 1986; Widdowson, 2007; Bremer, 2010). Fragmentation of quartz grains is an essential process in lateritic weathering, as emphasised by Brimhall et al. (1992) based on the results of microscopic and physicochemical studies. Earlier studies by Claisse (1972) concluded that quartz grains extracted from alterites were more soluble than other quartz grains when treated with strong acids.

Another common feature in hard laterite is the presence of quartz grains with fissures filled with gibbsite (Fig. 8A to C) or hematite (Fig. 8D), described as 'runiquartz' (Eswaran et al., 1975) (see also Kaloga, 1976; Achyuthan, 2004; Marcelino et al., 2010, 2018). The presence of runiquartz grains, which are very resistant, usually points to a lateritic origin of the coarse fraction of the soil in which they occur (Stoops, 1989).

In thin sections, fragments of quartz grains often show optical continuity (Fig. 8C), expressed by simultaneous extinction, which points to *in situ* alteration (Chauvel et al., 1983). Quartz dissolution may also be revealed by the presence of circumgranular voids (Muller et al., 1980/1981), resulting in so-called floating quartz grains (Fig. 9). These quartz grains show deeply etched surfaces in SEM images. Gibbsite, generally with palisadic fabric, sometimes crystallises in the created contact voids, which may eventually produce complete infillings of mouldic voids after total dissolution of the quartz grain (Delvigne & Boulangé, 1974; Schmidt-Lorenz, 1974a, 1974b). SEM observations of quartz grains have also shown that grains surrounded by hematite exhibit a powdery surface, interpreted as amorphous or microcrystalline silica; quartz grains in a goethitic

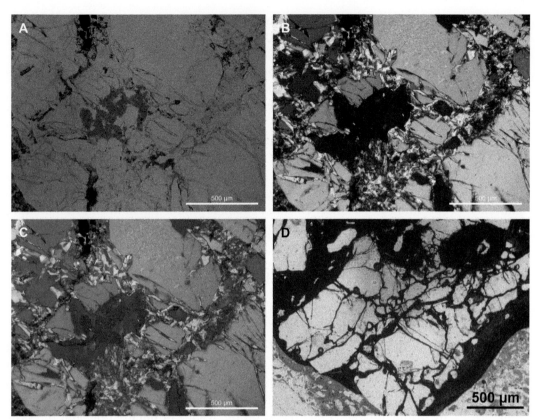

FIGURE 8 (A) Strongly corroded quartz grains with crystallisation of coarse gibbsite in cracks, in a laterite profile (DR Congo) (PPL). (B) The same field in XPL. (C) The same field in XPL with 1 λ retardation plate, illustrating the optical continuity of the quartz fragments. (D) Runiquartz grain with fissures partly filled with iron oxides in a laterite profile (Burkina Faso) (PPL).

environment display polyhedral etch figures, without iron oxide infillings, suggesting that iron oxide precipitation took place before quartz dissolution (Tshidibi, 1985).

Other features that have occasionally been reported, including pyrite occurrence in buried laterites and the presence of non-biogenic leucophosphite $(KFe^{3+}_2(PO_4)_2(OH) \cdot 2H_2O)$ in hard laterite crusts (Tiessen et al., 1996), presumably formed after laterite development.

2.3 Gravel-Rich Layer

The gravel-rich layer in the upper part of laterite profiles is often allochthonous and then consists mainly of transported iron-rich nodules, occasionally mixed with other coarse components such as lithic fragments (e.g., Zinn & Bigham, 2016). Depending on the length of transport and the geological uniformity of the region, this material can be more or less heterogeneous, showing different degrees of development based on fabric and composition. Even when it is derived from a same parent material, fragments exhibiting

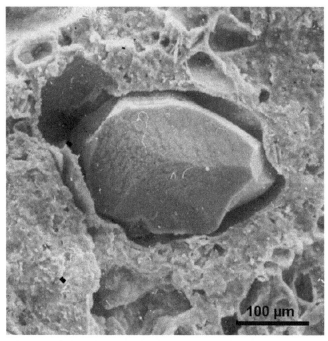

FIGURE 9 SEM image of a corroded sand-sized quartz grain, with an associated circumgranular void, in a hard laterite (Uganda).

different stages of evolution occur, as upslope profiles are truncated up to different depths. In many cases, fragments have undergone several cycles of cementation and transport. In the literature, numerous terms have been used to designate them, without any generally accepted definition, but they can all be described in terms of typic, concentric and nucleic nodules (Stoops, 2003).

Nodule-supported layers often have a conglomeratic aspect, with nodules as sharply delimited fabric units of different degrees of complexity and of various origins, as can be deduced from their internal fabric and composition (Fig. 10). The nodules are generally cemented, producing a hard continuous layer. This evolution corresponds to an absolute accumulation of iron, a decrease of the Fe_{ox}/Fe_{DCB} ratio and an increase in goethite content (Eswaran et al., 1981).

The most commonly observed nodule types are:

- *Typic nodules with saprolitic fabric*: saprolite fragments impregnated by iron oxides, whereby the degree of impregnation is related to water table position. The outlines and arrangement of minerals remain visible and weatherable minerals can be present. Boxwork pseudomorphs after ferromagnesian minerals can occur, pointing to residual iron accumulation.
- *Typic nodules with soil fabric*: soil material impregnated by iron oxides, often containing pedofeatures such as clay coatings. Their presence is an indication of truncation of iron oxide-impregnated soils in the surroundings.

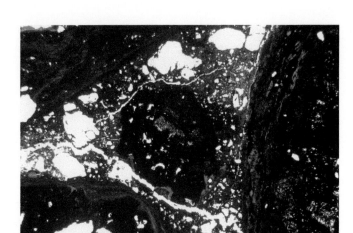

FIGURE 10 Rounded laterite fragments of different compositions and origins, cemented by goethite-impregnated material in a hard polygenetic laterite (upper terrace of the Niger River, Burkina Faso).

- *Typic nodules with plinthic cellular fabric* (see Section 2.1.2): rounded fragments of hardened plinthite, in different stages of development.
- *Concentric nodules*: nodules with concentric patterns, either aluminium- or iron-rich, of different genetic types. They often consist of small nucleus covered by several kaolinite-goethite coatings, which correspond to the pisolithic facies of Alexandre (2002). Concentric nodules containing charcoal fragments and iron oxide coatings have been also reported (Achyuthan & Fedoroff, 2008). Tangential cross-sections through kaolinite-goethite coatings can be mistaken for concentric nodules (see Section 2.2.3).
- *Complex nodules*: nodules consisting of fragments of older nodules, completely or partly surrounded by kaolinite-goethite coatings. They can include fragments of older kaolinite-goethite coatings and clearly have a polygenetic origin.
- *Manganiferous nodules*: manganese nodules have been reported for lateritic profiles in Western Africa (Beauvais & Nahon, 1985). Large nodules often have a core displaying a lithic fabric (including pseudomorphs after primary minerals), consisting of a mixture of cryptocrystalline todorkite ($(Na,Ca,K,Ba,Sr)_{1-x}$ $(Mn,Mg,Al)_6O_{12}\cdot3\text{-}4H_2O$), birnessite ($(Na,Ca,K)_{0.6}(Mn^{4+},Mn^{3+})_2O_4\cdot1.5H_2O$), lithiophorite and goethite. This core is surrounded by a compound coating with concentric layers dominated, from centre to nodule surface, by cryptocrystalline nsutite, cryptomelane and well-crystallised lithiophorite. In smaller nodules, the core is very small or absent and the dominant feature is the compound coating, similar in composition to the one described above, but in some cases with several

repetitions of the complete sequence. The Mn content decreases toward the surface of the nodule, whereas the Al content increases. The coatings are interpreted by the authors as having formed by centripetal growth. Similar coatings, but containing plant fragments as inclusion, were observed by Grandin and Perseil (1977) for samples from Ivory Coast. Compound nodules with an iron- and manganese-rich core, with birnessite and vernadite ((Mn,Fe,Ca,Na)(O,OH)$_2 \cdot n$H$_2$O) as the main manganese compounds, and surrounded by thick iron oxide coatings, were described in India by Achyuthan (2004). Their formation requires external input of manganese, which involves processes other than those resulting in laterite development by residual accumulation.

- *Sandy nodules*: nodules formed in sandy materials, often in alluvial deposits. They form as a result of lateral transport of iron, leading to the development of iron oxide nodules (Bocquier, 1973; Bourman et al., 1987) or continuous cemented layers (e.g., Denisoff, 1957; Alexandre, 2002). The nodules or layers first have a chitonic c/f-related distribution pattern, which evolves to close porphyric, with all packing voids filled by goethite. A similar fabric is often observed for Cenozoic ferruginous sandstones in Europe. Although this material as such is not lateritic, these goethite-impregnated sandy nodules frequently occur in allochthonous gravel-rich laterite layers. Leaching of the gravel-rich laterite layer, removing iron oxides and clay, can produce a material in which the nodules are transformed to entirely deferruginised earthy aggregates (Bellier et al., 1987).

In Oxisols of Malaysia, assumed tubular goethite occurs both within lateritic nodules and along their sides (Zauyah & Bisdom, 1981; Zauyah, 1983). SEM images show hollow tubes with a cross-section of 2-3 µm, but no mineralogical or chemical data are given. Similar tubular features were observed by Comerma et al. (1977) for petroplinthite from Venezuela.

In Mediterranean regions, plinthite or palaeoplinthite may be superposed on unrelated soil materials, such as Bt horizons. In thin sections, the plinthite-related features include iron oxide hypocoatings and impregnative nodules (Rellini et al., 2007).

Gravelly petroplinthite composed of nodules within a grey groundmass with speckled b-fabric were reported for the Ligurian coast in Italy (Rellini et al., 2015). The nodules are of various compositional types and have sharp boundaries, pointing to an allochthonous origin.

3. Bauxites and Bauxitic Laterites

As mentioned in the Introduction, deeply buried bauxites and karst bauxites are not considered in this review. In aluminium-rich laterites and bauxites, several types of gibbsite enrichment can be observed in thin sections.

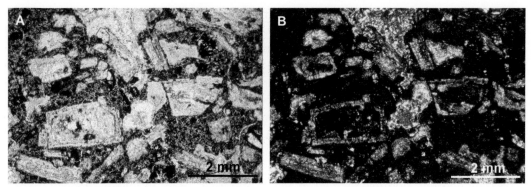

FIGURE 11 Gibbsite pseudomorphs after plagioclase crystals in strongly weathered andesitic material (Ujung Kulon Peninsula, Indonesia). (A) In PPL. (B) In XPL.

In the lower part of bauxitic laterite profiles, at the transition to the fresh rock, the fabric of the parent rock can be preserved (e.g., de Oliveira et al. 2013). Occasionally this phenomenon is observed throughout the profile (Boulangé et al., 1975). In the case of coarse-grained rocks, the outlines of mineral grains can be recognised (Fig. 11) and a boxwork pattern develops with empty cells, or with cells filled with cryptocrystalline gibbsite resulting from residual aluminium accumulation due to leaching of silica (Bocquier et al., 1983; Boulangé, 1983; Melfi & Carvalho, 1983; Groke et al., 1983). Also mouldic voids, developed as a result of leaching of more soluble elements, may be filled with coarse-crystalline gibbsite in the overlying layers, resulting in absolute aluminium accumulation (de Oliveira et al. 2013). This sometimes evolves to a continuous gibbsite groundmass around saprolitic nodules impregnated by iron oxides, explained by external input of aluminium (Fig. 12).

Relatively coarse-grained gibbsite (20-50 μm) (Fig. 12B), oriented normal to the sides of the original voids, has been described as the main constituent of the cement between

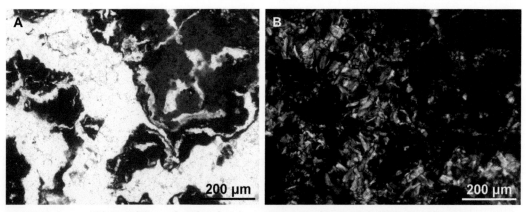

FIGURE 12 Coarse gibbsite infillings of voids in laterite on serpentine (Cameroon). (A) In PPL. (B) In XPL.

iron-rich aggregates in the lower or middle part of bauxite profiles (Lucas et al., 1989; Dedecker & Stoops, 1993). In this setting, it forms septa between lateritic or saprolitic nodules. The occurrence of coarse-grained gibbsite in depth has also been reported by other authors (e.g., Kotschoubey et al., 1989). The formation of gibbsite coatings has been attributed mainly to desilicification of illuvial kaolinite coatings (Boulangé et al., 1975; Boulangé, 1983, 1985).

Microcrystalline gibbsite is generally rather pure, but it sometimes contains inclusions of clayey material (Fig. 13). It has been observed at the transition between a cemented bauxite layer and its clay cover (Lucas et al., 1989; Dedecker & Stoops, 1993). It forms a continuous layer at the bottom, and a discontinuous nodular layer in the lower part of the cover, and it gradually disappears toward the upper part. The formation of microcrystalline gibbsite probably results from the removal of silica and iron from the originally iron-bearing kaolinitic groundmass (Schmidt-Lorenz, 1964). SEM studies by Bárdossy et al. (1978) have shown that gibbsite in strongly leached lateritic bauxites of highland areas is relatively coarse-grained (0.5-50 µm) and euhedral, whereas in lowland lateritic bauxites the gibbsite is finer (0.1-10 µm) and subhedral or anhedral, which is still coarser than the very fine-grained xenotopic gibbsite (0.05-0.5 µm) in karst bauxites. As a result of the destruction of the clay, the groundmass is transformed to microcrystalline gibbsitic material, iron is concentrated as hematite hypocoatings along planar voids (see Section 2.1.2) (Banerji, 1989; Varajão et al., 1989) and coarse-grained gibbsite develops in the available pore space as coatings and infillings of vughs, shrinkage cracks and circumgranular voids around dissolving quartz grains (Schmidt-Lorenz, 1974b).

Cryptocrystalline gibbsite is often observed in bauxitic material, especially in karst bauxites. It has the appearance of colloidal material, and it is the main component of nucleic or concentric bauxitic nodules (Fig. 14). The nucleus is a clastic mineral particle. The number of layers within the coating is typically between 2 and 10, but it may exceed 20, with a thickness of 1 to 100 µm and gradual to sharp boundaries (Bárdossy, 1982). Nodules with a diameter between 100 µm and 1 mm are the most common. They are

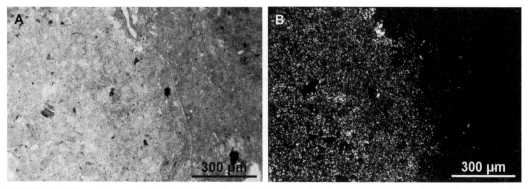

FIGURE 13 Diffuse microcrystalline gibbsite nodule (left) in a kaolinitic groundmass (right), in a bauxite profile (Trombetas, Brazil). (A) In PPL. (B) In XPL.

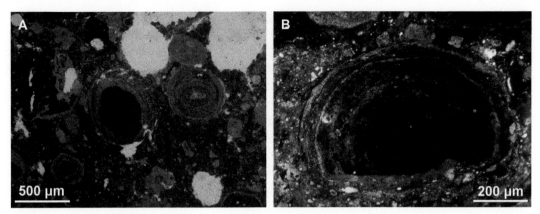

FIGURE 14 Gibbsite nodules in a bauxite deposit (Suriname). (A) Concentric nodules of microcrystalline gibbsite (PPL). (B) Fragment of a concentric nodule of microcrystalline gibbsite, covered by a continuous gibbsite layer, producing a new concentric nodule (PPL).

characterised by layers with sharp boundaries, the inner layers closely following the outlines of the nucleus but changing to a more rounded shape away from the centre. The layers are visible due to small changes in iron oxide content. Eccentrically concentric nodules ('spastoids'; Bárdossy, 1982) are common in some bauxites. The outer layer can enclose smaller nodules, engulfed by the faster growing larger nodule. The nodules are interpreted as authigenic diagenetic features (Bárdossy, 1982). In older deposits, with considerable overburden, the nodules may be flattened horizontally. Sometimes new layers surround fragmented nodules (Fig. 14B).

Gibbsitic nodules are characteristic of bauxitic hard laterites. In a homogeneous gibbsite-hematite groundmass, typic nodules are first formed by iron redistribution under hydromorphic conditions (Boulangé & Bocquier, 1983). Iron moves to cracks (cf. ferritic laterites), structural elements become more rounded and an internodular mass rich in boehmite (AlO(OH)) develops. Within the internodular mass, other nodules appear, with a centre depleted in iron oxides and a strongly developed iron oxide hypocoating. They are later covered by a coating of boehmite and recrystallised gibbsite. A repetition of this process results in the formation of concentric gibbsite nodules. Subsequent shrinkage of the nodules can lead to the formation of radial fissures, giving rise to concentric and septaric complex nodules.

Other gibbsite features include gravel layers composed of rounded gibbsite nodules (Dedecker & Stoops, 1999). Various types of gibbsite pedofeatures in laterites are illustrated by Delvigne (1998).

4. Conclusions

Based on available published data, it is not possible to give an unequivocal interpretation of each observed fabric unit in a laterite, isolated from the field context. Hematitic features seem rather related to residual accumulation of iron, whereas goethitic features commonly indicate an external input of iron, in a more porous system. Infillings with gibbsite indicate absolute accumulation of aluminium by vertical or lateral transport in solution or in the form of clay suspensions.

Compared with most other regolith materials, laterites and bauxites are characterised by relative coarse fabric units. This implies that the use of large thin sections is often required. To distinguish different fabrics in parts with high iron oxide content, sections should be as thin as possible, and OIL observations can provide useful complementary information. However, when the iron oxide content is high (>15-20%), the fabric of bauxites is difficult to observe in thin sections (Bárdossy & Pantó, 1971). X-ray computed topography seems a promising tool for studies of some aspects of the fabric of lateritic materials (Zinn et al., 2015).

A review of the extensive literature on laterites and bauxites makes it clear that interpretation of the copious data is difficult because of the lack of precise and complete descriptions, and the use of imprecise and generally undefined terminologies, both for field observations and in thin section studies. Published micromorphological interpretations are often based on a single profile, or a small number of related profiles, and often not sufficiently supported by other data. A genetic grouping of the observed features in a comprehensive way is not an easy task, as it may imply the acceptance of a general model of genesis and evolution.

The presence of mottles with fine plinthitic hematite forming hypocoatings along planar voids and the disaggregation and congruent dissolution of quartz grains are diagnostic for *in situ* lateritisation by relative iron accumulation. Absolute accumulation of iron results in the formation of goethite and goethite-kaolinite coatings around hematite/goethite nodules.

Micromorphological observations are an indispensable tool to understand the spatial and chronological relations between the various constituents of these often polygenetic materials. The information they provide pertains not only to the processes that took place but, more importantly, also to the hierarchy of those processes. Micromorphological studies are therefore in many cases necessary to be able to interpret the results of geochemical analysis.

References

Achyuthan, H., 2004. Paleopedology of ferricrete horizons around Chennai, Tamil Nadu, India. Revista Mexicana de Ciencias Geológicas 21, 133-143.

Achyuthan, H. & Fedoroff, N., 2008. Ferricretes in Tamil Nadu, Chennai, South-Eastern India: from landscape to micromorphology, genesis, and paleoenvironmental significance. In Kapur, S., Mermut, A. & Stoops, G. (eds.), New Trends in Soil Micromorphology. Springer-Verlag, Berlin, pp. 111-136.

Aleva, G.J.J., 1986. Classification of laterites and their textures. Geological Survey of India, Memoir 120, 8-28.

Aleva, G.J.J., 1987. Voids in laterites: suggestions for their description and classification. Travaux ICSOBA 16/17, 141-153.

Aleva, G.J.J. (ed.), 1994. Laterites. Concepts, Geology, Morphology and Chemistry. ISRIC, Wageningen, 154 p.

Alexander, L.T. & Cady, J.G., 1962. Genesis and Hardening of Laterite in Soils. SCS-USDA Technical Bulletin 1282, 90 p.

Alexandre, J., 2002. Les Cuirasses Latéritiques et Autres Formations Ferrugineuses Tropicales. Exemples du Haut Katanga Méridional. Annales du Musée Royal de l'Afrique Centrale, Sciences Géologiques 107, 118 p.

Alexandre, J. & Tshidibi, N.Y.B., 1985. Les enduits ferrugineux associés aux cuirasses latéritiques successives du Haut-Shaba. Nature, structure et mode de formation. In Alexandre, J. & Symoens, J.J. (eds.), Les Processus de Latéritisation. Académie Royale des Sciences d'Outre-Mer, Brussels, pp. 37-46.

Ambrosi, J.P., Nahon, D. & Herbillon, A.J., 1986. The epigenetic replacement of kaolinite by hematite in laterite. Petrographic evidence and the mechanisms involved. Geoderma 37, 283-294.

Banerji, P.K., 1989. East coast bauxite deposits of India revisited: some problems of bauxite and laterite genesis. Travaux ICSOBA 19, 331-340.

Bárdossy, G., 1982. Karts Bauxites. Bauxite Deposits on Carbonate Rocks. Akadémiai Kiadó, Budapest, 441 p.

Bárdossy, G. & Aleva, G.J.J., 1990. Lateritic Bauxites. Akadémiai Kiadó, Budapest, 624 p.

Bárdossy, G. & Nicolas, J., 1973. Proposition pour un terminologie des bauxites. Travaux ICSOBA 9, 99-104.

Bárdossy, G. & Pantó, G.Y., 1971. Investigation of Bauxites with the help of electron-probe. Tschermaks Mineralogische und. Petrografische Mitteilungen 15, 165-184.

Bárdossy, G., Csanády, A. & Csordás, A., 1978. Scanning electron microscope study of bauxites of different ages and origins. Clays and Clay Minerals 26, 245-262.

Beauvais, A. & Nahon, D., 1985. Nodules et pisolites de dégradation des profils d'altération manganésifères sous conditions latéritiques. Exemples de Côte d'Ivoire et du Gabon. Sciences Géologiques, Bulletin 38, 359-381.

Beauvais, A., Melfi, A., Nahon, D. & Trescases, J.J., 1987. Pétrologie du gisement latéritique manganésifère d'Azul (Brésil). Mineralium Deposita 22, 124-134.

Bellier, G., Humbel, F.X. & Lamouroux, M., 1987. Evolution actuelle d'une carapace ferrugineuse en forêt de Citeaux (Val-de-Saône). Cahiers ORSTOM, Série Pédologie 23, 27-42.

Bocquier, G., 1973. Génèse et Evolution de Deux Toposéquences de Sols Tropicaux du Tchad. Interprétation Biogéodynamique. Mémoires de l'ORSTOM n° 62. ORSTOM, Paris, 350 p.

Bocquier, G., Boulangé, B., Ildefonse, P., Nahon, D. & Muller, D., 1983. Transfers, accumulation modes, mineralogical transformations and complexity of historical development in lateritic profiles. In Melphi, A.J. & Carvalho, A. (eds.), Lateritisation Processes. Instituto Astronômico e Geofisico, University of São Paulo, pp. 331-337.

Boulangé, B., 1983. Aluminium concentration in a bauxite derived of granite (Ivory Coast): relative and absolute accumulations. Travaux ICSOBA 13, pp.109-116.

Boulangé, B., 1984. Les Formations Bauxitiques Latéritiques de Côte d'Ivoire: les Facies, leur Transformation, leur Distribution et l'Evolution du Modèle. Travaux et Documents de l'ORSTOM n° 175, ORSTOM, Paris, 341 p.

Boulangé, B., 1985. Les mécanismes de concentration de l'aluminium dans une bauxite sur granite: les accumulations relatives et absolues. In Alexandre, J. & Symoens, J.J. (eds.), Les Processus de Latéritisation. Académie Royale des Sciences d'Outre-Mer, Brussels, pp. 27-35.

Boulangé, B. & Bocquier, G., 1983. Le rôle du fer dans la formation des pisolithes alumineux au sein des cuirasses bauxitiques latéritiques. Sciences Géologiques, Mémoires 72, 29-44.

Boulangé, B., Paquet, H. & Bocquier, G., 1975. Le rôle de l'argile dans la migration et l'accumulation de l'alumine de certaines bauxites tropicales. Comptes Rendus de l'Académie des Sciences, Série D, 280, 2183-2186.

Bourman, R.P. & Ollier, C.D., 2002. A critique of the Schellmann definition and classification of 'laterite'. Catena 47, 117-131.

Bourman, R.P., Milnes, A.R. & Oades, J.M., 1987. Investigations of ferricretes and related ferruginous materials in parts of southern and eastern Australia. Zeitschrift für Geomorphologie 64, 1-24.

Bremer, H., 2010. Geoecology in the Tropics. Zeitschrift für Geomorphologie 54, Supplement Issue 1, 337 p.

Brimhall, G.H., Chadwick, O.A., Lewis, C.J., Compston, W., Williams, I.S., Danti, K.J., Dietrich, W.E., Power, M.E., Hendricks, D. & Bratt, J., 1992. Deformational mass transport and invasive processes in soil evolution. Science 255, 695-702.

Buchanan, F., 1807. A Journey from Madras through the Countries of Mysore, Canara and Malabar. Volume 2. East India Company, London, 556 p.

Charman, J.H., 1988. Laterite in Road Pavements. Construction Industry and Information Association, London, 71 p.

Chauvel, A., Boulet, R., Join, P. & Bocquier, G., 1983. Aluminum and iron oxy-hydroxide segregation in nodules of latosols developed on Tertiary sediments (Barreiras Group) near Manaus (Amazon Basin), Brazil. In Melphi, A.J. & Carvalho, A. (eds.), Lateritisation Processes. Instituto Astronômico e Geofisico, University of São Paulo, pp. 507-526.

Claisse, G. 1972. Etude sur la solubilisation du quartz en voie d'altération. Cahiers ORSTOM, Série Pédologie 10, 97-122.

Comerma, J.A., Eswaran, H. & Schwertmann, U., 1977. A study of plinthite and ironstone from Venezuela. Proceedings of the International Conference on Classification and Management of Tropical Soils. Kuala Lumpur, pp. 18-26.

Costantini, E.A.C. & Priori, S., 2007. Pedogenesis of plinthite during early Pliocene in the Mediterranean environment: case study of a buried paleosol at Podere Renieri, central Italy. Catena 71, 425-443.

de Barros Machado, A., 1983. The contribution of termites to the formation of laterites. In Melphi, A.J. & Carvalho, A. (eds.), Lateritisation Processes. Instituto Astronômico e Geofisico, University of São Paulo, pp. 261-270.

de Barros Machado, A., 1987. On the origin and age of the Steep Rock buckshot, Ontario, Canada. Chemical Geology 60, 337-349.

De Craene, A. & Laruelle, J., 1955. Genèse et altération des latosols équatoriaux et tropicaux humides. Bulletin Agricole du Congo Belge 46, 1113-1243.

Dedecker, D. & Stoops, G., 1993. Micromorphological and mineralogical study of a polygenetic laterite profile (Trombetas area, Brazil). Pedologie 43, 335-356.

Dedecker, D. & Stoops G., 1999. A morpho-synthetic system for the higher level description of micro-fabrics of bauxitic, kaolinitic soils. A first approximation. Catena 35, 317-325.

Delvigne, J.E., 1998. Atlas of Micromorphology of Mineral Alteration and Weathering. Canadian Mineralogist Special Publication 3, 495 p.

Delvigne, J. & Boulangé, B., 1974. Micromorphologie des hydroxydes d'aluminium dans les niveaux d'altération et dans les bauxites. In Rutherford, G.K. (ed.), Soil Microscopy. The Limestone Press, Kingston, pp. 665-681.

Denisoff, I., 1957. Un type particulier de concrétionnement en cuvette centrale congolaise. Pedologie 7, 119-123.

de Oliveira, F.S., Varajão, A.F.D.C., Varajão, C.A.C., Boulangé, B. & Soares, C.C.V., 2013. Mineralogical, micromorphological and geochemical evolution of the facies from the Bauxite deposit of Barro Alto, Central Brazil. Catena 105, 29-39.

Didier, P., Nahon, D., Fritz, B. & Tardy, Y., 1983. Activity of water as a geochemical controlling factor in ferricretes. A thermodynamic model in the system kaolinite Fe-oxihydroxide Fe-Al. Sciences Géologiques, Bulletin 71, 25-34.

Duchaufour, Ph., 1960. Précis de Pédologie. Masson & Cie, Paris, 437 p.

Eswaran, H. & Raghu Mohan, N.G., 1973. The microfabric of petroplinthite. Soil Science Society of America Proceedings 37, 79-82.

Eswaran, H., Sys, C. & Sousa, E.C., 1975. Plasma infusion. A pedological process of significance in the humid tropics. Anales de Edafología y Agrobiología 34, 665-674.

Eswaran, H., Comerma, J. & Sooryanaraynan, V., 1981. Scanning electron microscopic observations on the nature and distribution of iron minerals in plinthite and petroplinthite. In Lateritisation Processes. Proceedings of the International Seminar on Lateritisation Processes (Trivandrum, India). Balkema, Rotterdam, pp. 335-341.

Fedoroff, N., Courty, M.A., & Zhengtang Guo, 2010. Palaeosoils and relict soils. In Stoops, G., Marcelino, V. & Mees, F. (eds.), Interpretation of Micromorphological Features of Soils and Regoliths. Elsevier, Amsterdam, pp. 623-662.

Fedoroff, N., Courty, M.A., & Zhengtang Guo, 2018. Palaeosols and relict soils, a conceptual approach. In Stoops, G., Marcelino, V. & Mees, F. (eds.), Interpretation of Micromorphological Features of Soils and Regoliths. Second Edition. Elsevier, Amsterdam, pp. 821-862.

Fölster, H., 1971. Ferrallitische Böden aus Sauren Metamorphen Gesteine in den Feuchten und Wechselfeuchten Tropen Afrikas. Göttinger Bodenkundliche Berichte 20, 231 p.

Fölster, H., Kalk, E. & Moshrefi, N., 1971. Complex pedogenesis of ferrallitic savanna soils in south Sudan. Geoderma 6, 135-149.

Fritsch, E., Montes-Lauar, C.R., Boulet, R., Melfi, A.J., Balan, E. & Magat, P., 2002. Lateritic and redomorphic features in faulted landscape near Manaus. European Journal of Soil Science 53, 203-217.

Grandin, G. & Perseil, E.A., 1977. Le gisement de manganèse de Mokta (Côte d'Ivoire). Transformations minéralogiques des minerais par action météorique. Bulletin de la Société Géologique de France 19, 309-317.

Grandin, G. & Perseil, E.A., 1983. Les minéralisations manganésifères volcano-sédimentaires du Blafo-Guéto (Côte d'Ivoire) — paragenèses — altération climatique. Mineralium Deposita 18, 99-111.

Groke, M.C.T., Melfi, A.J. & Carvalho, A. 1983. Bauxitic alteration on basic and alkaline rocks in the state of São Paulo, Brazil, under tropic humid climate. In Melfi, A.J. & Carvalho, A. (eds.), Lateritisation Processes. Instituto Astronômico e Geofisico, University of São Paulo, pp. 236-250.

Hamilton, R., 1964. A short note on droplet-formation in ironcrusts. In Jongerius, A. (ed.), Soil Micromorphology. Elsevier, Amsterdam, pp. 277-278.

Harrison, J.B., 1933. The Katamorphism of Igneous Rocks under Humid Tropical Conditions. Imperial Bureau of Soil Science, Rothamsted Experimental Station, Harpenden, 79 p.

Horváth, Z., Varga, B. & Mindszenty, A., 2000. Micromorphological and chemical complexities of a lateritic profile from basalt (Jos Plateau, Central Nigeria). Chemical Geology 170, 81-93.

Humbert, R.P., 1948. The genesis of laterite. Soil Science 65, 281-290.

IUSS Working Group WRB, 2015. World Reference Base for Soil Resources 2014, Update 2015. World Soil Resources Reports No. 106. FAO, Rome, 192 p.

Kaloga, B., 1976. Contribution à l'étude du cuirassement: relations entre les gravillons ferrugineux et leurs matériaux d'emballage. Cahiers ORSTOM, Série Pédologie 14, 299-319.

Kotschoubey, B., Truckenbrodt, W. & de Queiroz Menezes, L.A., 1989. Polyphasic origin of the Ipixuna bauxite, northeastern state of Para, Brazil. Travaux ICSOBA 19, 105-113.

Kubiëna, W.L., 1954. Micromorphology of laterite formation in Rio Muni (Spanish Guinea). Transactions of the 5th International Congress of Soil Science, Léopoldville, Volume IV, pp. 77-84.

Lacroix, A., 1913. Les latérites de la Guinée et les produits d'altération qui leur sont associés. Nouvelles Archives du Musée National d'Histoire Naturelle, Paris, 5, 255-356.

Lambiv Dzemua, G., Mees, F., Stoops, G. & Van Ranst, E., 2011. Micromorphology, mineralogy and geochemistry of lateritic weathering over serpentinite in south-east Cameroon. Journal of African Earth Sciences 60, 38-48.

Lambiv Dzemua, G., Gleeson, S.A. & Schofield, P.F., 2013. Mineralogical characterization of the Nkamouna Co-Mn laterite ore, southeast Cameroon. Mineralium Deposita 48, 155-171.

Lelong, F., 1967. Sur les formations latéritiques de Guyane Française: 'manière d'être' de la kaolinite et de la gibbsite; origine des phyllites micacées. Comptes Rendus de l'Académie des Sciences 264, 2713-2716.

Lucas, Y., Kobilsek, B. & Chauvel, A., 1989. Structure, genesis and present evolution of Amazonian bauxites developed on sediments. Travaux ICSOBA 19, pp. 81-94.

Magaldi, D. & Bidini, D., 1991. Microscopic and submicroscopic characterization of a well developed plinthite in a buried Middle Pleistocene soil in Northern Tuscany (Italy). Quaderni di Scienza del Suolo 3, 31-44.

Maignien, R., 1966. Review of Research on Laterites. UNESCO, Paris, 148 p.

Marcelino, V., Stoops, G. & Schaefer, C.E.G.R., 2010. Oxic and related materials. In Stoops, G., Marcelino, V. & Mees, F. (eds.), Interpretation of Micromorphological Features of Soils and Regoliths. Elsevier, Amsterdam, pp. 305-327.

Marcelino, V., Schaefer, C.E.G.R. & Stoops, G., 2018. Oxic and related materials. In Stoops, G., Marcelino, V. & Mees, F. (eds.), Interpretation of Micromorphological Features of Soils and Regoliths. Second Edition. Elsevier, Amsterdam, pp. 663-689.

McCrea, A.F. & Gilkes, R.J., 1987. The microstructure of lateritic pallid zone. In Fedoroff, N., Bresson, L. M., & Courty M.A. (eds.), Micromorphology des Sols, Soil Micromorphology. AFES, Plaisir, pp. 501-506.

McCrea, A.F., Anand, R.R. & Gilkes, R.J., 1990. Mineralogical and physical properties of lateritic pallid zone materials developed from granite and dolerite. Geoderma 47, 33-58.

McFarlane, M.J., 1983. Laterites. In Goudie, A. & Pye, K. (eds.), Chemical Sediments and Geomorphology. Academic Press, New York, pp. 7-58.

Melfi, A.J. & Carvalho, A., 1983. Bauxitization of alkaline rocks in southern Brazil. Sciences Géologiques, Mémoires 72, 161-172.

Miedema, R., Jongmans, A.G. & Brinkman, R., 1987. The micromorphology of a typical catena from Sierra Leone, West Africa. In Fedoroff, N., Bresson, L.M. & Courty, M.A. (eds.), Micromorphologie des Sols, Soil Micromorphology. AFES, Plaisir, pp. 137-144.

Mindszenty, A., 1978. Tentative interpretation of the micromorphology of bauxitic laterites. Proceedings of the 4th International Congress ICSOBA. Volume 2. Bauxites. National Technical University of Athens, pp. 599-613.

Mindszenty, A., 1983. Some bauxitic textures and their genetic interpretation. Geologicky Zbornik - Geologica Carparthica 34, 665-675.

Muggler, C.C. & Buurman, P., 2000. Erosion, sedimentation and pedogenesis in a polygenetic oxisol sequence in Minas Gerais, Brazil. Catena 41, 3-17.

Muller, J.P. & Bocquier, G., 1987. Textural and mineralogical relationships between ferruginous nodules and surrounding clayey matrices in a laterite from Cameroon. In Schultz, L.G., van Olphen, H. & Mumpton, F. A. (eds.), Proceedings of the International Clay Conference. Denver, pp. 186-194.

Muller, D., Bocquier, G., Nahon, D. & Paquet, H., 1980/1981. Analyse des différenciations minéralogiques et structurales d'un sol ferrallitique à horizons nodulaires du Congo. Cahiers ORSTOM, Série Pédologie 18, 87-109.

Nahon, D.B., 1986., Evolution of iron crusts in tropical landscapes. In Colman, S.M. & Dethier, D.P. (eds.), Rates of Chemical Weathering of Rocks and Minerals. Academic Press, Orlando, pp. 169-191.

Nahon, D.B., 1991. Introduction to the Petrology of Soils and Chemical Weathering. John Wiley & Sons, New York, 313 p.

Nahon, D., Janot, C., Karpoff, A.M., Paquet, H. & Tardy, Y., 1977. Mineralogy, petrography and structures of iron crusts (ferricretes) developed on sandstones in the western part of Senegal. Geoderma 19, 263-277.

Nahon, D., Beauvais, A., Boeglin, J-L., Ducloux, J. & Nziengui-Mapangou, P., 1983 Manganite formation in the first stage of lateritic manganese ores in Africa. Chemical Geology 40, 25-42.

Nahon, D.B., Herbillon, A.J. & Beauvais, A., 1989. The epigenic replacement of kaolinite by lithiophorite in a manganese-lateritic profile, Brazil. Geoderma 44, 247-259.

Réfega, A.A.G., 1972. Micromorfologia de Formações Lateríticas. Contribuição para o Estudo da sua Génese. Universidade de Luanda, Nova Lisboa, 25 p.

Rellini, I., Trombino, L., Firpo, M. & Piccazzo, M., 2007. Geomorphological context of 'plinthitic pale-osols' in the Mediterranean region: examples from the coast of western Liguria (northern Italy). Cuaternario y Geomorfología 21, 27-40.

Rellini, I., Trombino, L., Carbone, C. & Firpo, M., 2015. Petroplinthite formation in a pedosedimentary sequence along a northern Mediterranean coast: from micromorphology to landscape evolution. Journal of Soils and Sediments 15, 1311-1328.

Rosolen, V., Lamotte, M., Boulet, R., Trichet, J., Rouer, O. & Melfi, A.J., 2002. Genesis of a mottled horizon by Fe-depletion within a laterite cover in the Amazon Basin. Comptes Rendus Geoscience 334, 187-195.

Rutherford, G.K., 1972. A study of the polycyclic nature of a lateritic soil profile from Guyana using micromorphology. In Kowalinski, S. & Drozd, J. (eds.), Soil Micromorphology. Panstwowe Wydawnicto Naukowe, Warsawa, pp. 387-398.

Schellmann, W., 1982. Eine neue Lateritdefinition. Geologische Jahrbuch D58, 32-47.

Schellmann, W., 1983. A new definition of laterite. Natural Resources and Development 18, 7-21.

Schmidt-Lorenz, R., 1964. Zur Mikromorphologie der Eisen- und Aluminiumoxydanreicherung beim Tonmineralabbau in Lateriten Keralas und Ceylons. In Jongerius, A. (ed.), Soil Micromorphology. Elsevier, Amsterdam, pp. 279-290.

Schmidt-Lorenz, R., 1974a. Lateritisiering - ein Sonderfall der Ferralitisierung. Mitteilungen Deutschen Bodenkundliche Gesellschaft 20, 68-79.

Schmidt-Lorenz, R., 1974b. Nachweis von Laterit-Spuren in paläopedogenem Material, aufgezeigt an Beispielen aus Mitteleuropa. Mitteilungen Deutschen Bodenkundliche Gesellschaft 20, 114-122.

Schmidt-Lorenz, R., 1980. Soil reddening through hematite from plinthitized saprolite. In Joseph, K.T. (ed.), Conference on Classification and Management of Tropical Soils. Malaysian Society of Soil Science, Kuala Lumpur, pp. 101-106.

Sivarajasingham, S., Alexander, L.T., Cady, J.G. & Cline, M.G., 1962. Laterite. Advances in Agronomy 14, 1-60.

Soil Survey Staff, 2014. Keys to Soil Taxonomy. 12th Edition. USDA-Natural Resources Conservation Service, Washington, 360 p.

Stoops, G., 1968. Micromorphology of some characteristic soils of the Lower Congo (Kinshasa). Pedologie 18, 110-149.

Stoops, G., 1970. Scanning electron microscopy applied to the micromorphological study of a laterite. Pedologie 20, 268-280.

Stoops, G., 1989. Relict properties in soils of humid tropical regions with special reference to Central Africa. In Bronger, A. & Catt, J.A. (eds.), Paleopedology. Nature and Application of Paleosols. Catena Supplement 16, pp. 95-106.

Stoops, G., 2003. Guidelines for Analysis and Description of Soil and Regolith Thin Sections. Soil Science Society of America, Madison, 184 p.

Stoops, G. & Schaefer, C.E.G.R., 2010. Pedoplasmation: formation of soil material. In Stoops, G., Marcelino, V. & Mees, F. (eds.), Interpretation of Micromorphological Features of Soils and Regoliths. Elsevier, Amsterdam, pp. 69-79.

Stoops, G. & Schaefer, C.E.G.R., 2018. Pedoplasmation: formation of soil material. In Stoops, G., Marcelino, V. & Mees, F. (eds.), Interpretation of Micromorphological Features of Soils and Regoliths. Second Edition. Elsevier, Amsterdam, pp. 59-71.

Stoops, G., Shi Guang Chun & Zauyah, S., 1990. Combined micromorphological and mineralogical study of a laterite profile on graphite sericite phyllite from Malacca (Malaysia). Bulletin de la Société Belge de Géologie 99, 79-92.

Stubendorff, U., 1986. The mechanical properties of tropical weathering products and their relationship to the mineralogical and chemical composition. Geologisches Jahrbuch C43, 3-137.

Suddhiprakarn, A & Kheoruenromne, I., 1994. Fabric features in laterite and plinthite layers in Ultisols in northeast Thailand. In Ringrose-Voase, A.J. & Humphreys, G.S. (eds.), Soil Micromorphology: Studies in Management and Genesis. Developments in Soil Science, Volume 22. Elsevier, Amsterdam, pp. 51-64.

Sys, C., 1968. Suggestions for the classification of tropical soils with lateritic materials in the American classification. Pedologie 18, 189-198.

Tardy, Y., 1992. Diversity and terminology of lateritic profiles. In Martini, I.P. & Chesworth, W. (eds.), Weathering, Soils and Paleosols. Elsevier, Amsterdam, pp. 379-405.

Tardy, Y., 1993. Pétrologie des Latérites et des Sols Tropicaux. Masson, Paris, 459 p.

Tiessen, H., LoMonaco, S., Ramirez, A., Santos, M.C.D. & Shang, C., 1996. Phosphate minerals in a lateritic crust from Venezuela. Biogeochemistry 34, 1-17.

Tshidibi, N.Y.B., 1985. Evolution du quartz au sein des cuirasses latéritiques et des sols ferrugineux. Geo-Eco-Trop 8, 93-110.

Valeton, I., 1972. Bauxites. Developments in Soil Science, Volume 1. Elsevier, Amsterdam, 226 p.

Valeton, I., 1973. Considerations for the description and nomenclature of bauxite. Travaux ICSOBA 9, 105-108.

Varajão, A.F.D.C., Boulangé, B. & Melfi, A.J., 1989. The petrologic evolution of the facies in the kaolinite and bauxite deposits of Vargem dos Óculos, Quadrilátero Ferrífero, Minas Gerais, Brazil. Travaux ICSOBA 19, 137-146.

Varentsov, I.M., 2013. Manganese Ores of Supergene Zone: Geochemistry of Formation. Springer, Dordrecht, 344 p.

Varghese, T. & Byju, G., 1993. Laterite Soils. Their Distribution, Characteristics, Classification and Management. STEG, Government of Kerala, Thiruvananthapuram, 116 p.

Widdowson, M., 2007. Laterite and ferricrete. In Nash, D.J. & McLaren, S.J. (eds.), Geochemical Sediments and Landscapes. Blackwell, Oxford, pp. 45-94.

Yakushev, V.M., 1968. Influence of termite activity on the development of laterite soil. Soviet Soil Science 1, 109-111.

Yongue-Fouateu, R., Ghogomu, R.T., Penaye, J., Ekodeck, G.E., Stendal, H. & Colin, F., 2006. Nickel and cobalt distribution in the laterites of the Lomié region, south-east Cameroon. Journal of African Earth Sciences 45, 33-47.

Zauyah S., 1983. Micromorphology of some lateritic soils in Malaysia. In Bullock, P. & Murphy, C.P. (eds.), Soil Micromorphology, Volume 2, Soil Genesis. AB Academic Publishers, Berkhamsted, pp. 667-674.

Zauyah, S. & Bisdom, E.B.A., 1983. SEM-EDXRA investigations of tubular features and iron nodules in lateritic soils from Malaysia. Geoderma 30, 219-232.

Zauyah, S., Schaefer, C.E.G.R. & Simas, F.N.B., 2010. Saprolites. In Stoops, G., Marcelino, V. & Mees, F. (eds.), Interpretation of Micromorphological Features of Soils and Regoliths. Elsevier, Amsterdam, pp. 49-68.

Zauyah, S., Schaefer, C.E.G.R. & Simas, F.N.B., 2018. Saprolites. In Stoops, G., Marcelino, V. & Mees, F. (eds.), Interpretation of Micromorphological Features of Soils and Regoliths. Second Edition. Elsevier, Amsterdam, pp. 37-57.

Zeese, R., Schwertmann, U., Tietz, G.F. & Jux, U., 1994. Mineralogy and stratigraphy of three deep laterite profiles of the Jos plateau (Central Nigeria). Catena 21, 195-214.

Zinn, Y.L. & Bigham, J.M., 2016. Pedogenic and lithogenic gravels as indicators of soil polygenesis in the Brazilian Cerrado. Soil Research 54, 440-450.

Zinn, Y.L., Carducci, C.E. & Araújo, M.A., 2015. Internal structure of a vermicular ironstone as determined by X-ray computed tomography scanning. Revista Brasileira de Ciência do Solo 39, 345-349.

Chapter 25

Regoliths and Soils on Volcanic Ash

Georges Stoops[1], Sergey Sedov[2], Sergei Shoba[3]

[1]*GHENT UNIVERSITY, GHENT, BELGIUM;*
[2]*UNIVERSIDAD NACIONAL AUTÓNOMA DE MÉXICO, MEXICO CITY, MÉXICO;*
[3]*LOMONOSOV MOSCOW STATE UNIVERSITY, MOSCOW, RUSSIA*

1. Introduction

Soils formed on volcanic sediments (pyroclastic material) have particular characteristics, which differentiate them from soils derived from other parent materials. During the initial stages of weathering, especially of the volcanic glass, extensive accumulations of nanocrystalline Si-Al components, such as allophane and imogolite, are produced. As a result, these soils have specific physical and chemical properties, with regard to for example pH-dependent surface charge, P retention, alkaline reaction with NaF, bulk density and thixotropy, as well as specific micromorphological characteristics. For young soils, this has resulted in their grouping at the highest taxonomic levels in international soil classification systems: the Andosol reference group in WRB (IUSS Working Group WRB, 2015) and the Andisol order in Soil Taxonomy (Soil Survey Staff, 2014). Some of these characteristics (e.g., undifferentiated b-fabric) remain detectable even in more developed soils. This is reflected by the prefix qualifier Andic added to the soil group name in WRB (IUSS Working Group WRB, 2015) and to the subgroup level name in Soil Taxonomy (Soil Survey Staff, 2014). Throughout this manuscript, only WRB terminology will be used, except where soil orders of the Soil Taxonomy system are cited as mentioned in the original publication.

The first micromorphological descriptions of soils on volcanic ash were probably made by Kubiëna (1956) and later by Frei (1964). The first systematic paper on Andosol micromorphology is by Kawai (1969). Spanish pedologists published a large number of micromorphological studies on soils from the Canary Islands in the 19 seventies and eighties (e.g., Benayas et al., 1974, 1978, 1980; Tejedor Salguero et al., 1975, 1978, 1979; Rodríguez-Rodríguez et al., 1979, 1980a, 1980b, 1988). The micromorphology of volcanic ash soils in Europe has been reviewed by Stoops (2007).

The general trend of pedogenic evolution on volcanic parent materials under humid climates consists of a transition of soils with amorphous fine compounds toward soils with crystalline clay. This crystallisation process creates the preconditions for clay illuviation so that Andosols evolve to soils with an argic horizon. In Europe such soils are often palaeosoils (e.g., Quantin et al., 1978; Tejedor Salguero et al., 1978).

The evolution of Andosols to soils with crystalline clay proceeds via transitional stages, known as 'intergrade' Andosols (Fernández-Caldas et al., 1982), in which some features of Andosols are preserved (thick A horizon, relatively high content of the oxalate-extractable Fe and Al), but which already include a high proportion of crystalline clay in the fine material (often halloysite).

Little information has been published on the micromorphology of Podzols on volcanic ash. Also Ferralsols, Vertisols and arid soils on volcanic ash have received little attention in the literature.

The micromorphology of subsurface indurated horizons that formed in pyroclastic materials or their derivates in volcanic regions of Central and South America have received some attention (Hidalgo et al., 1992; Oleschko et al., 1992; Poetsch & Arikas, 1997; Jongmans et al., 2000; Poetsch, 2004). They are known under various local names (tepetates, talpetates, cangahua, ñadis). In the WRB classification, they mostly fit the definition of either duric or fragic horizon (IUSS Working Group WRB, 2015). Originally they are located in the subsoil of moderately weathered soils (Luvisols, Vertisols, Cambisols) constituting their BC horizons, but they are frequently exposed along the surface due to erosion. The most important cementing agent is pedogenic amorphous silica (e.g., Miehlich, 1992; Poetsch & Arikas, 1997; see also Gutiérrez-Castorena & Effland, 2010; Gutiérrez-Castorena, 2018) but allophane and imogolite have been detected as well (Jongmans et al., 2000).

Not all soils with andic properties are formed on volcanic parent materials. Addition of volcanic ash to soils on non-volcanic substrates in the vicinity of active volcanoes is often sufficient to give rise to chemical, mineralogical and micromorphological properties similar to those of Andosols (e.g., Vingiani et al., 2014). It also can be at the origin of a melanic variant of the Mollic horizon. On the other hand, soils on volcanic formations can show non-andic properties as a result of a high input of other materials, e.g., of aeolian origin, as described for parts of the Canary Islands where dust derived from the Sahara is deposited in large quantities (von Suchodoletz et al., 2009). Soils with andic properties have been described for non-volcanic parent materials (e.g., shales), but no micromorphological research has been published.

This chapter gives a structured overview of the micromorphological characteristics of soils and regoliths formed on volcanic ash. It demonstrates that soils formed on tephra are quite different from those formed on other parent materials.

2. Microstructure

In an initial stage of soil formation, tephra have a coarse monic basic microstructure (Fig. 1A), which evolves to chitonic when fine material becomes available (see also Section 3.2). The c/f ratio varies between 15/1 to 20/1 as described for ashes of historic eruptions (e.g., Stoops et al., 2006; Stoops & Gérard, 2007), in boreal conditions (e.g., Sokolov, 1972; Simpson et al., 1999; Shoba et al., 2004; Stoops et al., 2008) and in arid environments (Stoops, unpublished data). When more fine material is present, fine aggregates form, leading to a chito-enaulic or enaulic basic microstructure, which evolves to a granular microstructure.

Andosols are characterised throughout the world by a well-separated granular microstructure (see Figs. 1B to D, and 2D), which changes to enaulic, chitonic and coarse monic basic microstructures with increasing depth. In young or coarse-textured soils, an enaulic basic microstructure is observed. The size of the granular peds ranges from 30 to 400 µm (e.g., Bech-Borras et al., 1977) and has been reported to increase with depth (Bech-Borras et al., 1977; Benayas et al., 1980; Zarei et al., 1997). In tropical Andosols, small granules (50 µm) are often coalescent, forming larger aggregates (1.5 mm) (Stoops, 1983). Bimodal granular microstructures have been recognised for surface horizons (Colombo et al., 2007). The peds usually show a uniform internal fabric, although concentric fabrics may be found in deeper layers or in transported materials (Fig. 1C) (Stoops, 1972, 2013; Stoops & Gérard, 2007). Biogenic pores are common, especially in the upper horizons. They occur in the form of channels and vughs, sometimes filled by loosely packed excrements.

The origin of the granular microstructure or the enaulic basic microstructure has not been clarified. The bimodal microstructures in A horizons, combined with granular infillings (Colombo et al., 2007), support the theory of a biological origin. However, the fact that granular microstructures have been observed for deeper parts of the profile and in soils with very restricted biological activity, such as soils on volcanic ash in boreal

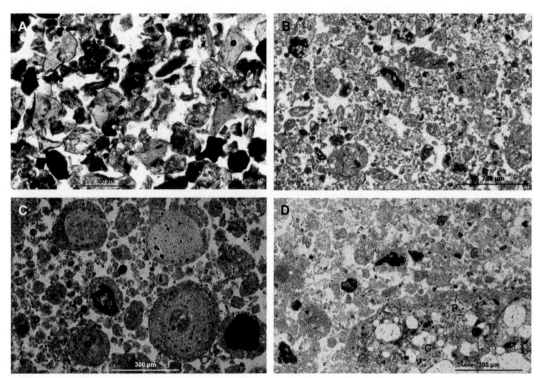

FIGURE 1 Microstructure of soils on volcanic ash. (A) Coarse monic basic microstructure, consisting of opaque grains and greenish-brown volcanic glass fragments (sideromelane) in the Bw horizon of an Andosol (Iceland) (PPL). (B) Granular microstructure in the Bw horizon of an Andosol (Iceland) (PPL). Note the coatings of micromass on the larger pyroclasts. (C) Granular microstructure with larger peds showing a well-developed concentric internal fabric in the C horizon of an Andosol (Galapagos Islands) (PPL). (D) Granular microstructure in a Calcic Gypsiorthid on volcanic ash (Raqqa, Syria) (PPL). Note presence of palagonite-like (orange) weathering (P) on glass fragment (G). The greyish micromass contains small calcite crystals (calcitic crystallitic b-fabric in XPL, not shown).

regions (Stoops et al., 2008) and on Santorini (Stoops et al. 2006; Stoops & Gérard, 2007), suggests that other processes may also be responsible. Dubroeucq et al. (2002) considered the granules as excrements of Enchytraeidae and nematodes. Shin and Stoops (1988) and Colombo et al. (2007) mentioned that the granular microstructure is better expressed in thin sections of air-dried material, than in those of samples for which acetone replacement was used. Observed microstructures can be partly an artifact caused by drying, especially in Hydrandepts (Chartres et al., 1985).

In the Bw horizon of Andosols the granular microstructure is often replaced by a blocky structure (e.g., Dubroeucq et al., 2002), combined with a moderately or weakly separated intrapedal granular microstructure. A (sub)angular blocky and locally massive microstructure has been reported to predominate in A horizons of imperfectly drained Andosols (Nieuwenhuyse et al., 1993).

Intergrades of Andosols to soils with crystalline clay display a specific set of micromorphological features, studied for a buried volcanic soil chronosequence in Central

Mexico (Sedov et al., 2003b). Their Ah horizons have a granular microstructure with abundant packing voids in some (subordinate) parts and a subangular blocky microstructure with fewer planes and channels in other parts.

In soils on volcanic ash with a Bt horizon, a granular microstructure is still dominant throughout the profile, although it is sometimes more complex than in the less developed soils. In red Mediterranean soils on lapilli on the Canary Islands, the microstructure is granular in the upper part of the profile, becoming more compact with depth due to compaction and agglomeration of the granules (Fedoroff & Rodriguez, 1978). A more compact microstructure in Bt or Bw horizons has also been reported in other studies (Goenadi & Tan, 1989).

In cold areas, soils on volcanic ash often display a platy or lenticular microstructure (Romans et al., 1980; Stoops & Gérard, 2007; Stoops et al., 2008) resulting from frost action (see Van Vliet-Lanoë, 2010; Van Vliet-Lanoë & Fox, 2018). The lenticular aggregates can preserve an intrapedal granular microstructure (Stoops et al., 2008).

In Ferralsols on volcanic substrates, documented for the Canary Islands, a less granular and more blocky microstructure than in Andosols is generally observed (Rodríguez-Rodríguez et al., 1988) and a porphyric c/f-related distribution pattern gradually appears (Tan & Goenadi, 1994). In relatively young Ferralsols a weak blocky microstructure prevails, becoming granular with increasing age (West et al., 1997).

In Vertisols derived from volcanic materials, a fine granular or fine blocky microstructure was observed for the A horizon, becoming angular blocky with centimetre-sized peds in the Bw horizon (Rodriguez Hernandez et al., 1979).

For soils on volcanic ash of arid areas, little information is available. Zarei et al. (1997), studying soils from the arid Fuerteventura island (eastern Canary Islands), stated that, depending on their texture, the surface horizons have an enaulic and chitonic c/f-related distribution pattern, grading to a granular microstructure, which gradually changes with depth to angular blocky and spongy microstructure, with common granular infillings. The same authors observed that the expression of the round peds increases with increasing abundance of amorphous material. In a calcic Gypsiorthid near Raqqa (Syria) (Stoops, 1980), the granular microstructure and micromass coatings around pyroclasts strongly resemble those in weakly developed soils as described before (Fig. 1D).

Paddy soils on young volcanic ash material show characteristics that are not related to the Andosol fabric, but rather to hydromorphic soils (see Lindbo et al., 2010; Vepraskas et al., 2018), except for the composition of the coarse fraction of the groundmass. The microstructure is blocky, or massive with channels, and the c/f-related distribution pattern is always porphyric (Stoops, 1984).

Duric or fragic horizons, such as tepetate, usually have a loamy texture with a porphyric c/f-related distribution pattern (Alfaro Sánchez et al., 1992). Pores are rare, presented mostly by thin, discontinuous, poorly interconnected fissures and very sparse channels. Jongmans et al. (2000) documented a chitonic c/f-related distribution pattern due to the presence of non-crystalline Al-silicate coatings.

According to Stoops and Gérard (2007) a relationship exists between bulk density and microstructure, bulk density being higher in soils with a coarse monic c/f-related distribution pattern, intermediate in soils with enaulic c/f-related distribution pattern and granular microstructures and lowest in soils with a blocky microstructure.

3. Groundmass

3.1 Coarse Material

3.1.1 Identification of Components

Many soil scientists still use local or obsolete terms, although an internationally adopted terminology is available (Schmid, 1981; Fisher & Schmincke, 1984a; Le Maitre et al., 2002). For our purpose the following criteria can be used to characterise pyroclasts: (i) grain size, (ii) vesicularity, (iii) crystallinity and texture, and (iv) chemical composition. The subdivision according to grain size is summarised in Table 1. Mainly grains smaller than 63 mm are of interest for microscopic studies. Subdivisions based on crystallinity and porosity are summarised in Table 2. Because the determination of the petrographic composition of volcanic rocks requires full mineralogical analyses for holocrystalline materials (see also Stoops & Mees, 2018), and chemical analyses for hypocrystalline and vitric materials, its discussion is beyond the scope of this chapter.

The identification and quantification of volcanic glass is of special importance for soil classification and genetic studies. Unaltered glass particles are optically isotropic. Si-rich volcanic glass (saturated, acid, rhyolitic) is colourless and has a negative optical relief ($n < 1.54$). Si-poor glass (neutral or basic, undersaturated, sideromelane) is darker and has a positive relief (see Fig. 1A). Tachylyte is a very dark variety of basaltic glass, opaque in thin sections (Fig. 2C). The shape of glass particles is very characteristic: they usually have an irregular outline with sharp angles and cavities in the form of 'embayments'. They can be highly porous, sometimes with a fibrous aspect, forming pumice (rhyolitic glass) (see Figs. 4B, and 6A and B), or scoria (basaltic glass) (see Fig. 7).

Table 1 Classification and nomenclature of pyroclasts and pyroclastic deposits based on clast size

Clast size (mm)	Pyroclast	Pyroclastic deposit[a]	
		Mainly unconsolidated (tephra)	Mainly consolidated (pyroclastic rock)
>63	Bomb, block	agglomerate; bed of blocks or bombs; block tephra	agglomerate; pyroclastic breccia
63-2	Lapillus	layer or bed of lapilli; lapilli tephra	lapilli tuff
2-1/16	Coarse ash grain	coarse ash	coarse (ash) tuff
<1/16	Fine ash grain (dust grain)	fine ash (dust)	fine (ash) tuff (dust tuff)

[a]Should contain at least 75 vol. % of volcanic material.
After Schmid (1981).

Table 2 Crystallinity and vesicularity of tephra

Term	Crystallinity
Holocrystalline	Only mineral grains
Hypocrystalline and mesocrystalline	Mineral grains and glass
Glass-bearing	<20% glass
Glass-rich	20-50% glass
Glassy	50-80% glass
Vitric or holohyaline	>80% glass

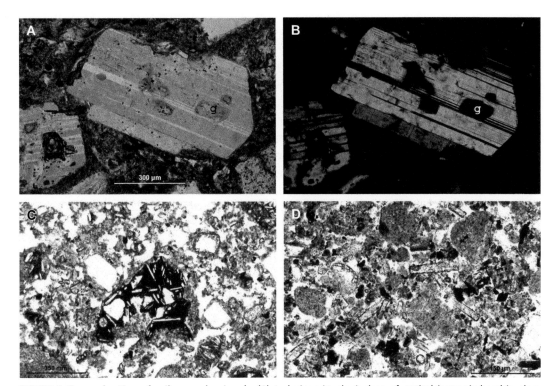

FIGURE 2 Coarse fraction of soils on volcanic ash. (A) Inclusions in plagioclase of optical isotropic basaltic glass (g) in an Andosol (Rwanda) (PPL), with partly crossed polarisers. (B) Same field in XPL. (C) Fragment of tachylytic rock with plagioclase laths and other coarse grains in an opaque matrix in the B horizon of an Andosol under pine-oak forest (Central Mexico) (PPL). (D) Phytoliths in the groundmass of an A horizon with granular microstructure in an Andosol under pine-oak forest (Central Mexico) (PPL).

A micromorphometric study of pumice gravel on the Canary Islands illustrates the large external surface ($10.7 \, cm^2 \, cm^{-3}$) and the even larger internal surface area ($25.0 \, cm^2 \, cm^{-3}$) of this material (Kress-Voltz, 1964, 1967). Glass can also be present as a component of rock fragments and as inclusions in volcanic minerals (Fig. 2A and B), or it can form coatings that cover those constituents. These features should be taken into account when quantifying volcanic glass and they are useful indicators of a volcanic origin of the coarse material in other soils.

A notable feature of many Andosols and other soils on volcanic ash is the high concentration of phytoliths in the silt fraction (e.g., Stoops, 1983; Chartres et al., 1985; Poetsch & Arikas, 1997; Stoops et al., 2007) (Fig. 2D) (see also Gutiérrez-Castorena & Effland, 2010; Gutiérrez-Castorena, 2018). They can account for 15% and more of the coarse silt fraction (Sedov et al., 2003a; Fehér et al., 2007), whereas in most other soil types they are below 1-2%, even in the topsoil (Kamanina, 1997). Phytoliths are the dominant component of the coarse fraction of soils on volcanic ash in Veracruz (Mexico) (Dubroeucq et al., 2002). High quantities of highly weatherable silicium-rich components, especially volcanic glass, probably liberate high concentrations of Si in the soil solution, providing a higher uptake of Si by plants and protecting phytoliths already present from dissolution (Vallejo Gómez et al., 2000). Their identification and quantification is not easy, as acid glass and phytoliths have similar optical properties. Moreover, the shape of volcanic glass particles often resembles that of phytoliths (e.g., Van Ranst et al. 2011). If microprobe analyses are not available, a reliable discrimination between these two components can be done by immersion of the separated sand and silt fractions in glycerine. This liquid has a refractive index of 1.48, which is higher than the maximum for phytoliths (1.47) (Piperno, 1988) and lower than the minimum for Si-rich glass (1.49) (Shoji et al., 1993). Checking whether the Becke line moves in or out the grain when the microscope tube is raised allows distinguishing between both materials in glycerine grain mounts.

3.1.2 Alteration Before Pedogenesis

Some pyroclasts show signs of alteration by hydrothermal, pneumatolitic and subaquatic processes, which frequently accompany volcanic eruptions and may affect pyroclasts before they are involved in pedogenesis. The products of this alteration are variable; a few important examples will be discussed below.

Feldspars sometimes enclose fine-grained muscovite or paragonite (sericite), which is liberated during further weathering, adding illitic material to the micromass. Fe-Mg silicates are often transformed to chlorite and/or smectite. Especially in andesitic and basaltic rock fragments, chloritic alteration products are commonly found (Fig. 3A). They form coatings and/or infillings of vacuoles, sometimes in combination with zeolites and/or carbonates. These chlorites can be a starting material for the later formation of 2:1 clay minerals, including smectites and vermiculites, in the regolith and the soil. Olivine is frequently altered, partially or totally, to iddingsite pseudomorphs, following a pellicular or irregular linear pattern. Although iddingsite behaves optically as a single crystal (yellowish- or reddish-brown, weakly pleochroic), it is in fact a mixture of cryptocrystalline goethite with phyllosilicates, such as smectite, chlorite, talc or mica (Delvigne et al., 1979).

Glass of pyroclastic ejecta derived from intermediate and Si-rich magmas is mostly transformed into clay minerals (cf. bentonite formation at a macroscopic scale).

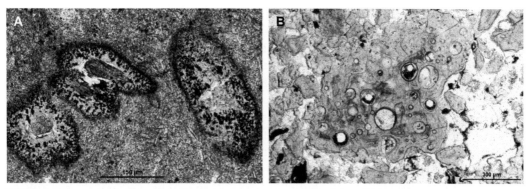

FIGURE 3 Non-pedogenic alteration of minerals in volcanic ash soils. (A) Hydrothermal alteration of pyroxenes: incomplete (right) and complete (left) pseudomorphs of smectite and iron oxides after pyroxene crystals in the C horizon of a buried truncated volcanic palaeosoil (Nevado de Toluca, Central Mexico) (PPL). (B) Sideromelane transformed to orange palagonite in tuff (Galapagos Islands) (PPL). Note alternating zones with different refractive indices. At the left upper side some greenish unaltered sideromelane grains are noticed.

Opalisation is restricted to very specific settings. Hydrothermal alteration of rhyolitic glass leads to its devitrification, forming fine-grained quartz, feldspar and cristobalite, resulting in optical anisotropy. In the case of acid sulphate hydrothermal alteration, kaolinite is formed as well (Wilson et al., 1997). Palagonite is the primary alteration product of sideromelane, formed by an interaction at ambient temperature with aqueous solutions (Fisher & Schmincke, 1984b; Stroncik & Schmincke, 2002). In thin sections it has a yellow, orange or reddish brown colour and refractive indices mostly below 1.54, diminishing with increasing water content. Two main palagonite varieties are recognised: (i) gel-palagonite, which is optically isotropic and usually crystal-free and concentrically banded (Fig. 3B), and (ii) fibro-palagonite, which is slightly to strongly optically anisotropic and either fibrous or granular in structure. Palagonite contains secondary minerals, including different kinds of clays (mainly smectites), zeolites, carbonate minerals, opal and Fe oxides. During the palagonitisation process some chemical elements are leached out, while others show relative gains, reflected by either a decrease or an increase of the refractive index of the alteromorph. Allophanic palagonite-like alteromorphs have been identified in volcanic soils by Gérard et al. (2007) (see also Section 3.1.4 and Section 5.2).

Correct identification of these alteration products is important for understanding the subsequent mineral transformations in soils. Hydrothermal phyllosilicates occurring as inclusions, being liberated from their 'host' volcanogenic minerals by weathering, could be a source of clay components for the soil fine material (Jongmans et al., 1994a, 1999; Sedov & Shoba, 1994).

3.1.3 Sedimentary Features

Tephra differ from many other parent materials not only by their mineralogical composition, but also by their layering, resulting from periodic ash deposition. This deposition sequence can be formed during a single eruption cycle, or as a result of eruptions separated in time. In the latter case pedogenesis may have influenced the older material. Profiles can therefore comprise several ash deposits, affected by successive phases of weathering, which greatly complicates the study of pedogenesis (Dalrymple, 1964).

A typical aspect of the soils from Iceland is the preservation of a fine microstratification (millimetre- to centimetre-sized), recognisable by differences in composition (e.g., light versus dark basaltic glass, mineral and organic constituents) or in grain size distribution (Simpson et al., 1999; Milek, 2006; Stoops et al., 2008). Elongated organ residues show a parallel, (sub)horizontal arrangement (Fig. 4A), alternating with thin ash layers, pointing to gradual sedimentation and very limited pedoturbation, especially bioturbation (Stoops et al., 2008). Similar characteristics were observed for volcanic ash soils under coniferous forests in a cool boreal climate (Shoba et al., 2004).

Rejuvenation of older soils by addition of younger ash fall is a common phenomenon in areas of active volcanism. It can be recognised by the presence of less-weathered coarse components in a more strongly weathered groundmass. Examples include the presence of fresh pumice fragments in a strongly weathered soil on andesite (Fig. 4B) (Fauzi & Stoops, 2004) and the occurrence of fresh minerals derived from a recent eruption in the surface horizon of an Andosol (Jongmans et al., 1994b). In extreme cases this can result in the so-called inverted weathering profiles, where soil material is less weathered at the top than at the bottom (Rodríguez-Rodríguez et al., 1988; Bakker et al., 1996; Solleiro-Rebolledo et al., 2003). Recent deposition of light-coloured acid ash can form a pale upper horizon, resembling an albic horizon.

In lower parts of the landscape or in depressions, accumulation of layers of colluvial volcanic material interrupted by cycles of pedogenesis can occur (e.g., Sauer & Zöller, 2006).

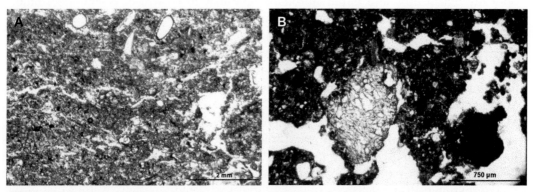

FIGURE 4 Sedimentary features in soils on volcanic ash. (A) Detail of alternation of mineral (upper part) and organic (lower part) deposits in the C horizon of an Andosol (Iceland) (PPL). Note horizontal arrangement of the few elongated organ residues. (B) Pumice fragment from a subrecent ash fall (Krakatau, 1883) incorporated in the A horizon of a strongly weathered volcanic ash soil (Western Java, Indonesia) (PPL).

A special case are tephra-rich tsunami deposits, but information on their micro-morphological characteristics is scarce. Some macroscopic and microscopic features have been described for tsunami deposits originating from the Santorini eruption in Greece, including the presence of reworked fragments of building stones and ceramics, marine shells and microfauna remains, as well as a multimodal chaotic fabric, and especially in thin sections, a sharp contact between fresh ash and soil-derived red-brown sediments (Bruins et al., 2008).

3.1.4 Evolution in Soils

In very weakly developed soils (Regosols), the coarse material shows practically no weathering in thin sections. The main alteration agents during the first stages of weathering of tephra are various species of lichens by excreting organic acids and by trapping dust and weathering products (Vingiani et al., 2013). Most of these components and organic matter were accumulated in the vacuoles of tephra fragments and gradually built a substrate for fungi or plants.

In Andosols, the coarse material mainly consists of weatherable volcanogenic minerals such as plagioclase, olivine, pyroxenes, oxi-hornblende, volcanic glass and opaque minerals (magnetite, ilmenite), as well as volcanic rock fragments. This material commonly shows evidence of weathering, ranging from superficial etching to partial dissolution (Fig. 5A), total congruent dissolution (e.g., hypersthene) or replacement by alteration products (Fig. 5C and D). Although olivine is considered to be more easily weathered than pyroxenes, the latter are often less resistant because of their prominent cleavage (Stoops, 2013). Olivine is often altered to iddingsite, which is a mixture of goethite and phyllosilicates (see Section 3.1.2). Iddingsite is very stable in soils and is often the only remaining lithogenic mineral, apart from inherited Fe or Fe-Ti oxides, that is observed for the most highly weathered Andosols (Fig. 5E and F) (Stoops, 2013). The most resistant minerals, especially in andesitic and basaltic tephra, are magnetite and/or ilmenite (e.g., Schwaighofer, 1976). The original position of tephra fragments that are completely weathered is often marked by clusters of these residual opaque minerals (Chartres et al., 1985; Jongmans et al., 1991). Increased fragmentation of vitreous material from bottom to top of the soils has been noticed (Bech-Borras et al., 1977).

In submarine volcanic glass, fine tubular microcavities that are a few micrometres wide and up to several hundred micrometres long can form by the action of hyphae and/or bacteria (Staudigel et al., 2008; McLoughlin et al., 2010). This pattern of alteration is not yet reported in published soil studies, because nobody looked for it, but it was observed for a Viking grave site on Iceland (Burns, pers. comm., 2016). Fission tracks and alpha ray recoil trails are a magnitude smaller than these biogenic features.

Weathering of small basalt fragments in the solum is often different from that of the underlying massive basalt substrate (Stoops, 1972). Leaching in the open soil system is stronger than in the less open saprolite system, resulting for instance in gibbsite formation.

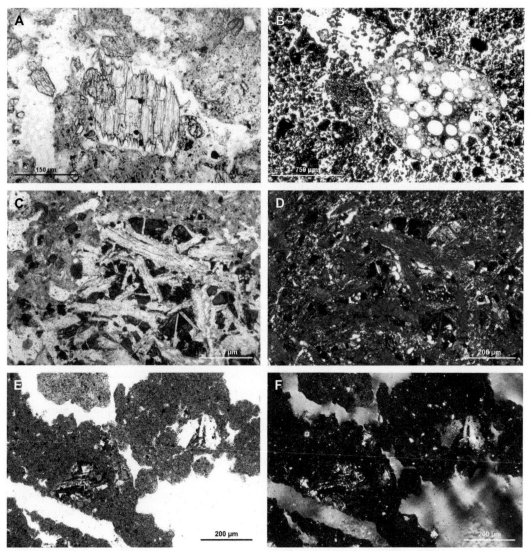

FIGURE 5 Weathering of coarse material in Andosols. (A) Congruent dissolution of pyroxene (note saw-edges) in the A horizon of a buried truncated volcanic palaeosoil (Nevado de Toluca, Central Mexico) (PPL). (B) Sideromelane scoria weathered to allophane with palagonite-like appearance in a groundmass with granular microstructure in Andosol (Rwanda) (PPL). (C) Weathering of plagioclase laths to isotropic colourless colloids, total disappearance of interstitial augite and preservation of reddish iddingsite grains, in a basalt fragment (B horizon of Andosol, Galápagos Islands) (PPL). (D) Same field in XPL. (E) Coarse fraction consisting exclusively of reddish iddingsite grains (in the groundmass and inside weathered rock fragment) and small magnetite grains, in the B horizon of a strongly weathered Andosol (Galápagos). (PPL) (F) Same field in XPL.

Sideromelane often weathers to a palagonite-like allophane alteromorph (Gérard et al., 2007) (see also Section 2.2.2). This material is characterised by a yellow, orange or orange-red colour (Figs. 1D, 3B and 5B), a negative relief, and often by the presence of concentric rings with slightly different refractive indices. Energy-dispersive microanalysis indicates a total loss of alkaline and alkaline earth elements, together with hydration. This type of pedogenic weathering has not been recognised in earlier studies, although some authors have previously postulated that palagonite-like substances in soils could be weathering products (Bech-Borras et al., 1977; Paluskova, 1988). As these alteromorphs quickly disintegrate (Rodríguez-Rodríguez et al., 1980a, 1980b), many Andosols containing basaltic vitric tephra are characterised by abundant yellow or orange fragments in the coarse fraction (Stoops, 2013). As described by Gebhardt et al. (1969), weathering of trachytic glass ($n = 1.510$) starts with a brownish allophane coating ($n = 1.600$), in contact with the sides of the glass particle, followed by weathering of the glass to kaolinite ($n = 1.557$) beneath the allophane coating. As weathering continues, kaolinite also forms within the vacuoles, and finally a complete kaolinitic alteromorph is formed. A similar sequence was observed by Ivanov et al. (2014), documented by EDS analysis of coatings on sand-sized volcanic glass particles in B horizons of associated Andosols and Podzols on the Commanders Islands. The inner coating was composed of allophane, ferrihydrite and organic matter, whereas organic matter was dominant over allophane in the middle coating and iron oxides were dominant in the outer coating, confirming the similarity between the studied Andosols and Podzols in terms of pedogenesis. Coatings have also been described for rhyolitic glass shards from the C horizon of Vitric Andosols in Veracruz, Mexico. The composition of these isotropic, limpid, non-laminated coatings was close to that of glass in deeper parts of the profile, becoming less Si-rich in higher parts (Dubroeucq et al., 1998).

Admixture of non-volcanic components due to aeolian or colluvial input is possible, and sometimes important (e.g., von Suchodoletz et al., 2009). Few data are available on differences in micromorphological characteristics between different Andosols. Tejedor Salguero et al. (1975, 1978, 1979) mentioned an increased weathering of the coarse material from Vitrandepts to Eutrandepts and further to Dystrandepts (Soil Survey Staff, 1975).

Although the formation of a Bt horizon is associated with a more advanced degree of weathering and many volcanogenic materials (especially volcanic glass) are already destroyed, the coarse material of soils with an argic horizon in many cases still contain silicates of volcanic origin, such as plagioclase and pyroxenes, as well as pyroclasts. The mineral grains and rock fragments are usually altered; sometimes illuvial clay invades the cracks and etch pits. Various types of alteromorphs are observed, in particular complete halloysite pseudomorphs after porous volcanic glass in Mexico (Sedov, unpublished data).

In Ferralsols of the Canary Islands the coarse material is strongly weathered and often only fragments impregnated by iron oxides remain visible. Volcanic glass alters in well-aerated conditions to a yellowish isotropic product with many fissures, followed by an intense rubification starting from the sides of the vacuoles, giving rise to dark red optically isotropic alteromorphs which are readily fragmented by pedoturbation (Rodríguez-Rodríguez et al., 1980b). Euhedral magnetite grains are commonly present throughout the profile (Schwaighofer, 1976; Rodríguez-Rodríguez et al., 1980a; Tejedor Salguero et al., 1984/1985). When all other minerals are dissolved, this can give a characteristic dotted aspect to strongly weathered soils on pyroclastic material, basalt or dolerite (Stoops, 2013). It is responsible for the unexpected high values of oxalate extractable iron, often exceeding the DCB-extractable iron (e.g., Algoe et al., 2012).

Coarse material of soils with duric horizon such as tepetate consists of angular to weakly rounded grains of volcanogenic minerals and often volcanic glass (Alfaro Sánchez et al., 1992). In most cases their weathering status is low except for horizons in more developed soils or palaeosoils (e.g., Rhodic Luvisols). Sometimes phytoliths are so abundant that they become one of the dominant components of the silt fraction (Solleiro-Rebolledo et al., 2003).

3.2   Micromass

The origin of the micromass of soils developed on volcanic material still needs further study. It can be part of the original tephra or derived from *in situ* weathering of the coarse fraction (e.g., Van Ranst et al., 2008). It can also be of aeolian origin, as suggested by the presence of non-volcanic components, such as fine muscovite flakes (e.g., Rodríguez-Rodríguez et al., 1988; Jahn et al., 1992). Also the liberation of hydrothermal phyllosilicates from their 'host' volcanogenic minerals by weathering could be a source of fine material (Jongmans et al., 1994a; Sedov & Shoba, 1994), as mentioned before.

In regoliths and Regosols, fine material is very rare and displays an undifferentiated b-fabric. It occurs as coatings on the coarse mineral grains (Sokolov, 1972; Schwaighofer, 1976; Bertrand & Fagell, 2008) (Figs. 1B and 6A). Internal hypocoatings of micromass

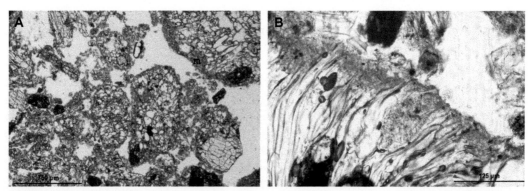

FIGURE 6 Micromass of soils on volcanic ash. (A) Pumice fragments surrounded by micromass coatings (m) and with internal hypocoatings in a shallow soil on historic volcanic ash deposits (Santorini, Greece) (PPL). (B) Detail of micromass internal hypocoating in pumice fragment. Note the micromass entering the open vacuoles. (Santorini, Greece) (PPL).

within pumice fragments, along their sides, were observed by Stoops et al. (2006), Colombo et al. (2007), Stoops (2007) and Stoops and Gérard (2007) (Fig. 6). These internal hypocoatings form by intrusion of micromass in near-surface vacuoles of the fragments (Fig. 6B). Weathering probably contributes to the opening of these vacuoles. Moreover, fine roots often penetrate in the pores of the pumice fragments, where nutrients are available (Sanborn, 2010). The presence of micromass coatings and internal hypocoatings is of great importance for the evaluation of the fertility of these soils, as they mainly occur on rock fragments that are larger than the fine earth fraction and therefore are removed for routine physicochemical analyses, leading to wrong information on the composition and fertility of the soil. The coatings of fine material should not be mistaken for those that form around rock fragments during the eruption (called armed pyroclasts by petrographers, aggregates of coarse ash surrounded by much finer ash) and are preserved during soil formation (Colombo et al., 2007). Similar micromass coatings are observed for non-volcanic soils subject to frost phenomena and explained as stress features (see Van Vliet-Lanoë, 2010; Van Vliet-Lanoë & Fox, 2018).

In Andosols, the micromass generally has an undifferentiated b-fabric, as already stated by Kawai (1969), because of the predominance of amorphous or cryptocrystalline colloidal components (allophane, imogolite, ferrihydrite). Pain (1971) proposed to use an undifferentiated b-fabric as a criterion for the recognition of volcanic ash parent material. The submicroscopic structure of these components is at nanometre scale and is usually studied with transmission electron microscopy (TEM) (e.g., Dubroeucq et al., 2002). Variable quantities of dark humus pigment are present in the fine material. Humus forms very stable complexes with allophanes and persists in buried Andosols of Pleistocene age (Sedov et al., 2003a). The Bw horizons of Andosols have higher clay content than the A horizons and sometimes contain rounded aggregates of different colour, pointing to a colluvial origin (derived from erosion of more evolved soils) rather than pedogenesis *in situ* (Stoops, 2007). Also Paluskova (1988) mentioned the frequent presence of rounded polymictic aggregates in fossil soil sediments on Santorini.

In intergrades to more evolved soils, some areas inside blocky peds are depleted of humus pigment (Sedov et al., 2003b). In soils where allophane is gradually replaced by crystalline clays, the micromass generally presents an undifferentiated b-fabric, even when halloysite is abundant, but stipple-speckled b-fabrics have been described for soils containing kaolinite or illite (e.g., Remmelzwaal, 1978; Fernández-Caldas et al., 1982; Malucelli et al., 1999). A transition of undifferentiated b-fabric in granular allophanic horizons to speckled b-fabric in more weathered blocky material has been described for soils on ash near Pompeii, Italy (Vogel et al., 2016). A similar evolution from undifferentiated to speckled and even weakly striated b-fabric was observed for a chronosequence on Papua New Guinea (Chartres et al., 1985). The b-fabric was more developed with depth, which was attributed to an increase in cryptocrystalline gibbsite in the clay fraction, as confirmed by XRD analysis. The lack of striated b-fabric was explained by the absence of frequent cycles of wetting and drying (Chartres et al, 1985).

In Ferralsols, the b-fabric becomes generally stipple-speckled to weakly striated in a first stage, but finally changes again to undifferentiated, with a cloudy limpidity, typical for oxic or ferralic materials (Rodríguez-Rodríguez et al., 1988; West et al., 1997; see also Marcelino et al., 2010, 2018).

Vertisols derived from volcanic materials can show striated b-fabrics (e.g., Rodriguez Hernandez et al., 1979; Yerima et al., 1987).

Arid soils on andesitic material in southern Spain show striated b-fabrics, together with clay coatings, considered as features inherited from a more humid climate (Aguilar & Delgado, 1974). Calcitic crystallitic b-fabrics have also been observed (Stoops, 1980; Zarei et al., 1997).

The fine material of soils with duric horizon has a variable b-fabric. It is usually undifferentiated or stipple-speckled, sometimes poro- and granostriated (Alfaro Sánchez et al., 1992).

SEM images of fracture surfaces show that the micromass of allophane-rich soils consists of aggregates of small (5-500 µm) globules, in which allophane seems to encapsulate other clay minerals present, whereas the micromass of halloysitic soil materials consists of randomly oriented tubules (Eswaran, 1983).

4. Organic Material

A common feature of most soils on volcanic ash, especially Andosols, is the abundance of fungal hyphae and sklerotia (Fig. 7). They are responsible for a specific humus component, rich in chitine-derived substances (Nierop & Buurman, 2007).

In Regosols, the type and abundance of organic material is mainly determined by climate. It ranges from occurrences of a few roots and hyphae in the dry soils of Santorini (Stoops et al., 2006, 2007; Stoops & Gérard, 2007) to abundant coarse plant residues and fungal hyphae at 4500 m altitude on the Orizaba volcano (Mexico) (Dubroeucq et al., 2002). In the latter example, this is combined with high activity of the mesofauna, as shown by the large excrements containing both organic and mineral fragments. Abundant faunal excrements were also observed for alpine meadow soils of Mexican volcanic highlands (Dubroeucq et al., 2002). A close association between fine roots and pumice grains has been observed, whereby root development appears to be favoured by the vesicular fabric of the grains and their capacity to retain water (De León Gonzales et al., 2007; Sanborn, 2010).

Within the spectrum of climates suitable for Andosol formation, higher humidity is observed to cause greater humus accumulation (Quantin, 1985; Miehlich, 1991), resulting in stronger staining of the fine material with dark humus pigment.

In cool boreal climates under coniferous forest, fragments of partially decomposed plant tissues (raw humus), as well as fungal mycelia, are abundant in the surface horizon, whereas dark organic pigment (colloidal humus) is limited (Shoba et al., 2004). Signs of faunal activity, such as excrements and litter fragmentation, are very scarce.

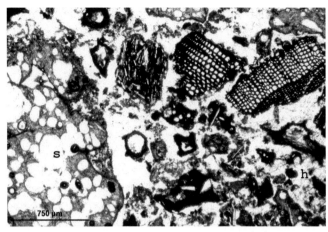

FIGURE 7 Charcoal fragments and sideromelane (s) scoria in Andosol (Tenerife, Canary Islands) (PPL). Note the presence of fungal hyphae (h) in the right lower corner.

Fragments of organ and tissue residues are rather few and found mostly in the surface horizon of Andosols. Charcoal particles are common, not only in the topsoil, but also throughout the profile, including the C horizon (Fig. 7). They originate from the burning of vegetation, which is quite common in regions with volcanic activity.

Frei (1964) stated that fine organic matter coats all coarse grains in Andosols of Ecuador. Humification of organic material is the least advanced in Vitrandepts, compared to Eutrandepts and Dystrandepts (Soil Survey Staff, 1975), and the former also contain numerous fungal hyphae and sklerotia (Tejedor Salguero et al., 1975, 1978, 1979). Dubroeucq et al. (2002) described the presence of organic gel-like substances in the lower part of Andosols on the Mexican highlands, which they considered to be a result of lateral downslope movements. Andosols in the humid tropics have in general a darker and more abundant micromass than in temperate soils, as well as abundant fungal hyphae and phytoliths (Stoops, unpublished data).

5. Pedofeatures

5.1 Illuvial Clay Coatings

Coatings and infillings evidencing illuviation of fine colloidal materials are usually not present, as the major part of the nanocrystalline fine material is coagulated and fixed in rather stable granular aggregates.

In Andosols intermediate to soils with crystalline clays, few (much less than 1%) but clear thin depositional clay coatings are located mostly in channels, but also on walls of planar voids and even in packing voids within areas with a granular structure of the Ah horizon (Sedov et al., 2003b). The coatings are optically isotropic or anisotropic with very low interference colours and mostly without lamination. Their colour varies from yellow to bright brown due to colouration by organic pigments or iron oxides.

Illuvial coatings composed of a mixture of clay and silt, as observed for soils of Tenerife, Canary Islands, are probably related to rapid melting of snow along exposed slopes (Benayas et al., 1974). Strong illuviation, up to 25% by volume, was noticed for Luvisols of the Canary Islands with preservation of a granular microstructure (500-1000 μm) (Rodríguez-Rodríguez et al., 1979, 1980a). For a soil chronosequence formed on Late Quaternary rhyolite tephra, Bakker et al. (1996) reported abundant pumice fragments and low clay contents for the youngest soils, and clay illuviation, clay formation, gleying and a high degree of weathering of primary minerals in the other end-member. Jongmans et al. (1991), studying soil formation on terraces containing volcanic components, concluded that the development of textural pedofeatures (and striated b-fabrics) occurs in soils on terraces that are at least 0.5 million years old. In general, phytoliths are rather frequent in the coarse material of these Bt horizons, in contrast with illuvial horizons on non-volcanic parent rocks. As the majority of phytoliths occurs dispersed in the groundmass and not as part of the illuvial features, they are not accumulated in the Bt horizon by illuviation process. Both the granular aggregates and phytoliths are probably relic features, inherited from an earlier stage.

Illuvial coatings of oriented clay were also described by Stoops (1972) and Goenadi and Tan (1989) for Ultisols on volcanic tephra (Fig. 8A and B). Also soils with duric horizon may contain clay coatings (Oleschko, 1990; Oleschko et al., 1992; Fig. 8C). One example are thick clay coatings composed of halloysite and cristobalite, with higher cristobalite content in the indurated zones (Dubroeucq, 1991). Coatings and infillings of impure clay with poor orientation and abundant silt particles are supposed to be generated by overland flows with a high concentration of suspended matter, associated with Pleistocene eruptions. In these cases, induration may result from structure collapse following water saturation, together with sedimentation from the surface flow (Solleiro-Rebolledo et al., 2003).

5.2 Authigenic Clay Coatings

Already in the initial stage of weathering of volcanic glass, thin isotropic coatings are formed around glass fragments (see Section 3.1.4). Nanocrystalline Si-Al-rich coatings and infillings have been described for Andosols of humid tropical regions (Jongmans et al., 1994b, 1995). They are colourless, optically isotropic and non-laminated in unweathered rock fragments, becoming microlaminated and stained by Fe oxides in higher parts of the profile. These coatings and infillings are considered to be the product of *in situ* coprecipitation of Al and Si, forming allophane. They are subsequently transformed by desilicification, initially near the voids, first to imogolite (SEM-EDS analysis) and later to very fine-grained gibbsite, giving the coatings a slightly anisotropic aspect in XPL.

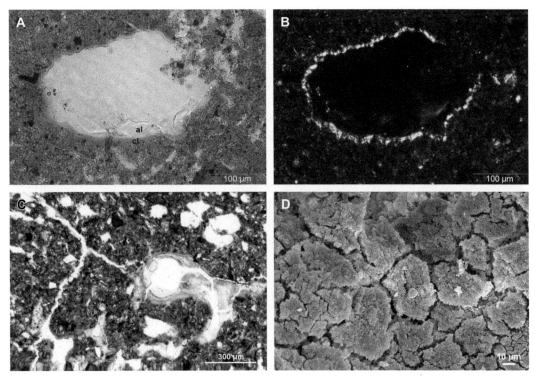

FIGURE 8 Pedofeatures in soils on volcanic ash. (A) Limpid, colourless allophane coating (al) superimposed on yellowish crystalline clay coating (cl) in the C horizon of Luvisol (Isla Santa Cruz, Galapagos) (PPL). (B) Same field in XPL. Note the absence of interference colours in the allophane coating. (C) Clay coatings in channels in a duric horizon (tepetate) under a Luvisol (Santa Catarina, Morelos, Mexico) (PPL). (D) Organo-mineral coating on volcanic mineral grain in a Vitric Andosol (Kamchatka, Russia) (SEM). Note network of fissures.

Jongmans et al. (1996) reported coatings in surface horizons on slopes of active volcanoes, consisting of a mixture of secondary opal-A and ash particles of different sizes. These workers concluded that the formation of such coatings is a common but transient phenomenon under strongly acidic environmental conditions that favour rapid weathering. Absence of a vegetation cover and of fresh ash additions are prerequisites for their formation.

Isotropic coatings, observed for the lower part of the profiles of some Andosols from Costa Rica, are thought to develop by formation of allophane from dissolved Si and Al present in percolating soil solutions (Jongmans et al., 1995).

In Podzols on coarse volcanic material, pale yellow coatings of isotropic clay occur in deeper parts of the Bw horizon (De Coninck & Rijkse, 1986).

For palaeosoils (Luvisols) of the Allier terraces (France), Jongmans et al. (1994c) described coatings that are partly optically isotropic and partly anisotropic, formed by *in situ* precipitation of allophane, which subsequently transforms to unoriented smectitic clays.

Similar observations were made by Gérard et al. (2007) for modern soils. The rather low interference colours and occasional optical isotropism of clay components both in the groundmass and in the pedofeatures is related to the presence of nanocrystalline components, the relatively low birefringence of minerals of the kaolinite group and, in case of halloysite dominance, the tubular or globular particle shape that does not promote formation of oriented (face-to-face) clay domains. Optically isotropic clay coatings in soils with an optically isotropic groundmass were also considered diagnostic of volcanic ash parent material by Dalrymple (1964).

Limpid, colourless, optically isotropic coatings, overlying smectitic clay coatings were reported for a Hapludult (Stoops, 1972, 2013; Eswaran et al., 1973), illustrating the polygenetic character of the soil due to later ashfalls (Fig. 8A and B). Optically isotropic clay coatings in Ferralsols on Java, showing anisotropic zones, were considered by Buurman & Jongmans (1987) as *in situ* precipitated allophane coatings recrystallising under the influence of desiccation to halloysite, or formed as a result of more recent ashfall.

Poetsch and Arikas (1997) and Poetsch (2004), using phase contrast microscopy for sections thinner than the standard thickness (20-30 µm), could identify opaline films juxtaposed on clay coatings in some indurated horizons. A similar observation was reported by Creutzberg et al. (1990). Some authors (Dubroeucq, 1992; Hidalgo et al., 1992) stated that the amorphous silica is closely associated with halloysite particles in the micromass and clay pedofeatures of these soils. It cannot be distinguished by optical methods and should be studied by means of electron microscopy and selective extractions.

5.3   Other Pedofeatures

For Andosols of the alpine meadows of tropical highlands, Dubroeucq et al. (2002) described dark organic coatings in channels in the lower part of the profile. The authors attributed this feature to the vertical and lateral migration of organic gels together with silica and allophane during the rainy season.

Nieuwenhuyse et al. (1993) mentioned the presence of iron oxide hypocoatings in Haplaquands. They were also observed for paddy soils from the Philippines (Stoops, 1984).

Rounded grains of opal or allophane-opal (50-700 µm, exceptionally 1600 µm) were described by von Buch (1967). The grains are clear or turbid, white or yellowish, with very low stress interference colours and a flow fabric. The refractive index ranges from 1.40 to 1.48. They are probably derived from the fragmentation of opaline coatings and infillings as they are soft in the lower horizons and become hard toward the surface. Similar features are frequently observed in thin sections of volcanic ash soils (Stoops, 2013). Another type of fragments are sand- and silt-sized composite particles consisting of mineral fragments, phytoliths and groundmass material cemented by hard allophane-like material, which are supposed to have formed irreversibly during dry periods

(Riezebos & Lustenhouwer, 1983). These granules are not destroyed during grain size analysis and behave as coherent particles during erosion.

Coatings and nodules of fine-grained gibbsite are observed under conditions of strong leaching (Kawai, 1969), especially in the middle and the lower part of profiles (Stoops, 1972; Eswaran et al., 1973; Morrás, 1976).

Infillings with granular aggregates in ant galleries and fungus chambers were described by Eschenbrenner (1994).

Only limited information is available on the presence of placic horizons in soils on volcanic ash. Pinheiro & Rodríguez-Rodríguez (1994) and Pinheiro et al. (2004) described placic horizons occurring in Placudands and Placaquands (Soil Survey Staff, 2014) of the Terceira Island (Azores). The groundmass consists of pyroclasts and a brownish, speckled or limpid clay with an undifferentiated b-fabric, resulting in a chitonic or enaulic c/f-related distribution pattern. The placic horizon is formed by a superposition of a red to dark red, rather limpid, optically isotropic, locally globular material, forming a dense porphyric layer with a lamellar fabric on top, and a diffuse transition to the lower parts (see also Wilson & Righi, 2010; Van Ranst et al., 2018).

In duric horizons on volcanic ash, various pedofeatures can be present, depending on the pedogenic environment. Besides clay coatings (Fig. 8C), they include ferruginous nodules in more humid soil types (Solleiro-Rebolledo et al., 2003) and carbonates in drier areas (Fedoroff et al., 1994), as well as components derived from palaeosoils such as fragments of Bt horizons or fragments of clay coatings. Some of these features contribute to the cementation of the horizon (Quantin, 1992).

Horizons beneath stratified litter in Kamchatka show a combination of features, typical for Andosols and for the spodic horizon of Podzols (Shoba et al., 2004). Pyroclastic grains are covered by thick optically isotropic micromass coatings with a bright yellow-brown colour, often dissected by a network of desiccation fissures that are clearly seen in SEM images (Fig. 8D). Compound coatings of allophane and organic matter surrounding volcanic ash grains were observed by Ivanov et al. (2014) (see Section 3.1.4). For the Bs horizons of Podzols on coarse materials in New Zealand, De Coninck and Rijkse (1986) observed aggregates of loosely packed pumice fragments and fine pellets to coexist with loose infillings and denser aggregates with a yellowish micromass, coated by monomorphic organic material. The authors concluded that the pellets are the result of local decomposition of the root system.

In a calcic Gypsiorthid described by Stoops (1980 and unpublished data), calcite penetrates in the vacuoles of the pyroclasts. Several of these pyroclasts show remains of limpid, yellowish, optically anisotropic coatings of strongly oriented clay, pointing to an earlier phase of pedogenesis under different conditions, whereas calcite pendants result from present-day soil formation. Sauer and Zöller (2006) described orthic, disorthic and anorthic micritic carbonate nodules for a complex profile on Lanzarote (Canary Islands). The morphology of calcic and petrocalcic horizons in arid soils on volcanic ash is similar to that of these horizons in soils on other parent materials (Jahn et al., 1985; Jahn & Stahr, 1994).

6. Conclusions

Studies of topo-, chrono- and climasequences of soils on volcanic ash (e.g., Stoops, 1972; Fernández-Caldas et al., 1982; Chartres et al., 1985; Dubroeucq et al., 1998; Stoops, 2013) clearly show a general trend: already in an early stage of pedogenesis a granular microstructure develops, and this persists until a blocky microstructure appears in more developed soils (e.g., Luvisols). The amount of mineral grains and glass particles that were part of the parent material rapidly diminishes till only iddingsite, magnetite and/or ilmenite remain in andesitic or basaltic tephra. The b-fabric is undifferentiated and evolves only slowly to weakly speckled in more developed soils. This evolution seems to occur more or less in parallel with an evolution of allophane to halloysite, and subsequently to cryptocrystalline gibbsite. Striated b-fabrics only form when hydromorphic conditions interfere. As stated by Chesworth (1973), all materials evolve to a residual system, and time nullifies the influence of parent materials. From a micromorphological point of view, not all fabric elements of the parent material or early alteration stages seem to be obliterated, as for instance b-fabric remains undifferentiated or weakly speckled as long as no hydromorphism occurs.

The literature contain many examples of the use of chemical and/or physical methods to disentangle the complex geological and pedological history of a profile by for instance differences in grain size and trace element content (Buurman et al., 2004). Also mineralogical studies of the coarse fraction, using grain mounts, are commonly used for this purpose. Interdisciplinary studies (Arnalds et al., 2007) show that in most cases micromorphological investigations yield the clearest results. Compared to grain mount studies, thin sections have the advantage that minerals and rock fragments of medium and coarse sand size can be identified and that even larger grains can be studied (Stoops & Van Driessche, 2007).

In studies of volcanic materials little effort has been made to distinguish materials derived from direct ashfall and those transported and recycled by aeolian activity, colluvial processes (e.g., lahar formation) or tsunamis. In addition, it is not always clear whether the described soils developed on tephra or on lava flows. A major contribution of micromorphology has been to clarify the complexities of accumulation patterns, including occurrences of standstill phases (accumulation of organic matter and phytoliths), water-born sedimentation and aeolian deposition.

Insufficient attention has been given to vertical petrographic heterogeneity when studying pedogenesis in soils on volcanic ash. In some cases the fabric of Bw horizons seems to be the result of colluvial input derived from more developed soils, rather than the result of a pedogenic process (Stoops, 2007). This is not necessarily detected by other research methods, such as chemical and mineralogical analyses.

Information on soils on volcanic ash in arid environments is very limited, apart from some data for palaeosoils. More research is necessary.

Several issues related to the micromorphology of soils formed on volcanic material have still to be clarified. The most prominent questions are (i) whether the origin of the typical granular microstructure is related to biological activity or physicochemical processes, or both; (ii) whether fine material that appears during early stages of soil development formed *in situ* by weathering or fragmentation, or whether it represents an aeolian input; (iii) whether differences exist between the micromorphological characteristics of allophanic Andosols (dominated by non-crystalline Al silicates) and non-allophanic Andosols (dominated by organometallic complexes, e.g., on shales); and (iv) to what extent lateral translocation of materials between pedons is important, as suggested by Chen et al. (1999).

Acknowledgements

The authors want to thank Dr. A. Jongmans, Dr. H. Mücher and Dr. I. Simpson for their useful corrections and suggestions.

References

Aguilar, J. & Delgado, M., 1974. Micromorphological study of soils developed on andesitic rocks in oriental Andalucia (Spain). In Rutherford, G.K. (ed.), Soil Microscopy. The Limestone Press, Kingston, pp. 281-295.

Alfaro Sánchez, G., Oleschko, K. & Meza Sánchez, M., 1992. Rasgos micromorfológicos de los tepetates de Hueypoxtla (Estado de México). Terra 10, 253-257.

Algoe, C., Stoops, G., Vandenberghe, R.E. & Van Ranst, E., 2012. Selective dissolution of Fe-Ti oxides - Extractable iron as a criterion for andic properties revisited. Catena 92, 49-54.

Arnalds, O., Bartoli, F., Buurman, P., Garcia-Rodeja, E., Oskarsson, H. & Stoops, G. (eds.), 2007. Soils of Volcanic Regions of Europe. Springer-Verlag, Berlin, 644 p.

Bakker, L., Lowe, D.J. & Jongmans, A.G., 1996. A micromorphological study of pedogenic processes in an evolutionary soil sequence formed on Late Quaternary rhyolitic tephra deposits, North Island, New Zealand. Quaternary International 34-36, 249-261.

Bech-Borras, J., Fedoroff, N. & Sole, A., 1977. Etude des andosols d'Olot (Gerona, Espagne). 3e partie: micromorphologie. Cahiers ORSTOM, Série Pédologie 15, 381-390.

Benayas, J., Alonso, J. & Fernández Caldas, E., 1974. Effect of the ecological environment on the micromorphology and mineralogy of Andosols (Tenerife Island). In Rutherford, G.K. (ed.), Soil Microscopy. The Limestone Press, Kingston, pp. 306-319.

Benayas, J., Fernández Caldas, E. & Tejedor Salguero, M.L., 1978. Estudio micromorfológico de Vitrandepts (I. Tenerife). Anales de Edafologia y Agrobiologia 37, 295-302.

Benayas, J., Fernández Caldas, E., Tejedor Salguero, M.L. & Rodríguez Rodríguez, A., 1980. Características micromorfológicas de los suelos de una climasecuencia de la vertiente meridional de la Isla de Tenerife. Anales de Edafologia y Agrobiologia 39, 51-74.

Bertrand, S. & Fagell, N., 2008. Nature and deposition of andosol parent material in south-central Chili (36-42°S). Catena 73, 10-22.

Bruins, H.J., MacGillivray, J.A., Synolakis, C.E., Benjamini, C., Keller, J., Kisch, H.J., Klügel, A. & van der Plicht, J., 2008. Geoarchaeological tsunami deposits at Palaikastro (Crete) and the Late Minoan IA eruption of Santorini. Journal of Archaeological Science 35, 191-212.

Buurman, P. & Jongmans, A.G., 1987. Amorphous clay coatings in a lowland Oxisol and other andesitic soils of West Java, Indonesia. Pemberitaan Penelitaan Tanah dan Pupuk 7, 31-40.

Buurman, P., Garcia Rodeja, E., Martinez Cortizas, A. & van Doesburg, J.D.J., 2004. Stratification of parent material in European volcanic and related soils studied by laser-diffraction grain-sizing and chemical analysis. Catena 56, 127-144.

Chartres, C.J., Wood, A. & Pain, C.F., 1985. The development of micromorphological features in relation to some mineralogical and chemical properties of volcanic ash soils in highland Papua New Guinea. Australian Journal of Soil Research 23, 339-354.

Chen, Z.S., Asio, V.B. & Yi, D.F., 1999. Characteristics and genesis of volcanic soils along a toposequence under subtropical climate in Taiwan. Soil Science 164, 510-525.

Chesworth, W., 1973. The residua system of chemical weathering. Journal of Soil Science 24, 69-81.

Colombo, C., Palumbo, G., Sellitto, V.M., Terribile, F., Gérard, M. & Stoops, G., 2007. Characteristics and genesis of volcanic ash soils in Southern Central Italy: Phlegreaen Fields (Campania) and Vico Lake (Latium). In Arnalds, O., Bartoli, F., Buurman, P., Garcia-Rodeja, E., Oskarsson, H. & Stoops, G. (eds.), Soils of Volcanic Regions of Europe. Springer-Verlag, Berlin, pp. 197-230.

Creutzberg, D., Kauffman, J.H., Bridges, E.M. & Guillermo Del Posso, G.M., 1990. Micromorphology of 'cangahua': a cemented subsurface horizon in soils from Ecuador. In Douglas, L.A. (ed.), Soil Micromorphology: A Basic and Applied Science. Developments in Soil Science, Volume 19. Elsevier, Amsterdam, pp. 367-372.

Dalrymple, J.B., 1964. The application of soil micromorphology to the recognition and interpretation of fossil soils in volcanic ash deposits from the North Island, New Zealand. In Jongerius, A. (ed.), Soil Micromorphology. Elsevier, Amsterdam, pp. 339-350.

De Coninck, F. & Rijkse, W.C., 1986. Micromorphology of three podzols in coarse volcanic material (N. Island, New Zealand). Transactions of the 13th Congress of the International Society of Soil Science, Hamburg, pp. 1549-1550.

De León-González, F., Gutiérrez-Castorena, M.C., González-Chávez, M.C.A. & Castillo-Juárez, H., 2007. Root-aggregation in a pumiceous sandy soil. Geoderma 142, 308-317.

Delvigne, J., Bisdom, E.B.A., Sleeman, J. & Stoops, G., 1979. Olivines, their pseudomorphs and secondary products. Pedologie 29, 247-309.

Dubroeucq, D., 1991. Les 'tepetates' de Xalapa, Veracruz (Mexico): une induration pédologique dans des sols d'origine volcanique. Microscopie et minéralogie des indurations. Cahiers. ORSTOM, Série Pédologie 26, 235-242.

Dubroeucq, D., 1992. Los tepetates de la región de Xalapa, Veracruz (México): un endurecimiento de origen pedológico. Terra 10, 233-240.

Dubroeucq, D., Geissert, D. & Quantin, P., 1998. Weathering and soil forming processes under semi-arid conditions in two Mexican volcanic ash soils. Geoderma 86, 99-122.

Dubroeucq, D., Geissert, D., Barois, I. & Ledru, M.P., 2002. Biological and mineralogical features of Andisols in the Mexican volcanic highlands. Catena 49, 183-202.

Eschenbrenner, V., 1994. The influence of fungus-cultivating ants (Hymenoptera, Forrnicidae, Attini) on the morphology of andosols in Martinique. In Ringrose-Voase, A. & Humphreys, G.S. (eds.), Soil Micromorphology: Studies in Management and Genesis. Developments in Soil Science, Volume 22. Elsevier, Amsterdam, pp. 405-410.

Eswaran, H., 1983. Characterization of domains with the scanning electron microscope. Pedologie 33, 41-54.

Eswaran, H., Stoops, G. & De Paepe, P., 1973. A contribution to the study of soil formation on Isla Santa Cruz, Galápagos. Pedologie 23, 100-122.

Fauzi, A.I. & Stoops, G., 2004. Influence of Krakatau ash fall on pedogenesis in West Java. Example of a toposequence in the Honje Mountains, Ujung Kulon Peninsula. Catena 56, 45-66.

Fedoroff, N. & Rodriguez, A., 1978. Comparaison micromorphologique des sols rouges des Iles Canaries et du basin mediterraneen. In Delgado, M. (ed.), Soil Micromorphology. University of Granada, pp. 867-928.

Fedoroff, N., Courty, M.A., Lacroix, E. & Oleschko, K., 1994. Calcitic accretion on indurated volcanic materials (example of tepetates, Altiplano, Mexico). Transactions of the 15th World Congress of Soil Science, Acapulco, Volume 6a, pp. 460-473.

Fehér, O., Langohr, R., Füleky, G. & Jakab, S., 2007. Late Glacial-Holocene genesis of Andosols from the Seaca-Tãtarca (South-Gurghiu Mountains, Romania). European Journal of Soil Science 58, 405-418.

Fernández-Caldas, E., Tejedor Salguero, M.L. & Quantin, P., 1982. Suelos de Regions Volcánicas Tenerife, Islas Canarias. Coleccion Viera y Clavijo, Universidad de La Laguna, CSIC, 250 p.

Fisher, R.V. & Schmincke, H.U., 1984a. Pyroclastic Rocks. Springer-Verlag, Berlin, 472 p.

Fisher, R.V. & Schmincke, H.U., 1984b. Alteration of volcanic glass. In Fisher, R.V. & Schmincke, H.U. (eds.), Pyroclastic Rocks. Springer-Verlag, Berlin, pp. 312-345.

Frei, E., 1964. Micromorphology of some tropical mountain soils. In Jongerius, A. (ed.), Soil Micromorphology. Elsevier, Amsterdam, pp. 307-312.

Gebhardt, H., Hugenroth, P. & Meyer, B., 1969. Pedochemische Verwitterung und Mineral-Umwandlung im Trachyt-Bims, Trachyt-Tuff und in den Tuff-Mischsedimenten der Laacher Eruptionsphase. Göttinger Bodenkundliche Berichte 11, 1-83.

Gérard, M., Caquineau, S., Pinheiro, J. & Stoops, G., 2007. Weathering and allophane neoformation in soils on volcanic ash from the Azores. European Journal of Soil Science 58, 496-515.

Goenadi, D. & Tan, K.H., 1989. Mineralogy and micromorphology of soils from volcanic tuffs in the humid tropics. Soil Science Society of America Journal 53, 1907-1911.

Gutiérrez-Castorena, Ma. del C. & Effland, W.R., 2010. Pedogenic and biogenic siliceous features. In Stoops, G., Marcelino, V. & Mees, F. (eds.), Interpretation of Micromorphological Features of Soils and Regoliths. Elsevier, Amsterdam, pp. 471-496.

Gutiérrez-Castorena, Ma. del C., 2018. Pedogenic siliceous features. In Stoops, G., Marcelino, V. & Mees, F. (eds.), Interpretation of Micromorphological Features of Soils and Regoliths. Second Edition. Elsevier, Amsterdam, pp. 127-155.

Hidalgo, M.C., Quantin, P. & Zebrowski, C., 1992. La cementación de los tepetates: estudio de la silicificación. Terra 10, 192-201.

IUSS Working Group WRB, 2015. World Reference Base for Soil Resources 2014, Update 2015. World Soil Resources Reports No. 106. FAO, Rome, 192 p.

Ivanov, A., Shoba, S. & Krasilnikov, P., 2014. A pedogeographical view of volcanic soils under cold humid conditions: the Commander Islands. Geoderma 235, 48-58.

Jahn, R. & Stahr, K., 1994. Formation of petrocalcic horizons in soils from basic pyroclastics under the semiarid climate of Lanzarote (Spain). Transactions of the 15th World Congress of Soil Science, Acapulco, Volume 6a, pp. 474-480.

Jahn, R., Gudmundsson, T. & Stahr, K., 1985. Carbonatisation as a soil forming process on soils from basic pyroclastic fall deposits on the Island of Lanzarote, Spain. In Fernández-Caldas, E. & Yaalon, D. (eds.), Volcanic Soils. Catena Supplement 7, 87-97.

Jahn, R., Zarei, M. & Stahr, K., 1992. Development of andic soil properties and of clay minerals in the semiarid climate of Lanzarote (Spain). Mineralogica et Petrographica Acta 35A, 193-201.

Jongmans, A.G., Feijtel, T.C.J., Miedema, R., van Breemen, N. & Veldcamp, A., 1991. Soil formation in a Quaternary terrace sequence of the Allier, Limagne, France. Macro- and micromorphology, particle size distribution, chemistry. Geoderma 49, 215-239.

Jongmans, A.G., van Oort, F., Nieuwenhuyse, A., Buurman, P., Jaunet, A.M. & van Doesburg, J.D.J., 1994a. Inheritance of 2:1 phyllosilicates in Costa Rican Andosols. Soil Science Society of America Journal 58, 494-501.

Jongmans, A.G., van Oort, F., Buurman, P., Jaunet, A.M. & van Doesburg, J.D.J., 1994b. Morphology, chemistry and mineralogy of isotropic aluminosilicate coatings in a Guadeloupe Andisol. Soil Science Society of America Journal 58, 501-507.

Jongmans, A.G., van Oort, F., Buurman, P. & Jaunet, A.M., 1994c. Micromorphology and submicroscopy of isotropic and anisotropic Al/Si coatings in a Quaternary Allier terrace, (France). In Ringrose-Voase, A. & Humphreys, G.S. (eds.), Soil Micromorphology: Studies in Management and Genesis. Developments in Soil Science, Volume 22. Elsevier, Amsterdam, pp. 285-291.

Jongmans, A.G., Verburg, P., Nieuwenhuyse, A. & Van Oort, F., 1995. Allophane, imogolite, and gibbsite in coatings in a Costa Rican Andisol. Geoderma 64, 327-342.

Jongmans, A.G., Mulder, J., Groenestein, K. & Buurman, P., 1996. Soil surface coatings at Costa Rican recently active volcanoes. Soil Science Society of America Journal 60, 1871-1880.

Jongmans, A.G., van Oort, F., Denaix, L. & Jaunet, A.M., 1999. Mineral micro- and nano-variability revealed by combined micromorphology and in situ submicroscopy. Catena 35, 259-279.

Jongmans, A.G., Denaix, L., van Oort, F. & Nieuwenhuyse, A., 2000. Induration of soil horizons by allophane and imogolite in Costa Rican volcanic soils. Soil Science Society of America Journal 64, 254-262.

Kamanina, I.Z., 1997. Phytoliths data analysis of soils of different landscape zones. In Pinilla, A., Juan-Tresserras, J. & Machado, J. (eds.), First European Meeting on Phytolith Research. Madrid. Monografías del Centro de Ciencias Medioambientales 4, 23-32.

Kawai, K., 1969. Micromorphological studies of Andosols in Japan. Bulletin of the National Institute of Agricultural Sciences (Japan) 20, 77-154.

Kress-Voltz, M., 1964. Gefüge und Strukturuntersuchungen an vulkanogenen Edaphoiden. In Jongerius, A. (ed.), Soil Micromorphology. Elsevier, Amsterdam, pp. 139-150.

Kress-Voltz, M., 1967. Mikromorphologische und mikromorphometrische Untersuchungen an vulkani-schem Material und dessen praktische Bedeutung. In Kubiëna, W.L. (ed.), Die Mikromorphometrische Bodenanalyse. Ferdinand Enke Verlag, Stuttgart, pp. 92-101.

Kubiëna, W.L., 1956. Materialien zur Geschichte der Bodenbildung auf der Westkanaren. Comptes Rendus du VI Congrès International de Science du Sol, Paris, Volume E, pp. 241-246.

Le Maitre, R.W., Streckeisen, A., Zanettin, B., Le Bas, M.J., Bonin, B., Bateman, P., Bellieni, G., Dudek, A., Efremova, S., Keller, J., Lameyre, J., Sabine, P.A., Schmid, R., Sorensen, H. & Woolley, A.R. (eds.), 2002. Igneous Rocks. A Classification and Glossary of Terms. Recommendations of the International Union of Geological Sciences Subcommission on the Systematics of Igneous Rocks. Cambridge University Press, Cambridge, 252 p.

Lindbo, D.L., Stolt, M.H. & Vepraskas, M.J., 2010. Redoximorphic features. In Stoops, G., Marcelino, V. & Mees, F. (eds.), Interpretation of Micromorphological Features of Soils and Regoliths. Elsevier, Amsterdam, pp. 129-147.

Malucelli, F., Terribile, F. & Colombo, C., 1999. Mineralogy, micromorphology and chemical analysis of andosols on the Island of São Miguel (Azores). Geoderma 88, 73-98.

Marcelino, V., Stoops, G. & Schaefer, C.E.G.R., 2010. Oxic and related materials. In Stoops, G., Marcelino, V. & Mees, F. (eds.), Interpretation of Micromorphological Features of Soils and Regoliths. Elsevier, Amsterdam, pp. 305-327.

Marcelino, V., Schaefer, C.E.G.R. & Stoops, G., 2018. Oxic and related materials. In Stoops, G., Marcelino, V. & Mees, F. (eds.), Interpretation of Micromorphological Features of Soils and Regoliths. Second Edition. Elsevier, Amsterdam, pp. 663-689.

McLoughlin, N., Staudigel, H., Furnes, H., Eickmann, B. & Ivarsson, M., 2010. Mechanisms of micro-tunneling in rock substrates: distinguishing endolithic biosignatures from abiotic microtunnels. Geobiology 8, 245-255.

Miehlich, G., 1991. Chronosequences of Volcanic Ash Soils. Hamburger Bodenkundliche Arbeiten 15, 207 p.

Miehlich, G., 1992. Formation and properties of tepetate in the central highlands of México. Terra 10, 137-144.

Milek, K.B., 2006. Aðalstraeti, Reykjavik, 2001. Geoarchaeological report on the deposits within the house of the soils immediately pre- and post-dating its occupation. In Roberts, H.M. (ed.), Excavations at Aðalstraeti 2003. Fornleifastofnun Islands. Reykjavik, pp. 73-114.

Morrás, H.J.M., 1976. Gibbsite glaebules of a soil profile from Santa Cruz Island, Galapagos (Ecuador). Pedologie 26, 91-96.

Nierop, K.G.J. & Buurman, P., 2007. Thermally assisted hydrolysis and methylation of organic matter in two allophanic volcanic ash soils from the Azores Islands. In Arnalds, O., Bartoli, F., Buurman, P., Garcia-Rodeja, E., Oskarsson, H. & Stoops, G. (eds.), Soils of Volcanic Regions of Europe. Springer-Verlag, Berlin, pp. 411-422.

Nieuwenhuyse, A., Jongmans, A.G. & van Breemen, N., 1993. Andisol formation in a Holocene beach ridge plain under the humid tropical climate of the Atlantic coast of Costa Rica. Geoderma 57, 423-442.

Oleschko, K., 1990. Cementing agents morphology and its relation to the nature of 'Tepetates'. In Douglas, L. A. (ed.), Soil Micromorphology: A Basic and Applied Science. Developments in Soil Science, Volume 19. Elsevier, Amsterdam, pp. 381-386.

Oleschko, K., Zebrowski, C., Quantin, P. & Fedoroff, N., 1992. Patrones micromorfológicos de organización de arcillas en tepetates (México). Terra 10, 183-191.

Pain, L.F., 1971. Micromorphology of soils developed from volcanic ash and river alluvium in the Kohada Valley, Northern District, Papua. Journal of Soil Science 22, 275-280.

Paluskova, K., 1988. Fossile Verwittererungshorizonte in Vulkaniten der Inselgruppe Santorin (Kykladen, Griechenland). Mitteilungen Geologisches Paläontologisches Institut Universtät Hamburg 67, 145-289.

Pinheiro, J. & Rodríguez-Rodríguez, A., 1994. Micromorphology of placic horizons of Andosols of the Azores. Transactions of the 15th World Congress of Soil Science, Acapulco, Volume 6b, pp. 222-223.

Pinheiro, J., Tejedor Salguero, M. & Rodriguez, A., 2004. Genesis of placic horizons in Andisols from Terceira Island, Azores − Portugal. Catena 56, 85-94.

Piperno, D.R., 1988. Phytolith Analysis: An Archaeological and Geological Perspective. Academic Press, San Diego, 280 p.

Poetsch, T., 2004. Forms and dynamics of silica gel in a tuff-dominated soil complex: results of micromorphological studies in the central highlands of Mexico. Revista Mexicana de Ciencias Geológicas 21, 195-201.

Poetsch, T. & Arikas, K., 1997. The micromorphological appearance of free silica in some soils of volcanic origin in central Mexico. In Zebrowski, C., Quantin, P. & Trujillo, G. (eds.), Suelos Volcánicos Endurecidos. Impressora Polar, Quito, pp. 56-64.

Quantin, P., 1985. Characteristics of the Vanuatu Andosols. In Fernández-Caldas, E. & Yaalon, D.Y. (eds.), Volcanic Soils. Catena Supplement 7, 99-105.

Quantin, P., 1992. L'induration des matériaux volcaniques pyroclastiques en Amerique Latine: processus géologiques et pédologiques. Terra 10, 24-33.

Quantin, P., Fernández-Caldas, E. & Tejedor Salguero, M.L., 1978. Séquence climatique des sols récents de la region septentrionale de Tenerife (Iles Canaries). Cahiers ORSTOM, Série Pédologie 16, 397-412.

Remmelzwaal, A., 1978. Soil Genesis and Quaternary Landscape Development in the Thyrrhenian Coastal Area of South-Central Italy. PhD Dissertation, University of Amsterdam, 309 p.

Riezebos, P.A. & Lustenhouwer, W.J., 1983. Characteristics and significance of composite particles derived from a Colombian andosol profile. Geoderma 30, 195-217.

Rodriguez Hernandez, C.M., Fernández-Caldas, E., Fedoroff, N. & Quantin, P., 1979. Les Vertisols des Iles Canaries occidentales. Etude physico-chimique, minéralogique et micromorphologique. Pedologie 29, 71-107.

Rodríguez-Rodríguez, A., Tejedor Salguero, M.L. & Fernández-Caldas, E., 1979. Suelos fersialíticos sobre lapillis basálticos. II Características micromorfológicas. Interpretation y classificacion. Anales de Edafologia y Agrobiologia 38, 1945-1950.

Rodríguez-Rodríguez, A., Fedoroff, N., Tejedor Salguero, M.L. & Fernández-Caldas, E., 1980a. Suelos fersialíticos sobre cenizas volcanicas. III Características micromorfológicas. Interpretation y classificacion. Anales de Edafologia y Agrobiologia 39, 37-49.

Rodríguez-Rodríguez, A., Fedoroff, N., Tejedor Salguero, M.L. & Fernández-Caldas, E., 1980b. Observaciones preliminares sobre la alteracion en los suelos fersialíticos sobre materiales volcanicos (Islas Canarias). Anales de Edafologia y Agrobiologia 39, 1923-1940.

Rodríguez-Rodríguez, A., Jimenez Mendoza, C.C. & Tejedor Salguero, M.L., 1988. Micromorfologia de los suelos ferralíticos en las Islas Canarias. Anales de Edafologia y Agrobiologia 47, 409-430.

Romans, J.C.C., Robertson, L. & Dent, D.L., 1980. The micromorphology of young soils from South-East Iceland. Geografiska Analer 62A, 93-103.

Sanborn, P., 2010. Soil formation on supraglacial tephra deposits, Klutlan Glacier, Yukon Territory. Canadian Journal of Soil Science 90, 611-618.

Sauer, D. & Zöller, L., 2006. Mikromorphologie der Paläoböden der Profile Femés und Guatiza, Lanzarote. In Zöller, L. & von Suchodoletz, H. (eds.), Östliche Kanareninseln – Natur, Mensch, Umweltprobleme. Bayreuther Geographische Arbeiten 27, pp. 105-130.

Schmid, R., 1981. Descriptive nomenclature and classification of pyroclastic deposits and fragments: recommendations of the IUGS Subcommission on the Systematics of Igneous Rocks. Geology 9, 41-43.

Schwaighofer, B., 1976. Mineralogisch-chemische Untersuchungen bei der Verwitterung pyroklastische Gesteine auf Teneriffe (Kanarische Inseln). Geoderma 16, 285-315.

Sedov, S.N. & Shoba, S.A., 1994. Types of pedogenesis on basic rocks in boreal regions. Transactions of 15th World Congress of Soil Science, Acapulco, Volume 6b, pp. 84-85.

Sedov, S., Solleiro-Rebolledo, E., Morales-Puente, P., Arias-Herrería, A., Vallejo-Gómez, E. & Jasso-Castañeda, C., 2003a. Mineral and organic components of the buried paleosols of the Nevado de Toluca, central Mexico as indicators of paleoenvironments and soil evolution. Quaternary International 106/107, 169-184.

Sedov, S.N., Solleiro-Rebolledo, E. & Gama-Castro, J.E., 2003b. Andosol to Luvisol evolution in Central Mexico: timing, mechanisms and environmental setting. Catena 54, 495-513.

Shin, J.S. & Stoops, G., 1988. Composition and genesis of volcanic ash soils in Jeju Island. I. Physico-chemical and macro-micromorphological properties. Journal of the Mineralogical Society of Korea 1, 32-39.

Shoba, S., Targulian, V., Sedov, S., Sakarov, A. & Zacharichina, L., 2004. Pedogenesis and weathering on tephra: climate and time dependency. In Kapur, S., Akça, E., Montanarella, L., Ozturk, A. & Mermut, A. (eds.), Extended Abstracts of the 12th International Working Meeting on Soil Micromorphology, Adana, pp. 37-39.

Shoji, S., Nanzyo, M. & Dahlgren, R.A., 1993. Volcanic Ash Soils. Genesis, Properties and Utilization. Elsevier, Amsterdam, 288 p.

Simpson, I.A., Milek, K.B. & Gudmundsson, G., 1999. A reinterpretation of the Great Pit at Hofstadir, Iceland, using sediment thin section micromorphology. Geoarchaeology 14, 511-530.

Soil Survey Staff, 1975. Soil Taxonomy. A Basic System of Soil Classification for Making and Interpreting Soil Surveys. 1st edition. Agriculture Handbook 436. USDA, NRCS, Washington, 754 p.

Soil Survey Staff, 2014. Keys to Soil Taxonomy. 12th Edition. USDA-Natural Resources Conservation Service, Washington, 360 p.

Sokolov, I., 1972. Weathering of volcano-clastic deposits in conditions of cold humid climate. In Kowalinski, S. & Drozd, J. (eds.), Soil Micromorphology. Panstwowe Wydawnicto Naukowe, Warsawa, pp. 513-518.

Solleiro-Rebolledo, E., Sedov, S., Gama-Castro, J.E., Flores-Román, D. & Escamilla-Sarabia, G., 2003. Paleosol-sedimentary sequences of the Glacis de Buenavista, central Mexico: interaction of Late Quaternary pedogenesis and volcanic sedimentation. Quaternary International 106-107,185-201.

Staudigel, H., Furnes, H., McLoughlin.N., Banerjee, N.R., Connell, L.B. & Templeton, A., 2008. 3.5 Billion years of glass bioalteration: volcanic rocks as a basis for microbial life? Earth-Science Reviews 89, 156-176.

Stoops, G., 1972. Micromorphology of some important soils of Isla Santa Cruz (Galápagos). In Kowalinski, S. & Drozd, J. (eds.), Soil Micromorphology. Panstwowe Wydawnicto Naukowe, Warsawa, pp. 407-420.

Stoops, G., 1980. Micromorphological descriptions. Tour Guide, Soil Classification Workshop, ICOMMORT, ACSAD, Damascus, pp. 56-57.

Stoops, G., 1983. Mineralogy and micromorphology of some andisols of Rwanda. In Beinroth, F.H., Neel, H. & Eswaran, H. (eds.), Proceedings of the Fourth International Soil Classification Workshop, Rwanda, June 1981. Part 1, Papers. ABOS-AGCD, Agricultural Editions 4, pp. 150-164.

Stoops, G., 1984. Micromorphological descriptions of some wetland soils in the Philippines. In Workshop on Wetland Soils, Field Book. International Rice Research Institute, Los Baños, pp.125-150.

Stoops, G., 2007. Micromorphology of soils derived from volcanic ash in Europe. A review and synthesis. European Journal Soil Science 58, 356-377.

Stoops, G., 2013. A micromorphological evaluation of pedogenesis on Isla Santa Cruz (Galápagos). Spanish Journal of Soil Science 3, 14-37.

Stoops, G. & Gérard, M., 2007. Micromorphology. In Arnalds, O., Bartoli, F., Buurman, P., Garcia-Rodeja, E., Oskarsson, H. & Stoops, G. (eds.), Soils of Volcanic Regions of Europe. Springer-Verlag, Berlin, pp. 129-140.

Stoops, G. & Mees, F., 2018. Groundmass composition and fabric. In Stoops, G., Marcelino, V. & Mees, F. (eds.), Interpretation of Micromorphological Features of Soils and Regoliths. Second Edition. Elsevier, Amsterdam, pp. 73-125.

Stoops, G. & Van Driessche, A., 2007. Mineralogy of the sand fraction – results and problems. In Arnalds, O., Bartoli, F., Buurman, P., Garcia-Rodeja, E., Oskarsson, H. & Stoops, G. (eds.), Soils of Volcanic Regions of Europe. Springer-Verlag, Berlin, pp. 141-153.

Stoops, G., Valvoulidou, E. & Monteiro, F., 2006. Micropedology, mineralogy and biology of soils derived from volcanic ash on Santorini (Greece). Geophysical Research Abstracts, Volume 8, EGU06-A-10388.

Stoops, G., FitzPatrick, E.A. & Gérard, M., 2007. Micromorphological descriptions of thin sections of volcanic ash soils of COST-622 reference profiles. In Arnalds, O., Bartoli, F., Buurman, P., Garcia-Rodeja, E., Oskarsson, H. & Stoops, G. (eds.), Soils of Volcanic Regions of Europe. Springer-Verlag, Berlin, 58 p. (CD-ROM).

Stoops, G., Gérard, M. & Arnalds, O., 2008. A micromorphological study of Andosol genesis in Iceland. In Kapur, S., Mermut, A. & Stoops, G (eds.), New Trends in Micromorphology. Springer-Verlag, Berlin, pp 67-90.

Stroncik, N.A. & Schmincke, H.U., 2002. Palagonite – a review. International Journal of Earth Sciences 91, 680-697.

Tan, K.H. & Goenadi, D.H., 1994. The normal related distribution pattern of soils developed in volcanic ash in the humid tropics. In Ringrose-Voase, A & Humphreys, G.S. (eds.), Soil Micromorphology: Studies in Management and Genesis. Developments in Soil Science, Volume 22. Elsevier, Amsterdam, pp. 343-352.

Tejedor Salguero, M.L., Benayas, J. & Fernández-Caldas, E., 1975. Estudio físico-quìmico y micro-morfológico de intergrados andosol-tierra parda oligotrófica, en un perfil complejo. Anales de Edafologia y Agrobiologia 34, 813-828.

Tejedor Salguero, M.L., Benayas, J. & Fernández-Caldas, E., 1978. Physicochemical and micromorpho-logical study of climatic and chronologic sequences of soils in Tenerife (Canary Islands). In Delgado, M. (ed.), Soil Micromorphology. Department of Edaphology, University of Granada, pp. 631-652.

Tejedor Salguero, M.L., Quantin, P. & Fernández-Caldas, E., 1979. Séquence climatique des sols anciens de la region septentrionale de Tenerife (Iles Canaries). Cahiers ORSTOM, Série Pédologie 17, 119-127.

Tejedor Salguero, M.L., Garcia-Lopez, L. & Fernández-Caldas, E, 1984/1985. Les sols ferrallitiques des Iles Canaries (Espagne). Cahiers ORSTOM, Série Pédologie 21, 109-116.

Vallejo Gómez, E., Sedov, S.N., Oleschko, K. & Shoba, S.A., 2000. Phytoliths in surface and buried Andosols of central Mexico: morphological variety, rates of accumulation and weathering. In Vrydaghs, L. & Degraeve, A. (eds.), Man and the (Palaeo)environment. The Phytolith Evidence. 3rd International Meeting on Phytolith Research, Brussels, pp. 17.

Van Ranst, E., Utami, S.R., Verdoodt, A. & Qafoku, N.P., 2008. Mineralogy of a perudic Andosol in central Java, Indonesia. Geoderma 144, 379-386.

Van Ranst, E., Dumon, M., Tolossa, A.R., Cornelis, J.T., Stoops, G., Vandenberghe, R.E. & Deckers, J., 2011. Revisiting ferrolysis processes in the formation of Planosols for rationalizing the soils with stagnic properties in WRB. Geoderma 163, 265-274.

Van Ranst, E., Wilson, M.A. & Righi, D., 2018. Spodic materials. In Stoops, G., Marcelino, V. & Mees, F. (eds.), Interpretation of Micromorphological Features of Soils and Regoliths. Second Edition. Elsevier, Amsterdam, pp. 633-662.

Van Vliet-Lanoë, B., 2010. Frost action. In Stoops, G., Marcelino, V. & Mees, F. (eds.), Interpretation of Micromorphological Features of Soils and Regoliths. Elsevier, Amsterdam, pp. 81-108.

Van Vliet-Lanoë, B. & Fox, C., 2018. Frost action. In Stoops, G., Marcelino, V. & Mees, F. (eds.), Interpretation of Micromorphological Features of Soils and Regoliths. Second Edition. Elsevier, Amsterdam, pp. 575-603.

Vepraskas, M.J., Lindbo, D.L. & Stolt, M.H., 2018. Redoximorphic features. In Stoops, G., Marcelino, V. & Mees, F. (eds.), Interpretation of Micromorphological Features of Soils and Regoliths. Second Edition. Elsevier, Amsterdam, pp. 425-445.

Vingiani, S., Terribile, F. & Adamo, P., 2013. Weathering and particle entrapment at the rock–lichen interface in Italian volcanic environments. Geoderma 207/208, 244-255.

Vingiani, S., Scarciglia, F., Mileti, F.A., Donato, B. & Terribile, F., 2014. Occurrence and origin of soils with andic properties in Calabria (southern Italy). Geoderma 232/234, 500-516.

Vogel, S., Märker, M., Rellini, I., Hoelzmann, P., Wulf, S., Robinson, M., Steinhübel, L., Di Maio, G., Imperatore, C., Kastenmeier, P., Liebmann, L., Esposito, D. & Seiler, F., 2016. From a stratigraphic sequence to a landscape evolution model: Late Pleistocene and Holocene volcanism, soil formation and land use in the shade of Mount Vesuvius (Italy). Quaternary International 394, 155-179.

von Buch, M.W., 1967. Mikromorphologische Untersuchungen von Strukturelementen und Kieselsäurebildungen in älteren vulkanischen Böden der Collipulli-Serie, Frontera, Südchile. Geoderma 1, 249-276.

von Suchodoletz, H., Kühn, P., Hambach, U., Dietze, M., Zöller, L. & Faust, D., 2009. Loess-like and palaeosol sediments from Lanzarote (Canary Islands/Spain) - indicators of palaeoenvironmental change during the Late Quaternary. Palaeogeography, Palaeoclimatology, Palaeoecology 278, 71-87.

West, L.T., Lowrence, K.S., Dayot, A.A., Tomas, L.M. & Yeck, R.D., 1997. Micromorphology and soil development as indicators of ash age on Mindinao, the Philippines. In Shoba, S., Gerasimova, M. & Miedema, R. (eds.), Soil Micromorphology: Studies on Soil Diversity, Diagnostics, Dynamics. Moscow, Wageningen, pp. 335-343.

Wilson, M.A. & Righi, D., 2010. Spodic materials. In Stoops, G., Marcelino, V. & Mees, F. (eds.), Interpretation of Micromorphological Features of Soils and Regoliths. Elsevier, Amsterdam, pp. 251-273.

Wilson, M.A., Rodman, A.W., White, G.N., Thoma, D.P. & Shovic, H.F., 1997. Acid sulfate hydrothermal soil development from rhyolite flow and tuff: Yellowstone National Park, Wyoming, USA. In Shoba, S., Gerasimova, M. & Miedema, R. (eds.), Soil Micromorphology: Studies on Soil Diversity, Diagnostics, Dynamics. Moscow, Wageningen, pp. 219-231.

Yerima, B.P.K., Wilding, L.P., Calhoun, F.G. & Hallmark, C.T., 1987. Volcanic ash-influenced Vertisols and associated Mollisols of El Salvador. Physical, chemical, and morphological properties. Soil Science Society of America Journal 51, 699-708.

Zarei, M., Jahn, R. & Stahr, K., 1997. Entwicklung der Mikromorphologie von Böden in einer Chronosequenz aus Vulkaniten in Beziehung zum Mineralneubildingsprozess. In Stahr, K. (ed.), Mikromorphologische Methoden in der Bodenkunde. Hohenheimer Bodenkundliche Hefte 40, 179-208.

Chapter 26

Anthropogenic Features

W. Paul Adderley, Clare A. Wilson, Ian A. Simpson, Donald A. Davidson

UNIVERSITY OF STIRLING, STIRLING, SCOTLAND, UNITED KINGDOM

CHAPTER OUTLINE

Interpretation of Micromorphological Features of Soils and Regoliths. https://doi.org/10.1016/B978-0-444-63522-8.00026-7

1. Introduction

Humans have had an influence on pedogenic processes to a varying extent throughout the world, either through direct intervention or by indirect, long-range, action. Examples of direct interventions include disturbance of soil by tillage, management of standing vegetation using fire and addition of manures. Such interventions may be both intensive and rapid. In contrast, indirect human actions include long-term changes to vegetation and land cover, human-induced atmospheric changes affecting aerosol depositions to soils and human-induced climate changes altering biological, chemical and physical soil processes. These indirect actions are cumulative and long term in their impact on soils and regoliths. The understanding that the nature of the impact of these direct and indirect actions on soils can be recognised has led to debate about the role of soil use in the geological concept of the Anthropocene, as defined by Crutzen (2002). Some authors consider anthropogenic soils as a possible marker for the onset of this period (Certini & Scalenghe, 2011), with direct human actions on soils forming part of the wider debate about how to rationalise a definition of the Anthropocene as a geological epoch (Waters et al., 2014).

The role of humans as a factor in soil formation has been considered in many soil classification exercises (Bidwell & Hole, 1965; Bridges, 1978; Amundsen & Jenny, 1991; Dudal et al., 2002). In many of the major classification systems used today, a diverse set of anthropogenic influences have been considered during development of the classification but not always adopted. As a trend, direct human influences appear to be increasingly understood within the concept of soils as a key environmental resource (Bullock & Gregory, 1991) and, from this perspective, are now fully considered in soil classification systems (Dudal, 1990). Interest in urban soils has also been raised by the influential textbook by Fanning and Fanning (1989). Coupled with the growing recognition of anthropogenic impacts on agricultural soils, human impact has also become recognised for soils in urban environments (e.g., Schleuss et al., 1998), and urban soils are now defined in soil classification systems (Effland & Pouyat, 1997). Combined, this activity can be seen to have led to the development of definitions for Anthrosols and Technosols in the World Reference Base for Soil Resources (Nachtergaele, 2003; IUSS Working Group WRB, 2006, 2015).

A major question immediately raised in discussion of anthropogenic soil features is whether certain micromorphological features can be the result of anthropogenic influences or can be considered as natural. This problem has long been of concern in soil classification systems (Manil, 1959). In micromorphological studies, it has elicited discussion on the relative impact of individual factors of soil formation (Dalrymple, 1962) and on equifinality of processes acting in geoarchaeological contexts (Goldberg & Macphail, 2006). With this array of potential topics, and to set limits to what is considered 'anthropogenic', only those features that have arisen through direct human interventions are considered in this chapter.

Direct interventions are typically those related to agriculture through tillage, irrigation and addition of materials to increase soil fertility. Tillage encompasses all direct mechanical interventions that occur during agricultural cultivation processes. This includes surface disturbances through spade and hoe tillage by hand and by utilising a source of traction, ploughing, harrowing and rolling. Modern highly mechanised agricultural practices include subsoiling (deep-ripping) and mole draining, whereby the surface soil may remain relatively intact but may greatly modify soil porosity and related hydrological properties at depth. Primary tillage processes generally aim to loosen and mix the soil, whereas secondary tillage processes tend to shape plant beds and produce a finer soil structure. Spade cultivation varies markedly, and anthropological studies in the north Atlantic region (Fenton, 1970) have captured a remarkable variety of forms including the pronounced bed forms seen on the Faroe Islands (av Skarði, 1970). In modern agricultural systems, tillage usually includes the repeated passage of machinery over the soil surface, and the compound effects of these actions therefore also need to be considered.

Many of these direct interventions can, therefore, be directly linked to the definitions of various diagnostic horizons of anthrosols in soil classification systems. In the World Reference Base, these are (i) hortic horizons, formed through intensive fertilisation and/ or deep cultivation over a long period, (ii) irrigic and hydragric horizons, formed by irrigation practices, (iii) plaggic horizons formed by addition of turfs over a long period, and (iv) terric horizons, formed by deposition of mineral materials (IUSS Working Group WRB, 2015).

In modern urbanised environments, direct human interventions may be more pronounced, with soil properties conditioned by non-natural substrates such as metalliferous residues, in soils that are classified as Technosols (IUSS Working Group WRB, 2015). With such soils, there is also a nuanced distinction between soils that are created and then left alone and those that are under continuous or regular impact through land management, typically the application of wastes. These soils may strongly contrast seminatural soils, as Technosols may be the result of repeated additions of mineral materials. Differences in porosity and infiltration have been reported for such soils (Paradelo & Barral, 2013).

Other direct interventions that can be recognised as anthropogenic features based on soil micromorphology include the practice of burning vegetation *in situ* as fire-clearance husbandry (see Steensberg, 1993), clearing standing vegetation for grazing (Mallik & FitzPatrick, 1996), clearance for settlement of forest areas (Arroyo-Kalin, 2010, 2012; Brancier et al., 2014) and addition of ash materials as fertiliser (Simpson et al., 2002; Adderley et al., 2006).

In developing better understandings of anthropogenic features, soil micromorphological studies have an increasing role to play by informing the debate on soil quality and landscape management (Mermut & Eswaran, 2001), as one of the several soil-based proxies allowing palaeoenvironmental reconstruction (Terwilliger et al., 2013) and studies on soils inherited by farming communities (Simpson, 1997). In doing so, a need to understand differences in spatial or geographic distribution and between various anthropogenic practices across a range of geographic regions has become increasingly important. Coupled with these spatial elements, changes through time are likely to occur. This raises the possibility of different, or confounded, origins of discrete features. In considering anthropogenic features over time, there is an obvious link with studies of archaeological soil and sediment materials (Macphail & Goldberg, 2010, 2018) and buried soil horizons (Cremaschi et al., 2018). As such, these may be subject to additional forces other than typical soil-forming processes, both natural and anthropogenic.

The effects of anthropogenic interventions in soil development as recognised through soil micromorphology are dependent not only on the nature, intensity and duration of the intervention but also on the original properties of the soil. This makes identification of micromorphological features associated with specific anthropogenic processes particularly difficult and complex. To identify the imprint of a range of processes, understanding of past land use may become essential for establishing a secure interpretation. In approaching the interpretation of anthropogenic features found in soils and regoliths through conventional soil micromorphological techniques, a description of each set of diagnostic micromorphological features is required. Through comparative analysis ('differential diagnosis'; Kubiëna, 1970), individual anthropogenic processes can be recognised. Such an analysis should follow the general methodology established by Bullock et al. (1985) and updated by Stoops (2003).

2. Microstructure

2.1 Agricultural Soils

2.1.1 Tillage

The immediate effects of individual tillage operations on the soil are typically considered to be short-lived (Mackie-Dawson et al., 1989) and can be usefully considered separately from the more general effects of sustained periods of intensive cultivation. Such contrasts have been investigated in many different trials, typically utilising repeated mechanical intervention to monitor both its impact and the recovery from such impacts (Bottinelli et al., 2014). Typically, these studies have focussed on processes associated with compaction. The most obvious effect of compaction is a change in pore size and increased bulk density. Both Brewer (1964) and Greenland (1977) defined pore size classifications, and the latter associated pore size with soil water potential. Micromorphometric approaches using computer-based image analysis have been adopted to quantify porosity and water movement differences, as done in a key study by Bouma et al. (1979). More recently, Pagliai et al. (2003) have extended this approach to include shape criteria in the classification of macropores. The classifications proposed by Brewer, Greenland and Pagliai are all still commonly used for this purpose today.

Reported effects observed through conventional optical micromorphology analysis of tillage operations include an immediate increase in porosity, as well as groundmass heterogeneity, associated with subsoiling (Grevers & de Jong, 1992), deep ploughing and mole drainage (Borchert, 1967) and the use of deep-ripping implements (Pagliai et al., 2004). Francis et al. (1988) found that the total porosity of ploughed soil increased but that this increase was dominated by unstable macropores. Borchert (1967) observed bridging between aggregates 4 years after cultivation. In general, total microporosity appears to remain unchanged or to decrease because of compaction caused by farm vehicle traffic (Servadio et al., 2005). After ploughing, a change in microporosity from vertical channels to planar voids has also been noted (Francis et al., 1988; Stoops et al., 1988; Lamande et al., 2003), but channels will next reappear (Mackie-Dawson et al., 1989). Aggregates in ploughed soils tend to be blocky or prismatic, as opposed to the crumb microstructure found in grassland soils.

Pedality may decrease in degree of development because of repeated tillage, to the point where soils may become apedal (Jongerius, 1983). The temporal framework of these studies is important, as the use of mechanised tillage practices over long periods typically shows a decrease in total porosity compared with the use of other land-use practices (Domżł et al., 1993). This may not be apparent in short-term experimental studies.

Tillage, particularly of clay-rich soils, can result in the development of compacted plough pans with a platy microstructure at, and below, the base of the Ap horizon, resulting from shear forces applied to soil constituents at the base of the plough layer (Jongerius, 1983; Pagliai et al., 2003). The texture of the soil will influence the relative strength of such plough pans, with weaker pan development in sandy soils (Kooistra et al., 1984).

The size and morphology of soil aggregates has been used as an indicator of tillage (Soares et al., 2005; Usai, 2005). It has been demonstrated experimentally that incorporation of exogenous organic materials has a significant effect on the stability of aggregates of different sizes (Six et al., 2004; Kristiansen et al., 2006). Therefore, interpretation of tillage must also consider anthropogenic amendment or fertilisation of the soil, reflected in the nature of the groundmass (Fig. 1). The effects of cultivation and the resilience of the formed features appear to be affected by the nature of the soil; by the type, seasonality, and frequency of tillage operations and by the concentration of soil amendments (see Wilson et al., 2002).

Many of the structural changes in soils following tillage are seasonal in nature (Hall, 1994), but abandoned agricultural soils may demonstrate the stability of these microstructures over longer periods (Gebhardt, 1993). However, high levels of bioturbation in old agricultural soils have been considered to explain the total loss of cultivation-related structural features and the development of crumb microstructures in fields abandoned 200 and 40 years ago (Davidson, 2002). Similarly, experimental tillage studies suggest that bioturbation effects may become dominant within a decade (Boersma & Kooistra, 1994). Even limited levels of bioturbation have been shown to alter the growth pattern of mycorrhiza, which, in turn, has an impact on soil microstructure (Johnson et al., 2001; Staddon & Fitter, 2001).

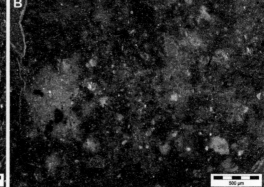

FIGURE 1 Heterogeneous groundmass due to incorporation of volcanic ash (a), in a pre-Hispanic buried agricultural soil (Hacienda Zuleta, Ecuador). (A) PPL. (B) OIL (see Wilson et al., 2002 for more detailed information on site context).

2.1.2   Burning of Soils and Above-ground Vegetation

The practice of clearing vegetation by managed fire regimes is adopted for many different purposes worldwide. This includes traditional heathland clearance (Dodgshon & Olsson, 2006), 'slash-and-burn' rotational agricultural systems and partial clearance of above-ground vegetation in plantations (Sant'anna et al., 2009).

Although fire clearance is still practised in many agricultural contexts worldwide, some burning practices are less common today than in past times. One such practice is the production of charcoal, which was very common in Europe until the use of coal became widespread in the 18th and 19th centuries. Such charcoal production can give rise to pronounced anthropogenically modified soils in European forests containing charcoal, burned soil fragments and exotic components (e.g., diatoms, phytoliths) (Gebhardt, 2007). With the efficiency of the charcoal production increased by the continuous reuse of charcoal production sites, tightly clustered spatial areas with intense anthropogenic soil modification may be found as a legacy of this practice (Gebhardt, 2007).

Experimental studies of changes in micromorphological features through burning typically consider a small set of experimental factors, including temperature and duration of the fire. As part of an experiment that contrasted different burning frequencies, Phillips et al. (2000) also reported strong differences in the type and abundance of micromorphological features linked to changes in bioturbation. It is clear that to develop an understanding of soil burning and addition of burnt materials requires consideration of temporal factors beyond the duration and intensity of a single fire event.

In an early study of colours resultant from soil burning, Mathieu and Stoops (1972) linked these shifts to the mineralogical composition of iron oxides. Many subsequent studies (see Deák et al., 2017) have examined similar phenomena to develop interpretations of past land use. Although these experimental studies, typically using strongly contrasting conditions, are useful in examination of a well-defined burning event, distinctions between natural fires and human-induced fires within a specific landscape context are much less pronounced. To address this important distinction, recent developments in stable isotope analysis of pyrogenic carbon (Ascough et al., 2011; Bird & Ascough, 2012) may prove to be a useful means of validating micromorphological assessment.

The direct effects of fire on soil are dependent on the texture and mineralogical composition of the soil and on the intensity, oxygen regime and temperature of the fire. Courty et al. (1989) suggest that these conditions may result in planar voids in coarse-textured soils, along with disaggregation of surface aggregates and the development of vesicular voids. Intraaggregate planar voids were also noted for soils affected by fires of moderate intensity (350-450°C), whereas at higher temperatures fragmentation of aggregates and the development of a granular structure through subsequent biological activity have been reported across soils of various textures (Phillips et al., 2000; Mallol et al., 2007).

2.1.3 Irrigation

The most commonly reported effects of irrigation on soil microstructure are disaggregation of aggregates and a decrease in porosity. For example, Prikhod'ko (2002) found that irrigation of Chernozems led to disintegration of mineral and organic aggregates. This structural degradation was most pronounced in clay-rich soils. Infilling of pores by disaggregated materials also led to a decrease in porosity. However, in some other instances, such as Solonetz soils and vertic Chernozems, irrigation has been noted to increase faunal activity, resulting in a crumb or granular microstructure and in increased porosity (Prikhod'ko, 2002; Tursina, 2002). The alkalinity of irrigation water is also an issue, as it affects the dispersivity of the soil materials. Tursina (2002) reports that irrigation using highly alkaline waters of a lake in Ukraine caused considerable loss of structure and formation of a surface crust, in contrast to sewage waters that caused little change in porosity. Bresson and Moran (2004) found that flood irrigation, in particular, causes collapse of soil structure and the development of hard-setting soil crusts (see also Pagliai & Stoops, 2010; Williams et al., 2018).

Wet rice cultivation produces distinctive soil profiles with an anthraquic surface horizon underlain by a hydragric horizon (IUSS Working Group WRB, 2015). Typically, these soils are apedal or have weakly developed crumb or blocky microstructures, and they can have a high porosity dominated by channels (Arimura, 1976; Miura et al., 1992). At the base of the anthraquic horizon, a plough pan with a platy microstructure frequently develops (Gong et al., 1998). The hydragric horizon is typically also apedal, but with a significantly lower channel- and vugh-dominated porosity (Fig. 2) compared with the anthraquic horizon (Hseu & Chen, 1997). Seasonality of changes in microstructure in wet cultivation systems is poorly researched to date, but the importance of shrinkage in the development of structural properties of paddy field soil has been demonstrated (Janssen et al., 2006).

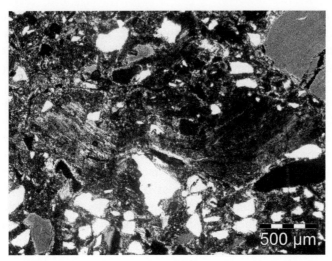

FIGURE 2 Typical aspect of a hydragric horizon, with an apedal microstructure, few channels and vughs, an orthic iron oxide nodule with diffuse boundaries and a strial b-fabric in part of the groundmass (paddy soils, Anuradhapura, Sri Lanka) (XPL).

2.1.4   Other Practices

Agricultural practices other than tillage and burning may be recognised to directly influence soil microstructure development (e.g., Dalrymple & Jim, 1984; Ricks Presley et al., 2004), but because tillage operations are usually also involved, it can be hard to isolate the features that those other practices produce. The reported effects of manuring and liming on soil microstructure are largely explained by increased bioturbation. This can result in the development of crumb microstructure, increased soil porosity, particularly microporosity, and the development of channels (Pagliai et al., 2004; Grieve et al., 2005). Where the addition of inorganic materials results in vertical accretion of topsoils, compaction lower in the profile can produce apedal intervals, as with deep plaggen soils (Meharg et al., 2006).

Less well understood are the effects on microstructure and aggregate development of combined manuring and tillage. Most agricultural systems rely on the use of managed fertiliser inputs, which in traditional farming systems are organic manures. In many circumstances, their use may cause no visible increase in soil organic matter content because of its rapid decomposition or uptake by crops. Barratt (1967, 1970) examined the effect of various fertiliser inputs to soils under pasture in long-term field experiments, revealing distinct differences between treatments with respect to both humus form and associated microstructures. These distinctive soil fabrics are often the result of changes in the nature and distribution of faecal materials.

A comparative assessment of orchard soils showed that microstructural features related to earthworm activity are different for plots where different agricultural practices were applied and that a relative absence of earthworms led ultimately to compaction of the uppermost horizons (Jongmans et al., 2003). Pulleman et al. (2005), in field experiments considering the effect of earthworm activity on soil microstructure, have demonstrated that clear differences emerge due to the combined effects of tillage and manuring, with clay enrichment of aggregates rich in organic matter. These were found to be common in soils under pasture but not when the same soil was subject to conventional tillage.

Where soils are extensively reworked by soil animals, infillings with crescent-like fabric features may result from welding together of earthworm casts (Bal, 1973). In this and other experimental studies (Spring, 2003; Davidson et al., 2004), individual faecal materials from other soil animals have been identified as soil aggregates. The tracks left by the movement of these animals may result in small-scale compaction or bioturbation. An experimental study of earthworm activity by Pérès et al. (1998) allowed surface porosity, granular aggregate formation and earthworm abundance to be correlated with organic amendment treatments.

Both the chemical and physical properties of the amending materials are of importance. Considering the former, excessive use of peat and turf materials as an alternative to farmyard manure may lead to soil acidification, which, in turn, may alter the microfabric through suppression of faunal activity (Collins & Coyle, 1980). Soil organic matter 'recalcitrance', a concept criticised by Bayeve and Laba (2015), is increasingly being clarified and understood through combinations of micromorphology and nuclear magnetic resonance techniques. In addition to the role of soil animals and microbiota in the physical comminution of organic matter (e.g., Chaparro et al., 2012), the consideration of soil enzyme processes (e.g., Min et al., 2015), the understanding of the role of phenolic compounds in inhibiting these enzymes in anthropogenic soils (e.g., Esiana, 2015) and the interrelationship between such processes and microfabric (e.g., Baldock & Skjemstad, 2000) are allowing a new understanding of organic matter transformations in soils to develop.

2.2   Technosols and Urban Soils

Because of the great diversity of waste materials and the large variations in intensity of their impact, identifying specific structural features associated with urban soils is more difficult than for many agricultural soils. The diversity of recorded microstructures thereby reflects the diversity of deposits and processes. For example, areas of land that are affected by trampling, such as recreational footpaths or animal tracks, may be characterised by massive or platy microstructures, compaction and low porosity, whereas some Technosols may be granular with frequent channels, depending on the level of biological activity (Giani et al., 2004; Vissac, 2005). Compaction in deep sequences of anthropogenic soils typically results in low porosities, apedal microstructures and a lack of horizonation (Davidson et al., 2006).

3. Groundmass

The groundmass has important interpretative value in anthropogenic soils, besides allowing to establish which soil-forming processes were dominant before the onset of human activity (see also Stoops & Mees, 2018). Imprints of such later activity may also be expressed in the groundmass through deliberate or accidental inputs and through reorganisation of mineral, biogenic and organic materials. Where anthropogenic soils result from additions of materials rather than disturbance, there can be a distinctive effect on particle size distribution (Davidson et al., 2007). These instances may also result in differences in c/f ratio between those soils and unamended surrounding soils. Therefore, this ratio itself may have interpretative value. This effect has been noted in connection with the addition of lime materials to irrigated agricultural soils, which resulted in the presence of various carbonates in surface horizons and in modified patterns of aggregation (van Oort et al., 2008). Furthermore, the alteration of hydrological properties and aeration through compaction or other anthropogenic processes can result in significant physicochemical changes. These changes may affect the stability of individual components, resulting in their differential alteration and hence changing the c/f ratio.

3.1 Coarse Mineral Components

Mineral artefacts that end up in a soil because of waste disposal, land reclamation and manuring are among the most widespread and resilient features of anthropogenic intervention. These artefacts may remain relatively inert in the soil or act as a stimulant, inhibitor or modifier of ongoing soil processes, such as biological turnover by soil animals (Jones et al., 1994).

3.1.1 Agricultural Soils

Anthropogenic coarse mineral components in agricultural soils include amending materials such as inorganic residues from burning of vegetation, inorganic fertilisers and lime and deliberately added exogenous soil materials. The latter include artefacts (e.g., pottery, glass) as well as animal and plant remains (e.g., bone, wood), as commonly described in archaeological studies (Macphail & Goldberg, 2010, 2018). Soils with distinct concentrations of animal and plant remains include black-earth soils from Amazonia (Arroyo-Kalin, 2010, 2012), French Guiana (Brancier et al., 2014), and West Africa (Fairhead & Leach, 2008). In contexts where large quantities of wood ash and plant ash are deliberately applied to increase soil fertility, instrumental colour analysis of optical microscopy images may allow identification of various ash materials (Simpson et al., 2002; Adderley et al., 2006). Similarly, with advances in sampling and analysis techniques for features in soil sections, such as microdrilling (van Oort et al., 1994; Denaix et al., 1999) and laser ablation ICP-MS (Bruneau et al., 2002), previously intractable questions relating to the effects of anthropogenic processes can be traced.

Mixing of different surface horizons and mixing of soil materials and exogenous materials are two tillage-related processes, most readily interpreted by analysis of thin sections (Miedema, 1997). Besides mixing of surface organic horizons with mineral subsurface horizons, tillage may also result in incorporation of natural surface deposits, such as coarse volcanic ash, creating banded patterns (Wilson et al., 2002). Prolonged tillage may result in significant changes in soil texture of the surface horizons of soils, through mixing of materials or through differential mineral weathering (Pang et al., 2006). The redistribution of residues of biogenic silica thin section has been used to describe different tillage disturbances (Golyeva, 2001; Clarke, 2003). Similarly, calcitic spheroids have been used as an indicator of past earthworm activity (Jongmans et al., 2001). In these instances, the relative distribution of this type of residues, which also include sponge spicules and diatoms, provides information on the origin of soil materials. However, the use of these residues as an indicator of tillage requires careful consideration of both their seasonal production and their short- and long-term preservation.

3.1.2 Technosols

Technosols are largely defined by the presence of significant quantities of artefactual materials in the coarse fraction. In urban soils, the range of inclusions that may be encountered is vast, and there may be huge vertical and lateral variations in their nature and abundance. Technic materials may include redeposited stone, mineral soil and sediment fragments, brick, cement, concrete, mortar and other construction debris, as well as bone, shell, asphalt, glass, slag and metallic objects (Alexandrovskaya & Alexandrovskiy, 2000; Prokof'eva et al., 2001).

3.2 Micromass

As previously discussed, additions of mineral materials to soils may occur as a result of land management. In addition to the changes that such materials may present in respect of the c/f ratio, distinct alterations of the limpidity and b-fabric of the micromass may occur (Wilson et al., 2002; Stoops, 2003). Reorganisation of the fabric through increased soil fauna activity can lead to the creation of areas with an undifferentiated b-fabric. Such areas may also result from incorporation of organic materials through diffusion of colloidal organic matter and illuviation of substances in solution (Babel, 1975). These changes have been reported as the development of bands with darker micromass (Fig. 3) and of parallel striated b-fabrics linked to compaction and shear forces. Changes in b-fabric and the appearance of a dotted limpidity can be interpreted as the incorporation of small fragments of carbonised materials. The presence of a crystallitic b-fabric in calcareous Chernozems has been linked to irrigation, resulting in dissolution and recrystallisation of carbonates (Tursina, 2002).

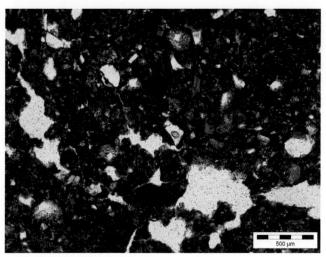

FIGURE 3 Pre-Hispanic buried agricultural soil, characterised by a dark micromass (Hacienda Zuleta, Ecuador) (PPL).

4. Organic Components

The observation of organic components and of their interaction with mineral components is essential in understanding anthropogenic impacts on soil organisation and the resulting changes in physical, chemical and biological properties. It is often difficult to generalise observations from past studies because the imprint found in the soil groundmass is determined by the combination of processes and soil properties, which are unique and case-specific. For example, contrasting results have been obtained for the effect of anthropogenic activity on organic matter content of the groundmass. Studies of modern and ancient cultivated soils in Yemen revealed an increase in all classes of plant residues in irrigated soils (Verba et al., 1995), and high organic matter contents have also been recorded for Mediterranean terrace soils (Darwish & Zurayk, 1997), buried cultivated soils in Ecuador (Wilson et al., 2002) and anthraquic paddy soil horizons (Miura et al., 1992; Gong et al., 1998), presumably as a result of manuring. However, Jongerius (1983) and Simpson et al. (1998) report a decline in total organic matter content, following cultivation.

Observations of organic components are of interest in the study of anthropogenic soils, because they may reveal chemical imprints of anthropogenic actions, as well as mechanical stress from physical or biological perturbation. The imprint of organic components such as roots may influence the groundmass. This may not be limited to the physical extent of the isolated component seen in thin sections. FitzPatrick (1993), for instance, discusses effects of roots that extend far into the groundmass. Such patterns may be the result of either active or former root systems. With respect to identification of palaeosoils, which commonly show superimposed relict fabric features (Kemp, 1999), the recognition of features recording the former presence of crop plant species may be of particular concern. In a review of biophysical interactions at the root-soil interface, Young (1998) has demonstrated that observable changes, including soil structural alterations, can extend up to 10 mm from the root surface. The so-called 'terra preta' or 'black-earth' soils, first identified as a marker of pre-Colombian human activity in Amazonia (Denevan, 2003), are characterised by high concentrations of charcoal, bone fragments and other charred residues compared with neighbouring undisturbed Oxisols (McCann et al., 2001; Woods et al., 2006; Arroyo-Kalin, 2010). The nature of the amending materials has been of principal interest; however, other groundmass features have also been reported. For example, comminuted charcoal may enter the soil and be incorporated into the groundmass through bioturbation. Likewise, highly comminuted bone materials have been reported for some black-earth sites (Schaefer et al., 2004) in Amazonia, but this anthropogenic signature is lacking at other localities (Brancier et al., 2014).

5. Pedofeatures

The impact of anthropogenic activity on soils and regoliths is most frequently interpreted through consideration of relationships between distinct pedofeatures and groundmass features. Studies of pedofeature development can be based on thin section observations following field-scale experimentation, with suitable controls such that the genesis of observed features can be attributed to human activity. Although the range of anthropogen-derived pedofeatures is vast (Miedema et al., 1994), several key features have become established in the canon of soil micromorphology. Such features include those indicative of large-scale field practices, such as cultivation, irrigation and fertilisation.

5.1 Large-Scale Field Practices

5.1.1 Tillage

In most temperate soils, the primary impacts of tillage at the profile scale are compaction and a change in soil pH (Babel, 1975). These changes will affect biological processes, and they will result in the development of a range of pedofeatures, as described below. Regardless of the context, associating tillage operations with specific pedofeatures requires careful interpretation because not all pedofeatures present will have formed as a result of those activities. Because tillage is likely to be done repeatedly, pedofeatures specific to surface horizons in arable farming systems, including excremental features (Davidson, 2002) (Fig. 4) and dusty clay coatings (agricutans; Jongerius, 1970), can only give an indication of the environment at the time of formation. The presence or absence of dusty clay coatings and clay coatings darkened by organic matter has been discussed extensively as an indicator of tillage activity, reflecting successional patterns of deposition (e.g., Usai, 2001; Deák et al., 2017). Illuviation features including dusty coatings resultant from ploughing have also been used as an indicator of past agricultural practice (Jongerius, 1983). However, the validity of using the occurrence of

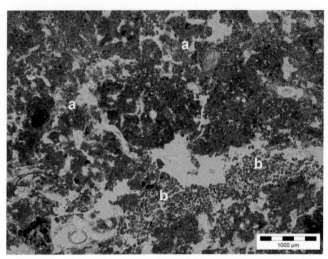

FIGURE 4 Mixed excrements of earthworms (a) and springtails (*Collembola* sp.) (b), in old cultivated soils (Sourhope, Scotland) (PPL, blue dyed resin).

coarse textural coatings as a diagnostic tool for anthropogenic activity has been questioned by the finding that such features can also develop through natural processes (Dalrymple & Theocharopoulos, 1987; see also Kühn et al., 2018). Wilson et al. (2002) found abundant mammilate excrements, infillings with bow-like patterns, silty clay coatings (Fig. 5) and amorphous iron/manganese oxide nodules and depletion pedofeatures. Pedofeatures that have been used as an indicator in studies of tillage include

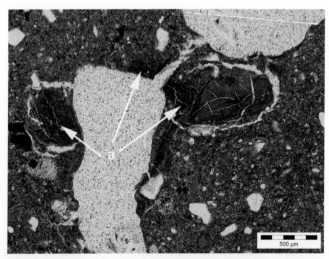

FIGURE 5 Clay coatings and infillings (a), in soils from a pre-Hispanic camellón-ridged field system buried beneath volcanic ash deposits dated to AD 1280 (Hacienda Zuleta, Ecuador) (PPL) (see Wilson et al., 2002).

organic matter accumulations in pores and excremental pedofeatures infilling relict faunal channels (Bal, 1973; Kleyer & Babel, 1984). Lima et al. (2002) and Ruivo et al. (2003) both cite instances of illuvial clay occurring in faunal burrows.

5.1.2   Fertilisation

Studies of pedofeatures related to fertiliser use are particularly valuable in examining land management practices over long periods and have been extensively applied in historical and archaeological studies (Davidson & Carter, 1998; Carter, 2001; Usai, 2001; Simpson et al., 2002; Adderley et al., 2006). Where the materials applied are organic, such as farmyard manures, fuel ashes, sewage sludges and industrial wastes, various pedo-features may result, including carbonate coatings and illuvial clay coatings (Lima et al., 2002; Xiubin et al., 2002; van Oort et al., 2008). Discrete pedofeatures may appear following biological or chemical alteration of exogenous amending materials. A special case of organic amendment is the use of other soil materials to improve soil fertility, as in plaggen soils (Fig. 6) and hortic soils (Pape, 1970; Spek et al., 2003; Davidson et al., 2004; Adderley et al., 2006). Typically, these soils contain excremental (see Fig. 4) and textural pedofeatures. In tropical contexts, Schaefer et al. (2004) have noted clear channel infillings of B horizon materials in A horizons, and vice versa, as well as abundant excrement pedofeatures producing a distinctive crumb structure at the soil surface.

5.1.3   Irrigation and Wet Cultivation

Irrigation of fields can lead to the development of a range of pedofeatures, distinct from those resulting from associated tillage. In soils already physically disturbed by tillage, irrigation can result in the formation of textural coatings such as clay coatings preserved

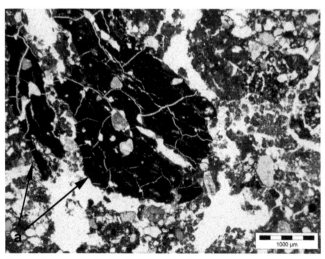

FIGURE 6 Fragments of amorphous uncarbonised materials rich in organic matter, containing some quartz grains and rhyolite fragments (a); the materials are probably turf residues from plaggen-type manuring, in intensively managed soils (Kailyard areas, Papa Sour, Shetland Isles) (PPL) (see Adderley et al., 2006).

in iron oxide nodules (e.g., Brinkman et al., 1973) and clay infillings in the pore space between microaggregates in the plough layer and lower horizons (e.g., Verba et al., 1995, 2002; Prikhod'ko, 2002; Tursina, 2002; Soares et al., 2005). However, irrigation water composition, inherent soil properties and the timing and nature of other tillage and manuring operations have a strong influence on the nature, development and persistence of textural features. For irrigated saline soils, Verba et al. (1995, 2002) note the formation of carbonate-clay coatings, and for chernozemic soils, Tursina (2002) describes coatings that are predominantly dark and rich in organic matter. The effects of irrigation on the oxidation state of soils, particularly cultivated soils with degraded structure and clay-illuvial horizons, are responsible for the common occurrence of depletion and impregnative iron oxide pedofeatures, including nodules and hypocoatings. Their development may be promoted by incorporation of organic matter at depth through ploughing. Costantini et al. (2006) report an increase in abundance of micritic calcite nodules with increasing depth. The frequency of iron oxide impregnation pedofeatures may also increase with proximity to any plough pan present.

In irrigated calcareous or saline soils, various pedofeatures may develop through dissolution and reprecipitation, such as impregnative calcitic features, including calcite nodules along void walls and the sides of sand grains, as well as carbonate depletion features (e.g., Prikhod'ko, 2002; Verba et al., 2002; Costantini et al., 2006). In the context of long-term historic irrigation of soils derived from calcareous alluvium, the formation of calcite nodules was observed, but translocation of fine materials was notably absent (El-Khatib & Stoops, 1987).

In soils under wet rice cultivation, impregnation and depletion pedofeatures of iron and manganese oxides are common features (Miura et al., 1992; Hseu & Chen, 1997; Gong et al., 1998; Zhang & Gong, 2003), throughout both the anthraquic and hydragric horizons. Alternating wetting and drying, together with physical disturbance through ploughing, may also result in the formation of textural clay and silt pedofeatures. Miura et al. (1992) found that such textural features are less common in wet cultivation soils than in dry upland rice cultivation systems. They also note that extinction lines observed for the clay coatings tend to be diffuse, which has been ascribed to the aquic moisture regime of these soils.

5.2 Technosols and Urban Soil Contexts

In urban contexts, pedofeatures have been related to soil sealing and soil disturbance during construction of paths, roads or buildings (International Committee for Anthropogenic Soils, 2007). The resulting pedofeatures include iron oxide depletion and impregnation features resulting from anoxic conditions below sealed surfaces (Prokof'eva et al., 2001; Vissac, 2005). Soils in urban contexts typically exhibit raised pH because of additions of calcareous building materials and other waste materials (Schleuss et al., 1998; Zhao et al., 2007). As a result of the addition of calcareous, sulphidic or phosphatic waste materials, intrusive crystallitic pedofeatures of calcite, gypsum, vivianite and other minerals can be present (Alexandrovskaya & Alexandrovskiy, 2000). Prokof'eva et al. (2001) report that compared with local undisturbed Chernozems, soils in Moscow have a high carbonate content, containing carbonates in the form of crystals, crystal intergrowths, spherulites, nodules and coatings.

6. Conclusions

Although there are many challenges to developing, through micromorphology, an understanding of soils and regoliths that have been influenced by direct human interventions, the application of micromorphology to Anthrosols and Technosols may lead to better understanding of their development and may allow improvement of their future management. Identification of the nature and intensity of human interventions, which is typically expected from micromorphologists in studies of Anthrosols and Technosols, is confounded by many process-led and land management issues.

The use of soil micromorphology has been demonstrated to help in risk assessment of metal-polluted soils of various types (e.g., Labanowski et al., 2007; van Oort et al., 2008). Seasonal changes, and the different rates of change for natural and anthropogenic processes operating in a single soil profile, create many localised phenomena and localised alterations of different soil constituents. This poses a major challenge to sampling design when extensive landscapes are considered. Despite such difficulties, there is a growing need to interpret anthropogenic features in soils and regoliths to deepen our understanding of the sustainability of both present-day and future land use.

Finally, considering past land use and conceptualising the Anthropocene as a geological epoch, soil micromorphology of anthropogenic features has an important role in developing a clearer understanding of the historical onset of land management.

References

Adderley, W.P., Simpson, I.A. & Davidson, D.A., 2006. Historic landscape management: a validation of quantitative soil thin-section analyses. Journal of Archaeological Science 33, 320-334.

Alexandrovskaya, E.I. & Alexandrovskiy, A.L., 2000. History of the cultural layer in Moscow and accumulation of anthropogenic substances in it. Catena 41, 249-259.

Amundsen, R. & Jenny, H., 1991. The place of humans in the state factor theory of ecosystems and their soils. Soil Science 151, 99-109.

Arimura, S., 1976. Studies on micromorphological properties of red soils and paddy soils in Northwestern Kyusyu. Pedologist 20, 140-151.

Arroyo-Kalin, M., 2010. The Amazonian formative: crop domestication and anthropogenic soils. Diversity 2, 473-504.

Arroyo-Kalin, M., 2012. Slash-burn-and-churn: landscape history and crop cultivation in pre-Columbian Amazonia. Quaternary International 249, 4-18.

Ascough, P.L., Bird, M.I., Francis, S.M., Thornton, B., Midwood, A.J., Scott, A.C. & Apperley, D., 2011. Variability in oxidative degradation of charcoal: influence of production conditions and environmental exposure. Geochimica et Cosmochimica Acta 75, 2361-2378.

av Skarði, J., 1970. Faroese cultivating and peat spades. In Gailey, A. & Fenton, A. (Eds.), The Spade in Northern and Atlantic Europe. Ulster Folk Museum, Belfast, pp. 67-73.

Babel, U., 1975. Micromorphology in soil organic matter. In Gieseking, J.E. (ed.), Soil Components. Volume 1. Organic Components. Springer, New York, pp. 369-473.

Bal, L., 1973. Micromorphological Analysis of Soils. Lower Levels in the Organisation of Soil Organic Matter. Soil Survey Paper 6, Netherlands Soil Survey Institute, Wageningen, 174 p.

Baldock, J.A. & Skjemstad, J.O., 2000. Role of the soil matrix in protecting natural materials against biological attack. Organic Geochemistry 31, 697-710.

Barratt, B.C., 1967. Differences in humus forms and their microfabrics induced by long-term topdressings in hayfields. Geoderma 1, 209-227.

Barratt, B.C., 1970. Effect of long-term fertilizer top dressing in hayfield on humus forms and their micromorphology. AGRI Digest 21, 11-18.

Bayeve, P.C. & Laba, M., 2015. Moving away from the geostatistical lamppost: why, where, and how does the spatial heterogeneity of soils matter? Ecological Modelling 298, 24-38.

Bidwell, O.W. & Hole, F.D., 1965. Man as a factor of soil formation. Soil Science 99, 65-72.

Bird, M.I., & Ascough, P.L., 2012. Isotopes in pyrogenic carbon: a review. Organic Geochemistry 42, 1529-1539.

Boersma, O.H. & Kooistra, M.J., 1994. Differences in soils structure of silt loam Typic Fluvaquents under various agricultural management practices. Agriculture, Ecosystems & Environment 51, 21-42.

Borchert, H., 1967. Untersuchungen der Bodengefügeveränderung von meliorierten Böden. Geoderma 1, 371-390.

Bottinelli, N., Hallaire, V., Goutal, N., Bonnaud., P. & Ranger, J., 2014. Impact of heavy traffic on soil macroporeosity of two silty forest soils: initial effect and short-term recovery. Geoderma 217/218, 10-17.

Bouma, J., Jongerius, A. & Schoonderbeek, D., 1979. Calculation of saturated hydraulic conductivity of some pedal clay soils using micromorphometric data. Soil Science Society of America Journal 43, 261-264.

Brancier, J., Cammas, C., Todisco, D. & Fouache, E., 2014. A micromorphological assessment on anthropogenic features in pre-Columbian French Guiana Dark Soils (FGDS): first results. Zeitschrift für Geomorphologie 58, Supplementary Issues 2, 109-139.

Bresson, L.M. & Moran, C.J., 2004. Micromorphological study of slumping in a hardsetting seedbed under various wetting conditions. Geoderma 118, 277-288.

Brewer, R., 1964. Fabric and Mineral Analysis of Soils. John Wiley and Sons, New York, 470 p.

Bridges, E.M., 1978. Interaction of soil and mankind in Britain. Journal of Soil Science 29, 125-139.

Brinkman, R., Jongmans, A.G., Miedema, R. & Maaskant, P., 1973. Clay decomposition in seasonally wet, acid soils: micromorphological, chemical and mineralogical evidence from individual argillans. Geoderma 10, 259-270.

Bruneau, P.M.C., Ostle, N., Davidson, D.A., Grieve, I.C. & Fallick, A., 2002. Determination of rhizosphere 13C pulse signals in soil thin sections by Laser Ablation Isotope Ratio Mass Spectromery (LA-IRMS). Rapid Communications in Mass Spectrometry 16, 2190-2194.

Bullock, P., & Gregory, P.J., 1991 Soils: a neglected resource in urban areas. In Bullock, P., & Gregory, P.J. (eds.), Soils in the Urban Environment. Blackwell, Oxford, pp 1-5.

Bullock, P., Fedoroff, N., Jongerius A., Stoops, G., Tursina, T. & Babel, U., 1985. Handbook for Soil Thin Section Description. Waine Research Publications, Wolverhampton, 152 p.

Carter, S.P., 2001. A reassessment of the origins of the St Andrews 'garden soil'. Tayside and Fife Archaeological Journal 7, 87-97.

Certini, G. & Scalenghe, R., 2011. Anthropogenic soils are the golden spikes for the Anthropocene. The Holocene 21, 1269-1274.

Chaparro, J.M., Sheflin, A.M., Manter, D.K., & Vivanco, J.M., 2012. Manipulating the soil microbiome to increase soil health and plant fertility. Biology and Fertility of Soils 48, 489-499.

Clarke, J., 2003. The occurrence and significance of biogenic opal in the regolith. Earth-Science Reviews 60, 175-194.

Collins, J.F. & Coyle, E., 1980. Long-term changes in soil macro- and micromorphological properties under the influence of peat debris. Journal of Soil Science 31, 547-558.

Costantini, E.A.C., Lessovaia, S. & Vodyanitskii, Y.U., 2006. Using the analysis of iron and iron oxides in paleosols (TEM, geochemistry and iron forms) for the assessment of present and past pedogenesis. Quaternary International 156/157, 200-211.

Courty, M.A., Goldberg, P. & Macphail, R., 1989. Soils and Micromorphology in Archaeology. Cambridge University Press, Cambridge, 344 p.

Cremaschi, M., Trombino, L. & Zerboni, A. 2018. Palaeosoils and relict soils, a systematic review. In Stoops, G., Marcelino, V. & Mees, F. (eds.), Interpretation of Micromorphological Features of Soils and Regoliths. Second Edition. Elsevier, Amsterdam, pp. 863-894.

Crutzen, P.J., 2002. Geology of mankind. Nature 415, 23.

Dalrymple, J.B., 1962. Some micromorphological implications of time as a soil forming factor, illustrated from sites in south-eastern England. Zeitschrift für Pflanzenernährung, Düngung und Bodenkunde 98, 232-239.

Dalrymple, J.B. & Jim, C.Y., 1984. Experimental study of soil microfabrics induced by isotropic stresses of wetting and drying. Geoderma 34, 43-68.

Dalrymple, J.B. & Theocharopoulos, S.P., 1987. Intrapedal cutans. Lateral differences in their properties and their spatial clustering. Geoderma 41, 149-180.

Darwish, T.M. & Zurayk, R.A., 1997. Distribution and nature of Red Mediterranean soils in Lebanon along an altitudinal sequence. Catena 28, 191-202.

Davidson, D.A., 2002. Bioturbation in old arable soils: quantitative evidence from soil micromorphology. Journal of Archaeological Science 29, 1247-1253.

Davidson, D.A. & Carter, S.P., 1998. Micromorphological evidence of past agricultural practices in cultivated soils: the impact of a traditional agricultural system on soils in Papa Stour, Shetland. Journal of Archaeological Science 25, 827-838.

Davidson, D.A., Bruneau, P.M.C., Grieve, I.C. & Wilson, C.A., 2004. Micromorphological assessment of the effect of liming on faunal excrement in an upland grassland soil. Applied Soil Ecology 26, 169-177.

Davidson, D.A., Dercon, G., Stewart, M. & Watson, F., 2006. The legacy of past urban waste disposal on local soils. Journal of Archaeological Science 33, 778-783.

Davidson, D.A., Dercon, G., Simpson, I.A., Dalsgaard, K., Spek, T. & Plant, D.A., 2007. The identification and significance of inputs to anthrosols in north-west Europe. Atti della Società Toscana di Scienze Naturali, Memorie Serie A, 112, 79-83.

Deák, J., Gebhart, A., Lewis, H., Usai, M.R. & Lee, H., 2017. Soils disturbed by vegetation clearance and tillage. In Nicosia, C. & Stoops, G. (eds), Archaeological Soil and Sediment Micromorphology. John Wiley & Sons Ltd, Chichester, pp. 233-264.

Denaix, L., Van Oort, F., Pernes, M. & Jongmans, A.G., 1999. Transmission X-Ray diffraction of undisturbed soil microfabrics obtained by microdrilling in thin sections. Clays and Clay Minerals 47, 637-646.

Denevan, W.M., 2003. The native population of Amazonia in 1492 reconsidered. Revista de Indias 62, 175-188.

Dodgshon, R.A. & Olsson, G.A., 2006. Heather moorland in the Scottish Highlands: the history of a cultural landscape, 1600-1880. Journal of Historical Geography 32, 21-37.

Domżł, H., Hodara, J., Słowińska-Jurkiewicz, A., & Turski, R., 1993. The effects of agricultural use on the structure and physical properties of three soil types. Soil and Tillage Research 27, 365-382.

Dudal, R., 1990. An International Reference Base for soil classification (IRB). Transactions of the 14th International Congress of Soil Science, Volume V, Kyoto, pp 38-43.

Dudal, R., Nachtergaele, F. & Purnell, M., 2002. The human factor of soil formation. Proceedings of the 17th World Congress of Soil Sciences, Bangkok, Thailand. Paper no. 93, 8 p.

Effland, W.R. & Pouyat, R.V., 1997. The genesis, classification, and mapping of soils in urban areas. Urban Ecosystems, 1, 217-228.

El-Khatib, B. & Stoops, G., 1987. Micromophological characteristics of the Oasis-soils from Syria. In Fedoroff, N., Bresson, L.M. & Courty, M.A. (eds.), Micromorphologie des Sols, Soil Micromorphology. AFES, Plaisir, pp. 207-211.

Esiana, B.O.I., 2015. The Long-Term Dynamics of Soil Organic Carbon in the Anthropogenic Soils of Scotland's Medieval Urban Landscape. PhD Dissertation, University of Stirling, 245 p.

Fairhead, J. & Leach, M., 2008. Amazonian dark earths in Africa? In Woods, W. (ed.), Terra Preta Nova: A Tribute to Wim Sombroek. Springer-Verlag, Berlin, pp. 265-278.

Fanning, D.S. & Fanning, M.C.B., 1989. Soil: Morphology, Genesis, and Classification. Wiley, New York, 416 p.

Fenton, A., 1970. Paring and burning. In Gailey, A. & Fenton, A. (eds), The Spade in Northern and Atlantic Europe. Ulster Folk Museum, Belfast, pp. 155-193.

FitzPatrick, E.A., 1993. Soil Microscopy and Micromorphology. John Wiley & Sons, Chichester, 304 p.

Francis, G.S., Cameron, K.C. & Kemp, R.A., 1988. A comparison of soil porosity and solute leaching after six years of direct drilling or conventional cultivation. Australian Journal of Soil Research 26, 637-649.

Gebhardt, A., 1993. Micromorphological evidence of soil deterioration since the mid-Holocene at archaeological sites in Brittany, France. The Holocene 3, 333-341.

Gebhardt, A., 2007. Impact of charcoal production activities on soil profiles: the micromorphological point of view. ArchaeoSciences 31, 127-136.

Giani, L., Chertov, O., Gebhardt, C., Kalinina, O., Nadporozhskaya, M. & Tolkdorf-Lienemann, E., 2004. Plagganthrepts in northwest Russia? Genesis, properties and classification. Geoderma 121, 113-122.

Goldberg, P., & Macphail, R.I., 2006. Practical and Theoretical Geoarchaeology. Blackwell, Oxford, 455 p.

Golyeva, A., 2001. Biomorphic analysis as a part of soil morphological investigations. Catena 43, 217-230.

Gong, Z., Zhang, G. & Luo, G., 1998. New soil horizons formed by anthropogenic activities. Transactions of the 16th World Congress of Soil Science, Montpellier, Symposium 16, Paper no. 1430, 11 p.

Greenland, D.J., 1977. Soil damage by intensive arable cultivation: temporary or permament? Philosophical Transactions of the Royal Society of London A 281, 193-208.

Grevers, M.C.J. & de Jong, E., 1992. Soil-structure and crop yield over a 5-year period following subsoiling Solonetzic and Chernozemic soils in Saskatchewan. Canadian Journal of Soil Science 73, 81-91.

Grieve, I.C., Davidson, D.A. & Bruneau, P.M.C., 2005. Effects of liming on void space and aggregation in an upland grassland soil. Geoderma 125, 39-48.

Hall, N.W., 1994. Soil structure transformation over the growing season – a micromorphological approach. In Ringrose-Voase, A.J. & Humphreys, G.S. (eds.), Soil Micromorphology: Studies in Management and Genesis. Developments in Soil Science, Volume 22. Elsevier, Amsterdam, pp 659-667.

Hseu, Z.Y. & Chen, Z.S., 1997. Microscopic redoximorphic features in an alfisol with plinthite and anthraquic flooding conditions in Taiwan. In Shoba, S., Gerasimova, M. & Miedema, R. (eds.), Soil Micromorphology: Studies on Soil Diversity, Diagnostics, Dynamics. Moscow, Wageningen, pp. 255-270.

International Committee for Anthropogenic Soils, 2007. Anthropogenic Soils CD-ROM. Ver 2.0. USDANRCS, National Soil Survey Center, Lincoln.

IUSS Working Group WRB, 2006. World Reference Base for Soil Resources. 2nd Edition. World Soil Resources Reports No. 103. FAO, Rome, 128 p.

IUSS Working Group WRB, 2015. World Reference Base for Soil Resources 2014, Update 2015. World Soil Resources Reports No. 106. FAO, Rome, 192 p.

Janssen, I., Peng, X.H. & Horn, R., 2006. Physical soil properties of paddy fields as a function of cultivation history and texture. Soil Management for Sustainability 38, 446-455.

Johnson, D., Leake, J.R. & Read, D.J., 2001. Novel in-growth core system enables functional studies of grassland mycorrhizal mycelial networks. New Phytologist 152, 555-562.

Jones, C.G., Lawton, J.H. & Shachak, M., 1994. Organisms as ecosystem engineers. Oikos 69, 373-386.

Jongerius, A., 1970. Some morphological aspects of regrouping phenomena in Dutch soils. Geoderma 4, 311-331.

Jongerius, A., 1983. Micromorphology in agriculture. In Bullock, P. & Murphy, C.P. (eds.), Soil Micromorphology. Volume 1. Techniques and Applications. AB Academic Publishers, Berkhamstead, pp. 111-138.

Jongmans, A.G., Pulleman, M.M. & Marinissen, J.C.Y., 2001. Soil structure and earthworm activity in a marine silt loam under pasture versus arable land. Biology and Fertility of Soils 33, 279-285.

Jongmans, A.G., Pulleman, M.M., Balabane, M., van Oort, F. & Marinissen, J.C.Y., 2003. Soil structure and characteristics of organic matter in two orchards differing in earthworm activity. Applied Soil Ecology 24, 219-232.

Kemp, R.A., 1999. Micromorphology of loess-paleosol sequences: a record of paleoenvironmental change. Catena 35, 179-196.

Kleyer, M. & Babel, U., 1984. Gefügebildung durch Bodentiere in 'konventionell' und 'biologisch' bewirtschafteten Ackerböden. Zeitschrift für Pflanzenernährung, Düngung und Bodenkunde 147, 98-109.

Kooistra, M.J., Bouma, J., Boersma, O.H. & Jager, A., 1984. Physical and morphological characterisation of undisturbed and disturbed ploughpans in a sandy loam soil. Soil and Tillage Research 4, 405-417.

Kristiansen, S.M., Schjønning, P., Thomsen, I.K., Olesen, J.E., Kristensen, K. & Christensen, B.T., 2006. Similarity of differently sized macro-aggregates in arable soils of different texture. Geoderma 137, 147-154.

Kubiëna, W.L., 1970. Micromorphological Features of Soil Geography. Rutgers University Press, New Brunswick, 256 p.

Kühn, P., Aguilar, J., Miedema, R. & Bronnikova, M., 2018. Textural Pedofeatures and Related Horizons. In Stoops, G., Marcelino, V. & Mees, F. (eds.), Interpretation of Micromorphological Features of Soils and Regoliths. Second Edition. Elsevier, Amsterdam, pp. 377-423.

Labanowski, J., Sebastia, J., Foy, E., Jongmans, T., Lamy, I. & van Oort, F., 2007. Fate of metal-associated POM in a soil under arable land use contaminated by metallurgical fallout in northern France. Environmental Pollution 149, 59-69.

Lamande, M., Hallaire, V., Curmi, P., Peres, G. & Cluzeau, D., 2003. Changes of pore morphology, infiltration and earthworm community in a loamy soil under different agricultural managements. Catena 54, 637-649.

Lima, H.L. Schaefer, C.E.R., Mello, J.W.V., Gilkes, R.J. & Ker, J.C., 2002. Pedogenesis and pre-Colombian land use of 'Terra Preta Anthrosols' ('Indian Black Earth') of Western Amazonia. Geoderma 110, 1-17.

Mackie-Dawson, L.A., Mullins, C.E., Goss, M.J., Court, M.N. & FitzPatrick, E.A., 1989. Seasonal changes in the structure of clay soils in relation to soil management and crop type: II. Effects of cultivation and cropping at Compton Beauchamp. Journal of Soil Science 40, 283-292.

Macphail, R.I. & Goldberg, P., 2010. Archaeological materials. In Stoops, G., Marcelino, V. & Mees, F. (eds.), Interpretation of Micromorphological Features of Soils and Regoliths. Elsevier, Amsterdam, pp. 589-622.

Macphail, R.I. & Goldberg, P., 2018. Archaeological materials. In Stoops, G., Marcelino, V. & Mees, F. (eds.), Interpretation of Micromorphological Features of Soils and Regoliths. Second Edition. Elsevier, Amsterdam, pp. 779-819.

Mallik, A.U. & FitzPatrick, E.A., 1996. Thin section studies of Calluna heathland soils subject to prescribed burning. Soil Use and Management 12, 143-149.

Mallol, C., Marlowe, F.W., Wood, B.M. & Porter, C.C., 2007. Earth, wind, and fire: ethnoarchaeological signals of Hadza fires. Journal of Archaeological Science 34, 2035-2052.

Manil, G., 1959. General considerations on the problem of soil classification. Journal of Soil Science 10, 5-13.

Mathieu, C. & Stoops, G., 1972. Observations pétrographiques sur la paroi d'un four à chaux Carolingien creusé en sol limoneaux. Archéologie Médiévale 2, 347-355.

McCann, J.M., Woods, W.I. & Meyer, D.W., 2001. Organic matter and anthrosols in Amazonia: interpreting the Amerindian legacy. In Rees, R.M., Ball, B.C., Campbell, C.D. & Watson, C.A. (eds.), Proceedings of the BSSS Conference on Sustainable Management of Soil Organic Matter. CABI, Wallingford, pp. 180-190.

Meharg, A.A., Deacon, C., Edwards, K.J., Donaldson, M., Davidson, D.A., Spring, C., Scrimgeour, C.M., Feldman, J. & Rabb, A., 2006. Ancient manuring practices pollute arable soils at the St Kilda World Heritage Site, Scottish North Atlantic. Chemosphere, 64, 1818-1828.

Mermut, A.R. & Eswaran, H., 2001. Some major developments in soil science since the mid-1960s. Geoderma 100, 403-426.

Miedema, R., 1997. Applications of micromorphology of relevance to agronomy. Advances in Agronomy 59, 119-169.

Miedema, R., Chartres, C.J., Courty, M.A., McSweeney, K., Oleshko, K. & Rabenhorst, M.C., 1994. Soil micromorphology: towards an analytical and quantitative tool for assessing anthropogenic influences on soils. Transactions of the 15th World Congress of Soil Science, Acapulco, Volume 1, pp. 143-162.

Min, K., Freeman, C., Kang, H., & Choi, S., 2015. The regulation by phenolic compounds of soil organic matter dynamics under a changing environment. BioMed Research International 15, 1-11.

Miura, K., Tulaphitak, T. & Kyuma, K., 1992. Pedogenetic studies on some selected soils in Northeast Thailand. 2. Micromorphological characteristics. Soil Science and Plant Nutrition 38, 495-503.

Nachtergaele, F.O.F., 2003. The future of the FAO legend and the FAO/UNESCO Soil Map of the World. In Eswaran, H., Rice, T., Ahrens, R. & Stewart, B.A. (eds.), Soil Classification. A Global Desk Reference. CRC Press, Boca Raton, pp. 147-157.

Pagliai, M. & Stoops, G., 2010. Physical and biological surface crusts. In Stoops, G., Marcelino, V. & Mees, F. (eds.), Interpretation of Micromorphological Features of Soils and Regoliths. Elsevier, Amsterdam, pp. 419-440.

Pagliai, M., Marsili, A., Servadio, P., Vignozzi, N. & Pellegrini, S., 2003. Changes in some physical properties of a clay soil in Central Italy following the passage of rubber tracked and wheeled tractors of medium power. Soil and Tillage Research 73, 119-129.

Pagliai, M., Vignozzi, N. & Pellegrini, S., 2004. Soil structure and the effect of management practices. Soil and Tillage Research 79, 131-143.

Pang, J., Hu, X., Huang, C. & Xu, Z., 2006. Micromorphological features of old cultivated and modern soils in Guanzhong areas, Shaanxi Province, China. Agricultural Sciences in China 5, 691-699.

Pape, J.C., 1970. Plaggen soils in The Netherlands. Geoderma 4, 229-255.

Paradelo, R. & Barral, M.T., 2013. Influence of organic matter and texture on the compactability of Technosols. Catena 110, 95-99.

Pérès, G., Cluzeau, D., Curmi, P. & Hallaire, V., 1998. Earthworms activity and soil structure changes due to organic enrichments in vineyard systems. Biology and Fertility of Soils 27, 417-424.

Phillips, D.H., Foss, J.E., Buckner, E.R., Evans, R.M. & FitzPatrick, E.A., 2000. Response of surface horizons in an oak forest to prescribed burning. Soil Science Society of America Journal 64, 754-760.

Prikhod'ko, V.E., 2002. Micromorphological diagnostics of the transformation of soil properties in steppe and semidesert soils upon their irrigation in the lower Volga region. Eurasian Soil Science 35, 588-598.

Prokof'eva, T.V., Sedov, S.N., Stroganova, M.N. & Kazdym, A.A., 2001. An experience of the micromorphological diagnostics of urban soils. Eurasian Soil Science 34, 783-792.

Pulleman, M.M., Six, J., van Breemen, N. & Jongmans, A.G., 2005. Soil organic matter distribution and microaggregate characteristics as affected by agricultural management and earthworm activity. European Journal of Soil Science 56, 453-467.

Ricks Presley, D., Ransom, M.D., Kluitenberg, G.J. & Finnell, P.R., 2004. Effects of thirty years of irrigation on the genesis and morphology of two semiarid soils in Kansas. Soil Science Society of America Journal 68, 1916-1926.

Ruivo, M.P., Arroyo-Kalin, M.A., Schaefer, C.E.R., Costi, H.T., De Souza Arcanjo, S.H., Lima, H.N., Pulleman, M.M. & Creutzberg, D., 2003. The use of soil micromorphology for the study of the formation and properties of Amazonian dark earths. In Lehmann, J., Kern, D.C., Glaser, B. & Woods, W.I. (eds.), Amazonian Dark Earths: Origin, Properties, Management. Kluwer, Dordrecht, pp. 243-254.

Sant'anna, S.A.C., Fernandes, M.F., Ivo, W.M.P.M., Costa J.L.S., 2009. Evaluation of soil quality indicators in sugarcane management in sandy loam soil. Pedosphere 19, 312-322.

Schaefer, C.E.G.R., Lima, H.N., Gilkes, R.J. & Mello, J.W.V., 2004. Micromorphology and electron microprobe analysis of phosphorus and potassium forms of an Indian Black Earth (IBE) Anthrosol from Western Amazonia. Australian Journal of Soil Research 42, 401-409.

Schleuss, U., Wu, Q. & Blume, H.P., 1998. Variability of soils in urban and periurban areas in Northern Germany. Catena 33, 255-270.

Servadio, P., Marsili, A., Vignozzi, N., Pellegrini, S. & Pagliai, M., 2005. Effects on some soil qualities in central Italy following the passage of four wheel drive tractor fitted with single and dual tires. Soil and Tillage Research 84, 87-100.

Simpson, I.A., 1997. Relict properties of anthropogenic deep top soils as indicators of infield management in Marwick, West Mainland Orkney. Journal of Archaeological Science 24, 365-380.

Simpson, I.A., Bryant, R.G., & Tveraabak, U., 1998. Relict soils and early arable land management in Lofoten, Norway. Journal of Archaeological Science 25, 1185-1198.

Simpson, I.A., Adderley, W.P., Guðmunsson, G., Hallsdóttir, M., Sigurgeirsson, M.A. & Snaesdóttir, M., 2002. Soil limitations to agrarian land production in pre-modern Iceland. Human Ecology 30, 423-443.

Six, J., Bossuyt, H., Degryze, S. & Denef, K., 2004. A history of research on the link between (micro) aggregates, soil biota and soil organic matter dynamics. Soil and Tillage Research 79, 7-31.

Soares, J.L.N, Espindola, C.R. & Pereira, W.L.M., 2005. Physical properties of soils under intensive agricultural management. Scientia Agricola 62, 165-172.

Spek, T., Groenman-van Waateringe, W., Kooistra, M. & Bakker, L., 2003. Formation and land use history of Celtic Fields in North-West Europe – An interdisciplinary case study at Ziejen, The Netherlands. European Journal of Archaeology 6, 141-173.

Spring, C.A., 2003. The Effects of Earthworms on Soil Structure in an Upland Grassland. PhD Dissertation, University of Stirling, 251 p.

Staddon, P.L. & Fitter, A.H., 2001. The differential vitality of intraradical mycorrhizal structures and its implications. Soil Biology and Biochemistry 33, 129-132.

Steensberg, A., 1993. Fire-clearance Husbandry: Traditional Techniques Throughout the World. Poul Kristensen, Herning, 239 p.

Stoops, G., 2003. Guidelines for Analysis and Description of Soil and Regolith Thin Sections. Soil Science Society of America, Madison, 184 p.

Stoops, G. & Mees, F., 2018. Groundmass composition and fabric. In Stoops, G., Marcelino, V. & Mees, F. (eds.), Interpretation of Micromorphological Features of Soils and Regoliths. Second Edition. Elsevier, Amsterdam, pp. 73-125.

Stoops, G., Mathieu, C., Maryam, A. & Gholamreza, K., 1988. Micromorphometric aspects of transformations of the macroporosity in irrigated soils. Egyptian Journal of Soil Science 28, 339-348.

Terwilliger, V.J., Eshetu, Z., Disnar. J., Jacob. J.R., Adderley, W.P., Huang, Y., Alexandre, M. & Fogel, M.L., 2013. Environmental changes and the rise and fall of civilizations in the northern Horn of Africa: an approach combining δD analyses of land-plant derived fatty acids with multiple proxies in soil. Geochimica et Cosmochimica Acta 111, 140-161.

Tursina, T.V., 2002. Micromorphological methods in the analysis of pedogenetic problems. Eurasian Soil Science 35, 777-791.

Usai, M.R., 2001. Textural features and pre-Hadrian's Wall ploughed Paleosols at Stanwix, Carlisle, Cumbria, U.K. Journal of Archaeological Science 28, 541-553.

Usai, M.R., 2005. Textural pedofeatures as tools to diagnose past cultivation – a controlled experiment. In Smith, D.N., Brickley, M.B. & Smith, W. (eds.), Fertile Ground. Papers in Honour of Susan Limbrey. Oxbow Books, Oxford, pp. 162-164.

van Oort, F., Jongmans, A.G. & Jaunet, A.M., 1994. The progression from optical light microscopy to transmission electron microscopy in the study of soils. Clay Minerals 29, 247-254.

van Oort, F., Jongmans, A.G., Lamy, I., Baize, D. & Chevallier, P., 2008. Impacts of long-term waste-water irrigation on the development of sandy Luvisols: consequences for metal pollutant distributions. European Journal of Soil Science 59, 925-938.

Verba, M.P., Al-Kasiri A.S., Goncharova, N.A. & Chizhikova, N.P., 1995. Impact of irrigation on micromorphological features of desert soils of Hadramout Vallet (Yemen). Eurasian Soil Science 27, 108-124.

Verba, M., Aidabekova, L. & Yamnova, I., 2002. Micromorphological and mineralogical peculiarities of old-irrigated soils in Aral Sea Basin. Proceedings of the 17th World Congress of Soil Sciences, Bangkok, Paper no. 1267, 2 p.

Vissac, C., 2005. Study of a historical garden soil at the Grand-Pressigny site (Indre-et-Loire, France): evidence of landscape management. Journal of Cultural Heritage 6, 61-67.

Waters, C.N., Zalasiewicz, J.N., Williams, M., Ellis, M.A. & Snelling, A.M., 2014. A stratigraphical basis for the Anthropocene? Geological Society, London, Special Publications 395, 1-21.

Williams, A., Pagliai, M. & Stoops, G., 2018. Physical and biological surface crusts. In Stoops, G., Marcelino, V. & Mees, F. (eds.), Interpretation of Micromorphological Features of Soils and Regoliths. Second Edition. Elsevier, Amsterdam, pp. 539-574.

Wilson, C.A., Simpson, I.A. & Currie, E.J., 2002. Soil management in Pre-Hispanic raised field systems: micromorphological evidence from Hacienda Zuleta, Ecuador. Geoarchaeology 17, 261-283.

Woods, W.I., Falcão, N.P.S. & Teixeira, W.G., 2006. Biochar trials aim to enrich soil for smallholders. Nature 443, 144.

Xiubin, H, Keli, T., Jungliang, T. & Matthews, J.A., 2002. Paleopedological investigation of three agricultural loess soils on the loess plateau of China. Soil Science 167, 478-491.

Young, I.M., 1998 Biophysical interactions at the root-soil interface: a review. Journal of Agricultural Science 130, 1-7.

Zhang, G.L. & Gong, Z.T., 2003. Pedogenic evolution of paddy soils in different soil landscapes. Geoderma 115, 15-29.

Zhao, Y.G., Zhang, G.L., Zepp, H. & Yang, J.L., 2007. Establishing a spatial grouping base for surface soil properties along urban-rural gradient: a case study in Nanjing, China. Catena 69, 74-81.

Chapter 27

Archaeological Materials

Richard I. Macphail[1], Paul Goldberg[2, 3]

[1]UNIVERSITY COLLEGE LONDON, LONDON, UNITED KINGDOM;
[2]BOSTON UNIVERSITY, BOSTON, MA, UNITED STATES;
[3]UNIVERSITY OF WOLLONGONG, WOLLONGONG, NSW, AUSTRALIA

1. Introduction

This chapter deals with the use of soil micromorphology for examining archaeological materials, to understand their formation and the manner in which they enter the archaeological record. This knowledge can then be applied to reconstruct past human

Interpretation of Micromorphological Features of Soils and Regoliths. https://doi.org/10.1016/B978-0-444-63522-8.00027-9

technologies and activities, especially when combined with microfacies analysis (Courty, 2001). Included in the discussion are natural soils and sediments employed in constructions that have been little transformed. These elements should be readily recognisable to workers experienced in natural soils and sediments. In contrast, archaeological materials found in occupation areas often have origins and formation processes completely outside the experience of most micropedologists and petrologists. This article is therefore partly meant to familiarise them with these somewhat atypical objects in thin section. Lastly, manufactured materials are often familiar to material scientists, such as archaeometallurgists, but it is important that these materials, and their weathered transformations, are readily recognised more widely in thin section. The variety of archaeological materials being reported on has grown exponentially since the 1980s (Courty et al., 1989) and as a result only a restricted coverage of the subject is offered here. Soil micromorphological descriptions and supportive data are detailed in referenced articles and in the figure captions.

2. Natural Soils and Sediments Employed in Construction

Natural materials, such as turf, brickearth, loess, alluvium and till, can be employed in their raw state or be mixed with various mineral and plant tempers to produce adobe materials, such as daub and mud brick (see Fig. 3C). Various raw, little-transformed and burned variants will be examined here.

When encountered in their raw state, natural soils and sediments used as constructional materials may be difficult to discern. Thin sections can show the occurrence of a natural material which is not present in a local natural soil or sediment and which is thus anomalous or exotic. The archaeological context may provide key information (e.g., mound or tumulus, floor or wall within a structure), but sometimes the context is not clear, and the presence of soils and sediments as a constructional material may only be recognised on the basis of soil horizons or sediments being in anomalous positions. For example, the occurrence of Bt horizon material along with textural pedofeatures (e.g., clay coatings) that are not oriented according to the present 'way-up' position may indicate that a soil slab has been used for construction. When soils are employed, different horizons can be used. The use of Ah, A2, Bt and Ctk horizons of Luvisols have been reported, alongside turves from podzols (Evans, 1957; Fenton, 1968; Macphail, 1987, 2003). However, field and archaeological explanations may well be wrong, and it is the chief role of soil micromorphology to correct these interpretations (see Fig. 3A).

2.1 Turf

Turf (topsoil) has been employed ubiquitously to construct mounds and ramparts in temperate regions (Van Nest, et al., 2001; Goldberg & Macphail, 2006). It can provide important environmental information on past landscapes, especially when combined with dating, palynology and the study of land snails (Alexandrovskiy & Chichagova, 1998; also note pioneer studies by Dimbleby (1962), Evans (1972) and reports by I. Cornwall reviewed by Macphail (1987)) (Fig. 1). Turf has specific characteristics according to the edaphic conditions under which it formed and the types of organic forms, such as mull, moder and mor. Excrements of soil mesofauna and preservation of organic matter reflect these conditions (Babel, 1975; Bal, 1982; Nys et al., 1987; see also Gerasimova & Lebedeva-Verba, 2010; Stolt & Lindbo, 2010; Gerasimova & Lebedeva, 2018; Ismail-Meyer et al., 2018; Kooistra & Pulleman, 2018). For example, at some poorly drained locations, grass litter decays slowly at the soil surface as thin microlaminated layers of leaves and excrements forming a laminated mull (Barrat, 1964); the initial construction of both Hadrian's Wall (~AD 122) near Carlisle, United Kingdom, and the Viking Age Gokstad Ship Burial Mound near Sandefjord, Vestfold, Norway, employed laminated mull turves, with sedge grassland being the turf source at Gokstad (Macphail et al., 2013).

The presence of turf or buried turf (Carter, 1990) can be recognised from its humic and biological character in the form of pure organic matter, mineral soil with high organic matter content organised as organo-mineral excrements and sometimes a preponderance of organic excrements of all sizes, with structures and excrements becoming generally finer up-profile; note, however, that the broad organic excrements of comminuters may most commonly occur in surface organic horizons. Plant residues and relict roots may also be expected to be more numerous in an upward direction, thus pointing to the 'right-way-up' for the turf (Fig. 1).

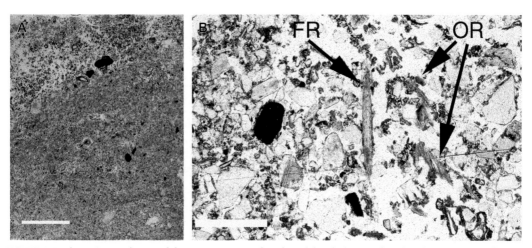

FIGURE 1 Turf. *In situ* modern turf from upper part of turf roof (Bagböle, Umeå University Experimental Farm, north Sweden; see Viklund, 1998; Goldberg & Macphail, 2006). (A) Acidic grassland mull horizon with living grass roots in right-way up position, very abundant thin organic and organo-mineral excrements in the uppermost part and the presence of wood charcoal providing evidence of past land use history (conifer forest clearance or management) (NL, scale bar length 2 mm). (B) Detail of same sample showing fresh roots (FR) from grass currently growing on the turf roof, and old roots (OR) showing 'browning' that are probably relicts of original pasture soil used to construct the turf roof; thin and sometimes coalesced organic excrements occur between sand grains (PPL, scale bar length 1 mm).

During mound or rampart construction, any type of turf can be employed. Romans and Robertson (1983) suggested that the 5.5-m-high Neolithic mound at Strathallan, Scotland, was constructed of poorly stable agricultural topsoil; this construction led to disaggregation and the formation of very abundant textural pedofeatures (dusty clay void coatings and silty clay infillings) (Barclay, 1983; Macphail et al., 1987). Both grazed grassland and woodland turf were collected to construct the putative mound at Romano-British Folly Lane, St Albans, whereas a series of Bronze Age mounds at West Heath were constructed from acid heathland turf developed on podzols (Scaife & Macphail, 1983; Macphail et al., 1998; see also Macphail et al., 2003).

Turf may be laid 'right-way-up', inverted or face to face. The experimental grassland mull turf roof at Umeå, north Sweden, was constructed with the bottom turf facing downwards and the upper ('living') turf facing upwards (Cruise & Macphail, 2000; Goldberg & Macphail, 2006) (Fig. 1). Modern Icelandic turf roofs have been investigated by Milek (2006). Turf roofs cannot be made out of moder or mor horizons because these are prone to rapid oxidation and decay.

The remains of (burned) grass turf wall material from the estimated 1-m-thick and 1.5-m-high walls from Iron Age Denmark were identified by a microaggregated fabric and highly humic aspect (Nørnberg & Courty, 1985). Similarly, turf fragments in a sunken building were found to contain organo-mineral and amorphous organic matter; minerogenic lenses within turf fragments are relict of rapid tephra fallout (Simpson et al., 1999). Viking house turf walls have also been studied to investigate contemporary soils (Milek, 2004).

Turf in mounds and buried turf (*in situ* topsoil) undergo various transformations. Changes related to a decrease in pH have been reported (Macphail, 1993; Crowther et al., 1996). In field experiments, base-rich turf became compact as organo-mineral excrements became coalesced, and a more acidic mesofauna producing thin bacillocylinders became dominant (Crowther et al., 1996). At Neolithic Easton Down, Wiltshire, the base-rich but decalcified turf that had formed on chalk seems to have been transformed in exactly the same way (Macphail, 1993). On loess over chalk at the turf rampart of Neolithic Belle Tout, East Sussex, no individual excrements are visible, and only the compact remains of a totally biologically reworked microfabric with relict fabric features of burrowing are present in a soil with a superimposed accommodating angular blocky microstructure. In acid sandy soils, compaction and 'ageing' can lead to changes in c/f-related distribution pattern (from enaulic progressively to chitonic, gefuric and porphyric), a decrease in interference colours of the plant material and preferential preservation of poorly decomposable organic materials (lignified cells) and charcoal (Scaife & Macphail, 1983; Macphail et al., 2003). In other field experiments and at a Bronze Age site, the base of the turf core displays a predominantly porphyric c/f-related distribution pattern and well-preserved organic matter, presumably due to the maintenance of soil wetness. Organic matter and turf layers may become mineralised through postdepositional movement of sesquioxides (Runia, 1988; Macphail et al., 2003). The chemical and micromorphological transformation of barrow turf into spodic pyrophosphate and dithionite Fe- and Al-enriched monomorphic material was reported by Fisher and Macphail (1985) (see also Vepraskas et al., 2018). Mounds can develop iron/manganese oxide pans because of localised waterlogging, which tends to accentuate turf structures within them, as well as the contact between the mound and the underlying soil; at the Gokstad Mound mobile phosphate had formed concentrations of vivianite in litter-rich turf layers (Bouma et al.,1990; Macphail et al., 1998, 2013). Fragments of turf, which retain characteristic biological microfabrics such as thin burrows (passage features) and excrements, are sometimes common components in the fills of sunken feature houses (e.g., *Grubenhäuser*) because of collapsed turf walls and roofs constructed from turf (Simpson et al., 1999; Macphail et al., 2006a; Milek, 2006).

2.2 Ground-Raising Constructional Materials

Constructions often include ground-raising and surface-sealing activities (Blume, 1989). Raw materials derived from local soils and underlying bedrock are often employed (see Figs. 2 to 4). In northwest Europe, for example, this material is commonly derived from the typical Luvisols formed on loess and from aeolian silt with alluvial admixtures of sand (brickearth). In the Balkans and in the Near and Middle East, tells include mud bricks derived from local alluvial soils (Goldberg, 1979; Rosen, 1986; Stoops & Nijs, 1986; Matthews et al., 1996; Haită, 2003; Love, 2012; Macphail et al., 2017a). The presence of rounded clay clasts and allochthonous carbonate nodules can be indicative of an alluvial soil origin (Goldberg, 1979; see also Nodarou et al., 2008). From Roman, medieval to modern times, these materials were employed in urban areas for levelling ground ahead of construction (Macphail, 1994). Sometimes burned daub from razed buildings was also used. Mesopotamian tells also developed through incidental and deliberate ground-raising and the accumulation of decayed mud brick (Friesem et al., 2011). This was brought about by dumps of all kinds of anthropogenic materials, such as ashes and building debris, which were also employed as 'constructional packing' below constructed plaster floors and matted surfaces (Matthews & Postgate, 1994; Matthews, 1995; Matthews et al., 1996) (see Fig. 8A). At Phoenician Tel Dor, Israel, however, ground-raising associated with manufactured floors and animal penning episodes was achieved through refuse accumulation associated with occupation (Shahack-Gross et al., 2005). Microdebris included wood ash, bone, charcoal, ceramics, constructional materials and abundant phytoliths. There have been similar findings at other Middle Eastern sites (Stoops & Nijs, 1986; Boivin & French, 1997/1998; Matthews et al., 1998; Gur-Arieh et al., 2014).

Mounds constructed of soil include medieval mottes in Europe (e.g., Gebhardt & Langohr, 1999) and Mississippian Culture mounds in the United States (Sherwood, 2006). Nevertheless, mound-raising may involve both the use of coherent soil slabs and simply dumping of soil, regolith and rock fragments in so-called basket loads. In these, the various soil materials become mixed as soil clasts and form a heterogeneous layer. In this case, the once-open voids are commonly filled with dusty to impure clay that is associated with textural intercalations whose texture is governed by the original grain size of the soil, all because of slaking of disturbed soil. This phenomenon was also noted by Romans and Robertson (1983) for barrow constructions and is similar to disturbed tree-throw hollow fills and to soils transported, deposited and modified under freeze-thaw conditions (Mücher, 1974; Macphail & Goldberg 1990; Macphail, 1999). The presence of the aforementioned pedofeatures, which are anomalous for a natural soil profile, may help distinguish between buried natural soils and ground-raising that uses soil. Lastly, construction layers employing dumped soils may develop reddish colours in the field (Sherwood, 2006), because iron-stained clay is concentrated in clay infillings and fine intercalations, but it remains conjectural whether such red layers were constructed intentionally (Courty et al., 1989).

2.3   Floors, Surfaces and Walls

Floors have been constructed from various coherent loamy and clayey deposits that are local to a site, producing so-called clay floors (Rentzel, 2011). In coastal sandy areas, estuarine silty clay (Macphail, 1990) and till (Nørnberg & Courty, 1985) have been used to produce stable living surfaces. Local wetland sediments, often rich in phytoliths and diatoms, have been imported to produce daub and clay floors, for example, in Iron Age and Migration Period (c. AD 0 to 600) houses at Jarlsberg, Vestfold and Åker gård, Hedmark, Norway (Viklund et al., 2013; McGraw, personal communication); at the unusually well-preserved site of Åker gård, byre areas (see Section 3.2.1) had earth floors strengthened by being pebbled, while relict periglacial silt loams were favoured for hearth constructions. The artificiality of such floors is inferred from the occasional inclusion of anthropogenic materials (e.g., charcoal, pottery, burned daub, bone and, even rarely, coprolites) and various iron, iron phosphate and calcium phosphate staining of their surfaces and voids (hypocoatings). The construction of clay floors was not the case, however, in most Viking and medieval sites in Iceland (e.g., 12th-century Reykholt Farm in Borgarfjörður, Iceland; Sveinbjarnardóttir et al., 2007). In north-west Europe, brickearth and loess soils were employed to produce floor slabs of natural material (Figs. 2 and 3). This included soil from both the pale A2 horizon and the more clay-rich Bt and Ctk horizon; the latter may include relict calcitic pedofeatures. Such types of deposits are specifically built and represent truly constructed floors. They are, however, often found alternating with dark coloured, charcoal-rich occupation accumulations, which are not constructed floors per se but represent beaten or trampled material and are unlikely to have been deposited on purpose (Fig. 2). Constructed clay floors thus mark the renewal of a clean living surface (see also Matthews et al., 1996). At Reykholt Farm, this was achieved by creating thin clay plastered floors that alternate with occupation deposits (Sveinbjarnardóttir et al., 2007). The difference between a constructed floor and overlying beaten floor deposits can be noted in the field and in thin section as a knife-sharp,

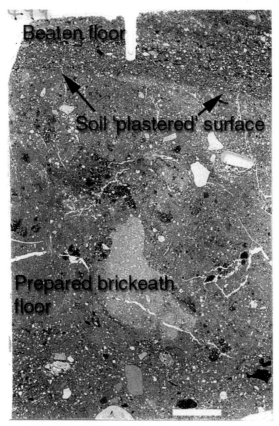

FIGURE 2 Earth floors. Medieval floor and beaten floor sequence (Spitalfields Hospital, London, sample M499, layer 4850 over layer 4900; Museum of London Archaeological Service; Goldberg & Macphail, 2006). The brickearth floor is a prepared mixture composed of brickearth, 'soil' and anthropogenic inclusions, with an apparent 'white soil' plastered surface employing iron-poor soil (probable A2 horizon); a beaten floor has formed over this by trampling; the brickearth floor is affected by some iron phosphate staining compared with the beaten floor which is rich in various forms of phosphate (e.g., fine bone and coprolites), organic matter and burned material (ashes, charcoal, burned soil) (NL, scale bar length 1 cm).

horizontal boundary, the occupation floor deposits showing fine sorting and horizontal orientation of materials such as bone, eggshell and mollusc shell fragments. The degree of lamination reflects the degree of humidity, the level of protection from the elements and the nature of any floor coverings (Gé et al., 1993; Cammas, 1994; Courty et al., 1994; Cammas et al., 1996; Matthews et al., 1997; Macphail et al., 2007). It has to be kept in mind, however, that floors were often swept, and sweepings were dumped elsewhere, and that 'occupation' deposits in fact result from 'disuse', squatter occupation or building decay (Macphail, 1994; Cowan, 2003; Galinié, personal communication; Sheldon, personal communication). Such deposits do not show the fine laminations as found in

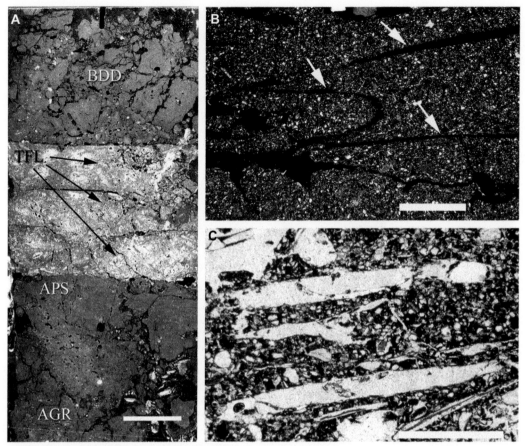

FIGURE 3 Floors and adobe. Middle Neolithic earth- and rock-based building material, and ground-raising and construction materials (Middle Neolithic — Yangshao, Huizui, Henan Province, China). (A) Adobe ground-raising deposits (AGR) and plant-tempered adobe preparation surface (APS), both composed of local loess, covered by a floor constructed from tufa (tufa floor layer, TFL), covered by burned daub debris (BDD) (NL, scale bar length 2 cm). (B) Plant-tempered mud-plastered loess forming a floor preparation surface (*arrows* indicate void pseudomorphs after plant remains); the dense character of the groundmass is due to soil slaking caused by the mud-plastering process; patches of microfabric with high interference colours result from the formation of textural intercalations caused by internal slaking creating a striated b-fabric (XPL, scale bar length 1 mm). (C) Mud brick (Iron Age, Tell Yoqneam, Israel). Earth-based, dark brown mud brick with typical large voids with elongated angular shapes that are moulds of plant remains (PPL, scale bar length 1 mm).

beaten floors and are also more likely to be markedly contaminated by latrine waste, for example. The use of brickearth slabs to form walls within timber-framed buildings in Roman London has long been documented, and one intact fallen lime plaster-coated brickearth wall at 1st to 2nd century Southwark, London, sealed a short-lived disuse occupation deposit within an abandoned building shell that included, in addition to brickearth fragments, biomixed coarse and fine charcoal, bone, eggshell, amorphous organic matter and earthworm excrements (Cowan, 2003).

Constructed floors may show iron oxide and iron phosphate staining and associated vivianite formation (see Karkanas & Goldberg, 2018) because of contamination from liquid waste, but overall they are less phosphate-rich than the beaten occupation accumulations. In addition, the surface of soil slabs in constructed floors is sometimes 'mud plastered', which is recognisable because of the evidence of slaking, which concentrates infillings with clay and intercalations in the uppermost few millimetres. This plastering may have been carried out to produce a dark red colour, caused by the development of clay void coatings and intercalations where the iron content of the soil is concentrated in the clay, or simply to seal a surface. This type of surface plastering has been noted at various sites (Boivin, 1999; Matthews et al., 2000; Macphail & Crowther, 2007) (Figs. 2 to 4); it has been argued, however, that fresh plaster was not necessarily always employed and that fragmented material may have been reused (Karkanas & Van de Moortel, 2014). Sealing is also recognised for floors constructed from calcareous till (Chalky Boulder Clay) at the 12th-century wheat barn, Cressing Temple, Essex, and at Iron Age Uppåkra, Scania, Sweden, which has similar geology (Macphail, 1995; Macphail et al., 2017a). Roman sunken floor buildings on chalk had their subterranean floors sealed with rammed fine chalk (Bennett et al., 2008). This practice is expedient because minor recrystallisation of calcite produces a semicemented material without the use of lime. In the United Kingdom, chalky 'clunch' walls and a chalky daub called 'cob' have similar characteristics.

Floors were produced from brickearth, loess or till, which was often mixed with dung and other organic matter, including a plant temper such as straw (Karkanas, 2007). Through time, this organic matter decomposed, which leaves straight-sided voids pseudomorphic after straw (see Figs. 3 to 5). In some cases, phytoliths or iron and manganese oxide stains can be observed.

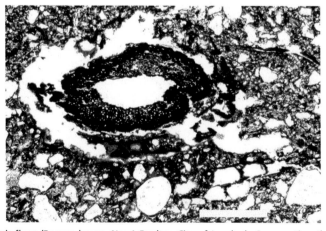

FIGURE 4 Burned daub floor (Roman house, No. 1 Poultry, City of London). Cross section through charred straw, with associated blackened humic staining and shrinkage voids (PPL, scale bar length 1 mm).

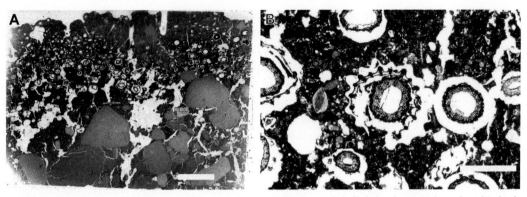

FIGURE 5 Clay oven. Experimental oven constructed from Cretaceous red chalk (a local material employed in both Roman and Saxon times at West Heslerton, North Yorkshire, UK) and straw. (A) Layered structure, with a lower rubefied layer which has been most affected by heating, a straw-tempered middle layer and an upper surface-sealing mud layer (NL, scale bar length 1 cm). (B) Detail of the straw-tempered layer, showing blackened and charred straw, as well as shrinkage voids (PPL, scale bar length 1 mm).

Building stone in constructions is outside the remit of this chapter, but when studying constructions, an understanding of petrographic features of rocks that were or could have been employed is obviously very useful. For example, Jurassic Oolitic limestone can be found within hydraulic mortar used in the construction of an aqueduct near Basel, Switzerland (Rentzel, 1998) (see Fig. 21), while Maya lime plaster floors were tempered with local siliceous and sponge-spicule-rich mud flat sediments at Marco Gonzalez, Belize, during the Early Classic to Late Classic (c. AD 250 to 700/760; Graham et al., 2017; Macphail et al., 2017b) (Fig. 8A). The study of Roman building materials at Pessinus, Turkey, included the characterisation of local limestone types and marble as background to the investigation of lime plasters (Stoops, 1984a). Soil micromorphology had to be employed to identify the use of quarried slabs of tufa as flooring material, which in the field had been mistakenly identified as manufactured lime plaster floors (Macphail & Crowther, 2007) (Fig. 3). In contrast to plaster, tufa is characterised by the blackened plant remains (Babel, 1975; Stoops, 2003) and by calcitic pseudomorphs after plant tissues (see Karkanas, 2007). Phytolith layers from oxidised stabling deposits, for example, have also been misidentified in the field as plaster floors because of their pale colour (Shahack-Gross et al., 2005; Albert et al., 2008).

The fills of sunken feature houses (e.g., *Grubenhäuser*) (Simpson et al., 1999; Milek, 2006; Wegener, 2009) can be modelled as having a soil fabric dominated by mesofauna excrements. In First Nation Canadian pit houses, anthropogenic materials including burned residues and coprolites are strongly mixed with earth-based and organic roofing debris by biological activity (Goldberg, 2000). This soil material can be either natural when the locality had been little altered by human activity or strongly anthropogenic. In the latter case, when settlements are long-established, evidence includes very abundant fine charcoal and rubefied mineral grains, abundant and articulated phytoliths, rare to occasional dung fragments and residues, and coprolites (Macphail et al., 2006b;

Macphail, 2016a). In sunken feature house fills, local silt or sandy loam hand-formed into loom weights during early medieval times can be found as anomalous fragments of clay or silt loam materials (Macphail et al., 2006b). Loom weights can be common inclusions in archaeological fills associated with textile manufacturing.

Mud brick, daub and adobe are similar earth-based building materials, which can be recognised from their composition, void patterns (if plant-tempered), inclusions (charcoal, dung, soil and sediment clasts) and, in some instances, the presence of textural and fabric pedofeatures or the occurrence of slaking if material is puddled or poured as a slurry into a mould, all of which can produce fabric heterogeneity from the employment of different soil materials and clayey intercalations merging into dusty clay void coatings where the voids are closed vughs or even vesicles (Levine et al., 2004; Goldberg & Macphail, 2006; Nodarou et al., 2008). When burned, these building materials preserve well, but they can be modified by weathering and reworked by colluvial transport (Goldberg, 1979) or totally transformed into soil through pedogenic processes (Goldberg, 1979; Macphail, 1994; Friesem et al., 2014a, 2014b).

Burned variants of the above-listed constructional materials often occur at archaeological sites (Kruger, 2015; Forget & Shahack-Gross, 2016). The chief characteristics caused by heating are blackening of once-humic areas (iron oxides are not strongly transformed because of localised reducing conditions or are incompletely burned) and reddening caused by the formation of hematite (oxidising conditions), best seen in OIL (Goldberg & Macphail, 2006). Another effect of heating is the formation of a crack structure, because of the loss of water by clay (Courty et al., 1989) (Figs. 4 and 5). In the case of experimental daub made from loess, a major change in macroscopic colour from hue 7.5 YR to 5 YR took place between 400 and 500°C (Dammers & Joergensen, 1996). The inclusion of organic materials produced grey, reduced areas. Examples from an experimental straw-tempered oven and a Roman straw-tempered brickearth floor demonstrate how the straw becomes charred and the surrounding matrix shrinks through desiccation (Figs. 5 and 6). Clearly, the temperature affecting the floor was not

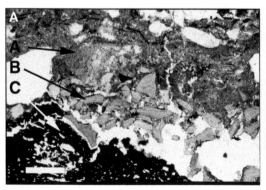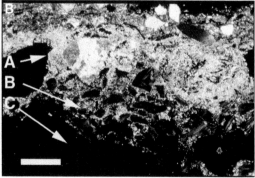

FIGURE 6 Hearth ash (Brean Down, Somerset, UK). *In situ* Bronze Age hearth within a round house. (A) Charcoal (c), ash (a) and burned bone (b) layers (PPL, scale bar length 0.5 mm). (B) Same field in XPL, illustrating high interference colours of calcite ash and modified birefringence of charred bone.

excessive, as the major loss of organic matter in daub tempered by barley straw occurs between 200 and 300°C, with most organic material lost at 400°C (Dammers & Joergensen, 1996; see also Berna et al., 2007) (see Fig. 8B). At an early Neolithic site in Hungary, burned daub and burned plant-tempered daub constitute the most abundant evidence of constructions (Crowther, 2003, 2007; Macphail, 2003; Carneiro & Mateiciucová, 2007). Soil micromorphological and mineralogical analyses at Tel Dor reveal that surfaces constructed from local calcareous sandstone were heated to produce a plaster floor (Shahack-Gross et al., 2005); heated plaster floors also characterised salt working levels at Late Classic Marco Gonzalez, Belize (Graham et al., 2017) (Fig. 8A).

2.4   Organic Floor Coverings

Floors and surfaces were sometimes covered with organic materials, which include turfs (Milek, 2006), mats and, more conjecturally, skins (Courty, 2001; Courty et al., 1994; Macphail et al., 1997). These organic floor coverings occur in Palaeolithic to medieval sediments, where 'mats' are characterised by thin, millimetre-thick layers of articulated phytoliths of Poaceae (grasses, reeds, straw) or they consist of identifiable plant remains, as at the Middle Stone Age of Sibudu Cave, KwaZulu-Natal, South Africa (Goldberg et al., 2009b). As found in Viking Age Trondheim and Oslo, Norway, under waterlogged conditions, centimetre-thick accumulations of organic coverings can be preserved on wooden floors, which under more strongly oxidising conditions may be in part preserved by iron phosphate impregnation (Macphail & Goldberg, 2017, pp. 377-379). In certain cases, these surfaces are associated with horizontal planar voids and iron and phosphate staining of the associated fine sediments as a result of trampling and of the mat acting as a hydraulic barrier, a phenomenon that also affects many clay floors and other floor make-ups (Matthews et al., 1996; Macphail et al., 1997; Macphail, 2005; Milek, 2006). Voids between the woven mat fibres can be partially filled with thin organo-mineral excrements from both the buried soil and overlying trampled soil (Macphail & Cruise, 2001). Beaten floor deposits (Gé et al., 1993; Cammas, 1994; Cammas et al., 1996; Milek, 1997; Macphail et al., 2004, 2006b, 2007; Sveinbjarnardóttir et al., 2007) (see Fig. 2) that can form on floors and matted surfaces have been already briefly mentioned in Section 2.3. Some of the anthropogenic materials within them, and which help characterise both *in situ* use of space and local activities, will be dealt with in the following text (e.g., domestic/hearth, stabling and industrial waste origins) (Cammas, 1994; Matthews et al., 1997; Banerjea et al., 2015).

3.   Waste Materials

Human occupation produces concentrations of all kinds of organic and inorganic waste (Courty et al., 1989, 1994; Goldberg & Macphail, 2006). These are found in middens, structures, or pits or simply concentrated around hunter-gatherer campsites. Additionally, these occurrences are the focus of scavenging by animals that leave their own waste.

3.1   Inorganic Waste Materials

In addition to fuel residues (charcoal, clinker from coal, and charred peat, turf and dung) that generally can be found in and around occupation areas (Carter, 1998; Milek, 2005; Milek & French, 2007), inorganic waste materials were commonly employed in northern regions for soil amelioration, as for example the manured soils at Papa Stour, Shetland (Adderley et al., 2006). Hearths normally produce calcitic ash (Wattez & Courty, 1987; Courty et al., 1989; Shahack-Gross et al., 2004a, 2014), often in layers alternating with wood charcoal beds, the latter recording incomplete combustion, whereby the layered structure can be preserved in conditions of rapid burial (Fig. 7) (Macphail, 1990). In the last case, ethnoarchaeological observations of different Hadza hearth types in Tanzania revealed that differently functioning hearths were both disturbed and weathered by natural agencies, with people reusing hearths and scooping out ashes, for example; there have also been experiments to replicate Middle Palaeolithic hearth fires (Mallol et al., 2007, 2013). The presence of vesicular porous char in hearths indicates the burning of flesh or animal fat (Goldberg et al., 2009b; Villagran et al., 2013; Mentzer, 2014). Although often converted by weathering into general poorly diagnostic micritic calcite masses (Courty et al., 1989; Gur-Arieh et al., 2014; Shahack-Gross et al., 2014), some wood ash can, on occasion, be preserved as coarse lozenge-shaped crystals, which are pseudomorphs after calcium oxalate crystals that formed inside the wood cells (Franceschi & Horner, 1980; Brochier, 1983; Wattez & Courty, 1987; Karkanas et al., 2007; Mentzer, 2014; Polo-Díaz et al., 2016), and termed POCC, an acronym of 'pseudomorphose d'oxalate de calcium en calcite' (Brochier & Thinon, 2003). If temperatures exceed 600°C, these pseudomorphs are converted into lime (CaO), which in turn is transformed to undiagnostic micritic calcite by hydration and carbonation (Brochier & Thinon, 2003). Bark and leaves produce greater amounts of calcium oxalates than wood (Franceschi & Horner, 1980; Brochier & Thinon, 2003), and calcium oxalates are abundant where ashes are

FIGURE 7 Grass ash (Early Iron Age site, Maiden Castle ditch, Dorchester, UK). (A) Burned Poaceae remains from cereal processing and dung, rich in phytoliths, including articulated phytolith sheets; note humic staining which may indicate the presence of burned dung (PPL, scale bar length 100 μm). (B) Same field in XPL, showing scattered and clustered fine calcite ash, small calcite spherulites suggesting dung may be a component, and cereal material (charred seeds).

derived mainly from stabling of animals foddered on woodland 'leaf hay'. Ash from grass, as well as from cereal processing and dung, produces finer crystals and an abundance of associated phytoliths, some of which may have melted to produce vesicular slags (Macphail & Cruise, 2001; Milek, 2005; Weiner, 2010) that can be enriched in phosphate (Fig. 7). These slags are autofluorescent in blue light, possibly because of a carbonate-substituted hydroxylapatite component (Weiner, 2010; Karkanas, personal communication). Burned phytoliths have a refractive index greater than 1.440 (Albert et al., 2008). Some siliceous burned monocotyledonous plant pseudomorphs are spherical in shape, or oblate when more strongly heated; these partially melted siliceous plant remains have seemingly vesicular voids which are simply relicts of natural vessels within stems (Macphail et al., 2012). Ash from cereal processing is often mixed with charred chaff (Macphail, 1991, 2000). Weathering, reuse of hearths and disturbance by digging scavengers can produce isolated and granular aggregates of ash. Some hearths produce burned bone (Karkanas et al., 2007), which appears to be more resistant to acid soil weathering than unburned bone, possibly because of alteration of the phosphate into a more stable, but unknown form than apatite (Berna, personal communication). Burned stones from hearths or burned rock mounds sometimes have residual ash cemented to their surfaces as a testimony to their hearth origin. Clearly, the accurate micromorphological and mineralogical identification of burned bone and hearth deposits in early human sites is crucial (Goldberg et al., 2001; Karkanas et al., 2007; Mentzer, 2014).

Burned food waste typically includes charred bone (rubefied, with lowered interference colours and lowered autofluorescence), calcined bone (whitish, optically isotropic and non-autofluorescent under blue light), burned eggshell (Fig. 9), large burned mollusc shells and charred cereal grains or seeds.

Burned soil and constructional materials that only show reddening may be of domestic hearth origin. Strongly burned soil showing a loss of birefringence of the aluminium-silicate minerals (as in some potsherds) and quartz (and melted 'dewdrop'-shaped and bubbled vesicular quartz, with a loss of birefringence mirroring loss of crystalline structure) is more typical of industrial hearths when very high temperatures are reached (>850°C, up to 1713°C) (Courty et al., 1989; Guélat et al., 1998; Berna et al., 2007; see also Mathieu & Stoops, 1972; Angelucci, 2008) (Fig. 8B).

Experiments employing a combination of open fires and furnaces with bellows showed kaolinite was lost and smectite suffered dehydroxylation and partial vitrification at 500-700°C, quartz was transformed at 1000-1200°C to a glaze-like (glassy) phase and to cristobalite at 1300°C (Berna et al., 2007). Salt working typically employs low-temperature fires to boil brine, sometimes using salt-rich coastal sediments. Hearth surfaces (Fig. 8A) and briquetage become rubefied because of this processing, but development of a high-temperature glazed surface has also been observed for briquetage used as bellows plates (Macphail et al., 2012). The 'redhills' of Essex, United Kingdom, and salt working levels at Marco Gonzalez have developed from the rapid accumulation of burned tidal flat sediments used to extract brine (Graham et al., 2017; Macphail et al., 2017b) (Fig. 8A).

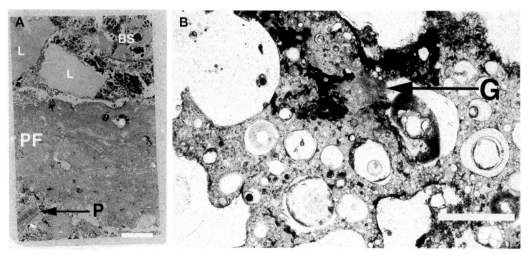

FIGURE 8 Industrial activities (salt-working hearth floor at Marco Gonzalez, Belize, and furnace waste from Åmot øvre, Stokke, Vestfold, Norway). (A) Flatbed scan of Late Classic Maya lime plaster floor (PF). This shows a sequence of a lime floor constructed over debris which includes pottery fragments (P; probable Coconut Walk ware used in salt working). The lime floor surface is weakly rubefied from the use of small fires to heat brine; ash and other fuel waste (charcoal) occurs above, where there is a dump of limestone (L) and burned sediments (BS) used in brine extraction (Graham et al., 2017; Macphail et al., 2017b) (NL, scale bar length is 1 cm). (B) Vesicular, fused fine sands and partially melted sands (quartz and feldspars which have lost or partially lost their birefringence), with small area of silicate glass formation (G) as identified by micro-FTIR. Iron staining suggests an association with iron-working (Viklund et al., 2013; F. Berna, personal communication) (PPL, scale bar length 1 mm).

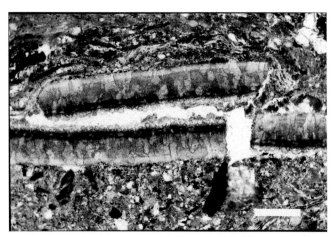

FIGURE 9 Burned eggshell (early medieval London Guildhall, City of London, UK), showing horizontal alignment, typical edge crystals and blackening produced by burning (PPL, scale bar length 0.5 mm).

Different activities can be implied from inclusions of leather and non-ferrous metal materials (craftwork), bark and wood (wood working or construction), exotic igneous rock fragments (grindstones) and exotic limestone and sandstone associated with lime plasters and mortar (construction). Thus, predominantly domestic and industrial spaces and areas within structures that are employed to house domesticated animals can be differentiated on the basis of the associated archaeological materials (Cammas et al., 1996; Macphail et al., 1997, 2004, 2007; Shahack-Gross et al., 2004a, 2005; Milek, 2006; Banerjea et al., 2015).

3.2 Organic Waste Materials

Animal and human faecal waste is a source of phosphate and organic matter. Although bat guano (Shahack-Gross et al., 2004b) and bone-rich bird guano remains are typical of some natural cave sediments (Macphail & Goldberg, 1999; Goldberg & Macphail, 2012), scavenging by birds can be similarly recorded from occupation areas and cremation- and excarnation-associated deposits where bodies were exposed (Macphail & Crowther, 2008; Macphail & Goldberg, 2017, pp. 487-489; see also Angelucci 2008).

Dung of herbivores (cattle, sheep, goats, horses) can be ubiquitous: in structures, fields and trackways where finely fragmented residues can occur alongside phosphate-enriched inwashed clays and neoformed iron, calcium or phosphate nodules (Engelmark & Linderholm, 2008; Macphail, 2011; Macphail et al., 2017a); manured fields may include articulated phytoliths that can be identified and counted in thin section (Devos et al., 2013). Herbivore dung can normally be differentiated from that of omnivores (pigs, humans) and carnivores (dogs, hyenas) (Courty et al., 1989, 1994; Horwitz & Goldberg, 1989; Macphail, 2000; Goldberg et al., 2009a).

Oxidised dung remains, especially from ovicaprids, may be rich in 'faecal spherulites', formed of calcite or monohydrocalcite (Brochier, 1983; Brochier et al., 1992; Canti, 1999; Shahack-Gross et al., 2004a, 2004b; Shahack-Gross, 2011). However, not all calcite spherulites (<20 μm) found in archaeological sediments are necessarily of faecal origin. Calcite spherulites have been experimentally obtained from recrystallisation of *Tamarix aphylla* wood ash (Shahack-Gross & Finkelstein, 2008). Clearly, the simple presence of calcite spherulites is inconclusive evidence for stabling or animal stocking. In tell sites, individual spherulite occurrences can be ubiquitous, and only where they are concentrated in high numbers do they have significance. Even in this case, it is their organisation within the soil microfabric and relationship with other constituents that allow any proper interpretation of their occurrence (*in situ* and microstratified stabling layers with identifiable dung remains, dumps, background herbivore presence, trampled concentrations) (Binder et al., 1993; Macphail et al., 1997, 2017a; Matthews et al., 1997; Sordoillet, 1997, 2009; Boschian & Montagnari-Kokelji, 2000; Shahack-Gross et al., 2003, 2004a, 2004b; Angelucci et al., 2009).

3.2.1 *Herbivore Dung*

Herbivore dung dominated by Poaceae remains (grasses and cereals) is not universal, and both ovicaprids and cattle were stalled or overwintered on a diet of woodland 'leaf hay' during prehistory in Swiss lake village sites (e.g., fir) and in Mediterranean rock shelters (e.g., oak). In the latter case, ashed stabling deposits record layered deposits of dark brown twig wood 'stable floor' layers and grey layers of ashed dung and fodder that were first fully identified at the Arene Candide (Italy) and Pendimoun (France) caves (Courty et al., 1991; Binder et al., 1993; Macphail & Goldberg, 1995). Subsequent investigations of numerous Neolithic animal management sites (e.g., France, Italy, Spain and Switzerland) have confirmed these findings (Brochier, 1996; Macphail et al., 1997; Sordoillet, 1997; Boschian & Montagnari-Kokelji, 2000; Akeret & Rentzel, 2001; Angelucci et al., 2009; Polo-Díaz et al., 2016). Calcium oxalates, their pseudomorphs and burned (blackened) residues are abundant, while phytoliths are rare. Charred twig wood, leaves and bark, and calcite ash pseudomorphic after the original wood cell structure are also visible, alongside ashed herbivore coprolites. It can also be noted that dung residues can contribute to grass ash (see Fig. 7); in some cases, recognisable fragments of ashed dung pellets can be preserved.

Herbivore dung is dominated by poorly digested plant tissues and small amounts of amorphous organic matter and organic phosphate staining that is normally pale to dark brown under PPL and blackish brown under OIL. Cattle often produce dung with microlayered 'long' plant fragments, whereas sheep/goats produce rounded pellets with stained margins and a convoluted structure formed from 'short' plant fragments reflecting how these animals differently process a grass and cereal diet; herbivores are efficient at digesting cellulose and hence this material suffers a loss of birefringence as part of the humification process (Courty et al., 1994; Akeret & Rentzell, 2001; Macphail et al., 2004). Experimental and ethnoarchaeological stabling and animal compounds from the United Kingdom and Africa have demonstrated these differences (Macphail & Goldberg, 1995; Shahack-Gross et al., 2003, 2008; Shahack-Gross, 2011).

Stabling features include thin layers of phytolith-dominated dung residues resulting from stabling within buildings, which had at first been misidentified as plaster layers before a thin section study was carried out (Shahack-Gross et al., 2005). Compacted dung enriched in liquid animal waste can form phosphate-enriched layers and even cemented crusts (Heathcote, 2002), and in this process carbonates are partially transformed to Ca-phosphates (Shahack-Gross et al., 2003) on the floors of stables (e.g., as carbonate-substituted hydroxylapatite; Macphail et al., 2004, 2006b). These crusts have a finely laminated character because of the horizontal compaction of grass stems (and cereal remains) and the long, articulated phytoliths that characterise these plant remains; as humification continues, these phytoliths become more visible but in such crusts remain articulated as >1-mm-size lengths (Macphail & Cruise, 2001). In some cases, they occur within a finely dotted yellowish brown cement that is autofluorescent under blue light when hydroxylapatite is present. How these stabling features preserve relates to the type of dung, whether it was burned or not, and overall environmental conditions and ageing (Figs. 10 and 11).

At Atzmaut rock shelter in the Negev Desert (Rosen et al., 2005; Macphail & Crowther, 2008), recent goat dung is dominated by fragmented pellets and plant fragments, with organic material being more dominant than phosphate, whereas compressed, aged and

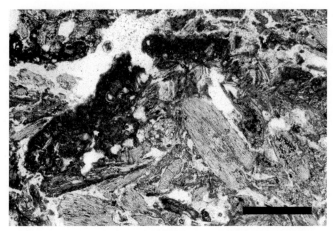

FIGURE 10 Recent goat dung stabling level (Atzmaut rock shelter, Mizpe Ramon, Negev Desert, Israel), showing plant tissue remains from fragmented goat pellets and dark brown humified amorphous organic matter (PPL, scale bar length 1 mm).

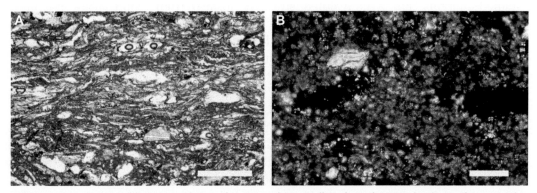

FIGURE 11 Bronze Age stabling accumulation of compacted goat pellets, from the same locality as the material in Figure 10. (A) Layered deposit mainly composed of humified amorphous organic matter, phytoliths and humified monocotyledonous plant fragments (optically isotropic or very low birefringence), and with some scattered silt and fine sand mineral grains (PPL, scale bar length 1 mm). (B) Detail showing a concentration of small calcitic faecal spherulites that are only partially masked by amorphous organic matter and are responsible for the overall optical anisotropy of the layer (XPL, scale bar length 100 μm).

oxidised Bronze Age stable floor layers are phosphate enriched, and interference colours are moderately high because faecal spherulites are no longer obscured by organic matter. Matthews et al. (1997) report interbedded lenses of dung pellet fragments and digested plant remains from stables. At 1st millennium BC Lattes, southern France, pure, bedded dung of horses or cattle is believed to have accumulated under humid exterior conditions, whereas dung bedded with quartz grains accumulated in dry conditions (Cammas, 1994).

Both layered dung from cattle and fragments of ovicaprid dung are reported from a Swiss Gallo-Roman stable at Brig Gliss VS, but here the stabling deposits were accidentally burned *in situ* (Guélat et al., 1998). This is not an uncommon occurrence where dung layers become generally strongly 'blackened' and 'browned' (see also Macphail et al., 2004). A burning temperature of c. 900°C was estimated from the presence of vitrified phytoliths at Brig Gliss VS (Guélat et al., 1998).

Dung fragments can also be preserved by iron, and iron and manganese mineralisation in which poor but still recognisable pseudomorphs are formed in intermittently waterlogged conditions (Macphail et al., 1998) or they can be completely preserved by anaerobic conditions (e.g., Akeret & Rentzel, 2001; Ismail-Meyer & Rentzel, 2004). Phosphate-cemented dung or stable floor fragments preserve well and occur in middens and manured soils; dung itself has been employed as constructional material, for example, to line wattle walls.

In an experimental manured plot in northern Sweden, characterised by a high proportion of organic phosphate, recognisable herbivore manure occurs as fragments (4 mm) of compact humified layered plant fragments within an amorphous organic matter matrix that resembles stable floor crust material (Macphail et al., 2004, 2006b; Goldberg & Macphail, 2006). Similar fragments occur in experimentally manured fields at Butser Ancient Farm and in Iron Age to Viking Period cultivated soils in Denmark, Norway and Sweden, and these still retain high proportions of their phosphate as organic phosphate from these dung inputs; dung fragments also occur in a well fill associated with one a mixed farming system (Engelmark & Linderholm, 1996; Goldberg & Macphail, 2006; Viklund et al., 2013). Such manuring raised the level of biological activity compared with the natural soil, usually manifesting itself as an increase in the amount and size of organic and organo-mineral excrements compared with a previous dominance of extremely thin organic excrements. In present-day northern Norway, when virgin and cultivated soils were treated with cattle slurry, the cultivated sandy soil developed an increase in fine material content (Sveistrup et al., 1995). The surface of the virgin soil became compacted, however, and examples of slurry coatings and panning were observed for all soils. In archaeological soils, such modern slurry coatings need to be identified as contamination (Viklund et al., 2013). On the other hand, phosphatic void coatings due to seepage from cess pits can be penecontemporaneous (Macphail, 2016b; Macphail & Goldberg, 2017, pp. 458-461).

3.2.2 *Pig Dung*

Pig dung may contain both large amounts of plant tissues and amorphous organic matter and organic phosphate, along with calcium oxalates from partially digested plant material and ingested silt (Courty et al., 1989, 1994). The latter is sometimes also present in herbivore dung and found intercalated in laminated stabling crusts. Pigs fed on a husk-rich cereal feed at 'West Stow Anglo-Saxon Village', Suffolk, United Kingdom, produced a moderately phosphate-enriched trampled surface soil in their enclosure where cereal husks occur as horizontally oriented inclusions; the immediately underlying soil contains neoformed amorphous Fe-P-Ca nodules (Macphail & Crowther, 2011). The pig pasture at West Stow was churned by pigs forming hetero-geneous soils; such mixing and microcolluviation was also reported from the 'medieval' farm at Lann-Gouh, Brittany, France (Gebhardt, 1995). Faecal spherulites have also been found in fresh wild boar and pig excrements (Brochier, 1996; Canti, 1999) (Figs. 12 and 13).

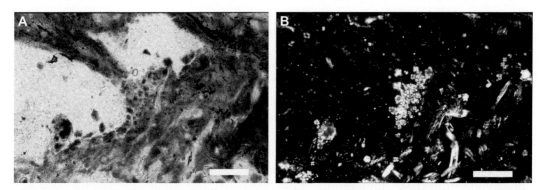

FIGURE 12 Fresh pig dung (wild boar) (Northern Apennines, Italy). (A) Excrement likely containing root fragments and biogenic mineral material (PPL, scale bar length 100 μm). (B) Same field in XPL, showing relict food material, such as root fragments, as birefringent cellulose; biogenic crystals occur as single calcitic faecal spherulites and as aggregates of probable calcium oxalates.

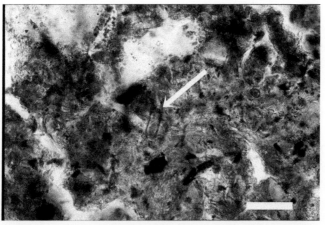

FIGURE 13 Pig dung (coprolite) (Late Bronze Age/Early Iron Age Potterne, Wiltshire, UK) essentially composed of amorphous organic matter and organic phosphate and containing soil diatoms (*arrow*) such as reasonably well-preserved *Hantzschia amphioxys* and *Navicula mutica* (aquatic species indicating drinking from ponds; Nigel Cameron, University College London, personal communication) (PPL, scale bar length 50 μm).

3.2.3 Dog Coprolites

Dogs and other canids scavenge and produce coprolites that can contain coarse bone fragments, including fish bone and fins (Matsui et al., 1996), as well as quartz silt that becomes embedded in a presumed hydroxylapatite cement (Fig. 14). This cement is commonly a fine dotted blackish grey (PPL) material, with undifferentiated b-fabric, white in OIL and strongly autofluorescent under blue light (Courty et al., 1989; Macphail, 2000). As reported for Ca-phosphate-cemented hyena coprolites, it is believed that trapped intestinal gas in dog coprolites can produce vesicles and that ingested fur or wool leaves narrow curved voids (Horwitz & Goldberg, 1989; Lewis, 1997; Larkin et al., 2000; Macphail & Goldberg, 2012). Where dogs scavenged sheep carcasses as evidenced by bone gnawing, there are wool-size curved voids in the coprolites; pollen analysis on both thin sections and prepared pollen samples shows that much cereal material was eaten, possibly from scavenged human faeces (Scaife, 2000; Cruise, personal communication). It can be noted here that when bone is digested by dogs (and by humans and birds, for example) it can become leached, losing birefringence and autofluorescence; when embedded in guano and/or cess, however, the edges of bone fragments may become phosphatised and display high autofluorescence (Courty et al., 1989; Macphail & Goldberg, 1999; Macphail & Crowther, 2008).

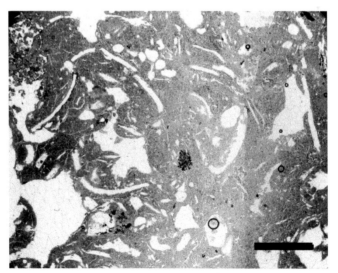

FIGURE 14 Dog coprolite (Middle Saxon West Heslerton, North Yorkshire) showing typical dark greyish groundmass, with numerous curved voids pseudomorphic after fur or wool (dogs gnawed sheep carcasses at this site) (PPL, scale bar length 1 mm).

3.2.4 Human Coprolites

Human coprolites, as identified through nematode egg analysis (Kenward & Hall, 1995), may have a presumed hydroxylapatite matrix similar to that of dog coprolites but differ by normally being a yellowish colour (Courty et al., 1989; Goldberg et al., 2009a; Macphail, 2016b). They also often have a different content reflecting their omnivorous diet (both bone and plant remains occur), and they do not typically contain much clastic mineral material or coarse bone. As such, they are far more likely to include plant food fragments, such as legume testa and cereal material, including articulated phytolith-rich bran; thin sections of mummified intestinal contents found solely leguminous seed cases in some Chilean mummies (Macphail, 2016b) (Figs. 15 and 16). Here the cellulose of the seed cases is still birefringent, unlike the humified cellulose found in herbivore dung — herbivores being better adapted to digest cellulose compared with humans. Body stains and probable intestinal remains that autofluoresce in blue light have been found in an Iron Age grave in Norway where bands of vivianite can also mirror wooden coffin remains; both amorphous yellow (Ca-P) and reddish brown (Fe-Ca-P) body stains occur in the pelvic area of a warrior found in a boat grave near the Gokstad Mound (Macphail et al., 2013; Viklund et al., 2013).

Latrines contain mineralised cess, again with a probable carbonate hydroxylapatite composition (Fig. 17). Such waste can contain embedded cereal material of dietary origin, as well as sphagnum moss that was most likely used as toilet paper. Fine fragments of this type of latrine waste often occur as a regular component of Roman and Medieval beaten floors in the United Kingdom. Dumped toilet waste and seepage from

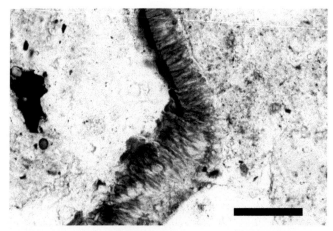

FIGURE 15 Human coprolite (Viking Age Coppergate site, York), identification confirmed by nematode egg studies (Andrew Jones, York Trust). Detail of legume testa as a food remain within the apatite groundmass; the hook-like tissue cells making up the testa are diagnostic (Ann Butler, Institute of Archaeology, University College London, personal communication) (PPL, scale bar length 25 μm).

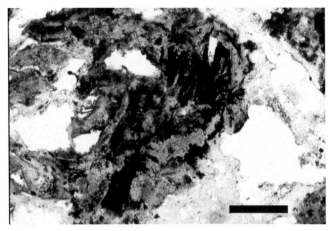

FIGURE 16 Human coprolite (Middle Saxon settlement site Maiden Lane, London, UK), identification confirmed by parasite egg study (Claire de Rouffignac, Museum of London Archaeological Service, personal communication). The sample exhibits voids from trapped gas; dark black and reddish brown colours relate to iron and manganese oxide staining associated with unidentifiable but likely food plant traces embedded within the coprolite (PPL, scale bar length 1 mm).

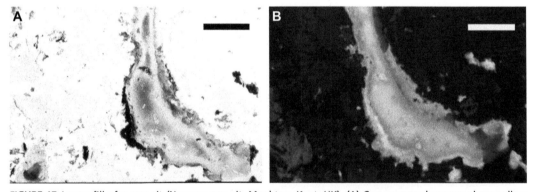

FIGURE 17 Lower fill of a cess pit (Norman cess pit, Monkton, Kent, UK). (A) Cess preserved as amorphous yellow material (probably calcium phosphate); mineralised cess (formed in aerobic conditions) is yellow, mainly optically isotropic although some birefringent materials can also be present in addition to probable hydroxylapatite (PPL, scale bar length 1 mm). (B) Same field in BLF, showing strong autofluorescence of the material.

cess pits, like modern day sewage sludge, produces yellow amorphous infillings and nodules of Fe-Ca-phosphates, sometimes with neoformation of vivianite (Macphail, 2016b; see Karkanas & Goldberg, 2018). It is common to find phosphate-embedded charcoal, both in latrine waste and in fields, implying that such material — which may also include embedded ash — was used as a 'night soil' fertiliser worldwide (Henning & Macphail, 2004; Goldberg & Macphail, 2006).

4. Manufactured Materials

The study of manufactured materials in archaeology is commonly thought of as being a specialist field. However, archaeological sites often include fragments of pottery and other ceramic materials (e.g., roof tiles, pipes, kilns). These materials need to be distinguished routinely from natural rocks or debris, which provides clues to local industries (see Berna et al., 2007). In complex societies, the types of materials (e.g., lime and gypsum plasters), and how they have been applied (e.g., single coat vs several layers), can be indicative of the status of buildings or rooms and the associated social standing of the former occupants. The petrology of pottery has been well studied (Rice, 1987), and handmade pottery can be distinguished from wheel-made ceramics by, at low magnification comparing air-void patterns, general coarse particle distribution and general aspects of the groundmass, and at high magnification examining the morphology and arrangement of clay domains formed by the packing of clay- and silt-size particles and overall birefringence fabrics (Courty et al., 1989; Roux & Courty, 1998; Quinn, 2013). Neolithic pottery has commonly been manufactured using organic tempers and local soils and sediments (Spataro, 2002).

4.1 Stone Tools

Stone tools are probably the earliest worked material to be found in archaeological sites. Flint or chert is composed of microcrystalline quartz or chalcedony and is a common example of a natural rock material that flakes well to produce sharp-edged tools (Angelucci, 2010). Flint, which can occur in a variety of colours and lustres in the field, is normally colourless in thin section (Fig. 18), unless stained by iron oxides. It may occur as an exotic rock because of human importation. Other silicate-dominated rocks used for tools are quartzite and various volcanic rocks, such as rhyolites, basalts, or obsidian.

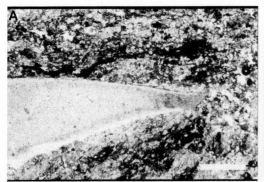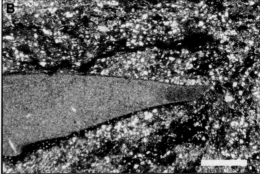

FIGURE 18 Manufactured stone tools (Lower Palaeolithic Boxgrove, West Sussex, UK). (A) Edge of colourless flint flake of anthropogenic origin, present in once-humic (now replaced by iron oxides) silty colluvium (PPL, scale bar length 1 mm). (B) Same field in XPL, showing low-order interference colours of microcrystalline quartz.

When burned, colourless flint can become greyish brown (PPL) (white in OIL), and cracks appear along with so-called pot-lidding of small hemispherical spalls that lack striking platforms (Andrefsky, 1998). Heating can also cause a change in texture resulting in a reduction in crystal size. Fire-altered cherts can be found with other heat-altered rocks in middens and burned rock mounds (e.g., Goldberg & Guy, 1996).

4.2   Plasters and Mortars

Lime- and gypsum-based plasters and mortar have been studied worldwide. Plaster floors are formed from layers of grey to greyish brown (PPL) micritic lime plaster that often includes unburned limestone components (Goren & Goldberg, 1991) and fine organic fragments, sometimes along with fine burned mineral material and charcoal (Matthews et al., 1996) (Fig. 19). Plaster is normally whitish in OIL. A comparison between experimental lime plaster and natural calcareous sediments (Karkanas, 2007) indicates that the most promising features for identifying lime are transitional textures of partially carbonised slaked lime (composed of poorly crystallised portlandite and mixtures of cryptocrystalline calcite) that can be observed in the lime lumps and the binding matrix cement. Well-formed calcitic groundmass and shrinkage cracks are also possible indicators of lime plasters (Karkanas, 2007). Fine charcoal and reddened burned mineral inclusions may also suggest the presence of burned lime.

Chalk, limestone and, in coastal areas where limestone is unavailable, shells have been burned to produce lime. For example, burned conch shell, of presumed lime-making origin, occurs in some New World Maya lime plasters on the island site of

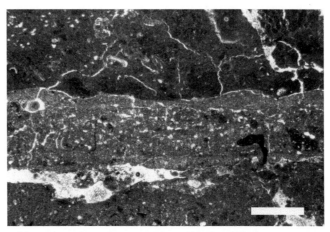

FIGURE 19 Manufactured lime plaster (Pre-Pottery Neolithic Yiftah El, Israel). Layers of very fine and sometime pure lime plaster, applied as cover of a floor; note very thin layers of finer plaster that cover a coarser layer in the middle and traces of fine organic matter from the manufacturing and mixing process (PPL, scale bar length 0.5 mm).

Marco Gonzalez, Belize (Macphail et al., 2017a; Graham et al., 2017); tempering material included high proportions of unburned shell, pottery fragments and siliceous marine tidal flat sediments, which had sometimes been burned. In Roman structures, a typical coarse/fine ratio for a plaster, or matrix within a coarse tempered mortar, is 60/40 (Stoops, 1984b; Macphail, 2003).

Plaster often shows very thin layers associated with the plastering process, carried out, for example, on brickearth walls or over mortared wall surfaces, ahead of painting (Pye, 2000/2001) (Fig. 20). Babylonian red floors had, in some cases, a 20-mm-thick supporting 'white' layer composed of a microcrystalline calcite cement and a temper of well-sorted (1-3 mm) coarse sand to fine gravel, composed of quartz and feldspar, below a c. 6-mm-thick red surface layer. The latter is characterised by a similar temper, but the matrix is formed from microcrystalline calcite and hematite dust (Stoops & Stoops, 1994). Multiple plaster floors are considered as evidence of both ritual activity and level of social status, but contemporary floors in different domestic structures may show a varying composition, which was probably determined by the individual households (Karkanas & Efstratiou, 2009).

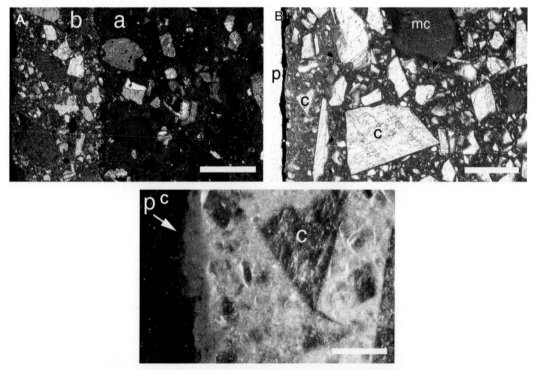

FIGURE 20 Roman wall lime plaster (The House of Amaranthus, Pompeii). (A) Reference section through fine wall plaster layers, including an inner fine plaster (a, with sand-size temper derived from volcanic material) and an outer pure lime wall surface plaster (b, with unburned calcite inclusions from the manufacturing process) (XPL, scale bar length 1 mm). (B) Opaque painted surface (p), covering the pure lime wall plaster surface, with inclusions of coarse-grained calcite (c) and micritic carbonate aggregates (mc) (PPL, scale bar length 0.5 mm). (C) Detail of paint layer, which from SEM-EDS seems to be iron-oxide based (OIL, scale bar length 100 μm).

Fragments of plaster and mortar are common in the Roman to early Medieval dark earth of European cities and as weathering features in Maya sites of the New World (Macphail, 1994; Cammas, 2004; Straulino et al., 2013; Borderie et al., 2014; Graham et al., 2017) and within fills of tells (Matthews et al., 1996; Boivin & French, 1997/1998). In Europe, mortar may contain brick fragments and volcanic rock clasts, which may imply that they are hydraulic mortars because these are a source of neoformed aluminium silicates that cement this mortar type rather than simply being cemented by neoformed calcite. Similarly, it has been suggested that the Maya may also have used volcanic materials to produce pozzolanic plasters in mainland Belize (Villaseñor & Graham, 2010). However, the identification of neoformed aluminium silicates in hydraulic mortar requires X-ray diffraction analyses (e.g., Rentzel, 1998). In addition, relics of calcium of aluminium hydrates are amorphous, featureless and optically isotropic in thin section (St John et al., 1998).

4.3 Metal Working

Hammerscale, iron slag and silt to coarse sand-size vesicular iron spheroids are evidence of industrial activities and may occur alongside fuel ash waste and high-temperature melted and fused soil; in the last, minerals may be altered, with quartz and feldspar, for example, losing their birefringence (Viklund et al., 2013, see Section 3.1) (see Fig. 8B). Opaque in transmitted light, hammerscale has a layered metallic edge, visible in OIL and SEM-BSE images, comprising an outer hematite layer and inner magnetite and wüstite (FeO) layers in some examples (Kresten & Hjärthner-Holdar, 2001). Iron slag is very dark (PPL) and has a typical vesicular void pattern. It includes optically isotropic areas, dark materials showing dendritic patterns (e.g., wüstite) and iron silicates such as fayalite $(Fe^{2+}_2(SiO_4))$; such patterns are strongly visible in SEM-BSE images (Kresten & Hjärthner-Holdar, 2001; Macphail, 2003). Iron slag and hammerscale also often occur as poorly preserved nodular material because of postdepositional gleying. Thin, angular iron flakes also occur in kitchen debris from the use of iron pots, and these can also be differentiated from iron-pan fragments by a metallic lustre under OIL, as confirmed by EDS testing (Macphail et al., 2016).

Non-ferrous metal and glass working also produce metal droplets ('prills') and 'glassy' furnace slags (Merkel, personal communication). Quantitative SEM-EDS and microprobe analyses are often necessary alongside optical microscopy. For example, petrographic analysis of a Bronze Age pit fill showed colourless to grey cassiterite aggregates (SnO_2), with characteristic high relief, high birefringence and extinction parallel to cleavage (Macphail & Crowther, 2008); blue-green aggregates in Roman London were identified as copper and tin working waste and associated corrosion products. Presumed bell-making activities in medieval Magdeburg also produced blue-green stained bronze droplets (Goldberg & Macphail, 2006).

A Cu-Pb alloy was found in trampled floor spreads at a medieval hospital, alongside strongly burned sediments (Fig. 21). In thin section, it was found to be a mainly opaque

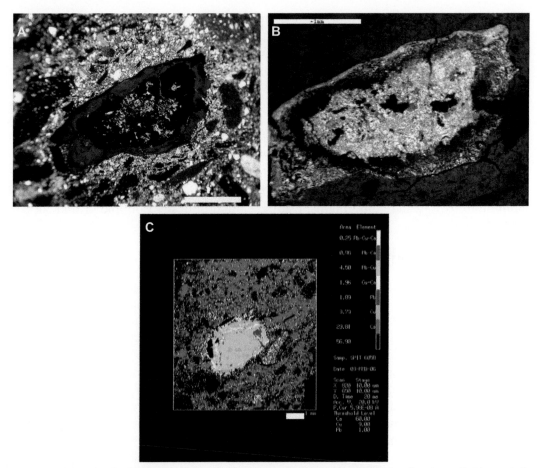

FIGURE 21 Metal droplet from non-ferrous metal working, within ash-rich deposits (medieval Spitalfields Hospital, sample M605B, London; Museum of London Archaeological Service). (A) Metal droplet including some crystals of unknown composition; note high-order interference colours of the ash (XPL, scale bar length 1 mm). (B) Backscattered electron image of same field; the corrosion rim is less bright than the unaltered interior because it still includes heavier elements (Pb). (C) Element map for the same field (rotated), showing that the metal droplet is composed of a Pb-Cu alloy (centre) with areas of pure Cu and Pb, and the corrosion rim consists of Cu- and Ca-phases, indicating that corrosion products are mixed into the Ca-dominated ash occupation deposit.

material (PPL), with elsewhere generally very low interference colours, but with bright yellow and reddish colours in OIL. On the other hand, it was possible with the help of specialist archaeometallurgists to identify probable corroding lead fragments in Roman ashy dumps where pure lead is opaque, with a blackish lustre under OIL (Borderie et al., 2014). A coating of 'red' lead oxide is also opaque but reddish under OIL, while corrosion staining of the surrounding calcitic ashes also produced grey (PPL), high-order interference colours and white (OIL) lead carbonate.

5. Conclusions

An archaeological material is any matter that has been utilised or produced by humans or that formed as a by-product of their activities. Soil micromorphology, using optical microscopy and associated techniques, is one of the best methods for characterising and identifying archaeological materials, especially when they occur as microscopic fragments at archaeological sites. In fact, although bulk and microfossil studies can provide useful supporting information, often the data from these techniques can only really be properly understood when soil micromorphology is available. For example, some phytoliths can be understood better when unfragmented in thin section, while the exact nature of phosphate-enrichment in archaeological contexts can often be identified, as produced by *in situ* bone, ash or dung concentrations, or which only result from secondary phosphate accumulations. In addition, the identification of such microartefacts is an essential new tool for characterising archaeological contexts.

Archaeological materials include natural mineral and organic materials that have been used in their raw state (dung, flint, turf, tufa, wood) or as little-transformed material (earth-floors and walls, adobe and mud brick, leather), as well as fully manufactured products (iron, lime plaster, metal alloys). Other archaeological matter include waste from occupation, such as dung and mineralised coprolites from domestic animals, human waste (coprolites, cess), food (bone, cereal grains, eggshell, mollusc shell) and hearth remains (ash, charcoal, briquetage). Sites also produce strongly burned residual remains, fused ash, slag, vitrified ceramics, burned bone and flint, for example. As research continues, the number of identified archaeological materials continues to increase and permits an ever deeper understanding of different cultures and human activities through time. Clearly, it is also required to comprehend the nature of the archaeological context and type of deposit in which these materials occur. Such strategies can provide a fuller understanding of past human activities operating at the site.

Acknowledgements

The authors thank the anonymous referees, Panagiotis Karkanas and the editors for their comments and gratefully acknowledge the specialists who supplied data supporting and complementing soil micromorphological identifications and interpretations, namely Francesco Berna, Jan Bill, Marie-Agnès Courty, John Crowther, Gill Cruise, Tom Gregory, Takis Karkanas, Johan Linderholm, John Merkel, Kevin Reeves, Thilo Rehren, Ruth Shahack-Gross, Karin Viklund, Luc Vrydaghs, and the many members of the Archaeological Soil Micromorphology Working Group who have contributed so much to this discipline over the last 25 years. Because of a lack of space most of the supporting chemical and microfossil data for this chapter have had to be omitted. The authors also gratefully acknowledge long-term funding by various North American, British and European government agencies, funding bodies and archaeological companies. R.I. Macphail specifically thanks The Leverhulme Trust for support with this project.

References

Adderley, W.P., Simpson, I.A. & Davidson, D.A., 2006. Historic landscape management: a validation of quantitative soil thin-section analyses. Journal of Archaeological Science 33, 320-334.

Akeret, Ö. & Rentzel, P., 2001. Micromorphology and plant macrofossil analysis of cattle dung from the Neolithic lake shore settlement of Arbon Bleiche 3. Geoarchaeology 16, 687-700.

Albert, R.M., Shahack-Gross, R., Cabanes, D., Gilboa, A., Lev-Yadun, S., Portillo, M., Sharon, I., Boaretto, E. & Weiner, S., 2008. Phytolith-rich layers from the Late Bronze and Iron Ages at Tel Dor (Israel): mode of formation and archaeological significance. Journal of Archaeological Science 35, 57-75.

Alexandrovskiy, A.L. & Chichagova, O.A., 1998. Radiocarbon age of Holocene paleosols of the East European forest-steppe zone. Catena 34, 197-207.

Andrefsky, W., 1998. Lithics. Cambridge Manuals in Archaeology. Cambridge University Press, Cambridge, 258 p.

Angelucci, D.E., 2008. Geoarchaeological insights from a Roman age incineration feature (ustrinum) at Enconsta de Sant'Ana (Lisbon, Portugal). Journal of Archaeological Science, 35, 2624-2633.

Angelucci, D.E., 2010. The recognition and description of knapped lithic artifacts in thin section. Geoarchaeology 25, 220-232.

Angelucci, D.E., Boschian, G., Fontanals, M., Pedrotti, A. & Vergès, J.M., 2009. Shepherds and karst: the use of caves and rock shelters in the Mediterranean region during the Neolithic. World Archaeology 41, 191-214.

Babel, U., 1975. Micromorphology of soil organic matter. In Giesking, J.E. (ed.), Soil Components. Volume 1: Organic Components. Springer-Verlag, New York, pp. 369-473.

Bal, L., 1982. Zoological Ripening of Soils. Agricultural Research Report n° 850. Centre for Agricultural Publishing and Documentation, Wageningen, 365 p.

Banerjea, R.Y., Bell, M., Matthews, W. & Brown, A., 2015. Applications of micromorphology to understanding activity areas and site formation processes in experimental hut floors. Archaeological and Anthropological Sciences 7, 89-112.

Barclay, G.T., 1983. Sites of the third millennium BC to the first millennium AD at North Mains, Strathallan, Perthshire. Proceedings of the Society of Antiquities of Scotland 113, 122-281.

Barrat, B.C., 1964, A classification of humus forms and microfabrics in temperate grasslands. Journal of Soil Science 15, 342-356.

Bennett, P., Clark, P., Hicks, A., Rady, J. & Riddler, I., 2008. At the Great Crossroads. Prehistoric, Roman and Medieval Discoveries on the Isle of Thanet 1994-95. Canterbury Archaeological Trust Occasional Papers No 4. Canterbury Archaeological Trust Ltd, Canterbury, 366 p.

Berna, F., Behar, A., Shahack-Gross, R., Berg, J., Boaretto, E., Gilboa, A., Sharon, I., Shalev, S., Shilstein, S., Yahalom-Mack, N., Zorn, J.R. & Weiner, S., 2007. Sediments exposed to high temperatures: reconstructing pyrotechnological processes in Late Bronze Age and Iron Age Strata at Tel Dor (Israel). Journal of Archaeological Science 34, 358-373.

Binder, D., Brochier, J.E., Duday, H., Helmer, D., Marinval, P., Thiebault, S. & Wattez, J., 1993. L'abri Pendimoun à Castellar (Alpes-Maritimes): nouvelles données sur le complex culturel de la imprimée dans son contexte stratigraphique. Gallia Préhistoire 35, 177-251.

Blume, H.P., 1989. Classification of soils in urban agglomerations. Catena 16, 269-275.

Boivin, N.L., 1999. Life rythms and floor sequences: excavating time in rural Rajasthan and Neolithic Çatalhöyük. World Archaeology 31, 367-388.

Boivin, N.L. & French, C.I.A., 1997/1998. New questions and answers in the micromorphology of the occupation deposits at the Souks Site, Beirut. Berytus 43, 181-210.

Borderie, Q., Fondrillon, M., Nicosia, C., Devos, Y. & Macphail, R.I., 2014. Bilan des recherches et nouveaux éclairages sur les terres noires: des processus complexes de stratification aux modalités d'occupation des espaces urbains. In Lorans, E. (ed.), Archéologie de l'Espace Urbain – Partie II. CTHS, Tours, pp. 213-223.

Bouma, J., Fox, C.A. & Miedema, R., 1990. Micromorphology of hydromorphic soils: applications for soil genesis and land evaluation. In Douglas, L.A. (ed.), Soil Micromorphology: A Basic and Applied Science. Developments in Soil Science, Volume 19. Elsevier, Amsterdam, pp. 257-278.

Boschian, G. & Montagnari-Kokelji, E., 2000. Prehistoric shepherds and caves in the Trieste Karst (northeastern Italy). Geoarchaeology 15, 331-371.

Brochier, J.E., 1983. Bergeries et feux de bois néolithiques dans le Midi de la France, caractérisation et incidence sur la raisonnement sédimentologique. Quärtar 33/34, 181-193.

Brochier, J.E., 1996. Feuilles ou fumiers? Observations sur le rôle des poussières sphérolitiques dans l'interprétation des dépôts archéologiques holocènes. Anthropozoologica 24, 19-30.

Brochier, J.E. & Thinon, M., 2003. Calcite crystals, starch grains aggregates or … POCC? Comment on 'calcite crystals inside archaeological plant tissues'. Journal of Archaeological Science 30, 1211-1214.

Brochier, J.E., Villa, P. & Giacomarra, M., 1992. Shepherds and sediments: geo-ethnoarchaeology of pastoral sites. Journal of Anthropological Archaeology 11, 47-102.

Cammas, C., 1994. Approche micromorphologique de la stratigraphie urbaine à Lattes: premiers résultats. Lattara 7, 181-202.

Cammas, C., 2004. Les 'terres noires' urbaines du Nord de la France: première typologie pédo-sédimentaire. In Verslype, L. & Brulet, R. (eds), Terres Noires – Dark Earth. Université Catholique de Louvain, Louvain-la-Neuve, pp. 43-55.

Cammas, C., Wattez, J. & Courty, M.A., 1996. L'enregistrement sédimentaire des modes d'occupation de l'espace. In Castelletti, L. & Cremaschi, M. (eds.)., Proceedings of the 13th International Congress of Prehistoric and Protohistoric Sciences, Volume 3, Paleoecology. ABACO, Forli, pp. 81-86.

Canti, M.G., 1999. The production and preservation of faecal spherulites: animals, environment and taphonomy. Journal of Archaeological Science 26, 251-258.

Carneiro, A. & Mateiciucová, I., 2007. Daub fragments and the question of structures. In Whittle, A. (ed.), The Early Neolithic on the Great Hungarian Plain: Investigations of the Körös Culture Site of Ecsegfalva 23, County Békés, Volume I. Institute of Archaeology, Budapest, pp. 255-288.

Carter, S.P., 1990. The stratification and taphonomy of shells in calcareous soils: implications for landsnail analysis in archaeology. Journal of Archaeological Science 17, 495-507.

Carter, S., 1998. The use of peat and other organic sediments as fuel in northern Scotland: identifications derived from soil thin sections. In Mills, C.M. & Coles, G. (eds.), Life on the Edge: Human Settlement and Marginality. Oxbow Books, Oxford, pp. 99-104.

Courty, M.A., 2001. Microfacies analysis assisting archaeological stratigraphy. In Goldberg, P., Holliday, V.T. & Ferring, C.R. (eds.), Earth Sciences and Archaeology. Kluwer, New York, pp. 205-239.

Courty, M.A., Goldberg, P. & Macphail, R.I., 1989. Soils and Micromorphology in Archaeology. Cambridge Manuals in Archaeology. Cambridge University Press, Cambridge, 344 p.

Courty, M.A., Macphail, R.I. & Wattez, J., 1991. Soil micromorphological indicators of pastoralism: with special reference to Arene Candide, Finale Ligure, Italy. Revista di Studi Liguri 57, 127-150.

Courty, M.A., Goldberg, P. & Macphail, R.I., 1994. Ancient people – lifestyles and cultural patterns. In Wilding, L. & Oleshko, K. (eds.), Micromorphological indicators of anthropogenic effects on soils. Proceeding of the Symposium of Subcommission B, 15th International Conference of Soil Science, Acapulco, pp. 250-269.

Cowan, C. (ed.), 2003. Urban Development in North-west Roman Southwark: Excavations 1974-90, Monograph 16. MOLAS, London, 209 p.

Crowther, J., 2003. Potential magnetic susceptibility and fractional conversion studies of archaeological soils and sediments. Archaeometry 45, 685-701.

Crowther, J., 2007. Chemical and magnetic properties of soils and pit fills. In Whittle, A. (ed.), The Early Neolithic on the Great Hungarian Plain: Investigations of the Körös Culture Site of Ecsegfalva 23, County Békés, Volume I. Institute of Archaeology, Budapest, pp. 227-254.

Crowther, J., Macphail, R.I. & Cruise, G.M., 1996. Short-term burial change in a humic rendzina, Overton Down Experimental Earthwork, Wiltshire, England. Geoarchaeology 11, 95-117.

Cruise, G.M. & Macphail, R.I., 2000. Microstratigraphical signatures of experimental rural occupation deposits and archaeological sites. In Roskams, S. (ed.), Interpreting Stratigraphy. University of York, York, pp. 183-191.

Dammers, K. & Joergensen, R.G., 1996, Progressive loss of carbon and nitrogen from simulated daub on heating. Journal of Archaeological Science 23, 639-648.

Devos, Y., Nicosia, C., Vrydaghs, L. & Modrie, S., 2013. Studying urban stratigraphy: Dark Earth and a microstratified sequence on the site of the Court of Hoogstraeten (Brussels, Belgium). Integrating archaeopedology and phytolith analysis. Quaternary International 315, 147-166.

Dimbleby, G.W., 1962. The Development of British Heathlands and their Soils. Oxford Forestry Memoir No. 23. Clarendon Press, Oxford, 121 p.

Engelmark, R. & Linderholm, J., 1996. Prehistoric land management and cultivation. A soil chemical study. In Mejdahl, V. & Siemen, P. (eds.), Proceedings of the 6th Nordic Conference on the Application of Scientific Methods in Archaeology. Arkaeologiske Rapporter 1, Esbjerg Museum, Esbjerg, pp. 315-322.

Engelmark, R., & Linderholm, J., 2008. Environmental Archaeology. Man and Landscape – A Dynamic Interrelation. The Öresund Fixed Link Project [in Swedish]. Kulturmilö, Malmö, 92 p.

Evans, E.E., 1957. Irish Folk Ways. Routledge and Kegan Paul, London, 324 p.

Evans, J.G., 1972. Land Snails in Archaeology. Seminar Press, London, 436 p.

Fenton, A., 1968. Alternating stone and turf – an obsolete building practice. Folk Life 6, 94-103.

Fisher, P.F. & Macphail, R.I., 1985. Studies of archaeological soils and deposits by micromorphological techniques. In Fieller, N.R.J., Gilbertson, D.D. & Ralph, N.G.A. (eds.), Palaeoenvironmental Investigations: Research Design, Methods and Data Analysis. British Archaeological Reports International Series 258, pp. 92-112.

Forget, M.C.L. & Shahack-Gross, R., 2016. How long does it take to burn down an ancient Near Eastern city? The study of experimentally heated mud-bricks. Antiquity 90, 1213-1225.

Franceschi, V.R. & Horner, H.T., 1980. Calcium oxalate crystals in plants. Botanical Review 46, 361-427.

Friesem, D., Boaretto, E., Eliyahu-Behar, A. & Shahack-Gross, R., 2011. Degradation of mud brick houses in an arid environment: a geoarchaeological model. Journal of Archaeological Science 38, 1135-1147.

Friesem, D.E., Karkanas, P., Tsartsidou, G. & Shahack-Gross, R., 2014a. Sedimentary processes involved in mud brick degradation in temperate environments: a micromorphological approach in an ethnoarchaeological context in northern Greece. Journal of Archaeological Science 41, 556-567.

Friesem, D.E., Tsartsidou, G., Karkanas, P. & Shahack-Gross, R., 2014b. Where are the roofs? A geo-ethnoarchaeological study of mud brick structures and their collapse processes, focusing on the identification of roofs. Archaeological and Anthropological Sciences 6, 73-92.

Gé, T., Courty, M.A., Matthews, W. & Wattez, J., 1993. Sedimentary formation processes of occupation surfaces. In Goldberg, P., Nash, D.T. & Petraglia, M.D. (eds.), Formation Proceses in Archaeological Contexts. Monographs in World Archaeology 17. Prehistory Press, Madison, pp. 149-163.

Gebhardt, A., 1995. Soil micromorphological data from traditional and experimental agriculture. In Barham, A.J. & Macphail, R.I. (eds.), Archaeological Sediments and Soils: Analysis, Interpretation and Management. Institute of Archaeology, London, pp. 25-40.

Gebhardt, A. & Langohr, R., 1999. Micromorphological study of construction materials and living floors in the medieval motte of Werken (West Flanders, Belgium). Geoarchaeology 14, 595-620.

Gerasimova, M. & Lebedeva-Verba, M., 2010. Topsoils - mollic, takyric and yermic horizons. In Stoops, G., Marcelino, V. & Mees, F. (eds.), Interpretation of Micromorphological Features of Soils and Regoliths. Elsevier, Amsterdam, pp. 351-368.

Gerasimova, M. & Lebedeva, M., 2018. Organo-mineral surface horizons. In Stoops, G., Marcelino, V. & Mees, F. (eds.), Interpretation of Micromorphological Features of Soils and Regoliths. Second Edition. Elsevier, Amsterdam, pp. 513-538.

Goldberg, P., 1979. Geology of Late Bronze Age mudbrick from Tel Lachish. Tel Aviv 6, 60-71.

Goldberg, P., 2000. Micromorphological aspects of site formation at Keatley Creek. In Hayden, B. (ed.), The Ancient Past of Keatley Creek. Archaeology Press, Simon Fraser University, Burnaby, pp. 79-95.

Goldberg, P. & Guy, J., 1996. Micromorphological observations of selected rock ovens, Wilson-Leonard site, Central Texas. In Castelletti, L. & Cremaschi, M. (eds.), Proceedings of the 13th International Congress of Prehistoric and Protohistoric Sciences, Volume 3, Paleoecology. ABACO, Forli, pp. 115-122.

Goldberg, P. & Macphail, R.I., 2006. Practical and Theoretical Geoarchaeology. Blackwell Publishing, Oxford, 455 p.

Goldberg, P. & Macphail, R.I., 2012. Gorham's Cave sediment micromorphology. In Barton, R.N.E., Stringer, C. & Finlayson, C. (eds.), Neanderthals in Context. A report of the 1995-1998 Excavations at Gorham's and Vanguard Caves, Gibraltar. Monograph 75, Oxford University School of Archaeology, Oxford, pp. 50-61.

Goldberg, P., Weiner, S., Bar-Yosef, O., Xu, Q. & Liu, J., 2001, Site formation processes at Zhoukoudian, China. Journal of Human Evolution 41, 483-530.

Goldberg, P., Berna, F. & Macphail, R.I., 2009a. Comment on 'DNA from Pre-Clovis Human Coprolites in Oregon, North America'. Science 325, 148-c.

Goldberg, P., Miller, C.E., Schiegl, S., Ligouis, B., Berna, F., Conard, N.J. & Wadley, L., 2009b. Bedding, hearths, and site maintenance in the Middle Stone Age of Sibudu Cave, KwaZulu-Natal, South Africa. Archaeological and Anthropological Sciences 1, 95-122.

Goren, Y. & Goldberg, P., 1991. Petrographic thin sections and the development of Neolithic plaster production in Northern Israel. Journal of Field Archaeology 18, 131-140.

Graham, E., Macphail, R., Turner, S., Crowther, J., Stegemann, J., Arroyo-Kalin, M., Duncan, L., Whittet, R., Rosique, C. & Austin, P., 2017. The Marco Gonzalez Maya site, Ambergris Caye, Belize: assessing the impact of human activities by examining diachronic processes at the local scale. Quaternary International 437, 115-142.

Guélat, M., Paccolat, O. & Rentzel, P., 1998. Une étable Gallo-Romaine à Brigue-Glis VS, Waldmatte. Evidences archéologiques et micromorphologiques. Annuaire de la Société Suisse de Préhistoire et d'Archéologie 81, 171-182.

Gur-Arieh, S., Shahack-Gross, R., Maeir, A.M., Lehmann, G., Hitchcock, L.A. & Boaretto, E., 2014. The taphonomy and preservation of wood and dung ashes found in archaeological cooking installations: case studies from Iron Age Israel. Journal of Archaeological Science 46, 50-67.

Haită, C., 2003. Micromorphology. Inhabited space disposition and uses. Analysis of an occupation zone placed outside the dwellings. In Popovici, D. (ed.), Archaeological Pluridisciplinary Researches at Borduşani-Popină, Pluridisciplinary Researches Series VI., National Museum of Romanian History, Bucharest, pp. 51-74.

Heathcote, J.L., 2002. An Investigation of the Pedosedimentary Characteristics of Deposits Associated with Managed Livestock. PhD Dissertation, University College London, 266 p.

Henning, J. & Macphail, R.I., 2004., Das karolingische Oppidum Büraburg: archälogische und mikromorphologische Studien zur Funktion einer frümittelaterlichen Bergbefestigung in Nordhessen. In Hänsel, B. (ed.), Parerga Praehistorica. Jubiläumsschrift zur Prähistorischen Archäologie 15 Jahre UPA, Band 100. Verlag Dr Rudolf Habelt GmbH, Bonn, pp. 221-252.

Horwitz, L.K. & Goldberg, P., 1989. A study of Pleistocene and Holocene hyaena coprolites. Journal of Archaeological Science 16, 71-94.

Ismail-Meyer, K. & Rentzel, P., 2004. Mikromorphologische Untersuchung der Schichtabfolge. In Jacomet, S., Leuzinger, U. & Schibler, J. (eds.), Die jungsteinzeitliche Seeufersiedlung Arbon, Bleiche 3, Umwelt und Wirtschaft. Archäologie im Thurgau, Band 12. Archäologie im Thurgau, Band 12. Departement für Erziehung und Kultur des Kantons Thurgau, Frauenfeld, pp. 66-80.

Ismail-Meyer, K., Stolt, M. & Lindbo, D., 2018. Soil organic matter. In Stoops, G., Marcelino, V. & Mees, F. (eds.), Interpretation of Micromorphological Features of Soils and Regoliths. Second Edition. Elsevier, Amsterdam, pp. 471-512.

Karkanas, P., 2007. Identification of lime plaster in prehistory using petrographic methods: a review and reconsideration of the data on the basis of experimental and case studies. Geoarchaeology 22, 775-796.

Karkanas, P. & Efstratiou, N., 2009. Floor sequences in Neolithic Makri, Greece: micromorphology reveals cycles of renovation. Antiquity 83, 955-967.

Karkanas, P. & Goldberg, P., 2018. Phosphatic features. In Stoops, G., Marcelino, V. & Mees, F. (eds.), Interpretation of Micromorphological Features of Soils and Regoliths. Second Edition. Elsevier, Amsterdam, pp. 323-346.

Karkanas, P. & Van de Moortel, A., 2014. Micromorphological analysis of sediments at the Bronze Age site of Mitrou, central Greece: patterns of floor construction and maintenance. Journal of Archaeological Science 43,198-213.

Karkanas, P., Shahack-Gross, R., Ayalon, A., Barkai, R., Bar-Matthews, M., Frumkin, A., Gopher, A. & Stiner, M., 2007. Evidence for habitual use of fire at the end of the Lower Paleolithic: site formation processes at Qesem Cave, Israel. Journal of Human Evolution 53, 197-212.

Kenward, H.K. & Hall, A.R., 1995. Biological Evidence from Anglo-Scandinavian Deposits at 16-22 Coppergate. York Archaeological Trust, York, 797 p.

Kooistra, M.J. & Pulleman, M.M., 2018. Features related to faunal activity. In Stoops, G., Marcelino, V. & Mees, F. (eds.), Interpretation of Micromorphological Features of Soils and Regoliths. Second Edition. Elsevier, Amsterdam, pp. 447-469.

Kresten, P. & Hjärthner-Holdar, E., 2001. Analyses of the Swedish ancient iron reference slag W-25:R. Historical Metallurgy 35, 48-51.

Kruger, R.P., 2015. A burning question or, some half-baked ideas: patterns of sintered daub creation and dispersal in a modern wattle and daub structure and their implications for archaeological interpretation. Journal of Archaeological Method and Theory 22, 883-912.

Larkin, N., Alexander, J. & Lewis, M.D., 2000. Using experimental studies of recent faecal material to examine hyena coprolites from the West Runton Freshwater Bed, Norfolk, UK. Journal of Archaeological Science 27, 19-31.

Levine, M.N., Arthur A., Joyce, A.A. & Goldberg, P., 2004. Earthen Mound Construction at Río Viejo on the Pacific Coast of Oaxaca, Mexico. Poster presented at the 69th Annual Meeting of the Society for American Archaeology, Montreal.

Lewis, M., 1997. An Analysis of the Early Middle Pleistocene Hyena Fauna Coprolites from Boxgrove, Sussex. MSc Dissertation, University College London, 347 p.

Love, S., 2012. The geoarchaeology of mudbricks in architecture: a methodological study from Çatalhöyük, Turkey. Geoarchaeology 27, 140-156.

Macphail, R.I., 1987. A review of soil science in archaeology in England. In Keeley, H.C.M. (ed.), Environmental Archaeology: A Regional Review Vol. II. Historic Buildings & Monuments Commission for England, London, pp. 332-379.

Macphail, R.I., 1990. Soil history and micromorphology. In Bell, M. (ed.), Brean Down Excavations 1983-1987. English Heritage, London, pp. 187-196.

Macphail, R.I., 1991. The archaeological soils and sediments. In Sharples, N.M. (ed.), Maiden Castle: Excavations and Field Survey 1985-6. English Heritage, London, pp. 106-118.

Macphail, R.I., 1993. Soil micromorphology. In Whittle, A., Rouse, A.J. & Evans, J.G. (eds.), A Neolithic Downland Monument in its Environment L: Excavations at the Easton Down Long Barrow, Bishops Canning, North Wiltshire. Proceedings of the Prehistoric Society 59, 218-219, 234-235.

Macphail, R.I., 1994. The reworking of urban stratigraphy by human and natural processes. In Hall, A.R. & Kenward, H.K. (eds.), Urban-Rural Connexions: Perspectives from Environmental Archaeology. Oxbow Books, Oxford, pp. 13-43.

Macphail, R.I., 1995. Cressing Temple (AD 1250) Wheat Barn: full assesment of soil micromorphology. Unpublished report, Essex County Field Archaeology Unit, Braintree, 8 p.

Macphail, R.I., 1999. Sediment micromorphology. In Roberts, M.B. & Parfitt, S.A. (eds.), Boxgrove. A Middle Pleistocene hominid site at Eartham Quarry, Boxgrove, West Sussex. English Heritage, London, pp. 118-148.

Macphail, R.I., 2000. Soils and microstratigraphy: a soil micromorphological and micro-chemical approach. In Lawson, A.J. (ed.), Potterne 1982-5: Animal Husbandry in Later Prehistoric Wiltshire. Wessex Archaeology Report Vol. 17. Wessex Archaeology, Salisbury, pp. 47-70.

Macphail, R.I., 2003. Industrial activities – some suggested microstratigraphic signatures: ochre, building materials and iron-working. In Wiltshire, P.E.J. & Murphy, P. (eds.), The Environmental Archaeology of Industry. Oxbow Books, Oxford, pp. 94-106.

Macphail, R.I., 2005. Soil micromorphology and chemistry. In Shelley, A. (ed.), Dragon Hall, King Street, Norwich: Excavation and Survey of a Late Medieval Merchant's Trading Complex. East Anglian Archaeology, Norwich, pp. 175-178.

Macphail, R.I., 2011. Micromorphological analysis of road construction sediments. In Malim, T. & Hayes, L. (eds.), An Engineered Iron Age Road, Associated Roman Use (Margary Route 64), and Bronze Age Activity Recorded at Sharpstone Hill, 2009. Transactions of the Shropshire Archaeological and Historical Society 85, 53-55.

Macphail, R.I., 2016a. House pits & Grubenhausen. In Gilbert, A.S. (ed.), Encyclopedia of Geoarchaeology. Springer Scientific, Dordrecht, pp. 425-432.

Macphail, R.I., 2016b. Privies and latrines. In Gilbert, A.S. (ed.), Encyclopedia of Geoarchaeology. Springer Scientific, Dordrecht, pp. 682-687.

Macphail, R.I. & Crowther, J., 2007. Soil micromorphology, chemistry and magnetic susceptibility studies at Huizui (Yiluo region, Henan Province, northern China), with special focus on a typical Yangshao floor sequence. Bulletin of the Indo-Pacific Prehistory Association 27, 93-113.

Macphail, R.I. & Crowther, J., 2008. Illustrations from soil micromorphology and complementary investigations. In Thiemeyer, H. (ed.), Archaeological Soil Micromorphology – Contributions tot the Archaeological Soil Micromorphology Working Group Meeting 3rd to 5th April 2008, Frankfurt. Frankfurter Geowissenschaftlichen Arbeiten, Serie D, Band 30, pp. 81-87.

Macphail, R.I. & Crowther, J., 2011. Experimental pig husbandry: soil studies from West Stow Anglo-Saxon Village, Suffolk, UK. Antiquity Project Gallery, Antiquity 85, 330.

Macphail, R.I. & Cruise, G.M., 2001. The soil micromorphologist as team player: a multianalytical approach to the study of European microstratigraphy. In Goldberg, P., Holliday, V. & Ferring, R. (eds.), Earth Science and Archaeology. Kluwer Academic/Plenum Publishers, New York, pp. 241-267.

Macphail, R.I. & Goldberg, P., 1990. The micromorphology of tree subsoil hollows: their significance to soil science and archaeology. In Douglas, L.A. (ed.), Soil Micromorphology: A Basic and Applied Science. Developments in Soil Science, Volume 19. Elsevier, Amsterdam, pp. 425-429.

Macphail, R.I. & Goldberg, P., 1995. Recent advances in micromorphological interpretations of soils and sediments from archaeological sites. In Barham, A.J. & Macphail, R.I. (eds.), Archaeological Sediments and Soils: Analysis, Interpretation and Management. Institute of Archaeology, London, pp. 1-24.

Macphail, R.I. & Goldberg, P., 1999. The soil micromorphological investigation of Westbury Cave. In Andrews, P., Cook, J., Currant, A. & Stringer, C. (eds.), Westbury Cave. The Natural History Museum Excavations 1976-1984. Western Academic & Specialist Press, Bristol, pp. 59-86.

Macphail, R.I. & Goldberg, P., 2012. Soil Micromorphology of Gibraltar coprolites. In Barton, R.N.E. (ed.), Gibraltar Neanderthals in Context: A Report of the 1995-98 Excavations at Gorham's & Vanguards Caves, Gibraltar. Oxford University School of Archaeology, Monograph 75, Oxford University Press, Oxford, pp. 240-243.

Macphail, R.I. & Goldberg, P., 2017. Applied Soils and Micromorphology in Archaeology. Cambridge University Press, Cambridge, 630 p.

Macphail, R., Romans, J.C.C. & Robertson, L., 1987. The application of micromorphology to the understanding of Holocene soil development in the British Isles; with special reference to early cultivation. In Fedoroff, N., Bresson, L.M. & Courty, M.A. (eds.), Micromorphologie des Sols, Soil Micromorphology. AFES, Plaisir, pp. 647-656.

Macphail, R.I., Courty, M.A., Hather, J. & Wattez, J., 1997. The soil micromorphological evidence of domestic occupation and stabling activities. In Maggi, R. (ed.), Arene Candide: A Functional and Environmental Assessment of the Holocene Sequence (Excavations Bernabò Brea-Cardini 1940-50). Memorie dell'Istituto Italiano di Paleontologia Umana, Roma, pp. 53-88.

Macphail, R.I., Cruise, G.M., Mellalieu, S.J. & Niblett, R., 1998. Micromorphological interpretation of a 'turf-filled' funerary shaft at St. Albans, United Kingdom. Geoarchaeology 13, 617-644.

Macphail, R.I., Crowther, J., Acott, T.G., Bell, M.G. & Cruise, G.M., 2003. The experimental earthwork at Wareham, Dorset after 33 years: changes to the buried LFH and Ah horizon. Journal of Archaeological Science 30, 77-93.

Macphail, R.I., Cruise, G.M., Allen, M.J., Linderholm, J. & Reynolds, P., 2004. Archaeological soil and pollen analysis of experimental floor deposits, with special reference to Butser Ancient Farm, Hampshire, UK. Journal of Archaeological Science 31, 175-191.

Macphail, R.I., Linderholm, J. & Karlsson, N., 2006a. Scanian pithouses; interpreting fills of grubenhäuser: examples from England and Sweden. In Engelmark, R. & Linderholm, J. (eds.), Proceedings of the 8th Nordic Conference on the Application of Scientific Methods in Archaeology. Umeå University, pp. 119-127.

Macphail, R.I., Cruise, G.M., Allen, M.J. & Linderholm, J., 2006b. A rebuttal of the views expressed in 'Problems of unscientific method and approach in archaeological soil and pollen analysis of experimental floor deposits; with special reference to Butser Ancient Farm, Hampshire, UK by R.I. Macphail, G.M. Cruise, M. Allen, J. Linderholm and P. Reynolds' by Matthew Canti, Stephen Carter, Donald Davidson and Susan Limbrey. Journal of Archaeological Science, 33, 299-305.

Macphail, R.I., Crowther, J. & Cruise, G.M., 2007, Micromorphology and post-Roman town research: the examples of London and Magdeburg. In Henning, J. (ed.), Post-Roman Towns and Trade in Europe, Byzantium and the Near-East. New Methods of Structural, Comparative and Scientific Methods in Archaeology. Walter de Gruyter & Co., Berlin, pp. 303-317.

Macphail, R.I., Crowther, J., & Berna, F., 2012. Soil Micromorphology, Microchemistry, Chemistry, Magnetic Susceptibility and FTIR. In Biddulph, E., Foreman, S., Stafford, E., Stansbie, D. & Nicholson, R. (eds.), London Gateway. Iron Age and Roman salt making in the Thames Estuary; Excavations at Stanford Wharf Nature Reserve, Essex. Oxford Archaeology Monograph No. 18. Oxford Archaeology, Oxford, 193 p.

Macphail, R.I., Bill, J., Cannell, R., Linderholm, J. & Rødsrud, C.L., 2013. Integrated microstratigraphic investigations of coastal archaeological soils and sediments in Norway: the Gokstad ship burial mound and its environs including the Viking harbour settlement of Heimdaljordet, Vestfold. Quaternary International 315, 131-146.

Macphail, R.I., Linderholm, J. & Eriksson, S., 2016. Kaupang 2015 (Kaupangveien 224), Vestfold, Norway: Soil Micromorphology, Chemistry and Magnetic Susceptibility Studies (Report for KHM, UiO). Institute of Archaeology, University College London, 33 p.

Macphail, R.I., Bill, J., Crowther, J., Haiță, C., Linderholm, J., Popovici, D. & Rødsrud, C.L., 2017a. European ancient settlements – a guide to their composition and morphology based on soil micromorphology and associated geoarchaeological techniques; introducing the contrasting sites of Chalcolithic Borduşani-Popină, Borcea River, Romania and the Viking Age Heimdaljordet, Vestfold, Norway. Quaternary International 460, 30-47

Macphail, R.I., Graham, E., Crowther, J. & Turner, S., 2017b. Marco Gonzalez, Ambergris Caye, Belize: a geoarchaeological record of ground raising associated with surface soil formation and the presence of a Dark Earth. Journal of Archaeological Science 77, 35-51.

Mallol, C., Marlowe, F.W., Wood, B.M. & Porter, C.C., 2007. Earth, wind, and fire: ethnoarchaeological signals of Hadza fires. Journal of Archaeological Science 34, 2035-2052.

Mallol, C., Hernández, C.M., Cabanes, D., Machado, J., Sistiaga, A., Pérez, L. & Galván, B., 2013. Human actions performed on simple combustion structures: an experimental approach to the study of Middle Palaeolithic fire. Quaternary International 315, 3-15.

Mathieu, C. & Stoops, G., 1972. Observations pétrographiques sur la paroi d'un four à chaux caroligien creusé en sol limoneux. Archéologie Médiévale 2, 347-354.

Matsui, A., Hiraya, R., Mijaji, A. & Macphail, R.I., 1996. Availability of soil micromorphology in archaeology in Japan [*in Japanese*]. Archaeology Society of Japan 38, 149-152.

Matthews, W., 1995. Micromorphological characterisation and interpretation of occupation deposits and microstratigraphic sequences at Abu Salabikh, Southern Iraq. In Barham, A.J. & Macphail, R.I. (eds.), Archaeological Sediments and Soils: Analysis, Interpretation and Management. Institute of Archaeology, London, pp. 41-74.

Matthews, W. & Postgate, J.N., 1994. The imprint of living in a Mesopotamian city: questions and answers. In Luff, R. & Rowley-Conwy, P. (eds.), Whither Environmental Archaeology? Oxbow Monograph 38. Oxbow Books, Oxford, pp. 171-212.

Matthews, W., French, C.A.I., Lawrence, T. & Cutler, D., 1996. Multiple surfaces: the micromorphology. In Hodder, I. (ed.), On the Surface: Çatalhöyük 1993-95. McDonald Institute for Archaeological Research and British Institute of Archaeology at Ankara, Cambridge, pp. 301-342.

Matthews, W., French, C.A.I., Lawrence, T., Cutler, D.F. & Jones, M.K., 1997. Microstratigraphic traces of site formation processes and human activity. World Archaeology 29, 281-308.

Matthews, W., French, C.A.I., Lawrence, T., Cutler, D.F. & Jones, M.K., 1998. Microstratigraphy and micromorphology of depositional sequences. In Oates, D., Oates, J. & McDonald, H. (eds.), Excavations at Tell Brak. Volume I, The Mitani and Old Babylonian Periods. McDonald Institute Monograph, McDonald Institute for Archaeological Research, Cambridge, pp. 135-140.

Matthews, W., Hastorf, C.A. & Ergenekon, B., 2000. Ethnoarchaeology: studies in local villages aimed at understanding aspects of the Neolithic site. In Hodder, I. (ed.), Towards Reflexive Method in Archaeology: The Example at Çatalhöyük. McDonald Institute for Archaeological Research and British Institute of Archaeology at Ankara, pp. 177-188.

Mentzer, S.M., 2014. Microarchaeological approaches to the identification and interpretation of combustion features in prehistoric archaeological sites. Journal of Archaeological Method and Theory 21, 616-668.

Milek, K., 1997. Soil micromorphology and the medieval urban environment: examples from Ely and Peterborough, England. In De Boe, G. & Verhaeghe, F. (eds.), Environment and Subsistence in Medieval Europe: Papers from the 'Medieval Brugge 1997' Conference, Volume 9. Institute for the Archaeological Heritage, Zellik, pp. 155-168.

Milek, K., 2004. Aðalstræti, Reykjavík, 2001: geoarchaeological report on deposits within the house and the soils immediately pre- and post-dating its occupation. In Roberts, H. (ed.), Excavations at Aðalstræti, 2003, Fornleifastofnun Íslands, Reykavík, pp. 73-114.

Milek, K., 2005. Soil micromorphology. In Sharples, N.M. (ed.), A Norse Farmstead in the Outer Hebrides: Excavations at Mound 3, Bornais, South Uist. Oxbow Books, Oxford, pp. 98-104.

Milek, K., 2006. Houses and Households in Early Icelandic Society: Geoarchaeology and the Interpretation of Social Space. PhD Dissertation, University of Cambridge. 413 p.

Milek, K. & French, C., 2007. Soils and Sediments in the Settlement and Harbour at Kaupang. In Skre, D. (ed.), Kaupang in Skiringssal. Aarhus University Press, Aarhus, pp. 321-360.

Mücher, H.J., 1974. Micromorphology of slope deposits: the necessity of a classification. In Rutherford, G.K. (ed.), Soil Microscopy. The Limestone Press, Kingston, pp. 553-556.

Nodarou, E., Frederick, C. & Hein, A., 2008. Another (mud)brick in the wall: scientific analysis of Bronze Age earthen construction materials from East Crete. Journal of Archaeological Science 35, 2997-3015.

Nørnberg, P., & Courty, M.A., 1985. Standard geological methods used on archaeological problems In Edgren, T. & Jungner, H. (eds.), Proceedings of the Third Nordic Conference on the Application of Scientific Methods in Archaeology, ISKOS 5. Finnish Antiquarian Society, Helsinki, pp. 107-118.

Nys, C., Bullock, P. & Nys, A., 1987. Micromorphological and physical properties of a soil under three different species of trees. In Fedoroff, N., Bresson, L.M. & Courty, M.A. (eds.), Micromorphologie des Sols. Soil Micromorphology. AFES, Plaisir, pp. 459-464.

Polo-Díaz, A., Alonso Eguíluz, M., Ruiz, M., Pérez, S., Mújika, J., Albert, R.M. & Fernández Eraso, J., 2016. Management of residues and natural resources at San Cristóbal rock-shelter: contribution to the characterisation of chalcolithic agropastoral groups in the Iberian Peninsula. Quaternary International 414, 202-225.

Pye, E., 2000/2001. Wall painting in the Roman empire: colour, design and technology. Archaeology International 4, 24-27.

Quinn, P.S., 2013. Ceramic Petrography. The Interpretation of Archaeological Pottery and Related Artefacts in Thin Section. Archeopress, Oxford, 234 p.

Rentzel, P., 1998. Ausgewählte Grubenstrukturen aus spätlatènezeitlichen Fundstelle Basel-Gasfabrik: Geoarchäologische interpretation der Grubenfüllungen. Jahresbericht der Archäologischen Bodenforschung des Kantons Basel-Stadt 1995, pp. 35-79.

Rentzel, P., 2011. Spuren der Nutzung im Mithraeum von Biesheim − Mikromorphologische Untersuchungen. In Reddé, M. (ed.) Oedenburg Volume 2. L'Agglomération Civile et les Sanctuaires. 2 − Matériel et Etudes. Monographien des Römisch-Germanischen Zentralmuseums, Mainz, pp. 248-255.

Rice, P.M., 1987. Pottery Analysis: A Sourcebook. University of Chicago Press, Chicago, 559 p.

Romans, J.C.C. & Robertson, L., 1983. The general effects of early agriculture on soil. In Maxwell, G.S. (ed.), The Impact of Aerial Reconnaissance on Archaeology. Research Report No. 49, Council for British Archaeology, London, pp. 136-141.

Rosen, A.M., 1986. Cities of Clay. The Geoarchaeology of Tells. University of Chicago Press, Chicago, 167 p.

Rosen, S.A., Savinetsky, A.B., Plakht, J., Kisseleva, N.K., Khassanov, B.F., Pereladov, A.M. & Haiman, M., 2005. Dung in the desert: preliminary results from the Negev Ecology Project. Current Anthropology 46, 317-327.

Roux, V. & Courty, M.A., 1998. Identification of wheel-fashioning methods: technological analysis of 4th–3rd millennium BC oriental ceramics. Journal of Archaeological Science 25, 747-763.

Runia, L.T., 1988. So-called secondary podzolisation in barrows. In Groenman-van Waateringe, W. & Robinson, M. (eds.), Man-Made Soils. British Archaeological Reports, International Series 410. Archaeopress, Oxford, pp. 129-142.

Scaife, R.G., 2000. Coprolites: microfossil examination and potential. In Lawson, A.J. (ed.), Potterne 1982-5: Animal Husbandry in Later Prehsitoric Wiltshire. Wessex Archaeology Report Vol. 17. Wessex Archaeology, Salisbury, pp. 71-72.

Scaife, R.G. & Macphail, R.I., 1983. The post-Devensian development of heathland soils and vegetation. In Burnham, P. (ed.), Soils of the Heathlands and Chalklands. South-East Soils Discussion Group, Wye, pp. 70-99.

Shahack-Gross, R., 2011. Herbivorous livestock dung: formation, taphonomy, methods for identification, and archaeological significance. Journal of Archaeological Science 38, 205-218.

Shahack-Gross, R. & Finkelstein, I., 2008. Subsistence practices in an arid environment: a geo-archaeological investigation in an Iron Age site, the Negev Highlands, Israel. Journal of Archaeological Science 35, 965-982.

Shahack-Gross, R., Marshall, F. & Weiner, S., 2003. Geo-ethnoarchaeology of pastoral sites: the identification of livestock enclosures in abandoned Maasai settlements. Journal of Archaeological Science 30, 439-459.

Shahack-Gross, R., Marshall, F., Ryan, K. & Weiner, S., 2004a. Reconstruction of spatial organisation in abandoned Maasai settlements: implications for site structure in Pastoral Neolithic of East Africa. Journal of Archaeological Science 31, 1395-1411.

Shahack-Gross, R., Berna, F., Karkanas, P. & Weiner, S., 2004b. Bat guano and preservation of archaeological remains in cave sites. Journal of Archaeological Science 31, 1259-1272.

Shahack-Gross, R., Albert, R.M., Gilboa, A., Nagar-Hillman, O., Sharon, I. & Weiner, S., 2005. Geoarchaeology in an urban context: the uses of space in a Phoenician monumental building at Tel Dor (Israel). Journal of Archaeological Science 32, 1417-1431.

Shahack-Gross, R., Simons, A. & Ambrose, S.H., 2008. Identification of pastoral sites using stable nitrogen and carbon isotopes from bulk sediment samples: a case study in modern and archaeological pastoral settlements in Kenya. Journal of Archaeological Science 35, 983-990.

Shahack-Gross, R., Berna, F., Karkanas, P., Lemorini, C., Gopher, A. & Barkai, R., 2014. Evidence for the repeated use of a central hearth at Middle Pleistocene (300 ky ago) Qesem Cave, Israel. Journal of Archaeological Science 44, 12-21.

Sherwood, S., 2006. The geoarchaeological study of Mound A, Shiloh Indian Mounds National Historic Landmark, Hardin County, Tennessee. Geological Society of America Abstracts with Programs 38, p. 391.

Simpson, I.A., Milek, K.B. & Gudmundsson, G., 1999. A reinterpretation of the Great Pit at Hofstadir, Iceland using sediment thin section micromorphology. Geoarchaeology 14, 511-530.

Sordoillet, D., 1997. Formation des dépôts archéologiques en grotte: la Grotte du Gardon (Ain) durant le Néolithique. In Bravard, J.F. & Prestreau, M. (eds.), Dynamique du Paysage. Entretiens de Géoarchéologie. Documents d'Archéologie en Rhône-Alpes No. 15. Ministère de la Culture, Lyon, pp. 39-57.

Sordoillet, D., 2009. Géoarchéologie de Sites Préhistoriques: Le Gardon (Ain), Montou (Pyrénées-Orientales) et Saint-Alban (Isère). Éditions de la Maison des Sciences de l'Homme, Paris, 188 p.

Spataro, M., 2002. The First Farming Communities of the Adriatic: Pottery Production and Circulation in the Early and Middle Neolithic. Quaderni della Società per la Preistoria e Protostoria della Regione Friuli-Venezia Giulla, Trieste, No. 9, 255 p.

St John, D.A., Poole, A.B. & Sims, I., 1998. Concrete Petrography. A Handbook of Investigation Techniques. Arnold, London, 474 p.

Stolt, M.H. & Lindbo, D.L., 2010. Soil organic matter. In Stoops, G., Marcelino, V. & Mees, F. (eds.), Interpretation of Micromorphological Features of Soils and Regoliths. Elsevier, Amsterdam, pp. 369-396.

Stoops, G., 1984a. The environmental physiography of Pessinus in function of the study of the archaeological stratigraphy and natural building materials. Dissertationes Archaeologicae Gardenses 22, 38-50.

Stoops, G., 1984b. Petrographic study of mortar and plaster samples. Dissertationes Archaeologicae Gardenses 22, 164-170.

Stoops, G., 2003. Guidelines for Analysis and Description of Soil and Regolith Thin Sections. Soil Science Society of America, Madison, 184 p.

Stoops, G. & Nijs, R., 1986. Micromorphological characteristics of some tell materials from Mesopotamia. Pedologie, 36, 329-336.

Stoops, G.S. & Stoops, G.J., 1994. Petrographic study of red floor fragments from the palaces at Babylon and Susa. Mesopotamian History and Environment. Occasional Publications, Vol. 2. Cinquante-Deux Réflexions sur le Proche-Orient Ancien. Peeters, Leuven, pp. 477-486.

Straulino, L., Sedov, S., Michelet, D. & Balanzario, S., 2013. Weathering of carbonated materials in ancient Maya constructions (Río Bec and Dzibanché): limestone and stucco deterioration patterns. Quaternary International 315, 87-100.

Sveinbjarnardóttir, G., Erlendsson, E., Vickers, K., McGovern, T.H., Milek, K.B., Edwards, K.J., Simpson, D.D.A. & Cook, G., 2007. The palaeoecology of a high status Icelandic farm. Environmental Archaeology 12, 187-206.

Sveistrup, T., Marcelino, V. & Stoops, G., 1995. Effects of slurry application on the microstructure of the surface layers of soils from northern Norway. Norwegian Journal of Agricultural Sciences 9, 1-13.

Van Nest, J., Charles, D.K., Buikstra, J.E. & Asch, D.L., 2001. Sod blocks in Illinois Hopewell mounds. American Antiquity 66, 633-650.

Vepraskas, M.J., Lindbo, D.L. & Stolt, M.H., 2018. Redoximorphic features. In Stoops, G., Marcelino, V. & Mees, F. (eds.), Interpretation of Micromorphological Features of Soils and Regoliths. Second Edition. Elsevier, Amsterdam, pp. 425-445.

Viklund, K., 1998. Cereals, Weeds and Crop Processing in Iron Age Sweden. Methodological and Interpretive Aspects of Archaeobotanical Evidence. Archaeology and Environment 14, Umeå University, 192 p.

Viklund, K., Linderholm, J. & Macphail, R.I., 2013. Integrated palaeoenvironmental study: micro- and macrofossil analysis and geoarchaeology (soil chemistry, magnetic susceptibility and micromorphology). In Gerpe, L.E. (ed.), E18-prosjektet Gulli-Langåker. Oppsummering og Arkeometriske Analyser, Bind 3, Fagbokforlaget, Bergen, pp. 25-83.

Villagran, X.S., Schaefer, C.E.G.R., & Ligouis, B., 2013. Living in the cold: geoarchaeology of sealing sites from Byers Peninsula (Livinston Island, Antarctica). Quaternary International 315, 184-199.

Villaseñor, I. & Graham, E., 2010. The use of volcanic materials for the manufacture of pozzolanic plasters in the Maya lowlands: a preliminary report. Journal of Archaeological Science 37, 1339-1347.

Wattez, J. & Courty, M.A., 1987. Morphology of ash of some plant materials. In Fedoroff, N., Bresson, L.M. & Courty, M.A. (eds.), Micromorphologie des sols. Soil micromorphology. AFES, Plaisir, pp. 677-683.

Wegener, O., 2009. Soil micromorphological investigations on trampling floors in pit houses (Grubenhäuser) of the deserted medieval town Marsleben (Saxony-Anhalt). In Thiemeyer, H. (ed.), Archaeological Soil Micromorphology. Frankfurter Geowissenschaftlichen Arbeiten, Serie D, Band 30, pp. 133-141.

Weiner, S., 2010. Microarchaeology. Beyond the Visible Archaeological Record. Cambridge University Press, Cambridge, 414 p.

Chapter 28

Palaeosoils and Relict Soils: A Conceptual Approach

Nicolas Fedoroff[1,†], Marie-Agnès Courty[2], Zhengtang Guo[3]

[1]AGROTECH PARIS, PARIS, FRANCE (DECEASED);
[2]UNIVERSIDAD ROVIRA Y VIRGILI, TARRAGONA, SPAIN;
[3]CHINESE ACADEMY OF SCIENCES, BEIJING, CHINA

CHAPTER OUTLINE

†Deceased February 14, 2013.

Interpretation of Micromorphological Features of Soils and Regoliths. https://doi.org/10.1016/B978-0-444-63522-8.00028-0

1. Introduction

Soils have played an important role in Earth history since 3000 million years ago (Retallack, 1990; Retallack & Mindszenty, 1994). Primitive soils, which formed by weathering, appeared before vegetation developed (e.g., Driese et al., 2007). Later soils formed in conditions with different atmosphere compositions than those existing at present and with a different impact of cosmic events (e.g., Jones & Bo Lim, 2000).

Definitions of palaeosoils vary widely. For Morrison (1977), palaeosoils are soils of obvious antiquity, whereas Butzer (1971) defines them as ancient soils. For other authors, they are soils formed in a landscape of the past (Ruhe, 1965; Yaalon, 1971) or in a past environment (Yaalon, 1983). In this chapter, the term 'palaeosoil' is restricted to buried soils of any age, for which functioning was totally or partially inhibited by burial. We apply it also to truncated soils, even if just a thin basal interval is preserved. Relict soils are defined as soils belonging to the present-day soil cover but which have characteristics inherited from the past, resulting from different environmental conditions than those occurring today.

The aims and history of palaeopedology differ from those of pedology (Fedoroff & Courty, 2002). Palaeosoils have been approached with different objectives. Many authors (e.g., Kukla & An Zhisheng, 1989) considered palaeosoils as stratigraphic markers in continental sequences, without attempting to analyze them, whereas others considered some layers in such sequences as palaeosoils just because of their colour. Analysis and interpretation of palaeosoils with methods used in pedology, including soil micromorphology, began at a relatively late stage (e.g., Fedoroff & Goldberg, 1982). The palaeo-environmental significance of palaeosoils has only recently been taken into account (e.g., Fedoroff et al., 1990; Kemp, 1999; Felix-Henningsen & Mauz, 2004; Ferraro et al., 2004). Proxy indicators for defining climatic characteristics under which palaeosoils have developed have been used by some authors (e.g., Wang & Follmer, 1998; Retallack, 2005a).

Soil micromorphology is at present commonly applied in investigations of palaeosoils and relict soils. Mücher and Morozova (1983) have presented a foundation for the micromorphological description of palaeosoils and related sediments, whereas Fedoroff et al. (1990) drafted keys for the recognition of environmental changes in the past. Most papers deal with Quaternary palaeosoils, especially those formed during the last inter-glacial, the last glacial cycle and the Holocene (e.g., Kemp et al., 1994; Cremaschi & Trombino, 1998; Kemp, 1998; Srivastava & Parkash, 2002), but pre-Quaternary palaeosoils have also been studied (e.g., Tate & Retallack, 1995; McCarthy et al., 1998; Driese et al., 2007). Micromorphological analyses of palaeosoils often show differences with present-day soil functioning. Elementary attributes of pedogenic facies of the past are similar to present ones, but they can be different by showing a higher degree of development or by the presence of assemblages of pedofeatures unknown in modern soils. Pedofeatures specific of pedogenesis of the past are rare. Micromorphological studies reveal also that almost all palaeosoils were affected by processes of *in situ* reworking, erosion, transport, deposition and allochthonous aggradation (e.g., Retallack, 2005b; Sephton et al., 2005), which are often underestimated or not even recognised by other observations.

Examination in thin sections of well-developed soils, especially those dating back to the Pleistocene, nearly always reveals the presence of relict characteristics (e.g., Fedoroff, 1997). In some relict soils, such as Ferralsols, the history of the soil is almost totally wiped out, but in others, such as calcretes and ferricretes (e.g., Achyuthan & Fedoroff, 2008), the history is preserved to a large extent.

In this chapter, an interdisciplinary approach is used, dealing with analysis of sets of interacting entities and the interactions within those systems (systems analysis). Such an approach is commonly utilised by sedimentary petrologists (e.g., Humbert, 1976a, 1976b).

2. Methodology

2.1 Recognition of Palaeosoils and Relict Soils

A first task when investigating continental sequences is to decide whether the layers defined in the field as palaeosoils are soils that formed *in situ*, transported soil materials (pedosediments), or something else (e.g., McCarthy, 2002). *In situ* soils are characterised in thin sections by a continuous pedogenic facies recognised by at least one of the following features: (i) undisturbed features resulting from soil biological activity, such as passage features and channels with root residues or excrements; (ii) a pedogenic microstructure; (iii) a pedogenic b-fabric; or (iv) one or more types of undisturbed pedofeatures. In contrast, pedosediments are recognised by (i) an absence of *in situ* biogenic features, but the common presence of tissue and charcoal fragments; (ii) a massive microstructure, and occasionally a structure dominated by packing of rounded aggregates; or (iii) sedimentary features (see also Mücher et al., 2010). Some layers which seem to be free of pedogenic characteristics in the field can show a pedogenic facies in thin sections. For example, the typic Late Pleistocene loess layers of the Loess Plateau of China consist entirely of passage features and various types of excrements (Guo, 1990). Consequently, this examination must be extended to underlying and overlying horizons or layers to characterise variations in pedogenic facies, identify other possible types of facies, compare groundmass characteristics and identify features related to erosion and aggradation.

Palaeosoils are very rarely preserved as complete and undisturbed profiles. Some discontinuities are easy to identify in the field, such as truncations, stone lines and the superimposition of allochthonous materials on pedogenic horizons. The field diagnosis can be considerably improved by examination in thin sections, which can reveal discontinuities in pedofeatures, as well as features due to erosion, *in situ* reworking (see Fig. 15), mass transportation, or minor aggradation (e.g., accumulation of aeolian dust). It also contributes to the determination of which *in situ* horizon or horizons are part of the investigated palaeosoil, on the basis of its pedogenic facies.

Palaeosoils commonly alternate with sediments, as in loess sequences (e.g., Kukla & An Zhisheng, 1989) or in deltas and alluvial plains (Srivastava & Parkash, 2002), creating palaeosoil-sediment sequences termed 'pedocomplexes' (e.g., Morrison, 1977; Feng & Wang, 2005). The palaeosoils can be (i) juxtaposed, i.e., lying in a close vertical succession, without penetration of pedofeatures of the overlying into the underlying palaeosoil; (ii) superimposed, i.e., with penetration of pedofeatures of the overlying into the underlying palaeosoil; or (iii) cumulic, i.e., palaeosoils in which a type of pedofeature (in general a textural pedofeature) occurs without significant change in a thick homogeneous horizon (Fig. 1) (e.g., Yin & Guo, 2006). An example of a well-studied pedocomplex is the marine isotope stage 5 (MIS 5) pedocomplex, also known as the Eemian pedocomplex, that is present throughout the loess belt of the northern hemisphere, characterisation of which includes micromorphological analyses (e.g., Fedoroff & Goldberg, 1982; Guo, 1990; Kemp et al., 1995, 1997, 2001; Stremme, 1998; Bronger, 2003; Chen et al., 2003; Chlachula et al., 2004). In the most complete sections, this pedocomplex consists of three juxtaposed palaeosoils, but in some regions they merge into a single soil profile in which no boundaries or transitions are recognised, even in thin sections (e.g., Xifeng loess section, China; Guo, 1990). Pedocomplexes are also frequent in karstic dolines, for which some micromorphological data are available (Boulet et al., 1986; Kühn & Hilgers, 2005).

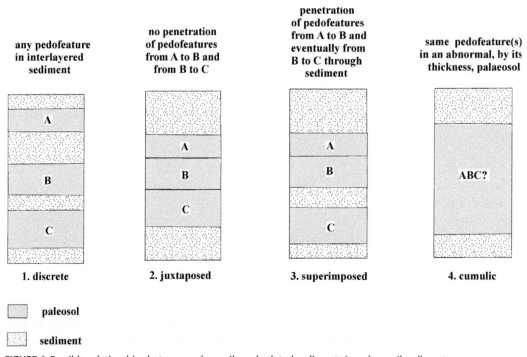

FIGURE 1 Possible relationships between palaeosoils and related sediments in palaeosoil-sediment sequences.

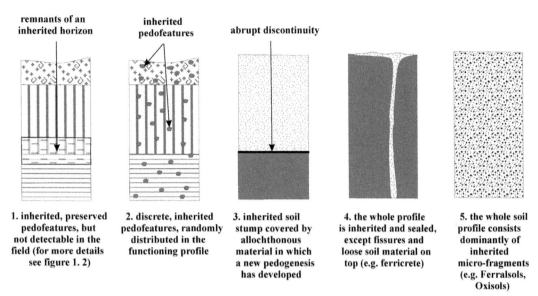

FIGURE 2 Types of distribution of inherited attributes in relict soils.

Relict soils can be subdivided into (i) soils with relict characteristics, which cannot be detected in the field; (ii) soils with discrete, randomly distributed inherited pedofeatures; (iii) soils with an ancient truncated base whose age can reach millions of years, covered by partially or entirely allochthonous materials; (iv) totally relict soils, except for the presence of fissures and channels in which water circulates and in which living organisms are active; and (v) soils consisting of microfragmented relict soil materials, e.g., small fragments of ferruginous features as in Ferralsols (Fig. 2).

Relict or inherited characteristics can be (i) discrete pedofeatures characterised by sharp boundaries and by a relict groundmass, embedded in a functioning groundmass; (ii) a disturbed or fragmented pedogenic facies embedded in a different soil material; or (iii) an undisturbed pedogenic facies which is unrelated to present-day soil-forming conditions, e.g., a platy microstructure due to ice lensing in a deep horizon of a mid-latitude soil. The structure of the parent material can be preserved in the truncated base (isalterite; Delvigne, 1998), or it can be reworked but without major aggradation of allochthonous material (alloterite; Delvigne, 1998). An example of deciphering relict soils can be found in Achyuthan and Fedoroff (2008).

2.2   Reconstruction of History

When reconstructing the history of soils of the past and deciphering their environmental significance, one has to bear in mind a few basic concepts: (i) the evolution of soils at geologic timescales was discontinuous; (ii) the memory of soils is palimpsest-like; and (iii) as a consequence of discontinuous soil evolution, pedofeatures and other attributes of soils exposed for long periods are organised according to hierarchies.

2.2.1 Discontinuous Soil Evolution

The theory of biorhexistasy, based on field observations, describes climatic conditions necessary for periods favourable for soil development, called biostasy, separated by episodes of soil erosion, called rhexistasy (Erhart, 1956). The concept of the pedogenic phase, which is derived from the biostasy phase, was formulated later (Fedoroff & Courty, 2002, 2005). A pedogenic phase is defined as a period of soil development induced by an invariant soil-forming process, characterised by a unique pedogenic facies (Fig. 3). For example, a Luvisol with a Bt horizon characterised by microlaminated clay coatings and infillings or, as illustrated in Figure 4A, a Kastanozem with a completely bioturbated B horizon characterised by passage features and infillings with

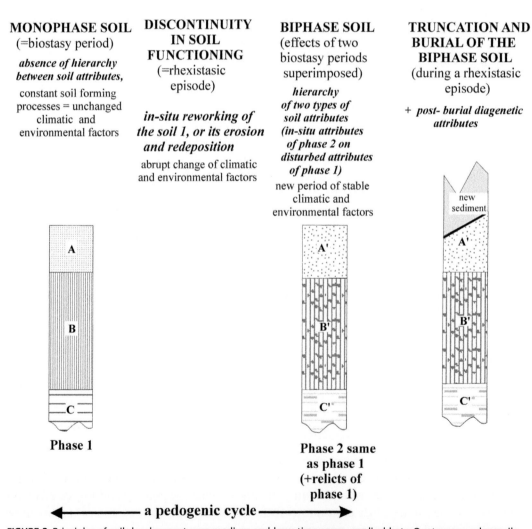

FIGURE 3 Principles of soil development over medium and long time spans, applicable to Quaternary palaeosoils.

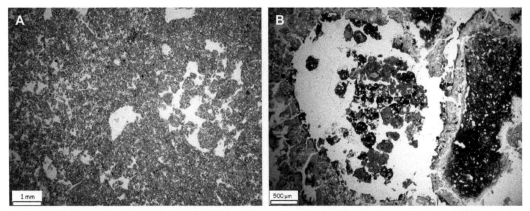

FIGURE 4 (A) Monophase palaeosoil, with moderate rubefaction, totally churned by soil fauna, with abundant passage features and aged fecal pellets (Kastanozem, S1 palaeosoil, Loess Plateau, China, Xifeng section) (PPL). The soil is interpreted to have developed during a period with a stable, rather warm, semiarid climate, under a steppic vegetation (MIS 5). (B) Polyphase soil showing the following hierarchy: (i) monophase micritic carbonate nodule (right), dissolved along its sides; (ii) yellowish grey clayey illuvial features, partially integrated in the groundmass; (iii) dark brown clay coatings along voids; and (iv) loose infilling with brown clayey aggregates and black aggregates, whereby the latter seem to postdate the former (Pampean soil on loess, Azul area, Argentina) (PPL). The sequence of events is as follows: (i) loess deposition; (ii) calcite accretion in the form of discrete features; and (iii) clay illuviation and partial dissolution of the micritic nodule. The brown clayey aggregates probably belong to the clay illuviation phase, whereas the black aggregates are related to an event requiring further investigation.

excrements, must be considered as monophase soils. A rhexistasic episode following a pedogenic phase can be characterised by, for example, loess aggradation or *in situ* soil reworking. This is usually followed by a pedogenic phase similar to the preceding phase of soil formation. A cycle consists of a pedogenic phase and a rhexistasic episode (Fig. 3). In Chinese loess, such cycles are repeated many times (Kukla & An Zhisheng, 1989). An example of polyphase evolution is given in Figure 4B. Cycles also exist in relict soils, for which they are more difficult to identify. They can be impossible to detect, as in most Ferralsols (see Fig. 2). A biostasy period should correspond to a unique pedogenic phase, but biphase and even three-phase palaeosoils without any rhexistasic attribute are quite common. For example, two successive pedogenic phases are recognised in calcareous vertic palaeosoils of Madeira, specifically Vertisol development followed by carbonate enrichment (Goodfriend et al., 1996). Duration and intensity of pedogenic phases can be unequal, some corresponding to extremely strongly developed profiles, for instance, relict soils on high terraces of the Rhone valley (Bornand, 1978) and palaeosoils assigned to MIS 14-15 in China (Guo, 1990). In contrast to the latter, palaeosoils of MIS 11, 9 and 7 in China are moderately to weakly developed.

According to the classical model (Jenny, 1941; Yaalon, 1983), soils develop toward a steady state, balanced with existing factors of soil formation. If the evolution of palaeosoils and relict soils is considered to be discontinuous, this concept must be regarded with caution. Some relict soils seem to have reached a steady state, but close investigation at microscopic scales shows that these soils have been affected to some extent by erosion, transport and sediment aggradation during rhexistasic episodes. For instance, the upper horizons of Ferralsols in northeastern Argentina have been affected by complete reworking during episodes of aeolian activity (Iriondo & Kröhling, 1997). The notion of threshold, referring to a fluctuation in soil development resulting from internal factors (e.g., Ewing et al., 2006), must also be considered with caution, as it probably reflects a modification of environmental parameters in some cases.

2.2.2 Soil Memory

Targulian and Sokolova (1996) and Targulian and Goryachkin (2004) defined 'soil memory' as an assemblage of persistent pedogenic solid-phase properties, in the form of a palimpsest-like type of memory. This concept of soil memory for palaeosoils and relict soils can be enlarged, including the following items (Fedoroff & Courty, 2005):

- assemblage of persistent pedogenic properties, which includes pedofeatures and microstructure; these properties have to be regrouped into pedogenic facies (if more than one is present), and a hierarchy has to be established between them;
- measurement of soil magnetic susceptibility and other magnetic properties; soil magnetic minerals are too fine-grained to be identified by optical microscopy, but they have been observed using electron microscopy (e.g., Perkins, 1996; Grimley & Arruda, 2007); most investigations of magnetic susceptibility of palaeosoils are conducted without using microscopic techniques (e.g., Maher et al, 2003);
- observation of vegetation and fauna remnants such as pollen grains, phytoliths, charcoal fragments, shells and insect cuticles; these remnants have to be separated from the soil mass rather than studied in thin sections;
- nature of the coarse mineral fraction (abundance, distribution, sorting, degree of weathering, external morphology) and its variation through the profile; quartz and zircon can be used for detecting depletion and aggradation (e.g., Brimhall et al., 1991); and
- identification of possible sedimentary features and features caused by soil reworking (Mücher & Morozova, 1983; Mücher et al., 2010).

The palimpsest-like type of soil memory refers to the notion that the memory is constantly renewed, even when the soil is buried (e.g., decay of organic components), but that a part of the pedogenic assemblage is preserved through time. Soil micromorphology is the most efficient method for identification and reconstruction of assemblages which are partly erased, degraded or superimposed by secondary constituents. The following principles can be considered:

- The effects of soil functioning, and consequently of environmental events, are maximal at the soil surface and at near-surface levels, and they decrease sharply with depth. In deep horizons the soil memory is incomplete and selective. For example, superimposed silty and clayey coatings and infillings at the top of a Bt horizon tend to merge progressively into microlaminated features at the bottom of this horizon, and they are replaced by coatings without microlamination in deep C horizons (Targulian et al., 1974). Consequently, the accuracy of the registration of climatic fluctuations tends to decrease with depth.

- The soil memory is renewed rapidly in surface and near-surface horizons, but it is preserved for a long time in deeper horizons, especially in horizons which are out of reach of soil mesofauna and macrofauna. Ice lens microstructures dating from the Late Pleistocene can be observed in Luvisols on loess at depths of 70–80 cm in the Paris basin (Fedoroff & Courty, 2005).

- The depth of occurrence of a pedofeature within a profile varies with the type of soil-forming processes, climate factors and related soil water regimes. Biogenic processes are concentrated near the soil surface, whereas clay illuviation usually occurs at intermediate depth in B horizons, and effects of hydrolysis are detected in the weathering zone.

- The resistance of pedofeatures and other soil attributes to ageing is very variable. The most susceptible to ageing are excrement features, followed in order by silty textural features, microlaminated clay coatings or infillings, and finally nodules rich in iron oxides (Fedoroff & Courty, 2002). Continuous cemented horizons (duricrusts), such as calcrete, ferricrete, gypcrete and silcrete, have played a specific role in conservation of relict soils because of their resistance to erosion.

Stable and humid climatic phases tend to be overemphasised in the soil memory, whereas the registration of shorter and drier phases can be totally absent. During humid phases, water percolating regularly in great amounts favours illuviation to great depth. For instance, thick well-developed relict Luvisols on Middle Pleistocene terraces in the Rhone valley are characterised by thick red clay coatings and infillings distributed throughout an argillic horizon with a thickness of a few metres, whereas registration of juxtaposed or superimposed later pedological phases is minimal (Bornand, 1978).

In polyphase palaeosoils and relict soils, the memory of the first phase is erased to a variable extent during the following phase. In a first approximation, the degree of memory obliteration is correlated with the duration of soil functioning. Various cases of soil memory preservation are presented in Figure 2. In this figure, the soil memory is best preserved in case 4, it is not easily detectable in case 1, and the relevant features are too fine for observation by optical microscopy in case 5 (microfragmentation).

Micromorphological analysis of the memory of a palaeosoil at the level of a thin section consists of the following steps:

- identifying all pedofeatures and other soil attributes present, using, for instance, the system of Bullock et al. (1985) and Stoops (2003), separating autochthonous and allochthonous features, grouping them into types, and characterising the assemblage of each type;

- grouping features into pedogenic facies, which correspond to the total of all pedo-features and other soil attributes occurring simultaneously under constant environmental conditions;
- examining the groundmass to detect features caused by erosion or sedimentation; and
- establishing a hierarchy between the different pedogenic facies, including erosional and sedimentary features if necessary.

2.2.3   Systems Analysis of Polygenetic Palaeosoils and Soils

Systems analysis can be applied for deciphering and interpreting polygenetic palaeosoils and soils at microscopic levels. The procedure is the following: (i) identify all features (pedogenic as well as sedimentary); (ii) regroup similar features in facies; (iii) order the facies in sequences of events (establishment of a hierarchy; Humbert, 1976a); and (iv) interpret from a pedological-sedimentological viewpoint each facies as well as the transition between facies. In palaeosoils and relict soils, a hierarchy can exist between all types of pedofeatures (e.g., textural and crystallitic; see Fig. 4B) and other attributes such as features resulting from erosion and sediments aggradation. Each type of pedofeature is considered to define a pedogenic phase, a period during which climate parameters are supposed to have been constant. A hierarchy must first be established at the level of the thin section and then extended to the palaeosoil as a whole and eventually to the catena.

In relict soils, establishment of hierarchies requires at first a separation of inherited features from those formed in present-day conditions throughout the profile. Inherited features can then be subdivided into simple and complex (e.g., pisoliths; see Fig. 11) and into *in situ*, reworked and transported. Features resulting from erosion and aggradation of allochthonous materials must be identified. If they are present, their relationships with pedofeatures have to be determined. Pedofeatures can be pre-, syn- or posterosional and pre-, syn- or post-aggradational.

3. Common Types of Hierarchies

We present here the most common hierarchies of textural, ferruginous and calcitic features and the related facies. Examples of siliceous features and facies can be found in Summerfield (1983), Milnes et al. (1991), Milnes and Thiry (1992), Thiry (1999) and Poetsch (2004). Inherited gypsum crystals and crystal intergrowths, and less frequently barite and celestite occurrences, can be present in palaeosoils and relict soils, but they do not form complex assemblages in which hierarchies can be recognised, except if they are associated with calcitic features.

Partially or totally dissolved gypsum crystals and crystal intergrowths, identified by their preserved external forms, are common in palaeosoils of arid and semiarid regions (Guo, 1990; Sullivan & Koppi, 1993). For more soluble salts, the lifetime of their occurrence is generally too short to be preserved as part of a hierarchy (e.g., Hamdi-Aissa, 2001).

3.1   Textural Features

The following types of occurrences involving one or two kinds of textural features are the most common:

- only one type of textural feature (e.g., microlaminated clay coatings), the absence of hierarchy (Fig. 5, case 1); in general some of these features are deformed or burrowed by soil fauna (e.g., small fragments of clay coatings incorporated in excrements), whereby the two processes (illuviation and faunal activity) are synchronous; such a facies characterise monophase soils;

- juxtaposed and concordant features (e.g., parallel lamination in coatings) (Fig. 5, case 2), which implies that two different illuvial phases occurred successively, without erosion or another abrupt event separating stages with different environmental conditions (e.g., replacement of a conifer forest by deciduous vegetation); such a facies characterises biphase soils;

- juxtaposed but discordant features (e.g., non-parallel lamination in coatings or cross-bedding) (Fig. 5, case 3), which means that a moderate interruption occurred between the different illuvial phases, during a rhexistasic episode; such a facies also characterises biphase soils;

- unrelated features (e.g., different types of clay coatings in different sets of voids) (Fig. 5, case 4), which implies that moderate soil disturbance took place between the two illuvial phases;

- one type of feature occurring in the form of fragments dispersed in the groundmass (clay coatings fragments), whereas an undisturbed feature (clay coatings) covers the sides of voids (Fig. 5, case 5) or coats the fragmented feature, both meaning that Luvisol development was followed by disruption of the Luvisol (e.g., by mass transportation), followed in turn by a new illuvial phase, similar to the first.

Hierarchies of three or more kinds of textural features (Fig. 7A) also occur quite frequently.

The hierarchy of textural features which is established in the horizon appearing in the field as the palaeosoil or its key horizon should be compared with the hierarchy in the underlying and overlying horizons or layers (Fig. 6). Relationships between illuvial facies in these horizons or layers, within a single profile, can be for instance:

- Textural features (clay coatings) that are undisturbed in the key horizon (Fig. 6, case 1) occur as fragments in the overlying horizon (Fig. 6, case 2), which means that the palaeosoil was truncated and then covered by reworked pedogenic material derived from the palaeosoil.

- A textural feature (clay coating) both occurs in the key horizon and penetrates into the underlying horizon (Fig. 6, case 3), which has to be interpreted as an illuvial phase able to penetrate the underlying horizon.

Monophase pedogenesis

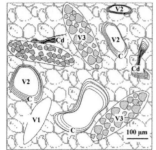

1

Absence of hierarchy between pedofeatures (clay illuvial pedogenic facies + simultaneous faunal activity)

C   microlaminated clay coating
Cd microlaminated clay coating deformed and/or burrowed by the soil fauna
V1 bare void
V2 coated void
V3 void infilled by faecal pellets

C1 type 1 of textural features
C2 type 2 of textural features
V1 coated void
V2 bare void

Biphase pedogenesis (two concordant phases)

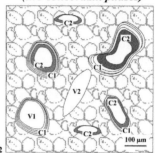

2

Hierarchy between two types of juxtaposed and concordant textural features (two different phases of illuviation which occurred successively)

Biphase pedogenesis (two discordant phases)

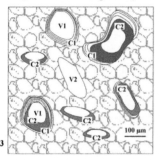

3

Hierarchy between two types of juxtaposed, but discordant textural features (two different phases of illuviation separated by a rhexistasic? event)

C1 type 1 of textural features
C2 type 2 of textural features
V1 coated void
V2 bare void
V3 second set of voids

Biphase pedogenesis (a cycle if C1 and C2 are similar)

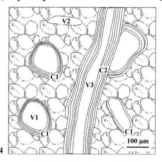

4

as 3, but the discordance is clearly expressed

Biphase pedogenesis (a cycle if C1f and C2 are similar)

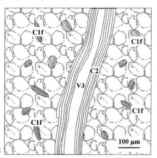

5

as 3 and 4, but the first type of textural features is fragmented indicating an in-situ reworking or a mass transport (the rhexistasic episode is clearly identified)

C1f fragment of type 1textural features
C2 type 2 textural features
V3 second set of voids

FIGURE 5 Types of hierarchy involving one or two sets of textural features, observed for the Bt3 horizon of Luvisols in Western Europe, developed during the Early Holocene (case 1), the Pleistocene-Holocene transition (cases 2 and 3), and the Late Pleistocene (cases 4 and 5).

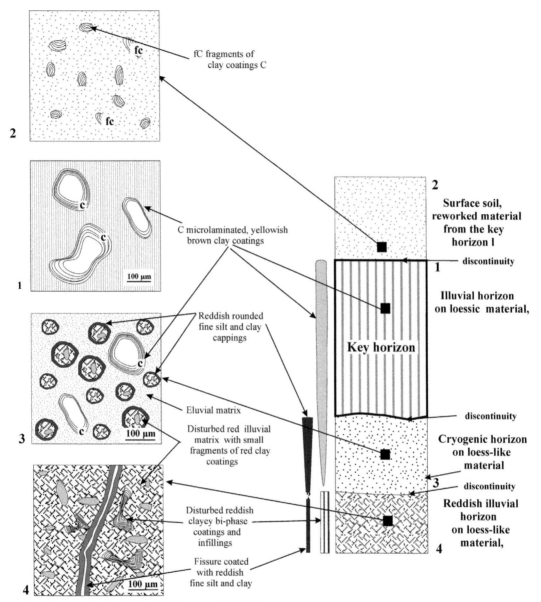

FIGURE 6 Example of relationships between illuvial facies in a sequence of superimposed deeply altered loessic materials and palaeosoils (cf. Fig. 1, case 3), as observed for Late Pleistocene sequences in the Po valley, Italy (see also Cremaschi, 1991).

- Rounded fine silt and clay cappings (Fig. 7B), discordant relative to the clay coatings, are present in the underlying horizon (Fig. 6, case 3); they record a cold phase which has induced permafrost conditions.
- Three or more types of textural features, usually discordant and overlapping, are observed in lower horizons as a result of successive pedogenic phases (Fig. 6, case

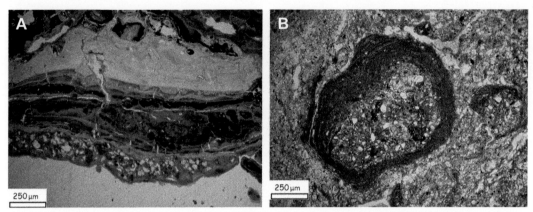

FIGURE 7 Hierarchy of textural features. (A) Layered deposit showing the following hierarchy, from the ferruginous nodule to the void: (i) whitish grey clayey mass; (ii) irregular, red clayey laminae with embedded reddish brown aggregates; (iii) reddish and whitish microlaminae; (iv) packing of reddish brown aggregates; and (v) red clay microlaminae (nodular horizon of a laterite profile, Dogon lowlands, Mali) (PPL). The whitish grey mass corresponds to a high groundwater stand, probably in relation with a humid climate, whereas the following layer indicates a well-drained soil and the third one corresponds to a new phase of high groundwater levels; the reddish brown aggregates again suggest a well-drained soil, whereas the quartz grains, especially abundant splinters, indicate an aeolian episode; the red clay laminae correspond to a well-drained soil in a stable landscape. (B) Rounded silt capping, around an aggregate embedded in a groundmass affected by clay eluviation, from a layer just above a fragmented argic horizon (see Fig. 15) (Po valley, northern Italy) (PPL). The feature is inherited from a cold episode.

4, 7a, 8). Fine silt and clay coatings in fissures occur as an extension in another form of cappings that are present at higher levels, whereas disturbed reddish clayey coatings and infillings belong to two illuvial phases. Such assemblages of textural features commonly occur in loess that is strongly affected by soil development, e.g., in northern Italy, where loess aggradation was moderate and where climatic factors were favourable for illuviation even during cold periods (Cremaschi, 1991).

Different textural feature types organised according to a well-expressed hierarchy can merge with greater depth into a single type. Such assemblage can be observed in relict soils developed on deeply karstified limestone (Atalay, 1997; Fedoroff, 1997). One type of textural feature can be present throughout a very thick horizon (e.g., Yin & Guo, 2006).

3.2 Ferruginous Features

Horizons characterised by ferruginous nodules, pisoliths and continuous crusts (ferricretes or laterites), corresponding to ferric, plinthic, petroplinthic and pisoplinthic diagnostic horizons of the World Reference Base (IUSS Working Group WRB, 2006) appear to consist in thin sections of a great variety of ferruginous features and facies assembled according to various hierarchies (Delvigne, 1998; Stoops & Marcelino, 2010), which are mainly inherited. The main types are described below.

A first type is iron oxide impregnations of an autochthonous material, which are either continuous or discontinuous in the form of mottles (Fig. 9, case 1.1). The coarse material can consist of gravel that represents the primary ferricrete (Fig. 9, case 1.2). A hierarchy is usually absent in both types. These forms should be considered as inherited, based on their location in the profile and their relationships with present-day ground-water levels. Discontinuous and continuous cementation result from *in situ* accretion of iron oxides, which occurred during a period of high but seasonally fluctuating ground-water rich in chelates (Schwertmann, 1985). Both types have to be considered as monophase plinthite, or as monophase petroplinthite if it is petrified. The host material is supposed to be free from ferruginous grains and gravels.

Ferruginous nodules characterised by sharp boundaries and a random distribution, in a different host material (Fig. 9, case 2), indicate severe reworking or erosion of a plinthitic soil, of which the fine fraction has been reworked or carried away and the mottles have become rounded. In this way, the soil has lost its less resistant parts while the most resistant constituents were preserved in the form of nodules. The nodules in such a horizon are thus inherited. In nodular facies that formed *in situ*, nodules are uniform in morphology and size, whereas in the case of transported material the facies consists of nodules of different types and dimensions.

Nodular horizons (Fig. 8) are frequently characterised by two and even more types of nodules between which a precise hierarchy cannot be established. Red, black and brown nodules have been described for a nodular horizon of a relict Plinthosol in Youth Island, Cuba (Gonzalez, 1991). The red nodules impregnate the same clayey material as red mottles that formed *in situ* at the base of the profile. From bottom to top of the profile, reddening of the nodules increases, their boundaries become sharper, their size decreases, and their shape becomes rounded. The red nodules undoubtedly belong to the *in situ* profile, or they may have been transported from a similar profile higher on the slope. The black nodules are opaque, masking the impregnated material. They appear to consist of pure oxides, and they contain abundant polyconcave vughs, partly coated by gibbsite crystals. Black nodules do not exist at present in any soils of the study area. Consequently, it is assumed that they were formed in Plinthosols, developed during a period with very high rainfall, favouring alteration. These Plinthosols were later completely eroded, whereby the resistant black nodules were preserved and scattered on the soil surface. The brown nodules are in fact pisoliths, as discussed below.

Ferruginous impregnation of the interstitial material of a nodular facies, usually in the form of bridges between the nodules (Fig. 9, case 3), has to be considered as a two-phase ferricrete. A rise of a fluctuating water table in a nodular horizon is responsible for such secondary ferruginous aggradation. Fragments of this type of two-phase ferricrete can be embedded in a new ferruginous impregnation (Fig. 9, case 4), producing a three-phase ferricrete.

Pisoliths, which are common in surface horizons in the tropics (Gonzalez, 1991; Achyuthan & Fedoroff, 2008), consist of a nucleus and a cortex (Delvigne, 1998), which have to be considered as being a part of a hierarchy (Figs. 10 and 11). The nature of the

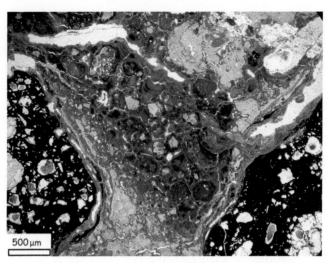

500 µm

FIGURE 8 Hierarchy of ferruginous features. Monophase, homogeneous, black, opaque nodules, with strongly weathered quartz grains (partly dissolved); the material between nodules comprises subrounded aggregates of whitish grey clayey material and dark red clayey aggregates with silt-sized quartz grains, and it also includes three generations of juxtaposed, cross-bedded clay coatings (Dogon lowlands, Mali) (PPL). The nodules were formed as part of a continuous or semicontinuous ferricrete, during a single phase characterised by a fluctuating water table; the ferricrete was subsequently disjointed and only the nodules were preserved; the sequence of events recorded for the material between nodules, is comparable to the one described for Fig. 7A.

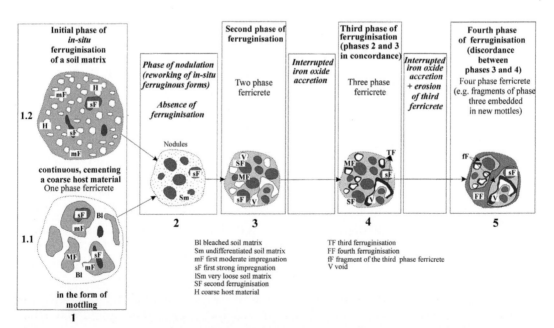

Bl bleached soil matrix
Sm undifferentiated soil matrix
mF first moderate impregnation
sF first strong impregnation
lSm very loose soil matrix
SF second ferruginisation
H coarse host material

TF third ferruginisation
FF fourth ferruginisation
fF fragment of the third phase ferricrete
V void

FIGURE 9 Schematised sequence of the evolution of a ferricrete (see also Achyuthan & Fedoroff, 2008).

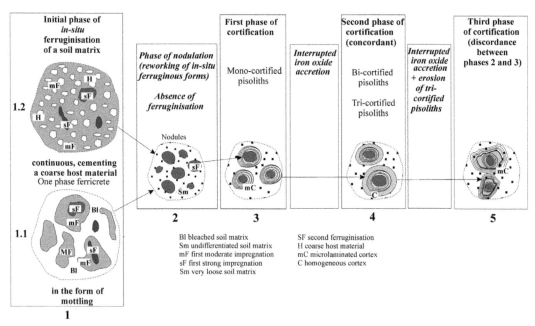

FIGURE 10 Schematised sequence of the evolution of pisoliths (see also Achyuthan & Fedoroff, 2008).

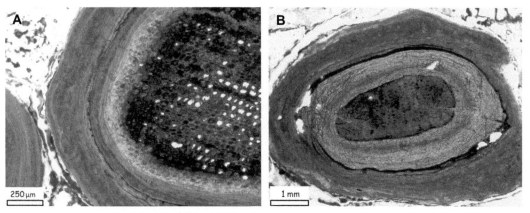

FIGURE 11 Ferruginous pisoliths (Youth island, Cuba). (A) Biphase pisolith with a ferruginised charcoal core and a microlaminated cortex, corresponding to two different phases of iron oxide accretion (PPL). (B) Four-phase pisolith, consisting of a red ferruginised charcoal core and a three-member cortex composed of two microlaminated members separated by a dark brown quasi-opaque layer (PPL). The pisolith records four phases of iron oxide accretion, whereby the second and fourth phase appear to be similar.

nucleus varies widely. Examples include the following: (i) nodules originating from mottles present in the mottled clay horizon of the investigated profile (Gonzalez, 1991) (Fig. 10, case 1); (ii) ferruginised fragments of a weathered bedrock, e.g., in ferricretes of southeastern Deccan with nuclei of weathered charnockite that occurs in outcrops at a distance of 20 km (Achuythan & Fedoroff, 2008); (iii) fragments of soil horizons; and (iv) charcoal fragments. Charcoal as pisoliths nucleus is quite common (Fig. 11). Gonzalez (1991) mentions them as occurring in considerable abundance in Youth Island, Cuba, and they have also been observed in West Africa (Eschenbrenner, 1987; Delvigne, 1998) and southeastern Deccan (Achuythan & Fedoroff, 2008). These charcoal fragments indicate wildfires, which can affect periodically wet zones (Goldammer & Seibert, 1989) and which could correspond to abrupt events (Kennett et al., 2008). The cortex of pisoliths has no relation with the nucleus. It is generally laminated or microlaminated (Fig. 10, case 3 and 4). Clay particles are commonly accreted together with the iron oxides. The cortex frequently consists of two or more types of laminae of different colours, e.g., red and dark brown, which may be concordant or discordant (Fig. 10, case 4 and 5). Cortification probably occurs in a porous surface horizon during a high groundwater stand, whereby ferruginous nodules act as nuclei for bacterial precipitation of iron oxides from water rich in chelates (Emerson & Revsbech, 1994) and clay particles are trapped in the oxides (see also Stoops & Marcelino, 2010). Alternation of red (hematite) and dark brown (goethite) laminae probably results from fluctuations from warm humid to cooler humid periods (Berner, 1969; Bondeulle & Muller, 1988). Cross-bedding indicates erosive episodes during which the soft materials were eroded while the pisoliths were reworked or transported (Fig. 10, case 5). A final phase of ferruginous bridging can cement the pisoliths. Very few radiometric dates have been obtained for the cortex of ferruginous pisoliths, but accretion is probably a slow process (e.g., 0.01-0.02 µm/year; Bernal et al., 2006).

3.3   Calcitic Facies

Buried or relict calcic horizons and petrocalcic horizons (IUSS Working Group WRB, 2006) occur in geological series from the Precambrian (Melezhik et al., 2004; Lewis et al., 2008) to the Holocene. Petrocalcic horizons or calcretes are described in numerous publications (e.g., Milnes, 1992; Nash & Smith, 2003; McLaren, 2004) (see also Durand et al., 2010). Micromorphological investigations have shown the common complexity of most of these horizons (e.g., Candy et al., 2003). Here we propose a classification of calcitic features and facies based on the concept of hierarchy.

Three main primary types of calcitic features occurring in soils can be distinguished (Fedoroff & Courty, 1994): (i) discrete calcitic features of biogenic origin; (ii) sparitic to micritic crystallisations in the form of nodules of various sizes, merging eventually in continuous horizons in the vadose zone, as a result of evaporation of water saturated with respect to calcite; and (iii) lamellar surface crusts. Like other soil-forming processes, accretion of calcite was discontinuous at geological timescales. Discontinuities result from variations in water table depth, intensity of evaporation, soil water composition and intensity of calcitic dust aggradation. The most common types of hierarchy of calcitic features and facies in palaeosoils and relict soils are as follows (Fig. 12):

- Absence of hierarchy, with the calcitic facies consisting of simple features which can be of biogenic origin and/or consist of sparitic or micritic crystallisations

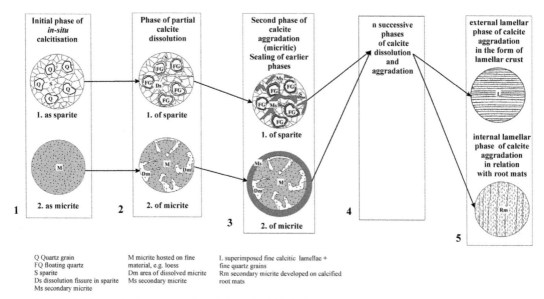

Q Quartz grain
FQ floating quartz
S sparite
Ds dissolution fissure in sparite
Ms secondary micrite

M micrite hosted on fine
material, e.g. loess
Dm area of dissolved micrite
Ms secondary micrite

L superimposed fine calcitic lamellae +
fine quartz grains
Rm secondary micrite developed on calcified
root mats

FIGURE 12 Schematised sequence of evolution of calcitic facies.

(Fig. 12, case 1). Such facies must be considered as monophase. The common secondary micritisation of primary biogenic forms should be considered as belonging to the same phase as the initial calcitic aggradation (Fedoroff et al., 1994). Such primary facies are common in palaeosoils of the Loess Plateau of China, where discrete biogenic features or nodules of various sizes are monophase (Guo, 1990). The nodules locally merge into continuous monophase calcrete.

- A two-phase hierarchy, consisting of calcitic aggradation followed by partial or total dissolution (Fig. 12, case 2). Calcite dissolution produces various features. The most common are (i) partial diffuse dissolution of a sparitic groundmass expressed by floating detrital grains (e.g., Robinson et al., 2002), (ii) dissolution of a calcitic groundmass along fissures, which can be filled by illuvial clay, (iii) scoriaceous hard nodules (Guo, 1990), characterised by dissolution along the edges of the nodule, (iv) partial dissolution of soft nodules, resulting in a less pronounced crystallitic b-fabric, and (v) complete dissolution of nodules, whereby the external form is preserved (mouldic pores) (Fig. 13). Examples of two-phase features can be found in palaeosoils of the Loess Plateau of China (Guo, 1990).

- A three-phase hierarchy, consisting of a two-phase hierarchy as described above, followed by a new phase of calcitic aggradation, most frequently as micritic calcite (Fig. 12, case 3). This new phase can be biogenic (e.g., calcified root mats) or non-biogenic (e.g., sparitic cement).

Polygenetic calcitic assemblages are different from assemblages in ferricretes because after a few phases of calcite aggradation and dissolution the calcitic groundmass

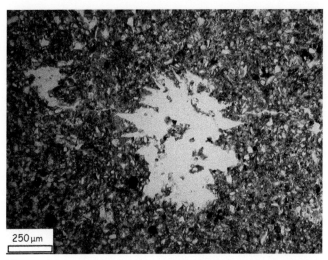

FIGURE 13 mouldic voids, pseudomorphic after gypsum crystal intergrowths, in the upper part of an undisturbed Kastanozem (Loess Plateau of China, Xifeng section) (PPL). The mineral probably crystallised during the transition from MIS 5 to MIS 4 (see Fig. 20), and it was later dissolved.

becomes petrified and sealed for water penetration (e.g., Wright et al., 1993; Durand et al., 2006). Further phases of calcification can occur along the top of the horizon and on walls of vertical fissures in the form of lamellar crusts. Each lamella of a surface crust is characterised by a dense, micritic groundmass in which one can recognise residues of cyanobacteria, green algae and fungi (Wright et al., 1996), as well as detrital silt-sized quartz, frequently in the form of splinters (Smalley & Vita-Finzi, 1968). Between the massive calcrete and the lamellar crust, rounded calcitic aggregates are frequently observed. Development of these lamellar crusts can be interpreted to begin with erosion of the friable upper horizon of the calcitic soil down to the petrified horizon, during a rhexistasic episode. Rounded calcitic aggregates are probably remnants of this erosion, which was mainly aeolian. A biogenic surface crust subsequently colonised the eroded soil, protecting it from further erosion. Biogenic filaments trapped aeolian dust from which carbonates were derived and reprecipitated, cementing the crust (Verrecchia et al., 1995). When a lamella was hardened, a new one developed on top. Lamellar crusts are assumed to have developed during periods that were arid and probably windy (Stahr et al., 2000).

Internal lamellar crusts can be related to climate fluctuations. During a wetter period, the petrified horizon was dissolved along fissures, whereas during the following arid period, calcium carbonate, probably supplied by aeolian dusts, was in excess and precipitated in various forms. The alternation of the wet and arid sequence may be responsible for the lamellar structure (Alonso-Zarza, 1999).

4. Reworked Materials

Soil reworking, soil erosion, transportation and redeposition of soil materials, and allochthonous sediment aggradation appear to be underestimated in palaeopedology. Careful analysis of palaeosoils and relict soils in thin sections shows that soil covers of the past were frequently, to some extent, affected by these processes (e.g., Kemp et al., 2004). Exceptions include palaeosoils buried by man (e.g., soils under burial mounds) and by tephra.

4.1   *In Situ* Soil Reworking and Mass Transportation

A continuum exists between *in situ* reworked soils and mass-transported soil materials (Lacerda, 2007) (Fig. 14, case 1). Both are frequently difficult to recognise and distinguish in the field. In thin sections, both are recognised by one or more of the following characteristics: (i) a massive microstructure with variable abundance of closed poly-concave vughs, frequently grading to vesicles; (ii) a granular microstructure with rounded to subrounded aggregates which are not of excremental origin; (iii) a more or less homogenised groundmass in which the colur of the parent soil is preserved; (iv) the presence of fragmented pedofeatures; and (v) the presence of clayey to silty clay intercalations or of thin unlaminated clay coatings in vesicles. Ferruginous and calcitic pedofeatures tend be rounded with sharp boundaries. Fragments of clay textural features are angular to subangular, with little or no deformation, the latter resulting in the preservation of their original aspect in XPL (see also Mücher et al., 2010).

Thin section analyses contribute to the identification of the cause of disruption of the parent soil. Soil can be disrupted by water saturation, inducing soil material dispersion and leading to the collapse of the soil structure. Completely dispersed soil material having reached thixotropy is characterised by a homogenised groundmass and by an apedal microstructure, with occluded polyconcave vughs or vesicles. In partially dispersed soil material, rounded aggregates inherited from the parent soil are embedded in dispersed material. The thixotropic state may occur in moving groundwater, in which case textural intercalations would indicate groundwater movement along selected pathways through the groundmass. A thixotropic state also occurs above a permafrost and even above a zone of seasonal deep frost (Fig. 15). Cryoturbation results in structure collapse and local displacements (e.g., Sanborn et al., 2006). Earthquakes can induce structure collapse when the soil or the subsoil is water-saturated (e.g., Wolf et al., 2006). Disruption of palaeosoils can also result from a violent airburst due to the impact of a cosmic bolide (Courty et al., 2008). In thin sections, such disruption can be expressed as loose packing of angular to subangular aggregates of various sizes, initially described as frost-shattered soil (Guo, 1990).

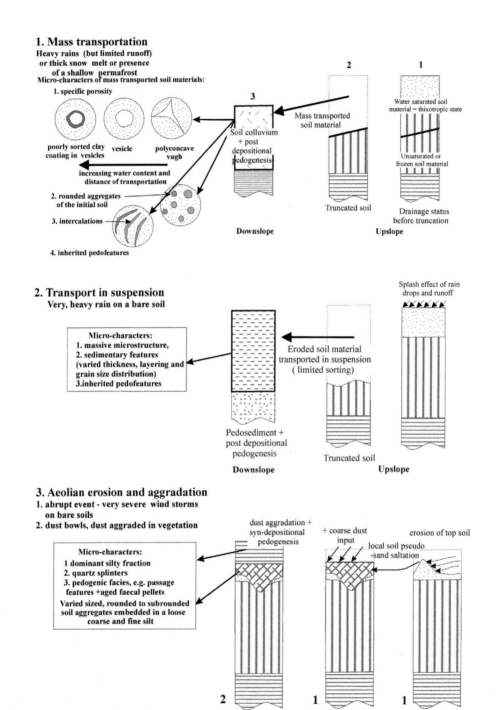

FIGURE 14 Macromorphological and micromorphological characteristics of a mass-transported soil, soil materials transported in suspension, and a soil affected by aeolian erosion and aggradation.

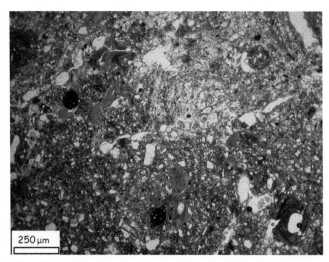

FIGURE 15 Disturbed Chromic Luvisol, showing three types of reddish illuvial features: (i) small rounded fragments, randomly distributed in the reddish groundmass; (ii) weakly disturbed large infillings; and (iii) undisturbed clay and fine silt infillings (Po valley, northern Italy) (PPL). The fragmented facies is considered to be the result of deep frost action; illuviation has gone through different phases (i) development of a Chromic Luvisol, of which the small rounded red fragments are remnants; (ii) churning of the Luvisol; (iii) a new illuvial phase, recorded by the weakly disturbed large infillings; and (iv) another illuvial phase, during which silt grains were also translocated.

Mass-transported water-saturated soil material can be distinguished from an *in situ* reworked soil by (i) rounded aggregates belonging either to deep soil horizons or to the parent material (clay balls); (ii) a greater homogenisation of the groundmass in which randomly distributed, sorted fragments of pedofeatures can be present; and (iii) the presence of dusty, poorly oriented clay coatings in vesicles. The latter result from deposition of water-suspended clays in residual voids after the mass-transported material has been stabilised. They indicate mass transportation in a rather liquid form, as in present-day desert mudflows. Clay balls (Fig. 16) are torn from the dry floor on which the mass-transported material has slid down. If transportation results from collapse, the soil material consists of angular to subangular aggregates derived from various horizons of the eroded soil and in some cases from the underlying material (Rust & Nanson, 1989).

Soil micromorphology has revealed that lower horizons of deep polyphase soils are often characterised by clay fragments (see Fig. 5, case 5), whose abundance varies from scattered occurrences ($\sim$1%) to high concentrations in materials composed of an almost pure accumulation of fragments. They exist from midlatitudes to the tropics. In the Rambouillet forest of the Paris Basin, such horizons are C horizons of Gleyic Luvisols developed over a thin loess cover that was deposited during various episodes of the last and penultimate glacial cycles. These loess layers, which are strongly affected by several phases of pedogenesis dominated by clay illuviation (Fedoroff, 1968), were either

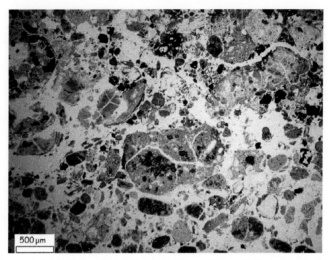

FIGURE 16 Pedosediment derived from a Vertisol, in the form of rounded aggregates (clay balls) mixed with volcanic fragments, with associated opaque aggregates (Sangiran dome, Java, Indonesia) (PPL). The clay balls result from heavy rain, affecting a Vertisol fragmented by desiccation; the opaque aggregates are probably heated soil material.

reworked or mass-transported, as a result of deep soil freezing or permafrost conditions during the glacial maximal. Such horizons have also been reported for Red Mediterranean soils (Chromic Luvisols), especially those overlying calcretes (Mücher et al., 1972), and for some tropical soils (e.g., Boulet et al., 1986) (Fig. 17). This suggests that soil covers were periodically destabilised by drastic climatic events, such as a drought followed by heavy rains. Clay-with-flints (argiles à silex), which occurs in patches over a large area in southern England and the Paris basin, consists, to a some extent, of chalk dissolution residues and mainly of Cenozoic sediments deeply affected by soil-forming processes (Pepper, 1973; Laignel et al., 1998). Pedological aspects of these formations have only rarely been investigated (Stoops & Mathieu, 1970; Thorez et al., 1971).

4.2 Transport in Suspension

Soil materials transported in suspension and deposited as pedosediments are easily distinguished in thin sections from mass-transported materials by a massive microstructure, a groundmass whose colour is close to that of the parent soil, and horizontal sedimentary layering with layers of variable thickness and texture (Figs. 14, case 2 and 18). Inherited pedofeatures can be present, but in lesser abundance than in mass-transported soil materials. Ferruginous nodules, because of their resistance to both

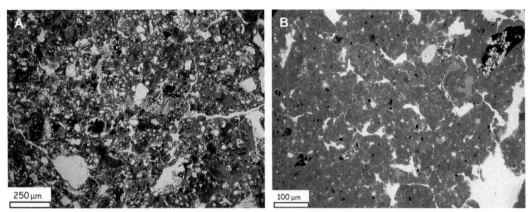

FIGURE 17 Disturbed facies, at the bottom of the transition from a weathered zone to a ferralic horizon (Misiones province, Argentina) (PPL). (A) Soil consisting of coalescent aggregates derived from (i) dark red coatings and infillings, (ii) ferruginised weathered basalt, and (iii) whitish grey clayey material. This occurrence of a disturbed facies, in the tropics, cannot be considered as a result of a deep frost, in contrast to occurrences in midlatitudes (see Fig. 15). (B) Partially homogenised groundmass, comprising (i) aggregates with varying degree of compaction, (ii) an aggregate with red clay infillings, (iii) fragments of ferruginous features, (iv) weathered bedrock fragments, and (v) quartz grains, partly in the form of splinters. This facies is considered to be an aeolian soil-sediment that originates from ferralic soil developed on weathered basalts (see also Iriondo & Kröhling, 2007); the undisturbed red clay infillings are postdepositional.

physical disaggregation and chemical weathering, can help in identifying the source of the pedosediment. Flood-suspended particles can penetrate inside coarse deposits within river beds, where they are deposited in packing voids.

4.3   Aeolian Processes

The widespread occurrence of features resulting from aeolian processes shows that wind action on the soil cover is quite common at present and was episodically very strong in the past. These features occur very widely, including occurrences in the humid tropics (e.g., Iriondo & Kröhling, 1997, 2007) (see Fig. 17B). Quartz exoscopy (e.g., Le Ribault, 1977), using scanning electron microscopy, is one of the methods that contribute to identifying aeolian episodes in palaeosoils and relict soils.

Well-sorted coarse silt in which quartz splinters are common is a good indicator for aeolian dust aggradation (Smalley & Vita-Finzi, 1968). Thin section observations have shown that many soil covers in the past were affected by minor dust aggradation that was not sufficiently predominant to create typical loess. Such minor aggradations have been observed for interglacial palaeosoils (e.g., Guo, 1990) and in areas adjacent to loess deposits, but they also exist in areas where loess is absent, such as the humid tropics (e.g., Berger et al., 1994).

Some clay coatings and infillings present in non-Luvisol-related soils have been attributed to aeolian dust (Brimhall & Lewis, 1992), for example, abundant thick clay infillings in ferricretes of the Ilgorn Plateau, southwestern Australia. The question of fine dust penetration into soils is not resolved. For instance, it is still unclear how the fine dust from the Sahara that is deposited along the northern side of the Mediterranean Sea is incorporated into soils.

Various forms of calcite, gypsum and more soluble minerals are common in palaeosoils and relict soils of arid to subhumid areas. When occurring as isolated grains, their external morphology can result from aeolian processes, e.g., rounded grains produced by wind winnowing. After their deposition on the soil surface, most of these mineral occurrences are dissolved, followed by transport of ions in solution and by precipitation at depth.

The aeolian origin of lunette dunes is well known (Cooke et al., 1993). They are characterised by packing of rounded aggregates or crystals (e.g., gypsum crystals) that originate from the adjacent wind-deflated sabkha (playa) and morphology of which can be more or less altered by postdepositional soil-forming processes (Hachicha et al., 1987). Lunette-like aeolian features can also exist without being associated with a sabkha. Along the coast of southern Israel, there is a south-to-north sequence from typical loess, over loess containing rounded soil aggregates, to Vertisols (Rognon et al., 1987). In many tropical soils, thin sections reveal characteristics of aeolian origin, for instance, the presence of silt-sized quartz grains, some of them in the form of splinters, in Ferralsols developed on large basaltic plateaus in northern Argentina (Iriondo & Kröhling, 1997) (see Fig. 17B). The parent material of the entire soil profile can be aeolian in origin, as in the Sahel of western Africa (Coudé-Gaussen, 1987; McTainsh et al., 1997) and in southern India (Achyuthan & Fedoroff, 2008). The aeolian origin of these relict tropical soils ('sols ferrugineux tropicaux' in western Africa; Bertrand, 1998) is indicated in thin sections by one or more of the following characteristics: (i) homogeneity of the groundmass throughout the profile; (ii) rounded aggregates, more or less altered by postdepositional soil-forming processes; (iii) association of predominantly oxic characteristics with calcitic features; and (iv) inherited rounded ferruginous features that are randomly distributed (Achyuthan & Fedoroff, 2008).

5. Palaeoenvironmental Significance

Many papers have demonstrated a close relationship between climate and soil type, but opinions diverge about climate impact on the evolution of past soil covers. Soil micromorphology has contributed considerably to this debate for Quaternary palaeosoils (e.g., Fedoroff et al., 1990; Kemp, 2001; Kemp et al., 2001; Felix-Henningsen & Mauz, 2004) as well as for older formations (e.g., Driese & Ober, 2005).

Deciphering the environmental significance of palaeosoils and relict soils is usually based on comparison with modern analogues. Some palaeosoils can be interpreted adequately by applying this concept. For instance, Holocene Luvisols can be used as

FIGURE 18 Layered silt and silty clay crust, crossed by a passage feature (left). The coarser grained layers have elongated horizontal voids, due to ice lensing (Pampean pleniglacial loess, Tortugas quarry, Argentina) (PPL). The crust consists of loess reworked by runoff, deposited in the form of a sedimentary surface crust.

modern analogues for interglacial palaeo-Luvisols, both being characterised by micro-laminated clay coatings and infillings. Palaeosoils developed during late Pleistocene interglacials (e.g., MIS 5) in the Chinese Loess Plateau region are typic Luvisols (Guo, 1990).

Palaeosoils and related formations frequently present pedofeature assemblages and facies for which no modern analogues are known. For instance, large channels (a few centimetres in diameter) in soils on loess that developed during glacial cycles in northern Italy are filled, from bottom to top, by (i) cross-bedded microlaminated clay and fine silt (Fig. 19, case 4.1); (ii) massive coarse clay (Fig. 19, case 4.2); (iii) massive coarse silt with embedded silt-sized fragments of clay-illuvial features, discordantly covering dusty clay (Fig. 19, case 4.3); and (iv) a fine layer of clay and fine silt (Fig. 19, case 4.4). Similar channels were observed in Périgord, southwestern France (Sellami, 1999). The coarse silt with embedded clayey fragments could result from an abrupt event such as a Heinrich event or the Younger Dryas. Fully satisfactory interpretations for such features are still lacking. Cross-bedded microlaminated clay and fine silt coatings occur presently in soils covered by a thick snow cover melting rapidly in spring (Fedoroff et al., 1981).

Collisions with extraterrestrial bolides have been suggested to have had a regional or global impact on soil covers (Bunopas et al., 2001). Ufnar et al. (2001) described features related to the K/T impact in Cretaceous palaeosoils. Courty et al. (2008), using soil

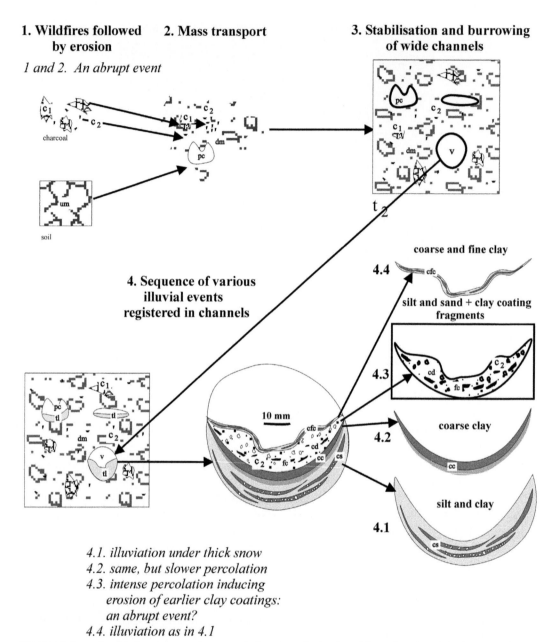

FIGURE 19 Example of facies, unknown in presently forming soils, in altered loess deposits of the last glacial cycle (Po valley, Italy).

micromorphology, have proposed a method for deciphering the effects of cosmic impacts on soils. Such impacts in soils are identified by one or more of the following characteristics: (i) groundmass disruption; (ii) heated fragments; (iii) charcoal fragments; and (iv) allochtonuous materials such as microtektites, glass shards and bituminous components. Four tektite-strewn areas resulting from such collisions are known in for the Cenozoic and Quaternary (McCall, 2001), but pedological aspects have not been investigated.

Heated fragments of palaeosoils and *in situ* heated palaeosoils have been recognised, for instance, as subangular fragments and ellipsoidal aggregates in Java, where they are related to an occurrence of tektites (Courty et al., 2007) (see Fig. 16). In this study, heating, up to a few hundred degrees Celsius, is recognised in thin sections by (i) a complete loss of birefringence of the originally smectitic groundmass; (ii) an opaque aspect of the groundmass (Courty et al., 1989); and (iii) compaction of ellipsoidal aggregates probably by heating. In this case, heating is thought to result of a blast of hot air resulting from a meteorite impact. Heat-transformed palaeosoils have also been described for volcanic regions (Usai, 1996).

Charcoal fragments that cannot be the result of human activity are common in palaeosoils (Van Vliet-Lanoë, 1976) and relict soils. Their abundance is underestimated, or their presence is not even recognised, when micromorphological investigations are not carried out. They can be in the form of (i) randomly distributed or aligned fragments in reworked soil material; (ii) nodules consisting of ferruginised charcoal fragments (see Fig. 11); or (iii) silty textural features containing abundant small charcoal fragments. Charcoal fragments are at least partly related to wildfires associated with discontinuities in soil development (rhexistasic episodes). Some extensive wildfires, such as those of the Younger Dryas, may have been caused by cosmic impacts (Kennett et al., 2008).

Layers of tephra are common in palaeosoils and related sediments, but they have only been used as a stratigraphic tool. The impact of major volcanic eruptions such as the 74,000 ka Toba eruption (Rampino & Self, 1993) on soils is unknown.

6. Transitions in Palaeosoil Sequences and Their Significance

Various types of transitions exist in palaeosoil sequences, which have to be distinguished from pedogenic horizon transitions. They can be in the form of planar abrupt truncations, or the discontinuity can be irregular, wavy or in the form of tongues. The transition can also be very progressive, even undetectable in thin sections, although radiometric dates indicate a gap of thousands to hundred thousand of years between levels separated by only a few centimetres. Transitions are frequently emphasised by stone lines, especially in the tropics. To interpret the transition in palaeosoil sequences, the underlying and overlying layers have to be compared (see Fig. 17). Earthquakes can also be responsible for the presence of discontinuities (e.g., Wolf et al., 2006).

The best documented example of transition is the passage from the last interglacial pedocomplex to the pleniglacial loess (MIS 5 to MIS 4) as it occurs in the Loess Plateau of China (Guo, 1990; Porter & An Zhisheng, 1995). This transition is characterised in thin sections by a layer with a coarse disrupted assemblage (Fig. 20A), consisting of abundant small rounded aggregates derived from the underlying Kastanozem (see Fig. 4A), randomly embedded in a coarse loessic material rich in detrital calcite grains

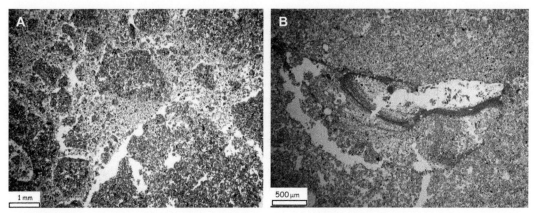

FIGURE 20 Facies recognised for the transition from the last interglacial pedocomplex (MIS 5) to the pleniglacial loess (MIS 4) (Loess Plateau of China, Xifeng section). (A) Fragments of a Kastanozem, embedded in a coarse-grained loose calcareous loess, containing rounded aggregates of the Kastanozem (PPL). This facies cannot be due to a deep frost and has been interpreted as the result of an airburst. (B) Blackish brown coating probably corresponding to an episode of wildfires followed by heavy rains (PPL).

(Guo, 1990). Below this layer and apparently related to it, infillings composed of blackish brown massive fine silt rich in small charcoal fragments are recognised (Fig. 20B), as well as mouldic voids after barite or gypsum crystal intergrowths (see Fig. 13). Calcitic features in the form of micritic infillings, micritic hypocoatings and sparitic infillings composed of rounded crystals in channels are present throughout the loess cover, the transition layer and underlying Kastanozem. A grain size increase occurs within the transition layer and just above in the pleniglacial loess (Porter & An Zhisheng, 1995), together with an increase in mica and heavy mineral content. A tentative reconstruction of the sequence of events, based on the hierarchy of attributes, could be the following: (i) aeolian erosion of the top of the Kastanozem, indicated by small rounded aggregates, associated with coarse loess deposition resulting from exceptional dust storms, more severe than those responsible for the formation of the main loess deposit; (ii) disruption of the top of upper Kastanozem, possibly resulting from an air blast, which shattered the soil to some centimetres depth; (iii) an episode of wildfires followed by heavy rains, which is registered by the blackish brown infillings; (iv) accumulation of gypsum or barite together with coarse loess, followed by dissolution during a wetter episode and by precipitation as crystal intergrowths in the Kastanozem, which were later dissolved during a wetter period of unknown age; and (v) development of calcitic features, apparently related to deposition of the main loess deposit, independent from the air blast. A similar type of transition from MIS 5 to MIS 4 in loess and palaeosoil sequences has been observed in the Rhine valley (Rousseau et al., 2002) and in Pampean loess sections (Fig. 21) (Rosario, Argentina). In the latter study, rounded fragments of a Luvisol, probably belonging to the MIS 5 pedocomplex, are embedded in a coarse loess (Kemp et al., 2004).

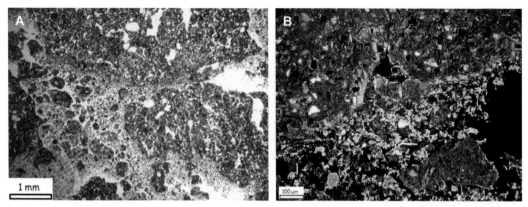

FIGURE 21 Facies recognised for the transition from a well-developed Luvisol to a pleniglacial loess (Pampean loess, Tortugas quarry, Argentina). (A) Fragments of an argillic horizon characterised by thin pale brown clay coatings, embedded in a coarse-grained loose loess. The facies is comparable to that recognised for the MIS 5 to MIS 4 transition in China, but here the preexisting soil is a Luvisol. (PPL) (B). Detail of the previous showing the occurrence of acicular and rod-shaped calcite crystals (XPL).

7. Reconstruction of the History of Relict Soils

Reconstructing the history of relict soils requires first the identification of present-day pedogenic attributes, then detection of all relict attributes from the soil surface down to the unaltered parent material and finally the establishment of hierarchies between functioning and inherited attributes starting from the bottom of the profile. The palaeoenvironmental significance of each of these events should be considered at the end.

This is illustrated by the example of a Haplic Luvisol developed in loess deposited during MIS 2 in the Paris Basin, which, like many soils of soil covers of midlatitudes, contains *in situ* relict features (Fig. 22). In this soil, undisturbed relict characteristics are recognised in the B3t and C horizons, whereas fragments of clay coatings occur in the B2t horizon (Fedoroff, 1968). In thin sections (Fig. 22, case 1), the groundmass of the B3t horizon is a non-calcareous loess with local zones of ice lensing, crossed by abundant channels. The most common channels (V1) are coated by a dark brown, rather dusty, moderately well-oriented, massive clay coatings, juxtaposed concordantly by yellowish brown, well-oriented, translucent, microlaminated clay coatings. A few channels (V2) are lined only by massive coatings, and others (V3) only by microlaminated coatings. Some channels (V4) are uncoated, and others (V5) are filled with excrements which can also be present in V2 and V3 channels. The hierarchy of features is from the groundmass to the middle of channels: (i) zones of ice lensing; (ii) dark brown massive fine silt and clay coatings; and (iii) yellowish brown microlaminated clay coatings. Observations for channels V2, V3 and V4 are apparently not in agreement with this hierarchy. The explanation could be that channels V2 became occluded when sedimentation of yellowish brown translucent clay started, whereas channels V3 developed after the

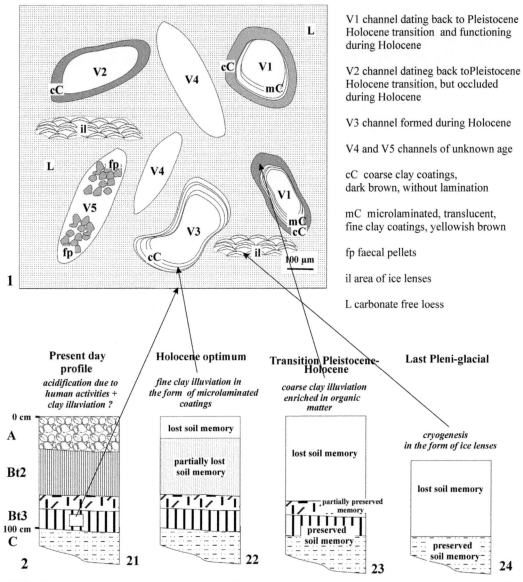

FIGURE 22 Reconstruction of the history of a Haplic Luvisol on loess, based on the micromorphology of its B3t horizon.

formation of the massive coatings. The history of this horizon could be described as the following sequence of events: (i) development of ice lenses during or immediately after loess aggradation; (ii) a first pedogenic phase characterised by dark brown massive fine silt and clay coatings; and (iii) a second pedogenic phase characterised by yellowish brown microlaminated clay coatings. Ice lenses indicate deep frost affecting a soil saturated with water. The dark brown coatings (cC) indicate a fundamental change in environmental conditions, which switched from an environment with very strong winds and a dusty atmosphere during loess sedimentation to biostasic conditions. The dark brown colour is due to organic matter, which was translocated together with the clay and fine silt. For this period, we can assume that the vegetation was dominated by conifers, a hypothesis based on palynological data for the MIS 2 to 1 transition in northwestern Europe. The weak sorting of fine silt and clay in these coatings and the absence of layering indicate rapid percolation in spring during snow melt. The microlaminated clay coatings are typical of Luvisols formed during the Holocene. The excrements cannot be included in this hierarchy, being formed during both illuvial phases. Figure 22 (case 2) is an attempt to reconstruct the soil profile at its main stages of development, by extrapolation of results obtained for the B3t horizon.

8. Dating Palaeosoil Development

Soil micromorphology contributes only indirectly to dating of palaeosoil development. Soil micromorphology cannot replace radiometric dates and stratigraphic correlations, but the lifetime of a polyphase palaeosoil can be estimated from the number and nature of pedogenic phases detected. For instance, in loess-palaeosoil sequences, the illuvial facies developed during the last interglacial (MIS 5e) formed over ~10,000 years (e.g., Hearty et al., 2007). The duration of the exposure of palaeosoils interstratified in continental deltaic formations has been estimated from the extent of development of pedofeatures (Srivastava & Parkash, 2002). In this study, passage features and weak iron oxide staining are considered to indicate a few years of exposure, whereas a moderately well-developed illuvial facies is assumed to require a few thousands of years to develop, by comparison with Early Holocene and MIS 5e illuvial facies. Estimation of the lifetime of relict soils is less certain as they are not included in a stratigraphic sequence and their different facies usually have different ages.

Soil micromorphology can also contribute to dating palaeosoils development by supplying radiometric dating laboratories with samples of specific microscopic features instead of bulk samples. Only a few papers combining microscopic analysis and radiometric datings have been published. Courty et al. (1994) integrated geochemical microanalysis and micromorphology to decipher the genesis of calcitic pendants in Spitsbergen, whereas Bernal et al. (2006) proposed a radiometric method for measuring the rate of accretion of ferruginous pisolith cortexes. OSL (optically stimulated luminescence) dating was also utilised in combination with microscopic investigations on an Argentinean loess-palaeosoil sequence by Kemp et al. (2003), while Kühn & Hilgers (2005) used the same approach in southern Taunus (Germany). AMS (accelerator mass spectrometry) radiocarbon dating requires only a very small amount of carbon and could be applied to small charcoal fragments that are frequently present in palaeosoils (Carcaillet, 2001).

9. Diagenesis

Diagenesis of recent palaeosoils (Middle Pleistocene to Holocene) usually only concerns biological activity and organic matter (Chichagova, 1995). As soon as a soil is buried, the soil fauna feeding on plant residues disappears. For instance, earthworm populations may disappear, followed by the consumption of earthworm excrements by mites, replacing them with mite excrements. Fresh organic fragments in a buried soil are progressively humified and tend to disappear, whereas charcoal fragments are preserved, especially if they are ferruginised. In much older palaeosoils, buried by less than a few hundred metres and not affected by tectonic movements or thermal activity, microscopic characteristics are weakly altered or unaltered by diagenesis (Retallack, 1991). In lithified palaeosoils, changes are more important, but pedogenic characteristics can still be recognised in thin sections (Retallack & Wright, 1990).

Groundwater can act as a major pedogenic factor when it occurs near the soil surface, but it can also induce diagenetic processes in buried soils, similar to near-surface processes (Gibling & Rust, 1992). Development of diagenetic attributes depends on the rate of groundwater circulation, redox conditions and ionic concentration. Textural intercalations can occur at great depth where groundwater moves along faults. Soluble salts may be either dissolved or precipitated. Some iron and manganese oxide impregnations in palaeosoils, unrelated to other pedofeatures, are undoubtedly of diagenetic origin.

Pressure resulting from deep burial combined with circulating groundwater leads to collapse of voids and ultimately to the development of a massive microstructure (Sheldon & Retallack, 2001).

10. Conclusions

Soil micromorphology has largely contributed to deciphering of palaeosoils and relict soils, but its potential is far from exhausted. In the future, systematic coupling of thin section analysis by polarised light microscopy with submicroscopic and microanalytical techniques will boost soil micromorphology in this field (Courty et al., 2008). An integration of soil micromorphology with magnetic measurements (e.g., Tsatskin et al., 2001, 2006; Maher et al., 2003), radiometric dating (e.g., Bernal et al., 2006) and stable isotope analysis (Courty et al., 1994) is also important.

References

Achyuthan, H. & Fedoroff, N., 2008. Ferricretes in Tamil Nadu, Chennai, South-Eastern India: from landscape to micromorphology, genesis, and paleoenvironmental significance. In Kapur, S., Mermut, A. & Stoops, G. (eds.), New Trends in Soil Micromorphology. Springer-Verlag, Berlin, pp. 111-136.

Alonso-Zarza, A.M., 1999. Initial stages of laminar calcrete formation by roots: examples from the Neogene of central Spain. Sedimentary Geology 126, 177-191.

Atalay, I., 1997. Red Mediterranean soils in some karstic regions of Taurus mountains, Turkey. Catena, 28, 247-260.

Berger, W.H., Yasuda, M.K., Bickert, T., Wefer, G. & Takayama, T., 1994. Quaternary time scale for the Ontong Java Plateau: Milankovitch template for Ocean Drilling Program Site 806. Geology 22, 463-467.

Bernal, J.P., Eggins, E., McCulloch, M.T., Grün, R. & Eggleton, R.A., 2006. Dating of chemical weathering processes by *in situ* measurement of U-series disequilibria in supergene Fe-oxy/hydroxides using LA-MC-ICPMS. Chemical Geology 235, 76-94.

Berner, R.A. 1969. Goethite stability and the origin of red beds. Geochimica et Cosmochimica Acta 33, 267-273.

Bertrand, R., 1998. Du Sahel à la Forêt Tropicale: Clés de Lecture des Sols dans les Paysages Ouest-africains. CIRAD, Montpellier, 272 p.

Boudeulle, M. & Muller, J.P., 1988. Structural characteristics of hematite and goethite and their relationships with kaolinite in a laterite from Cameroon. A TEM study. Bulletin de Minéralogie 111, 149-166.

Bornand, M., 1978. Altération des Matériaux Fluvio-glaciaires, Genèse et Evolution des Sols sur Terrasses Quaternaires dans la Moyenne Vallée du Rhône. Thèse d'État, Université des Sciences et Techniques du Languedoc, Montpellier, 363 p.

Boulet, R., Lucas, Y. & Lamouroux, M., 1986. Organización tridimensional de la cobertura pedológica: ejemplo del estudio de una dolina y sus immediaciones en la región de Quivican — Cuba. In Sol et Eau: Résumés des Actes du Séminaire. ORSTOM, Paris, pp. 81-133.

Brimhall, G.H. & Lewis, C., 1992. Bauxitic and lateritic ores. In Nierenberg, W.A. (ed.), Encyclopedia of Earth System Science, Volume 1. Academic Press, San Diego, pp. 321-335.

Brimhall, G.H, Lewis, C., Ford, C., Bratt, J., Taylor, G. & Warin, O., 1991. Quantitative geochemical approach to pedogenesis: importance of parent material reduction, volumetric expansion, and aeolian influx in laterization. Geoderma 51, 51-91.

Bronger, A., 2003. Correlation of loess-paleosol sequences in East and Central Asia with SE Central Europe: towards a continental Quaternary pedostratigraphy and paleoclimatic history. Quaternary International 106/107, 11-31.

Bullock, P., Fedoroff, N., Jongerius, A., Stoops, G., Tursina, T. & Babel, U., 1985. Handbook for Soil Thin Section Description. Waine Research Publications, Wolverhampton, 152 p.

Bunopas, S., Vella, P., Hada, S., Fontaine, H., Burrett, C., Haines, P., Khositanont, S. & Howard, K.T., 2001. Australasian tektites and catastrophic products enclosed in impact ejecta horizon from the Buntharik impact event in Thailand. Gondwana Research 4, 586.

Butzer, K.W., 1971. Environment and Archaeology. Second Edition. Aldine Atherton, Chicago, 703 p.

Candy, I., Black, S., Sellwood, B.W. & Rowan, J.S., 2003. Calcrete profile development in Quaternary alluvial sequences, southeast Spain: implications for using calcretes as a basis for landform chronologies. Earth Surface Processes and Landforms 28, 169-185.

Carcaillet, C., 2001. Brassages particuliers dans des sols mis en évidence à l'aide de datations au ^{14}C par AMS. Comptes Rendus de l'Académie des Sciences 332, 21-28.

Chen, F.H., Qiang, M.R., Feng, Z.D., Wang, H.B. & Bloemental, J., 2003. Stable East Asian monsoon climate during the Last Interglacial (Eemian) indicated by paleosol S1 in the western part of the Chinese Loess Plateau. Global and Planetary Change 36, 171-179.

Chichagova, O.A., 1995. Composition, properties and radiocarbon age of humus in paleosols. GeoJournal 36, 207-212.

Chlachula, J., Kemp, R.A., Jessen, C.A., Palmer, A.P. & Toms, P.S., 2004. Landscape development in response to climatic change during Oxygen Isotope Stage 5 in the southern Siberian loess region. Boreas 62, 60-75.

Cooke, R., Warren, A. & Goudie, A., 1993. Desert Geomorphology. UCL Press, London, 526 p.

Coudé-Gaussen, G., 1987. The perisaharan loess: sedimentological characterization and paleoclimatical significance. GeoJournal 15, 177-183.

Courty, M.A., Goldberg, P. & Macphail, R., 1989. Soils and Micromorphology in Archaeology. Cambridge Manuals in Archaeology. Cambridge University Press, Cambridge, 344 p.

Courty, M.A., Marlin, C., Dever, L., Tremblay, P. & Vachier, P., 1994. The properties, genesis and environmental significance of calcitic pendants from the High Arctic (Spitsbergen). Geoderma 61, 71-102.

Courty, M.A., Brasseur, B. & Fedoroff, N., 2007. The soil record of instantaneous processes linked to cosmic events and related consequences. Geophysical Research Abstracts 9, 10859.

Courty, M.A., Crisci, A., Fedoroff, M., Grice, K., Greenwood, P., Mermoux, M., Smith, D. & Thiemens, M., 2008. Regional expression of the widespread disruption of soil-landscapes by the 4 kyr BP impact-linked dust event using pedo-sedimentary micro-fabrics. In Kapur, S., Mermut, A. & Stoops, G. (eds.), New Trends in Soil Micromorphology. Springer-Verlag, Berlin, pp. 211-236.

Cremaschi, M. (ed.), 1991. Loess Aeolian Deposits and Related Paleosols in the Mediterranean Region. Quaternary International, Volume 5, 137 p.

Cremaschi, M. & Trombino, L., 1998. The paleoclimatic significance of paleosols in Southern Fezzan (Libyan Sahara): morphological and micromorphological aspects. Catena 34, 131-156.

Delvigne, J.E., 1998. Atlas of Micromorphology of Mineral Alteration and Weathering. Canadian Mineralogist Special Publication 3, 495 p.

Driese, S.G. & Ober, E.G., 2005. Paleopedologic and paleohydrologic records of precipitation seasonality from Early Pennsylvanian 'Underclay' paleosols, U.S.A. Journal of Sedimentary Research 75, 997-1010.

Driese, S.G., Medaris, L.G., Runkel, A.G. & Langford, R.P., 2007. Differenciating pedogenesis from diagenesis in early paleoweathering surfaces formed on granitic composition parent materials. Journal of Geology 115, 387-416.

Durand, N., Gunnell, Y., Curmi, P. & Ahmad, S.M., 2006. Pathways of calcrete development on weathered silicate rocks in Tamil Nadu, India: mineralogy, chemistry and paleoenvironmental implications. Sedimentary Geology 192, 1-18.

Durand, N., Monger, H.C. & Canti, M.G., 2010. Calcium carbonate features. In Stoops, G., Marcelino, V. & Mees, F. (eds.), Interpretation of Micromorphological Features of Soils and Regoliths. Elsevier, Amsterdam, pp. 149-194.

Emerson, D. & Revsbech, N.P., 1994. Investigation of an iron-oxidizing microbial mat community located near Aarhus, Denmark: field studies. Applied Environnenmental Microbiology 60, 4022-4031.

Erhart, H., 1956. La Genèse des Sols en tant que Phénomène Géologique. Esquisse d'une Théorie Géologique et Géochimique. Biostasie et Rhésistasie. Masson, Paris, 90 p.

Eschenbrenner, V., 1987. Les Glébules des Sols de Côte d'Ivoire. Nature et Origine en Milieu Ferrallitique. Modalités de leur Concentration. Rôle des Termites. PhD Dissertation, Université de Bourgogne, 498 + 282 p.

Ewing, S.E., Sutter, B., Owen, J., Nishiizumi, K., Sharp, W., Cliff, S.S., Perry, K., Dietrich, W., McKay, C.P. & Admunson, R., 2006. A threshold in soil formation at Earth's arid-hyperarid transition. Geochimica et Cosmochimica Acta 70, 5293-5322.

Fedoroff, N., 1968. Genèse et morphologie des sols à horizon B textural en France atlantique. Science du Sol 1, 29-65.

Fedoroff, N., 1997. Clay illuviation in Red Mediterranean soils. Catena 28, 171-198.

Fedoroff, N. & Courty, M.A., 1994. Organisation du sol aux échelles microscopiques. In Bonnneau, M. & Souchier, B. (eds.), Pédologie, 2. Constituants et Propriétés du Sol, Second Edition. Masson, Paris, pp. 349-375.

Fedoroff, N. & Courty, M.A., 2002. Paléosols et sol reliques. In Miskovsky, J.C. (ed.), Géologie de la Préhistoire. Géopré, Presses Universitaires de Perpignan, pp. 277-316.

Fedoroff, N. & Courty, M.A., 2005. Les Paléosols et sols archéologiques: mémoire des climats et des Hommes. In Girard, M.C., Walter, C., Remy, J.C., Berthelin, J. & Morel, J.L. (eds.), Sols et Environnement. Dunod, Paris, pp. 151-186.

Fedoroff, N. & Goldberg, P., 1982. Comparative micromorphology of two late Pleistocene paleosols (in the Paris Basin). Catena 9, 227-251.

Fedoroff, N., de Kimpe, C.R., Page, F. & Bourbeau, G., 1981. Essai d'interprétation des transferts sous forme figurée dans les podzols du Québec méridional à partir de l'étude micromorphologique des profils. Geoderma 26, 25-45.

Fedoroff, N., Courty, M.A. & Thompson, M.L., 1990. Micromorphological evidence of paleoenvironmental change in pleistocene and holocene paleosols. In Douglas, L.A. (ed.), Soil Micromorphology: A Basic and Applied Science. Developments in Soil Science, Volume 19. Elsevier, Amsterdam, pp. 653-665.

Fedoroff, N., Courty, M.A., Lacroix, E. & Oleschko, K., 1994. Calcitic accretion on indurated volcanic materials (example of tepetates, Altiplano, Mexico). Transactions of the 15th World Congress of Soil Science, Acapulco, Mexico, Volume 6a, pp. 460-473.

Felix-Henningsen, P. & Mauz, B., 2004. Paleoenvironmental significance of soils on ancient dunes of the northern Sahel and southern Sahara of Chad. Die Erde 135, 321-340.

Feng, Z.D. & Wang, H.B., 2005. Pedostratigraphy and carbonate accumulation in the last interglacial pedocomplex of the Chinese loess plateau. Soil Science Society of America Journal 69, 1094-1101.

Ferraro, F., Terhorst, B., Ottner, F. & Cremaschi, M., 2004. Val Sorda: an upper Pleistocene loess-paleosol sequence in northeastern Italy. Revista Mexicana de Ciencias Geológicas 21, 30-44.

Goldammer, J.G. & Seibert, B., 1989. Natural rain forest fires in Eastern Borneo during the Pleistocene and Holocene. Naturwissenschaften 76, 518-520.

Gibling, M.R. & Rust, B.P., 1992. Silica-cemented paleosols (ganisters) in the Pennsylvanian Waddens Cove Formation, Nova Scotia, Canada. In Wolf, K.H. & Chilingarian, G.V. (eds.), Diagenesis, III. Developments in Sedimentology, Volume 47. Elsevier, Amsterdam, pp. 621-655.

Gonzalez, J.E., 1991. Genesis y Funcionamentos de los Principales Tipos de Suelos de la Isla de la Juventud. PhD Dissertation, Ministerio de Agricultura, Instituto de Suelos, La Habana, 234 p.

Goodfriend, G.A., Cameron, R.A.D., Cook, L.M., Courty, M.A. & Fedoroff, N., 1996. The Quaternary aeolian sequence of Madeira: stratigraphy, chronology and paleoenvironmental interpretation. Palaeogeography, Palaeoclimatology, Palaeoecology 120, 195-234.

Grimley, D.A & Arruda, N.K., 2007. Observations of magnetite dissolution in poorly drained soils. Soil Science 172, 968-982.

Guo, Z., 1990. Succession des Paléosols et des Loess du Centre-ouest de la Chine. Approche Micromorphologique. PhD Dissertation, Université de Paris VI, 226 p.

Hachicha, M., Stoops, G. & M'hiri, A., 1987. Aspects micromorphologiques de l'évolution des sols de lunettes argileuses en Tunisie. In Fedoroff, N., Bresson, L.M. & Courty, M.A. (eds.), Micromorphologie des Sols, Soil Micromorphology. AFES, Plaisir Paris, pp 193-197.

Hamdi-Aissa, B., 2001. Le fonctionnement actuel et passé de sols du Nord Sahara (cuvette d' Ouargla). Approches micromorphologique, géochimique et minéralogique et organisation spatiale. PhD Dissertation, Institut National Agronomique Paris-Grignon, Paris, 308 p.

Hearty, P.J., Hollin, J.T., Neumann, A.C., O'Leary, M.J. & McCulloch, M., 2007. Global sea-level fluctuations during the Last Interglaciation (MIS 5e). Quaternary Science Reviews 26, 2090-2112.

Humbert, L. 1976a. Eléments de Pétrographie Dynamique des Systèmes Calcaires. Tome 1. Description Macroscopique et Microscopique, Diagenèse, Applications. Editions Technip, Paris, 213 p.

Humbert, L. 1976b. Eléments de Pétrographie Dynamique des Systèmes Calcaires. Tome 2. Atlas Photographique. Editions Technip, Paris, 200 p.

Iriondo, M. & Kröhling, D.M., 1997. The tropical loess. In An Zhisheng & Zhou Weijian (eds.), Quaternary Geology. Proceedings of the 30th International Geological Congress, Beijing,Volume 21. VSP, Utrecht, pp. 61-77.

Iriondo, M.H. & Kröhling, D.M., 2007. Non-classical types of loess. Sedimentary Geology 202, 352-368.

IUSS Working Group WRB, 2006. World Reference Base for Soil Resources. 2nd Edition. World Soil Resources Reports No. 103. FAO, Rome, 128 p.

Jenny, H., 1941. Factors of Soil Formation: A System of Quantitative Pedology. Mc-Graw Hill, New York, 281 p.

Jones, T.P. & Bo Lim, 2000. Extraterrestrial impacts and wildfires Palaeogeography, Palaeoclimatology, Palaeoecology 164, 57-66.

Kemp, R.A., 1998. Role of micromorphology in paleopedological research. Quaternary International 51/52, 133-141.

Kemp, R.A., 1999. Micromorphology of loess-paleosol sequences: a record of paleoenvironmental change. Catena 35, 179-196.

Kemp, R.A., 2001. Pedogenic modification of loess: significance for palaeoclimatic reconstructions, Earth-Science Reviews 54, 145-156.

Kemp, R.A., Jerz, H., Grottenthaler, W. & Preece, R.C., 1994. Pedosedimentary fabrics of soils within loess and colluvium in southern England and Germany. In Ringrose-Voase, A.J. & Humphreys, G.S. (eds.), Soil Micromorphology: Studies in Management and Genesis. Developments in Soil Science, Volume 22. Elsevier, Amsterdam, pp. 207-219.

Kemp, R.A., Derbyshire, E., Meng Xingmin, Chen Fahu & Pan Baotian, 1995. Pedosedimentary reconstruction of a thick loess-sequence near Lanzhou in north-central China. Quaternary Research 43, 30-45.

Kemp, R.A., Derbyshire, E. & Meng Xingmin, 1997. Micromorphological variation of the S1 paleosol across northwest China. Catena 31, 77-90.

Kemp, R.A., Derbyshire, E. & Meng Xingmin, 2001. A high resolution micromorphological record of changing landscapes and climates on the western Loess plateau of China during oxygen isotope stage 5. Palaeogeography, Palaeoclimatology, Palaeoecololology 170, 157-169.

Kemp, R.A., Toms, P.S., Sayago, J.M., Derbyshire, E., King, M. & Wagoner, L., 2003. Micromorphology and OSL dating of the basal part of the loess-paleosol sequence at La Mesada in Tucumán province, Northwest Argentina. Quaternary International 106/107, 111-117.

Kemp, R.A., Toms, P.S., King, M. & Kröhling, D.M., 2004. The pedosedimentary evolution and chronology of Tortugas, a Late Quaternary type-site of the northern Pampa, Argentina. Quaternary International 114, 101-112.

Kennett, D.J., Kennett, J.P., West, G.J., Erlandson, J.M., Johnson, J.R., Hendy, I.L., West, A., Culleton, B.J., Jones, T.L. & Stafford, T.W., 2008. Wildfire and abrupt ecosystem disruption on California's Northern Channel Islands at the Ållerød-Younger Dryas boundary (13.0-12.9 ka). Quaternary Science Reviews 27, 2530-2545.

Kühn, P. & Hilgers, A., 2005. Reconstruction of multiphase Late Glacial/Holocene soil formation by integrated luminescence dating and micromorphology – a case study from the southern Taunus Foreland, Germany. Geophysical Research Abstracts 7, 00005.

Kukla, G. & An Zhisheng, 1989. Loess stratigraphy in Central China. Palaeogeography, Palaeoclimatology, Palaeoecololology 72, 203-225.

Lacerda, W.A., 2007. Landslide initiation in saprolite and colluvium in southern Brazil: field and laboratory observations. Geomorphology 87, 104-119.

Laignel, B., Quesnel, F., Lecoustumer, M-N. & Meyer, R., 1998. Variabilité du cortège argileux des formations résiduelles à silex de l'Ouest du bassin de Paris. Comptes Rendus de l'Académie des Sciences 326, 467-472.

Le Ribault, L., 1977. L'Exoscopie des Quartz. Editions Masson, Paris, 200 p.

Lewis, B.D., Williams, N.D. & Greenberg, J.K., 2008. Unusual calcrete formation on Precambrian marble from the Merelani Tanzanite district, Simanjiro, Tanzania. Geological Society of America Abstracts with Programs, Volume 40, no. 5, p. 75.

Maher, B.A., Alekseev, A. & Alekseeva, T., 2003. Magnetic mineralogy of soils across the Russian Steppe: climatic dependence of pedogenic magnetite formation. Palaeogeography, Palaeoclimatology, Palaeoecolology 201, 321-341.

McCall, G.J.H., 2001. Tektites in the geological record. Bulletin of Canadian Petroleum Geology 50, 158-177.

McCarthy, P.J., 2002. Micromorphology and development of interfluve paleosols: a case study from the Cenomanian Dunvegan Formation, NE British Columbia, Canada. Bulletin of Canadian Petroleum Geology 50, 158-177.

McCarthy, P.J., Martini, I.P. & Leckie, D.A., 1998. Use of micromorphology for palaeoenvironmental interpretation of complex alluvial palaeosols: an example from the Mill Creek Formation (Albian), southwestern Alberta, Canada. Palaeogeography, Palaeoclimatology, Palaeoecology 143, 87-110.

McLaren, S., 2004. Characteristics, evolution and distribution of Quaternary channel calcretes, Southern Jordan. Earth Surface Processes and Landforms 29, 1487-1507.

McTainsh, G.H., Nickling, G. & Lynch, A.W., 1997. Dust deposition and particle size in Mali, West Africa. Catena 29, 307-322.

Melezhik, V.A., Fallick, A.E. & Grillo, S.M., 2004. Subaerial exposure surfaces in a Palaeoproterozoic ^{13}C-rich dolostone sequence from the Pechenga Greenstone Belt: palaeoenvironmental and isotopic implications for the 2330-2060 Ma global isotope excursion of ^{13}C/^{12}C. Precambrian Research 133, 75-103.

Milnes, A.R., 1992. Calcretes. In Martini, I.P. & Chesworth, W. (eds.), Weathering, Soils and Paleosols. Developments in Earth Surface Processes, Volume 2. Elsevier, Amsterdam, pp. 309-347.

Milnes, A.R. & Thiry, M., 1992. Silcretes. In Martini, I.P. & Chesworth, W. (eds.). Weathering, Soils and Paleosols. Developments in Earth Surface Processes, Volume 2. Elsevier, Amsterdam, pp. 349-378.

Milnes. A.R., Wright, M.J. & Thiry, M., 1991. Silica accumulation in saprolites and soils in South Australia. In Nettleton, W.D. (ed.), Occurrence, Characteristics and Genesis of Carbonate, Gypsum and Silica Accumulations in Soils. Soil Science Society of America Special Publication 26, SSSA, Madison, pp. 121-149.

Morrison, R.B., 1977. Quaternary soil stratigraphy, concepts, methods and problems. In Mahaney, W.C. (ed.), Quaternary Soils. Geo-Abstracts, Norwich, pp. 77-108.

Mücher, H.J. & Morozova, T.D., 1983. The application of soil micromorphology in Quaternary geology and geomorphology. In Bullock, P. & Murphy, C.P. (eds.), Soil Micromorphology. Volume 1. Techniques and Applications. AB Academic Publishers, Berkhamsted, pp. 151-194.

Mücher, H.J., Carballas, T., Guitián Ojea, F., Jungerius, P.D., Kroonenberg, S.B. & Villar, M.C., 1972. Micromorphological analysis of effects of alternating phases of landscape stability and instability on two soil profiles in Galicia, N.W. Spain. Geoderma 8, 241-266.

Mücher, H., van Steijn, H. & Kwaad, F., 2010. Colluvial and mass wasting deposits. In Stoops, G., Marcelino, V. & Mees, F. (eds.), Interpretation of Micromorphological Features of Soils and Regoliths. Elsevier, Amsterdam, pp. 37-48.

Nash, D.J. & Smith, R.F., 2003. Properties and development of channel calcretes in a mountain catchment, Tabernas Basin, southeast Spain. Geomorphology 50, 227-250.

Pepper, D.M., 1973. A comparison of the 'Argile à Silex' of northern France with the 'Clay-with-Flints' of southern England. Proceedings of the Geologists' Association 84, 331-352.

Perkins, A.M., 1996. Observations under electron microscopy of magnetic minerals extracted from speleothems. Earth and Planetary Sciences Letters 139, 281-289.

Poetsch, T., 2004. Forms and dynamics of silica gel in a tuff-dominated soil complex: results of micromorphological studies in the central highlands of Mexico. Revista Mexicana de Ciencias Geológicas 21, 195-201.

Porter, S.C. & An Zhisheng, 1995. Correlation between climate events in the North Atlantic and China during the last glaciation. Nature 375, 305-308.

Rampino, M.R. & Self, S., 1993. Climate volcanism feedback and the Toba eruption of 74,000 years ago. Quaternary Research 40, 269-280.

Retallack, G.J., 1990. Soils of the Past: An Introduction to Paleopedology. Unwin Heyman, London, 520 p.

Retallack, G.J., 1991. Untangling the effects of burial alteration and ancient soil formation. Annual Review of Earth and Planetary Sciences 19,183-206.

Retallack, G.J., 2005a. Pedogenic carbonate proxies for amount and seasonality of precipitation in paleosols. Geology 33, 333-336.

Retallack, G.J., 2005b. Earliest Triassic claystone breccias and soil-erosion crisis. Journal of Sedimentary Research 75, 679-695.

Retallack, G.J. & Mindszenty, A., 1994. Well preserved late Precambrian paleosols from Northwest Scotland. Journal of Sedimentary Research 64, 264-281.

Retallack, G.J., & Wright, V.P., 1990. Micromorphology of lithified paleosols. In Douglas, L.A. (ed.), Soil Micromorphology: A Basic and Applied Science. Developments in Soil Science, Volume 19. Elsevier, Amsterdam, pp. 641-652.

Robinson, S.A., Andrews, J.E., Hesselbo, S.P., Raddley, J.D., Dennis, P.F., Harding, I.C. & Allen, P., 2002. Atmospheric pCO_2 and depositional environment from stable-isotope geochemistry of calcrete nodules (Barremian, Lower Cretaceous, Wealden Beds, England). Journal of the Geological Society 159, 215-224.

Rognon, P., Coudé-Gaussen, G., Fedoroff, N. & Goldberg, P., 1987. Micromorphology of loess in the norther Negev (Israel). In Fedoroff, N., Bresson, L.M. & Courty, M.A. (eds.), Micromorphologie des Sols, Soil Micromorphology. AFES, Plaisir, pp. 631-638.

Rousseau, D.D., Antoine, P., Hatté, C., Lang, A., Zöller, L., Fontugne, M., Ben Othman, D., Luck, J.M., Moine, O., Labonne, M., Bentaleb, I. & Jolly, D., 2002. Abrupt millennial climatic changes from Nussloch (Germany) Upper Weichselian aeolian records during the Last Glaciation. Quaternary Science Reviews 21, 1577-1582.

Ruhe, R.V., 1965. Quaternary paleopedology. In Wright, H.E. & Frey, D.G. (eds.), The Quaternary of the United States. Princeton University Press, Princeton, pp. 755-764.

Rust, B.R. & Nanson, G.C., 1989. Bedload transport of mud as pedogenic aggregates in modern and ancient rivers. Sedimentology 36, 291-306.

Sanborn, P.T., Scott Smith, C.A., Froese, D.G., Zazula, D.G. & Westgate, J.A., 2006. Full-glacial paleosols in perennially frozen loess sequences, Klondike goldfields, Yukon Territory, Canada. Quaternary Research 66, 147-157.

Schwertmann, U., 1985. The effect of pedogenic environments on iron oxides minerals. Advances in Soil Science 1, 172-200.

Sellami, F., 1999. Reconstitution de la Dynamique des Sols des Terrasses de la Dordogne à Creysse (Bergeracois). Impact sur la Conservation des Sites en Plein Air. PhD Dissertation, Institut National Agronomique Paris-Grignon, Paris, 255 p.

Sephton, M.A., Looy, C.V., Brinkhuis, H., Wignall, P.B., de Leeuw, J.W. & Visscher, H., 2005. Catastrophic soil erosion during the end-Permian biotic crisis. Geology 33, 941-944.

Sheldon, N.D. & Retallack, G.J., 2001. Equation for compaction of paleosols due to burial. Geology 29, 233-247.

Smalley, I.J. & Vita-Finzi, C., 1968. The formation of fine particles in sandy deserts and the nature of 'desert' loess. Journal of Sedimentary Research 38, 766-774.

Srivastava, P. & Parkash, B., 2002. Polygenetic soils of the north-central part of the Gangetic Plains: a micromorphological approach. Catena 46, 43-259.

Stahr, K., Kühn, J., Trommler, J., Papenfuss, K.H., Zarei, M. & Singer, A., 2000. Palygorskite-cemented crusts (palycretes) in Southern Portugal. Australian Journal of Soil Research 38, 169-188.

Stoops, G., 2003. Guidelines for Analysis and Description of Soil and Regolith Thin Sections. Soil Science Society of America, Madison, 184 p.

Stoops, G. & Marcelino, V., 2010. Lateritic and bauxitic materials. In Stoops, G., Marcelino, V. & Mees, F. (eds.), Interpretation of Micromorphological Features of Soils and Regoliths. Elsevier, Amsterdam, pp. 329-350.

Stoops, G. & Mathieu, C., 1970. Aspects micromorphologiques des argiles à silex de Thiérarche. Science du Sol 2, 103-116.

Stremme, H.E., 1998. Correlation of Quaternary pedostratigraphy from western to eastern Europe. Catena 34, 105-112.

Sullivan, L.A. & Koppi, A.J., 1993 Barite pseudomorphs after lenticular gypsum in a buried soil from central Australia. Australian Journal of Soil Research 31, 393-396.

Summerfield, M.A., 1983. Petrography and diagenesis of silcrete from the Kalahari Basin and Cape coastal zone, southern Africa. Journal of Sedimentary Research 53, 895-909.

Targulian, V.O. & Goryachkin, S.V., 2004. Soil memory: types of record, carriers, hierarchy and diversity. Revista Mexicana de Ciencias Geológicas 21, 1-8.

Targulian, V.O. & Sokolova, T.A., 1996. Soil as a biotic/abiotic natural system: a reactor, memory and regulator of biospheric interactions. Eurasian Soil Science 29, 30-38.

Targulian, V.O., Birina, A.G., Kulikov, A.V., Sokolova, T.A. & Tselischeva, L.K., 1974. Arrangement, Composition and Genesis of Sod-pale-podzolic Soil Derived from Mantle Loam. Morphological Investigation. Nauka, Moscow, 55 p.

Tate, T.A. & Retallack, G.J., 1995. Thin sections of paleosols. Journal of Sedimentary Research 65, 579-580.

Thiry, M., 1999. Diversity of continental silicification features: examples from the Cenozoic deposits in the Paris Basin and neighbouring basement. In Thiry, M. & Simon-Coinçon, R. (eds.), Palaeoweathering, Palaeosurfaces and Related Continental Deposits. International Association of Sedimentologists Special Publication 27, pp. 87-128.

Thorez, J., Bullock, P., Catt, J.A. & Weir, A.H., 1971. The petrography and origin of deposits silling solution pipes in the chalk near South Mimms, Hertfordshire. Geological Magazine 108, 413-423.

Tsatskin, A., Heller, F., Gendler, T.S., Virina, E.I., Spassov, S., Du Pasquier, J., Hus, J., Hailwood, E.A., Bagin, V.I. & Faustov, S.S., 2001. A new scheme of terrestrial paleoclimate evolution during the last 1.5 Ma in the western Black Sea region: integration of soil studies and loess magnetism. Physics and Chemistry of the Earth 26, 911-916.

Tsatskin, A., Gendler, T.S. & Heller, F., 2006. Identification of multiphase paleosols in aeolian sequences by micromorphology, enviromagnetism and Mössbauer spectroscopy. Geophysical Research Abstracts 8, 05025.

Ufnar, D.F., González, L.A., Ludvigson, G.A., Brenner, R.L. & Witzke, B.J., 2001. Stratigraphic implications of meteoric sphaerosiderite $\delta^{18}O$ values in paleosols of the Cretaceous (Albian) Boulder Creek Formation, NE British Columbia Foothills, Canada. Journal of Sedimentary Research 71, 1017-1038.

Usai, M.R., 1996. Paleosol Interpretation. Micromorphological and Pedological Studies. William Sessions Limited, York, 168 p.

Van Vliet-Lanoë, B., 1976. Corrélation entre la présence de charbon de bois au sommet des paléosols et les dégradations climatiques. Pédologie 26, 97.

Verrecchia, E., Yair, A., Kidron, G.J. & Verrecchia, K., 1995. Physical-properties of the psammophile cryptogamic crust and their consequences to the water regime of sandy soils, north-western Negev Desert, Israel. Journal of Arid Environments 29, 427-437.

Wang, H. & Follmer, L.R., 1998. Proxy of monsoon seasonally in carbon isotopes from paleosols of the southern Chinese Loess Plateau. Geology 26, 987-990.

Wolf, L.W., Tuttle, M.P., Browning, S. & Park, S., 2006. Geophysical surveys of earthquake-induced liquefaction deposits in the New Madrid seismic zone. Geophysics 71, 223-230.

Wright, V.P., Turner, M.S., Andrews, J.E. & Spiro B., 1993. Morphology and significance of super-mature calcretes from the Upper Old Red Sandstone of Scotland. Journal of the Geological Society 150, 871-883.

Wright, V.P., Beck, V.H. & Sanz-Montero, M.E., 1996. Spherulites in calcrete laminar crusts: Biogenic $CaCO_3$ precipitation as a major contributor to crust formation: discussion. Journal of Sedimentary Research 66, 1040-1041.

Yaalon, D.H., 1971. Soil forming processes in time and space. In Yaalon, D.H. (ed.), Paleopedology: Origin, Nature and Dating of Paleosols. International Soil Science Society and Israel University Press, Jerusalem, pp. 29-39.

Yaalon, D.H., 1983. Climate, time, and soil development. In Wilding, L.P., Smeck, N.E. & Hall, G.F. (eds.), Pedogenesis and Soil Taxonomy. I. Concepts and Interactions. Developments in Soil Science, Volume 11A. Elsevier, Amsterdam, pp. 233-251.

Yin, Q. & Guo, Z., 2006. Mid-pleistocene vermiculated red soils in southern China as an indication of unusually strengthened East Asian monsoon. Chinese Science Bulletin 51, 213-220.

Chapter 29

Palaeosoils and Relict Soils: A Systematic Review

Mauro Cremaschi, Luca Trombino, Andrea Zerboni

UNIVERSITÀ DEGLI STUDI DI MILANO, MILANO, ITALY

CHAPTER OUTLINE

1. Introduction

The efficacy of micropedology in palaeopedological studies is widely accepted (Stoops, 2009, 2014), but no clear consensus exists for the definition of palaeosoils (Catt, 1986). Suggested definitions have variably emphasised soil development in a past environment (e.g., Yaalon, 1971; Ferrari & Magaldi, 1976) and pedogenesis that began in the past (e.g.,

Interpretation of Micromorphological Features of Soils and Regoliths. https://doi.org/10.1016/B978-0-444-63522-8.00029-2

Bos & Sevink, 1975; Duchaufour, 1983; Fedoroff et al., 2010, 2018). Palaeosoil definitions are also complicated by the common incompleteness of palaeosoil profiles and by poor applicability of classification systems developed for modern soils (Zerboni et al., 2011). Moreover, new advances in palaeopedological research, for instance, the concepts of pedosedimentary sequences and pedostratigraphic levels (Costantini & Priori, 2007) or three-dimensional studies of past landscapes (Costantini et al., 2009), have to be taken into account. In view of these difficulties in proposing definitions and subdivision of palaeosoils, this chapter is limited to presenting an overview of various types of formations recording pedogenic processes that occurred or started in the past. In the following paragraphs, various palaeosoil categories are discussed, starting with buried soils (i.e., palaeosoils separated from the present-day surface), followed by palaeosoils that are not separated from the surface (i.e., relict soils, polygenetic soils), and by some specific examples (i.e., reworked palaeosoils fragments, rock coatings).

2. Buried Soils

Buried soils (Morozova, 1963; Catt, 1986) are palaeosoils covered by a sediment interval that is sufficiently thick to isolate them from present-day pedogenic processes. When the buried soil is not completely isolated from surface processes, features that developed before and after burial need to be distinguished. A further challenge can be the study of intervals comprising a series of closely spaced buried soils (pedocomplexes) (Catt, 1988; Van Vliet-Lanoë, 1990; Bronger & Sedov, 2003).

Most micromorphological studies of buried soils concern Quaternary formations, including the pioneering work of Smolikova (1967, 1978) and Bronger (1972, 1978) in Central Europe, as well as many studies that were conducted as part of archaeological investigations (see Courty et al., 1989). However, many studies dealing with older occurrences, from Precambrian to Cainozoic, are also available (see Section 2.6).

2.1   Buried Soils in Loess

The better-known examples of Late Tertiary and Quaternary buried soils are the loess/palaeosoils sequences of Eastern Europe, Central Asia, Argentina, the US Great Plains region, North Africa and Israel, and especially, the Loess Plateau of Central China, on which numerous studies have focused.

Thin section studies have been applied to the thick and generally weakly weathered loess-bearing pedostratigraphic sequences of Central and Eastern Europe, allowing to distinguish palaeopedons and to characterise pedogenic overprints on aeolian sediments (Konecka-Betley, 1994). Diagnostic horizons of fossil soils of different ages were studied for several sections of Upper Pleistocene loess distributed throughout Poland. The occurrence at the microscale of abundant clay translocation, the mobilisation of organic matter, the growth of large calcite crystals and the neoformation of Mn/Fe oxide nodules in the micromass permitted to document the occurrence of five interstadial soils with argillic, cambic and mollic diagnostic horizons (Konecka-Betley, 1994). Calcitic pedogenic features were studied for a loess-palaeosoil pedocomplex near Kursk (Western Russia), coupling micropedology and SEM (Kovda et al., 2009). The authors noted a high variability of calcareous features at macro-, micro- and submicroscopic levels and attributed it to several wet-dry and cold-warm stages. They identified, for instance, the occurrence of pseudomycelium in modern chernozems, whereas in the buried pedocomplexes these features are progressively more leached as a consequence of warmer conditions, resulting in a greater abundance of calcite coatings along ped faces (as short thick rods) and intrapedal voids (as needles and rods). The same study illustrates that calcite nodules found in modern chernozems and in the oldest pedocomplex vary significantly; in the most ancient horizons, the central part of micritic nodules is strongly impregnated by Fe oxides.

Palaeosoils in loess sequences of the China Loess Plateau and Central Asia have been investigated by thin section studies to reconstruct past environmental changes (e.g., Derbyshire, 1995; Guo et al., 1996; Kemp & Derbyshire, 1998). Kemp et al. (1999) demonstrated the high sensitivity of micromorphology in detecting palaeoenvironmental changes within the rapidly aggrading landscape of the Chinese Loess Plateau. Derbyshire et al. (1995) concluded that micromorphology is a more powerful technique if combined with magnetic susceptibility, particle size and geochemical analyses, allowing identification of a palaeoclimatic gradient in palaeosoil evolution along the Loess Plateau. In the Jiuzhoutai sequence, they identified some key features correlated to specific environmental settings, such as lenticular gypsum crystals recording restricted leaching in semiarid conditions, successive layers with relative depletion and enrichment in fine particles pointing to cryogenic processes, and both loose to welded excrements and zones of associated calcite depletion-accumulation indicating wetter environmental conditions.

Guo et al. (2008) studied a series of early Miocene and Quaternary buried soils on the Loess Plateau in China to infer changes in monsoon intensity. The Miocene palaeosoils have a greater abundance of clay pedofeatures than the more recent ones, in the form of coatings and infillings with high interference colours. The difference in abundance records stronger soil development, suggesting significantly stronger and/or more persistent summer monsoons. These soils, moreover, show a higher degree of incorporation of illuvial clay into the groundmass through alternating shrink and swell (vertic) processes, typical for climates with contrasting seasons, thus suggesting strong seasonality in northern China in the early Miocene.

The micromorphological characteristics of complex S5 (Zhisheng et al., 1987; Zhao, 2003) are of special significance in terms of Pleistocene palaeoclimate, being formed under woodland vegetation during an interval corresponding to a climatic optimum. Moreover, micropedological studies evidenced that the pedocomplex preserves several distinct layers consisting of the upper Bts horizon, as well as some horizons showing desiccation cracks, and a basal horizon with $CaCO_3$ nodules (Zhao, 2003). Formation of pedogenic carbonate features in the lower horizons, caused by dissolution of calcite and limited vertical leaching combined with a low biological activity (recorded by weakly developed excremental features in the topsoil), points to more arid conditions. In more humid parts of the Plateau, pedosequences are extensively bioturbated and can be almost completely decalcified in the upper horizons, combined with formation of carbonate coatings and needle-fibre calcite in the underlying parent material, or older palaeosoils. The comparison of different bioclimate indicators, based on mineralogy, micromorphology and palynology, for a Holocene palaeosoil of the Ansai loess section from China, highlighted that oriented clay coatings developed by mechanical translocation of clay, whereas the argillic horizon developed by strong weathering and subsequent eluviation-illuviation of clay over a vertical distance of about 60 cm. This evidence emphasises that in a single palaeosoil in loess not only features of a period of humid and warm climate are present, but also the imprint of several different bioclimatic environments can be distinguished (He et al., 2008).

Loess sequences from Tadjikistan contain numerous palaeosoils, whose genetic classification is difficult because almost all buried soils are truncated as well as affected by recalcification by carbonates derived from the overlying loess. For these sequences, micromorphology allowed distinguishing between inherited and pedogenic carbonates and identification of the occurrence of clay illuviation (Bronger et al., 1995), providing detailed palaeoclimatic information, especially for the Brunhes chron (Bronger et al., 1998). The profiles contain silt to sand-sized grains that are irregularly distributed in the groundmass, contrasting with pedogenic micritic and acicular calcite occurring in the lower layers, resulting from leaching of the overlying loess. The intensity and duration of pedogenesis is highlighted by the degree of clay illuviation and incorporation of clay coatings in the groundmass.

Micropedological studies of loess and related wind-blown deposits have been carried out at a variety of locations and helped to (i) identify occasional input/reworking of parent materials by non-aeolian processes (e.g., occurrence of sedimentary structures associated with water flow along ephemeral land surfaces); (ii) distinguish the presence of weakly developed soil horizons (i.e., cumulic soils); and (iii) quantify the contribution of specific pedogenic processes in shaping buried palaeosoils (as discussed above), especially in those case where many processes were superimposed (Kemp, 1999). The above-described processes may be superimposed to weaker pedogenic processes such as frost action: recorded by platy microstructures or banded fabrics (Kemp, 1999). In many other cases, thin section studies have been successfully applied to pedosedimentary sequences in Argentina (Amiotti et al., 2001; Kemp et al., 2004), central Alaska (Josephs, 2010), Israel (Rognon et al., 1987) and Lanzarote (von Suchodoletz et al., 2009).

2.2 Buried Soils in Sandy Aeolian Sediments

Buried soils of roughly the same age as palaeosoils in loess sequences occur as intercalations in coastal aeolian deposits. Aeolianites (or kurkars in the eastern Mediterranean regions) are littoral aeolian calcarenite deposits that record reworking of

FIGURE 1 Degraded clay coatings, with associated Fe/Mn oxide hypocoatings, in a palaeosoil in aeolianite deposits (Elba Island, Italy). (A) PPL. (B) XPL.

vast quantities of shallow-marine biogenic carbonate sediment into coastal dunes (Brooke, 2001). Different cycles of aeolianite deposition and palaeosoil development along the coasts of the Elba Island (Tuscany, central Italy) were studied by Cremaschi and Trombino (1998a). Clay coatings (Fig. 1), iron-manganese oxide nodules and a strongly rubified micromass that characterise the truncated palaeosoils are also typical of present-day Red Mediterranean Soils.

Detailed micromorphological studies of aeolianites from Israel evidenced that soil formation is not only constrained by climatic factors but also by continuous aeolian clay (dust) input with relatively low and variable time rates. This is suggested by degraded clay coatings, well-expressed striated b-fabrics in leached horizons and dissolution-affected fragments of calcareous aeolianite inherited from the bedrock, found in the Bk horizons (Tsatskin et al., 2009). Palaeosoils corresponding to living floors buried by aeolian sediments are more common in pre-Holocene contexts (Tsatskin & Ronen, 1999; Cremaschi et al., 2015). For instance, a complex pedosequence of aeolianites and interlayered palaeosoils from the coast of Northern Israel records several stages of development of discrete palaeosoils within a Mousterian pedocomplex, marked by episodes of aggradation and accelerated dust input under different environmental conditions (Tsatskin & Ronen, 1999). In this case, common micritic coatings (disrupted by faunal churning) and iron oxide segregations suggest decalcification and rubification in the lowermost palaeosoil intervals. The more complex juxtaposition of pedofeatures in the overlying interval indicates that clay illuviation predates calcite illuviation and recrystallisation. Tsatskin and Ronen (1999) also identified a palaeo-Vertisol that in thin section presents strongly developed shrink swell-related microfabric, common Mn oxide coatings and calcitic pseudomorphs after land snail shells. Finally, the occurrence of colluvial microlayers and various ferric accumulations (associated with gleying) in the uppermost soil interval suggests the occurrence of colluvial and aeolian processes associated with climatic instability.

2.3   Buried Soils in Glacial Sediments

Buried soils can be found in the context of Quaternary glacial and periglacial deposits (e.g., moraine fronts, fluvioglacial sediments, colluvial deposits). In the Val Sorda (Northern Italy), lodgement till of the Last Glacial Maximum advance of the Garda Glacier covers a thick and well-dated palaeosoil/loess pedosequence, including a cumulic chernozem overlying a less weathered loess body and a reddish palaeosoil developed on till (i.e., Alfisol) (Cremaschi, 1987; Ferraro et al., 2004; Ferraro, 2009). The common reddish clay coatings preserved in the red palaeosoils at the bottom of the loess sequence suggest intense pedogenesis, which took place during Marine Isotopic Stage (MIS) 5 (Cremaschi & Zerboni, 2016). The upper part of the sequence shows less intense pedogenesis and continuous dust input, which is indicated by pedogenic features that developed during loess deposition: finely fragmented charcoal, calcified root cells and organic coatings intermixed with aeolian sediments. The over-consolidation of the pedosequence, the obliteration of the porosity and, in the uppermost part of the sequence, the deformation of layered calcareous sediment by glaciotectonic stress (Fig. 2) are related to overthrust by a thick ice sheet (at least 500-700 m) during the Last Glacial Maximum.

Perennially frozen loess deposits in the Klondike goldfields (Alaska) highlight the occurrence and the properties of palaeosoils formed in full-glacial environments during MIS 2 and 4 (Sanborn et al., 2006). Buried A horizons present a banded to spongy microstructure, the latter with partially decomposed root residues and locally abundant organic matter accumulation associated with a weak granular microstructure. In the C horizons, laminations with particle-size sorting are common. The microstructures are similar to those in modern soils that contained ice lenses under perennially or seasonally frozen conditions (Cryosols), but the same features also resemble those of modern soils of temperate and boreal grasslands, in which organic matter accumulates as root

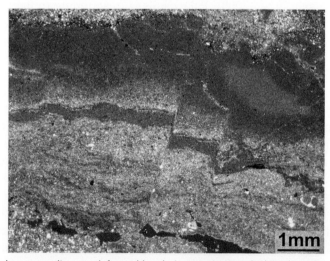

FIGURE 2 Layered calcareous sediments, deformed by glaciotectonic stress (Val Sorda pedosequence, Italy) (XPL).

detritus. Sanborn et al. (2006) conclude that these characteristics of the palaeosoils indicate affinities with soils of steppe and tundra environments, indicating a diversity of local environments in full-glacial landscapes.

2.4 Buried Soils in Alluvial Sediments

The migrating pattern of large rivers and the continuous input of sediments from overbank sedimentation due to seasonal or occasional floods can bury stable surfaces and the associated soils (e.g., Limbrey, 1993; Zárate et al., 2000; Hoffmann et al., 2009).

Holocene alluvial deposits in the overbank facies of the Po River plain at the edge of the Apennine alluvial fans in Northern Italy are systematically intercalated with weakly developed Atlantic and Subboreal buried soils (ranging from Regosols to Cambisols), characterised by carbonate leaching and slight hydromorphism (Cremaschi & Nicosia, 2012; Cremaschi et al., 2015). Presence of finely comminuted charcoal (Fig. 3A), redistributed calcitic wood ash, dusty clay coatings (Fig. 3B) and phosphatic aggregates records anthropogenic activities from the Late Neolithic through Chalcolithic to early Bronze Age.

Along an alluvial landscape of the Pampas (southern Argentina), Zárate et al. (2000) identified a pedosequence including horizons with a high concentration of bioclastic particles (diatoms, shells and humic acid-stained material), levels dominated by non-bioclastic coarse particles with a preferred alignment of clay domains (stipple-speckled or granostriated b-fabric) and levels strongly affected by carbonate precipitation (calcitic-crystallitic b-fabric), suggesting alternating phases of pedogenic and sedimentary inputs. A decrease in clastic sedimentation and then reestablishment of fluvial aggradation with a dominance of bioclastic sedimentation mark the early-Holocene shift from a dry subhumid to humid climate. This was followed by increasing aeolian inputs (loess) during the mid-Holocene as the climate reverted to dry subhumid, and temporary subaerial exposure of the sediments led to palaeosoil development (Zárate et al., 2000).

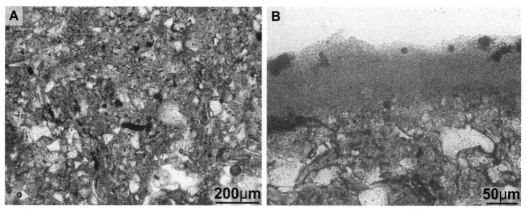

FIGURE 3 Features recording evidence of agricultural activity, in a Late Neolithic/Chalcolithic palaeosoil in alluvial fan deposits (Po Plain, Italy). (A) Finely comminuted charcoal (PPL). (B) Dusty clay coating (PPL).

A dark organic layer buried by Oligocene estuarine sediments in the area of Leuven (Belgium) has traditionally been interpreted as the remnant of Cenozoic oil seepage. A micromorphological study by Van Herreweghe et al. (2003) shows a strong accumulation of organic matter (dark bands) occurring as coatings with a characteristic polygonal cracking pattern and as bridges between quartz grains, corresponding to features common in present-day hydromorphic podzols of temperate and tropical regions (see Van Ranst et al., 2018). The buried palaeosoil can therefore be interpreted as a giant buried spodic horizon.

2.5   Buried Soils in Colluvial Deposits

Cover beds (or colluvial deposits) can set apart pedological bodies that formed during phases of geomorphological stability in contexts that are overall dominated by slope dynamics. They can be considered as 'geomorphologically' buried palaeosoils.

A Late Quaternary, widespread, complex sequence of colluvium and intercalated palaeosoils from South Africa has been investigated by Botha and Fedoroff (1995), who identified clay coatings and iron oxide hypocoatings along channels as expression of the dominant soil-forming processes, pointing to conditions of seasonally impeded drainage and strong biological activity. Fragments of red limpid clay coatings occurring in voids, and infillings composed of red colluvium, suggest short-distance transport of material derived from eroded soil profiles upslope. In the same study, homogenous, typic impregnative nodules are interpreted as detrital petroplinthic pedorelicts derived from ferruginous mottles.

Soils located above the present-day treeline in the Northern Apennines (Italy) can be subdivided into an upper unit, composed of the superficial soil of colluvial origin, and a lower unit, constituted by the underlying more developed palaeosoil (Compostella et al., 2014). The colluvial unit is characterised by a grain-dominated microstructure, a dominance of weakly weathered claystone fragments in the coarse fraction and the presence of subrounded soil aggregates from dismantled older horizons, whereas the lower unit has a blocky microstructure, common channels, a dominance of moderately to strongly weathered sandstone fragments in the coarse fraction and three generations of clay coatings, which are, respectively, microlaminated, non-laminated and dusty. A similar profile was observed by Cremaschi et al. (1984a) at the Bagioletto site (northern Apennine), where a debris flow, activated by periglacial processes in the Little Ice Age, covers a palaeosoil consisting of a Bt horizon with Fe-rich microlaminated clay coatings in voids (Fig. 4). This palaeosoil includes Mesolithic and Chalcolithic archaeological material, situating its formation in the Boreal to the Subboreal periods.

2.6   Pre-Cainozoic Buried Soils

Palaeosoils that are part of the geological record are often lithified and testify significant, and sometimes abrupt, climate changes. They were buried by subsequent sedimentary events and became lithified after burial. In some cases, they have been affected by

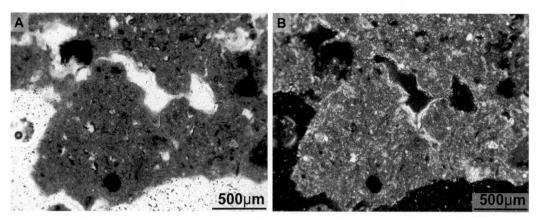

FIGURE 4 Clay illuviation features in buried palaeosoil beneath colluvial deposits (Bagioletto, northern Apennine, Italy). (A) (PPL). (B) XPL.

heating, metamorphism and/or deformation, which tends to mask pedogenic features, as demonstrated by an extensive study of Precambrian palaeosoils in Canada (Gall, 1992). Microscopic evidence of diagenetic mechanical and geochemical overprinting includes compaction, development of veins that cross-cut pedogenic features, and widespread alkali and alkaline earth element metasomatism (Gall, 1992). Soils can also be covered by basalt flows or tephra (e.g., Benayas et al., 1987; Wieder et al., 1990; Sheldon, 2003; Marques et al., 2014), isolating them from surface processes (Sheldon, 2006). A common issue in the interpretation of palaeosoils in the geological record is to make a distinction between factors responsible for pedogenesis, acting on the fresh parent material (i.e., clay illuviation, nodule formation), and those related to early to late diagenetic processes (i.e., calcite recrystallisation, dolomitisation) (Teruggi & Andreis, 1971; Retallack & Wright, 1990; Retallack, 1991, 1992; Srivastava & Sauer, 2014). Pedogenesis and diagenesis may produce analogous effects, but the spatial relationships of microscopic features, as well as analysis of features related to specific processes, usually allow attributing each feature either to pedogenesis or to postburial processes. Specific lithofacies corresponding to buried soils can be important stratigraphic markers, allowing long-distance correlation between outcrops (e.g., Berra et al., 2010).

Buried palaeosoils that are part of the geological record are widespread in a number of specific sedimentary rock sequences, for instance, in the sandstones interlayered between limestones belonging to platform depositional systems. In coastal settings, sediments can record abrupt environmental changes, corresponding to important sea-level drops. The latter can be responsible for emersion and the onset of karstification and weathering (e.g., Berra et al., 2010; Meister et al., 2013). At several localities of the Dolomites, the Norian Dolomia Principale Formation is capped or interlayered by (polycyclic) discontinuous Terra Rossa-type palaeosoils, up to 10 m thick. These Terra Rossa-type palaeosoils display a reddish micromass (consisting of diagenetic dolomitic microsparite), dark clay infillings and coatings (Fig. 5A), brownish ferruginous intrapedal

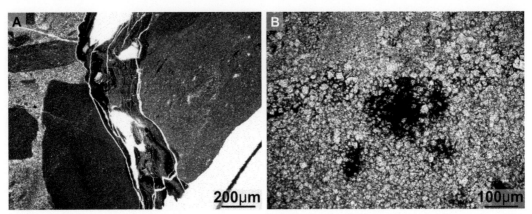

FIGURE 5 Residual pedofeatures in Terra Rossa-type palaeosoils developed in limestone (Southern Alps, Italy). (A) Coarse clay infilling of a fracture (Malga Flavona Member) (PPL). (B) Ferruginous nodule in dolomitic limestone (Malga Flavona Member) (PPL).

nodules (Fig. 5B), and traces of bioturbation (root casts), pointing to a warm and wet tropical pedoclimate (Trombino et al., 2006; Berra et al., 2016). Well-preserved palaeosoils interlayered in the Triassic Calcare Rosso limestone from the southern Italian Alps are attributed to interplay between extensive karstification and pedogenesis after emersion (Berra et al., 2011; Vola & Jadoul, 2014). These Terra Rossa palaeosoils have a dark red marly calcareous matrix, with greenish mottles when micaceous minerals are prevalent, or with reddish-yellow to brownish-yellow mottles when dolomite predominates (Binda, 2011; Vola & Jadoul, 2014). In a few cases, they contain quartz grains, vaguely outlined soil aggregates and calcitic root structures; these features possibly originated after pedogenesis under warm and wet environmental conditions.

Examples of palaeosoils in non-calcareous sedimentary successions include lithified palaeosoils contained within overbank sediments that have been described for the Himalayan Foreland (Srivastava et al., 2013; Srivastava & Sauer, 2014). Despite diagenetic alteration, evidence of palaeopedogenic processes is preserved in the form of striated b-fabrics (because of pedogenic slickenside development), dense impure micrite nodules with diffuse boundaries, root channels and thick microlaminated iron-rich clay coatings. These horizons were interpreted as ferruginous palaeosoils developed during the Oligocene under tropical weathering. In another micromorphological study of Himalayan Foreland deposits, the well-developed red colour of palaeosoils was mainly attributed to burial diagenesis, rather than to weathering and soil development under tropical climatic conditions (Singh et al., 2009).

Sandy to muddy lithified fluvial sequences can preserve evidence of former pedogenesis. For instance, palaeosoils in the Maghra El-Bahari Formation (Upper Cretaceous-Lower Tertiary, Eastern Desert of Egypt) display well-oriented clay coatings and infillings that are incorporated into the micromass by vertic processes, as well as palygorskite fibres, indicating pedogenesis under arid conditions, and small iron oxide nodules (pointing to hydromorphic soil conditions) (Wanas & Abu El-Hassan, 2006). The occurrence of pedogenic spheroidal ferruginous nodules and pyrite in the groundmass of palaeosoils of the Paddy Member of the Albian Peace River Formation (Western Canada) was explained by Ufnar et al. (2005) by hydromorphic conditions and the production of sulphides by sulphate-reducing bacteria in the soil water. Micromorphological studies of palaeosoil sandstone deposits have been performed at several localities. For instance, a palaeosoil from the Lower Triassic Bunter Sandstone of Germany (the Karneol-Horizont), consisting of a lower polygenetic calcrete with dolomite and massive silica replacement of carbonates and an upper interval showing vertic features was described by Martins and Pfefferkorn (1988). In England, the Old Red Sandstone Formation preserves micromorphological evidence of calcrete formation, thus suggesting pedogenesis under arid environmental conditions during the Permian (Allen, 1986).

Buried palaeosoils on igneous and metamorphic rocks are well preserved in some cases. A red palaeosoil on Lower Palaeozoic low-grade metamorphic rocks of the Brabant Massif (Belgium), covered by Cretaceous marine sediments, could be identified as a truncated ferralitic saprolite, based on preserved lithic fabrics and kaolinite alteromorphs (Stoops, 1992; Mees & Stoops, 1999). In intrabasaltic palaeosoils at the Giant's Causeway of Ireland, small segregated areas of iron oxides surrounded by green and grey zones point to former reducing conditions (Phillips et al., 2010).

3. Relict Soils

Relict soils are surface soils showing pedogenic features that formed in past pedoclimatic conditions, different from present-day conditions (Ruellan, 1971). Such palaeofeatures are most easily recognised when they record a stronger degree of soil development or weathering than present-day local soils; therefore they are not significantly affected by current environmental conditions. A particular but rare type of relict soils are exhumed soils, which are formerly buried soils that have been brought to surface by erosion.

Relict features related to ice segregation processes, in the form of lenticular platy microstructures and fragmented clay coatings, have been described by Kühn (2003) and Kühn et al. (2006) for Bt horizons of Pleistocene and Holocene Luvisols and Albeluvisols of North Germany, allowing to attribute the initial phases of pedogenesis to Late Glacial periglacial conditions. Systematic micromorphological investigations of soil profiles with sand wedges revealed the occurrence of numerous rounded fragments of limpid clay coatings within frost-related lenticular microplates (Kühn, 2003), whose disruption occurred during the Late Glacial. The co-occurrence of fragmented and intact clay coatings confirms the incidence of subsequent Holocene clay illuviation, allowing reconstruction of a sequence of multiple phases of clay illuviation.

Late Tertiary to the Holocene relict soils, both exhumed and not exhumed, have been described for the Central Sahara by Cremaschi and Trombino (1998b). Reddish, undifferentiated to striated micromass, clay coatings (and fragments of coatings) and

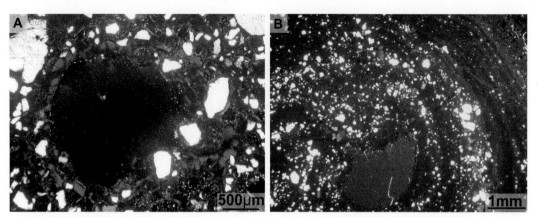

FIGURE 6 Pedofeatures in relict palaeosoils of the Central Sahara. (A) Clay coating (XPL). (B) Detail of a typic, Fe-rich nodule with thick kaolinite-goethite coatings enclosing few quartz grains, remaining after dismantling of a pre-Quaternary lateritic soil (XPL).

amorphous iron oxide nodules point to pedogenesis during warm and humid phases (Fig. 6A). In the same region, superimposed relict B horizons were found to record several humid phases promoting pedogenesis, as shown by the degree of mineral grain weathering (including quartz), reddening of the micromass, striated b-fabric, and the occurrence of ferruginous nodules and clay coatings (Zerboni et al., 2011), In this study, micropedological indicators suggest that throughout the Quaternary the degree of expression of pedogenic features requiring water availability progressively decreased in the region. In the groundmass, rubification and iron oxide nodule development took place during the earliest weathering phases, whereas the most recent phases are mainly characterised by pedoturbation and calcite mobilisation.

Relict soils within pockets at the top of the limestone-bearing Rohri Hills (Sind, Pakistan) contain abundant Fe/Mn oxide nodules (lateritic concretions) in a clayey micromass, revealing pedogenesis under more humid conditions than those existing at present (Biagi & Cremaschi, 1988). The same Fe/Mn oxide nodules are also scattered along the hilltop surface, suggesting dismantling of the original tropical soil.

In the Sahara, where today erosion is the main surface process (Swezey, 2001), little evidence of *in situ* palaeosoils is preserved, although other features related to old weathering phases may have survived. On the top of the Tadrart Acacus Massif typic, Fe-rich nodules with thick kaolinite-goethite coatings including few quartz grains (Fig. 6B) were observed as part of the clastic fraction of the desert pavement. They were interpreted as the sole remnant of pre-Quaternary lateritic soils, from which the groundmass had been completely removed (Zerboni et al., 2015a).

In the Ligurian Alps (Northern Italy), a representative profile comprises (i) a lower interval with an angular blocky microstructure, strongly weathered coarse mineral grains including runiquartz (Eswaran et al., 1975; see also Marcelino et al., 2018; Stoops &

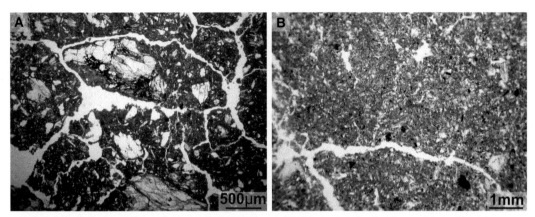

FIGURE 7 Micromorphological features of a profile comprising a palaeosoil and a relict soil (Ligurian Alps, Italy). (A) Groundmass of the strongly weathered palaeosoil, recording subtropical to tropical conditions (PPL). (B) Groundmass of the surface relict soil, developed in the overlying loess (PPL).

Marcelino, 2018), a reddish micromass (Fig. 7A), two generations of clay coatings and infillings (i.e., limpid reddish clay coatings and dense infillings, and thin yellowish clay coatings) and few alteromorphic iron oxide nodules; and (ii) the overlying soil in the loess cover, with a granular microstructure, unweathered minerals, yellow micromass (Fig. 7B), few clay coatings (i.e., microlaminated yellowish clay coatings and infillings, with sharp extinction lines in XPL), and orthic and anorthic iron oxide nodules (Rellini et al., 2009). The micromorphological characteristics of the lower interval are considered to be comparable to those of strongly weathered present-day subtropical to tropical soils, whereas those in the upper interval show features comparable to Late Glacial interstadial soils of central Europe.

On the slopes of a limestone plateau 400 m a.s.l. in Northern Italy, Trombino and Ferraro (2002) identified a soil profile composed of two different pedological units: a Terra Rossa relict fersiallitic palaeosoil with reddish micromass and striated b-fabric (due to incorporation of clay coatings), overlain by a thin soil developed from loess, with yellowish micromass and crystallitic b-fabric. Terra Rossa palaeosoils on the calcareous plateaus of Western Europe are considered to be examples of relict palaeosoils (Duchaufour, 1983). They developed during the Late Tertiary and the Pleistocene, persisting today in a completely different pedoclimatic regime.

4. Polygenetic Soils

Polygenetic soils (Bos & Sevink, 1975) developed in the course of multiple periods of pedogenesis that were not separated by periods of significant erosion or sedimentation. This can result in superposition of features that formed in contrasting environmental settings, which can be distinguished by thin section studies (Tursina, 2009).

4.1   Polygenetic Soil on Late Pleistocene Loess

The sequences of Late Pleistocene palaeosoils in the Paris Basin (Fedoroff & Goldberg, 1982) are a paradigmatic example of polygenetic soils on loess. Different pedogenic cycles are represented by (i) yellowish red, laminated clay coatings formed in a phase of clay illuviation in the last interglacial, (ii) hydromorphic degradation related to cooling at the beginning of the last glacial phase (not based on micromorphological features), (iii) reddish brown, dusty, laminated clay coatings formed by periglacial degradation during the Last Glacial Maximum, and (iv) various stages of weak clay illuviation (yellow laminated clay coatings) in the Holocene.

Different generations of pure clay coatings, alternating with coarse clay to fine sand coatings, were observed for the Bagaggera profile (Northern Italy) (Fig. 8), suggesting alternating periods of soil development in temperate climate and periods marked by structure degradation in cooler climatic conditions (Cremaschi et al., 1984b). The interpretation of the micropedological observations is supported by luminescence dating of the sequence and the occurrence of Mousterian and Aurignacian lithics within the profile, allowing age determination of distinct pedogenetic phases (Cremaschi et al., 1990).

In the Ghiardo profile, at the southern edge of the Po Plain, a different type of evolution is revealed by micromorphological analysis (Cremaschi et al., 2015), with occurrences of small charcoal particles, and granular and platy microstructures induced by frost (Cremaschi & Van Vliet-Lanoë, 1990), which are superimposed by features recording strong vertic behaviour in the form of clay coating fragments (Fig. 9A) and striated b-fabric (Fig. 9B). Textural pedofeatures are widespread in the E horizon; layered coatings and infillings with alternating dusty and silty clay confirm that the topsoil was exposed for some time and reworked in response to Chalcolithic deforestation.

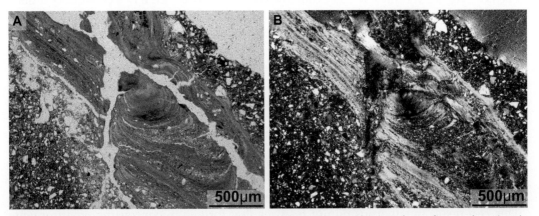

FIGURE 8 Compound coating composed of alternating pure clay coatings and coarse clay to fine sand coatings in a polygenetic soil (Bagaggera loess sequence, northern Italy). (A) PPL. (B) XPL.

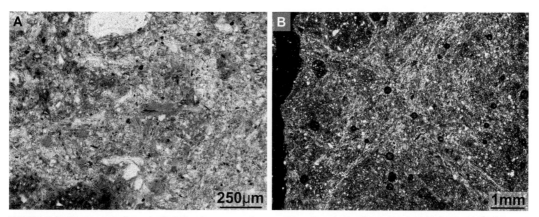

FIGURE 9 Features recording vertic behaviour, superimposed on preexisting features, in a polygenetic palaeosoil (Ghiardo profile, northern Italy). (A) Fragments of clay coatings (PPL). (B) Striated b-fabric (XPL).

A Mid-Upper Pleistocene pedosedimentary sequence in the central Po Plain (Monte Netto, Northern Italy), consists of several loess layers showing different degrees of weathering (Zerboni et al., 2015b). The lower part of the sequence comprises a strongly weathered (rubified) palaeosoil developed on colluvial sediments (Mid-Pleistocene age), covered by three different loess intervals showing different degrees of pedogenesis, expressed by microstructure type (ranging from massive to subangular blocky), micromass colour, and, mainly, pedofeature characteristics: (i) thickness, colour and grain-size of textural pedofeatures (i.e., yellowish to orange clay coatings and fragments of coatings); (ii) type and density of amorphous pedofeatures (i.e., iron and manganese oxide nodules, with both sharp and diffuse boundaries, and Fe/Mn oxide hypocoatings); and (iii) presence of iron oxide-depleted areas in the groundmass. The polygenetic nature of the Monte Netto pedosequence is testified by the juxtaposition or superimposition of the these pedofeatures, with, for instance (from oldest to most recent), stratigraphic superimposition of fragmented red clay coatings, yellow layered clay coatings, orange and limpid clay infillings, and yellowish brown dusty clay coating, allowing recognition of several subsequent phases of illuviation, with decreasing intensity (Fig. 10).

4.2   Vetusols

Vetusols (Cremaschi, 1987), also called archaeosoils (Hubschman, 1975), are thick polygenetic soils related to ancient surfaces, developed under a particular and specific set of climate-controlled pedogenic processes (e.g., decarbonatation, mineral weathering, clay illuviation, rubification) persisting for a long time (at least since the Lower-Middle Pleistocene), whereby their formation is controlled by geomorphological surface stability. Vetusols seem to be less sensitive to the Quaternary climate changes than other types of palaeosoils (Cremaschi & Busacca, 1994; Busacca & Cremaschi, 1998).

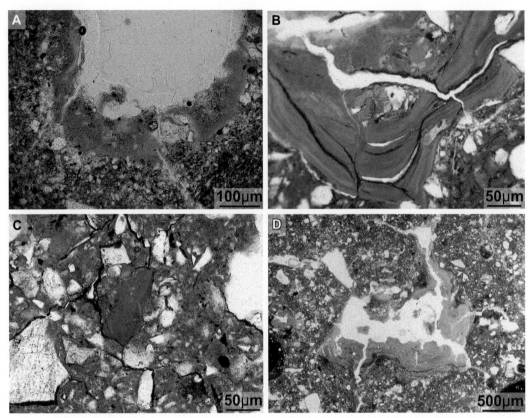

FIGURE 10 Diverse textural pedofeatures in a polygenetic soil (Monte Netto, northern Italy). (A) Yellowish brown dusty clay coating (PPL). (B) Orange, limpid clay infilling (PPL). (C) Fragment of red clay coating (PPL). (D) Layered yellowish clay coating (PPL).

The palaeosoil's chronosequence on the terraced Trebbia River alluvial fans (Northern Italy) includes five members, dating from the Early/Middle Pleistocene to the Early Holocene. The different members of the chronosequence, from youngest to oldest, are characterised by increasing thickness of the weathering profiles, as well as increasing complexity and number of their horizons, mineralogical maturity and number of identifiable loess covers (Busacca & Cremaschi, 1998). The presence of several generations of clay coatings (Fig. 11), often juxtaposed to coarse/dusty coatings (see Kühn et al., 2018), and of reddish sand-sized aggregates rich in kaolinite and hematite (i.e., pseudosands; Cremaschi & Sevink, 1987) allows correlation between the oldest member of the chronosequence and relict palaeosoils of the region.

The Collecchio palaeosoil, a Ferretto-type palaeosoil in Northern Italy, displays a deeply weathered ferralitic soil characterised by abundant clay coatings and common large Mn-bearing and ferruginous concentric nodules as well as diffuse iron oxide impregnations of the micromass, pointing to a plinthitic horizon, developed under past

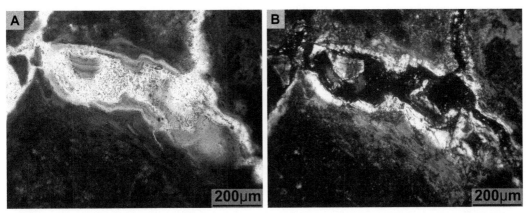

FIGURE 11 Juxtaposed clay coatings of different generations in a Vetusol (Val Trebbia, northern Italy). (A) PPL. (B) XPL.

tropical environmental conditions (Ferrari & Magaldi, 1968). The so-called Ferretto palaeosoil was therefore interpreted as a marker for the long Mindel-Riss interglacial, but more recent studies have identified it as a Vetusol (Billard, 1995).

4.3   Other Occurrences

In polygenetic soils along stable surfaces in the Gangetic Plains (India), clay illuviation was more intense during the most recent of several recognised pedogenic phases, as indicated by enrichment of the groundmass of clay intercalations and strongly bire-fringent microlaminated clay pedofeatures. Clay pedofeatures that formed during earlier stages are degraded, in the form of bleaching, fragmentation, loss of preferred orientation and development of a coarse speckled appearance (Srivastava & Parkash, 2002). Moreover, in the older phase, the formation of irregularly shaped carbonate nodules with inclusions of groundmass material suggests the occurrence of a more recent dry phase, followed by dissolution-reprecipitation and partial to complete removal of carbonates from the soil during wetter conditions. In the Granada Basin (SE Spain), discontinuous sedimentation and soil development episodes during the Pleistocene were governed by pulses of tectonic uplift, whereby the main pedogenic processes included leaching of carbonates, clay illuviation, and rubification, but the degree of development of the weathering horizons varied over time (Ortiz et al., 2002). Those soils, formed on Early Pleistocene surface show the strongest development, but after their formation they were partially truncated and recalcified, thus resulting in polygenetic soils. In the most recent soils, relative thickness of clay coatings, increasing progressively in more ancient soils, confirms the occurrence of a long pedogenic phase, whereas the occurrence of frost-fragmented clay coatings is related to the occurrence of cold episodes following Bt horizon formation.

In a palaeopedological study of deposits of the Quaternary intramontane basin of the Val d'Agri (Southern Italy), Zembo et al. (2012) described several different spatial relationships between pedogenic features, corresponding to temporal relationships among pedogenetic processes identified, indicative of highly variable climatic conditions and/or water regime: (i) sharply delimited nucleic concentric ferruginous nodules with juxtaposed amorphous iron oxide hypocoatings; (ii) compound juxtaposed reddish clay coatings and overlying limpid yellowish clay coatings, and (in other samples) reddish and limpid yellowish clay coatings; (iii) different generations of sparite coatings juxtaposed to micritic nodules, or acicular calcite crystals locally juxtaposed to calcite coatings and infillings; and (iv) anorthic iron and manganese oxide nodules with silt and clay coating. These observations allowed identifying six pedogenic cycles characterised by different soil-forming processes acting under specific palaeoenvironmental conditions, from old to young: (i) warm-humid conditions in a subtropical-tropical climate (Chromic Luvisol development phase); (ii) moderately semiarid environment (Petric Calcisol development phase); (iii) warm conditions with seasonal water deficit (second Chromic Luvisol development phase); (iv) increasing seasonality and/or moderate aridity (Gleic Vertisol development phase); (v) humid-temperate climatic conditions with forest vegetation (third Chromic Luvisol development phase); and (vi) decline of vegetation cover during climate deterioration (Calcic Luvisol development phase).

Examples of palaeosoils recording polycyclic pedogenesis include occurrences in Western Liguria (Northern Italy), showing compound juxtaposed clay coatings (Rellini et al., 2007) (Fig. 12). A younger generation of yellow limpid clay coatings is characterised by sharp or diffuse extinction lines, pointing to weak Holocene pedogenesis under a temperate-humid climate. It is juxtaposed to an older generation of reddish dusty clay coatings, showing a stronger degree of assimilation into the groundmass. The latter are similar to those found in deeply weathered tropical soils and are not compatible with the present climate. For the same area, Rellini et al. (2015) described another example of polycyclic pedogenesis, based on compound juxtaposed but discordant clay coatings. The younger generation is represented by yellowish-grey dusty clay coatings and/or

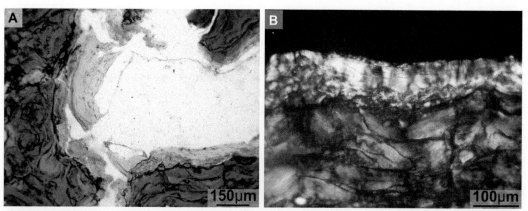

FIGURE 12 Juxtaposed yellowish and reddish clay coatings in petroplintite (Liguria, Italy). (A) Overview (PPL). (B) Detail (XPL).

dense infillings, pertaining to a Holocene pedogenesis, whereas thin reddish brown limpid clay coatings and dense infillings, without lamination but with sharp extinction lines, are observed as an older generation, indicative of a strong illuvial phase that happened during the Middle Pleistocene. The discordance between both generations of clay coatings suggests an interruption of pedogenesis.

5. Reworked Soil Material

The presence of reworked soil material, either as soil horizon fragments or in more dispersed form, demonstrates the former occurrence of soils that formed in past environmental conditions, and it can indicate the existence of palaeosurfaces that have been dismantled by erosion.

Mroczek (2013) demonstrated the usefulness of reworked soil fragments in discriminating between undisturbed and recycled loess (Smalley, 1972; Pécsi & Richter, 1996) and thus to establish a relationship between the youngest loess cover (and related palaeosoils) and older loess deposits. Deformed clay infillings and rounded fragments of clay coatings partly with undulating extinction bands, and rounded Fe and Mn/Fe oxide nodules are inherited from older, pre-altered sediments.

Rellini et al. (2015) reported the presence of lateritic soils fragments (see also Marcelino et al., 2018; Stoops & Marcelino, 2018) in a coastal pedosedimentary sequence developed in NW Italy during the Early Quaternary. The fragments are Fe oxide-impregnated nodules (Fig. 13), often containing runiquartz and strongly weathered mineral grains. They record the former existence upslope of strongly weathered palaeosoils that are now completely dismantled. Moreover, these lateritic nodules are responsible for iron oxide enrichment and hardening of soil horizons, resulting in the development of a petroplinthic horizon (IUSS Working Group WRB, 2015), in a context that should not have been favourable for its formation. Strongly weathered reworked soil fragments, described as Fe-rich amorphous nodules (see Fig. 6), have also been

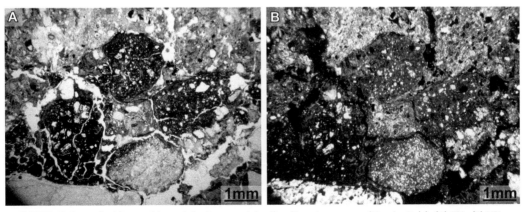

FIGURE 13 Reworked iron oxide nodules in a coastal soil-sediment sequence (Liguria, Italy). (A) PPL. (B) XPL.

identified in other environmental contexts, for instance, by Zerboni et al. (2011) in palaeosoils from the Libyan Sahara. They suggest the presence of former ferrugineous soils that developed during the Middle Pleistocene, and which were subsequently truncated, reworked and incorporated into younger soils.

Reworked soil fragments generally help in identifying phases of strong weathering that occurred in the past, but in some cases, they also help in identifying phases of less intense soil development. Compostella et al. (2014) identified subrounded to subangular soil aggregates (Fig. 14) in soil profiles located above the present-day treeline at about 1800 m a.s.l. in the Northern Apennines (Italy), which seem to originate from dismantled palaeosoils characterised by strong mineral alteration and clay illuviation, which are today inactive in the area. The abundance of these fragments of pre-altered soil material could explain the apparent (at field description level) strong development of the horizons in which they are found.

Reworked soil fragments can be found also in infillings of many kinds of sedimentary traps, including sinkholes and karst pockets. The latter include occurrences in Late Pleistocene aeolianites of the Yucatán peninsula, for which reworked fragments of clay coatings and redoximorphic features were observed, pointing to continuous pedogenesis in humid conditions, lasting several thousand years during the Early-Middle Holocene. Thick recalcified pedosediments in the same pockets indicate redeposition of soil-derived material during a late phase of landscape change that could be related to ancient Maya land-use (Cabadas-Báez et al., 2010). In the pedogenic fill of a sinkhole in Central Syria, rounded aggregates of carbonate-free soil material occur (Fig. 15), attesting the former presence of a currently dismantled fersiallitic soil (Trombino, 2007). Reworked soil fragments, including anorthic Fe/Mn oxide concretions, found in the uppermost horizon of loess sequences in northern Italy also suggest colluvial reworking of wind-blown sediments and their reorganisation promoted by human-induced deforestation (Cremaschi et al., 2015).

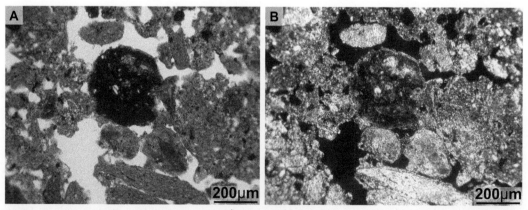

FIGURE 14 Subrounded to rounded soil aggregate, in a profile above the treeline, interpreted as being derived from a relict forest soil (northern Apennines, Italy). (A) PPL. (B) XPL.

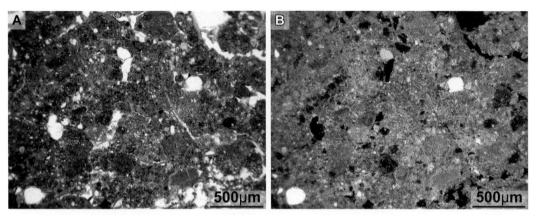

FIGURE 15 Rounded aggregates of reddish non-calcareous soil material and fragments of clay coatings, in calcareous sinkhole deposits (central Syria). (A) PPL. (B) XPL.

6. Weathering Surfaces and Rock Coatings

The first step of soil development on rock substrates is alteration along subaerially exposed bare rock surfaces (Blume et al., 2016). Weathered surfaces and coatings on rock outcrops (Dorn, 1996) can therefore be considered as palaeosoils, recording palaeo-environmental information. Weathering rinds represent the initial step of the decay of the rock outcrop, whereas rock coatings are surface accretions, related to weathering. Thin section studies of weathering surfaces and rinds have mainly focussed on weathering products and salt accumulation, and on microorganisms and their by-products, aimed at understanding the nature and rates of weathering processes, to interpret their geomorphological and palaeoclimatic significance (e.g., Dorn et al., 1992; Dorn, 1996; Viles & Goudie, 2007; André et al., 2008; Liu & Broecker, 2008; Favero-Longo et al., 2011; Dorn et al., 2013a, 2013b; Goldsmith et al., 2014; Marszałek et al., 2014). Micromorphology may help the investigation of various cavernous surface features (e.g., tafoni, honeycomb structures), analysing the loss or accumulation of mineral and organic constituents (Guglielmin et al., 2005; Mol & Viles, 2012; Dorn et al., 2013a). Because most authors have used different terminologies for their descriptions, Zerboni (2008) suggested the use of the concepts and terminology proposed by Stoops (2003) for sake of uniformity.

Rock varnish is a dark brown to black coating that in arid and semiarid lands covers stable subaerially exposed rock surfaces (Dorn & Oberlander, 1981; Dorn, 1996, 2009). It consists mainly of clay minerals and Mn/Fe oxides (Potter & Rossman, 1977), but phosphates, sulphates, carbonates and organic particles can also be present (Dorn et al., 1989; Cremaschi, 1996; Zerboni, 2008). At the microscopic scale, rock varnish is generally layered, with each layer corresponding to specific environmental settings. In Mn oxide-bearing rock varnish of the Sahara (Zerboni, 2008), the basal layer consists of laminated red clay coatings and infillings, occupying the space between quartz grains (Fig. 16),

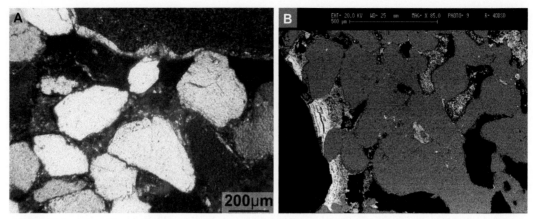

FIGURE 16 Rock varnish on sandstone. (A) Rock varnish comprising, from base to top, clay infillings of packing pores, followed by a dense opaque Mn/Fe oxide-rich layer, followed by a thin clay coating (Central Sahara) (XPL). (B) SEM image of a similar feature.

formed by clay illuviation under wet environmental conditions. The overlying layer developed during progressively more arid conditions that promoted the formation of manganese oxides, which impregnated the previously deposited clay (see also Vepraskas et al., 2018). The resulting impregnation is generally dense and diffuse, becoming less intense toward the bottom. Finally, thin dusty coatings form the outermost layer, corresponding to deposition of dust under arid climate conditions. Rock varnish has been described also for tropical regions as a product of past environmental conditions. In southern Katanga (DR Congo), residual lateritic gravel on exposed surfaces is covered by varnish, whose colour and fabric are different for different planation levels (Alexandre, 2002). Also in arid regions rock varnish is a good indicator of the relative chronology of geomorphological units (Dorn, 1988; Hooke & Dorn, 1992). In Antarctica, rock varnish developed on sandstone is more complex than on granitic rocks. It consists of alternating layers of dusty silica-rich glass, rich in Al and alkali elements (Na, Ca, K), and dark opaque jarosite-bearing layers (Zerboni & Guglielmin, 2016). The silica- and jarosite-bearing laminae can be very thin and sometimes have a stromatolithic aspect (Fig. 17). On the external part of rocks, they form continuous coatings, whereas in the inner part of the sandstone they occur as coatings and infillings along intergranular voids.

Mn/Fe-bearing crusts and coatings representing the classical Mn-rich varnish are thicker and enclose quartz grains, commonly with pellicular coatings on grains that may have a fibrous structure, resulting from biomineralisation of former fungal hyphae replaced by Mn/Fe oxides (Mergelov et al., 2012; Zerboni & Guglielmin, 2016). Voids within rock surface deposits can be filled with gypsum, formed from components that are commonly available in local aerosols and they can also contain biogenic oxalate accumulations as by-products of algal and fungal metabolism.

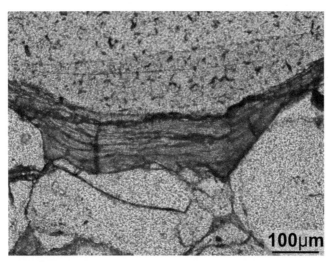

FIGURE 17 Complex rock varnish on sandstone, composed of alternating layers of pale siliceous material and dark fine-grained jarosite (Antarctica) (PPL).

7. Palaeosoil Development Indices

Several authors have proposed micropedological indices to quantify the degree of palaeosoil development. The construction of micropedological indexes requires a good understanding of soil micromorphology and the possibility to analyse a large set of slides to reach statistically validated assumptions. One example is the Micromorphological Soil Development Index (MISODI) (Magaldi & Tallini, 2000), proposed to assess soil development and degree of weathering of palaeosoils of Central Italy. This index considers microstructure types, b-fabric patterns, frequency and thickness of clay and iron oxide coatings, frequency and size of iron nodules, and degree of alteration of coarse mineral grains. Based on this index, Khormali et al. (2003) proposed the MISECA (Micromorphological Index of Soil Evolution in highly Calcareous arid to semiarid Conditions) index of soil evolution for highly calcareous materials in arid and semiarid environments. It takes into account roughly the same features as the MISODI index, with addition of the frequency of calcite depletion features. The S_{BC} index (Xiaomin et al., 1994) is another micropedological indicator suggesting arid or subhumid environments. In this case, the ratio between the volume of channels (and excrements) and that of calcite grains that are part of the groundmass is considered to be sensitive to environmental conditions, with subhumid environments represented by the palaeosoil intervals, separated by interval recording more arid episodes. Finally, even if they were not conceived expressly for palaeosoils, indices expressing aspects of clay illuviation (Miedema & Slager, 1972; see also Kühn et al., 2018) can be used to compare the degree of development of palaeosoils comprising Bt horizons (Rellini et al., 2015).

8. Conclusions

The micromorphological study of palaeosoils is an important tool to investigate the processes that promoted soil formation and weathering in the past (e.g., Fedoroff, 1967, 1971; Kemp, 1998, 1999), allowing to obtain information about palaeoclimates and palaeoenvironments. One prerequisite to obtain a reliable palaeoclimatic reconstruction is that micropedological characteristics of present-day analogous soils have to be well known and understood.

Although micromorphology has been applied for several decades in palaeosoil interpretation (Kubiëna, 1963), deducing palaeoenvironmental and palaeoclimatic information from soils remains a challenging process, requiring careful analysis of soil characteristics. Pedogenic features are the result of the concomitant action of many processes, whereby climate plays a fundamental role, but also topography, vegetation and parent material are important, and, particularly in palaeosoil studies, a careful consideration of the factor time is mandatory. Only if the effect of factors other than climate can be considered to be minimal or invariant, it will be possible to infer the true role of past climate in the formation of palaeosoils.

Micropedological investigation also offers a fundamental tool to examine pedo-complexes and pedosedimentary sequences, in which palaeosoils alternate with un-weathered (or very weakly weathered) sediments (e.g., Fedoroff & Goldberg, 1982; Kemp et al., 1995; Kemp, 1999; Kemp & Zárate, 2000; Mroczek, 2013; Zerboni et al., 2015b).

From a theoretical point of view, no micromorphological feature can be considered by itself to be indicative of past processes: all the features that can be observed in thin section can record past pedogenic phases, if they are not in equilibrium with present-day pedoenvironmental conditions, in terms of both climate and landscape position.

References

Alexandre, J., 2002. Les Cuirasses Latéritiques et Autres Formations Ferrugineuses Tropicales. Exemples du Haut Katanga Méridional. Annales du Musée Royal de l'Afrique Centrale, Sciences Géologiques 107, 118 p.

Allen, J.R.L., 1986. Pedogenic calcretes in the Old Red Sandstone facies (Late Silurian-Early Carboniferous) of the Anglo-Welsh area, southern Britain. In Wright, V.P. (ed.), Palaeosols: Their Recognition and Interpretation. Blackwell Scientific, Oxford, pp. 58-86.

Amiotti, N., Blanco, M. del C. & Sanchez, L.F., 2001. Complex pedogenesis related to differential aeolian sedimentation in microenvironments of the southern part of the semiarid region of Argentina. Catena 43, 137-156.

André, M.F., Hall, K., Bertran, P. & Arocena, J., 2008. Stone runs in the Falkland Islands: periglacial or tropical? Geomorphology 95, 524-543.

Benayas, J., Barragan, E., Galvan, J., Hernando, J., Palomar, M.L. & Roquero, C., 1987. Microscopy and chemical composition of paleosols affected by the heating of basaltic flows ('almagres') in the island of Tenerife, Spain. In Fedoroff, N., Bresson, L.M. & Courty, M.A. (eds.), Micromorphologie des Soils, Soil Micromorphology. AFES, Plaisir, pp. 591-596.

Berra, F., Jadoul, F. & Anelli, A., 2010. Environmental control on the end of the Dolomia Principale/Hauptdolomit depositional system in the central Alps: coupling sea-level and climate changes. Palaeogeography, Palaeoclimatology, Palaeoecology 290, 138-150.

Berra, F., Jadoul, F., Binda, M. & Lanfranchi, A., 2011. Large-scale progradation, demise and rebirth of a high-relief carbonate platform (Triassic, Lombardy Southern Alps, Italy). Sedimentary Geology 239, 48-63.

Berra, F., Carminati, E., Jadoul, F. & Binda, M., 2016. Does compaction-induced subsidence control accommodation space at the top of prograding carbonate platforms? Constraints from the numerical modelling of the Triassic Esino Limestone (Southern Alps, Italy). Marine and Petroleum Geology 78, 621-635.

Biagi, P. & Cremaschi, M., 1988. The early paleolithic sites of the Rohri Hills (Sind, Pakistan) and their environmental significance. World Archaeology 19, 421-433.

Billard, A., 1995. Le mythe du 'grand interglaciaire' Mindel-Riss d'après l'étude des sols du Nord de Turin (Italie). Antropozoikum 22, 5-62.

Binda, M., 2011. Facies Distribution of a Rimmed Carbonate Platform and Overlying Regressive Carbonates: the Esino Limestone and Calcare Rosso Facies in the Central Southern Alps (Lombardy). PhD Dissertation, Università degli Studi di Milano, 267 p.

Blume, H.P., Brümmer, G.W., Fleige, H., Horn, R., Kandeler, E., Kögel-Knabner, I., Kretzschmar, R., Stahr, K. & Wilke, B.M., 2016. Scheffer/Schachtschabel: Soil Science. Springer, Heidelberg, 618 p.

Bos, R.H.G. & Sevink, J., 1975. Introduction of gradational and pedomorphic features in descriptions of soils. Journal of Soil Science 26, 1365-2389.

Botha, G.A. & Fedoroff, N., 1995. Palaeosols in Late Quaternary colluvium, northern KwaZulu-Natal, South Africa. Journal of African Earth Sciences 21, 291-311.

Bronger, A., 1972. Zur Mikromorphologie und Genese von Paläoböden auf Löss im Karpatenbecken. In Kowalinski, S. & Drozd, J. (eds.), Soil Micromorphology. Panstwowe Wydawnicto Naukowe, Warsawa, pp. 607-616.

Bronger, A., 1978. Climatic sequences of steppe soils from Eastern Europe and the USA with emphasis on the genesis of the 'Argillic Horizon'. Catena 5, 33-51.

Bronger, A. & Sedov, S.N., 2003. Vetusols and paleosols: natural versus man-induced environmental change in the Atlantic coastal region of Morocco. Quaternary International 106/107, 33-60.

Bronger, A., Winter, R., Derevjanko, O. & Aldag, S., 1995. Loess-palaeosol-sequences in Tadjikistan as a palaeoclimatic record of the Quaternary in Central Asia. In Derbyshire, E. (ed.), Wind Blown Sediments in the Quaternary Record. Quaternary Proceedings Volume 4, John Wiley & Sons, Chichester, pp. 69-81.

Bronger, A., Winter, R. & Heinkele, T., 1998. Pleistocene climatic history of East and Central Asia based on paleopedological indicators in loess-paleosol sequences. Catena 34, 1-17.

Brooke, B., 2001. The distribution of carbonate eolianites. Earth-Science Reviews 55, 135-164.

Busacca, A. & Cremaschi, M., 1998. The role of time versus climate in the formation of deep soils of the Apennine fringe of the Po valley, Italy. Quaternary International 51/52, 95-107.

Cabadas-Báez, H., Solleiro-Rebolledo, E., Sedov, S., Pi-Puig, T. & Gama-Castro, J., 2010. Pedosediments of karstic sinkholes in the eolianites of NE Yucatán: a record of Late Quaternary soil development, geomorphic processes and landscape stability. Geomorphology 122, 323-337.

Catt, J.A., 1986. Soils and Quaternary Geology. Clarendon Press, Oxford, 267 p.

Catt, J.A., 1988. Soils of the Plio-Pleistocene: do they distinguish types of interglacial? Philosophical Transactions of the Royal Society of London B 318, 539-557.

Compostella, C., Mariani, G.S. & Trombino, L., 2014. Holocene environmental history at the treeline in the Northern Apennines, Italy: a micromorphological approach. The Holocene 24, 393-404.

Costantini, E.A.C. & Priori, S., 2007. Pedogenesis of plinthite during early Pliocene in the Mediterranean environment: case study of a buried paleosol at Podere Renieri, central Italy. Catena 71, 425-443.

Costantini, E.A.C., Makeev, A. & Sauer, D., 2009. Recent developments and new frontiers in paleo-pedology. Quaternary International 209, 1-5.

Courty, M.A., Goldberg, P. & Macphail, R.I., 1989. Soils and Micromorphology in Archaeology. Cambridge Manuals in Archaeology. Cambridge University Press, Cambridge, 344 p.

Cremaschi, M., 1987. Paleosols and Vetusols in the Central Po Plain (Northern Italy): A Study in Quaternary Geology and Soil Development. Unicopli, Milano, 316 p.

Cremaschi, M., 1996. The rock varnish in the Messak Sattafet (Fezzan, Libyan Sahara), age, archaeo-logical context, and palaeoenvironmental implication. Geoarchaeology 11, 393-421.

Cremaschi, M., & Busacca, A., 1994. Deep soils on stable or slowly aggrading surfaces: time versus climate as soil-forming factors. Geografia Fisica e Dinamica Quaternaria 171, 87-95.

Cremaschi, M. & Nicosia, C., 2012. Sub-Boreal aggradation along the Apennine margin of the Central Po Plain: geomorphological and geoarchaeological aspects. Géomorphologie: Relief, Processus, Environment 18, 155-174.

Cremaschi, M. & Sevink, J., 1987. Micromorphology of paleosol chronosequences on gravelly sediments in Northern Italy. In Fedoroff, N., Bresson, L.M. & Courty M.A. (eds.), Micromorphoilogie des Sols, Soil Micromorphology. AFES, Plaisir, pp. 577-584.

Cremaschi, M. & Trombino, L., 1998a. Eolianites, sea level changes and paleowinds in the Elba island (Central Italy) during Late Pleistocene. In Busacca, A.J., (ed.), Dust Aerosols, Loess Soils and Global Change. Washington State University College of Agriculture and Home Economics, Pullman, pp. 131-134.

Cremaschi, M. & Trombino, L., 1998b. The palaeoclimatic significance of palaeosols in Southern Fezzan (Libyan Sahara): morphological and micromorphological aspects. Catena 34, 131-156.

Cremaschi, M. & Van Vliet-Lanoë, B., 1990. Traces of frost activity and ice segregation in Pleistocene loess deposits and till of northern Italy: deep seasonal freezing or permafrost? Quaternary International 5, 39-48.

Cremaschi, M. & Zerboni, A., 2016. La questione dell'apparato morenico del Garda alla luce delle recenti ricerche sul Torrion della Val Sorda. Geologia Insubrica 12/1, 103.

Cremaschi, M., Biagi, P., Castelletti, L., Leoni, L., Accorsi, C., Mazzanti, M. & Rodolfi, G., 1984a. Il sito mesolitico di Monte Bagioletto, nel quadro delle variazioni ambientali oloceniche dell'Appennino Tosco - Emiliano. Emilia Preromana, 9, 11-46.

Cremaschi, M., Orombelli, G. & Salloway, J.C., 1984b. Quaternary stratigraphy and soil development at the southern border of the Central Alps, the Bagaggera sequence. Rivista Italiana di Paleontologia Stratigrafica 90, 565-603.

Cremaschi, M., Fedoroff, N., Guerreschi, A., Huxtable, J., Colombi, N., Castelletti, L. & Maspero, A., 1990. Sedimentary and pedological processes in the Upper Pleistocene loess of northern Italy. The Bagaggera sequence. Quaternary International 5, 23-38.

Cremaschi, M., Zerboni, A., Nicosia, C., Negrino, F., Rodnight, H. & Spötl, C., 2015. Age, soil-forming processes, and archaeology of the loess deposits at the Apennine margin of the Po Plain (northern Italy). New insights from the Ghiardo area. Quaternary International 376, 173-188.

Derbyshire, E., 1995. Aeolian sediments in the Quaternary record: an introduction. Quaternary Science Reviews 14, 641-643.

Derbyshire, E., Kemp, T.R. & Meng, X., 1995. Variations in loess and palaeosol properties as indicators of palaeoclimatic gradients across the Loess Plateau of North China. Quaternary Science Reviews 14, 681-697.

Dorn, R.I., 1988. A rock varnish interpretation of alluvial-fan development in Death Valley, California. National Geographic Research 4, 56-73.

Dorn, R.I., 1996. Rock Coatings. Developments in Earth Surface Processes, Volume 6, Elsevier, Amsterdam, 429 p.

Dorn, R.I., 2009. Rock varnish and its use to study climatic change in geomorphic settings. In Parsons, A. J. & Abrahams, A.D. (eds.), Geomorphology of Desert Environments. Second edition, Springer, Dordrecht, pp. 657-673.

Dorn, R.I. & Oberlander, T.M., 1981. Rock varnish origïn, characteristics, and usage. Zeitschrift für Geomorphologie 25, 420-436.

Dorn, R.I., Jull, A.J.T., Donahue, D.J., Linick, T.W. & Toolin, L.J., 1989. Accelerator mass spectrometry radiocarbon dating of rock varnish. Geological Society of America Bulletin 101, 1363-1372.

Dorn, R.I., Krinsley, D.H., Liu, T., Anderson, S., Clark, J., Cahill, T.A., & Gill, T.E., 1992. Manganese-rich rock varnish does occur in Antarctica. Chemical Geology 99, 289-298.

Dorn, R.I., Gordon, S.J., Allen, D., Cerveny, N., Dixon, J.C., Groom, K.M., Hall, K., Harrison, E., Mol, L., Paradise, T.R., Summer, P., Thompson, T. & Turkington, A.V., 2013a. The role of fieldwork in rock decay research: case studies from the fringe. Geomorphology 200, 59-74.

Dorn, R.I., Krinsley, D.H., Langworthy, K.A., Ditto, J. & Thompson, T.J., 2013b. The influence of mineral detritus on rock varnish formation. Aeolian Research 10, 61-76.

Duchaufour, Ph., 1983. Pédologie. Tome I. Pédogenèse et Classification. Masson, Paris, 491 p.

Eswaran, H., Sys, C., & Sousa, E.C., 1975. Plasma infusion. A pedological process of significance in the humid tropics. Anales de Edafología y Agrobiología. 34, 665-674.

Fang Xiaomin, Li Jijun, Derbyshire, E., Fitzpatrick, E.A. & Kemp, R.A., 1994. Micromorphology of the Beiyuan loess-paleosol sequence in Gansu Province, China: geomorphological and paleoenvironmental significance. Palaeogeography, Palaeoclimatology, Palaeoecology 111, 289-303.

Favero-Longo, S.E., Gazzano, C. & Girlanda, M., 2011. Physical and chemical deterioration of silicate and carbonate rocks by meristematic microcolonial fungi and endolithic lichens (Chaetothyriomycetidae). Geomicrobiology Journal 28, 732-744.

Fedoroff, N., 1967. Un example d'application de la micromorphologie à l'étude des paleosols. Bulletin de l'Association Francaise pur l'Etude du Sol 3, 193-209.

Fedoroff, N., 1971. The usefulness of micropedology in paleopedology. In Yaalon, D.H. (ed.), Paleopedology: Origin, Nature and Dating of Paleosols. International Soil Science Society and Israel University Press, Jerusalem, pp. 159-160.

Fedoroff, N. & Goldberg, P., 1982. Comparative micromorphology of two late Pleistocene paleosols (in the Paris Basin). Catena 9, 227-251.

Fedoroff, N., Courty, M.A. & Zhengtang Guo, 2010. Palaeosols and relict soils. In Stoops, G., Marcelino, V. & Mees, F. (eds.), Interpretation of Micromorphological Features of Soils and Regoliths, Elsevier, Amsterdam. pp. 623-662.

Fedoroff, N., Courty, M.A. & Zhengtang Guo, 2018. Palaeosols and relict soils, a conceptual approach. In Stoops, G., Marcelino, V. & Mees, F. (eds.), Interpretation of Micromorphological Features of Soils and Regoliths. Second Edition. Elsevier, Amsterdam, pp. 821-862.

Ferrari, G.A. & Magaldi, D., 1968. I paleosuoli di Collecchio (Parma) ed. il loro significato (Quarternario continentale padono) − Nota 1. L'Ateneo Parmense. Acta Naturalia 4, 57-92.

Ferrari, G. & Magaldi, D., 1976. Il problema del Loess. Studio interdisciplinare del rilievo di Trino Vercellese. Gruppo di Studio del Quaternario Padano, Quaderno 3, 34-39.

Ferraro, F., 2009. Age, sedimentation, and soil formation in the Val Sorda loess sequence, Northern Italy. Quaternary International 204, 54-64.

Ferraro, F., Terhorst, B., Ottner, F. & Cremaschi, M., 2004. Val Sorda: an upper Pleistocene loess-paleosol sequence in northeastern Italy. Revista Mexicana de Ciencias Geológicas 21, 30-47.

Gall, Q., 1992. Precambrian paleosols in Canada. Canadian Journal of Earth Sciences 29, 2530-2536.

Goldsmith, Y., Stein, M. & Enzel., Y., 2014. From dust to varnish: geochemical constraints on rock varnish formation in the Negev Desert, Israel. Geochimica et Cosmochimica Acta 126, 97-111.

Guglielmin, M., Cannone, N., Strini, A. & Lewkowicz, A.C., 2005. Biotic and abiotic processes on granite weathering landforms in a cryotic environment, Northern Victoria Land, Antarctica. Permafrost and Periglacial Processes 16, 69-85.

Guo, Z.T., Fedoroff, N. & Liu, D.S., 1996. Micromorphology of the loess-paleosol sequence of the last 130 ka in China and palaeoclimatic events. Science in China, Series D 39, 468-477.

Guo, Z.T., Sun, B., Zhang, Z.S., Peng, S.Z., Xiao, G.Q., Ge, J.Y., Hao, Q.Z., Qiao, Y.S., & Liang, M.Y., 2008. A major reorganization of Asian climate by the early Miocene. Climate of the Past 4, 153-174.

He, X., Bao, Y., Hua, L. & Tang, K., 2008. Clay illuviation in a Holocene palaeosol sequence in the Chinese Loess Plateau. In Kapur, S., Mermut, A. & Stoops, G. (eds.), New Trends in Soil Micromorphology. Springer-Verlag, Berlin, pp. 237-252.

Hoffmann, T., Erkens, G., Gerlach, R., Klostermann, J. & Lang, A., 2009. Trends and controls of Holocene floodplain sedimentation in the Rhine catchment. Catena 77, 96-106.

Hooke, R. & Dorn, R.I., 1992. Segmentation of alluvial fans in Death Valley, California: new insights from surface exposure dating and laboratory modelling. Earth and Surface Processes and Landforms 17, 557-574.

Hubschman, J., 1975. Terrefort molassique et terrasses récentes de la région toulousaine. Bulletin AFEQ 34, 125-136.

IUSS Working Group WRB, 2015. World Reference Base for Soil Resources 2014, Update 2015. World Soil Resources Reports No. 106. FAO, Rome, 192 p.

Josephs, R.L., 2010. Micromorphology of an early Holocene loess-paleosol sequence, Central Alaska, U.S.A. Arctic, Antarctic, and Alpine Research 42, 67-75.

Kemp, R.A., 1998. Role of micromorphology in paleopedological research. Quaternary International 51/52, 133-141.

Kemp, R.A., 1999. Micromorphology of loess-paleosol sequences: a record of paleoenvironmental change. Catena 35, 179-196.

Kemp, R.A. & Derbyshire, E., 1998. The loess soils of China as records of climatic change. European Journal of Soil Science 49, 525-539.

Kemp, R.A. & Zárate, M.A., 2000. Pliocene pedosedimentary cycles in the southern Pampas, Argentina. Sedimentology 47, 3-14.

Kemp, R.A., Derbyshire, E., Meng Xingmin, Fahu, C. & Baotian, P., 1995. Pedosedimentary reconstruction of thick loessepaleosol sequence near Lanzhou in North Central China. Quaternary Research 43, 30-45.

Kemp, R.A., Derbyshire, E. & Meng Xingmin, 1999. Comparison of proxy records of Late Pleistocene climate change from a high-resolution loess-palaeosol sequence in north-central China. Journal of Quaternary Science 14, 91-96.

Kemp, R.A., King, M., Toms, P., Derbyshire, E., Sayago, J.M. & Collantes, M.M., 2004. Pedosedimentary development of part of a Late Quaternary loess-palaeosol sequence in northwest Argentina. Journal of Quaternary Science 19, 567-576.

Khormali, F., Abtahi, A., Mahmoodi, S. & Stoops, G., 2003. Argillic horizon development in calcareous soils of arid and semiarid regions of southern Iran. Catena 53, 273-301.

Konecka-Betley, K., 1994. Fossil soils of late Pleistocene developed from loesses. Roczniki Gleboznawcze 45, 55-62.

Kovda, I., Sycheva, S., Lebedeva, M. & Inozemtzev, S., 2009. Variability of carbonate pedofeatures in a loess-paleosol sequence and their use for paleoreconstructions. Journal of Mountain Science 6, 155-161.

Kubiëna, W.L., 1963. Paleosoils as indicators of paleoclimates. Arid Zone Research 20, 207-208.

Kühn, P., 2003. Micromorphology and Late Glacial/Holocene genesis of Luvisols in Mecklenburg-Vorpommern (NE-Germany). Catena 54, 537-555.

Kühn, P., Billwitz, K., Bauriegel, A., Kühn, D. & Eckelmann, W., 2006. Distribution and genesis of Fahlerden (Albeluvisols) in Germany. Journal of Plant Nutrition and Soil Science 169, 420-433.

Kühn, P., Aguilar, J., Miedema, R. & Bronnikova, M., 2018. Textural Pedofeatures and Related Horizons. In Stoops, G., Marcelino, V. & Mees, F. (eds.), Interpretation of Micromorphological Features of Soils and Regoliths. Second Edition. Elsevier, Amsterdam. pp. 377-423.

Limbrey, S., 1993. Micromorphological studies of buried soils and alluvial deposits in a Wiltshire river valley. In Needham, S. (ed.), Alluvial Archaeology in Britain. Oxbow Monograph 27, Oxbow Books, Oxford, pp. 53-64.

Liu, T. & Broecker, W.S., 2008. Rock varnish microlamination dating of Late Quaternary geomorphic features in the drylands of western USA. Geomorphology 93, 501-523.

Magaldi, D. & Tallini, M., 2000. A micromorphological index of soil development for the Quaternary geology research. Catena 41, 261-276.

Marcelino, V., Schaefer, C.E.G.R. & Stoops, G., 2018. Oxic and related materials. In Stoops, G., Marcelino, V. & Mees, F. (eds.), Interpretation of Micromorphological Features of Soils and Regoliths. Second Edition. Elsevier, Amsterdam. pp. 663-689.

Marques, R., Prudêncio, M.I., Waerenborgh, J.C., Rocha, F., Dias, M.I., Ruiz, F., Ferreira da Silva, E., Abad, M. & Muñoz, A.M., 2014. Origin of reddening in a paleosol buried by lava flows in Fogo island (Cape Verde). Journal of African Earth Sciences 96, 60-70.

Marszałek, M., Alexandrowicz, Z. & Rzepa, G., 2014. Composition of weathering crusts on sandstones from natural outcrops and architectonic elements in an urban environment. Environmental Science And Pollution Research 21, 14023-14036.

Martins, U.P. & Pfefferkorn, H.W., 1988. Genetic interpretation of a lower Triassic paleosol complex based on soil micromorphology. Palaeogeography Palaeoclimatology Palaeoecology 64, 1-14.

Mees, F. & Stoops, G., 1999. Palaeoweathering of Lower Palaeozoic rocks of the Brabant Massif, Belgium: a mineralogical and petrographical analysis. Geological Journal 34, 349-367.

Meister, P., Mckenzie, J.A., Bernasconi, S.M. & Brack, P., 2013. Dolomite formation in the shallow seas of the Alpine Triassic. Sedimentology 60, 270-291.

Miedema, R. & Slager, S., 1972. Micromorphological quantification of clay illuviation. Journal of Soil Science 23, 309-314.

Mergelov, N.S., Goryachkin, S.V., Shorkunov, I.G., Zazovskaya, E.P. & Cherkinsky, A.E., 2012. Endolithic pedogenesis and rock varnish on massive crystalline rocks in East Antarctica. Eurasian Soil Science 45, 901-917.

Mol, L. & Viles, H., 2012. The role of rock surface hardness and internal moisture in tafoni development in sandstone. Earth Surface Processes and Landforms 37, 301-314.

Morozova, T.D., 1963. Micromorphological study of buried soils. Soviet Soil Science 6, 852-857.

Mroczek, P., 2013. Recycled loesses — a micromorphological approach to the determination of local source areas of Weichselian loess. Quaternary International 296, 241-250.

Ortiz, I., Simón, M., Dorronsoro, C., Martín, F. & García, I., 2002. Soil evolution over the Quaternary period in a Mediterranean climate (SE Spain). Catena 48, 131-148.

Pécsi, M. & Richter, G., 1996. Löss: Herkunft, Gliederung, Landschaften. Zeitschrift für Geomorphologie, Supplementband 98, 391 p.

Phillips, D.H., Smith, B.J., Russell, M.I. & McAlister, J.J., 2010. Micromorphology and mineralogy of intra-basaltic palaeosols at Giant's Causeway, Northern Ireland. Geochimica et Cosmochimica Acta 74, A815.

Potter, R.M. & Rossman, G.R., 1977. Desert varnish: the importance of clay minerals. Science 196, 1446-1448.

Rellini, I., Trombino, L., Firpo, M. & Piccazzo, M., 2007. Geomorphological context of 'plinthitic pale-osols' in the Mediterranean region: examples from the coast of western Liguria (northern Italy). Cuaternario y Geomorfología 21, 27-40.

Rellini, I., Trombino, L., Firpo, M. & Rossi, P.M., 2009. Extending westward the loess basin between the Alps and the Mediterranean region: micromorphological and mineralogical evidences from the slope of the Ligurian Alps (northern Italy). Geografia Fisica e Dinamica Quaternaria 32, 103-116.

Rellini, I., Trombino, L., Carbone, C. & Firpo, M., 2015. Petroplinthite formation in a pedosedimentary sequence along northern Mediterranean coast: from micromorphology to landscape evolution. Journal of Soils and Sediments 15, 1311-1328.

Retallack, G.J., 1991. Untangling the effects of burial alteration and ancient soil formation. Annual Review of Earth and Planetary Sciences 19, 183-206.

Retallack, G.J., 1992. How to find a Precambrian paleosol. In Schidlowski, M., Golubic, S., Kimberley, M.M., McKirdy Sr., D.M. & Trudinger, P.A. (eds.), Early Organic Evolution. Implications for Mineral and Energy Resources. Springer, Berlin, pp. 16-30.

Retallack, G.J., & Wright, V.P., 1990. Micromorphology of lithified paleosols. In Douglas, L.A. (ed.), Soil Micromorphology: A Basic and Applied Science. Developments in Soil Science, Volume 19. Elsevier, Amsterdam, pp. 641-652.

Rognon, P., Coudé-Gaussen, G., Fedoroff, N. & Goldberg, P., 1987. Micromorphology of loess in the norther Negev (Israel). In Fedoroff, N., Bresson, L.M. & Courty, M.A. (eds.), Micromorphologie des Sols, Soil Micromorphology. AFES, Plaisir, pp. 631-638.

Ruellan, A., 1971. L'histoire des sols, quelques problèmes de définition et d'interprétation. Cahier ORSTOM, Série Pédologie 9, 335-344.

Sanborn, P.T., Scott Smith, C.A., Froese, D.G., Zazula, D.G. & Westgate, J.A., 2006. Full-glacial paleosols in perennially frozen loess sequences, Klondike goldfields, Yukon Territory, Canada. Quaternary Research 66, 147-157.

Sheldon, N.D., 2003. Pedogenesis and geochemical alteration of Columbia River Basalts, Picture Gorge, Oregon. Geological Society of America Bulletin 115, 1377-1387.

Sheldon, N.D., 2006. Using paleosols of the Picture Gorge Basalt to reconstruct the middle Miocene climatic optimum. PaleoBios 26, 27-36.

Singh, S., Parkash, B. & Awasthi, A.K., 2009. Origin of red color of the Lower Siwalik palaeosols: a micromorphological approach. Journal of Mountain Science 6, 147-154.

Smalley, I.J., 1972. The interaction of great rivers and large deposits of primary loess. Transactions of the New York Academy of Sciences 34, 534-542.

Smolikova, L., 1967. Polygenese der fossilen Lössböden der Tschechoslowakei im Lichte mikromor-phologischer Untersuchungen. Geoderma 1, 315-324.

Smolikova, L., 1978. Bedeutung der Bodenmikromorphologie für die Datierung archäologischer Horizonte. In Delgado, M. (ed.), Soil Micromorphology. University of Granada, pp. 1199-1222.

Srivastava, P. & Parkash, B., 2002. Polygenetic soils of the north-central part of the Gangetic Plains: a micromorphological approach. Catena 46, 243-259.

Srivastava, P. & Sauer, D., 2014. Thin-section analysis of lithified paleosols from Dagshai Formation of the Himalayan Foreland: identification of paleopedogenic features and diagenetic overprinting and implications for paleoenvironmental reconstruction. Catena 112, 86-98.

Srivastava, P., Patel, S., Singh, N., Jamir, T., Kumar, N., Aruche, M. & Patel, R.C., 2013. Early Oligocene paleosols of the Dagshai Formation, India: a record of the oldest tropical weathering in the Himalayan foreland. Sedimentary Geology 294, 142-156.

Stoops, G., 1992. Micromorphological study of pre-Cretaceous weathering in the Brabant Massif (Belgium). In Schmitt, J.M. & Gall, Q. (eds.), Mineralogical and Geochemical Records of Paleoweathering. ENSMP, Mémoires des Sciences de la Terre 18, 69-84.

Stoops, G., 2003. Guidelines for analysis and description of soil and regolith thin sections. Soil Science Society of America, Madison, 184 p.

Stoops, G., 2009. Seventy years 'Micropedology' 1938-2008. The past and future. Journal of Mountain Science 6, 101-106.

Stoops, G., 2014. The 'fabric' of soil micromorphological research in the 20th century — a bibliometric analysis. Geoderma 2013, 193-202.

Stoops, G. & Marcelino, V., 2018. Lateritic and bauxitic materials. In Stoops, G., Marcelino, V. & Mees, F. (eds.), Interpretation of Micromorphological Features of Soils and Regoliths. Second Edition. Elsevier, Amsterdam. pp. 691-720.

Swezey, C., 2001. Eolian sediment responses to late Quaternary climate changes: temporal and spatial patterns in the Sahara. Palaeogeography, Palaeoclimatology, Palaeoecology 167, 119-155.

Teruggi, M.E. & Andreis, R.R., 1971. Micromorphological recognition of paleosolic features in sediments and sedimentary rocks. In Yaalon, D.H. (ed.), Paleopedology: Origin, Nature and Dating of Paleosols. International Soil Science Society and Israel University Press, Jerusalem, pp. 161-172.

Trombino, L., 2007. Micromorphological reconstruction of the archaeological land use and palae-oenvironment of Tell Mishrifeh: evidence from the sinkhole south of the site. In Morandi Bonacossi, D. (ed.), Urban and Natural Landscapes of an Ancient Syrian Capital. Settlement and Environment at Tell Mishrifeh/Qatna and in Central-Western Syria. Forum Editrice, Udine, pp. 115-122.

Trombino, L. & Ferraro, F., 2002. Paleosuoli tipo Terra Rossa al margine prealpino. Il caso di studio del Monte Casto (Val Sabbia — Brescia). Il Quaternario 15, 131-140.

Trombino, L., Anelli, A., Berra, F. & Jadoul, F., 2006. Integration of micro- and nano-morphological analyses to reconstruct the depositional and post-depositional evolution of paleosols in the Norian succession of the southern Alps (northern Italy). Geophysical Research Abstracts 8, 05764.

Tsatskin, A. & Ronen, A., 1999. Micromorphology of Mousterian paleosol in aeolianites at the site Habonim, Israel. Catena 34, 365-384.

Tsatskin, A., Gendler, T.S., Heller, F., Dekman, I. & Frey, G.L., 2009. Towards understanding paleosols in Southern Levantine eolianites: integration of micromorphology, environmental magnetism and mineralogy. Journal of Mountain Science 6, 113-124.

Tursina, T.V., 2009. Methodology for the diagnostics of soil polygenesis on the basis of macro- and micromorphological studies. Journal of Mountain Science 6, 125-131.

Ufnar, D.F., González, L.A., Ludvigson, G.A., Brenner, R.L., Witzke, B.J. & Leckie, D., 2005. Reconstructing a mid-Cretaceous landscape from paleosols in Western Canada. Journal of Sedimentary Research 75, 984-996.

Van Herreweghe, S., Deckers, S., De Coninck, F., Merckx, R. & Gullentops, F., 2003. The paleosol in the Kerkom sands near Pellenberg (Belgium) revisited. Netherlands Journal of Geosciences 82, 149-159.

Van Ranst, E., Wilson, M. & Righi, D., 2018. Spodic materials. In Stoops, G., Marcelino, V. & Mees, F. (eds.), Interpretation of Micromorphological Features of Soils and Regoliths. Second Edition Elsevier, Amsterdam. pp. 633-662.

Van Vliet-Lanoë, B., 1990. Le pédocomplexe de Warneton, Ou en est-on ? Bilan paléopédologique et micromorphologique. Quaternaire 1, 65-75.

Vepraskas, M.J., Lindbo, D.L. & Stolt, M.H. & 2018. Redoximorphic features. In Stoops, G., Marcelino, V. & Mees, F. (eds.), Interpretation of Micromorphological Features of Soils and Regoliths. Second Edition Elsevier, Amsterdam. pp. 425-445.

Viles, H.A. & Goudie, A.S., 2007. Rapid salt weathering in the coastal Namib desert: implications for landscape development. Geomorphology 85, 49-62.

Vola, G. & Jadoul, F., 2014. Applied stratigraphy and carbonate petrography of the Arabescato Orobico dimension stone from the Bergamasc Alps (Calcare Rosso, Italy). Italian Journal of Geoscience 133, 294-314.

von Suchodoletz, H., Kühn, P., Hambach, U., Dietze, M., Zöller, L. & Faust, D., 2009. Loess-like and palaeosol sediments from Lanzarote (Canary Islands/Spain) — indicators of palaeoenvironmental change during the Late Quaternary. Palaeogeography, Palaeoclimatology, Palaeoecology 278, 71-87.

Wanas, H.A. & Abu El-Hassan, M.M., 2006. Paleosols of the Upper Cretaceous-Lower Tertiary Maghra El-Bahari Formation in the northeastern portion of the Eastern Desert, Egypt: their recognition and geological significance. Sedimentary Geology 183, 243-259.

Wieder, M., Singer, A. & Gvirtzman, G., 1990. Micromorphological study of deep buried Jurassic basalt-derived paleosols from northern Israel. In Douglas, L.A. (ed.), Soil Micromorphology: A Basic and Applied Science. Developments in Soil Science, Volume 19. Elsevier, Amsterdam, pp. 697-703.

Yaalon, D.H., 1971. Soil forming processes in time and space. In Yaalon, D.H. (ed.), Paleopedology: Origin, Nature and Dating of Paleosols. International Soil Science Society and Israel University Press, Jerusalem, pp. 29-39.

Zárate, M., Kemp, R.A., Espinosa, M. & Ferrero, L., 2000. Pedosedimentary and palaeoenvironmental significance of a Holocene alluvial sequence in the southern Pampas, Argentina. The Holocene 10, 481-488.

Zembo, I., Trombino, L., Bersezio, R., Felletti, F. & Dapiaggi, M., 2012. Climatic and tectonic controls on pedogenesis and landscape evolution in a quaternary intramontane basin (Val d'Agri basin, southern Apennines, Italy). Journal of Sedimentary Research 82, 283-309.

Zerboni, A., 2008. Holocene rock varnish on the Messak plateau (Libyan Sahara): chronology of weathering processes. Geomorphology 102, 640-651.

Zerboni, A., Guglielmin, M., 2016. Complex rock varnish from the Dry Valleys (Antarctica) suggests the interaction of biochemical weathering and dust accretion in a frontier critical zone. Goldschmidt 2016, Abstract n° 2983.

Zerboni, A., Trombino, L. & Cremaschi, M., 2011. Micromorphological approach to polycyclic pedo-genesis on the Messak Settafet plateau (central Sahara): formative processes and palae-oenvironmental significance. Geomorphology 125, 319-335.

Zerboni, A., Perego, A. & Cremaschi, M., 2015a. Geomorphological map of the Tadrart Acacus massif and the Erg Uan Kasa (Libyan Central Sahara). Journal of Maps 11, 772-787.

Zerboni, A., Trombino, L., Frigerio, C., Livio, F., Berlusconi, A., Michetti, A.M., Spötl, C. & Rodnight, H., 2015b. A loess-paleosols sequence at Monte Netto (northern Italy) records upper Pleistocene climatic changes in the central Po Plain. Journal of Soil and Sediments 15, 1329-1350.

Zhao, J., 2003. Paleoenvironmental significance of a paleosol complex in Chinese loess. Soil Science 168, 63-72.

Zhisheng, A., Tungsheng, L., Yizhi, Z., Fuqing, S. & Zhongly, D., 1987. The paleosol complex S5 in the China Loess Plateau — a record of climatic optimum during the last 1.2 Ma. GeoJournal 15, 141-143.

Chapter 30

Micromorphological Features and Their Relation to Processes and Classification: General Guidelines and Overview

Georges Stoops[1], Vera Marcelino[1], Florias Mees[2]

[1]*GHENT UNIVERSITY, GHENT, BELGIUM;*
[2]*ROYAL MUSEUM FOR CENTRAL AFRICA, TERVUREN, BELGIUM*

CHAPTER OUTLINE

Interpretation of Micromorphological Features of Soils and Regoliths. https://doi.org/10.1016/B978-0-444-63522-8.00030-9

1. Introduction

The aim of micromorphology is to contribute to the understanding of soils and regoliths and their identification, genesis and relative chronology. The first step in this process is identification of the components and analysis of the fabric, followed by its description. The next, most important step is the interpretation of the observations. This chapter intends to be a help in reaching this goal. The first part comprises general guidelines for interpretation. The second part provides a guide, in table format, to micromorphological features discussed in the various chapters of this book.

In many cases, the study of a single soil thin section is not sufficient for understanding all genetic processes. In practice, the interpretation of micromorphological features mostly requires comparative studies between different horizons of a same profile, between different profiles and between samples collected during different periods (e.g., before and after cultivation), or between natural and experimental materials.

When using the tables, the reader should realise that in some cases the absence of a feature may be more characteristic than its presence, which cannot be fully outlined in a table.

1.1 Criteria for *In Situ* Formation and Relative Chronology

For interpretation of genetic processes in soils and regoliths, it is important to determine which features were formed *in situ* and which were inherited from the parent material. Regarding genetic processes, it is also important to state a chronological order to understand the sequence of processes, even if the soil was interpreted as monogenetic in the field.

The following rules can help in distinguishing between *in situ* and inherited features:
- Impregnative features (e.g., nodules, hypocoatings) with diffuse boundaries and composition and fabric of the coarse material that are similar to those of the surrounding groundmass are formed *in situ*; however, one should take into account

that (i) dilution of the coarse material can occur, as a result of crystallisation pressure expulsing grains, in the case of crystalline impregnative materials, and (ii) impregnative nodules can acquire sharp boundaries in vertic and other pedoturbated materials (disorthic nodules).

- Nodules in which the nature and arrangement of the coarse material are different from those in the surrounding groundmass are (i) inherited from the parent material (e.g., colluvial material), or (ii) formed during an earlier stage of soil development, provided the difference in texture and/or mineralogical composition can be explained by a more advanced stage of weathering of the surrounding groundmass (e.g., weatherable minerals present in the nodule, but absent in the groundmass).
- Features related to surfaces or voids (coatings, hypocoatings, quasicoatings) always postdate surface of void formation; coatings associated with the surface of voids are generally formed *in situ*, which is not always the case for coatings covering the surface of grains (e.g., iron oxide coatings on transported sand grains or pebbles).
- Infillings postdate void formation and are therefore generally formed *in situ*; passage features are also considered as *in situ*.
- Intercalations are always considered as having formed *in situ*, although their pedogenic nature is not always evident.
- Euhedral crystals of minerals known to form in soil environments, especially of more soluble minerals, are supposed to be formed *in situ*.

Monogenetic soils are very rare, except for weakly developed profiles. Already during formation of a mature soil, considered as monogenetic, different phases occur, which may have different micromorphological imprints. If a change in climate or geomorphological position takes place, additional overprinting of features occurs. One of the important tasks of micromorphology is to clarify the relative chronology of micromorphological features, and thus of pedogenic processes. In many cases, it is the only possible method to disentangle the complex history of a soil (Cremaschi et al., 2018; Fedoroff et al., 2018). However, setting up the relative chronology of two or more features is not always easy. The following simple rules may help: (i) when two coatings, infillings or intercalations intersect, the continuous one is the youngest; (ii) in the case of juxtaposed coatings, the one immediately related to the covered surface is the oldest, and the overlying coating is the youngest; (iii) features enclosed by other ones are older than the surrounding features; and (iv) features crossing others and continuing in the surrounding groundmass postdate the crossed feature (e.g., infillings of planar voids crossing a nodule).

1.2   Relicts of the Parent Material

A soil obviously bears characteristics of its parent material. These are best preserved and most pronounced in the coarse fraction (see also Stoops & Mees, 2018). Some authors propose that the degree of expression of these characteristics gradually diminishes, ultimately resulting in a similar soil material for different parent materials (Chesworth, 1973). Because the coarse material is mainly, but not exclusively, inherited from the parent material, it is anyway our most important source of information about the latter.

Changes occurring during saprolite or soil development are an indication of the degree of weathering.

1.2.1 Nature of the Parent Material

For identification of the parent material, the shape, relative size, absolute size and mineralogical composition of the grains are important.

A rounded shape points to transported grains and indicates that the profile has developed in sedimentary deposits (e.g., alluvium, weathered sandstone). Angular grains, on the contrary, are typical for *in situ* weathering of igneous or metamorphic rocks, or for some glacial sedimentary parent materials.

Both the absolute size of the grains and their sorting can give information about the genesis of sedimentary materials. For example, aeolian deposits have a good sorting, whereas fan deposits are poorly sorted. Material derived from *in situ* weathering generally shows poor sorting.

For the interpretation of the mineralogical composition of the coarse fraction, the reader is referred to handbooks on petrography (see Stoops & Mees, 2018). Special attention should be given to specific details visible in thin sections such as the following:

- Colour: e.g., minerals such as biotite, hornblende and tourmaline can have different colours depending on their origin; differences in UV or blue light fluorescence or cathodoluminescence can also point to different origins
- Inclusions: e.g., inclusions of volcanic glass in feldspar grains point to a volcanic origin (Stoops et al., 2018); different inclusions in quartz grains (e.g., rutile needles, mica flakes) are indicative of different origins; fissured quartz grains with hematite or gibbsite infillings point to a lateritic origin (Marcelino et al., 2018; Stoops & Marcelino, 2018)
- Zoning: e.g., zoning in feldspar and pyroxenes generally excludes a metamorphic origin
- Deformation features: e.g., wavy extinction, especially in quartz grains, points to a stress deformation of the source rock (e.g., metamorphic rocks)

The abrupt appearance or disappearance of a mineral species in a profile, which cannot be explained by weathering, generally points to a lithological discontinuity. In a homogeneous profile, the same minerals should occur throughout the profile in the same proportions, as far as weathering did not change relative mineral abundance.

1.2.2 Degree of Weathering

For determining the degree of weathering, it is important to look both to the individual grains and to the proportions of different mineral species.

Weathering of individual grains can be expressed by their shape (e.g., denticulate pyroxene or amphibole crystals) and/or by their optical properties (e.g., change of colour, relief and interference colours of biotite or chlorite). For more details the reader is referred to the excellent atlas by Delvigne (1998) (see also Stoops & Mees, 2018). Hydrothermal weathering or diagenesis of the parent rock can be misleading in studies of weathering in surface conditions. Changes in proportions of different minerals (e.g., feldspar/quartz ratio) can also give indications on the occurrence and degree of weathering.

The composition of the coarse material does not necessarily reflect only the nature of the rock from which it is derived, but also the influence of later additions, such as aeolian materials (loess, volcanic ash) and anthropogenic deposits (e.g., coal, pottery fragments, chert).

In the case of alluvial or colluvial material (Mücher et al., 2018; Cremaschi et al., 2018), part of the coarse fraction can be of pedogenic origin, for example, reworked iron oxide or carbonate nodules.

The fine material of the soil is in most cases so strongly influenced by pedogenic processes that it is generally no longer possible to deduce information on the parent material from the study of the micromass in thin sections.

2. Guide to Features

In the following part, an effort has been made to relate the most common micromorphological features, as reported in this book, to processes, soil horizons and types of soil and regolith materials. The indicated horizon type does not strictly correspond to a diagnostic horizon. For instance, micromorphological characteristics of an oxic horizon may be reported, even when the cation exchange capacity of the soil is too high to qualify for the oxic diagnostic horizon according to Soil Taxonomy, or morphological vertic properties may be present in materials that do not meet the climatic requirements for Vertisols.

The overviews presented in these tables are neither systematic nor exclusive, and some features therefore appear in more than one section. Not all features mentioned in the various chapters of the book could be listed in these tables.

2.1   Microstructure and c/f-Related Distribution Patterns

Gravelly, sandy and silty materials without clay are characterised by the presence of microscopically visible, inherent microporosity (simple packing pores), absent in both loamy and clayey materials. This large inherent porosity may give rise to special features in coarse sand and gravel, such as pendants and cappings. Even at a macroscopic scale, it is obvious that the main factor determining the expression of pedogenic processes is the texture of the parent material, e.g., sandy Vertisols or clayey Spodosols do not exist. On a microscopic scale, the influence of texture is still more evident, e.g., the micromorphological expression of an argillic horizon will be quite different in sandy soils and in loamy or clayey materials. Therefore it is useful to make a distinction between materials of different textural classes. For gravelly, sandy or silty materials, basic microstructures, named according to their c/f-related distribution pattern, are used (Stoops, 2003). Intergrades between the basic microstructures may occur, as already proposed by Stoops & Jongerius (1975), and mixtures are commonly observed.

2.1.1   Basic Microstructures

For pure gravelly, sandy or silty materials, coarse monic can be considered as an initial stage. Addition of fine material by weathering or by external input leads first to enaulic, chitonic or gefuric types, depending on composition, and finally to porphyric. When mass transfer occurs from the coarse to the fine fraction through weathering, transformation to single-spaced, double-spaced or open porphyric can take place (Chadwick & Nettleton, 1994). Illuviation of fine material, combined with thorough pedoturbation, can also gradually give rise to single-spaced to open porphyric patterns.

The evolution of coarse monic to porphyric comprises successively monic, gefuric, chitonic and porphyric when cohesive forces dominate, which is the case when the fine fraction is composed of covalently bonded silica, iron, aluminium and organic matter. The series is monic, enaulic, gefuric-chitonic and porphyric when adhesive forces prevail, which occurs when ionically bonded salts are dominant, in arid environments.

Basic related distribution patterns of silty or loamy materials are difficult to describe in thin sections in transmitted light, as overlapping of grains is quite common.

Gradual transitions between enaulic basic microstructures and granular microstructures (Stoops, 2008) can be observed in soils. Chitonic and gefuric basic microstructures are practically inexistent in surface horizons, and spongy microstructures are exceptional in all profile positions.

Because microstructure descriptions refer to the soil as a whole, including pedofeatures, the 'coarse material' and 'fine material' columns in the tables for basic microstructure types include not only coarse and fine mineral components of the groundmass but also fabric elements such as organic material, clay coatings and gypsum crystals.

In the following tables, the term 'various' as coarse material description is used to indicate a composition ranging from quartz-free to quartz-rich. 'Resistant only' refers to coarse material composed essentially of quartz, with minor amounts of highly stable minerals such as anatase, rutile, tourmaline, zircon and certain opaque minerals. The term 'mainly resistant' refers to quartz with some relatively stable but weatherable minerals such as muscovite and microcline.

Basic microstructure — coarse monic c/f-related distribution pattern

Coarse material	Fabric	Occurrence	Chapter	Figures
Various	Various, including banded arrangement (with sorting, grading)	Sandy or silty parent materials, including layered deposits unaffected by pedogenesis	—	—
Various	Random	Disruptional microlayer of surface crusts	Williams et al. (2018)	—
Mainly resistant, with organic residues	Random	A and E horizons of Spodosols	Van Ranst et al. (2018)	V9
Volcanic components	Random, banded	Fresh volcanic ash and lower parts of Andosols	Stoops et al. (2018)	S1A
Lenticular gypsum crystals	Random, or crescent fabric	Hypergypsic horizons	Poch et al. (2018)	—

Basic microstructure — enaulic c/f-related distribution pattern

Coarse material	Fine material	Occurrence	Chapter	Figures
Various	Dark; undifferentiated b-fabric	Surface horizons	—	—
Various	Organo-mineral material	Sediments with high organic matter content	Ismail-Meyer et al. (2018)	—
Various	Black, excrements	Moder	Ismail-Meyer et al. (2018)	—
Various	Fine mineral aggregates	Sedimentary crusts	Williams et al. (2018)	—
Mainly resistant	Dark polymorphic organic matter; undifferentiated b-fabric	Friable spodic horizons	Van Ranst et al. (2018)	V3
Resistant only	Yellowish/reddish brown; undifferentiated b-fabric	Sandy oxic materials	Marcelino et al. (2018)	M4
Volcanic	Yellow, brown; undifferentiated b-fabric	Young soils on volcanic ash	Stoops et al. (2018)	—

Basic microstructure — chitonic c/f-related distribution pattern

Coarse material	Fine material	Occurrence	Chapter	Figures
Various	Mainly limpid clay, with continuous orientation	Bt horizons	Kühn et al. (2018)	K12
Various	Laminated	Slow water flow colluvial deposits	Mücher et al. (2018)	—
Various	Optically isotropic colourless material	Duric and fragic horizons	Gutiérrez-Castorena (2018)	—
Mainly resistant	Pure monomorphic organic matter to optically isotropic iron oxides	Bh or Bir horizons of Spodosols; placic horizons	Van Ranst et al. (2018)	V1, V7
Resistant only	Optically isotropic brown to reddish material	Sandy oxic material	Marcelino et al. (2018)	M4
Volcanic	Optically isotropic material	A horizons of young volcanic ash soils	Stoops et al. (2018)	S6A

Basic microstructure — gefuric c/f-related distribution pattern

Coarse material	Fine material	Occurrence	Chapter	Figures
Various	Fine clay, with continuous orientation, concave and convex forms	Bt horizons in sandy material	Kühn et al. (2018)	K4, K12
Various	Coarse clay and silt	Laminated colluvial deposits	Mücher et al. (2018)	—
Various	Banded clay	Sedimentary crusts	Williams et al. (2018)	—

Continued

Basic microstructure — gefuric c/f-related distribution pattern — cont'd

Coarse material	Fine material	Occurrence	Chapter	Figures
Various	Pedogenic silica	Silcretes	Gutiérrez-Castorena (2018)	—
Mainly resistant	Dark monomorphic organic matter	Friable spodic horizons	Van Ranst et al. (2018)	—
Mainly resistant, with organic residues	Organic fine material	A or E horizons of Spodosols	Van Ranst et al. (2018)	—

2.1.2 Granular Microstructure

The granular microstructure is defined as being composed of small spherical aggregates, but this definition is often not strictly followed. In many cases, it is also used to describe small equant aggregates with an irregular or angular to subangular shape, as can be deduced from published photographs. In some cases, fine crumb microstructures are described as granular. A correct description of granular aggregates would help understanding the genesis of this microstructure type, which is still under discussion (see Kovda & Mermut, 2018; Marcelino et al., 2018; Stoops et al., 2018), both biological and physical processes being proposed.

Coarse material	Fine material	Occurrence	Chapter	Figures
Various	Various	Cultivated soils	Adderley et al. (2018)	—
Various	Various	Surface horizon of Vertisols	Kovda & Mermut (2018)	—
Various	Greyish, brownish, dark; undifferentiated b-fabric	Surface horizons of soils other than Vertisols	Gerasimova & Lebedeva (2018)	GL6, GL10
Various, fine silt size	Various	Upper part of frost-affected soils	Van Vliet-Lanoë & Fox (2018)	VF1E, VF1F
Various	Various	Pedosediments, aeolian deposits	Fedoroff et al. (2018)	—
Various	Various	Faunal features	Kooistra & Pulleman (2018)	—
Mainly resistant	Organic	Surface horizons of Spodosols	Van Ranst et al. (2018)	—
Resistant only	Yellowish/reddish brown; undifferentiated b-fabric	Oxic materials	Marcelino et al. (2018)	M1, M2
Resistant only	Yellowish/reddish brown, undifferentiated b-fabric	Internodular infilling in lateritic gravel	Stoops & Marcelino (2018)	—
Volcanic	Yellow, brown; undifferentiated b-fabric	Young soils on volcanic ash	Stoops et al. (2018)	S1B-S1D, S2D, S5A
None	Fine organic	Moder	Ismail-Meyer et al. (2018)	—

2.1.3 Crumb Microstructure

Coarse material	Fine material	Occurrence	Chapter	Figures
Various	Dark, speckled or dotted; undifferentiated b-fabric	Surface horizons, especially of Mollisols, Vertisols; cultivated soils	Gerasimova & Lebedeva (2018); Kovda & Mermut (2018); Adderley et al. (2018)	GL1
Mainly resistant	Dark brown polymorphic organic matter; undifferentiated b-fabric	Bh horizons of Spodosols (rare)	Van Ranst et al. (2018)	V4

2.1.4 Blocky Microstructure

Blocky microstructures are not characteristic of specific soils, but their presence excludes materials such as spodic and andic. The large dimensions of the peds often make it impossible to observe them in thin sections. Angular blocky microstructures seen in thin sections are often artefacts resulting from drying of the material before impregnation.

Coarse material	Fine material	Occurrence	Chapter	Figures
Various	Various	Frost-affected soils	Van Vliet-Lanoë & Fox (2018)	VF1C, VF1D
Various	Various	Subsurface horizons of Vertisols	Kovda & Mermut (2018)	KM1, KM2
Various	Various (moderate to high content)	Subsurface horizons of various other soil types	—	—

2.1.5 Platy Microstructure

Platy microstructures can be the result of external or internal oriented pressures, or they can be related to a banded arrangement of particles. They only form when sufficient fine material is present.

Coarse material	Fine material	Occurrence	Chapter	Figures
Various	Various	Frost-affected soils and sediments	Van Vliet-Lanoë & Fox (2018)	VF7A, VF15D
Various	Various	Sedimentary surface crusts	Williams et al. (2018)	W7
Various	Various	Intervals with yermic or takyric properties	Gerasimova & Lebedeva (2018)	
Various	Various	Frost-affected mollic and chernic horizons	Gerasimova & Lebedeva (2018)	GL4
Various	Various	Frost-affected clay-eluvial horizons	Kühn et al. (2018)	—
Various	Various	Plough pans	Adderley et al. (2018)	—
Various	Various	Base of anthraquic horizons in paddy soils	Adderley et al. (2018)	—

Continued

— cont'd

Coarse material	Fine material	Occurrence	Chapter	Figures
Various	Various	Solifluction materials	Mücher et al. (2018)	—
Various	Various (high content)	Degraded Vertisols	Kovda & Mermut (2018)	—
Various	Calcareous	Lamellar calcareous crusts	Durand et al. (2018); Fedoroff et al. (2018)	—
Organic	Fine organic material	Sedimentary organic layers; ripening peat	Ismail-Meyer et al. (2018)	—
Various	Various	Animal tracks, footpaths	Adderley et al. (2018)	—

2.1.6   Lenticular Microstructure

Coarse material	Fine material	Occurrence	Chapter	Figures
Various (low content)	Various	Frost-affected soils	Van Vliet-Lanoë & Fox (2018); Gerasimova & Lebedeva (2018); Fedoroff et al. (2018)	VF1A, VF1B, VF6
Various	Various	Intervals with yermic properties	Gerasimova & Lebedeva (2018)	GL17
Various	Clay	Frost-affected E horizons	Kühn et al. (2018)	—

2.1.7   Spongy Microstructure

Coarse material	Fine material	Occurrence	Chapter	Figures
Various	Organo-mineral material	Mollic horizons	Gerasimova & Lebedeva (2018)	GL7
Mainly resistant	Polymorphic organic matter	Lower part of spodic horizons (rare)	Van Ranst et al. (2018)	—
Phosphatic material	Phosphatic material	Guano	Karkanas & Goldberg (2018)	KG8

2.1.8   Vesicular Microstructure

Coarse material	Fine material	Occurrence	Chapter	Figures
Various (low content)	Various	Natural surface crusts with yermic or takyric properties	Gerasimova & Lebedeva (2018); Williams et al. (2018)	GL13, GL15; W4
Various (low content)	Various	Irrigation crusts	Williams et al. (2018)	W6
Various (low content)	Various	Surface layer affected by fire	Adderley et al. (2018)	—
Various (high content)	Various	Frost-affected soils	Van Vliet-Lanoë & Fox (2018)	VF6B, VF8
Various	Various	Mass-transported water-saturated soil material	Fedoroff et al. (2018)	—

2.1.9   Vughy Microstructure

Coarse material	Fine material	Occurrence	Chapter	Figures
Various	Various	Soils with initially granular microstructure, following compaction (star-shaped vughs)	—	—
Various	Various	Frost-affected soils (star-shaped vughs, associated with blocky microstructure)	Van Vliet-Lanoë & Fox (2018)	—
Various	Various	*In situ* reworked or transported soil material (polyconcave vughs)	Fedoroff et al. (2018)	—

2.1.10   Channel Microstructure

Coarse material	Fine material	Occurrence	Chapter	Figures
Various	Various	Soils and sediments with biological activity, mainly in subsurface horizons	Kooistra & Pulleman (2018)	—
Various	Various	Agricultural soils	Adderley et al. (2018)	—
Weathered minerals (saprolitic fabric)	Weathering products	Soil materials undergoing pedoplasmation	Stoops & Schaefer (2018)	—

2.1.11   Massive Microstructure

Coarse material	Fine material	Occurrence	Chapter	Figures
Various (low content)	Various	Sediments unaffected by wetting and drying, freezing and thawing, or biological activity	—	—
Various (low content)	Various	Pedosediments	Fedoroff et al. (2018)	—

2.2   Groundmass

2.2.1   Heterogeneity

Heterogeneity of the groundmass, as expressed by variations in colour, limpidity or composition, is an important indication of some processes. Regoliths, such as saprolites on coarse-grained rocks, have an inherent heterogeneity at a microscopic scale.

Nature	Occurrence	Chapter	Figures
Groundmass containing material derived from other parts of the same profile	Soils affected by bioturbation; soils with anthropogenic disturbance (subsoiling, digging); Vertisols	Kooistra & Pulleman (2018); Gerasimova & Lebedeva (2018); Adderley et al. (2018); Kovda & Mermut (2018)	GL1E; KM5, KM6

Continued

— cont'd

Nature	Occurrence	Chapter	Figures
Idem, with also anthropogenic elements	Cultivated soils	Adderley et al. (2018)	A1
Groundmass containing rounded fragments of materials with different groundmass types	Mass movement deposits	Mücher et al. (2018)	—
Groundmass containing angular sedimentary clasts	Landslide deposits	Mücher et al. (2018); Stoops et al. (2018)	M4
Groundmass containing allochthonous elements, often showing (sub) horizontal layering and compaction	Archaeological living floors	Macphail & Goldberg (2018)	MG2
Groundmass containing angular clasts and limpid clayey material, partly as fragments of coatings	Collapsed weathering products of dissolved limestone	Stoops & Schaefer (2018)	SS9
Partly preserved saprolitic fabric	Saprolites modified partly by pedoplasmation	Stoops & Schaefer (2018)	SS1, SS7

2.2.2 Special Fabric Types

Nature	Occurrence	Chapter	Figures
Banded fabric, parallel orientation of elongated particles; strial b-fabric	Sedimented or translocated material, transported soil material	Mücher et al. (2018); Fedoroff et al. (2018)	F18
Idem (often with grading), in surface positions	Surface crusts, seals, drain fillings	Williams et al. (2018)	W7, W8, W9
Banded fabric, with volcanic constituents	Material formed by successive ash falls	Stoops et al. (2018)	
Banded fabric and striated b-fabric below the surface	Soils subjected to shear forces during ploughing	Adderley et al. (2018)	—
Laminar fabric, parallel to surface	Laminated colluvium	Mücher et al. (2018)	M1
Laminar fabric, with calcitic-crystallitic b-fabric	Laminar crusts on top of calcrete or bedrock	Durand et al. (2018)	D6
Rock fragments with vertical orientation	Frost-affected soils	Van Vliet-Lanoë & Fox (2018)	VF10
Crescent fabrics (passage features)	Mollic horizons; cultivated soils; hologypsic layers; oxic materials	Gerasimova & Lebedeva (2018); Adderley et al. (2018); Poch et al. (2018); Marcelino et al. (2018)	P5C; M3
Crescent fabric in saprolitic fabric	Soil material undergoing pedoplasmation	Stoops & Schaefer (2018)	SS4A
Lithic fabric, with secondary minerals	Saprolites	Zauyah et al. (2018)	Z1-Z5

2.2.3   Coarse Material

The composition of the coarse fraction reflects the nature of the parent material for soils, or the source material for sediments. In the next table, only some specific types are mentioned (see also Stoops & Mees, 2018).

Nature	Occurrence	Chapter	Figures
Resistant only (quartz with pitted and corroded surfaces, fragmented quartz grains, runiquartz)	Oxic and lateritic materials	Marcelino et al. (2018); Stoops & Marcelino (2018); Gutiérrez-Castorena (2018)	M4, M6
Volcanic components and their alteration products	Volcanic ash soils	Stoops et al. (2018)	S1A, S1D, S2A-S2C, S3B, S4B, S5B
Calcite grains and limestone fragments	Lower horizons of temperate and (semi)arid soils on calcareous parent material	Durand et al. (2018)	–
Phytoliths	(Buried) surface horizons; Andosols; Vertisols; archaeological deposits	Kaczorek et al. (2018); Gerasimova & Lebedeva (2018); Stoops et al. (2018); Kovda & Mermut (2018); Macphail & Goldberg (2018); Adderley et al. (2018)	K18-K4; S2D; KM7; MG8
Interconnected phytoliths	Soils with *in situ* organic matter decomposition	Kaczorek et al. (2018); Macphail & Goldberg (2018)	K3E, K3F
Vesicular glass slags, derived from phytoliths	Heated material	Macphail & Goldberg (2018); Kaczorek et al. (2018)	MG9; K4
Diatoms, sponge spicules	Marine, lacustrine or fluvial sediments; subaquatic or continuously wet materials, anmoor	Kaczorek et al. (2018)	K5-K8
Shells and shell fragments	Various; weakly developed soils on marine or lacustrine sediments; archaeological deposits	Durand et al. (2018); Macphail & Goldberg (2018)	D1, D2, D17; MG10
Plant ashes	Soils with anthropogenic influences; archaeological deposits (black earth, fire place remnants)	Adderley et al. (2018); Macphail & Goldberg (2018)	MG7-MG9
Fragments of bone or burned bone	Cave sediments, archaeological materials, carnivore excrements	Karkanas & Goldberg (2018); Adderley et al. (2018); Macphail & Goldberg (2018)	KG1; MG7
Fish bones, in phosphatic groundmass	Guano	Karkanas & Goldberg (2018)	–
Fragments of concrete, mortar, brick, pottery, slag, asphalt	Urban soils, 'black earth', Technosols	Adderley et al. (2018); Macphail & Goldberg (2018)	–
Stone tools	Archaeological deposits	Macphail & Goldberg (2018)	MG19

2.2.4 Micromass

The following table contains examples taken from various chapters. More information is presented in the chapter dealing specifically with groundmass characteristics (Stoops & Mees, 2018).

Nature	Occurrence	Chapter	Figures
Dark, dotted; undifferentiated or weakly speckled b-fabric	Surface horizons	Gerasimova & Lebedeva (2018)	—
Brown to black; mainly undifferentiated b-fabric	Spodic horizons, placic horizons	Van Ranst et al. (2018)	V11
Optically isotropic, autofluorescent	Al-rich spodic horizons	Van Ranst et al. (2018)	—
Grey, yellow, brown, speckled; undifferentiated b-fabric; volcanic coarse components	Andosols	Stoops et al. (2018)	—
Fine-grained; strial b-fabric	Clayey sediments; sedimentary surface crusts, drain infillings	Kovda & Mermut (2018); Williams et al. (2018)	W10
Yellow, red, speckled; undifferentiated b-fabric, seldom weak circular, crescent or porostriated	Oxic materials	Marcelino et al. (2018)	M1-M3
Mostly dark brown; strongly expressed striated b-fabric, possibly masked by calcitic-crystallitic b-fabric	Vertic materials	Kovda & Mermut (2018); Kühn et al. (2018); Cremaschi et al. (2018)	KM1, KM5, KM6, KM8, KM10B; C9B
Calcitic-crystallitic b-fabric	Carbonate-rich soil material, with inherited or pedogenic carbonates	Durand et al. (2018); Gerasimova & Lebedeva (2018); Kovda & Mermut (2018)	D3; GL8
Striated b-fabrics (mainly grano- and porostriated)	Frost-affected soils	Van Vliet-Lanoë & Fox (2018)	VF9B
Galaxy and striated b-fabric	Landslide and debris flow deposits; glacial sediments	Mücher et al. (2018)	M5
Partly opaque, undifferentiated b-fabric	Heated soil materials	Fedoroff et al. (2018)	—
Greyish; undifferentiated or crystallitic b-fabric; autofluorescent	Phosphate-rich materials	Karkanas & Goldberg (2018)	KG5
Micromass depleted in iron oxides; often well expressed striated b-fabric	Soils with long-term water saturation	Vepraskas et al. (2018)	V4

2.3 Organic Material

It is not the aim to give a detailed overview of the various occurrences of organic matter, which is covered in a specific chapter (Ismail-Meyer et al., 2018), that also includes a table summarising the characteristics of organic tissue types.

2.3.1 Coarse to Fine Organic Material

Nature	Occurrence	Chapter	Figures
Tissue residues, mineral-deficient excrements, fine organic material	F and H layer of moder	Ismail-Meyer et al. (2018)	I8A, I8B
Fragmented plant remains, well preserved, few excrements	F layer of mor	Ismail-Meyer et al. (2018)	—
Bark and cork remains only, well preserved	H layer of mor	Ismail-Meyer et al. (2018)	—
Coarse plant residues, light yellowish to reddish brown or dark brown	Fibric horizon	Ismail-Meyer et al. (2018)	I10A-I10D
Coarse plant residues, dark, loosely packed, porous, with dopplerite as pore infilling	Hemic horizon	Ismail-Meyer et al. (2018)	I10G, I10H
More or less recognisable organic debris, common iron-sulphides, common pseudomorphs after organic matter	Sapropel	Ismail-Meyer et al. (2018)	—
Parallel layers of various plant residues or organic fine material	Sedimentary organic lenses	Ismail-Meyer et al. (2018)	—
Rare small plant fragments; aquatic animal excrements	Anmoor	Ismail-Meyer et al. (2018)	—
Black brown fine organic matter, few dark plant residues; excrements can be present	Sapric horizon	Ismail-Meyer et al. (2018)	I11C, I11D
Amorphous organic matter, with horizontal cracks, consisting of excrements, diatoms and residues of aquatic plants	Gyttja	Ismail-Meyer et al. (2018)	—
Limpid yellowish organic material with small amounts of moss residues	Dy	Ismail-Meyer et al. (2018)	—

2.3.2 Organo-mineral Material

Nature	Occurrence	Chapter	Figures
Dark organic excrements with undifferentiated b-fabric, as aggregates between sand grains	A horizons below mor	Ismail-Meyer et al. (2018)	—
Brown to dark brown organic pigment, associated with fine mineral material, dotted limpidity, undifferentiated b-fabric, rarely speckled, plant residues	Surface horizons, especially mollic horizons and A horizons of cultivated soils; mull	Ismail-Meyer et al. (2018); Gerasimova & Lebedeva (2018); Adderley et al. (2018)	GL1-GL3, GL5, GL6, GL9, GL10; I8C, I8D; A3
Polymorphic aggregates between sand grains (enaulic c/f-related distribution pattern)	Friable spodic horizons	Van Ranst et al. (2018)	V3, V4
Monomorphic material between sand grains (gefuric, chitonic or close porphyric c/f-related distribution pattern)	Spodic horizons	Van Ranst et al. (2018)	V6, V7

Continued

— cont'd

Nature	Occurrence	Chapter	Figures
Layers composed of algae, cyanobacteria, lichens, mosses, with entrapped mineral grains	Biological surface crusts	Williams et al. (2018)	W12-W14
Charcoal fragments	Soils affected by bioturbation or tillage, archaeological material, volcanic ash soils, soils affected by wildfires	Adderley et al. (2018); Macphail & Goldberg (2018); Ismail-Meyer et al. (2018); Stoops et al. (2018); Fedoroff et al. (2018)	I7A, I7B, S7
Straw fragments or hair	Archaeological construction materials	Macphail & Goldberg (2018)	—

2.4 Pedofeatures

Coatings and infillings are considered here as one group, as they are closely associated, coatings gradually evolving to infillings. Moreover, tangential sections through coatings may resemble infillings in thin sections.

2.4.1 Textural Coatings and Infillings

Nature	Occurrence	Chapter	Figures
Limpid fine clay, yellowish/reddish, with continuous orientation, length slow optical orientation	Bt horizons of Luvisols, final phase of clay illuviation; some duripans in volcanic soils	Kühn et al. (2018); Van Vliet-Lanoë & Fox (2018); Fedoroff et al. (2018); Cremaschi et al. (2018); Stoops et al. (2018)	K1-K4, K14; VF7B; F7A, F8; C4, C6A, C8, C10A, C10D; S8A-S8C
Limpid fine clay with striated b-fabric	Illuvial features subjected to shearing (Vertisols, hydromorphic soils)	Kovda & Mermut (2018); Kühn et al. (2018)	K13
Moderate to well oriented with optically isotropic patches, yellowish/greyish fine clay with enclosed fine quartz and opaque particles	Soils affected by hydromorphism	Kühn et al. (2018)	—
Thick, yellowish grey, dotted, moderate continuous orientation, no pronounced lamination	Natric horizons	Kühn et al. (2018)	—
Dusty, greyish, moderately sorted, weak parallel orientation of clay, along base of horizontal channels	E or upper Bt horizons of soils with tillage; colluvial material	Kühn et al. (2018); Adderley et al. (2018); Mücher et al. (2018); Cremaschi et al. (2018)	K5, K6; C3B, C10A
Unsorted silt and coarse clay, weakly oriented particles	Flooded soils; soils with percolation of melt water	Kühn et al. (2018); Van Vliet-Lanoë & Fox (2018)	VF13
Thin clayey to fine sandy coatings, with strong continuous orientation, along horizontal surface	Skin seals (surface crust)	Williams et al. (2018)	—

— cont'd

Nature	Occurrence	Chapter	Figures
Silt and sand infillings, in planar voids	Solonchaks; takyric surface horizons	Kühn et al. (2018); Gerasimova & Lebedeva (2018)	GL11
Thick whitish clay coatings with continuous orientation, or silty clay with weak orientation	Glossic tongues with secondary illuviation	Kühn et al. (2018)	—
Silty-clayey cappings or coatings, on plates, lenses, grains	Frost-affected soils	Van Vliet-Lanoë & Fox (2018); Kühn et al. (2018)	VF6, VF9A, VF12
Fine material plastered to the walls of channels	Materials affected by specific faunal activity (e.g., termites)	Kooistra & Pulleman (2018); Marcelino et al. (2018)	KP14; M8
White kaolinitic clay coatings, with superposed granular hematite hypocoatings	Plinthitic and lateritic materials	Stoops & Marcelino (2018)	SM2
Yellowish to reddish brown speckled coatings, with diffuse lamination, speckled extinction pattern (goethite-kaolinite coatings)	Lateritic materials	Stoops & Marcelino (2018)	SM3-SM5
Fragments of illuvial coatings	Transported material; frost-affected soils; soils with bioturbation; soils with crystallisation of pedogenic minerals	Mücher et al. (2018); Kühn et al. (2018); Fedoroff et al. (2018); Cremaschi et al. (2018); Mees (2018); Van Vliet-Lanoë & Fox (2018); Kooistra & Pulleman (2018)	K7, K8, K10, K15; F15, F17A; C9A, C10C; M3D; VF14
Fragments of clay coatings, with kink-band fabric	Vertisols	Kovda & Mermut (2018)	—
Fragments of clay coatings forming a layer	Weathering residues of limestone; lower part of polyphase soils	Stoops & Schaefer (2018); Fedoroff et al. (2018)	SS9
Clay depletion hypocoatings	Soils with long-term water saturation	Vepraskas et al. (2018)	—
Micromass-depleted hypocoatings, along horizontal planes	Washed-out microlayer (surface crust)	Williams et al. (2018)	—
Carbonate-rich clay coatings	Irrigation crusts	Adderley et al. (2018); Williams et al. (2018)	—

2.4.2   *Iron and Manganese Oxides*

Nature	Occurrence	Chapter	Figures
Manganese oxide nodules	Materials subjected to short periods of water saturation	Vepraskas et al. (2018)	—
Iron/manganese oxide orthic nodules and hypocoatings	Materials subjected to longer periods of water saturation	Vepraskas et al. (2018); Adderley et al. (2018)	V2; A2 L3D-3E V3A-V3B

Continued

— cont'd

Nature	Occurrence	Chapter	Figures
Iron oxide coatings and hypocoatings along root channels, in a greyish groundmass	Materials drained after long period of water saturation	Vepraskas et al. (2018); Adderley et al. (2018)	
Iron/manganese oxide hypocoatings and disorthic nodules	Vertisols, water-saturated during part of the year	Kovda & Mermut (2018)	KM10A, KM10B
Concentric iron and manganese nodules, often with associated granostriation	Soils with alternating wetting and drying, vertic materials	Vepraskas et al. (2018); Kovda & Mermut (2018)	KM10C
Dendritic iron/manganese oxide nodules	Clayey materials	Kovda & Mermut (2018)	KM10D
Iron depletion hypocoatings and associated iron oxide quasicoatings	Soils with long periods of water saturation	Vepraskas et al. (2018)	V4
Optically isotropic brown-orange coatings or fibrous goethite coatings, in conducting voids	Hydromorphic soils, paddy soils	Vepraskas et al. (2018)	—
Iron oxide nodules	Near plough pans	Adderley et al. (2018)	—
Impregnative iron/manganese oxides in material with lithic fabric	Material in first stage of weathering, in oxidising conditions	Zauyah et al. (2018)	Z3, Z4, Z7
Optically isotropic, limpid, orange ferruginous material, as coatings in sandy materials	Bir horizons of Spodosols, placic horizons	Van Ranst et al. (2018)	—
Iron oxide coatings, hypo- and quasicoatings	Sulphuric materials with weathering of jarosite	Mees & Stoops (2018)	MS4
Iron oxide pseudomorphs after pyrite	Oxidised sulphidic materials	Mees & Stoops (2018)	—
Fine-grained hematite hypocoatings and impregnative nodules, in a kaolinitic groundmass	Plinthitic lateritic materials	Stoops & Marcelino (2018)	SM2, SM3
Coatings of coarse-crystalline goethite, oriented perpendicular to pore walls	Lateritic materials	Stoops & Marcelino (2018)	SM5
Goethite-kaolinite coatings, with speckled limpidity, diffuse lamination, speckled extinction pattern	Lateritic materials	Stoops & Marcelino (2018); Fedoroff et al. (2018)	SM6; F11
Manganese oxide coatings on hematitic or goethitic material	Lateritic materials	Stoops & Marcelino (2018)	—
Complex polygenetic hematite-goethite nodules	Lateritic materials	Stoops & Marcelino (2018); Fedoroff et al. (2018)	SM9
Concentric nodules with complex goethite/hematite/gibbsite coatings	Lateritic and bauxitic materials	Stoops & Marcelino (2018); Fedoroff et al. (2018)	F11

2.4.3 Calcium Carbonates

Nature	Occurrence	Chapter	Figures
Calcite coatings; orthic calcite nodules, micritic to sparitic	Soils in semiarid to temperate regions	Durand et al. (2018)	D4, D7
Disorthic rounded calcite nodules, often with iron-manganese oxide impregnation	Vertic materials	Kovda & Mermut (2018)	KM11, KM12

— cont'd

Nature	Occurrence	Chapter	Figures
Calcite pendants	Stony temperate to arid soils	Durand et al. (2018)	D5
Needle-fibre calcite infillings	Vadose settings, all latitudes	Durand et al. (2018)	D8
Coarse-grained calcite cement	Groundwater deposits	Durand et al. (2018)	D9
Concentric nodules (pisoliths)	Polyphase pedogenic (and sedimentary) material	Durand et al. (2018)	D11
Calcified roots and filaments	Various semiarid soil types	Durand et al. (2018)	D12-D14
Coarse-crystalline radial aggregates	Soils with earthworm activity, partly promoted by cultivation	Durand et al. (2018); Ismail-Meyer et al. (2018)	D15; I9A, I9B
Spherulitic aggregates	Dung deposits	Durand et al. (2018); Macphail & Goldberg (2018);	D16; MG12-13
Laminar calcareous crusts	Top of calcrete or bedrock	Durand et al. (2018)	D6
Calcite hypo- and quasicoatings	Arid to ustic soils, with fluctuating water table	Durand et al. (2018)	D7
Calcite depletion hypocoatings	Calcareous materials with leaching along pores	Mees (2018)	M1A, M1B
Partially dissolved calcareous features	Soils subjected to a change from dry to humid conditions	Fedoroff et al. (2018)	—
Cryptocrystalline material, partly with banded fabric; lime plaster and mortar	Archaeological materials, urban soils	Macphail & Goldberg (2018)	MG20, MG21

2.4.4   Gypsum

Nature	Occurrence	Chapter	Figures
Infillings of channels, mainly loose, seldom dense xenotopic	Gypsiferous soils	Poch et al. (2018); Kühn et al. (2018)	P5; K11
Lenticular crystals, crystal intergrowths, nodules	Gypsiferous soils	Poch et al. (2018); Kovda & Mermut (2018)	P8; KM6
Abundant loose, generally lenticular crystals, mostly with many passage features	Gypsic horizons	Poch et al. (2018)	P1A, P5C
Massive accumulations with xenotopic or subidiotopic fabric	Petrogypsic horizons	Poch et al. (2018)	P3
Gypsum pendants	Gravelly materials	Poch et al. (2018)	—
Accumulation of gypsum crystals and rounded clay aggregates	Aeolian gypsarenites	Poch et al. (2018)	P4
Coatings, infillings and clusters consisting of prismatic or tabular crystals	Sulphuric materials	Mees & Stoops (2018)	MS5

2.4.5   Pedogenic Silica and Silicates

Nature	Occurrence	Chapter	Figures
Opal coatings, infillings, nodules, pendants	Silcretes, duripans	Gutiérrez-Castorena (2018)	G3, G4, G6, G7
Micro- to macrocrystalline quartz coatings and nodules	Silcretes	Gutiérrez-Castorena (2018)	G1, G5
Chalcedony coatings, infillings and aggregates	Volcanic ash soils, silcretes, duripans, silicified sandstone	Gutiérrez-Castorena (2018)	G2
Cristobalite coatings	Volcanic materials, some duripans	Gutiérrez-Castorena (2018)	—
Tridymite coatings	Volcanic materials, silcretes	Gutiérrez-Castorena (2018)	—
Silicified roots	Silcretes, lacustrine sediments, duripans; acid sulphate soils	Gutiérrez-Castorena (2018); Mees & Stoops (2018)	—
Silica hypo- and quasicoatings, nodules	Acid sulphate soils	Mees & Stoops (2018)	—
Palygorskite-sepiolite coatings, hypocoatings, nodules	Semiarid soils	Mees (2018)	M1–M3
Zeolite crystals, infillings, coatings	Soils and regoliths with volcanic material, alkaline saline soils	Mees (2018)	M5, M6
Sodium silicates	Soils and regoliths with volcanic material, alkaline saline soils	Mees (2018)	M8

2.4.6   Other Minerals

Nature	Occurrence	Chapter	Figures
Calcium oxalate (whewelite), partly as phytoliths	Surface materials with lichens or fungi	Durand et al. (2018)	D21-D23
Pyrite grains, framboids, clusters, often associated with organic matter	Sulphidic materials	Mees & Stoops (2018)	MS1, M2
Weathered pyrite grains, including goethite pseudomorphs	Sulphuric materials	Mees & Stoops (2018)	—
Jarosite coatings, hypo- and quasicoatings, nodules	Sulphuric materials	Mees & Stoops (2018)	MS3
Bassanite, anhydrite	Gypsiferous soils, in hyperarid conditions	Mees & Tursina (2018)	MT4
Celestite	Mainly gypsiferous soils	Mees & Tursina (2018); Mees (2018)	MT6B
Barite	Various soil types (rare)	Mees & Tursina (2018); Vepraskas et al. (2018)	MT8
Halite	Saline soils	Mees & Tursina (2018)	MT1
Highly soluble sulphate minerals	Saline soils	Mees & Tursina (2018)	MT2-MT4, MT9
Aluminium-bearing and associated sulphate minerals	Oxidised sulphide-bearing materials	Mees & Stoops (2018)	—
Sodium carbonate minerals	Alkaline saline soils	Mees & Tursina (2018)	—

— cont'd

Nature	Occurrence	Chapter	Figures
Siderite	Bog ore, strongly hydromorphic soils	Vepraskas et al. (2018)	—
Phosphatic nodules, phosphatic excrements	Cave deposits; archaeological deposits	Karkanas & Goldberg (2018)	KG6, KG10, KG12
Vivianite nodules, infillings, crystal intergrowths	Waterlogged, peaty soils and archaeological deposits	Karkanas & Goldberg (2018); Vepraskas et al. (2018); Adderley et al. (2018)	KG16, KG17
Coarse-grained gibbsite coatings, infillings, nodules	Lateritic material; middle part of bauxite profiles	Stoops & Marcelino (2018); Stoops & Schaefer (2018)	SM12; SS3
Microcrystalline gibbsite coatings, infillings, nodules	Transition between cemented bauxite and cover layer	Stoops & Marcelino (2018)	—
Cryptocrystalline gibbsite coatings, infillings, nodules	Bauxitic material	Stoops & Marcelino (2018)	SM13
Concentric gibbsite nodules	Bauxitic material	Stoops & Marcelino (2018)	SM14
Gibbsite alteromorphs, with preservation of lithic fabric	Strongly leached saprolites or soils	Stoops & Marcelino (2018); Stoops et al. (2018)	SM11

References

Adderley, W.P., Wilson, C.A., Simpson, I.A. & Davidson, D.A., 2018. Anthropogenic features. In Stoops, G., Marcelino, V. & Mees, F. (eds.), Interpretation of Micromorphological Features of Soils and Regoliths. Second Edition. Elsevier, Amsterdam, pp. 753-777.

Chadwick, O.A. & Nettleton, W.D., 1994. Quantitative relationships between net volume change and fabric properties during soil evolution. In Ringrose-Voase, A.J. & Humphreys, G.S. (eds.), Soil Micromorphology: Studies in Management and Genesis. Developments in Soil Science, Volume 22. Elsevier, Amsterdam, pp. 353-360.

Chesworth, W., 1973. The parent rock effect in the genesis of soil. Geoderma 10, 215-225.

Cremaschi, M., Trombino, L. & Zerboni, A., 2018. Palaeosols and relict soils, a systematic review. In Stoops, G., Marcelino, V. & Mees, F. (eds.), Interpretation of Micromorphological Features of Soils and Regoliths. Second Edition. Elsevier, Amsterdam, pp. 863-894.

Delvigne, J.E., 1998. Atlas of Micromorphology of Mineral Alteration and Weathering. Canadian Mineralogist, Special Publication 3, 495 p.

Durand, N., Monger, H.C. & Canti, M.G., 2018. Calcium carbonate features. In Stoops, G., Marcelino, V. & Mees, F. (eds.), Interpretation of Micromorphological Features of Soils and Regoliths. Second Edition. Elsevier, Amsterdam, pp. 205-258.

Fedoroff, N., Courty, M.A. & Zhengtang Guo, 2018. Palaeosols and relict soils, a conceptual approach. In Stoops, G., Marcelino, V. & Mees, F. (eds.), Interpretation of Micromorphological Features of Soils and Regoliths. Second Edition. Elsevier, Amsterdam, pp. 821-862.

Gerasimova, M. & Lebedeva, M., 2018. Organo-mineral surface horizons. In Stoops, G., Marcelino, V. & Mees, F. (eds.), Interpretation of Micromorphological Features of Soils and Regoliths. Second Edition. Elsevier, Amsterdam, pp. 513-538.

Gutiérrez-Castorena, Ma.del C., 2018. Pedogenic siliceous features. In Stoops, G., Marcelino, V. & Mees, F. (eds.), Interpretation of Micromorphological Features of Soils and Regoliths. Second Edition. Elsevier, Amsterdam, pp. 127-155.

Ismail-Meyer, K., Stolt, M. & Lindbo, D., 2018. Soil organic matter. In Stoops, G., Marcelino, V. & Mees, F. (eds.), Interpretation of Micromorphological Features of Soils and Regoliths. Second Edition. Elsevier, Amsterdam, pp. 471-512.

Kaczorek, D., Vrydaghs, L., Devos, Y., Pető, Á. & Effland, W.R., 2018. Biogenic siliceous features. In Stoops, G., Marcelino, V. & Mees, F. (eds.), Interpretation of Micromorphological Features of Soils and Regoliths. Second Edition. Elsevier, Amsterdam, pp. 157-176.

Karkanas, P. & Goldberg, P., 2018. Phosphatic features. In Stoops, G., Marcelino, V. & Mees, F. (eds.), Interpretation of Micromorphological Features of Soils and Regoliths. Second Edition. Elsevier, Amsterdam, pp. 323-346.

Kooistra, M.J. & Pulleman, M.M., 2018. Features related to faunal activity. In Stoops, G., Marcelino, V. & Mees, F. (eds.), Interpretation of Micromorphological Features of Soils and Regoliths. Second Edition. Elsevier, Amsterdam, pp. 447-469.

Kovda, I. & Mermut, A., 2018. Vertic features. In Stoops, G., Marcelino, V. & Mees, F. (eds.), Interpretation of Micromorphological Features of Soils and Regoliths. Second Edition. Elsevier, Amsterdam, pp. 605-632.

Kühn, P., Aguilar, J., Miedema, R. & Bronnikova, M., 2018. Textural Pedofeatures and Related Horizons. In Stoops, G., Marcelino, V. & Mees, F. (eds.), Interpretation of Micromorphological Features of Soils and Regoliths. Second Edition. Elsevier, Amsterdam, pp. 377-423.

Macphail, R.I. & Goldberg, P., 2018. Archaeological materials. In Stoops, G., Marcelino, V. & Mees, F. (eds.), Interpretation of Micromorphological Features of Soils and Regoliths. Second Edition. Elsevier, Amsterdam, pp. 779-819.

Marcelino, V., Schaefer, C.E.G.R. & Stoops, G., 2018. Oxic and related materials. In Stoops, G., Marcelino, V. & Mees, F. (eds.), Interpretation of Micromorphological Features of Soils and Regoliths. Second Edition. Elsevier, Amsterdam, pp. 663-689.

Mees, F., 2018. Authigenic silicate minerals − sepiolite-palygorskite, zeolites and sodium silicates. In Stoops, G., Marcelino, V. & Mees, F. (eds.), Interpretation of Micromorphological Features of Soils and Regoliths. Second Edition. Elsevier, Amsterdam, pp. 177-203.

Mees, F. & Stoops, G., 2018. Sulphidic and sulphuric materials. In Stoops, G., Marcelino, V. & Mees, F. (eds.), Interpretation of Micromorphological Features of Soils and Regoliths. Second Edition. Elsevier, Amsterdam, pp. 347-376.

Mees, F. & Tursina, T.V., 2018. Salt minerals in saline soils and salt crusts. In Stoops, G., Marcelino, V. & Mees, F. (eds.), Interpretation of Micromorphological Features of Soils and Regoliths. Second Edition. Elsevier, Amsterdam, pp. 289-321.

Mücher, H., van Steijn, H. & Kwaad, F., 2018. Colluvial and mass wasting deposits. In Stoops, G., Marcelino, V. & Mees, F. (eds.), Interpretation of Micromorphological Features of Soils and Regoliths. Second Edition. Elsevier, Amsterdam, pp. 21-36.

Poch, R.M., Artieda, O. & Lebedeva, M., 2018. Gypsic features. In Stoops, G., Marcelino, V. & Mees, F. (eds.), Interpretation of Micromorphological Features of Soils and Regoliths. Second Edition. Elsevier, Amsterdam, pp. 259-287.

Stoops, G., 2003. Guidelines for Analysis and Description of Soil and Regolith Thin Sections. Soil Science Society of America, Madison, 184 p.

Stoops, G., 2008. Micromorphology. In Chesworth, W. (ed.), Encyclopedia of Soil Science. Springer, Dordrecht, pp. 458-466.

Stoops, G. & Jongerius, A., 1975. Proposal for a micromorphological classification of soil materials. I. A classification of the related distributions of fine and coarse particles. Geoderma 13, 189-199.

Stoops, G. & Marcelino, V., 2018. Lateritic and bauxitic materials. In Stoops, G., Marcelino, V. & Mees, F. (eds.), Interpretation of Micromorphological Features of Soils and Regoliths. Second Edition. Elsevier, Amsterdam, pp. 691-720.

Stoops, G. & Mees, F., 2018. Groundmass composition and fabric. In Stoops, G., Marcelino, V. & Mees, F. (eds.), Interpretation of Micromorphological Features of Soils and Regoliths. Second Edition. Elsevier, Amsterdam, pp. 73-125.

Stoops, G. & Schaefer, C.E.G.R., 2018. Pedoplasmation: formation of soil material. In Stoops, G., Marcelino, V. & Mees, F. (eds.), Interpretation of Micromorphological Features of Soils and Regoliths. Second Edition. Elsevier, Amsterdam, pp. 59-71.

Stoops, G., Sedov, S. & Shoba, S., 2018. Regoliths and soils on volcanic ash. In Stoops, G., Marcelino, V. & Mees, F. (eds.), Interpretation of Micromorphological Features of Soils and Regoliths. Second Edition. Elsevier, Amsterdam, pp. 721-751.

Van Vliet-Lanoë, B. & Fox, C., 2018. Frost action. In Stoops, G., Marcelino, V. & Mees, F. (eds.), Interpretation of Micromorphological Features of Soils and Regoliths. Second Edition. Elsevier, Amsterdam, pp. 575-603.

Van Ranst, E., Wilson, M.A. & Righi, D., 2018. Spodic materials. In Stoops, G., Marcelino, V. & Mees, F. (eds.), Interpretation of Micromorphological Features of Soils and Regoliths. Second Edition. Elsevier, Amsterdam, pp. 633-662.

Vepraskas, M.J., Lindbo, D.L. & Stolt, M.H., 2018. Redoximorphic features. In Stoops, G., Marcelino, V. & Mees, F. (eds.), Interpretation of Micromorphological Features of Soils and Regoliths. Second Edition. Elsevier, Amsterdam, pp. 425-445.

Williams, A.J., Pagliai, M. & Stoops, G., 2018. Physical and biological surface crusts and seals. In Stoops, G., Marcelino, V. & Mees, F. (eds.), Interpretation of Micromorphological Features of Soils and Regoliths. Second Edition. Elsevier, Amsterdam, pp. 539-574.

Zauyah, S., Schaefer, C.E.G.R. & Simas, F.N.B., 2018. Saprolites. In Stoops, G., Marcelino, V. & Mees, F. (eds.), Interpretation of Micromorphological Features of Soils and Regoliths. Second Edition. Elsevier, Amsterdam, pp. 37-57.

Author Index

'Note: Page numbers followed by "f" indicate figures and "t" indicate tables.'

Subject Index

'*Note*: Page numbers followed by "f" indicate figures and "t" indicate tables.'

CPI Antony Rowe
Chippenham, UK
2018-10-16 23:23